# CEREBRAL CORTEX

## Volume 14

## Neurodegenerative and Age-Related Changes in Structure and Function of Cerebral Cortex

# CEREBRAL CORTEX

Series Editors: Edward G. Jones and Alan Peters



---

Volume 1: Cellular Components of the Cerebral Cortex

Volume 2: Functional Properties of Cortical Cells

Volume 3: Visual Cortex

Volume 4: Association and Auditory Cortices

Volume 5: Sensory-Motor Areas and Aspects of Cortical Connectivity

Volume 6: Further Aspects of Cortical Function, Including Hippocampus

Volume 7: Development and Maturation of Cerebral Cortex

Volume 8A: Comparative Structure and Evolution of Cerebral Cortex, Part I

Volume 8B: Comparative Structure and Evolution of Cerebral Cortex, Part II

Volume 9: Normal and Altered States of Function

Volume 10: Primary Visual Cortex in Primates

Volume 11: The Barrel Cortex of Rodents

Volume 12: Extrastriate Cortex in Primates

Volume 13: Models of Cortical Circuits

Volume 14: Neurodegenerative and Age-Related Changes in Structure and Function of Cerebral Cortex

Series Editors: Edward G. Jones and Alan Peters

# CEREBRAL CORTEX

## Volume 14
## Neurodegenerative and Age-Related Changes in Structure and Function of Cerebral Cortex

Edited by

ALAN PETERS

*Boston University School of Medicine*
*Boston, Massachusetts*

and

JOHN H. MORRISON

*Mount Sinai Medical Center*
*New York, New York*

Kluwer Academic / Plenum Publishers
New York, Boston, Dordrecht, London, Moscow

The Library of Congress cataloged the first volume of this title as follows:

Main entry under title:

Cerebral cortex.

Includes bibliographical references and index.
Contents: v. 1. Cellular components of the cerebral cortex.
1. Cerebral cortex—Collected works. I. Peters, Alan, 1929– II. Jones, Edward G., 1939– . [DNLM: 1. Cerebral cortex—Anatomy and histology. 2. Cerebral cortex—Physiology. WL 307 C4136]

QP383.C45 1984 612'.825 84-1982

ISBN 0-306-45966-3


233 Spring Street, New York, N.Y. 10013

10 9 8 7 6 5 4 3 2 1

A C.I.P. record for this book is available from the Library of Congress.



Printed in the United States of America

Sir John Carew Eccles (1903–1997)

John Eccles died peacefully at his home near Locarno on May 2, 1997. During the course of a long life, he dominated a whole era of neuroscience, not only through his works but also through the large number of neuroscientists that he trained and who went on to become leaders in their field. Trained by Charles Sherrington at Oxford, Eccles was in many ways the living embodiment of the Sherringtonian tradition in neuropsychology. His life in science was one from which many drew inspiration but one that few could hope to emulate. A skillful technician and a tireless experimenter, he could intellectually and physically exhaust whole relays of much younger persons. At the peak of his career, his scientific output was prodigious and groundbreaking.

During his highly productive years at Canberra, which became the Mecca to which virtually every neurophysiologist of a generation was drawn, Eccles was richly blessed with resources of material and personnel. But it was not always so: the first intracellular recordings from motoneurons, which led ultimately to the Nobel Prize, were begun before moving to Canberra, under conditions that were far from optimal and that would have made a lesser person blanch. Eccles deserved his place in the sun.

A formidable debater, Eccles never shirked the task of vigorously promoting his current ideas, yet he dropped them with a generous spirit when evidence to the contrary was marshaled against them. Indeed, his tergiversations often served to highlight an incremental advance in scientific understanding.

By the time he joined the editorial advisory board of *Cerebral Cortex* in 1984, his experimental career had been over for some years, Canberra and its successor laboratories at Chicago and Buffalo being long past. But he was still in full command of his intellectual powers, knew everyone who was doing anything significant in the field, and offered us many suggestions that were helpful in ensuring the success of our venture. Not content with being a mere adviser, he contributed to Volume 2 with a theory of the operation of the neocortex. The chapter ranged widely, from anatomical modularity through the role of calcium in long-term synaptic plasticity to models of memory and cognition, stressing the links between memory and cognitive experience and reflecting that "all experiencing is tinged with remembering." It was a style of writing that few scientists can command today, but typical of the man behind the pen.

John Eccles's passing presages the end of an era in neuroscience, an era in which the foundations of modern cellular neurophysiology were laid. His was not the sole contribution by any means, but such was the force of his personality, the power of his intellect, and his influence on a generation of neuroscientists, that, like one of similar attributes in the early seminal era of physiology, Johannes Müller, he was truly one of its colossi.

Edward G. Jones
*Davis, California*

# Contributors


**John R. Absher** Cingulum NeuroSciences Institute, Winston-Salem, North Carolina 27101; and Departments of Physiology, Pharmacology, and Neurology, Wake Forest University School of Medicine, Winston-Salem, North Carolina 27157-1083

**Marilyn S. Albert** Departments of Psychiatry and Neurology, Massachusetts General Hospital, Harvard Medical School, Boston, Massachusetts 02118

**Amy F. T. Arnsten** Section of Neurobiology, Yale Medical School, New Haven, Connecticut 06510-8001

**Gene J. Blatt** Department of Anatomy and Neurobiology, Boston University School of Medicine, Boston, Massachusetts 02118

**Constantin Bouras** Department of Psychiatry, HUG Belle-Idée, University of Geneva School of Medicine, CH-1225 Geneva, Switzerland; and Kastor Neurobiology of Aging Laboratories, Mount Sinai School of Medicine, New York, New York 10029

**Eva Braak** Department of Anatomy, J. W. Goethe University, D-60590 Frankfurt, Germany

**Heiko Braak** Department of Anatomy, J. W. Goethe University, D-60590 Frankfurt, Germany

**Peter B. Crino** Department of Neurology, University of Pennsylvania School of Medicine, Philadelphia, Pennsylvania 19104

**James H. Eberwine** Departments of Pharmacology and Psychiatry, University of Pennsylvania School of Medicine, Philadelphia, Pennsylvania 19104

**Lindsay A. Farrer** Departments of Medicine (Genetics Program) and Neurology, Boston University School of Medicine, and Department of Epidemiology and Biostatistics, Boston University School of Public Health, Boston, Massachusetts 02118

**Fred H. Gage** Laboratory of Genetics, The Salk Institute, La Jolla, California 92037

**James W. Geddes** Sanders-Brown Center on Aging and Department of Anatomy and Neurobiology, University of Kentucky, Lexington, Kentucky 40536-0230

**Stephen D. Ginsberg** The Center for Neurodegenerative Disease Research, Department of Pathology and Laboratory Medicine, University of Pennsylvania School of Medicine, Philadelphia, Pennsylvania 19104-4283

**James G. Herndon** Yerkes Regional Primate Research Center, Emory University, Atlanta, Georgia 30322

**Asao Hirano** Division of Neuropathology, Department of Pathology, Montefiore Medical Center, Bronx, New York 10467

**Patrick R. Hof** Kastor Neurobiology of Aging Laboratories, Fishberg Research Center for Neurobiology and Department of Geriatrics and Adult Development, and Department of Ophthalmology, Mount Sinai School of Medicine, New York, New York 10029

**Thomas L. Kemper** Departments of Pathology, Anatomy and Neurobiology, and Neurology, Boston University School of Medicine, Boston, Massachusetts 02118

**Ronald J. Killiany** Department of Neuroanatomy and Neurobiology, Boston University School of Medicine, Boston, Massachusetts 02118

**J. H. Kordower** Research Center for Brain Repair, Department of Neurological Sciences, Rush Alzheimer's Disease Center, Rush Presbyterian-St. Luke's Medical Center, Chicago, Illinois 60612

**Virginia M.-Y. Lee** The Center for Neurodegenerative Disease Research, Department of Pathology and Laboratory Medicine, University of Pennsylvania School of Medicine, Philadelphia, Pennsylvania 19104-4283

| | |
|---|---|
| **Alex Martin** | National Institute of Mental Health, Bethesda, Maryland 20892-1366 |
| **Eliezer Masliah** | Departments of Neurosciences and Pathology, University of California, San Diego, La Jolla, California 92093-0624 |
| **John H. Morrison** | Kastor Neurobiology of Aging Laboratories, Fishberg Research Center for Neurobiology and Department of Geriatrics and Adult Development, Mount Sinai School of Medicine, New York, New York 10029 |
| **Mark B. Moss** | Department of Anatomy and Neurobiology, Boston University School of Medicine, Boston, Massachusetts 02118 |
| **E. J. Mufson** | Research Center for Brain Repair, Department of Neurological Sciences, Rush Alzheimer's Disease Center, Rush Presbyterian-St. Luke's Medical Center, Chicago, Illinois 60612 |
| **Imaharu Nakano** | Department of Neurology, Jichi Medical School, Tochigi-ken, 329-0498 Japan |
| **Timothy J. Nicholson** | Department of Anatomy and Neurobiology, Boston University School of Medicine, Boston, Massachusetts 02118 |
| **Alan Peters** | Department of Anatomy and Neurobiology, Boston University School of Medicine, Boston, Massachusetts 02118 |
| **Daniel A. Peterson** | Laboratory of Genetics, The Salk Institute, La Jolla, California 92037; *present address:* Department of Neuroscience, Finch University of Health Sciences, The Chicago Medical School, North Chicago, Illinois 60064-3095 |
| **Douglas L. Rosene** | Department of Anatomy and Neurobiology, Boston University School of Medicine, Boston, Massachusetts 02118; and Yerkes Regional Primate Research Center, Emory University, Atlanta, Georgia 30322 |
| **M. Luise Schmidt** | The Center for Neurodegenerative Disease Research, Department of Pathology and Laboratory Medicine, University of Pennsylvania School of Medicine, Philadelphia, Pennsylvania 19104-4283 |
| **David P. Salmon** | Department of Neurosciences, University of California, San Diego, La Jolla, California 92093-0948 |

**D. Larry Sparks** Sun Health Research Institute, Sun City, Arizona 85351; and Division of Medical Examiner's Service, Kentucky Justice Cabinet, Frankfort, Kentucky 40601

**John Q. Trojanowski** The Center for Neurodegenerative Disease Research, Department of Pathology and Laboratory Medicine, University of Pennsylvania School of Medicine, Philadelphia, Pennsylvania 19104-4283

**Brent A. Vogt** Cingulum NeuroSciences Institute, Winston-Salem, North Carolina 27101; and Departments of Physiology, Pharmacology, and Neurology, Wake Forest University School of Medicine, Winston-Salem, North Carolina, 27157-1083

**Leslie J. Vogt** Cingulum NeuroSciences Institute, Winston-Salem, North Carolina 27101; and Departments of Physiology, Pharmacology, and Neurology, Wake Forest University School of Medicine, Winston-Salem, North Carolina 27157-1083

**Kent E. Vrana** Cingulum NeuroSciences Institute, Winston-Salem, North Carolina 27101; and Departments of Physiology, Pharmacology, and Neurology, Wake Forest University School of Medicine, Winston-Salem, North Carolina 27157-1083

**Mark J. West** Department of Neurobiology, University of Aarhus, 8000 Aarhus C, Denmark

# Preface

This volume of *Cerebral Cortex* is dedicated to Sir John Eccles, who was an active member of the advisory board for the series until his death in May 1997. His input as to what topics should be covered in future volumes of this series will be sorely missed.

The present volume is concerned with neurodegenerative disorders and age-related changes in the structure and function of the cerebral cortex, a topic that has attracted increasing interest as longevity and the number of aged individuals in the population increase. Although much of the research on the neurodegenerative effects of aging has been centered on Alzheimer's disease, most of the aging population will not be afflicted by this disease. They will, however, be affected by the consequences of normal aging, so the first few chapters of this volume are concerned with that topic.

Chapter 1, by Marilyn S. Albert and Mark B. Moss, gives an account of the cognitive changes that accompany normal human aging. Chapter 2, by Mark B. Moss, Ronald J. Killiany, and James G. Herndon, describes the cognitive decline that occurs in the rhesus monkey, a primate that provides a very useful model for ascertaining the effects of normal aging because rhesus monkeys apparently do not suffer from Alzheimer's disease. Some of the structural changes that take place in the cerebral cortex as a consequence of normal aging are described by Alan Peters in Chapter 3, and in both this chapter and in Chapter 4 by Mark J. West, it is noted that neuronal loss during normal aging is less widespread than originally thought. There are, however, structural and chemical alterations that do occur during normal aging, and one of these, discussed by Amy F. T. Arnsten in Chapter 5, is alteration in the catecholamines that are crucial to the cognitive functions of prefrontal cortex in normal aging. The changes are not confined to neurotransmitters, however, and in Chapter 6, Douglas L. Rosene and Timothy J. Nicholson examine concomitant alterations in neurotransmitter receptors.

The role of a different kind of chemical mediation, brought about by trophic factors, is discussed in Chapter 7, by Daniel A. Peterson and Fred H. Gage, in which they discuss endogenous brain factors that could prevent injured neurons from dying, as displayed in experimental models of injury. Chapter 8, by Patrick R. Hof,

Constantin Bouras, and John H. Morrison, provides a bridge between the effects on the cortex of normal aging and those of dementing disorders. They describe how some populations of neurons seem more selectively vulnerable to the effects of aging and disease than do others. In Chapter 9 Gene J. Blatt and Thomas L. Kemper present a detailed analysis of the anatomy of dementias, in which they conclude that the various, diverse dementias share a common involvement of the frontal lobes, with differing patterns of involvement of the other cortical lobes and of subcortical areas. In Chapter 10, Thomas L. Kemper further discusses the involvement of subcortical nuclei in normal aging and in degenerative changes of the brain. Chapter 11, the final chapter in this section, by Imaharu Nakano and Asao Hirano, describes the ultrastructural features of altered cells and inclusions found in dementing illnesses.

Before presenting the series of chapters focusing on Alzheimer's disease, there is an interesting contribution in Chapter 12 by Lindsay A. Farrer that deals with a different aspect of neurodegenerative disease: the detection of genetic modifiers through an analysis of clinical and genotypal information, so that ultimately the genetic bases of disorders can be determined and more-effective therapies developed.

The first of the chapters dealing with Alzheimer's disease, Chapter 13 by Marilyn S. Albert and Mark B. Moss, gives an account of the early behavioral changes that typify the disease. This chapter makes the point that even in the early stages, Alzheimer's patients show a dramatic decline in their ability to recall information after brief delay intervals, a difficulty the authors attribute to changes in entorhinal cortex. Indeed, it is the allocortex that seems to be first affected in Alzheimer's disease. A detailed account of the stages of progression in Alzheimer's disease is given in an extensive review of its pathology in Chapter 14, by Heiko Braak and Eva Braak. The correlations between pathology and dementia are also considered in Chapter 15, by Eliezer Masliah and David P. Salmon, but they emphasize alterations occurring at the synapses, and especially in the presynaptic terminals, where molecules associated with amyloid may lead to abnormal synaptic functioning. Chapter 16, by Brent A. Vogt, Alex Martin, Kent E. Vrana, John R. Absher, Leslie J. Vogt, and Patrick R. Hof, further emphasizes that Alzheimer's disease is a heterogeneous disease, possibly with subtypes. For example, not all patients display the same cognitive dysfunctions, and although neurodegeneration and atrophy occur consistently in medial temporal areas, there is substantial variation in the degree to which the prefrontal, parietotemporal, and occipital cortices are affected.

The molecular pathology of Alzheimer's disease is further examined in Chapter 17 by Stephen D. Ginsberg, M. Luise Schmidt, Peter B. Crino, James H. Eberwine, Virginia M.-Y. Lee, and John Q. Trojanowski, which looks at the genetic risk factors associated with the disease and the bases of the structural abnormalities. As these authors point out, understanding such abnormalities is crucial for elucidating the mechanisms that underlie neurodegeneration, and a more thorough understanding of these mechanisms will lead to new therapies for prevention or treatment. One possible mechanism underlying the neuropathology of Alzheimer's disease is the excitotoxic effect of glutamate, particularly with respect to the large pyramidal cells of the cortex, considered by James W. Geddes in Chapter 18, along with the possibility that changes in glutamate receptors account for the well-recognized cognitive deficits. Another possibility, explored in Chapter 19 by E. J. Mufson and J. H. Kordower, is that alterations in the expression of nerve growth factor could underlie the pathophysiology of the cholinergic basal forebrain neurons, which are known to degenerate in Alzheimer's disease. It has been suggested that this degeneration contributes to the cognitive decline, and these authors have shown decreased nerve growth factor activity in the cholinergic neurons, even though the levels are not altered in their target sites.

Chapter 20, the final chapter in this volume, by D. Larry Sparks, gives an account of vascular complications and alterations in Alzheimer's disease and discusses indications that heart disease and hypertension might contribute to conditions that predispose subjects to Alzheimer's disease.

Taken as a whole, these contributions give a picture of the cognitive changes and related pathology that occur in both normal aging and in dementias, with a particular focus on Alzheimer's disease. Some of the factors that bring about these alterations are also considered. The hope is that these factors will eventually be well enough understood to at least retard the onset of the overwhelming cognitive decline facing Alzheimer's disease patients, and perhaps even to prevent the more common though less devastating impairment that often accompanies normal aging.

Alan Peters
John H. Morrison

*Boston and New York*

# Contents

*Chapter 9*

**The Anatomy of Dementias**

*Gene J. Blatt and Thomas L. Kemper*

*Chapter 10*

**Age-Related Changes in Subcortical Nuclei that Project to the Cerebral Cortex**

*Thomas L. Kemper*



*Chapter 11*

**Ultrastructural Changes in Dementing Illnesses**

*Imaharu Nakano and Asao Hirano*

*Chapter 12*

**Locating Genetic Modifiers for Inherited Neurodegenerative Diseases**

*Lindsay A. Farrer*



*Chapter 13*

**Early Features of Alzheimer's Disease**

*Maryilyn S. Albert and Mark B. Moss*



*Chapter 14*

**Temporal Sequence of Alzheimer's Disease-Related Pathology**

*Heiko Braak and Eva Braak*

*Chapter 15*

**Neuropathological Correlates of Dementia in Alzheimer's Disease**

*Eliezer Masliah and David P. Salmon*



*Chapter 16*

**Multifocal Cortical Neurodegeneration in Alzheimer's Disease**

*Brent A. Vogt, Alex Martin, Kent E. Vrana, John R. Absher, Leslie J. Vogt, and Patrick R. Hof*

*Chapter 20*

**Vascular-Related and Mediated Alterations in Alzheimer's Disease**

*D. Larry Sparks*

# CEREBRAL CORTEX

Volume 14

Neurodegenerative and Age-Related Changes in Structure and Function of Cerebral Cortex

# 1

# Cognitive Profiles of Normal Human Aging

MARILYN S. ALBERT and MARK B. MOSS

## 1. Introduction

Age-related changes in cognition are evident in a number of cognitive domains. However, recent evidence indicates that it is not the result of widespread neuronal loss, as previously thought. Instead it appears that other neurobiologic factors are primarily responsible for age-related changes in cognition, including molecular changes within memory circuits and focal neuronal loss within subcortical nuclei that project widely to cortex.

This chapter begins with an overview of our current knowledge of age-related cognitive changes within six domains: memory, executive function, visuospatial processing, language, attention, and general intelligence. The emphasis is on the age-related changes in cognition that one sees in the absence of clinical disease. With respect to the human studies, a particular emphasis is placed on excluding subjects in the early stages of Alzheimer's disease (AD). Wherever possible, studies are also cited wherein the impact of medical diseases common in the elderly (e.g., hypertension, respiratory or cardiac disease, vitamin deficiency), has also been excluded, since these diseases may impair cognition as well. Individuals free of clinical disease can be of heuristic value, although nonrepresentative of the average

MARILYN S. ALBERT • Departments of Psychiatry and Neurology, Massachusetts General Hospital, Harvard Medical School, Boston, Massachusetts 02118. MARK B. MOSS • Department of Anatomy and Neurobiology, Boston University School of Medicine, Boston, Massachusetts 02118.
*Cerebral Cortex,* Volume 14, edited by Peters and Morrison. Kluwer Academic/Plenum Publishers, New York, 1999.

aging individual. Examining them permits differentiation of disease-related changes from those related to age.

There follows an overview of age-related changes in brain structure and function, with an emphasis on neuroimaging and morphological findings in humans. Wherever possible, comparisons are drawn between human data and recent findings from nonhuman primate models of human aging. These comparisons are instructive for a number of reasons. First, if the pattern of performance within a cognitive domain is similar in both humans and monkeys, the likelihood that the findings are accurate is strengthened. Second, an examination of carefully selected autopsy tissue in monkeys and humans can likewise serve to address, and potentially reinforce, similar conclusions. Third, carefully controlled lesion studies in monkeys can shed light on potential structure–function relationships in humans that cannot easily be evaluated by other means.

## 2. Domains of Cognitive Function

Although one could fractionate cognitive function into an almost unlimited number of components, most investigators in the field of neuropsychology view higher cortical function as composed of a relatively small number of major categories. For the purposes of this chapter, we discuss cognitive function within the following six domains: memory, executive function, visuospatial processing, language, attention, and general intelligence. This chapter will discuss all six, but will emphasize memory, executive function, and visuospatial processing because these have been extensively examined in both humans and monkeys.

## 3. Memory

Workers in the field of memory have accepted the conclusion that memory is not a unitary phenomenon, and most models of memory function hypothesize that memory consists of a series of specific yet interactive stories (e.g., Waugh and Norman, 1965; Tulving, 1972).

### 3.1. Types of Memory

*Sensory memory.* Sensory memory, or registration, represents the earliest stage of information processing. It is modality-specific (i.e., visual, auditory, tactile), highly unstable, and characterized by rapid decay.

*Primary/immediate memory.* The component of the memory system referred to as primary or immediate memory permits one to hold spans of auditory and/or visual information for relatively long periods of time by active rehearsal. The ability to concentrate on, rehearse, and recall a span of digits, words, or visual features is perhaps the best example of this capacity. Any disruption to the rehearsal process results in the information being lost from immediate memory. Experiments by

Petersen and Petersen (1959) have demonstrated that normal subjects forget a significant proportion of new information in less than one minute when distractions are present. The amount of information that immediate memory can store is limited to about five to seven items, as mentioned earlier.

In recent years a number of researchers have begun to include in this aspect of memory the concept of "working memory" (e.g., Baddeley, 1992). This theoretical construct identifies a system for the temporary holding and manipulation of information over a range of cognitive tasks. It also emphasizes the fact that there must be a dissociation between verbal domains within immediate or working memory (i.e., the "phonological loop" vs the "visuospatial scratchpad") that can be differentially affected by disease.

*Secondary memory.* To be retained over a long period of time, information from immediate memory must be assembled into multimodal units to be placed in storage. Storage of information by the memory system appears to take place differentially. As early as 1949, Hebb postulated that two processes were necessary for the brain to retain information. The first process, analogous to what we have termed recent memory, required the continual reverberation of a neural circuit (Hebb, 1949). The second process, equivalent to secondary memory, required an actual structural change in the neural pattern of the central nervous system.

*Explicit versus implicit secondary memory.* Over the past few years, accumulating evidence has suggested that memory functions are composed of at least two distinct types of memory, explicit and implicit. *Explicit memory* pertains to information that is acquired because a person makes a conscious effort to learn and retain it. *Implicit memory* pertains to information acquired without conscious effort. Implicit memory, therefore, is said to refer to "knowing how," or to "procedural" or "habit" memory (Ryle, 1949; Hirsh, 1974; Cohen and Squire, 1980) and is accessible primarily through performing a task or engaging a skill in which the knowledge is embedded (Squire, 1986).

### 3.2. Explicit Memory Performance and Aging

There is considerable information to indicate that changes in explicit sensory memory are minimal with age. For example, the time necessary to identify a single letter does not change significantly between the late teens and the early 70s. When seven-letter strings are used, the rate of letter identification increases with age by a factor of 1.3 (Cerella, Poon, and Fozard, 1982). A partial report paradigm showed that letter identification latencies change minimally with age when adjusted for the loss of visual discrimination experienced by the elderly. These and other data indicate that there is a minimal decline in sensory memory with age (see Craik, 1977; Poon, 1985, for detailed reviews).

Primary explicit memory also shows little decline with age. Most studies report no significant age differences in digit-span forward (Drachman and Leavitt, 1972), no age differences in word span (Talland, 1967), and only moderate differences in letter span (Botwinick and Storandt, 1974). Older subjects show as much of a recency effect (i.e., retrieval of the last few items on a list in a word list learning task) as younger subjects (Raymond, 1971).

However, there are substantial changes in explicit secondary memory, in contrast to the minimal age changes in sensory and primary memory (see Craik, 1977; Poon, 1985 for reviews). The age at which changes in secondary memory occur depends upon the methods that are used to test the memory store. Difficult explicit memory tasks (e.g., delayed recall) demonstrate statistically significant differences by subjects in their 50s, in comparison to younger individuals (Albert *et al.,* 1987a). Age decrements are greater on recall than recognition tasks. This is true whether words or pictures are used. Cueing during encoding or retrieval also alters the appearance of an age decline. Cueing at both encoding and retrieval produces the smallest age differences, whereas no cueing at either stage of the task produces the largest differences (Craik *et al.,* 1987). However, even with cued recall and recognition, there are often declines. Rabinowitz (1986) reported a 33% age-related decrement in cued recall and an 11% age-related decrement in recognition when comparing young and old subjects (mean age 19 vs 68).

Fig. 1 shows the performance of subjects across the age range on delayed recall of two lengthy paragraphs. That is, subjects are read the two paragraphs. Immediately after hearing each one, and then again after 20 minutes, subjects are asked to state what they can recall of the paragraphs.

A close examination of these data indicates that the older individuals are not more rapidly forgetting what they learned, but rather they are taking longer to learn it. For example, if one compares the difference between immediate and delayed recall over the life span, there are no statistically significant age differences (Petersen *et al.,* 1992). Thus, if one allows older subjects to learn material well (i.e., to the point where few errors are made), they do not forget what they have learned more rapidly than the young. However, if not given the time to learn material to the same level of proficiency as younger individuals, the average older person, after a delay, will retain less information than younger persons.

However, there is considerable variability among older subjects on tasks of this sort. There are numerous healthy older subjects who have test scores that overlap

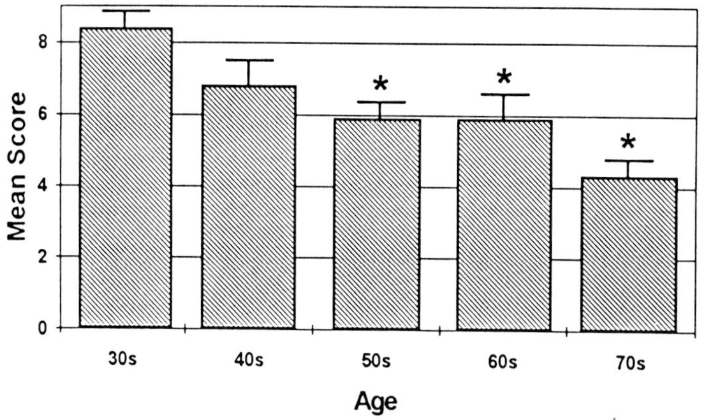


**Figure 1.** Delayed-recall performance of subjects 30 to 80 years old. The subjects are asked to report what they remember of two lengthy paragraphs after a 20-minute delay. There is a statistically significant difference between the decade groups, beginning with the subjects 50 and older. Reprinted from Peterson *et al.,* 1992, with kind permission of Elsevier Science, NL, Sara Burgerhartstraat 25,1055 KV, Amsterdam, The Netherlands.

those of subjects many years younger than themselves (e.g., about one-third of healthy 70-year-old humans have delayed-recall scores that overlap those of 30 year olds [equated for education]).

Similar findings have emerged from studies in monkeys. Since free recall cannot be easily tested in monkeys, considerable effort has gone into developing a memory task that uses recognition and also determines the quantity of information aging monkeys can retain across varying delay intervals. The delayed nonmatching-to-sample tasks (DNMS) is the most widely used method for assessing recognition memory in the nonhuman primate (Bachevalier and Mishkin, 1989; Murray and Mishkin, 1984, 1986; Mishkin, 1978; Mahut *et al.,* 1982; Zola-Morgan and Squire, 1985). This task relies upon a two-alternative forced-choice paradigm wherein the monkey is required to determine which of two objects was recently presented. By using the delayed nonmatching-to-sample task, many studies have shown that aged monkeys are impaired at learning the nonmatching principle but are, at best, only mildly impaired across the delay conditions of the test (Moss *et al.,* 1988; Arnsten and Goldman-Rakic, 1990; Bachevalier *et al.,* 1991; Presty *et al.,* 1987; Rapp and Amaral, 1989). Thus, like humans, the old monkeys take longer to learn something new but do not appear to forget this information more rapidly over lengthening delays than do younger monkeys.

Another task, the delayed-recognition-span test (DRST) (Rehbein, 1985; Moss, 1983), has demonstrated similar findings. This test requires a monkey to identify a novel stimulus from an increasing array of previously presented stimuli. The task has been administered with two types of conditions: a spatial version and an object version. In both conditions the monkey first sees a board with 18 positions (food wells) and a stimulus object on one position. A second stimulus is added while the board is obscured from view, and the animal is required to point to the new stimulus. On the first trial of the spatial version, the animal's task is to indicate which of two disks is occupying the new position on the board (i.e., a nonmatching-to-position task). On the first trial of the object version, the animal's task is to indicate which of two objects on the board is the new one (i.e., in essence, a nonmatching-to-sample task). However, since new disks/objects are then added in series to the previous ones on the board, the task becomes increasingly difficult. For example, young adult monkeys often achieve a span of five or greater before committing an error.

Monkeys aged 19–24 years (equivalent to humans aged approximately 57–72) are impaired on the spatial version of the DRST but not on the color version (Killiany *et al.,* submitted). This finding is similar to the declines in performance on delayed recall seen in middle-aged humans. Monkeys aged 25–29 years (equivalent to humans aged approximately 75–87) are impaired relative to young adult monkeys aged 5–7 years (equivalent to humans aged approximately 15–21) under both conditions of the task (Moss *et al.,* 1997). Monkeys over the age of 29 (equivalent to humans over the age of 87) are also impaired on this task (Moss *et al.,* this volume; Arnsten, this volume). This suggests that the performance of monkeys on the spatial version of the DRST may be functionally equivalent to the performance of humans on difficult delayed-recall tasks. In addition, the difference between performance on the DNMS and the spatial condition of the DRST is similar to what one sees between recall and recognition paradigms in aging humans.

Longitudinal assessment of memory using the DRST has also been examined

(Moss, 1993; see also Moss *et al.*, this volume). Young adult (5–7 years) and aged (23–26 years) monkeys were assessed on the spatial-recognition-span test after an interval of 4 years. Two of the three young animals showed a slight decline in performance, and one showed a slight improvement. In the aged group, two animals showed a very slight decline (as with the young) but the remaining two showed marked declines which, in one case, were severe. Thus, the longitudinal findings confirm the general conclusions of the cross-sectional data.

Like aging humans, aging monkeys also show considerable variability in performance. Among the oldest animals, some perform within the range of younger animals, even though the mean performance of the aged group declines significantly. For example, on DNMS, the number of trials-to-criterion ranges from 50 to 220 and the number of errors ranges from 29 to 60. Among the older monkeys, the number of trials-to-criterion ranges from 200 to 516 and the number of errors ranges from 50 to 115.

It should be noted that there is a long-standing and large body of research indicating that aged rodents, compared to young, are impaired on a wide variety of spatial memory tasks (e.g., Woodruff-Pak, 1990; Barnes, 1988). Old rodents, like old monkeys, show considerable variability in performance.

Lesion studies in monkeys indicate that the medial temporal region of the brain and/or diencephalic midline structures are essential for normal memory. A lesion of the hippocampus and the amygdala produces a profound defect in a monkey's ability to learn the nonmatching-to-sample principle on the DNMS and to retain correct information over delays of increasing length (e.g., Zola-Morgan and Squire, 1985). Likewise, damage to the mediodorsal nucleus of the thalamus produces a striking memory deficit on the DNMS (e.g., Aggleton and Mishkin, 1983). The impairment is present in the monkeys' ability to learn information (e.g., 140 trials-to-criterion for controls versus 315 for monkeys with mediodorsal thalamic lesions) and to retain the information at long delays (e.g., 80% at 10 min for controls vs 60% for lesioned monkeys). It is noteworthy that the deficits following these lesions are more severe than those seen in age-related declines.

Lesion studies in monkeys also indicate that the degree of memory loss following lesions of this type is related to the combination of structures damaged and also to the location of the lesion within the critical structures. For example, combined damage to the hippocampus and amygdala produces a more severe deficit than a lesion to either structure alone (e.g., Zola-Morgan and Squire, 1985). Similarly, a lesion in the posterior mediodorsal nucleus produces a more severe memory defect than one in the anterior region (Aggleton and Mishkin, 1983). Thus, age-related changes likely produce alterations within this memory network, but the location and extent are not yet clear.

Moreover, several neurotransmitters and hormones can have a modulatory effect on memory performance. For example, the basal forebrain is the primary source of neocortical acetylcholine (Mesulam and Van Hoesen, 1976), and the locus coeruleus is the major source of neocortical norepinephrine (Moore and Bloom, 1979), both of which modulate memory performance. Lesions of the three nuclei within the basal forebrain (the nucleus basalis of Meynert, the medial septal nucleus, and the diagonal band) produce a deficit of DNMS performance in young monkeys, which recovers over time (Aigner *et al.*, 1991). It has been hypothesized that recovery is related to compensatory responses in other neuromodulatory mechanisms; however, if these are also compromised, recovery might not occur. Thus, to determine the cause(s) of age-related memory deficits in monkeys, each aspect of the memory network must be evaluated in carefully selected animals, and comparisons must be made to human data. Given the difficulty of this task, the answers are not available.

### 3.3. Implicit Memory Performance and Aging

Age-related changes in implicit memory have been examined in a variety of ways. One method uses a "priming" technique in which learning is inferred by examining the change in items retained after a single exposure to a set of items (Graf and Schacter, 1985). For example, if one has been asked to read an inverted word once, the speed with which one will read that word a second time increases. Thus, learning is inferred by the change in speed. Commonly used priming paradigms include stem-completion and partial-word identification. The majority of studies indicate that there is no significant decline in priming ability with advancing age (see Graf, 1990; Howard, 1988, for reviews).

Skill learning is another paradigm used to examine implicit memory. Learning is inferred by examining the change in amount retained after multiple exposures to a procedure (e.g., rotary pursuit, reading inverted words, partial-word identification) rather than the change in recall of a specific item. Most of the skill-learning studies have examined the improvement in visual skills over time, such as reading inverted words or identifying fragmented pictures. The results published to date indicate that age-related impairments in skill learning appear to depend upon whether or not there is a primary difficulty in processing the information on which training occurs. For example, if inverted words are exposed for a longer duration in the older subjects than in the younger ones, age-related differences are minimal; or, if words are degraded by 37% rather than by 50%, age-related differences are again minimal (Hashtroudi *et al.,* 1991). There are little data on motor skill learning, but those that exist suggest that it is impaired with age (e.g., Harrington and Haaland, 1992).

Classical conditioning, on the other hand, shows significant declines with age (Solomon *et al.,* 1989). This is typically examined by observing the change in response to stimulus-1 that comes after coupling stimulus-1 and stimulus-2 together (e.g., eye-blink in response to a tone, after the tone has been coupled with a puff of air to the eye). It has been hypothesized that these declines are related to alterations in neurotransmitter levels in subcortical nuclei (discussed further below).

In monkeys, initial evidence suggests that procedural learning, at least with respect to motor skill learning, is only minimally affected by age. The performance of aged monkeys on a timed task, in which a piece of hard candy is moved along a metal form with a varying number and angle of curves, did not differ significantly either in latency or trials-to-criterion, as compared with young adult monkeys (Ronald, 1992).

Lesion studies in monkeys and findings in humans with brain damage indicate that the brain regions responsible for implicit memory are different from those responsible for explicit memory. Monkeys with lesions of the medial temporal lobe (i.e., hippocampus and amygdala) perform in the normal range on motor skill tasks, including the hard-candy task described earlier (Zola-Morgan and Squire,

1984). Likewise, humans with extensive lesions of the medial temporal lobe that produce a dramatic explicit memory deficit can learn new motor skills (e.g., Corkin, 1968). By contrast, monkeys with lesions of the nucleus accumbens are impaired on the hard-candy task but not on tests of explicit memory, including the DNMS task (Killiany and Mahut, 1992). The nucleus accumbens is continuous with the caudate and putamen in the monkey; thus, these findings implicate the striatum in motor skill learning. On the basis of data from patients with Huntington's disease, it has also been hypothesized that the striatal network is involved in motor skill learning (Heindel *et al.,* 1989). Since an impairment in motor skill learning with age has been reported, its origin may be related to alterations in the striatum.

## 4. Executive Function

The complex set of abilities sometimes referred to as "executive functions" include concurrent manipulation of information (e.g., cognitive flexibility), concept formation, and cue-directed behavior. The wide variety of abilities sometimes included under the term "executive function" is therefore striking.

Tests evaluating concept formation and set-shifting uniformly show significant changes with age, primarily when subjects are in their late 60s or 70s. For example, the similarities subtest of the Wechsler Adult Intelligence Scale (WAIS), which asks subjects to identify how two objects (e.g., a table and a chair) are similar to each other, is the subtest on the verbal scale of the WAIS that shows the greatest decline with age (Heaton *et al.,* 1986). Education appears to be a modifier of this decline in that subjects with fewer years of education demonstrate declines at younger ages; however, all subjects in the oldest group (mean age 68 years), regardless of educational level, show significant declines in performance.

Series-completion tests also show substantial age declines. These tests generally require the subject to examine a series of letters or numbers and determine the rule that governed the sequencing of the items. Cross-sectional and longitudinal data demonstrate age-related declines on tasks of this sort (e.g., Lachman and Jelalian, 1984; Schaie, 1983).

Proverb-interpretation tests, which require the subject to provide the general meaning of a proverb (e.g., "barking dogs seldom bite"), also demonstrate age-related declines (Albert *et al.,* 1990). This is true whether subjects are asked to provide the meaning of the proverb themselves or are given alternatives from among which to choose. Similarly, set-shifting tasks, such as the visual–verbal test (in which subjects are asked to look at a series of cards and indicate how three of the four objects on each card are alike in one way, and then how three of the objects are alike in another way also show substantial age-related declines. These changes appear to be related to the fact that older subjects have difficulty switching from one abstract answer to another (i.e., they tend to get the first item in the set correct but the second one wrong). Slowness in establishing mental set (i.e., getting the first item in the set wrong but the second one right) and failure to establish set (i.e., getting both items in the set wrong) did not increase differentially with age (Albert *et al.,* 1990).

Changes in executive function in monkeys have been examined primarily by using reversal-learning paradigms (Bartus *et al.,* 1979; Rapp, 1990; Lai *et al.,* 1995). Reversal learning involves responding to a change in reinforcement contingencies by first "unlearning" or breaking the initial stimulus–reinforcement bond, then acquiring, or "shifting," to a new one. In this way, reversal learning can be considered a measure of executive function. Data have accumulated to suggest that aged monkeys have difficulty in reversing established stimulus–reward contingencies, particularly when based on spatial location. Thus, compared to young adult monkeys, aged monkeys as a group are impaired on spatial-reversal, but not on object-reversal, learning. Moreover, aged monkeys tend to make more perseverative responses than do young adults on both spatial-reversal and object-reversal learning.

Lesion studies in monkeys indicate that reversal-learning ability is related to frontal lobe function. Pohl (1973) administered object- and spatial- ("place") reversal tasks to monkeys that underwent surgical ablation of the dorsolateral prefrontal region. Compared to monkeys with lesions of either the inferior temporal, or the parietal, cortex, monkeys with frontal lesions were markedly impaired on spatial-, but not object-reversal, learning. Classically, damage to the dorsolateral prefrontal cortex (DLPC) results in a spatial working-memory deficit demonstrated by monkeys on spatial delayed-response performances. Like the spatial delayed response, the spatial-reversal test may preferentially engage DLPC function because (1) internal versus external informative stimuli serve as cues for the correct response, and (2) this informative stimulus must be held "on line" for brief periods of time and continuously updated. Thus, the age-related changes seen in spatial-reversal learning suggest that prefrontal regions are altered with age and are responsible for executive function changes.

Direct behavioral support implicating the prefrontal region in age-related change is provided by Rapp and Amaral's finding that the spatial delayed-response performance by aged monkeys was impaired relative to that of young adults, and that spatial-reversal learning showed larger age-related declines than did performance on delayed nonmatching-to-sample tests (Rapp and Amaral, 1991). Bachevalier *et al.* (1991) corroborated these results and reported declines on spatial delayed-reversal tasks in earlier age groups than those for delayed nonmatching-to-sample tasks in a cross-sectional primate study.

## 5. Visuospatial Function

Visuospatial function is characterized by the ability to produce and recognize figures and to form relationships among spatial locations. Specific visual functions include the ability to recognize familiar faces, the ability to copy or match objects or pictures, and the ability to translate spatial elements from one mode to another. Translation of mirror-image spatial arrangements into self-oriented positions is an example of a task that requires intact and efficient visuospatial abilities.

Visuospatial ability therefore can be assessed by (1) constructional tasks, such as the assembly of blocks, sticks, or puzzles; (2) drawing tasks that involve copying; or (3) matching tasks that require the subject to identify pictures with similar elements.

The most complex three-dimensional construction task in common use is the block design subtest of the WAIS. The subject is presented with a two-dimensional drawing in red and white of a target design, and a set of blocks (some sides of the blocks are all red, some sides are all white, and some sides are half red and half white). The subject is asked to arrange the three-dimensional blocks so that they mimic the two-dimensional design. To receive credit the subject must assemble the blocks correctly within a specified time limit. This task shows substantial declines with age (Doppelt and Wallace, 1955).

Performance on figure-drawing tasks is also affected by age. Older subjects are impaired, compared with the young, in depicting and perceiving three-dimensional drawings. Plude *et al.* (1986) asked a group of young and old adults (mean ages 21 and 67, respectively, and equated for static visual acuity) to draw a cube to command. The cubes were then rated by 10 independent raters, with an interrater reliability of 0.98. The drawings of the young adults were rated as significantly better than those of the old. The older subjects were also less accurate than the young in judging the adequacy of cube drawings that were distorted to varying degrees. The elderly were less accurate than the young in discriminating between distorted and undistorted cubes. They were, however, equally able to copy a cube when landmarks were provided regarding the size of the lines. Comparable reports of the depiction and perception of two-dimensional drawings are not available.

Spatial abilities in monkeys also show change with age and, as with humans, the presence or degree of impairment appears related to the demand characteristics of the task. Lai *et al.* (1995) showed that aged monkeys were impaired on a spatial task that required the ability to use allocentric rather than egocentric spatial cues. Monkeys were first required to learn a simple two-choice spatial-discrimination task by displacing, on a small table top, one of two identical plaques set either to the left or to the right of the animal, to a 90% learning criterion. Once the criterion was reached, each animal was wheeled to the other side of the table so that, with respect to the animals' body position, the previously correct plaque was now reversed. Thus, an animal that learned first to displace the plaque to its left now had to displace the plaque to its right. Unlike young monkeys, aged monkeys evidenced a strong tendency to choose the plaque that retained the same location with reference to body position (egocentric cue) rather than to choose the plaque that retained the same location with reference to other external cues in the environment (allocentric cues). There is a significant impairment in spatial-reversal learning performance as a function of aging. In an increasing fashion, early-aged (19–24 years), advanced-aged (25–29 years), and oldest-old monkeys (30 years and older) were impaired when reward contingencies based on spatial position were reversed. Early-aged monkeys were impaired only on the initial reversals, whereas the oldest-old were impaired on all reversals. In contrast, there was no significant difference between young monkeys and any aged monkey group in initially learning to choose the left or the right plaque.

## 6. Language

Linguistic ability is thought to encompass at least four domains: phonological, lexical, syntactic, and semantic. It was previously assumed that linguistic ability is preserved into very old age, primarily because performance on the vocabulary subtest of the WAIS, the best general estimate of verbal intelligence, is well maintained until individuals are in their 80s (Schaie, 1983). However, within the last decade, a number of studies have shown that some aspects of linguistic knowledge decline with age, although not until relatively late in life (i.e., >70).

*Phonology.* Phonologic knowledge refers to use of the sounds of language and the rules for their combination. Phonologic capabilities are well preserved with age (Bayles and Kaszniak, 1987).

*Lexicon.* Psycholinguists distinguish between the lexical representation, that is, the name of an item, and semantic representation, that is, the meaning of a word (Clark and Clark, 1977). The lexicon of healthy older individuals appears to be intact, as are the semantic relationships of the lexicon.

*Syntax.* Syntactic knowledge refers to the ability to meaningfully combine words. A large number of studies have shown that age has little effect on syntax. Obler *et al.* (1985) found that syntactic forms that were difficult for older individuals were also difficult for younger ones.

*Semantic knowledge.* Older individuals appear to have difficulty with the semantic aspects of word retrieval. Several groups of investigators have reported that scores on a test of confrontation-naming decrease with age (Borod *et al.,* 1980; LaBarge *et al.,* 1986; Albert *et al.,* 1987b). However, declines in naming ability do not become statistically significant until subjects are in their 70s (Albert *et al.,* 1987b). When subjects could not correctly name an item, the most common error they committed was semantic in nature, that is, they produced semantically related associates, circumlocutions, and nominalizations. The nature of these semantic errors suggests that older individuals have a great deal of knowledge about the target word. For example, a semantically related associated ("dice" for "dominoes") can only be produced if the subject comprehends the general category associated with the stimulus item.

Verbal fluency also assesses semantic ability. In a verbal fluency task, a subject is asked to name as many examples of a category (e.g., animals or vegetables) as possible in specified period of time (e.g., 1 min) or as many words as possible beginning with a particular letter (e.g., *F*) within a specified period of time. Several studies report a decline in verbal fluency with age (Obler *et al.,* 1985; Albert *et al.,* 1987b). These changes also occur relatively late in the life span (>70). Thus, semantic linguistic ability appears to change with advancing age, while other aspects of linguistic ability are relatively well preserved.

## 7. Attention

The concept of attention has been discussed since the time of William James (1890). It is often thought to encompass at least three interrelated aspects: sustained attention (vigilance), selective attention (extracting relevant from irrelevant information), and attentional capacity (total attentional resources available to an individual) (see Parasuraman and Davies, 1984; Hasher and Zacks, 1979, for reviews). Some investigators also include within the area of attention the complex set of abilities sometimes referred to as "executive functions". For the purposes of the present discussion, the term "attention" will be used to describe sustained attention only. Executive function ability is discussed in another section of this chapter.

Sustained attention assesses the individual's ability to focus on and perform a simple task without losing track of the task objective. Memory demands are minimized in tests of sustained attention by limiting the information that must be remembered to material that falls within a person's immediate memory span (i.e., $7 \pm 2$). Digit-span forward is the most commonly used test of sustained attention and is included in many tests of general intelligence (e.g., WAIS-R and Wechsler Memory Scale-Revised [WMS-R]). Visual or auditory continuous-performance tasks that require the individual to identify a repeating letter (e.g., *A*) or a repeated letter sequence (*I* before *X*) (Mirsky, 1978) are other common means of evaluating sustained attention.

Many studies have demonstrated that sustained attention is well preserved with age. A wide variety of studies suggest that there is less than one standard deviation (SD) of change between 20 and 80 years of age (e.g., Spitz, 1972).

It is only recently that tests of sustained attention have been directly assessed in studies of aged monkeys. Adapting attentional tasks from those used in humans by Posner (Posner, 1980; Posner and Cohen, 1984), Baxter and Voytko (1996) showed that aged monkeys were as proficient as young adult monkeys at shifting attention to a peripherally cued spatial location. In addition, if one assumes that sustained attention is a prerequisite for the ability to learn simultaneous-discrimination tasks, results from this testing paradigm can be also used to address this issue. Aged monkeys are as efficient as young adults in their ability to solve two-choice discrimination problems using either pattern (Moss *et al.*, 1988) or spatial (Lai *et al.*, 1995) stimuli. Thus it seems reasonable to conclude that, as in humans, aging in monkeys does not impair simple or sustained attentional processes.

## 8. General Intelligence

Although intelligence tests measure most of the abilities previously discussed, they do so in a complex way. Intelligence tests were designed to predict with a reasonable degree of certainty how a person would function in an academic environment, not to provide a complete assessment of cognitive function. Thus, intelligence tests do not assess all aspects of cognitive ability. For example, the Wechsler Adult Intelligence Scale does not include an evaluation of memory. In addition, IQ tests do not assess cognitive abilities in relative isolation from one another. Many of the tasks require a complex interaction of abilities and often depend upon speed for an adequate level of performance. Nevertheless, intelligence testing has been one of the most explored topics in the psychology of aging.

There is widespread agreement that intelligence test performance declines with age. There has, however, been considerable debate about when such declines occur and about their magnitude. The age at which declines are observed depends upon the methodology employed. There is some consensus that relatively little decline in performance occurs before people are in their 50s (Schaie and Labouvie-Vief, 1974). After this age, results differ depending upon whether cross-sectional or longitudinal testing designs are employed. The cross-sectional method shows declines of 1 SD or more beginning about age 60 (Doppelt and Wallace, 1955; Schaie, 1983); over the age of 70, scores drop sharply. The longitudinal method shows declines beginning in the late 60s. Both methods find substantial declines after individuals reach their mid-70s. Thus, the major difference between cross-sectional and longitudinal investigations is observed between subjects in their late 50s and those in their early 60s. In this age range, the cross-sectional method shows greater age declines than the longitudinal method.



## 9. Age-Related Changes in Brain Structure and Function

There is substantial evidence that as people age, they show significant increases in cerebrospinal fluid (CSF) and decreases in brain tissue (Stafford *et al.*, 1988; Zatz *et al.*, 1982); that is, there is increasing atrophy with age. This alteration becomes statistically significant when subjects are in their 70s (Stafford *et al.*, 1988).

However, recent studies in humans suggest that the decrease in brain tissue is primarily a decrease in white matter (Guttmann *et al.*, in press; see Figure 2). Decreases in gray matter are only marginally statistically significant; approximately 48% of brain tissue consists of gray matter among both 30 year olds and 70 year olds (Guttmann *et al.*, in press). These findings are consistent with recent data in hu-

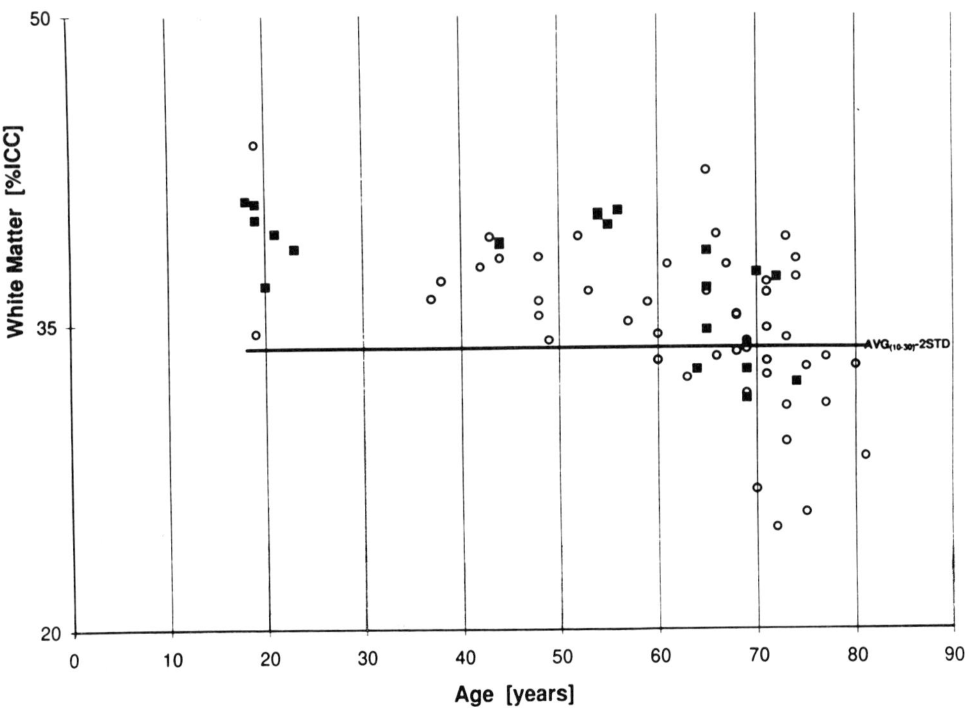


**Figure 2.** White matter volume as a percent of intracranial cavity volume (%ICC) for male (filled squares) and female (open circles) across the age range. A horizontal line is drawn 2 SD below the mean for the 20-year-old subjects. Note that only subjects 60 and older fall below the horizontal line, graphically demonstrating differences in the volume of white matter with advancing age. From Guttman *et al.*, in press.

mans (Haug *et al.,* 1984; Terry *et al.,* 1987; Leuba and Garey, 1989), indicating that, with advancing age, neuronal loss in the cortex is either not significant or not as extensive as pre-1984 reports had suggested (Brody, 1955, 1970; Colon, 1972; Shefer, 1973; Henderson *et al.,* 1980; Anderson *et al.,* 1983; Pakkenberg and Gunderson, 1997). There are comparable data in monkeys. A recent study of MRIs in young and old monkeys demonstrates significant decreases in white matter volume with age, but minimal decreases in gray matter volume (Rosene *et al.,* submitted).

Neuronal counts of numerous brain regions in young and old monkeys also demonstrate minimal neuronal cortical loss with age, including the striate cortex (Vincent *et al.,* 1989), motor cortex (Tigges *et al.,* 1992), frontal cortex (Peters *et al.,* 1994), and entorhinal cortex (Amaral, 1993; Gomez-Isla *et al.,* 1996). (It should be noted that these findings do not exclude focal changes with age in small areas of gray matter, such as selected areas of cortex or small subcortical nuclei, as discussed further below.)

Data in monkeys and humans also indicate that neuronal loss is highly selective within the hippocampal formation. For example, whereas the subiculum shows a significant age-related loss in humans and a similar trend in monkeys, the CA1, CA2, and CA3 fields show no evidence of an age-related neuronal loss (Amaral, 1993; Rosene, 1993; West *et al.,* 1994; Gomez-Isla *et al.,* 1996). Equivalent data have recently been reported in rodents, where it was shown that even in the subset of animals with declines on a memory tasks, there was no decrease in the number of neurons in the various hippocampal subfields (Rapp and Gallagher, 1996). (See Morrison and Hof, 1997, for a more detailed discussion of these issues.)

A recent positron-emission tomography (PET) study in humans also suggests that the hippocampal formation is more highly functional with age than previously thought (Schacter *et al.,* 1996). In this investigation, young and elderly subjects attempted to learn two types of word lists. The testing conditions were such that high levels of recall were achieved with one set of word lists and low levels of recall were achieved with the other. Analyses of the PET data from both the young and older adults revealed blood flow increases in the hippocampal formation in comparisons wherein the high-recall condition was emphasized; bilateral blood flow increases were seen in the High Recall minus baseline comparison; and right unilateral increases were seen in the High Recall minus Low Recall comparison. The pattern of blood flow increases appeared similar in both the young and older subjects—in both *z*-score units and location of change—in the high-recall condition. This suggests that the features of recollection indexed by hippocampal activity operate in a similar manner in old and young.

Nevertheless, the older subjects recalled less than the younger subjects. Comparisons involving the low-recall conditions highlighted this. Most striking was the difference between the groups in the Low Recall minus Baseline comparison, which produced blood flow increases bilaterally in the anterior frontal lobe (centering on area 10) for the young subjects but not for the older ones. By contrast, the older subjects demonstrated unilateral blood flow increases in the right posterior frontal lobe (centering on area 46) and the right motor area (centering on areas 4 and 6). These differences in frontal activation between the young and older adults may reflect differences in retrieval strategies when subjects are attempting to recall information that is not well learned. (It should be noted that some investigators use the term "working memory" to refer to the on-line strategies employed to store and retrieve information and, by this definition, it would appear that older individuals have a deficit in "working memory.")

It seems unlikely, based on the reports of preserved neuronal populations in the cortex cited above, that these frontal-activation differences between young and old are the result of substantial cortical neuronal loss. Age-related alterations in subcortical nuclei that project to the cortex seem to be the more-likely explanation for these findings, because there is substantial neuronal loss in selected subcortical regions involved in the neurotransmitter systems important for memory function, such as the basal forebrain and the locus coeruleus (e.g., Chan-Palay and Asan, 1989; Rosene, 1993). For example, in humans and monkeys there is an approximately 50% neuronal loss with age in the basal forebrain and a 35–40% loss in the locus coeruleus and dorsal raphe (Kemper, 1993). This compares with an approximate loss of 5% in CA1 of the hippocampus.

In addition, there is an age-related decrease in dopaminergic binding sites in the caudate nucleus (Severson *et al.*, 1982) and the substantia nigra, and a loss of neurons in the substantia nigra of about 6% per year (McGeer *et al.*, 1977). This loss of dopamine is thought to be responsible for many neurological symptoms that occur more often with age, such as decreased arm-swing and increased rigidity (Odenheimer *et al.*, 1994). Changes in dopamine levels may also cause age-related changes in cognitive flexibility. This is suggested by the fact that patients with Parkinson's disease (a disorder associated with a loss of cells in the substantia nigra and a severe decline in dopamine levels) have cognitive deficits that have been variably described as "mental inflexibility" (Lees and Smith, 1983), a disorder of the "shifting attitude" (Cools *et al.*, 1984), an "instability of cognitive set" (Flowers and Robertson, 1985), and a difficulty with "set formation, maintenance, and shifting" (Taylor *et al.*, 1987). An age-related functional loss of dopamine has been demonstrated in monkeys (Arnsten *et al.*, 1994), making it possible to study this hypothesis in nonhuman primates. Moreover, although neuronal loss appears minimal in the hippocampus with age, recent reports suggest that molecular changes within the memory circuits are likely to contribute to performance declines in memory with advancing age. A shift in *N*-methyl-D-aspartate (NMDA) receptor expression has been found in the perforant pathway of monkeys, in the absence of significant changes in other glutamate receptors and in the absence of structural degeneration in this brain circuit critical for normal memory (Gazzaley *et al.*, 1996). This change in NMDA expression could be related to alterations in long-term potentiation (LTP) reported in rodents (Barnes, 1994). It is also consistent with the finding of decreased "place cell" stability in aged rodents. Place cells in the hippocampus permit rodents to maintain a "spatial map" of their environment. In old rodents, these cells appear to function normally in familiar environments but are "unstable" (in old rodents compared with young ones) in new environments (Barnes *et al.*, 1997).

It seems reasonable to hypothesize that molecular changes such as those described above may be responsible for the fact that older individuals (human and monkey) take longer on average to learn something new, but once material is well learned, there is no difference in retaining information over delays. This may also explain why, in the PET study cited above, hippocampal activations appear equivalent between young and elderly in the high-recall conditions (i.e., when items were well learned).

## 10. Conclusion

Significant age-related changes in mental abilities occur in both healthy humans and monkeys. The age at which these changes are evident differs with the ability being evaluated and the method of evaluation. There is considerable variability among both elderly humans and elderly monkeys; some subjects show minimal changes with age and others show substantial declines.

There is increasing evidence that these alterations in ability are related to changes in brain structure and function. However, the commonly held notion that widespread neuronal death is responsible for these declines appears unwarranted. Rather, neuronal loss in specific subcortical nuclei that project to cortex, combined with molecular changes within the structurally intact memory circuits, appears to be responsible for age-related declines in cognition. The ability to alter these cognitive changes, and/or the brain changes associated with them, remains to be determined.

## 11. References

Aggleton, J., and Mishkin, M., 1983, Memory impairments following restricted medial thalamic lesions in monkeys, *Exp. Brain Res.* **21:**199–209.

Aigner, T., Mitchell, S., Aggleton, J., DeLong, M., Struble, R., Price, D., Wenk, G., Pettigrew, K., and Mishkin, M., 1991, Transient impairment of recognition memory following ibotenic-acid lesions of the basal forebrain in macaques, *Exp. Brain Res.* **86:**18–26.

Albert, M., Duffy, F., and Naeser, M., 1987a, Non-linear changes in cognition with age and neurophysiological correlates, *Can. J. Psychol.* **41:**141–157.

Albert, M., Heller, H., and Milbert, W., 1987b, Changes in naming ability with age, *Psychol. Aging* **41:**141–157.

Albert, M., Wolfe, J., and Lafleche, G., 1990, Differences in abstraction ability with age, *Psychol. Aging* **5:**94–100.

Amaral, D., 1993, Morphological analyses of the brains of behaviorally characterized aged nonhuman primates, *Neurobiol. Aging* **14:**671–672.

Anderson, J., Hubbard, B., Coghill, G., and Slidders, W., 1983, The effect of advanced old age on the neurone content of the cerebral cortex: Observations with an automatic image analyser point counting method, *J. Neurol. Sci.* **58:**233–244.

Arnsten, A., and Goldman-Rakic, P., 1990, Analysis of alpha-2 adrenergic agonist effects on the delayed non-matching-to-sample performance of aged rhesus monkeys, *Neurobiol. Aging* **11:**583–590.

Arnsten, A., Cai, J., Steere, J., and Goldman-Rakic, P., 1994, Dopamine D2 receptor mechanisms in the cognitive performance of young adult and aged monkeys, *Psychopharmacology (Berlin)* **116:**143–151.

Bachevalier, J., and Mishkin, M., 1989, Mnemonic and neuropathological effects of occluding the posterior cerebral artery in *Macaca mulatta, Neuropsychologia* **27:**83–105.

Bachevalier, J., Landis, L., Walker, M., Brickson, M., Mishkin, M., Price, D., and Cork, L., 1991, Aged monkeys exhibit behavioral deficits indicative of widespread cerebral dysfunction, *Neurobiol. Aging* **12:**99–111.

Baddeley, A., 1992, Working memory, *Science* **255:**556–559.

Barnes, C., 1988, Aging and physiology of spatial memory, *Neurobiol. Aging* **9:**663–668.

Barnes, C., 1994, Normal aging: Regionally specific changes in hippocampal synaptic transmission, *Trends Neurosci.* **17:**13.

Barnes, C., Suster, N., Shen, J., and McNaughton, B., 1997, Multistability of cognitive maps in the hippocampus of old rats, *Nature* **388:**272–275.

Bartus, R., Dean, R., and Fleming, D., 1979, Aging in the rhesus monkey: Effects on visual discrimination learning and reversal learning, *J. Gerontol.* **34:**209–219.

Bayles, K., and Kaszniak, A., 1987, *Communication and Cognition in Normal Aging and Dementia,* Little Brown, Boston.

Baxter, M. G., and Voytko, M. L., 1996, Spatial orienting of attention in adult and aged rhesus monkeys, *Behav. Neurosci.* **110:**898–904.

Borod, J., Goodglass, H., and Kaplan, E., 1980, Normative data on the Boston Diagnostic Aphasia Examination, parietal lobe battery, and Boston Naming Test, *J. Clin. Neuropsychol.* **2:**209–215.

Botwinick, J., and Storandt, M., 1974, *Memory, Related Functions, and Age,* Charles C. Thomas, Springfield, Ill.

Brody, H., 1970, Structural changes in the aging nervous system, *Interdiscipl. Top. Gerontol.* **7:**9–21.

Brody, H., 1955, Organization of cerebral cortex. III. A study of aging in the human cerebral cortex, *J. Comp. Neurol.* **102:**511–556.

Cerella, J., Poon, L., and Fozard, J., 1982, Age and iconic read-out, *J. Gerontol.* **37:**197–202.

Chan-Palay, V., and Asan, E., 1989, Quantitation of catecholamine neurons in the locus coeruleus in human brains of normal young and older adults in depression, *J. Comp. Neurol.* **287:**357–372.

Clark, H., and Clark, E., 1977, *Psychology and Language: An Introduction to Psycholinguistics,* Harcourt, Brace, Jovanovich, New York.

Cohen, N., and Squire, L., 1980, Preserved learning and retention of pattern analyzing skill in amnesia: Dissociation of knowing how and knowing that, *Science* **210:**207–210.

Colon, E., 1972, The elderly brain: A quantitative analysis of the cerebral cortex in two cases, *Psychiat. Neurol. Neurochir.* **75:**261–270.

Cools, A., Van Den Bercken, J., Horstink, M., Van Spaendonck, and Berger, H., 1984, Cognitive and motor shifting aptitude disorder in Parkinson's disease, *J. Neurol. Neurosurg. Psychiat.* **4:**443–453.

Corkin, S., 1968, Acquisition of motor skill after bilateral medial temporal lobe excision, *Neuropsychologia* **6:**225–265.

Craik, F., 1977, Age differences in human memory, in: *Handbook of the Psychology of Aging* (J. Birren and K. Schaie, eds.), Van Nostrand Reinhold, New York, pp. 384–420.

Craik, F., Byrd, M., and Swanson, J., 1987, Patterns of memory loss in three elderly samples, *Psychol. Aging* **2:**79–86.

Doppelt, J., and Wallace, W., 1955, Standardization of the Wechsler Adult Intelligence Scale for older persons, *J. Abnorm. Soc. Psychol.* **51:**312–330.

Drachman, D., and Leavitt, J., 1972, Memory impairment in the aged: Storage versus retrieval deficit, *J. Exp. Psychol.* **93:**302–308.

Flowers, K., and Robertson, C., 1985, The effect of Parkinson's disease on the ability to maintain mental set, *J. Neurol. Neurosurg. Psychiat.* **48:**517–529.

Gazzaley, A., Siegel, S., Kordower, J., Mufson, E., and Morrison, J., 1996, Circuit-specific alterations of NMDA receptors in the hippocampus of aged monkeys, *Proc. Natl. Acad. Sci. USA* **93:**3121–3125.

Gomez-Isla, T., Price, J., McKeel, D., Morris, J., Growdon, J., and Hyman, B., 1996, Profound loss of layer II entorhinal cortex neurons occurs in very mild Alzheimer's disease, *J. Neurosci.* **16:**4491–4500.

Graf, P., 1990, Life-span changes in implicit and explicit memory, *Bull. Psychonom. Soc.* **28:**353–358.

Graf, P., and Schacter, D., 1985, Implicit and explicit memory for new associations in normal and amnesic subjects, *J. Exp. Psychol.* **11:**501–518.

Guttmann, C., Jolesz, F., Kikinis, R., Killiany, R., Moss, M., Sandor, T., and Albert, M., White matter changes with normal aging, *Neurology,* in press.

Harrington, D., and Haaland, K., 1992, Skill learning in the elderly: Diminished implicit and explicit memory for a motor sequence, *Psychol. Aging* **7:**425–434.

Hasher, L., and Zacks, R., 1979, Automatic and effortful processes in memory, *J. Exp. Psychol.* **108:**356–388.

Hashtroudi, S., Chrosniak, L., and Schwartz, B., 1991, Effects of aging on priming and skill learning, *Psychol. Aging* **6:**605–615.

Haug, H., 1984, Macroscopic and microscopic morphometry of the human brain and cortex: A survey in the light of new results, *Brain Pathol.* **1:**123–149.

Heaton, R., Grant, I., and Matthews, C., 1986, Differences in neuropsychological function test performance associated with age, education, and sex, in: *Neuropsychological Assessment of Neuropsychiatric Disorders* (I. Grant and K. Adams, eds.), Oxford, New York, pp. 100–120.

Hebb, D. O., 1949, The Organization of Behavior, Wiley, New York.

Heindel, W., Salmon, D., Shults, C., Wallicke, P., and Butters, N., 1989, Neuropsychological evidence for multiple implicit memory systems: A comparison of Alzheimer's, Huntington's, and Parkinson's disease, *J. Neurosci.* **9:**582–587.

Henderson, G., Tomlinson, B., and Gibson, P., 1980, Cell counts in human cerebral cortex in normal adults throughout life, using an image analyzing computer, *J. Neurol. Sci.* **46:**113–136.

Hirsh, R., 1974, The hippocampus and contextual retrieval of information from memory: A theory, *Behav. Biol.* **12:**421–444.

Howard, D., 1988, Implicit and explicit assessment of cognitive aging, in: *Cognitive Development in Adulthood* (C. Brainerd and M. Howe, eds.), Springer-Verlag, New York, pp. 3–37.

James, W., 1890, *The Principles of Psychology,* Holt, New York.

Kemper, T., 1993, The relationship of cerebral cortical changes to nuclei in the brainstem, *Neurobiol. Aging* **14:**659–660.

Killiany, R., and Mahut, M., 1992, Lesions of the nucleus accumbens in adult rhesus monkeys result in a deficit of motor learning but not in S-R associative learning or memory, *Soc. Neurosci. Abstr.* **18:**1063.

Killiany, R., Moss, M., Herndon, J., Lai, Z., and Herndon, J., Recognition memory function in early aged rhesus monkeys, submitted.

LaBarge, E., Edwards, D., and Knesevich, J., 1986, Performance of normal elderly on the Boston Naming Test, *Brain Lang.* **27:**380–384.

Lachman, M., and Jelalian, E., 1984, Self-efficacy and attributions for intellectual performance in young and elderly adults, *J. Gerontol.* **39:**577–582.

Lai, Z., Moss, M., Killiany, R., and Rosene, D., 1995, Executive system dysfunction in the aged monkeys: Spatial and object reversal learning, *Neurobiol. Aging* **16:**947–954.

Lees, A., and Smith, E., 1983, Cognitive deficits in the early stages of Parkinson's disease, *Brain* **106:**257–270.

Leuba, G., and Garey, L., 1989, Comparison of neuronal and glial numerical density in primary and secondary visual cortex, *Exp. Brain Res.* **77:**31–38.

Mahut, H., Zola-Morgan, S., and Moss, M., 1982, Hippocampal resections impair associative learning and recognition memory in the monkey, *J. Neurosci.* **2:**1214–1229.

McGeer, P., McGeer, E., and Suzuki, J., 1977, Aging and extrapyramidal function, *Arch. Neurol.* **34:**33–35.

Mesulam, M.-M., and Van Hoesen, G., 1976, Acetylcholinesterase-rich projections from the basal forebrain of the rhesus monkey to neocortex, *Brain Res.* **109:**152–157.

Mirsky, A., 1978, Attention: A neuropsychological perspective, in: *Education and the Brain: Seventy-seventh Yearbook of the Study of Aging,* Part II. (J. Chall and A. Mirsky, eds.), Chicago University Press.

Mishkin, M., 1978, Memory in monkeys severely impaired by combined, but not separate, removal of amygdala and hippocampus, *Nature* **273:**297–298.

Moore, R., and Bloom, F., 1979, Central catecholamine neuron systems: Anatomy and physiology of the dopamine systems, *Annu. Rev. Neurosci.* **2:**113–168.

Morrison, J., and Hof, P., 1997, Life and death of neurons in the aging brain, *Science* **278:**412–419.

Moss, M., 1983, Assessment of memory in amnesic and dementia patients: Adaptation of behavioral tests used with non-human primates, *INS Bull.* **5:**15.

Moss, M., 1993, The longitudinal assessment of recognition memory in aged rhesus monkeys, *Neurobiol. Aging* **14:**635–636.

Moss, M., Rosene, D., and Peters, A., 1988, Effects of aging on visual recognition memory in the rhesus monkey, *Neurobiol. Aging* **9:**495–502.

Moss, M. B., Killiany, R. J., Lai, Z. C., Rosene, D. L., and Herndon, J., 1997, Recognition memory span in rhesus monkeys of advanced age, *Neurobiol. Aging* **18:**13–19.

Murray, E., and Mishkin, M., 1984, Severe tactual as well as visual memory deficits follow combined removal of the amygdala and hippocampus in monkeys, *J. Neurosci.* **4:**2565–2580.

Murray, E., and Mishkin, M., 1986, Visual recognition in monkeys following rhinal cortical ablations combined with either amygdalectomy or hippocampectomy, *J. Neurosci.* **6:**1991–2003.

Obler, L., Nicholas, M., Albert, M. L., and Woodward, S., 1985, On comprehension across the adult life span, *Cortex* **21:**273–280.

Odenheimer, G., Funkenstein, H., Beckett, L., Chown, M., Pilgrim, D., Evans, D., and Albert, M., 1994, Comparison of neurologic changes in successfully aging persons vs the total aging population, *Arch. Neurol.* **51:**573–580.

Pakkenberg, B., and Gundersen, H., 1997, Neocortical neuron number in humans: Effect of sex and age, *J. Comp. Neurol.* **384:**312–320.

Parasuraman, R., and Davies, R., 1984, *Varieties of Attention,* Academic, New York.

Peters, A., Josephson, K., and Vicent, S., 1991, Effects of aging on the neuroglial cells and pericytes within area 17 of the rhesus monkey *(Macaca mulatta), Anat. Rec.* **229:**384–398.

Peters, A., Leahu, D., Moss, M., and McNally, K., 1994, The effects of aging on area 46 of the frontal cortex of the rhesus monkey, *Cereb. Cortex* **6:**621–635.

Petersen, L., and Petersen, M., 1959, Short term retention of individual items, *J. Exp. Psychol.* **91:**341–343.

Petersen, R., Smith, G., Kokmen, E., Ivnik, R., and Tangalos, E., 1992, Memory function in normal aging, *Neurology* **42:**396–401.

Plude, D., Milberg, W., and Cerella, J., 1986, Age differences in depicting and perceiving tridimensionality in simple line drawings, *Exp. Aging Res.* **12:**221–225.

Pohl, W., 1973, Dissociation of spatial discrimination deficits following frontal and parietal lesions in monkeys, *J. Comp. Physiol. Psych.* **82:**227–239.

Poon, L., 1985, Differences in human memory with aging, in: *Handbook of the Psychology of Aging* (J. Birren and K. Schaie, eds.), Van Nostrand Reinhold, New York, pp. 427–462.

Posner, M. I., 1980, Orienting of attention, *Q. J. Exp. Psychol.* **32:**3–25.

Posner, M. I., and Cohen, Y., 1984, Components of visual orienting, in: *Attention and Performance: Control of Language Processes* (H. Bouma and D. G. Bowhuis, eds.), Lawrence Erlbaum, New Jersey, pp. 531–556.

Presty, S., Bachevalier, J., Walker, L., Struble, R., Price, D., Mishkin, M., and Cork, L., 1987, Age differences in recognition memory of the rhesus monkey *(Macaca mulatta), Neurobiol. Aging* **8:**435–440.

Rabinowitz, J., 1986, Priming in episodic memory, *J. Gerontol.* **41:**204–213.

Rapp, P., 1990, Visual discrimination and reversal learning in the aged monkey *(Macaca mulatta), Behav. Neurosci.* **104:**876–888.

Rapp, P., and Amaral, D., 1989, Evidence for task-dependent memory dysfunction in the aged monkey, *J. Neurosci.* **9:**3568–3576.

Rapp, P., and Amaral, D., 1991, Recognition memory deficits in a subpopulation of aged monkeys resemble the effects of medial temporal lobe damage, *Neurobiol. Aging* **12:**481–486.

Rapp, P., and Gallagher, M., 1996, Preserved neuron number in the hippocampus of aged rats with spatial learning deficits, *Proc. Natl. Acad. Sci.* **93:**9926–9930.

Raymond, B., 1971, Free recall among the aged, *Psychol. Rep.* **29:**1179–1182.

Rehbein, L., 1983, Long-term effects of early hippocampectomy in the monkey, *Ph.D. thesis,* Northeastern University.

Ronald, P., 1992, Motor skill learning, performance, and handedness in aged rhesus monkeys, *M.A. thesis,* Boston University School of Medicine.

Rosene, D., 1993, Comparing age-related changes in the basal forebrain and hippocampus of the rhesus monkey, *Neurobiol. Aging* **14:**669–670.

Rosene, D., Lai, Z., Killiany, R., Moss, M., Jolesz, D., Sandor, T., and Albert, M., Age-related loss of white matter with preservation of gray matter in the forebrain of the rhesus monkey: An MRI study, *Neurobiol. Aging,* submitted.

Ryle, G., 1949, *The Concept of Mind,* Hitchinson, London.

Schacter, D., Savage, C., Alpert, N., Rauch, S., and Albert, M., 1996, The role of hippocampus and frontal cortex in age-related memory changes: A PET study, *NeuroReport* **7:**1165–1169.

Schaie, K., 1983, The Seattle longitudinal study: A 21-year exploration of psychometric intelligence in adulthood, in: *Longitudinal Studies of Adult Psychological Development* (K. Schaie, ed.), Guilford, New York, pp. 64–135.

Schaie, K., and Labouvie-Vief, G., 1974, Generational vs ontogenetic changes in adult cognitive behavior: A fourteen year cross-sequential study, *Dev. Psychol.* **10:**305–320.

Severson, J., Marcusson, J., Winblad, B., and Finch, C., 1982, Age-correlated loss of dopaminergic binding sites in human basal ganglia, *J. Neurochem.* **39:**1623–1631.

Shefer, V., 1973, Absolute number of neurons and thickness of the cerebral cortex during aging, senile and vascular dementia, and Pick's and Alzheimer's diseases, *Neurosci. Behav. Physiol.* **6:**319–324.

Solomon, P., Pomerleau, D., Bennett, L., James, J., and Morse, D., 1989, Acquisition of the classically conditioned eyeblink response in humans over the lifespan, *Psychol. Aging* **4:**34–41.

Spitz, H., 1972, Note on immediate memory for digits: Invariance over the years, *Psychol. Bull.* **78:**183–185.

Squire, L., 1986, Mechanisms of memory, *Science* **232:**1612–1619.

Stafford, J., Albert, M., Naeser, M., Sandor, T., and Garvey, A., 1988, Age-related differences in computed tomographic scan measurements, *Arch. Neurol.* **45:**405–419.

Talland, G., 1967, Age and the immediate memory span, *Gerontologist* **7:**4–9.

Taylor, A., Saint-Cyr, J , and Lang, A., 1987, Parkinson's disease: Cognitive changes in relation to treatment response, *Brain* **110:**35–51.

Terry, R., Deteresa, R., and Hansen, L., 1987, Neocortical cell counts in normal human adult aging, *Ann. Neurol.* **21:**530–539.

Tigges, J., Gordon, T., McClure, H., Hall, E., and Peters, A., 1988, Survival rate and life span of rhesus monkeys at the Yerkes Regional Primate Research Center, *Am. J. Primatol.* **15:**263–273.

Tigges, J., Herndon, J., and Peters, A., 1992, Neuronal population of area 4 during life span of rhesus monkeys, *Neurobiol. Aging* **11:**201–208.

Tulving, E., 1972, Episodic and semantic memory, in: *Organization of Memory* (E. Tulving and W. Donaldson, eds.), Academic, New York, pp. 381–403.

Vincent, S., Peters, A., and Tigges, J., 1989, Effects of aging on neurons within area 17 of rhesus monkey cerebral cortex, *Anat. Rec.* **223:**329–341.

Waugh, N., and Norman, D., 1965, Primary memory, *Psychol. Rev.* **72:**89–104.

West, M., Coleman, P., Flood, D., and Troncoso, J., 1994, Differences in the pattern of hippocampal neuronal loss in normal aging and Alzheimer's disease, *Lancet* **344:**769–772.

Woodruff-Pak, D., 1990, Neurobiological substrates of classical conditioning across the life span, *Ann. NY Acad. Sci.* **608:**150–173.

Zatz, L., Jernigan, T., and Ahumada, A., 1982, Changes in computed cranial tomography with aging: Intracranial fluid volume, *Am. J. Neurorad.* **3:**1–11.

Zola-Morgan, S., and Squire, L., 1984, Preserved learning in monkeys with medial temporal lesions: Sparing of motor and cognitive skills, *J. Neurosci.* **4:**1072–1085.

Zola-Morgan, S., and Squire, L., 1985, Medial temporal lesions in monkeys impair memory in a variety of tasks sensitive to amnesia, *Behav. Neurosci.* **17:**558–564.

# 2

# Age-Related Cognitive Decline in the Rhesus Monkey

MARK B. MOSS, RONALD J. KILLIANY,
and JAMES G. HERNDON

## 1. Introduction

It is only in the last 25 years that significant strides have been taken toward the development of primate models of normal human aging. Together with advances in the fields of molecular biology, neuroimaging, and behavioral neuroscience, these models have emerged as potentially important components in our understanding of the age-related cognitive decline in humans (Peters *et al.*, 1996). This is not to say that studies using other species, particularly rodents, have provided a wealth of information on the neurobiological substrates of aging. However, the rich behavioral repertoire of the monkey, particularly the monkey's remarkable ability to perform a variety of complex short-term memory tasks, many administered to normal humans in geriatric research studies, has made this species particularly suited to assess cognitive function at a level not possible in nonprimates.

This chapter will review, from a human neuropsychological perspective, earlier findings and recent advances in our understanding of cognitive decline that occurs

---

MARK B. MOSS and RONALD J. KILLIANY • Department of Neuroanatomy and Neurobiology, Boston University School of Medicine, Boston, Massachusetts 02118. JAMES G. HERNDON • Yerkes Regional Primate Research Center, Emory University, Atlanta, Georgia 30322.

*Cerebral Cortex,* Volume 14, edited by Peters and Morrison. Kluwer Academic/Plenum Publishers, New York, 1999.

in the aging monkey. In large part, we will rely on data obtained in our multidisciplinary research program exploring the neural substrates of cognitive decline in the aged monkey. Within each cognitive domain, where information is available, age-related changes are considered for three age groups within the range of aged monkeys (19–33+ years). Finally, a brief review of longitudinal changes in aged monkeys and an overview of the possible neurobiological base to account for age-related cognitive decline are provided, a topic more extensively discussed in three other chapters in this volume (Arnsten; Rosene; and Peters and Nicholson).

## 2. Life Span of the Rhesus Monkey

A study from our program examining the survival rates of 763 monkeys kept at two locations at Yerkes Regional Primate Research Center (Tigges *et al.,* 1988) showed that the maximum life span of the rhesus monkey extends into the fourth decade, but that only 50% of monkeys reached 16 years of age, and only about 25% reached 25 years of age. This study, together with earlier data from various population studies (Dyke *et al.,* 1986; Davis, 1985; Bowden and Williams, 1984) suggests that at best, only 30–40% of rhesus monkeys attain an age of 20 years, no more than 25% survive to 25 years of age, and very few animals survive to 30 years of age and then only under ideal conditions of individual housing and medical care. In summary, the typical adult life span of the rhesus monkey may be considered to extend from 5–30 years. On this basis, we have concluded that monkeys more than 20 years of age can be considered old. Within the "aged" range, taking into account the estimated ratio of monkey-to-human years (approximately 1 to 3, we consider monkeys 19–23 "early-aged" animals, monkeys 24–28 "advanced-aged" animals, and monkeys 29 and older "oldest-old" animals.

## 3. Cognitive Function

### 3.1. Attention

Although attention may be considered the "gateway" to higher cortical processing, and several aspects of attention have been shown to be preserved in normal human aging (Greenwood *et al.,* 1993; Hartly *et al.,* 1990), few studies have assessed directly this function in aged monkeys. Davis (1978) demonstrated that aged monkeys were more likely to evidence distraction in initial learning, and several studies (see section below), by virtue of unimpaired performance on simple two-choice discrimination-learning tasks, have inferred normal attentional processing in aged monkeys. Recently, by adapting attentional tasks from those used in humans by Posner (Posner, 1988; Posner and Cohen, 1984), Baxter and Voytko (1996) have shown that aged monkeys are as proficient as young adult monkeys at shifting attention to a peripherally-cued spatial location. Thus, initial studies suggest that attentional processes in aged monkeys, as in aged humans, are spared.

### 3.2. Learning and Memory

#### 3.2.1. Simple Discrimination Learning

Learning in aged monkeys has traditionally been assessed by the use of simple two-choice discrimination tasks. In a typical paradigm, monkeys are rewarded for choosing the designated correct stimulus of a stimulus pair (objects, colors, patterns, or stimulus positions) presented simultaneously over a series of trials to a learning criterion of 90% accuracy. In most studies, it has been shown that aged monkeys, as a group, do not evidence any significant difficulty in simple discrimination learning (Lai *et al.,* 1995a; Bachevalier *et al.,* 1991; Arnsten and Goldman-Rakic, 1990; Rapp, 1990; Bartus *et al.,* 1979), though in one study (Voytko, 1993) it was found that a group of aged monkeys evidenced significant impairment in learning object discriminations. In an earlier study from our program (Moss *et al.,* 1988), we found no significant difference in trials or errors to learning criterion between a group of six young adult and a group of six advanced-aged monkeys on three different simple discrimination tasks (brightness, hue, and pattern). In a later study, we found that a group of oldest-old monkeys were impaired in learning a simple object discrimination (Moss *et al.,* 1997), suggesting that simple discrimination learning is preserved until the end of the life span.

#### 3.2.2. Concurrent-Discrimination Learning

Concurrent discrimination is an aspect of learning that has been assessed both in monkeys (Malamut *et al.,* 1984; Moss *et al.,* 1981; Cowey and Gross, 1970) and in humans (Oscar-Berman and Zola-Morgan, 1980). Concurrent-discrimination tasks assess the ability to learn associations between reward and given stimulus objects in the presence of proactive and retroactive interference. In an earlier study, Bartus and Dean (1979) showed that aged monkeys were impaired on a concurrent-discrimination task, leading to the suggestion that aged monkeys are particularly sensitive to interference. In one form of the task, 20 pairs of objects are presented in succession daily until the animal performs all 20 discriminations with no errors in one session (Bachevalier *et al.,* 1991). A significant difference was found in performance of this task between a group of oldest-old monkeys and a group of young adult control animals. In contrast, the same investigators found that a group of early-aged and advanced-aged monkeys were, as a group, unimpaired on the task. Recent findings from our program (Moss *et al.,* unpublished) support the findings of Bachevalier *et al.* (1991) that oldest-old monkeys are impaired on concurrent learning. Preliminary data suggest that impairment on concurrent learning also may characterize monkeys of early and advanced age. Taken together, these results suggest that the capacity to simultaneously learn multiple object–reward associations declines with age. In fact, this capacity deteriorates to such an extent that the monkeys in the oldest-old group evidenced impairment at learning simple pair discriminations (see section above).

#### 3.2.3. Delayed Nonmatching-to-Sample

Normal human aging is associated with a decline in several domains of cognitive function, among which short-term memory is perhaps the earliest and most prominent (Albert and Moss, 1996). Several aspects of short-term memory commonly assessed in aged humans are recall of word lists, of aurally presented short narratives, and of visually presented designs on visual or spatial locations. In some instances, recognition paradigms, in which the relevant stimuli are presented in a "yes/no" or forced-choice format, are used in conjunction with recall tasks. In monkeys, because of the difficulty in creating a true free-recall test condition, memory assessment is limited primarily to recognition paradigms. One test commonly used to assess visual-recognition memory in the monkey is the delayed nonmatching-to-sample (DNMS) task (Bachevalier *et al.,* 1991; Rapp and Amaral, 1991; Arnsten and Goldman-Rakic, 1990; Rapp and Amaral, 1989; Moss *et al.,* 1988; Presty *et al.,* 1987). This is a benchmark visual-recognition-memory task that has been used by several laboratories to assess memory in monkeys with selective telencephalic and diencephalic lesions (Zola-Morgan and Squire, 1985; Mahut *et al.,* 1982; Mishkin, 1978; Gaffan, 1974) and resembles, in many respects, clinical tests that are used to assess memory function in patients with a variety of neurological disorders (Albert and Moss, 1996). The DNMS task requires an animal to identify a novel object from a previously presented familiar one, with increasing delays interposed between the two presentation trials. On the trial-unique paradigm, a different pair of objects is used for each trial. The task can be made more difficult by increasing the delay between presentation of the sample and the recognition trials (delay condition) or by increasing the number of samples prior to pairing them with novel objects in the recognition trial (list condition). As a group, aged monkeys evidence impairment on the acquisition and performance of this task, although several studies have reported normal to near-normal performance by individual aged monkeys (Bachevalier *et al.,* 1991; Rapp and Amaral, 1989, 1991; Moss *et al.,* 1988; Presty *et al.,* 1987). Nevertheless, several studies spanning the last ten years have collectively revealed a significant relationship between performance on the DNMS task and aging across the entire adult age range in monkeys.

In a recent study of recognition memory from our program using the DNMS task (Killiany *et al.,* 1997), we found that early-aged monkeys, as a group, were not impaired with respect to young adult monkeys in the acquisition of the task. However, they were impaired on the delay and the list performance measures, a pattern consistent with findings from an earlier study by Presty *et al.* (1987). Together with earlier findings from Bartus and colleagues (1979), who demonstrated that rhesus monkeys estimated to be in the early-aged range (18–21 yrs) are impaired on tasks of reversal learning and delayed response (Bartus *et al.,* 1979, 1978), these data support the view that at least some aspects of learning and memory function may evidence decline as early as the later part of the second decade in the rhesus monkey.

On closer inspection of the performance of the early-aged monkey on the delay and list conditions of the DNMS in our study, monkeys showed a significant impairment when delays greater than 30 seconds were interposed between the sample and the recognition trials, or when list lengths of 3, 5, or 10 items were used. At first glance, this appeared at variance with the study by Presty *et al.* (1987), who found a significant overall age effect on a combined performance measure consisting of three delays (30, 60, and 120 sec) and three list lengths (3, 5, and 10 items). *A posteriori* paired comparisons in their study did not show a significant impairment in the group of early-aged animals. It is plausible that our finding of a significant impairment by early-aged monkeys was related to the increased demands of two additional longer delays (300 sec and 600 sec). Indeed, early-aged monkeys in our study evidenced a marked decline in performance accuracy with delays of 300 or 600 seconds, and analyses revealed a direct relationship between performance and age. On the list condition, the performance of the animals in the early-aged group was significantly impaired with respect to the performance of the young adult animals. Taken together, the data from the study by Presty *et al.* (1987) and our study lead to two conclusions. First, the impairment on the performance conditions of the DNMS by early-aged monkeys is relatively mild and is evident only under conditions of relatively high memory demands. Second, as pointed out by Rapp and Amaral (1991), there does not appear to be any significant relationship between rate of acquisition and accuracy on the performance conditions on DNMS. This adds support to the view that the two phases of the task may require different cognitive abilities and that they may, in turn, be differentially affected by age.

When we assessed recognition memory in a group of monkeys of advanced age (25–29 years), a different pattern of impairment was seen (Killiany *et al.,* 1997). Only one of the six monkeys in the advanced-aged group reached learning criterion as efficiently as young adult monkeys, a percentage significantly lower than that seen in our study with early-aged monkeys (Killiany *et al.,* 1997). On each of the four delay conditions of the DNMS, the accuracy of advanced-aged animals was significantly worse than that of the young adult animals. Though the performance of both groups was significantly affected by the increased delays, the advanced-aged group was not affected disproportionately. Each group evidenced about a 15% decrement in performance when delays were extended from 30 to 300 seconds. The accuracy of performance for the advanced-aged group declined significantly relative to the young adult group with lists of 3 or 5, but not 10, items. However, no significant interaction between age and lists was found, suggesting that the monkeys of advanced age were not disproportionately affected by the increased demand on memory.

Since the list condition of the DNMS requires the recognition of objects that are serially presented, performance on specific item positions in a list can be affected by the presence of proactive and retroactive interference effects. Evidence from earlier studies has suggested that monkeys with hippocampal ablations may be disproportionately sensitive to interference (Mahut *et al.,* 1982; Moss *et al.,* 1981) and that in normal human aging and certain age-related dementias (Kemper, 1994; Hyman *et al.,* 1984), the integrity of the hippocampus may be compromised. To see if such effects may have contributed to the impairment of monkeys in the advanced-aged group on this aspect of the recognition task, we assessed the pattern of errors made on each item for all list lengths (Moss *et al.,* 1988). Specifically, the number of errors committed by each monkey at the first, middle, and last list positions of the 3-, 5-, and 10-object lists was calculated and analyzed for changes in performance as a function of list position. No significant differences in performance among item positions were obtained for either group at list lengths of 3 and 5 objects. Similarly, for the list of 10 objects, the young adult group showed no significant difference in performance between the first and middle positions, but the younger monkeys had significantly improved performance between the middle and the last positions. In contrast, the performance on the middle positions by monkeys in the advanced-aged group did not differ significantly from performance on the first or last positions. In fact, four of the six monkeys in this group committed more errors at the last than at the middle position. These findings suggested that monkeys of advanced age failed to show enhancement of learning for items at the end of a list, that is, a recency effect. Together with findings from earlier studies (Bartus and Dean, 1979), this suggests that advanced-aged animals may be disproportionately sensitive to interference effects.

With regard to the behavioral assessment of the oldest-old rhesus monkeys, few data are available. In a large cohort of aged monkeys studied by Bachevalier *et al.* (1991), the oldest group, consisting of four monkeys aged 28–31 years (originally assessed 2 years earlier), evidenced a pattern of deficits that differed from the other two aged groups. In another study assessing performance of aged monkeys from 22 to 33 years on the DNMS task, Rapp and Amaral (1991) noted that a 33-year-old monkey, the oldest monkey in the study, was the only animal that could not reach learning criterion, even within 3000 trials, as compared to the 200–400 trials required by the other monkeys in the study to master their task.

Recently, we assessed a group of five monkeys between the ages of 30 and 35. As a group and individually, these oldest-old monkeys were markedly impaired relative to young adults on acquisition on the DNMS task. Monkeys in this oldest group required a mean number of trials twice that required by monkeys of advanced age. Further appreciation for the extent of this deficit comes from a comparison of DNMS performance by the oldest-old monkeys with that of a group of 25 young adult monkeys as part of our study on age-related patterns of cognitive performance (Herndon *et al.,* 1997). Based on the number of errors committed before reaching learning criterion, the oldest-old group was nearly 14 standard deviation units (SD) above the mean of the young adult group, whereas the SD for advanced-aged monkeys (also impaired relative to young adults) was only 4.7 SD above the mean of the young adult group. Taken together, these findings suggest that survival beyond advanced age may be associated with a marked and precipitous decline in the ability to learn new operational cognitive rules, a pattern now emerging in studies of normal human aging (Howieson *et al.,* 1993).

In contrast to their severe impairment on acquisition of DNMS, the oldest-old monkeys evidenced only a moderate impairment on a 2-minute delay condition in the performance phase of the DNMS task, obtaining a mean accuracy of 77%. Though significantly lower than that achieved by young adult monkeys (89%), accuracy was comparable to that of monkeys of advanced age (79%). Presty *et al.* (1987) reported similar levels of accuracy (82%) on DNMS in five monkeys in the range of 25–29 years (mean age of 27) on a 2-minute delay on the DNMS task. At a delay of 10 minutes, the performance of the oldest-old fell to a mean accuracy of 69% as compared with 85% by young adults. This finding suggests that oldest-old monkeys may forget more rapidly than do younger animals. The observation of a disproportionate memory decline as delays increase has also been noted in a subset of younger-aged (mean age of 24; "early" and "advanced" aged as these terms are used in this chapter) monkeys by Rapp and Amaral (1991). Based on evidence from lesion studies in monkeys, these authors ascribed this pattern of performance to compromise of medial temporal lobe structures. The extent to which these structures have been morphologically or neurochemically compromised in the aged monkey, however, remains unknown.

On the basis of our findings and also those of others (Presty *et al.,* 1987; Rapp and Amaral, 1991), we believe that in the monkey there is a significant but mild age-related deficit in recognition memory. This deficit appears at a relatively early age (about the third decade of life) and plateaus sometime around the middle of that decade. No additional decline in recognition memory appears to take place with successful aging. The ability to learn new strategies, on the other hand, seems to change at a different rate and on a different course. This deficit does not appear to become significant until the more advanced age (about the middle of the third decade), and performance declines apparently continue into the oldest-old age. A similar behavioral profile has been reported in aging humans. A study of 161 healthy subjects aged 62 to 100 revealed that performance on learning or acquisition paradigms declines uniformly with age, whereas performance on tasks of delayed recall stays relatively stable, when adjusted for the amount of material initially learned (Petersen *et al.,* 1992).

### 3.2.4. Delayed Recognition Span

The DNMS task relies upon a two-alternative, forced-choice paradigm in which the monkey is required to discriminate, after a delay, which of two objects was previously presented. This task employs a win–shift or nonmatch strategy, and it measures the rate of forgetting by imposing various delays between the sample presentation and the recognition trial. However, it is limited to two-choice recognition (see Fig. 1). To characterize recognition memory within spatial as well as nonspatial stimulus domains, and to assess the memory capacity or "load" of aged monkeys, we developed a task that uses the same win–shift strategy as the DNMS—the delayed recognition span task (DRST) (Moss, 1983; Rehbein and Mahut, 1983). The DRST employs a nonmatching paradigm but requires the monkey to identify a novel stimulus among an increasing array of previously presented stimuli, using either spatial or nonspatial stimulus cues. It has been used in monkeys to assess

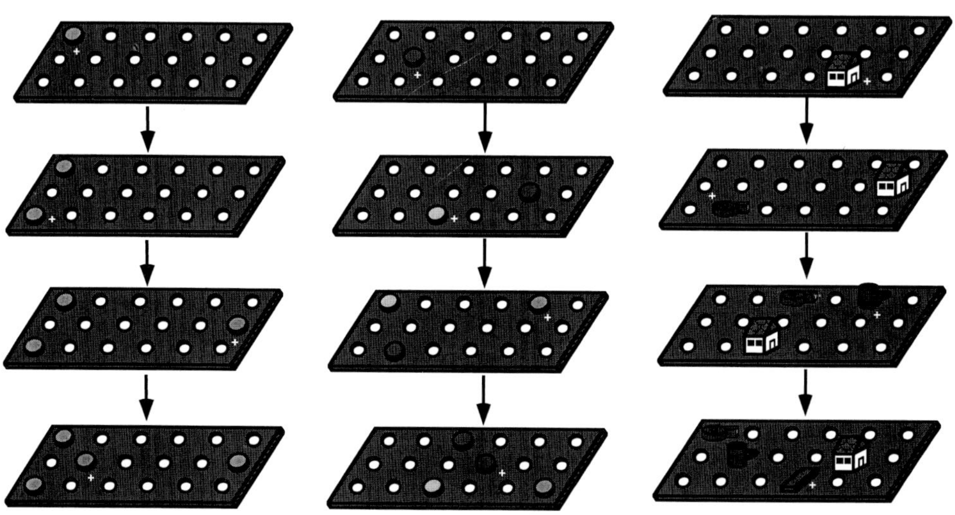

**Figure 1.** Schematic showing sequence of four presentations in one trial of the spatial condition of the delayed-recognition-span task. + denotes baited well.

memory function following damage to the hippocampus in infants (Killiany and Mahut, 1991; Rehbein and Mahut, 1983) and adults (Beason *et al.*, 1990) and, in humans, to differentiate patterns of memory dysfunction in a variety of neurologic disorders (Martin *et al.*, 1995; Lange *et al.*, 1992; Salmon *et al.*, 1989; Moss *et al.*, 1986). In addition, it has been used for assessing memory function in normal human aging (Inouye *et al.*, 1993). The DRST can be administered using different classes of stimulus material to help characterize recognition memory deficits across several stimulus domains.

We have administered the DRST using the spatial, color, and object conditions with our monkeys. For the spatial condition, 15 identical plain brown discs are used as stimuli; for the color condition, 15 different colors are used; and for the object condition, 15 junk objects are used. The color and the object conditions are administered in a fashion similar to that of the spatial condition, except that the position of the previously correct stimuli is changed in a pseudorandom fashion so that the animal has to identify the new stimulus based on visual cues, without the use of spatial cues.

Relative to young adults, early-aged monkeys were significantly impaired on both the spatial and the object components of the DRST. Moreover, a linear relationship was observed between age and degree of impairment. In contrast to their performance on the spatial and object conditions of the DRST, early-aged monkeys, as a group, were not significantly impaired on the color condition. This finding was surprising, because early-aged monkeys were impaired in the other nonspatial span condition using three-dimensional, random junk objects.

It is unclear why the degree of difficulty on nonspatial recognition span by early-aged monkeys would be related to the nature of the stimuli. One explanation is that aged monkeys may benefit when the number of stimulus dimensions to be processed and recognized is limited, as is the case with color stimuli. However, because aged monkeys have little difficulty learning object-discrimination tasks (Bachevalier *et al.*, 1991; Rapp, 1990; Moss *et al.*, 1988; Bartus *et al.*, 1979), it is hard to support this notion and, indeed, further investigation is required.

In a recent study from our program (Moss *et al.*,1997), monkeys of advanced age given the DRST obtained lower recognition spans across both spatial and color stimulus classes than did young adult monkeys. On average, the advanced-aged monkeys retained only two-thirds the amount of information as that retained by the young adult monkeys. Parallel studies using the delayed-recognition-span test with human subjects have also shown a significant decline in recognition-memory span in normal aging, as well as in various amnesic and dementia states (Inouye *et al.*, 1993; Moss *et al.*, 1986).

In view of recent findings that, in humans, the oldest-old may undergo cognitive decline different in extent and pattern from that of younger aged groups (Howieson *et al.*, 1993), we decided to evaluate a group of oldest-old rhesus monkeys to see if the pattern of cognitive impairment on the DRST differed from that of monkeys of early or advanced age. We found that the oldest-old monkeys were impaired relative to young adults on the DRST-spatial using either repeated or unique series. However, their mean span (1.83) did not differ significantly from that of monkeys of advanced age (2.00). In addition, three of the four monkeys evidenced improvement with repeated trials, an effect that we have also noted in advanced-age (Moss *et al.*, 1997) and early-aged (Killiany *et al.*, 1997) monkeys.

These findings together suggest that the age-related decline seen in visuospatial memory function in aging monkeys may not accelerate over the latter part of the life span.

On the object condition of the DRST, oldest-old monkeys were significantly impaired relative to young adults for both the repeated and the trial-unique series. Examination of findings on this task in our study of early-aged monkeys (Killiany *et al.*, 1997) indicates no significant different in performance between the two age groups, and initial data for the object recognition span in monkeys of advanced age suggests a range of performance similar to that of the oldest-old. These findings suggest that, as with spatial recognition span, age-related decline in recognition span for three-dimensional objects may not change beyond advanced age.

Some features of the DRST are worth considering in the context of spatial memory function in aged monkeys. For this task, the number of spatial locations on the test board is limited to 18. Over the course of the task, the number of times a disc appears over any one of the 18 positions is nearly equivalent. Hence, the memory demand of the task shifts from being based on novelty of the stimulus (as in the trial-unique DNMS) to being based on the temporal order of the stimuli presented within a trial, as in the DNMS using repeated objects. In this regard, the spatial DRST has some elements in common with the delayed-response task, in which remembering temporal factors (the last stimulus baited) is key to successful performance. Thus, it is not surprising that, like delayed response (Bachevalier *et al.*, 1991; Arnsten and Goldman-Rakic, 1990; Rapp and Amaral, 1989; Walker *et al.*, 1988; Davis *et al.*, 1982; Marriott and Abelson, 1980; Bartus *et al.*, 1978), the spatial condition of the DRST appears to be relatively sensitive to aging in the monkey. Indeed, monkeys of advanced age (25–29 years) evidence further impairment in spatial recognition span (Moss *et al.*, 1997), which appears to be most severe in oldest-old monkeys (Moss *et al.*, 1997).

One other aspect of the DRST to be considered is the series of repeated trials that were embedded within each condition of the task. Both repeated and unique sequences were included in the delayed-recognition-span test as a means of assessing the ability to learn and recall under both stimulus conditions within the same task. Hebb (1961) found that in humans, the recall of an arbitrary sequence of highly familiar information, such as digits, improved when the same series was embedded repeatedly within presentations of random sequences. Similarly, we found that both the aged and the young adult monkeys in our program performed better on an embedded repeated sequence than they did on the unique sequences, although the performance level of the aged animals was still significantly lower than that of young adults. These findings suggest that aged monkeys, like young adults, can benefit from repetition, but the improvement is modest. Because all monkeys achieved their highest overall span on the object condition, it may be that their ability to improve performance with repetition has reached asymptotic levels toward a maximum memory capacity.

While there is general agreement in the literature regarding the ability of aged monkeys to learn two-choice visual discrimination tasks efficiently, discrepancies exist when considering the presence and extent of impairment on tasks of visual recognition memory. As mentioned above, many early-aged, advanced-aged, and oldest-old monkeys evidence only mild impairment on the DNMS, particularly on the performance component of the task. Rapp and Amaral (1989), for example, demonstrated only a modest difference in the performance of aged and young monkeys across delays as long as 22 hours. In contrast, other studies have assessed recognition memory in the aging rhesus monkey with the delayed-response task and found a robust impairment (Bachevalier *et al.,* 1991; Arnsten and Goldman-Rakic, 1990; Rapp and Amaral, 1989; Walker *et al.,* 1988; Davis *et al.,* 1982; Marriott and Abelson, 1980; Bartus *et al.,* 1978). This discrepancy in performance between the DNMS and delayed-response tasks has been addressed directly by Rapp and Amaral (1991), who concluded that temporal order and spatial features were the likely candidates accounting for the differing effects.

The impairment in recognition memory, and in particular, in spatial recognition, may be related to several factors. First, the DRST requires the animal to keep track of an increasing stimulus array during a trial (i.e., increasing "memory load"). In addition to increasing demand on memory capacity, this likely increases the amount of task interference, an effect that may be sensitive to aging (Moss *et al.,* 1988; Bartus and Dean, 1979). Second, as with the delayed-response task, the spatial condition requires the animal to identify the new stimulus on the basis of spatial position and temporal order. Tasks that rely primarily on these cues have been shown to produce marked impairments in aged monkeys (Bachevalier *et al.,* 1991; Anderson *et al.,* 1993; Lai *et al.,* 1995a).

In view of these differential age effects on learning and memory function, it is tempting to conclude that age-related cognitive declines in learning and memory function result from two separate processes. Thus, the capacity for short-term storage of information, although impaired with aging, reaches asymptotic decline in advanced age. In contrast, the capacity for learning new rules may undergo progressive decline throughout the latter part of the life span. Whether this decline in learning new rules extends to other forms of learning (i.e., associative learning) remains to be seen. Further, it is unclear whether the differential effects of aging on new learning and memory load or storage can be functionally dissociated in the pathogenesis of normal aging.

### 3.2.5. Delayed Response

The delayed-response task (DR) is a classic behavioral task known to be sensitive to frontal cortical system damage, as first demonstrated in the 1930s with Jacobson's (1935) studies in the chimpanzee. Over the last fifty years, several investigators have confirmed and extended these early findings in a variety of species, including the monkey (Goldman and Rosvold, 1970; Iversen, 1970; Butter *et al.,* 1968; Gross, 1963; Battig *et al.,* 1960; Mishkin, 1957). Battig *et al.* (1960) and later Goldman and Rosvold (1972) demonstrated an impairment on DR in monkeys with lesions of the caudate nucleus. Similarly, an impairment on DR was found following lesions of the hippocampal formation (Zola-Morgan *et al.,* 1989), but as reported in the same study, did not extend to lesions of the amygdala. Perhaps the most robust effect on DR performance is that following damage to the prefrontal cortex. Goldman and Rosvold (1970) behaviorally fractionated the frontal cortical regions and identified the dorsolateral surface as a critical locus. This includes the same region in which Peters *et al.* (1994) have identified a significant correlation between thinning of layer I cortex and cognitive impairment in aged monkeys. In humans, the DR task has been used to elucidate the nature of memory dysfunction in patients with alcoholic Korsakoff's syndrome (Oscar-Berman *et al.*, 1992), patients with schizophrenia (Park *et al.*, 1995a; Park and Holzman, 1992; Raine *et al.*, 1992), and their relatives (Park *et al.*, 1995b). Of particular interest, deficits on the DR task have been recently reported in intermediate-aged adults (Oscar-Berman *et al.*, 1991), but little has been reported in more advanced-aged subjects.

In aged monkeys, the DR task has been used by several investigators (Bachevalier *et al.*, 1991; Rapp and Amaral, 1989; Walker *et al.*, 1988; Arnsten and Goldman-Rakic, 1985; Davis *et al.*, 1982; Marriott and Abelson, 1980; Bartus *et al.*, 1978; Medin and Davis, 1974) to characterize memory dysfunction. The findings indicate that aged monkeys evidence increasing degrees of impairment with longer delays, though in many instances they are impaired even on delays as short as five seconds (Walker *et al.*, 1988; Arnsten and Goldman-Rakic, 1985). Further, aged monkeys appear to be disproportionately affected by increased delays, a pattern that in contrast to that seen on the DNMS (e.g., Rapp and Amaral, 1989; Moss *et al.*, 1988; Presty *et al.*, 1987). This suggests that the DR task may be more sensitive to age-related cognitive decline, raising the possibility that frontal cortical systems may be disproportionately vulnerable to the aging process.

### 3.3. Executive System Function

Although there is an impressive literature on age-related decline in memory with normal aging and age-related disease, the number of studies aimed at the assessment of executive function is more limited. Executive system functions are a set of integrative cognitive skills that are vital to performing higher-level activities of daily living. They encompass abilities such as cognitive flexibility, cognitive tracking, set maintenance, abstraction, divided attention, insight, and social judgment. Cognitive flexibility refers to the capacity to change a response pattern with changing reinforcement contingencies. Cognitive tracking refers to the capacity to keep track of or correctly identify correct response–reward associations among frequently alternating contingencies. Set maintenance refers to the ability to maintain a given mode of responding. Abstraction refers to the capacity to identify a common element among stimuli that appear to differ along several dimensions. Divided attention refers to the capacity to attend to two or more stimulus sources simultaneously. In large measure, many of these functions share the requirement of inhibiting interfering responses.

Executive functions govern a person's capacity to solve ongoing problems of how, when, and what to do in an environmentally adaptive manner. Many patients have sustained considerable loss of verbal, visuospatial, or memory skills without significant changes in independent living (Lezak, 1995). However, impairments of executive function tend to affect all aspects of daily living skills, making independent living difficult or even impossible. These types of impairments have been described in a variety of ways such as mental inflexibility (Lees and Smith, 1983), disorder of shifting attitude (Cools *et al.*, 1984), and dysexecutive syndrome (Baddeley, 1986).

In normal aged humans, a variety of studies have been reported showing impairments on tasks of executive system function (Albert *et al.*, 1990; Albert *et al.*, 1987; Lezak, 1995; Talland, 1961). One of the most consistent findings is the decrement in performance on tasks of divided attention (Craik and Simon, 1980; Craik, 1971; Brinley and Fichter, 1970; Broadbent and Gregory, 1965). Studies in normal adults have also reported age-related deficits in abstraction (Mack and Carlson, 1978; Bromley, 1957). Caution in interpreting some of these results was raised by a later study (Hess and Slaughter, 1986) showing that deficits in abstraction may be exacerbated by tasks that make heavy demands on memory and attention. However, a recent study revealed impairments on two tasks of abstraction and set shifting that had low memory and attentional demands in optimally healthy aged subjects (Albert *et al.*, 1990), suggesting that impairment in abstraction is not merely the result of underlying deficits in memory or attention.

Evidence from the neuropsychological, neuropathological, and neurological literature suggests that the executive functions are mediated by the prefrontal cortical areas (Shimamura *et al.*, 1991; Stuss, 1991; Albert, 1988; Damasio, 1985; Stuss and Benson, 1986). However, the relationship of the age-related decline in this cognitive domain to specific neuropathological and neurochemical alterations is unclear.

In nonhuman primates, studies of frontal lobe function date back to the 1930s, with Jacobson's (1935) classical studies of the delayed-response task in the chimpanzee. Over the last fifty years, several investigators have confirmed and extended these early findings in a variety of species, including the monkey (Goldman and Rosvold, 1970; Iversen, 1970; Butter *et al.*, 1968; Gross, 1963; Mishkin, 1957). Three tests that were found to be sensitive, but not specific, to lateral prefrontal cortical damage were the delayed-response task, a test of visuospatial memory (Battig *et al.*, 1960); the conditional position response test, a task of conditional learning (Goldman and Rosvold, 1970); and place- or spatial-reversal learning, a test of cognitive flexibility (Pohl, 1973; Goldman *et al.*, 1971). Of these three tasks, reversal-learning appears to be the most sensitive in discriminating between various types of cortical damage (Pohl, 1973). It should be noted that the prefrontal cortex in the monkey is clearly not unitary in function, but rather can be functionally subdivided into at least two major behaviorally relevant subdivisions: one located on the ventral or orbital surface, the other on the dorsolateral convexity. Several studies have been performed to demonstrate the behavioral dissociation between these two loci (Bachevalier and Mishkin, 1986; Pohl 1973; Goldman *et al.*, 1970; Mishkin *et al.*, 1969; Mishkin, 1964; Butter *et al.*, 1963), but more work is needed to fully understand their function.

Only a limited number of studies have been performed in aged monkeys using reversal-learning tasks, and these have generated apparently contradictory findings. Thus, Bartus *et al.* (1979) reported an age-related impairment on both pattern and color reversals, while Rapp (1990) reported no impairment on either pattern or object reversal. We reported, in a preliminary study (Lai *et al.*, 1995a), an age-related impairment on spatial but not object reversals.

While these results seem confusing at first glance, two factors stand out that may account for them. The first is task complexity. Bartus *et al.* (1979) used a task that displayed nine stimuli simultaneously in a 3-by-3 matrix, whereas Rapp (1990) and Lai *et al.* (1995a) used two-choice paradigms. Chance performance in the 3-by-3 design (11%) is much lower than in the two-choice design (50%) and hence makes the task more difficult. With this factor in mind, we can speculate that aged monkeys are not impaired on a relatively easy form of the pattern-reversal task, but they exhibit robust impairments on a more difficult form of the task.

The second factor to consider is the age of the monkeys being studied. In these previous studies, Bartus *et al.* (1979) examined monkeys 18 to 22 years of age, Rapp (1990) examined monkeys (without known birth dates) thought to be about 23 to 27 years of age, and Lai *et al.* (1995a) examined monkeys 20 to 28 years of age. Because it is clear from the work of Bachevalier *et al.* (1991), and from the previous work we have presented in this chapter, that monkeys over the age of 19 are not a homogeneous group, it may be that individual differences account for differences in results on executive function tests.

We have compared the performance early-aged (n = 11), advanced-aged (n = 9) and oldest-old monkeys (n = 5) to young adult animals on spatial and object reversals (Lai *et al.*, 1995a, and unpublished data). For spatial reversals, in increasing severity, early-aged, advanced-aged, and oldest-old monkeys were impaired on the task. Early-aged monkeys were impaired only on the first reversal, and the oldest-old were impaired on all three reversals. For object-reversal learning, only the monkeys in the oldest-old group evidenced a significant impairment relative to young adults.

Taken together with results of previous reports, these findings suggest that aging is associated with a progressive impairment in executive-system function in the nonhuman primate, a finding paralleling that of normal human aging (see Albert and Moss, this volume).

### 3.4. Motor Skills

Motor skills are a complex set of activities encompassing voluntary and involuntary regulation, integration, proprioception, dexterity, and both gross and fine muscle strength. Such skills are evident both actively in an organism's responses and passively in its posture or appearance.

Motor skills govern people's capacity to adapt, change, control, and respond to their surroundings. Impairment of motor skills have the tendency to interfere with, or alter, everyday life. These types of impairments have been described in a variety of ways, such as paralysis, apraxia, ataxia, motor weakness, and motor slowing (Lezak, 1995).

In normal aged humans, studies have been conducted on a wide variety of motor skills, ranging from the ability to track movement with a finger (Carey *et al.*, 1994), to functional abilities such as brushing teeth (Felder *et al.*, 1994), to even highly complex skills such as playing golf (Over and Thomas, 1995). Overall, these studies and many others have found a decline in both speed and accuracy with age (Rabbit 1979; Fozard and Thomas, 1975; Loveless and Sanford, 1974; Surwillow, 1968; Birren and Botwinick, 1955).

Many investigators have also noted an increased amount of variability in the data of the aged human subjects. This may be due in part to the varying degree of physical fitness. It has been established that healthy adults, particularly those deemed fit in terms of cardiovascular health, have faster reaction times than those who are sedentary or unfit (Light, 1978; Spirduso, 1975; Abrahams and Birren, 1973; Botwinick and Thompson, 1968). It has even been reported that some aged individuals show little to no loss of motor function when their performance is compared with that of young adult subjects; however, in a cross-sectional study of this type (Rowe and Kahn, 1987), it is possible that aged subjects may have been inadvertently selected in a biased fashion, which would have tended to exclude impaired individuals.

Evidence from the neuropsychological, neuropathological, electrophysiological, and neurological literature makes clear that motor functions depend upon many levels of neural activity, requiring the integration of cortex, subcortex, cerebellum, brainstem, and peripheral nervous system (Asanuma, 1973; Ferrier, 1875). However, the relationship of the age-related changes in this system to specific neuropathological and neurochemical alterations remains unclear.

With respect to aged monkeys, only a limited number of studies has been done. Two of these studies measured reaction time while the monkeys were performing a memory task (Bachevalier *et al.,* 1991; Davis, 1978) and found that monkeys, like humans, evidenced a significant increase in reaction time with age. A third study (Baxter and Voytko, 1996), designed to assess the most expeditious response made by the monkey, failed to find an age effect. These results suggest that there may be an overall motor slowing with age in the monkey, but that the capacity to muster a rapid response on demand remains relatively intact.

Our studies of motor system function in the aged monkey have focused on tasks of fine motor control and motor skill. For example, we assessed the performance of young adult monkeys, early-aged monkeys, advanced-aged monkeys, and oldest-old monkeys on both types of tasks (Ronald, 1994). We used the task developed by Kuypers (Lawrence and Kuypers, 1968) to assess fine motor control. The apparatus for this task is a testing board with a series of wells of decreasing size. The monkey is simply required to obtain the food reward from each well. We have also used a motor skill task (Davis *et al.,* 1954) to assess the performance of our aged monkeys. This task required the monkey to guide a hard candy along a bent wire to obtain the candy as a reward. Response latency, errors, and success rate were recorded. Overall, we have not observed any significant group differences in performance on either task. A few individual monkeys in the oldest-old group, however, performed worse on both tasks than any of the younger monkeys in terms of response speed and error number.

A more thorough assessment of motor skill was conducted by another group of investigators (Brickson *et al.,* 1987). They used the same hard-candy paradigm but utilized more difficult combinations of bends in the wire. They found no significant differences in performance on relatively easy portions of the task. On the difficult portion, however, they found a significant group difference in performance, with monkeys in the early-aged, advanced-aged and oldest-old ranges requiring significantly more time to solve it than monkeys of young adult age required.

These studies suggest that in the monkey, as in the human, there is an age-related decline in response speed and accuracy. This decline can be mitigated with exercise, practice, and optimal general health. As the tasks become more complex, however, the deleterious effects of age become more pronounced.

## 4. Performance Patterns

As reviewed above, behavioral studies over the past 20 years have clearly demonstrated the existence of age-related cognitive dysfunction in monkeys, and have laid the groundwork for the eventual discovery of thc neurobiological bases of such declines. We strongly felt that progress toward this latter goal would be facilitated by characterization of normal levels of performance for healthy young adults and by determination of age-specific *prevalences* of rates of cognitive impairment. Further, a single overall index of cognitive ability would provide another measure by which to characterize the degree of age-related impairment along a continuum from normal to severe. Such an overall index of cognitive ability would be a useful supplement to performance measures from individual tasks, particularly in epidemiological studies, where each individual must be rated as impaired or unimpaired in order to examine the relationship of risk factors to later cognitive decline.

We addressed this issue (Herndon *et al.*, 1997) by examining the performance of young rhesus monkeys on each item in a battery of six measures of cognitive performance and by comparing the performance of aged monkeys (as distributed in the three age groups described above) with normal levels of performance. To derive a summary measure of cognitive ability, we next employed Principal Components Analysis (PCA). PCA is a statistical technique yielding a linear transformation of a group of variables into a single number that accounts for the maximum possible proportion of the variance in all of the test battery items. Finally, we described age-specific rates of global impairment among aged monkeys as determined by this single, overall measure.

We used 53 monkeys in the analysis, and three measures of cognitive function (DNMS acquisition, DMNS delay, and DRST-spatial condition) were available. For 30 of these animals, three additional measures of performance, for a total of six, were available (the color condition of DRST, and spatial- and object-reversal learning). Our strategy was first, to derive a summary score from the performance of the smaller group of monkeys on all six tests, and second, to perform an essentially equivalent analysis on the full group, for which at least three independent tests were available. For the 30 monkeys measured on all six, we were able to compare the results of the two analyses. Because the summary score based on only three tests was highly correlated ($r = 0.93$) with the summary score derived from all six tests, we concluded that an adequate description of global cognitive performance could be derived from the smaller set of three. The formula for this summary measure, Cognitive Performance Index (CPI), was $y = 0.55a + 0.6b + 0.6c$, where $a$ is the number of errors required to attain criterion in DNMS (multiplied by $-1$ to provide uniformity of direction), $b$ is the percentage of correct responses during 100 trials of DNMS performance with 120 seconds delay, and $c$ is the DRST-spatial recognition span. We defined cognitive impairment as performance on the CPI below the range displayed by monkeys less than 15 years old. We then found that 50% of our early-aged animals ($n = 12$), 87% of our advanced-aged animals ($n = 15$), and 100% of our oldest-old group ($n = 4$) were impaired according to this criterion.

This global deficit can best be understood in terms of the individual cognitive tasks that make up the measure. A summary of age-specific deficits on individual tasks is presented in Table 1.

It is clear from this table that the pattern of impairment was not uniform among age groups across tasks. For the early-aged animals, prevalence of impairment ranged from zero for the DRST-color condition to more than two-thirds for the DRST-spatial condition. From the early-aged group to the oldest-old group, the

**Table I. Age-Specific Prevalence (%) of Monkeys Severely Impaired (Below Minimum Performance of Young Monkeys) on Each of Six Cognitive Tasks, According to Age Group.** ***No.*** **Indicates Number of Impaired/Number Tested**

| | DNMS Acquisition | | DNMS 120-Sec Delay | | DRST-Spatial | | DRST-Color | | Spatial Reversal | | Object Reversal | |
|---|---|---|---|---|---|---|---|---|---|---|---|---|
| Age (years) | No. | Prevalence (CI) | No. | Prevalence (CI) | No. | Prevalence (CI) | No. | Prevalence (CI) | No. | Prevalence (CI) | No. | Prevalence (CI) |
| <15 | 25 | | 24 | | 18 | | 15 | | 18 | | 25 | |
| 19–23 | 12 | 25[a] | 12 | 42[b] | 12 | 67[b] | 10 | 0 | 11 | 45[b] | 11 | 27[a] |
| | | (10–38) | | (21–73) | | (43–91) | | (0–31) | | (1–19) | | (11–61) |
| 24–28 | 16 | 56[b] | 16 | 63[b] | 15 | 63[b] | 8 | 0.25 | 5 | 80[b] | 5 | 40[a] |
| | | (35–81) | | (41–85) | | (43–91) | | (8–66) | | (48–100) | | (29–95) |
| 29+ | 4 | 100[b] | 3 | 67[b] | 3 | 100[b] | 1 | 100 | 3 | 100[b] | 3 | 33 |
| | | (41–100) | | (30–100) | | (0–100) | | (3–100) | | (30–100) | | (30–100) |
| $\chi^2$ | 6.9[d] | | 1.2 | | 0.5 | | 6.7[d] | | 3.6[c] | | 0.1 | |

Exact 95% confidence intervals (CI) for each prevalence estimate given in parentheses.
[a]Different from animals <15 years of age, $p < 0.05$, Fisher's exact test.
[b]Different from animals <15 years of age, $p < 0.01$, Fisher's exact test.
[c]$\chi^2$ for trend, $p < 0.06$ (young group not included in trend calculation).
[d]$\chi^2$ for trend, $p < 0.01$ (young group not included in trend calculation).

prevalence of impairment increased for DNMS-acquisition, DRST-color, and spatial-reversal tasks. For the remaining three tasks, the prevalence of impairment exhibited in early old age appears to have been asymptotical.

One finding from this analysis is that the prevalence of impairment on the DNMS acquisition and on the performance of the nonmatching task shows markedly different patterns. Thus, in DNMS acquisition, relatively few monkeys in the early-aged group (25%) can be considered impaired, but this proportion increases to 100% over the remainder of the life span. In contrast, performance of DNMS under the 120-second-delay condition appears more severely impaired among the early-aged animals than does the DNMS acquisition.

Of the tests in our battery, those in which the monkey was required to make a response based upon spatial, rather than pattern or color, characteristics reveal the most severe impairment in the early-aged animals, as seen by comparing the performance of the 19–23-year age group (early-aged) on the DRST-spatial and on the spatial-reversal task with the performance on the nonspatial conditions of these same two tasks. Thus, while spatial reversal and DRST-spatial were both characterized by very high prevalence of impairment in early-aged animals, rates of impairment were relatively low in the DRST-color condition (0%) and in the object-reversal task (27%). In identifying a particular sensitivity to aging in tasks requiring memory of location in space, these findings parallel earlier results obtained by several investigators with the delayed-response task, which also requires the subjects to remember information about spatial position (Bachevalier *et al.,* 1991; Arnsten and Goldman-Rakic, 1990; Rapp and Amaral, 1989; Walker *et al.,* 1988; Davis *et al.,* 1982; Marriott and Abelson, 1980; Bartus *et al.,* 1978).

The analysis of cognitive impairment on a task-by-task basis emphasizes that age-related decline in cognitive ability is both widespread and multidimensional, that must not be ignored if structure–function relationships in this decline are to be identified. The CPI should not be viewed as a substitute for individual measures of performance, however, but rather as a supplementary index useful in classifying animals in much the same way that composite measures are used to characterize individual humans in terms of cognitive impairment.

## 5. Longitudinal Studies

In a recent study described above (Moss *et al.,* 1997), we compared the performance of four young adults (5–7 years) to advanced-aged (25–29 years) monkeys on both the spatial and color conditions of the DRST. We found that aged monkeys were impaired relative to young adults on both conditions of the task, a finding consistent with the data from normal aged humans (Moss *et al.,* 1986). We assessed the spatial- and color-recognition spans in this same group of animals 1 year after their initial assessment. Performance in the young adult group was quite consistent. On the spatial span, three of the four animals showed no change in performance, and the fourth evidenced a decline of less than 10%. For the color span, again three animals in the groups showed no change in performance over the 1-year interval. The fourth animal evidenced slightly improved performance.

In the group of five aged monkeys, we observed a very different pattern of performance for the two tasks than that seen in the young adults. On the spatial condition, two animals showed no change in performance, one showed a slight improvement, one showed a mild loss, and the last evidenced a marked loss. In contrast to their performance on the spatial condition, none of the animals showed an appreciable loss in recognition span on the color condition. In fact, two showed a slight improvement.

We have also had the opportunity to longitudinally assess recognition-memory function in a different cohort of young and aged monkeys. We tested a group of three young adult (5–7 years) and four advanced-aged (23–26 years) monkeys 4 years after initial assessment on the spatial-recognition-span test. The pattern of findings in the initial assessment paralleled closely that of the study mentioned previously, with the aged animals obtaining a significantly lower recognition span (mean of 2.1) than that of the young adults (mean of 2.6). Four years later, two of the three animals in the young adult group showed a slight decline in performance, whereas one showed a slight improvement. In the aged group, a more striking dichotomous performance was noted. As in the young adult group, two animals showed very slight decline, and the remaining two showed marked decline, in one case severe. We were intrigued by two aspects of the findings. First, of the seven animals tested, only one showed a severe decline in function over the 4-year interval. Second, and perhaps more important, there was minimal loss of function in two of the aged animals, including one that was 25 years old at the initial testing.

To better understand the pattern of performance by the two markedly impaired aged monkeys, we did an error analysis, which revealed that, unlike the monkeys that did not significantly decline over the 4 years, the two markedly impaired animals tended to make a greater proportion of their errors by selecting the last previously presented disk. Whether this effect represents enhanced perseverative tendencies, or is related more to serial-order effects, remains to be evaluated.

In summary, longitudinal assessment of recognition memory in the rhesus monkey reveals that in aged animals, unlike in young adults, performance is quite variable. Some animals clearly show a marked decline, even within a 1-year interval, while others remain virtually unchanged even after 4 years. It appears, therefore, that like aged humans, aged monkeys are characterized by individual differences. While it is important to understand the neurobiological basis for the marked decline observed in some animals, it is perhaps more important to understand why some animals appear to age successfully. As we continue our studies, adding neuroimaging, neuroanatomical, and neurochemical data, we hope to better identify the factors that reflect or support successful, as opposed to unsuccessful aging in the monkey.

## 6. Neurobiological Considerations

The neurobiological basis of age-related cognitive dysfunction is not well understood, but has become an increasing focus of study in recent years. By relating the pattern of behavioral performance to findings from neurobiologic studies of aged monkeys, several authors have identified a variety of changes that may play a role. Structures or neural systems implicated in these deficits include the medial temporal lobe/hippocampal system (Moss *et al.*, 1988; Rapp and Amaral, 1991) and prefrontal cortices (Bartus, 1979; Peters *et al.*, 1994; Lai *et al.*, 1995; Peters *et al.*, 1997).

With regard to the hippocamps, previous studies have identified error patterns on memory tasks in aged monkeys that parallel those seen in lesion studies in young adult monkeys. Rapp and Amaral (1991) identified a subset of animals that were disproportionately impaired on the delay and the list conditions of the DNMS task, a pattern consistent with medial temporal lobe damage in young monkeys (Bachevalier *et al.*, 1991; Zola-Morgan *et al.*, 1989; Bachevalier and Mishkin, 1986; Murray and Mishkin, 1986; Zola-Morgan and Squire, 1986, 1985; Murray and Mishkin, 1984; Mahut *et al.*, 1982; Mishkin, 1978). In an earlier study (Moss *et al.*, 1988), we identified a pattern of errors on the list condition of the DNMS by aged monkeys that is also characteristic of monkeys with hippocampal damage (Mahut *et al.*, 1982).

It is tempting to consider aging as a mild functional lesion of the hippocampus, a notion advanced by several investigators in the field (Poon, 1985; Struble *et al.*, 1985). It is of interest, however, that neuron loss does not appear to occur in the aged monkey hippocampus. At least two groups (Rosene, 1993; West *et al.*, 1993) have provided initial evidence that no age-related cell loss occurs in the hippocampal CA1 subfield or subiculum of the monkey, and recent evidence suggests only a selective and mild loss of synapses in the dentate gyrus (Tigges *et al.*, in press). Similarly, Amaral (1993) has reported no cell loss in the hippocampus or the entorhinal cortex. Yet West (1993) has reported an age-related neuronal loss in subfield CA4 and in the subiculum in the hippocampus of normal aged humans. Thus, it remains to be determined to what extent the hippocampus undergoes age-related cell loss in primates (See chapters by Rosene and Nicholson; West, this volume).

Possibly more-subtle "sublethal" changes may be occurring in the hippocampus that do not manifest the same degree of impairment as those found after near-total removal of this structure. For example, recent work in our laboratory has shown that the hippocampus in monkeys undergoes volume loss with age and, moreover, that this loss appears related to cognitive decline (Lai *et al.*, 1995b; Rosene *et al.*, submitted). When combined with the finding of Tigges *et al.* (in press) that the number of synapses in the hippocampus does not change with age, one is left to conclude the volume loss is attributable to other changes in cell morphology such as shrinkage, dendritic regression or, as has been suggested recently, damage and loss of white matter (Rosene *et al.*, submitted).

The growing body of evidence of impairment on tasks of spatial memory and those requiring retention of temporal order (Rapp and Amaral, 1989) implicates the prefrontal cortex as another major neuropathological locus in aged monkeys, a view advanced earlier by Bartus and his colleagues (Bartus, 1979; Bartus and Dean, 1979). Adding support to this notion are findings that aged monkeys are impaired on tasks of executive system function (Lai *et al.*, 1995a; Bachevalier 1993), a cognitive domain that includes abilities likely mediated by prefrontal cortical areas (Bachevalier and Mishkin, 1986; Goldman-Rakic and Brown, 1981). Similarly, the impairments in performance that aged monkeys evidence on delayed-response (Bachevalier *et al.*, 1991; Rapp and Amaral, 1989; Walker *et al.*, 1988; Arnsten and Goldman-Rakic, 1985; Marriott and Abelson, 1980; Bartus *et al.*, 1978; Medin *et al.*, 1974) and reversal-learning tasks (Lai *et al.*, 1995) have been historically associated with prefrontal cortical damage in both monkeys and humans (Oscar-Berman *et al.*, 1991; Goldman and Rosvold, 1970; Milner, 1963; Battig *et al.*, 1960). Consistent with this view, an analysis of the errors made by early- and advanced-aged monkeys showed that they made significantly more perseverative errors on the DRST and reversal tasks than did young adults. However, the extent to which frontal cortical system pathology may underlie the deficits observed in aged monkeys remains unclear.

Studies are converging to suggest that there is no significant age-related loss of neurons in area 46 of the prefrontal cortex (Peters *et al.*, 1994), area 4 of the frontal cortex (Tigges *et al.*, 1990) nor in other cortical regions (Amaral, 1993; Peters and Sethares, 1993; Rosene, 1993; West *et al.*, 1993; Vincent *et al.*, 1989; see chapter by Peters, this volume). Though neurons in area 40 of the prefrontal cortex do not undergo age-related cell loss, the white matter in this region has been shown to undergo marked alteration with age (Peters *et al.*, 1994). Moreover, a recent study by these same investigators revealed a loss of synapses and a reduction in thickness of layer I in area 46 in a group of aged rhesus monkeys as compared with a group of young adults (Peters *et al.*, 1997).

Age-related amyloid plaques are more likely to develop in the prefrontal and temporal cortices than in more-caudal regions of the brain (Struble *et al.*, 1985). Nevertheless, there is apparently no relationship between cognitive impairment and plaque density (Slone *et al.*, 1997), suggesting that age-related accumulation of amyloid in plaques is not a significant pathogenic feature of normal aging in the monkey. Support for this has recently been obtained in a study of amyloid plaques that indicates that the monkey has mainly the less-toxic Aβ1-40, whereas in humans, the more-toxic Aβ1-42 predominates (Gearing *et al.*, in press).

Despite evidence of a preserved number of neurons in prefrontal cortex, reduction in endogenous concentrations of dopamine in this area have been reported (Goldman-Rakic and Brown, 1981). Loss of serotonin binding sites with age has also been reported in both the frontal cortex and the hippocampus (Marcusson *et al.*, 1984). This is particularly interesting, because neuronal loss has been demonstrated in the brainstem nucleus raphe dorsalis and the nucleus centralis superior of aged monkeys (Kemper, 1993), both of which are serotonergic and project mainly to the forebrain. Additional evidence of subcortical neuron loss within the cholinergic basal forebrain nuclei is conflicting. Rosene (1993) reported an age-related decline in the number of AChE-positive neurons in all three major basal forebrain nuclei—the medial septum, the diagonal band, and the nucleus basalis—and Stroessner-Johnson *et al.* (1992) reported age-related loss of ChAT-positive neurons in the medial septum. However, Voytko *et al.* (1995) recently reported no age-related cell loss in the nucleus basalis using an antibody to the p75 low-affinity nerve growth factor receptor. Clearly, the possible contribution of cortically projecting subcortical-site neuron loss to age-related cognitive decline needs further investigation.

One intriguing change recently noted in both humans and monkeys, and one particularly relevant to the present study, is an apparent loss of white matter, but not gray matter, that occurs with age (Rosene *et al.*, submitted; Lai *et al.*, 1995b; Albert, 1993). This is supported by a recent neuropathological study of area 46 in aged monkeys that confirms the presence of myelin degeneration in the deep layers of cortex and underlying white matter (Peters *et al.,* 1994). There has also been some preliminary indication that loss of white matter volume determined by MRI may correlate positively and significantly with cognitive decline (Lai *et al.,* 1995b). Given the nature of the cognitive deficits found with aging, there is likely a disruption of the functional integrity of prefrontal cortical systems and medial temporal lobe structures such as the hippocampus. This is made more intriguing in the context of myelin degeneration, which would compromise communication within and between both areas.

The results of the present investigation add to the evidence that rhesus monkeys undergo cognitive declines, beginning at a relatively early age, a finding that parallels the observations made in normal human aging (Albert and Moss, 1996). Initial findings from anatomical studies raise the possibility that this decline may be related more to a deterioration of cortical information processing due to dystrophic myelin or altered neuronal function than to a cortical neuronal cell loss.

## 7. Conclusions

Evidence has accumulated that monkeys, like normal aged humans, evidence age-related cognitive decline and that the pattern of this decline changes with age. Further, as in humans, the changes that occur with normal aging in monkeys are apparently not universal. Some monkeys age quite successfully, while others age unsuccessfully. One of the primary challenges in researching the effects of normal aging is determining the neurobiological basis of these differing patterns of age-related change.

ACKNOWLEDGMENTS. This research was supported by NIH grants PO1-AG00001 and R55-AG12610 from the National Institute on Aging and by NIH grant RR-00165 from the National Center for Research Resources to the Yerkes Regional Primate Research Center. Boston University School of Medicine and the Yerkes Regional Primate Research Center are both fully accredited by the American Association for Accreditation of Animal Laboratory Care. We wish to thank our colleagues, Drs. Douglas Rosene, Alan Peters, Tom Kemper, Julie Sandell, Johannes Tigges, Carmella Abraham, and Bradley Hyman, for their contributions to this body of work, and also Jessica Berman, Beverly Duryea, Claudia Fitzgerald, and Tara Moore for their valuable assistance with this project.

## 8. References

Abrahams, J. P., and Birren, B. E., 1973, Reaction time as a function of age and behavioral predisposition to coronary heart disease, *J. Gerontol.* **28**:471–478.

Albert, M. S., and Moss, M. B., 1996, Neuropsychology of aging: Findings in humans and monkeys, in: *Handbook of the Biology of Aging, 4th Edition* (E. L. Schneider and J. W. Rowe, eds., J. H. Morrison, section ed.), Academic, San Diego, pp. 217–233.

Albert, M. S., 1988, Cognitive function, in: *Geriatric Neuropsychology* (M. S. Albert and M. B. Moss, eds.), Guilford, New York, pp. 33–53.

Albert, M. S., 1993, Neuropsychological and neurophysiological changes in healthy adult humans across the age range, *Neurobiol. Aging* **14:**623–625.

Albert, M. S., Duffy, F. H., and Naeser, M. A., 1987, Nonlinear changes in cognition and their neurophysiological correlates, *Can. J. Psychol.* **41:**141–157.

Albert, M. S., Wolfe, J., and Lafleche, G., 1990, Differences in abstraction ability with age, *Psychol. and Aging* **5:**94–100.

Amaral, D. G., 1993, Morphological analyses of the brains of behaviorally characterized aged nonhuman primates, *Neurobiol. Aging* **14:**671–672.

Anderson, J. R., Anthouard, M., de Monte, M., and Kempt, J., 1993, Differences in performance of young and old monkeys on a visuospatial memory task, *Q. J. Exper. Psychol.* **46B:**391–398.

Arnsten, A. F. T., and Goldman-Rakic, P. S., 1990, Analysis of alpha-2 adrenergic agonist effects on the delayed nonmatching to sample performance of aged rhesus monkeys, *Neurobiol. Aging* **11:**583–590.

Arnsten, A. F. T., and Goldman-Rakic, P. S., 1985, α-2 adrenergic mechanisms in prefrontal cortex associated with cognitive decline in aged nonhuman primates, *Science* **230:**1273–1276.

Asanuma, H., 1973, Cerebral cortical control of movement, *Physiologist* **16:**143–166.

Bachevalier, J., 1993, Behavioral changes in aged rhesus monkeys, *Neurobiol. Aging* **14:**619–621.

Bachevalier, J., and Mishkin, M., 1986, Visual recognition impairment follows ventromedial, but not dorsolateral, prefrontal lesions in monkeys, *Behav. Brain Res.* **20:**249–261.

Bachevalier, J., Landis, L. S., Walker, L. C., Brickson, M., Mishkin, M., Price, D. L., and Cork, L. C., 1991, Aged monkeys exhibit behavioral deficits indicative of widespread cerebral dysfunction, *Neurobiol. Aging* **12:**99–111.

Baddeley, A., 1986, *Working memory,* Oxford University Press, Oxford.

Bartus, R. T., 1979, Effects of aging on visual memory, sensory processing, and discrimination learning in a nonhuman primate, in: *Sensory Systems and Communication in the Elderly* (J. M. Ordy and K. Brizzee, eds.), Raven, New York, pp. 137–145.

Bartus, R. T., and Dean, R. L., 1979, Recent memory in aged non-human primates: Hypersensitivity to visual interference during retention, *Exp. Aging Res.* **5:**385–400.

Bartus, R. T., Fleming, D., and Johnson, H. R., 1978, Aging in the rhesus monkey: Debilitating effects on short-term memory, *J. Gerontol.* **34:**209–219.

Bartus, R. T., Dean, R. L., and Fleming, D. L., 1979, Aging in the rhesus monkey: Effects on visual discrimination learning and reversal learning, *J. Gerontol.* **34:**209–219.

Battig, K., Rosvold, H. E., and Mishkin, M., 1960, Comparison of the effects of frontal and caudate lesions on delayed response and alternation in monkeys, *J. Comp. Physiol. Psychol.* **53:**400–440.

Baxter, M. G., and Voytko, M. L., 1996, Spatial orienting of attention in adult and aged rhesus monkeys, *Behav. Neurosci.* **110:**898–904.

Beason, L., Moss, M. B., and Rosene, D. L., 1990, Effects of entorhinal, parahippocampal, or basal forebrain lesions on recognition memory in the monkey, *Soc. Neurosci. Abstr.* **16:**725.

Birren, J. E., and Botwinick, J., 1955, Speed of response as a function of perceptual difficulty and age, *J. Gerontol.* **10:**433–436.

Botwinick, J., and Thompson, L. W., 1968, Age differences in reaction time: An artifact? *Gerontologist* **8:**25–28.

Bowden, D. M., and Williams, D. D., 1984, Aging, in: *Advances in Veterinary Science and Comparative Medicine,* Vol. 28, Academic, New York, pp. 305–341.

Brickson, M., Bachevalier, J., Watermeier, L. S., Walker, L. C., Struble, R. G., Price, D. L., Mishkin, M., and Cork, L. C., 1987, Performance of aged rhesus monkeys on a visuospatial task, *Soc. Neurosci. Abstr.* **13:**1627.

Brinley, J., and Fichter, J., 1970, Performance deficits in the elderly in relation to memory load and set, *Gerontologist* **25:**30–35.

Broadbent, D. E., and Gregory, M., 1965, Some confirmatory results on age differences in memory for simultaneous stimulation, *Br. J. Psychol.* **56:**77–80.

Bromley, D., 1957, Effects of age on intellectual output, *J. Gerontol.* **12:**318–323.

Butter, C. M., Mishkin, M., and Rosvold, H. E., 1963, Conditioning and extinction of a food-rewarded response after selective ablations of frontal cortex in rhesus monkeys, *Exp. Neurol.* **7:**65–75.

Butter, C. M., Mishkin, M., and Mirsky, A. F., 1968, Emotional responses toward humans in monkeys with selective frontal lesions, *Physiol. and Behav.* **3:**213–215.

Carey, J. R., Bogard, C. L., King, B. A., and Suman, V. J., 1994, Finger-movement tracking scores in healthy subjects, *Percept. Motor Skills* **79:**563–576.

Cools, A. R., Van Den, J. H. L., Bercken, M. W. I., Horstink, K., Van Spaendonck, P. M., and Berger, P. M., 1984, Cognitive and motor shifting aptitude disorder in Parkinson's disease, *J. Neurol. Neurosurg. Psychiatry* **4:**443–453.
Cowey, A., and Gross, C. G., 1970, Effects of foveal prestriate and inferotemporal lesions on visual discrimination by rhesus monkeys, *Exp. Brain Res.* **11:**128–144.
Craik, F. I. M., 1971, Age differences in recognition memory, *Q. J. Exp. Psychol.* **23:**316–323.
Craik, F. I. M., and Simon, E., 1980, Age differences in memory: The role of attention and depth of processing, in: *New Directions in Memory and Aging* (L. W. Poon, J. L. Fozard, L. S. Cermak, and L. W. Thompson, eds.), Lawrence Erlbaum, New Jersey.
Damasio, A. R., 1985, The frontal lobes, in: *Clinical Neuropsychology* (K. M. Heilman and E. Valenstein, eds.), Oxford University Press, New York, pp. 339–376.
Davis, R. T., 1978, Old monkey behavior, *Exp. Gerontol.* **13:**237–250.
Davis, R. T., 1985, Introduction, in: *Behavior and Pathology of Aging in Rhesus Monkeys* (R. T. Davis and C. W. Leathers, eds.), Alan R. Liss, New York, pp. 1–9.
Davis, R. T., McDowell, A. A., Deter, C. W., and Stelle, J. P., 1954, Performance of rhesus monkeys on selected laboratory tasks presented before and after a large single dose of whole-body X radiation, *Exp. Gerontol.* **4:**20–26.
Davis, R. T., Bennet, C. L., and Weisenburger, R. P., 1982, Repeated measurements of forgetting by rhesus monkeys *(Macaca mulatta), Percept. Motor Skills* **55:**703–709.
Dyke, B., Gage, T. B., Mamelka, P. M., Goy, R. W., and Stone, W. H., 1986, A demographic analysis of the Wisconsin Regional Primate Center rhesus colony, 1962–1982, *Am. J. Primatol.* **10:**257–269.
Felder, R., Reveal, M., Lemon, S., and Brown, C., 1994, Testing toothbrushing ability of elderly patients, *Spec. Care Dentist.* **14:**153–157.
Ferrier, D., 1875, Experiments of the brain of monkeys, I, *Proc. R. Soc. London* **23:**409–430.
Fozard, J. L., and Thomas, J. C., 1975, Psychology of aging: Basic findings and some psychiatric applications, in: *Modern Perspectives in the Psychiatry of Old Age* (J. G. Howells, ed.), Brunner/Hassel, New York.
Gaffan, D., 1974, Recognition impaired and association intact in the memory of monkeys after transection of the fornix, *J. Comp. Physiol. Psychol.* **86:**1100–1109.
Gearing, M., Tigges, J., Mori, H., and Mirra, S. S., Alpha beta is a major form of beta-amyloid in nonhuman primates, *Neurobiol. Aging,* in press.
Goldman, P. S., and Rosvold, H. E., 1970, Localization of function within the dorsolateral prefrontal cortex of the rhesus monkey, *Exp. Neurol.* **27:**291–304.
Goldman, P. S., and Rosvold, H. E., 1972, The effects of selective caudate lesions in infant and juvenile rhesus monkeys, *Brain Res.* **43:**53–66.
Goldman, P. S., Rosvold, H. E., and Mishkin, M., 1970, Selective sparing of function following prefrontal lobectomy in infant monkeys, *Exp. Neurol.* **29:**221–226.
Goldman, P. S., Rosvold, H. E., Vest, B., and Galkin, T. W., 1971, Analysis of the delayed-alternation deficit produced by dorsolateral prefrontal lesions in the rhesus monkey, *J. Comp. Physiol. Psychol.* **77:**212–220.
Goldman-Rakic, P. S., and Brown, R. M., 1981, Regional changes in monoamines in cerebral cortex and subcortical structures of aging rhesus monkeys, *Neuroscience* **6:**177–178.
Greenwood, P. M., Parasuraman, R., and Hazby, J. V., 1993, Changes in visuospatial attention over the adult lifespan, *Neuropsychologia* **31:**471–485.
Gross, C. G., 1963, Comparison of the effects of partial and total lateral frontal lesions on test performance by monkeys, *J. Comp. Physiol. Psychol.* **56:**41–47.
Hartly, A. A., Kieley, J. M., and Slabach, E. H., 1990, Age differences and similarities in the effects of cues and prompts, *J. Exp. Psychol.* **16:**523–537.
Hebb, D. O., 1961, Distinctive features of learning in the higher animal, in: *Brain Mechanisms and Learning* (J. F. Delafresnay, ed.), Blackwell, Oxford, pp. 58–72.
Herndon, J., Moss, M. B., Killiany, R. J., and Rosene, D. L., 1997, Patterns of cognitive decline in early, advanced, and oldest of the old aged rhesus monkeys, *Behav. Brain Res.* **87:**25–34.
Hess, T. M., and Slaughter, S. L., 1986, Aging effects on prototype abstraction and concept identification, *J. Gerontol.* **41:**214–221.
Howieson, D. B., Holm, L. A., Kaye, J. A., Oken, B. S., and Howieson, J., 1993, Neurologic function in the optimally healthy oldest old, *Neurology* **43:**1882–1886.
Hyman, B. T., Van Hoesen, G. W., Demasio, A. R., and Barns, C. L., 1984, Alzheimer's disease: Cell specific pathology isolates the hippocampal formation, *Science* **225:**1168–1170.

Inouye, S. K., Albert, M. S., Mohs, R., Sun, K., and Berkman, L. F., 1993, Cognitive performance in a high-functioning community-dwelling elderly population, *J. Gerontol.* **48:**146–151.
Iversen, S. D., 1970, Interference and inferotemporal memory deficits, *Brain Res.* **19:**277–289.
Jacobson, C. F., 1935, Functions of the frontal association are in primates, *Arch. Neurol. Psychiatry* **33:**558–569.
Kemper, T. L., 1993, The relationship of cerebral cortical changes to nuclei in the brainstem, *Neurobiol. Aging* **14:**659–660.
Kemper, T. L., 1994, Neuroanatomical and neuropathological changes during aging and dementia, in: *Clinical Neurology and Aging* (M. L. Albert and J. E. Knoefel, eds.), Oxford University Press, New York, pp. 3–67.
Killiany, R. J., Moss, M. B., Rosene, D. L., Herndon, J., and Lai, Z. C., 1997, Recognition memory function in early aged rhesus monkeys, *Psychobiology,* submitted.
Killiany, R. J., and Mahut, H., 1991, Ontogenetic development of recognition memory span in rhesus macaques, *IBRO Abstr.* **165:**312.
Lai, Z. C., Moss, M. B., Killiany, R. J., Rosene, D. L., and Herndon, J. G., 1995a, Executive system dysfunction in the aged monkey: Spatial and object reversal learning, *Neurobiol. Aging* **16:**947–95.
Lai, Z. C., Rosene, D. L., Killiany, R. J., Pugliese, D., Albert, M. S., and Moss, M. B., 1995b, Age-related changes in the brain of the rhesus monkey: MRI changes in white matter but not gray matter, *Soc. Neurosci. Abstr.* **21:**156.
Lange, K. W., Robbins, T. W., Marsden, C. D., James, M., Owen, A. M., and Paul, G. M., 1992, L-dopa withdrawal in Parkinson's disease selectively impairs cognitive performance in tests sensitive to frontal lobe function, *Psychopharmacology* **107:**394–400.
Lawrence, D. G., and Kuypers, H. G. J. M., 1968, The functional organization of the motor system in the monkey. I. The effects of bilateral pyramidal lesions, *Brain* **91:**1–18.
Lees, A. J., and Smith, E., 1983, Cognitive deficits in the early stages of Parkinson's disease, *Brain* **106:**257–270.
Lezak, M. D., 1995, *Neuropsychological Assessment,* 3rd Ed., Oxford University Press, New York.
Light, K. C., 1978, Effects of mild cardiovascular and cerebrovascular disorders on serial reaction time performance, *Exp. Aging Res.* **4:**3–22.
Loveless, N. E., and Sanford, A. J., 1974, Effects of age on contingent negative variation and preparatory set in a reaction time task, *J. Gerontol.* **29:**52–63.
Mack, J. L., and Carlson, N. J., 1978, Conceptual deficits and aging: The category test, *Percept. Motor Skills* **46:**123–128.
Mahut, H., Zola-Morgan, S., and Moss, M. B., 1982, Hippocampal resections impair associative learning and recognition memory in the monkey, *J. Neurosci.* **2:**1214–1229.
Malamut, B. L., Saunders, R. C., and Mishkin, M., 1984, Monkeys with combined amygdalo-hippocampal lesions succeed in object discrimination learning despite 24-hour intertrial intervals, *Behav. Neurosci.* **98:**759–769.
Marcusson, J. O., Morgan, D. G., Winblad, B., and Finch, C. E., 1984, Serotonin-2 binding sites in human frontal cortex and hippocampus: Selective loss of S-2A sites with age, *Brain Res.* **311:**51–56.
Marriott, J. G., and Abelson, J. S., 1980, Age differences in short-term memory of test-sophisticated rhesus monkeys, *Age* **3:**7–9.
Martin, E. M., Pitrak, D. L., Pursell, K. J., Mullane, K. M., and Novak, R. M., 1995, Delayed recognition memory span in HIV-1 infection, *J. Int. Neuropsychol. Soc.* **1:**575–580.
Medin, D. L., and Davis, R. T., 1974, in: *Behavior of Non-Human Primates* (A. M. Shrier and F. Stollmitz, eds.), Academic, New York, pp. 1–47.
Milner, B., 1963, Effects of different brain lesions on card sorting, *Arch. Neurol.* **9:**90–100.
Mishkin, M., 1957, Effects of small frontal lesions on delayed alternation in monkeys, *J. Neurophysiol.* **20:**615–622.
Mishkin, M., 1964, Perseveration of central sets after frontal lesions in monkeys, in: *The Frontal Granular Cortex and Behavior* (J. M. Warren and K. Akert, eds.), McGraw-Hill, New York, pp. 219–241.
Mishkin, M., 1978, Memory in monkeys severely impaired by combined, but not by separate, removal of amygdala and hippocampus, *Nature* **273:**297–298.
Mishkin, M., Vest, B., Waxler, M., and Rosvold, H. E., 1969, A re-examination of the effects of frontal lesions on object alternation, *Neuropsychologia* **7:**357–363.
Moss, M. B., 1983, Assessment of memory in amnesic and dementia patients: Adaptation of behavioral tests used with non-human primates, *INS Bull.* **3:**10–17.

Moss, M. B., Mahut, H., and Zola-Morgan, S., 1981, Concurrent discrimination learning of monkeys after hippocampal, entorhinal, or fornix lesions, *J. Neurosci.* **1:**227–240.

Moss, M. B., Albert, M. S., Butters, N., and Payne, M., 1986, Differential patterns of memory loss among patients with Alzheimer's disease, Huntington's disease, and alcoholic Korsakoff's syndrome, *Arch. Neurol.* **43:**239–246.

Moss, M. B., Rosene, D. L., and Peters, A., 1988, Effects of aging on visual recognition memory in the rhesus monkey, *Neurobiol. Aging* **9:**495–502.

Moss, M. B., Killiany, R. J., Lai, Z. C., Rosene, D. L., and Herndon, J. G., 1997, Recognition memory span in rhesus monkeys of advanced age, *Neurobiol. Aging* **18:**13–19.

Murray, E. A., and Mishkin, M., 1984, Severe tactual as well as visual memory deficits follow combined removal of the amygdala and hippocampus in monkeys, *J. Neurosci.* **4:**2565–2580.

Murray, E. A., and Mishkin, M., 1986, Visual recognition in monkeys following rhinal cortical ablations combined with either amygdalectomy or hippocampectomy, *J. Neurosci.* **6:**1991–2003.

Oscar-Berman, M., and Zola-Morgan, S., 1980, Comparative neuropsychology and Korsakoff's syndrome. II. Two-choice visual discrimination learning, *Neuropsychologia* **18:**513–525.

Oscar-Berman, M., McNamara, P., and Freedman, M., 1991, Delayed-response tasks: Parallels between experimental ablation studies and findings in patients with frontal lesions, in: *Frontal Lobe Function and Dysfunction* (H. S. Levin, H. M. Eisenbery, and A. L. Benton, eds.), Oxford University Press, New York, pp. 230–255.

Oscar-Berman, M., Hutner, N., and Bonner, T. R., 1992, Visual and auditory spatial and nonspatial delayed-response performance by Korsakoff and non-Korsakoff alcoholic and aging individuals, *Behav. Neurosci.* **106:**613–622.

Over, R., and Thomas, P., 1995, Age and skilled psychomotor performance: A comparison of younger and older golfers, *Int. J. Aging Hum. Dev.* **41:**1–12.

Park, S., and Holzman, P. S., 1992, Schizophrenics show spatial working memory deficits, *Arch. Gen. Psychiatry* **49:**975–982.

Park, D., Holzman, P. S., and Lenzenweger, M. F., 1995a, Individual differences in spatial working memory in relation to schizophrenia, *J. Abnorm. Psychol.* **104:**355–363.

Park, D., Holzman, P. S., and Goldman-Rakic, P. S., 1995b, Spatial working memory deficits in the relatives of schizophrenics, *Arch. Gen. Psychiatry* **52:**821–828.

Peters, A., Sethares, C., and Moss, M. B., 1998, The effects of aging on layer 1 in area 46 of prefrontal cortex in the rhesus monkey, *Cerebral Cortex* **8:**671–684.

Peters, A., and Sethares, C., 1993, Aging and the Meynert cells in rhesus monkey primary visual cortex, *Anat. Rec.* **236:**721–729.

Peters, A., Gruia-Leahu, D., Moss, M. B., and McNally, K., 1994, The effects of aging on area 46 of the frontal cortex of the rhesus monkey, *Cereb. Cortex* **6:**621–635.

Peters, A., Rosene, D. L., Moss, M. B., Kemper, T. L., Abraham, C. R., Tigges, J., and Albert, M. S., 1996, Neurobiological bases of age-related cognitive decline in the rhesus monkey, *J. Neuropathol. Exp. Neurol.* **55:**861–874.

Petersen, R. C., Smith, G., Kokmen, E., Ivnik, R. J., and Tangalos, E. G., 1992, Memory function in normal aging, *Neurology* **42:**396–401.

Pohl, W. G., 1973, Dissociation of spatial discrimination deficits following frontal and parietal lesions in monkey, *J. Comp. Physiol. Psychol.* **82:**227–239.

Poon, L. W., 1985, Differences in human memory with aging, in: *Handbook of the Psychology of Aging* (J. E. Dirren and K. W. Schaie, eds.), Van Nostrand Reinhold, New York, pp. 427–462.

Posner, M. I., 1988, Structures and functions of selective attention, in: *Master Lectures in Clinical Neuropsychology* (T. Boll and B. Bryant, eds.), American Psychological Association, Washington, D.C., pp. 173–202.

Posner, M. I., and Cohen, Y., 1984, Components of performance, in: *Attention and Performance* (H. Bouma and D. Bowhuis, eds.), Lawrence Erlbaum, Hillsdale, NJ, pp. 531–556.

Presty, S. K., Bachevalier, J., Walker, L. C., Struble, R. G., Price, D. L., Mishkin, M., and Cork, L. C., 1987, Age differences in recognition memory of the rhesus monkey *(Macaca mulatta), Neurobiol. Aging* **8:**435–440.

Rabbit, P. M. A., 1979, How old and young subjects monitor and control responses for accuracy and speed, *Br. J. Psychol.* **70:**305–311.

Raine, A., Lencz, T., Reynolds, G. P., Harrison, G., Sheard, C., Medley, I., Reynolds, L. M., and Cooper, J. E., 1992, An evaluation of structural and functional prefrontal deficits in schizophrenia: MRI and neuropsychological measures, *Psychiatry Res.* **45:**123–137.

Rapp, P. R., 1990, Visual discrimination and reversal learning in the aged monkey *(Macaca mulatta), Behav. Neurosci.* **104:**876–888.

Rapp, P. R., and Amaral, D. G., 1989, Evidence for a task-dependent memory dysfunction in the aged monkey, *J. Neurosci.* **9:**3568–3576.

Rapp, P. R., and Amaral, D. G., 1991, Recognition memory deficits in a subpopulation of aged monkeys resemble the effects of medial temporal lobe damage, *Neurobiol. Aging* **12:**481–486.

Rehbein, L., and Mahut, H., 1983, Long-term deficits in associative and spatial recognition memory after early hippocampal damage in monkeys, *Soc. Neurosci. Abstr.* **9:**639.

Ronald, P., 1994, Motor skill learning, performance, and handedness in aged rhesus monkeys, master's thesis, Department of Anatomy, Boston University School of Medicine.

Rosene, D. L., 1993, Comparing age-related changes in the basal forebrain and hippocampus of the rhesus monkey, *Neurobiol. Aging* **14:**669–670.

Rosene, D. L., Lai, Z. C., Killiany, R. J., Moss, M. B., Jolesz, F., and Albert, M. S., Age related loss of white matter with preservation of gray matter in the forebrain of the rhesus monkey: An MRI study, *Neurobiol. Aging,* submitted.

Rosene, D. L., Moss, M. B., Berman, J. D., Williams, K. A., and Killiany, R. J., Preservation of neuron number and size in the hippocampal formation of the aging rhesus monkey, *Neurobiol. Aging,* submitted.

Rowe, J. W., and Kahn, R., 1987, Human aging: Useful vs. successful, *Science* **237:**143–149.

Salmon, D. P., Granholm, E., McCullough, D., Butters, N., and Grant, I., 1989, Recognition memory span in mild and moderately demented patients with Alzheimer's disease, *J. Clin. Exp. Neuropsych.* **4:**429–443.

Shimamura, A. P., Janowsky, J. S., and Squire, L. R., 1991, What is the role of frontal lobe damage in memory disorders? in: *Frontal Lobe Function and Dysfunction* (H. S. Levin, H. M. Eisenberg, and A. L. Benton, eds.), Oxford University Press, Oxford, pp. 173–198.

Sloane, J., Pietropaolo, M., Rosene, D. L., Moss, M. B., Peters, A., Kemper, T. L., and Abraham, C. R., Lack of correlation between plaque burden and cognition in the aged monkey, in press.

Spirduso, W. W., 1975, Reaction and movement time as a function of age and physical activity level, *J. Gerontol.* **35:**850–865.

Stroessner-Johnson, H. M., Rapp, P. R., and Amaral, D. G., 1992, Cholinergic cell loss and hypertrophy in the medial septal nucleus of the behaviorally characterized aged rhesus monkey, *J. Neurosci.* **12:**1936–1944.

Struble, R. J., Price, D. L., Jr., Cork, L. C., and Price, D. L., 1985, Senile plaques in cortex of aged normal monkeys, *Brain Res.* **361:**267–275.

Stuss, D. T., 1991, Interference effects on memory function in posteukotomy patients: An attentional perspective, in: *Frontal Lobe Function and Dysfunction* (H. S. Levin, H. M. Eisenberg, and A. L. Benton, eds.), Oxford University Press, Oxford, pp. 157–172.

Stuss, D. T., and Benson, D. F., 1986, *The Frontal Lobes,* Raven, New York.

Surwillo, W. W., 1968, Timing of behavior in senescence and the role of the central nervous system, in: *Aging and Behavior* (G. A. Talland, ed.), Academic, New York.

Talland, G. A., 1961, Effects of aging on the formation of sequential and spatial concepts, *Percept. Motor Skills* **13:**210.

Tigges, J., Gordon, T. P., McClure, H. M., Hall, E. C., and Peters, A., 1988, Survival rate and life span of rhesus monkeys at the Yerkes Regional Primate Research Center, *Am. J. Primatol.* **15:**263–273.

Tigges, J., Herndon, J. G., and Peters, A., 1990, Neuronal population of area 4 during life span of the rhesus monkey, *Neurobiol. Aging* **11:**201–208.

Tigges, J., Herndon, J. C., and Rosene, D. L., Preservation into old age of synaptic number and size in the supragranular layer of the dentate gyrus in rhesus monkeys, *Anat. Rec.,* in press.

Vincent, A. E., Peters, A., and Tigges, J., 1989, Effects of aging on the neurons within area 17 of rhesus monkey cerebral cortex, *Anat. Rec.* **223:**329–341.

Voytko, M. L, 1993, Cognitive changes during normal aging in monkeys assessed with an automated test apparatus, *Neurobiol. Aging* **14:**643–644.

Voytko, M. L., Sukhov, R. C., Walker, L. C., Breckler, S. J., Price, D. L., and Koliatson, V E., 1995, Neuronal number and size are preserved in the nucleus basalis of aged rhesus monkeys, *Dementia* **6:**131–141.

Walker, L. C., Kitt, C. A., Struble, R. G., Wagster, M. V., Price, D. L., and Cork, L. C., 1988, The neural basis of memory decline in aged monkeys, *Neurobiol. Aging* **9:**657–666.

West, M. J., 1993, Regionally specific loss of neurons in the aging human hippocampus, *Neurobiol. Aging* **14**:287–293.
West, M. J., Amaral, D. G., and Rapp, P. R., 1993, Preserved hippocampal cell number in aged monkeys with recognition memory deficits, *Soc. Neurosci. Abstr.* **19**:599.
Zola-Morgan, S., and Squire, L. R., 1985, Medial temporal lesions in monkeys impair memory on a variety of tasks sensitive to human amnesia, *Behav. Neurosci.* **99**:22–38.
Zola-Morgan, S., and Squire, L. R., 198impairment in monkeys following lesions limited to the hippocampus, *Behav. Neurosci.* **100**:155–160.
Zola-Morgan, S., Squire, L. R., and Amaral, D. G., 1989, Lesions of the hippocampal formation, but not lesions of the fornix or the mammillary nuclei, produce long-lasting memory impairment in monkeys, *J. Neurosci.* **9**:898–913.

3

# Normal Aging in the Cerebral Cortex of Primates

ALAN PETERS

## 1. Introduction

Studies of the neurobiological basis of normal aging have gained impetus in recent years, because it has become clear that even normal aging is accompanied by a decline in several aspects of cognitive function. The need to understand the underlying basis for cognitive decline with age is made even more important by the fact that our population is gradually shifting toward one that contains a greater proportion of older individuals. And at the same time as medical knowledge and treatment of life-threatening problems, such as heart attacks, diabetes and cancer, improves, life expectancy is also being prolonged. Yet there has been little emphasis on the changes that occur in the nervous system during the normal aging process. Rather, the emphasis has been on the changes that occur as a consequence of Alzheimer's disease, because the behavioral changes induced by Alzheimer's disease profoundly affect the lives of Alzheimer's patients and their families, and because the alterations in the structure of the brain—such as the presence of neuritic plaques, neurofibrillary tangles, and amyloid deposits—are very obvious. The structural changes resulting from normal age-related cognitive decline are much less obvious,

ALAN PETERS • Department of Anatomy and Neurobiology, Boston University School of Medicine, Boston, Massachusetts 02118.

*Cerebral Cortex,* Volume 14, edited by Peters and Morrison. Kluwer Academic/Plenum Publishers, New York, 1999.

and normal aging has much less social impact, for it is accepted as being "normal." But normal aging changes are important to understand both in their own right and because they constitute the baseline upon which the changes brought about by Alzheimer's disease are superimposed. For example, it is necessary to answer questions such as: Is the distribution of plaques during normal aging similar to, or different from, that in patients who have Alzheimer's disease? Are neurons lost from the cerebral cortex during normal aging, or does severe cortical neuronal loss occur only in Alzheimer's disease?

One of the difficulties in studying the effects of normal aging in humans is to know whether the tissue being examined is from normal individuals or from individuals who showed signs of dementia. Few humans have been studied behaviorally for long enough before death to have accurate data about their cognitive state. In addition, the inevitable postmortem delays make it extremely difficult to obtain human tissue that is well preserved.

One solution is to develop suitable models of normal human aging. A commonly used model is that of the aging laboratory rat or mouse. These rodent models provide the advantages of stable gene pools, large litters, and relatively short lives. For example, the average laboratory rat has a maximum life span of about 3 years, and some investigators consider even a 15-month-old rat to be aged. Yet if rats are kept active and prevented from becoming obese, they can live for up to 5 years (West *et al.*, 1984; Peters *et al.*, 1987). The main problem with laboratory rats, however, is the difficulty of devising meaningful behavioral tasks to reveal age-related cognitive declines. A much better model is the nonhuman primate, which has a longer life span and can be behaviorally tested, even using some cognitive tasks adapted from human neuropsychological test batteries (e.g., Albert and Moss, 1984; Mishkin, 1993; Zola-Morgan and Squire, 1985). Such tests show that the primary deficits that occur with age affect short-term memory and executive system function (see chapter by Moss, Killiany, and Herndon, this volume, for further details). Moreover, the monkey brain more closely resembles that of the human, and the monkey model offers the potential for developing testable treatments and interventions to decelerate or reverse normal cognitive decline.

In our studies, we have been using the rhesus monkey (*Macaca mulatta*) as a model for determining the behavioral and neurobiological changes that can occur as a consequence of normal aging (see Peters *et al.*, 1996). However, before these studies could begin, we needed to determine the life span of the rhesus monkey, since this was not accurately known. Our assessment (Tigges *et al.*, 1988), using data available at Yerkes Regional Primate Research Center, showed that only 50% of monkeys reach an age of 16 years. One quarter reach 25 years, and the records show that the oldest known monkey, a male, died at age 35. Similar data have been obtained in other studies (e.g., Dyke *et al.*, 1986; Davis, 1985; Bowden and Williams, 1984; Smith, 1982), but the point is that monkeys over 20 years of age can be considered old, and to assess the effects of aging, they can be compared with individuals 5 or 6 years old, since monkeys become sexually mature at 4 to 5 years of age.

Although this review focuses on how normal aging affects the monkey cerebral cortex, in some areas the data are very sparse. In many instances there is much more information available about the human brain, so wherever possible reference will also be made to that information. Also, the studies on nonhuman primates are intended as guides to determine what factors underlie, or account for, normal age-related cognitive decline in humans.

## 2. Neurons in the Cerebral Cortex

One of the most publicized changes reported in the normally aging human brain is neuron loss. Although it now seems that reports of large neuronal losses from cerebral cortex are probably erroneous, the concept is so ingrained that anyone showing signs of memory loss is said to be "losing his/her neurons." The long-standing belief that neurons are lost from the human cortex during aging appears to originate in the work of Brody (1955), who counted neurons in sections from the precentral, postcentral, and superior temporal gyri and occipital cortex of human brains from birth to age 95. Brody concluded that although the postcentral gyrus is minimally affected, the other three areas showed significant neuron loss. This loss was reported to be as much as 32% to 49%, mainly from layers II to IV. Brody (1970) also found a loss of some 48% of neurons from the superior temporal gyrus between 50 and 90 years of age, and Henderson *et al.* (1980), using a much larger sample of brains, reported a progressive loss of neurons, 38% to 45%, between the years of 16 and 95 in a number of cortical areas. Using a similar computerized imaging system to examine the brains of 19 nondemented persons aged 69 to 95 years, Anderson *et al.* (1983) also reported neuronal loss with age and suggested that after 70 years, there is a cell loss from neocortex amounting to about 1% of the neurons per year. These authors pointed out, however, that some of the apparent loss might be due to neuronal shrinkage. Other investigators have also reported a loss of neurons from the cortex during aging (e.g., Colon, 1972; Shefer, 1973; Devaney and Johnson, 1980). The obvious conclusion drawn from these studies was that there is significant neuronal loss during normal aging, and it was reasonable to suggest this as the underlying cause of any normal cognitive decline.

In contrast to these results, Cragg (1975) concluded that there is no age-related neuron loss from the human cerebral cortex, and he suggested that the earlier studies had not taken into account the variable tissue shrinkage that occurs when paraffin-wax-embedded sections are used. Thus, using frozen sections, Cragg (1975) found no loss of neurons from the frontal and temporal gyri. Cragg's conclusion was supported by the results of Terry *et al.* (1987), who examined the brains of 51 normal individuals between 24 and 100 years of age. Using automatic counting methods, they examined the midfrontal, superior temporal, and inferior temporal areas, and concluded that the number of neurons and the neuronal density are unchanged with age. However, Terry *et al.* (1987) also observed a decrease in size of some of the larger neurons with age, and because this appeared to be coupled with an increase in the number of smaller neurons, they suggested that some of the larger neurons are not lost with age, but rather, they shrink. A similar conclusion emerges from a study of the effects of age on the sizes of neurons in precentral gyrus carried out by Schutz and Hunziker (1980). They concluded that there is no difference in neuronal size when groups of subjects aged 19 to 44 are compared with groups aged 65 to 74, but that by 85 to 94, the neuronal perikarya have shrunk.

Another extensive study of human brains was carried out by Haug and his colleagues (Haug *et al.*, 1984; Haug, 1985; Haug and Eggers, 1991). Haug *et al.* (1984) examined the visual cortex and area 11 of the frontal cortex in 78 brains between 18 and 111 years of age. In both areas they found that neuron density does not decrease with age. Further, in area 11 they determined that neuron size remains constant up to age 60, although the neurons display a small decrease in volume beyond that age. In area 17, on the other hand, there is no decrease in the size of neurons with age. A subsequent report on the results of examining five cortical areas in 120 brains (Haug, 1985) reaches basically the same conclusion, namely, that there is no loss of neurons from the cortex during normal aging. Haug (1985) also suggests that during fixation, tissue from younger brains shrinks more than tissue from older brains, so that when sections from the cortices of younger and older brains are compared, the younger brains appear to have a greater neuronal density. In their studies of the primary visual cortex in a series of human brains, Leuba and Garey (1987) and Leuba and Krasik (1994a and b) also concluded that there is no neuron loss during normal aging, and they pointed out the large variation in total neuronal number that can occur among individual brains: they found the total number of neurons in area 17 to vary between 90 and 200 million.

Braak and Braak (1986) have also carried out a detailed examination of the neurons in area 11 of the frontal cortex of 18 human brains, and they conclude that in normal aging there is only a slight overall loss of neurons with age. They suggest that this loss occurs because, in the normally aged brains of patients ranging from 88 to 96 years, there is a smaller percentage of nonpyramidal cells than in younger brains. This conclusion has not been confirmed by others, although it is now relatively easy to investigate it further by using antibodies to γ-aminobutyric acid (GABA), a selective label for nonpyramidal cells.

What emerges from these studies is that, as better-preserved material is examined and greater care taken to ensure that the brains being studied are indeed normal, the evidence that neurons are lost during normal aging grows weaker. Investigators have also become more aware of the pitfalls associated with various methods of determining neuron numbers.

On a much smaller scale, findings on whether there is a neuron loss from the cortex of aging nonhuman primates have shown a similar pattern to that of the human studies. Starting in the 1970s, Brizzee and his colleagues (Brizzee, 1973; Brizzee *et al.*, 1975, 1980) compared the sensorimotor and frontal cortices of rhesus monkeys 4 to 7 years with brains from monkeys 18 to 20 years. They found that though the thickness of the cortex did not change with age, there was a significant decrease in the packing density of neurons. This decrease was pronounced in the middle layers, leading them to conclude that with age comes a loss of cortical neurons in the monkey brain. Later, Brizzee *et al.* (1980) examined the gyri in the frontal cortex that border the principal salcus and found lower numbers of neurons encountered through the thickness of the cortex in older monkeys, again suggesting a loss of neurons with age.

These were the only studies of the effects of aging on the rhesus monkey cerebral cortex prior to our own, wherein we compared the number of neurons present in the cortices of young monkeys 4 to 6 years old with those in monkeys over 24 years old. We took tissue from behaviorally tested monkeys whose brains were perfused with mixtures of paraformaldehyde and glutaraldehyde to ensure that the tissue was optimally fixed. We embedded the cortical tissue in plastic, then counted the number of neuronal profiles displaying their nuclei in 250-μm-wide-by 1-μm-thick sections passing through the depth of the cortex. The sections are aligned vertically, as judged by the fact that the section plane passes parallel to the lengths of the apical dendrites of the pyramidal cells, and, provided that the sizes of the counting objects—the neuronal nuclei—do not alter with age, which they do not, this method provides a comparative assessment of neuron number beneath a given cortical surface area. From an examination of both the primary visual cortex (Vincent *et al.*, 1989; Peters *et al.*, 1996) and area 46 of the frontal cortex (Peters *et al.*, 1994), it was concluded that there is no change in the overall thickness of the cortex with age and no indication of a significant neuronal loss. The same tissue has also been used to examine the fine structure of the neurons, and, with the exception of a few neurons in layer I (discussed later) there is no indication that the neuronal cell bodies in the older animals are undergoing any regressive changes other than the accumulation of some lipofuscin, as shown in the pyramidal cell illustrated in Fig. 1. Interestingly, as in the human cerebral cortex (Braak and Braak, 1986), lipofuscin is more evident in the perikarya of *non*pyramidal cells than in the pyramidal cells.

More recently, Kim *et al.* (1997) have also examined the effects of aging on the number and size of neurons in the cytochrome oxidase (CO) blobs within the supragranular layers of primary visual cortex in the rhesus monkey. They cut sections in the tangential plane parallel to the surface of the cortex, then visualized the CO blobs histochemically. Comparing the cortices of four young monkeys 5 to 12 years with four monkeys 24 to 27 years, Kim *et al.* (1997) found no age-related difference in the number of CO blobs per unit area of cortex, no difference in the soma size of neurons in the blobs, and no differences in the neuron density in the blobs. In earlier studies of the retinae and lateral geniculate bodies of these same monkeys (Ahmad and Spear, 1993; Spear *et al.*, 1994; Kim *et al.*, 1960), no statistically significant differences were found between the young and old monkeys in the numbers of retinal ganglion cells, nor of parvocellular neurons of the lateral geniculate bodies that provide inputs to the neurons in the cortex CO blobs. Consequently, these authors suggest that in terms of neuron number, the retinogeniculo-striate pathway in rhesus monkeys seem relatively unaffected by age.

A similar approach to ours in determining the effects of aging on neuronal number in primary visual cortex (Vincent *et al.*, 1989) has been used to examine the motor cortex (Tigges *et al.*, 1990), but in this case tissue was taken from monkeys ranging in age from 1 day to 35 years, and frozen sections were used. As in visual cortex, no age-associated loss of neurons from motor cortex was discovered. The large Betz cells in layer V of this cortex were examined separately, and although there was some indication that their cell bodies shrink slightly with age, there was no suggestion of a neuronal loss. The large Betz cells do, however, accumulate masses of lipofuscin with age. This begins at about 5 years and continues to such an extent that, in some of the neurons, lipofuscin actually displaces the nucleus to one side of the cell body.

Terry *et al.* (1987) also pointed out that in the human cortex, there is an apparent decrease in the size of the large neurons with age similar to that in the monkey Betz cells. To further investigate the effects of age on large cortical neurons, we have examined the large solitary Meynert cells in layer VI of the monkey primary visual cortex (Peters and Sethares, 1993) and found no indication that these neurons are lost with age, or that they become smaller. Furthermore, their perikarya show little sign of aging, because they accumulate such small amounts of lipofuscin that the fine perikaryal structure of the neurons from young monkeys is hardly distinguishable from that of old monkeys. Another difference is that some

Betz cells in older monkeys develop perikaryal inclusions consisting of parallel sheets of filaments (Tigges, 1992), whereas such inclusions have not been encountered within Meynert cells. Thus, not all neurons are similarly affected by age. Why the Betz cells shrink, accumulate masses of lipofuscin, and develop filamentous inclusions as they age, while the Meynert cells change so little, is unknown. It may

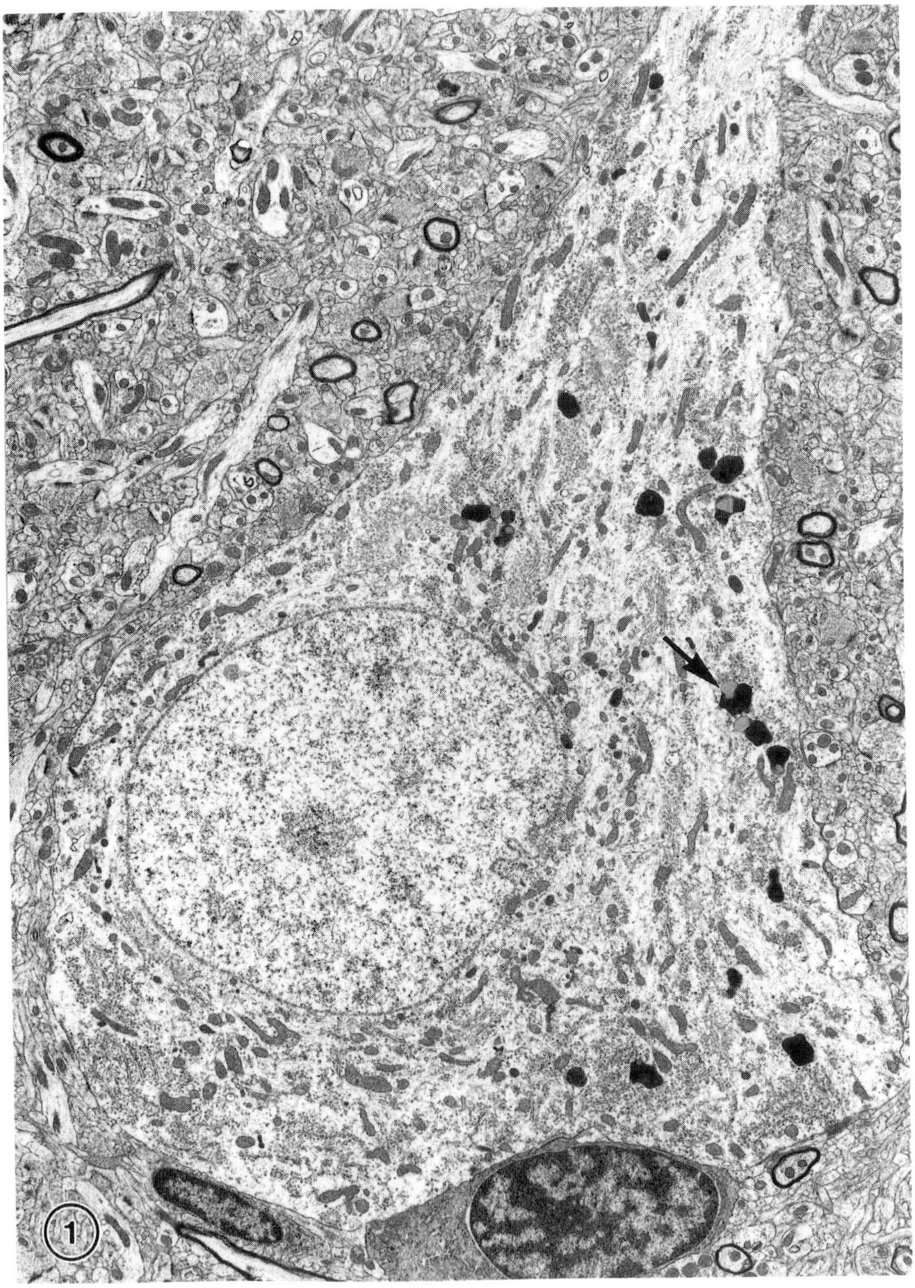


**Figure 1.** Large pyramidal cell from area 46 of a 25-year-old monkey. As is typical, the cell body of this neuron shows no signs of degeneration and contains only a small amount of lipofuscin (*arrow*). × 5,000.

reflect a difference in the metabolism of the two neuronal types, with the Meynert cells being more able to sustain their level of metabolic activity throughout life.

It should be pointed out that neurofibrillary tangles do not occur in the perikarya of neurons in the aging monkey brain, and, apart from the inclusions in Betz cells, in our fine-structural studies of the aged cortex, we have encountered no bundles of filamentous material in the neurons. However, Hof and Morrison (1995) have shown that in normal monkey cortex, many of the large neurons label with an antibody to nonphosphorylated neurofilament protein (SMI32). They suggest that in the human cortex, similar neurons may be vulnerable to degenerative processes leading to dementia (Hof *et al.,* 1990; Hof and Morrison, 1990; see chapter by Hof, Bouras, and Morrison, this volume), but this is not the case in the monkey cortex, apparently.

Our conclusion, based on examination of three neocortical areas from the brains of rhesus monkeys, is that there is no indication of a significant neuronal loss with age. Rosene *et al.* (unpublished data) have come to a similar conclusion with respect to the hippocampus of the rhesus monkey, because neuronal counts reveal no age-related change in neuronal density within either the subiculum or the CA1 subfield.

In a recent study of the effects of aging on the neurons and neuroglial cells of the rat parietal cortex, Peinado *et al.* (1997) have also concluded that there is no loss of neurons from this cortical area with age, although there may be an age-related decrease in the size of some neurons.

For another perspective on the effects of aging on the cerebral cortex, including that of rodents, see the review by Flood and Coleman (1988), which contains some useful data tables.

## 3. Variation in the Sizes of Cortical Areas

It can be argued that our strip-counting approach to evaluating possible neuronal loss from the visual, motor, and prefrontal cortices of the rhesus monkey is only valid if the cortex volume does not change with age, since our results show only that there is no loss of neurons from beneath a unit area of cortical surface. If the overall volume, or the areal extent of a piece of the cortex, decreases with age, this would result in an overall loss of neurons. To examine this, we determined the surface area and the volume of the primary visual cortex (area 17) in the brains of 14 rhesus monkeys, and for the purposes of analysis the monkeys were separated into two sets, one set between 4 and 12 years old, the other over 25 (Peters *et al.,* 1997). Cortex thickness and number of neurons displaying nuclei were determined in 250-μm-wide- by-1-μm-thick sections; the surface area and volume of area 17 in each monkey was determined using 30-μm-thick, Nissl-stained serial sections. In these monkeys, the thickness of area 17 varied between 1.26 mm and 1.78 mm, with a mean depth of 1.55 mm; the number of neuronal profiles with nuclei varied between 416 and 522; the surface area of area 17 varied between 702 $mm^2$ and 1480 $mm^2$, with a mean value of 956 $mm^2$; and the volume varied between 1,190 $mm^3$ and 2,503 $mm^3$, with a mean value of 1,792 $mm^3$. Despite these variations, no significant differences were found between values for the sets of young and old monkeys, indicating that there is no loss of either neurons or volume from area 17 with age.

The most surprising result was the large size variation in area 17 among individual monkeys. However, examination of the literature showed that others have recorded similar large differences in the extent of area 17 among individual monkeys. For example, Van Essen *et al.* (1984) measured the surface area of the primary visual cortex in the brains of 24 *Macaca fascicularis* and obtained values from 690 $mm^2$ to 1560 $mm^2$, with a mean value of 1,195 $mm^2$. In another study Purves and LaMantia (1993) examined the striate cortex of neonatal and young adult male rhesus monkeys, and their data from a sample of 11 young adult monkeys show the surface area to vary between 885 $mm^2$ and 1,346 $mm^2$, with a mean value of 1,069 $mm^2$. In these three studies, the mean values obtained for the surface area of primary visual cortex are similar, about 1,000 $mm^2$, although among individual brains there can be a two-fold difference in the area of primary visual cortex. Purves and LaMantia (1993) also give the weights of the brains they examined, and interestingly, although there is a tendency for larger brains to have the most extensive primary visual cortices, there is no exact correlation between the two values.

Similar differences among individuals also occur in the primary visual cortices of humans. For example, Stensaas *et al.* (1974) examined 52 human brains obtained at autopsy and found that the surface area of the smallest striate cortex was 1,284 $mm^2$ and the largest was 3,702 $mm^2$, a three-fold difference among individuals. If the volumes of striate cortices from only brains over 25 years old are considered, a similar variation can also be seen in data generated by Murphy (1985). In a more recent study, Leuba and Krafsik (1994a) measured the volume of area 17 in human brains from individuals between 66 and 93 years of age and found the volume of area 17 to vary between 3,250 $mm^3$ and 6,380 $mm^3$, again showing a three-fold difference. In another study, Leuba and Krafsik (1994b) measured the surface area and volume of area 17 of the human cortex from midgestation until old age, and recorded a large variation in the dimensions of primary visual cortex in brains from mature individuals. Leuba and Kraftsik (1994b) indicate that there might be a slight decrease in the volume of area 17 during aging, even though they detected no change in the number of neurons.

Although we found no similar data on the amount of cortex devoted to a given function for other areas of the cortex, we assume that variations similar to those occurring in primary visual cortex do occur in other cortical areas. In both humans and nonhuman primates, there is likely to be an enormous normal variation in the numbers of neurons in any specific functional area of the cerebral cortex. Consequently, even if the total numbers of neurons within a given cortical area are calculated from both young and old brains, it would be necessary to examine large numbers of brains to ascertain if there is a small overall age-related neuron loss. On the other hand, a large neuronal loss, such as can occur in Alzheimer's disease, is easily detected. For example, Leuba and Krafsik (1994a) were able to show that although there is no difference in the volume of area 17 between normal and Alzheimer's-diseased brains, the latter have an average of 33% fewer neurons beneath 1 $mm^2$ of cortical surface, a difference that is readily apparent. On the other hand, in a recent study using newer methods to calculate neuronal numbers, Regeur *et al.* (1994) concluded that there is no overall loss of neurons even from the brains of patients with Alzheimer's disease.

Another indication that there is no loss of cortical neurons with age comes from studies of changes in total cortical volume. Although it is commonly agreed that there is a reduction in total cerebral volume with age, it is not always recognized that this is not due to a loss of gray matter. It is due to a loss of white matter. For example, Haug and Eggers (1991) found that when brains from humans 20 to 40 years of age are compared with those between 75 and 85 years of age, there is a reduction of 6% in the volume of the entire brain, and a reduction of 3% in the volume of the entire gray matter cortex. But the reduction in the volume of white matter is 11%. Similarly, on the basis of their morphometric analyses of the brains of 19 nondemented persons aged 69 to 95 years, Anderson *et al.* (1983) determined that between 70 and 85 years of age the reduction in the volume of the cerebral cortex is only 2–3%, while the reduction in white matter is about 11%. Also Albert (1993) reports that on the basis of MRI scans from 70 carefully screened human subjects between 30 to 80 years old, there is no decrease in the volume of gray matter, but there is a decrease in the amount of white matter. And more recently, Double *et al.* (1996) have used point-counting to show that although there is a small decline in the total volume of the normal human brain with age, amounting to about 2 ml per year, it is due to a change in the amount of white matter, since the volume of gray matter is unaltered. These authors point out that frontal white matter is particularly affected. It is also interesting to note that Leuchter *et al.* (1994), using MRI, find that periventricular lesions of white matter are relatively common in elderly people, and they suggest that such lesions could well lead to disconnections between brain regions.

Results such as these support the view that there is no extensive loss of cortical neurons during aging (see Peters *et al.*, 1998). They point instead to changes in the white matter. White matter has not been extensively examined in aging studies. However, in a review of the effects of aging on the human brain, Kemper (1994) points out that when myelin-stained preparations of older brains are compared with those from younger brains, the staining of the myelin is much paler in the older brains, indicating a loss of myelin. Also, Lintl and Braak (1983) report that when they examined the striate cortex of normal individuals aged 18 to 96 years, from the third decade onward they found the amount of myelin in the stripe of Gennari to be reduced. The stripe of Gennari essentially contains intracortical fibers, so this study suggests that some intracortical fibers are being lost with age.

At present there are no comparable publications that relate to the effects of aging on the monkey brain, but Rosene and Moss (unpublished data) have begun to obtain MRI scans from the brains of behaviorally tested rhesus monkeys, and are finding that when the brains of young monkeys (5 to 14 years) are compared to those of elderly monkeys (20 to 32 years), there is a significant age-related decline in the overall volume of the forebrain, just as in humans. This is due to a statistically significant decrease in the volume of the white matter, accompanied by an increase in the volume of the ventricles. The gray matter is largely unaffected. Moreover, a comparison of these data with behavioral data obtained from the same monkeys indicates a correlation between the extent of white-matter loss and the overall decline in cognitive function.

Whether the reported age-related decrease in the volume of white matter is due primarily to a loss of fibers, or to degeneration of myelin sheaths, or to both, is not yet known. But as shown later in this review, there is a very evident breakdown in the integrity of myelin in the gray and white matter of the cerebral hemispheres of old monkeys.

## 5. Effects of Age on Dendrites

Even though there appears to be no significant loss of neurons from the cerebral cortex during aging, this does not mean that the dendrites and axons are unaffected. Unfortunately, there is little information available about the effects of aging on these neuritic processes, because the only way to determine if they are lost during aging is to visualize entire neurons in young and old animals and then to make a comparison between them. And even that is fraught with problems, due to the extensive variation in the morphology of individual neurons. One method to visualize the dendritic trees of neurons is through use of the Golgi method. Using this method, early studies of aging human material suggested that dendrites atrophy with normal aging (e.g., Scheibel *et al.*, 1975), although some of apparent atrophy may have been due to suboptimal fixation of the tissue. In another study, Buell and Coleman (1981) sampled neurons from the parahippocampal gyrus of human brains using the Golgi–Cox method of neuronal impregnation, and in the old brains they encountered both normal neurons and others with atrophied dendritic trees. Upon measuring the dendritic trees of the pyramidal neurons in layer II, they concluded that in normal aged individuals, these neurons had longer and more branched dendrites than in the brains of either young individuals or those with senile dementia. Buell and Coleman (1981) suggest that in the normally aging cortex, there are some regressing neurons whose dendritic trees are dying back, and to compensate for this, the dendritic trees of other, surviving, neurons expand. Nakamura *et al.* (1985) have also reported that the basal dendritic trees of pyramidal neurons in layers III and IV of human motor cortex are reduced during normal aging.

Clearly some of the results obtained using human material can be criticized because the tissue is not optimally preserved, so that some of the reported alterations may be postmortem artifacts. However, using tissue from the brains from monkeys aged 7 to 28, fixed by vascular perfusion, Cupp and Uemura (1980) examined Golgi-impregnated pyramidal neurons from layers III and IV of the superior frontal gyrus. They concluded that until 20 years of age, there is a continued increase in length and branching of the apical dendrites of pyramidal cells, but no increase in length of basal dendrites, even though they continue to branch. Both sets of dendrites lost both length and branches by 27 to 28 years of age. Subsequently Uemura (1980) reported that these dendrites also lose spines.

In our studies of the effects of aging on the cerebral cortex of the rhesus monkey, many of the cortices were also examined by electron microscopy. In aging cortices some dendrites with reduced numbers of mitochondria and microtubules are encountered occasionally throughout the neuropil of the lower layers, more commonly in the upper layers, and frequently in layer I. These dendrites are often swollen and instead of the normal organelles, their cytoplasm contains vacuoles and membranous inclusions; it seems likely that the alterations in their cytoplasm indicate that these dendrites are dystrophic. An example of such a dendrite is illustrated in Fig. 2. This shows the apical dendrite of a pyramidal cell in which the cytoplasmic matrix and the microtubules have broken down, leaving a space that contains membranous bodies. Such altered dendrites are commonly observed in layer I of both visual (Vincent *et al.*, 1989) and frontal (Peters *et al.*, 1994) cortex in old monkey, as shown in Fig. 3, and because the majority of dendrites in layer I belong to the apical tufts of pyramidal cells, it is assumed that the dendritic branches in many of these tufts are undergoing regression. This would fit with the conclusions reached by Cupp and Uemura (1980), but a more complete study is necessary before definite statements can be made.

What causes the dendrites in the upper layers—and especially in layer I of the cortices of old monkeys—to swell is not known, but it is worth noting that infusion of *N*-methyl D-aspartate into the lateral ventricles of rat brains produces a number of changes, among them a swelling of some dendrites in the cerebral cortex. Dietrich *et al.* (1992) suggest that this swelling is due to the activation of the NMDA receptors, which brings about a disruption of dendritic membranes. Whether similar changes occur during the normal aging process is worth investigating.

## 6. Effects of Age on Layer I

Because layer I is the only layer in the monkey cerebral cortex in which obvious age changes are consistently observed, we now digress and consider these changes.

In old monkeys it is common to find that the thickness of layer I is reduced when compared with that of young monkeys. Although more data are needed, our preliminary analysis indicates that in area 17 the extent of the thinning of layer I correlates with age, while in area 46 of frontal cortex it correlates as well with the cognitive decline exhibited by individual monkeys (see Peters *et al.*, 1996). As pointed out in the previous section, at least some of this thinning can be attributed to degeneration of the apical dendrites of pyramidal cells, for it is common to observe dendrites with profoundly altered cytoplasm in layer I of old monkeys (Fig. 3). These particular dendrites may also be growing shorter as layer I becomes thinner with age.

In addition to the changes in the dendrites, the glial limiting membrane on the outside of layer I is also noticeably thicker in many of the old monkeys (Fig. 4). In young monkeys the glial limiting membrane is usually formed by two or three layers of astrocytic processes, but in the old monkeys it may be several times thicker, and the astrocytic processes contain more cytoplasmic filaments. Since it is known that astrocytes can respond to degenerative alterations in the brain by forming scars, this thickening of the glial limiting membrane is possibly in response to the degeneration of dendritic branches in the apical tufts of the pyramidal cells, and perhaps also to a loss of some of the inputs to layer I.

The other interesting point is that layer I is the only layer in which we have observed morphological changes in neuronal perikarya. In old monkeys it is often observed that the cytoplasm of layer I neurons are sparse in organelles, and are vacuolated or contain membrane-bound vacuoles (Fig. 5). On the other hand, in

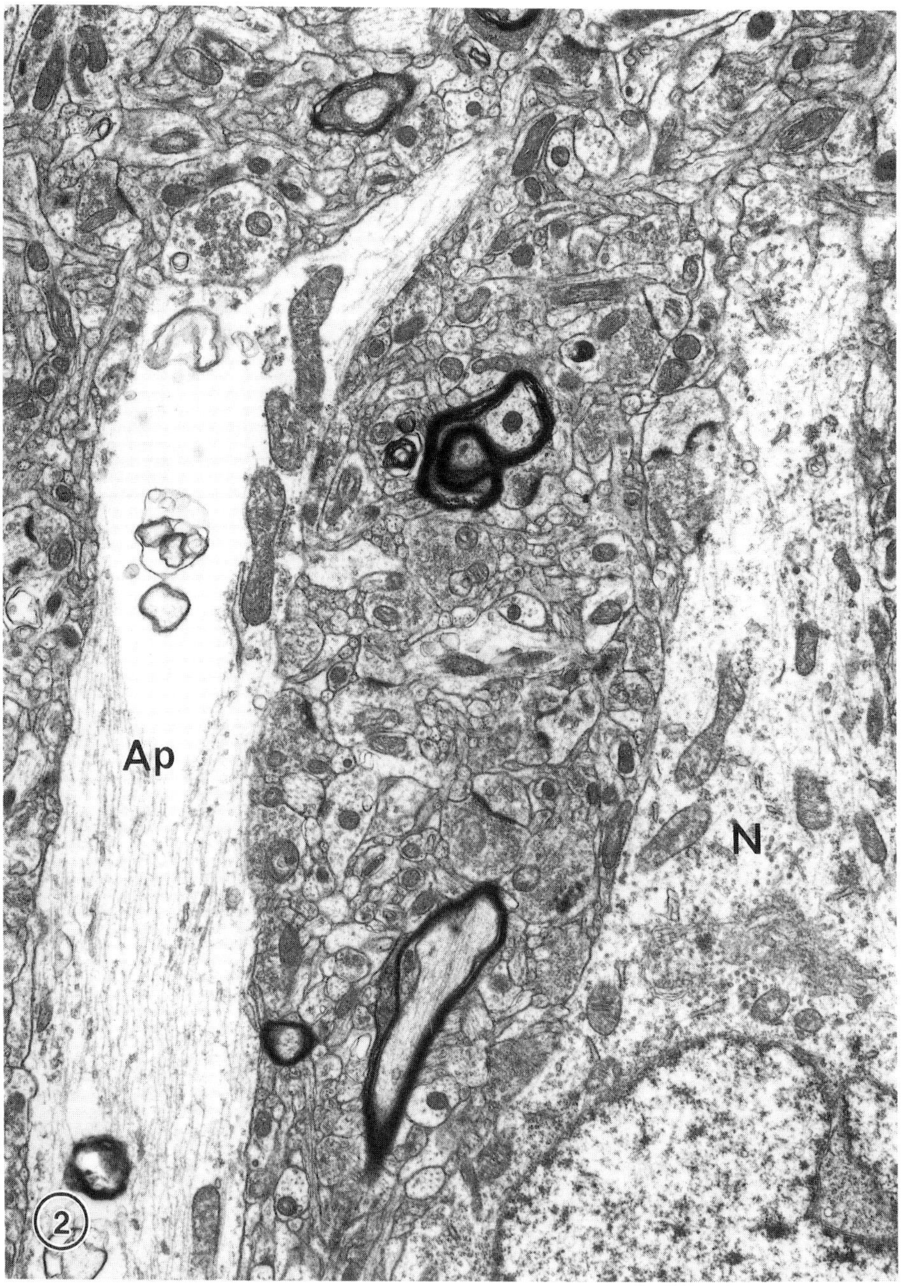


**Figure 2.** Layer II, area 46, of a 26-year-old monkey. At the bifurcation site of the apical dendrite (*Ap*) on the left the organelles have broken down, so that cytoplasm is vacuolated and contains membranous inclusions. In contrast, the pyramidal cell (*N*) on the right shows no signs of degeneration, but between them is a myelinated axon whose sheath appears abnormal. × 10,000.

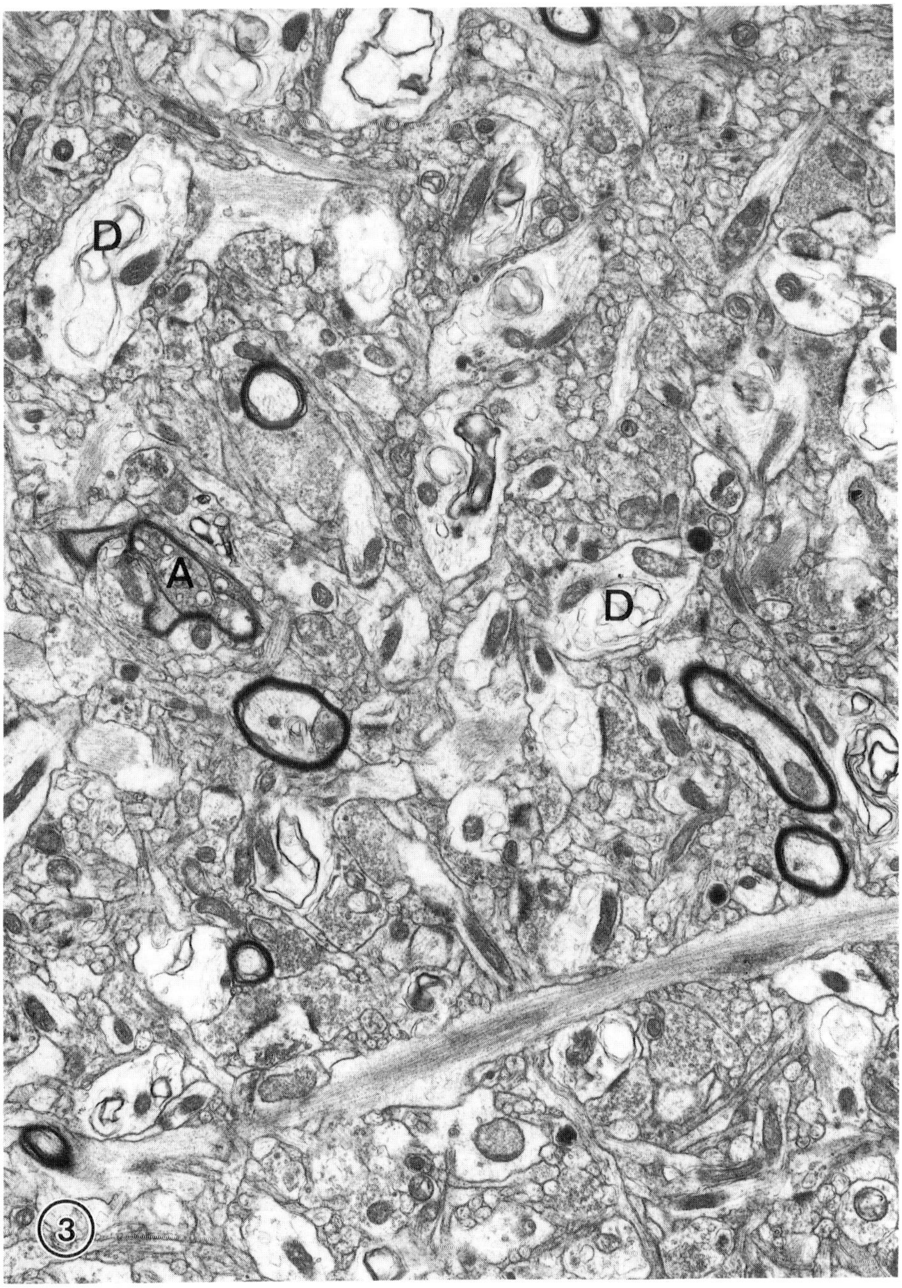


**Figure 3.** Neuropil of layer I, area 46, of a 26-year-old monkey. Many of the dendrites (*D*) appear to be degenerating, since their cytoplasm contains vacuoles and membranous whorls. Some of the myelinated axons (*A*) also show degenerative changes, while the majority of unmyelinated axons appear normal. × 10,000.

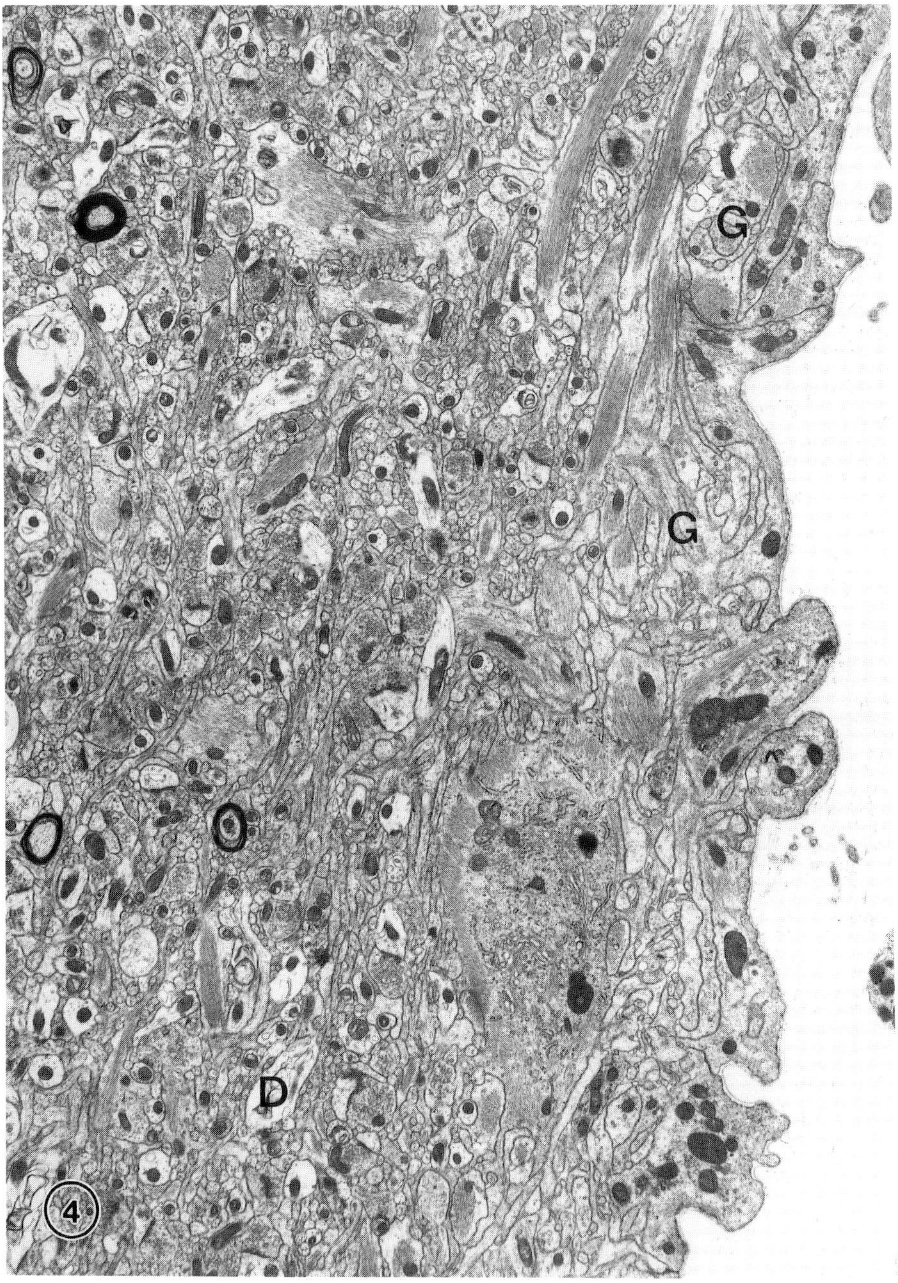


**Figure 4.** The glial limiting membrane and the outer portion of layer I, area 46, of a 26-year-old monkey. Compared to that of a younger monkey, the glial limiting membrane (*G*) is thickened. It consists of the layered processes of astrocytes containing thick filament bundles. There is some degeneration of dendrites (*D*), but less than in the field shown in Fig. 3. × 6,000.

marked contrast to the dendrites and cell bodies, most of the profiles of unmyelinated axons and axon terminals that account for a large proportion of the neuropil of layer I show little sign of degenerative changes, although axons with degenerating sheaths are not uncommon (see Fig. 3). No detailed analysis has been done, but it seems likely that the overall thinning of layer I in old monkeys must lead to some loss of synapses, and some of these may be derived from extrinsic sources. For example, Beal *et al.* (1991) have shown that with age comes a reduction in the level of choline acetyltransferase in monkeys. A reduction in the level of this enzyme, which is involved in the formation of acetylcholine, may indicate a loss of cholinergic input to the cortex, much of which terminates in layer I (Mrzljak *et al.*, 1995). This suggests that there may be a loss of the cholinergic afferent input to layer I, an input largely derived from the basal nucleus of Meynert. The available data concerned with age-related changes in this nucleus, however, do not provide a

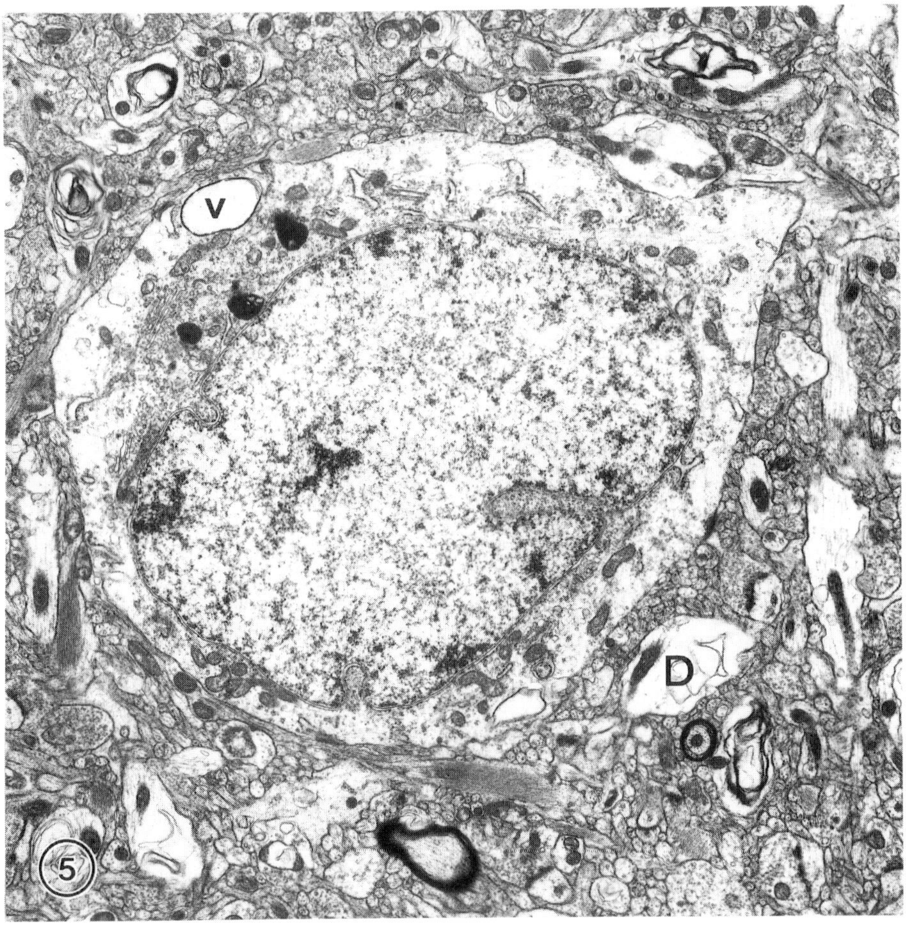


**Figure 5.** A neuron in layer I, frontal cortex, of a 26-year-old monkey. In the periphery of the perikaryon, the neuron's cytoplasm is pale and contains few organelles. That and the presence of the membrane-bound vacuole (*v*) suggests an early stage of degeneration. Note the degenerating dendrites (*D*) in the surrounding neuropil. × 8,000.

clear picture. For example, Amaral (1993) reported that there is no age-related loss of neurons from this nucleus in aging monkeys, whereas Rosene (1993) reported that, on the basis of acetycholinesterase staining, there is a loss of neurons throughout the magnocellular nuclei. In a contribution to a previous volume in this series, Vogt (1991) proposed that because of its rich cholinergic innervation, layer I may play a basic role in attention setting. If that is true, layer I would be in a prime position to regulate information processing in the cortex, and so a loss of cholinergic synapses would be expected to have a pronounced effect on memory.

There is also a significant dopaminergic input to layer I derived from neurons in the substantia nigra, and preliminary studies on the monkey indicate that this nucleus loses neurons with age (Siddiqi *et al.,* 1995). Beal *et al.* (1991) have also demonstrated an age-related reduction in the levels of dopamine, and, using monkeys 2 to 18 old, Goldman-Rakic and Brown (1981) showed that with age there is a reduction in the levels of dopamine in some areas of the cortex, most notably in prefrontal cortex, where the reduction is over 50%.

Another important input to layer I is from other cortical areas: for example, the feedback fibers from area 18 to area 17 terminate in layer I (e.g., Rockland, 1994). These afferents are large-diameter myelinated axons that contribute to the horizontal plexus of myelinated fibers in layer Ib, and it is worth noting that, in myelin-stained preparations of area 17, when compared with young monkeys, the staining intensity of this fiber plexus is diminished in old monkeys (Peters, unpublished data).

Obviously, at present any theories about the potential role of layer I in normal age-related memory loss can only be based on conjecture, but it is the only layer of the cortex in the rhesus monkey to show pronounced age-related morphological changes.

## 7. Age-Related Changes in Synapses

There is very little information about the effects of normal aging on the synapses of the primate cerebral cortex, and almost all of it relates to the human brain and, other than the study by Masliah *et al.* (1993), all of the published studies have employed electron microscopy to determine synaptic densities. Masliah *et al.* (1993) used antibodies to synaptophysin to label the axon terminals, which were then counted by light microscopy.

The first study of the effects of aging on the synapses in the human cortex was that of Cragg (1975), who found no significant decline in the synaptic density in frontal cortex between 16 and 72 years of age. A similar result was obtained by Huttenlocher (1979), who used the method of phosphotungstic acid staining to reveal the postsynaptic densities. Hutterlocher (1979) examined the synapses in layer III of the middle frontal gyrus in 21 normal brains, ranging in age from newborn to 90 years. The synaptic density throughout adult life (16 to 72 years) was found to be about $11 \times 10^8$ synapses per $mm^3$ of tissue, and it changed little with age. Gibson (1983) obtained similar values for the concentration of synapses in the neuropil of the frontal and temporal regions of postmortem brains between 16 and

90 years old, and determined that although there is a decrease of about 20% in the number of synapses in frontal cortex, there is no loss from temporal cortex. Masliah *et al.* (1983) also found that compared with younger brains, those from individuals over 60 years of age had an average of 20% fewer labeled presynaptic terminals. Thus there might be some loss of synapses from the frontal cortex of the human brain in old age, but most of the tissue has been taken postmortem, in which structural preservation is less than optimal. In other parts of the cortex, such as the temporal lobe and the striate cortex, where Huttenlocher and de Courten (1987) ascertained the number of synapses to be relatively unchanged in adult brains until 71 years of age, the synapse loss may be even less.

One other study of synaptic changes with age in the human brain is that of Adams (1987), who examined layer I in the precentral and postcentral cortex. In contrast to our findings in the macaque, Adams (1987) found no age-related change in the thickness of layer I, and found a decrease only in the precentral gyrus, due primarily to the decrease in the numbers of asymmetric axospinous synapses, presumably reflecting a loss of spines from the postsynaptic dendrites. Adams (1987) also found that the synapse loss was accompanied by an increase in the mean length of the postsynaptic density profiles, reflecting an increase in the size of the synaptic junction. DeKosky and Scheff (1990) found that the significant loss of synapses in the frontal lobes of Alzheimer's patients is also accompanied by a compensatory increase in the dimensions of synaptic junctions.

For monkeys, the information is even more sparse. Uemura (1980), using ethanolic phostotungstic acid to label synapses in the prefrontal cortex of rhesus monkeys, found a loss of synapses in older monkeys, paralleling the loss of dendritic spines described earlier by Cupp and Uemura (1980). The loss of synapses was of the order of 20%, similar to that described by some investigators examining the human prefrontal cortex. In contrast, Tigges *et al.* (1992) were unable to detect any age change in the structure and number of synapses on the perikarya of Betz cells in monkey motor cortex (area 4), and Zecevic *et al.* (1989) found no change in the frequency of symmetric synapses in the neuropil of the motor cortex of rhesus monkeys aged 3 to 20 years.

The only other studies that appear to have been carried out on the effects of aging on synapses in the nonhuman primate brain are in the hippocampus. Tigges *et al.* (1995) examined the synapses in the outer third of the dentate gyrus and found no change in the total number of synapses with age beyond a 3% decrease in the numbers of synapses involving dendritic shafts. And in this same cortex, Tigges *et al.* (1996, unpublished) have been unable to detect any age-related loss of synapses in the inner third of the molecular layer. These data agree with those generated by Gazzaley *et al.* (1996), who using an antibody to synaptophysin to compare synaptic densities in the outer and inner molecular layers of the hippocampus in young and old monkeys, found no change in total synaptic density with age.

More studies are needed to determine if there is a loss of synapses from cerebral cortex of human and nonhuman primates in normal aging, and whether any such loss is specific for certain cortical areas or global. But based on the available evidence, the tentative conclusion is that, except perhaps for the frontal cortex, normal aging brings only a small decrease in the number of synapses.

## 8. Effects of Aging on Axons and Their Sheaths

The tentative conclusion that there may be only a small loss of synapses with age does not seem to fit with the observed white-matter loss, because such an observation implies an age-related axon loss. It is possible that axons are not being lost, but that instead their sheaths are breaking down and that this is what leads to the diminution in the white-matter volume with age. This latter possibility agrees with observations by Kemper (1994) and by Lintl and Braak (1983) that there is a diminution in the staining intensity of white matter with age. A breakdown of myelin sheaths has been reported by Vincent *et al.* (1989) and by Peters *et al.* (1994) in their studies of the effects of aging on the striate and frontal cortices of rhesus monkeys. Some myelin sheaths in both the gray matter and the underlying white mater show degenerative age changes, and these changes take various forms (Fig. 6). The most common change is a splitting of the myelin lamellae so that adjacent turns of the sheath become separated. Sometimes the space produced by the split is occupied by a dense cytoplasm; other times the sheath may enclose dark cytoplasm containing lysosomes and lamellae of myelin. But as shown in Fig. 7, even in these examples it is not uncommon for an apparently intact axon to be present inside the outer sheath. The source of the dark cytoplasm in altered sheaths is unknown. It could originate from a microglial cell, because these cells are known to phagocyte myelin (e.g., Mosley and Cuzner, 1996), or it could be from the oligodendrocyte forming the sheath; both neuroglial cell types have a dark cytoplasm. Other myelin sheaths show blobs that can also contain dense cytoplasm; yet others are very swollen and they surround vacuoles (Fig. 6,v), sometimes with the axon still intact but pushed to one side of the swollen lumen enclosed by the sheath. It is also not uncommon in the cortices of old monkeys to find axons occupying sheaths that are excessively large, so that the axon occupies only one side of the large lumen. In other cases axons are surrounded by complex myelin sheaths that seem to consist of one sheath inside another.

In most cases it seems to be the myelin that is affected or breaking down, because the enclosed axon is usually intact (Fig. 7). Only few degenerating axons have been encountered in the cerebral hemispheres of old monkeys.

Obviously, these kinds of changes need to be further investigated, because it is conceivable that a breakdown of the myelin sheaths of some axons could lead to changes in conduction rates in some neuronal circuits. As a preliminary step to determine if the breakdown in myelin could play a role in age-related behavioral deficits, Peters *et al.* (1994) made a subjective assessment of the amount of degeneration of myelin in the white matter beneath the frontal cortices of a series of old monkeys. A preliminary assessment based on four monkeys suggested a possible correlation between the ranking of the monkeys in terms of the extent of the degeneration of myelin and their performance of the delayed nonmatching-to-sample test, suggesting in turn that, because age-related cognitive changes in normal aging are unlikely to be due to a loss of neurons, and perhaps not of synapses, they might instead be due to a breakdown of myelin sheaths of axons that form connections between neurons. Such a breakdown of myelin would affect the conduction rates of those axons, which would be expected to result in a disruption of the timing in neuronal circuits.

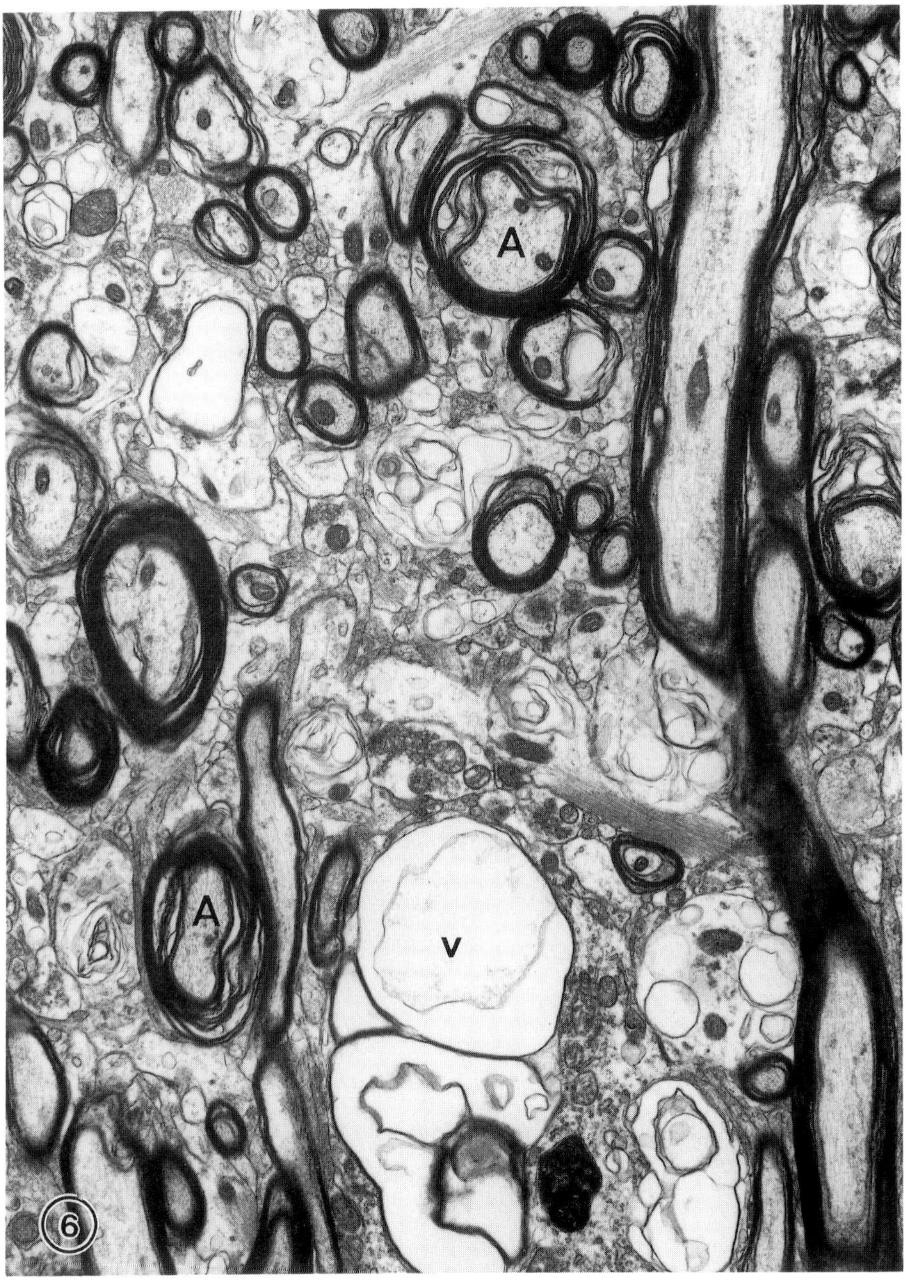


**Figure 6.** Layer VI, area 46, of a 26-year-old monkey shows myelin degeneration. Though some of the axons have normal sheaths, those of other axons (*A*), with split and empty vacuoles (*v*) apparently surrounded by myelin, can occur throughout the white matter. × 10,000.

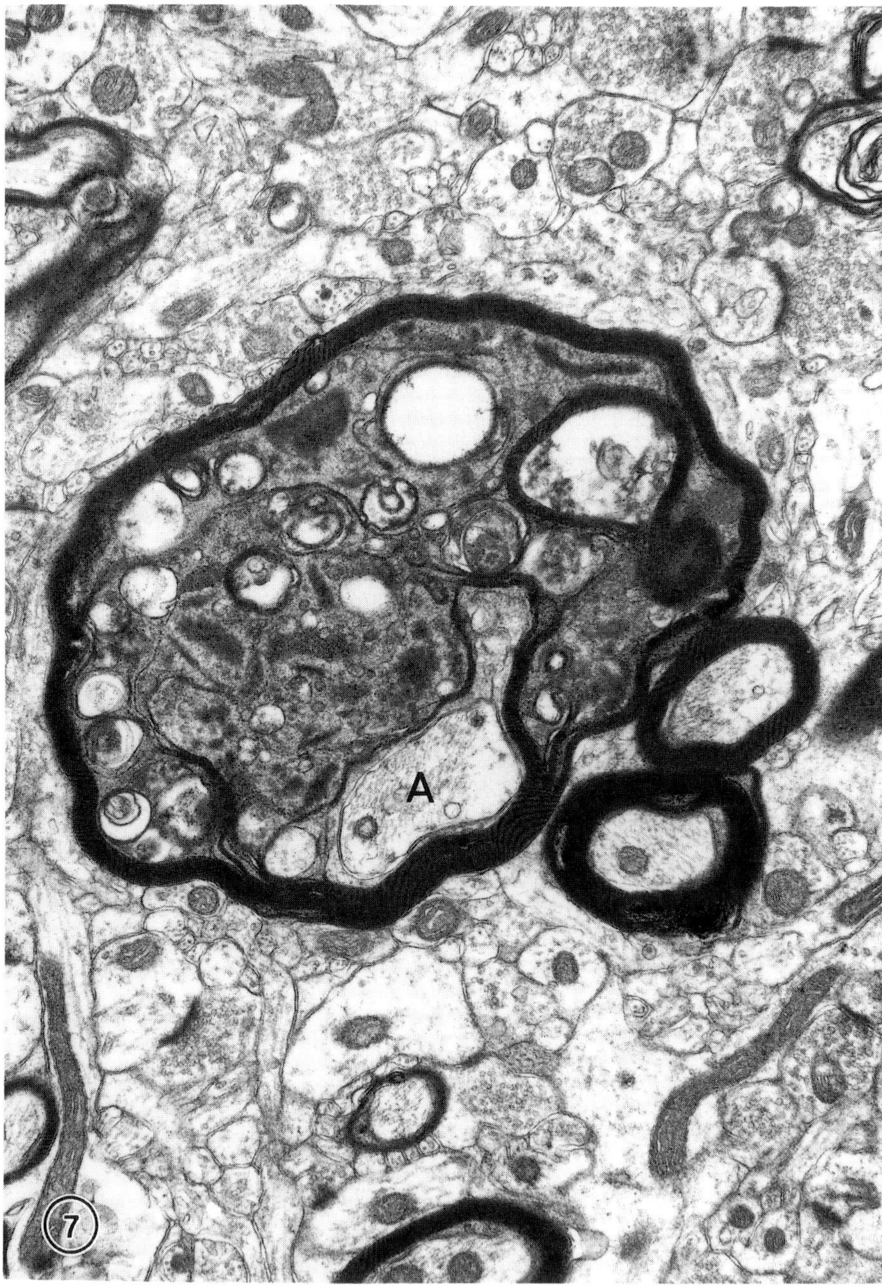


**Figure 7.** A large degenerating, myelinated fiber in layer VI, area 46, of a 25-year-old monkey. This degenerating fiber surrounds a dark cytoplasm that may be derived from either a microglial cell or from the oligodendrocyte forming the sheath. There are also some myelin lamellae, and to one side is a relatively intact axon (*A*). Note that the two adjacent myelinated fibers appear normal. × 10,000.

## 9. Effects of Aging on Neuroglial Cells

There are few studies on the effects of normal aging on the neuroglial cell population in the cerebral cortex. In their study of normal aging of the frontal and visual cortices of the human brain, Haug *et al.* (1984) recorded no significant numerical change in the neuroglial cell population. The same conclusion was reached by Leuba and Garey (1987) in their examination of areas 17 and 18, although Leuba and Krafsik (1994a) did find an increase in the numbers of glial cells in area 17, but not in area 18, in cases of Alzheimer's disease. In contrast to the above results, Devaney and Johnson (1980) concluded that there is a substantial age-related increase in the glial population density in the human striate cortex, and Terry *et al.* (1987) also found the number of glial cells to increase with age. Again, the true situation is not obvious.

For the monkey, numerical data are even more sparse and seem confined to one study by Peters *et al.* (1991) in which comparisons in the numbers of neuroglial cells were determined by counting the profiles of neuroglial cells with nuclei in thin sections through the striate cortex. This approach, though unlikely to produce very accurate numerical data, has the advantage of identifying the various types of neuroglial cells with certainty. The outcome suggested that the total number of neuroglial cells does increase slightly with age, because of an increase in the numbers of microglial and oligodendroglial cells. Thus, when young and old monkeys were compared, the numbers of oligodendroglial cell profiles increased by about 35% and the numbers of microglial cells by about 50%. The numbers of astrocytes remained stable, which means that in young monkeys, the oligodendroglial cells accounted for 35% of the total neuroglial cell population and, in old monkeys, for 40%. For microglial cells, the values were 7.6% and 9.4%.

Numbers aside, there are obvious changes in the morphology of the neuroglial cell population with age. All three types of neuroglial cells accumulate cytoplasmic inclusions in normal aged rhesus monkeys. The inclusions are so obvious that it is better to look at the neuroglial cells than the neurons to determine if a piece of cortex is from a young or old monkey. The kinds of inclusions seen in the cytoplasm of the three cell types are typical of the particular glial cell type.

The microglial cells are generally assumed to be the principal phagocytes of the central nervous system, which is probably why this neuroglial type shows the greatest numerical increase in old monkeys. Microglial cells also contain the most heterogeneous types of inclusions, from dark granular inclusions, to dark inclusions that appear laminated, to other inclusions that are pale and have a frothy appearance. Sometimes, as illustrated in Fig. 8, the inclusions are so massive that they fill the cytoplasm; as a result, the cytoplasm may be confined to a thin rim surrounding the inclusions, with the nucleus being pushed to one side of the perikaryon. Unfortunately, even though these phagocytozed contents are so prominent, their appearance in the cortex offers few clues to their source, so it is not known what kinds of elements from aging cortex neuropil are being phagocytozed by the microglial cells. In the optic nerve, on the other hand, we have encountered microglial cells that obviously contain broken-down myelin.

The appearance of the phagocytozed material within the cytoplasm of the astrocytes also gives no clues to its origin (Fig. 9). In the aging cortex, the material is

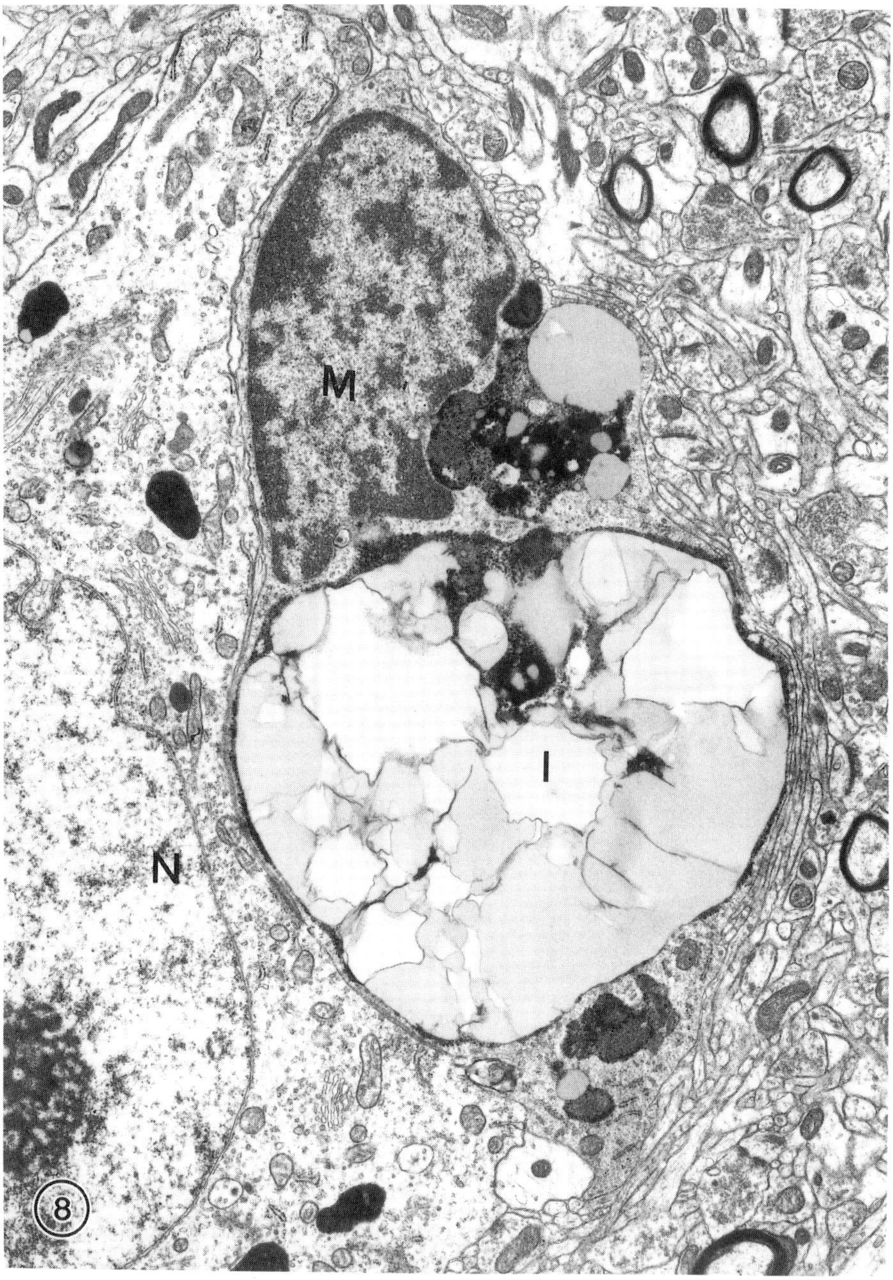


**Figure 8.** A microglial cell (*M*) lying next to a neuron (*N*) in layer IV, area 46, of a 26-year-old monkey. The microglial cell has the typical dark nucleus containing two inclusions. The larger inclusion (*I*) appears to contain a lipid material, while the contents of the smaller inclusion look more heterogeneous. The inclusions are so large that the cytoplasm of the microglial cells forms only a thin covering around them. × 10,000.

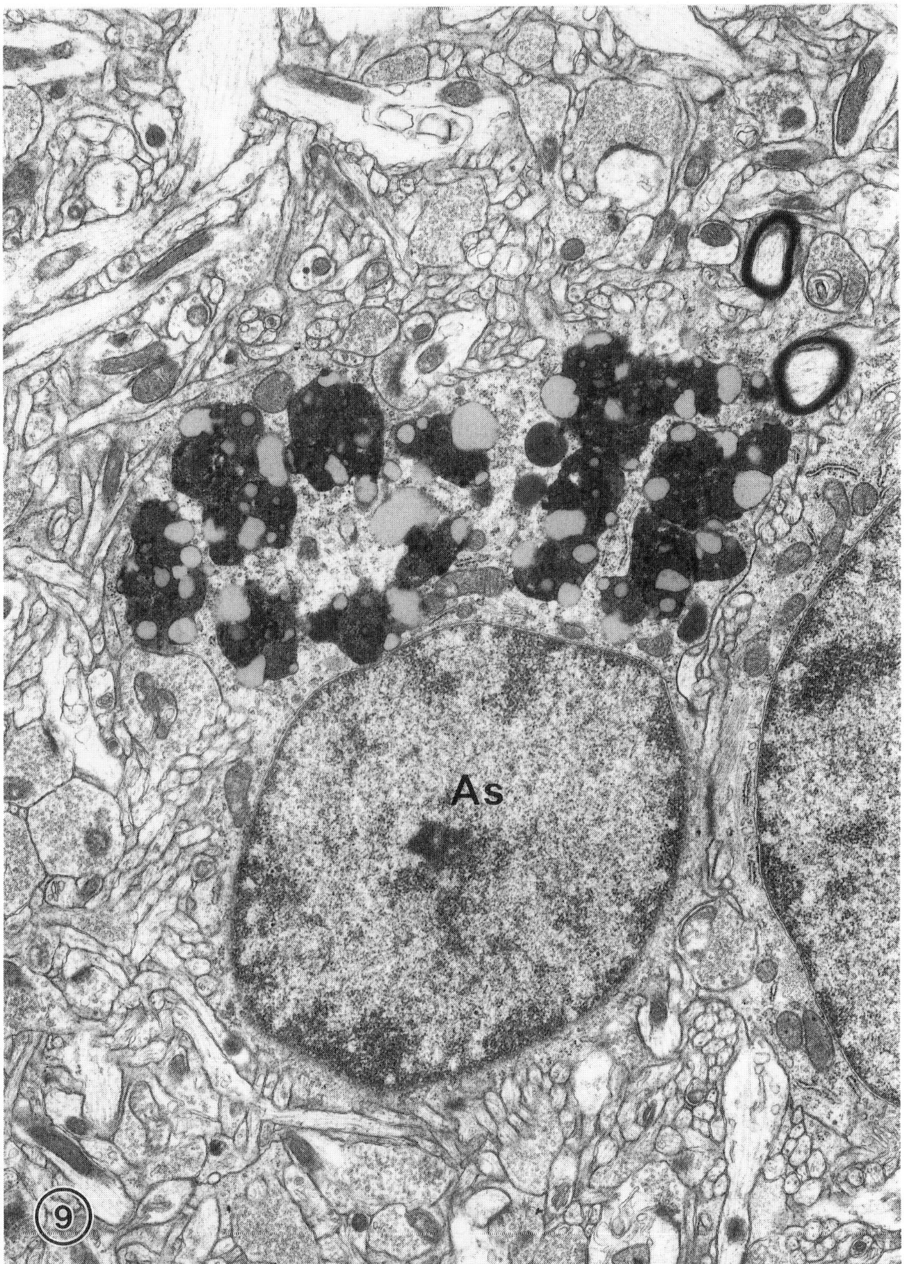


**Figure 9.** An astrocyte (*As*) from layer II, area 46, of a 25-year-old monkey. The cytoplasm to one side of the nucleus is filled with dark inclusions. × 10,000.

membrane-bound and resembles lipofuscin. It generally consists of two components, one with a dark, granular appearance and, embedded within it, the other component, pale and more homogeneous and sometimes having the appearance of bubbles embedded in the dark matrix. In the old brains, some of the astrocytic processes passing through the neuropil also seem to have an increased concentration of the typical glial intermediate filaments in their cytoplasm. This is certainly true of the astrocytic processes extending out toward and contributing to, the thickened outer glial limiting membrane in the older cortices (Fig. 4).

The oligodendrocyte is the only cell type that we have studied in detail in the cerebral cortices of old monkeys. This is because of their role in the formation of the myelin sheaths, which frequently show degenerative changes in the central nervous system. In the cortex of areas 46 and 17 from old monkeys, many of the oligodendrocytes have developed swellings along their processes (Peters, 1996). These swellings are filled with dark inclusions that resemble age pigment, and similar inclusions can also occur within the perikarya (Fig. 10). Furthermore, when the Perl's stain for the detection of iron-containing compounds is used to visualize oligodendrocytes, it is seen that in younger monkeys oligodendrocytes usually occur individually, whereas in the cortices of older monkeys it is not uncommon to find the cell bodies in groups (Fig. 10), or rows. These groups, or rows, are often situated close to capillaries, and tight junctions may exist between the adjacent oligodendroglial cell bodies.

The groups of oligodendrocytes in the cortices of older monkeys have likely been produced by cell division, although it is generally assumed that population of oligodendrocytes is stable, because there is little evidence that these cells undergo an extensive proliferation in the mature normal brain, although they will proliferate in some plaques in cases of multiple sclerosis (see Ludwin, 1995; Norton, 1996). On the other hand, there is some evidence that the numbers of oligodendrocytes do increase in the cortices of old monkeys (Peters *et al.*, 1991). And clearly, the oligodendrocytes are being affected by the aging process, as evidenced by the accumulation of dense inclusions in their processes and perikarya. The source of these inclusions is presumably related to the breakdown of the myelin sheaths in the cortices and white matter of old monkeys.

Is the breakdown of the myelin sheaths in older monkeys due to an inability of the oligodendrocytes to maintain their sheaths or is something like an immune response causing the myelin to break down and so bring about alterations in the oligodendrocytes? In the twitcher mouse, an animal model for globoid leukodystrophy, in which there is a disintegration of myelin sheaths, LeVine and Torres (1992) reported that some of the oligodendrocytes have large swellings adjacent to their myelin sheaths, suggesting that the material in these swellings is derived from the breakdown of the myelin, and that it moves along the processes of the oligodendrocytes towards the cell bodies to produce the dark inclusions within the perikarya. On the other hand, Mitrovic *et al.* (1994) suggest that microglial cell-derived nitric oxide plays an important role in multiple sclerosis and other inflammatory diseases affecting myelin, because it damages oligodendrocytes. Mitrovic *et al.* (1994) have shown that in tissue cultures, oligodendrocytes are the neuroglial cells most susceptible to the effects of nitric oxide, which ultimately results in their death. The effects of normal aging on oligodendrocytes and their myelin sheaths needs to be investigated further, since myelin breakdown is one of the most obvious changes in the aging brain.

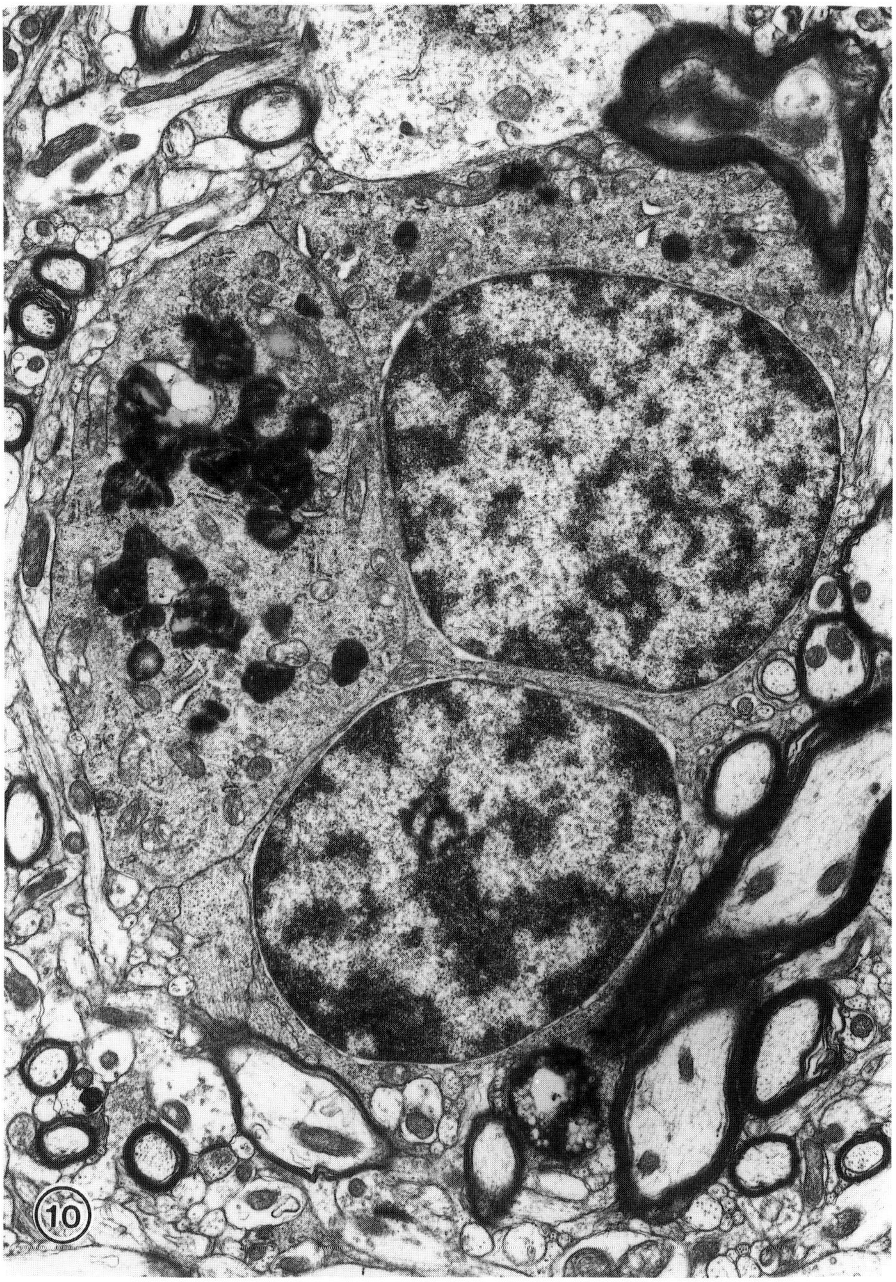


**Figure 10.** A group of three oligodendrocytes in area 17 of a 28-year-old monkey. Two of the oligodendrocytes have been sectioned through their nuclei, the third through a cell pole containing dense inclusions. × 10,000.

## 10. Neuritic Plaques and Amyloid

In humans, the memory loss associated with Alzheimer's disease has often been attributed to the formation and presence of neuritic plaques containing amyloid. Plaques do occur in the cortex of the aged monkey, as described initially by Wisniewski *et al.* (1973), who examined four rhesus monkeys. The exact ages of the monkeys were unknown, but three of them were believed to be between 16 and 18 years, and one between 22 and 23 years. The oldest monkey and one of the younger monkeys had plaques, and after examining the brains of a large number of monkeys, Price *et al.* (1983) determined that plaques first appear in monkeys between 17 and 20 years of age (see also Uno, 1993). Yet, although most older monkeys develop some plaques in their cortices, monkeys do not commonly develop the numbers of plaques that characterize Alzheimer's disease, they never seem to show the same loss of neurons, and they do not evidence the magnitude of behavioral deficits indicative of Alzheimer's disease.

To determine the distribution of plaques in monkeys, Heilbronner and Kemper (1990) examined 55 cytoarchitectural areas in the cerebral cortices of three monkeys over 25 years of age. They found the plaques to have a predilection for the frontal areas and for the primary somatosensory cortex. Lower densities of plaques were present in the amygdala and insula, and in the cingulate, temporal, occipital, and parietal association areas; the lowest densities occurred in the hippocampus and in the primary visual and auditory areas. In an earlier study, Struble *et al.* (1985) also reported that the highest density of plaques in the monkey brain is in the temporal lobe and prefrontal cortex, with few plaques in the occipital lobe and hippocampus, a distribution quite different from that in Alzheimer's disease in humans, wherein plaques are most concentrated in the hippocampus, amygdala, entorhinal cortex, and the temporal and parietal lobes (e.g., Kemper, 1994).

A more recent study measured plaque density in the brains of 14 behaviorally tested rhesus monkeys between 5 and 30 years old (Sloane *et al.*, 1997). Plaques were present in four of the brains, from monkeys aged 25, 29, and 30 years of age, and they were most evident in the frontal lobes, as shown in the study by Heilbronner and Kemper (1990). Plaque density does tend to increase with age, as it does in normally aging humans (e.g., Mackenzie, 1994). The density of plaques was then compared with the monkey's behavioral performance and, although both cognitive dysfunction and plaque number increased with age, no correlation was found between the frequency of plaques and the extent of behavioral dysfunction (Sloane *et al.*, 1996; unpublished). This agrees with previous reports that plaque distribution and density in monkeys does not account for their cognitive deficits (Struble *et al.*, 1985; Cork, 1993).

The known facts about the structure of senile plaques were reviewed in an earlier volume in this series (Peters, 1991), so the same ground will not be covered here. The structure of plaques in monkeys seems very similar to that in humans, and obtaining well-preserved material from the monkey cortex would offer an excellent opportunity to carry out an in-depth evaluation of plaque structures and components; no useful, detailed studies of the fine structure have so far been carried out beyond that of Wisniewski *et al.* (1973). With the present availability of antibodies to structural components of plaques and to neurotransmitters, it is an opportune time to carry out a combined chemical and structural analysis of their components and to determine what kinds of neurons contribute to plaques, and which of their processes provide neurites to them.

The deposition of β-amyloid (Aβ) in plaques and blood vessels in aged monkey brain is similar to that seen in humans (e.g., Walker *et al.,* 1990; Struble *et al.,* 1985; Cork *et al.,* 1990; Kanemura *et al.,* 1996), but it has recently become evident that, in contrast to Alzheimer's disease, in which the earliest and more-prominent form of Aβ in senile plaques is $A\beta_{42}$, in the monkey $A\beta_{40}$ is the predominant form (Gearing *et al.,* 1996). A similar result has been obtained by Kanemura *et al.* (1996), who used antibodies to distinguish between the two types of amyloid, and concluded that the proportion of $A\beta_{40}$ to $A\beta_{42}$ is much greater in the monkey brain than in the human brain. This may mean that the processing of amyloid is different in monkeys than in humans with Alzheimer's disease, and may help to provide some clues about how to alleviate the human disease.

## 11. Conclusions

In both monkey and human, there is little evidence for the concept that cognitive decline in normal aging can be attributed to a loss of neurons. Neither does there appear to be a correlation between the accumulation of neuritic plaques and normal cognitive decline. Also, it is unlikely that the changes in astrocytes and microglial cells represent anything other than the normal response of those cells to damage, resulting in their activation to undertake phagocytosis. Of course, that leads to the question of what they are phagocytozing, a question as yet unanswered. The changes in the oligodendrocytes are accompanied by an age-related breakdown in the integrity of myelin sheaths.

There may be withdrawal in the lengths of dendrites of some neurons, especially in those that extend into layer I, and in the monkey there is evidence that, although the thinning of layer I of area 17 correlates only with aging, in area 46 of frontal cortex the thinning correlates not only with aging but with cognitive decline. What instigates the thinning is not yet known, but there is a suggestion that thinning and the morphological alterations that accompany it may be linked to a loss of neurons in nuclei, such as the substantia nigra and the basal forebrain, that project to the neocortex.

There is an overall loss of white matter, but not of gray matter, from the cerebral hemispheres with age, and in monkeys at least, the overall reduction in white matter volume correlates with cognitive decline. The loss of white matter may be due not to extensive axonal degeneration, but to a breakdown of myelin sheaths. More work needs to be done to establish a firm link between cognitive impairment and the breakdown in the integrity of myelin sheaths, and investigation into the cause of the myelin breakdown is needed as well. The breakdown in the integrity of myelin would obviously reduce the rate of conduction along fibers (e.g., Ritchie, 1984; Waxman *et al.,* 1995), and if only a proportion of the fibers are affected, this would lead to a disruption and loss of coherence in the timing of synaptic sequences within neuronal circuits.

ACKNOWLEDGMENTS. This work was supported by Program Project grant 2PO1 AG00001 from the National Institute on Aging, and by grant RR-00165 from the Division of Research Resources to Yerkes Regional Primate Research Center. I wish to thank Karen McNally for her participation in these studies and for the preparation of the illustrations.

## 12. References

Ahmad, A., and Spear, P. D., 1993, Effects of aging on the size, density, and numbers of rhesus monkey lateral geniculate neurons, *J. Comp. Neurol.* **334:**631–643.

Adams, I., 1987, Comparison of synaptic changes in the precentral and postcentral cerebral cortex of aging humans: A quantitative ultrastructural study, *Neurobiol. Aging* **8:**203–212.

Albert, M. S., 1993, Neuropsychological and neurophysiological changes in healthy adult humans across the age range, *Neurobiol. Aging* **14:**623–625.

Albert, M. S., and Moss, M. B., 1984, The assessment of memory disorders in patients with Alzheimer's disease, in: *Neuropsychology of Memory* (L. Squire and N. Butters, eds.), Guilford, New York, pp. 236–246.

Amaral, D. D., 1993, Morphological analyses of the brains of behaviorally characterized aged nonhuman primates, *Neurobiol. Aging* **14:**671–672.

Anderson, J. M., Hubbard, B. M., Coghill, G. R., and Slidders, W., 1983, Neuropsychological changes in healthy adults across the age range, *Neurobiol. Aging* **14:**623–625.

Beal, M. F., Walker, L. C., Storey, E., Segar, L., Price, D. L., and Cork, L. C., 1991, Neurotransmitters in neocortex of aged rhesus monkey, *Neurobiol. Aging* **12:**407–412.

Bowden, D. M., and Williams, D. D., 1984, Advances in veterinary science and comparative medicine, in: *Aging*, Academic, New York, pp. 305–341.

Braak, H., and Braak, E., 1986, Ratio of pyramidal cells versus non-pyramidal cells in the human frontal isocortex and changes in ratio with aging and Alzheimer's disease, *Prog. Brain Res.* **70:**185–212.

Brizzee, K. R., 1973, Quantitative studies on aging in cerebral cortex of rhesus monkey and albino rat, with notes on the effects of prolonged low-dose ionizing radiation in the rat, *Prog. Brain Res.* **40:**141–160.

Brizzee, K. R., Klara, P., and Johnson, J. E., 1975, Changes in microanatomy, neurocytology, and fine structure with aging, in: *Neurobiology of Aging* (J. M. Ordy and K. R. Brizzee, eds.), Plenum, New York, pp. 574–594.

Brizzee, K. R., Ordy, J. M., and Bartus, R. T., 1980, Localization of cellular changes within multimodal sensory regions in aged monkey brain: Possible implications of age related cognitive loss, *Neurobiol. Aging* **1:**45–52.

Brody, H., 1955, Organization of cerebral cortex. III. A study of aging in the human cerebral cortex, *J. Comp. Neurol.* **102:**511–516.

Brody, H., 1970, Structural changes in the aging nervous system, *Interdiscipl. Top. Gerontol.* **7:**9–21.

Buell, S. J., and Coleman, P. D., 1981, Quantitative evidence for selective dendritic growth in normal human aging but not in senile dementia, *Brain Res.* **214:**23–41.

Colon, E. J., 1972, The elderly brain: A quantitative analysis of the cerebral cortex in two cases, *Psychiat. Neurol. Neurochir.* **75:**261–270.

Cork, L. C., 1993, Plaques in prefrontal cortex of aged, behaviorally tested monkeys: Incidence, distribution, and relationship to task performance, *Neurobiol. Aging* **14:**675–676.

Cork, L. C., Masters, C., Beyreuther, K., and Price, D. L., 1990, Development of senile plaques: Relationship of neuronal abnormalities and amyloid deposits, *Am. J. Pathol.* **137:**1383–1392.

Cragg, B. G., 1975, The density of synapses and neurons in normal, mentally defective, and aging human brains, *Brain* **98:**81–90.

Cupp, C. P., and Uemura, E., 1980, Age-related changes in prefrontal cortex of *Macaca mulatta:* Quantitative analysis of dendritic branching patterns, *Exp. Neurol.* **69:**143–163.

Davis, R. T., 1985, Introduction, in: *Behavior and Pathology of Aging in Rhesus Monkeys* (R. T. Davis and C. W. Leathers, eds.), Alan R. Liss, New York.

DeKosky, S. T., and Scheff, S. W., 1990, Synapse loss in frontal cortex biopsies in Alzheimer's disease: Correlation with cognitive severity, *Ann. Neurol.* **27:**457–464.

Devaney, K. O., and Johnson, H. A., 1980, Neuron loss in the aging visual cortex of man, *J. Gerontol.* **35:**836–841.

Dietrich, W. D., Halley, M., Alonso, O., Globus, M. Y.-T., and Busto, R., 1992, Intraventricular infusion of *N*-methyl-D-aspartate. II. Acute neuronal consequences, *Acta Neuropathol.* **84:**630–637.

Double, K. L., Halliday, G. M., Kril, J. J., Harasty, J. A., Cullen, K., Brooks, W. S., Creasey, H., and Broe, G. A., 1996, Topography of brain atrophy during normal aging and Alzheimer's disease, *Neurobiol. Aging* **17:**513–521.

Dyke, B., Gage, T. B., Mamelka, P. M., Goy, W., and Stone, W. H., 1986, A demographic analysis of the Wisconsin Regional Primate Center rhesus colony (1962–1982), *Am. J. Primatol.* **10:**257–269.

Flood, D. G., and Coleman, P. D., 1988, Neuron numbers and sizes in aging brain: Comparisons of human, monkey, and rodent data, *Neurobiol. Aging* **9:**453–463.

Gazzaley, A. H., Siegel, S. J., Kordower, J. H., Mufson, E. J., and Morrison, J. H., 1996, Circuit-specific alterations of *N*-methyl-D-aspartate receptor subunit 1 in the dentate gyrus of aged monkeys, *Proc. Natl. Acad. Sci. USA* **93:**3121–3125.

Gearing, M., Tigges, J., Mori, H., and Mira, S. S., 1996, $A\beta_{40}$ is a major form of β-amyloid in nonhuman primates, *Neurobiol. Aging* **17:**903–908.

Gibson, P. H., 1983, EM study of the number of cortical synapses in the brains of aging people and people with Alzheimer-type dementias, *Acta Neuropathol. (Berlin)* **62:**127–133.

Goldman-Rakic, P. S., and Brown, R. M., 1981, Regional changes of monoamines in cerebral cortex and subcortical structures of aging rhesus monkeys, *Neurosci.* **6:**177–178.

Haug, H., 1985, Are neurons of the human cerebral cortex really lost during aging? A morphometric examination, in: *Senile Dementia of the Alzheimer Type* (J. Traber and W. H. Gispen, eds.), Springer-Verlag, Berlin, pp. 150–156.

Haug, H., and Eggers, R., 1991, Morphometry of the human cortex cerebri and corpus striatum during aging, *Neurobiol. Aging* **12:**336–338.

Haug, H., Kühl, S., Mecke, E., and Wasner, K., 1984, The significance of morphometric procedures in the investigation of age changes in cytoarchitectonic structures of human brain, *J. Hirnforsch.* **25:**353–374.

Heilbronner, P. L., and Kemper, T. L., 1990, The cytoarchitectonic distribution of senile plaques in three aged monkeys, *Acta Neuropathol.* **81:**60–65.

Henderson, G., Tomlinson, B. C., and Gibson, P. H., 1980, Cell counts in human cerebral cortex in normal adults throughout life, using an image analysis computer, *J. Neurol. Sci.* **46:**113–136.

Hof, P. R., Cox, K., and Morrison, J. H., 1990, Quantitative analysis of a vulnerable subset of pyramidal neurons in Alzheimer's disease. I. Superior and inferior temporal cortex, *J. Comp. Neurol.* **301:** 44–54.

Hof, P. R., and Morrison, J. H., 1990, Quantitative analysis of a vulnerable subset of pyramidal neurons in Alzheimer's disease. II. Primary and secondary visual cortex, *J. Comp. Neurol.* **310:**55–64.

Hof, P. F., and Morrison, J. H., 1995, Neurofilament protein defines regional patterns of cortical organization in the macaque monkey visual system: A quantitative immunohistochemical analysis, *J. Comp. Neurol.* **352:**161–186.

Hutterlocher, P. R., 1979, Synaptic density in human frontal cortex: Developmental changes and effects of aging, *Brain Res.* **163:**195–205.

Hutterlocher, P. R., and de Courten, C., 1987, The development of synapses in striate cortex of man, *Hum. Neurobiol.* **6:**1–9.

Kemper, T. L., 1994, Neuroanatomical and neuropathological changes during aging and dementia, in: *Clinical Neurology and Aging* (M. L. Albert and J. E. Knoefel, eds.), Oxford University Press, New York, pp. 3–67.

Kim, C. B. Y., Tom, B. W., and Spear, P. D., 1996, Effects of aging on the densities, numbers, and sizes of retinal gangion cells in rhesus monkey, *Neurobiol. Aging* **17:**431–438.

Kim, C. B. Y., Pier, L. P., and Spear, P. D., 1997, Effects of aging on numbers and sizes of neurons in histochemically defined subregions of monkey striate cortex, *Anat. Rec.* **247:**119–128.

Kanemura, K., Iwatsubo, T., and Ihara, Y., 1996, Comparable amyloid β-protein (Aβ) 42 (43) and Aβ40 deposition in the aged monkey brain, *Neurosci. Lett.* **214:**196–198.

Leuba, G., and Garey, L. J., 1987, Comparison of neuronal and glial numerical density in the developing and aging human visual cortex, *Hum. Neurobiol.* **6:**11–18.

Leuba, G., and Krafsik, R., 1994a, Visual cortex in Alzheimer's disease: Occurrence of neuronal death and glial proliferation, and correlation with pathological landmarks, *Neurobiol. Aging* **15:**29–43.

Leuba, G., and Krafsik, R., 1994b, Changes in volume, surface estimate, three-dimensional shape, and total numbers of neurons of the human primary visual cortex from midgestation until old age, *Anat. Embryol.* **190:**351–366.

Leuchter, A. F., Dunkin, J. J., Lufkin, R. B., Anzia, Y., Cook, I., and Newton, T. F., 1994, Effect of white matter disease on functional connections in the aging brain, *J. Neurol. Neurosurg. Psychiatry* **57:**1347–135.

LeVine, S. M., and Torres, M. V., 1992, Morphological features of degenerating oligodendrocytes in twitcher mice, *Brain Res.* **587:**348–352.

Lintl, P., and Braak, H., 1983, Loss of intracortical myelinated fibers: A distinctive age-related alteration in the human striate area, *Acta Neuropathol. (Berlin)* **61:**178–182.

Ludwin, S. K., 1995, Pathology of the myelin sheath, in: *The Axon* (S. G. Waxman, J. D. F. Koscsis, and P. K. Stys, eds.), Oxford University Press, New York, pp. 412–437.

Mackenzie, I. R. A., 1994, Seline plaques do not progressively accumulate with normal aging, *Acta Neuropathol.* **87:**520–525.

Masliah, E., Mallory, M., Hansen, L., DeTeresa, R., and Terry, R. D., 1993, Quantitative synaptic alterations in the human neocortex during normal aging, *Neurology* **43:**192–197.

Mishkin, M., 1993, Neural circuitry underlying behavioral deficits in aging, *Neurobiol. Aging* **14:**615–617.

Mitrovic, B., Ignarro, L. J., Montrestruque, S., Smoll, A., and Merrill, J. E., 1994, Nitric oxide as a potential pathological mechanism in demyelination: Its effects on primary glial cells *in vitro, Neuroscience* **61:**575–585.

Mosley, K., and Cuzner, M. L., 1996, Receptor-mediated phagocytosis of myelin by macrophages and microglia: Effects of oponization and receptor blocking agents, *Neurochem. Res.* **21:**481–487.

Mrzljak, L., Pappy, M., Leranth, C., and Goldman-Rakic, P. S., 1995, Cholinergic synaptic circuitry in the macaque prefrontal cortex, *J. Comp. Neurol.* **357:**603–617.

Murphy, G. M., 1985, Volumetric asymmetry in the human striate cortex, *Exp. Neurol.* **88:**288–230.

Nakamura, S., Akiguchi, I., Kameyama, M., and Mizuno, N., 1985, Age-related changes of pyramidal cell basal dendrites in layers III and V of human motor cortex: A quantitative Golgi study, *Acta Neuropathol. (Berlin)* **65:**281–284.

Norton, W. T., 1996, Do oligodendrocytes divide? *Neurochem. Res.* **21:**495–503.

Peinado, M. A., Quesada, A., Pedrosa, J. A., Martinez, M., Esteban, F. J., del Moral, M. L., and Peinado, J. M., 1997, Light microscopic quantification of morphological changes during aging in neurons and glia of rat parietal cortex, *Anat. Rec.* **247:**420–425.

Peters, A., 1991, Aging in monkey cerebral cortex, in: *Normal and Altered States of Function* (A. Peters and E. G. Jones, eds.), *Cerebral Cortex,* Vol. 9, Plenum, New York, pp. 485–510.

Peters, A., 1996, Age-related changes in oligodendrocytes in monkey cerebral cortex, *J. Comp. Neurol.* **371:**153–263.

Peters, A., Harriman, K. M., and West, C. D., 1987, The effect of increased longevity, produced by dietary restriction, on the neuronal population of area 17 in rat cerebral cortex, *Neurobiol. Aging* **8:**7–20.

Peters, A., Josephson, K., and Vincent, S. L., 1991, Effects of aging on the neuroglial cells and pericytes within area 17 of the rhesus monkey cerebral cortex, *Anat. Rec.* **229:**384–398.

Peters, A., Leahu, D., Moss, M. B., and McNally, K. J., 1994, The effects of aging on area 46 of the frontal cortex of the rhesus monkey, *Cereb. Cortex* **6:**621–635.

Peters, A., Morrison, J. H., Rosene, D. L., and Hyman, B. T., 1998, Are neurons lost from the primate cerebral cortex during normal aging? *Cerebral Cortex* **8:**295–300.

Peters, A., Nigro, N. J., and McNally, K. J., 1997, A further evaluation of the effect of age on striate cortex of the rhesus monkey, *Neurobiol. Aging* **18:**29–36.

Peters, A., Rosene, D. L., Moss, M. B., Kemper, T. L., Abraham, C. R., Tigges, J., and Albert, M. S., 1996, Neurobiological bases of age-related cognitive decline in the rhesus monkey, *J. Neuropathol. Exp. Neurol.* **55:**861–784.

Peters, A., and Sethares, C., 1993, Aging and the Meynert cells in rhesus monkey primary visual cortex, *Anat. Rec.* **236:**721–729.

Price, D. L., Whitehouse, P. J., Struble, R. g., Price, D. J., Cork, L. C., Hedreen, J. C., and Kitt, C. A., 1983, Basal forebrain cholinergic neurons and neuritic plaques in primate brain, in: *Biological Aspects of Alzheimer's Disease* (R. Katzman, ed.), Cold Spring Harbor Laboratory, Cold Spring Harbor, pp. 65–78.

Purves, D., and LaMantia, A., 1993, Development of blobs in the visual cortex of macaques, *J. Comp. Neurol.* **334:**169–175.

Regeur, L., Badsberge, J. G., Pakkenberg, H., Evans, S. M., and Pakkenberg, B., 1994, No global neocortical nerve cell loss in brains from patients with senile dementia of Alzheimer's type, *Neurobiol. Aging* **15:**347–352.

Ritchie, J. M., 1984, Pathophysiology of conduction in demyelinated nerve fibers, in: *Myelin* (P. Morrell, ed.), Plenum, New York, pp. 337–367.

Rockland, K. S., 1994, The organization of feedback connections from V2 (18) to V1 (17), in: *Primary Visual Cortex of Primates* (A. Peters and E. G. Jones, eds.), *Cerebral Cortex,* Vol. 10, Plenum, New York, pp. 261–300.

Rosene, D. L., 1993, Comparing age-related changes in the basal forebrain and hippocampus of the rhesus monkey, *Neurobiol. Aging* **14:**669–670.

Scheibel, M. E., Lindsay, R. D., Tomiyasu, U., and Scheibel, A. B., 1975, Progressive dendritic changes in aging human cortex, *Exp. Neurol.* **47:**392–403.

Schutz, U., and Hunziker, O., 1980, Comparative studies of neuronal perikaryal size and shape in the aging cerebral cortex, *J. Gerontol.* **35:**483–491.

Shefer, V. F., 1973, Absolute number of neurons and thickness of cerebral cortex during aging, senile and vascular dementia, Pick's and Alzheimer's diseases, *Neurobiol. Behav. Physiol.* **6:**319–324.

Siddiqi, Z. A., Kemper, T. L., and Blatt, G. J., 1994, Age-related neuronal loss in the substantia nigra of rhesus monkey, *Soc. Neurosci. Abstr.* **20:**49.

Siddiqi, Z. A., Kemper, T. L., Rosene, D. L., and Feldman, M. L., 1995, Age-related changes in the rhesus monkey: Neuronal loss in substantia nigra, pars compacta, and ventral tegmental area, *Soc. Neurosci. Abstr.* **21:**1565.

Sloane, J. A., Pietropaolo, H. F., Rosene, D. L., Moss, M. B., Peters, A., Kemper, T., and Abraham, C. R., 1997, Lack of correlation between plaque burden and cognition in the aged monkey, *Acta Neuropathol.* **94:**471–478.

Smith, D. G., 1982, A comparison of the demographic structure and growth of free ranging and captive groups of rhesus monkeys *(Macaca mulatta), Primates* **23:**24–30.

Spear, P. D., Moore, R. J., Kim, C. B. Y., Xue, J.-T., and Tumosa, N., 1984, Effects of aging on the primate visual system: Spatial and temporal processing by lateral geniculate neurons in young adult and old rhesus monkeys, *J. Neurophysiol.* **72:**402–420.

Stensaas, S. S., Eddington, D. K., and Dobelle, W. H., 1974, The topography and variability of the primary visual cortex in man, *J. Neurosurg.* **40:**747–755.

Struble, R. G., Price, D. L., Cork, L. C., and Price, D. L., 1985, Senile plaques in cortex of aged normal monkeys, *Brain Res.* **361:**267–275.

Terry, R. D., DeTeresa, R., and Hansen, L. A., 1987, Neocortical cell counts in normal human adult aging, *Ann. Neurol.* **21:**530–539.

Tigges, J., 1992, Novel inclusion bodies in Betz cells of cortical area 4 of aged rhesus monkeys, *Anat. Rec.* **233:**162–169.

Tigges, J., Gordon, T. P., McClure, H. M., Hall, E. C., and Peters, A., 1988, Survival rate and life span of the rhesus monkey, *Am. J. Primatol.* **15:**263–273.

Tigges, J., Herndon, J. G., and Peters, A., 1990, Neuronal population of area 4 during life span of the rhesus monkey, *Neurobiol. Aging* **11:**210–208.

Tigges, J., Herndon, J. G., and Peters, A., 1992, Axon terminals on Betz cell somata of area 4 in rhesus monkeys throughout adulthood, *Anat. Rec.* **232:**305–315.

Tigges, J., Herndon, J. G., and Rosene, D. L., 1995, Mild age-related changes in the gyrus in rhesus monkeys, *Acta Anat.* **153:**39–48.

Tigges, J., Herndon, J. G., and Rosene, D. L., 1996, Preservation into old age of synaptic number and size in the supragranular layer of the gyrus in rhesus monkeys, *Acta Anat.* **157:**63–72.

Uemura, E., 1980, Age-related changes in prefrontal cortex of *Macaca mulatta:* Synaptic density, *Exp. Neurol.* **69:**164–172.

Uno, H., 1993, The incidence of senile plaques and multiple infarction in aged macaque brain, *Neurobiol. Aging* **14:**673–674.

Van Essen, D. C., Newsome, W. T., and Maunsell, J. H. R., 1984, The visual field representation in striate cortex of the macaque monkey: Asymmetries, anisotropies, and individual variability, *Vision Res.* **24:**429–487.

Vincent, S. L., Peters, A., and Tigges, J., 1989, Effects of aging on the neurons within area 17 of rhesus monkey cerebral cortex, *Anat. Rec.* **223:**329–341.

Vogt, B. A., 1991, The role of layer 1 in cortical function, in: *Normal and Altered States of Function* (A. Peters and E. G. Jones, eds.), *Cerebral Cortex,* Vol. 9, Plenum, New York, pp. 199–222.
Walker, L. C., Masters, C., Beyreuther, K., and Price, L. C., 1990, Amyloid in the brains of aged squirrel monkeys, *Acta Neuropathol.* **80:**381–387.
Waxman, S. G., Kocsis, J. D. F., and Black, J. A., 1995, Pathophysiology of demyelinated axons, in: *The Axon: Structure, Function, and Pathophysiology* (S. G. Waxman, J. D. Kocsis, and P. K. Stys, eds.), Oxford University Press, New York and Oxford, pp. 438–461.
West, C. D., Volicer, L., and Vaughan, D. W., 1984, Aging in the food restricted rat: Body temperature, receptor function, and morphological changes in the brain, in: *Nutritional Intervention in the Aging Process* (H. J. Armbrecht, J. M. Prendergast, and R. M. Coe, eds.), Springer-Verlag, New York, pp. 111–137.
Wisniewski, H. M., Ghetti, B., and Terry, R. D., 1973, Neuritic (senile) plaques and filamentous changes in aged rhesus monkeys, *J. Neuropathol. Exp. Neurol.* **32:**566–584.
Zecevic, N., Bourgeois, J.-P., and Rakic, P., 1989, Changes in synaptic density in motor cortex of rhesus monkeys during fetal and postnatal life, *Develop. Brain Res.* **50:**11–32.
Zola-Morgan, S., and Squire, L. R., 1985, Medial temporal lesions in monkeys impair memory on a variety of tasks sensitive to human amnesia, *Behav. Neurosci.* **99:**22–34.

# 4

# Age-Related Neuronal Loss in the Cerebral Cortex

MARK J. WEST

## 1. Background

### 1.1. Historical Confusion

Age-related neuronal loss in the cerebral cortex has been the subject of numerous studies aimed at identifying the structural basis of senescent decline in neural function. Because cortical neurons are not generated or replaced during adult life, an evaluation of neuronal loss represents a simple means for quantifying the cumulative effects of the neurodegenerative aging process. As a single parameter, neuron number represents one fundamental expression of the information processing capacity of the tissue, that is, the number of integrative entities with separate output channels. It also provides a quantitative, structural framework in which changes in other parameters, such a synapse, receptor, and dendrite number, can more rigorously be evaluated.

Ten years ago, the vast literature in this field was reviewed by Coleman and Flood (Coleman and Flood, 1987; Flood and Coleman, 1988). They concluded that data in the studies on age-related neuron loss, up to that time, were "rife with confusion and seemingly contradictory data" (Coleman and Flood, 1987, p. 539)

MARK J. WEST • Department of Neurobiology, University of Aarhus, 8000 Aarhus C, Denmark.
*Cerebral Cortex,* Volume 14, edited by Peters and Morrison. Kluwer Academic/Plenum Publishers, New York, 1999.

and the issue of age-related neuron loss "swirling in controversy" (Coleman and Flood, 1987, p. 521). Although secular trends and subject selection (in human studies) and species and strain differences (in animal studies) undoubtedly contributed to this state of affairs, a significant part of this confusion was related to a lack of appreciation of the problems associated with the methods available at that time for counting neurons and to misconceptions as to what type of data constituted evidence of neuron loss (West, 1994; Coggeshall and Lenkan, 1996).

### 1.2. A More Clear but Still Incomplete View

In the intervening years, a new generation of design-based stereological methods using unbiased counting (Sterio, 1984; Gundersen, 1986) and sampling techniques (Gundersen and Jensen, 1987) has been developed for estimating the total number of neurons in delineated regions of the brain (West, 1993a). Discussions of the manner in which these techniques can be used to count neurons (Coggeshall, 1992; West, 1993a) have led to a more widespread appreciation of the fact that measures of neuronal density cannot be used by themselves to draw assumption-free conclusions about neuronal loss* (Swaab and Uylings, 1987) and that quantitative evaluations of neuronal loss can only be based on determinations or estimates of total neuron number (West, 1994). This development is notable in that most of the studies reviewed by Coleman and Flood were based on neuronal density data obtained with model-based stereological methods. It is also now widely accepted that unbiased estimation procedures†, involving the direct counting of neurons with tools, such as the disector, and unbiased sampling techniques that ensure that all parts of the region of interest have an equal probability of being sampled, can produce reliable, reproducible estimates of total neuron number (Saper, 1996; West and Coleman, 1996). Aging studies employing these techniques have led to a more clear, but still incomplete, view of the role that age-related neuronal loss plays in senescent functional decline (Wickelgren, 1996a,b). This view can be summarized in the following manner: (1) there is accumulating evidence that regional and layer-specific—rather than global—changes are the rule; (2) interspecies differences in age-related neuronal loss are significant and they confound models of human aging; (3) age-related neuronal loss, in both humans and animals, is not as

*The reason that numerical density cannot by itself be used as an expression of either neuron number or neuronal loss is that it is the ratio of the total number of objects (neurons) to the volume of the region under consideration. Differences in numerical densities can result from changes in the number of neurons, the volume of the region, or both. As a consequence, comparisons of densities cannot be used to make assumption-free statements about either neuronal loss or changes in the volume of the region. The same is true for numerical densities of synapses.

†Here, "unbiased" means that the average of repeated estimates made with a particular unbiased sampling scheme will approach the true value as the number of repetitions increases. Model-based methods (e.g., Abercrombie, 1946), which involve assumptions about the size, shape, or orientation of the objects being counted, will produce unbiased estimates if the assumptions are true. However, the means of the repeated estimates made with model-based systems will systematically deviate from the true value, i.e., be biased, if the assumptions are not true. Herein lies the problem with model-based methods, for it is extremely difficult (if not impossible) to make assumptions about the size, shape, and orientation of neurons or other biological entities that are appropriate for all parts of the individual, all individuals in a group, and different groups.

common as one would have been led to believe from the literature that existed on the topic ten years ago (see Peters, this volume).



## 2. Recent Studies with Improved Methods

A number of recent studies of age-related neuronal loss have employed unbiased stereological principles. In a recent paper, Pakkenberg and Gundersen (1997) reported a slight (10%), though significant, age-related reduction in the number of neurons in the human cerebral cortex. This finding should be contrasted with the landmark report of 20–50% losses in many cortical areas based on density measurements (Brody, 1955) which influenced thinking about the global loss of cortical neurons for four decades. Haug (1984) later described how density changes of the magnitude reported in earlier studies could be explained by his observation that young brains shrink more than old brains when fixed and that age-related changes in the numerical density of neurons need not involve neuronal loss. It has also become increasingly apparent that the earlier studies may have been confounded by the inclusion of early cases of Alzheimer's disease in the samples of allegedly normal aged subjects used in these studies (Evans *et al.*, 1989). The slight, though significant, global losses reported recently on the basis of estimates of total neuron number are not totally unanticipated, however. There has been accumulating evidence that specific cortical areas, such as the primary visual cortex (area 17) of monkeys (Peters *et al.*, 1997) and the superior temporal sulcus and entorhinal cortex (area 28) of humans (Gomez-Isla *et al.*, 1996, 1997) do not lose neurons with age and that age-related losses appear to be confined to specific regions, subregions, and laminae of the central nervous system (West, 1993b).

One cortical region of the human brain in which there is unbiased stereological data indicating age-related neuronal loss is the hippocampal region. In studies of the relationship between age and the total number of neurons in specific subdivisions of the hippocampus, we have detected significant losses in the dentate hilus (ca. 30%) and subiculum (ca. 50%), but no evidence of losses in the immediately adjacent dentate gyrus, CA3-2, or CA1 regions (West, 1993b; West *et al.*, 1994) (Fig. 1). In view of the role of the hippocampus in relational memory processes and the role of the neurons in the largely unidirectional flow of transformations of polysensory information through the subdivisions of hippocampal region (Witter, 1993), these regionally specific losses provide a potential structural basis for the observed degradation of specific cognitive processes during later life (Albert and Moss, 1996). At present, however, there is no direct evidence of a relationship between neuron number and cognitive status in normal aged subjects.

## 3. Differences in Neuronal Loss Patterns: Normal Aging and Alzheimer's Disease

In a number of regions of the CNS, the pattern of normal age-related neuronal loss is different from that associated with Alzheimer's disease (AD). West *et al.*

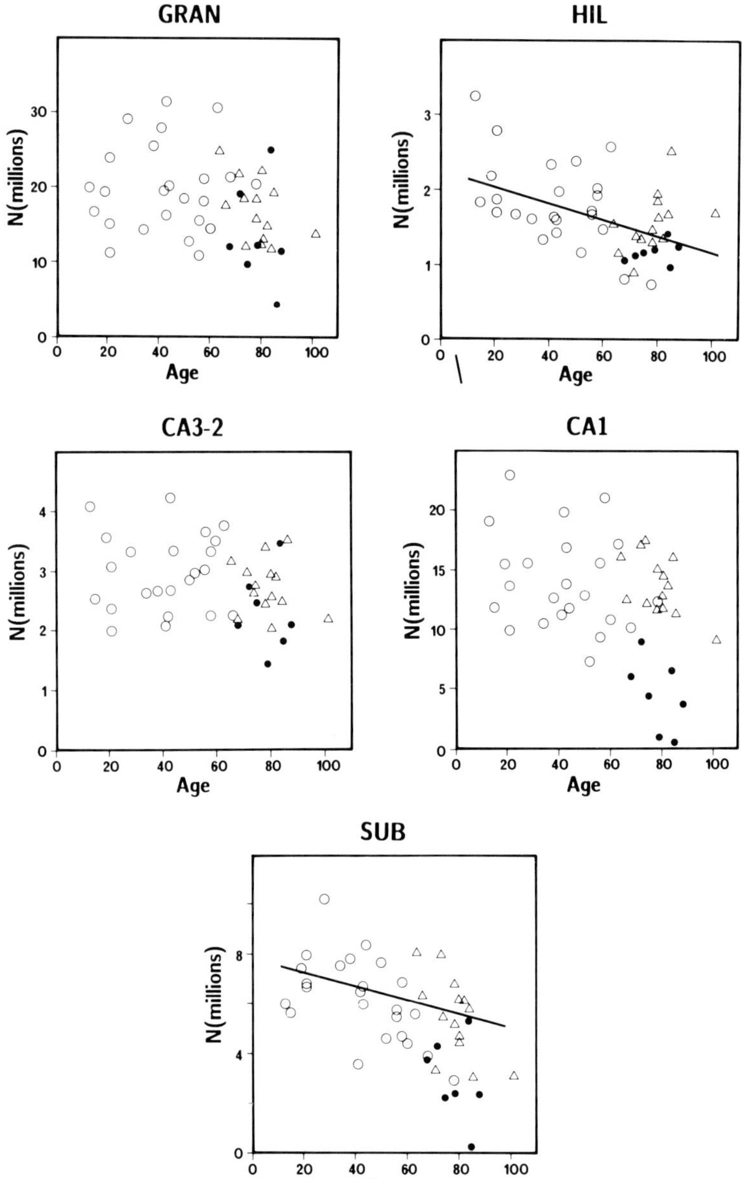


**Figure 1.** The total number of neurons (N) in the subdivisions of the hippocampal region plotted as a function of age. Individuals in the normal aging group, ages 13 to 101 years, as shown as open circles and open triangles, those in the Alzheimer's group as filled circles, and those in the normal age-matched AD control group as open triangles. Regressions of neuron-number-vs-age for the normal aging group that had slopes significantly different from zero are indicated by lines. Only HIL and SUB showed evidence of normal age-related neuronal loss. The magnitude of these losses is summarized in (Fig. 2). (GRAN: dentate granule cells; HIL: dentate hilar cells; CA3-2 and CA1: pyramidal cells of hippocampus proper; SUB: neurons in the deep and superficial layers of subiculum) From West *et al.* (1994) with permission. © The Lancet Ltd., 1994.

(1994) demonstrated that the total number of neurons in the CA1 region is not altered during normal aging, but is severely reduced in Alzheimer's patients by about 70% (Figs. 1 and 2). Similarly, Mouton *et al.* (1994) and Ohm *et al.* (1997) have demonstrated the lack of age-related losses in the locus coeruleus and significant reductions in Alzheimer's patients. Gomez-Isla *et al.* (1996) have described the absence of age-related neuronal changes in the individual layers of entorhinal cortex and the presence of a massive loss of neurons in layer II in very early-stage Alzheimer's patients. These data lend credence to the emerging view that (1) the neurodegenerative processes that characterize normal aging are qualitatively different from those that accompany Alzheimer's disease, and (2) Alzheimer's disease should be considered as neither accelerated aging nor an inevitable consequence of the normal aging process.

## 4. Interspecies Differences

Studies of age-related neuronal loss, carried out in experimental animals and based on unbiased estimates of the total neuron number, indicate that there are interspecies differences, or at least differences between humans and other species. Data from studies of the hippocampal region of laboratory rats provides a good example of this. In contrast to the normal age-related losses of dentate hilar and subicular neurons in humans (described above), there is no evidence of a change in the total number of neurons in any of the individual subdivisions of the hippocampal region of groups of Wistar (Rasmussen *et al.*, 1996) and Long-Evans rats (Rapp and Callagher, 1996) from 2 to 28 months of age. This conclusion should be contrasted with those drawn from analyses performed with previously available model-based techniques, in which age-related losses are described (Wickelgren,

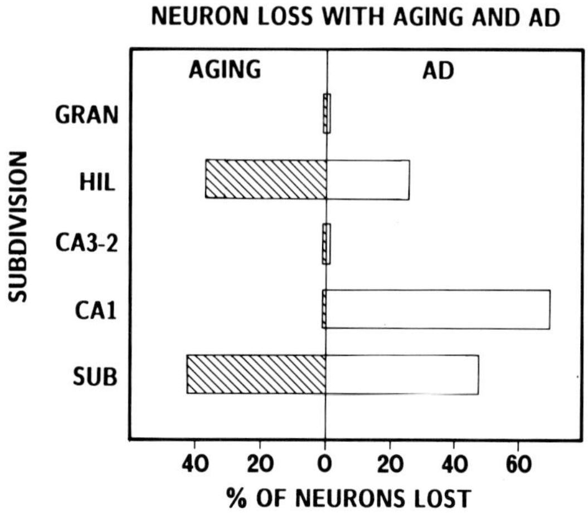


**Figure 2.** Percentage of neurons lost due to aging and Alzheimer's disease. Cross-hatched bars represent the percentage of neurons lost due to normal aging, calculated from the end points (youngest and oldest subjects) of the regressions of neuron number and age that had slopes significantly different from 0, i.e., only in GRAN and SUB (see Fig. 1). The open bars represent the percentage of neurons lost due to Alzheimer's disease in each subdivision, calculated from the differences in the means of the Alzheimer's group (filled circles, Fig. 1) and the normal age-matched AD controls (open triangles, Fig. 1) that were shown to be significantly different, i.e., only in HIL, CA1, and SUB. (GRAN: dentate granule cells; HIL: dentate hilar cells; CA3-2 and CA1: pyramidal cells of hippocampus proper; SUB: neurons in the deep and superficial layers of subiculum) From West *et al.* (1994) with permission. © The Lancet Ltd., 1994.

1996b). Similarly, preliminary studies of the hippocampal region of rhesus monkeys using modern stereological methods suggest that there is no age-related neuronal loss in the hippocampal subdivisions of this species up to 30–40 years of age (West *et al.,* 1993). These data lead to the conclusion that there are important differences between humans and other species with regard to age-related neuronal loss in homologous brain regions and raise questions about the comparability of aging processes in different species and the appropriateness of animal models for studying human aging.

## 5. Factors other than Neuronal Loss

There has yet to be demonstrated a correlation between neuron number in the entire brain (or any of its subdivisions) and performance in humans. Attempts have been made to identify correlations of this type in experimental animals. In the rat studies referred to above, there was, in spite of considerable variability in scores on tests designed to test the performance of the hippocampus, no demonstrable relationship between performance and the number of neurons (Rasmussen *et al.,* 1996; Rapp and Gallagher, 1996). A recent study examining the relationship between rearing conditions and the number of dentate granule cells in adult mice (Kempermann *et al.,* 1997) reported correlations between rearing conditions and dentate granule cell number and between rearing conditions and performance on hippocampal tasks, but not between performance and the number of granule cells. These findings underline the fact that there are significant individual differences in neuron number (West, 1993b; Peters *et al.,* 1997) and the fact that the functional capacity or performance of a neural system is also related to factors other than neuron number. Candidates for the latter include alterations in the neurons themselves, such as changes in the size, number, and type of synapses (Geinisman *et al.,* 1992); the length and shape of the dendrites (Flood, 1993; Buell and Coleman, 1979); the receptor properties of the cellular membranes (Gazzaley *et al.,* 1996; Wilson and Tonegawa, 1997); and changes in glia, including oligodendrocytes (Meier-Ruge *et al.,* 1992; Peters *et al.,* 1994). To date, only numerical density data is available for age-related synaptic changes, and few rigorous conclusions can be drawn about the absolute magnitude of age-related synaptic loss. Reports of age-related changes in the density of synapses (Masliah *et al.,* 1993) indicate that synapse changes play an important role. Recently published detailed descriptions of stereological protocols for making unbiased estimates of total synapse number (Geinisman *et al.,* 1996) can be expected to provide data on the absolute size of age-related synapse loss in the near future.

## 6. The Emerging View and Its Implications

The emerging view that age-related, cortical, neuronal loss is less widespread than previously thought (Wickelgren, 1996b) has implications for future studies of aging and for the development of strategies to prevent senescent decline in cortical functions. It is becoming increasingly apparent that factors other than neuronal loss play significant (perhaps primary) roles in age-related changes in functional capacity. In the absence of global age-related neuronal loss, the possibilities for developing successful strategies to reduce senescent functional decline are enhanced. In contrast to frank neuronal loss, age-related changes in the other parameters (e.g., dendrites, synapses, receptors, mitochondria, membranes) are more likely to be amenable to restoration. Neurons that die are lost forever and restorative strategies are limited. This need not be the case for changes in the numbers of receptors or synapses if the cells are viable. Today, the restoration of surviving neurons with trophic substances (Mattson and Scheff, 1994; Martinez-Serrano *et al.*, 1995) appears more feasible than the replacement of neurons lost from specific neural circuits. Another implication of this general conclusion is that the neuronal losses that do occur are regional and lamina-specific. This in turn implies that the conditions responsible for the losses in specific regions are not necessarily the same. Identifying the regions in which there are age-related neuronal losses is a necessary first step toward identifying the factors responsible for the losses and developing strategies for preventing these losses.

## 7. References

Abercrombie, M., 1946, Estimation of nuclear populations from microtome sections, *Anat. Rec.* **94:**239–247.

Albert, M. S., and Moss, M. B., 1996, Neuropsychology of aging: Findings in humans and monkeys, in: *Handbook of the Biology of Aging* (E. L. Schneider and J. W. Rowe, eds., H. J. Morrison, section ed.), Academic, San Diego, pp. 217–233.

Brody, H., 1955, Organization of the cerebral cortex. III. A study of aging in the human cerebral cortex, *J. Comp. Neurol.* **102:**511-556.

Buell, S. J., and Coleman, P. D. 1979, Dendritic growth in the aged human brain and failure of growth in senile dementia, *Science* **206:**854.

Coggeshall, R. E., 1992, A consideration of neural counting methods, *Trends Neurosci.* **15:**9–13.

Coggeshall, R. E., and Lenkan, H. A., 1996, Methods for determining numbers of cells and synapses: A case for more uniform standards of review, *J. Comp. Neurol.* **364:**6–15.

Coleman, P. D., and Flood, D. G., 1987, Neuron numbers and dendritic extent in normal aging and Alzheimer's disease, *Neurobiol. Aging* **8:**521–545.

Evans, D. A., Funkelstein, H. H., Albert, M. S., Scherr, P. A., Cook, N. R., Chown, M. J., Hebert, L. E., Hennekens, C. H., and Taylor, J. O., 1989, Prevalence of Alzheimer's disease in a community population of older persons, *J. Am. Med. Assoc.* **262:**2551–2556.

Flood, D. G., 1993, Critical issues in the analysis of dendritic extent in aging humans, primates, and rodents, *Neurobiol. Aging* **14:**649–654.

Flood, D. G., and Coleman, P. D., 1988, Neuron numbers and sizes in aging brain: Comparisons of human, monkey, and rodent data, *Neurobiol. Aging* **9:**453–463.

Gazzaley, A. H., Siegel, S. J., Kordower, J. H., Mufson, E. J., and Morrison, J. H., 1996, Circuit-specific alterations of *N*-methyl-D-aspartate receptor subunit 1 in the dentate gyrus of aged monkeys, *Proc. Natl. Acad. Sci. USA* **93:**3121–3125.

Geinisman, Y., de Toledo-Morrell, L., Morrell, F., Persina, I. S., and Rossi, M., 1992, Age-related loss of axospinous synapses formed by two afferent systems in the rat dentate gyrus as revealed by the unbiased stereological disector technique, *Hippocampus* **2:**437–444.

Geinisman, Y., Gundersen, H. J. G., Van der Zee, E., and West, M. J., 1996, Unbiased stereological estimation of the total number of synapses in a brain region, *J. Neurocytol.* **25:**805–819.

Gómez-Isla, T., Price, J. L., McKeel, D. W., Morris, J. C., Growdon, J. H., and Hyman, B. T., 1996,

Profound loss of layer II entorhinal cortex neurons occurs in very mild Alzheimer's disease, *J. Neurosci.* **16:**4491–4500.
Gundersen, H. J. G., 1986, Stereology of arbitrary particles, *J. Microsc.* **143:**3–45.
Gundersen, H. J. G., and Jensen, E. B., 1987, The efficiency of systematic sampling in stereology and its prediction, *J. Microsc.* **147:**229–263.
Haug, H., 1984, Macroscopic and microscopic morphometry of the human brain and cortex: A survey in the light of new results, *Brain Pathol.* **1:**123–149.
Kempermann, G., Kuhn, H. G., and Gage, F. H., 1997, More hippocampal neurons in adult mice living in an enriched environment, *Nature* **386:**493–495.
Martinez-Serrano, A., Fisher, W., and Bjorklund, A., 1995, Reversal of age-dependent cognitive impairments and cholinergic neuron atrophy by NGF-secreting neural progenitors grafted to the basal forebrain, *Neuron* **15:**473–484.
Masliah, E., Mallory, M., Hansen, L., DeTeresa, R., and Terry, R. D., 1993, Quantitative synaptic alterations in the human neocortex during normal aging, *Neurology* **43:**192–197.
Mattson, M. P., and Scheff, S. W., 1994, Endogenous neuroprotection factors and traumatic brain injury: Mechanisms of action and implications for therapy, *J. Neurotrauma* **11:**3–33.
Meier-Ruge, W., Ulrich, J., Bruhlman, M., and Meier, E., 1992, Age-related white matter atrophy in the human brain, *Ann. N.Y. Acad. Sci.* **673:**260–269.
Mouton, P. R., Pakkenberg, B., Gundersen, H. J. G., and Price, D. L., 1994, Absolute number and size of pigmented locus coeruleus neurons in young and aged individuals, *J. Chem. Neuroanat.* **7:**185–109.
Ohm, T. G., Busch, C. H., and Bohl, J., 1997, Unbiased estimation of neuronal numbers in human nucleus coeruleus during aging, *Neurobiol. Aging* **18:**393–399.
Pakkenberg, B., and Gundersen, H. J. G., 1997, Neocortical neuron number in humans: Effect of sex and age, *J. Comp. Neurol.* **384:**312–320.
Peters, A., Leahu, D., Mass, M. B., and McNally, K. J., 1994, The effects of aging on area 46 of the frontal cortex of the rhesus monkey, *Cereb. Cortex* **4:**621–635.
Peters, A., Nigro, N. J., and McNally, K. J., 1997, A further evaluation of the effect of age on striate cortex of the rhesus monkey, *Neurobiol. Aging* **18:**29–37.
Rapp, P. R., and Gallagher, M., 1996, Preserved neuron number in the hippocampus of aged rats with spatial learning deficits, *Proc. Natl. Acad. Sci. USA* **93:**9926–9930.
Rasmussen, T., Schleimann, T., Sorensen, J. C., Zimmer, J., and West, M. J., 1996, Memory-impaired aged rats: No loss of principal hippocampal and subicular neurons, *Neurobiol. Aging* **17:**1143–1147.
Saper, C. P., 1996, Any way you cut it: A new journal policy for the use of unbiased counting methods, *J. Comp. Neurol.* **364:**5.
Stereo, D. C., 1984, The unbiased estimation of number and sizes of arbitrary particles using the disector, *J. Microsc.* **134:**127–136.
Swaab, D. F., and Uylings, H. B. M., 1987, Density Measures: Parameters to avoid, *Neurobiol. Aging* **8:**574–576.
West, M. J., 1993a, New stereological methods for counting neurons, *Neurobiol. Aging* **44:**275–285.
West, M. J., 1993b, Regionally specific loss of neurons in the aging human hippocampus, *Neurobiol. Aging* **14:**287–293.
West, M. J., 1994, Advances in the study of age-related neuron loss, *Semin. Neurosci.* **6:**403–411.
West, M. J., and Coleman, P. D., 1996, How to count, *Neurobiol. Aging* **17:**503.
West, M. J., Amaral, D. G., and Rapp, P. R., 1993, Preserved hippocampal cell number in aged monkeys with recognition memory deficits, *Soc. Neurosci. Abstr.* **19(1):**599.
West, M. J., Coleman, P. D., Flood, D. G., and Troncoso, J. C., 1994, Differences in the pattern of hippocampal neuronal loss in normal aging and Alzheimer's disease, *Lancet* **344:**769–772.
Wickelgren, I., 1996a, For the cortex, neuron loss may be less than thought, *Science* **273:**48–50.
Wickelgren, I., 1996b, Is hippocampal cell death a myth? *Science* **271:**1229–1230.
Wilson, M. A., and Tonegawa, S., 1997, Synaptic plasticity, place cells and spatial memory: Study with second generation knockouts, *Trends Neurosci.* **20:**102–106.
Witter, M. P., 1993, Organization of the entorhinal-hippocampal system: A review of current anatomical data, *Hippocampus* **3:**33–44.

# 5

# Age-Related Cognitive Deficits and Neurotransmitters

## The Role of Catecholamine Mechanisms in Prefrontal Cortical Cognitive Decline

AMY F. T. ARNSTEN

## 1. Introduction

How do changes in neurotransmitters over the life span alter higher cortical function? More specifically, how might degeneration of monoamine systems projecting to cortex from brainstem contribute to age-related cognitive decline? These important questions have often been asked using nonhuman primate models, as monkeys have highly developed neocortices and show age-related changes comparable to those observed in humans. Aged monkeys and humans exhibit similar patterns of cognitive deficits, and similar neuropathological changes with advancing age. The aged monkey model has been especially important for examining age-related changes that cannot be observed reliably or ethically in humans. In particular, the monkey model obviates the problems of using human postmortem tissue, where

---

AMY F. T. ARNSTEN • Section of Neurobiology, Yale Medical School, New Haven, Connecticut 06510-8001.

*Cerebral Cortex,* Volume 14, edited by Peters and Morrison. Kluwer Academic/Plenum Publishers, New York, 1999.

illness can confound changes in transmitter levels, and where biochemical measures of low concentrations of cortical monoamines are often obscured by postmortem deterioration. Furthermore, monkeys do not appear to develop Alzheimer's disease (AD); thus, early cases of AD do not confound the study of normal aging as they can in human experiments. Finally, aged monkeys provide an opportunity for intensive pharmacological studies not possible in humans. Studies of aged monkeys have revealed a number of factors that may contribute to age-related cognitive decline. The current chapter will highlight the importance of catecholamine depletion to the loss of working-memory abilities subserved by the prefrontal cortex (PFC).

## 2. Pattern of Cognitive Deficits with Normal Aging

This topic is dealt with in greater detail in chapters in this volume by Moss and Rosene and by Albert. The following is intended as a brief review.

### 2.1. Monkeys

Although aging is a varied and complex process, decline in the cognitive functions of the prefrontal cortex (PFC) is consistently observed in aged nonhuman primates. Bartus (1979) was the first to note the similarity in cognitive deficits between aged monkeys and young monkeys with lesions to the PFC. For example, marked deficits on spatial working-memory tasks such as delayed response emerge early in the aging process (Bachevalier *et al.,* 1991; Bartus *et al.,* 1978; Rapp and Amaral, 1989; Walker *et al.,* 1988). These deficits are also apparent on the spatial memory-span task (see Moss and Rosene, this volume). As with prefrontal cortical lesioned monkeys, working-memory deficits in aged animals are particularly pronounced in the presence of distracting stimuli (Bartus and Dean, 1979) or conditions of high proactive interference (e.g., Rapp and Amaral, 1989). Aged monkeys, like PFC-lesioned monkeys, are also impaired on tasks where behavioral inhibition is critical; for example, they show impairment on reversals of previously learned discrimination problems (Bartus *et al.,* 1979; Rapp, 1990). In contrast, aged monkeys are not impaired on a test of covert orienting of visuospatial attention subserved by the parietal cortex (M. Voytko, personal communication), although they are impaired on a route-finding task linked to area 7 (Walker *et al.,* 1988). Recognition-memory abilities dependent upon the medial temporal lobe and also feature-discrimination learning dependent upon the inferior temporal cortex are impaired only in a subset of elderly monkeys (Bartus *et al.,* 1979; Rapp and Amaral, 1992), although modest deficits on recognition-memory tasks have been reported for very old monkeys (Moss *et al.,* 1988; Presty *et al.,* 1987).

### 2.2. Humans

Aged humans also show signs of PFC dysfunction (reviewed in Hochanadel and Kaplan, 1984; West, 1996), including an increased susceptibility to interference (e.g., Hoyer *et al.,* 1979). Chao, Knight, and colleagues have shown that aged subjects exhibit electrophysiological changes during tests of working memory, attention, and behavioral inhibition that resemble changes observed in patients with PFC lesions (Chao and Knight, 1997a,b; Nielsen-Bohlman and Knight, 1995). Consistent with PFC dysfunction, Davis *et al.* (1990) observed progressive deficits on the Wisconsin Card Sorting and Stroop tasks with increasing age, whereas recognition memory remained quite stable across the life span (Davis *et al.,* 1990). As with aged monkeys, elderly human subjects did not show deficits on the Posner Task of covert attentional shifting, suggesting that the visuospatial attentional abilities of the parietal lobe are less affected by the aging process (Greenwood *et al.,* 1993; Robinson and Kertzman, 1990). Thus the cognitive abilities of the PFC are consistently affected by the aging process in both human and nonhuman primates, whereas cognitive functions dependent upon more-posterior cortical regions are altered with lesser severity or consistency. This pattern contrasts with that observed in AD, where temporal and posterior cortical functions are often affected earlier and more severely than are PFC functions (Haxby *et al.,* 1985; Haxby *et al.,* 1990; Small *et al.,* 1989).

## 3. Pattern of Neuropathological and Neurochemical Changes

### 3.1. Neuropathological Changes

What structural changes may underlie the loss of PFC cognitive function with normal aging? Careful examination of the dorsolateral PFC in monkeys has shown no age-related loss of cortical cells, although there are changes in dendritic morphology and striking changes in axonal myelination (see Peters, this volume). Similarly, histological examination of the hippocampal formation in monkeys has revealed little change in cell number or integrity, although small changes have been observed in human hippocampus that may or may not be linked to early AD (see West, this volume). In contrast to preserved number of cells in cortex, the monoamine-containing nuclei in the brainstem that project to cortex show consistent degenerative changes. As reviewed in Kemper (this volume), there is approximately a 30–40% loss of neurons over the life span in the raphe (serotonin, or 5HT), locus coeruleus (norepinephrine, or NE) and substantia nigra/ventral tegmental area (dopamine, or DA) nuclei. There is also a loss of locus coeruleus (LC) cells in aged rodents (e.g., Leslie *et al.,* 1985), monkeys (Sladek and Sladek, 1979), and humans (e.g., German *et al.,* 1988; Iversen *et al.,* 1983; Vijayashankar and Brody, 1979). Cholinergic cell loss from the basal forebrain with normal aging is more controversial, and careful studies in aged monkeys have shown small (Stroessner-Johnson *et al.,* 1992) or no (Voytko *et al.,* 1995) changes over the life span. These differences may result from the varying methodologies used to identify cholinergic cells.

### 3.2. Neurochemical Changes

Consistent with the degeneration of monoamine cell groups over the life span, neurochemical studies of monkey cortex have revealed significant decreases in several monoamines and their associated precursors and/or metabolites. These findings are summarized in Figure 1.

Biochemical analysis of aged monkey cortex has shown a marked loss of DA from the PFC, and depletion is evident quite early in the aging process (Goldman-Rakic and Brown, 1981; Wenk *et al.*, 1989). The DA metabolite HVA is also markedly depleted from the PFC with age (Wenk *et al.*, 1989), and similar changes in DA and its metabolite have been observed in aged rat PFC (Luine *et al.*, 1990). In monkeys, DA loss with age is also prominent in the temporal pole (Goldman-Rakic and Brown, 1981), but other cortical and subcortical areas are more modestly affected (Goldman-Rakic and Brown, 1981; Wenk *et al.*, 1989). Levels of the catecholamine precursor l-dopa are also markedly reduced in the PFC, temporal pole, and surrounding cortical areas (Goldman-Rakic and Brown, 1981), suggesting that turnover of both DA and NE would be impaired with age. The very low levels of NE and DA in cortex have precluded accurate measurement of NE and DA levels from human postmortem tissue. Levels are also too low to be detected using current imaging techniques, which have visualized an approximately 40% loss of DA uptake sites in the striatum of living individuals over the life span (e.g., van Dyck *et al.*, 1995). Imaging studies in humans have also revealed a significant loss of D1 receptors from cortex in elderly humans (de Keyser *et al.*, 1990; Suhara *et al.*, 1991). Thus, the DA system is affected by the aging process at multiple levels.

Norepinephrine is not significantly reduced with age in monkey PFC (Goldman-Rakic and Brown, 1981; Wenk *et al.*, 1989), although rat studies have found a decrease in PFC NE with age (Luine *et al.*, 1990) and further suggest that NE turnover can be decreased even when NE levels are normal (Hadfield and Milio,

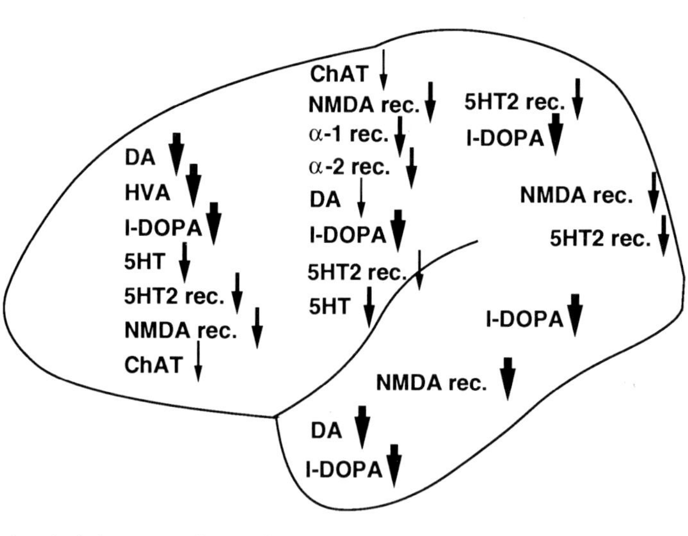


**Figure 1.** Biochemical changes in the aged monkey cortex. The greatest number of findings are in the frontal lobe, the most extensively studied region. The width of the arrow roughly represents the degree of age-related decrease. Both Goldman-Rakic and Brown (1981) and Wenk *et al.* (1989) agree that dopamine is greatly depleted from the PFC with age. DA, dopamine; HVA, DA metabolite; l-DOPA, catecholamine precursor; 5HT, serotonin, 5HT2 rec., serotonin-2 receptors; NMDA rec., *N*-methyl-D-aspartate receptors; ChAT, choline acetyltransferase; α-1 rec., α-1 adrenergic receptors; α-2 rec., α-2 adrenergic receptors. Data from Goldman-Rakic and Brown (1981); Wenk *et al.* (1989); Wenk *et al.* (1991); and Bigham and Lidow (1995). Additional discussion can be found in Rosene, this volume.

1990). As stated above, the large decrease with advancing age in the NE precursor l-dopa in monkey cortex encourages this possibility, as does the finding of reduced NE in the LC of aged monkeys (Sladek and Sladek, 1979). A small, age-related decline in α-1 and α-2 adrenergic receptors has been observed in sensorimotor cortices (areas 1 and 4), while β-receptors actually show a small increase with age in these same regions (Bigham and Lidow, 1995). Adrenergic receptors are less affected in the PFC (area 46), where there is only a small decrease in α-2 receptors in layer I (Bigham and Lidow, 1995). This loss of α-2 receptors may reflect loss of presynaptic sites with age. In humans, immunocytochemical analysis of catecholamine synthetic enzymes has shown a prominent loss of NE fibers from the PFC with advancing age (Gaspar *et al.*, 1989; Gaspar *et al.*, 1991).

Receptors 5HT, 5HT1, and 5HT2 appear to decrease in monkey cortex with advancing age (Bigham and Lidow, 1995; Goldman-Rakic and Brown, 1981; Wenk *et al.*, 1989), particularly in older animals (Wenk *et al.*, 1989). Although there are discrepancies in the literature regarding the specific cortical areas affected, there is general agreement that aging alters these systems. However, serotonin loss is not as marked as loss of DA in the PFC (Goldman-Rakic and Brown, 1981; Wenk *et al.*, 1989).

The cholinergic system has been a major focus of aging research. Studies of cholinergic markers in monkey cortex, however, have often observed variable results, due in part to the large individual variation in results from young animals (e.g., Wenk *et al.*, 1989). Modest decreases in choline acetyltransferase (ChAT) have been observed in the frontal pole and motor cortex with age (Wenk *et al.*, 1989), but these were not replicated (Wenk *et al.*, 1991). Age-related changes in cholinergic markers are further discussed in Rosene and Nicholson (this volume).

Why might DA in the PFC be especially vulnerable to the aging process? The DA neurons projecting to the PFC in rodents are known to have many special properties: They fire faster, they have fewer autoreceptors regulating cell firing and DA synthesis, and they are uniquely sensitive to stress (Deutch and Roth, 1990; Thierry *et al.*, 1976). Thus, exposure to mild (e.g., psychological) stress selectively increases DA turnover in the PFC, but not in other cortical or subcortical DA fields. The effects of stress are blocked by pretreatment with anxiolytics such as lorazepam and mimicked by anxiogenic beta carbolines such as FG7142 (Roth *et al.*, 1988). This sensitivity to stress is thought to arise in part form differential regulation of PFC-projecting DA neurons (see Deutch and Roth, 1990, for review) and may contribute to a preferential "burnout" of the meso-PFC DA system over the life span. The NE and 5HT systems are also very reactive to stress in rodents (Bliss *et al.*, 1968; Inoue *et al.*, 1993; Weiss *et al.*, 1981), although in the PFC, they are not as reactive as DA (Goldstein *et al.*, 1996). This stress sensitivity may also render these systems more susceptible to degenerative changes.

## 4. Neurotransmitter Modulation of the Prefrontal Cortex

Although age-related changes in neuromodulators have been documented in the monkey PFC, are these findings relevant to age-related PFC cognitive deficits? To address this question, we first must understand the role of these substances in the functioning of the PFC in young animals. Our current understanding of DA, NE, 5HT, and acetylcholine (ACh) mechanisms influencing PFC spatial working memory function is summarized in Fig. 2.

### 4.1. Dopamine

It has been known for almost 20 years that DA depletion in the monkey PFC profoundly impairs spatial working-memory performance (Brozoski *et al.,* 1979), a finding replicated in rats (Bubser and Schmidt, 1990; Simon, 1981) and marmosets (Roberts *et al.,* 1994). More recently, impairment has been found with acute infusion of D1 receptor antagonists (SCH23390, SCH39166) or D1/D2 receptor antagonists (haloperidol, *cis*-flupenthixol) into the PFC of monkeys (Sawaguchi and Goldman-Rakic, 1991; Sawaguchi and Goldman-Rakic, 1994) and rats (Seamans *et al.,* 1998). These effects are delay-dependent; no impairment was found following the zero-second-delay control condition (Sawaguchi and Goldman-Rakic, 1991; Sawaguchi and Goldman-Rakic, 1994). In contrast, selective D2 receptor antagonists (sulpiride, raclopride) were without effect, indicating that DA has its beneficial effects on PFC function through actions at the D1 receptor family (Sawaguchi and Goldman-Rakic, 1991; Sawaguchi and Goldman-Rakic, 1994; Seamans *et al.,* 1995). Consistent with D1 receptor actions, systemic administration of D1 receptor antagonist SCH23390 in monkeys (Arnsten *et al.,* 1994) and rats (Murphy *et al.,* 1996a) similarly impairs spatial working-memory performance. As with intra-PFC infusions, no impairment was found following zero-second-delay control trials (Arnsten *et al.,* 1994). This is consistent with specific working-memory impairment rather than nonspecific (e.g., motor or motivational) changes in behavior. Spatial working-memory impairment has also been observed following systemic treatment with low, presynaptic doses of D2 agonist quinpirole, which inhibits DA release (Arnsten *et al.,* 1995). Consistent with presynaptic drug actions, this response was not evident in young monkeys experimentally depleted of catecholamines with

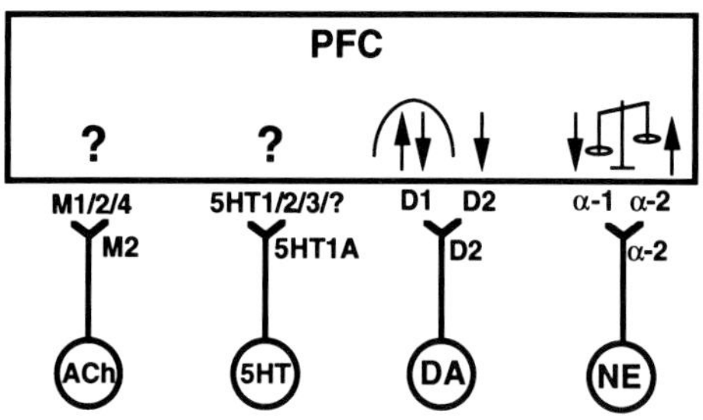


**Figure 2.** Known neuromodulatory influences on PFC cognitive function. Acetylcholine (Ach): Although it is likely that cholinergic mechanisms play an important role in PFC circuitry, findings remain conflicted. Systemic scopolamine administration impairs delayed-response performance (e.g., Bartus, 1982), whereas nucleus basalis lesions do not (Voytko *et al.,* 1995). Serotonin (5HT) depletion (Brozoski *et al.,* 1979) and 5HT3 receptor antagonists (Arnsten *et al.,* 1997) show no effect on spatial working-memory performance. Dopamine (DA) has a critical influence on PFC working memory function. Either too little or too much D1 receptor stimulation impairs PFC memory function (inverted "U"); beneficial D1 influences are represented by the upward arrow, detrimental actions by the downward arrow. In contrast, recent evidence suggests that DA actions at the D2 family of DA receptors have only detrimental effects on working-memory abilities (downward arrow). Norepinephrine (NE) has an important beneficial influence through actions at postsynaptic α-2 adrenergic PFC receptors (upward arrow), whereas NE actions at α-1 adrenergic receptors impair PFC function (downward arrow). The scales symbolize a balance between NE actions at α-1 vs. α-2 adrenergic receptors.

chronic reserpine treatment (Arnsten *et al.,* 1995). Electrophysiological findings suggest mechanisms whereby DA may have its beneficial effects on PFC circuits. Iontophoresis of DA has been found to enhance cue- and delay-related firing of PFC pyramidal cells in monkeys performing a spatial working-memory task (Sawaguchi *et al.,* 1988). More recently, *in vitro* studies of the rodent PFC have shown that an optimal level of D1 receptor stimulation sharpens NMDA signals arriving on apical dendrites and facilitates signal transfer from dendrite to soma (Yang and Seamans, 1996).

Although DA has essential, beneficial actions in the PFC, excessive DA receptor stimulation is detrimental to PFC function (Arnsten *et al.,* 1994; Arnsten and Goldman-Rakic, 1990b; Arnsten and Goldman-Rakic, 1998; Murphy *et al.,* 1996a; Murphy *et al.,* 1994; Verma and Moghaddam, 1996; Williams and Goldman-Rakic, 1995; Zahrt *et al.,* 1997). Increasing endogenous DA release in the PFC with mild stress exposure impairs spatial working-memory function in rats and monkeys (Arnsten and Goldman-Rakic, 1990b; Arnsten and Goldman-Rakic, 1998; Murphy *et al.,* 1996a; Murphy *et al.,* 1994). Impairment in performance correlates with increased DA turnover in the rodent PFC (Murphy *et al.,* 1996a; Murphy *et al.,* 1994), and performance is protected by pretreatment with agents that either prevent the rise in PFC DA turnover (Murphy *et al.,* 1996b), or block DA receptors (Arnsten and Goldman-Rakic, 1990b; Arnsten and Goldman-Rakic, 1998; Murphy *et al.,* 1996a; Murphy *et al.,* 1994). Interestingly, both D1 and D2 receptor-selective agents are effective in this regard, suggesting that DA has detrimental actions at both the D1 and D2 receptor families. Impaired spatial working-memory performance was also observed when the noncompetitive NMDA receptor antagonist ketamine was used to increase endogenous DA turnover in the rodent PFC (Verma and Moghaddam, 1996). The D1 receptor actions were not apparent in this study, although this may have been due to NMDA receptor blockade masking D1 modulatory actions. The importance of D1 receptor actions are underscored, however, by findings with D1 receptor agonists. Although very low doses of D1 agonists can improve working-memory performance in monkeys, higher doses can impair performance (Arnsten *et al.,* 1994; Cai and Arnsten, 1997). Indeed, infusion of the D1 receptor agonist SKF81297 into the rat PFC produces a dose-dependent impairment in spatial working-memory performance that can be blocked by D1 receptor antagonist pretreatment (Zahrt *et al.,* 1997). Low concentrations of D1 receptor antagonists have also been shown to enhance the memory-related firing of PFC pyramidal cells in monkeys performing a delayed-response task (Williams and Goldman-Rakic, 1995). Thus, either too little or too much stimulation of D1 receptors is detrimental to PFC function, as shown in Fig. 2.

### 4.2. Norepinephrine

Although most research has focused on DA mechanisms affecting the PFC, accumulating evidence indicates that NE has an equally important influence on PFC cognitive functioning (see Arnsten *et al.,* 1996, for review). Norepinephrine has marked, beneficial effects on PFC cognitive function through actions at postsynaptic, $\alpha$-2 receptors in the PFC. Thus, infusion of $\alpha$-2 antagonist yohimbine into the PFC produced a delay-dependent impairment in spatial working-memory performance in young monkeys (Li and Mei, 1994). Performance was not impaired by infusion of α-1 receptor antagonist prazosin or β-receptor antagonist propranolol, consistent with α-2 receptor mechanisms. Systemic administration studies show that yohimbine's effects are dose-dependent: low yohimbine doses improve working-memory performance through presynaptic actions, enhancing NE release and increasing endogenous stimulation of postsynaptic α-2 receptors, whereas higher yohimbine doses may block postsynaptic receptors and impair performance (Arnsten and Cai, 1993). Conversely, α-2 receptor stimulation with an agonist such as clonidine or guanfacine improves spatial working-memory function (Arnsten and Cai, 1993; Franowicz and Arnsten, 1998). Either systemic administration (Steere and Arnsten, 1994) or PFC infusion (Steere *et al.,* 1996) of guanfacine improves spatial working-memory performance in young monkeys. Consistent with drug actions at α-2 receptors localized *postsynaptic* to NE terminals, α-2 agonists are more, rather than less, effective in monkeys with NE depletion in the PFC (Arnsten and Goldman-Rakic, 1985) or global catecholamine loss (Cai *et al.,* 1993). Preliminary electrophysiological data suggest that α-2 agonists improve PFC function by enhancing delay-related activity of PFC pyramidal neurons in monkeys performing a delayed-response task (Li and Mei, 1994), whereas yohimbine suppresses memory-related activity (Bao-Ming Li, personal communication; Sawaguchi, 1998).

The importance of NE to PFC function is underscored by the findings that α-2 adrenergic agonists can improve PFC cognitive deficits in patients with cognitive disorders. Clonidine has been shown to improve, for example, performance on the Trails B, Word Fluency, and Stroop tasks in patients with schizophrenia (Fields *et al.,* 1988) or Korsakoff's syndrome (Mair and McEntee, 1986; Moffoot *et al.,* 1994). In support of a PFC mechanism, SPECT imaging studies of Korsakoff's patients have shown that clonidine's beneficial effects on the Word Fluency correlated with increased regional cerebral blood flow in the left PFC, the cortical region most affiliated with performance of this task (Moffoot *et al.,* 1994). Clonidine's beneficial effects in Korsakoff's patients appear to result from actions at postsynaptic receptors, as the drug has been most effective in those patients with greatest NE loss (McEntee and Mair, 1990). Finally, the symptoms of attention deficit hyperactivity disorder (ADHD) are thought to involve PFC dysfunction (Barkley *et al.,* 1992; Benson, 1991; Chelune *et al.,* 1986; Gorenstein and Mammato, 1989; Loge *et al.,* 1989; Mattes, 1980; Shue and Douglas, 1992), and clonidine and guanfacine have been helpful in this disorder (Chappell *et al.,* 1995; Horrigan and Barnhill, 1995; Hunt *et al.,* 1995; Hunt *et al.,* 1985). In particular, guanfacine has produced significant improvement on the Continuous Performance Task, a test of sustained attention and impulse inhibition that relies on the PFC (Chappell *et al.,* 1995). Thus, NE appears to have important beneficial effects on PFC function through α-2 receptors in humans just as it has in monkeys.

Most recently, data suggests that NE may have detrimental effects on PFC function via actions at α-1 adrenergic receptors. Infusion of selective α-1 adrenergic agonist phenylephrine into rat (Arnsten *et al.,* 1999) or monkey (Steere *et al.,* 1996) PFC induces spatial working-memory deficits that can be reverse by co-infusion of an α-1 receptor antagonist. These preliminary results suggest that NE may have opposing roles at postsynaptic α-1 versus α-2 receptors in the PFC, just as it does in the hypothalamus regulating ingestive behavior (Davies and Wellman, 1992; Leibowitz, 1988) and in the thalamus regulating arousal (Buzsaki *et al.,* 1991). Preliminary data comparing the binding of [$^3$H]-NE to recombinant human α adrenoceptors suggests that NE has higher affinity for the α-2A subtype than for either the α-1A or α-1D subtypes (J. P. Hieble, personal communication). Thus, α-2 mechanisms may predominate when basal NE release is moderate (e.g., during normal, attentive waking) and PFC function is optimal, whereas α-1 mechanisms may predominate under conditions of higher levels of NE release (e.g., during stress), contributing to PFC cognitive impairment.

### 4.3. Serotonin

There is little evidence that serotonin (5HT) alters dorsolateral PFC function. Depletion of 5HT from the dorsolateral PFC had no effect on spatial working-memory performance in monkeys (Brozoski *et al.,* 1979). In rats, depletion of serotonin or serotonin antagonist treatment has often been shown to improve reference memory tasks that depend upon posterior cortical or hippocampal function, but these treatments generally do not alter performance of working-memory tasks (Altman and Normile, 1986, 1988; Altman *et al.,* 1990; Normile and Altman, 1988). In aged monkeys, 5HT3 receptor antagonists have no effect on spatial working-memory performance, although these agents do improve performance of a visual discrimination task that depends upon the inferior temporal cortex (Arnsten *et al.,* 1997). These findings are consistent with rodent studies showing improvement on reference memory tasks with 5HT receptor blockade or depletion, and they suggest that 5HT may impede posterior cortical function while having little effect on PFC function. Further experiments with more-selective 5HT receptor agents will, however, be necessary for better understanding how 5HT mechanisms may influence PFC function.

### 4.4. Acetylcholine

Although studies of systemically administered cholinergic antagonists would suggest an important role for acetylcholine (ACh) in the PFC, few studies have addressed this issue explicitly in primates. Bartus's pioneering research demonstrated that the muscarinic antagonist scopolamine could induce working-memory deficits in young monkeys (Bartus, 1982; Bartus and Johnson, 1976), and similar learning and memory deficits were observed in humans treated with scopolamine (e.g., Drachman, 1981). These studies suggest that blocking the cholinergic input from the basal forebrain to the PFC results in marked PFC dysfunction. This interpretation is generally supported by lesion studies in lower primates, where lesions to the basal forebrain impair reversal learning abilities associated with the orbital prefrontal cortex (Ridley *et al.,* 1985; Roberts *et al.,* 1992). More recent studies in higher primates, however, have shown that lesions of the basal forebrain have surprisingly little effect on learning and memory performance (Aigner *et al.,* 1987; Voytko *et al.,* 1994), although they do increase susceptibility to scopolamine treatment (ibid). In their landmark study, *Voytko et al.* (1994) produced extensive lesions of the medial septum, diagonal band of Broca, and nucleus basalis of Meynert in cynomolgus monkeys, and they observed little effect on performance of the delayed-response, delayed nonmatching-to-sample, visual discrimination, spatial discrimination, or discrimination-reversal tasks, even with the difficult long delay conditions. The lack of effect on the delayed nonmatching-to-sample task was particularly surprising, because researchers have assumed that cholinergic influence from the basal forebrain is critical to the proper functioning of the medial temporal lobe. These results suggest that the marked deficits observed with scopolamine on memory tasks may result from its actions on other cholinergic systems, such as, profound effects on thalamic physiology and thus on thalamocortical activation. However, more-direct studies will be needed to define the role of cholinergic mechanisms in the PFC (e.g., examining the effects of scopolamine infusion into the PFC of monkeys performing a delayed-response task) and other cortical areas. It is noteworthy that the basal forebrain lesion produced by Voytko *et al.* (1994) did impair performance on the Posner task of visuospatial attention. As this task relies on the integrity of the parietal association cortex (Posner *et al.*, 1984), these data suggest that cholinergic mechanisms may be especially important for parietal cortical function.

There has been little or no research on the neurochemical needs of non-PFC cortical regions in higher primates, and this absence of data hinders our ability to understand the ramifications of age-related transmitter loss in non-PFC cortices. For example, Goldman-Rakic and Brown (1981) observed a marked DA depletion in the temporal pole with age, yet we have no idea as to the possible functional consequences of this loss. Similarly, we have little data to help us understand the possible functional ramifications of age-related cortical serotonin loss. It is likely that different cortical areas have different chemical needs; for example, cognitive functions of the PFC are improved by α-2 adrenergic agonists but not by 5HT3 receptor antagonists, whereas cognitive functions reliant on the inferior temporal cortex show the converse pattern of response (Arnsten *et al.*, 1997). Understanding the varying neurochemical needs of distinct cortical regions remains an important arena for future research.

## 5. The Cholinergic Hypothesis of Age-Related Cognitive Decline

The cholinergic hypothesis of age-related cognitive decline was first described by Bartus and colleagues (Bartus *et al.*, 1982), based on Bartus's pioneering research showing that the muscarinic antagonist scopolamine could induce working-memory deficits in young monkeys resembling naturally developed impairment in aged monkeys (Bartus, 1982; Bartus and Johnson, 1976). Similar learning and memory deficits were observed in humans treated with scopolamine (e.g., Drachman, 1981), leading to the idea that cholinergic loss underlay cognitive decline in normal aging. The marked degeneration of the basal forebrain cholinergic system in AD (Whitehouse *et al.*, 1982) was thus thought to produce the prominent deficits in learning and memory fundamental to this disease, whereas more-subtle changes in cholinergic function with normal aging were thought to underlie the more benign cognitive changes accompanying normal aging. Research in aged or lesioned rodents has, in general, supported this view, although even this literature remains controversial (see Fibiger, 1991, for review). However, research in higher primates with more developed neocortices has provided a more complex picture. Studies of aged rhesus monkeys have revealed only small changes in the cholinergic projections to cortex. Age-related changes in the nucleus basalis and medial septum are subtle and sometimes controversial (Stroessner-Johnson *et al.*, 1992; Voytko *et al.*, 1995; and see Kemper, this volume), and there is only a small decline in cholinergic uptake sites in cortex (Wenk *et al.*, 1989; see Rosene, this volume). Although it was hoped that age-related deficits on the delayed nonmatching-to-sample task would be related to loss of cholinergic neurons in the basal forebrain, a more subtle relationship was observed: deficits were related to a size increase in a portion of nucleus basalis neurons (Stroessner-Johnson *et al.*, 1992). Furthermore, indirectly acting cholinergic agonists such as physostigmine and tacrine have only small and inconsistent effects on working-memory performance in aged monkeys (Bartus and Dean, 1988). Finally, as described above, lesions to the basal forebrain have surprisingly little effect on learning and memory performance in young monkeys (Aigner *et al.*, 1987; Voytko *et al.*, 1994). Thus, although changes in the cholinergic system may contribute to age-related cognitive deficits, they are unlikely to play a primary role.

## 6. Functional Evidence for PFC Catecholamine Loss with Age

### 6.1. Dopamine Loss

Does DA depletion from the PFC with advancing age have functional significance? Data from aged rats and monkeys suggest that it does. For example, in aged rats, impairment on a spatial working-memory task correlated best with loss of DA metabolites from the PFC (Luine *et al.*, 1990). In monkeys, functional evidence for DA loss can be observed using treatments that block D1 receptors (SCH23390) or inhibit DA release (quinpirole). As can be seen in Fig. 3, the D1 receptor antagonist SCH23390 produces a dose-related impairment in delayed-response performance in young adult monkeys (Arnsten *et al.*, 1994). In contrast, the same does of SCH23390 have no effect on working-memory performance in aged monkeys. However, the sedative effects of SCH23390 are comparable in young versus aged monkeys (score of "I"—quieter than usual—following 0.1 mg/kg treatment), suggesting that the age-related loss of the drug's mnemonic effects is not simply due to altered kinetics or brain penetration in the aged animals, but is instead consistent

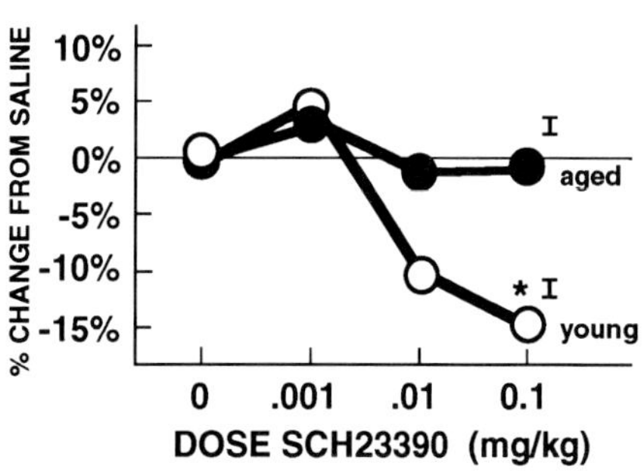


**Figure 3.** Effects of the D1 DA receptor antagonist SCH23390 on delayed-response performance and sedation ratings in young vs. aged monkeys. Results represent mean + SEM change from performance on saline vehicle. Young, but not aged, monkeys show a dose-related impairment in delayed-response performance; both groups show similar changes in sedation, consistent with decreased endogenous D1 receptor stimulation in the PFC of the aged animals. (*, significantly different from performance on saline; sedation rating of "I": calmer/quieter than usual.) Figure adapted from Arnsten *et al.* (1994).

with there being little endogenous DA stimulation of D1 receptors in the PFC of aged monkeys, and thus little effect of D1 receptor blockade. A similar loss of activity can be observed with D2 agonist treatment (Arnsten *et al.*, 1995). D2 agonists such as quinpirole inhibit DA release in the PFC as well as in other brain areas through actions at presynaptic D2 or D3 receptors (Plantje *et al.*, 1987). In young adult monkeys, low doses of quinpirole impair delayed-response performance, whereas higher doses improve it (Fig. 4A, heavy line, solid circles). The impairment

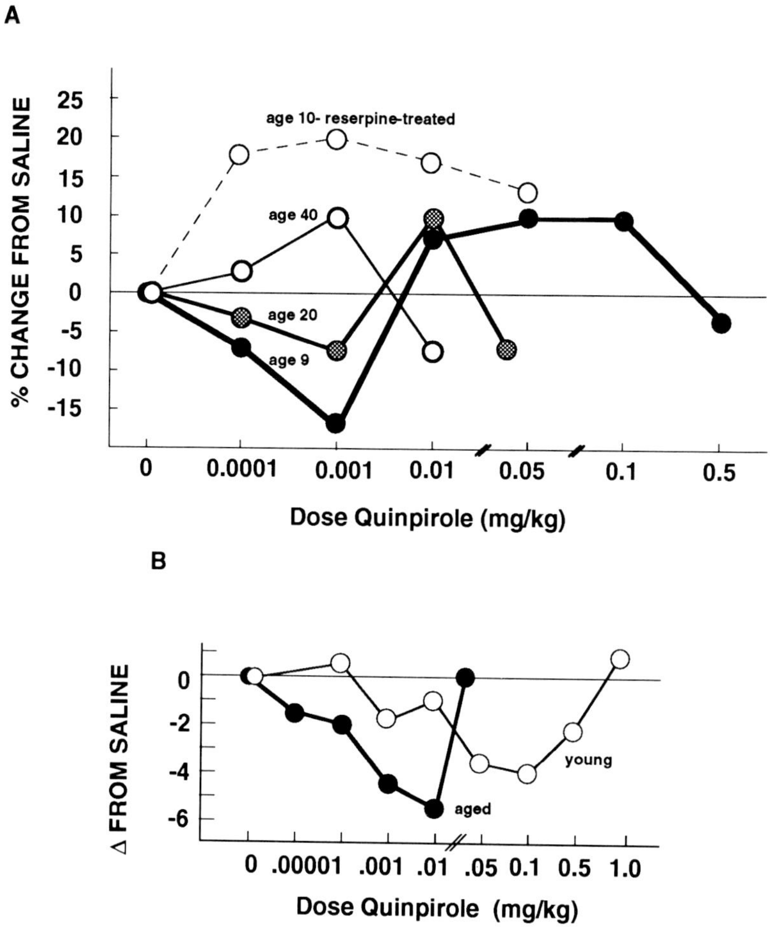


**Figure 4.** The effects of the D2 DA agonist quinpirole on delayed-response vs. fine motor performance in young vs. aged monkeys. (A) The effects of quinpirole on delayed-response performance. Young adult monkeys (e.g., age 9 years, solid line) show a biphasic response, with performance impaired at low doses and improved at higher doses, the result of drug actions at pre- vs. postsynaptic receptors, respectively. Only improvement is seen in young monkeys with DA depletion following chronic reserpine treatment (dashed line). Age causes the dose–response curve to resemble that of reserpine-treated animals: low dose impairment diminishes and is replaced by unilateral improvement in the oldest animals, consistent with age-related DA depletion. (B) Impaired fine motor performance with low dose quinpirole treatment remains in aged animals. This response appears to be presynaptic as well, as reserpine-treated animals do not exhibit fine motor impairment with quinpirole treatment (not shown), consistent with relatively preserved DA levels with age in the motor cortex as compared with the PFC. Figures adapted from Arnsten *et al.* (1995).

at low doses results from actions at presynaptic D2/D3 receptors, as these effects are no longer evident following chronic DA depletion with reserpine treatment (Fig. 4A, dashed line, open circles). In contrast, the enhancing effects of quinpirole result from actions at postsynaptic receptors, as these effects remain and are exaggerated in reserpine-treated monkeys. With advancing age, monkeys come to resemble reserpine-treated animals: There is a progressive loss of quinpirole's ability to impair delayed-response performance, and instead, there is emerging improvement, in the DA-depleted, reserpine-treated animals (Fig. 4A). Indeed, there is a significant correlation between estimated age and the change in the quinpirole response (Arnsten *et al.,* 1995). In contrast to quinpirole's effects on delayed-response performance, there is no loss of quinpirole's effects on fine motor performance (Fig. 4B), a task that relies on the skills of the motor cortices. These behavioral results are consistent with biochemical studies showing that DA loss is greater from the PFC than from the motor cortices in aged monkeys (Goldman-Rakic and Brown, 1981).

Given the marked depletion of DA from the PFC and the role of D1 DA receptors in supporting PFC function, it is natural to ask whether D1 agonists can improve PFC cognitive function in aged monkeys. Studies with four different D1 agonists (SKF38393, dihydrexidine, SKF81297, and A77636) have all shown some beneficial response in aged monkeys (Arnsten *et al.,* 1994; Cai and Arnsten, 1997). Aged monkeys, like reserpine-treated monkeys, are improved by the partial D1 agonist SKF38393, which has no effect on normal young animals (Arnsten *et al.,* 1994). Full D1 agonists are also able to improve performance; however, as seen in Figure 5, these agents often produce an inverted "U"-like dose-response curve, whereby the aged monkeys are impaired by higher doses. Indeed, aged monkeys may be particularly sensitive to the detrimental actions of D1 agonists (Arnsten *et al.,* 1994). These detrimental actions may limit the use of D1 agonists for the treatment of age-related cognitive decline. It is interesting to speculate that the increased susceptibility of aged animals to D1 detrimental actions may result from a loss of spines with advancing age (Arnsten, 1997). If dendritic spines are a major site for DA's beneficial neuromodulatory actions (Smiley and Goldman-Rakic, 1993), then loss of spines and changes in dendritic morphology with age may diminish the substrate for DA's beneficial actions in the PFC. This imbalance between DA's beneficial versus detrimental actions may contribute to age-related loss of PFC function.

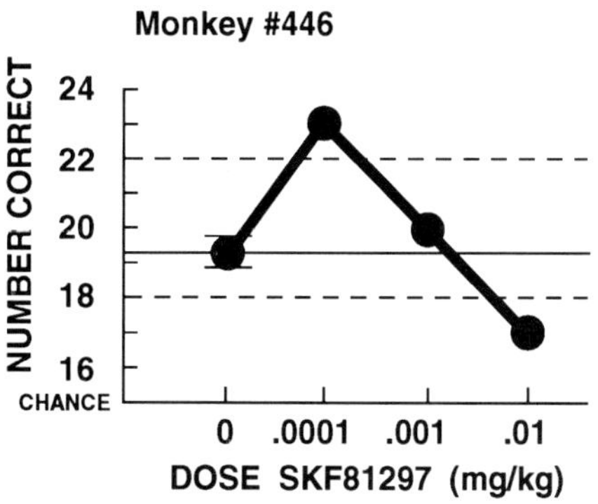


**Figure 5.** Response of aged monkey #446 to D1 DA agonist SKF81297. As with other D1 agonists, aged monkeys exhibit the inverted "U" dose–response curve: delayed-response performance improves following low doses and is impaired following higher doses of SKF81297. The horizontal line indicates mean response to saline vehicle; the dashed lines indicate the range of performance following saline vehicle treatment. Figure adapted from Cai and Arnsten, 1997.

### 6.2. Norepinephrine Loss

Degeneration of the noradrenergic system may also contribute to age-related loss of PFC function. Just as we can track changes in the DA system with altered responses to a D2 receptor agonist, changes in response to α-2 adrenergic drugs allow us to observe age-related changes on the NE system. Thus, as the NE cell is compromised with age, we observe fewer effects of α-2 agents at presynaptic receptors and more effects at postsynaptic receptors. For example, aged monkeys, like reserpine-treated young monkeys (Cai *et al.,* 1993), show marked improvement with α-2 agonists such as clonidine (Arnsten and Goldman-Rakic, 1985). As can be seen in Fig. 6 (A and B), these beneficial effects become more prominent with increasing age (Arnsten and Cai, 1993). Similarly, aged monkeys are improved by low doses of the α-2 agonist guanfacine, whereas young monkeys are not (Franowicz and Arnsten, 1998). Conversely, young monkeys are improved by very low doses of yohimbine, which increase NE release and promote endogenous stimulation of postsynaptic α-2 receptors (Arnsten and Cai, 1993). With advancing age, these beneficial effects of very low dose yohimbine treatment are no longer apparent (ibid, Arnsten and Cai, 1993) (Fig. 6, A and C), consistent with endogenous NE depletion in the aged animals.

Extensive evidence indicates that the beneficial effects of α-2 agonists in aged monkeys result from actions at α-2 receptors in the PFC. Infusion of α-2 agonists into the PFC of aged monkeys (Arnsten, 1997) or aged rats (Tanila *et al.,* 1996) improves spatial working-memory performance. Furthermore, α-2 agonists selectively improve performance of PFC tasks. Thus, they improve performance of delayed-response (Arnsten *et al.,* 1988), delayed-alternation (Arnsten and Goldman-Rakic, 1985), delayed matching-to-sample (repeated stimuli) (Jackson and Buccafusco, 1991) and discrimination-reversal (Franowicz and Arnsten, 1997) tasks, especially under distracting conditions that challenge the PFC (Arnsten and Contant, 1992; Jackson and Buccafusco, 1991), but they produce little improvement on performance of visual discrimination acquisition (Arnsten and Goldman-Rakic, 1985; Franowicz and Arnsten, 1997), Morris water maze (Sirviö *et al.,* 1991), or delayed nonmatching-to-sample (trial-unique stimuli) (Arnsten and Goldman-Rakic, 1990a) tasks that do not depend on PFC circuits.

Interestingly, the NE projection from the LC to the PFC is reciprocal; the dorsolateral and dorsomedial PFC may provide the sole cortical inputs to the LC (Arnsten and Goldman-Rakic, 1984). These anatomical findings suggest an interesting relationship between the PFC and LC, whereby PFC dysfunction may impair regulation of the LC and vice versa. Consistent with this suggestion, evidence indicates that PFC lesions result in disinhibition of LC firing (Sara and Herve-Minvielle, 1995), a state that decreases LC response to task-relevant stimuli (Kubiak *et al.,* 1995). Thus, in aged individuals a vicious cycle may develop in which LC degeneration impairs PFC function, which further disrupts LC function, and so forth.

To develop pharmacological treatments for age-related cognitive deficits in the elderly, it is important to understand the receptor subtype mechanisms underlying the beneficial effects of α-2 agonists in aged monkeys. Accumulating evidence is consistent with an α-2A pharmacological profile. Three α-2 receptor subtypes have been cloned in the primate: α-2A, α-2B, and α-2C, whose genes reside on chromosomes 10, 2, and 4, respectively (Kobilka *et al.,* 1987; Lomasney *et al.,* 1990; Regan *et al.,* 1988). Low doses of agonists with higher affinity for the α-2A subtype such as guanfacine and UK-14304 are able to improve delayed-response performance without hypotensive and sedative side effects, whereas agonists such as clonidine and B-HT920 with high affinity for the α-2B and α-2C receptors improve delayed-response performance at higher doses but also induce marked hypotension and sedation (Arnsten *et al.*, 1988; Arnsten and Leslie, 1991). Furthermore, the im-

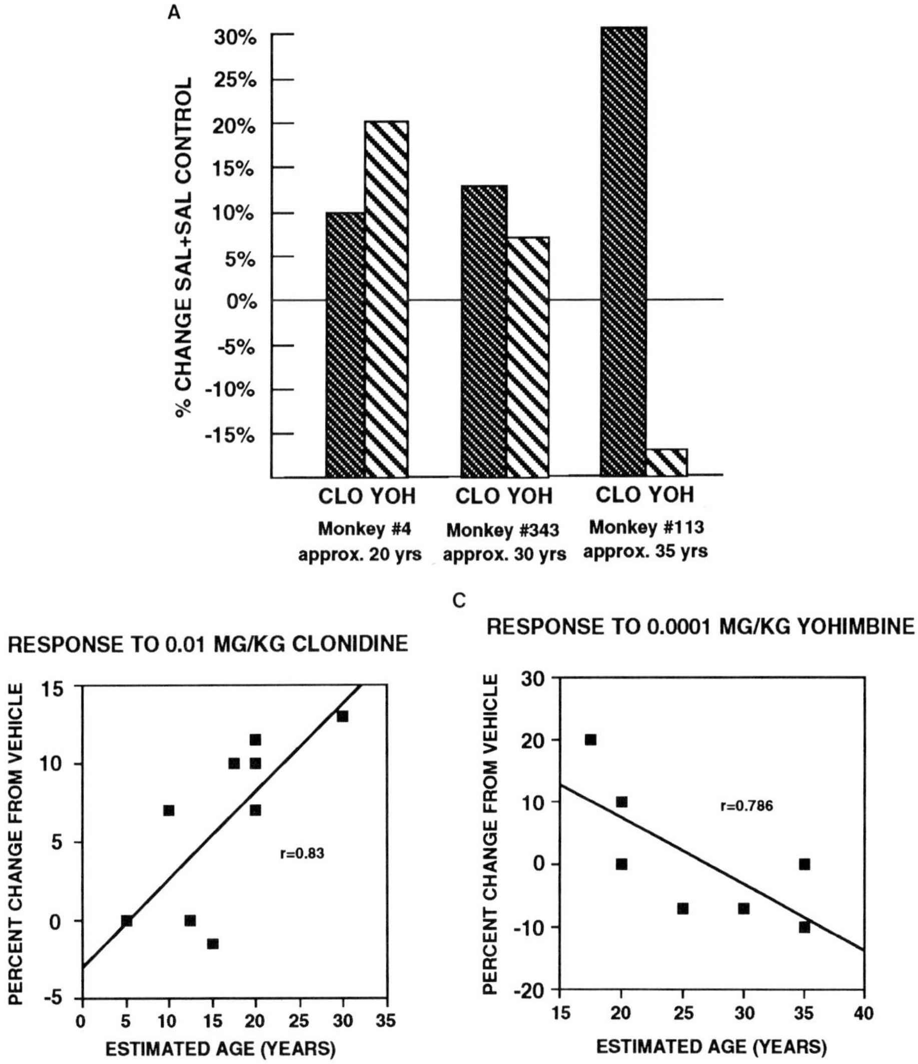


**Figure 6.** (A) The effects of the α-2 adrenergic agonist clonidine and the α-2 adrenergic antagonist yohimbine on the delayed-response performance of three representative aged monkeys spanning approximately 20–35 years. Results represent change from performance following vehicle (saline + saline) treatment. With age the beneficial effects of clonidine at postsynaptic receptors increase, whereas those of yohimbine at presynaptic receptors diminish. CLO 0.01 mg/kg clonidine; YOH 0.0001 mg/kg yohimbine. (B) Changes in response to clonidine (0.01 mg/kg) over the life span. Age correlates significantly with improvement on the delayed-response task with clonidine treatment, consistent with increasing drug actions at postsynaptic α-2 adrenergic receptors over the life span. (C) Conversely, the beneficial effects of low-dose yohimbine treatment decrease with age, consistent with age-related NE depletion. Adapted from Arnsten and Cai (1993).

provement in delayed-response performance can be reversed by yohimbine, which has high affinity for α-2A receptors, but not by prazosin, which has high affinity for α-2B and α-2C receptors in addition to its α-1-blocking properties (Arnsten and Goldman-Rakic, 1985). More recently, we have observed that the α-2C antagonist MK912 is similarly ineffective in blocking the beneficial effects of low-dose guanfacine treatment in aged monkeys (unpublished results). Interestingly, the α-2A receptor is most prevalent in the PFC, and electronmicrographic studies have demonstrated postsynaptic localization of α-2A immunoreactivity in the monkey PFC (Aoki *et al.,* 1994).

Based on this research in monkeys, the α-2A agonist guanfacine is now being investigated for use in ADHD. As described above, open trials indicate that guanfacine is highly effective in children and adolescents with ADHD (Chappell *et al.,* 1995; Horrigan and Barnhill, 1995; Hunt *et al.,* 1995), improving attention regulation and decreasing impulsivity with few side effects. With increasing age, however, guanfacine's side effects become more problematic (R. Hunt, personal communication), which limits its use in the elderly. Although guanfacine is currently the most selective agent for the α-2A site, it still has significant affinity for the α-2B and C subtypes. Thus, the development of more-selective α-2A agonists may be helpful in treating PFC dysfunction in the elderly.

## 7. Summary

Although many aging studies have focused on changes in the cholinergic system, catecholamines are critical to PFC function, and changes in catecholamines appear to contribute to the loss of PFC cognitive functions with normal aging. The loss of an optimal catecholamine environment in the PFC may exacerbate other age-related changes (e.g., altered dendritic morphology, loss of white matter), thus contributing to deterioration of higher cortical function.

ACKNOWLEDGMENTS. This work was supported by PHS grant AG06036.

## 8. References

Aigner, T. G., Mitchell, S. J., Aggleton, J. P., DeLong, M. R., Struble, R. G., Price, D. L., and Wenk, G. L., 1987, Effects of physostigmine and scopolamine on recognition memory in monkeys with ibotinic-acid lesions of the nucleus basalis of Meynert, *Psychopharmacology* **92:**292–300.

Altman, H. J., and Normile, H. J., 1986, Enhancement of the memory of a previously learned aversive habit following pretest administration of a variety of serotonergic antagonists in mice, *Psychopharmacology* **90:**24–27.

Altman, H. J., and Normile, H. J., 1988, What is the nature of the role of the serotonergic nervous system in learning and memory: Prospects for development of effective treatment strategy for senile dementia, *Neurobiol. Aging* **9:**627–638.

Altman, H. J., Normile, H. J., Galloway, M. P., Ramirez, A., and Azmitia, E. C., 1990, Enhanced spatial learning in rats following 5, 7-DHT-induced serotonergic deafferentation of the hippocampus, *Brain Res.* **518:**61–66.

Aoki, C., Go, C.-G., Venkatesan, C., and Kurose, H., 1994, Perikaryal and synaptic localization of alpha-2A-adrenergic receptor-like immunoreactivity, *Brain Res.* **650:**181–204.

Arnsten, A. F. T., 1997, Catecholamine regulation of the prefrontal cortex, *J. Psychopharm.* **11:**151–162.

Arnsten, A. F. T., and Goldman-Rakic, P. S., 1984, Selective prefrontal cortical projections to the region of the locus coeruleus and raphe nuclei in the rhesus monkey, *Brain Res.* **306:**9–18.

Arnsten, A. F. T., and Goldman-Rakic, P. S., 1985, Alpha-2 adrenergic mechanisms in prefrontal cortex associated with cognitive decline in aged nonhuman primates, *Science* **230:**1273–1276.

Arnsten, A. F. T., and Goldman-Rakic, P. S., 1990a, Analysis of alpha-2 adrenergic agonist effects on the delayed nonmatch-to-sample performance of aged rhesus monkeys, *Neurobiol. Aging* **11:**583–590.

Arnsten, A. F. T., and Goldman-Rakic, P. S., 1990b, Stress impairs prefrontal cortex cognitive function in monkeys: Role of dopamine, *Soc. Neurosci. Abstr.* **16:**164.

Arnsten, A. F. T., and Goldman-Rakic, P. S., 1998, Noise stress impairs prefrontal cortical cognitive function in monkeys: Evidence for a hyperdopaminergic mechanism, *Arch. Gen. Psychiatry* **55:**362–369.

Arnsten, A. F. T., and Leslie, F. M., 1991, Behavioral and receptor binding analysis of the alpha-2 adrenergic agonist, UK-14304 (5 bromo-6 [2-imidazoline-2-yl amino] quinoxaline): Evidence for cognitive enhancement at an alpha-2 adrenoceptor subtype, *Neuropharmacology* **30:**1279–1289.

Arnsten, A. F. T., and Contant, T. A., 1992, Alpha-2 adrenergic agonists decrease distractibility in aged monkeys performing a delayed response task, *Psychopharmacology* **108:**159–169.

Arnsten, A. F. T., and Cai, J. X., 1993, Post-synaptic alpha-2 receptor stimulation improves working memory in aged monkeys: Indirect effects of yohimbine vs. direct effects of clonidine, *Neurobiol. Aging* **14:**597–603.

Arnsten, A. F. T., Cai, J. X., and Goldman-Rakic, P. S., 1988, The alpha-2 adrenergic agonist guanfacine improves memory in aged monkeys without sedative or hypotensive side effects, *J. Neurosci.* **8:**4287–4298.

Arnsten, A. F. T., Cai, J. X., Murphy, B. L., and Goldman-Rakic, P. S., 1994, Dopamine D1 receptor mechanisms in the cognitive performance of young adult and aged monkeys, *Psychopharmacology* **116:**143–151.

Arnsten, A. F. T., Cai, J. X., Steere, J. C., and Goldman-Rakic, P. S., 1995, Dopamine D2 receptor mechanisms contribute to age-related cognitive decline: The effects of quinpirole on memory and motor performance in monkeys, *J. Neurosci.* **15:**3429–3429.

Arnsten, A. F. T., Lin, C. H., van Dyck, C. H., and Stanhope, K. J., 1997, The effects of 5-$HT_3$ receptor antagonists on cognitive performance in aged monkeys, *Neurobiol. Aging* **18:**21–28.

Arnsten, A. F. T., Mathew, R., Ubriani, R., Taylor, J. R., and Li, B.-M., 1999, Alpha-1 noradrenergic receptor stimulation impairs prefrontal cortical cognitive function, *Biol. Psychiatry* **45:**26–31.

Arnsten, A. F. T., Steere, J. C., and Hunt, R. D., 1996, The contribution of α-2 noradrenergic mechanisms to prefrontal cortical cognitive function: Potential significance to attention deficit hyperactivity disorder, *Arch. Gen. Psychiatry* **53:**448–455.

Bachevalier, J., Landis, L. S., Walker, L. C., Brickson, M., Mishkin, M., Price, D. L., and Cork, L. C., 1991, Aged monkeys exhibit behavioral deficits indicative of widespread cerebral dysfunction, *Neurobiol. Aging* **12:**99–111.

Barkley, R. A., Grodzinsky, G., and DuPaul, G. J., 1992, Frontal lobe functions in attention deficit disorder with and without hyperactivity: A review and research report, *J. Abnormal Child Psych.* **20:**163–188.

Bartus, R. T., 1979, Effects of aging on visual memory, sensory processing, and discrimination learning in a nonhuman primate, in: *Aging, Vol. 10* (J. M. Ordy and K. Brizzee, eds.), Raven, New York, pp. 85–114.

Bartus, R. T., 1982, Effects of cholinergic agents on learning and memory in animal models of aging, in: *Aging, Vol. 19* (S. Corkin, J. H. Growden, K. L. Davis, E. Usdin, and R. J. Wurtman, eds.), Raven, New York, pp. 271–280.

Bartus, R. T., and Johnson, H. R., 1976, Short-term memory in the rhesus monkey: Disruption from the anti-cholinergic scopolamine, *Pharmacol. Biochem. Behav.* **5:**39–46.

Bartus, R. T., and Dean, R. L., 1979, Recent memory in aged non-human primates: Hypersensitivity to visual interference during retention, *Exp. Aging Res.* **5:**385–400.

Bartus, R. T., and Dean, R. L., 1988, Tetrahydroaminoacridine, 3,4 diaminopyridine and physostigmine: Direct comparisons of effects on memory in aged primates, *Neurobiol. Aging* **9:**351–356.

Bartus, R. T., Fleming, D., and Johnson, H. R., 1978, Aging in the rhesus monkey: Debilitating effects on short-term memory, *J. Gerontol.* **33:**858–871.

Bartus, R. T., Dean, R. L., and Fleming, D. L., 1979, Aging in the rhesus monkey: Effects on visual discrimination learning and reversal learning, *J. Gerontol.* **34:**209–219.

Bartus, R. T., Dean, R. L., Beer, B., and Lippa, A. S., 1982, The cholinergic hypothesis of geriatric memory dysfunction, *Science* **217:**408–417.

Benson, D. F., 1991, The role of frontal dysfunction in attention deficit hyperactivity disorder, *J. Child Neurol.* **6:**S9–S12.

Bigham, M. H., and Lidow, M. S., 1995, Adrenergic and serotonergic receptors in aged monkey cortex, *Neurobiol. Aging* **16:**91–104.

Bliss, E. L., Ailion, J., and Zwaniger, J., 1968, Metabolism of norepinephrine, serotonin, and dopamine in rat brain with stress, *J. Pharmacol. Exp. Ther.* **164:**122–134.

Brozoski, T., Brown, R. M., Rosvold, H. E., and Goldman, P. S., 1979, Cognitive deficit caused by regional depletion of dopamine in prefrontal cortex of rhesus monkey, *Science* **205:**929–931.

Bubser, M., and Schmidt, W., 1990, 6-OHDA lesion of the rat prefrontal cortex increases locomotor activity, impairs acquisition of delayed alternation tasks, but does not affect uninterrupted tasks in the radial maze, *Behav. Brain Res.* **37:**157–168.

Buzsaki, G., Kennedy, B., Solt, V. B., and Ziegler, M., 1991, Noradrenergic control of thalamic oscillation: The role of alpha-2 receptors, *Eur. J. Neurol.* **3:**222–229.

Cai, J. X., and Arnsten, A. F. T., 1997, Dose-dependent effects of the dopamine D1 receptor agonists A77636 or SKF81297 on spatial working memory in aged monkeys, *J. Pharmacol. Exp. Ther.* **282:**1–7.

Cai, J. X., Ma, Y., Xu, L., and Hu, X., 1993, Reserpine impairs spatial working memory performance in monkeys: Reversal by the alpha-2 adrenergic agonist clonidine, *Brain Res.* **614:**191–196.

Chao, L. L., and Knight, R. T., 1997a, Age-related prefrontal alterations during auditory memory, *Neurobiol. Aging* **18:**87–95.

Chao, L. L., and Knight, R. T., 1997b, Prefrontal deficits in attention and inhibitory control with aging, *Cereb. Cortex* **7:**63–69.

Chappell, P. B., Riddle, M. A., Scahill, L., Lynch, K. A., Schultz, R., Arnsten, A., Leckman, J. F., and Cohen, D. J., 1995, Guanfacine treatment of comorbid attention deficit hyperactivity disorder and Tourette's Syndrome: Preliminary clinical experience, *J. Am. Acad. Child. Adol. Psychiatry* **34:**1140–1146.

Chelune, G. J., Ferguson, W., Koon, R., and Dickey, T. O., 1986, Frontal lobe disinhibition in attention deficit disorder, *Child Psych. Hum. Dev.* **16:**221–234.

Davies, B. T., and Wellman, P. J., 1992, Effects on ingestive behavior in rats of the alpha-1-adrenoreceptor agonist cirazoline, *Eur. J. Pharmacol.* **210:**11–16.

Davis, H. P., Cohen, A., Gandy, M., Colombo, P., Van Dusseldorp, G., Simolke, and Romano, J., 1990, Lexical priming deficits as a function of age, *Behav. Neurosci.* **104:**286–295.

de Keyser, J., De Backer, J.-P., Vauquelin, G., and Ebinger, G., 1990, The effect of aging on the D1 dopamine receptors in human frontal cortex, *Brain Res.* **528:**308–310.

Deutch, A. Y., and Roth, R. H., 1990, The determinants of stress-induced activation of the prefrontal cortical dopamine system, *Prog. Brain Res.* **85:**367–403.

Drachman, D. A., 1981, The cholinergic system, memory, and aging, in: *Aging, Vol. 17* (S. J. Enna, T. Samorajski, and B. Beer, eds.), Raven, New York, pp. 255–268.

Fibiger, H. C., 1991, Cholinergic mechanisms in learning, memory, and dementia: A review of recent evidence, *Trends Neurosci.* **14:**220–223.

Fields, R. B., Van Kammen, D. P., Peters, J. L., Rosen, J., Van Kammen, W. B., Nugent, A., Stipetic, M., and Linnoila, M., 1988, Clonidine improves memory function in schizophrenia independently from change in psychosis, *Schizophrenia Res.* **1:**417–423.

Franowicz, J. C. S., and Arnsten, A. F. T., 1997, The α-2A noradrenergic agonist, guanfacine, improves visual object discrimination reversal performance in rhesus monkeys, *Behav. Neurosci.* **111:**1–9.

Franowicz, J. C. S., and Arnsten, A. F. T., 1998, The α-2A noradrenergic agonist, guanfacine, improves delayed response performance in young adult rhesus monkeys, *Psychopharmacology* **136:**8–14.

Gaspar, P., Berger, B., Febvret, A., Vigny, A., and Henry, J. P., 1989, Catecholamine innervation of the human cerebral cortex as revealed by comparative immunohistochemistry of tyrosine hydroxylase and dopamine-beta-hydroxylase, *J. Comp. Neurol.* **279:**249–271.

Gaspar, P., Duyckaerts, C., Alvarez, C., Javoy-Agid, F., and Berger, B., 1991, Alterations of dopaminergic and noradrenergic innervations in motor cortex in Parkinson's disease, *Ann. Neurol.* **30:**365–374.

German, D. C., Walker, B. S., Manaye, K., Smith, W. K., Woodward, D. J., and North, A. J., 1988, The human locus coeruleus: Computer reconstruction of the cellular distribution, *J. Neurosci.* **8:**1776–1788.

Goldman-Rakic, P. S., and Brown, R. M., 1981, Regional changes of monoamines in cerebral cortex and subcortical structures of aging rhesus monkeys, *Neuroscience* **6:**177–187.

Goldstein, L. E., Rasmusson, A. M., Bunney, S. B., and Roth, R. H., 1996, Role of the amygdala in the coordination of behavioral, neuroendocrine, and prefrontal cortical monoamine responses to psychological stress in the rat, *J. Neurosci.* **16:**4787–4798.

Gorenstein, E. E., and Mammato, C. A., 1989, Performance of inattentive-overactive children on selected measures of prefrontal-type function, *J. Clin. Psychol.* **45:**619–632.

Greenwood, P. M., Parasuraman, R., and Haxby, J. V., 1993, Changes in visuospatial attention over the adult life span, *Neuropsychologia* **31:**471–485.

Hadfield, M. G., and Milio, C., 1990, Regional brain monoamine levels and utilization in middle-aged rats, *Life Sci.* **46:**295–299.

Haxby, J. V., Duara, R., Grady, C. L., Cutler, N. R., and Rapoport, S. I., 1985, Relations between neuropsychological and cerebral metabolic asymmetries in early Alzheimer's disease, *J. Cereb. Blood Flow Metab.* **5:**193–200.

Haxby, J. V., Grady, C. L., Koss, E., Horwitz, B., Heston, L., Schapiro, M., Friedland, R. P., and Rapoport, S. I., 1990, Longitudinal study of cerebral metabolic asymmetries and associated neuropsychological patterns in early dementia of the Alzheimer type, *Arch. Neurol.* **47:**753–760.

Hochanadel, G., and Kaplan, E., 1984, Neuropsychology of normal aging, in: *Clinical Neurology of Aging* (M. L. Albert, ed.), Oxford University Press, New York, pp. 231–244.

Horrigan, J. P., and Barnhill, L. J., 1995, Guanfacine for treatment of attention-deficit-hyperactivity disorder in boys, *J. Child Adol. Psychopharm.* **5:**215–223.

Hoyer, W. J., Rebok, G. W., and Sved, S. M., 1979, Effects of varying irrelevant information on adult age differences in problem solving, *J. Gerontol.* **34:**553–560.

Hunt, R. D., Mindera, R. B., and Cohen, D. J., 1985, Clonidine benefits children with attention deficit disorder and hyperactivity: Reports of a double-blind placebo-crossover therapeutic trial, *J. Am. Acad. Child Adol. Psychiatry* **24:**617–629.

Hunt, R. D., Arnsten, A. F. T., and Asbell, M. D., 1995, An open trial of guanfacine in the treatment of attention deficit hyperactivity disorder, *J. Am. Acad. Child Adol. Psychiatry* **34:**50–54.

Inoue, T., Koyama, T., and Yamashita, I., 1993, Effect of conditioned fear and stress on serotonin metabolism in the rat brain, *Pharmacol. Biochem. Behav.* **44:**371–374.

Iversen, L. L., Rossor, M. N., Reynolds, G. P., Hills, R., Roth, M., Mountjoy, C. Q., Foote, S. L., Morrison, J. H., and Bloom, F. E., 1983, Loss of pigmented dopamine beta hydroxylase positive cells from locus coeruleus in senile dementia of Alzheimer's type, *Neurosci. Lett.* **39:**95–100.

Jackson, W. J., and Buccafusco, J. J., 1991, Clonidine enhances delayed matching-to-sample performance by young and aged monkeys, *Pharmacol. Biochem. Behav.* **39:**79–84.

Kobilka, B. K., Matsui, H., Kobilka, T. S., Yang-Feng, T. L., Francke, U., Caron, M. G., Lefkowitz, R. J., and Regan, J. W., 1987, Cloning, sequencing, and expression of the gene encoding for the human platelet alpha-2-adrenergic receptor, *Science* **238:**650–656.

Kubiak, P., Ivanova, S., Rajkowski, J., and Aston-Jones, G., 1995, Contrasting responses of monkey locus coeruleus neurons to conditioned vs. unconditioned stimuli: Relationship to arousal, performance, and vigilance, *Soc. Neurosci. Abstr.* **21:**2092.

Leibowitz, S. F., 1988, Hypothalamic paraventricular nucleus: Interaction between alpha-2 noradrenergic system and circulating hormones and nutrients in relation to energy balance, *Neurosci. Biobehav. Rev.* **12:**101–106.

Leslie, F. M., Loughlin, S. E., Sternberg, D. B., Mcgaugh, J. L., Young, L. E., and Zornetzer, S. F., 1985, Noradrenergic changes in senescent memory loss, *Brain Res.* **359:**292–299.

Li, B.-M., and Mei, Z.-T., 1994, Delayed response deficit induced by local injection of the α-2 adrenergic antagonist yohimbine into the dorsolateral prefrontal cortex in young adult monkeys, *Behav. Neural. Biol.* **62:**134–139.

Loge, D. V., Staton, R. D., and Beatty, W. W., 1989, Performance of children with ADHD on tests sensitive to frontal lobe dysfunction, *J. Am. Acad. Child Adol. Psychiatry* **29:**540–545.

Lomasney, J. W., Lorenz, W., Allen, L. F., King, K., Regan, J. W., Yang-Feng, T. L., Caron, M., and Lefkowitz, R. J., 1990, Expansion of the alpha-2-adrenergic family: Cloning and characterization of a human alpha-2-adrenergic receptor subtype, the gene for which is located on chromosome 2, *Proc. Nat. Acad. Sci. USA* **87:**5094–5098.

Luine, V., Bowling, D., and Hearns, M., 1990, Spatial memory deficits in aged rats: Contributions of monoaminergic systems, *Brain Res.* **537:**271–278.

Mair, R. G., and McEntree, W. J., 1986, Cognitive enhancement in Korsakoff's psychosis by clonidine: A comparison with l-dopa and ephedrine, *Psychopharmacology* **88:**374–380.

Mattes, J. A., 1980, The role of frontal lobe dysfunction in childhood hyperkinesis, *Comp. Psychiatry* **21:**358–369.

McEntree, W. J., and Mair, R. G., 1990, The Korsakoff syndrome: A neurochemical perspective, *Trends Neurosci.* **13:**340–344.

Moffoot, A., O'Carroll, R. E., Murray, C., Dougall, N., Ebmeier, K., and Goodwin,G. M., 1994, Clonidine infusion increases uptake of Tc-exametazime in anterior cingulate cortex in Korsakoff's psychosis, *Psychol. Med.* **24:**53–61.

Moss, M. B., Rosene, D. L., and Peters, A., 1988, Effects of aging on visual recognition memory in the rhesus monkey, *Neurobiol. Aging* **9:**495–502.

Murphy, B. L., Roth, R. H., and Arnsten, A. F. T., 1994, The effects of FG7142 on prefrontal cortical dopamine and spatial working memory in rat and monkey, *Soc. Neurosci. Abstr.* **20:**1018.

Murphy, B. L., Arnsten, A. F. T., Goldman-Rakic, P. S., and Roth, R. H., 1996a, Increased dopamine turnover in the prefrontal cortex impairs spatial working memory performance in rats and monkeys, *Proc. Natl. Acad. Sci. USA* **93:**1325–1329.

Murphy, B. L., Arnsten, A. F. T., Jentsch, J. D., and Roth, R. H., 1996b, Dopamine and spatial working memory in rats and monkeys: Pharmacological reversal of stress-induced impairment, *J. Neurosci.* **16:**7768–7775.

Nielsen-Bohlman, L., and Knight, R. T., 1995, Prefrontal alterations during memory processing in aging, *Cereb. Cortex* **5:**541–549.

Normile, H. J., and Altman, H. J., 1988, Enhanced passive avoidance retention following posttrain serotonergic receptor antagonist administration in middle-aged and aged rats, *Neurobiol. Aging* **9:**377–382.

Plantje, J. F., Steinbush, H. W. M., Schipper, J., Dijcks, F. A., Verheijden, P. F. H. M., and Stoof, J. C., 1987, D-2 dopamine receptors regulate the release of [$^3$H]dopamine in rat cortical regions showing dopamine immunoreactive fibers, *Neuroscience* **20:**157–168.

Posner, M. I., Walker, J. A., Friedrich, F. J., and Rafal, R. D., 1984, Effects of parietal injury on covert orienting of visual attention, *J. Neurosci.* **4:**1863–1874.

Presty, S. K., Bachevalier, J., Walker, L. C., Struble, R. G., Price, D. L., Mishkin, M., and Cork, L. C., 1987, Age differences in recognition memory of the rhesus monkey *(Macaca mulatta), Neurobiol. Aging* **8:**435–440.

Rapp, P. R., 1990, Visual discrimination and reversal learning in the aged monkey *(Macaca mulatta), Behav. Neurosci.* **104:**876–884.

Rapp, P. R., and Amaral, D. G., 1989, Evidence for task-dependent memory dysfunction in the aged monkey, *J. Neurosci.* **9:**3568–3576.

Rapp, P. R., and Amaral, D. G., 1992, Individual differences in the cognitive and neurobiological consequences of normal aging, *Trends Neurosci.* **15:**340–345.

Regan, J. W., Kobilka, T. S., Yang-Feng, T. L., Caron, M. G., Lefkowitz, R. J., and Kobilka, B. K., 1988, Cloning and expression of a human kidney cDNA for an alpha-2 adrenergic receptor subtype, *Proc. Natl. Acad. Sci. USA* **85:**6301–6305.

Roberts, A. C., Salvia, M. A., Wilkinson, L. S., Collins, P., Muir, J. L., Everitt, B. J., and Robbins, T. W., 1994, 6-Hydroxydopamine lesions of the prefrontal cortex in monkeys enhance performance on an analog of the Wisconsin Card Sort Test: Possible interactions with subcortical dopamine, *Neuroscience* **14:**2531–2544.

Robinson, D. L., and Kertzman, C., 1990, Visuospatial attention: Effects of age, gender, and spatial reference, *Neuropsychologia* **28:**291–301.

Roth, R. H., Tam, S.-Y., Ida, Y., Yang, J.-X., and Deutch, A. Y., 1988, Stress and the mesocorticolimbic dopamine systems, *Ann. N.Y. Acad. Sci.* **537:**138–147.

Sara, S. J., and Herve-Minvielle, A., 1995, Inhibitory influence of frontal cortex on locus coeruleus, *Proc. Natl. Acad. Sci. USA* **92:**6032–6036.

Sawaguchi, T., 1998, Attenuation of delay-period activity of monkey prefrontal cortical neurons by an α-2 adrenergic antagonist during an oculomotor delayed-response task, *J. Neurophysiol.* **80:**2200–2205.

Sawaguchi, T., and Goldman-Rakic, P. S., 1991, D1 dopamine receptors in prefrontal cortex: Involvement in working memory, *Science* **251:**947–950.

Sawaguchi, T., and Goldman-Rakic, P. S., 1994, The role of D1-dopamine receptors in working memory:

Local injections of dopamine antagonists into the prefrontal cortex of rhesus monkeys performing an oculomotor delayed response task, *J. Neurophysiol.* **71:**515–528.

Sawaguchi, T., Matsumura, M., and Kubota, K., 1988, Dopamine enhances the neuronal activity of spatial short-term memory task in the primate prefrontal cortex, *Neurosci. Res.* **5:**465–473.

Seamans, J. K., Floresco, S. B., and Phillips, A. G., 1998, D1 receptor modulation of hippocampal-prefrontal cortical circuits integrating spatial memory with executive functions in the rat, *J. Neurosci.* **18:**1613–1621.

Shue, K. L., and Douglas, V. I., 1992, Attention deficit hyperactivity disorder and the frontal lobe syndrome, *Brain Cogn.* **20:**104–124.

Simon, H., 1981, Dopaminergic A10 neurons and the frontal system, *J. Physiol.* **77:**81–95.

Sirviö, J., Riekkinen, P., Vajanti, I., Koivisto, E., and Riekkinen, P. J., 1991, The effects of guanfacine, alpha-2 agonist, on the performance of young and aged rats in spatial navigation task, *Behav. Neural Biol.* **56:**101–107.

Sladek, J. R., and Sladek, C. D., 1979, Relative quantification of monoamine histofluorescence in young and old nonhuman primates, in: *Parkinson's Disease II* (C. E. Finch, D. E. Potter, and A. D. Kenny, eds.), Plenum, New York, pp. 231–239.

Small, G. W., Kuhl, D. E., Riege, W. H., Fujikawa, D. G., Ashford, J. W., Metter, E. J., and Mazziotta, J. C., 1989, Cerebral glucose metabolic patterns in Alzheimer's disease, *Arch. Gen. Psychiatry* **46:**527–532.

Smiley, J. F., and Goldman-Rakic, P. S., 1993, Heterogeneous targets of dopamine synapses in monkey prefrontal cortex demonstrated by serial section electron microscopy: A laminar analysis using the silver-enhanced diaminobenzidine sulfide (SIDS) immunolabeling technique, *Cereb. Cortex* **3:**233–238.

Steere, J. C., Li, B.-M., Jentsch, J. D., Mathews, R., Taylor, J. R., and Arnsten, A. F. T., 1996, Alpha-1 noradrenergic stimulation impairs, while alpha-2 stimulation improves, prefrontal cortex spatial working memory function, *Soc. Neurosci. Abstr.* **22:**1126.

Stroessner-Johnson, H. M., Rapp, P. R., and Amaral, D. G., 1992, Cholinergic cell loss and hypertrophy in the medial septal nucleus of the behaviorally characterized aged monkey, *J. Neurosci.* **12:**1936–1944.

Suhara, T., Fukuda, H., Inoue, O., Itoh, T., Suzuki, K., Yamasaki, T., and Tateno, Y., 1991, Age-related changes in human D1 dopamine receptors measured by positron emission tomography, *Psychopharmacology* **103:**41–45.

Tanila, H., Rama, P., and Carlson, S., 1996, The effects of prefrontal intracortical microinjections of an alpha-2 agonist, alpha-2 antagonist, and lidocaine on the delayed alternation performance of aged rats, *Brain Res. Bull.* **40:**117–119.

Thierry, A. M., Tassin, J. P., Blanc, G., and Glowinski, J., 1976, Selective activation of the mesocortical DA system by stress, *Nature* **263:**242–244.

van Dyck, C. H., Seibyl, J. P., Malison, R. T., Wallace, E., Zoghbi, S. S., Zea-Ponce, Y., Baldwin, R. M., Charney, D. S., Hoffer, P. B., and Innis, R. B., 1995, Age-related decline in dopamine transporter binding in human striatum with [$^{123}$I]β-CIT SPECT, *J. Nucl. Med.* **36:**1175–1181.

Verma, A., and Moghaddam, B., 1996, NMDA receptor antagonists impair prefrontal cortex function as assessed via spatial delayed alternation performance in rats: Modulation by dopamine, *J. Neurosci.* **16:**373–379.

Vijayashankar, N., and Brody, H., 1979, A quantitative study of the pigmented neurons in the nuclei locus coeruleus and subcoeruleus in man as related to aging, *J. Neuropathol. Exp. Neurol.* **38:**490–497.

Voytko, M. L., Olton, D. S., Richardson, R. T., Gorman, L. K., Tobin, J. T., and Price, D. L., 1994, Basal forebrain lesions in monkeys disrupt attention but not learning and memory, *J. Neurosci.* **14:**167–186.

Voytko, M. L., Sukhov, R. R., Walker, L. C., Breckler, S. J., Price, D. L., and Koliatsos, V. E., 1995, Neuronal number and size are preserved in the nucleus basalis of aged rhesus monkeys, *Dementia* **6:**131–141.

Walker, L. C., Kitt, C. A., Struble, R. G., Wagster, M. V., Price, D. L., and Cork, L. C., 1988, The neural basis of memory decline in aged monkeys, *Neurobiol. Aging* **9:**657–666.

Weiss, J. M., Goodman, P. A., Losito, B. G., Corrigan, S., Charry, J. M., and Bailey, W. H., 1981, Behavioral depression produced by an uncontrollable stressor: Relationship to norephinephrine, dopamine, and serotonin levels in various regions of the rat brain, *Brain Res. Rev.* **3:**167–205.

Wenk, G. L., Pierce, D. J., Struble, R. G., Price, D. L., and Cork, L. C., 1989, Age-related changes in multiple neurotransmitter systems in the monkey brain, *Neurobiol. Aging* **10:**11–19.

Wenk, G. L., Walker, L. C., Price, D. L., and Cork, L. C., 1991, Loss of NMDA, but not GABA-A, binding in the brains of aged rats and monkeys, *Neurobiol. Aging* **12:**93–98.

West, R. L., 1996, An application of prefrontal cortex function theory to cognitive aging, *Psychol. Bull.* **120**:272–292.

Whitehouse, P. J., Price, D. L., Struble, R. G., Clark, A. W., Coyle, J. T., and DeLong, M. R., 1982, Alzheimer's disease and senile dementia: Loss of neurons in the basal forebrain, *Science* **215**:1237–1239.

Williams, G. V., and Goldman-Rakic, P. S., 1995, Blockade of dopamine D1 receptors enhances memory fields of prefrontal neurons in primate cerebral cortex, *Nature* **376**:572–575.

Yang, C. R., and Seamans, J. K., 1996, Dopamine D1 receptor actions in layers V-VI rat prefrontal cortex neurons *in vitro:* Modulation of dendritic-somatic signal integration, *J. Neurosci.* **16**:1922–1935.

Zahrt, J., Taylor, J. R., Mathew, R. G., and Arnsten, A. F. T., 1997, Supranormal stimulation of dopamine D1 receptors in the rodent prefrontal cortex impairs spatial working memory performance, *J. Neurosci.* **17**:8528–8535.

# 6

# Neurotransmitter Receptor Changes in the Hippocampus and Cerebral Cortex in Normal Aging

DOUGLAS L. ROSENE and
TIMOTHY J. NICHOLSON

## 1. Introduction

As humans grow older, they all experience changes in their basic physiological systems. Although some of these changes can be debilitating or fatal, the quality of life during aging depends to a great extent on the functional status of the central nervous system. Over the past three decades, increasingly well-controlled and well-documented studies in humans and experimental animals have shown a variety of cognitive changes, including impairments in both memory and executive system function. In contrast to age-related dementias like Alzheimer's disease, these im-

DOUGLAS L. ROSENE • Department of Anatomy and Neurobiology, Boston University School of Medicine, Boston, Massachusetts 02118; and Yerkes Regional Primate Research Center, Emory University, Atlanta, Georgia 30322. TIMOTHY J. NICHOLSON • Department of Anatomy and Neurobiology, Boston University School of Medicine, Boston, Massachusetts 02118.

*Cerebral Cortex,* Volume 14, edited by Peters and Morrison. Kluwer Academic/Plenum Publishers, New York, 1999.

pairments are relatively mild, though they increase in severity with age and can become quite troublesome (see Moss and Albert, this volume). It seems likely that these changes in cognitive function result from changes localized in the cerebral cortex (including the hippocampal formation and other parts of the limbic system) or related parts of the forebrain (thalamus, basal ganglia, amygdala, basal forebrain). Investigations of the forebrain in both aged humans and experimental animals have uncovered a variety of age-related biological changes, including the appearance of amyloid plaques (e.g. Selkoe *et al.*, 1987), loss of neurons (e.g. Brody, 1955; Brizzee *et al.*, 1980), loss of myelin (e.g., Kemper, 1994), loss of synapses (e.g. Geinisman *et al.*, 1992), decreases in neurotransmitter levels (e.g. Wenk *et al.*, 1989), loss of neurotransmitter receptors (e.g. Wagster *et al.*, 1990), alterations in mitochondrial energy metabolism (e.g. Wallace, 1995), and changes in neurophysiological responsivity (e.g. Barnes, 1994; Tanila *et al.*, 1997).

Though investigators and clinicians differ as to which of these biological changes is the major causal factor underlying age-related cognitive impairments, the common wisdom has been that they are all related to the hallmark of normal aging: widespread, age-related loss of neurons.

In 1988 Flood and Coleman reviewed the evidence for widespread, ubiquitous neuron loss in normal aging and concluded that it was equivocal. They suggested that neuron loss in normal aging is regionally specific and limited in extent, a suggestion quite compatible with the heterogeneous distribution of many of the biological changes listed above. Over the past decade, however, a growing body of evidence has suggested that if a loss of neurons does occur in the forebrain, it must be minor. Thus Haug (1985) demonstrated that in humans, reports of decreased neuronal density in aging are likely to be an artifact of greater shrinkage of brain tissue from young individuals during histological processing. In the intervening years, methods of assessing neuron loss have improved greatly as a result of the methodological studies of the Danish group (e.g. West, 1993). Their techniques—the disector and the fractionator—have certainly increased the accuracy of neuron number estimates, but their major contribution has been to emphasize the use of systematic random sampling from the entire reference volume of the structure under consideration. Using these new techniques in both rats and monkeys, where control over postmortem delay, fixation, and tissue processing is far superior to that which can be achieved in humans, numerous investigators have reported a relative preservation of neurons in the cerebral cortex (Tigges *et al.*, 1990; Stroessner-Johnson *et al.*, 1992; Amaral, 1993; Rosene, 1993; Peters *et al.*, 1998; Rapp and Gallagher, 1996; Gazzaley *et al.*, 1997; Morrison and Hof, 1997). In a recent review, Peters *et al.* (1998) concluded that there is no strong evidence for neuron loss anywhere in the cerebral cortex or hippocampus and that if it does occur, it must be less than 10%. Because between-subject biological variability in cortical volume and neuron numbers in both humans and monkeys has been shown to exceed 100% for visual cortex (Leuba and Kraftsik, 1994; Peters *et al.*, 1998), it seems unlikely that such a modest loss of neurons, even if it occurred ubiquitously throughout the cerebral cortex, could account for the level of cognitive deficits reported in normal aging. Recently, Morrison and Hof (1997) suggested that age-related cognitive dysfunction must result from functional compromise of many or all of the neurons of the forebrain. Although many of the age-related changes cited above become candidates for such a functional change, Gazzaley *et al.* (1996) recently reported loss of a specific subunit for the glutamatergic NMDA receptor in the outer molecular layer of the hippocampal dentate gyrus without any apparent age-related alteration of the dendrites, synapses, or entorhinal neurons that innervate this territory. Because neuron loss apparently does not occur in this system, or to any great degree elsewhere, this and other age-related biological changes must be reexamined in light of preserved neuron numbers.

This chapter will review some of the age-related changes in neurotransmitter receptors that have been reported in the brain during normal aging. Although the monoamine (see Arnsten, this volume, for review), neuropeptide, and steroid receptor systems (e.g. Bhatnagar *et al.,* 1997) are clearly important and undergo significant age-related changes, this chapter will focus on three classical neurotransmitter systems: cholinergic, glutamatergic, and GABAergic. We will also summarize some of our own laboratory observations of changes in these three systems in normal-aging rhesus monkeys and discuss some of the possible mechanisms that could underlie these changes.

## 2. Technical Considerations

Although it is beyond the scope of this review to analyze all of the technical factors that influence findings on age-related changes in neurotransmitter receptors, such factors can dramatically affect the interpretation of the reported changes. The following four factors need to be considered.

First, receptors can be studied in two ways: with radioactively labeled ligands, which bind to the receptor with varying degrees of affinity and specificity for receptor subtypes, or with antibodies that recognize an epitope of the receptor or one of its subunits (e.g., Gazzaley *et al.,* 1996). Because ligands typically bind to the receptor's active extracellular domain, they typically identify functional receptors that are fully formed and inserted into the membrane. In contrast, some antibodies to receptors or subunits may bind to receptors or receptor subunits that are not fully formed and/or inserted into the membrane. Hence observations derived from these two methods may not be equivalent, because they look at receptors in different states.

Second, radioactively labeled ligands can be used to assess receptors in two ways: with membrane preparations, in which brain tissue is homogenized and processed as dissociated membrane fractions, or with on-the-slide receptor autoradiography, in which ligand is applied to tissue sections (Young and Kuhar, 1979a). In the former instance, the radioactivity demonstrating binding of ligand to the receptor is assessed in a liquid scintillation counter, whereas in the latter case the tissue sections are either apposed to autoradiographic film or coated with autoradiographic emulsion. If apposed to film together with radioisotope standards, the binding on tissue sections can be assessed quantitatively using image analysis systems. This yields estimates of binding in femtomoles of receptor per milligram of tissue, which can be compared with those from membrane homogenates. If coated with emulsion, microscopic evaluation and grain-counting can assess the sections quantitively within the experiment, but the results cannot be compared directly with either film or membrane preparation results.

Third, differences in the apparent density of receptors when assessed with any of the radioactively labeled ligand methods can be due to a change in receptor numbers, a change in receptor affinity, or a combination of the two. It is important to determine which of these pertains, because the likely causal mechanisms can be quite different. For example, if the decrease in binding is due to a reduction in receptor numbers, one might consider whether this is due to a loss of membranes into which the receptor is inserted (e.g., dendritic atrophy) or to a down-regulation of receptor production. Alternatively, a decrease due to a change in affinity would suggest either that the receptor is being constructed from different combinations of subunits due to altered gene expression (e.g., Milbrandt *et al.*, 1997), or that it has been altered through changes in second-messenger systems to which it is coupled (e.g., Joseph *et al.*, 1993). To determine whether decreased binding density is due to change in receptor number or change in receptor affinity, a saturation study can be conducted using different concentrations of the radioactively labeled ligand (Rosenthal, 1967). But unless this is done, single-concentration studies must be interpreted with caution. An excellent discussion of this and related issues of whether the ligand is binding to a single site or to multiple sites is provided by Keen and MacDermot (1993).

Finally, both postmortem delay (in human tissue) and the use of fixatives (in animal tissue) can affect apparent binding densities. Typically, tissue used for ligand-binding studies is frozen without fixation, whereas tissue for immunocytochemical studies is fixed with aldehydes, either by immersion (human tissue) or by perfusion-fixation (animal tissue). Human brain tissue is usually removed as soon as possible after death, then frozen without fixation on dry ice and stored at −80°C until cut. Even though efforts are always made to obtain the tissue promptly, postmortem delays before freezing are usually measured in hours. Most neurotransmitter receptors are quite robust and appear to survive the attack of endogenous proteolytic enzymes during these hours, but still this variable must be explicitly stated, and one should consider whether postmortem effects in young and old might be different. It is usually possible to freeze animal tissue within minutes of death to stop degradation of receptors and optimize binding, but some investigators recommend perfusing with a low concentration of fixative instead (e.g., Lidow, *et al.*, 1989). In any case, one must be very cautious in making comparisons between studies of human tissue frozen after a delay of hours, animal tissue frozen after a delay of minutes, and animal tissue fixed by perfusion. The same caveat applies to receptor studies that use antibodies to a receptor or its subunits. In this case human tissue is typically fixed by the immersion of tissue pieces removed after the same postmortem delay typical of frozen tissue, but with the added variable that immersion fixation proceeds slowly and will vary at different depths within the tissue because of slow diffusion rates. Even in animal studies, perfusion-fixation is a notoriously difficult variable to control; hence one must be cautious in comparing results of animal studies in different laboratories and especially cautious in making comparisons between the results with animal tissues and those with immersion-fixed human tissues.

## 3. Age-Related Changes in Cholinergic Receptors

The cholinergic system has long been a focus of studies on aging, particularly following the publication by Bartus *et al.* (1982) of the hypothesized cholinergic basis for age-related cognitive dysfunction. They asserted that age-related loss of cholinergic basal forebrain neurons led to a loss of cholinergic input to the cerebral cortex and limbic system and a consequent disruption of memory function. Though such a loss of cholinergic neurons from the nucleus basalis is well documented in Alzheimer's disease, still, as with earlier reports of cortical neuronal loss, recent studies raise serious doubt about whether any significant number of these neurons die during normal aging (e.g., Voytko *et al.*, 1995; Stroessner-Johnson, 1992). Nevertheless, there is evidence of age-related loss of various cholinergic markers in the forebrain suggesting a deafferentation, and hence investigators have carefully examined these receptors to determine if there are any consequent age-related changes.

Using the monkey as a model, Wenk *et al.* (1989) took tissue samples from 12 different regions of the cerebral cortex as well as from the caudate and putamen in 23 monkeys aged from 2 to 37 years. They performed *in vitro* ligand binding assays on membrane homogenates for four different neurotransmitter receptors, including the muscarinic M1 and nicotinic cholinergic receptors. Whereas a biochemical assay indicated a statistically significant decrease in the cholinergic synthetic enzyme choline acetyltransferase (ChAT) in the frontal pole, the lower levels of ChAT observed with age in other regions did not reach statistical significance. Similarly, using tissue homogenates from these regions, neither M1 nor nicotinic receptor binding showed any statistically significant reductions with age. Examination of their graphs however, indicates that as with ChAT levels, there was a consistent, age-related reduction in levels of M1 and nicotinic binding across all areas of the brain, but the variability was too high for statistical significancc. In a subscquent study from the same group, Wagster *et al.* (1990) reported no change in M1 receptor density but did find a significant decrease in M2 receptors in the temporal lobe of aged monkeys. In contrast, Vannucchi and Goldman-Rakic (1991) examined M1 binding in the frontal lobe of aged rhesus monkeys and found a statistically significant decrease. They conducted a full-saturation study of M1 binding and concluded that the decrease resulted from a decrease in receptor affinity, not from a decrease in receptor number (except in their oldest subject, which was about 30 years of age). They suggest that affinity change may precede actual loss of receptor number. It is interesting to speculate that if a change in receptor affinity results from changes in the type or proportion of their protein subunits due to altered gene regulation, this might be the first manifestation of the age-related changes that ultimately result in the loss of functional receptors reported in other studies.

Araujo *et al.* (1990), for example, examined nicotinic and muscarinic receptors in membrane homogenates, assessing both number (Bmax) and affinity (Kd) in 27-month-old Long-Evans rats as compared with 3-month-old rats. They found no age-related changes in affinity, but reported statistically significant reductions in the number of both receptors. For the nicotinic receptors, there was a significant decrease in numbers in the cerebral cortex, striatum, and hippocampus, but not in the thalamus. For muscarinic receptors, there was a statistically significant decrease in M2 receptors in the hippocampus, cerebral cortex, striatum, and thalamus. Similarly, they found decreases in M1 receptor numbers in the cerebral cortex and thalamus, but reduction in the hippocampus was not significant, and receptor numbers in the striatum were unchanged. In contrast to these diverse changes in the forebrain, none of the receptors showed age-related changes in the cerebellum. Using on-the-slide ligand binding instead of membrane preparations, Beigon *et al.*

(1989) compared M1 and M2 binding in Long-Evans rats that were young adults (6 months old) and aged (28 months old). They examined 63 different areas of the brain and reported that although binding densities were not affected by age in most areas, statistically significant reductions were observed in a few cortical regions and in the striatum. The greatest reduction in M2 receptor density occurred in the cholinergic nuclei of the basal forebrain, where reductions of nearly 40% were observed. Together these two studies provide reasonable evidence for age-related reductions in cholinergic receptor numbers, particularly in the forebrain. Even though they used an inbred strain of rats, however, there was still enough variability that it cannot yet be concluded that there is a generalized age-related reduction in forebrain cholinergic receptors. An interesting question posed by these results is whether between-subject variability might reflect the loss of receptors in only a subpopulation of the aged subjects. If so, this variability might be reduced if levels of cholinergic receptor binding were compared not with age, but with the degree of cognitive impairment. In other words, would the most cognitively impaired subjects have the lowest levels of cholinergic markers?

Two recent studies in rats used this approach, assessing cognitive performance with the Morris water maze. This task is thought to assess visuospatial mapping and is known to be sensitive to hippocampal damage in rats. Thus Aubert *et al.* (1995) tested aged Long-Evans rats (24–25 months) and divided them into cognitively impaired and cognitively spared groups based upon comparisons with young adults (6 months). Subsequently, the brains were evaluated for M1, M2, and nicotinic receptor binding using pirenzepine, AFDX-384, and cytisine binding, respectively. Although there were no significant differences in either M1 or nicotinic binding densities, M2 receptors showed a variety of statistically significant changes, ranging from decreases in prefrontal, orbitofrontal, cingulate, and insular cortices of old unimpaired rats relative to young adults to increases in these same areas when old impaired rats were compared with the old unimpaired rats. Surprisingly, in the latter comparison the only statistically significant change in the hippocampal formation in impaired relative to unimpaired rats was a higher M2 binding density in the molecular layer of the dentate gyrus. This study suggests a variability within the aged population, but instead of the cognitively impaired subjects having the greatest decreases in receptor density, they have the smallest.

In contrast, Smith *et al.* (1995) performed an almost identical study, using the Morris water maze to rate the cognitive performance of aged Long-Evans rats (24–25 months) and comparing binding for M1, M2, and nicotinic receptors in these versus young adult rats (6 months). Like Aubert *et al.* (1995), Smith *et al.* (1995) reported that there was no difference in nicotinic or M1 binding between young and old, but instead of the decrease in M2 binding in frontal cortices reported by Aubert *et al.* (1995), they observed a significant age-related decrease in M2 binding density only in the septal area and diagonal band. Their behavioral analysis also yielded different results. Rather than dividing their aged rats into impaired and unimpaired groups for group analyses, Smith *et al.* (1995) performed an analysis within the aged group correlating behavioral performance with receptor binding density. This analysis produced a result similar to Aubert *et al.* (1995) in that higher levels of M1 binding in the hippocampus correlated with poorer behavioral performance, despite the overall lack of effect on M1 binding. From these studies on rats, it appears that there are generally small and variably significant decreases in cholinergic receptor densities with age. The surprising finding is that within the aged subjects, *higher* binding densities (though still below the average of young adults) correlate with greater behavioral impairment, suggesting that age-related behavioral impairments, at least on the Morris water maze, result from either some other, undiscovered age-related dysfunction that correlates with relative sparing of cholinergic receptor densities, or from an interaction of cholinergic receptor changes and other variables. Despite these complications, the two studies point to the value of measuring age-related changes in both cognitive function and biological variables in the same subjects.

Another approach to studying cholinergic receptors is to use molecular methods to examine levels of the message for receptor proteins. Blake *et al.* (1991) compared muscarinic receptor densities and mRNA levels for the m1, m3, and m4 muscarinic receptor subtypes on adjacent tissue sections from young (5–6 months) and old (24–25 months) Wistar rats. They reported a significant decrease in ligand binding in most areas of the cortex but not in the hippocampus. In contrast, they found a significant reduction in mRNA for the m1 message only in the olfactory tubercle. Wang *et al.* (1992) compared mRNA levels for the m1 and m2 muscarinic receptor messages in brain homogenates from 12 young (5–8 months) and 16 old (24–28 months) Fischer 344 rats. They reported no statistically significant age-related decrease in mRNA levels, except for a decrease in m2 levels in the hypothalamus. These authors suggested that differences in receptors reported in other studies might thus be due to post-translational processing rather than the levels of mRNA. This is an interesting and plausible possibility, but it is also interesting that for both m1 and m2, mRNA levels were consistently lower in all other brain regions reported (hippocampus, frontal cortex, parietal cortex, and striatum for m1; thalamus, hypothalamus midbrain, and cerebellum for m2). In contrast, Birtsch *et al.* (1997) used non-isotopic (digoxigenin-UTP) *in situ* hybridization to compare the percentage of cortical neurons expressing the α-4-1 and α-5 subunits of the nicotinic receptor in Wistar rats at 3, 24, and 33 months of age. They reported that there was an age-related decrease in the number of neurons expressing α-4-1 message but not in the number expressing α-5. Hence, like the ligand binding studies, examination of messenger RNA levels suggests that there are age-related reductions but does not give a compelling picture of generalized or ubiquitous age-related losses. And it remains unclear how mRNA expression relates to assessments of receptors using ligands.

## 4. Age-Related Changes in GABAergic Receptors

The main inhibitory neurotransmitter in the brain is gamma-aminobutyric acid (GABA), and the $GABA_A$ receptor subtype has been the most studied. This receptor is of particular interest because the effect of GABA is modulated by the interaction of various ligands with various modulatory sites present on this same receptor, including alcohol, barbiturate anesthetics, and benzodiazepines (see Biggio and Costa, 1988, for a review). The benzodiazepines interact with a specific benzodiazepine (BZD) modulatory binding site that is present on many, but not all, of the $GABA_A$ receptors (Corda *et al.*, 1989), and they are able to potentiate the effect of GABA. Because of the importance of GABAergic inhibitory mechanisms for the function of all systems in the brain, both the $GABA_A$ receptor and its benzodiazepine modulatory site have been examined for age-related changes.

In their study of multiple neurotransmitter systems in the forebrain of the aging rhesus monkey, Wenk *et al.* (1989) reported that benzodiazepine binding was unchanged with age, although a nonsignificant reduction was observed in a few areas. Although the density of binding to the BZD modulatory site on the $GABA_A$ receptor suggests that the $GABA_A$ receptors are unchanged with age, there is abundant evidence from studies of the inferior colliculus that age-related changes in $GABA_A$ binding do occur in this brainstem site. Thus Milbrandt *et al.* (1996) reported a significant decrease in $GABA_A$ binding between 2 and 3 months of age, using GABA as the ligand in rat inferior colliculus. Yet they found no further age-related change at either 18–20 months or at 23–26 months of age. In addition, they looked at muscimol binding to the $GABA_A$ receptor and flunitrazepam binding to the benzodiazepine modulatory site on the $GABA_A$ receptor, but for these ligands they compared only binding densities between 3 and 26 months of age, which yielded no reported change in either muscimol or flunitrazepam binding. Although there were no differences in binding using single concentrations of GABA, muscimol, or flunitrazepam (for the BZD site), in a pharmacological experiment examining the modulation of the picrotoxin binding site on the $GABA_A$ receptor, these investigators reported that a given concentration of GABA inhibited binding of a ligand to the picrotoxin site more effectively in aged rats than in young rats. They speculate that this could mean either an age-related enhancement of receptor response to GABA or an enhancement of receptor desensitization by GABA. In either case their results suggest an early age-related decrease in $GABA_A$ receptors (although they did not do a saturation study to verify this), followed by an alteration in the functional composition of the receptor.

In contrast, Gutierrez *et al.* (1994) conducted a saturation study of muscimol binding to the $GABA_A$ receptor using membrane preparations and reported an age-related decrease in receptor numbers between 2 and 24–30 months of age. They also used antibodies to specific receptor subunits and *in situ* hybridization for the message for these same subunits. Both techniques demonstrated age-related decreases in message and in protein that were consistent with an age-related loss of $GABA_A$ receptors.

In a follow-up study, Milbrandt *et al.* (1997) examined levels of mRNA for different $GABA_A$ receptor subunits in 3-month-old as compared with 26-month-old rats. Using *in situ* hybridization, they reported that in the inferior colliculus of aged rats there was a statistically significant increase in the $\gamma$-1 subunit, as well as an increase in $\alpha$-2 that did not reach significance, with little or no decrease in the other subunits. So whereas Gutierrez *et al.* (1994) found a loss of some subunits comparing 2-month-old with 26-month-old rats, Milbrandt *et al.* (1997) found an *increase* in two subunits and no decrease in any other subunits. Milbrandt *et al.* (1997) suggest that the increase in these two subunits may be important, because they assemble together into $GABA_A$ receptors that show enhanced sensitivity to GABA. They also report that these changes in subunit expression in the inferior colliculus are regionally specific rather than ubiquitous within all of the subnuclei. Ruano *et al.* (1991) reported similar findings of regional specificity in age-related alterations of the $GABA_A$ receptor complex. They found an age-related increase in

GABA modulation of ligand binding to the BZD binding site on the $GABA_A$ receptor complex in the hippocampus, but not in the cerebral cortex (Ruano *et al.*, 1993). This suggests that there may be a variety of age-related compensatory changes in the $GABA_A$ receptor complex in different parts of the brain, and that such functional alterations may be mediated by alterations in the expression of message (mRNA) for different GABA receptor subunits (Rabow *et al.*, 1995).

The disparate results from these studies raise a number of questions about defining "aging," both for a given species and for a given system. With regard to age, one must consider when it is appropriate to establish a "young adult" baseline; it is common to regard a 3-month-old rat as a young adult, but what exactly is the status of a 2-month-old rat? Is this a juvenile, and the change from 2 to 3 months the final stage of early development, or is this an "early" young adult? Also, the auditory system shows age-related changes very early in life, so what is the proper age for establishing the "young adult" baseline here, and how does this relate to age-related functional decrements in audition? Although Milbrandt *et al.* (1996) reported a loss of GABA binding from 2 to 3 months of age but no further loss from 3 to 26 months, as well as no loss of BZD or muscimol binding from 3 to 26 months, they did not compare 2-month-old to 26-month-old rats on muscimol or BZD binding, the very comparison wherein Gutierrez *et al.* (1994) reported a receptor number loss.

In summary, there appears to be evidence of significant age-related change in the $GABA_A$ receptor complex, suggesting that there may be an early loss of these receptors followed by compensatory alterations in their composition and function. Alternatively, there may be little or no loss, but rather a progressive change in receptor subunit composition, reflected as either losses or increases depending upon the ligand or subunit examined and the ages studied.

## 5. Age-Related Changes in Glutamatergic Receptors

Glutamate is the principal excitatory neurotransmitter in the central nervous system, and reductions in glutamate are well documented in Alzheimer's disease, where forebrain neurons that use this neurotransmitter are clearly lost (e.g., Hyman et al., 1984). Of the several subtypes of glutamatergic receptors, the *N*-methyl–D-aspartate (NMDA) receptor has received the most interest, because it appears to be a necessary part of long-term potentiation—a neurophysiological model of the type of neuronal plasticity thought to underlie normal learning (Collingridge and Bliss, 1987).

Wenk *et al.* (1989) took tissue samples from 12 different areas of monkey cerebral cortex, as well as from caudate and putamen. These were analyzed using tissue homogenates from these regions for glutamate binding, using one concentration of $^{3}$H-glutamate. Although between-subject variance was quite high, there was a statistically significant increase in glutamate binding in the parietal lobe and no significant loss in any of the other areas that were examined. In contrast to this apparent stability and peculiar increase, Magnusson (1995) undertook an extensive study of the NMDA glutamate receptor subtype using six different ligands to assess the glutamate (NMDA/aspartate) binding site, the glycine modulatory site, and the channel protein where $Mg^{2+}$ ions form a block that is removed only with depolarization. Using young and old C57Bl/6 mice and *in vitro* on-the-slide autoradiography, Magnusson reported ubiquitous reductions in binding throughout the forebrain, especially in the frontal lobe and hippocampal formation. Interestingly, though all the ligands showed age-related reductions in binding, the reductions were greatest for the glutamate site, intermediate for binding to the channel protein (e.g., MK-801), and smallest for the glycine site. Because saturation studies were not done, it is unclear if these reductions are due to changes in receptor number or receptor affinity. Magnusson (1995), however, speculates that the differential size of the reductions seen with different ligands to different sites suggests that the NMDA receptor composition from its various subunits may be altered.

In marked contrast to these ubiquitous results with multiple ligands, Shimada *et al.* (1997) recently looked at the NMDA receptor in aging rats using ligands to both the channel protein and the glycine binding site. Instead of autoradiography, they used a membrane preparation and performed a full-saturation binding study. They found no age-related changes in the NMDA receptor. One might suspect a species difference between the mice studied by Magnusson (1995) and the rats studied by Shimada *et al.* (1997), but a recent study in monkeys obtained results more compatible with those of Magnusson.

Gazzaley *et al.* (1996) used antibodies that recognize three different sets of ionotropic glutamate receptor subunits: NMDAR1, GluR2/3, and Glu5/6/7. They examined the molecular layer of the dentate gyrus in old (24–32 years) and young adult (5–9 years) cynomolgus and rhesus monkeys using these antibodies, as well as antibodies to microtubule-associated protein-2 (MAP2) to label dendrites and an antibody to synaptophysin as a marker for presynaptic terminals. Using immunofluoresence and quantitative confocal microscopy, they found a decrease in immunofluoresence only for the NMDAR1 receptor subunit. Interestingly, this was localized to the distal two-thirds of the granule cell dendrites, where the glutamatergic entorhinal afferents impinge and where NMDA receptors are especially dense. Gazzaley *et al.* (1996) argue that this reduction is not a simple consequence of dendritic atrophy because there was no reduction in MAP2 immunofluoresence, nor was it a result of simple deafferentation, because there was no change in synaptophysin immunoreactivity. Although they cannot exclude the possibility that NMDA receptors in other forebrain areas are also affected, they argue that this dendritic reduction in one glutamatergic receptor subunit reflects a circuit-specific, age-related alteration that might account for age-related cognitive impairments. In any case, this subunit-specific change in the monkey hippocampus is compatible with Magnusson's suggestion that reductions in NMDA receptor binding might reflect an alteration in receptor subunits rather than a complete loss of the receptors.

## 6. Age-Related Changes in the Monkey Temporal Lobe

In our laboratory we examined seven different neurotransmitter receptor subtypes using *in vitro* autoradiography on adjacent sections through the hippocampal formation, amygdala, entorhinal cortex, and adjacent temporal lobe neocortex of

**Table I. Ligand Assay Procedures**

| Receptor | Ligand | Concen. | Buffer[a] | Displacer[b] | Reference |
|---|---|---|---|---|---|
| $M_1$ | [$^3$H]-PZ | 10nM | Krebs, Low ions | AT 10μM | Hammer *et al.*, 1980 |
| $M_2$ | [$^3$H]-OXO | 1nM | 10mM HEPES + Tris | AT 10μM | Vogt & Burns, 1988 |
| $Ch_{uptake}$ | [$^3$H]-HC-3 | 10nM | 10mM glycylglycine | HC3 1μM | Sandberg & Coyle, 1985 |
| $GABA_A$ | [$^3$H]-MUS | 5nM | 0.05 Tris-Citrate | GABA 1μM | Anke *et al.*, 1992 |
| BZD | [$^3$H]-FLU | 1nM | 50mM tris-HCl | CLON 1μM | Young & Kuhar, 1979b |
| NMDA | [$^3$H]-MK801 | 7nM | 20mM HEPES-K + 5mM | KET 200μM | Subramaniam & McGonigle, 1991 |
| Kainate | [$^3$H]-KA | 15nM | 50mM Tris-Citrate | KA 500μM | Monaghan *et al.*, 1986 |

[a]Appropriate ions are included in each assay buffer to maximize specific binding (see Refs.).
[b]Sections incubated with ligand plus displacer represent 'nonspecific' binding.
[c]Abbreviations: AT, atropine; BZD, benzodiazepine; Ch, choline; CLON, clonazepam; FLU, flunitrazepam; GABA, Gamma-aminobutyric acid; HC-3, Hemicholinium-3; KA, kainic acid; KET, ketanserin; MUS, muscimol; NMDA, *N*-methyl-D-aspartate; OXO, oxotremorine; PZ, pirenzepine.

behaviorally tested young and old rhesus monkeys. The binding densities observed in six young monkeys ranging from 5 (when monkeys reach sexual maturity) to 10 years of age were compared with those from seven old monkeys ranging from 23 to 31 years of age and representing the last quarter of the rhesus monkey life span (Tigges *et al.*, 1988). All were perfused with ice-cold Krebs buffer, the brains immediately removed and frozen at −75°C and stored at −80°C until cut into interrupted series of 15-μm-thick sections on a cryostat. *In vitro* on-the-slide ligand binding assays were then run on series of sections using tritium-labeled ligands according to the published protocols summarized in Table 1. After apposition to tritium-sensitive film, binding was analyzed using the Inquiry computer-based densitometry system. These images were analyzed in detail for each ligand by hippocampal subfield, amygdalar nucleus, and cortical architectonic area and layer. For ligands where single-concentration studies showed significant age-related changes in binding, additional sections were processed in a saturation study to determine if the changes were the result of a change in the number of receptors and/or a change in the affinity of the receptor for ligand. A summary of these studies, with the data collapsed into the binding for the entire hippocampal formation (Hip), amygdala (Amy), and the entorhinal and neocortical areas of the temporal lobe together (TCtx), is presented in Figure 1. Asterisks mark statistically significant ($p \leq 0.05$) group differences between young and old subjects.

For the M1 cholinergic receptor, we found significant reductions in binding in both the hippocampal formation and the amygdala. The somewhat smaller reduction in the temporal lobe cortices did not reach significance. In contrast, for the M2 receptor we found statistically significant reductions in all three areas. Saturation studies for both M1 and M2 receptors indicated that these changes were due to reductions in number, with no change in affinity.

Using hemicholinium to assay the high-affinity choline uptake (HACU) site found mainly on cholinergic terminals as a rough assay of innervation density, we observed a reduction in HACU binding in both the hippocampus and the temporal lobe cortices (sections for the amygdala were not available), but these were statis-

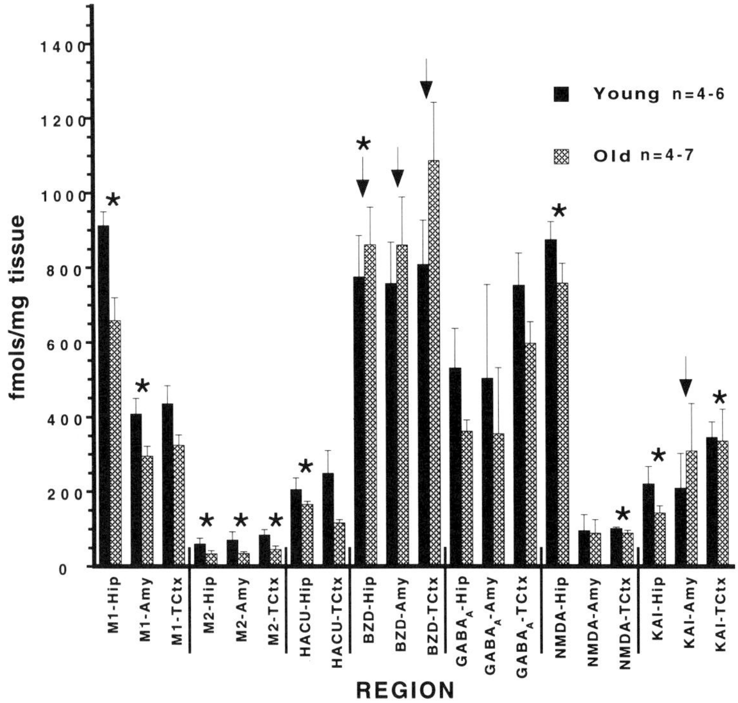


**Figure 1.** This figure presents a summary of the results of *in vitro*, on-the-slide ligand-binding studies of the temporal lobe of young (5–10 years of age) and old (23–31 years of age) rhesus monkeys. The ligand binding procedures are summarized in Table 1. For each region indicated, samples were taken from digitized images by lamina within subfields of the hippocampal formation and architectonic areas of the temporal lobe cortex as well as within nuclei of the amygdala, but for this analysis samples were averaged over the regions indicated. For each region the binding indicated is derived from the single concentration assay of Table 1. Asterisks indicate regions where there is a significant age-related difference between young and old ($p \leq 0.05$) and arrows indicate subfields where there is an increase in binding in the old relative to the young.

tically significant only in the hippocampal formation. This is consistent with the loss of M2 binding that is thought to be largely localized to presynaptic cholinergic terminals where M2 sites function as autoreceptors, and suggests a partial cholinergic deafferentation. Because lesion data suggest that M1 receptors will up-regulate—at least initially—after major deafferentation, perhaps some of the inconsistency among studies of cholinergic receptors results from a progressive but slower loss of M1 receptors that is masked, or even appears as an increase, in M1 binding due to the compensatory up-regulation of M1 sites in response to cholinergic deafferentation.

In contrast to cholinergic receptors, $GABA_A$ receptors were not significantly changed with age in the hippocampus, amygdala, or temporal lobe cortices, even though they were consistently lower. Yet the BZD binding site (a modulatory site on the $GABA_A$ receptor complex) showed increased binding in the aged monkeys relative to young adults. This increase was seen in all areas and reached statistical significance for the hippocampal formation. Subsequent saturation studies demonstrated that these changes were due to altered numbers of receptors and not a change in affinity. Because $GABA_A$ receptors can express either high- or low-affinity

BZD binding sites (e.g., Corda *et al.,* 1989), this suggests that the normal complement of $GABA_A$ receptors is altered in the aged brain, and thus a higher proportion of the reduced number of receptors now express the more-potent BZD site. Because the BZD site, when occupied, enhances the action of a given endogenous level of GABA on the $GABA_A$ receptor, it may be that the increase in BZD binding is a compensatory change in response to loss of the normal complement of $GABA_A$ receptors.

For the excitatory amino acid system, we examined both the NMDA and kainate receptor subtypes. As shown in Figure 1, there was a statistically significant binding reduction in both the hippocampal formation and in the temporal lobe cortices, but not in the amygdala. The same pattern was observed for the kainate receptor subtype: statistically significant reductions in the hippocampus and temporal lobe cortices, no significant difference in the amygdala (although there was actually a higher binding density in the elderly). Subsequent saturation studies revealed reduced kainate binding was due to a decreased number of receptors, whereas reduced NMDA binding was due to a decrease in affinity. The latter observation suggests that the age-related loss of NMDA binding may result from alterations in the construction of the receptor from its various subunits. This speculation can be addressed by using the techniques of molecular neurobiology to determine if there are changes in the message for any of these subunits.

## 7. Possible Mechanisms of Receptor Changes

Results summarized above as well as our own study suggest that there are indeed age-related alterations in these major neurotransmitter receptor systems. The general pattern indicates an age-related reduction in receptor binding for many of these receptor subtypes, but it is apparently variable between subjects and across different parts of the brain. It is less clear whether these binding reductions are due to a loss of receptors (i.e., a decrease in receptor numbers) or a reduction in the affinity of the receptor for ligand, because most of the studies have used only single concentrations of ligand. At least some of the binding reductions may be due to changes in receptor subunit composition, and these changes may be highly localized. With such diverse results, it is impossible to draw any firm conclusions about the overall effect of aging on neurotransmitter receptors. It is time, however, to begin seriously investigating mechanisms that may underlie age-related binding changes. Of the many possible mechanisms, at least four seem relatively plausible and worthy of consideration.

### 7.1. Membrane Loss

The simplest explanation for a reduction in receptor binding is a reduction in receptor numbers as a result of the loss of the neuronal membranes into which they are normally inserted. This could result from the atrophy of dendrites, where most postsynaptic receptors are located, or from atrophy of the terminal arborization of axons and/or loss of synaptic terminals, where many "presynaptic" autoreceptors are localized. Of course, if neurons die, loss of both dendrites and axons would surely follow. Even though neurons do not seem to be lost from the cerebral cortex (e.g., Peters *et al.,* 1996; Morrison and Hof, 1997), they are lost from brainstem areas, including those that project to the cerebral cortex (e.g., Kemper *et al.,* 1997; Kemper, this volume). Nevertheless, in most studies of neurotransmitter receptors, it is not at all clear if atrophy of dendrites or axons occurs independent of neuron loss. The recent study of Gazzaley *et al.* (1997) using antibodies to a dendritic marker (MAP2) and a synaptic marker (synaptophysin) is an important first step in addressing this possible mechanism, although arguments about the appropriateness of these two antibodies can be made.

### 7.2. Functional or Metabolic Down-Regulation

Another possible mechanism of binding reduction is a functional or metabolically mediated receptor down-regulation. In this scenario aging neurons would produce fewer neurotransmitter receptors, either in response to decreased demand due to reduced inputs of a given neurotransmitter (even though the synapses are still present and intact) or due to a general reduction in neuronal energy metabolism, glucose regulation, and blood flow. If this occurred, neurons would make and/or insert fewer receptors into their membrane, resulting in an overall reduction in receptor density.

### 7.3. Masking of Loss by Up-Regulation

Just as down-regulation of receptor numbers can occur, up-regulation of receptor numbers can also, most notably in response to acute, single-stage lesions. Although such lesions are not typical of the kind of slow, progressive changes that characterize normal aging, they do offer some insight into possible responses. Thus it may be that if a progressive deafferentation is occurring due to loss of axon-terminal arborizations (or simply due to reduced synaptic activity for a given neurotransmitter) postsynaptic receptors up-regulate as a compensatory response. Depending upon whether this same receptor is lost as a result of its partial presynaptic localization on degenerating axon arborizations or because of a parallel process of dendritic atrophy, such a compensatory up-regulation could be expressed as an increase in receptor density, or the postsynaptic increase could mask the presynaptic loss. In either case, studies of receptor density must carefully consider the possibility of such compensatory up-regulation, because both an observed increase an an increase that exactly compensates for (i.e., masks) a reduction in another neuronal compartment will produce a functional derangement of the system, because appropriate receptor densities are no longer present where they normally would be. That such changes may affect behavior is suggested by the observations of Smith *et al.* (1995) and Aubert *et al.* (1995).

### 7.4. Changes in Affinity and Gene Regulation of Subunits

An alternative to overall up-regulation or down-regulation of entire receptors is a change in the composition of receptor subunits. Different subunits can combine in different ways that confer different functional properties on the receptor (e.g., Rabow *et al.*, 1995). Consequently, compensatory responses could consist of alterations in the message for different receptor subtypes, with the resulting altered composition changing the functional character of the receptor. Consider, for example, our observed increase in benzodiazepine receptor numbers while the number of $GABA_A$ receptors are either unchanged or reduced. Perhaps in this case the subunit composition of the $GABA_A$ receptor is changed, leading to a higher proportion of the $GABA_A$ receptors that express a benzodiazepine binding site as the total number of $GABA_A$ receptors decreases slightly due to dendritic atrophy. Similarly, altered subunit composition of the NMDA receptor might result in the reduced ligand affinity of this receptor, which we observed, or in the altered presence of one critical subunit in the distal dendritic compartment observed by Gazzaley *et al.* (1996).

## 8. Summary and Conclusions

Age-related changes in neurotransmitter systems are probably an important factor underlying age-related changes in cognitive function. These changes are likely widespread, occurring in multiple neurotransmitter receptor systems but not expressing equally in all receptor subtypes or in all regions examined. Hence changes in neurotransmitter receptors are probably not a primary effect of normal aging. Instead, their heterogeneity suggests that there may be multiple causal mechanisms expressed differently in different parts of the brain. Although many specific causal mechanisms are plausible and the exact expression may reflect localized compensatory mechanisms, future investigations addressing receptor changes at the level of subunit gene regulation could provide further useful insights into how these different receptor-change mechanisms are expressed in normal aging.

ACKNOWLEDGMENTS. We wish to thank Elizabeth Johnson, Beverly Duryea, Claudia Fitzgerald, and Ramy Rizkalla for technical assistance with the ligand binding studies reported here, and Adrienne Sudanowicz and Brett Palana for assistance in preparation of the manuscript. These investigations were supported in part by NIH grant P01-AG00001 from the National Institute on Aging to Boston University, and grant RR00165 from the National Center for Research Resources to the Yerkes Regional Primate Research Center. Both the Yerkes Center and the Laboratory Animal Science Center at Boston University are fully accredited by the AAALAC.

## 9. References

Amaral, D. G., 1993, Morphological analysis of the brains of behaviorally characterized aged nonhuman primates, *Neurobiol. Aging* **14:**671–672.

Anke, M. M., Kulkulka, K. M., McAvoy, K. J., and Rolcosz, N. C., 1992, Regional distribution and kinetics of three sites on the $GABA_A$ receptor: Lack of effect of portacaval shunting, *J. Cereb. Blood Flow Metab.* **12:**334–346.

Araujo, D. M., Lapchak, P. A., Meaney, M. J., Collier, B., and Quirion, R., 1990, Effects of aging on nicotinic and muscarinic autoreceptor function in the rat brain: Relationship to presynaptic cholinergic markers and binding sites, *J. Neurosci.* **10:**3069–3078.

Aubert, I., Rowe, W., Meaney, M. J., Gauthier, S., and Quirion, R., 1995, Cholinergic markers in aged cognitively impaired Long-Evans rats, *Neuroscience* **67:**277–292.

Barnes, C. A., 1994, Normal aging: Regionally specific changes in hippocampal synaptic transmission, *Trends Neurosci.* **17(1):**13–18.

Bartus, R. T., Dean, R. L., and Fleming, D. L., 1979, Aging in the rhesus monkey: Effects on visual discrimination learning and reversal learning, *J. Gerontol.* **34:**209–219.

Bartus, R. T., Dean, R. L., Beer, B., and Lippa, A. S., 1982, The cholinergic hypothesis of geriatric memory dysfunction, *Science* **217:**408–417.

Biegon, A., Hanau, M., Greenberger, V., and Segal, M., 1989, Aging and brain cholinergic muscarinic receptor subtypes: An autoradiographic study in the rat, *Neurobiol. Aging* **10:**305–310.

Bhatnagar, M., Cintra, A., Chadi, G., Lindberg, J., Oitzl, M., De Kloet, E. R., Moller, A., Agnati, L. F., and Fuxe, K., 1997, Neurochemical changes in the hippocampus of the brown Norway rat during aging, *Neurobiol. Aging* **18:**319–327.

Blake, M. J., Appel, N. M., Joseph, J. A., Stagg, C. A., Anson, M., De Souza, E. B., and Roth, G. S., 1991, Muscarinic acetylcholine receptor subtype mRNA expression and ligand binding in the aged rat forebrain, *Neurobiol. Aging* **12:**193–199.

Biggio, G., and Costa, E., 1988, Chloride channels and their modulation by neurotransmitters and drugs, in: *Advances in Biochemical Psychopharmacology*, Vol. 45, Raven, New York.

Birtsch, C., Wevers, A., Traber, J., Maelicke, A., Bloch, W., and Schroder, H., 1997, Expression of α4-1 and α5 nicotinic cholinoceptor mRNA in the aging rat cerebral cortex, *Neurobiol. Aging* **18:**335–342.

Brizzee, K. R., Ordy, J. M., and Bartus, R., 1980, Localization of cellular changes within multimodal sensory regions in aged monkey brain: Possible implications for age-related cognitive loss, *Neurobiol. Aging* **1:**45–52.

Brody, H., 1955, Organization of cerebral cortex. III. A study of aging in the human cerebral cortex, *J. Comp. Neurol.* **102:**511–556.

Collingridge, G. L., and Bliss, T. V. P., 1987, NMDA receptors: their role in long-term potentiation, *Trends Neurosci.* **10:**288–293.

Corda, M. G., Giorgi, O., Longoni, B., Ongini, E., Pesce, G., Cruciani, R., and Biggio, G., 1989, Functional coupling of $GABA_A$ receptors and benzodiazepine recognition site subtypes in the spinal cord of the rat, *Eur. J. Pharmacol.* **169:**205–213.

Costa, E., and Guidotti, A., 1979, Molecular mechanism in the receptor action of benzodiazepines, *Annu. Rev. Pharmacol. Toxicol.* **19:**531–545.

Flood, D. G., and Coleman, P. D., 1988, Neuron numbers and sizes in aging brain: Comparisons of human, monkey, and rodent data, *Neurobiol. Aging* **9:**453–63.

Flynn, D. D., and Mash, D. C., 1986, Characterization of L-[3H]nicotine binding in human cerebral cortex: Comparison between Alzheimer's disease and the normal, *J. Neurochem.* **47:**1948–1954.

Gazzaley, A. H., Siegel, S. J., Kordower, J. H., Mufson, E. J., and Morrison, J. H., 1996, Circuit-specific alterations of *N*-methyl-D-aspartate receptor subunit 1 in the dentate gyrus of aged monkeys, *Proc. Natl. Acad. Sci. USA* **93:**3121–3125.

Gazzaley, A. H., Thakker, M. M., Hof, P. R., and Morrison, J. H., 1997, Preserved number of entorhinal cortex layer II neurons in aged macaque monkeys, *Neurobiol. Aging* **18:**549–554.

Geinisman, Y., DeToledo-Morrell, L., Morrell, F., Persina, I. S., and Rossi, M., 1992, Age-related loss of axospinous synapses formed by two afferent systems in the rat dentate gyrus as revealed by the unbiased sterological dissector technique, *Hippocampus* **2:**432–444.

Gutierrez, A., Khan, Z. U., Morris, S. J., and De Blas, A. L., 1994, Age-related decrease of GABAA receptor subunits and glutamic acid decarboxylase in the rat inferior colliculus, *J. Neurosci.* **14:**7469–7477.

Hammer, R., Berrie, C. P., Birdsall, N. J. M., Burgen, A. S. V., and Holme, E. C., 1980, Pirenzepine distinguishes between different subclasses of muscarinic receptors, *Nature* **283:**90–92.

Haug, H., 1985, Are neurons of the human cerebral cortex really lost during aging? A morphometric examination, in: *Senile Dementia of the Alzheimer Type* (J. Traber and W. H. Gispen, eds.), Springer-Verlag, Berlin, pp. 150–163.

Hyman, B. T., Van Hoesen, G. W., and Damasio, A. R., 1984, Cell-specific pathology isolates the hippocampal formation in Alzheimer's disease, *Science* **223:**121–122.

Keen, M., and MacDermot, J., 1993, Analysis of receptors by radioligand binding, in: *Receptor Autoradiography Principles and Practice,* (J. Wharton and J. M. Polak, eds.), Oxford University Press, London, pp. 23–55.

Joseph, J. A., Cutler, R., and Roth, G. S., 1993, Changes in G protein-mediated signal transduction in aging and Alzheimer's disease, *Ann. NY Acad. Sci.* **695:**42–45.

Kemper, T. L., Moss, M. B., Rosene, D. L., and Killiany, R. J., 1997, Age-related neuronal loss in the nucleus centralis superior of the rhesus monkey, *Acta Neuropathol.* **94:**124–130.

Kemper, T. L., 1994, Neuroanatomical and neuropathological changes during aging and dementia, in: *Clinical Neurology and Aging* (M. L. Albert and J. E. Knoefel, eds.), Oxford University Press, New York, pp. 3–67.

Leuba, G., and Kraftsik, R., 1994, Changes in volume, surface estimate, three-dimensional shape and total number of neurons of the human primary visual cortex from midgestation until old age, *Anat. Embryol.* **190:**351–366.

Lidow, M. S., Goldman-Rakic, P. S., Gallager, D. W., and Rakic, P., 1989, Quantitative autoradiographic mapping of serotonin 5-HT1 and 5-HT2 receptors and uptake sites in the neocortex of the rhesus monkey, *J. Comp. Neurol.* **280:**27–42.

Magnusson, K. R., 1995, Differential effects of aging on binding sites of activated NMDA receptor complex in mice, *Mech. Aging Develop.* **84:**227–243.

Milbrandt, J. C., Albin, R. L., Turgeon, S. M., and Caspary, D. M., 1996, GABAA receptor binding in the aging rat inferior colliculus, *Neuroscience* **73:**449–458.

Milbrandt, J. C., Hunter, C., and Caspary, D. M., 1997, Alterations of GABAA receptor subunit mRNA levels in the aging Fischer 344 rat inferior colliculus, *J. Comp. Neurol.* **379:**455–465.

Monaghan, D. T., Nguyen, L., and Cotman, C. W., 1986, The distribution of [$^3$H]kainate binding sites in primate hippocampus is similar to the distribution of both Ca2+-sensitive and Ca2+-insensitive [$^3$H]kainate binding sites in rat hippocampus, *Neurochem. Res.* **11:**1073–1082.

Morrison, J. H., and Hof, P. R., 1997, Life and death of neurons in the aging brain, *Science* **278:**412–419.

Peters, A., Morrison, J. H., Rosene, D. L., and Hyman, B. T., 1998, Are neurons lost from the cerebral cortex during normal aging? *Cereb. Cortex,* **8:**295–300.

Peters, A., Rosene, D. L., Moss, M. B., Kemper, T. L., Abraham, C. R., Tigges, J., and Albert, M. S., 1996, Neurobiological bases of age-related cognitive decline in the rhesus monkey, *J. Neuropathol. Exp. Neurol.* **55:**861–874.

Rabow, L. E., Russek, S. J., and Farb, D. H., 1995, From ion currents to genomic analysis: Recent advances in $GABA_A$ receptor research, *Synapse* **21:**189–274.

Rapp, P. R., and Gallagher, M., 1996, Preserved neuron number in the hippocampus of aged rats with spatial learning deficits, *Proc. Natl. Acad Sci. USA* **93:**9926–9930.

Rosene, D. L., 1993, Comparing age-related changes in the basal forebrain and hippocampus of the rhesus monkey, *Neurobiol. Aging* **14:**669–670.

Rosenthal, H. E., 1967, A graphic method for the determination and presentation of binding parameters in a complex system, *Anal. Biochem.* **20:**525–532.

Ruano, D., Cano, J., Machado, A., and Vitorica, J., 1991, Pharmacological characterization of $GABA_A$/-benzodiazepine receptor in rat hippocampus during aging, *J. Pharmacol. Exp. Ther.* **256:**902–908.

Ruano, D., Machado, A., and Vitorica, J., 1993, Absence of modifications of the pharmacological properties of the $GABA_A$ receptor complex during aging, as assessed in 3- and 24-month-old rat cerebral cortex, *Eur. J. Pharmacol.* **246:**81–87.

Sandberg, K., and J. T. Coyle, 1985, Characterization of [$^3$H]hemicholinium-3 binding associated with neuronal choline uptake sites in rat brain membranes, *Brain Res.* **348:**321–330.

Selkoe, D. J., Bell, D. S., Podlisny, M. B., Price, D. L., and Cork, L. C., 1987, Conservation of brain amyloid proteins in aged mammals and humans with Alzheimer's disease, *Science* **235:**873–877.

Shimada, A., Mukhin, A., Ingram, D. K., and London, E. D., 1997, *N*-methyl-D-aspartate receptor binding in brains of rats at different ages, *Neurobiol. Aging* **18:**329–333.

Smith, T. D., Gallagher, M., and Leslie, F. M., 1995, Cholinergic binding sites in rat brain: Analysis by age and cognitive status, *Neurobiol. Aging* **16:**161–173.

Stroessner-Johnson, H. M., Rapp, P. R., and Amaral, D. G., 1992, Cholinergic cell loss and hypertrophy in the medial septal nucleus of the behaviorally characterized aged rhesus monkey, *J. Neurosci.* **12:**1936–1944.

Subramaniam, S., and McGonigle, P., 1991, Quantitative autoradiographic characterization of the binding of (+)-5-methyl-10,11-dihydro-5H-dibenzo[a,d]cyclohepten-5,10-imine ([$^3$H]MK-801) in rat brain: Regional effects of polyamines, *J. Pharm. Exp. Ther.* **256:**811–819.

Tanila, H., Sipila, P., Shapiro, M., and Eichenbaum, H., 1997, Brain Aging: Impaired coding of novel environmental cues, *J. Neurosci.* **17:**5167–5174.

Tigges, J., Gordon, T. P., McClure, H. M., Hall, E. C., Peters, A., 1988, Survival rate and life span of rhesus monkeys at the Yerkes Regional Primate Research Center, *Am. J. Primatol.* **15:**263–273.

Tigges, J., Herndon, J. G., and Peters, A., 1990, Neuronal population of area 4 during the life span of the rhesus monkey, *Neurobiol. Aging* **11:**201–208.

Vannucchi, M. G., and Goldman-Rakic, P. S., 1991, Age-dependent decrease in the affinity of muscarinic M1 receptors in neocortex of rhesus monkeys, *Proc. Natl. Acad. Sci. USA* **88:**11475–11479.

Vogt, B. A., and Burns, D. L., 1988, Experimental localization of muscarinic receptor subtypes to cingulate cortical afferents and neurons, *J. Neurosci.* **8:**643–652.

Voytko, M. L., Sukhov, R. R., Walker, L. C., Breckler, S. J., Price, D. L., Koliatsos, V. E., 1995, Neuronal number and size are preserved in the nucleus basalis of aged rhesus monkeys, *Dementia* **6:**131–141.

Wagster, M. V., Whitehouse, P. J., Walker, L. C., Kellar, K. J., and Price, D. L., 1990, Laminar organization of age-related loss of cholinergic receptors in temporal neocortex of rhesus monkey, *J. Neurosci.* **51:**2879–2885.

Wallace, D. C., 1995, Mitochondrial DNA mutations, in: *Human Disease and Aging: Molecular Aspects of Aging* (K. Esser and G. M. Martin, eds.), Wiley, New York, pp. 163–177.

Wang, S.-Z., Shu, S.-Z., Joseph, J. A., and El Fakahany, E. E., 1992, Comparison of the level of mRNA encoding m1 and m2 muscarinic receptors in brains of young and aged arts, *Neurosci. Lett.* **145:**149–152.

Wenk, G. L., Pierce, D. J., Struble, R. G., Price, D. L., and Cork, L. C., 1989, Age-related changes in multiple neurotransmitter systems in the monkey brain, *Neurobiol. Aging* **10:**11–19.

West, M. J., 1993, New stereological methods for counting neurons, *Neurobiol. Aging* **14:**275–285.

Young, S. W., III, and Kuhar, M. J., 1979a, A new method for receptor autoradiography: [$^3$H]opioid receptors in rat brain, *Brain Res.* **179:**255–270.

Young, S. W., III, and Kuhar, M. J., 1979b, Autoradiographic localization of benzodiazepine receptors in the brains of humans and animals, *Nature* **280:**393–395.

7

# Trophic Factors in Experimental Models of Adult Central Nervous System Injury

DANIEL A. PETERSON
and FRED H. GAGE

## 1. Introduction

Unlike cells in most organ systems, neurons of the adult mammalian central nervous system (CNS) are exquisitely vulnerable to injury and alterations in their environment. Furthermore, those central neurons that survive a perturbation in their environment exhibit only limited regeneration to reestablish functional connectivity. Such observations led the pioneering neuroscientist Santiago Ramon y Cajal to state, "Once development was ended, the founts of growth and regeneration of the axons and dendrites dried up irrevocably. In adult centers, the nerve paths are something fixed and immutable; everything may die, nothing may be regenerated" (Cajal, 1928). The specialization within the CNS exacerbates the vulnerability of individual neurons by causing system-wide failures as a result of loss of specialized

DANIEL A. PETERSON and FRED H. GAGE • Laboratory of Genetics, The Salk Institute, La Jolla, California 92037. *Present address for D. A. P.:* Department of Neuroscience, Finch University of Health Sciences, The Chicago Medical School, North Chicago, Illinois 60064-3095.

*Cerebral Cortex,* Volume 14, edited by Peters and Morrison. Kluwer Academic/Plenum Publishers, New York, 1999.

neurons. For example, the selective vulnerability of dopaminergic projection neurons in the human substantia nigra leads to the movement disorders characterized as Parkinson's disease, affecting the entire organism. The mechanisms of neurodegeneration have been extensively studied in the hope of elucidating crucial steps in the sequence of neuronal death. One approach has sought the identity of endogenous survival factors whose presence could support injured adult neurons and prevent their death and the corresponding loss of function.

### 1.1. The Trophic Hypothesis

In contrast to the adult CNS, the developing CNS exhibits dynamic proliferation and migration of neurons, extensive outgrowth and arborization of neuronal processes, and remodeling of individual neurons and groups of neurons to form the specialized functional regions recognized in the adult CNS. The different spatial and temporal expressions of a number of substances within the developing CNS were observed to correlate with the survival and outgrowth of neurons, suggesting a regulatory role for these substances, which are called *trophic factors*. The subsequent developmentally regulated death of neurons, resulting in a thinning-out of neuronal populations, was hypothesized to result from competition among these neurons for a limited amount of trophic factor. This concept was formalized as the "trophic hypothesis" (Korsching, 1993). Support for the trophic hypothesis came from studies showing that the addition of these trophic factors in a controlled tissue culture environment could promote the survival and outgrowth of developing central neurons and suggested that the availability of trophic factors was the limiting factor for the survival of developing central neuronal populations. Mature, differentiated neurons of the adult CNS have proved difficult to culture, so most evidence has been derived from studies using cultured embryonic or early postnatal CNS neurons. Although primarily a theory of neuronal development, the premise of the trophic hypothesis suggests that trophic factors could be applied to maintain adult neurons and improve their ability to survive injury.

Various modalities have been proposed by which trophic factors act on neurons (Fig. 1). As conceptualized in the trophic hypothesis, trophic influence on developing neurons is target-derived. Evidence for this modality arises from the accumulation of trophic material in the target tissue following axotomizing lesion, from loss of projection neurons following removal of target tissue, and from retrograde transport of labeled trophic factor by projection neurons. As discussed later, both nerve growth factor (NGF) and brain-derived neurotrophic factor (BDNF) appear to exert their influence through the target-derived trophic modality. The regional coexistence of trophic factors and their receptors suggests another modality of trophic influence, wherein a cell is influenced by neighboring neurons, glia, or axonal-ensheathing glia. This *paracrine trophism* has been suggested by observations of trophic influence on local projection neurons and by evidence of glial-mediated trophic influence. Examples of trophic factors suggested to work through a paracrine modality include fibroblast growth factor-2 (FGF2), neurotrophin-3 (NT3), and ciliary neurotrophic factor (CNFT). These trophic factors are discussed in more detail later. The production of a trophic factor by neurons which themselves express the receptor for that trophic factor led to the formulation of the *autocrine trophic modality,* wherein the factor produced by a neurons is available to act upon the neuron that produced it, in addition to possibly acting in a paracrine modality on neighboring neurons expressing the receptor. Neither FGF2 or CNFT is secreted by classical secretory pathways, raising the possibility of another modality, *intracrine trophism,* where both production and signaling of the receptor occur intracellularly (Logan, 1990). Finally, neurons may be influenced in a trophic fashion by their input. *Afferent-derived trophism* may result from activity modulated by neurotransmitters directly, as suggested by the presence of receptors in the postsynaptic density (Wu *et al.,* 1996).

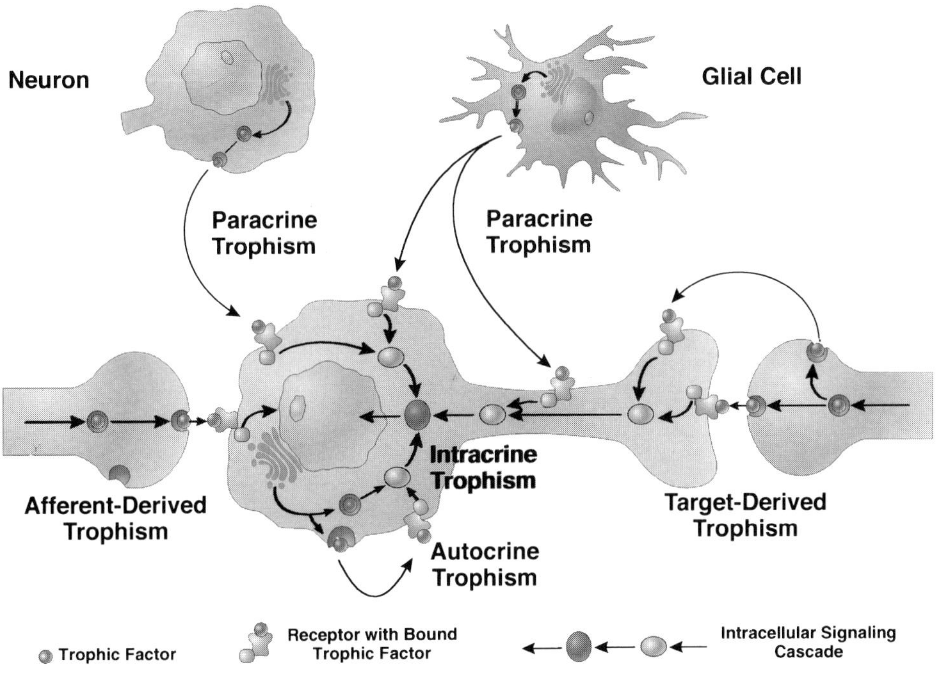


**Figure 1.** There are a number of ways in which trophic factors may act upon neurons. The trophic hypothesis of developmental regulation of neuronal cell death is based upon evidence that trophic factors may be supplied by target tissue acting upon receptors located on the axon terminals (Target-Derived Trophism). Depending upon the type of receptor activated, the trophic signal is conducted to the nucleus (arrows) through a cascade of signaling pathways (ovals). The presence of neuronal receptors for locally expressed trophic factors suggests another modality, whereby neurons also receive trophic support from neighboring neurons and glia (Paracrine Trophism). Neurons that both produce a trophic factor and express the receptor for that factor argue for another modality, wherein a neuron self-stimulates by producing trophic factors that activate receptors on their own surface (Autocrine Trophism) or act internally without surface presentation (Intracrine Trophism). A neuron may also receive trophic influence from distant neurons through their axonal projections (Afferent-Derived Trophism), through an interaction of traditional trophic factors and neurotransmitter release.

## 1.2. Trophic Factor Diversity

The number of substances for which trophic effects have been ascribed has grown extensively from the first described trophic factor, nerve growth factor (NGF), to include a wide variety of extracellular substances found in the CNS. These include factors related to NGF, referred to as the *neurotrophins* (BDNF, NT3, NT4/5, and NT6), CNFT, leukemia inhibitory factor (LIF), mitogenic or growth factors such as epidermal growth factor (EGF) and the fibroblast growth factors (FGFs), glial-derived neurotrophic factor (GDNF), gangliosides, classic cytokines such as IL-1 or TGFα, steroids, and even neurotransmitters. Furthermore, there is a vast array of synthetic peptides, receptor agonists, and receptor antagonists that have been shown to protect neurons from injury in various paradigms and have been referred to as trophic factors in that regard.

Despite the variety of trophic factors reported to have an effect on central neurons, not all neuronal populations are equally responsive to trophic factors. The trophic factor signal to the cell is mediated by a specific cell-surface receptor, and the presence of identified receptors is used to predict whether a given neuronal population will be responsive. We will focus on the cerebral cortex and limbic structures of the adult mammalian CNS and on those trophic substances that have been studied *in vivo* using experimental injury models or within which the presence of receptors suggests that a particular trophic substance might be neuroprotective. As a result, many important injury models investigating subcortical structures, such as those dealing with the dopaminergic nigrostriatal projection, will not be discussed in detail. For example, GDNF is a potent survival factor for dopaminergic mesencephalic neurons (Kearns and Gash, 1995; Sauer *et al.*, 1995) and spinal motor neurons (Li *et al.*, 1995). Glial-derived neurotrophic factor mRNA has been detected in adult hippocampus and cortex (Springer *et al.*, 1994) and is up-regulated in response to seizure activity in these structures (Humpel *et al.*, 1994; Schmidt-Kastner *et al.*, 1994), suggesting a possible role in injury responsiveness. However, because no experiments as of this writing, have directly examined the role of GDNF in cortex or hippocampus, this trophic factor will not be addressed. Other reviews may be consulted for information on trophic factors in injury models involving subcortical structures (Lapchak, 1996; Sendtner *et al.*, 1996; Lindsay *et al.*, 1994).

## 1.3. Trophic Influence Assessment *in Vivo*

Though more difficult to implement and interpret than *in vitro* studies, due to the large number of possible interactions with the surrounding environment, *in vivo* studies provide crucial evidence as to whether the hypotheses tested in carefully controlled *in vitro* conditions reflect the responsiveness of the injured adult CNS to trophic factor administration. Experimental *in vivo* administration of trophic factors permits assessment of their therapeutic effect on functional recovery of injured adult CNS. Despite strong evidence for the trophic hypothesis, it is not clear to what extent adult CNS neurons *in vivo* require the support of trophic factors to remain viable. Gene-targeting approaches have been applied to this question, producing mixed results. The main drawback with respect to elucidating adult trophic requirements using gene targeting is the influence on the entire developmental period rather than just on the adult animal. The application of conditional mutations (discussed below) may provide results that can be more clearly interpreted. In addition, recent evidence reveals that trophic factors participate in modulating rapid physiological events in neuronal plasticity and signaling (Berninger and Poo, 1996; Thoenen, 1995), suggesting that their physiological role in the adult may differ from the developmental period.

The ultimate therapeutic goal of experimental studies using trophic factors is to successfully prevent neuronal death, promote neurite regeneration, and produce a reinnervation that reestablishes the neuronal circuitry to provide a restoration of CNS function. *In vitro* studies have greatly increased our understanding of the mechanisms of trophic factor receptor activation and function in individual neurons; however, such reduced preparations are limited in their ability to reveal trophic factor influence on the survival and connectivity of neuronal ensembles. By investigating changes in the expression of trophic factors and their receptors *in vivo* following experimental injury and by delivery of hypothesized trophic factors to injured tissue, it is possible to assess the correspondence between the characteristics of a trophic factor in a controlled *in vitro* environment and its possible therapeutic role *in vivo*.

We will also examine the rationale for experimental studies using trophic factors. Expression of trophic factors around the neuronal cell bodies or in the region to which their axons project or elevation of a factor in response to injury might indicate responsiveness by a given neuronal population. An important assumption of studies where trophic factors have been added to injured neuronal populations, but one that is often not explicitly stated, is the hypothesis that neuronal populations sensitive to a particular trophic factor during the developmental period will also be responsive to that trophic factor in the adult CNS. It is therefore important to establish whether the neuronal population being investigated expresses known receptors for the proposed trophic factor, either constitutively or as a result of injury, prior to administering that trophic factor in the experimental model.

## 2. Experimental Approaches

### 2.1. Injury Models

The type of injury inflicted upon the adult CNS is generally intended to reflect the type of trauma evidenced by a disease process or clinically relevant insult. In this way, the extent of response by the adult CNS can be used to evaluate a proposed therapeutic intervention. One common approach is to perform an axotomizing lesion where a fiber pathway is transected so that the projection neurons and possibly the target neurons (if not directly removed by the lesion) undergo necrotic cell death. The goals of this injury model are to assess the trophic effect of candidate substances and to determine their neurite-promoting effect as an indicator of regenerative capacity. Because specific neuronal projections can be targeted by this approach, axotomizing lesion models lend themselves to precise neuroanatomical analysis and potential behavioral assessment for recovery of function. The axotomy may be performed by the stereotaxic insertion of a cannula from which a retractable knife blade is extended to sever a discrete anatomical pathway. For more extensive projection pathways it is often necessary to perform an aspirative lesion under direct visual control to encompass the projection. One advantage of the aspirative lesion is the resulting cavity, which provides a receptacle for the placement of a graft of trophic-factor-producing cells (Fig. 2).

An alternative to the invasive axotomizing lesion is a chemical lesion using an injected excitotoxic compound, such as an excitatory amino acid or its receptor agonist, that either kills neurons in the region of infusion or exposes axon terminals in the region of infusion to the toxin, thus retrogradely affecting the cell body. The excitotoxic lesion model also lends itself to neuroanatomical and behavioral

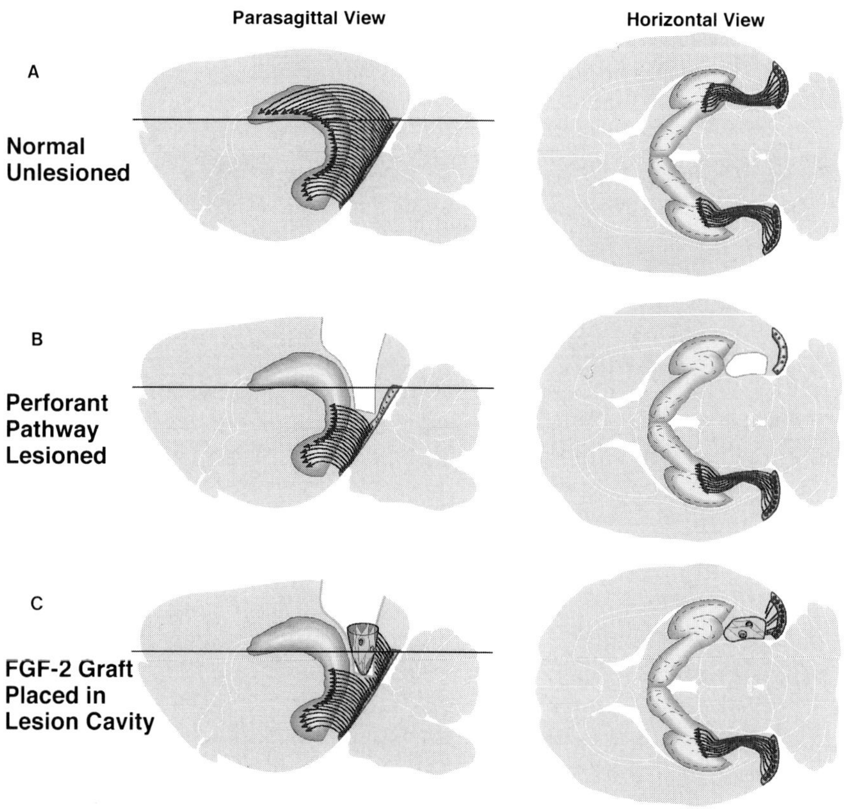


**Figure 2.** A schematic illustrating an axotomizing lesion followed by a solid graft of trophic factor-producing cells. The line through the parasagittal view of a rat brain (left column) indicates the dorsoventral level at which the horizontal view (right column) was taken. (A) The entorhinal cortex on the caudal pole of the cerebral cortex sends axonal projections, termed "perforant pathways," to the hippocampus. (B) A lesion aspirates away the overlying retrosplenial cortex and axotomizes perforant pathway projections to dorsal hippocampus, causing the neurons to die. (C) The lesion cavity can be filled with a "plug" of trophic factor-secreting fibroblasts (see also Fig. 3) that rescue the entorhinal neurons, providing a substrate for putative regeneration of perforant pathway projections to hippocampus.

analysis. Chemical lesions have the advantage of producing minimal anatomical disturbance, often leaving unharmed fibers passing through the injection region. However, control over the degree of injected excitotoxin diffusion may result in considerable variation of degeneration between conditions.

Ischemic insults have also been employed as an experimental model of stroke by transiently occluding either local blood flow as a model of focal ischemia or more-complete vascular occlusion as a model of global ischemia. By altering the duration and severity of the ischemic episode, the temporal response and ultimate vulnerability of neuronal populations can be studied in the presence or absence of proposed trophic factors. Occlusion models have the great advantage of closely corresponding to actual traumatic response from stroke or cardiac arrest, and therefore data from experimental intervention can be readily applied to therapeutic efforts.

Gene targeting approaches may offer potential adult CNS experimental injury models. Although traditional gene targeting addresses the developmental role of trophic factors and their receptors, recent advances in conditional mutations using site-specific recombinases raise the possibility of generating a transgenic animal whose genetic "lesion" is not made manifest until the animal reaches adulthood, when it can then be invoked in precise anatomical regions by injection of site-specific recombinases (Kilby *et al.*, 1993). Development of these strategies will offer an unprecedented level of control over the precise molecular mechanisms involved in neuronal injury and response to trophic factor availability.

### 2.2. Trophic Factor Delivery Systems

Trophic factors have been introduced to the adult CNS in a variety of ways. A common approach provides the trophic factor protein directly to the CNS, as in some early studies where Gelfoam packing was soaked in the trophic factor before its introduction to the lesion cavity. More frequently, trophic factor protein has been infused into the cerebrospinal fluid to bathe the brain or spinal cord surface. This infusion can be either intraventricular or intrathecal. Alternatively, intraparenchymal infusion can be used to target protein delivery within the CNS. Infusions can be either periodic, using a chronically implanted cannula, or constant, using a closed cannula system connected to an osmotic minipump. With the commercial production of recombinant trophic factors, the reproducibility and comparability of direct infusion studies have been enhanced. It still remains difficult, however, to determine the extent of spread of infused protein and to demonstrate that it remains unbound and biologically active where detected.

Recognizing the abundance of trophic factors produced in the developing brain, a number of studies during the 1970s and 1980s investigated intracerebral grafting of fetal brain tissue, both to provide a source of trophic factors and as a potential neuronal replacement therapy should the fetal neurons functionally integrate into the adult CNS (Björklund 1991; Gage and Björklund, 1987; Clarke *et al.*, 1986; Gage *et al.*, 1984; Gage *et al.*, 1983). Because of the reliance on specific fetal aging and precise microdissection of brain regions for grafting, it became difficult to account for variability between studies. Analysis of these studies was also compromised by the inability to control for the availability of only a specific trophic factor from the fetal graft. Furthermore, ethical considerations concerning the use of fetal tissue constrained the transfer of experimental results into therapeutic strategies.

With the advent of improved expression vectors, it became possible to genetically engineer somatic cells *in vitro* to express the transgene of interest (Karpati *et al.,* 1996; Ray *et al.,* 1995a; Rosenberg *et al.,* 1988). The ability to graft intracerebrally a known population of genetically modified somatic cells, typically fibroblasts or astrocytes, thus producing reliable quantities of a specific trophic factor made it possible to conduct more tightly controlled experiments to test the efficacy of trophic factors. This *ex vivo* gene therapy approach has provided a robust and reliable method for trophic factor delivery to experimental models of adult CNS injury (Fig. 3). *Ex vivo* gene therapy has the advantage that somatic cells can be taken from an individual, infected in culture to express the transgene for the

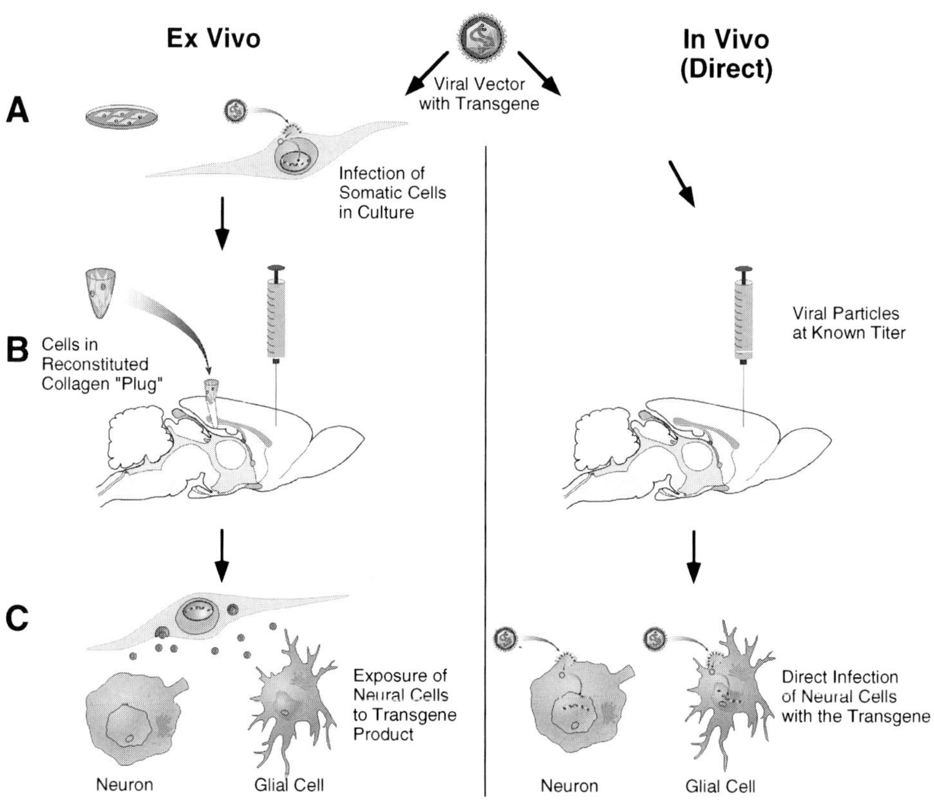


**Figure 3.** Basic scheme of gene therapy approaches for the delivery of trophic factors to the adult CNS. After construction, a viral vector containing the transgene to be expressed can be delivered to the adult CNS using two approaches, *ex vivo* gene therapy (left column) or *in vivo* gene therapy (right column). With *ex vivo,* the first step (A) involves infection of somatic cells (typically fibroblasts or astrocytes) in culture. Once cells are infected, and selected for their expression of the transgene construct, the cells can be grafted (B). Grafting takes place via intraparenchymal injection of suspended cells or by a "plug" of cells in reconstituted collagen that fill the lesion cavity and provide both expression of the transgene product and a physical bridge for extension of regenerative growth. Once grafted, the transgene product is made available to nearby neural cells (C). With *in vivo* gene therapy, the viral particles are directly injected (B) at a titer offering the best balance between infectivity and viral toxicity. The injected viral particles can infect neural cells in the vicinity of the injection (C), which in turn can express the transgene.

desired trophic factor, and then harvested and placed into the CNS of the same individual as an autologous graft. This approach avoids many of the ethical concerns of fetal grafts while simultaneously providing better control of trophic factor-delivery. Furthermore, cells producing different trophic factors could be combined in controlled ratios to provide a trophic factor cocktail to investigate their synergistic effects.

Neurons are difficult to infect with a transgene expression vector since they are postmitotic, which limits the opportunity for stable integration. Recently, vectors have been developed that can stably infect postmitotic cells (Karpati *et al.,* 1996; Naldini *et al.,* 1996). The use of adenovirus, adeno-associated virus (AAV), herpes-simplex virus (HSV), and the HIV-associated lentiviral vector raises the possibility of direct *in vivo* gene therapy (Fig. 3). The efficiency, selectivity, and toxicity of vectors for direct *in vivo* gene therapy of the CNS are currently areas of intense investigation, and are beyond the scope of this review (see Blömer *et al.,* 1997). Direct infection of neural tissue may provide the experimental advantage that neurons could be infected to express the proposed trophic substance in advance of the lesion (Fig. 4). Success in such experiments raises the possibility of preventative therapeutic approaches against neurodegenerative disease.

Finally, recent work on conditionally targeted gene mutations provides the potential for precise spatial and temporal expression of a targeted "knock-in" through the introduction of site-specific recombinases (Brooks *et al.,* 1997; Kilby *et*

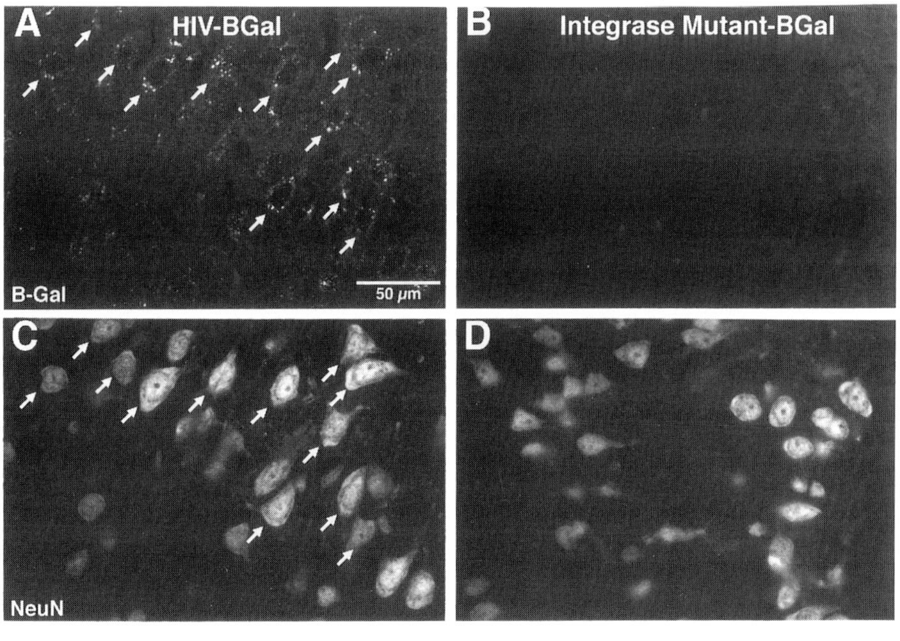


**Figure 4.** Lentiviral vector containing *Lac Z* as a marker transgene was injected into the dorsal hippocampus of an adult rat. The entorhinal cortex was examined six weeks later for evidence of retrograde transport and integration into the projection neurons of layer II as determined by immunofluorescent labeling of antibodies against the transgene product β-Galactosidase (B-Gal). Neurons (arrows), but not adjacent glial cells, of layer II expressed β-Gal in animals receiving the lentiviral vector, but not in control animals receiving an integrase-deficient lentiviral vector, illustrating that direct *in vivo* gene delivery may effectively protect projection neurons. See Naldini *et al.* (1996) for details.

al., 1993). This approach is distinct from the previous discussion of conditional mutants in that, rather than using the conditional mutation to produce a lesion, the conditional mutation turns on the expression of the introduced gene to test the efficacy of its gene product on neuroprotection and/or regeneration. For the time investment of generating a transgenic animal carrying the conditional mutation, it will be possible to precisely test hypotheses regarding trophic factor neuroprotection by autocrine expression or by expressing receptors or intracellular signaling components proposed to be the mechanism of trophic factor action.

## 3. Nerve Growth Factor

### 3.1. Nerve Growth Factor and Its Receptors

Nerve growth factor (NGF), a secreted polypeptide found throughout the body and within the CNS, was the first described neurotrophic factor and is currently the most thoroughly characterized (Levi-Montalcini, 1987). Its effect is mediated through the low-affinity p75 neurotrophin receptor, which binds all neurotrophins with similar affinity, and the high-affinity receptor tyrosine kinase trkA (Bothwell, 1995). The role of p75 in NGF stimulation is controversial at present since trkA, but not p75, is sufficient for NGF signaling (Ibáñez *et al.*, 1992; Loeb *et al.*, 1991). Furthermore, recent evidence suggests that, in certain cellular contexts, p75 may signal apoptotic pathways in the absence of NGF (Barrett and Georgiou, 1996; Frade *et al.*, 1996; Casaccia-Bonnefil *et al.*, 1996). Unfortunately, the disruption of the gene encoding for p75 by homologous recombination has produced conflicting reports concerning the survival of NGF-responsive central neurons during development (Peterson *et al.*, 1997; Peterson *et al.*, 1999; Van der Zee *et al.*, 1996). Further work will be required to clarify the role of p75 (Segal and Greenberg, 1996).

### 3.2. *In Vitro* Activity

NGF has been shown to be neuroprotective in a target-dependent manner on sympathetic and some sensory neurons (Levi-Montalcini, 1987). It acts as a survival factor for cultured basal forebrain cholinergic neurons (Svendsen *et al.*, 1994; Gähwiler *et al.*, 1990; Hsiang *et al.*, 1989; Hatanaka and Tsukui, 1986). Expression of a cholinergic phenotype may not be a reliable indicator of sensitivity to NGF administration, however, because prelabeling of septal projection neurons prior to culture revealed that NGF promoted the survival of all back-labeled septal neurons irrespective of phenotype (Arimatsu *et al.*, 1989). Furthermore, NGF enhances cholinergic expression in septal cultures (Hefti *et al.*, 1985), arguing for caution in assessing survival effects of trophic factor addition without a reliable estimate of changes of the total neuronal population. In addition to its neuroprotective role, NGF is reported to regulate neurotransmitter expression (Fusco *et al.*, 1989; Vantini *et al.*, 1989; Hefti *et al.*, 1985; Greene *et al.*, 1984; Gnahn *et al.*, 1983) and presynaptic neurotransmitter release (Knipper *et al.*, 1994a,b), function in activity-dependent neuronal plasticity (Berninger and Poo, 1996; Thoenen, 1995), and modulate neuroimmune interactions and tissue inflammation (Levi-Montalcini *et al.*, 1996).

### 3.3. *In Vivo* Activity

Several lines of evidence suggest that NGF may play a neurotrophic role in the adult CNS. High levels of both NGF mRNA and protein have been detected in the hippocampus and cerebral cortex (Shelton and Reichardt, 1986; Whittemore *et al.*, 1986). Some neurons that project to these regions have been shown to take up and retrogradely transport NGF (Seiler and Schwab, 1984; Schwab *et al.*, 1979). Furthermore, these neurons, which are primarily cholinergic (Korsching *et al.*, 1985), express both the low- and high-affinity NGF receptors (Kawaja and Gage, 1991a; Batchelor *et al.*, 1989). However, striatal cholinergic neurons, which do not project to these NGF-rich regions but do express the high-affinity trkA receptor in adult rats, also show a biological response to NGF *in vivo* (Gage *et al.*, 1989; Hagg *et al.*, 1989a).

NGF levels fluctuate in response to lesion. NGF increases in the hippocampus following fimbria-fornix lesion (Gasser *et al.*, 1986; Korsching *et al.*, 1986; Weskamp *et al.*, 1986; Collins and Crutcher, 1985), stimulation (Rocamora *et al.*, 1992; Ernfors *et al.*, 1991), or ischemic insult (Takeda *et al.*, 1993; Lindvall *et al.*, 1992). In contrast to the fimbria-fornix lesion, NGF protein in the hippocampus does not increase after perforant pathway lesion (Fagan *et al.*, 1997), and NGF mRNA does not increase in response to either lesion. Therefore, it appears that the increased NGF levels reflect an accumulation of NGF normally taken up and transported to the medial septum.

While NGF levels increase in the hippocampus following an axotomizing lesion of the fimbria-fornix, the cholinergic projection neurons of the medial septum undergo a massive cell death (Gage *et al.*, 1986). Concurrent intracerebroventricular infusion of NGF (Fischer and Björklund, 1991; Hagg *et al.*, 1989b; Kromer, 1987; Hefti, 1986; Williams *et al.*, 1986) or lesion cavity packing with NGF-treated Gelfoam (Otto *et al.*, 1989) prevented cell death in this population. Rescue was also achieved with fetal hippocampal grafts placed into the lesion cavity, which had the additional benefit of providing a physical bridge upon which regenerating septal projections could grow (Tuszynski *et al.*, 1990). Recently, septal cholinergic neurons have been rescued by infusion of antibodies to trkA, which act as specific agonists for the high-affinity NGF receptor (Lucidi-Phillipi *et al.*, 1996). The specificity of NGF effect *in vivo* appears to require expression of the low- or high-affinity receptors, because cells that do not express these receptors are not protected by NGF-producing grafts (Peterson *et al.*, 1996).

In addition to neuroprotection, NGF also exerts a tropic influence on adult cholinergic neurons in both intact and lesioned CNS (Chalmers *et al.*, 1996; Lucidi-Phillipi *et al.*, 1996; Kawaja *et al.*, 1992). Autologous grafts of fibroblasts genetically modified to express NGF into the fimbria-fornix lesion cavity rescued septal cholinergic neurons as described above for NGF infusion (Gage *et al.*, 1991; Stromberg *et al.*, 1990; Rosenberg *et al.*, 1988; Gage *et al.*, 1987), and they also promoted a regenerative response of cholinergic fibers into the fibroblast graft (Kawaja *et al.*, 1992). Examination of the fine structure of cholinergic ingrowth to NGF grafts revealed neurite association with the surface membrane of astrocytes whose invasion of the graft had preceded neurite ingrowth (Chalmers *et al.*, 1996; Kawaja and Gage, 1991b). This neurite–astrocyte association was correlated with the expression of cell adhesion molecules whose expression was up-regulated in response to NGF (Chalmers *et al.*, 1996). Cholinergic fibers of lesioned septal neurons could not only enter NGF-producing grafts placed in the fimbria fornix lesion cavity, but were also observed to transverse the entire NGF graft and enter the hippocampus (Eagle *et al.*, 1995). Upon entering the hippocampus, the regenerating septal cholinergic fibers innervated the appropriate layers of the dentate gyrus and displaced sympathetic fibers that had sprouted into these regions. Behavioral improvement on tasks assessing hippocampal performance in spatial memory suggested that the morphological reconnectivity through the NGF-bridging graft reestablished functional septohippocampal projections (Eagle *et al.*, 1995).

### 3.4. Clinical Experience with Nerve Growth Factor Administration

Atrophy of cholinergic projection neurons is one of the characteristic features of Alzheimer's disease. The observation of enhanced cholinergic phenotype and protection of septal neurons following NGF delivery in the *in vitro* and *in vivo* experimental models reviewed above suggested that NGF administration may reverse the cholinergic deficit in Alzheimer patients (Olson, 1993; Phelps *et al.*, 1989). Systemic administration of NGF resulted in localized hyperalgesia (Petty *et al.*, 1994), whereas patients receiving intracerebroventricular administration of NGF reported back pain and weight loss (Olson, 1993; Olson *et al.*, 1992). The intracerebroventricular administration of NGF produced mild improvement in a variety of physiological measures and psychological tests in Alzheimer patients (Olson, 1993; Olson *et al.*, 1992). However, intracerebroventricular administration of NGF has recently been shown to initiate sprouting of sensory and sympathetic fibers into the CNS, a warning to exercise caution in evaluating therapeutic administration of NGF by this route (Winkler *et al.*, 1997; see section 8.3). Further clinical trials will be needed to evaluate the safety and efficacy of NGF administration in reversing neurodegenerative disease.

## 4. Other Neurotrophins

### 4.1. Neurotrophins and Their Receptors

Investigation of peripheral sensory neurons revealed a heterogeneity in responsiveness to NGF with identification of subpopulations that were NGF-insensitive. Such observations prompted searches for alternative trophic substances for these subpopulations, leading to the characterization and cloning of brain-derived neurotrophic factor or BDNF (Ernfors *et al.*, 1990a; Leibrock *et al.*, 1989; Barde *et al.*, 1978). BDNF has structural similarity to NGF, with a 50% homology of its amino acid sequence. This structural similarity suggested the existence of additional related factors, which led to molecular studies that cloned a third member of this family of NGF-related proteins, called neurotrophin-3 or NT3 (Hohn *et al.*, 1990; Jones and Reichardt, 1990; Kaisho *et al.*, 1990; Maisonpierre *et al.*, 1990; Rosenthal *et al.*, 1990). A fourth factor was cloned from *Xenopus laevis* and called NT4 (Hallböök *et al.*, 1991). A fifth factor with structural homology to NGF was cloned and called NT5 (Berkemeier *et al.*, 1991) although because it has also been described as a mammalian homologue to the amphibian NT4 (Ip *et al.*, 1992a), it is commonly referred to as NT4/5. A sixth member, called NT6, has recently been cloned from

**Table I. Neurotrophin-Receptor Relationships**

| Trophic factor | Low-affinity receptor | High-affinity receptor(s) |
|---|---|---|
| NGF | p75 | trkA |
| BDNF | p75 | trkB |
| NT-3 | p75 | trkC (± trkB) |
| NT-4/5 | p75 | trkB |

teleosts, but no mammalian homologue has yet been described (Götz *et al.,* 1994). The members of this family of proteins with structural similarity to NGF are called neurotrophins.

The receptors to which the neurotrophins bind with high affinity are also receptor tyrosine kinases, with the original receptor for NGF being termed trkA. As mentioned previously, the low-affinity receptor p75 binds to all members of the neurotrophin family equally. The structural homology of the neurotrophins to NGF suggested that their receptors might also bear a similarity to trkA. Investigation of this resulted in the cloning of two receptors, trkB (Middlemas *et al.,* 1991; Klein *et al.,* 1990a) and trkC (Cordon-Cardo *et al.,* 1991; Lamballe *et al.,* 1991). BDNF, NT4/5, and to a lesser degree, NT3 activate trkB, whereas only NT3 activates trkC (Ip *et al.,* 1992a; Berkmeier *et al.,* 1991; Lamballe *et al.,* 1991; Soppet *et al.,* 1991). The relationship of the factors and their receptors is summarized in Table I (see Bothwell, 1995, for a detailed review).

The expression of mRNA for BDNF is widespread throughout the adult CNS, including midbrain dopaminergic neurons (Gall *et al.,* 1992) and olfactory bulb (Guthrie and Gall, 1991), but it is most pronounced in the cortical and hippocampal neurons (Hofer *et al.,* 1990; Phillips *et al.,* 1990; Wetmore *et al.,* 1990). The expression of the trkB receptor in neurons of the basal forebrain (Klein *et al.,* 1990b), combined with evidence of retrograde transport of $^{125}$I-BDNF to these neurons (DiStefano *et al.,* 1992), suggests that BDNF may act through the classical target-derived neurotrophic model (Wetmore *et al.,* 1991; Phillips *et al.,* 1990). This model for BDNF is additionally supported by recent evidence from Conner *et al.* (1997) describing the accumulation of BDNF immunoreactivity within the projection neurons of the basal forebrain and its subsequent loss following removal of the target region. The expression of the high-affinity trkB receptor for BDNF and NT4/5 is widespread throughout the CNS including, in addition to the basal forebrain, cerebral cortex, hippocampus, striatum, substantia nigra, cerebellar Purkinje cells, brainstem and spinal motor neurons, and brainstem sensory nuclei (Yan *et al.,* 1997). The distribution of NT3 and trkC is less widespread and has been localized to the hippocampus (Merlio *et al.,* 1992; Ernfors *et al.,* 1990b). Although mRNA for NT4/5 is present at low levels throughout the CNS, its precise cellular distribution has not yet been determined (Timmusk *et al.,* 1993; Ip *et al.,* 1992a).

## 4.2. *In Vitro* Activity

In addition to early reports of its survival effect on peripheral sensory neurons (Davies *et al.,* 1986), BDNF has also been shown to promote survival and differentiation of cultured neurons from basal forebrain (Nonomura and Hatanaka, 1992; Knüsel *et al.*, 1991), ventral mesencephalon (Hyman *et al.*, 1994; Spina *et al.*, 1992; Hyman *et al.*, 1991; Knüsel *et al.*, 1991), cerebellum (Larkfors *et al.*, 1996), and cortex (Ghosh *et al.*, 1994). However, BDNF failed to support the survival of basal forebrain neurons dependent upon NGF following its withdrawal (Kew and Sofroniew, 1997). Fetal hippocampal neuronal survival is promoted by BDNF (Lowenstein and Arsenault, 1996a; Pringle *et al.*, 1996; Mattson *et al.*, 1995; Cheng and Mattson, 1994), including protection from excitotoxic stress (however, see Marsh and Palfrey, 1996). All of these neuronal populations have been shown to retrogradely transport BDNF (Conner *et al.*, 1997; DiStefano *et al.*, 1992), suggesting that additional neuronal populations that retrogradely transport BDNF, such as the thalamus and entorhinal cortex, and that have not been assessed *in vitro,* may also prove sensitive to BDNF.

Of the other neurotrophins, NT3 reportedly supports the survival and differentiation of cultured fetal central neurons, including those from hippocampus (Cheng and Mattson, 1994; Ip *et al.*, 1993a; Ohsawa *et al.*, 1993), cerebral cortex (Widmer *et al.*, 1992), cerebellum (Larkfors *et al.*, 1996; Lindholm *et al.*, 1993), and spinal cord (Wong *et al.*, 1993). However, it has also been reported to be ineffective in promoting the survival of basal forebrain (Nonomura *et al.*, 1995) and hippocampal neurons in culture (Lowenstein and Arsenault, 1996b; Marsh and Palfrey, 1996; Knüsel *et al.*, 1991), and it failed to protect cultured hippocampal neurons from injury (Pringle *et al.*, 1996). NT4/5 has been reported to be a survival factor for cultured basal forebrain (Nonomura *et al.*, 1995), cerebellar (Larkfors *et al.*, 1996), and ventral mesencephalic neurons (Hyman *et al.*, 1994; Hynes *et al.*, 1994; Widmer and Hefti, 1994a,b).

### 4.3. *In Vivo* Activity

Neurotrophins play a critical role in the developmental survival of sensory or motor neurons, as determined by the extensive cell loss seen in animals with targeted disruption of the genes encoding for BDNF (Ernfors *et al.*, 1994; Jones *et al.*, 1994), trkB (Klein *et al.*, 1993), or NT3 (Wilkinson *et al.*, 1996; Fariñas *et al.*, 1994), with homozygous null mutants in each case demonstrating a short postnatal survival (see Snider, 1994, for review). Among the neurotrophin mutant reports, only Jones *et al.* (1994) reported cortical and hippocampal changes, finding a reduction in the number of calcium-binding protein-immunoreactive neurons in these structures.

Neurotrophin levels fluctuate in response to activity levels or neuronal damage (see Isackson, 1995; Wetmore and Olson, 1993, for reviews). Experimental induction of seizure activity in the basal forebrain (Isackson *et al.*, 1991), cortex (Ernfors *et al.*, 1991), and hippocampus (Rocamora *et al.*, 1992; Ballarin *et al.*, 1991; Ernfors *et al.*, 1991) results in increased BDNF expression in both adult and neonatal rats (Kornblum *et al.*, 1997). Hippocampal BDNF levels also increased as a result of an ischemic episode (Lindvall *et al.*, 1992). Removal of entorhinal afferents had no effect on BDNF or trkB levels within the hippocampus, arguing against a role for afferent modulation (Lapchak *et al.*, 1993). However, afferent modulation of glutamate release in hippocampal area CA1 has been recently reported to correlate with increased BDNF and trkB mRNA, suggesting that there is indeed an afferent role for neurotrophin regulation (Falkenberg *et al.*, 1996). The detection of neurotrophin receptors in the postsynaptic density (Wu *et al.*, 1996) further supports an afferent role for neurotrophin regulation.

In contrast to the elevation of BDNF, levels of NT3 were not induced following hippocampal damage (Ballarin *et al.*, 1991; Ernfors *et al.*, 1991); in fact, they are reported to decrease (Lindvall *et al.*, 1992; Rocamora *et al.*, 1992). However, long-term elevation of NT3 mRNA has been detected in the hippocampus following administration of an NMDA agonist (Rocamora *et al.*, 1993, 1994) or induction of stress (Smith *et al.*, 1995).

Neurotrophins have also been investigated *in vivo* for their efficacy in reducing death following experimental injury. Infusions of BDNF have been shown to protect spinal cord motor neurons from axotomizing lesion (Novikova *et al.*, 1996; Friedman *et al.*, 1995), at least those of a cholinergic phenotype (Tuszynski *et al.*, 1996). BDNF was reported to be neuroprotective in the striatum of rats following 1-methyl-4-phenylpyridinium (MPP+) injection when assessed by functional magnetic resonance imaging (Kirschner *et al.*, 1996), but a similar paradigm delivering BDNF-expressing fibroblasts had no effect on striatal lesion volume produced by an NMDA agonist (Frim *et al.*, 1993). BDNF failed to rescue either lesioned tyrosine hydroxylase-(TH)-positive nigral neurons (Lucidi-Phillipi *et al.*, 1995; Knüsel *et al.*, 1992) or fetal ventral mesencephalon grafted to the striatum (Yurek *et al.*, 1996; Sauer *et al.*, 1993). These studies did note, however, an increase in TH-positive neuron size, TH-immunoreactive sprouting (with intranigral grafts), or dopaminergic function (Yoshimoto *et al.*, 1995) in animals receiving BDNF. Delivery of BDNF may serve to stimulate non-dopaminergic nigral neurons such as peptidergic and GABAergic systems (Arenas *et al.*, 1996).

Damage to corticospinal neurons can result in up-regulation of the neurotrophin receptors trkB and trkC, with subsequent rescue of these neurons via intraparenchymal delivery of BDNF or NT3 (Giehl and Tetzlaff, 1996). Within cortical or limbic structures, most studies have examined the effect of BDNF on cholinergic neurons of the basal forebrain following a lesion of the fimbria-fornix pathway, which axotomizes their projections to the hippocampus. Compared to the strong rescue of lesioned cholinergic neurons by NGF, BDNF was either less effective than NGF (Morse *et al.*, 1993; Widmer *et al.*, 1993; Knüsel *et al.*, 1992) or did not significantly rescue the cholinergic neurons (Lucidi-Phillipi *et al.*, 1995; Koliatsos *et al.*, 1994). Interestingly, the numbers of neurons expressing the low-affinity p75 receptor was greater than those expressing ChAT, leading to speculation that these cells responded to injury by first down-regulating ChAT, then later down-regulating p75 (Widmer *et al.*, 1993). Sofroniew *et al.* (1993) have reported that cholinergic basal forebrain neurons atrophy in response to lesion but do not die, suggesting the presence of an undetected, nonexpressing cholinergic population. Alternatively, BDNF may induce p75 expression in this population (Lucidi-Phillipi *et al.*, 1995; Koliatsos *et al.*, 1994; Sauer *et al.*, 1993; Knüsel *et al.*, 1992). Unfortunately, none of these studies used stereological methods to establish the total number of neurons within the basal forebrain or the absolute number of each phenotype, making it impossible to address the issue of cell death compared to phenotype regulation from these reports. BDNF has also been reported as a survival factor for cortical serotonergic neurons in a chemical lesion model (Mamounas *et al.*, 1995). In addition to its neuroprotective role, BDNF promotes neurite sprouting *in vivo* for central dopaminergic (Yurek *et al.*, 1996; Lucidi-Phillipi *et al.*, 1995) and serotonergic (Mamounas *et al.*, 1995) neurons, retinal ganglion cells (Sawai *et al.*, 1996; Mansour-Robaey *et al.*, 1994), and in the spinal cord (Xu *et al.*, 1995).

Despite the variety of central neuronal types responding to NT3 *in vitro*, the results of *in vivo* injury experiments have been disappointing in terms of supporting a survival role for NT3. This ineffectiveness is in agreement with some of the *in vitro* studies discussed above and consistent with the lack of change, or even downregulation, of *in vivo* NT3 expression in some of the experimental models discussed above. NT3 failed to be neuroprotective for *in vivo* injury models for central cholinergic (Tuszynski *et al.*, 1996; Koliatsos *et al.*, 1994), dopaminergic (Kirschner *et al.*, 1996), or serotonergic (Mamounas *et al.*, 1995) neurons. However, NT3 does appear to support the survival *in vivo* of some cortical neurons (Giehl and Tetzlaff, 1996) and to promote neurite sprouting in the spinal cord (Senut *et al.*, 1995; Xu *et al.*, 1995; Schnell *et al.*, 1994).

Both NT4/5 and BDNF act through the trkB receptor, and several studies assessing BDNF effects on injury (discussed above) found NT4/5 infusion similarly effective (Arenas *et al.*, 1996; Tuszynski *et al.*, 1996; Friedman *et al.*, 1995). Infusion of NT4/5 has also effectively protected calbindin-immunopositive neurons in substantia nigra (Alexi and Hefti, 1996) and striatum (Alexi *et al.*, 1997) following excitotoxic injury, although the large induction of calbindin labeling observed in the nigra, in excess of control levels, argues for a stereological study to ascertain total cell numbers. NT4/5 has also been found to be effective in protecting against injury to the cortex and limbic system by preserving the cholinergic phenotype of lesioned septal neurons more effectively than BDNF (Alderson *et al.*, 1996), preventing apoptosis in dentate granule cells (Qiao *et al.*, 1996) and reducing cortical infarct volume with NT4/5 pretreatment followed by middle cerebral artery occlusion (Chan *et al.*, 1996).

### 4.4. Clinical Experience with Neurotrophin Administration

Compared to NGF, the neurotrophins have produced weaker *in vivo* experimental evidence for neuroprotective effects on cortical and limbic central neurons. As a result, clinical trials for BDNF and NT3 have focused on the responsiveness of peripheral neuropathies (not reviewed here) and the subcutaneous administration of BDNF for rescue of spinal motor neurons in patients with amyotrophic lateral sclerosis (ALS). While earlier phases of the trial showed tolerance to BDNF, the Phase 3 trial was abandoned due to a lack of demonstrated clinical efficacy in these ALS patients, although intrathecal administration of BDNF will continue to be pursued (Amgen, Inc., and Regeneron Pharmaceuticals, Inc., press release dated January 10, 1997, available through the Internet at *www.businesswire.com/cnn/regn.htm*). Unlike with NGF, whose protective effect is confined to cholinergic neurons, the experimental evidence for protection of cortical neurons by BDNF (Giehl and Tetzlaff, 1996; Mamounas *et al.*, 1995) and NT4/5 (Chan *et al.*, 1996; Qiao *et al.*, 1996) suggests that these factors are suitable candidates for clinical trials of cortical injury. However, initiation of further clinical trials of neurotrophin administration to reverse central degeneration will likely require the stimulation of additional evidence using animal models of experimental injury.

## 5. Ciliary Neurotrophic Factor

Although named for its ability to rescue ciliary ganglion neurons *in vitro* (Barbin *et al.*, 1984), CNFT has broad effects on central neurons in culture, including survival, phenotypic expression, neuroprotection, and physiological function (Stoop and Poo, 1995; Skaper *et al.*, 1992; Ip *et al.*, 1991; Magal *et al.*, 1991a,b). When applied to cultures of fetal hippocampal neurons, CNFT promoted the survival of GABAergic, cholinergic, and calbindin-immunopositive neurons (Ip *et al.*, 1991). Alternatively, CNFT may have differentially enhanced the expression of these phenotypes in surviving neurons, as has been reported for p75 following addition of CNFT to central neuronal cultures (Kew and Sofroniew, 1995; Magal *et al.*, 1991a). Development and differentiation of a variety of neuronal and nonneuronal cell types are influenced by CNFT in culture, but the extremely low levels of CNFT during development and the absence of discernible effect in a null mutant mouse during development suggest that the major influence of CNFT occurs during the postnatal period (Masu *et al.*, 1993; see Sendtner *et al.*, 1994 for review). However, programmed cell death of sensory neurons was reduced in a transgenic mouse overexpressing CNFT (Tolosano *et al.*, 1996), arguing that CNFT level, not the responsiveness of cells to CNFT, is what is important during development. Although not secreted by classical secretory pathways, CNFT is released by astrocytes in culture (Smith *et al.*, 1996; Kamiguchi *et al.*, 1995), and presentation of CNFT stimulates astrocyte synthesis of glial fibrillary acidic protein (Kahn *et al.*, 1997). CNFT signals through a cell-surface heterotrimer receptor whose signaling pathway is similar to the cytokine IL6 (Sariola *et al.*, 1994; Ip *et al.*, 1992b; Davis *et al.*, 1991).

In the adult rodent CNS, CNFT has been detected in astrocytes of the optic tract, olfactory bulb, and spinal cord white matter (Dobrea *et al.*, 1992; Stockli *et al.*, 1991) and in neurons located in the facial nucleus, dentate gyrus, olfactory bulb, basal forebrain, locus coeruleus, cerebral cortex, substantia nigra, hippocampus, and Purkinje cells (Henderson *et al.*, 1994). The essential receptor component for CNFT (CNFTR$\alpha$) has been localized to neurons in olfactory bulb, hippocampus, and cerebral cortex (Lee *et al.*, 1997; Kirsch and Hofmann, 1994), and to a variety of central motor neurons (MacLennan *et al.*, 1996). Interestingly, adult septal neurons *in vivo* have not been shown to express CNFTR$\alpha$ (Lee *et al.*, 1997a; Ip *et al.*, 1993c), although fetal septal neurons are responsive to CNFT applied *in vitro.* Recently, Kordower *et al.* (1997) reported that CNFTR$\alpha$ could be located in substantia nigra, Purkinje cells, hippocampus, and cerebral cortex of a nonhuman primate, and in a variety of central motor neurons. *In vivo,* CNFT mRNA is elevated following mechanical trauma (Asada *et al.*, 1995; Ip *et al.*, 1993b). Transient elevation of mRNA for both CNFT and CNFTR$\alpha$ has been reported in astrocytes of the outer molecular layer of the hippocampal dentate gyrus following unilateral entorhinal cortex lesion (Lee *et al.*, 1997b).

*In vivo* administration of CNFT to adult motor neurons has proved to be neuroprotective in a number of injury models (Tan *et al.*, 1996; Sagot *et al.*, 1995; Sendtner *et al.*, 1990, 1992) although not all reports have found CNFT to be effective, perhaps due to the very short half-life of CNFT protein (Clatterbuck *et al.*, 1994). In other motor systems, CNFT administration has also been reported to be neuroprotective following injury to the substantia nigra (Hagg and Varon, 1993) and striatum (Anderson *et al.*, 1996; Emerich *et al.*, 1996), including the nonhuman primate striatum (Emerich *et al.*, 1997). Despite the absence of CNFTRα in adult septum, Hagg *et al.* (1992) reported that CNFT rescued both cholinergic and noncholinergic septal neurons following fimbria-fornix transection while also up-regulating the expression of p75 above control levels. Axotomized thalamic projection neurons, which do express CNFTRα, also could be rescued by CNFT infusion (Clatterbuck *et al.*, 1993). Post-ischemic infusion of CNFT has been found to reduce cortical infarction and to be neuroprotective for thalamic and hippocampal neurons (Kumon *et al.*, 1996; Wen *et al.*, 1995), although others found the neuroprotection in hippocampus to be nonsignificant (Ogata *et al.*, 1996). Cortical neurons involved in motor activity have also been rescued by CNFT administration following retrograde degeneration due to axotomy (Dale *et al.*, 1995) or excitotoxicity (Emerich *et al.*, 1997).

The neuroprotective effect of CNFT on central motor neurons suggested that it might prove useful as a therapeutic agent for treatment of the progressive degenerative disease of spinal motor neurons, amyotrophic lateral sclerosis (ALS). Clinical trials assessing subcutaneous injections of CNFT have so far been disappointing due both to the dosage-related systemic side effects and to the lack of effect on the progression of ALS (Miller *et al.*, 1996a,b; ALS CNTF Treatment Study Group, 1995a,b; 1996). A possible explanation is that an insufficient amount of CNFT reaches these spinal motor neurons due to dilution of the systemic injection compounded by the short half-life of the protein. Phase I clinical trials administering CNFT via intrathecal infusion (Penn *et al.*, 1997) and *ex vivo* gene therapy using polymer-encapsulated genetically modified cells grafted intrathecally (Aebischer *et al.*, 1996) reported few neurological and no systemic side effects. However, the therapeutic value of CNFT for treatment of ALS with this mode of delivery has yet to be demonstrated. In addition, CNFT can lead to gliosis in animals models (Clatterbuck *et al.*, 1996; Levison *et al.*, 1996; Winter *et al.*, 1995); thus, histological assessment of intrathecal presentation of CNFT in these clinical trials will be needed to determine possible complicating responses.

## 6. Fibroblast Growth Factors

### 6.1. Fibroblast Growth Factors and Their Receptors

The fibroblast growth factors (FGFs) are a family of related peptides that consists, at present, of 10 members. Acidic and basic FGFs were the first members to be studied (Jaye *et al.*, 1986; Esch *et al.*, 1985; Gospodarowicz, 1974) and based upon their prevalence in the adult CNS (Bohlen *et al.*, 1985a,b), were quickly determined to have neurotrophic effects on CNS neurons *in vitro* (Walicke 1988a; Walicke and Baird 1988; Unsicker *et al.*, 1987; Morrison *et al.*, 1986; Walicke *et al.*,

1986). As additional members of the family were discovered, a numerical nomenclature was adopted where acidic FGF was designated FGF1, basic FGF was designated FGF2, and later members were assigned sequential numbers (Baird 1994).

Like the neurotrophins, the FGFs bind to both a low-affinity and a high-affinity receptor. The low-affinity receptor is a heparan sulfate proteoglycan (Kiefer *et al.*, 1990). The role of the low-affinity receptor is not completely understood, but it appears that the proteoglycan serves to deliver FGF to its high-affinity receptor (Ornitz *et al.*, 1992; Rapraeger *et al.*, 1991), although the low-affinity receptor may not be necessary for FGF activation on some cells (Allen and Maher, 1993). Currently, there are four high affinity receptors (FGFR1–4) identified that are transmembrane tyrosine kinase receptors of the immunoglobulin IgG superfamily whose variable extracellular IgG domains bind different members of the FGF family (Johnson and Williams, 1993).

Unlike the neurotrophins, FGF1, FGF2, and FGF9 lack the typical N-terminal signal sequence for secretion. This lack raises the question of how these FGFs are presented extracellularly. Whereas fibroblasts infected to express the transgene for FGF2 in its native, cell-associated form show biological activity both *in vitro* (Ray *et al.*, 1995b) and *in vivo* (Peterson *et al.*, 1996; Takayama *et al.*, 1995), parallel experiments with secreted FGF2 containing a signal sequence (NGF pre–pro sequence) upstream of the FGF2 gene failed to have biological activity. These FGFs may be released by a novel secretory pathway or act through some other mechanism (Florkiewicz *et al.*, 1995; Mignatti *et al.*, 1992; Logan, 1990).

### 6.2. *In Vitro* Activity

FGFs are potent mitogens for a variety of mesodermal and neuroectodermal-derived cells. Most investigation in the CNS, however, has focused upon their neuroprotective influence, and nearly all of these studies have examined the function of FGF2 (but see the later section on adult neurogenesis). Survival of neurons was supported by addition of FGF2 to cultures of hippocampal (Matsuda *et al.*, 1990; Walicke 1988a; Walicke *et al.*, 1986), cerebral cortical (Matsuda *et al.*, 1990; Walicke 1988a; Morrison *et al.*, 1986), septal (Matsuda *et al.*, 1990; Grothe *et al.*, 1989; Walicke 1988a), striatal (Matsuda *et al.*, 1990; Walicke 1988a), hypothalamic (Ishikawa *et al.*, 1992), and dopaminergic ventral mesencephalic (Fawcett *et al.*, 1995; Silani *et al.*, 1994; Mayer *et al.*, 1993; Matsuda *et al.*, 1990) neurons from embryonic and early postnatal CNS. FGF2 action as a survival factor was determined to result directly from its effect on neurons, and was found not to be due to glial cell mediation (Unsicker *et al.*, 1992; Walicke and Baird, 1988). Subsequent studies determined that FGF2 served not only as a survival factor but was neuroprotective against glutamate or NMDA-mediated excitotoxic insult (Louis *et al.*, 1993; Skaper *et al.*, 1993; Freese *et al.*, 1992; Mattson and Rychlik, 1990), glucose deprivation (Cheng *et al.*, 1993; Louis *et al.*, 1993; Mattson *et al.*, 1993a; Maiese *et al.*, 1993; Cheng and Mattson, 1992), and oxidative stress (Enokido *et al.*, 1992). The neuroprotective function of FGF2 is hypothesized to be mediated through regulation of intracellular calcium levels and prevention of intracellular accumulation of radical oxygen species leading to excitotoxic cell death (Peterson *et al.*, 1996; Mattson and Cheng, 1993).

## 6.3. *In Vivo* Activity

FGFs play an important role in development (Yamaguchhi and Rossant, 1995). Targeted disruption of FGFR1 results in embryonic death between days seven and nine for homozygous animals (Yamaguchi *et al.*, 1994; Deng *et al.*, 1994). Spontaneous deletion of FGFR1 in humans results in one form of Pfeiffer's syndrome characterized by craniofacial and limb abnormalities (Muenke *et al.*, 1994). Increasing evidence suggests that FGF receptors may play their important role in development by providing intracellular signaling for extracellular adhesion molecules (Mason, 1994).

FGF2 has been detected throughout the adult CNS in astrocytes, occasionally in microglia, and in specific neuronal populations, including the cingulate cortex and the hippocampal CA2 subfield (Gonzalez *et al.*, 1995; Gómez-Pinilla *et al.*, 1992; Hanneken and Baird, 1992; Matsuyama *et al.*, 1992; Woodward *et al.*, 1992; Pettmann *et al.*, 1986). The high-affinity FGF receptors also show widespread distribution in the adult CNS (Fagan *et al.*, 1997; Peterson *et al.*, 1996; Gonzalez *et al.*, 1995; Yazaki *et al.*, 1994). Although earlier studies reported that neurons primarily expressed FGFR1 (*flg*) and that FGFR2 (*bek*) was primarily expressed by fiber tract glia presumed to be oligodendrocytes (Asai *et al.*, 1993; Wanaka *et al.*, 1990), it was determined later that FGFR1 is expressed by both neurons and astrocytes and that oligodendrocytes express FGFR1 only in myelin sheaths associated with nerve fibers (Gonzalez *et al.*, 1995). Expression of FGF2 and FGF receptors is increased in response to physical trauma (Logan *et al.*, 1992; Finklestein *et al.*, 1988), axotomizing lesion (Fagan *et al.*, 1997; Gómez-Pinilla *et al.*, 1992), ischemia (Liu and Chen, 1994; Takami *et al.*, 1992), and seizure activity (Gall *et al.*, 1994; Riva *et al.*, 1994; van der Wal *et al.*, 1994; Humpel *et al.*, 1993) where FGF2 expression was also reported in dentate granule neurons and hippocampal CA1 neurons (Bugra *et al.*, 1994). Changes in the expression of FGF2 are also associated with neurodegenerative disorders such as Alzheimer's (Cotman and Gómez-Pinilla, 1991; Stopa *et al.*, 1990), Parkinson's (Tooyama *et al.*, 1993a), and Huntington's diseases (Tooyama *et al.*, 1993b), although the role of FGF2 in these conditions remains unclear.

Based upon *in vitro* evidence for neuroprotection by FGF2 and the up-regulation of its expression following CNS injury, efforts have been made to supply FGF2 to experimental models of adult CNS injury. Initial *in vivo* studies utilized the previously established fimbria-fornix lesion model where NGF had been used and found that infused FGF2 could also protect axotomized septal neurons (Otto *et al.*, 1989; Anderson *et al.*, 1988). A similar neuroprotective effect of infused FGF2 was found for preganglionic spinal cord neurons following target removal by adrenomedullectomy (Blottner and Baumgarten, 1992a,b).

FGF2 has been tested in other subcortical structures using experimental models of ischemia and excitotoxicity. Neuroprotection by FGF2 following an ischemic episode has been reported for thalamus (Yamada *et al.*, 1991) and striatum (Nozaki *et al.*, 1993). Intraparenchymal infusion of a single dose of FGF2 into hippocampal subfield CA1 prior to ischemic damage failed to protect these neurons (Hara *et al.*, 1991). However, a later study using continuous intraventricular infusion of FGF2 beginning two hours before the ischemic episode and continuing for four days did find that FGF2 protected hippocampal CA1 neurons (Nakata *et al.*, 1993). Because of the proposed excitotoxic mechanism of neuronal death caused by acute ischemic episodes, FGF2 has been assessed for its effect on chemical lesions of the striatum, where it has also been shown to be neuroprotective (Kirschner *et al.*, 1995; Date *et al.*, 1993; Frim *et al.*, 1993; Nozaki *et al.*, 1993; Otto and Unsicker, 1990). FGF2 was effective even when administered systemically (Kirschner *et al.*, 1995; Nozaki *et al.*, 1993), suggesting its therapeutic administration as a possible treatment for stroke. Broad administration of FGF2 may lead to confounds, however, when interpreting its therapeutic value, due to undesired effects on other structures both within the CNS and peripherally (see section 8.3). Furthermore, the efficacy of FGF2 administration *in vivo* in the studies discussed above was typically analyzed at a simple level, assessing, for example, two-dimensional cell counts from single sections, macroscopic estimation of lesion area, or estimation of immunopositive fiber density. Future studies should undertake to evaluate neuroprotection based upon stereological estimates of total cell number to account for parametric changes in lesioned and trophic factor-treated tissue.

Administration of FGF2 has also been investigated for rescuing cortical neurons. Cummings *et al.* (1992) performed a knife-cut axotomizing the projection neurons of entorhinal cortex layer II and observed a percentage loss of these neurons compared to the contralateral side. The cell loss could be prevented by intraventricular infusion of FGF2.

Examination of growth factor receptors expressed by entorhinal layer II (ECL2) neurons revealed that, although these neurons expressed the FGFR1 receptor for FGF2, they did not express either the low-affinity p75 or high-affinity trkA receptors for NGF (Peterson *et al.*, 1996). The responsiveness of lesioned ECL2 neurons to trophic factor rescue was then assessed by placing grafts of fibroblasts that had been genetically engineered to express either NGF or FGF2 in the lesion cavity to provide a localized supply of these factors (see also Fig. 2). Neuronal survival was assessed using unbiased stereology (Peterson *et al.*, 1994 and 1996). As predicted from the absence of receptors, NGF failed to prevent the neuronal loss that follows perforant pathway lesion. However, animals that received grafts of the FGF2 expressing fibroblasts showed that the ECL2 neurons were rescued (Peterson *et al.*, 1996) (Fig. 5).

Although there is substantial cell loss in ECL2 after perforant pathway lesion, a subpopulation remains (Fig. 5). Though ECL2 consists primarily of glutamatergic projection neurons, we analyzed subpopulations of ECL2 to elucidate phenotypic differences that might distinguish those ECL2 neurons that survived lesion from those that were vulnerable to it. Layer II of the entorhinal cortex is characterized by a specific laminar distribution of the calcium-binding protein calbindin-$D_{28K}$ (Fig. 6). These calbindin-positive ECL2 neurons have extensive dendritic ramifications into layer I and have exons that descend through layer III and coalesce along layer IV. Back labeling with the hippocampal injections of the retrograde tracer fluorogold demonstrate that these calbindin-positive ECL2 neurons project to the hippocampus. Calbindin colocalizes with the glutamatergic projection neurons so that about half of the glutamatergic neurons coexpress calbindin (Figs. 7A, B, arrows) and about half do not (Figs. 7A, B, arrowheads). When neuronal loss of ECL2 neurons following lesion was determined, it was found that only the calbindin-negative glutamatergic neurons were vulnerable, whereas the population of calbindin-positive glutamatergic neurons remained unchanged (Fig. 7).

To elucidate a mechanism for the selective vulnerability of ECL2 neurons, further examination was undertaken, which revealed that, in addition to using glutamate as their neurotransmitter, these neurons received glutamatergic input as

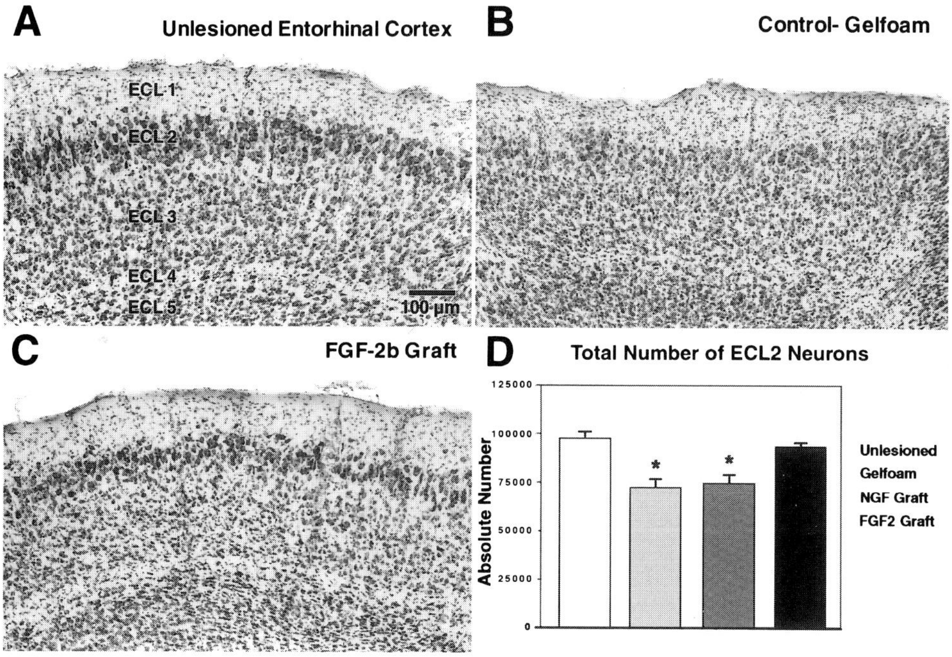


**Figure 5.** (A) Projection neurons in entorhinal cortex layer II (ECL2) form a distinct neuron lamina in the entorhinal cortex, with neuron-sparse molecular layer I (ECL1) above and the less densely packed layer III neurons (ECL3) below. (B) Following an aspirative lesion of the perforant pathway (see Fig. 2) the lesion cavity is packed only with Gelfoam and ELC2 neurons undergo degeneration. (C) A graft of FGF2-producing fibroblasts in the lesion cavity rescues the ECL2 neurons. (D) Quantitation using unbiased stereology confirms the significant ELC2 neuron loss post-lesion. It also confirms the ability of FGF2, but not NGF, to rescue these neurons. Reproduced from Peterson *et al.* (1996) with permission.

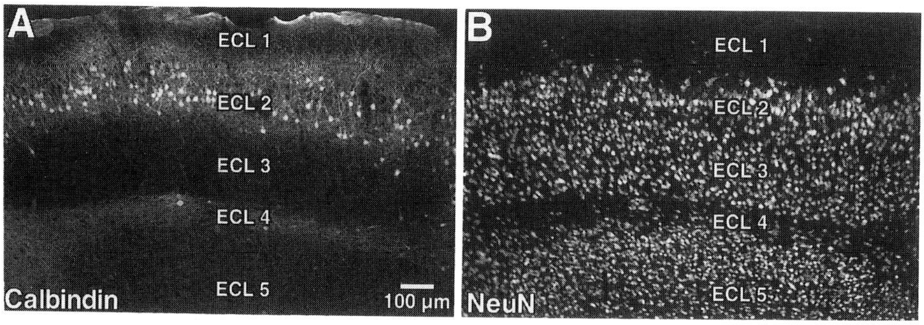


**Figure 6.** (A) Analysis of the immunofluorescent phenotype of entorhinal neurons reveals a specific layer II distribution of calbindin-positive cells. Comparison of the distribution of NeuN-immunoreactive neurons in the same section (B) demonstrates that calbindin-positive cells constituted a subset of ECL2 neurons. Reproduced from Peterson *et al.* (1996) with permission.

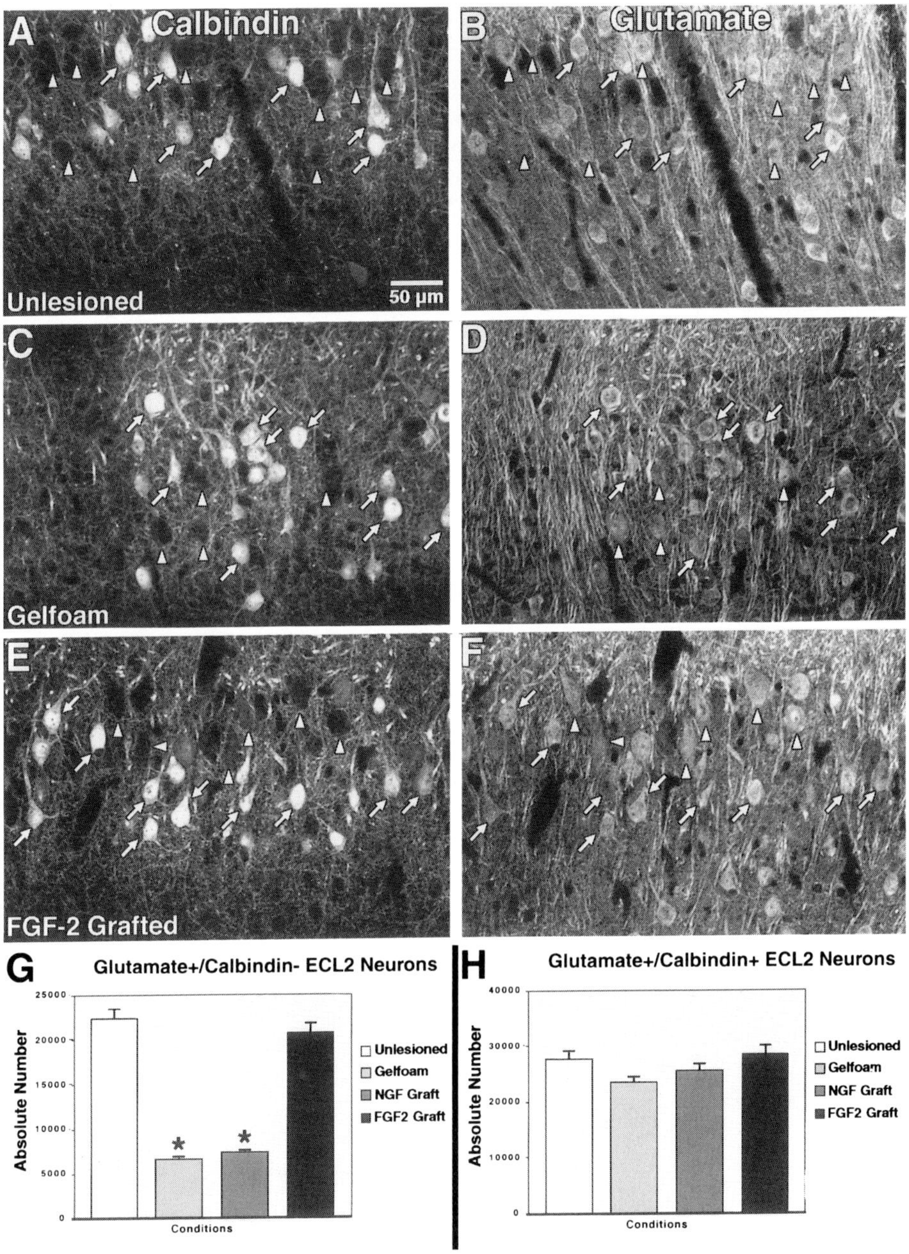


**Figure 7.** Examination of colocalization of calbindin-positive neurons in intact, unlesioned ECL2 (A) with the predominant neurotransmitter phenotype of ECL2 neurons, glutamate (B), demonstrates that calbindin-positive neurons were glutamatergic (arrows) but that some glutamatergic cells did not express calbindin (arrowheads). Analysis of phenotypes in perforant pathway lesioned animals (C, D) showed that the loss in ECL2 neurons occurred among the population of glutamatergic neurons that did not express calbindin (arrowheads), whereas the glutamatergic/calbindin-positive neurons (arrows) appeared unchanged. Animals grafted with FGF2-expressing fibroblasts (E, F) showed both glutamatergic/calbindin-positive neurons (arrows) and glutamatergic/calbindin-negative neurons (arrowheads) at levels comparable to the unlesioned animal (B, F). Stereological analysis of the glutamatergic/calbindin-negative population (G; arrowheads in above panels) as compared with the glutamatergic/calbindin-positive population (H; arrows in above panels) reveals that the calbindin-negative glutamatergic neurons in ECL2 are particularly vulnerable to the axotomizing lesion and sensitive to rescue by FGF2-, but not NGF-expressing grafts. From Peterson *et al.* (1996) with permission.

evidenced by their expression of NMDA receptors. Glutamate is the major effector of oxidative stress in the brain (see Geddes, this volume). This observation suggests a hypothesis whereby excessive stimulation of NMDA receptors by an acute, lesion-precipitated rise of extracellular glutamate (Graham *et al.*, 1990; Katayama *et al.*, 1990) could result in a massive influx of calcium. Those ECL2 neurons that coexpress calbindin are able to withstand this excitotoxic onslaught of calcium, whereas those that do not express calbindin are unable to buffer the excessive intracellular calcium and undergo oxidative stress culminating in their death (Iacopino *et al.*, 1992).

The ability of FGF2 to prevent loss of the vulnerable calbindin-negative population of ECL2 neurons implicates this growth factor in the prevention of excitotoxicity. One possible mechanism for this neuroprotective role has been suggested by Mattson *et al.* (1993b), where FGF2 was found to selectively suppress an NMDA receptor subunit, resulting in the inability of glutamate to generate calcium influx. Further work is needed to clarify whether interruption of the excitotoxic cascade is the mechanism by which FGF2 is neuroprotective for cortical neurons.

In addition to its neuroprotective role, there is evidence from *in vitro* studies that FGF2 promotes neurite outgrowth from central neurons, including cerebral cortex (Morrison *et al.*, 1986), cerebellum (Hatten *et al.*, 1988), and hippocampus (Lowenstein and Arsenault, 1996a; Mattson *et al.*, 1989; Walicke 1988b; Walicke *et al.*, 1986). Both FGF2 and FGF1 elicited neurite outgrowth when added to hypothalamic cultures, but FGF2 had more than 10 times the potency of FGF1 (Ishikawa *et al.*, 1992). In an elegant study, Woodhams and Atkinson (1996) demonstrated that regenerative outgrowth from slice cultures of entorhinal cortex could be elicited by FGF1. In contrast, we found only limited regenerative growth from the entorhinal cortex into FGF2 grafts *in vivo* after eight weeks (unpublished observation). In subcortical regions, other forms of FGF, such as FGF4 in the nigrostriatal system (Brecknell *et al.*, 1996) and FGF1 in the spinal cord (Cheng *et al.*, 1996), have been reported to elicit neurite outgrowth *in vivo* (see Grothe and Wewetzer, 1996, for a review of subcortical structures). In addition to regeneration, sprouting of intact cholinergic fibers into entorhinal projection zones of the dentate gyrus denervated by a perforant pathway lesion is also dependent upon FGF2 expression (Fagan *et al.*, 1997). The widespread expression of fibroblast growth factors and the mounting evidence of their biological effects on neuronal protection and stimulation of sprouting and regeneration suggest that the FGFs may hold potential for therapeutic administration in treatment of neuronal injury. Unfortunately, clinical trials evaluating systemic FGF2 adminstraton for neuroprotection against stroke have been terminated due to an unacceptable risk-to-benefit ratio (Scios and Wyeth-Ayerst Laboratories press release, July 21, 1998).

## 7. Trophic Factor Administration and Adult Neurogenesis

### 7.1. *In Vitro* Evidence

Although numerous attempts have been made to culture neurons from the adult CNS, most have had limited success. A number of investigators began studying the effect of growth factor administration upon cultures of adult CNS tissue and found that with high levels of FGF (Ray *et al.,* 1993; Richards *et al.,* 1992) or EGF (Reynolds and Weiss, 1992), a homogeneous population of cells emerged that could be expanded in culture under the continued presence of growth factors (see Ray *et al.,* 1997, for review). Although we originally found only limited neuronal differentiation within cultures from the adult hippocampus (Ray *et al.,* 1993), subsequent investigation (Palmer *et al.,* 1995) revealed that withdrawal of growth factors followed by the addition of forskolin, retinoic acid, or thyroid hormone led to a substantial increase in the number of neuronal phenotypes. Addition of the neurotrophins BDNF and NT3 to expanded adult cultures that were growth-arrested by withdrawal of FGF2 both increased neuronal frequency and promoted neurite outgrowth. These studies suggest that growth factors, though ineffective at rescuing adult central neurons in culture, are able to stimulate a population of neuronal progenitor cells resident in the adult CNS that, under appropriate conditions, can differentiate down a neuronal lineage.

### 7.2. *In Vivo* Evidence

Investigation into the location, kinetics, migration, and differentiation of neuronal precursors *in vivo* has revealed that the long-recognized germinal regions of the adult brain—the subventricular zone and the subgranular zone of the dentate gyrus—contribute to the neuronal populations of the olfactory bulb (Lois and Alvarez-Buylla, 1993; Luskin, 1993) and dentate gyrus (Kuhn *et al.,* 1996; Cameron *et al.,* 1993), respectively. Neurogenesis in the dentate gyrus is a regulated event, with increased neurogenesis following removal of excitatory input (Cameron *et al.,* 1995; Gould *et al.,* 1994) or environmental enrichment (Kempermann *et al.,* 1997), and decreased neurogenesis resulting from biological or psychological stress (Gould *et al.,* 1997; Cameron and Gould, 1994). Proliferation in the germinal zones can also be increased by intraventricular infusion of growth factors such as FGF2 (Kuhn *et al.,* 1997) (Fig. 8). Such proliferation in germinal zones may be an unanticipated side effect of system-wide delivery of trophic factors for therapeutic purposes.

These studies, in conjunction with the *in vitro* evidence of growth factor expansion and differentiation of cultured adult neuronal progenitors, suggest that it might be possible to recruit the endogenous adult neuronal progenitor population to replace losses in certain regions depleted of neurons following injury. Alternatively, it might be possible to graft expanded cultures of adult neuronal progenitors to the adult CNS to repopulate damaged regions. We reported that adult hippocampal progenitor cells, cultured long-term in the presence of FGF2 and expanded in culture, were able to integrate into the adult dentate gyrus following engraftment and to express a neuronal phenotype (Gage *et al.,* 1995). Furthermore, we recently reported that engraftment of the same long-term cultures of adult hippocampal progenitors was able to integrate into the olfactory bulb and differentiate into neurons, including TH-containing neurons, a phenotype not found in the adult hippocampus (Suhonen *et al.,* 1996). These studies suggest that the adult CNS retains adequate differentiation cues to enable growth-factor-stimulated neuronal progenitors to follow a neuronal lineage. Further studies will be

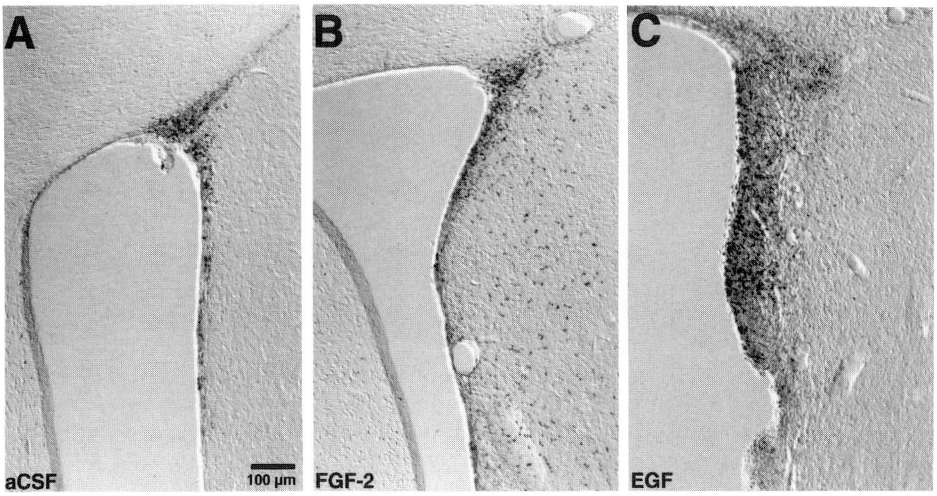


**Figure 8.** Widespread distribution of growth factors in adult CNS to protect specific neuronal populations may have unanticipated effects. Animals receiving intraventricular infusion of artificial cerebrospinal fluid (A; aCSF), FGF2 (B), or epidermal growth factor (C; EGF) were injected with the thymidine analog bromodeoxyuridine (BrdU) to label newly generated cells. Growth factor infusion stimulated cell proliferation (labeled with antibody against BrdU) in the subventricular zone of the anterior lateral ventricle walls, which contribute to adult neurogenesis. See Kuhn *et al.* (1997) for details.

necessary to explore the extent to which the recruitment or engraftment of adult neuronal progenitors may provide a complementary therapeutic approach to trophic-factor-mediated neuronal rescue and regenerative outgrowth promotion.

## 8. Conclusions

The biological activity of trophic substances has been intensely investigated in recent years, resulting in the increase of identified trophic factors and a greater understanding of the mechanisms of activity of previously identified factors. Although the adult CNS is highly vulnerable to injury, experimental models administering trophic factors suggest that they may have therapeutic potential. Despite the promise of the experimental studies described above, pitfalls in the analysis of *in vivo* material can contribute to unsubstantiated conclusions about the effects of trophic factor administration. Three interpretational hazards will be discussed: reliability of phenotypic expression, technical considerations in quantification of tissue sections, and possible side effects of broad trophic factor administration with relevance to therapeutic use of trophic factors.

### 8.1. Reliability of Phenotypic Expression

Trophic factor administration on central neuron survival *in vitro* and *in vivo* is usually assessed on the number of neurons expressing the characteristic phenotype of the neuronal population being investigated. For example, the success or lack of success of NGF administration in promoting septal neuronal survival is determined from the number of cholinergic neurons observed with and without NGF administration. Implicit in this interpretation is that phenotypic expression—the cholinergic phenotype, in this example—is equivalent to neuronal survival. The possibility also exists that NGF may rescue a greater number of neurons that simply fail to express the cholinergic phenotype. For instance, in the intact basal forebrain, there is a high level of correspondence between the expression of ChAT and p75 (Batchelor *et al.,* 1989). However, the expression of p75, but not ChAT, is reported to increase in the lesioned basal forebrain following administration of BDNF (Widmer *et al.,* 1993). Administration of CNFT reportedly increased the number of p75-expressing neurons above the unlesioned control level (Kew and Sofroniew, 1995; Hagg *et al.,* 1992). Similarly, administration of NT4/5 reportedly increased calbindin expression above control levels in the substantia nigra (Alexi and Hefti, 1996).

The possible error associated with the assumption that phenotypic expression equals survival is accentuated in the case of a presumed negative effect. That a trophic factor apparently fails to rescue a neuronal population based upon the absence of phenotypic expression may reflect simply a temporal absence of expression. Following lesion, ChAT expression in basal forebrain neurons can be lost preceeding neuronal death (Fischer and Björklund, 1991; O'Brien *et al.,* 1990) or even without neuronal death (Butterworth and Dragunow, 1996; Sofroniew *et al.,* 1993; Naumann *et al.,* 1992). It is not clear if the studies reporting a discordance between p75 and ChAT expression reflect an enhanced survival of a p75-positive/ChAT-negative population, the induction of p75 expression in a novel population of neurons, a temporal dissociation between expression of p75 and ChAT such that ChAT expression is lost prior to p75 expression, or a temporal dissociation with the failure of p75-positive neurons to express ChAT in the temporal interval examined. Although prelabeling projection neurons with dye prior to lesion is helpful in addressing this question (O'Brien *et al.,* 1990), the strength of the conclusions is limited by the inability to control for the detectable uptake of dye by all projection neurons.

## 8.2. Caveats Regarding Quantification of Tissue Sections

Because of the problems with accurate determination of cell markers discussed above, the only way to address the problem of phenotype expression and neuronal survival is to make a quantitative assessment based upon absolute numbers. This provides a reference against which to measure relative changes in detection of a marker. In tissue culture studies using monolayer cultures, the estimation of absolute numbers is relatively straightforward using a counterstain or other morphological criteria. The subset of the vessel's culture surface is sampled in a systematic random fashion, and the numerical results are expressed as an area fraction (mean number per area of the counting frame). Determination of the entire culture surface area can then be obtained by a geometric calculation using the radius (if round) or the length and width (if rectangular) or using the surface area value provided by the manufacturer. Multiplication of the area fraction by the total area yields an estimate of the total number.

In the case of *in vivo* studies of trophic factor effect on neuronal survival, the estimation of absolute number, though more complex, is crucial for a correct interpretation of the results. The need to estimate absolute numbers cannot be overemphasized in these studies, because there is usually some physical perturbation to the system to provoke the injury against which the trophic factor effect is to be measured. Depending upon the type of lesion used, there will be dimensional changes in the tissue region being assessed. These changes might manifest themselves as swelling or shrinkage, or as a differential effect of both, depending upon the surrounding physical boundaries of fiber tracts, ventricles, etc. The effect of dimensional changes upon two-dimensional estimates (number per area) can be drastic. For example, if there was considerable edema in the area being counted on the lesioned side compared to the equivalent area on the intact side, the increased interstitial space on the lesioned side would produce a lower number per unit area, when in fact the total number corrected for swelling may be equivalent to the intact side. Such a result might be incorrectly interpreted as there being no trophic factor effect even though the trophic factor may have rescued all neurons. (Of course, this presumes also an accurate estimate of neuronal loss caused by lesion against which to assess the effect of trophic factor administration!)

Because it is exceptionally difficult to model these dimensional changes, the efficient alternative is to apply unbiased stereology. Although a discussion of this technical approach is beyond the scope of this review (see Coggeshall and Lekan, 1996, and West, 1993, for reviews; see Peterson *et al.*, 1994, 1996, and 1997, and West, this volume, for examples), the approach of unbiased stereology can be summarized as making three-dimensional cell counts throughout the entire region, yielding an estimate of total cell number. The ability to produce estimates of total number means that the true extent of lesion effect and trophic factor administration can be assessed for each phenotype and for total neuronal number within the defined region.

## 8.3. Broad Effects of Trophic Factor Administration

The intervention of trophic factors in the processes of neuronal death and atrophy in the experimental models reviewed above gives hope that trophic factors may prove to be appropriate therapeutic agents for CNS injury and disease in patients. A primary difficulty in extrapolating the results from experimental studies to clinical trials is in determining the optimal method for delivery of the trophic factor. The permeability of the blood–brain barrier prevents or reduces the availability of many peripherally administered trophic factors in the CNS at an effective concentration. As a result, experimental studies have relied on the direct delivery strategies discussed previously (section 2.2). However, direct delivery strategies may produce an unacceptable degree of invasiveness for therapeutic purposes and may cause collateral injury to adjacent brain structures that can be tolerated in an experimental animal study but not in a patient. As a result, most clinical trials of trophic factors have utilized system-wide administration using either peripheral injection or delivery to the cerebrospinal fluid.

A danger of system-wide administration is development of side effects from trophic factor activity outside the CNS, such as stimulation of the peripheral nervous system leading to hyperalgesia, fever, weight loss, angiogenesis, and so on. In addition, the distribution of trophic factors within the CNS is not limited, and attention also must be given to the effect of chronic trophic factor administration on areas of the CNS outside of the region of intended action. For example, NGF administration has been proposed as a therapeutic strategy to stimulate cholinergic neurons in Alzheimer's disease patients, but in a recent study examining the effect of chronic intracerebroventricular infusion of NGF in rats, Winkler *et al.* (1997) reported the development of a substantial subpial Schwann cell hyperplasia with sprouting of sensory and sympathetic neurites in the medulla oblongata and spinal cord, which disappeared over time following cessation of NGF administration. This NGF-induced hyperplasia argues for the elucidation of more appropriate spatial and temporal parameters for delivery of NGF.

Trophic factors intended to stimulate or rescue a certain population of CNS neurons may not only act on CNS regions other than those anticipated, but their effect upon the CNS may also be different than intended. Delivery of FGF2 has been found to promote the survival of central neurons (Peterson *et al.*, 1996; Anderson *et al.*, 1988) and minimize CNS injury following an ischemic episode (Kirschner *et al.*, 1995; Nozaki *et al.*, 1993). Such evidence has prompted interest in clinical trials of FGF2 administration to protect against neuronal death; however, FGF2 also stimulates neural progenitor cells in the adult CNS (Ray *et al.*, 1993; see section 7). Intracerebroventricular infusion of FGF2 stimulated the proliferation of endogenous neural progenitor cells (Fig. 8) (Kuhn *et al.*, 1997), as did peripheral injection (Tao *et al.*, 1996). These studies suggest that delivery of trophic factors be carefully limited to minimize unanticipated or undesired effects of the trophic factors on other CNS regions or cell types that might result from system-wide administration.

Exogenous administration of naturally produced trophic factors in experimental models of adult CNS injury reveals that neurons can be protected when receiving supplemental amounts of trophic factors. Trophic factors are appropriate for a given population of neurons when those neurons are chronically exposed to it in the adult, such as septal projection neurons and NGF (Hefti *et al.*, 1986; Williams *et al.*, 1986), or when the factor is locally elevated in response to injury, such as cortical neurons and FGF2 (Fagan *et al.*, 1997; Peterson *et al.*, 1996). Within a given population of neurons, the trophic influence is evident in receptor-expressing subpopulations that either up-regulate their phenotypic expression in response to the trophic factor (Alexi and Hefti, 1996; Koliatsos *et al.*, 1994) or are selectively vulnerable to the injury (Peterson *et al.*, 1996). The successful rescue and stimulation of injured central neurons in experimental animal models provides a rationale for initiation of clinical trials for some trophic factors to determine their therapeutic value in patients.

ACKNOWLEDGMENTS. The authors thank J. Ray and M. L. Gage for helpful comments, and H. G. Kuhn and U. Blömer for access to unpublished material. The work conducted in the authors' laboratory was supported by NINDS P01-NS28121 and by a T. L. L. Temple Foundation/Alzheimer's Association Award and an American Federation for Aging Research Grant to D. A. P.

## 9. References

Aebischer, P., Schluep, M., Deglon, N., Joseph, J. M., Hirt, L., Heyd, B., Goddard, M., Hammang, J. P., Zurn, A. D., Kato, A. C., Regli, F., and Baetge, E. E., 1996, Intrathecal delivery of CNTF using encapsulated genetically modified xenogeneic cells in amyotrophic lateral sclerosis patients, *Nature Med.* **2:**696–699.

Alderson, R. F., Wiegand, S. J., Anderson, K. D., Cai, N., Cho, J. Y., Lindsay, R. M., and Altar, C. A., 1996, Neurotrophin-4/5 maintains the cholinergic phenotype of axotomized septal neurons, *Eur. J. Neurosci.* **8:**282–290.

Alexi, T., and Hefti, F., 1996, Neurotrophin-4/5 selectively protects nigral calbindin-containing neurons in rats with medial forebrain bundle transections, *Neuroscience* **72:**911–921.

Alexi, T., Venero, J. L., and Hefti, F., 1997, Protective effects of neurotrophin-4/5 and transforming growth factor-alpha on striatal neuronal phenotypic degeneration after excitotoxic lesioning with quinolinic acid, *Neuroscience* **78:**73–86.

Allen, L. E., and Maher, P. A., 1993, Expression of basic fibroblast growth factor and its receptor in an invasive bladder carcinoma cell line, *J. Cell Physiol.* **155:**368–375.

ALS CNTF Treatment Study (ACTS) Phase I–II Study Group, 1995a, The pharmacokinetics of subcutaneously administered recombinant human ciliary neurotrophic factor (rHCNTF) in patients with amyotrophic lateral sclerosis: Relation to parameters of the acute-phase response, *Clin. Neuropharmacol.* **18:**500–514.

ALS CNTF Treatment Study (ACTS) Phase I–II Study Group, 1995b, A phase I study of recombinant human ciliary neurotrophic factor (rHCNTF) in patients with amyotrophic lateral sclerosis, *Clin. Neuropharmacol.* **18:**515–532.

ALS CNTF Treatment Study Group, 1996, A double-blind placebo-controlled clinical trial of subcutaneous recombinant human ciliary neurotrophic factor (rHCNTF) in amyotrophic lateral sclerosis, *Neurology* **46:**1244–1249.

Anderson, K. J., Dam, D., Lee, S., and Cotman, C. W., 1988, Basic fibroblast growth factor prevents death of lesioned cholinergic neurons *in vivo, Nature* **322:**360–361.

Anderson, K. D., Panayotatos, N., Corcoran, T. L., Lindsay, R. M., and Wiegand, S. J., 1996, Ciliary neurotrophic factor protects striatal output neurons in an animal model of Huntington's disease, *Proc. Natl. Acad. Sci. USA* **93:**7346–7351.

Arenas, E., and Akerud, P., Wong, V., Boylan, C., Persson, H., Lindsay, R. M., and Altar, C. A., 1996, Effects of BDNF and NT-4/5 on striatonigral neuropeptides or nigral GABA neurons *in vivo, Eur. J. Neurosci.* **8:**1707–1717.

Arimatsu, Y., Miyamoto, M., Tsukui, H., and Hatanaka, H., 1989, Nerve growth factor promotes survival of retrogradely labeled hippocampus-projecting neurons in the rat basal forebrain *in vitro, Dev. Brain Res.* **45:**297–301.

Asada, H., Ip, N. Y., Pan, L., Razack, N., Parfitt, M. M., and Plunkett, R. J., 1995, Time course of ciliary neurotrophic factor mRNA expression is coincident with the presence of protoplasmic astrocytes in traumatized rat striatum, *J. Neurosci. Res.* **40:**22–30.

Asai, T., Wanaka, A., Kato, H., Masana, Y., Seo, M., and Tohyama, M., 1993, Differential expression of two members of FGF receptor gene family, FGFR-1 and FGFR-2 mRNA, in the adult rat central nervous system, *Mol. Brain Res.* **17:**174–178.

Baird, A., 1994, Fibroblast growth factors: Activities and significance of non-neurotrophin neurotrophic growth factors, *Curr. Opin. Neurobiol.* **4:**78–86.

Ballarin, M., Ernfors, P., Lindefors, N., and Persson, H., 1991, Hippocampal damage and kainic acid injection induce a rapid increase in mRNA for BDNF and NGF in the rat brain, *Exp. Neurol.* **114:** 35–43.

Barbin, G., Manthorpe, M., and Varon, S., 1984, Purification of the chick eye ciliary neuronotrophic factor, *J. Neurochem.* **43:**1468–1478.

Barde, Y. A., Lindsay, R. M., Monard, D., and Thoenen, H., 1978, New factor released by cultured glioma cells supporting survival and growth of sensory neurones, *Nature* **274:**818.

Barrett, G. L., and Georgiou, A., 1996, The low-affinity nerve growth factor receptor p75(NGFr) mediates death of PC12 cells after nerve growth factor withdrawal, *J. Neurosci. Res.* **45:**117–128.

Batchelor, P. E., Armstrong, D. M., Blaker, S. N., and Gage, F. H., 1989, Nerve growth factor receptor and choline acetyltransferase colocalization in neurons within the rat forebrain: Response to fimbria-fornix transection, *J. Comp. Neurol.* **284:**187–204.

Berkemeier, L. R., Winslow, J. W., Kaplan, D. R., Nikolics, K., Goeddel, D. V., and Rosenthal, A., 1991, Neurotrophin-5: A novel neurotrophic factor that activates trk and trkB, *Neuron* **7:**857–866.

Berninger, B., and Poo, M. M., 1996, Fast actions of neurotrophic factors, *Curr. Opin. Neurobiol.* **6:**324–330.

Björklund, A., 1991, Neural transplantation–an experimental tool with clinical possibilities, *Trends Neurosci.* **14:**319–322.

Blömer, U., Naldini, L., Kafri, T., Trono, D., Verma, I. M., and Gage, F. H., 1997, Highly efficient and sustained gene transfer in adult neurons with a lentiviral vector, *J. Virol.* **71:**6641–6649.

Blottner, D., and Baumgarten, H. G., 1992a, Basic fibroblast growth factor prevents neuronal death and atrophy of retrogradely labeled preganglionic neurons *in vivo, Exp. Neurol.* **118:**35–46.

Blottner, D., and Baumgarten, H. G., 1992b, Insulin-like growth factor-I counteracts bFGF-induced survival of nitric oxide synthase (NOS)-positive spinal cord neurons after target-lesion *in vivo, J. Neurosci. Res.* **32:**471–480.

Bohlen, P., Esch, F., Baird, A., and Gospodarowicz, D., 1985a, Acidic fibroblast growth factor (FGF) from bovine brain: Amino-terminal sequence and comparison with basic FGF, *EMBO J.* **4:**1951–1956.

Bohlen, P., Esch, F., Baird, A., Jones, K. L., and Gospodarowicz, D., 1985b, Human brain fibroblast growth factor: Isolation and partial chemical characterization, *FEBS Lett.* **185:**177–781.

Bothwell, M., 1995, Functional interactions of neurotrophins and neurotrophin receptors, *Annu. Rev. Neurosci.* **18:**223–253.

Brecknell, J. E., Du, J. S., Muir, E., Fidler, P. S., Hlavin, M. L., Dunnett, S. B., and Fawcett, J. W., 1996, Bridge grafts of fibroblast growth factor-4-secreting schwannoma cells promote functional axonal regeneration in the nigrostriatal pathway of the adult rat, *Neuroscience* **74:**775–784.

Brooks, A. I., Muhkerjee, B., Panahian, N., Cory-Slechta, D., and Federoff, H. J., 1997, Nerve growth factor somatic mosaicism produced by herpes virus-directed expression of *cre* recombinase, *Nature Biotechnol.* **15:**57–62.

Bugra, K., Pollard, H., Charton, G., Moreau, J., Ben-Ari, Y., and Khrestchatisky, M., 1994, aFGF, bFGF, and *flg* mRNAs show distinct patterns of induction in the hippocampus following kainate-induced seizures, *Eur. J. Neurosci.* **6:**58–66.

Butterworth, N. J., and Dragunow, M., 1996, Medial septal cholinergic neurons express c-jun but do not undergo DNA fragmentation after fornix-fimbria transections, *Mol. Brain Res.* **43:**1–12.

Cajal, S., Ramon y, 1928, *Degeneration and Regeneration of the Nervous System,* Hafner, New York.

Cameron, H. A., and Gould, E., 1994, Adult neurogenesis is regulated by adrenal steroids in the dentate gyrus, *Neuroscience* **61:**203–209.

Cameron, H. A., McEwen, B. S., and Gould, E., 1995, Regulation of adult neurogenesis by excitatory input and NMDA receptor activation in the dentate gyrus, *J. Neurosci.* **15:**4687–4692.

Cameron, H. A., Woolley, C. S., McEwen, B. S., and Gould, E., 1993, Differentiation of newly born neurons and glia in the dentate gyrus of the adult rat, *Neuroscience* **56:**337–344.

Casaccia-Bonnefil, P., Carter, B. D., Dobrowsky, R. T., and Chao, M. V., 1996, Death of oligodendrocytes mediated by the interaction of nerve growth factor with its receptor p75, *Nature* **383:**716–719.

Chalmers, G. R., Peterson, D. A., and Gage, F. H., 1996, Sprouting adult CNS cholinergic axons express NILE and associate with astrocytic surfaces expressing neural cell adhesion molecule, *J. Comp. Neurol.* **371:**287–299.

Chan, K. M., Lam, D. T., Pong, K., Widmer, H. R., and Hefti, F., 1996, Neurotrophin-4/5 treatment reduces infarct size in rats with middle cerebral artery occlusion, *Neurochem. Res.* **21:**763–767.

Cheng, B., and Mattson, M. P., 1992, Glucose deprivation elicits neurofibrillary tangle-like antigenic changes in hippocampal neurons: Prevention by NGF and bFGF, *Exp. Neurol.* **117:**114–123.

Cheng, B., and Mattson, M. P., 1994, NT-3 and BDNF protect CNS neurons against metabolic/excitotoxic insults, *Brain Res.* **640:**56–67.

Cheng, B., McMahon, D. G., and Mattson, M. P., 1993, Modulation of calcium current, intracellular calcium levels, and cell survival by glucose deprivation and growth factors in hippocampal neurons, *Brain Res.* **607:**275–285.

Cheng, H., Cao, Y., and Olson, L., 1996, Spinal cord repair in adult paraplegic rats: Partial restoration of hind limb function, *Science* **273:**510–513.

Clarke, D. J., Gage, F. H., Nilsson, O. G., and Björklund, A., 1986, Grafted septal neurons form cholinergic synaptic connections in the dentate gyrus of behaviorally impaired aged rats, *J. Comp. Neurol.* **252:**483–492.

Clatterbuck, R. E., Price, D. L., and Koliatsos, V. E., 1993, Ciliary neurotrophic factor prevents retrograde neuronal death in the adult central nervous system, *Proc. Natl. Acad. Sci. USA* **90:**2222–2226.

Clatterbuck, R. E., Price, D. L., and Koliatsos, V. E., 1994, Further characterization of the effects of brain-derived neurotrophic factor and ciliary neurotrophic factor on axotomized neonatal and adult mammalian motor neurons, *J. Comp. Neurol.* **342:**45–56.

Clatterbuck, R. E., Price, D. L., and Koliatsos, V. E., 1996, Ciliary neurotrophic factor stimulates the expression of glial fibrillary acidic protein by brain astrocytes *in vivo*, *J. Comp. Neurol.* **369:**543–551.

Coggeshall, R. E., and Lekan, H. A., 1996, Methods for determining numbers of cells and synapses: A case for more uniform standards of review, *J. Comp. Neurol.* **364:**6–15.

Collins, F., and Crutcher, K. A., 1985, Neurotrophic activity in the adult rat hippocampal formation: Regional distribution and increase after septal lesion, *J. Neurosci.* **5:**2809–2814.

Conner, J. M., Lauterborn, J. C., Yan, Q., Gall, C. M., and Varon, S., 1997, Distribution of brain-derived neurotrophic factor (BDNF) protein and mRNA in the normal adult rat CNS: Evidence for anterograde axonal transport, *J. Neurosci.* **17:**2295–2313.

Cordon-Cardo, C., Tapley, P., Jing, S. Q., Nanduri, V., O'Rourke, E., Lamballe, F., Kovary, K., Klein, R., Jones, K. R., Reichardt, L. F., and Barbacid, M., 1991, The trk tyrosine protein kinase mediates the mitogenic properties of nerve growth factor and neurotrophin-3, *Cell* **66:**173–183.

Cotman, C. W., and Gomez-Pinilla, F., 1991, Basic fibroblast growth factor in the mature brain and its possible role in Alzheimer's disease, *Ann. N.Y. Acad. Sci.* **638:**221–231.

Cummings, B. J., Yee, G. J., and Cotman, C. W., 1992, bFGF promotes the survival of entorhinal layer-II neurons after perforant path axotomy, *Brain Res.* **591:**271–276.

Dale, S. M., Kuang, R. Z., Wei, X., and Varon, S., 1995, Corticospinal motor neurons in the adult rat: Degeneration after intracortical axotomy and protection by ciliary neurotrophic factor (CNTF), *Exp. Neurol.* **135:**67–73.

Date, I., Yoshimoto, Y., Imaoka, T., Miyoshi, Y., Gohda, Y., Furuta, T., Asari, S., and Ohmoto, T., 1993, Enhanced recovery of the nigrostriatal dopaminergic system in MPTP-treated mice following intrastriatal injection of basic fibroblast growth factor in relation to aging, *Brain Res.* **621:**150–154.

Davies, A. M., Thoenen, H., and Barde, Y. A., 1986, The response of chick sensory neurons to brain-derived neurotrophic factor, *J. Neurosci.* **6:**1897–1904.

Davis, S., Aldrich, T. H., Valenzuela, D. M., Wong, V. V., Furth, M. E., Squinto, S. P., and Yancopoulos, G. D., 1991, The receptor for ciliary neurotrophic factor, *Science* **253:**59–63.

Deng, C. X., Wynshaw-Boris, A., Shen, M. M., Daugherty, C., Ornitz, D. M., and Leder, P., 1994, Murine FGFR-1 is required for early postimplantation growth and axial organization, *Genes Dev.* **8:**3045–3057.

DiStefano, P. S., Friedman, B., Radziejewski, C., Alexander, C., Boland, P., Schick, C. M., Lindsay, R. M., and Wiegand, S. J., 1992, The neurotrophins BDNF, NT-3, and NGF display distinct patterns of retrograde axonal transport in peripheral and central neurons, *Neuron* **8:**983–993.

Dobrea, G. M., Unnerstall, J. R., and Rao, M. S., 1992, The expression of CNTF message and immunoreactivity in the central and peripheral nervous system of the rat, *Dev. Brain Res.* **66:**209–219.

Eagle, K. S., Chalmers, G. R., Clary, D. O., and Gage, F. H., 1995, Axonal regeneration and limited functional recovery following hippocampal deafferentation, *J. Comp. Neurol.* **363:**377–388.

Emerich, D. F., Lindner, M. D., Winn, S. R., Chen, E. Y., Frydel, B. R., and Kordower, J. H., 1996, Implants of encapsulated human CNTF-producing fibroblasts prevent behavioral deficits and striatal degeneration in a rodent model of Huntington's disease, *J. Neurosci.* **16:**5168–5181.

Emerich, D. F., Winn, S. R., Hantraye, P. M., Peschanski, M., Chen, E. Y., Chu, Y., McDermott, P., Baetge, E. E., and Kordower, J. H., 1997, Protective effect of encapsulated cells producing neurotrophic factor CNTF in a monkey model of Huntington's disease, *Nature* **386:**395–399.

Enokido, Y., Akaneya, Y., Niinobe, M., Mikoshiba, K., and Hatanaka, H., 1992, Basic fibroblast growth factor rescues CNS neurons from cell death caused by high oxygen atmosphere in culture, *Brain Res.* **599:**261–271.

Ernfors, P., Ibáñez, C. F., Ebendal, T., Olson, L., and Persson, H., 1990a, Molecular cloning and neurotrophic activities of a protein with structural similarities to nerve growth factor: Developmental and topographical expression in the brain, *Proc. Natl. Acad. Sci. USA* **87:**5454–5458.

Ernfors, P., Wetmore, C., Olson, L., and Persson, H., 1990b, Identification of cells in rat brain and peripheral tissues expressing mRNA for members of the nerve growth factor family, *Neuron* **5:**511–526.

Ernfors, P., Bengzon, J., Kokaia, Z., Persson, H., and Lindvall, O., 1991, Increased levels of messenger RNAs for neurotrophic factors in the brain during kindling epileptogenesis, *Neuron* **7:**165–176.

Ernfors, P., Lee, K. F., and Jaenisch, R., 1994, Mice lacking brain-derived neurotrophic factor develop with sensory deficits, *Nature* **368:**147–150.

Esch, F., Baird, A., Ling, N., Ueno, N., Hill, F., Denoroy, L., Klepper, R., Gospodarowicz, D., Bohlen, P., and Guillemin, R., 1985, Primary structure of bovine pituitary basic fibroblast growth factor (FGF) and comparison with the amino-terminal sequence of bovine brain acidic FGF, *Proc. Natl. Acad. Sci. USA* **82:**6507–6511.

Fagan, A. M., Suhr, S. T., Lucidi-Phillipi, C. A., Peterson, D. A., Holtzman, D. M., and Gage, F. H., 1997, Endogenous FGF-2 is important for cholinergic sprouting in the denervated hippocampus, *J. Neurosci.* **17:**2499–2511.

Falkenberg, T., Lindefors, N., Camilli, F., Metsis, M., and Ungerstedt, U., 1996, Glutamate release correlates with brain-derived neurotrophic factor and trkB mRNA expression in the CA1 region of rat hippocampus, *Mol. Brain Res.* **42:**317–327.

Fariñas, I., Jones, K. R., Backus, C., Wang, X. Y., and Reichardt, L. F., 1994, Severe sensory and sympathetic deficits in mice lacking neurotrophin-3, *Nature* **369:**658–661.

Fawcett, J. W., Barker, R. A., and Dunnett, S. B., 1995, Dopaminergic neuronal survival and the effects of bFGF in explant, three dimensional, and monolayer cultures of embryonic rat ventral mesencephalon, *Exp. Brain Res.* **106:**275–282.

Finklestein, S. P., Apostolides, P. J., Caday, C. G., Prooser, J., Philips, M. F., and Flagsbrun, M. 1988, Increased basic fibroblast growth factor (bFGF) immunoreactivity at the site of focal brain wounds, *Brain Res.* **460:**253–259.

Fischer, W., and Björklund, A., 1991, Loss of AChE- and NGFr-labeling precedes neuronal death of axotomized septal-diagonal band neurons: Reversal by intraventricular NGF infusion, *Exp. Neurol.* **113:**93–108.

Florkiewicz, R. Z., Majack, R. A., Buechler, R. D., and Florkiewicz, E., 1995, Quantitative export of FGF-2 occurs through an alternative, energy-dependent, non-ER/Golgi pathway, *J. Cell Physiol.* **162:**388–399.

Frade, J. M., Rodriguez-Tebar, A., and Barde, Y. A., 1996, Induction of cell death by endogenous nerve growth factor through its p75 receptor, *Nature* **383:**166–168.

Freese, A., Finklestein, S. P., and DiFiglia, M., 1992, Basic fibroblast growth factor protects striatal neurons *in vitro* from NMDA-receptor mediated excitotoxicity, *Brain Res.* **575:**351–355.

Friedman, B., Kleinfeld, D., Ip, N. Y., Verge, V. M., Moulton, R., Boland, P., Zlotchenko, E., Lindsay, R. M., and Liu, L., 1995, BDNF and NT-4/5 exert neurotrophic influences on injured adult spinal motor neurons, *J. Neurosci.* **15:**1044–1056.

Frim, D. M., Uhler, T. A., Short, M. P., Ezzedine, Z. D., Klagsbrun, M., Breakefield, X. O., and Isacson, O., 1993, Effects of biologically delivered NGF, BDNF, and bFGF on striatal excitotoxic lesions, *NeuroReport* **4:**367–370.

Fusco, M., Oderfeld-Nowak, B., Vantini, G., Schiavo, N., Gradkowska, M., Zaremba, M., and Leon, A., 1989, Nerve growth factor affects uninjured, adult rat septohippocampal cholinergic neurons, *Neuroscience* **33:**47–52.

Gage, F. H., and Björklund, A., 1987, Trophic and growth-regulating mechanisms in the central nervous system monitored by intracerebral neural transplants, in: *Selective Neuronal Death* (G. Bock and M. O'Conner, eds.), John Wiley and Sons, Chichester, England, pp. 143–159.

Gage, F. H., Dunnett, S. B., Stenevi, U., and Björklund, A., 1983, Aged rats: Recovery of motor impairments by intrastriatal nigral grafts, *Science* **221:**966–969.

Gage, F. H., Björklund, A., Stenevi, U., Dunnett, S. B., and Kelly, P. A. T., 1984, Intrahippocampal septal grafts ameliorate learning impairments in aged rats, *Science* **225:**533–536.

Gage, F. H., Wictorin, K., Fischer, W., Williams, L. R., Varon, S., and Björklund, A., 1986, Retrograde cell changes in medial septum and diagonal band following fimbria-fornix transection: Quantitative temporal analysis, *Neuroscience* **19:**241–255.

Gage, F. H., Wolff, J. A., Rosenberg, M. B., Xu, L., Yee, J. K., Shults, C., and Friedmann, T., 1987, Grafting genetically modified cells to the brain: Possibilities for the future, *Neuroscience* **23:**795–807.

Gage, F. H., Batchelor, P., Chen, K. S., Chin, D., Higgins, G. A., Koh, S., Deputy, S., Rosenberg, M. B., Fischer, W., and Björklund, A., 1989, NGF receptor reexpression and NGF-mediated cholinergic neuronal hypertrophy in the damaged adult neostriatum, *Neuron* **2:**1177–1184.

Gage, F. H., Kawaja, M. D., and Fisher, L. J., 1991, Genetically modified cells: Applications for intracerebral grafting, *Trends Neurosci.* **14:**328–333.

Gage, F. H., Coates, P. W., Palmer, T. D., Kuhn, H. G., Fisher, L. J., Suhonen, J. O., Peterson, D. A., Suhr, S. T., and Ray, J., 1995, Survival and differentiation of adult neuronal progenitor cells transplanted to the adult brain, *Proc. Natl. Acad. Sci. USA* **92:**11879–11883.

Gähwiler, B. H., Rietschin, L., Knopfel, T., and Enz, A., 1990, Continuous presence of nerve growth factor is required for maintenance of cholinergic septal neurons in organotypic slice cultures, *Neuroscience* **36:**27–31.

Gall, C. M., Gold, S. J., Isackson, P. J., and Seroogy, K. B., 1992, Brain-derived neurotrophic factor and neurotrophin-3 mRNAs are expressed in ventral midbrain regions containing dopaminergic neurons, *Mol. Cell. Neurosci.* **3**:56–63.

Gall, C. M., Berschauer, R., and Isackson, P. J., 1994, Seizures increase basic fibroblast growth factor mRNA in adult rat forebrain neurons and glia, *Mol. Brain Res.* **21**:190–205.

Gasser, U. E., Weskamp, G., Otten, U., and Dravid, A. R., 1986, Time course of the elevation of nerve growth factor (NGF) content in the hippocampus and septum following lesions of the septohippocampal pathway in rats, *Brain Res.* **376**:351–356.

Ghosh, A., Carnahan, J., and Greenberg, M. E., 1994, Requirement for BDNF in activity-dependent survival of cortical neurons, *Science* **263**:1618–1623.

Giehl, K. M., and Tetzlaff, W., 1996, BDNF and NT-3, but not NGF, prevent axotomy-induced death of rat corticospinal neurons *in vivo, Eur. J. Neurosci.* **8**:1167–1175.

Gnahn, H., Hefti, F., Heumann, R., Schwab, M. E., and Thoenen, H., 1983, NGF-mediated increase of choline acetyltransferase (ChAT) in the neonatal rat forebrain: Evidence for a physiological role of NGF in the brain? *Brain Res.* **285**:45–52.

Gómez-Pinilla, F., Lee, J. W., and Cotman, C. W., 1992, Basic FGF in adult rat brain: Cellular distribution and response to entorhinal lesion and fimbria-fornix transection. *J. Neurosci.* **12**:345–355.

Gonzalez, A. M., Berry, M., Maher, P. A., Logan, A., and Baird, A., 1995, A comprehensive analysis of the distribution of FGF-2 and FGFR1 in the rat brain, *Brain Res.* **701**:201–226.

Gospodarowicz, D., 1974, Localisation of a fibroblast growth factor and its effect alone and with hydrocortisone on 3T3 cell growth, *Nature* **249**:123–127.

Götz, R., Köster, R., Winkler, C., Raulf, F., Lottspeich, F., Schartl, M., and Thoenen, H., 1994, Neurotrophin-6 is a new member of the nerve growth factor family, *Nature* **372**:266–269.

Gould, E., Cameron, H. A., and McEwen, B. S., 1994, Blockade of NMDA receptors increases cell death and birth in the developing rat dentate gyrus, *J. Comp. Neurol.* **340**:551–565.

Gould, E., McEwen, B. S., Tanapat, P., Galea, L. A. M., and Fuchs, E., 1997, Neurogenesis in the dentate gyrus of the adult tree shrew is regulated by psychosocial stress and NMDA receptor activation, *J. Neurosci.* **17**:2492–2498.

Graham, S. H., Shiraishi, K., Panter, S. S., Simon, R. P., and Faden, A. I., 1990, Changes in extracellular amino acid neurotransmitters produced by focal cerebral ischemia, *Neurosci. Lett.* **110**:124–130.

Greene, L. A., Seeley, P. J., Rukenstein, A., DiPiazza, M., and Howard, A., 1984, Rapid activation of tyrosine hydroxylase in response to nerve growth factor, *J. Neurochem.* **42**:1728–1734.

Grothe, C., and Wewetzer, K., 1996, Fibroblast growth factor and its implications for developing and regenerating neurons, *Int. J. Dev. Biol.* **40**:403–410.

Grothe, C., Otto, D., and Unsicker, K., 1989, Basic fibroblast growth factor promotes *in vitro* survival and cholinergic development of rat septal neurons: Comparison with the effects of nerve growth factor, *Neuroscience* **31**:649–661.

Guthrie, K. M., and Gall, C. M., 1991, Differential expression of mRNAs for the NGF family of neurotrophic factors in the adult rat central olfactory system, *J. Comp. Neurol.* **313**:95–102.

Hagg, T., and Varon, S., 1993, Ciliary neurotrophic factor prevents degeneration of adult rat substantia nigra dopaminergic neurons *in vivo, Proc. Natl. Acad. Sci. USA* **90**:6315–6319.

Hagg, T., Hagg, F., Vahlsing, H. L., Manthorpe, M., and Varon, S., 1989a, Nerve growth factor effects on cholinergic neurons of neostriatum and nucleus accumbens in the adult rat, *Neuroscience* **30**:95–103.

Hagg, T., Fass-Holmes, B., Vahlsing, H. L., Manthorpe, M., Conner, J. M., and Varon, S., 1989b, Nerve growth factor (NGF) reverses axotomy-induced decreases in choline acetyltransferase, NGF receptor, and size of medial septum cholinergic neurons, *Brain Res.* **505**:29–38.

Hagg, T., Quon, D., Higaki, J., and Varon, S., 1992, Ciliary neurotrophic factor prevents neuronal degeneration and promotes low affinity NGF receptor expression in the adult rat CNS, *Neuron* **8**:145–158.

Hallböök, F., Ibáñez, C. F., and Persson, H., 1991, Evolutionary studies of the nerve growth factor family reveal a novel member abundantly expressed in Xenopus ovary, *Neuron* **6**:845–858.

Hanneken, A., and Baird, A., 1992, Immunolocalization of basic fibroblast growth factor: Dependence on antibody type and tissue fixation, *Exp. Eye Res.* **54**:1011–1014.

Hara, H., Onodera, H., Kawagoe, J., and Kogure, K., 1991, Failure of basic fibroblast growth factor to prevent postischemic neuronal damage in the rat, *Eur. J. Pharmacol.* **209**:195–198.

Hatanaka, H., and Tsukui, H., 1986, Differential effects of nerve-growth factor and glioma-conditioned medium on neurons cultured from various regions of fetal rat central nervous system, *Dev. Brain Res.* **30:**47–56.

Hatten, M. E., Lynch, M., Rydel, R. E., Sanchez, J., Joseph-Silverstein, J., Moscatelli, D., and Rifkin, D. B., 1988, *In vitro* neurite extension by granule neurons is dependent upon astroglial-derived fibroblast growth factor, *Dev. Biol.* **125:**280–289.

Hefti, F., 1986, Nerve growth factor (NGF) promotes survival of septal cholinergic neurons after fimbrial transections, *J. Neurosci.* **6:**2155–2162.

Hefti, F., Hartikka, J., Eckenstein, F., Gnahn, H., Heumann, R., and Schwab, M., 1985, Nerve growth factor increases choline acetyltransferase but not survival or fiber outgrowth of cultured fetal septal cholinergic neurons, *Neuroscience* **14:**55–68.

Henderson, J. T., Seniuk, N. A., and Roder, J. C., 1994, Localization of CNTF immunoreactivity to neurons and astroglia in the CNS, *Mol. Brain Res.* **22:**151–165.

Hofer, M., Pagliusi, S. R., Hohn, A., Leibrock, J., and Barde, Y. A., 1990, Regional distribution of brain-derived neurotrophic factor mRNA in the adult mouse brain, *EMBO J.* **9:**2459–2464.

Hohn, A., Leibrock, J., Bailey, K., and Barde, Y. A., 1990, Identification and characterization of a novel member of the nerve growth factor/brain-derived neurotrophic factor family, *Nature* **344:**339–341.

Hsiang, J., Heller, A., Hoffmann, P. C., Mobley, W. C., and Wainer, B. H., 1989, The effects of nerve growth factor on the development of septal cholinergic neurons in reaggregate cell cultures, *Neuroscience* **29:**209–223.

Humpel, C., Lippoldt, A., Chadi, G., Ganten, D., Olson, L., and Fuxe, K., 1993, Fast and widespread increase of basic fibroblast growth factor messenger RNA and protein in the forebrain after kainate-induced seizures, *Neuroscience* **57:**913–922.

Humpel, C., Hoffer, B., Stromberg, I., Bektesh, S., Collins, F., and Olson, L., 1994, Neurons of the hippocampal formation express glial cell line-derived neurotrophic factor messenger RNA in response to kainate-induced excitation, *Neuroscience* **59:**791–795.

Hyman, C., Hofer, M., Barde, Y. A., Juhasz, M., Yancopoulos, G. D., Squinto, S. P., and Lindsay, R. M., 1991, BDNF is a neurotrophic factor for dopaminergic neurons of the substantia nigra, *Nature* **350:**230–232.

Hyman, C., Juhasz, M., Jackson, C., Wright, P., Ip, N. Y., and Lindsay, R. M., 1994, Overlapping and distinct actions of the neurotrophins BDNF, NT-3, and NT-4/5 on cultured dopaminergic and GABAergic neurons of the ventral mesencephalon, *J. Neurosci.* **14:**335–347.

Hynes, M. A., Poulsen, K., Armanini, M., Berkemeier, L., Phillips, H., and Rosenthal, A., 1994, Neurotrophin-4/5 is a survival factor for embryonic midbrain dopaminergic neurons in enriched cultures, *J. Neurosci. Res.* **37:**144–154.

Iacopino, A., Christakos, S., German, D., Sonsalla, P. K., and Altar, C. A., 1992, Calbindin-D28K-containing neurons in animal models of neurodegeneration: Possible protection from excitotoxicity, *Mol. Brain Res.* **13:**251–261.

Ibáñez, C. F., Ebendal, T., Barbany, G., Murray-Rust, J., Blundell, T. L., and Persson, H., 1992, Disruption of the low affinity receptor-binding site in NGF allows neuronal survival and differentiation by binding to the trk gene product, *Cell* **69:**329–341.

Ip, N.Y., Li, Y. P., van de Stadt, I., Panayotatos, N., Alderson, R. F., and Lindsay, R. M., 1991, Ciliary neurotrophic factor enhances neuronal survival in embryonic rat hippocampal cultures, *J. Neurosci.* **11:**3124–3134.

Ip, N. Y., Ibáñez, C. F., Nye, S. H., McClain, J., Jones, P. F., Gies, D. R., Belluscio, I., Le Beau, M. M., Espinosa, R., III, Squinto, S. P., Persson, H., and Yancopoulos, G. D., 1992a, Mammalian neurotrophin-4: Structure, chromosomal localization, tissue distribution, and receptor specificity, *Proc. Natl. Acad. Sci. USA* **89:**3060–3064.

Ip, N. Y., Nye, S. H., Boulton, T. G., Davis, S., Taga, T., Li, Y., Birren, S. J., Yasukawa, K., Kishimoto, T., Anderson, D. J., Stahl, N., and Yancopoulos, G. D., 1992b, CNTF and LIF act on neuronal cells via shared signaling pathways that involve the IL-6 signal transducing receptor component gp130, *Cell* **69:**1121–1132.

Ip, N. Y., Li, Y., Yancopoulos, G. D., and Lindsay, R. M., 1993a, Cultured hippocampal neurons show responses to BDNF, NT-3, and NT-4, but not NGF, *J. Neurosci.* **13:**3394–3405.

Ip, N. Y., Wiegand, S. J., Morse, J., and Rudge, J. S., 1993b, Injury-induced regulation of ciliary neurotrophic factor mRNA in the adult rat brain, *Eur. J. Neurosci.* **5:**25–33.

Ip, N. Y., McClain, J., Barrezueta, N. X., Aldrich, T. H., Pan, L., Li, Y., Wiegand, S. J., Friedman, B., Davis, S., and Yancopoulos, G. D., 1993c, The alpha component of the CNTF receptor is required for signaling and defines potential CNTF targets in the adult and during development, *Neuron* **10:**89–102.

Isackson, P. J., 1995, Trophic factor response to neuronal stimuli or injury, *Curr. Opin. Neurobiol.* **5:**350–357.

Isackson, P. J., Huntsman, M. M., Murray, K. D., and Gall, C. M., 1991, BDNF mRNA expression is increased in adult rat forebrain after limbic seizures—temporal patterns of induction distinct from NGF, *Neuron* **6:**937–948.

Ishikawa, K., Ohe, Y., Okutomi, Y., Takeuchi, T., and Suzuki, M., 1992, Neurotrophic effects of fibroblast growth factors on peptide-containing neurons in culture from postnatal rat hypothalamus, *Neuroendocrinology* **55:**193–198.

Jaye, M., Howk, R., Burgess, W., Ricca, G. A., Chiu, I. M., Ravera, M. W., O'Brien, S. J., Modi, W. S., Maciag, T., and Drohan, W. N., 1986, Human endothelial cell growth factor: Cloning, nucleotide sequence, and chromosome localization, *Science* **233:**541–545.

Johnson, D. E., and Williams, L. T., 1993, Structural and functional diversity in the FGF receptor multigene family, *Adv. Cancer Res.* **60:**1–41.

Jones, K. R., and Reichardt, L. F., 1990, Molecular cloning of a human gene that is a member of the nerve growth factor family, *Proc. Natl. Acad. Sci. USA* **87:**8060–8064.

Jones, K. R., Farinas, I., Backus, C., and Reichardt, L. F., 1994, Targeted disruption of the BDNF gene perturbs brain and sensory neuron development but not motor neuron development, *Cell* **76:**989–999.

Kahn, M. A., Huang, C. J., Caruso, A., Barresi, V., Nazarian, R., Condorelli, D. F., and de Vellis, J., 1997, Ciliary neurotrophic factor activates JAK/Stat signal transduction cascade and induces transcriptional expression of glial fibrillary acidic protein in glial cells, *J. Neurochem.* **68:**1413–1423.

Kaisho, Y., Yoshimura, K., and Nakahama, K., 1990, Cloning and expression of a cDNA encoding a novel human neurotrophic factor, *FEBS Lett.* **266:**187–191.

Kamiguchi, H., Yoshida, K., Sagoh, M., Sasaki, H., Inaba, M., Wakamoto, H., Otani, M., and Toya, S., 1995, Release of ciliary neurotrophic factor from cultured astrocytes and its modulation by cytokines, *Neurochem. Res.* **20:**1187–1193.

Karpati, G., Lochmuller, H., Nalbantoglu, J., and Durham, H., 1996, The principles of gene therapy for the nervous system, *Trends Neurosci.* **19:**49–54.

Katayama, Y., Becker, D. P., Tamura, T., and Hovda, D. A., 1990, Massive increases in extracellular potassium and the indiscriminate release of glutamate following concussive brain injury, *J. Neurosurg.* **73:**889–900.

Kawaja, M. D., and Gage, F. H., 1991a, Nerve growth factor receptor immunoreactivity in the rat septohippocampal pathway: A light and electron microscope investigation, *J. Comp. Neurol.* **307:**517–529.

Kawaja, M. D., and Gage, F. H., 1991b, Reactive astrocytes are substrates for the growth of adult CNS axons in the presence of elevated levels of nerve growth factor, *Neuron* **7:**1019–1030.

Kawaja, M. D., Rosenberg, M. B., Yoshida, K., and Gage, F. H., 1992, Somatic gene transfer of nerve growth factor promotes the survival of axotomized septal neurons and the regeneration of their axons in adult rats, *J. Neurosci.* **12:**2849–2864.

Kearns, C. M., and Gash, D. M., 1995, GDNF protects nigral dopamine neurons against 6-hydroxydopamine *in vivo*, *Brain Res.* **672:**104–111.

Kempermann, G., Kuhn, H. G., and Gage, F. H., 1997, More hippocampal neurons in adult mice living in an enriched environment, *Nature* **386:**493–495.

Kew, J. N., Sofroniew, M. V., 1995, Ciliary neurotrophic factor supports p75NGFR-immunoreactive non-cholinergic, but not cholinergic, developing septal neurons *in vitro*, *Neuroscience* **66:**793–804.

Kew, J., and Sofroniew, M., 1997, Brain-derived neurotrophic factor, acidic and basic fibroblast growth factors, insulin-like growth factor-I, and various antioxidants do not prevent the apoptotic death of developing septal cholinergic neurons following nerve growth factor withdrawal *in vitro*, *Neuroscience* **76:**809–820.

Kiefer, M. C., Stephans, J. C., Crawford, K., Okino, K., and Barr, P. J., 1990, Ligand-affinity cloning and structure of a cell surface heparan sulfate proteoglycan that binds basic fibroblast growth factor, *Proc. Natl. Acad. Sci. USA* **87:**6985–6989.

Kilby, N. J., Snaith, M. R., and Murray, J. A., 1993, Site-specific recombinases: Tools for genome engineering, *Trends Genet.* **9:**413–421.

Kirsch, M., and Hofmann, H. D., 1994, Expression of ciliary neurotrophic factor receptor mRNA and protein in the early postnatal and adult rat nervous system, *Neurosci. Lett.* **180:**163–166.

Kirschner, P. B., Henshaw, R., Weise, J., Trubetskoy, V., Finklestein, S., Schulz, J. B., and Beal, M. F., 1995, Basic fibroblast growth factor protects against excitotoxicity and chemical hypoxia in both neonatal and adult rats, *J. Cereb. Blood Flow Metab.* **15:**619–623.

Kirschner, P. B., Jenkins, B. G., Schulz, J. B., Finkelstein, S. P., Matthews, R. T., Rosen, B. R., and Beal, M. F., 1996, NGF, BDNF, and NT-5, but not NT-3, protect against MPP+ toxicity and oxidative stress in neonatal animals, *Brain Res.* **713:**178–185.

Klein, R., Conway, D., Parada, L. F., and Barbacid, M., 1990a, The trkB tyrosine protein kinase gene codes for a second neurogenic receptor that lacks the catalytic kinase domain, *Cell* **61:**647–656.

Klein, R., Martin-Zanca, D., Barbacid, M., and Parada, L. F., 1990b, Expression of the tyrosine kinase receptor gene trkB is confined to the murine embryonic and adult nervous system, *Development* **109:**845–850.

Klein, R., Smeyne, R. J., Wurst, W., Long, L. K., Auerbach, B. A., Joyner, A. L., and Barbacid, M., 1993, Targeted disruption of the trkB neurotrophin receptor gene results in nervous system lesions and neonatal death, *Cell* **75:**113–122.

Knipper, M., da Penha Berzaghi, M., Blochl, A., Breer, H., Thoenen, H., and Lindholm, D., 1994a, Positive feedback between acetylcholine and the neurotrophins nerve growth factor and brain-derived neurotrophic factor in the rat hippocampus, *Eur. J. Neurosci.* **6:**668–671.

Knipper, M., Leung, L. S., Zhao, D., and Rylett, R. J., 1994b, Short-term modulation of glutamatergic synapses in adult rat hippocampus by NGF, *NeuroReport* **5:**2433–2436.

Knüsel, B., Winslow, J. W., Rosenthal, A., Burton, L. E., Seid, D. P., Nikolics, K., and Hefti, F., 1991, Promotion of central cholinergic and dopaminergic neuron differentiation by brain-derived neurotrophic factor but not neurotrophin 3, *Proc. Natl. Acad. Sci. USA* **88:**961–965.

Knüsel, B., Beck, K. D., Winslow, J. W., Rosenthal, A., Burton, L. E., Widmer, H. R., Nikolics, K., and Hefti, F., 1992, Brain-derived neurotrophic factor administration protects basal forebrain cholinergic, but not nigral dopaminergic, neurons from degenerative changes after axotomy in the adult rat brain, *J. Neurosci.* **12:**4391–4402.

Koliatsos, V. E., Price, D. L., Gouras, G. K., Cayouette, M. H., Burton, L. E., and Winslow, J. W., 1994, Highly selective effects of nerve growth factor, brain-derived neurotrophic factor, and neurotrophin-3 on intact and injured basal forebrain magnocellular neurons, *J. Comp. Neurol.* **343:**247–262.

Kordower, J. H., Yaping-Chu, and MacLennan, A. J., 1997, Ciliary neurotrophic factor receptor alpha-immunoreactivity in the monkey central nervous system, *J. Comp. Neurol.* **377:**365–380.

Kornblum, H. I., Sankar, R., Shin, D. H., Wasterlain, C. G., and Gall, C. M., 1997, Induction of brain derived neurotrophic factor mRNA by seizures in neonatal and juvenile rat brain, *Mol. Brain Res.* **44:**219–228.

Korsching, S., 1993, The neurotrophic factor concept: A reexamination, *J. Neurosci.* **13:**2739–2748.

Korsching, S., Auburger, G., Heumann, R., Scott, J., and Thoenen, H., 1985, Levels of nerve growth factor and its mRNA in the central nervous system of the rat correlate with cholinergic innervation, *EMBO J.* **4:**1389–1393.

Korsching, S., Heumann, R., Thoenen, H., and Hefti, F., 1986, Cholinergic denervation of the rat hippocampus by fimbrial transection leads to a transient accumulation of nerve growth factor (NGF) without change in mRNANGF content, *Neurosci. Lett.* **66:**175–180.

Kromer, L. F., 1987, Nerve growth factor treatment after brain injury prevents neuronal death, *Science* **235:**214–216.

Kuhn, H. G., Dickinson-Anson, H., and Gage, F. H., 1996, Neurogenesis in the dentate gyrus of the adult rat: Age-related decrease of neuronal progenitor proliferation, *J. Neurosci.* **16:**2027–2033.

Kuhn, H. G., Winkler, J., Kempermann, G., Thal, L. J., and Gage, F. H., 1997, Epidermal growth factor and fibroblast growth factor-2 have different effects on neural progenitors in the adult rat brain, *J. Neurosci.* **17:**5820–5829.

Kumon, Y., Sakaki, S., Watanabe, H., Nakano, K., Ohta, S., Matsuda, S., Yoshimura, H., and Sakanaka, M., 1996, Ciliary neurotrophic factor attenuates spatial cognition impairment, cortical infarction, and thalamic degeneration in spontaneously hypertensive rats with focal cerebral ischemia, *Neurosci. Lett.* **206:**141–144.

Lamballe, F., Klein, R., and Barbacid, M., 1991, trkC, a new member of the trk family of tyrosine protein kinases, is a receptor for neurotrophin-3, *Cell* **66:**967–979.

Lapchak, P. A., 1996, Therapeutic potentials for glial cell line-derived neurotrophic factor (GDNF) based upon pharmacological activities in the CNS, *Rev. Neurosci.* **7:**165–176.

Lapchak, P. A., Araujo, D. M., and Hefti, F., 1993, BDNF and trkB messenger RNA expression in the rat hippocampus following entorhinal cortex lesions, *NeuroReport* **4:**191–194.

Larkfors, L., Lindsay, R. M., and Alderson, R. F., 1996, Characterization of the responses of Purkinje cells to neurotrophin treatment, *J. Neurochem.* **66:**1362–1373.

Lee, M. Y., Hofmann, H. D., and Kirsch, M., 1997a, Expression of ciliary neurotrophic factor receptor-alpha messenger RNA in neonatal and adult rat brain: An *in situ* hybridization study, *Neuroscience* **77:**233–246.

Lee, M. Y., Deller, T., Kirsch, M., Frotscher, M., and Hofmann, H. D., 1997b, Differential regulation of ciliary neurotrophic factor (CNTF) and CNTF receptor alpha expression in astrocytes and neurons of the fascia dentata after entorhinal cortex lesion, *J. Neurosci.* **17:**1137–1146.

Leibrock, J., Lottspeich, F., Hohn, A., Hofer, M., Hengerer, B., Masiakowski, P., Thoenen, H., and Barde, Y. A., 1989, Molecular cloning and expression of brain-derived neurotrophic factor, *Nature* **341:**149–152.

Levi-Montalcini, R., 1987, The nerve growth factor 35 years later, *Science* **237:**1154–1162.

Levi-Montalcini, R., Skaper, S. D., Dal Toso, R., Petrelli, L., and Leon, A., 1996, Nerve growth factor: From neurotrophin to neurokine, *Trends Neurosci.* **19:**514–520.

Levison, S. W., Ducceschi, M. H., Young, G. M., and Wood, T. L., 1996, Acute exposure to CNTF *in vivo* induces multiple components of reactive gliosis, *Exp. Neurol.* **141:**256–268.

Li, L., Wu, W., Lin, L. F., Lei, M., Oppenheim, R. W., and Houenou, L. J., 1995, Rescue of adult mouse motoneurons from injury-induced cell death by glial cell line-derived neurotrophic factor, *Proc. Natl. Acad. Sci. USA* **92:**9771–9775.

Lindholm, D., Castren, E., Tsoulfas, P., Kolbeck, R., Berzaghi, M., da P., Leingartner, A., Heisenberg, C. P., Tesarollo, L., Parada, L. F., and Thoenen, H., 1993, Neurotrophin-3 induced by tri-iodothyronine in cerebellar granule cells promotes Purkinje cell differentiation, *J. Cell Biol.* **122:**443–450.

Lindsay, R. M., Wiegand, S. J., Altar, C. A., and DiStefano, P. S., 1994, Neurotrophic factors: From molecule to man, *Trends Neurosci.* **17:**182–190.

Lindvall, O., Ernfors, P., Bengzon, J., Kokaia, Z., Smith, M. L., Siesjo, B. K., and Persson, H., 1992, Differential regulation of mRNAs for nerve growth factor, brain-derived neurotrophic factor, and neurotrophin 3 in the adult rat brain following cerebral ischemia and hypoglycemic coma, *Proc. Natl. Acad. Sci. USA* **89:**648–652.

Liu, H. M., and Chen, H. H., 1994, c-fos protein expression and ischemic changes in neurons vulnerable to ischemia/hypoxia, correlated with basic fibroblast growth factor immunoreactivity, *J. Neuropathol. Exp. Neurol.* **53:**598–605.

Loeb, D. M., Maragos, J., Martin-Zanca, D., Chao, M. V., Parada, L. F., and Greene, L. A., 1991, The trk proto-oncogene rescues NGF responsiveness in mutant NGF-nonresponsive PC12 cell lines, *Cell* **66:**961–966.

Logan, A., 1990, Intracrine regulation at the nucleus: A further mechanism of growth factor activity? *J. Endocrinol.* **125:**339–343.

Logan, A., Frautschy, S. A., Gonzalez, A. M., and Baird, A., 1992, A time course for the focal elevation of synthesis of basic fibroblast growth factor and one of its high-affinity receptors (*flg*) following a localized cortical brain injury, *J. Neurosci.* **12:**3828–3837.

Lois, C., and Alvarez-Buylla, A., 1993, Proliferating subventricular zone cells in the adult mammalian forebrain can differentiate into neurons and glia, *Proc. Natl. Acad. Sci. USA* **90:**2074–2077.

Louis, J. C., Magal, E., Gerdes, W., and Seifert, W., 1993, Survival-promoting and protein kinase C-regulating roles of basic FGF for hippocampal neurons exposed to phorbol ester, glutamate, and ischaemia-like conditions, *Eur. J. Neurosci.* **5:**1610–1621.

Lowenstein, D. H., and Arsenault, L., 1996a, Dentate granule cell layer collagen explant cultures: Spontaneous axonal growth and induction by brain-derived neurotrophic factor or basic fibroblast growth factor, *Neuroscience* **74:**1197–1208.

Lowenstein, D. H., and Arsenault, L., 1996b, The effects of growth factors on the survival and differentiation of cultured dentate gyrus neurons, *J. Neurosci.* **16:**1759–1769.

Lucidi-Phillipi, C. A., Gage, F. H., Shults, C. W., Jones, K. R., Reichardt, L. F., and Kang, U. J., 1995, Brain-derived neurotrophic factor-transduced fibroblasts: Production of BDNF and effects of grafting to the adult rat brain, *J. Comp. Neurol.* **354:**361–376.

Lucidi-Phillipi, C. A., Clary, D. O., Reichardt, L. F., and Gage, F. H., 1996, TrkA activation is sufficient to rescue axotomized cholinergic neurons, *Neuron* **16:**653–663.

Luskin, M. B., 1993, Restricted proliferation and migration of postnatally generated neurons derived from the forebrain subventricular zone, *Neuron* **11**:173–189.

MacLennan, A. J., Vinson, E. N., Marks, L., McLaurin, D. L., Pfeifer, M., and Lee, N., 1996, Immunohistochemical localization of ciliary neurotrophic factor receptor alpha expression in the rat nervous system, *J. Neurosci.* **16**:621–630.

Magal, E., Burnham, P., and Varon, S., 1991a, Effect of CNTF on low-affinity NGF receptor expression by cultured neurons from different rat brain regions, *J. Neurosci. Res.* **30**:560–566.

Magal, E., Burnham, P., and Varon, S., 1991b, Effects of ciliary neuronotrophic factor on rat spinal cord neurons *in vitro:* Survival and expression of choline acetyltransferase and low-affinity nerve growth factor receptors, *Dev. Brain Res.* **63**:141–150.

Maiese, K., Boniece, I., DeMeo, D., and Wagner, J. A., 1993, Peptide growth factors protect against ischemia in culture by preventing nitric oxide toxicity, *J. Neurosci.* **13**:3034–3040.

Maisonpierre, P. C., Belluscio, L., Squinto, S., Ip, N. Y., Furth, M. E., Lindsay, R. M., and Yancopoulos, G. D., 1990, Neurotrophin-3: A neurotrophic factor related to NGF and BDNF, *Science* **247**:1446–1451.

Mamounas, L. A., Blue, M. E., Siuciak, J. A., and Altar, C. A., 1995, Brain-derived neurotrophic factor promotes the survival and sprouting of serotonergic axons in rat brain, *J. Neurosci.* **15**:7929–7939.

Mansour-Robaey, S., Clarke, D. B., Wang, Y. C., Bray, G. M., and Aguayo, A. J., 1994, Effects of ocular injury and administration of brain-derived neurotrophic factor on survival and regrowth of axotomized retinal ganglion cells, *Proc. Natl. Acad. Sci. USA* **91**:1632–1636.

Marsh, H. N., and Palfrey, H. C., 1996, Neurotrophin-3 and brain-derived neurotrophic factor activate multiple signal transduction events but are not survival factors for hippocampal pyramidal neurons, *J. Neurochem.* **67**:952–963.

Mason, I., 1994, Cell signalling. Do adhesion molecules signal via FGF receptors? *Curr. Biol.* **4**:1158–1161.

Masu, Y., Wolf, E., Holtmann, B., Sendtner, M., Brem, G., and Thoenen, H., 1993, Disruption of the CNTF gene results in motor neuron degeneration, *Nature* **365**:27–32.

Matsuda, S., Saito, H., and Nishiyama, N., 1990, Effect of basic fibroblast growth factor on neurons cultured from various regions of postnatal rat brain, *Brain Res.* **520**:310–316.

Matsuyama, A., Iwata, H., Okumura, N., Yoshida, S., Imaizumi, K., Lee, Y., Shiraishi, S., and Shiosaka, S., 1992, Localization of basic fibroblast growth factor-like immunoreactivity in the rat brain, *Brain Res.* **587**:49–65.

Mattson, M. P., and Rychlik, B., 1990, Glia protect hippocampal neurons against excitatory amino acid-induced degeneration—Involvement of fibroblast growth factor, *Int. J. Dev. Neurosci.* **8**:399.

Mattson, M. P., and Cheng, B., 1993, Growth factors protect neurons against excitotoxic/ischemic damage by stabilizing calcium homeostasis, *Stroke* **24**:1136–1140.

Mattson, M. P., Murrain, M., Guthrie, P. B., and Kater, S. B., 1989, Fibroblast growth factor and glutamate: Opposing roles in the generation and degeneration of hippocampal neuroarchitecture, *J. Neurosci.* **9**:3728–3740.

Mattson, M. P., Zhang, Y., and Bose, S., 1993a, Growth factors prevent mitochondrial dysfunction, loss of calcium homeostasis, and cell injury, but not ATP depletion in hippocampal neurons deprived of glucose, *Exp. Neurol.* **121**:1–13.

Mattson, M. P., Kumar, K. N., Wang, H., Cheng, B., and Michaelis, E. K., 1993b, Basic FGF regulates the expression of a functional 71 kDa NMDA receptor protein that mediates calcium influx and neurotoxicity in hippocampal neurons, *J. Neurosci.* **13**:4575–4578.

Mattson, M. P., Lovell, M. A., Furukawa, K., and Markesbery, W. R., 1995, Neurotrophic factors attenuate glutamate-induced accumulation of peroxides, elevation of intracellular $Ca^{2+}$ concentration, and neurotoxicity and increase antioxidant enzyme activities in hippocampal neurons, *J. Neurochem.* **65**:1740–1751.

Mayer, E., Dunnett, S. B., Pellitteri, R., and Fawcett, J. W., 1993, Basic fibroblast growth factor promotes the survival of embryonic ventral mesencephalic dopaminergic neurons. I. Effects *in vitro, Neuroscience* **56**:379–388.

Merlio, J. P., Ernfors, P., Jaber, M., and Persson, H., 1992, Molecular cloning of rat trkC and distribution of cells expressing messenger RNAs for members of the trk family in the rat central nervous system, *Neuroscience* **51**:513–532.

Middlemas, D. S., Lindberg, R. A., and Hunter, T., 1991, trkB, a neural receptor protein-tyrosine kinase: Evidence for a full-length and two truncated receptors, *Mol. Cell Biol.* **11**:143–153.

Mignatti, P., Miromoto, T., and Rifkin, D. B., 1992, Basic fibroblast growth factor, a protein devoid of

secretory signal sequence, is released by cells via a pathway independent of the endoplasmic reticulum-Golgi complex, *J. Cell Physiol.* **151**:81–93.
Miller, R. G., Bryan, W. W., Dietz, M. A., Munsat, T. L., Petajan, J. H., Smith, S. A., and Goodpasture, J. C., 1996a, Toxicity and tolerability of recombinant human ciliary neurotrophic factor in patients with amyotrophic lateral sclerosis, *Neurology* **47**:1329–1331.
Miller, R. G., Petajan, J. H., Bryan, W. W., Armon, C., Barohn, R. J., Goodpasture, J. C., Hoagland, R. J., Parry, G. J., Ross, M. A., and Stromatt, S. C., 1996b, A placebo-controlled trial of recombinant human ciliary neurotrophic (rhCNTF) factor in amyotrophic lateral sclerosis, rhCNTF, ALS Study Group, *Ann. Neurol.* **39**:256–260.
Morrison, R. S., Sharma, A., de Vellis, J., and Bradshaw, R. A., 1986, Basic fibroblast growth factor supports the survival of cerebral cortical neurons in primary culture, *Proc. Natl. Acad. Sci. USA* **83**:7537–7541.
Morse, J. K., Wiegand, S. J., Anderson, K., You, Y., Cai, N., Carnahan, J., Miller, J., DiStefano, P. S., Altar, C. A., Lindsay, R. M., and Alderson, R. F., 1993, Brain-derived neurotrophic factor (BDNF) prevents the degeneration of medial septal cholinergic neurons following fimbria transection, *J. Neurosci.* **13**:4146–4156.
Muenke, M., Schell, U., Hehr, A., Robin, N. H., Losken, H. W., Schinzel, A., Pulleyn, L. J., Rutland, P., Reardon, W., Malcolm, S., and Winter, R. M., 1994, A common mutation in the fibroblast growth factor receptor 1 gene in Pfeiffer syndrome, *Nature Genet.* **8**:269–274.
Nakata, N., Kato, H., and Kogure, K., 1993, Protective effects of basic fibroblast growth factor against hippocampal neuronal damage following cerebral ischemia in the gerbil, *Brain Res.* **605**:354–356.
Naldini, L., Blomer, U., Gallay, P., Ory, D., Mulligan, R., Gage, F. H., Verma, I. M., and Trono, D., 1996, *In vivo* gene delivery and stable transduction of nondividing cells by a lentiviral vector, *Science* **272**:263–267.
Naumann, T., Peterson, G. M., and Frotscher, M., 1992, Fine structure of rat septohippocampal neurons. II. A time course analysis following axotomy, *J. Comp. Neurol.* **325**:219–242.
Nonomura, T., and Hatanaka, H., 1992, Neurotrophic effect of brain-derived neurotrophic factor on basal forebrain cholinergic neurons in culture from postnatal rats, *Neurosci. Res.* **14**:226–233.
Nonomura, T., Nishio, C., Lindsay, R. M., and Hatanaka, H., 1995, Cultured basal forebrain cholinergic neurons from postnatal rats show both overlapping and non-overlapping responses to the neurotrophins, *Brain Res.* **683**:129–139.
Novikova, L., Novikov, L., and Kellerth, J. O., 1996, Brain-derived neurotrophic factor reduces necrotic zone and supports neuronal survival after spinal cord hemisection in adult rats, *Neurosci. Lett.* **220**:203–206.
Nozaki, K., Finklestein, S. P., and Beal, M. F., 1993, Basic fibroblast growth factor protects against hypoxia-ischemia and NMDA neurotoxicity in neonatal rats, *J. Cereb. Blood Flow Metab.* **13**:221–228.
O'Brien, T. S., Svendsen, C. N., Isacson, O., and Sofroniew, M. V., 1990, Loss of true blue labelling from the medial septum following transection of the fimbria-fornix: Evidence for the death of cholinergic and non-cholinergic neurons, *Brain Res.* **508**:249–256.
Ogata, N., Ogata, K., Imhof, H. G., and Yonekawa, Y., 1996, Effect of CNTF on ischemic cell damage in rat hippocampus, *Acta Neurochir. (Wien)* **138**:580–583.
Ohsawa, F., Widmer, H. R., Knüsel, B., Denton, T. L., and Hefti, F., 1993, Response of embryonic rat hippocampal neurons in culture to neurotrophin-3, brain-derived neurotrophic factor, and basic fibroblast growth factor, *Neuroscience* **57**:67–77.
Olson, L., 1993, NGF and the treatment of Alzheimer's disease, *Exp. Neurol.* **124**:5–15.
Olson, L., Nordberg, A., von Holst, H., Backman, L., Ebendal, T., Alafuzoff, I., Amberla, K., Hartvig, P., Herlitz, A., Lilja, A., *et al.*, 1992, Nerve growth factor affects 11C-nicotine binding, blood flow, EEG, and verbal episodic memory in an Alzheimer patient (case report), *J. Neural Transm.* **4**:79–95.
Ornitz, D. M., Yayon, A., Flanagan, J. G., Svahn, C. M., Levi, E., and Leder, P., 1992, Heparin is required for cell-free binding of basic fibroblast growth factor to a soluble receptor and for mitogenesis in whole cells, *Mol. Cell Biol.* **12**:240–247.
Otto, D., and Unsicker, K., 1990, Basic FGF reverses chemical and morphological deficits in the nigrostriatal system of MPTP-treated mice, *J. Neurosci.* **10**:1912–1921.
Otto, D., Frotscher, M., and Unsicker, K., 1989, Basic fibroblast growth factor and nerve growth factor administered in gel foam rescue medial septal neurons after fimbria-fornix transection, *J. Neurosci. Res.* **22**:83–91.

Palmer, T. D., Ray, J., and Gage, F. H., 1995, FGF-2-responsive neuronal progenitors reside in proliferative and quiescent regions of the adult rodent brain, *Mol. Cell Neurosci.* **6:**474–486.

Penn, R. D., Kroin, J. S., York, M. M., and Cedarbaum, J. M., 1997, Intrathecal ciliary neurotrophic factor delivery for treatment of amyotrophic lateral sclerosis (phase I trial), *Neurosurgery* **40:**94–99.

Peterson, D. A., Lucidi-Phillipi, C. A., Eagle, K. A., and Gage, F. H., 1994, Perforant path damage results in progressive neuronal death and somal atrophy in layer II of entorhinal cortex and functional impairment with increasing postdamage age, *J. Neurosci.* **14:**6872–6885.

Peterson, D. A., Lucidi-Phillipi, C. A., Murphy, D. P., Ray, J., and Gage, F. H., 1996, Fibroblast growth factor-2 protects entorhinal layer II glutamatergic neurons from axotomy-induced death, *J. Neurosci.* **16:**886–898.

Peterson, D. A., Leppert, J. T., Lee, K. F., and Gage, F. H., 1998, Basal forebrain neuronal loss in mice lacking neurotrophin receptor p75, *Science* **277:**837–838.

Peterson, D. A., Dickinson-Anson, H. A., Leppert, J. T., Lee, K. F., and Gage, F. H., 1999, Central neuronal loss and behavioral impairment in mice lacking neurotrophin receptor p75, *J. Comp. Neurol.* **404:**1–20.

Pettmann, B., LaBourdette, G., Weibel, M., and Sesenbrenner, M., 1986, The brain fibroblast growth factor (bFGF) is localized in neurons, *Neurosci. Lett.* **68:**175–180.

Petty, B. G., Cornblath, D. R., Adornato, B. T., Chaudhry, V., Flexner, C., Wachsman, M., Sinicropi, D., Burton, L. E., and Peroutka, S. J., 1994, The effect of systemically administered recombinant human nerve growth factor in healthy human subjects, *Ann. Neurol.* **36:**244–246.

Phelps, C. H., Gage, F. H., Growdon, J. H., Hefti, F., Harbaugh, R., Johnston, M. V., Khachaturian, Z. S., Mobley, W. C., Price, D. L., Raskind, M., *et al.,* 1989, Potential use of nerve growth factor to treat Alzheimer's disease, *Neurobiol. Aging* **10:**205–207.

Phillips, H. S., Hains, J. M., Laramee, G. R., Rosenthal, A., and Winslow, J. W., 1990, Widespread expression of BDNF but not NT3 by target areas of basal forebrain cholinergic neurons, *Science* **250:**290–294.

Pringle, A. K., Sundstrom, L. E., Wilde, G. J. C., Williams, L. R., and Iannetti, F., 1996, Brain-derived neurotrophic factor, but not neurotrophin-3, prevents ischaemia-induced neuronal cell death in organotypic rat hippocampal slice cultures, *Neurosci. Lett.* **211:**203–206.

Qiao, X., Hughes, P. E., Venero, J. L., Dugich-Djordjevic, M. M., Nichols, N. R., Hefti, F., and Knüsel, B., 1996, NT-4/5 protects against adrenalectomy-induced apoptosis of rat hippocampal granule cells, *NeuroReport* **7:**682–686.

Rapraeger, A. C., Krufka, A., and Olwin, B. B., 1991, Requirement of heparan sulfate for bFGF-mediated fibroblast growth and myoblast differentiation, *Science* **252:**1705–1708.

Ray, J., Peterson, D. A., Schinstine, M., and Gage, F. H., 1993, Proliferation, differentiation, and long-term culture of primary hippocampal neurons, *Proc. Natl. Acad. Sci. USA* **90:**3602–3606.

Ray, J., Fisher, L. J., and Gage, F. H., 1995a, Implantation of genetically modified cells in the brain, in: *Somatic Gene Therapy* (P. L. Chang, ed.) CRC, Boca Raton, FL, pp. 161–182.

Ray, J., Hogg, J., Beutler, A. S., Takayama, H., Baird, A., and Gage, F. H., 1995b, Expression of biologically active basic fibroblast growth factor by genetically modified rat primary skin fibroblasts, *J. Neurochem.* **64:**503–513.

Ray, J., Palmer, T. D., Suhonen, J., Takahashi, J., and Gage, F. H., 1997, Neurogenesis in the adult brain: Lessons learned from the studies of progenitor cells from the embryonic and adult central nervous systems, in: *Isolation, Characterization, and Utilization of CNS Stem Cells* (F. H. Gage and Y. Christen, eds.) Springer-Verlag, Berlin, pp. 129–149.

Reynolds, B. A., and Weiss, S., 1992, Generation of neurons and astrocytes from isolated cells of the adult mammalian central nervous system, *Science* **255:**1707–1710.

Richards, L. J., Kilpatrick, T. J., and Bartlett, P. F., 1992, *De novo* generation of neuronal cells from the adult mouse brain, *Proc. Natl. Acad. Sci. USA* **89:**8591–8595.

Riva, M. A., Donati, E., Tascedda, F., Zolli, M., and Racagni, G., 1994, Short- and long-term induction of basic fibroblast growth factor gene expression in rat central nervous system following kainate injection, *Neuroscience* **59:**55–65.

Rocamora, N., Palacios, J. M., and Mengod, G., 1992, Limbic seizures induce a differential regulation of the expression of nerve growth factor, brain-derived neurotrophic factor, and neurotrophin-3 in the rat hippocampus, *Mol. Brain Res.* **13:**27–33.

Rocamora, N., Massieu, L., Boddeke, H. W., Mengod, G., and Palacios, J. M., 1993, Neuronal death and neurotrophin gene expression: Long-lasting stimulation of neurotrophin-3 messenger RNA in the degenerating CA1 and CA4 pyramidal cell layers, *Neuroscience* **53:**905–908.

Rocamora, N., Massieu, L., Boddeke, H. W., Palacios, J. M., and Mengod, G., 1994, Differential regulation of the expression of nerve growth factor, brain-derived neurotrophic factor, and neurotrophin-3 mRNAs in adult rat brain after intrahippocampal injection of quinolinic acid, *Mol. Brain Res.* **26:**89–98.

Rosenberg, M. B., Friedmann, T., Robertson, R. C., Tuszynski, M., Wolff, J. A., Breakefield, X. O., and Gage, F. H., 1988, Grafting of genetically modified cells to the damaged brain: Restorative effects of NGF gene expression, *Science* **242:**1575–1578.

Rosenthal, A., Goeddel, D. V., Nguyen, T., Lewis, M., Shih, A., Laramee, G. R., Nikolics, K., and Winslow, J. W., 1990, Primary structure and biological activity of a novel human neurotrophic factor, *Neuron* **4:**767–773.

Sagot, Y., Tan, S. A., Baetge, E., Schmalbruch, H., Kato, A. C., and Aebischer, P., 1995, Polymer encapsulated cell lines genetically engineered to release ciliary neurotrophic factor can slow down progressive motor neuronopathy in the mouse, *Eur. J. Neurosci.* **7:**1313–1322.

Sariola, H., Sainio, K., Arumae, U., and Saarma, M., 1994, Neurotrophins and ciliary neurotrophic factor: Their biology and pathology, *Ann. Med.* **26:**355–363.

Sauer, H., Fischer, W., Nikkhah, G., Wiegand, S. J., Brundin, P., Lindsay, R. M., and Björklund, A., 1993, Brain-derived neurotrophic factor enhances function rather than survival of intrastriatal dopamine cell-rich grafts, *Brain Res.* **626:**37–44.

Sauer, H., Rosenblad, C., and Björklund, A., 1995, Glial cell line-derived neurotrophic factor, but not transforming growth factor beta 3, prevents delayed degeneration of nigral dopaminergic neurons following striatal 6-hydroxydopamine lesion, *Proc. Natl. Acad. Sci. USA* **92:**8935–8939.

Sawai, H., Clarke, D. B., Kittlerova, P., Bray, G. M., and Aguayo, A. J., 1996, Brain-derived neurotrophic factor and neurotrophin-4/5 stimulate growth of axonal branches from regenerating retinal ganglion cells, *J. Neurosci.* **16:**3887–3894.

Schmidt-Kastner, R., Tomac, A., Hoffer, B., Bektesh, S., Rosenzweig, B., and Olson, L., 1994, Glial cell-line derived neurotrophic factor (GDNF) mRNA upregulation in striatum and cortical areas after pilocarpine-induced status epilepticus in rats, *Mol. Brain Res.* **26:**325–330.

Schnell, L., Schneider, R., Kolbeck, R., Barde, Y. A., and Schwab, M. E., 1994, Neurotrophin-3 enhances sprouting of corticospinal tract during development and after adult spinal cord lesion, *Nature* **367:**170–173.

Schwab, M. E., Otten, U., Agid, Y., and Thoenen, H., 1979, Nerve growth factor (NGF) in the rat CNS: Absence of specific retrograde axonal transport and tyrosine hydroxylase induction in locus coeruleus and substantia nigra, *Brain Res.* **168:**473–483.

Segal, R. A., and Greenberg, M. E., 1996, Intracellular signalling pathways activated by neurotrophic factors, *Annu. Rev. Neurosci.* **19:**463–489.

Seiler, M., and Schwab, M. E., 1984, Specific retrograde transport of nerve growth factor (NGF) from neocortex to nucleus basalis in the rat, *Brain Res.* **300:**33–39.

Sendtner, M., Kreutzberg, G. W., and Thoenen, H., 1990, Ciliary neurotrophic factor prevents the degeneration of motor neurons after axotomy, *Nature* **345:**440–441.

Sendtner, M., Schmalbruch, H., Stockli, K. A., Carroll, P., Kreutzberg, G. W., and Thoenen, H., 1992, Ciliary neurotrophic factor prevents degeneration of motor neurons in mouse mutant progressive motor neuronopathy, *Nature* **358:**502–504.

Sendtner, M., Carroll, P., Holtmann, B., Hughes, R. A., and Thoenen, H., 1994, Ciliary neurotrophic factor, *J. Neurobiol.* **25:**1436–1453.

Sendtner, M., Holtmann, B., and Hughes, R. A., 1996, The response of motoneurons to neurotrophins, *Neurochem. Res.* **21:**831–841.

Senut, M. C., Tuszynski, M. H., Raymon, H. K., Suhr, S. T., Liou, N. H., Jones, K. R., Reichardt, L. F., and Gage, F. H., 1995, Regional differences in responsiveness of adult CNS axons to grafts of cells expressing human neurotrophin 3, *Exp. Neurol.* **135:**36–55.

Shelton, D., and Reichardt, L., 1986, Studies on the expression of the beta nerve growth factor (NGF) gene in the central nervous system: Level and regional distribution of NGF mRNA suggest that NGF functions as a trophic factor for several distinct populations of neurons, *Proc. Natl. Acad. Sci. USA* **83:**2714–2718.

Silani, V., Mariani, D., Donato, F. M., Ghezzi, C., Mazzucchelli, F., Buscaglia, M., Pardi, G., and Scarlato,

G., 1994, Development of dopaminergic neurons in the human mesencephalon and *in vitro* effects of basic fibroblast growth factor treatment, *Exp. Neurol.* **128:**59–76.

Skaper, S. D., Negro, A., Dal Toso, R., and Facci, L., 1992, Recombinant human ciliary neurotrophic factor alters the threshold of hippocampal pyramidal neuron sensitivity to excitotoxin damage: Synergistic effects of monosialogangliosides, *J. Neurosci. Res.* **33:**330–337.

Skaper, S. D., Leon, A., and Facci, L., 1993, Basic fibroblast growth factor modulates sensitivity of cultured hippocampal pyramidal neurons to glutamate cytotoxicity: Interaction with ganglioside GM1, *Dev. Brain Res.* **71:**1–8.

Smith, G. M., Hale, J., Pasnikowski, E. M., Lindsay, R. M., Wong, V., and Rudge, J. S., 1996, Astrocytes infected with replication-defective adenovirus containing a secreted form of CNTF or NT3 show enhanced support of neuronal populations *in vitro, Exp. Neurol.* **139:**156–166.

Smith, M. A., Makino, S., Kvetnansky, R., and Post, R. M., 1995, Stress and glucocorticoids affect the expression of brain-derived neurotrophic factor and neurotrophin-3 mRNAs in the hippocampus, *J. Neurosci.* **15:**1768–1777.

Snider, W. D., 1994, Functions of the neurotrophins during nervous system development: What the knockouts are teaching us, *Cell* **77:**627–638.

Sofroniew, M. V., Cooper, J. D., Svendsen, C. N., Crossman, P., Ip, N. Y., Lindsay, R. M., Zafra, F., and Lindholm, D., 1993, Atrophy, but not death, of adult septal cholinergic neurons after ablation of target capacity to produce mRNAs for NGF, BDNF, and NT3, *J. Neurosci.* **13:**5263–5276.

Soppet, D., Escandon, E., Maragos, J., Middlemas, D. S., Reid, S. W., Blair, J., Burton, L. E., Stanton, B. R., Kaplan, D. R., Hunter, T., Nikolics, K., and Parada, L. F., 1991, The neurotrophic factors, brain-derived neurotrophic factor, and neurotrophin-3 are ligands for the trkB tyrosine kinase receptor, *Cell* **65:**895–903.

Spina, M. B., Squinto, S. P., Miller, J., Lindsay, R. M., and Hyman, C., 1992, Brain-derived neurotrophic factor protects dopamine neurons against 6-hydroxydopamine and *N*-methyl-4-phenylpyridinium ion toxicity: Involvement of the glutathione system, *J. Neurochem.* **59:**99–106.

Springer, J. E., Mu, X., Bergmann, L. W., and Trojanowski, J. Q., 1994, Expression of GDNF mRNA in rat and human nervous tissue, *Exp. Neurol.* **127:**167–170.

Stockli, K. A., Lillien, L. E., Naher-Noe, M., Breitfeld, G., Hughes, R. A., Raff, M. C., Thoenen, H., and Sendtner, M., 1991, Regional distribution, developmental changes, and cellular localization of CNTF-mRNA and protein in the rat brain, *J. Cell Biol.* **115:**447–459.

Stoop, R., and Poo, M. M., 1995, Potentiation of transmitter release by ciliary neurotrophic factor requires somatic signaling, *Science* **267:**695–699.

Stopa, E. G., Gonzalez, A. M., Chorsky, R., Corona, R. J., Alvarez, J., Bird, E. D., and Baird, A., 1990, Basic fibroblast growth factor in Alzheimer's disease, *Biochem. Biophys. Res. Commun.* **171:**690–696.

Stromberg, I., Wetmore, C. J., Ebendal, T., Ernfors, P., Persson, H., and Olson, L., 1990, Rescue of basal forebrain cholinergic neurons after implantation of genetically modified cells producing recombinant NGF, *J. Neurosci. Res.* **25:**405–411.

Suhonen, J. O., Peterson, D. A., Ray, J., and Gage, F. H., 1996, Differentiation of adult hippocampus-derived progenitors into olfactory neurons *in vivo, Nature* **383:**624–627.

Svendsen, C. N., Kew, J. N., Staley, K., and Sofroniew, M. V., 1994, Death of developing septal cholinergic neurons following NGF withdrawal *in vitro:* Protection by protein synthesis inhibition, *J. Neurosci.* **14:**75–87.

Takami, K., Iwane, M., Kiyota, Y., Miyamoto, M., Tsukuda, R., and Shiosaka, S., 1992, Increase of basic fibroblast growth factor immunoreactivity and its mRNA level in rat brain following transient forebrain ischemia, *Exp. Brain Res.* **90:**1–10.

Takayama, H., Ray, J., Raymon, H. K., Baird, A., Hogg, J., Fisher, L. J., and Gage, F. H., 1995, Basic fibroblast growth factor increases dopaminergic graft survival and function in a rat model of Parkinson's disease, *Nature Med.* **1:**53–58.

Takeda, A., Onodera, H., Sugimoto, A., Kogure, K., Obinata, M., and Shibahara, S., 1993, Coordinated expression of messenger RNAs for nerve growth factor, brain-derived neurotrophic factor, and neurotrophin-3 in the rat hippocampus following transient forebrain ischemia, *Neuroscience* **55:**23–31.

Tan, S. A., Deglon, N., Zurn, A. D., Baetge, E. E., Bamber, B., Kato, A. C., and Aebischer, P., 1996, Rescue of motoneurons from axotomy-induced cell death by polymer encapsulated cells genetically engineered to release CNTF, *Cell Transplant.* **5:**577–587.

Tao, Y., Black, I. B., and DiCicco-Bloom, E., 1996, Neurogenesis in neonatal rat brain is regulated by peripheral injection of basic fibroblast growth factor (bFGF), *J. Comp. Neurol.* **376:**653–663.

Thoenen, H., 1995, Neurotrophins and neuronal plasticity, *Science* **270:**593–598.

Timmusk, T., Belluardo, N., Metsis, M., and Persson, H., 1993, Widespread and developmentally regulated expression of neurotrophin-4 mRNA in rat brain and peripheral tissues, *Eur. J. Neurosci.* **5:**605–613.

Tolosano, E., Cutufia, M. A., Hirsch, E., Stefanuto, G., Voyron, S., Fasolo, A., Silengo, L., and Altruda, F., 1996, Ciliary neurotrophic factor constitutively expressed in the nervous system of transgenic mice protects embryonic dorsal root ganglion neurons from apoptosis, *Eur. J. Neurosci.* **8:**521–529.

Tooyama, I., Kawamata, T., Walker, D., Yamada, T., Hanai, K., Kimura, H., Iwane, M., Igarashi, K., McGeer, E. G., and McGeer, P. L., 1993a, Loss of basic fibroblast growth factor in substantia nigra neurons in Parkinson's disease, *Neurology* **43:**372–376.

Tooyama, I., Kremer, H. P., Hayden, M. R., Kimura, H., McGeer, E. G., and McGeer, P. L., 1993b, Acidic and basic fibroblast growth factor-like immunoreactivity in the striatum and midbrain in Huntington's disease, *Brain Res.* **610:**1–7.

Tuszynski, M. H., Buzsaki, G., and Gage, F. H., 1990, Nerve growth factor infusions combined with fetal hippocampal grafts enhance reconstruction of the lesioned septohippocampal projection, *Neuroscience* **36:**33–44.

Tuszynski, M. H., Mafong, E., and Meyer, S., 1996, Central infusions of brain-derived neurotrophic factor and neurotrophin-4/5, but not nerve growth factor and neurotrophin-3, prevent loss of the cholinergic phenotype in injured adult motor neurons, *Neuroscience* **71:**761–771.

Unsicker, K., Reichert-Preibsh, H., Schmidt, R., Pettmann, B., Labourdette, G., and Sensenbrenner, M., 1987, Astroglial and fibroblast growth factors have neurotrophic functions for cultured peripheral and central nervous system neurons, *Proc. Natl. Acad. Sci. USA* **84:**5459–5463.

Unsicker, K., Reichert-Preibsch, H., and Wewetzer, K., 1992, Stimulation of neuron survival by basic FGF and CNTF is a direct effect and not mediated by non-neuronal cells: Evidence from single cell cultures, *Dev. Brain Res.* **65:**285–288.

van der Wal, E. A., Gomez-Pinilla, F., and Cotman, C. W., 1994, Seizure-associated induction of basic fibroblast growth factor and its receptor in the rat brain, *Neuroscience* **60:**311–323.

Van der Zee, C. E. E. M., Ross, G. M., Riopelle, R. J., and Hagg, T., 1996, Survival of cholinergic forebrain neurons in developing p75NGFR-deficient mice, *Science* **274:**1729–1732.

Vantini, G., Schiavo, N., Di Martino, A., Polato, P., Triban, C., Callegaro, L., Toffano, G., and Leon, A., 1989, Evidence for a physiological role of nerve growth factor in the central nervous system of neonatal rats, *Neuron* **3:**267–273.

Walicke, P. A., 1988a, Basic and acidic fibroblast growth factors have trophic effects on neurons from multiple CNS regions, *J. Neurosci.* **8:**2618–2627.

Walicke, P. A., 1988b, Interactions between basic fibroblast growth factor (FGF) and glycosoaminoglycans in promoting neurite outgrowth, *Exp. Neurol.* **102:**144–148.

Walicke, P. A., and Baird, A., 1988, Neurotrophic effects of basic and acidic fibroblast growth factors are not mediated through glial cells, *Brain Res.* **468:**71–79.

Walicke, P., Cowan, W. M., Ueno, N., Baird, A., and Guillemin, R., 1986, Fibroblast growth factor promotes survival of dissociated hippocampal neurons and enhances neurite extension, *Proc. Natl. Acad. Sci. USA* **83:**3012–3016.

Wanaka, A., Johnson, E. M., Jr., and Milbrandt, J., 1990, Localization of FGF receptor mRNA in the adult rat central nervous system by *in situ* hybridization, *Neuron* **5:**267–281.

Wen, T. C., Matsuda, S., Yoshimura, H., Kawabe, T., and Sakanaka, M., 1995, Ciliary neurotrophic factor prevents ischemia-induced learning disability and neuronal loss in gerbils, *Neurosci. Lett.* **191:**55–58.

Weskamp, G., Gasser, U. E., Dravid, A. R., and Otten, U., 1986, Fimbria-fornix lesion increases nerve growth factor content in adult rat septum and hippocampus, *Neurosci. Lett.* **70:**121–126.

West, M. J., 1993, New stereological methods for counting neurons, *Neurobiol. Aging* **14:**275–285.

Wetmore, C., and Olson, L., 1993, Expression and regulation of neurotrophins and their receptors in hippocampal systems, *Hippocampus* **3:**171–182.

Wetmore, C., Ernfors, P., Persson, H., and Olson, L., 1990, Localization of brain-derived neurotrophic factor mRNA to neurons in the brain by *in situ* hybridization, *Exp. Neurol.* **109:**141–152.

Wetmore, C., Cao, Y. H., Pettersson, R. F., and Olson, L., 1991, Brain-derived neurotrophic factor: Subcellular compartmentalization and interneuronal transfer as visualized with anti-peptide antibodies, *Proc. Natl. Acad. Sci. USA* **88:**9843–9847.

Whittemore, S., Ebendal, T., Larkfors, L., Olson, L., Seiger, A., Stromberg, I., and Persson, H., 1986, Development and regional expression of beta nerve growth factor messenger RNA and protein in the rat central nervous system, *Proc. Natl. Acad. Sci. USA* **83:**817–821.

Widmer, H. R., and Hefti, F., 1994a, Neurotrophin-4/5 promotes survival and differentiation of rat striatal neurons developing in culture, *Eur. J. Neurosci.* **6:**1669–1679.

Widmer, H. R., and Hefti, F., 1994b, Stimulation of GABAergic neuron differentiation by NT-4/5 in cultures of rat cerebral cortex, *Dev. Brain Res.* **80:**279–284.

Widmer, H. R., Knüsel, B., and Hefti, F., 1992, Stimulation of phosphatidylinositol hydrolysis by brain-derived neurotrophic factor and neurotrophin-3 in rat cerebral cortical neurons developing in culture, *J. Neurochem.* **59:**2113–2124.

Widmer, H. R., Knüsel, B., and Hefti, F., 1993, BDNF protection of basal forebrain cholinergic neurons after axotomy: Complete protection of p75NGFR-positive cells, *NeuroReport* **4:**363–366.

Wilkinson, G. A., Fariñas, I., Backus, C., Yoshida, C. K., and Reichardt, L. F., 1996, Neurotrophin-3 is a survival factor *in vivo* for early mouse trigeminal neurons, *J. Neurosci.* **16:**7661–7669.

Williams, L. R., Varon, S., Peterson, G. M., Wictorin, K., Fischer, W., Björklund, A., and Gage, F. H., 1986, Continuous infusion of nerve growth factor prevents basal forebrain neuronal death after fimbria fornix transection, *Proc. Natl. Acad. Sci. USA* **83:**9231–9235.

Winkler, J., Ramirez, G. A., Kuhn, H. G., Peterson, D. A., Day-Lollini, P. A., Stewart, G. R., Tuszynski, M. H., Gage, F. H., and Thal, L. J., 1997, Reversible Schwann cell hyperplasia and sprouting of sensory and sympathetic neurites after intraventricular administration of nerve growth factor, *Ann. Neurol.* **41:**82–93.

Winter, C. G., Saotome, Y., Levison, S. W., and Hirsh, D., 1995, A role for ciliary neurotrophic factor as an inducer of reactive gliosis, the glial response to central nervous system injury, *Proc. Natl. Acad. Sci. USA* **92:**5865–5869.

Wong, V., Arriaga, R., Ip, N. Y., and Lindsay, R. M., 1993, The neurotrophins BDNF, NT-3, and NT-4/5, but not NGF, up-regulate the cholinergic phenotype of developing motor neurons, *Eur. J. Neurosci.* **5:**466–474.

Woodhams, P. L., and Atkinson, D. J., 1996, Regeneration of entorhino-dentate projections in organotypic slice cultures: Mode of axonal regrowth and effects of growth factors, *Exp. Neurol.* **140:**68–78.

Woodward, W. R., Nishi, R., Meshul, C. K., Williams, T. E., Coulombe, M., and Eckenstein, F. P., 1992, Nuclear and cytoplasmic localization of basic fibroblast growth factor in astrocytes and CA2 hippocampal neurons, *J. Neurosci.* **12:**142–152.

Wu, K., Xu, J. L., Suen, P. C., Levine, E., Huang, Y. Y., Mount, H. T. J., Lin, S. Y., and Black, I. B., 1996, Functional trkB neurotrophin receptors are intrinsic components of the adult brain postsynaptic density, *Mol. Brain Res.* **43:**286–290.

Xu, W. M., Guenard, V., Kleitman, N., Aebischer, P., and Bunge, M. B., 1995, A combination of BDNF and NT-3 promotes supraspinal axonal regeneration into Schwann cell grafts in adult rat thoracic spinal cord, *Exp. Neurol.* **134:**261–272.

Yamada, K., Kinoshita, A., Kohmura, E., Sakaguchi, T., Taguchi, J., Kataoka, K., and Hayakawa, T., 1991, Basic fibroblast growth factor prevents thalamic degeneration after cortical infarction, *J. Cereb. Blood Flow Metab.* **11:**472–478.

Yamaguchi, T. P., and Rossant, J., 1995, Fibroblast growth factors in mammalian development, *Curr. Opin. Genet. Dev.* **5:**485–491.

Yamaguchi, T. P., Harpal, K., Henkemeyer, M., and Rossant, J., 1994, FGFR-1 is required for embryonic growth and mesodermal patterning during mouse gastrulation, *Genes Dev.* **8:**3032–3044.

Yan, Q., Radeke, M. J., Matheson, C. R., Talvenheimo, J., Welcher, A. A., and Feinstein, S. C., 1997, Immunocytochemical localization of TrkB in the central nervous system of the adult rat, *J. Comp. Neurol.* **378:**135–157.

Yazaki, N., Hosoi, Y., Kawabata, K., Miyake, A., Minami, M., Satoh, M., Ohta, M., Kawasaki, T., and Itoh, N., 1994, Differential expression patterns of mRNAs for members of the fibroblast growth factor receptor family, *FGFR-1-FGFR-4,* in rat brain, *J. Neurosci. Res.* **37:**445–452.

Yoshimoto, Y., Lin, Q., Collier, T. J., Frim, D. M., Breakefield, X. O., and Bohn, M. C., 1995, Astrocytes retrovirally transduced with BDNF elicit behavioral improvement in a rat model of Parkinson's disease, *Brain Res.* **691:**25–36.

Yurek, D. M., Lu, W., Hipkens, S., and Wiegand, S. J., 1996, BDNF enhances the functional reinnervation of the striatum by grafted fetal dopamine neurons, *Exp. Neurol.* **137:**105–118.

# 8

# Cortical Neuropathology in Aging and Dementing Disorders

## Neuronal Typology, Connectivity, and Selective Vulnerability

PATRICK R. HOF, CONSTANTIN BOURAS, and JOHN H. MORRISON

## 1. Introduction

Alzheimer's disease (AD) represents the most common form of dementia in elderly individuals, affecting nearly 11% of the population over 65 years of age. Possibly up to 50% of those over 85 can be diagnosed clinically as having "probable AD" (Moss and Albert, 1988; Evans *et al.*, 1992). Clinical and neuropathological studies also indicate that cases with AD alone or in combination with other diseases account for

PATRICK R. HOF • Kastor Neurobiology of Aging Laboratories, Fishberg Research Center for Neurobiology and Department of Geriatrics and Adult Development, and Department of Ophthalmology, Mount Sinai School of Medicine, New York, New York 10029. CONSTANTIN BOURAS • Department of Psychiatry, HUG Belle-Idée, University of Geneva School of Medicine, CH-1225 Geneva, Switzerland; and Kastor Neurobiology of Aging Laboratories, Mount Sinai School of Medicine, New York, New York 10029. JOHN H. MORRISON • Kastor Neurobiology of Aging Laboratories, Fishberg Research Center for Neurobiology and Department of Geriatrics and Adult Development, Mount Sinai School of Medicine, New York, New York 10029.

*Cerebral Cortex,* Volume 14, edited by Peters and Morrison. Kluwer Academic/Plenum Publishers, New York, 1999.

more than 50% of all demented patients over 65 (Tomlinson *et al.*, 1968; Mortimer *et al.*, 1981; Fallet-Bianco *et al.*, 1990; Jorm, 1991; Copeland *et al.*, 1992; Fratiglioni *et al.*, 1991). Our understanding of the structural, molecular, and biochemical nature of AD has grown substantially in recent years, yet the causal pathogenetic events that lead to dementia are not known. Many hypotheses on the development of the typical lesions of AD have been proposed and, despite a considerable diversity of data, it appears that the structural and functional changes seen in AD isolate the cerebral cortex as the predominant substratum of the clinical symptomatology observed in demented patients (De Lacoste and White, 1993; Morrison, 1993; Hof and Morrison, 1994, 1996; Hof *et al.*, 1995a; Hof, 1997). When considering the pathology at regional and cellular levels, however, a generalized loss of cortical function appears unlikely, as AD involves certain populations of neurons displaying a specific regional and laminar distribution and specific connectivity patterns, whereas other neuron types remain largely unaffected (Morrison *et al.*, 1987; Morrison, 1993; Hof and Morrison, 1994, 1996; Hof *et al.*, 1995a; Gómez-Isla *et al.*, 1996, 1997; Bobinski *et al.*, 1997; Hof, 1997; Morrison and Hof, 1997). Thus, different degrees of neuronal vulnerability exist in AD that may be described in terms of morphologic and biochemical characteristics among distinct neuronal populations and sets of connections. In addition, the study of pathologic alterations associated with normal brain aging may offer clues as to the cellular mechanisms responsible for the development of dementia, because many neuronal populations that are devastated in early AD undergo mild but consistent changes in cognitively intact elderly individuals as well.

Neuroanatomical studies of the cellular organization and connectivity of the nonhuman primate brain have considerably expanded our knowledge of the organization of the major afferent and efferent systems of the cerebral cortex as well as its intrinsic circuitry. Such investigation have been particularly illuminating in regard to the organization of neocortical and hippocampal circuits, and the use of specific histochemical techniques has permitted detailed characterization of neuron typology and connectivity and of the neurochemical characteristics of identified circuits (Campbell and Morrison, 1989; Hof and Morrison, 1991, 1995; Hof *et al.*, 1990a, 1995b,c, 1996b, 1997a,b; Kisvárday *et al.*, 1990; Campbell *et al.*, 1991; Morrison and Hof, 1992, 1997; Nimchinsky *et al.*, 1996). In particular, certain neuronal types within the cerebral cortex and some of the major cortical afferent systems have now been characterized as to the presence of key structural proteins, neurotransmitters, and receptors. The human brain has become the object of intensive scrutiny with chemically specific methods, although there are obvious limitations in the application of experimental procedures to the human brain. Data from nonhuman primates, however, have fostered the development of organizational schemes for the cerebral cortex that relate various molecules of interest to specific groups of neurons or circuits (Carmichael and Price, 1994; Hof and Morrison, 1995; Hof *et al.*, 1995, 1996b, 1997a,b; Nimchinsky *et al.*, 1996), and data drawn from anatomical and functional imaging studies suggest that comparable cortical hierarchies exist in the human cerebral cortex (e.g., Clarke and Miklossy, 1990; Clarke, 1993, 1994, 1995; Miklossy, 1993; Watson *et al.*, 1993; Sereno *et al.*, 1995; Tootell *et al.*, 1995; Tootell and Taylor, 1995; DeYoe *et al.*, 1996). Also, cross-correlations among human and nonhuman primates are made possible by the fact that many of the histochemical techniques are very reliable in the postmortem human brain. Although inferring information on the organizational principles of the human brain directly from the macaque monkey brain should clearly be done with caution, such correlations have been useful in regard to the neuropathology of brain aging and AD, as well as to the normal functional anatomy of the human cerebral cortex.

Most of the clinicopathologic studies of normal brain aging and AD are based on relatively small patient samples (Tomlinson *et al.*, 1968, 1970; Katzman *et al.*, 1988; Dickson *et al.*, 1991; Morris *et al.*, 1991; J. L. Price *et al.*, 1991; Arriagada *et al.*, 1992a,b; Crystal *et al.*, 1982; Hof *et al.*, 1992b; Berg *et al.*, 1993; Bierer *et al.*, 1995). The study of large retrospective series is essential to evaluate precisely the age-related factors leading to lesion formation and to define the qualitative and quantitative differences in lesion distribution between elderly subjects without cognitive deficits, patients with moderate age-associated memory impairment, and patients with dementia typical of AD. Such analyses are crucial to determine whether AD is the end point of a continuum with normal brain aging (Giannakopoulos *et al.*, 1994b, 1997b). It is also important to examine the neuropathological changes in dementing illnesses other than AD, because the differences in distribution and structural composition of lesions in those disorders may provide important clues to the origin and progression of the pathology. In this chapter we review data on the relationship between the distribution of pathologic changes and the localization of specific elements of the cortical pathways affected by these alterations in both normal brain aging and AD, as well as in certain related dementing conditions. We also discuss the relationship between the neurochemical phenotype of a given neuron and its relative vulnerability or resistance to the degenerative process in view of the correlations with distribution of cellular pathologic changes, neurochemical characteristics related to vulnerability, and affected cortical circuits.

## 2. Neuropathological Evidence of Selective Neuronal Vulnerability in Aging and Alzheimer's Disease

The fact that many neuropsychiatric diseases involve the degeneration of identifiable cortical neuronal subsets with disease-specific pathological distribution patterns has been long suspected. As early as 1910, Joseph Shaw Bolton reported in his Goulstonian lectures that in the case of many psychiatric conditions, including senile dementia, clear clinicopathological correlations existed between the regional and laminar distribution of cellular changes observed in various cortical areas and the clinical symptomatology (Bolton, 1910). Later, Morel (1945) and Mutrux (1947) pointed to the fact that the projection layers of the neocortex (i.e., layers III and V) are selectively affected in AD, and that there is an increase in severity of pathological changes along cortical hierarchies, the primary sensory cortices being less affected than their target association areas. This pattern, since confirmed by many authors, indicates that the cellular alterations in aging and senile dementia are not ubiquitous or uniform, but take place in restricted, identifiable populations of cortical neurons, and that these cell-specific alterations are reflected by the local distribution patterns of histopathological AD markers (Morel and Wildi, 1952, 1955; Goodman, 1953; Mutrux, 1958; Wildi *et al.*, 1964; Jamada and Mehraein,

1968; Wildi and Dago-Akribi, 1968; Droz *et al.*, 1982; Kemper, 1984; Pearson *et al.*, 1985; Rogers and Morrison, 1985; Duyckaerts *et al.*, 1986; Esiri *et al.*, 1986; Lewis *et al.*, 1987; Hof *et al.*, 1989, 1990a,b, 1993a, 1997c; Hof and Morrison, 1990; Arnold *et al.*, 1991; Braak and Braak, 1991; Hof and Bouras, 1991; Bouras *et al.*, 1994; De Lacoste and White, 1993; Vogt *et al.*, this volume).

## 2.1. Neuropathological Changes

### 2.1.1. Macroscopic Changes

In cognitively intact elderly people, a mild degree brain atrophy is often observed in the mesial and inferior temporal lobe structures as well as in the frontal and parietal neocortex in comparison to younger individuals. The cause of this atrophy in intact people is poorly understood, but it contrasts sharply with overt AD, where gross changes have been described. In fact, recent studies using prospective magnetic resonance imaging in populations of intellectually preserved and mildly affected elders have demonstrated that even in the presence of very mild cognitive impairment, hippocampal volumes can be about 25% smaller than in younger individuals and intellectually intact elders, and that at-risk individuals from families of late-onset AD experience an 8% annual reduction in hippocampal volume. Hippocampal and parahippocampal volume reduction in individuals with minimal cognitive impairment also show region-specific patterns that permit us to distinguish between these two groups. The atrophy detected by imaging techniques can be considered a pathognomonic feature if clinical symptoms of dementia are present (Killiany *et al.*, 1993; Jobst *et al.*, 1994; Lehéricy *et al.*, 1994; Fox *et al.*, 1996; Convit *et al.*, 1997; Fama *et al.*, 1997). Shrinkage of the gyri is usually associated with a loss of brain weight. In AD the gyral atrophy is most often apparent in the frontal and temporal association areas, but the parietal lobe may also be severely affected. Cerebral atrophy is also seen on the ventral surface of the temporal lobe, and in particular in the parahippocampal gyrus. The paracentral gyri are less affected by atrophy than are the frontal and temporal areas, but in many cases they do display mild atrophy. No region of the cerebral cortex is totally spared, but the occipital areas are relatively preserved. There is a thickening of leptomeninges over the convexity and near the midline, and leptomeningeal blood vessels are often infiltrated with amyloid Aβ protein in aging (Surbeck, 1961; Hamano *et al.*, 1997). Additional gross changes include ventricular dilation, enlargement of the temporal horn, and pallor of the substantia nigra and locus coeruleus (see Terry *et al.*, 1981, 1994; Katzman and Terry, 1983; Mountjoy *et al.*, 1983; Kemper, 1984, for reviews).

### 2.1.2. Neurofibrillary Tangles and Senile Plaques

In addition to the macroscopic changes, both AD and normal brain aging are characterized by the presence of several cellular alterations throughout the cerebral cortex. In particular, the distribution and density of neurofibrillary tangles (NFTs) and senile plaques (SPs), the two classical hallmarks of AD, have been studied in great detail, and their presence in certain key cortical regions is required of the neuropathologic diagnosis of AD (Terry and Katzman, 1983; Mirra *et al.*, 1993). Neurofibrillary tangles represent the accumulation of abnormal components of the neuronal cytoskeleton that form paired helical filaments and are composed of intracellular argentophilic fibers that are also intensely labeled by histochemical stains such as thioflavin S (Kidd, 1963, 1964; Terry and Wisniewski, 1970; Brion, 1990; Vallet *et al.*, 1992; Vickers *et al.*, 1992, 1996; Vickers, 1997) (Fig. 1A). In the hippocampus and neocortex, NFTs occur in the perikaryon of large pyramidal neurons, and SPs are found throughout the cortical layers with a preferential distribution in layers II, III, and V of neocortical association regions (Dayan, 1970a,b; Shefer, 1972, 1976; Terry and Katzman, 1983; Kemper, 1984; Perry, 1986; Rogers and Morrison, 1985; Lewis *et al.*, 1987; Arnold *et al.*, 1991). End-stage NFTs, referred to as extracellular or ghost tangles, are also observed. Electron microscopy of NFTs reveals the presence of paired helical filaments formed by an apparent pair of strands, wound around one another with crossover repeats of approximately 75–80 nm, and widths ranging from 5 to 22 nm (Kidd, 1963, 1964; Terry and Wisniewski, 1970). A variety of cytoskeletal proteins, such as microtubule-associated protein tau, and the neurofilament triplet protein, as well as ubiquitin, are involved in paired helical filament formation (Brion, 1990; Trojanowski *et al.*, 1993b; Vickers *et al.*, 1996). The microtubule-associated protein tau is the main component of paired helical filaments. A total of six tau isoforms can be generated from a single tau gene, located on chromosome 17, by alternative splicing. The expression of tau is developmentally regulated, so that all six tau isoforms are found in the adult human brain, whereas only the smallest isoform, known as fetal tau, is present in the fetal human brain (Bramblett *et al.*, 1992; Matsuo *et al.*, 1994). Immunoblot

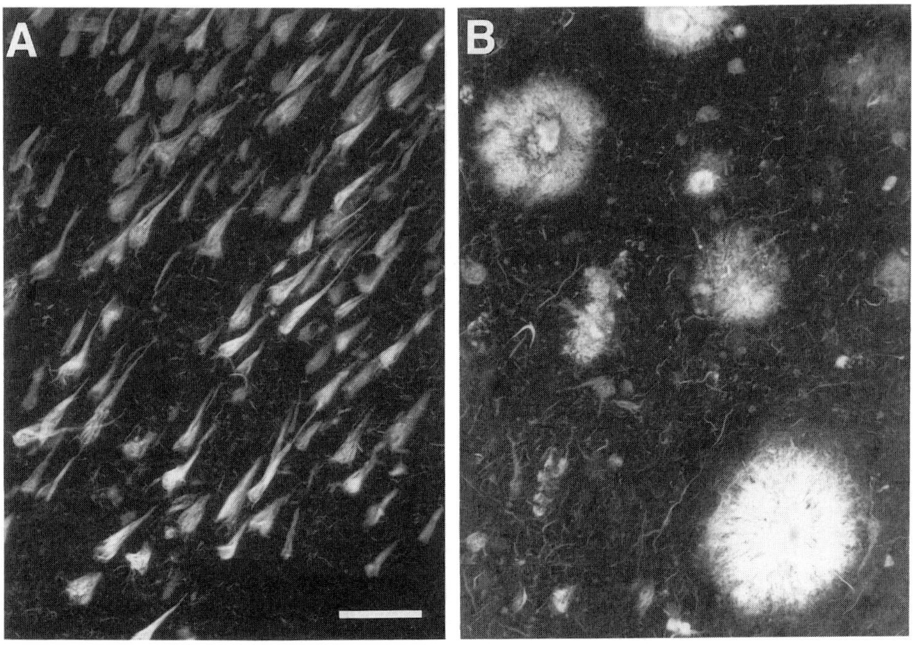


**Figure 1.** Examples of NFTs (A) and SPs (B) from the hippocampus of a severe AD case. Note the flame-shaped morphology of NFTs and the more variable features of SPs. In B a classical plaque with a central amyloid core and a corona of degenerating neurites, an isolated dense amyloid core, a predominantly neuritic plaque, and a few small diffuse deposits can be easily recognized. Thioflavine S staining using the method described by Vallet *et al.* (1992). Bar = 50 μm.

studies in homogenates of postmortem AD brain show that adsorbed anti-paired helical filaments recognize an abnormally phosphorylated tau protein triplet, known as tau 55, 64, 69 (Delacourte *et al.*, 1990; Goedert *et al.*, 1994; Buée-Scherrer *et al.*, 1996a). Recently, an additional 74 kDa tau isoform has been observed in AD brain extracts in high concentrations preferentially in young demented cases, suggesting the existence of age-related biochemical modifications of tau proteins (Sergeant *et al.*, 1997). Paired helical filament-tau isoforms are generally considered to be hyperphosphorylated compared with normal adult tau, but recent analyses have demonstrated that biopsy-derived normal adult human tau is phosphorylated *in vivo* at the same sites as those on paired helical filament tau (Bramblett *et al.*, 1992; Matsuo *et al.*, 1994). It is thus possible that a down-regulation of phosphatases in the AD brain induces the generation of hyperphosphorylated paired helical filament tau that does not bind microtubules and that aggregate as paired helical filaments in NFTs (see Ginsberg *et al.*, this volume).

Senile plaques (SPs) are composed of dystrophic neurites and glial elements with a central amyloid core (Terry and Wisniewski, 1970; Brion, 1990). The classic SP consists of amyloid Aβ protein deposits intensely labeled by thioflavine S (the plaque core), surrounded by degenerating axon terminals and dendritic segments (Fig. 1B). Some SPs display only an amyloid core without dystrophic neurites, whereas others are formed uniquely by dystrophic neurites and glial elements and are frequently referred to as neuritic plaques. Finally, diffuse amyloid deposits are usually observed, and are especially prominent in oldest-old AD patients (Hauw *et al.*, 1986). The major biochemical component of amyloid deposits is the Aβ protein, a polypeptide of 40–42 amino acid residues derived through proteolytic cleavage from a set of larger protein isoforms collectively referred to as the amyloid precursor protein (APP), that are encoded by a single gene on chromosome 21 (Brion, 1990). Three major isoforms of APP are generated from this gene by alternative splicing (APP 770, 751, 695). All three isoforms are found in the human brain (Robakis, 1994; Checler, 1995). Structurally, APP is a protein of 110–135 kDa, with a large extracellular N-terminal domain, a cell-surface domain, and a smaller intracellular C-terminal domain. The proteolytic cleavage of this molecule in the Aβ domain is catalyzed by a group of enzymes known as α, β, and γ secretases (Checler, 1995). It is thought that the cleavage of APP outside the Aβ domain is responsible for generating large C-terminal fragments containing the Aβ sequence that are possibly amyloidogenic. Two distinct species of Aβ protein with different C-terminals, Aβ40 and Aβ42(43), are deposited in the brains of patients with AD (Mann *et al.*, 1996). In addition, the composition of the Aβ protein found in the meninges in AD is only slightly different from that in the parenchymal deposits with predominance of Aβ40, whereas both Aβ40 and Aβ42 are found in normal aging (Hamano *et al.*, 1997). The mechanism of SP formation is, however, not fully understood. Classic SPs may develop from diffuse amyloid deposits, although no consistent colocalization of these two forms of SP has been observed. Alternatively, SPs may arise from amyloid in or near the wall of the cerebral blood vessels, but most SPs are not associated with amyloidotic microvessels. A third hypothesis involves the release of Aβ sequence from clusters of dystrophic neurites, with subsequent formation of diffuse Aβ deposits and classic SPs (see Terry *et al.*, 1994, for review).

Whereas AD is characterized by substantial amounts of NFTs and SPs in the hippocampus and association regions of the neocortex, these lesions may also occur in normal aging, but in much lower numbers (Dayan, 1970b). In fact, as will be discussed in greater detail, only certain hippocampal circuits are affected by

NFT formation, and no clear distribution and densities of SP patterns seem to exist in normal brain aging (Arriagada *et al.*, 1992a,b; Bouras *et al.*, 1993, 1994). In addition, rarer cases of AD with early onset (before 65 and as early as 30 years), many occurring in families and linked to mutations in the APP gene or in the presenilin-1 and -2 genes on chromosomes 14 and 1, respectively, exhibit a much more aggressive form of dementia than cases of late-onset AD. In these younger cases, the distribution of cortical atrophy is similar to that in late-onset AD but is much more severe, and NFTs and SPs are observed in much higher densities in association neocortex and hippocampus with these younger AD patients (Giannakopoulos *et al.*, 1996b).

### 2.1.3. Neuronal and Synaptic Loss in AD

Severe neuronal loss is consistently observed in the brains of demented patients and involves mainly the hippocampus and association neocortex rather than primary sensory and motor areas (Colon, 1972, 1973; Shefer, 1972; Ball, 1977; Terry *et al.*, 1981, 1987; Mountjoy *et al.*, 1983; Coleman and Flood, 1987; Hof *et al.*, 1990a; Davies *et al.*, 1992; Gómez-Isla *et al.*, 1996, 1997). Neuronal loss involves the largest pyramidal cells and is significantly correlated with the presence of NFTs in neocortical association areas (Mountjoy *et al.*, 1983; Bondareff *et al.*, 1989; Hof *et al.*, 1990a). In normal aging neuronal loss has long been considered to be substantial (Brody, 1955), but newer, more-rigorous methods to estimate neuron numbers have demonstrated that it is in fact minimal (Haug *et al.*, 1984; Coleman and Flood, 1987; Pakkenberg and Gundersen, 1997). In AD cases neuron loss is present only in the subiculum and in the dentate gyrus, showing that, in terms of neuronal death, aging and AD depend on mechanisms that affect nonequivalent populations of neurons, at least in the hippocampal subdivisions (West and Gundersen, 1990; West, 1993; West *et al.*, 1994). Estimates of neuron loss in the hippocampal formation of elderly humans varies among studies, however, and, although West and colleagues demonstrated a loss in the hilus and subiculum, a recent stereological analysis of total neuron numbers in various hippocampal subdivisions revealed correlation between age and neuron numbers in CA1 and subiculum, and severe neuron loss in the dentate gyrus and subiculum, only in AD cases (Simic *et al.*, 1997). These discrepancies may be due to slight differences in the quantification methods employed and may also reflect the substantial anatomical variability that exists among individual human brains.

Synapse loss may be another early marker of the dementing process. Decreases in synapse density, as measured by synapse number estimates or via synaptophysin immunoreactivity, are strongly correlated with the cognitive impairment as assessed by Blessed or Mini-Mental scores (DeKosky and Scheff, 1990; Terry *et al.*, 1991; Masliah *et al.*, 1991c, 1993, and this volume; Sze *et al.*, 1997), a correlation even stronger than that of NFTs to mental decline, indicating that synapse loss might be a more accurate neuropathologic reflection of the circuit disruption underlying dementia (Terry *et al.*, 1991).

### 2.1.4. Other Histopathological Changes

Other pathologic alterations commonly seen in AD brains include neuropil threads, granulovacuolar degeneration, diffuse amyloid deposits, and amyloid angiopathy. Neuropil threads are composed of degenerating elements, presumably of dendritic origin, from NFT-bearing neurons, and they contain paired helical filaments (Braak *et al.*, 1986; Braak and Braak, 1988). They seem to appear early in the course of the disease and are particularly prevalent in the hippocampus and entorhinal cortex (Braak and Braak, 1991; Vickers *et al.*, 1992, 1994b). Granulovacuolar degeneration affects principally the hippocampal neurons in the subiculum and CA1 fields (Tomlinson and Kitchener, 1972). The significance and etiopathogenesis of this pathologic profile has not been elucidated, but this lesion is fairly specific for a certain group of NFT-prone hippocampal neurons. In addition, granulovacuolar lesions share immunogenic characteristics with NFTs, in that they are stained by antibodies to the microtubule-associated protein tau (Dickson *et al.*, 1987b; Bondareff *et al.*, 1991). No clear correlation, however, between the density of granulovacuolar degenerations and NFTs has been established (Tomlinson *et al.*, 1968; Dayan, 1970a). The deposition of the Aβ amyloid protein in the form of large diffuse plaque-like structures without amyloid cores seems to be a common feature in both AD and normal brain aging, and is observed throughout the cerebral cortex and in some subcortical regions as well. In the cerebral cortex, these deposits, referred to as diffuse plaques (Clinton *et al.*, 1992), predominate in the parahippocampal gyrus and in the superficial layers of the neocortex, and they do not appear to be related to the progression of the dementia. The presence of amyloid deposition in the wall of cerebral blood vessels has been frequently associated with the occurrence of dementia (Scholz, 1938; Morel, 1943; Pantelakis, 1954; Surbeck, 1961), although it is not a consistent feature of the dementing process. In addition, such vascular pathology can be observed in many cortical regions, and its presence does not appear to be systematically correlated with a higher incidence of NFTs or SPs. It is worth noting that certain domains of the cerebral vascular bed are more affected by amyloid angiopathy than others, in particular the rich microvascular plexus in layer IVC of the primary visual cortex (Morel, 1943; Pantelakis, 1954; Surbeck, 1961; Buée *et al.*, 1994) (Fig. 2), yet these areas are not particularly prone to NFT formation or cell loss. Finally, the presence of plaque-like structures that lack Aβ immunoreactivity but are labeled by antibodies to a 100kDa molecule probably related to paired helical filaments has been reported in aged and AD brains (Schmidt *et al.*, 1997). Though the exact nature of the protein recognized by these antibodies is unclear, it has been proposed that these structures represent the deposition of a nonamyloid precursor of at least certain SP.

### 2.2. Distribution of Cortical Neurofibrillary Tangles and Senile Plaques

The distribution of pathological markers in the cerebral cortex of AD is neither random or generalized, and recent quantitative analyses of neuron loss and NFT and SP distribution in neocortex and hippocampus have found correlations between the localization of SPs and NFTs in specific cortical regions and layers, and in neurons furnishing subsets of long corticocortical as well as key hippocampal projections (Mutrux, 1947; Kemper, 1984; Pearson *et al.*, 1985; Rogers and Morrison, 1985; Duyckaerts *et al.*, 1986; Esiri *et al.*, 1986; Hyman *et al.*, 1986; Lewis *et al.*, 1987; Braak *et al.*, 1989; Hof *et al.*, 1990a; Hoff and Morrison, 1990; Arnold *et al.*, 1991; Morrison and Hof, 1997) (Figs. 3–5). Overall, the *pattern* of neuronal loss closely matches that of NFT distribution in the hippocampal formation and

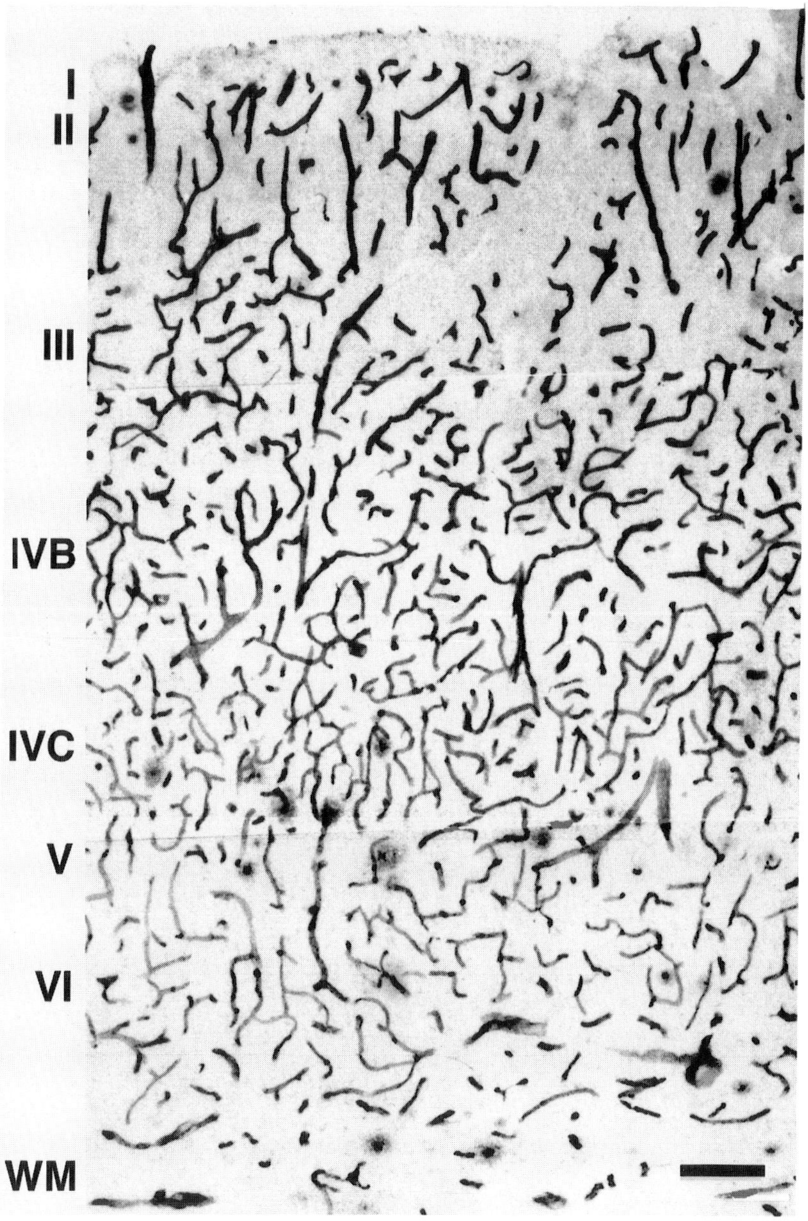


**Figure 2.** Distribution of cerebral microvessels in the primary visual cortex in an AD case. Note the increasing density of blood vessels in layer IVC and the presence in this layer of SPs and amyloid deposits in contact with the microvasculature (Buée *et al.*, 1992). Microvessels and amyloid deposits stained by an antibody to the heparan sulfate proteoglycan core protein. Bar = 100 μm.

neocortex (Ball, 1977; Terry *et al.*, 1981; Bondareff *et al.*, 1989; Hof *et al.*, 1990a; Vogt *et al.*, 1990). Recently, however, it has been reported that in certain regions of the cerebral cortex, NFT densities could not reliably account for the whole extent of neuron loss, and that differential laminar changes of this latter measure could represent a better index to assess the severity of the pathologic changes (Hof *et al.*,

1990a; Vogt *et al.,* 1990, 1992; Gómez-Isla *et al.,* 1996, 1997). Also, in absence of NFTs, the quantity of SPs does not correlate with the degree of cognitive impairment.

In the hippocampal formation, consistent patterns of change are observed, with some subfields severely devastated and other, adjacent regions almost completely unaltered. It is well known since the topographic study of Hirano and Zimmerman (1962) that layer II of the entorhinal cortex, the subiculum, and the CA1 field of the hippocampus represent particularly vulnerable sites that systematically display very high NFT densities in AD (Fig. 4). Layers III and V of the entorhinal cortex are less severely affected (Hyman *et al.,* 1984, 1986; Brady and Mufson,

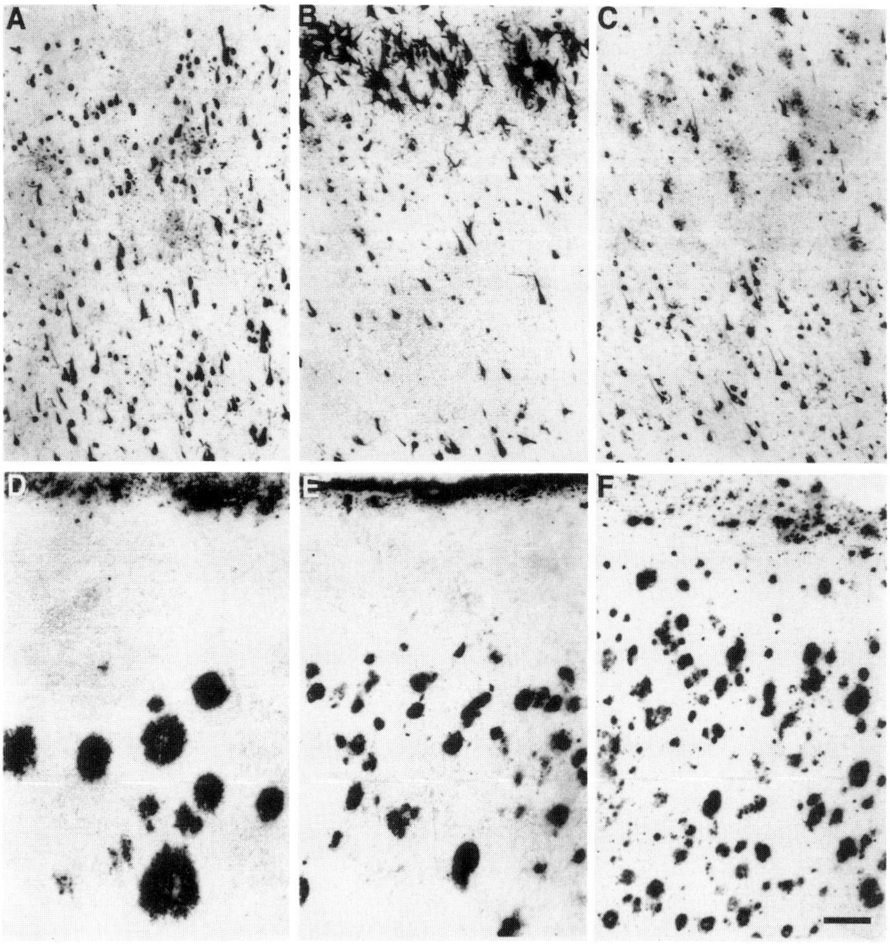


**Figure 3.** Regional patterns of NFT (A–C) and SP (D–F) distribution in the CA1 field of the hippocampus (A, D), layers I–III of the entorhinal cortex (B, E), and in layers III–V (C) and I–III (F) of the inferior temporal cortex (area 20) in an AD case. Note the high NFT densities in all of these regions. Layer II of the entorhinal cortex contains particularly high numbers of large NFTs (B), and there is a bilaminar pattern in layers III and V of area 20, with more NFTs in the deep layers (C). There are fewer SPs in the CA1 field compared with the other regions (D–F). Materials were stained with antibodies AD2 to hyperphosphorylated tau proteins (A–C) or 953 to amyloid Aβ protein. Bar = 50 μm.

1991; Vickers *et al.*, 1992). In the hippocampus, the most affected zones are the CA1 and subiculum, whereas the parasubiculum, CA2, CA3, and the dentate gyrus are much less affected. In addition, the area of transition between the allocortex and the temporal neocortex (the transentorhinal region) (Braak and Braak, 1985) displays a very specific pattern of degeneration in AD in that NFTs outline a gradual sweeping of layer II downward through the outer layers, terminating in the deep



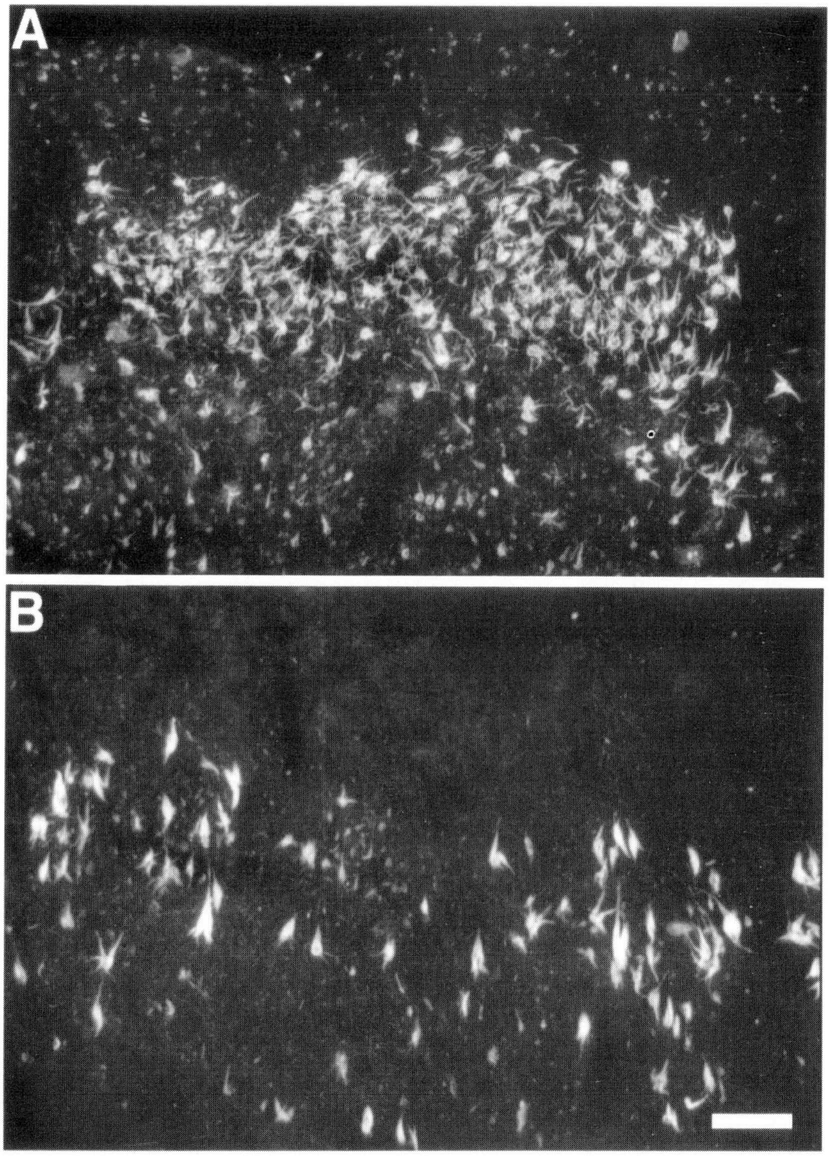


**Figure 4.** Examples of NFT distribution in layer II of the entorhinal cortex (A) and in the pyramidal cell layer of the CA1 field of the hippocampus (B) in a severe AD case, showing the massive involvement of these particular layers. Nearly all projection neurons in layer II of entorhinal cortex and a large number of pyramidal neurons in the CA1 field contain NFTs. Thioflavine S staining. Bar = 50 μm.

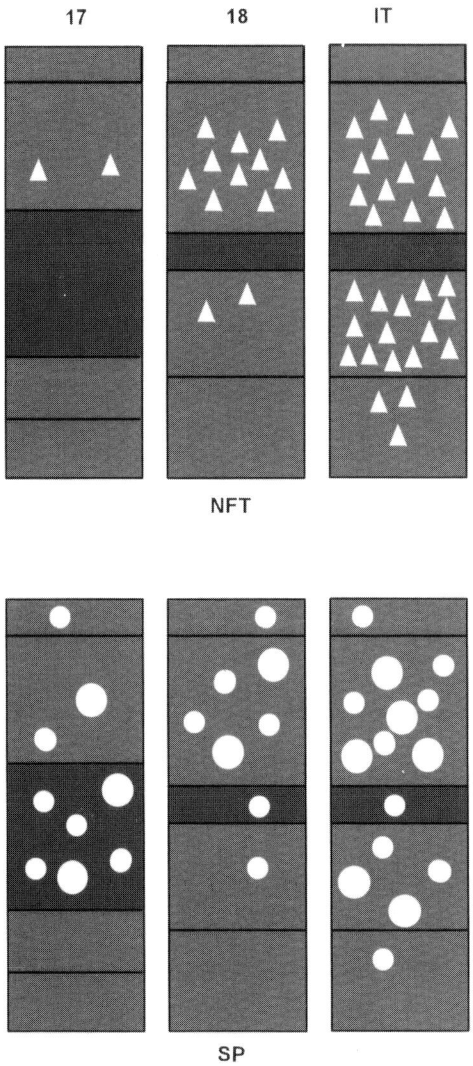


**Figure 5.** The laminar distribution of NFTs (triangles) and SPs (circles) follows the distribution of certain corticocortically projecting neurons along neocortical hierarchies. This schematic is based on three anatomic levels from the hierarchic organization of visual areas (Felleman and Van Essen, 1991; Ungerleider, 1995). In the primary visual cortex (area 17) only rare NFTs are seen in layer III that projects to the superficial layers of area 18. This corresponds to the presence of consistent SP densities in these layers in area 18. From the superficial layers of area 18, feedforward neurons projecting to higher hierarchical levels are involved, resulting in higher NFT densities. The terminal fields of the projections from area 18 in layers III and V/VI of area IT contain increasing SP densities. In area IT, NFTs are present in higher numbers not only in layer III but also in layer V and VI, indicating that feedback neurons are also involved by the degenerative process. This may be correlated to the consistently high SP densities in layers II–IV of area 17, which receives feedback projections from areas 18 and IT. Pearson *et al.* (1985); Rogers and Morrison (1985); Lewis *et al.* (1987); Hof *et al.* (1989, 1990b, 1997c); De Lacoste and White (1993); Morrison (1993).

portion of layer III. This site appears to be the earliest cortical region consistently affected during the progression of AD changes (Braak and Braak, 1991). The distribution of SPs in the hippocampal formation is somewhat variable, although certain zones, such as layer II of the entorhinal cortex, the molecular layer of the dentate gyrus, and the superficial layer of the subiculum (Hyman *et al.,* 1990), tend to contain high SP densities.

In the neocortex, NFTs reside primarily within layers III and V (Figs. 3, 5), with considerable variability in their regional densities. For instance, primary sensory and motor regions have far fewer NFTs than association areas (Pearson *et al.,* 1985; Esiri *et al.,* 1986; Lewis *et al.,* 1987; Hof and Morrison, 1990; Hof *et al.,* 1990a; Arnold *et al.,* 1991). Thus, when comparing primary (area 17) with secondary visual cortex (area 18), an approximately 20-fold increase in the number of NFTs is found in area 18, and inferior temporal cortex areas that probably correspond to visual association areas are characterized by a further doubling of the NFTs (Lewis *et al.*, 1987; Hof *et al.*, 1989, 1990b; De Lacoste and White, 1993) (Fig. 5). Similar differences are found between the primary auditory cortex and auditory association areas, which display a 10-fold increase in NFT density (Esiri *et al.*, 1986; Lewis *et al.*, 1987). There are also fundamental differences in laminar NFT distribution among neocortical areas. For instance, in areas 17 and 18, most of the NFTs are located in layer III, whereas in the inferior temporal cortex, only about 40% of the NFTs are in layer III. The auditory association regions display even greater laminar differences, with only 27% of the NFTs found in layers II and III (Lewis *et al.*, 1987). Generally, at the regional level, temporal cortex is first in NFT numbers, the frontal cortex second, the parietal cortex third, and the occipital cortex last (Arnold *et al.*, 1991); some areas of the cerebral cortex, such as the posterior cingulate cortex, exhibit large case-to-case discrepancies in NFT (and SP) counts (Vogt *et al.*, 1990, 1992, and this volume).

Senile plaques overall display a more even distribution among cortical areas than do NFTs, and minor differences in density are observed between closely related cortical areas, such as areas 17 and 18 (Lewis *et al.*, 1987; Hof and Morrison, 1990; Arnold *et al.*, 1991) (Fig. 5). SP distribution is far from being homogeneous, however (Wildi and Dago-Akribi, 1968), as the frontal and anterior cingulate cortex have been shown to contain more SPs than temporal regions (Rogers and Morrison, 1985). At early stages of dementia, SPs are not consistently observed in the temporal neocortex, as with NFTs, but may be first observed in the occipital cortex. In these regions, SPs are usually more numerous in layers II and III than in infragranular layers (Rogers and Morrison, 1985). Senile plaques also display characteristic laminar patterns in several areas of the cerebral cortex. In particular, they are abundant in layer IV of area 17, where they predominate in sublayer IVC and mark the border between this layer and layer V (Braak *et al.*, 1989; Beach and McGeer, 1992), as well as in the retrosplenial cortex, in the parvocellular zone of layer III (Braak *et al.*, 1992a). Although the significance of these subtle regional patterns is not fully elucidated, it is interesting to note that the high incidence of SPs in layer IVC of area 17 is frequently associated with the presence of amyloid angiopathy and a specialized microvasculature (Buée *et al.*, 1994) (Fig. 2), suggesting a possible relationship between these two lesions, at least in this particular region.

### 2.3. Correlations between Lesion Distribution and Specific Circuits

The distribution and density of NFTs as well as SPs in the cerebral cortex has been correlated with the degree of cognitive impairment in AD. Thus, there is a significant correlation between the mean SP frequency in several cortical regions and dementia severity assessed within six weeks of death using the Blessed Dementia Scale (Tomlinson *et al.*, 1970). Other quantitative studies of the association of SPs and NFTs with dementia severity in AD showed a stronger correlation with cortical NFTs, however (as well as synapse loss), than with SPs, particularly in the temporal neocortex, entorhinal cortex, and hippocampus (Wilcock *et al.*, 1982; Terry *et al.*, 1991; Arriagada *et al.*, 1992a; Bouras *et al.*, 1993, 1994). With the exception of certain visual areas, no strong correlation between NFT and SP severity exist among association neocortical areas (Bouras *et al.*, 1994). Moreover, synaptic damage and loss has been demonstrated in the neocortex of elderly nondemented individuals, suggesting an age-dependent mechanism for the loss of synapses in that region (Masliah *et al.*, 1993).

### 2.3.1. Early Involvement of the Projection from the Entorhinal Cortex to the Hippocampus

The presence of neuronal loss, NFTs, and SPs in the cerebral cortex of nondemented elderly individuals has been described by many authors (Brody, 1955; Wildi *et al.*, 1964; Tomlinson *et al.*, 1968; Wildi and Dago-Akribi, 1968; Dayan, 1970a; Ball, 1977; Droz *et al.*, 1982; Mountjoy *et al.*, 1983; Ulrich, 1985), and recent studies of cognitively intact elderly people have demonstrated that amyloid deposition and NFT formation are common features of normal brain aging and may be present in densities comparable to those observed in demented patients, but in restricted cortical territories (Crystal *et al.*, 1988; Katzman *et al.*, 1988; Hubbard *et al.*, 1990; Dickson *et al.*, 1991; J. L. Price *et al.*, 1991; Arriagada *et al.*, 1992a,b; Hof *et al.*, 1992b; Vickers *et al.*, 1992; Berg *et al.*, 1993; Bouras *et al.*, 1993, 1994; Giannakopoulos *et al.*, 1994b; Bierer *et al.*, 1995). For example, nondemented cases have been described that exhibit NFTs almost exclusively in the entorhinal cortex, with lower NFT counts in the CA1 and temporal neocortex, and numerous SPs in the temporal neocortex (Braak and Braak, 1990). Braak and Braak have shown that the transentorhinal cortex and layer II of the entorhinal cortex are consistently involved at the early stages of the degenerating process, and they have developed a neuropathologic staging scheme for dementia based on the progression of the alterations in the hippocampal formation (Braak and Braak, 1985, 1991). Their stages I to III exhibit mild involvement of the entorhinal cortex, supporting the notion that the medial temporal cortex is selectively and significantly affected before any clinical symptomatology is detected by neuropsychological testing. Thus, preclinical stages of dementia may be characterized by high NFT densities restricted to select areas of the medial and inferior temporal cortex, including hippocampus, entorhinal cortex, and perirhinal cortex, with only sparse NFTs in the other neocortical regions in cases presenting with very mild symptoms of cognitive decline (Hof *et al.*, 1992b; Bouras *et al.*, 1993, 1994; Bierer *et al.*, 1995) (Fig. 6). For example, the inferior temporal cortex may contain SPs at this stage, but it is practically devoid of NFTs, whereas it is severely affected by NFT formation in demented cases. This pattern of NFT distribution corresponds well to the minimal neuronal loss observed in the entorhinal cortex in cases with very mild cognitive impairment, whereas the presence of more severe dementing symptoms is correlated with neuronal loss in neocortical regions (Gómez-Isla *et al.*, 1996, 1997).

Macaque monkeys with restricted lesions of the hippocampus and amygdala display a sparing of learning skills, whereas lesions of the perirhinal and parahippocampal fields TF and TH result in severe memory disturbances, stressing the fact that these cortical fields are crucial elements of the memory system (Zola-Morgan and Squire, 1984; Zola-Morgan *et al.*, 1989, 1992; Squire and Zola-Morgan, 1991; Suzuki *et al.*, 1993). Therefore, substantial degeneration of presumed corticocortical circuits within the neocortex emerges as a necessary factor for the clinical expression of the dementia (Pearson *et al.*, 1985; Rogers and Morrison, 1985; Lewis *et al.*, 1987; Hof *et al.*, 1990a; Hof and Morrison, 1990). These observations imply that elderly individuals can maintain a high level of cognitive performance while sustaining significant compromise of hippocampal circuits, and that they may rely more on neocortical than on hippocampal circuits for memories essential in daily activities. In terms of cognitive performance, healthy elders may describe difficulties in learning and retrieving new information, although in reality the problem lies in the amount of information that the individuals can learn within a given period of time in comparison to younger subjects. The major difference with early AD patients is that after a delay healthy elders retain the new information, whereas very mildly impaired patients lose most of it (Albert, 1996; Albert and Moss, 1996),

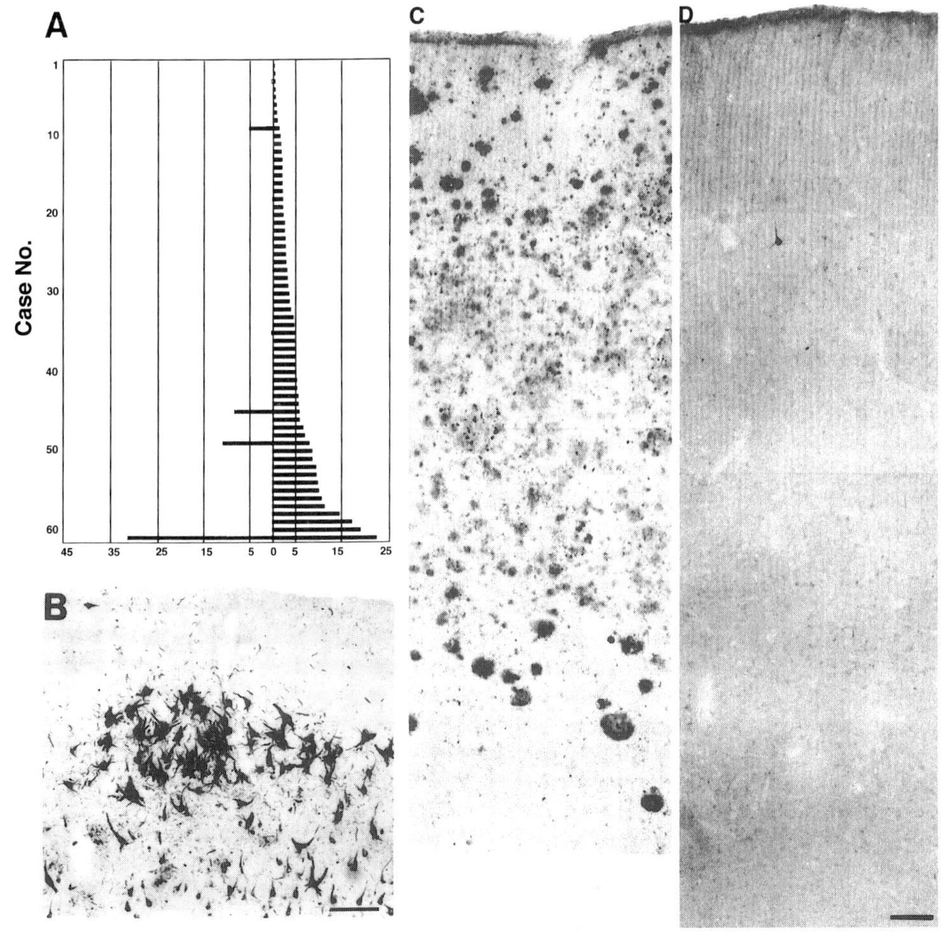


**Figure 6.** Distribution of pathologic changes in normal brain aging and possible preclinical stages of AD. (A) Densities (per mm$^2$) of NFTs (right side of panel) and SPs (left side of panel) in layer II of the entorhinal cortex in a series of 61 nondemented cases. *All* have NFT formation in this layer, whereas SPs are inconsistently observed. (B) Layer II of entorhinal cortex of case 60 (near bottom of graph in A); there are numerous NFTs. (C, D) In the temporal neocortex, very few NFTs are seen, mostly in layer III (D), whereas large amounts of amyloid deposition are present in some of these cases suggesting no correlation between amyloid deposition and clinical and neuropathological status (see Bouras *et al.*, 1993, 1994). Bars = 50 μm.

correlating not only with neuronal loss in the entorhinal cortex, but also with changes in volumetric indices within the temporal lobe, as revealed by neuroimaging studies (Albert, 1996; Fama *et al.*, 1997). Increases in the volume of the temporal horn of the lateral ventricle may selectively reflect loss of projections from the entorhinal cortex. Thus, normal aging is characterized by nearly intact cognitive abilities, despite the presence of scarce neurofibrillary pathology in the entorhinal cortex. It is therefore tempting to refer to such age-related changes as "asymptomatic AD-like neurofibrillary pathology."

Recent retrospective surveys of a sample of over 1100 nondemented autopsy cases show that the density and distribution of NFTs and SPs vary depending on the age of death and the cortical area studied (Bouras *et al.*, 1993, 1994; Giannakopoulos *et al.*, 1993, 1994a,b,c, 1995a,c, 1996a, 1997b; Hof *et al.*, 1995a, 1996a) (Table I). The entorhinal cortex is consistently involved by NFTs, while the number of cases with NFTs in the CA1 field increases age: low NFT densities (fewer than 10 per mm$^2$) are found in 2% of the patients younger than 70 years, but in 46% of the patients older than 95 years. In the inferior temporal cortex (area 20), low NFT densities are present in 17% of cases 65–69 years old and 43% of cases older than 95. However, the superior frontal cortex (area 9) and area 17 do not consistently contain NFTs in nondemented cases. In respect to SPs, the number of cases increases overall with age in the CA1 field as well as in neocortical areas 20, 9, and 17. In this nondemented population, the percentage of high-density SP cases is lower below 85 years, but in older age groups the number of patients presenting with moderate-to-high SP densities increases, particularly in patients 95 years and up. The differential cortical vulnerability to NFT and SP formation has also been extended to a subgroup of patients showing age-associated memory impairment (Crook *et al.*, 1986). In such cases, NFTs are always present in layer II of the entorhinal cortex, as well as in the CA1 field and temporal neocortex, but in consistently higher densities as compared with cognitively intact individuals (independent of age) whereas SPs are present in highly variable densities throughout the cerebral cortex (Bouras *et al.*, 1993, 1994). In cases with very mild cognitive changes, NFTs are rare in neocortical areas and, when present, are preferentially distributed in layer III of area 20 (Hof *et al.*, 1992a; Bouras *et al.*, 1993, 1994) (Fig. 7). It has been shown that in very mildly affected cases, NFTs in layer II of the entorhinal cortex correlate with a substantial decrease in neuron numbers, whereas in cognitively normal cases, the number of neurons remains constant from 60 to 90 years of age (Gómez-Isla *et al.*, 1996), indicating that in some individuals, the presence of NFTs in this region may remain asymptomatic throughout the entire life span.

These data indicate that both NFT and SP formation have age-related patterns and are differentially distributed within the cerebral cortex in elderly nondemented individuals, revealing the existence of a dynamic process during which lesion formation begins in the entorhinal cortex, and, in cases who develop AD, subsequently occurs throughout the CA1 field, the subiculum, area 20, and finally in other neocortical areas (Braak and Braak, 1991; Hof *et al.*, 1992a; Bouras *et al.*, 1993, 1994). Of note, substantial SP formation in neocortical association areas is frequently seen in both nondemented and age-associated memory-impairment cases, consistent with several reports that the early and generalized SP formation in the neocortex is not necessarily accompanied by a significant cognitive decline

(Crystal *et al.*, 1988; Katzman *et al.*, 1988; Dickson *et al.*, 1991; Morris *et al.*, 1991; J. L. Price *et al.*, 1991; Arriagada *et al.*, 1992a,b; Hof *et al.*, 1992b; Bouras *et al.*, 1993, 1994; Bierer *et al.*, 1995). Early NFT formation in medial and inferior temporal structures contrasts with the sparing of other cortical areas in both cognitively intact individuals and in patients with age-associated memory impairment, suggesting that damage of these areas correlates with memory impairment but is insufficient to cause the massive cognitive deterioration observed in AD. Therefore, moderate-to-severe involvement of the entorhinal cortex and CA1 field appears

**Table I. Density and Distribution of NFTs and SPs in the Cerebral Cortex of Elderly People[a]**

| | 65–69 | 70–74 | 75–79 | 80–84 | 85–89 | 90–94 | 95–105 |
|---|---|---|---|---|---|---|---|
| ***NFT*** | | | | | | | |
| *CA1* | | | | | | | |
| rare | 97.0 | 89.5 | 78.0 | 68.9 | 59.3 | 54.1 | 46.0 |
| moderate | 3.0 | 9.7 | 18.9 | 24.9 | 28.1 | 35.9 | 46.0 |
| severe | 0 | 0.8 | 3.1 | 6.2 | 12.6 | 10.0 | 8.0 |
| *Area 20* | | | | | | | |
| rare | 79.5 | 78.1 | 72.0 | 63.8 | 57.2 | 50.9 | 46.0 |
| moderate | 17.4 | 18.4 | 21.2 | 26.4 | 28.4 | 37.7 | 43.2 |
| severe | 3.1 | 3.5 | 6.8 | 9.8 | 14.4 | 11.4 | 10.8 |
| *Area 9* | | | | | | | |
| rare | 98.5 | 95.6 | 95.5 | 94.0 | 92.6 | 89.3 | 94.5 |
| moderate | 1.5 | 3.5 | 3.8 | 5.4 | 5.6 | 10.7 | 5.5 |
| severe | 0 | 0.9 | 0.7 | 0.6 | 1.8 | 0 | 0 |
| *Area 17* | | | | | | | |
| rare | 100 | 98.5 | 97.3 | 95.8 | 95.1 | 94.3 | 94.6 |
| moderate | 0 | 1.5 | 2.3 | 3.9 | 4.2 | 5.7 | 5.4 |
| severe | 0 | 0 | 0.4 | 0.3 | 0.7 | 0 | 0 |
| ***SP*** | | | | | | | |
| *CA1* | | | | | | | |
| rare | 92.3 | 76.3 | 66.3 | 56.0 | 51.2 | 50.4 | 46.0 |
| moderate | 6.2 | 15.8 | 21.2 | 31.4 | 29.5 | 28.9 | 27.0 |
| severe | 1.5 | 7.9 | 12.5 | 12.6 | 19.3 | 20.7 | 27.0 |
| *Area 20* | | | | | | | |
| rare | 80.0 | 64.9 | 56.1 | 44.6 | 37.5 | 36.5 | 29.7 |
| moderate | 13.9 | 21.1 | 19.3 | 28.1 | 25.3 | 24.5 | 27.0 |
| severe | 6.1 | 14.0 | 24.6 | 27.3 | 37.2 | 39.0 | 43.3 |
| *Area 9* | | | | | | | |
| rare | 80.0 | 63.2 | 58.0 | 47.0 | 38.6 | 38.4 | 37.8 |
| moderate | 13.9 | 23.7 | 23.1 | 27.5 | 29.5 | 27.7 | 24.3 |
| severe | 6.1 | 13.2 | 18.9 | 25.5 | 31.9 | 33.9 | 43.3 |
| *Area 17* | | | | | | | |
| rare | 84.6 | 71.9 | 63.5 | 49.7 | 39.7 | 41.3 | 40.5 |
| moderate | 7.7 | 17.5 | 19.4 | 25.9 | 25.0 | 29.3 | 29.7 |
| severe | 7.7 | 10.6 | 17.1 | 24.4 | 35.5 | 29.4 | 29.8 |

[a]Data represents percentage of cases in seven age groups, independent of clinical diagnosis. Over 1100 elderly cases were analyzed. Adapted from Giannakopoulos *et al.* (1994b).

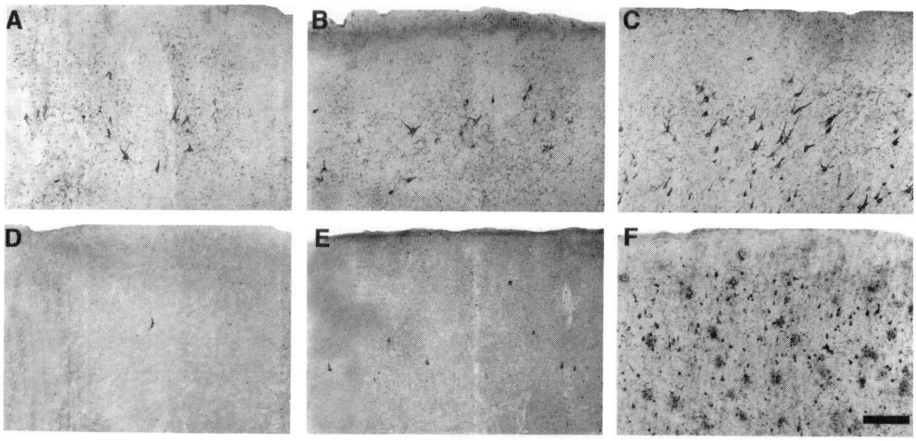


**Figure 7.** Neurofibrillary tangle distribution in layer II of the entorhinal cortex (A–C) and in layers II/III of area 20 (D–F) in normal aging (A, D), very mild cognitive impairment (B, E), and in AD (C, F). In both regions, there are higher NFT densities in cases with very mild impairment and in demented cases as compared with normal aging. The entorhinal cortex is significantly affected in normal aging, indicating that NFT formation in this region is a consistent feature of brain senescence, whereas cognitive impairment is linked to NFTs in neocortical regions. This does not imply a necessary continuum between aging and demented states. Adapted from Bouras *et al.* (1994)., by permission of Oxford University Press. Bar = 50 μm.

compatible with normal cognitive status or very mild cognitive impairment that still allows for independent daily living, whereas presence of the degenerative changes to *neocortical* areas is linked to clinical signs of dementia. In this respect, area 20 may represent an interface between affected and unaffected cortical regions, consistent with several reports suggesting that the presence of increasing NFT densities within adjacent cortical components of the medial and inferior aspects of the temporal cortex is a neuropathologically relevant marker of incipient dementia (Hubbard *et al.*, 1990; Hof *et al.*, 1992b; Bouras *et al.*, 1993). In a neuropathological survey of 70 nonselected cases in a general hospital, Hubbard and colleagues (1970) showed that severe NFT formation in the hippocampus and in area 20 with no involvement of other neocortical areas is not associated with overt clinical signs of dementia. Similarly, in a quantitative neuropathological evaluation of 61 nondemented elderly patients, we found a few apparently normal cases with much higher NFT densities in the hippocampal formation and in area 20, and proposed that the presence of NFTs in these regions could precede the development of clinical signs of dementia (Bouras *et al.*, 1993).

### 2.3.2. Involvement of Specific Cortical Circuits in AD

The distribution NFTs and SPs in the hippocampal formation displays a strong correlation with specific projections. Most of the pathways interconnecting the different anatomic components of hippocampal formation undergo NFT formation (Hyman *et al.*, 1990). The perforant pathway that originates in layer II of the entorhinal cortex is massively affected in AD (Hyman *et al.*, 1984, 1986), and end-stage NFTs in this layer are correlated with a high SP density in the molecular layer of the dentate gyrus, where the pathway terminates (Senut *et al.*, 1991). The degeneration of the perforant pathway does not seem to directly involve its target neurons, because reconstruction of Lucifer Yellow-injected granule cells has failed to demonstrate morphologic alterations of these neurons or the participation of their apical dendrites in SP formation in the molecular layer (Einstein *et al.*, 1994), although this does not preclude the possibility of more subtle molecular changes in their dendritic arbor (see below). High NFT density in layer V of the entorhinal cortex might correlate with the degeneration of connections to specific nuclei in the amygdala as well as in other limbic and neocortical areas (Hyman *et al.*, 1990). The projections from the hippocampus to the entorhinal cortex, amygdala, and neocortical areas originate from the CA1 field and the subiculum, which are consistently affected by NFTs in AD (Kalus *et al.*, 1989; Hyman *et al.*, 1990), and among these connections the distribution of SPs is often complementary to that of NFTs, supporting the notion that the regional pathologic changes in the hippocampal formation reflect the disconnection of intrinsic pathways and of projections to other cortical regions (Hirano and Zimmermann, 1962; Brady and Mufson, 1990; Hyman *et al.*, 1990).,

The regional and laminar variations in NFT and SP distribution in the neocortex of AD cases suggest that NFTs are closely associated with the presence of corticocortically projecting neurons. In the case of SPs, their distribution correlates with termination zones of some corticocortical projections, but other neuronal systems contribute to their formation (Pearson *et al.*, 1985; Morrison *et al.*, 1985; Rogers and Morrison, 1985; Duyckaerts *et al.*, 1986; Esiri *et al.*, 1986; Lewis *et al.*, 1987; Braak *et al.*, 1989; Hof and Morrison, 1990; Hof *et al.*, 1990a; Arnold *et al.*, 1991; De Lacoste and White, 1993) (Fig. 5). The neurons of origin of corticocortical pathways can be grouped into three distinct classes; this is reflected in their differential laminar distribution. These projections can also be differentiated by the distribution of their axonal terminals in target areas. Specifically, corticocortical projections can be classified as feedforward, feedback, and lateral (Felleman and Van Essen, 1991). Feedforward projections ascend within the hierarchy of a given functional system (i.e., from a primary sensory area to an association area), feedback projections descend the same hierarchy, and lateral projections link cortical regions that are at the same functional level within the cortical hierarchy. In this model, inferred from data obtained in nonhuman primates, feedforward connections originate preferentially from neurons located in the superficial layers and terminate in layer IV and the deep portion of layer III, feedback connections emanate predominantly from the deep layers and terminate in layers I and VI, and lateral connections arise from the deep layers and project to all cortical layers (Felleman and Van Essen, 1991). According to this scheme, the preferential distribution of NFTs in layers III and V suggests that elements of all three types are involved in the degenerating process (Fig. 5). The fact that layer V contains slightly higher NFT densities than the superficial layers in the association neocortex suggests that feedback and lateral projections may be more affected in AD than feedforward projections (Lewis *et al.*, 1987). The feedforward projections in areas 17 and 18 are clearly damaged: NFTs predominate in layer III, which contains most of the efferent corticocortical neurons in these areas. The majority of neurons projecting to the prefrontal cortex from the occipital and temporal cortex and from the occipital cortex to the temporal cortex are located in layer III, although the ratio of these neurons in the supragranular to the infragranular layers decreases as one progresses from the occipital to the inferior temporal cortex and to superior temporal regions (Barbas, 1986; Barbas and Blatt, 1995). This shift in laminar distribution is paralleled by the progressive displacement of NFTs from supragranular layers to infragranular layers that occurs in the occipitotemporal direction (Lewis *et al.,* 1987) (Fig. 5).

The less-specific regional and laminar distribution of SPs (Arnold *et al.,* 1991) suggests that they do not originate exclusively from the degeneration of the same circuit components as NFTs in a given cortical region. Although SP distribution clearly reflects the degeneration of projections from NFT-containing neurons in certain pathways (Pearson *et al.,* 1985; Rogers and Morrison, 1985; Lewis *et al.,* 1987; Senut *et al.,* 1991; Vickers *et al.,* 1996; Vickers, 1997), and although the slightly higher incidence of SPs in the supragranular layers and in layer IV in the association cortex reflects the involvement of specific feedforward visual and auditory projections (Lewis *et al.,* 1987), SP presence in a large variety of neurochemical markers, including classical neurotransmitters and neuropeptides, indicates that local circuit neurons and subcortical afferents may participate as well in their formation (Morrison *et al.,* 1985; Walker *et al.,* 1988).

### 2.3.3. Neuropathology of the Oldest-Old: Specific Patterns of Lesion Distribution in Brains from Centenarians

The fact that there is an age-related progression in the severity of neuropathological changes in brain aging may indicate that AD represents an unavoidable end point of the normal senescence process (Ohm, 1997). In this respect, it is relevant to consider what would be the upper limit of normal brain aging, and to what degree the neuropathological characteristics of AD in the "oldest-old" differ from those in the younger population (Hauw *et al.,* 1986; Mizutani and Shimada, 1990, 1992; Delaère *et al.,* 1993; Giannakopoulos *et al.,* 1993, 1994a,c, 1995a,c, 1996a, 1997b). Several reports on populations of centenarians and nonagenarians demonstrate that overall, the regional and laminar patterns of NFT and SP distribution in nondemented very old people show subtle differences as compared to distribution in younger nondemented cases (Bouras *et al.,* 1993, 1994; Giannakopoulos *et al.,* 1993, 1994a,c, 1995c, 1996a, 1997b). In cognitively intact centenarians, however, there is a more frequent involvement of areas 22 (superior temporal cortex) and 24 and 23 (anterior and posterior cingulate cortex), which contrasts with observations in nondemented younger cases (see above), suggesting a certain degree of differential cortical vulnerability after 90 years of age (Giannakopoulos *et al.,* 1996a). In fact, quantitative analyses have demonstrated several particularities in the distribution of AD lesions in this age group.

Unlike in younger cases, demented centenarians show much higher NFT densities in the anterior CA1 field of the hippocampus than do demented centenarians (Table II), suggesting that massive NFT formation in the anterior portion of the hippocampus is a marker of the development of AD in centenarians. In sharp contrast to findings in younger AD cases, NFT densities in the entorhinal cortex are comparable between demented and nondemented centenarians. In areas 7b and 39 (postero-inferior parietal and angular cortices), 22, 23, and 24, demented centenarians have significantly higher NFT densities compared with nondemented

**Table II. NFT and SP Densities in Hippocampus, Entorhinal Cortex, and Temporal Neocortex of Centenarians[a]**

| | NFT | | SP | |
|---|---|---|---|---|
| | ND | AD | ND | AD |
| CA1 | 34.0 ± 4.8 | 65.3 ± 9.0 | 2.5 ± 0.6 | 2.7 ± 1.0 |
| Subiculum | 17.6 ± 2.7 | 19.0 ± 3.0 | 1.8 ± 0.6 | 2.1 ± 0.5 |
| Entorhinal II | 35.5 ± 4.8 | 29.2 ± 6.1 | 4.5 ± 1.0 | 4.3 ± 1.0 |
| Entorhinal V | 16.5 ± 2.7 | 13.2 ± 3.5 | 1.5 ± 0.4 | 1.6 ± 0.6 |
| Area 20 II/III | 16.7 ± 2.4 | 20.8 ± 6.7 | 7.9 ± 1.6 | 8.0 ± 2.0 |
| Area 20 V/VI | 11.4 ± 1.9 | 15.0 ± 3.5 | 4.4 ± 1.1 | 3.4 ± 1.0 |

[a]Data represent mean regional and laminar NFT and SP densities per mm$^2$ ± SD in nondemented (ND) and AD centenarians. Adapted from Giannakopoulos *et al.* (1993).

centenarians, whereas there is no significant difference in areas 9 and 20, nor in areas 17 and 18 (Table III). In addition, when a distinction is made between cognitively intact centenarians and centenarians presenting with very mild cognitive impairment, a clear association is observed between increased NFT formation in *layers V and VI* of area 20 and the cognitive performance (Fig. 8) (Giannakopoulos *et al.,* 1996a). This differs from younger cases with age-associated memory impairment, who are characterized by increases in NFT counts in *layer III* of area 20 (Hof *et al.,* 1992b; Bouras *et al.,* 1993, 1994), implying that different subsets of corticocortical connections from the inferior temporal cortex are affected in the early stages of the dementing process in very old people. Another striking differ-

**Table III. NFT and SP Densities in Frontal, Parietal, Visual, and Cingulate Cortices of Centenarians[a]**

| | NFT | | SP | |
|---|---|---|---|---|
| | ND | AD | ND | AD |
| Area 9 II/III | 4.4 ± 1.5 | 6.9 ± 4.3 | 5.7 ± 1.7 | 7.8 ± 1.7 |
| Area 9 V/VI | 1.6 ± 0.6 | 4.6 ± 2.5 | 3.9 ± 1.1 | 4.4 ± 1.1 |
| Area 7 II/III | 1.2 ± 0.3 | 16.8 ± 5.2 | 13.2 ± 2.3 | 26.0 ± 3.0 |
| Area 7 V/VI | 0 | 14.4 ± 4.1 | 4.7 ± 1.0 | 9.9 ± 1.5 |
| Area 17 II/III | 0.1 ± 0.1 | 10.8 ± 2.5 | 10.0 ± 1.5 | 11.1 ± 1.5 |
| Area 17 IV–VI | 0 | 2.0 ± 1.0 | 11.3 ± 2.2 | 7.1 ± 1.0 |
| Area 24 II/III | 4.5 ± 1.0 | 23.0 ± 4.0 | 22.8 ± 3.2 | 33.8 ± 3.7 |
| Area 24 V/VI | 1.5 ± 0.5 | 11.7 ± 2.6 | 12.1 ± 3.1 | 12.1 ± 1.6 |
| Area 23 II/III | 6.2 ± 2.5 | 17.5 ± 6.4 | 13.9 ± 2.1 | 26.4 ± 2.9 |
| Area 23 V/VI | 0.7 ± 0.2 | 8.6 ± 2.4 | 4.4 ± 0.8 | 8.2 ± 0.9 |

[a]Data represent mean regional and laminar NFT and SP densities per mm$^2$ ± SD in nondemented (ND) and AD centenarians. Major differences in lesion densities between the two groups are observed in the parietal and cingulate areas. Adapted from Giannakopoulos *et al.* (1995c).

ence in demented centenarians is that there are more NFTs in the *superficial* layers of association cortex (Giannakopoulos *et al.*, 1994a,c, 1995c, 1996a), whereas in younger AD cases, NFTs tend to be present in slightly higher densities in the *infragranular* layers (Lewis *et al.*, 1987; Hof *et al.*,1990a), suggesting a greater involving of feedforward projections than of lateral and feedback projections in centenarians with AD. Also, NFT formation in the temporal neocortex may play a less prominent role in the clinical expression of AD in centenarians as compared with younger patients. Although some involvement of area 20 may account for very mild symptoms of cognitive impairment in this age group, overt clinical signs of AD in oldest-old individuals requires both high NFT densities in the CA1 field and a progression of the alterations to areas 7, 22, 23, and 24, which are not consistently affected in most of the other cases in the early stages of the dementing process (Hof *et al.*, 1992b; Giannakopoulos *et al.*, 1995c, 1996a). The particular NFT distribution in centenarians suggests a certain displacement of these lesions such that parietal and cingulate areas are more affected than is usually the case in AD, whereas superior frontal and inferior temporal association areas are relatively preserved. The exact clinical repercussion of these differences is, however, unknown, and detailed neuropsychological testing as well as functional imaging studies will be

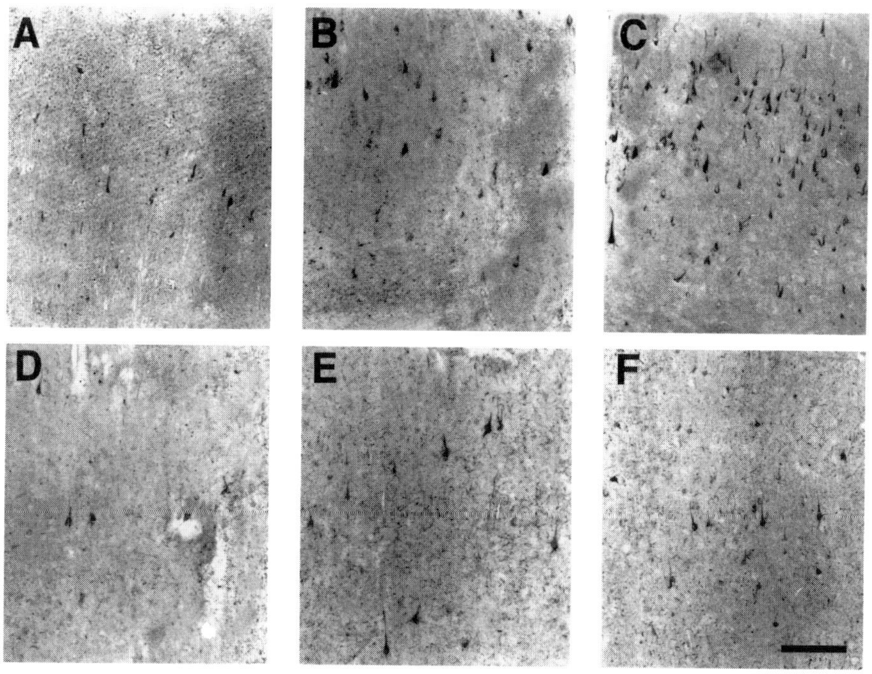


**Figure 8.** Neurofibrillary tangles in the CA1 field of hippocampus (A–C) and layers V/VI of area 20 (D–F) in a nondemented centenarian (A, D), a patient in his tenth decade with very mild cognitive impairment (B, E), and a demented centenarian (C, F). The CA1 field contains the highest NFT densities in the demented case. Also, there is an increase in NFT densities in layer V and VI of area 20 in the mildly affected and demented cases. These patterns contrast with those observed in a younger population of elderly individuals (Bouras *et al.*, 1993, 1994; Giannakopoulos *et al.*, 1993, 1996a). Adapted from Giannakopoulos *et al.* (1996a). Reproduced with permission from the Journal of Neuropathology and Experimental Neurology. Bar = 50 μm.

necessary to assess whether cortical functions that depend on cingulate and parietal areas are preferentially compromised in AD centenarians.

Several studies of centenarians indicate that, as with younger populations, there is no relationship between SP formation in the neocortex and the early stages of the dementing process (Delaère *et al.*, 1993; Giannakopoulos *et al.*, 1993). Evaluations of centenarians from separate cohorts drawn from psychiatric and geriatric hospitals, however, have revealed a clear correlation between SP densities in areas 7, 9, 20, 22, 23, and 24 and the severity of dementia (Fig. 8) (Giannakopoulos *et al.*, 1994a,c, 1995c, 1996a). These observations differ from studies of younger demented and nondemented cases which show no correlation between dementia severity and SP densities (Arriagada *et al.*, 1992a,b; Berg *et al.*, 1993; Bierer *et al.*, 1995) and imply that, although SP formation may not be etiologically linked to the early stages of cognitive impairment, it may represent a reliable pathological hallmark of AD severity in centenarians.

Besides these particularities in NFT and SP distribution, centenarians show a differential pattern of neuronal loss in the cerebral cortex in comparison to younger cohorts (Giannakopoulos *et al.*, 1996a). No statistically significant difference in neuron density is observed in the CA fields, the hilus of the dentate gyrus, or the subiculum between demented and nondemented centenarians. In contrast, demented centenarians have a 31% neuronal loss in layer II and 18% in layer V of the entorhinal cortex; 27% in layers II and III and 28% in layers V and VI of area 20; and 5% in layers II and III and 8% in layers V and VI of area 9, as compared with nondemented cases. The absence of neuronal loss in the CA fields and subiculum of demented centenarians contrasts with recent data on younger AD cohorts showing up to 68% neuron loss in the CA1 field and 47% in the subiculum (West *et al.*, 1994). These results suggest that substantial neuronal depletion in the CA1 field reflects a qualitative difference between normal aging and AD. Centenarians, however, display no significant neuronal loss overall in the CA1 field despite the presence of high NFT densities, indicating that these NFTs may be at the early stages of their formation and that neuron death has not yet occurred (Giannakopoulos *et al.*, 1996a). The neuronal loss observed in layer II and V of the entorhinal cortex and in areas 9 and 20 of demented centenarians parallels observations in younger demented individuals (Lippa *et al.*, 1992; O'Banion *et al.*, 1994; Gómez-Isla *et al.*, 1996, 1997) and implies that decreasing neuronal density in these areas is associated with AD severity and duration in all age groups. Neuronal loss in the entorhinal complex of centenarians is compatible with estimates of total neuron numbers in the entorhinal cortex of patients younger than 95 years, in whom neuronal losses of 60% in layer II of the entorhinal cortex take place in cases with very mild cognitive impairment and of 90% in severe AD cases (Gómez-Isla *et al.*, 1996). The extent of this loss in demented centenarians is lower, however, than in younger AD cases (Lippa *et al.*, 1992; Gómez-Isla *et al.*, 1996), suggesting that a mild decrease in neuron numbers in cortical regions such as the entorhinal cortex and areas 9 and 20 may be sufficient to induce a significant degree of cognitive failure after 95 years of age (Hansen *et al.*, 1988a). The apparent dissociation between NFT formation and neuron densities in demented centenarians suggests that neuronal loss unrelated to the presence of NFTs is the rule in the oldest-old, in contrast to AD cases under 85 years, where a correlation was documented (Gómez-Isla *et al.*, 1996; Bobinski *et al.*, 1997), and supports the notion that NFTs cannot account for all of the neuronal loss in AD (Hof *et al.*, 1990a; Gómez-Isla *et al.*, 1997).

From a clinical and epidemiological standpoint, it is important to note that these data from very old individuals strengthen the hypothesis that AD is not a continuum with normal brain aging, and that in centenarians the symptoms of dementia may differ considerably from those in younger patients (Rebeck *et al.*, 1994; Ritchie and Kildea, 1995; Sobel *et al.*, 1995; Giannakopoulos *et al.*, 1995a; Asada *et al.*, 1996; Lautenschlager *et al.*, 1996; Politoff and Monson, 1997). For example, dementia prevalence levels off from 95 years of age on, and no association between AD and apolipoprotein E allele $\epsilon4$ is observed in centenarians (Rebeck *et al.*, 1994; Ritchie and Kildea, 1995; Sobel *et al.*, 1995; Asada *et al.*, 1996; Lautenschlager *et al.*, 1996), suggesting that very old individuals are characterized by differential brain aging and susceptibility to the dementing process (Hauw *et al.*, 1986; Mizutani and Shimada, 1990, 1992; Delaère *et al.*, 1993; Giannakopoulos *et al.*, 1993, 1994a,b,c, 1995a,c, 1996a) and stressing the necessity of genetic and biochemical analyses to define further the characteristics of brain aging in this population.

### 2.3.4. Atypical AD Cases and Restricted Corticocortical Pathway Involvement

Whereas visual association cortices contain high densities of SPs and NFTs in AD, area 17 may contain numerous SP but relatively few NFTs, even at advanced stages of the disease (Pearson *et al.*, 1985; Rogers and Morrison, 1985; Lewis *et al.*, 1987; Braak *et al.*, 1989; Hof *et al.*, 1989, 1990b; Hof and Morrison, 1990; Arnold *et al.*, 1991). Total neuronal counts in area 17 do not differ between AD and control brains, unlike in frontal, parietal, and temporal areas (Mountjoy *et al.*, 1983), and area 17 is one of the last regions of the neocortex to be affected by NFT formation in the progression of AD (Arnold *et al.*, 1991). Neurofibrillary tangles are present in area 18, but their laminar distribution differs from that in temporal, parietal, and frontal cortices, as they are predominantly located within layer III and are far less numerous (Lewis *et al.*, 1987; Hof *et al.*, 1989, 1990b; Hof and Morrison, 1990). Thus, based on NFT distribution, areas 17 and 18 are relatively spared in AD as it usually presents, suggesting that the corticocortical outflow from area 17 to area 18 is far less affected than it is among temporal or parietal visual association areas early in the course of the disease (Hof and Morrison, 1990; Hof *et al.*, 1989, 1990a,b, 1993a, 1997c; Morrison *et al.*, 1997). Furthermore, the increasing degree of neuropathologic involvement along visual cortical hierarchies from occipital cortex to parietal and temporal cortex is consistent with many visual function deficits observed in AD patients, although not all modalities of visual function are similarly affected (Cogan, 1985; Hinton *et al.*, 1986; Neary and Snowden, 1987; Cronin-Golomb *et al.*, 1993; Curcio and Drucker, 1993; Kurylo *et al.*, 1994, 1996; Parasuraman *et al.*, 1995; Mentis *et al.*, 1996; Pietrini *et al.*, 1996).

In recent years, several studies in humans have shown that many AD patients present with prominent visual symptoms. Since SP and NFT formation represent a reliable correlate of the disconnection of certain corticocortical pathways, the identifiable neurological deficits observed in such AD cases may be directly linked to such disconnection. In these cases, the visual impairment affects specific aspects of visual function (Cogan, 1985; Benson *et al.*, 1988; Fletcher, 1994; Mackenzie Ross *et al.*, 1996). Several of the patients showing predominantly visual function impairment have symptoms comparable to those in 1909 by Hungarian physician Reszö Bálint—a complex visual syndrome characterized by cortical paralysis of visual fixation associated with optic ataxia and a severe disturbance of visual attention in patients with large bilateral watershed infarcts in the parieto-occipital region (Bálint, 1909; see Fletcher, 1994). The occurrence of a Bálint's-like syndrome in AD patients was first documented by Ernst Grünthal (1928) and Ferdinand Morel (1945) (Fig. 9). Most of the recently described AD cases with occipital alterations display a complex neurologic symptomatology and variable disturbances of visual function that include alexia, anomia, agraphia, and transcortical sensory aphasia, as well as Bálint's and Gerstmann's syndromes, with relatively good preservation of memory in the early stages of their illness. This symptomatology contrasts sharply with the memory impairment and temporospatial disorientation that is usually seen AD. The cortical atrophy and metabolic deficits, as shown in some of these atypical AD cases through functional imaging or at autopsy, predominate in the parieto-temporo-occipital junction and occipital cortex, and these cases are frequently referred to as posterior cortical atrophy (Magnani *et al.*, 1982; Nissen *et al.*, 1985; Benson *et al.*, 1988; Hof *et al.*, 1989, 1990a, 1993a, 1997c; Jacquet *et al.*, 1990; Mendez *et al.*, 1990a,b,c; Berthier *et al.*, 1991; Hof and Bouras, 1991; Graff-Radford *et al.*, 1993; Levine *et al.*, 1993; Victoroff *et al.*, 1994; Levy *et al.*, 1995; Mackenzie Ross *et al.*, 1996; Pietrini *et al.*, 1996). A particularly prominent feature in these cases is a severe deficit in motion perception and target tracing (Neary and Snowden, 1987; Sadun *et al.*, 1987; Fletcher and Sharpe, 1988; Katz and Rimmer, 1989; Kiyosawa *et al.*, 1989; Kuskowski *et al.*, 1989; Cronin-Golomb *et al.*, 1993; Kurylo *et al.*, 1994, 1996; Silverman *et al.*, 1994; Butter *et al.*, 1996; Furey-Kurkjian *et al.*, 1996; Mentis *et al.*, 1996). The clinical symptomatology of AD patients with posterior cortical atrophy suggests that subsets of cortical pathways linking primary visual occipital regions to posterior parietal and cingulate visual association cortex are affected in the

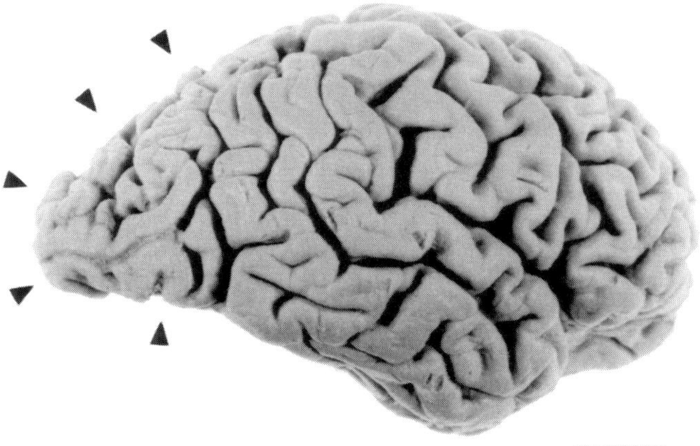

**Figure 9.** Lateral view of the right hemisphere in AD with posterior cortical atrophy presenting as a Bálint's syndrome (Hof *et al.*, 1993a). A generalized atrophy predominates on the occipital lobe (arrowheads). This patient was first reported by Ferdinand Morel (1945). From Hof *et al.* (1993a). © Springer-Verlag. Bar = 2 cm.

early stages of the dementia (Hof *et al.*, 1989, 1990a, 1993a, 1997c; Hof and Bouras, 1991).

Quantitatively, inordinately high densities of AD lesions are observed in the occipital regions of AD cases with posterior cortical atrophy and visuomotor impairment resembling Bálint's syndrome (Hof *et al.*, 1989, 1990b, 1993a, 1997c; Jacquet *et al.*, 1990; Mendez *et al.*, 1990a,b,c; Berthier *et al.*, 1991; Hof and Bouras, 1991; Graff-Radford *et al.*, 1993; Levine *et al.*, 1993; Victoroff *et al.*, 1994), with a gradient in NFT densities from area 17 to the visual association regions in the anterior occipital (area 19) and parietal cortices (area 7b) as well as in the posterior cingulate cortex (area 23). One patient, also reported to have prosopagnosia and apperceptive visual agnosia (Hof and Bouras, 1991), showed a comparable occipital displacement in the lesion distribution, with a preferential localization of NFTs and SPs in mid and inferior temporal cortex (areas 37, 20, and 21) instead of inferior parietal and posterior cingulate cortex, demonstrating specific involvement of the occipitotemporal visual pathway, consistent with the clinical symptomatology (Hof and Bouras, 1991). It is worth noting that impairment in constructional tasks (i.e., constructional apraxia) in AD is correlated neuropathologically with the presence of neuritic changes in the occipital cortex early in the course of the disease (Nielsen *et al.*, 1996). Similarly, a triad of visuospatial problems, agraphia, and impairment in bimanual tasks in AD appears to correlate with bilateral parietal atrophy and the preferential distribution of lesions in areas 17, 18, 20, and 23 (Mackenzie Ross *et al.*, 1996).

In comparison to AD cases with the usual dementia presentation, NFT densities in areas 17, 18, and 19 contain a much higher density in the cases with posterior cortical atrophy. Differences were most marked in the occipital areas and were predominant in areas 17 and 18, with increases over conventional AD cases as high as 35-fold (Hof *et al.*, 1989, 1990b, 1993a, 1997). Smaller differences from conventional AD exist in areas 19,7b/7m, and 23 (Fig. 10) (Table IV). Neurofibrillary tangle counts do not differ significantly in posterior cortical atrophy cases in areas 37, 20, and 21 as compared to AD cases, except in the case presenting with visual agnosia. In the prefrontal areas (areas 9, 45, and 46) of posterior cortical atrophy cases, NFT densities are consistently lower than in conventional AD cases, supporting the notion of lesion distribution in occipital regions that are usually not severely involved in AD (Hof *et al.*, 1989, 1990a,b, 1993a, 1997c; Hof and Morrison, 1990; Arnold *et al.*, 1991; Hof and Bouras, 1991). Senile plaques are also much more numerous in the occipital cortex of posterior cortical atrophy cases than is usual in AD cases. Depending on the area, up to 5-fold increases in SP density are found in the posterior cortical atrophy cases, the largest differences being observed in areas 17 and 18 (Table IV) (Fig. 11).

The distribution of NFTs and SPs in the posterior cortical atrophy cases suggests that specific elements of visual corticocortical circuits are severely disrupted. The high density of NFTs in layer III of area 17 indicates that the projection to area 18 is affected, and the high density of NFTs in layer III of areas 18 and 19 similarly suggests that feedforward projections to higher-level regions in the occipitoparietal junction are involved. It appears that specific portions of ventral and dorsal area 19 are also severely affected in the posterior cortical atrophy cases. When compared to cortical parcellation schemes obtained in macaque monkeys as well as from neuropathological and functional imaging studies in humans, it seems that these particular areas of the occipital cortex may represent the human equivalent of areas V4 and MT and areas in the intraparietal sulcus of monkeys (Clarke and Miklossy, 1990; Felleman and Van Essen, 1991; Clarke, 1993, 1994, 1995; Watson *et al.*, 1993; Miklossy, 1993; Hof and Morrison, 1995; McCarthy *et al.*, 1995; Orban *et al.*, 1995; Sereno *et al.*, 1995; Tootell *et al.*, 1995; Tootell and Taylor, 1995; De Yoe *et al.*, 1996; Engel *et al.*, 1997). A loss of Meynert cells in layer VI of area 17 is observed in posterior cortical atrophy cases with Bálint's-like syndrome, suggesting that long forward projections to the area corresponding in the human brain to area MT are involved (Hof *et al.*, 1989, 1990b), and this region consistently contains higher NFT and SP densities than the surrounding occipital fields (Hof *et al.*, 1997c). Furthermore, a recent study in human brains also demonstrated the presence of projec-

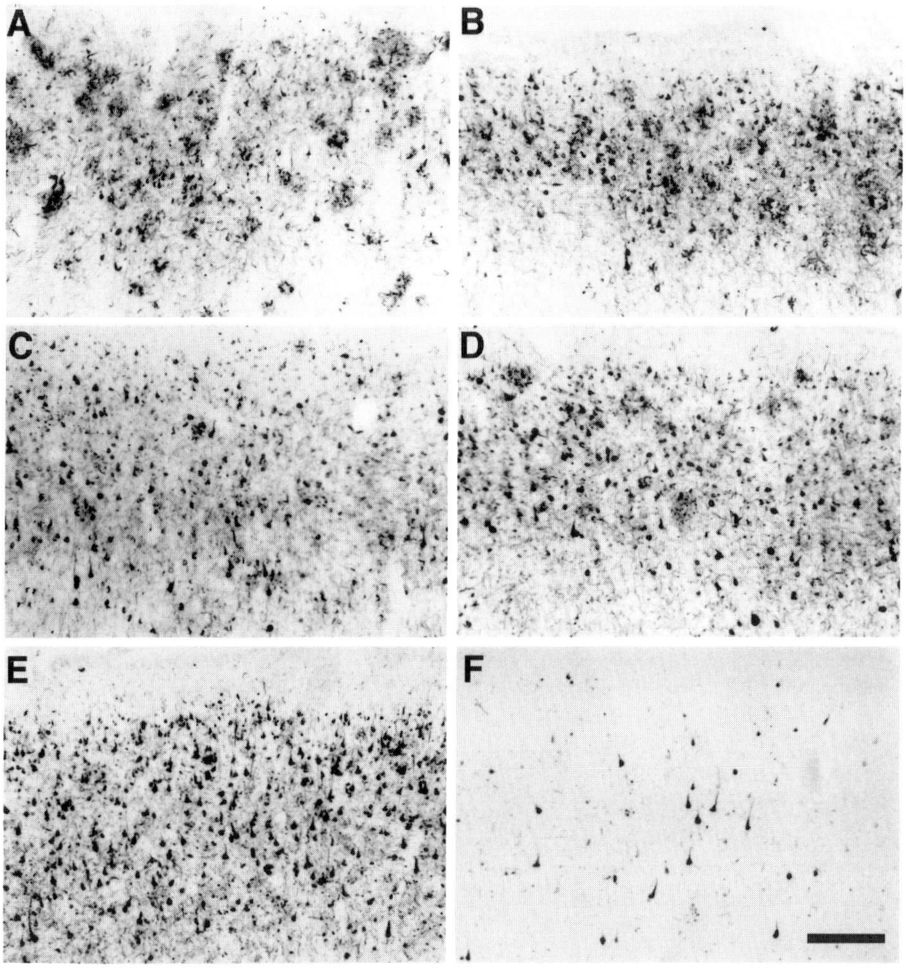


**Figure 10.** NFT distribution in layers II and III of area 17 (A), area 18 (B), dorsal area 19 (C), parietal area 7b (D), posterior cingulate area 23 (E), and prefrontal area 9 (F) in AD with posterior cortical atrophy. Note the progressive NFT increase and very high overall NFT densities in the occipital, parietal, and cingulate regions, whereas the prefrontal cortex has much lower NFT densities than are usually observed in AD. From Hof *et al.* (1993a). © Springer-Verlag. Bar = 50 μm.

tions from occipital cortex to dorsal posterior parietal cortex and angular gyrus (areas 7b and 39) that may correspond to some of the visual association areas located in the parietal cortex of the monkey that are specialized in visuomotor attention and visuospatial tasks (Miklossy, 1993). The existence in humans of a separate region involved in kinetic contour detection has also been reported (Orban *et al.*, 1995), and the infero-posterior and medial parietal areas 7b/7m that are severely affected in posterior cortical atrophy cases may include this region, which is distinct from area MT. The consistent and severe involvement of area 23 in the posterior cortical atrophy cases is particularly relevant, because this cortical area is an important component of the visuomotor system (Olson and Musil, 1992a,b; Olson *et al.*, 1993; Vogt *et al.*, 1997).

Thus, the high numbers of neurofibrillary tangles in layers V and VI of area 18 indicates that feedback projections to area 17 are also affected. The disruption of feedback projection (which terminate in both supra- and infragranular layers) is well correlated to the high SPs densities in all layers of area 17 in these cases. Also, increases in SP density in area 18 suggest an involvement of feedback projections from the area 19 complex and of feedforward projections from area 17. The same logic can be applied to the very high SP densities in areas 7b/7m and 23, possibly reflecting degeneration of feedforward connections from area 19 as well as to the prefrontal areas 9 and 46, which receive projections from the parietal visual areas (Wilson *et al.*, 1993). Taken together, these data support the involvement of cortical

**Table IV. NFT and SP Counts in the Neocortex in Posterior Cortical Atrophy**[a]

| Areas | Layers | NFTs | SPs |
|---|---|---|---|
| 17 (V1) | I/III | **80** (15) | **70** (35) |
| | IV | **1** (0) | **30** (25) |
| | V/VI | **30** (3) | **40** (20) |
| 18 (V2) | I/III | **150** (40) | **122** (80) |
| | IV/VI | **170** (60) | **90** (45) |
| 19 (V3–V4) | I/III | **210** (80) | **130** (90) |
| | IV/VI | **240** (90) | **100** (65) |
| MT (V5) | I/III | **400** (110) | **160** (80) |
| | IV/VI | **430** (130) | **135** (75) |
| 7m/7b | I/III | **230** (75) | **145** (90) |
| | IV/VI | **250** (95) | **90** (65) |
| 23 | I/III | **170** (45) | **140** (75) |
| | IV/VI | **170** (50) | **85** (75) |
| 20 (TEO) | I/III | **105** (220) | **80** (100) |
| | IV/VI | **130** (245) | **60** (85) |
| 9 | I/III | **45** (160) | **70** (110) |
| | IV/VI | **70** (175) | **50** (75) |

[a]Data represent the number of lesions under a 7-mm-wide cortical traverse in one AD case with posterior cortical atrophy (bold values) as compared with an AD case with the usual course of the disease (values in parentheses).

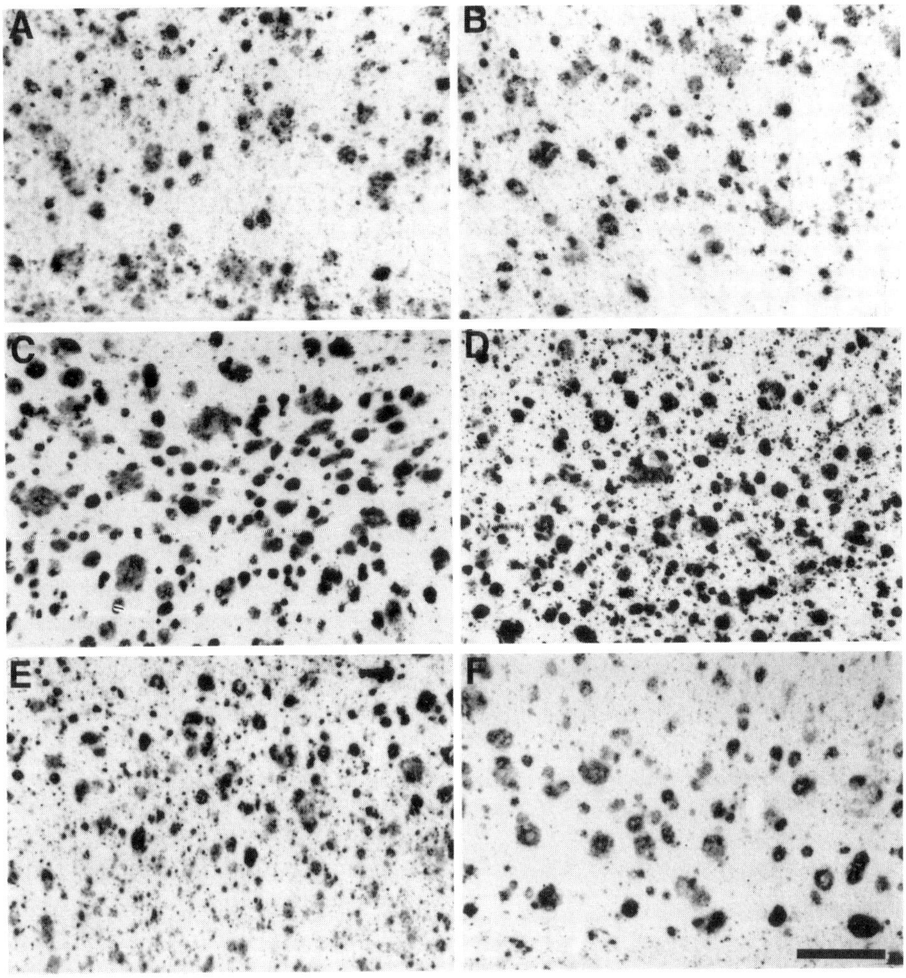


**Figure 11.** SP distribution in area 17 (A), area 18 (B), dorsal area 19 (C), parietal area 7b (D), posterior cingulate area 23 (E), and temporal area 20 (F) in AD with posterior cortical atrophy. Note the progressive SP increase and very high overall SP densities in the occipital, parietal, and cingulate regions, whereas the temporal cortex has much lower SP densities than the other regions. (A) shows layer III and IV of area 17; note the well-defined band of amyloid deposition in layer IVC. The other panels show layers II and III. From Hof *et al.* (1993a). © Springer-Verlag. Bar = 50 μm.

networks linking parietal, cingulate, and occipital regions (Vogt *et al.*, 1990, 1992, 1997; Minoshima *et al.*, 1994; Reiman *et al.*, 1996) to posterior cortical atrophy cases. Interestingly, positron-emission tomography studies have shown that posterior cingulate cortex and cinguloparietal cortex appear to be selectively and severely affected regions in the very early stages of AD, stressing the role of these areas in learning and memory impairment (Minoshima *et al.*, 1997).

Rare cases of AD presenting with somatosensory and motor deficits or parietal lobe syndrome have also been described (Crystal *et al.*, 1982; Jagust *et al.*, 1990; Golaz *et al.*, 1992; Mackenzie Ross *et al.*, 1996). In these cases too, the distribution of AD lesions was focal and more severe on sensory-motor cortex and parietal cortex.

Similarly, demented cases presenting with severe language impairment but relatively good preservation of praxis and gnosis even after several years of clinical evolution had a predominant left-hemisphere atrophy, whereas one case with marked right-hemisphere atrophy had a massive deficit in all of the cognitive functions early in the course of dementia (Giannakopoulos *et al.,* 1994d) (Fig. 12). In these asymmetric AD cases, NFTs and SPs were preferentially localized in the prefrontal, temporal, and posterior parietal cortices in both hemispheres, whereas the hippocampal formation displayed lower lesion densities than neocortical areas, and significantly higher lesion densities were found in the more atrophic side, with a regional topography that matched the clinical symptoms (Giannakopoulos *et al.,* 1994d). It is clear that certain cases of dementia do not meet the accepted clinical and neuropathologic criteria for the definition of AD, yet they show the same histopathologic features (Engel *et al.,* 1992; see Vogt *et al.,* this volume). These atypical AD cases may therefore represent examples of clinicopathological subtypes of AD that are likely to involve the disruption of subsets of corticocortical connections, consistent with findings that posterior cortical atrophy is clinically homogeneous but displays neuropathologic heterogeneity (Victoroff *et al.,* 1994; see also Vogt *et al.,* this volume). These findings stress the need for detailed and accurate neuropsychologic and functional testing, as well as for meticulous neuropathologic assessment of dementing syndromes.

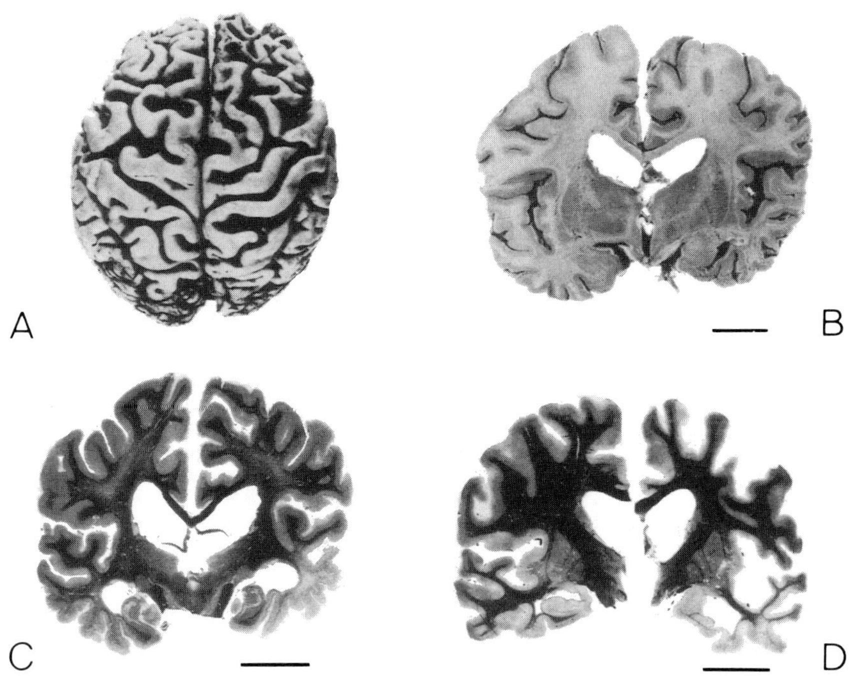


**Figure 12.** Asymmetric variant of AD. (A–D) Gross morphology (A, B) and coronal sections stained with the Luxol–Van Gieson method (C, D) through a brain with predominant right-hemisphere atrophy. Adapted from Bouras *et al.* (1990). Bars = 2 cm.

## 3. Morphologic and Neurochemical Correlates of Neuronal Vulnerability in AD

### 3.1. Neuronal Types and Development of Neurofibrillary Tangles

The fact that distinct components of the hippocampus and neocortex show differential degrees of involvement in aging parallels the observation that the progression of AD might be causally linked to differential neuronal vulnerability within select components of the cortical circuitry (Morrison *et al.*, 1987; Hof *et al.*, 1990a; Hof and Morrison, 1990; West and Gundersen, 1990; Vickers *et al.*, 1992, 1994b, 1996; Morrison, 1993; West, 1993; West *et al.*, 1994; Gómez-Isla *et al.*, 1996, 1997; Morrison and Hof, 1997; Simic *et al.*, 1997; Vickers, 1997). The transition from healthy neuron to NFT may offer essential clues as to the molecular and cellular correlates of vulnerability. Thus studies of the molecular and morphologic phenotype of specific neuronal classes in nondemented elderly individuals as well as in animal models may offer clues to the etiopathogenetic factors linked to the relative vulnerability of these neurons in dementia (see below). Although NFTs appear to be distributed throughout the cerebral cortex and in many subcortical structures, it is evident that not all neuron types are prone to NFT formation in AD (Braak and Braak, 1986; Hof *et al.*, 1990a; Gómez-Isla *et al.*, 1996, 1997). In the neocortex, the laminar distribution of NFTs indicates their strong association with pyramidal neurons. However, quantitative analyses using stereology have shown that the number of NFTs cannot account for the total population of pyramidal neurons, suggesting that only certain pyramidal subgroups are particularly prone to accumulating paired helical filaments (Giannakopoulos *et al.*, 1996a; Gómez-Isla *et al.*, 1997). Specifically, large pyramidal cells in layer III and V are the most affected cell class, whereas smaller pyramidal neurons in layers II, VI, and the upper part of layer III are more resistant to NFT formation. In contrast, the small pyramidal cells and spiny stellate cells in layer IV are remarkably resistant in AD, as are most of the classes of inhibitory interneurons. The fact that certain neurons are more vulnerable to NFT formation correlates with their morphology and connectivity: the neurons at risk are efferent cells that send long projections to other cortical regions and possibly to certain subcortical structures; all are large pyramidal neurons. The same reasoning can be applied to the hippocampus and entorhinal cortex: large numbers of pyramidal efferent neurons in layers II and V of the entorhinal cortex and in the CA1 field and subiculum are severely affected. Yet the large efferent neurons in the CA3 field and dentate hilus do not consistently form NFTs, and appear to be relatively resistant to degeneration in aging and in AD, suggesting that cellular characteristics other than strictly size and morphology are related to vulnerability or resistance to the degenerative process.

### 3.2. Abundance of Nonphosphorylated Neurofilament Protein in Subsets of Vulnerable Neurons

Neurofilament protein is a major component of the neuronal cytoskeleton. In macaque monkey and human cerebral cortex stained with antibodies to nonphosphorylated epitopes on the medium (168 kDa) and heavy (200 kDa) molecular-weight subunits of the neurofilament protein, only subpopulations of large pyramidal neurons that exhibit clear regional and laminar distribution patterns are labeled (Campbell and Morrison, 1989; Hof *et al.*, 1990a; Hof and Morrison, 1990, 1995; Hof and Nimchinsky, 1992; Hof *et al.*, 1995b; Nimchinsky *et al.*, 1995, 1996, 1997). The intensity of neurofilament protein immunoreactivity varies as a function of cell size, the largest neurons being the most intensely labeled, both in the perikaryon and the dendritic arbor (Campbell and Morrison, 1989) (Fig. 13). Neurofilament protein, as well as several other cytoskeletal proteins, has been implicated in NFT formation (Ksiezak-Reding *et al.*, 1987; Zhang *et al.*, 1989; Trojanowski *et al.*, 1993b), suggesting that neurons containing high levels of nonphosphorylated neurofilament protein are particularly prone to NFT formation.

Detailed quantitative descriptions of the morphological, laminar, and regional characteristics of neurofilament protein-containing neurons in specific prefrontal, cingulate, temporal, and occipital areas of both monkey and human cortex have revealed a substantial degree of regional heterogeneity in size, density, and laminar distribution (Campbell and Morrison, 1989; Hof *et al.*, 1990a; Hof and Morrison, 1990, 1995; Hof and Nimchinsky, 1992; Carmichael and Price, 1994; Hof *et al.*, 1995b; Nimchinsky *et al.*, 1995, 1996). In addition, the distribution of neurofilament protein-immunoreactive neurons matches closely that demonstrated for certain cortical efferent systems. In fact, extensive quantitative analyses of more than 30 corticocortical pathways in the monkey neocortex combining tract-tracing and immunohistochemistry have demonstrated that the neurofilament protein-containing neurons represent a subpopulation of corticocortically projecting neurons (Campbell *et al.*, 1991; Hof *et al.*, 1995c, 1996b, 1997a,b; Nimchinsky *et al.*, 1996; Morrison *et al.*, 1997). These studies revealed that the proportion of neurofilament protein-containing neurons varies substantially among corticocortical projections, with low numbers of immunoreactive neurons in the limbic and short corticocortical connections (0–30%), intermediate numbers in nonvisual callosal connections (30–45%), and very high numbers (60–100%) in long association pathways linking polymodal regions, cingulomotor pathways, and visual callosal connections (Campbell *et al.*, 1991; Hof *et al.*, 1995c, 1996b, 1997b; Nimchinsky *et al.*, 1996). Also, in anterior cingulate cortex, neurofilament protein-containing neurons are restricted to layer V, the origin of most long corticocortical projections from anterior cingulate cortex (Barbas, 1986; Hof *et al.*, 1995c). Area 9 in prefrontal cortex, as well as superior and inferior temporal cortex, has a high density of neurofilament protein-containing cells in III and V, and corticocortical projections emanate from III and V in these regions (Barbas, 1986; Hof *et al.*, 1995c). The correlation between the origin of long corticocortical projections and neurofilament protein-containing neurons is particularly striking in the primary visual cortex of both macaque monkey and human. In both, the large, strongly neurofilament protein-immunoreactive neurons are restricted to layer IVB and to the Meynert cell class of layer VI, both known as sites of origin of the projections from area V1 to area MT (Shipp and Zeki, 1989a; Felleman and Van Essen, 1991; Hof *et al.*, 1996b). Furthermore, in the visual system, there are higher proportions of neurofilament protein-immunoreactive neurons in the projections to the parietal cortex (up to 85%) as compared with the projections to the temporal cortex (up to 40%) (Hof *et al.*, 1996b).

These observations on the regional and laminar distribution of neurofilament

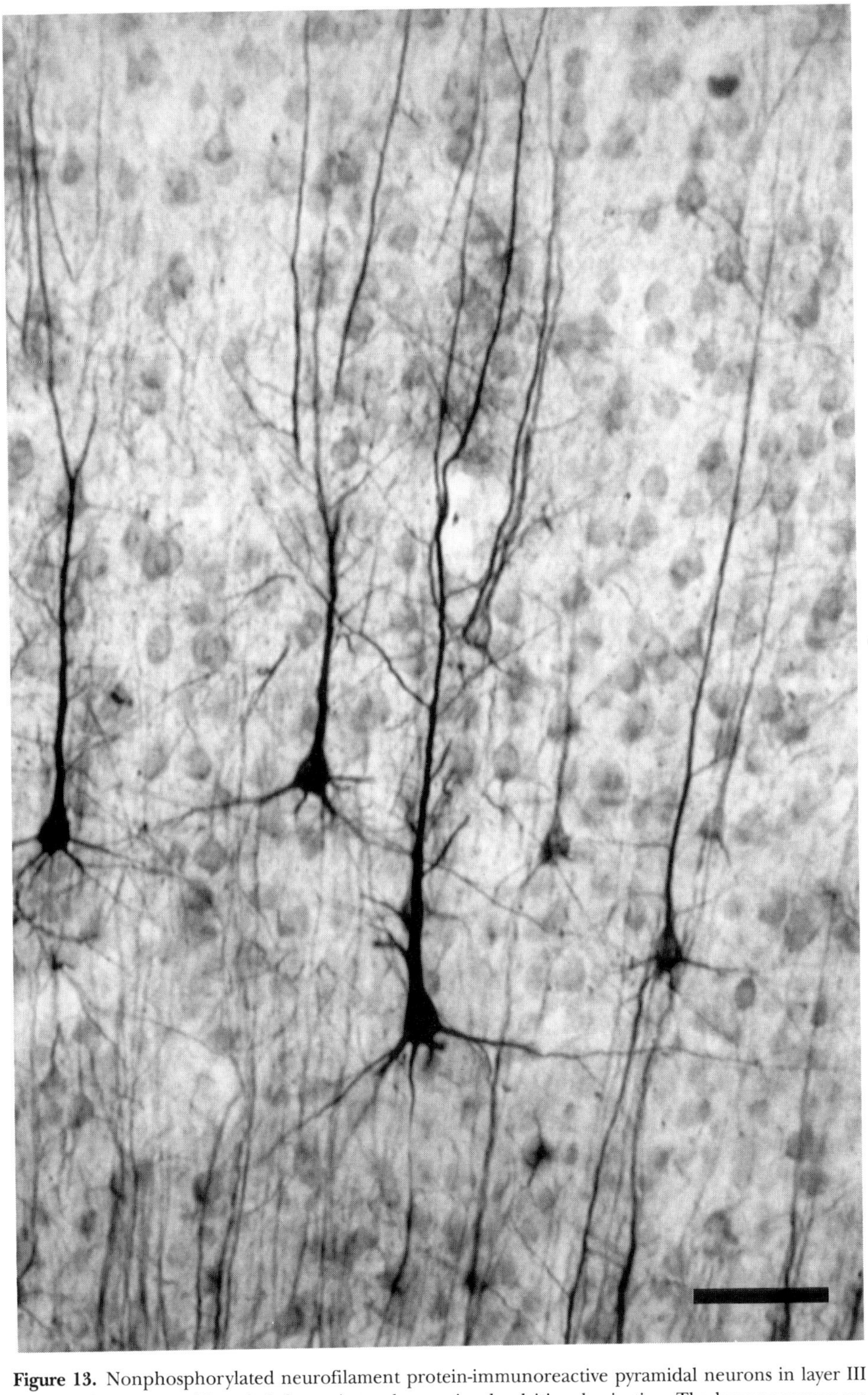

**Figure 13.** Nonphosphorylated neurofilament protein-immunoreactive pyramidal neurons in layer III of normal neocortex. Note their large size and extensive dendritic arborization. The larger neurons are more intensely stained (Campbell and Morrison, 1989). These neurons represent a relatively small population of the total pyramidal cells, as evidenced by the Giemsa counterstain. Immunolabeled with an antibody (SMI-32) to medium and large neurofilament protein subunits. Bar = 100 μm.

protein-containing neurons indicate that, depending on which cortical pathway is considered, they represent a significant portion of the neurons of origin of long corticocortical projections. We have also observed that in certain parts of the neocortical and hippocampus, these neurons are highly susceptible to cell death as well as to NFT formation in AD (Morrison *et al.,* 1987; Hof *et al.,* 1990a; Hof and Morrison, 1990; Vickers *et al.,* 1992, 1994b, 1996). The laminar distribution of neurofilament protein-containing neurons in visual areas of normal human cortex, such as Brodmann areas 18 and 20 and in prefrontal and anterior cingulate cortex, is comparable to the laminar distribution of NFTs (Fig. 14). Furthermore, there is an inverse correlation between the densities of NFTs and densities of neurofilament protein-immunoreactive neurons, in that cortical layers showing NFT densities in AD cases have dramatically decreased numbers of such neurons (Hof *et al.,* 1990a; Hof and Morrison, 1990a). A comparable situation exists in layers II and V of entorhinal cortex and in the pyramidal neurons of the subiculum, which have a very high density of neurofilament protein in healthy human brain, whereas they display a severe loss of neurofilament-protein-containing cells in AD (Morrison *et al.,* 1987; Vickers *et al.,* 1992, 1994b) (Fig. 15). These observations show that neurofilament protein-containing neurons are highly vulnerable in AD. In fact, quantitative analyses have demonstrated a severe loss of neurofilament protein-containing neurons in layers III and V in the inferior temporal and superior frontal cortex (Fig. 16). The severity of the loss correlates with the size of these neurons, in that the largest neurons (larger than 350 $\mu m^2$ of perikaryal cross-sectional area) are the most affected (up to 90% cell loss), and the intermediate size neurons (250–350 $\mu m^2$) are more susceptible to degeneration than are the smaller (less than 250 $\mu m^2$) immunoreactive neurons (Fig. 17) (Hof *et al.,* 1990a).

Although the primary visual cortex is relatively spared in AD, a certain degree of pathology exists in areas 17 and 18 that is reflected in restricted neuronal changes. Visual dysfunction is frequently observed in patients with AD and has been associated with dysfunction of cortical areas at higher levels of visual processing, such as regions in the temporal and parietal cortices (Cogan, 1985; Hinton *et al.,* 1986; Neary and Snowden, 1987; Cronin-Golomb *et al.,* 1993; Curcio and Drucker, 1993; Kurylo *et al.,* 1994, 1996; Parasuraman *et al.,* 1995; Mentis *et al.,* 1996; Pietrini *et al.,* 1996; Hof *et al.,* 1997c). Interestingly, the distribution and density patterns of neurofilament protein-containing neurons in the visual areas of the human occipital cortex differ substantially from the patterns observed in inferior temporal and prefrontal cortex (Campbell and Morrison, 1989; Hof *et al.,* 1990a; Hof and Morrison, 1990). In area 17 the neurofilament protein-immunoreactivity is confined to three kinds of cells: large, intensely stained neurofilament protein-containing neurons that are mainly in layer IVB; Meynert cells that border layers V and VI; and isolated cells in the deep portion of layers III and IVA. Small, lightly stained neurons are also present in layers III, V, and VI, whereas the large, darkly labeled neurons in area 18 predominate in the deep portion of layer III, with a few isolated in layer V. Areas 9 and 20 have consistently higher densities of intensely labeled, large neurons in deep layer III than does area 18, and in addition, have a much higher density of labeled neurons in layers V and VI than do both occipital regions (Campbell and Morrison, 1989; Hof *et al.,* 1990a; Hof and Morrison, 1990).

In contrast to the loss in areas 9 and 20, the neurofilament protein-containing neuron loss in area 17 and 18 is minimal (Table V). The loss is limited to large

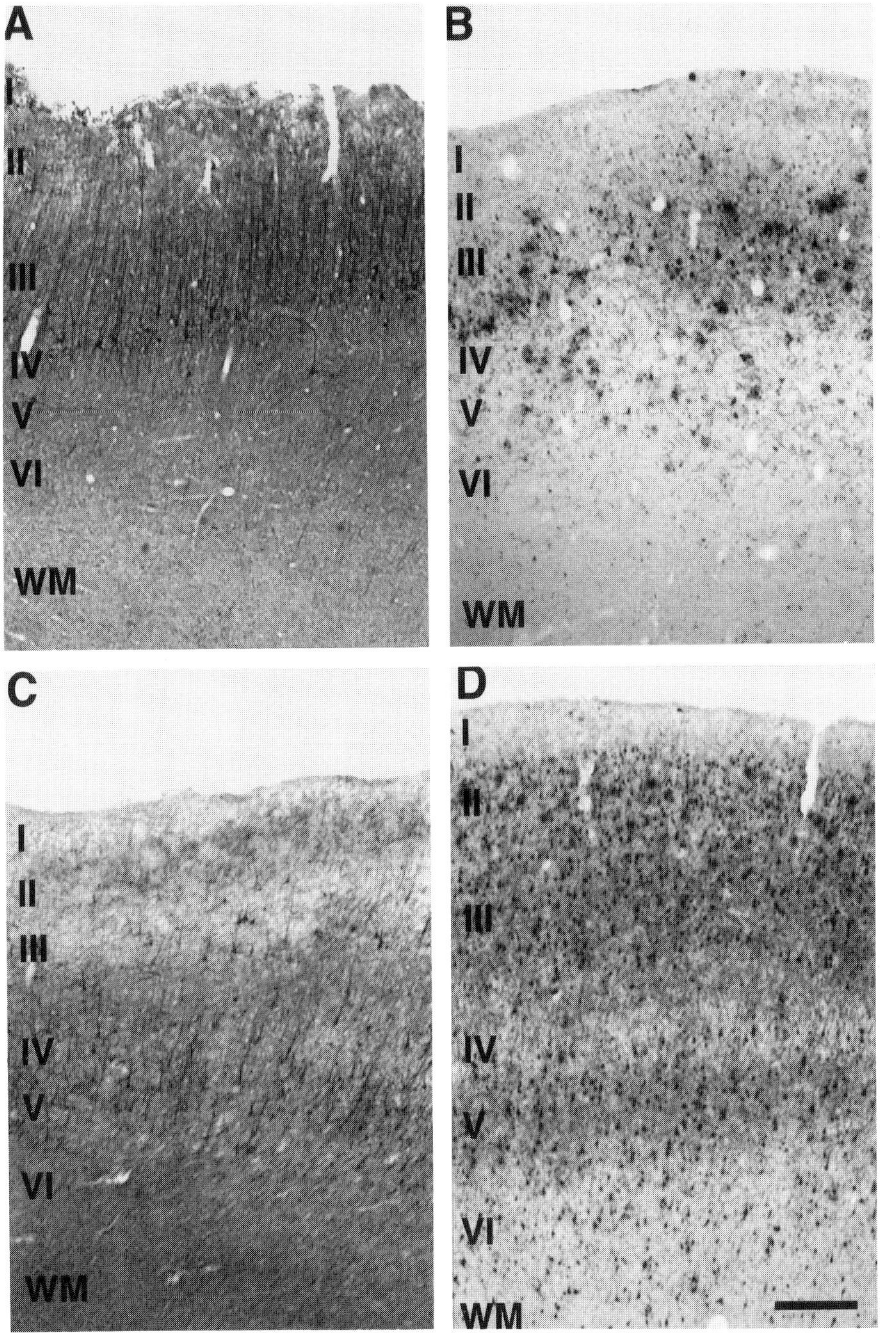


**Figure 14.** Correlative distribution of neurofilament protein-enriched neurons (A, C) and NFTs (B, D) in area 18 (A, B) and area 20 (C, D) of severe AD. In area 18 the neurons form a band in the deep portion of layer III and are moderately affected as compared with control subjects (Hof and Morrison, 1990). Their distribution matches that of NFTs (A, B). In the inferior temporal cortex, a severe loss of neurofilament protein-immunoreactive neurons in both layers III and V (C) correlates with the very high NFT densities (D). Bar = 100 μm.

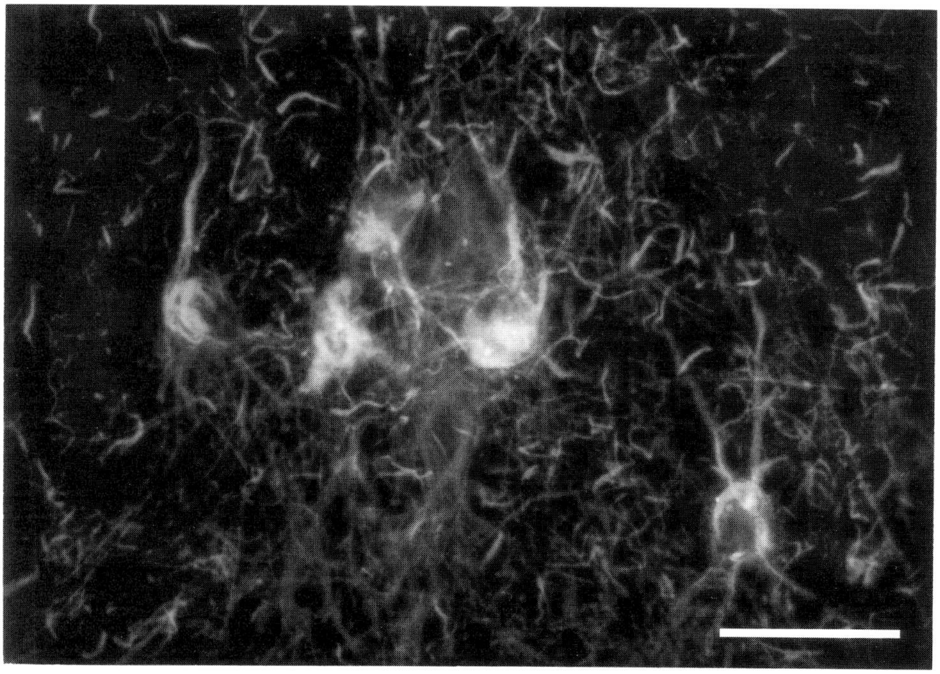

**Figure 15.** Double labeling of neurons in layer II of the entorhinal cortex with antibodies to nonphosphorylated neurofilament protein (light gray areas) and hyperphosphorylated tau proteins in AD with mild cognitive changes. All neurons contain some degree of accumulation of pathologic tau proteins coextensive with neurofilament protein immunoreactivity (bright zones in neurons). Note also the many tau-immunoreactive neuritic elements and neuropil threads. Bar = 50 μm.

neurons in layer IVB and to the Meynert cell class, and does not exceed 25% in these layers. In area 18, neurofilament protein-containing neuron loss occurs in the deep part of layer III and averages 18%. Thus, the extensive short corticocortical projection from area 17 to area 18 originating in layer III is presumably spared in AD, as the lightly immunoreactive cells in layers III of area 17 are only marginally affected in AD cases. In addition, the lightly labeled cells in layers V and VI in area 17 that probably project to a variety of subcortical sites are intact in AD (Hof and Morrison, 1990). The restricted vulnerability of the neurofilament protein-containing layer IVB and Meynert cells in AD is particularly interesting in light of data obtained from retrograde transport studies combined with immunohistochemistry in macaque monkeys, which indicate that all neurons in layers IVB and VI in area V1 that project to area MT contain neurofilament protein. In addition, in the monkey the projections from areas V2 and V3 to area MT originate primarily from layer III neurofilament protein-containing pyramidal cells (Hof *et al.*, 1996b), precisely the same location as the vast majority of large neurofilament protein-containing cells in human areas 18 and 19 (Hof and Morrison, 1990). Thus, although the projection from area 17 to area 18 may be intact, and the overall neuronal loss may not appear significant in area 17, two crucial projections involved in visuomotor skills are compromised in AD, namely the equivalents of macaque monkey projections from area V1 and V2 to MT (Mishkin *et al.*, 1983; Van Essen and Maunsell, 1983; DeYoe and Van Essen, 1988; Livingstone and Hubel, 1988; Zeki and Shipp,

1988; Shipp and Zeki, 1989a,b; Ungerleider, 1995; Hof *et al.*, 1996b). This apparently higher vulnerability of select components of the occipitoparietal projections may be why deficits in visuomotor abilities are relatively common in AD (Katz and Rimmer, 1989; Kiyosawa *et al.*, 1989; Kuskowski *et al.*, 1989; Fletcher, 1994; Parasuraman *et al.*, 1995; Mentis *et al.*, 1996), and it is worth noting that the region located in the occipito-temporal junction considered to correspond to area MT in the human contains generally large numbers of NFTs in AD cases and is characterized by high numbers of neurofilament protein-immunoreactive neurons organized in a very distinct laminar pattern (Adams *et al.*, 1996; Hof *et al.*, 1997c).

In addition to regional distribution and relationship to connectivity, NFT identified by thioflavine-S stain or immunohistochemistry using antibodies to neurofilament protein and to microtubule-associated protein tau have revealed dynamic cellular alterations in vulnerable neuronal populations during normal aging. For instance, layer II of the entorhinal cortex contains neurofilament protein-immunoreactive neurons that also display immunoreactivity to tau protein and thioflavine S-positive materials, suggesting the existence of transitional forms of NFTs (Vickers *et al.*, 1992) (Fig. 18A–C). In such cases, very rare NFTs are seen in the frontal cortex, which contains preserved neurofilament protein-immunoreactive neurons in layers III and V. In AD cases, however, most NFTs in the entorhinal cortex progress to an end-stage and are no longer immunoreactive to tau and neurofilament proteins, but are stained only with thioflavine S (Fig. 18D–F). At this

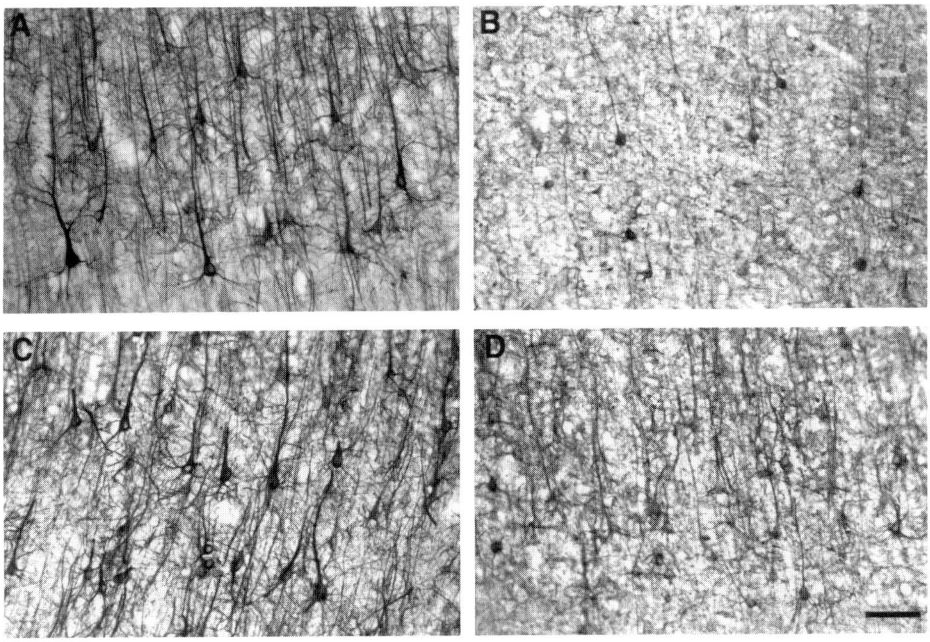

**Figure 16.** Involvement of pyramidal neurons in layer III of the superior frontal cortex (A, B) and inferior temporal cortex (C, D) in control cases (A, C) and severe AD (B, D). Note the dramatic neuronal loss and the changes in dendritic morphology and staining intensity in the remaining neurons in AD. Hof *et al.* (1990a). Bar = 120 μm.

stage, transitional NFT forms are observed in the frontal cortex, indicating a time-dependent process in the formation of NFTs in certain neurofilament protein-containing neurons (Vickers *et al.,* 1992) (Fig. 18G–I). Moreover, the presence of high levels of neurofilament protein appears to be a prerequisite for the NFT formation, because certain neurons, such as the pyramidal neurons of the CA1 field, that do not normally express detectable levels of this protein in young adults yet are prone to NFT formation in AD, begin to show increasing levels of immunoreactivity with age (Vickers *et al.,* 1994b).

Thus, one of the neurochemical characteristics of neurons vulnerable to NFT formation in AD appears to be the presence of high somatic and dendritic concentrations of nonphosphorylated neurofilament proteins. The degree to which corticocortically projecting cells are also neurofilament protein-immunoreactive neurons varies among corticocortical projections, and both the length of a given corticocortical projection and the functional nature of the interconnected regions

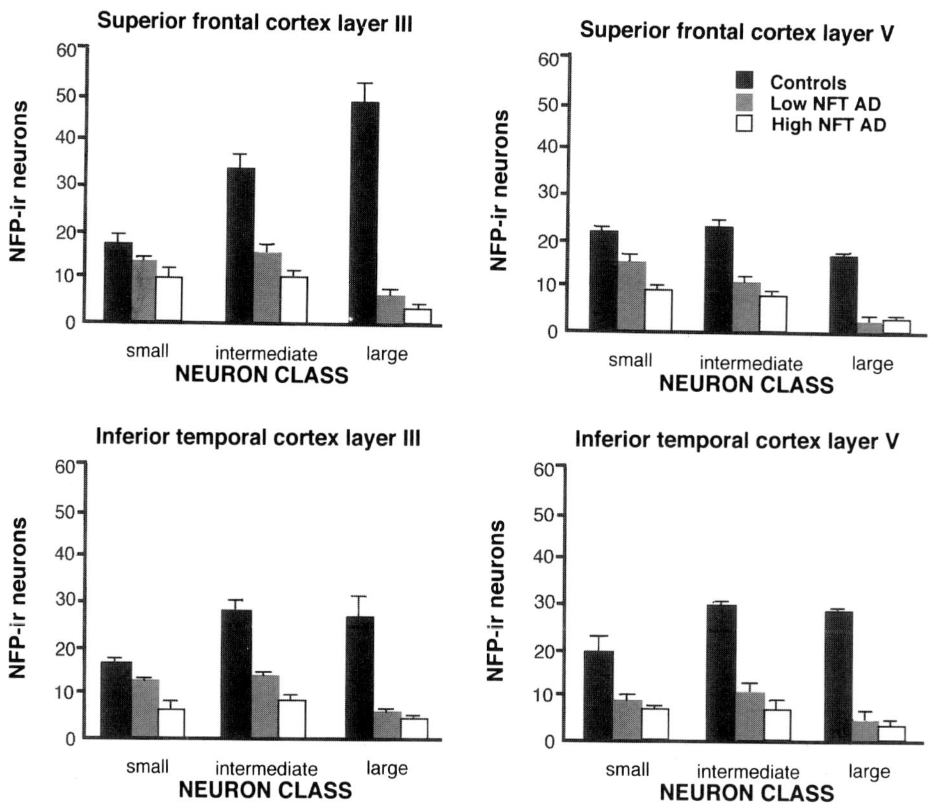


**Figure 17.** Quantitative analysis of neurofilament protein-immunoreactive (ir) neuron loss in layers III and V of the superior frontal and inferior temporal cortices in AD with low NFT densities (less than 30 per mm$^2$) or high NFT densities (more than 30 per mm$^2$) compared with control cases. Note the correlation between neuron loss and disease severity assessed by NFT counts. Neuron loss is more marked in the largest neurons and occurs even in cases with low NFT densities. Perikaryal areas were calculated at the time neurons densities were recorded and neurons were categorized as small (<250 $\mu m^2$ of perikaryal cross-sectional area), intermediate (250–350 $\mu m^2$) and large (>350 $\mu m^2$). Hof *et al.* (1990a).

**Table V. Neurofilament Protein-Immunoreactive Neuron Counts in Areas V1 and V2 of Nondemented Elderly and Alzheimer's Disease Cases[a]**

| | | Neuron counts | |
|---|---|---|---|
| | | Control cases | AD cases |
| *Area V1* | | | |
| Layers | I | 0.0 | 0.0 |
| | II | 0 | 0 |
| | IIIs[b] | 2.7 ± 0.5 | 2.6 ± 0.5 |
| | IIId[b] | 6.9 ± 0.5 | 7.3 ± 0.8 |
| | IVA | 0.2 ± 0.1 | 0.1 ± 0.1 |
| | IVB | 8.2 ± 0.7 | 6.2 ± 0.4 |
| | IVCα | 1.9 ± 0.3 | 1.9 ± 0.3 |
| | IVCβ | 0.2 ± 0.1 | 0 |
| | V | 4.6 ± 0.2 | 4.5 ± 0.2 |
| | VI | 7.4 ± 0.6 | 7.5 ± 0.8 |
| | M | 3.9 ± 0.5 | 2.9 ± 0.2 |
| *Area V2* | | | |
| Layers | I | 0 | 0 |
| | II | 0.1 ± 0.1 | 0 |
| | IIIs | 1.2 ± 0.2 | 0.8 ± 0.2 |
| | IIId | 23.0 ± 1.2 | 19.1 ± 0.9 |
| | IV | 1.1 ± 0.2 | 0.5 ± 0.1 |
| | V | 3.9 ± 0.5 | 3.3 ± 0.3 |
| | VI | 6.1 ± 0.4 | 4.9 ± 0.4 |

[a]Data represent neurofilament protein-immunoreactive reuron counts per mm of cortical traverse and are means ±SEM from five control and five AD brains. There are differences between AD cases and controls in layer IVB and the Meynert cells in area V1 and in the deep portion of layer III in area V2. M: Meynert cells; s and d refer to the superficial and deep portions of layer III, respectively. Adapted form Hof and Morrison (1990).
[b]Superficial layer III.
[c]Deep layer III.

may constitute determinant variables of the phenotype of these projection neurons (Campbell *et al.,* 1991; Hof *et al.,* 1995c, 1996b, 1997a,b; Nimchinsky *et al.,* 1996). Clearly, corticocortical projection cells of origin have a heterogeneous neurochemical phenotype, and additional analyses must be performed to describe more precisely the cortical connectivity with biochemical characteristics and to determine the features that confer selective vulnerability on a given subset of neurons in AD. It is, however, conceivable that a defined neuronal subpopulation, characterized by a set of neurochemical attributes and highly vulnerable in AD, can be correlated with specific corticocortical projections of known function.

### 3.3. Glutamate Receptor Implication

It has been proposed that disruption of glutamate metabolism and glutamate receptor-mediated excitotoxicity is one of the major mechanisms of neuron death in many neurodegenerative disorders, as well as in other pathological conditions

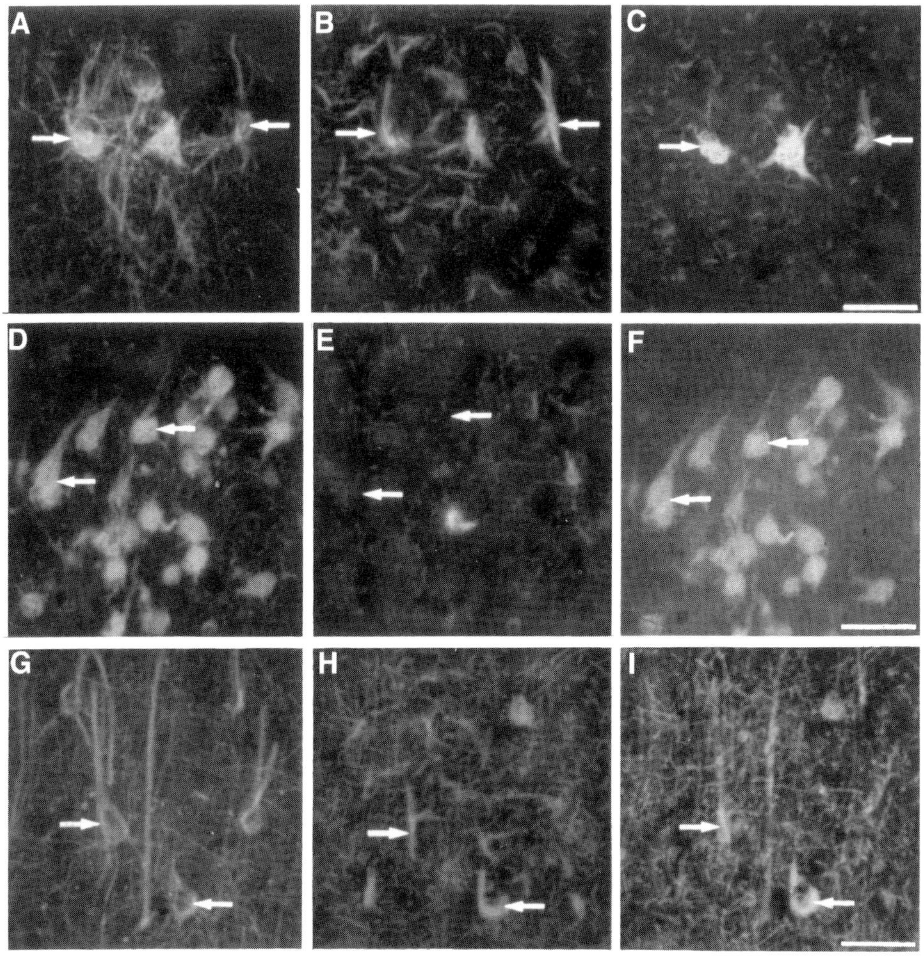


**Figure 18.** Dynamic changes in cytoskeletal components in AD as compared with normal aging. Multiple labeling of neurofilament protein (A, D, G, I), tau proteins (B, E, H, I), and thioflavine S (C, F) in layer II of the entorhinal cortex (A–F) and layer III of area 9 (G–I). An elderly nondemented case is shown in A–C and an AD case in D–I. Neurofilament and tau proteins are visible in the entorhinal complex in the control case (arrows; A, B) but not in the AD case, where only thioflavine S-stained NFTs remain (arrows; D–F). These NFTs have also lost their immunoreactivity to tau protein at this late stage (compare B and E). In AD, transitional neurons showing changes similar to those in the entorhinal complex of controls are seen in the frontal cortex (arrows; G–I). Adapted from Vickers *et al.* (1992) *Brain Res.* **594:**273–278, with kind permission of Elsevier Science—NL, Sara Burgerhartstraat 25, 1055 KV Amsterdam, The Netherlands. Bars = 30 μm (A–C, G–I) and 20 μm (D–F).

such as brain ischemia and epilepsy (Choi, 1988, 1990; Choi and Rothman, 1990; Olney, 1990; Farooqui and Horrocks, 1991; Meldrum and Garthwaite, 1991; Coyle and Puttfarcken, 1993; Francis *et al.,* 1993; Turski and Turski, 1993; DeFelipe *et al.,* 1994; Ellison, 1995; Lipton and Rosenberg, 1995; Blümcke *et al.,* 1996; Morrison *et al.,* 1996; Ottersen *et al.,* 1996; Obrenovitch and Urenjak, 1997). Glutamate receptor-mediated excitotoxicity presumably results from increased calcium flux leading to toxic intracellular concentrations of this ion. Although the *N*-methyl-D-aspartate (NMDA) receptor has been the main receptor subtype associated with calcium flux, it is now clear that both kainate and α-amino-3-hydroxy-5-methyl-4-isoxazole propionic acid-(AMPA)-kainate receptors also increase calcium flux (Sommer *et al.,* 1991; Verdoorn *et al.,* 1991; Gasic and Heinemann, 1992). Thus, the analysis of glutamate receptor profiles of cortical neurons and related circuits becomes a crucial feature when correlating identified neurons and circuits with the potential for vulnerability to degeneration through excitotoxicity, because the defining characteristics of a given glutamatergic circuit regarding ion fluxes depends on the subunit composition of the receptors dominating that circuit. Such comprehensive analyses allow for the integration of glutamate receptor characteristics into a comprehensive neurochemical profile of vulnerability, and have become possible as glutamate receptor subunit-specific antibodies have become available (for review see Huntley *et al.,* 1994b; Morrison *et al.,* 1996a).

### 3.3.1. Cellular Distribution of Glutamate Receptor Subunits: Neuronal Types and Cortical Circuits

Several studies have demonstrated that distinct pyramidal and nonpyramidal subtypes can be classified according to their glutamate receptor profiles (Huntley *et al.,* 1993, 1994a,b; Vickers *et al.,* 1993; Morrison *et al.,* 1996b; Vissavajjhala *et al.,* 1996). For instance, certain pyramidal neurons have AMPA subunits GluR2/3 and kainate subunits GluR5/6/7, whereas others have GluR2/3 but lack GluR5/6/7 (Huntley *et al.,* 1994b). In human cerebral cortex, NMDA and non-NMDA receptor are present in the vast majority of neurons, yet specific regional and cell class-specific patterns are visible (Blümcke *et al.,* 1994, 1996; Vickers *et al.,* 1995a; He *et al.,* 1996, 1997; Hof *et al.,* 1996c; Ong and Garey, 1996; Huntley *et al.,* 1997). Although all three classes of ionotropic glutamate receptors are present in hippocampal neurons, the neurons in the CA2-3 fields generally contain higher labeling intensities of GluR1 and GluR2/3 than do the CA1 field and hilus, and GluR4 labeling is rather sparse (Blümcke *et al.,* 1994, 1996; Eastwood *et al.,* 1994; Hyman *et al.,* 1994; Day *et al.,* 1995; Ikonomovic *et al.,* 1995; Hof *et al.,* 1996c). Also, in human neocortex GluR1 and GluR4 are present in apparently higher levels in nonpyramidal neurons, whereas pyramidal cells display a strong immunoreactivity to GluR2/3, GluR5/6/7 and NMDAR1 subunits (Vikers *et al.,* 1995a; He *et al.,* 1996, 1997; Hof *et al.,* 1996c; Ong and Garey, 1996; Huntley *et al.,* 1997) (Fig. 19). In addition, retrograde tracing studies in macaque monkeys combined with immunohistochemistry have shown that approximately 70% of the pyramidal neurons in layer III of prefrontal association cortex that project to the superior temporal cortex are GluR5/6/7-immunoreactive, whereas virtually none are in layer V (Huntley *et al.,* 1994b). Interestingly, this percentage in layer III is comparable to the proportion of neurofilament protein-containing neurons participating in this projection (Hof *et al.,* 1995b), and the distribution of GluR5/6/7 parallels that of neurofilament protein in feedforward visual projections from areas V1, V2, and V3 to areas V4 and MT (Hof *et al.,* 1997b) (Table VI). Comparably, neurons in layer V of the entorhinal cortex that project to the superior temporal cortex are all GluR5/6/7-immunoreactive (Good and Morrison, 1995). In contrast, the NMDA and AMPA receptors have a less selective distribution among corticocortical projections; they are present in nearly all of the corticocortically-projecting neurons (Hof *et al.,* 1997b). These data demonstrate a selective parcellation of kainate receptor subunits to subpopulations of long corticocortically projecting neurons, and, at least in some cases, a possible relationship to another component of the molecular phenotype of these circuit neurons.

Cortical GABAergic interneurons can be subdivided into largely non-overlapping subsets defined by the selective distribution of the three calcium-binding proteins, parvalbumin, calbindin, and calretinin (Hendry *et al.*, 1989; Celio, 1990; Hof and Nimchinsky, 1992; Hof *et al.*, 1993c). These subsets also display a substantial degree of heterogeneity in their content of NMDA and non-NMDA receptor subunits such that fewer calretinin neurons are immunoreactive for NMDAR1, GluR2, and GluR5/6/7 subunit proteins than are parvalbumin- and clabindin-containing neurons, which have a much higher representation of all three glutamate receptor classes (Vickers *et al.*, 1993; Huntley *et al.*, 1994a, 1997; Morrison *et al.*, 1996b; Vissavajjhala *et al.*, 1996) (Table VII).

The overall density of GluR5/6/7-immunoreactive neurons is higher in the association areas of prefrontal cortex as compared with the primary sensory areas (Huntley *et al.*, 1993), and—in prefrontal, temporal, motor, and sensory areas of

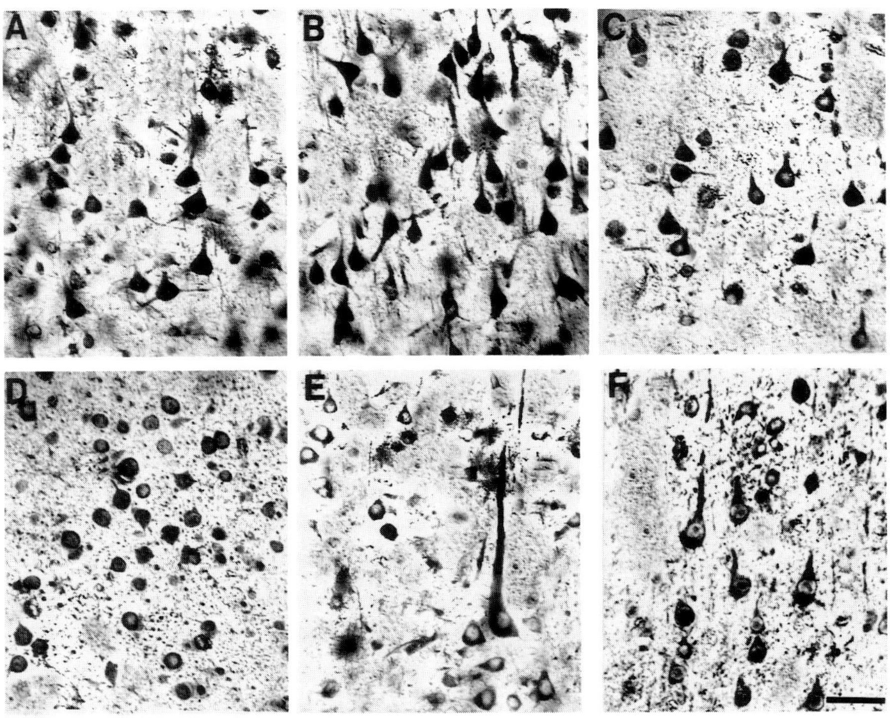


**Figure 19.** Distribution of immunolabeling for GluR2 and GluR5/6/7 in human neocortex (A, B). Immunoreactivity for GluR2 in layers III (A) and V (B) of area 7. Note the strong staining of pyramidal neurons. Apical dendrites are also visible (C, D). GluR5/6/7-immunoreactive neurons in layer III of area 7 (C) and in a tangential section through layer III of area 9 (D). There are many labeled perikarya as well as fibers cut in cross sections (D). (E, F) GluR5/6/7 immunoreactivity in large pyramidal neurons and dendritic elements in layer V of area 20, (#), and polymorphic neurons in layer VI in the same region (F). On F some glial elements are also labeled. Adapted from Vickers *et al.* (1995a), *Brain Res.* **594:**273–278, with kind permission of Elsevier Science—NL, Sara Burgerhartstraat 25, 1055 KV, Amsterdam, The Netherlands. Bar = 60 μm.

**Table VI. Distribution of Neurofilament Protein-Immunoreactive Neurons in Corticocortical Projections between Areas MT, V4, V1, and V2[a]**

| Projection/layer | NFP (%) | Projection/layer | NFP (%) |
|---|---|---|---|
| *Feedforward to MT* | | *Feedforward to V4* | |
| **V1 → MT** | | **V1 → V4** | |
| IVB | 100 | III | 25.1 |
| Meynert | 100 | | |
| **V2 → MT** | | **V2 → V4** | |
| III | 78.0 | III | 32.2 |
| V | 0 | V | 0 |
| **V3 → MT** | | **V3 → V4** | |
| III | 74.8 | III | 35.9 |
| V | 57.2 | V | 28.4 |
| *Feedback to V1* | | *Feedback to V2* | |
| **MT → V1** | | **MT → V2** | |
| II–III | 60.0 | II–III | 61.8 |
| V–VI | 79.1 | V–VI | 79.6 |
| **V4 → V1** | | **V4 → V2** | |
| II–III | 60.1 | II–III | 63.2 |
| V–VI | 80.1 | V–VI | 75.9 |
| **V3 → V1** | | **V3 → V2** | |
| II–III | 64.8 | II–III | 67.2 |
| V–VI | 72.8 | V–VI | 76.5 |

[a]Data represent mean laminar percentages of double-labeled neurons (i.e., retrogradely labeled and neurofilament protein-immunoreactive). Adapted from Hof *et al.* (1997a).

monkey neocortex—GluR5/6/7 subunit-immunoreactivity is higher in the pyramidal layers, which furnish and receive corticocortical, commissural, and local excitatory projections (Huntley *et al.,* 1994a). In contrast, NMDAR1-immunoreactive neurons and synapses are densely distributed throughout all layers of monkey and human neocortex (Huntley *et al.,* 1997), suggesting that the kainate class of glutamate receptor subunits plays a more selective role in mediating the activity of specific areas and pathways. Such differences in regional and laminar patterns of subunit-identified cells and synapses suggest a differential involvement of the subunits in mediating the activity of distinct excitatory pathways, and may be key

**Table VII. Colocalization of Glutamate Receptor Subunit Proteins with Calcium-Binding Proteins in the Neocortex[a]**

| | GluR2 | GluR5/6/7 | NMDAR1 |
|---|---|---|---|
| Parvalbumin | 95 | 50 | 50 |
| Calbindin | 90 | 30 | 50 |
| Calretinin | 60 | 1 | 6 |

[a]Data represent percentages of colocalization in calcium-binding, protein-containing interneurons. Adapted from Vickers *et al.* (1993); Morrison *et al.* (1996b); Vissavajjhala *et al.* (1996); and Huntley *et al.* (1997).

factors in determining the relative vulnerability of specific neuronal classes to neurodegeneration.

### 3.3.2. Glutamate Receptor Subtype-Specific Alterations in Aging

Age-related memory impairment in nonhuman primates occurs in the absence of overt dementia (Rapp and Amaral, 1992; Gallagher and Rapp, 1997; Moss *et al.*, 1997), and changes in the expression of certain glutamate receptors have been reported in the hippocampus of animal models of aging as well as in elderly humans (Armstrong *et al.*, 1994; García-Ladona *et al.*, 1994; Pagliusi *et al.*, 1994; Ikonomovic *et al.*, (1995). Additionally, the role of NMDA receptors and the integrity of the excitatory projections from the entorhinal cortex to the dentate gyrus (perforant path) are crucial factors in memory formation (Bliss and Lømo, 1973; Bliss and Collingridge, 1993; Morris *et al.*, 1986; Meunier *et al.*, 1993), and they represent a functional target to investigate potential shifts in glutamate receptor subunit distribution during aging. Recent analyses combining glutamate receptor subunit-specific antibodies and quantitative confocal laser scanning microscopy demonstrated that NMDAR1 immunofluorescence intensity within the dentate gyrus of aged macaque monkeys (older than 25 years) was decreased by about 30% in the outer portion of the dentate gyrus molecular layer in comparison to its intensity in adults and juvenile animals (Fig. 20A) (Gazzaley *et al.*, 1996a). This decrease in NMDAR1 immunofluorescence intensity occurs selectively within the distal dendrites of the dentate gyrus granule cells, which receive the perforant path input from the entorhinal cortex, but not in the proximal dendrites, which receive an intrinsic excitatory input from the dentate hilus as well as from noncortical structures (Gazzaley *et al.*, 1996a). Furthermore, AMPA (GluR2/3)

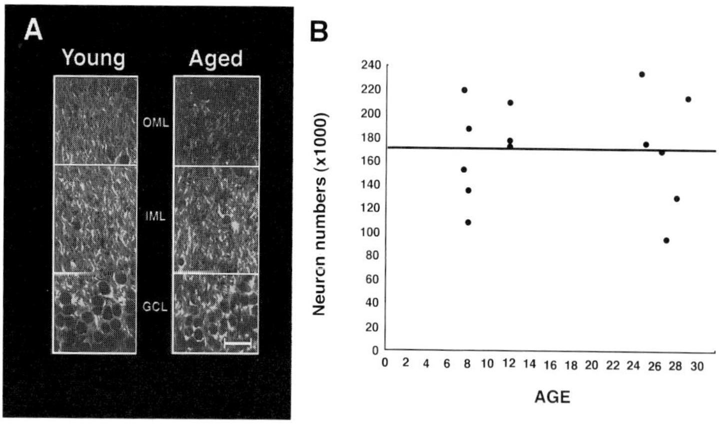


**Figure 20.** (A) Age-related changes in the perforant path terminal zone of the macaque monkey. Aged monkeys show a consistent decrease in staining intensity for the NMDAR1 subunit restricted to the outer molecular layer of the dentate gyrus (OML) as compared with young animals. No such changes are visible in the inner molecular (IML) and granule cell (GCL) layers (a pseudocolored confocal image using a glow scale re-creates relative staining intensities). There was no structural alteration of these dendrites. Adapted from Gazzaley *et al.* (1996b). Bar = 25 μm. (B) Stereologic estimates of total neuron numbers in layer II of the entorhinal cortex in the same animals. There is no neuron loss in aged monkeys as compared with young ones, demonstrating an NMDA receptor change without a structural degeneration of the perforant path. Gazzaley *et al.* (1996b, 1997).

and kainate (GluR5/6/7) subunits are not altered in the aged animals, demonstrating that the intradendritic changes in NMDAR1 immunofluorescence is specific to that class of glutamate receptors, even though all three classes of glutamate receptors are colocalized within these dendrites (Siegel *et al.*, 1994, 1995). Interestingly, total synaptic density as reflected by synaptophysin immunoreactivity, as well as the dendritic integrity demonstrated using microtubule-associated protein 2, show no different in staining intensity or dendritic area across the molecular layer in old animals as compared with young animals.

These findings suggest that circuit-specific shifts in NMDAR1 subunit concentration within segments of granule cell dendrites corresponding to sublaminar patterns of input segregation occurs in the dentate gyrus of aged monkeys without alterations in dendritic morphology and synapse density. Further evidence that these age-related changes are restricted to the expression of these receptors in the dentate granule cell dendrites is provided by recent stereological data showing no neuronal loss in layer II islands of the anterior portion of the entorhinal cortex, which provide the projection to the dentate gyrus, in the same group of macaque monkeys (Fig. 20B) (Gazzaley *et al.*, 1997). In these animals, total neuron counts in layer II average 167,700 neurons, whereas in the young adult and juvenile monkeys this value is 169,900 and 170,700, respectively (Gazzaley *et al.*, 1997). The intradendritic distribution of a neurotransmitter receptor appears, therefore, to be modifiable in specific circuits during aging, and such subtle changes may constitute the molecular and cellular substrate for age-associated memory impairment.

### 3.3.3. Implications for AD

With respect to the pathogenesis of AD, not only have glutamate receptors been implicated in neurotoxicity, but glutamate is thought to be the major neurotransmitter in corticocortical systems, because it has been found in pyramidal neurons in humans (Hyman *et al.*, 1987; Greenamyre and Young, 1989; Greenamyre and Maragos, 1993; Huntley *et al.*, 1994a, 1997; Vickers *et al.*, 1995a; Hof *et al.*, 1996c). Also, glutamate-containing neurons are prone to NFT formation (Hyman *et al.*, 1987, 1994; Kowall and Beal, 1991), and depletion of glutamate has been reported in the perforant pathway in AD cases (Greenamyre *et al.*, 1985; Hyman *et al.*, 1987; Greenamyre and Young, 1989; Dewar *et al.*, 1991; Greenamyre and Maragos, 1993). In AD several studies using autoradiography, immunocytochemistry, or *in situ* hybridization have revealed substantial changes in the distribution and densities of various glutamate receptor subunits in the human cerebral cortex (Dewar *et al.*, 1991; Carlson *et al.*, 1993; García-Ladona *et al.*, 1994; Greenamyre and Maragos, 1993). For instance, AMPA-selective subunits GluR1 and GluR2/3 are severely reduced in neocortex and entorhinal cortex of demented cases, but their density increases in hippocampal regions that are relatively spared in AD, such as dentate gyrus and CA2-3 fields (Greenamyre and Maragos, 1993; Armstrong *et al.*, 1994; García-Ladona *et al.*, 1994; Yasuda *et al.*, 1995; Armstrong and Ikonomovic, 1996). Binding to AMPA and NMDA receptors is not altered in area 17 in AD, but is significantly reduced at higher cortical levels in the visual system (Carlson *et al.*, 1993), correlating well with the notion that primary visual cortex is spared in AD and that there is a progressive increase in the severity of involvement of visual association areas (Lewis *et al.*, 1987; Hof and Morrison, 1990).

Neuronal alterations mediated via excitotoxicity are thought to result from abnormally high cytosolic levels of calcium that come from over-activation of certain calcium-permeable glutamate receptors, particularly NMDA receptors, and/or from secondary activation of voltage-gated calcium channels (Choi and Rothman, 1990; Szatowski and Atwell, 1994). Recent studies have shown that after ischemic brain injury, there is a down-regulation in the expression of the calcium-impermeable GluR2 subunit in the CA1 field of the hippocampus (Pellegrini-Giampietro *et al.,* 1992, 1994; Pollard *et al.,* 1993; Tsubokawa *et al.,* 1995). Such switches in expression may represent a cellular correlate of the toxic increase in calcium flux, and define subsets of pyramidal neurons particularly prone to degeneration. Interestingly, neurons in culture that possess high levels of calcium-permeable AMPA/kainate and NMDA receptors are highly sensitive to calcium load exposure, and such neurons are much more likely to express the calcium-permeable AMPA subunits GluR1 and GluR4 than the calcium-impermeable GluR2 subunit (Turetsky *et al.,* 1994; Yin *et al.,* 1994; Lu *et al.,* 1996).

Glutamate toxicity that is presumably mediated through calcium-activated processes has been linked with both degeneration and cytoskeletal alterations similar to those underlying NFT formation (De Boni and Crapper McLachlan, 1985; Sautière *et al.,* 1992; Mattson *et al.,* 1992; Yang *et al.,* 1995). Kainate induces excitotoxic damage in the rodent hippocampus in an age-related fashion, and exposure of primary neuronal cultures to glutamate increases immunoreactivity to hyperphosphorylated tau protein by over 200% (Sindou *et al.,* 1994; Kesslak *et al.,* 1995). Also, acute as well as chronic exposure to NMDA induces increases in immunostaining of phosphorylated tau protein at the serine 202 phosphorylation site (Couratier *et al.,* 1996). Furthermore, it has been demonstrated that motoneurons as well as cortical neurons in culture that are heavily stained by the same antibody to nonphosphorylated neurofilament protein used in our neuropathological studies of AD (see above) are also hypersensitive to AMPA/kainate receptor-mediated toxicity (Gottron *et al.,* 1995; Carriedo *et al.,* 1996). Studies of primate cortex employing antisera to kainate subunit proteins have demonstrated regional, laminar, and cellular distribution patterns that correlate with the distribution of neuropathologic profiles in AD, as the labeled neurons are primarily located in layers II, III, and V in the association cortex (Huntley *et al.,* 1994b; Morrison *et al.,* 1996). Double-labeling analyses have shown that neurofilament protein-immunoreactive neurons highly vulnerable to NFT formation in AD are consistently immunoreactive for both kainate subunits and AMPA/kainate subunits (Vickers *et al.,* 1993; Huntley *et al.,* 1994b; Hof *et al.,* 1997b). The glutamate receptor profile of these particular pyramidal neurons may play a role in interactions between glutamate excitotoxicity and cytoskeletal pathology similar to that described in tissue culture preparations. However, only about one-third of the pyramidal neurons that contain glutamate receptor subunits also contain neurofilament protein (Vickers *et al.,* 1993), correlating with the observation that more neurons are lost in AD than can be accounted for by NFT formation alone (Hof *et al.,* 1990a; Gómez-Isla *et al.,* 1997). In addition, certain neurons die through apoptotic mechanisms in AD and therefore do not undergo NFT formation (Su *et al.,* 1994; Cotman and Anderson, 1995; Vyas *et al.,* 1997). The distribution of DNA damage in the cerebral cortex shows region-specific patterns that do not correlate clearly with the distribution of NFTs and SPs, further suggesting that NFT formation and apoptotic and necrotic mechanisms of neuron death are evolving independently in AD (Lucassen *et al.*, 1997). Thus, it may be possible that neurons with high levels of both glutamate receptors and neurofilament protein degenerate through NFT formation, whereas pyramidal cells expressing low levels of the neurofilament triplet protein but having high levels of calcium-permeable glutamate receptor subunits degenerate without NFT formation, thus accounting for the additional neuronal loss. Clearly, the interplay between the molecular and morphological characteristics that determines the resistance or vulnerability of different types of neurons is highly complex and represents a key issue in our understanding of the mechanisms of neuronal degeneration in AD.

### 3.4. Other Proteins Associated with Neuronal Degeneration

In addition to neurofilament protein and other components of the cytoskeleton, many proteins have been found in association with neuronal degeneration in AD, indicating that NFT formation and other degenerative changes depend on a constellation of cellular mechanisms. Some of these proteins are found in a variety of pathological changes and are therefore not specific to AD. This is particularly the case for ubiquitin, which is found in NFTs, Pick bodies, and Lewy bodies (Love *et al.*, 1988, 1993). Several enzymes involved in the phosphorylation of various proteins are consistently observed in association with NFTs (Saitoh *et al.*, 1991). Protein phosphorylation may interfere with the specific function of proteins and represents an important mechanism by which neurons can adequately maintain their normal function and metabolism and integrate extracellular inputs. In this context, isoforms of protein kinase C have been found in degenerating neurons as well as in senile plaques and diffuse amyloid deposits (Masliah *et al.*, 1990). Protein kinase C alterations have also been found in the brains of nondemented individuals, indicating perhaps another marker of neuronal changes in the early stages of AD (Masliah *et al.*, 1991a). Also, pathological phosphorylation of tau protein has been demonstrated in aging, AD, and several other dementing illnesses, resulting in the deposition of certain abnormal isoforms of tau protein (Delacourte *et al.*, 1990, 1996; Flament *et al.*, 1990a,b; Vermersch *et al.*, 1992a,b, 1997; Trojanowski *et al.*, 1993b). Interestingly, tau protein is phosphorylated *in vitro* by type II calcium/calmodulin-dependent protein kinase, and this enzyme has been localized within hippocampal neurons that are prone to NFT formation in AD and copurifies on immunoblots with paired helical filaments, the major constituent of NFT (McKee *et al.*, 1990; Xiao *et al.*, 1996). Another protein kinase, casein kinase II, is present in high concentration in NFTs, and its overexpression precedes NFT formation in susceptible neurons (Iimoto *et al.*, 1989; Masliah *et al.*, 1992a).

Similarly, large hippocampal pyramidal neurons have been shown to contain higher levels of copper–zinc superoxide dismutase in AD, suggesting a possible modification in the metabolism of free radicals in these susceptible neurons (Delacourte *et al.*, 1988). The presence of nitrotyrosine as a product of the peroxynitrate reaction with superoxide dismutase and neurofilament protein has been reported in NFTs as well as in NFT-free neurons in AD (Good *et al.*, 1996; Smith *et al.*, 1997). With respect to the proteins involved in the generation of free radicals, several involved in the metabolism of iron, such as transferrin and lactotransferrin, have been reported in association with degenerating neurites, NFTs, and SPs, as well as with a variety of other neurodegenerative lesions (Connor *et al.*, 1992; Kawamata *et al.*, 1993; Leveugle *et al.*, 1994) (Fig. 21). Further evidence for the role of free radicals in neuronal degeneration is provided by the observation of the enzyme heme oxygenase-1, advanced Maillard reaction end products, and products of lipid peroxidation such as 4-hydroxy-2-nonenal, within NFT in several neuro-

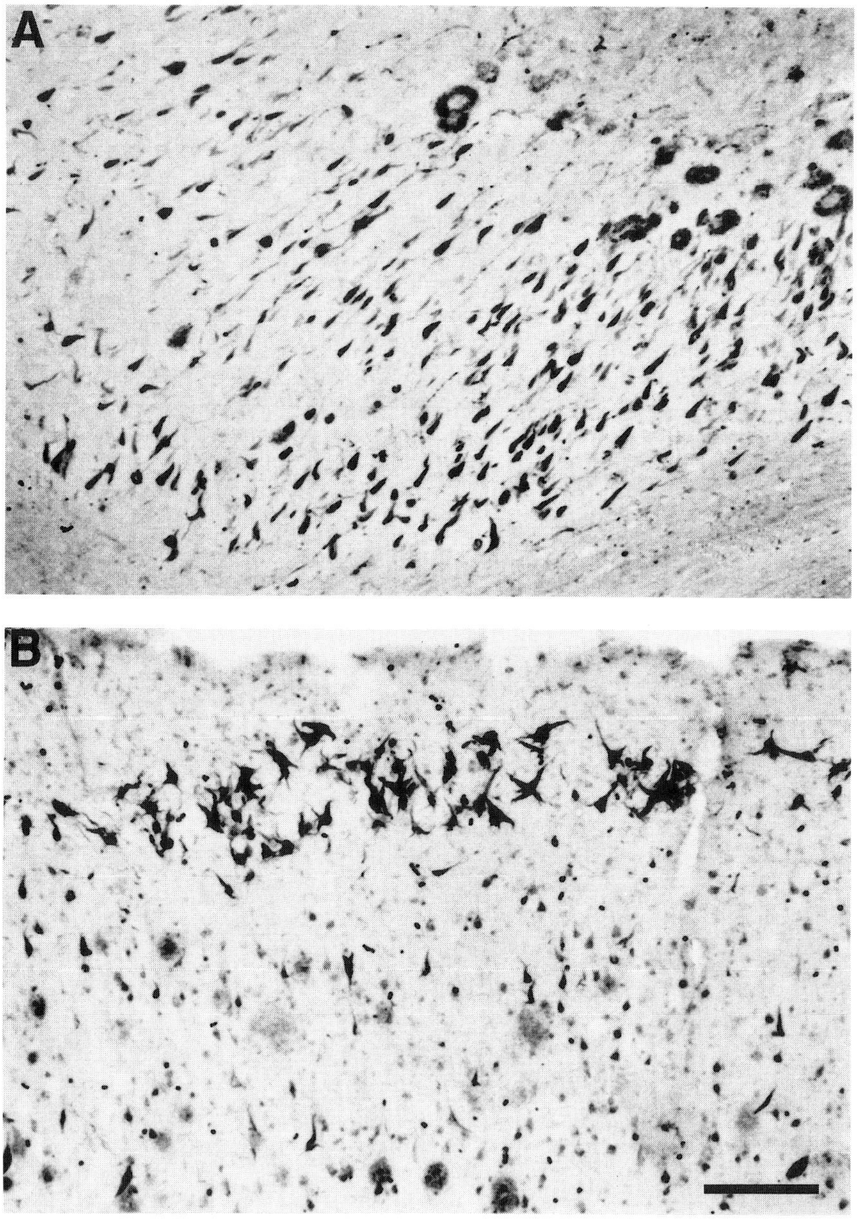


**Figure 21.** Lactotransferrin immunohistochemistry in AD. There is a strong staining of NFT-containing neurons in the pyramidal layer of the CA1 field (A) and in layer II of the entorhinal cortex (B). A few SPs and amyloid deposits are also labeled in (B). Adapted from Leveugle *et al.* (1994). Bar = 100 μm.

degenerative diseases (Smith *et al.*, 1994a,b; Yan *et al.*, 1994; Castellani *et al.*, 1995; Montine *et al.*, 1997a,b; Münch *et al.*, 1997). Interestingly, 4-hydroxy-2-nonenal forms irreversible pyrrole adducts when reacting with proteins, and such adducts are present in NFTs with a strong association with apolipoprotein E4 homozygosity but not in other neurodegenerative disorders, indicating a strong interaction between apolipoprotein E genotype and lipid peroxidation in the AD brain (Montine *et al.*, 1997a). In addition, a variety of proteins—mitogen-activated kinases; glycogen synthase kinases; thrombospondin; phospholipase C$\partial$; growth-associated protein 43; amyloid precursor protein-related compounds including $\alpha_1$-antichymotrypsin, and apolipoproteins E and J; presenilin-1; several classes of proteoglycans, and others—have been shown in NFT-containing neurons and SPs (Masliah *et al.*, 1991a,c, 1992b; Namba *et al.*, 1991; Shimohama *et al.*, 1991, 1995; Buée *et al.*, 1992, 1993, 1994, 1996; Gollin *et al.*, 1992; McGeer *et al.*, 1992; Su *et al.*, 1992; Trojanowski *et al.*, 1993a; McGeer and McGeer, 1995; Metzger *et al.*, 1996; Murphy *et al.*, 1996; Yamaguchi *et al.*, 1996; Giannakopoulos *et al.*, 1997a, 1998; Levey *et al.*, 1997; Weber *et al.*, 1997; Wisniewski *et al.*, 1997) (Fig. 22). Along the same lines, several molecules associated with inflammatory and immune functions, such as cytokines, complement proteins, and acute phase reactants, have been observed in the brain in AD and other neurodegenerative diseases (McGeer and McGeer, 1995). Recently, sequestration of cytoplasmic RNA species has also been found in SPs, NFTs, and neuropil threads in AD (Ginsberg *et al.*, 1997, and see this volume), demonstrating that the machinery for gene transcription and translation exists in AD lesions and suggesting that specific gene products may be involved in the local development of these lesions.

Finally, cholinesterase-containing large pyramidal neurons are severely affected in AD cases, and the enzyme has also been localized within SPs, NFTs, and neuropil threads (Heckers *et al.*, 1992; Gómez-Ramos *et al.*, 1992, 1994; Morán *et al.*, 1994). Many neuropeptides have been associated with degenerative neurons in AD. In particular, galanin neurons in the basal forebrain are particularly sensitive (Kowall and Beal, 1989; see also Mufson and Kordower, this volume), as are substance P-immunoreactive neurons in the neocortex (Bouras *et al.*, 1986, 1990; Quigley and Kowall, 1991). Various neuropeptide-containing fibers have also been observed in degenerating neurites within SPs (Morrison *et al.*, 1985; Bouras *et al.*, 1987, 1990; Walker *et al.*, 1988; Armstrong *et al.*, 1989). The exact significance of all of these neurochemical markers in the development of AD has not been elucidated; however, the convergent involvement of certain biochemical classes of molecules, such as glutamate receptors, cytoskeletal proteins, and certain metabolic or receptor proteins, may help to define select categories of vulnerable neurons and provide a cohesive neurochemical profile of vulnerability.

### 3.5. Morphologic and Molecular Characteristics of Resistant Neuron Subpopulations

Despite the diversity of molecules associated with vulnerability and the fact many neuronal classes are affected by pathological changes in dementing illnesses, not all neuronal categories are equally affected, and some identifiable subsets of neurons are highly resistant to the degenerative process of AD. Several classes of

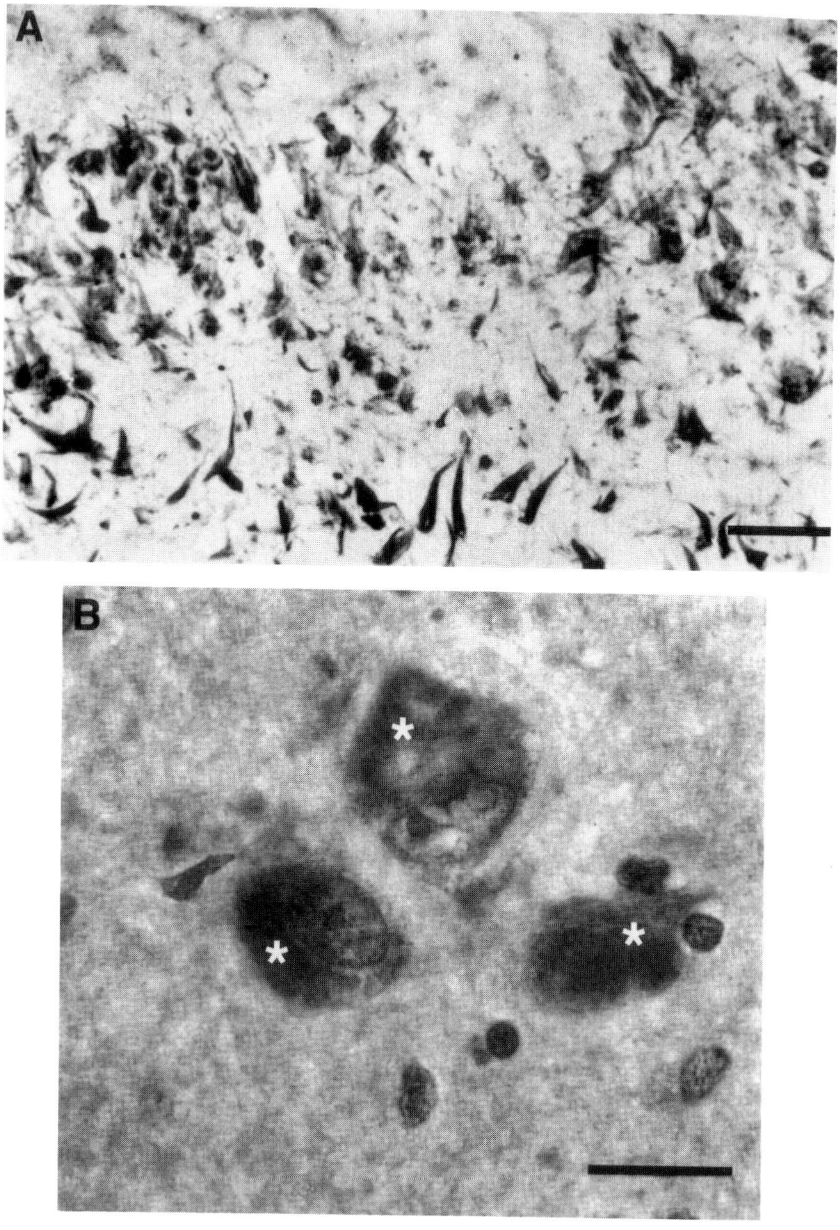


**Figure 22.** Cellular localization of apolipoprotein E (A), presenilin-1 (B) and thrombospondin (C) in AD. There is a strong labeling of NFT in layer II of entorhinal cortex with apolipoprotein E (Hof *et al.*, 1994b), and an intense immunoreactivity in neurons apparently free of NFT with presenilin-1 (Giannakopoulos *et al.*, 1997a) and thrombospondin (Buée *et al.*, 1992). Bars = 50 μm (A) and 20 μm (B, C).

molecules have been observed preferentially in resistant neurons. In particular, several calcium-binding proteins have been shown to represent useful markers for subsets of morphologically non-overlapping cortical interneurons that are remarkably resistant in neurodegenerative disease (Baimbridge *et al.*, 1992; Heizmann and Braun, 1992; Andressen *et al.*, 1993; Hof and Morrison, 1994; Iacopino *et al.*, 1994;

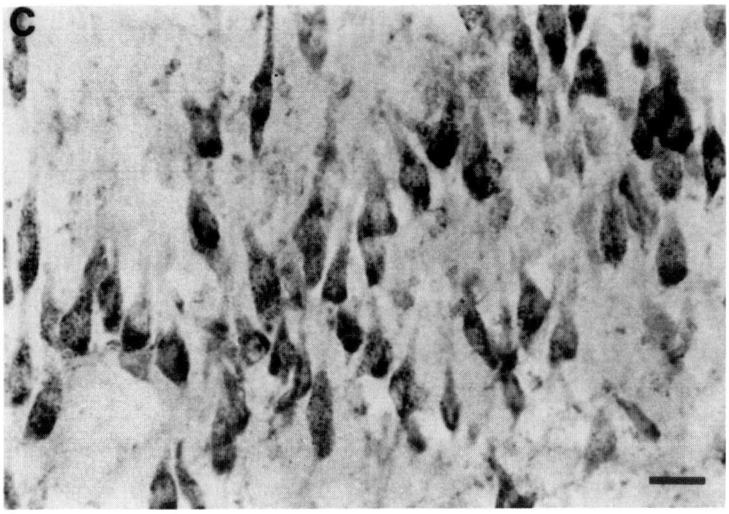

**Figure 22.** (*Continued*)

Sampson *et al.,* 1997). The major role of calcium-binding protein is considered to be intracellular calcium buffering and transport and the regulation of several enzymes (Andressen *et al.,* 1993). Interestingly, very few of these calcium-binding, protein-containing interneurons show neurofilament protein-immunoreactivity. The lack of significant levels of neurofilament protein and the rarity of NFT formation in these neurons has been proposed as a determinant of their resistance to the degenerative process of AD (Sampson *et al.,* 1997). In view of the impaired calcium homeostasis that occurs during neurodegeneration, it is possible that a high intracellular concentration of these protein prevents or protects against the deleterious effects of calcium metabolism imbalance and calcium-mediated excitotoxicity (Sloviter, 1989; Leranth and Ribak, 1991; Sloviter *et al.,* 1991; Heizmann and Braun, 1992; Andressen *et al.,* 1993; Hof and Morrison, 1994; Iacopino *et al.,* 1994; Pike and Cotman, 1995; Vogt Weisenhorn *et al.,* 1996). Among calcium-binding proteins, the distribution of parvalbumin, calbindin, and calretinin have been studied in detail in the human neocortex and hippocampus. In the neocortex these proteins subdivide the GABAergic interneurons into three major interneuron populations that together probably account for nearly all of the GABAergic interneurons (Andressen *et al.,* 1993; Hof *et al.,* 1993b). In the macaque monkey and human neocortex, these subclasses of calcium-binding proteins or GABA-containing neurons show a high degree of morphological specificity and correspond to well-defined elements of the local cortical microcircuitry: most parvalbumin-immunoreactive neurons correspond to basket and chandelier cells, calbindin-immunoreactive neurons represent mostly double bouquet cells, and calretinin is present in the bipolar and bitufted neurons (Andressen *et al.,* 1993; Hof *et al.,* 1993b; Condé *et al.,* 1994; Gabbott *et al.,* 1996a,b). Although these populations are largely non-overlapping, a small subgroup of chandelier neurons in the infragranular layers of the human temporal neocortex appears to contain both calbindin and parvalbumin (Del Río and DeFelipe, 1997a). In the hippocampal formation, a similar segregation of these proteins is observed among different types of inhibitory interneurons.

### 3.5.1. Parvalbumin

Most parvalbumin-immunoreactive cells in the neocortex are large-to-medium-sized multipolar neurons with a large cytoplasmic rim around the nucleus. Their axons are rarely observed, but terminal boutons in the form of baskets are numerous on the surface of neighboring pyramidal neurons. Moreover, parvalbumin-immunoreactive endings on the initial segment of pyramidal cell axons are commonly seen, indicating the presence of a large population of parvalbumin-containing chandelier cells (see Andressen *et al.*, 1993; Condé *et al.*, 1994; Gabbott *et al.*, 1996a,b; Nimchinsky *et al.*, 1997, for review). A much larger variety of morphological types of parvalbumin-immunoreactive interneurons has been found in the hippocampus (Braak *et al.*, 1991; Höchli *et al.*, 1991; Freund and Buzsáki, 1996).

In AD several studies have shown that the parvalbumin-containing neocortical interneurons display a well-preserved morphology, even in severe cases and suffer no significant cell loss (Ferrer *et al.*, 1991, 1993d; Hof *et al.*, 1991b; Fonseca *et al.*, 1993), although some studies have reported a substantial loss (Arai *et al.*, 1987; Satoh *et al.*, 1991). Quantitative analyses of the superior frontal cortex (area 9) and inferior temporal cortex (area 20) in AD cases have demonstrated that parvalbumin-containing neurons have similar cellular densities in layers II–IV and V/VI in AD and in control cases in both areas (Figs. 23 and 24). Also, when grouped by size, no difference in parvalbumin-containing neurons exists between AD brains and control brains, further demonstrating the integrity of these neurons in dementia. Parvalbumin-immunoreactive aberrant terminal sprouting, has however, been observed, as well as reduced numbers of terminal cartridges from chandelier cells, pointing to involvement of the parvalbumin-containing neuron axon terminals in the degenerative process. In contrast to neocortical regions, parvalbumin-immunoreactive neurons are affected in the hippocampus and entorhinal cortex in AD. In the CA1 in particular, there is a consistent loss of these neurons and a severe reduction in their dendritic fields (Chan-Palay *et al.*, 1991); a recent study has also demonstrated that in the entorhinal cortex, parvalbumin-immunoreactive neurons are affected not because of NFT formation, but because of a progressive vulnerability related to the severe loss of projection neurons (Solodkin *et al.*, 1996).

### 3.5.2. Calretinin

Calretinin-positive neurons are small-to-large-sized bipolar and double bouquet neurons with vertically oriented dendrites, and less frequently, small multipolar neurons (see Andressen *et al.*, 1993; Hof *et al.*, 1993b; Condé *et al.*, 1994; Gabbott *et al.*, 1996a,b; Nimchinsky *et al.*, 1997, for reviews). Some calretinin-containing neurons, particularly in the upper layers, have a dendritic arborization typical of double bouquet cells. The small bipolar calretinin-positive neurons appear to be concentrated in the upper layers. In primate materials the vast majority of calretinin-positive cells display the features of double bouquet and bipolar neurons (Fig. 25) and occur in both superficial and deep layers. Layer I contains large calretinin-containing neurons with horizontal processes, reminiscent of Cajal–Retzius cells. In addition, a population of calretinin-immunoreactive layer V pyramidal neurons has been observed in the human cortex, principally in the anterior cingulate regions, but also in the entorhinal cortex (Nimchinsky *et al.*, 1997). In the hippocampus two major populations of calretinin-containing interneurons have been described based on their morphology and local connectivity: spiny interneurons, which contact the dendrites of principal neurons and aspiny interneurons, whose role seems to be to specifically target other interneurons (Freund and Buzsáki, 1996). Recently, a population of typical bipolar calretinin-immu-

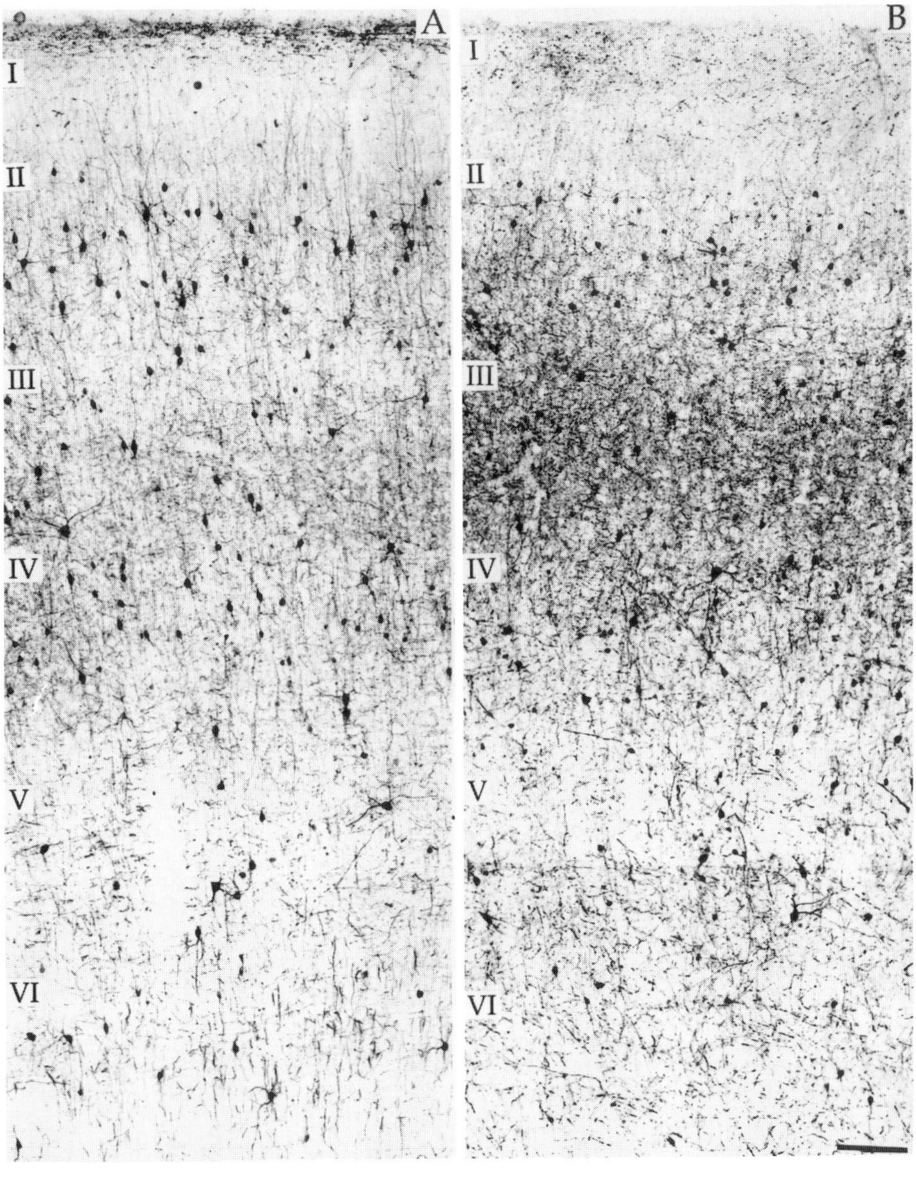


**Figure 23.** Distribution of parvalbumin-immunoreactive neurons in the superior frontal cortex of a control case (A) and an AD case (B). These interneurons are present in all neocortical layers but predominate in layers II–IV. In the AD case, there is a remarkable preservation of these neurons' number and morphology. From Hof *et al.* (1991b). Reproduced with permission from the *Journal of Neuropathology and Experimental Neurology*. Bar = 100 μm.

noreactive neurons in the human neocortex has been reported to also make dendrodendritic and dendrosomatic contacts with other interneurons containing calretinin or parvalbumin, but not calbindin (Del Río and DeFelipe, 1997b).

Quantitative analyses of areas 9 and 20 of both control and AD cases have shown only marginal differences between them in laminar neuron densities (Hof *et al.,* 1993c; Fonseca and Soriano, 1995). Moreover, the staining intensity of the labeled profiles is comparable in control and AD cases, and the calretinin-immunoreactive neurons displayed a preserved morphology in AD in that their somata and dendritic arborizations remain intact (Fig. 25). Like parvalbumin-immunoreactive interneurons in the hippocampal formation, calretinin-containing neurons show alterations in their dendritic trees in the absence of cell loss or association with NFTs or SPs (Brion and Résibois, 1994). This observation supports the notion

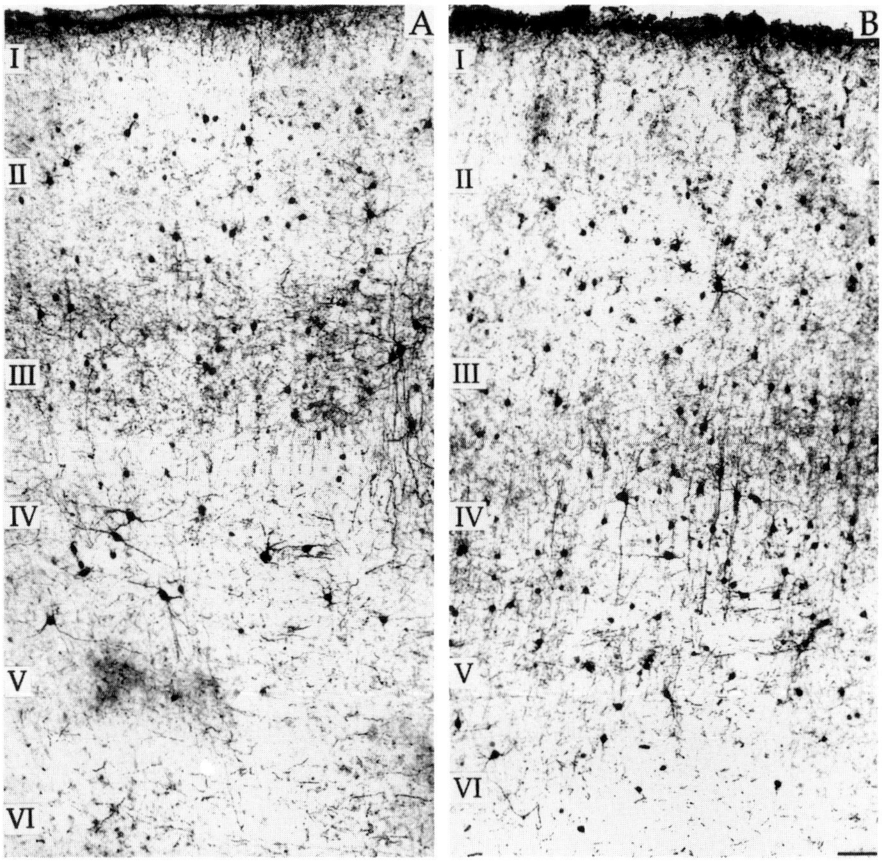


**Figure 24.** Distribution of parvalbumin-immunoreactive neurons in the inferior temporal cortex of a control case (A) and an AD case (B). Note distribution patterns are similar to Figure 23. This area has slightly lower densities of parvalbumin-immunoreactive neurons than does the superior frontal cortex. In the AD case, there is a remarkable preservation of these neurons' number and morphology. From Hof *et al.* (1991b). Reproduced with permission from the *Journal of Neuropathology and Experimental Neurology.* Bar = 100 μm.

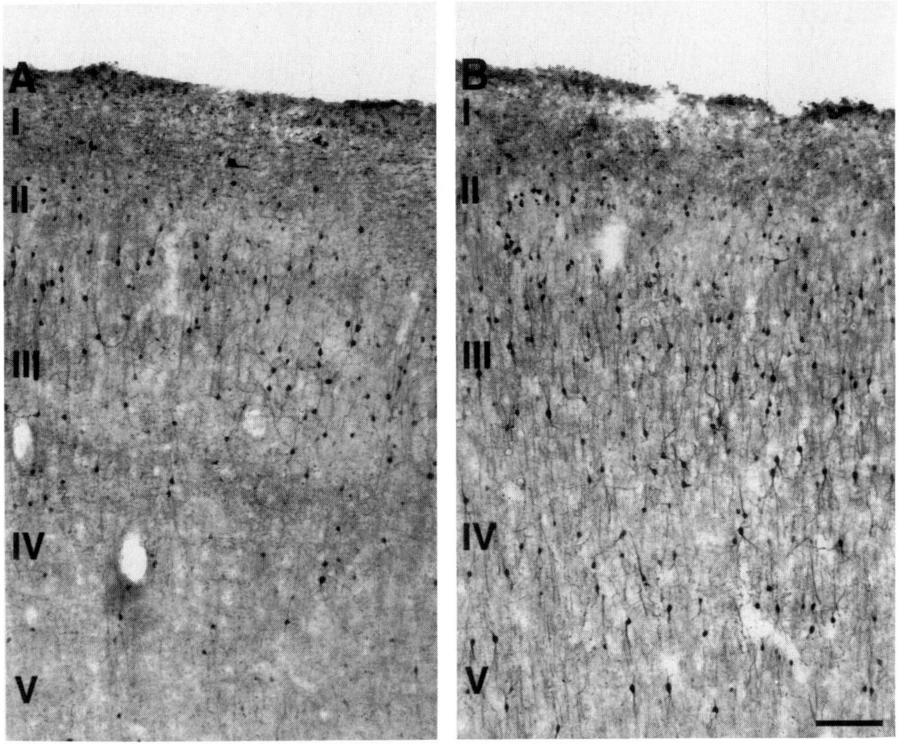


**Figure 25.** Distribution of calretinin-immunoreactive neurons in the superior frontal cortex of a control case (A) and an AD case (B). These bipolar/bitufted interneurons predominate in layers II–V. In the AD case, there is a remarkable preservation of these neurons' number and morphology. Bar = 100 μm.

that hippocampal interneurons are ultimately involved in AD, but only at the late stages, when there is a severe loss of projection neurons.

### 3.5.3. Calbindin

There are two major calbindin-containing neuron classes in the cerebral cortex: small pyramidal neurons and interneurons (Hendry *et al.*, 1989; Hof and Morrison, 1991; Baimbridge *et al.*, 1992; Hayes and Lewis, 1992; Andressen *et al.*, 1993; Hof *et al.*, 1993b; Condé *et al.*, 1994; Gabbott *et al.*, 1996a,b). Small pyramidal neurons are lightly immunostained and preferentially located in layer III; therefore, they are possibly a subset of corticocortical projection neurons (Hof and Morrison, 1991; Hayes and Lewis, 1992). Interneurons, on the other hand, are darkly immunostained and occur in layers II, III, V, and VI. Most interneurons show the characteristic dendritic arborization of double bouquet cells (Fig. 26). They are small-to-intermediate-sized and located predominantly in the layers II and III, with a sparser population in the deep layers. The coexistence of calbindin with NADPH-diaphorase (nitric oxide synthase) activity and somatostatin indicates that a subpopulation of calbindin neurons belongs to the Martinotti cell type. The calbindin-immunoreactive pyramidal neurons represent a relatively homogeneous popula-

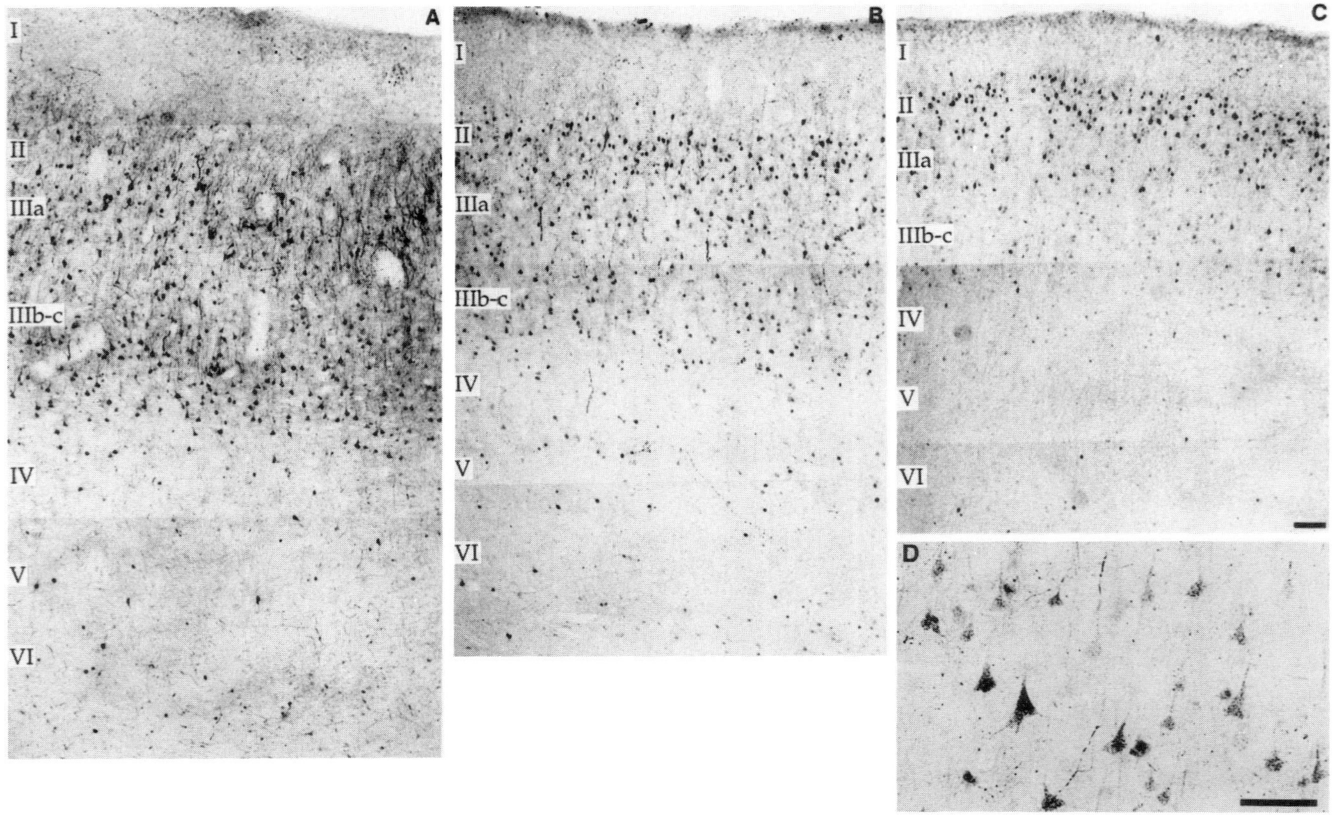


**Figure 26.** Distribution of calbindin-immunoreactive neurons in the superior frontal cortex of a control case (A), an AD case with mild cognitive impairment (B), and a severe AD case (C). There is a dense population of labeled interneurons in layers II and superficial III, whereas layers V and VI have fewer neurons (A). In the AD cases, there is a gradual loss of immunoreactivity in the dendritic arborizations of these neurons, but their numbers are retained (B, C). In layer III a population of small pyramidal neurons also contain calbindin (A, D). These neurons are progressively lost in AD. Adapted from Hof and Morrison (1991). Bars = 100 μm.

tion and are characterized by a light and punctate staining of the perikaryon (Hof and Morrison, 1991). Similarly, calbindin is observed in the interneurons, dentate granule cells, and principal cells in the hippocampal formation. Calbindin-containing interneurons in the hippocampus innervate the dendrites of principal neurons (Seress *et al.*, 1992; Freund and Buzsáki, 1996).

Whereas both parvalbumin- and calretinin-containing neurons are generally resistant to degeneration and show a well-preserved morphology and staining pattern in AD, the case of the calbindin-immunoreactive cells is more complex, because these interneurons display some degree of differential vulnerability (Fig. 26) (Table 8). The population of calbindin interneurons in layers II and III shows a variable degree of resistance to degeneration among AD cases (Ichimiya *et al.*, 1988; Hof and Morrison, 1991; Ferrer *et al.*, 1993c; Nishiyama *et al.*, 1993), and their loss appears related to dementia severity (Ferrer *et al.*, 1993c). Our quantitative analysis of the prefrontal cortex has shown that, whereas the interneurons in layers II and III are rather resistant in AD, the interneurons in layers V and VI, though less numerous than in layers II and III, are affected only in AD cases displaying high NFT densities. This could reflect differential connectivity patterns of interneurons located in layer II and the upper third of layer III from those located in layer V. For instance, layer V calbindin-containing cells may project to pyramidal cells that are more sensitive in AD, whereas the targets of neurons in layers II and III could be less vulnerable (Hof and Morrison, 1991) (Fig. 26, Table VIII). Somatostatin and calbindin are colocalized in some interneurons (DeFelipe *et al.*, 1989), and somatostatin-immunoreactive fibers are present in SPs and thus degenerate to some degree in AD (Morrison *et al.*, 1985; Bouras *et al.*, 1987, 1990). Layer III pyramidal neurons are thought to be a principal target of these interneurons (De Lima and Morrison, 1989). Therefore, somatostatinand calbindin-immunoreactive fibers are more vulnerable in AD than are the cell bodies of the interneurons that contain them. The differential vulnerability of the axonal portion of all three classes of calcium-binding, protein-containing interneurons may be linked to their synaptic relationships with the highly vulnerable pyramidal neurons in AD. In AD neocortex, a subpopulation of layer III pyramidal neurons that contain calbindin is highly sensitive to degeneration (Hof and Mor-

**Table VIII. Calbindin-Immunoreactive Neuron Counts in the Prefrontal Cortex (Area 9) in Control and AD Brains[a]**

| | Cortical layers | | |
|---|---|---|---|
| | II/IIIs[b] | IIId[c] | IV/VI |
| Control | 136 ± 9 | 193 ± 14 | 68 ± 7 |
| Mild AD | 141 ± 17 | 167 ± 12 | 64 ± 7 |
| Severe AD | 126 ± 8 | 79 ± 6 | 38 ± 5 |

[a]Data represent mean neuron counts per mm cortical traverse ±SEM. The decrease in layer IIId, already noticeable in the mild AD group, is largely due to the loss of pyramidal neurons containing calbindin; the interneurons are preserved. A subgroup of interneurons in the deep layers also appears vulnerable in severe AD. Adapted from Hof and Morrison (1991).

[b]Refers to superficial layer III.

[c]Refers to deep layer III.

rison, 1991). Thus, like the neurofilament protein-immunoreactive pyramidal neurons, the calbindin-containing pyramidal cells may represent a subset of corticocortical projections that is dramatically affected in AD (Fig. 26). This is further supported by the fact that neurofilament protein is colocalized with calbindin in a large number of neocortical pyramidal neurons (Hayes and Lewis, 1992).

### 3.5.4. Other Factors Conferring Resistance to Degeneration

Several neuropeptides have been associated with partial resistance to degeneration of cortical neurons in AD. Somatostatin neurons resistant to degeneration have been observed, and this resistance seems dependent upon the colocalization of neuropeptide Y and/or NADPH diaphorase (Gaspar *et al.*, 1989). The density of certain interneurons containing neuropeptide Y, somatostatin, and NADPH diaphorase is not decreased in AD neocortex and hippocampus, so although some of these neurons are morphologically altered, they are not eliminated (Chan-Palay, 1987; Kowall and Beal, 1988; Hyman *et al.*, 1992). These interneurons predominate in the deep layers of the neocortex, in the subcortical white matter, and in most fields of the hippocampus. The combination of factors such as high cytoplasmic levels of calcium-binding proteins and GABA with the morphological features of locally projecting interneurons may confer a heightened resistance on these cells. Another example of neurochemical specialization linked to resistance is that of corticotropin-releasing, factor-containing neurons that remain unaffected in AD (Kelly and Kowall, 1989). This neuropeptide is colocalized with parvalbumin in some chandelier cells (Lewis and Lund, 1990). Moreover, parvalbumin-immunoreactive neurons are resistant to degeneration in the cerebral cortex in Huntington's disease and Pick's disease (Cudkowicz and Kowall, 1990; Arai *et al.*, 1991). Conversely, we and others have recently noted that in Down's syndrome, there is a severe loss of all three calcium-binding, protein-immunoreactive neurons in the neocortex (Kobayashi *et al.*, 1990; Hof, unpublished). Possibly the loss of calcium-binding, protein-immunoreactive neurons in such cases is related to the structural abnormalities that occur during brain development in Down's syndrome.

Thus, the relative resistance or vulnerability of calcium-binding proteins appears to be highly variable and dependent upon multiple factors, suggesting that the AD pattern cannot be considered a reliable model of interneuron viability in all neurodegenerative diseases. Other examples of the differential vulnerability of these interneuronal subsets include Guamanian amyotrophic lateral sclerosis/parkinsonism–dementia complex, postencephalitic parkinsonism, and corticobasal degeneration (all are resistant; Hof *et al.*, 1994b; unpublished); frontal lobe dementia (calbindin-containing neurons are vulnerable, but parvalbumin-containing neurons are resistant) (Ferrer *et al.*, 1993b); and Creutzfeldt-Jakob disease (parvalbumin-containing neurons are vulnerable) (Ferrer *et al.*, 1993a). Together, these observations suggest that the relationships between etiopathogenesis and specific cellular vulnerability are highly complex at the level of specific neurochemical, morphological, and connectional features. The situation may be further complicated by the fact that calbindin gene expression is reduced not only in neurodegenerative disorders, but also during aging (Iacopino and Christakos, 1990), and that in aged rats, fewer calbindin- (but not parvalbumin-) containing neurons are observed in the CA1 field of the hippocampus in comparison to young animals (Potier *et al.*, 1994). Also, molecules other than calcium-binding proteins and neuropeptides have recently been observed to either confer a certain degree of resistance or to be present in neurons that are not prone to NFT formation. In particular, presenilin-1 and apolipoprotein J (clusterin) may protect against the development of NFTs (Giannakopoulos *et al.*, 1997a; Giannakopoulos *et al.*, 1998) (Fig. 22).

The consistent presence of the AMPA receptor subunit protein GluR2 on the vast majority calcium-binding, protein-containing interneurons that was recently demonstrated using a GluR2-specific antibody in rat and macaque monkey may also participate in protecting these neurons against degeneration, as GluR2 gates the calcium fluxes (Vissavajjhala *et al.*, 1996) (Table VII). Consistent with this observation is the fact that in neuronal cultures, GABAergic cells expressing GluR1 and GluR4, but only low levels of GluR2/3, are highly vulnerable to excitotoxic insults (Yin *et al.*, 1994). The relative representation of GluR2 in different neuronal subtypes is, however, perhaps more appropriately viewed at the synaptic level. Electrophysiological studies have demonstrated that pyramidal cells in CA1 possess AMPA receptors that have lower calcium permeability than GABAergic interneurons (Sommer *et al.*, 1991; Verdoorn *et al.*, 1991), suggesting that the AMPA receptors on GABAergic interneurons might have a lower representation of GluR2 than those on pyramidal cells. A recent quantitative immunogold double-labeling study showed that in fact, though GluR2 is often present at asymmetrical synapses on both neuronal classes, it is more abundant at synapses on pyramidal cells than on GABAergic interneurons (He *et al.*, 1998). This may seem counterintuitive given the higher resistance of interneurons to excitotoxic insults and degeneration relative to pyramidal neurons, but the GluR2 representation should be viewed in the context of the differential distribution of calcium-binding proteins. The CA1 pyramidal neurons do not contain any of the three calcium-binding proteins discussed above, and therefore would be highly vulnerable to influx of calcium through AMPA receptors in the event of a pathological down-regulation of GluR2, whereas the GABAergic neurons that contain the calcium-binding proteins would be less vulnerable in such a scenario. It would be interesting to know whether the small population of calretinin-immunoreactive neurons that does not contain GluR2 (Vissavajjhala *et al.*, 1996) corresponds to the neurons that establish dendrodentritic and dendrosomatic appositions on other interneurons (Freund and Buszáki, 1996; Del Río and DeFelipe, 1997b), and to what degree these particular neurons are affected under pathologic conditions. Clearly, a more comprehensive morphological and biochemical knowledge of cell typology and potential shifts in expression of key proteins is required before we can assess which critical features correlate with a specific neuronal population's relative vulnerability or resistance in neurodegenerative disorders.

## 4. Other Dementing Disorders: Evidence for Disease-Specific Involvement of Cortical Circuitry

The cellular alterations underlying degenerative brain diseases have been particularly well analyzed in AD, yet many of the changes included in the neuropathological diagnosis of AD are also present, with varying degrees of severity, in other dementing illnesses. For instance, diseases such as progressive supranuclear palsy (PSP), corticobasal degeneration (CBD), Pick's disease, Lewy body dementia (LDB), posttraumatic dementia (including dementia pugilistica), postencephalitic parkinsonism (PEP), and Guamanian amyotrophic lateral sclerosis/parkinsonism–dementia complex (ALS/PDC) are all characterized by the presence in several brain regions of neuronal and glial changes involving the cytoskeleton, as well as amyloid deposition in certain cases. In these "AD-related" conditions, however, NFT and amyloid deposition distribution differs from that observed in AD with respect to the regional and laminar localization of lesions. In addition, the same neurofilament protein-enriched neurons affected in AD are prone to degeneration in Huntington's disease, which involves mostly subcortical structures but invariably includes cognitive impairment in its progression as well (Cudkowicz and Kowall, 1990), whereas these neurons are mostly preserved in Guamanian cases with ALS/PDC. Interneurons that are not affected in AD show differential involvement in other types of dementia. Thus, the pathological changes at the cellular level as well as the overall distribution of lesions in these disorders indicate the existence of disease-specific patterns. Comparing the neuropathologic differences among these dementing disorders with the patterns observed in AD may provide important information about the etiopathogenetic mechanisms unique to each condition.

### 4.1. Down's Syndrome

Down's syndrome (DS) has long been regarded as a potential model to study the development and etiology of the pathological mechanisms leading to AD (Burger and Vogel, 1973; Zellweger, 1977; Oliver and Holland, 1986; Mann, 1988), because the progressive involvement of restricted cortical regions in the evolution of dementia may be paralleled by the development of pathologic alterations in DS. Downs syndrome is characterized by a typical phenotype that includes frequent cardiac abnormalities, impaired brain growth and maturation resulting in various degrees of mental retardation, appearance of age-related pathological changes early in life, and reduced life expectancy (Zellweger, 1977). Moreover, many patients with DS older than 50 years are demented, demonstrating the possible association of DS with AD (Mann, 1988). Functionally, the cerebral metabolic rate for glucose as measured by positron-emission tomography is significantly lower in old as compared with young DS patients and is quite similar to the regional distribution of hypometabolism observed in AD (Schapiro *et al.*, 1988; Pietrini *et al.*, 1997).

The presence of dementia in DS is correlated with low brain weight, frontocaudal shortening, variable hypoplasia of the frontal lobe, and characteristic atrophy of the superior temporal gyrus (Kemper, 1991). At the microscopic level, there is severe neuronal loss in the hippocampal formation and neocortex as well as in subcortical structures. Reduced dendritic arborization, decreased numbers of spines, spine atrophy, and anomalies in spine orientation have also been observed on pyramidal neurons in the hippocampus and neocortex of DS cases, as early as the fetal stages (Takashima *et al.*, 1981; Ferrer and Gullotta, 1990). Neurofibrillary tangles and SPs are plentiful in the cerebral cortex and subcortical regions of DS cases, the deposition of NFTs and amyloid occurs prior to severe neuronal loss, and amyloid deposits are commonly observed before the formation of NFTs and neuritic changes (Mann *et al.*, 1985, 1986; Mann and Esiri, 1989; Motte and Williams, 1989; Hof *et al.*, 1995d; Hyman *et al.*, 1995). Neurofibrillary tangles are first observed in layer II of entorhinal cortex and in the hippocampus, whereas they are present in the neocortex only in DS cases of 40 to 50 years (Fig. 27; Table IX). Neocortical regions in the temporal and frontal lobes display moderate-to-high lesion densities in DS cases older than 40 years, whereas primary sensory regions such as visual and auditory cortex are less severely affected. The laminar distribution of NFTs and SPs is similar to that in AD (Hof *et al.*, 1995d). Downs syndrome cases younger than 40 years exhibit fewer neuropathologic changes (Table IX).

The hippocampal formation, including entorhinal cortex and the transition zone in perirhinal cortex, contains the highest number of NFTs in DS cases (Fig. 27, Table IX), with layer II of the entorhinal cortex involved in all of the cases older than 35. The subiculum and CA1-3 of the hippocampus are severely affected in the DS cases older than 45 years. The perirhinal cortex contains the highest NFT densities in the oldest DS cases. In the neocortex NFT densities are generally lower than in the hippocampus and are consistently present only in association regions in cases older than 50 (Table IX). The inferior temporal cortex has higher NFT densities than the superior frontal cortex. Senile plaques and diffuse amyloid deposits are generally found throughout the cerebral cortex in DS cases older than 15 years and they are distributed more evenly than NFTs among the different cortical regions (Hof *et al.*, 1995d). Senile plaques are slightly less numerous in the hippocampal formation than in the neocortex. Unlike NFTs, SPs and amyloid deposits are numerous in the primary and secondary visual areas in DS, where their densities are comparable to those in the superior frontal and inferior temporal cortices (Fig. 27, Table X). Like the primary visual cortex, the primary auditory cortex has very low NFT numbers. The cingulate cortex shows lesion densities comparable to the adjacent frontal and parietal cortices.

There are strong correlations between the number of NFTs and SPs and ages of DS cases; the strongest correlations are between age and number of SPs in layers II/III and IV–VI of the primary and secondary visual cortex ($r = 0.79$-$0.88$; Hof *et al.*, 1995d). No such correlation exists for NFTs in the primary visual cortex ($r = 0.40$), and a low correlation is present in the secondary visual area ($r = 0.66$), reflecting the inconsistent NFT distribution in these regions. All of the other cortical regions have high correlation coefficient values for both NFTs and SPs ($r \geq 0.76$), with the strongest correlation for NFTs in layer II of the entorhinal cortex and in layers V and VI of inferior temporal cortex ($r = 0.84$). The highest correlation coefficient values for SPs outside of the visual cortex are in the perirhinal cortex ($r = 0.85$); weaker correlations are observed in the CA1-3 fields of the hippocampus ($r = 0.72$) and other cortical regions. Finally, even though the lesion distribution in DS is qualitatively comparable to that observed in AD, certain cortical regions exhibit quantitative age-related differences from AD cases such that very old DS cases have higher NFT densities in the perirhinal cortex, entorhinal cortex, and the subiculum in comparison to age-matched AD cases. Generally, AD cases also contain fewer SPs than do old DS cases.

The distribution of neuropathologic alterations in DS cerebral cortex indicates that the cortical structures consistently involved in AD and aging are comparably affected in DS. In particular, by the examination of cases of various ages, NFTs appear first in layer II of the entorhinal cortex in middle-aged DS cases, which

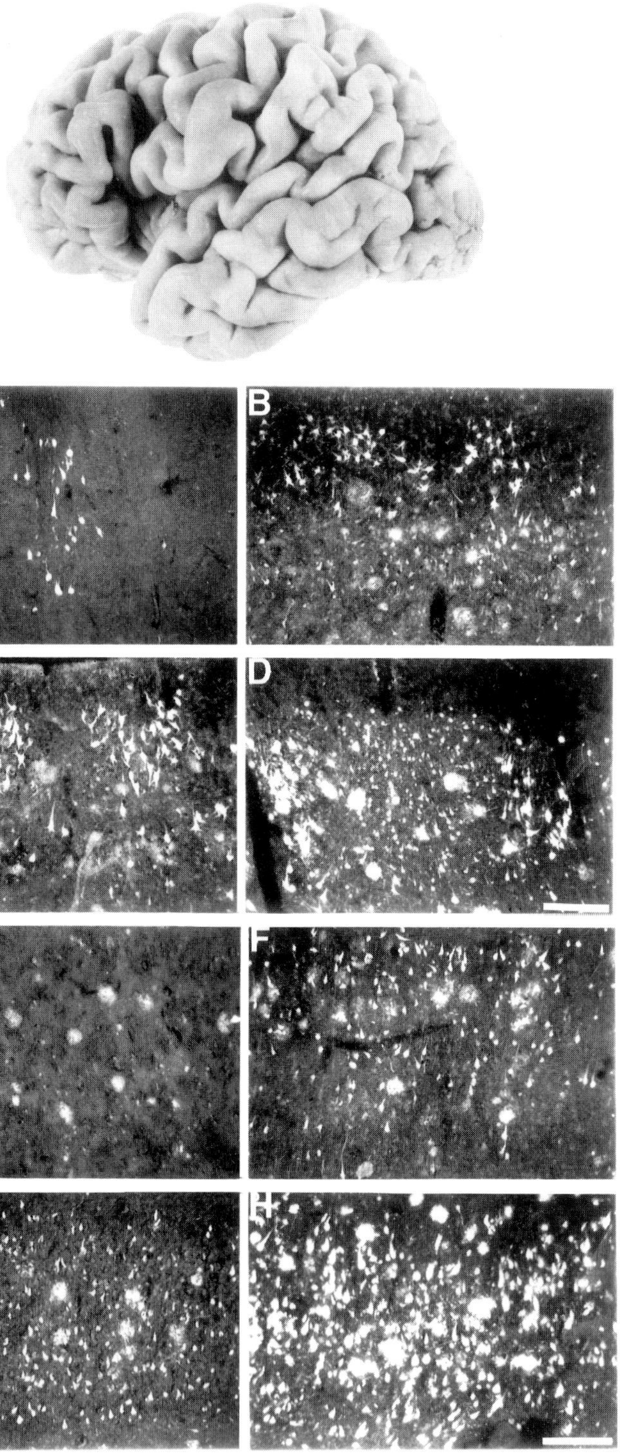
A
B
C
D
E
F
G
H

**Table IX. Distribution and Densities of NFTs in DS Cases**[a]

| Area | 40[b] | 44 | 54 | 54 | 55 | 65 | 74 |
|---|---|---|---|---|---|---|---|
| CA1-3 | 3.3 | 0.5 | 65.8 | 99.2 | 32.3 | 55.0 | 90.7 |
| Subiculum | 1.8 | 0 | 103.0 | 84.0 | 43.8 | 131.8 | 161.9 |
| Entorhinal | | | | | | | |
| Layers II | 50.0 | 1.5 | 36.0 | 53.7 | 60.2 | 94.5 | 109.7 |
| V | 0 | 0 | 50.8 | 40.0 | 67.0 | 93.0 | 93.6 |
| 35/36 | | | | | | | |
| Layers II/III | 5.5 | 0 | 38.0 | 79.8 | 64.8 | 99.5 | 134.5 |
| V/VI | 8.0 | 0 | 54.0 | 91.6 | 82.3 | 143.3 | 171.7 |
| 9 | | | | | | | |
| Layers II/III | 0 | 0.4 | 13.4 | 42.6 | 12.0 | 12.8 | 33.5 |
| V/VI | 0 | 0.8 | 21.6 | 54.6 | 32.8 | 33.0 | 51.5 |
| 20 | | | | | | | |
| Layers II/III | 0 | 0 | 20.6 | 59.0 | 24.5 | 29.5 | 65.3 |
| V/VI | 0 | 0 | 30.2 | 4.8 | 37.3 | 63.8 | 96.5 |
| 17 | | | | | | | |
| Layers II/III | 0 | 0 | 0 | 10.3 | 0 | rare | rare |
| V/VI | 0 | 0 | 0 | 4.4 | 0 | 0 | 0 |
| 18 | | | | | | | |
| Layers II/III | 0 | 0 | 3.8 | 27.0 | 20.0 | 19.4 | 22.5 |
| V/VI | 0 | 0 | 0 | 8.0 | 5.6 | 7.2 | 0 |

[a]Data represent mean NFT count per mm of cortical traverse. *Rare* indicates that only a few lesions per slide are present in that region. NFT density progressively increases with age, and NFTs are observed first in the entorhinal cortex. Neocortical Involvement is not seen before age 54. Adapted from Hof *et al.* (1995d).
[b]Age of DS cases.

parallels observations in AD and in normal aging (discussed above). Quantitative analysis of lesion densities revealed significant differences between AD and DS. Furthermore, amyloid deposition in the form of SPs or diffuse amyloid deposits is a major and early feature of DS that is observed prior to NFT formation in DS cerebral cortex (Wisniewski *et al.*, 1985; Giaccone *et al.*, 1989; Hof *et al.*, 1995d). Incidentally, SP densities might not be correlated with dementia in DS, because cases younger than 40 years have densities comparable to those in AD, yet generally without dementia. The development of pathologic lesions in DS may not be entirely similar to that seen in AD, although both conditions are likely to share several pathogenetic mechanisms. The time-course of NFT appearance in DS cortex shows that no considerable number of NFTs appears in the neocortex until about 50 years of age, despite already severe involvement of hippocampus and entorhinal cortex

←

**Figure 27.** Neuropathologic changes in DS brains (A–C, and E–G) compared to severe AD (D, H). In layer II of the entorhinal cortex (A–C) and CA1 field (E–G), there is a progressive increase of NFT and SP densities with age in DS. The DS cases were 40, 55, and 65 years old, respectively. NFTs are first to appear in the entorhinal cortex, while only SPs are in the CA1 field (A, E). The severity of pathologic changes is comparable for both diseases. Note the very striking macroscopic changes in gyral and sulcal patterns in the brain of a 44-year-old DS patient (top). Adapted from Hof *et al.* (1995d). Bars = 100 μm.

(Mann and Esiri, 1989; Hof *et al.*, 1995d; Hyman *et al.*, 1995). Thus dementia in DS may correlate with the involvement of comparable populations of neocortical pyramidal neurons, as in AD. Moreover, regional cerebral glucose metabolism reduced in neocortical regions in demented DS cases but not in cases younger than 40 years, and the reduction in cerebral glucose utilization correlates with the presence of neuropathological lesions in DS (Schapiro *et al.*, 1988, 1990, 1992).

At the molecular level, cytoskeletal abnormalities have been reported in DS, such as aberrant neurofilament protein expression in the first months of life, as well as alterations of axons labeled by an antibody to the heavy molecular weight neurofilament protein subunit, suggesting that abnormal regulation of cytoskeletal elements could be causally linked to the progression of dementia in DS patients (Plioplys, 1987). In addition, biochemical and immunohistochemical data indicate that a progressive incorporation of certain components of the cytoskeleton into NFTs takes place in DS in a fashion similar to that of AD, and subpopulations of neurons in DS reexpress developmentally regulated antigens (Wolozin *et al.*, 1988; Sparks and Hunsacker, 1992). Abnormally phosphorylated forms of the microtubule-associated protein tau are early markers for the NFT-prone population of neurons in DS, and tau-immunoreactive neurons have been shown to persist through life in DS (Sparks and Hunsacker, 1992; Hof *et al.*, 1995d). Comparative

**Table X. SP Distribution and Densities in DS Cases**[a]

| Area | 20[b] | 25 | 35 | 40 | 44 | 54 | 54 | 55 | 65 | 74 |
|---|---|---|---|---|---|---|---|---|---|---|
| CA1-3 | 4.0 | 3.8 | 5.6 | 8.8 | 4.0 | 50.6 | 18.4 | 13.8 | 12.5 | 40.8 |
| Subiculum | 10.4 | 8.4 | 10.4 | 7.5 | 11.3 | 31.4 | 21.6 | 16.5 | 17.5 | 52.3 |
| *Entorhinal* | | | | | | | | | | |
| Layers II/III | 7.2 | 14.8 | 19.2 | rare[c] | 23.0 | 44.2 | 26.2 | 23.0 | 30.5 | 37.4 |
| V/VI | 9.4 | 10.6 | 14.8 | rare | 16.0 | 32.6 | 21.0 | 12.3 | 27.0 | 29.0 |
| *35/36* | | | | | | | | | | |
| Layers II/III | 10.5 | 15.5 | 11.8 | rare | 18.3 | 38.6 | 47.6 | 23.8 | 39.0 | 55.4 |
| V/VI | 6.3 | 9.7 | 9.4 | rare | 14.3 | 29.6 | 35.8 | 17.5 | 20.0 | 31.0 |
| *9* | | | | | | | | | | |
| Layers II/III | 22.8 | 25.6 | 14.6 | 6.0 | 16.3 | 33.0 | 53.0 | 25.0 | 44.2 | 44.5 |
| V/VI | 16.8 | 16.0 | 9.0 | 4.5 | 9.8 | 22.2 | 33.8 | 14.8 | 28.8 | 18.5 |
| *20* | | | | | | | | | | |
| Layers II/III | 22.2 | 20.4 | 18.8 | 5.0 | 11.8 | 44.4 | 50.8 | 28.8 | 43.5 | 25.0 |
| V/VI | 12.4 | 13.8 | 10.8 | 2.3 | 8.0 | 26.4 | 33.0 | 11.8 | 23.0 | 16.5 |
| *17* | | | | | | | | | | |
| Layers II/III | 21.0 | 6.0 | 6.4 | 7.6 | 14.5 | 20.2 | 33.8 | 33.8 | 47.5 | 51.0 |
| V/VI | 8.6 | 3.8 | 9.0 | 9.0 | 12.5 | 14.6 | 32.4 | 21.5 | 38.3 | 52.3 |
| *18* | | | | | | | | | | |
| Layers II/III | 31.4 | 15.0 | 12.2 | 13.3 | 13.3 | 40.8 | 57.0 | 32.8 | 37.5 | 45.5 |
| V/VI | 22.8 | 10.2 | 7.2 | 7.7 | 6.5 | 20.0 | 31.6 | 21.8 | 26.3 | 28.2 |

[a]Data adapted from Hof *et al.*, 1995d showing SP densities per mm of cortical traverse. Note that SPs are observed earlier than NFTs (Table IX) and in some DS cases younger than 44, occur in all cortical areas investigated. Adapted form Hof *et al.* (1995d).
[b]Age of DS cases.
[c]*Rare* indicates that only a few lesions per slide are present.

studies between AD and DS have shown that the abnormal phosphorylation of tau in DS precedes NFT formation in affected neurons (Mann *et al.*, 1989; Flament *et al.*, 1990a; Hanger *et al.*, 1991). Additional factors may contribute to enhanced neuronal vulnerability and lead to development of dementia in DS. Down's syndrome is not due to a genetic mutational event, but is the consequence of a chromosomic aberration and resulting gene dosage effect. However, DS expression in the brain is highly variable among patients and does not necessarily result in a severe mental handicap and somatic malformations. Thus DS does not represent a uniform brain substrate, which translates into its variable expression of neuropathologic changes. For instance, overexpression of the gene encoding the amyloid Aβ protein and its precursor due to triplication of chromosome 21 may result in the early deposition of amyloid in the brain of these patients as well as elevated serum concentration of the amyloid precursor, and isolated SP amyloid cores from DS cases show morphological differences from AD, indicating that other mechanisms are involved in SP formation in DS (Allsop *et al.*, 1986; Rumble *et al.*, 1989). Hyman and colleagues have also demonstrated that inheritance of the apolipoprotein E4 genotype in DS cases may be another risk factor for increased amyloid deposition (Hyman *et al.*, 1995).

The triplication of chromosome 21 could also induce the overexpression of proteins other than amyloid that may participate in the accelerated degenerative process of DS. patients with DS exhibit an elevated activity of copper–zinc superoxide dismutase, an enzyme coded by a gene located in the 21q22 region of chromosome 21 (Minc-Golomb *et al.*, 1991), and the gene encoding the glutamate receptor subunit GluR5 is located in the vicinity of the copper–zinc superoxide dismutase gene at the chromosome 21 locus 21q21.1-22.1 (Eubanks *et al.*, 1993). These associations provide additional links between early neurodegenerative changes in DS and excitotoxic mechanisms in cortical neurons more sensitive to excitatory amino-acid-mediated toxicity, and they point further to the many similarities between the development of DS and the progression of cellular changes in the early stages of AD.

### 4.2. Guamanian Amyotrophic Lateral Sclerosis/Parkinsonism–Dementia Complex

Amyotrophic lateral sclerosis/parkinsonism–dementia complex (ALS/PDC) is highly prevalent in the native Chamorro population of the Marianas Islands in the Western Pacific (Zimmermann, 1945; Koerner, 1952; Arnold *et al.*, 1953; Tillema and Wijnberg, 1953; Mulder *et al.*, 1954; Hirano *et al.*, 1961a,b, 1966, 1967, 1968; Lessell *et al.*, 1962; Andrews, 1967; Brody *et al.*, 1971, 1975; Chen and Chase, 1986; Garruto and Yase, 1986; Hirano and Llena, 1986; Rodgers-Johnson *et al.*, 1986; Kurland, 1988; Hirano, 1990; Bailey-Wilson *et al.*, 1993; Lilienfeld *et al.*, 1994). Reportedly, pure ALS cases have slowly declined in recent years but inordinate numbers of patients with PD, ALS or ALS/PDC are still living on Guam. Prevalences of 23.5 per thousand of ALS and 57.4 for PD have been reported for people older than 55 years in the three southern villages of Umatac, Merizo, and Inarajan (Fig. 28). These figures are orders of magnitude higher than comparable values for Rochester, Minnesota (0.33 and 9.16 per thousand, respectively; Lilienfeld *et al.*,

1994). Most cases are found in geographical isolates located in southern Guam and on Rota, the two southernmost islands of the archipelago (Brody, 1971, 1975; Chen and Chase, 1986; Garruto and Yase, 1986; Kurland, 1988; Zhang *et al.*, 1990a,b, 1995, 1996; Lilienfeld *et al.*, 1994). The disease complex is referred to as lytico-bodig by the native population, but both PD and ALS are known to occur separately (Elizan *et al.*, 1966; Hirano *et al.*, 1967; Chen and Chase, 1986; Lilienfeld *et al.*, 1994). The etiopathogenetic mechanisms of ALS/PDC are not yet determined, although environmental neurotoxins have been implicated, as have genetic factors (Garruto and Yase, 1986; Crapper McLachlan *et al.*, 1988; Kurland, 1988; Kisby *et al.*, 1992; Bailey-Wilson *et al.*, 1993; Lilienfeld *et al.*, 1994). Incidence of the disease does not decrease among Chamorro migrants from Guam, but it seems that an exposure period on the island of at least two decades is necessary for the development of symptoms (Torres *et al.*, 1957; Eldridge *et al.*, 1969; Garruto *et al.*, 1980; Lilienfeld *et al.*, 1994). Genetic factors alone do not account for all of the cases, as several patients unrelated to the ethnic Chamorro background have been docu mented among the Japanese, Filipino, and Caucasian migrants to Guam (Elizan *et*

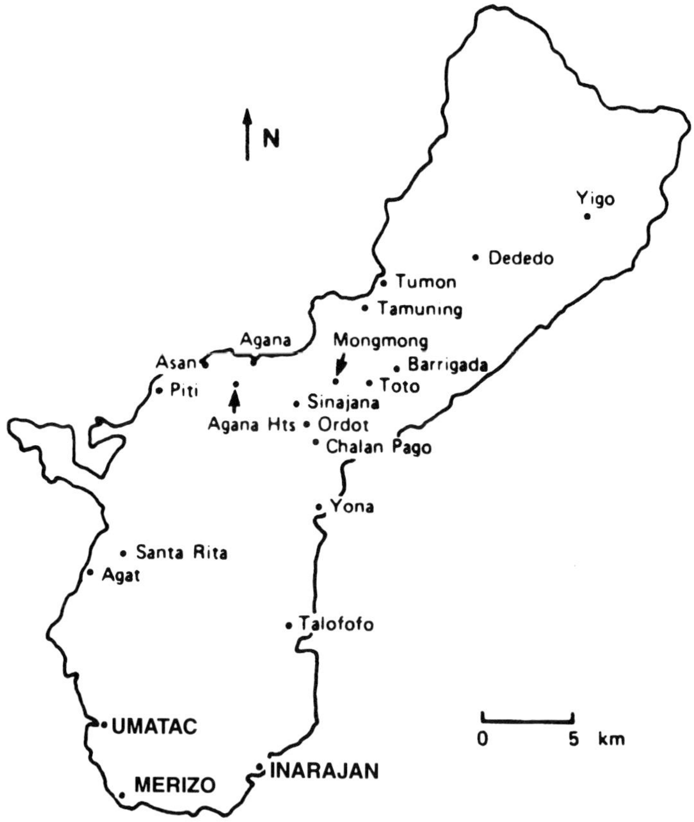


**Figure 28.** Map of Guam. The three Chamorro villages with the highest prevalence of ALS/PDC cases (Umatac, Merizo, and Inarajan) are all located in the southernmost part of the island. However, cases are observed throughout the island, not only in Chamorros but also in a minority of Filipino, Japanese, and Caucasian migrants.

al., 1966; Forno and O'Flanagan, 1973; Garruto *et al.*, 1981; Chen *et al.*, 1982; Lilienfeld *et al.*, 1994). In addition, several foci with high incidence of motor neuron disorders have been reported in other Western Pacific regions, such as the Kii peninsula in Japan and among isolated Auyu and Jakai populations of the Irian Jaya (West New Guinea) (Yase, 1970; Gajdusek, 1982; Garruto, 1991).

Whereas Guamanian ALS is clinically indistinguishable from sporadic (continental) ALS in the early stages, PD is characterized by an insidious progressive decline consisting of mental slowness and an extrapyramidal syndrome that includes bradykinesia, rigidity, and less frequently, tremor (Hirano *et al.*, 1961a; Chen and Chase, 1986). Neuropathological examination reveals severe cortical atrophy and neuronal loss in neocortex, hippocampus, and basal nucleus of Meynert, and depigmentation of the substantia nigra and locus coeruleus in cases with predominant PD (Hirano *et al.*, 1916b, 1996; Nakano and Hirano, 1983; Lilienfeld *et al.*, 1994), as well as severe loss of upper and lower motoneurons in cases with predominant ALS (Brody *et al.*, 1971; Hirano *et al.*, 1966, 1967; Lepore *et al.*, 1988). Severe NFT formation and granulovacuolar degeneration in the neocortex and hippocampus are consistently observed in both ALS and PD cases (Hirano *et al.*, 1961b, 1966, 1968; Lilienfeld *et al.*, 1994). Importantly, Guamanian ALS shows moderate-to-high densities of cortical NFTs, whereas NFTs are not commonly observed in sporadic ALS (Malamud *et al.*, 1961; Hirano *et al.*, 1967). The ultrastructure of NFTs in ALS/PDC is comparable to that observed in AD (Hirano *et al.*, 1967), and NFTs from Guamanian cases share certain biochemical features of NFTs from AD cases (Guiroy *et al.*, 1993; Shankar *et al.*, 1989; Ito *et al.*, 1991a,b; Hof *et al.*, 1994b; Buée-Scherrer *et al.*, 1995; Buée *et al.*, 1996; Mawal-Dewan *et al.*, 1996). In recent studies using Western blot analysis of tau protein, an abnormal triplet referred to as tau 55, 64, 69 (Delacourte *et al.*, 1990; Goedert *et al.*, 1994) was observed in all cortical areas in AD brains, whereas the Guamanian cases showed heterogeneity in the tau protein profile, in that subcortical structures have a pathological tau protein doublet (tau 64, 69), whereas the hippocampus and neocortex show the AD-related tau triplet (Buée-Scherrer *et al.*, 1995). The tau 64, 69 doublet is characteristic of PSP (Flament *et al.*, 1991; Buée-Scherrer *et al.*, 1995). Interestingly, vertical-gaze palsy has recently been found to occur with high prevalence among the Guamanian patients, suggesting that there may be some degree of overlap between these conditions (Lepore *et al.*, 1988). In fact, one PD case with vertical-gaze palsy from our series exhibited the tau 64, 69 doublet typical of progressive supranuclear palsy in neocortical samples (Buée-Scherrer *et al.*, 1995). Note that in PSP, neocortical NFTs have a laminar distribution comparable to that observed in ALS/PDC, with about twice to three times as many NFTs in layers II and III as in layers V and VI, in some cases (Hof *et al.*, 1992a; see below). Neuronal Hirano bodies and glial tau-immunoreactive hazy and coiled inclusions in astrocytes and oligodendrocytes have also been reported in ALS/PDC (Lilienfeld *et al.*, 1994; Oyanagi *et al.*, 1997).

Quantitative analyses of regional and laminar distribution of NFTs in the neocortex and hippocampal formation of ALS/PDC have revealed that in cases with predominant PD, the cellular localization of NFTs corresponds to a subpopulation of pyramidal neurons, with the laminar distribution in neocortical areas differing considerably from that in AD cases (Hof *et al.*, 1991c, 1994a,b). For instance, in association cortex of AD cases, NFTs are slightly more numerous in layers V and VI than in layers II and III (see above). Conversely, in Guamanian PD cases, 62–90% of

NFTs are observed in layer II and in the outer third of layer III, whereas infragranular layers display relatively modest densities of NFTs as compared with supragranular layers (Hof *et al.,* 1991c, 1994a,b). This difference in the laminar distribution of NFTs suggests that components of the neocortical circuitry other than those thought to be affected in AD may be involved in Guamanian cases, which may account for the differences between them in neuropsychological manifestations and the course of their dementing syndromes (Hof *et al.,* 1991c) (Fig. 29). This particular NFT distribution appears to be highly consistent in ALS/PDC, because it is conspicuous even in ALS and PD cases with low NFT counts, and it suggests that in these cases, NFT distribution reflects the localization of the neurons most severely affected by the degenerative process.

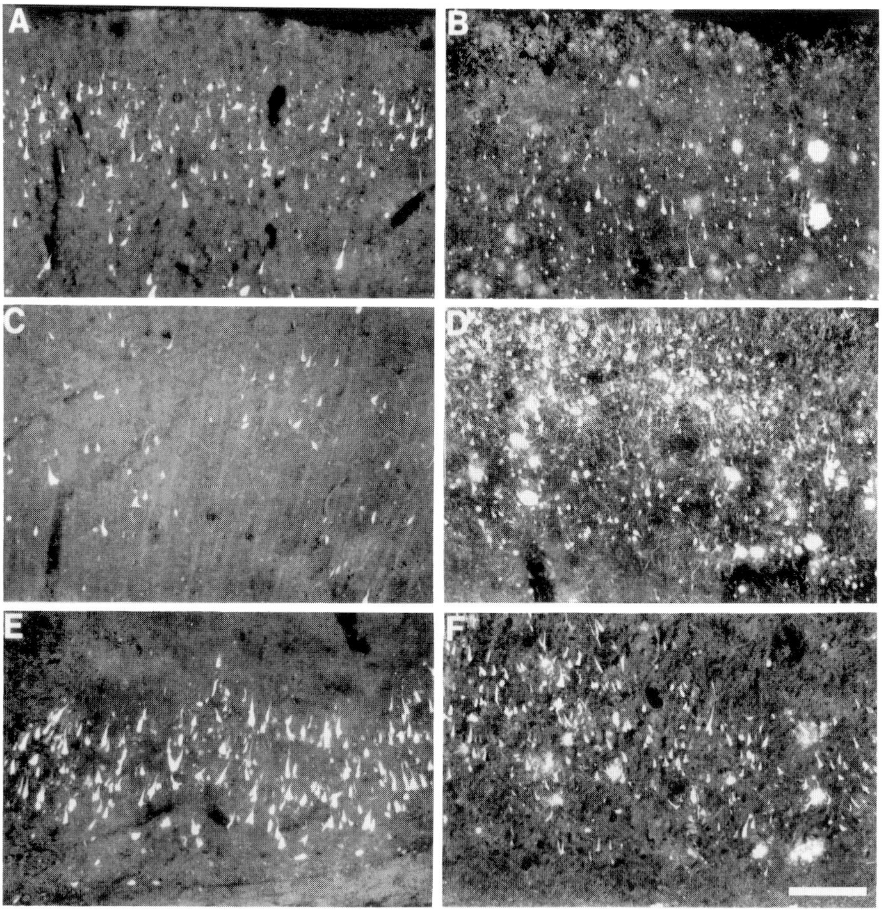


**Figure 29.** Comparison of NFT distribution in layers II and III of area 9 (A, B), in layers V and VI of area 21 (C, D), and in the CA1 field of the hippocampus (E, F) in a Guamanian PD case (A, C, E) and a Caucasian AD case (B, D, F). Note the inverse NFT distribution patterns in the neocortex and the absence of SPs in the Guamanian case. Thioflavine S staining. Reprinted from Hof *et al.* (1991c), *Brain Res.* **564:**306–313, with kind permission of Elsevier Science—NL, Sara Burgerhartstraat 25, 1055 KV, Amsterdam, The Netherlands. Bar = 50 μm.

In PD cases with low NFT densities, there are virtually no NFTs in the lower half of layer III. Also, there are larger differences between NFT counts in layers II/III and V/VI in Guamanian PD than there are in AD. The PD cases show up to 4.0-fold higher NFT counts in layers II/III as compared to layers V/VI in area 9 and up to 4.4-fold higher counts in area 21. In AD cases, differences in NFT counts between supra- and infragranular layers are less remarkable, with values up to 1.9-fold higher in layers V and VI of area 9 and 1.7-fold higher in area 21 (Hof *et al.*, 1991c) (Fig. 29, Table XI). Besides this major difference in the laminar NFT distribution pattern between AD and Guamanian PD cases, however, AD cases have higher overall NFT densities in the neocortex, although the range of NFT counts overlaps to a large degree between the two groups (Table XI). The ratios of NFT counts in layers II and III and total NFT counts across the entire cortical thickness in association cortex are much higher in PD than in AD. The mean ratios are 0.73 and 0.41 in PD and AD cases, respectively (Table XI) (Hof *et al.*, 1991c). In addition, Guamanian cases with predominant ALS exhibit a lesion distribution comparable to that found in cases with predominant PD, but with lower NFT numbers in the neocortex (Hof *et al.*, 1994a). The hippocampus in Guamanian ALS and PD cases contains more-consistently high NFT densities than in AD cases (Table XI; Fig. 30). In the hippocampus of both brain groups, a similar NFT distribution pattern is observed (Fig. 30) (Hof *et al.*, 1994a). Neurofibrillary tangles are present in the entire pyramidal layer of Ammon's horn, with very high densities in the CA1 field. Guamanian PD cases also contain higher NFT counts in CA1 than AD cases (Table XI).

Although the distribution of NFTs is qualitatively comparable in Guamanian ALS and PD cases, there are substantial quantitative differences in regional and laminar NFT densities between the two brain groups (Hof *et al.*, 1994a). Cases with predominant ALS symptomatology generally exhibit lower NFT densities than do

**Table XI. Comparison of NFT Densities in the Cerebral Cortex of Guamanian PD and AD Cases[a]**

| Area | Layer | PD | AD |
|---|---|---|---|
| CA1 | | 136 ± 20 | 85 ± 14 |
| Entorhinal | II/III | 110 ± 11 | 90 ± 6 |
| | V/VI | 50 ± 5 | 40 ± 4 |
| 21 | II/III | 42 ± 5 | 63 ± 4 |
| | V/VI | 15 ± 4 | 92 ± 5 |
| 9 | II/III | 17 ± 5 | 41 ± 7 |
| | V/VI | 9 ± 3 | 50 ± 9 |
| 4 | II/III | 9 ± 2 | 2 ± 1 |
| | V/VI | 4 ± 2 | 0 |
| 17 | II/III | 2 ± 2 | 0 |
| | V/VI | 0 | 0 |

[a]Data represent mean densities per mm cortical traverse and were obtained from 8 non-Caucasian PD cases from Guam and 5 Caucasian AD cases. Note the reverse laminar pattern of NFT distribution in the Guamanian cases as compared with AD cases, and the presence of NFTs in area 4 in PD. Adapted from Hof *et al.* (1991c, 1994a,b).

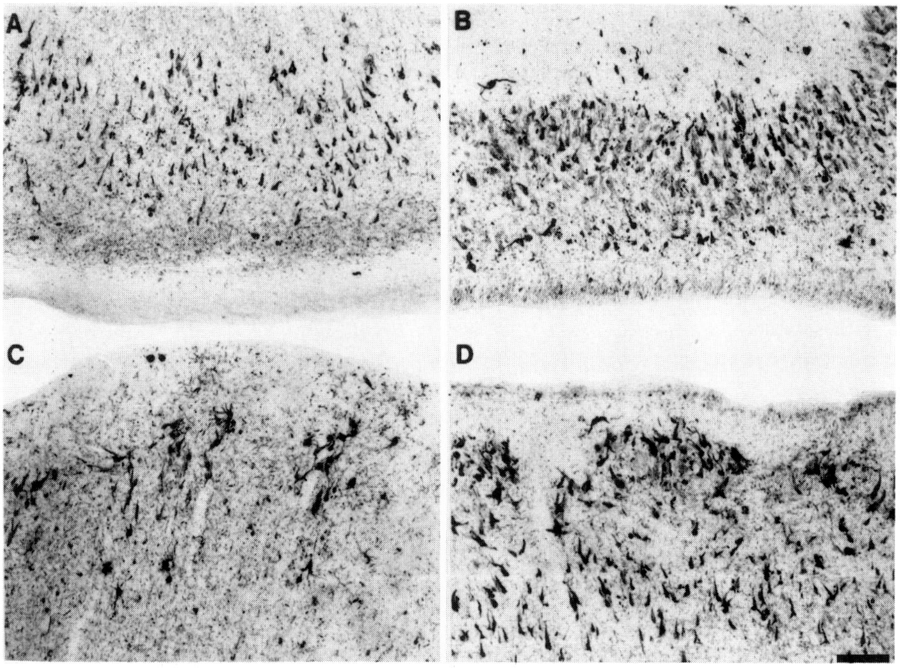


**Figure 30.** Comparison of NFT distribution in the CA1 field (A, B) and layer II of entorhinal cortex (C, D) in a Guamanian ALS case (A, C) and a Guamanian case with severe PD (B, D). The patterns are similar, but the PD case is more severely affected. From Hof *et al.* (1994a), *Brain Res.* **650:**107–116, with kind permission of Elsevier Science—NL, Sara Burgerhartstraat 25, 1055 KV, Amsterdam, The Netherlands. Bar = 65 μm.

cases with predominant PD (Fig. 31). On average, PD cases contain 2.2-fold higher NFT densities in the CA1 field of the hippocampus, and 2.3-fold higher in the subiculum, than ALS cases (Table XII; Fig. 31). Layer II of the entorhinal cortex is involved to comparable degrees in both brain groups, whereas layers III, V, and VI have 6.6-fold higher NFT densities in PD than in ALS (Table XII). The perirhinal cortex is more severely affected by NFT formation in the PD cases, with approximately 5.0-fold higher NFT counts than in ALS cases. Neocortical areas 21, 9, and 7b show less-marked differences, whereas the anterior and posterior cingulate cortex displays much higher NFT counts in the PD cases (up to 13.0-fold differences) (Table XII). Interestingly, area 4 exhibits more NFT in PD than in ALS, with differences as high as 2.3-fold in layers II and III and 3.3-fold in layers V and VI (Table XII). Ratios of lesion counts in the supragranular layers to total lesion counts across the full cortical thickness show no differences among cases in both groups in each neocortical area (Hof *et al.*, 1994a), with mean ratio values of 0.74 in ALS and 0.73 in PD. Finally, a striking difference between AD and Guamanian PD brains is that amyloid deposits, amyloid angiopathy, and SPs are not consistently observed in any of the neocortical and hippocampal areas investigated in ALS/PDC cases (Hof *et al.*, 1991c, 1994a,b; Lilienfeld *et al.*, 1994). Amyloid has been observed to aggregate around ghost NFTs in PD cases, and NFTs in Guamanian cases have also been shown to contain immunoreactivity for inflammatory response proteins similarly to those in AD (Schwab *et al.*, 1995, 1996).

The fact that layer II of the entorhinal cortex is the only region to show comparable lesion densities in both ALS and PD cases may be related to the particular vulnerability of these neurons at very early stages of the disease. It is possible that in ALS/PDC, these neurons are also highly sensitive to the neurodegenerative process, and that their select involvement in early stages of the disease could be reflected in the comparable amount of NFTs observed in PD and ALS. Conversely, the much higher NFT densities observed in PD cases in all other cortical regions, as well as in layers III and VI of the entorhinal cortex, may indicate that at the cellular level, the degenerative process develops more rapidly in PD than in ALS, although there are conflicting reports on the duration of the illness in cases with either predominant ALS or PD symptomatology (Elizan *et al.*, 1966; Rodgers-Johnson, 1986). In addition, there is clearly some heterogeneity among the Guamanian patients, because there is no correlation between severity of pathological changes and duration of disease (Hof *et al.*, 1991c, 1994a). It is not known whether the pyramidal neurons prone to NFT formation in PD cases are less vulnerable to a possibly common etiopathogenetic factor in the ALS cases. Guamanian cases with predominant ALS symptoms, however, generally remain cognitively intact until late in the course of their illness, as opposed to the clear impairment seen in the cases with predominant PD (Elizan *et al.*, 1966; Hirano *et al.*, 1968; Hudson, 1981). In

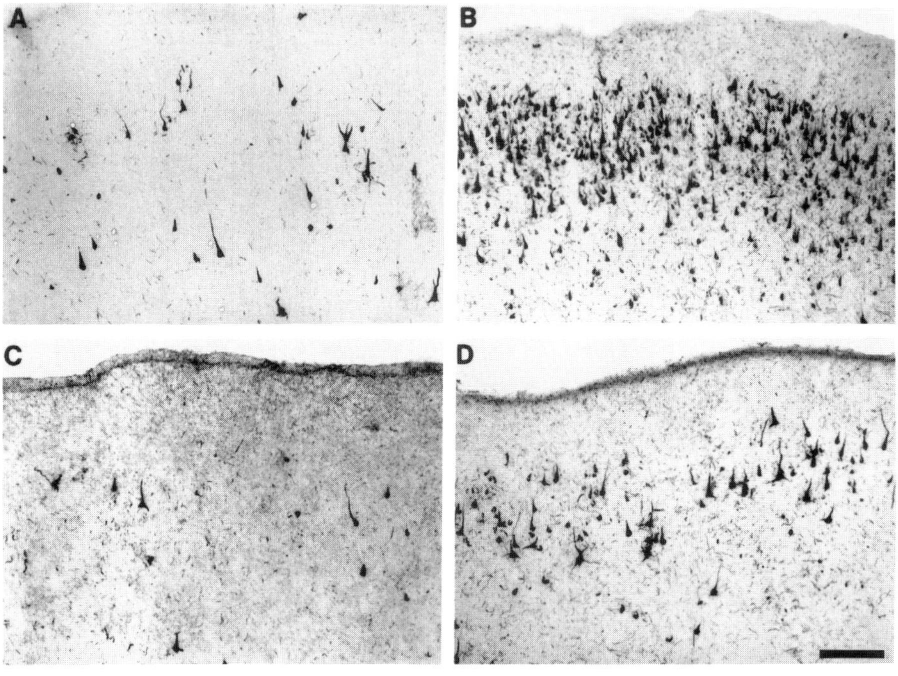


**Figure 31.** Comparison of NFT distribution in layers II and III of area 21 (A, B) and area 9 (C, D) in a Guamanian ALS case (A, C) and a Guamanian case with severe PD (B, D). As in the hippocampus and entorhinal cortex, NFT distribution patterns are similar, but the PD case is more severely affected. From Hof *et al.* (1994a), *Brain Res.* **650:**107–116, with kind permission of Elsevier Science—NL, Sara Burgerhartstraat 25, 1055 KV, Amsterdam, The Netherlands. Bar = 65 μm.

**Table XII. Comparison of NFT Densities in the Cerebral Cortex of Guamanian with Predominant ALS or Predominant PD Symptomatology[a]**

| Area | Layer | PD | AD |
|---|---|---|---|
| CA1 | | 56 ± 12 | 120 ± 15 |
| Subiculum | | 48 ± 10 | 130 ± 15 |
| Entorhinal | II/III | 74 ± 9 | 112 ± 13 |
| | V/VI | 15 ± 4 | 50 ± 10 |
| 35/36 | II/III | 20 ± 4 | 90 ± 11 |
| | V/VI | 7 ± 2 | 35 ± 5 |
| 21 | II/III | 15 ± 3 | 40 ± 4 |
| | V/VI | 6 ± 2 | 12 ± 3 |
| 9 | II/III | 7 ± 2 | 15 ± 2 |
| | V/VI | 4 ± 1 | 9 ± 2 |
| 7b | II/III | 2 ± 1 | 6 ± 2 |
| | V/VI | 1 ± 1 | 2 ± 1 |
| 24 | II/III | 3 ± 1 | 15 ± 3 |
| | V/VI | 0 | 7 ± 2 |
| 23 | II/III | 1 ± 1 | 8 ± 2 |
| | V/VI | 0 | 3 ± 1 |
| 4 | II/III | 5 ± 2 | 11 ± 3 |
| | V/VI | 2 ± 1 | 5 ± 2 |
| 17 | II/III | 0 | 2 ± 1 |
| | V/VI | 0 | 0 |
| 18 | II/III | 0 | 2 ± 1 |
| | V/VI | 0 | 1 ± 1 |

[a]Data represent mean densities per mm cortical traverse and were obtained from 9 ALS and 12 PD cases. Note that Guamanian ALS cases have consistently lower NFT densities as compared with PD cases. Adapted from Hof *et al.* (1994a,b).

Guamanian cases with predominant ALS, NFTs are not considered to be a common pathological change in the sporadic form (Malamud *et al.,* 1961; Hudson, 1981). However, Guamanian ALS cases display all of the usual accompanying pathologic and clinical characteristics of sporadic ALS (Malamud *et al.,* 1961; Hirano *et al.,* 1966, 1967; Brody *et al.,* 1971). It is therefore possible that the presence of NFTs in the brains of Guamanian ALS cases represents an overlap of the classical features of severe ALS with a relatively mild form of slowly progressive dementia comparable to that seen in PD cases.

Similarly, the differential distribution of NFTs in the neocortex of Guamanian and AD cases may account in part for the differences in the clinical presentation of dementia in these diseases. Guamanian patients display bradyphrenia rather than the temporospatial disorientation that is characteristic of AD. The memory and cognitive impairments of the Guamanian patients progress at a slower rate than is generally observed in AD. We have proposed that the pattern of NFT distribution in AD represents a widespread breakdown of corticocortical projections. In Guamanian ALS/PDC, the distribution of NFTs also matches corticocortical systems; however, their remarkable preference for layers II and III suggests that feedforward projections are more severely damaged than feedback projections (Hof *et al.,* 1991c, 1994a,b). Moreover, because in Guamanian cases, NFTs are densest in layer II and in the upper third of layer III, it is possible that a different population of feedforward projections generally spared in AD is affected, as well as the short-distance corticocortical projections. Similarly, the severe changes observed in layers V and VI in AD also indicate that feedback corticocortical projection neurons are more involved in AD cases than in Guamanian cases.

Immunohistochemical analysis of some Guamanian cases revealed that the patterns of neuronal vulnerability differ in AD and ALS/PDC (Hof *et al.,* 1994b). The population of neurofilament protein-immunoreactive pyramidal neurons proposed to furnish long corticocortical connections that are subject to degeneration in AD appeared to be mostly unaffected in ALS/PDC (Fig. 32) (Hof *et al.,* 1994b). This observation is well correlated with the preferential distribution of NFTs in the superficial position of layer III and in layer II of neocortical areas in the Guamanian cases. These differences could also be related to the fact that the molecular profile of NFTs in Guamanian cases is not entirely equivalent to that observed in AD cases (Buée-Scherrer *et al.,* 1995). In Guamanian ALS cases, the large neurofilament-positive cells in layer V of primary motor cortex (Betz cells) are dramatically affected, as they are in sporadic ALS cases (Nihei *et al.,* 1993; Hof *et al.,* 1994b). Finally, the populations of interneurons labeled by parvalbumin and calretinin that are unaffected in AD are also resistant to the degenerative process in ALS/PDC and sporadic ALS cases (Fig. 32) (Hof *et al.,* 1994b).

The pathogenetic mechanisms of Guamanian ALS/PDC have not yet been determined. High concentrations of aluminum in the island's soil as well as the ingestion of the toxin β-*N*-methylamino-L-alanine (BMAA) from the seeds of the cycad plant *Cycas circinalis* have been related to the development of the disease and tentatively implicated in NFT formation (Gibbs *et al.,* 1982; Perl *et al.,* 1982; Perl and Pendlebury, 1986; Garruto *et al.,* 1984; Crapper-McLachlan *et al.,* 1988; Kurland, 1988; Piccardo *et al.,* 1988; Kisby *et al.,* 1992), and the high levels of zinc—another potential neurotoxic ion—in the cycad plant flour used in traditional Chamorro cooking has been recently considered as another source of environmental hazard (Duncan *et al.,* 1992). Of the possible toxic agents, aluminum is known to be a potent neurotoxin and has also been implicated as an environmental factor in the pathogenesis of AD (Perl and Brody, 1980; Perl, 1985; Perl and Good, 1987). Note that the NFTS of ALS/PDC cases contain high concentrations of elemental aluminum in comparison to control cases (Perl *et al.,* 1982). Although AD and ALS/PDC share several pathologic features, the differences in NFT distribution and in neuronal vulnerability indicate that their pathogenetic process is unlikely to be entirely equivalent. Further analysis and comparison of selective neuronal vulnerability patterns as well as thorough reevaluation of the genetic and environmental background of Chamorro people will be necessary to elucidate the pathogenetic mechanisms specific to each disorder.

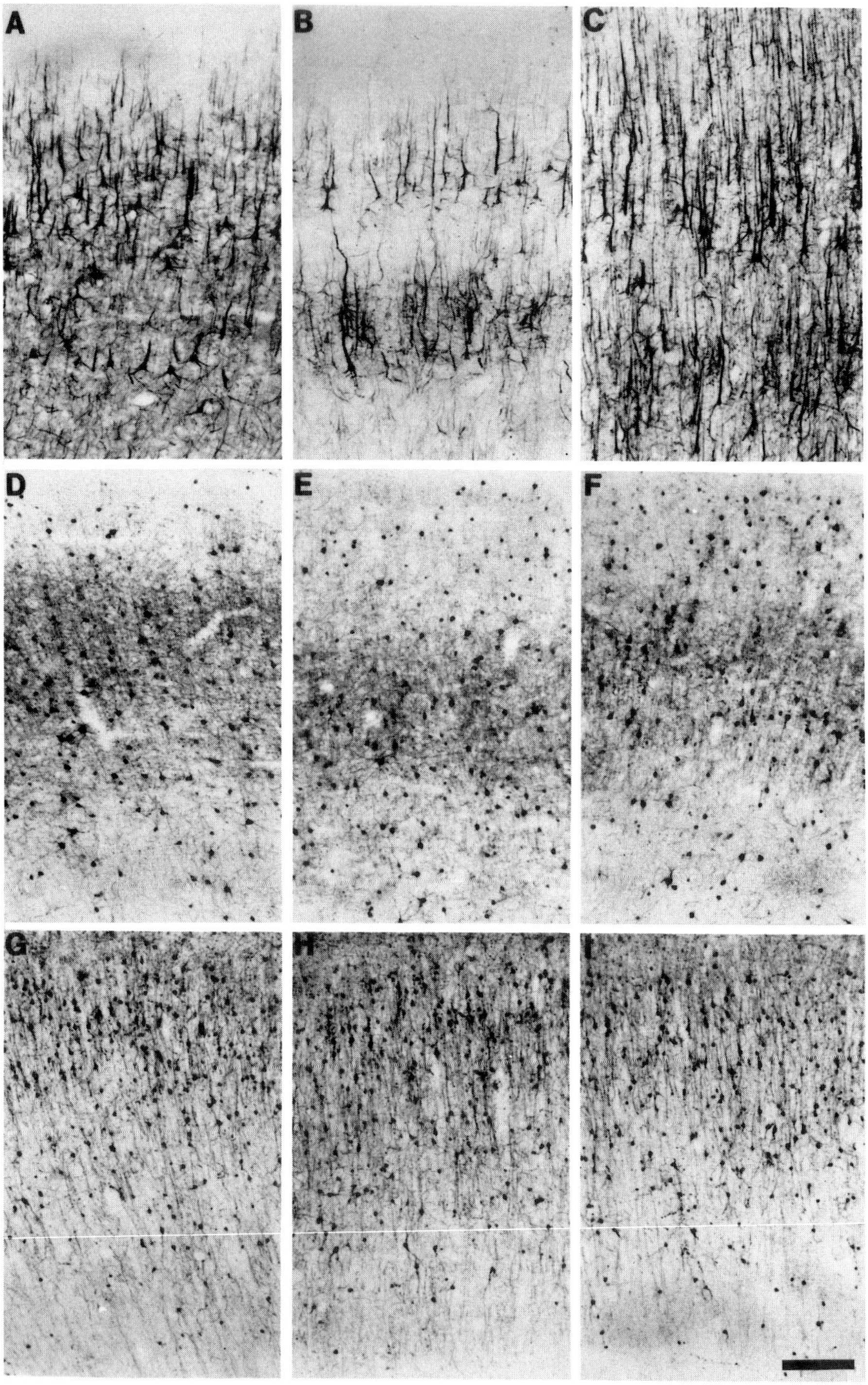


**Figure 32.** Immunoreactivity for neurofilament protein (A–C), parvalbumin (D–F), and calretinin (G–I) in a Guamanian ALS/PDC case (A, D, G), a Caucasian AD case (B, E, H), and a Caucasian control case (C, F, I). There is a loss of neurofilament protein-containing neurons in the AD case, but these neurons are relatively preserved in the Guamanian case, which parallels NFT distribution in both diseases. Interneurons containing parvalbumin and calretinin are not affected in Guamanian ALS/PDC. From Hof *et al.* (1994b). © Springer-Verlag. Bar = 100 μm.

### 4.3. Progressive Supranuclear Palsy and Corticobasal Degeneration

Progressive supranuclear palsy (PSP) and corticobasal degeneration (CBD) are progressively dementing illnesses, both of which combine a heterogeneous symptomatology of movement disorder and cognitive impairment, rendering their clinical and neuropathologic diagnosis difficult, as they share many characteristics with Lewy body dementia, Parkinson's disease associated with dementia, Pick's disease, and postencephalitic parkinsonism (Litvan *et al.*, 1996, 1997a,b; Schneider *et al.*, 1997). Thus, PSP and CBD are discussed together, as they share several neuropathological features. Progressive supranuclear palsy was originally recognized as a clinical and neuropathological entity by Steele, Richardson, and Olszewski (Steele *et al.*, 1964), although cases corresponding to its description were recorded in the beginning of this century (e.g., Posey, 1904; Alajouanine *et al.*, 1926). Progressive supranuclear palsy is characterized clinically by a supranuclear ophthalmoplegia of the vertical gaze; pseudobulbar palsy; dysarthria; facial, nuchal, and troncular muscular dystonia; and in most cases by a certain degree of dementia (Steele *et al.*, 1964; Behrman *et al.*, 1969; Constantinidis *et al.*, 1970; Dix *et al.*, 1971; Jellinger, 1971; Steele, 1972; Albert *et al.*, 1974; Probst and Dufresne, 1975; Litvan *et al.*, 1996, 1997b; Schneider *et al.*, 1997). Corticobasal degeneration is a slowly progressive, late-onset neurodegenerative disorder characterized by an asymmetric akinetic-rigid syndrome associated with cognitive deficits (apraxia and aphasia), extrapyramidal motor dysfunction (rigidity and dystonia), and moderate dementia late in the course of the disease (Rebeiz *et al.*, 1967, 1968; Paulus and Selim, 1990; Riley *et al.*, 1990; Rinne *et al.*, 1994; Litvan *et al.*, 1996, 1997b; Massman *et al.*, 1996; Schneider *et al.*, 1997).

Histopathologically, PSP is characterized by variable degree of NFT formation and neuronal loss in several subcortical structures such as globus pallidus, subthalamic nucleus, zona incerta, substantia nigra, nucleus basalis of Meynert, tegmentum, raphe, dentate nucleus of the cerebellum, and nuclei of the cranial nerves, including Westphal–Edinger nucleus and the nuclei of the III, IV, dorsal X, and XII cranial nerve. The nuclei of the V, VI, and VII cranial nerves are generally less affected (Steele *et al.*, 1964; Behrman *et al.*, 1969; Constantinidis *et al.*, 1970; Dix *et al.*, 1971; Jellinger, 1971; Probst and Dufresne, 1975; Akashi *et al.*, 1989; Tagliavini *et al.*, 1984). At the ultrastructural level, NFTs in PSP are different from those observed in AD and Guamanian ALS/PDC, and are composed of 13–16 nm straight tubules, although 22-nm twisted tubules have also been reported (Tellez-Nagel *et al.*, 1973; Tomonaga *et al.*, 1977; Bugiani *et al.*, 1979; Yagashita *et al.*, 1979; Ksiezak-Reding *et al.*, 1994). Classically, PSP is considered a model of the subcortical dementias (Albert *et al.*, 1974; D'Antona *et al.*, 1985), but several reports have demonstrated a substantial cortical involvement, pointing to the differences in NFT distribution between PSP and AD (Ishino and Otsuki, 1976; Akashi *et al.*, 1989; Hauw *et al.*, 1990; Arima *et al.*, 1992; Braak *et al.*, 1992b; Hof *et al.*, 1992a; Mori *et al.*, 1994; Hanihara *et al.*, 1995; Higuchi *et al.*, 1995; Verny *et al.*, 1996; Wakabayashi *et al.*, 1997). Most NFTs in PSP are observed in the hippocampal formation entorhinal and perirhinal cortex (Ishino and Otsuki, 1976; Braak *et al.*, 1992b; Hof *et al.*, 1992a). A striking feature of PSP is the presence of numerous NFTs within the granule cell layer of the dentate gyrus (Fig. 33*B, C*) (Arima *et al.*, 1992; Hof *et al.*, 1992a; Higuchi *et al.*, 1995; Wakabayashi *et al.*, 1997). The pyramidal layer of the

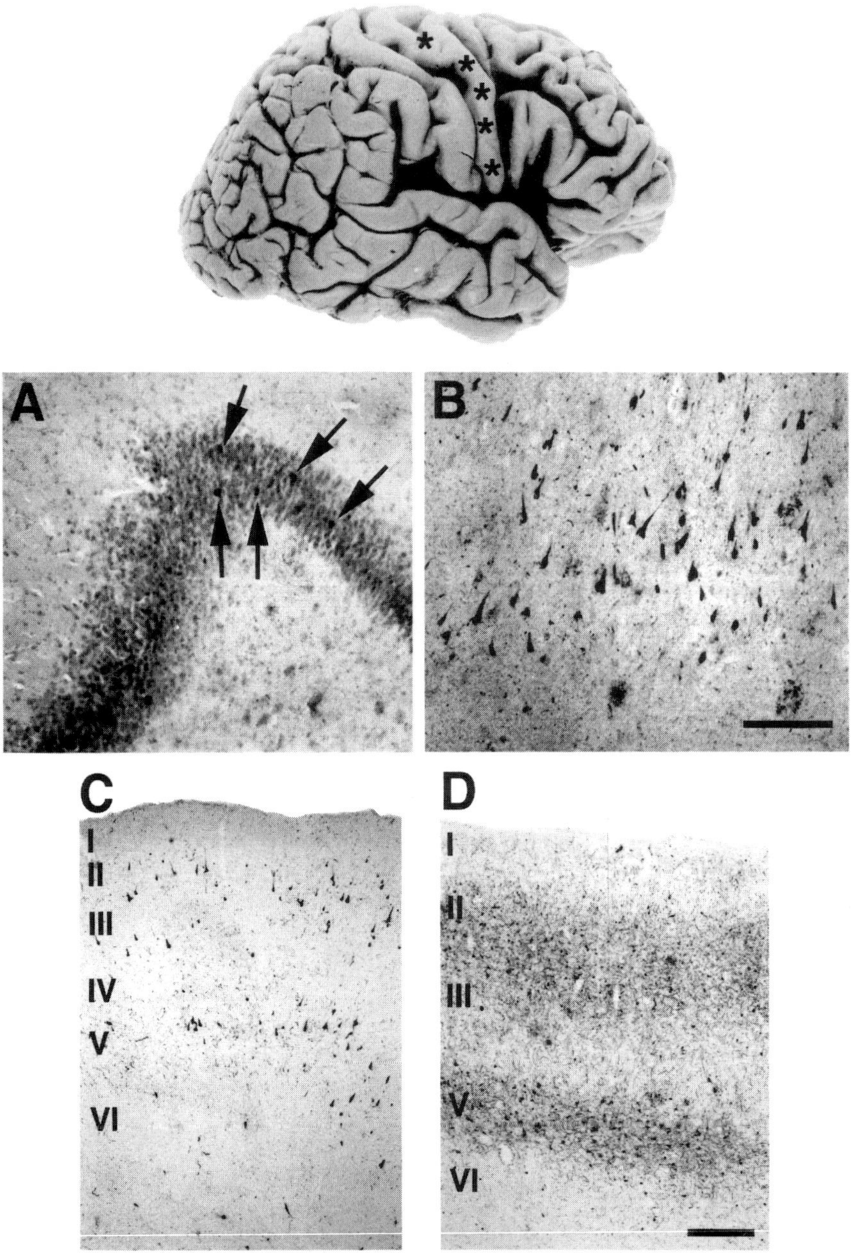


**Figure 33.** Distribution of NFT in PSP cerebral cortex. (A) Some granule cells in the dentate gyrus are consistently involved in NFT formation in PSP (arrows). (B) The CA1 field shows a high number of NFTs in certain cases, as well as neuritic plaques (C, D). In the neocortex, NFTs are predominantly distributed in the superficial layers (C, area 9); in the primary motor cortex, layer V is more severely affected (D). In PSP the primary motor cortex is consistently affected by tau-immunoreactive cellular alterations (D) and shows a marked atrophy macroscopically (top, asterisks). Adapted from Hof *et al.* (1992a). Bars = 100 μm.

hippocampus and subiculum contains moderate-to-high NFT densities. The entorhinal cortex displays a laminar pattern of NFT distribution comparable to that in AD, with numerous NFTs in layers II and V (Braak *et al.*, 1992b; Hof *et al.*, 1992a). In the neocortex, temporal lobe areas generally contain slightly more NFTs than prefrontal regions, with higher densities in layers II and III (Fig. 33*D*) (Hof *et al.*, 1992a). The occipital cortex is devoid of NFTs in PSP. Remarkably, high numbers of NFTs are present in the primary motor cortex mainly in layer V (Fig. 33*E*), and many PSP cases exhibit a marked bilateral atrophy of the precentral gyrus (Fig. 33*A*, *E*) (Hauw *et al.*, 1990; Hof *et al.*, 1992a; Daniel *et al.*, 1995). Senile plaques are not commonly observed in PSP. Other neuropathological features of PSP are the presence of ballooned neurons immunoreactive for αB-crystallin and phosphorylated neurofilament protein, and high densities of tau-immunoreactive neuronal threads and astrocytic tangles (Probst *et al.*, 1988; Higuchi *et al.*, 1995; Dickson *et al.*, 1986).

The pattern of NFT distribution observed in PSP is similar to that existing in Guamanian ALS/PDC (Hof *et al.*, 1991c, 1994a,b), although cortical NFT densities in our series of Guam ALS/PDC cases were generally higher than in PSP cases. A preferential distribution of NFT in layers II and III has also been reported in cases of posttraumatic dementia and PEP (see below). In addition, the presence of numerous NFTs in the granule cell layer of the dentate gyrus suggests that certain intrinsic elements of the hippocampal circuits are affected in PSP. It is possible, that certain neuronal populations are selectively affected in the cortex of PSP, with a vulnerability profile substantially different from that seen in AD (Hof *et al.*, 1992a). Although it is unclear how the NFT distribution is related to clinical PSP symptoms, the severe subcortical pathology characteristics of PSP is likely to play a major role in its progression, whereas in AD, dementia is probably more related to neocortical involvement. The repercussion of primary motor cortex involvement on motor symptoms in PSP patients is unknown, but the fact that NFTs occur principally in layers V and VI in PSP cases as opposed to layer III (which contains the majority of NFTs in the primary motor cortex in AD [Golaz *et al.*, 1992]) suggests selective damage to the output layers of the primary motor cortex in PSP. We have also reported that in Guamanian ALS/PDC, the primary motor cortex is severely affected (Hof *et al.*, 1994a,b), and a PSP-like syndrome has been documented clinically and neuropathologically in several Guamanian cases in the recent years (Lepore *et al.*, 1988). Thus PSP and ALS/PDC may share certain pathological features, including pathological changes in the motor cortex, preferential distribution of NFTs in the superficial layers of the neocortex, subcortical involvement, and scarcity of SPs.

Neuropathological examination of CBD cases shows many features comparable to PSP, such as gliosis, neuronal loss, achromatic swollen neurons, neuritic changes, NFTs, and glial tangles (Rebeiz *et al.*, 1967, 1968; Paulus and Selim, 1970; Ksiezak-Reding *et al.*, 1994; Buée-Scherrer *et al.*, 1996b; Bergeron *et al.*, 1997). In most cases high numbers of tau-immunoreactive threads, NFTs, and glial tangles are seen throughout the hippocampus and neocortex, but as in PSP, amyloid deposits are less frequent (Fig. 34). The highest numbers of NFTs are found in the hippocampal formation, whereas lower numbers were observed in primary motor, inferior temporal, and prefrontal cortices, and in subcortical structures (Fig. 34) (Buée-Scherrer *et al.*, 1996b). Occipital areas are devoid of NFTs. As in PSP, most CBD cases show severe changes in the primary motor cortex (Buée-Scherrer *et al.*, 1996b). Swollen achromatic neurons were positively stained by antibodies to hyper-

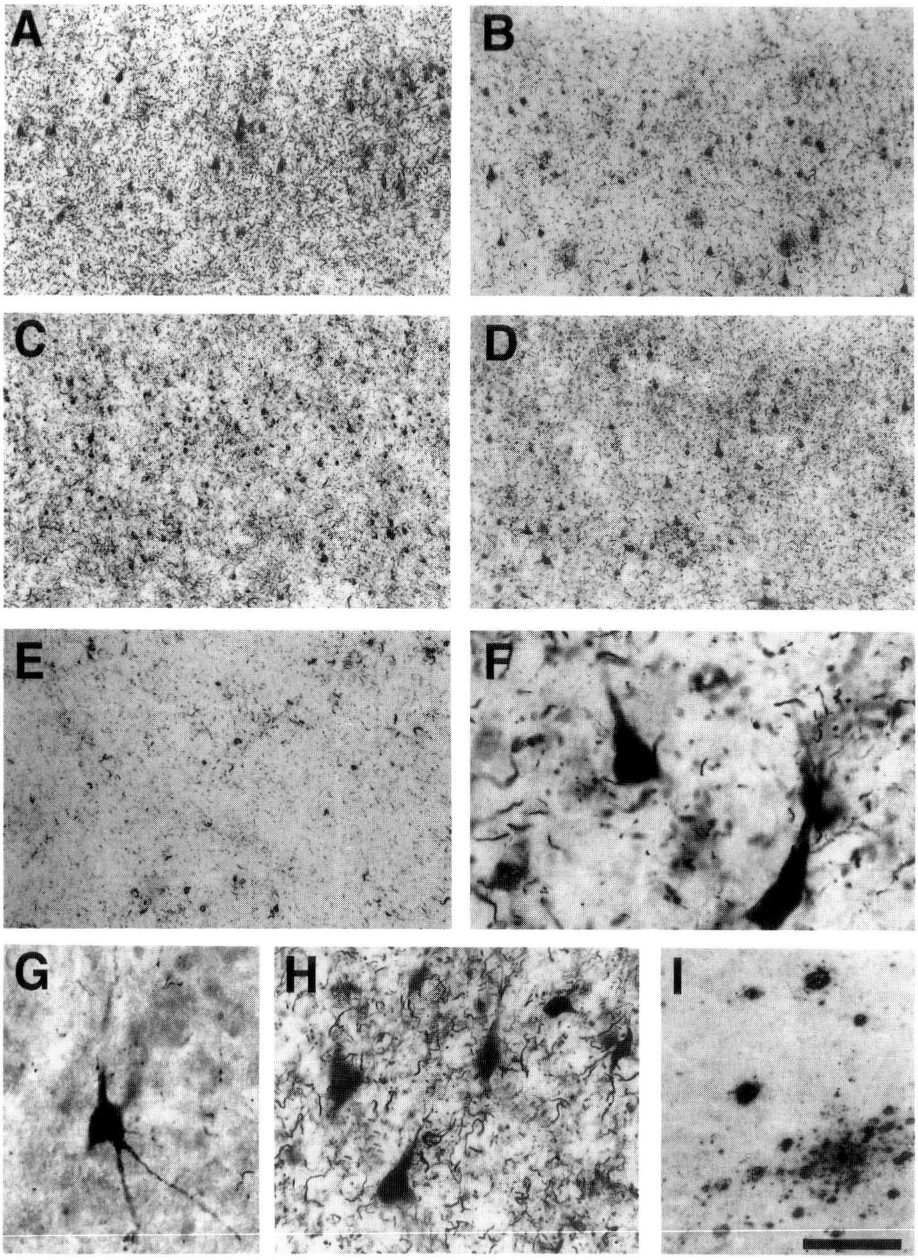


**Figure 34.** Neuropathologic profiles in CBD. (A–E) Distribution of hyperphosphorylated tau proteins in the CA1 field (A), area 20 (B), area 9 (C), primary motor cortex (D), and basal ganglia (E). There are high numbers of NFTs and degenerating neurites in these regions. Swollen achromatic neurons are labeled by anti-tau (F, H) and antiphosphorylated neurofilament protein antibodies (G) in neocortex. Rare amyloid deposits are seen in these cases (I). Adapted from Buée-Scherrer *et al.* (1996b). © Springer-Verlag. Bar (on I) = 100 μm (A–E, I); 50 μm (G, H); and 25 μm (F).

phosphorylated tau proteins and phosphorylated neurofilament protein, and are much more numerous than in PSP (Fig. 34F–H; Buée-Scherrer *et al.*, 1996b; Bergeron *et al.*, 1997). Very high densities of tau-immunoreactive degenerating neurites are present throughout the cerebral cortex. This neuritic pathology appears to be much more severe in CBD than in other neurodegenerative disorders. Glial tau-immunoreactive tangles involving both astrocytes and oligodendrocytes are found throughout the cerebral cortex (Dickson *et al.*, 1996; Takahashi *et al.*, 1996; Bergeron *et al.*, 1997).

The neuropathological findings in CBD indicate that the cytoskeletal alterations are different in this condition from those in AD and PSP (Ksiezak-Reding *et al.*, 1994; Uchihara *et al.*, 1994; Wakabayashi *et al.*, 1994b; Feany and Dickson, 1995; Buée-Scherrer *et al.*, 1996b; Dickson *et al.*, 1996; Takahashi *et al.*, 1996). Interestingly, homozygosity for the A0 allele of protein tau is highly prevalent in PSP but not in other related conditions (Conrad *et al.*, 1997), suggesting a possible direct involvement of tau in the development of PSP. Biochemical analyses have shown that both CBD and PSP brains contain a pathological tau 64, 69 doublet in all cortical regions analyzed thus far, except in the visual (Ksiezak-Reding *et al.*, 1994; Vermersch *et al.*, 1994; Feany *et al.*, 1995; Buée-Scherrer *et al.*, 1996b; Schmidt *et al.*, 1996). Feany and colleagues have recently shown that different tau isoforms in PSP (with exon 3) and in CBD (without exon 3) may produce the same tau doublet 64, 69 (Feany *et al.*, 1995). Thus, different degrees of phosphorylation may induce particular shifts in molecular weight and result in different tau profiles in these conditions. These specific tau profiles, including the tau triplet of AD, Parkinson's disease with dementia, and Guamanian ALS/PDC, the tau 64, 69 doublet of CBD and PSP, and the tau 55, 64 doublet of Pick's disease (Delacourte *et al.*, 1990, 1996; Vermersch *et al.*, 1993, 1994; Ksiezak-Reding *et al.*, 1994; Feany *et al.*, 1995; Buée-Scherrer *et al.*, 1995, 1996b; Mawal-Dewan *et al.*, 1996; and see below), may be related to the presence of predominant tau isoforms in the subset of neurons particularly vulnerable to a given disorder or to differential cell type-specific metabolism. In particular, the differences in tau expression in PSP and CBD as compared with AD may reflect the fact that the cognitive decline and neurologic symptoms observed in PSP and CBD evolve along different pathogenetic mechanisms than they do in AD.

### 4.4. Postencephalitic Parkinsonism

Neurofibrillary tangles are present in several subcortical structures in the brain of patients exposed to the pandemic of influenza in the years 1915–1930 and who later developed postencephalitic parkinsonism (PEP) (Duvoisin and Yahr, 1965; Martin, 1965; Ishii and Nakamura, 1981; Mitsuyama and Seyama, 1981; Geddes *et al.*, 1993; Cheyette and Cummings, 1995; Litvan *et al.*, 1996). The brain areas where NFTs have been described include basal ganglia, nucleus basalis of Meynert, hypothalamus, substantia nigra, locus coeruleus, raphe, periaqueductal gray matter, and reticular formation (Duvoisin and Yahr, 1965; Martin, 1965; Torvik and Meen, 1966; Ishii and Nakamura, 1981; Mitsuyama and Seyama, 1981; Geddes *et al.*, 1983; Litvan *et al.*, 1996). In addition, NFTs have been reported in many cases in the hippocampus and neocortex (Ishii and Nakamura, 1981; Hof *et al.*, 1992d; Buée-Scherrer *et al.*, 1997). Ultrastructurally, NFTs in PEP cases are similar to those found in AD (Ishii and Nakamura, 1981).

Dementia is not a common feature of PEP, and in most cases the clinical presentation is dominated by the extrapyramidal symptoms, which almost invariably develop after the acute phase of the disease (Duvoisin and Yahr, 1965; Rail *et al.*, 1981; Howard and Lees, 1987). We had the opportunity to study brains from patients born between 1892 and 1912 who had been affected by the pandemic of influenza in the years 1918–1924. The clinical presentation of these cases was consistent with the diagnosis of PEP and included episodes of oculogyric crises, somnolence, inversion of the day/night rhythm, mutism, a degree of psychosis, and development of a parkinsonian symptomatology including resting tremor and akinetorigid syndrome (Duvoisin and Yahr, 1965; Howard and Lees, 1987; Geddes *et al.*, 1993; Cheyette and Cummings, 1995; Litvan *et al.*, 1996). None of these patients was aphasic or apraxic and all were well oriented (Hof *et al.*, 1992d; Buée-Scherrer *et al.*, 1997). Neuropathologically, all of these cases had the typical features of PEP (Duvoisin and Yahr, 1965; Martin, 1965; Constantinidis and Tissot, 1967; Ishii and Nakamura, 1981; Mitsuyama and Seyama, 1981; Geddes *et al.*, 1993; Litvan *et al.*, 1996), with loss of pigmented neurons in the substantia nigra, presence of NFTs in the substantia nigra and several other brainstem nuclei, cell loss in the globus pallidus, severe gliosis, and absence of Lewy bodies.

Such PEP cases may show large differences in NFT densities among cortical regions (Hof *et al.*, 1992d). In the CA1 field of the hippocampus and subiculum, NFT counts of 3–40 per mm have been reported. The entorhinal cortex contains very high NFT densities in layers II and V (Fig. 35) (Hof *et al.*, 1992d). In the

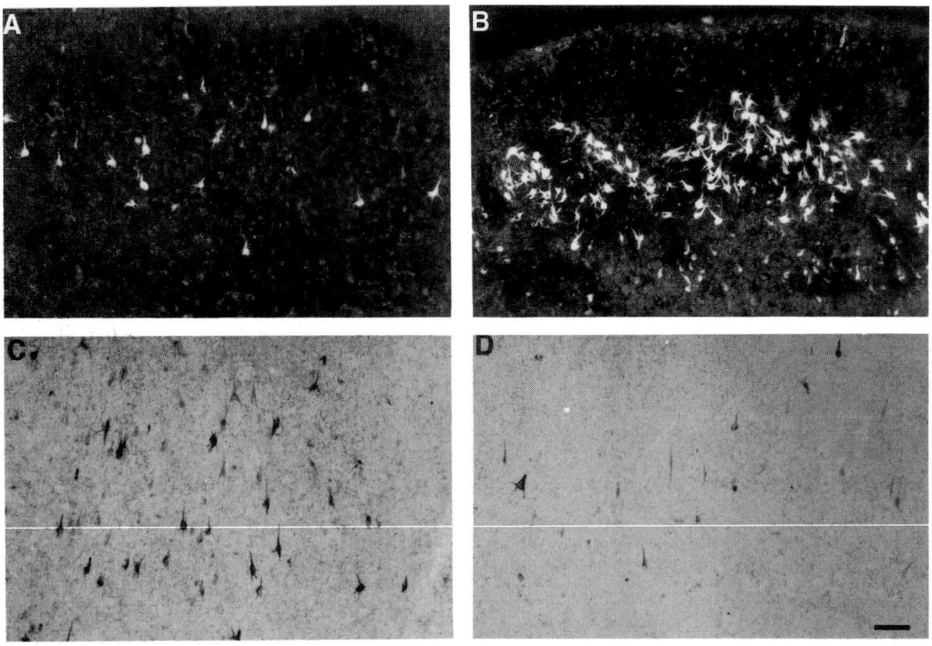


**Figure 35.** Distribution of NFT in PEP cerebral cortex. These lesions are consistently observed in the CA1 field of the hippocampus (A) and layer II of the entorhinal cortex (B). They are denser in the superficial layers (C) of the neocortex compared with the deep layers (D). Thioflavine S (A, B) or an antibody to hyperphosphorylated tau proteins (C, D). From Hof *et al.* (1992c). Bar = 40 μm.

temporal neocortex, NFTs are predominantly located in layers II and III, and are much less numerous in layers V and VI, with 36% more NFTs in the supragranular layers than in the infragranular layers (Fig. 35). Neurofibrillary tangles are rarely encountered in the prefrontal, parietal, and occipital cortices, indicating that some regions are preferentially affected by the degenerative process. All NFTs are stained by antibodies to hyperphosphorylated tau protein and ubiquitin, as in other neurodegenerative disorders (Hof *et al.*, 1992d; Wong *et al.*, 1996; Buée-Scherrer *et al.*, 1997). Moderate SP densities are present in the neocortex and hippocampus of PEP cases, where they predominate in the CA1 field and subiculum, and in supragranular layers of neocortex (Hof *et al.*, 1992d). SP formation has been also reported in association with NFTs (Wong *et al.*, 1996). In addition, numerous tau-immunoreactive neuritic elements are commonly observed in the neuropil and in the vicinity of SPs.

Biochemically, a triplet of hyperphosphorylated tau proteins similar to that observed in AD is observed in all PEP cases (tau 55, 64, 69), and the severity of NFT formation parallels the intensity of tau labeling in Western blots (Buée-Scherrer *et al.*, 1997). The tau doublets typical of PSP and CBD (see above) and Pick's disease are not observed in PEP cases. In PEP, the tau 55, 64, 69 triplet is found in cortical and subcortical brain regions, in contrast to AD, where this triplet is restricted to the hippocampal formation and association neocortex (Delacourte *et al.*, 1990; Buée-Scherrer *et al.*, 1995). The tau triplet is also present in brain areas usually spared in AD such as primary motor cortex and basal ganglia, and its regional distribution differs among PEP cases, suggesting some heterogeneity in the neurodegenerative process that may reflect the severity of the disease (Buée-Scherrer *et al.*, 1997).

These results demonstrate that NFTs and SPs can be common pathological features in the cerebral cortex of PEP cases, although their density is relatively low compared to AD cases. The distribution of lesions, in particular NFTs, in the cerebral cortex of PEP patients displays clear differences from that of AD. The laminar distribution of NFT in neocortical areas of PEP cases differs from that observed in AD; it is reminiscent of that observed in cases presenting with Guamanian ALS/PDC, PSP, and CBD (see above), as well as in posttraumatic encephalopathies (Hof *et al.*, 1991a, 1992c). Whereas dementia in AD is likely to be related to severe neocortical pathological changes, in PEP the severe involvement of subcortical structures is likely to play a major role in the progression of the disease, but it is not clear how the distribution of NFTs is related to the clinical presentation (Duvoisin and Yahr, 1965; Gibbs and Lees, 1987; Howard and Lees, 1987; Calne and Lees, 1988; Vieregge *et al.*, 1991). In addition, several studies have demonstrated that the progression of deterioration in PEP patients is not related to aging but rather to factors inherent in the disease itself (Gibbs and Lees, 1987; Calne and Lees, 1988; Vieregge *et al.*, 1991). Interestingly, tau-immunoreactive NFTs have been observed in the neocortex of an infant with subacute sclerosing panencephalitis, a measles-virus-related illness (Bancher *et al.*, 1996). The etiopathogenetic mechanisms of AD and PEP are likely to differ and involve distinct neuronal populations. Guamanian ALS/PDC exhibits the same electrophoretic tau profile as PEP, with a high heterogeneity in the regional distribution among cases (Buée-Scherrer *et al.*, 1995, 1997), suggesting comparable pathogenic mechanisms in both conditions. Both PEP and Guamanian ALS/PDC have been tentatively linked to external etiopathogenetic factors such as viruses and environmental toxins (Elizan and Casals, 1991; Hof *et al.*, 1992d; Geddes *et al.*, 1993; Lilienfeld *et al.*, 1994), and they are qualitatively characterized by a comparable NFT distribution pattern, suggesting that NFT formation in PEP may affect different cortical systems than in AD, and that severe cortical involvement in PEP is more common than previously thought.

### 4.5. Dementia Pugilistica and Posttraumatic Dementia

The occurrence of dementia and parkinsonian symptoms is a consistent sequelae of long-term cerebral injury in professional boxers, referred to as punch drunk syndrome or dementia pugilistica (Martland, 1928; Millspaugh, 1937; Roberts, 1969; Corsellis *et al.*, 1973). Its clinical features have been reviewed in detail by Roberts; they include poor performance on neuropsychological testing, inordinate aggressiveness, progressive social decline, and dementia, as well as cerebellar, pyramidal, and extrapyramidal symptoms (Roberts, 1969). The morphological changes observed in the brains of boxers are related to the acute and chronic effects of head trauma, which result in the formation of thrombi, aneurysms, subdural hematomas, intracerebral hemorrhages, edema, herniation, and massive ischemia (Casson *et al.*, 1984; Lampert and Hardman, 1984; Guterman and Smith, 1987; Jordan, 1987; Adams and Bruton, 1989). Chronic brain damage is characterized by the presence of generalized cerebral atrophy, enlargement of the ventricles, occurrence of septum cavum, loss of Purkinje cells and substantia nigra neurons, and formation of NFTs, SPs, and amyloid deposits (Brandenburg and Hallervorden, 1954; Grahmann and Ule, 1957; Roberts, 1969; Corsellis *et al.*, 1973; Lampert and Hardman, 1984; Roberts *et al.*, 1990a,b; Tokuda *et al.*, 1991; Hof *et al.*, 1992c; Geddes *et al.*, 1996; McKenzie *et al.*, 1996). An association between head trauma resulting in loss of consciousness and AD has also been demonstrated (French *et al.*, 1985; Mortimer *et al.*, 1985; Borenstein Graves *et al.*, 1990; Leininger *et al.*, 1990; Williams *et al.*, 1991), and cases of posttraumatic dementia following acute head injuries have been reported and are characterized by numerous SPs and amyloid deposits, and NFT formation in the cerebral cortex (Corsellis and Brierley, 1959; Rudelli *et al.*, 1982; Roberts *et al.*, 1990a,b, 1991, 1994; Clinton *et al.*, 1991; McKenzie *et al.*, 1994; Sherriff *et al.*, 1994; Graham *et al.*, 1995).

In dementia pugilistica, NFTs are preferentially distributed in layers II and upper III of neocortical areas, whereas layers V and VI are less affected (Hof *et al.*, 1992c) (Fig. 36). This distribution is similar to that observed in dementing illnesses other than AD and DS (see above). In the hippocampal formation, NFT distribution is comparable to that observed in AD cases, with numerous NFTs in the CA1 field, subiculum, and layers II and V of the entorhinal cortex (Fig. 36). A quantitative analysis has demonstrated that NFT density in the neocortex is generally higher

---

**Figure 36.** NFT distribution in the cerebral cortex of a dementia pugilistica case. This patient was a professional boxer with over 500 bouts in his career (Hof *et al.*, 1992d). Note the very high number of lesions in layer II of the entorhinal cortex (A) and in the superficial layers of the neocortex (area 20, B). Thioflavine S staining. Bar = 40 μm.

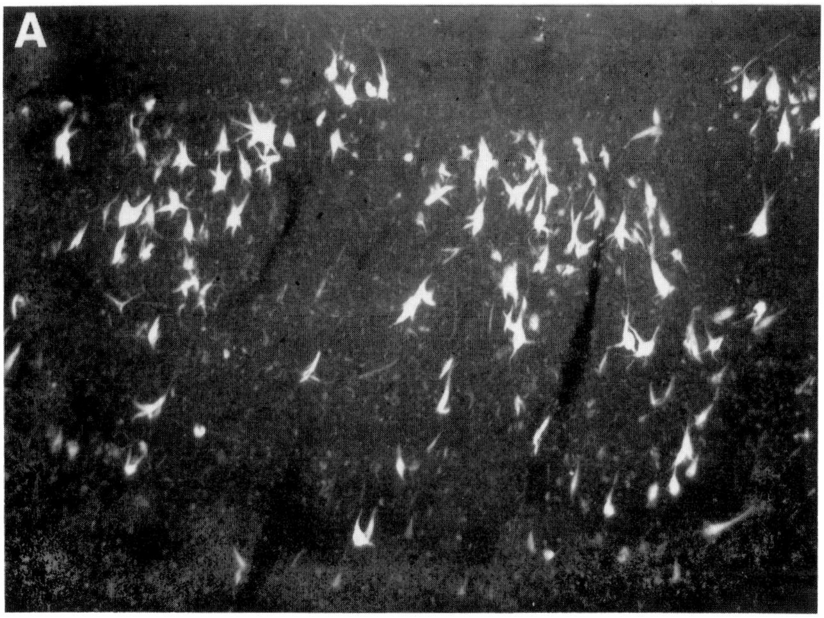
A

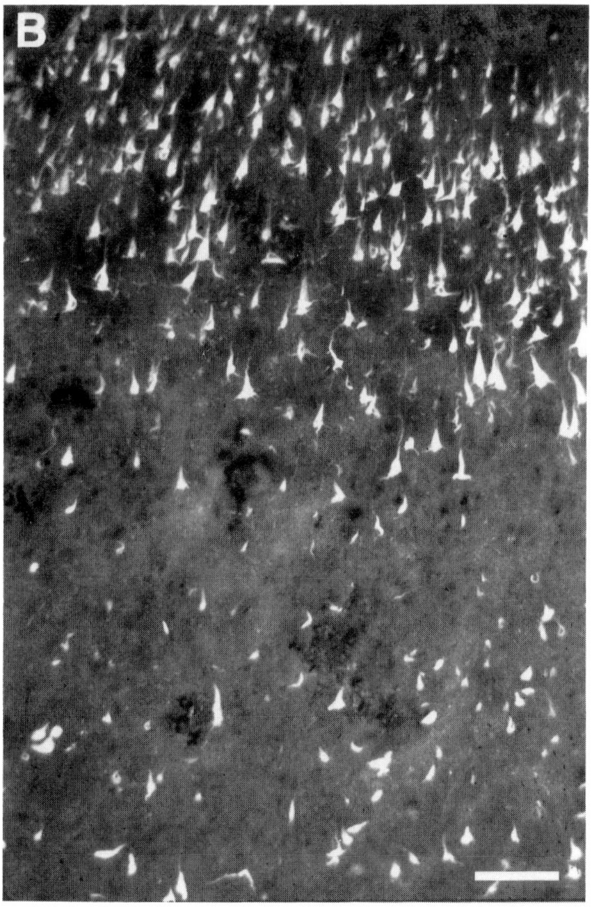
B

in dementia pugilistica than in AD cases, whereas the hippocampal formation shows comparable values (Hof *et al.,* 1992c). Temporal neocortical regions contain higher NFT densities than do parietal and frontal areas; cingulate, sensory, and motor cortices have much lower NFT densities. The highest neocortical NFT densities are observed in the perirhinal and inferior temporal cortex. NFTs in dementia pugilistica are strongly immunolabeled by an antibody to hyperphosphorylated tau proteins (Hof *et al.,* 1992c). It is thus possible that in dementia pugilistica, NFTs show biochemical modifications comparable to those observed in AD, consistent with neuropathological findings and analyses of NFT biochemistry in the dementia (Roberts, 1988; Dale *et al.,* 1991; Tokuda *et al.,* 1991; Hof *et al.,* 1992c; Geddes *et al.,* 1996; McKenzie *et al.,* 1996).

The presence of amyloid deposition and SP formation has, however, been reported in the brains of dementia pugilistica cases and in posttraumatic AD (Rudelli *et al.,* 1982; Roberts *et al.,* 1990a,b, 1994; Clinton *et al.,* 1991; Tokuda *et al.,* 1991; MacKenzie *et al.,* 1994, 1996; Sherriff *et al.,* 1994; Graham *et al.,* 1995). These studies also pointed to the fact that deposition of amyloid and amyloid precursor protein may occur rapidly following injury, in some cases within a few hours. Many of these cases present with extensive cortical amyloid deposition, but not every case displayed such changes (Roberts *et al.,* 1990a). In fact, staining for the amyloid Aβ protein shows highly variable SP numbers or amyloid deposits, and our study demonstrated that boxers' brains are nearly devoid of amyloid deposition (Hof *et al.,* 1992c), suggesting that it may not be a consistent feature in the pathogenesis of dementia pugilistica. In this respect too, dementia pugilistica is comparable to Guamanian ALS/PDC, PSP, and PEP in that fewer SPs and amyloid deposits overall are encountered in these conditions (see above). In one of our cases, a few NFTs also showed a strong immunoreactivity to amyloid (Hof *et al.,* 1992c). Similar observations of amyloid-containing NFTs have been reported in AD, Guamanian ALS/PDC, and dementia pugilistica cases (Allsop *et al.,* 1990; Ito *et al.,* 1991b; Yamaguchi *et al.,* 1991; Schwab *et al.,* 1995). Labeling of NFTs by antibodies to the amyloid Aβ protein occurs in extracellular NFTs and may represent secondary deposition of amyloid fibers (Allsop *et al.,* 1990; Yamaguchi *et al.,* 1991). Although this phenomenon appears not to be as common in dementia pugilistica as in AD and Guamanian cases, it is likely that comparable deposition mechanisms in NFTs exist in these conditions. The high density of NFTs without significant SP formation in PSP, PEP, and Guamanian cases, however, supports the notion that NFT formation and subsequent dementia can occur in the absence of amyloid deposition. In addition, variable amounts of neuropil threads have been observed in dementia pugilistica cases (Hof *et al.,* 1992c; McKenzie *et al.,* 1996), suggesting that neuritic pathology occurs to a limited degree in absence of severe amyloid deposition, and that the relative paucity of threads in dementia pugilistica constitutes another difference from the AD cases.

In the context of chronic head injury, we reported the case of a 24-year-old autistic woman suffering from a residual state of infantile autism and presenting with self-injury behavior that had included head-banging, eye-gouging, and self-biting since childhood (Hof *et al.,* 1991a). Her brain contained numerous neurofibrillary tangles in the perirhinal and entorhinal cortex, and NFTs were found in the amygdala and in the piriform and orbitofrontal cortex. In the neocortex, tangles were located in layers II and III, as in dementia pugilistica, and there were no neuritic plaques or amyloid deposits (Hof *et al.*, 1991a). NFTs in this case appear to be distributed in regions that are functionally related to the primate olfactory system (Tagaki, 1991), and we proposed that because the olfactory bulb and tracts were extremely atrophic in this case, it was possible that degeneration occurred through the olfactory pathway. It has been proposed that a route of entry for potentially toxic agents in the etiopathogenesis of lesions in dementia could be through a damaged olfactory system and would promote NFT formation in susceptible neurons (Perl and Good, 1987; Talamo *et al.*, 1989). This may be a particularly relevant hypothesis in the case of posttraumatic dementia and dementia pugilistica, in which repeated or acute head trauma may cause irreversible damage to the olfactory nerves. Thus, global NFT distribution and localization of neurons prone to NFT formation after head injury and in dementia pugilistica indicate that common mechanisms may take place in the pathogenesis of these conditions. The differential laminar NFT distribution in neocortex, however, suggests that the events leading to restricted neuronal vulnerability are not the same as those in AD, but that similar populations of neurons are involved in posttraumatic dementia and in dementing illnesses other than AD.

### 4.6. Pick's Disease

Pick's disease is a relatively rare form of dementia that usually affects individuals in their fifth and sixth decade (Pick, 1892; Onari and Spatz, 1926; Sjögren *et al.*, 1952; Lüers and Spatz, 1957; Constantinidis *et al.*, 1974; Jakob, 1979; Tissot *et al.*, 1985). Pick's disease is distinguishable from AD by several typical features, but it can be difficult to differentiate clinically from early-onset AD and other dementing illnesses (Sjögren *et al.*, 1952; Constantinidis *et al.*, 1974; Tissot *et al.*, 1985; Mendez *et al.*, 1993; Litvan *et al.*, 1997a). Early in the course of the disease, signs of frontal disinhibition, such as mood disturbances, irritability, and aggressiveness, are commonly associated with incoherent joviality, early loss of social awareness, aberrant behaviors, hyperorality, and bulimia, as well as verbal stereotypias, echolalia, and progressive language impoverishment resulting in mutism (Pick, 1892; Constantinidis *et al.*, 1974; Wechsler *et al.*, 1982; Tissot *et al.*, 1985; Mendez *et al.*, 1993).

Neuropathologically, Pick's disease is characterized by a prominent frontotemporal lobar atrophy, gliosis, severe neuronal loss, swollen neurons, and also neuronal inclusions, referred to as Pick bodies, in the cerebral cortex and some subcortical structures. The cortical atrophy is frequently asymmetric and predominates in the frontal and temporopolar regions, whereas the parietal, occipital, and posterior temporal cortices are rarely involved, although panencephalitic and parietal variants have been reported (Onari and Spatz, 1926; Neumann, 1949; Schenk and van Mansvelt, 1955; Lüers and Spatz, 1957; Jakob, 1960, 1961, 1979; Brion *et al.*, 1973; Constantinidis *et al.*, 1974; Cambier *et al.*, 1981; Shibayama *et al.*, 1983; Tissot *et al.*, 1985; Cochran *et al.*, 1994). In the affected neocortical regions, the atrophy is particularly visible in the supragranular layers (Schenk and van Mansvelt, 1955; Arnold *et al.*, 1994; Cochran *et al.*, 1994; Hof *et al.*, 1994c; Jellinger, 1994a). The hippocampal formation usually displays dramatic atrophy and very high densities of Pick bodies, particularly in the dentate gyrus. Pathologic changes are frequently encountered in subcortical regions, including the caudate nucleus, amygdala, nucleus basalis of Meynert, substantia nigra, locus coeruleus, and central gray, as well as in the olfactory bulb (Constantinidis *et al.*, 1974; Tissot *et al.*, 1985; Yoshimura *et al.*, 1989). Finally, recent studies throughout the layers of the atrophic cortex, as well as glial tangles, NFTs, and more rarely, SPs (Sinha *et al.*, 1993; Cochran *et al.*, 1994; Hof *et al.*, 1994c; Jellinger, 1994a; Bulée-Scherrer *et al.*, 1996b) (Fig. 37).

Pick bodies are the neuropathological hallmark of Pick's disease. They are spherical argentophilic inclusion bodies consisting of randomly distributed 10–20 mm-wide straight fibrils and constricted larger fibrils that show an approximately 160 nm periodicity between constrictions (Wisniewski *et al.*, 1972; Brion *et al.*, 1973; Munoz-Garcia and Ludwin, 1984). Pick bodies have been shown to share antigenic

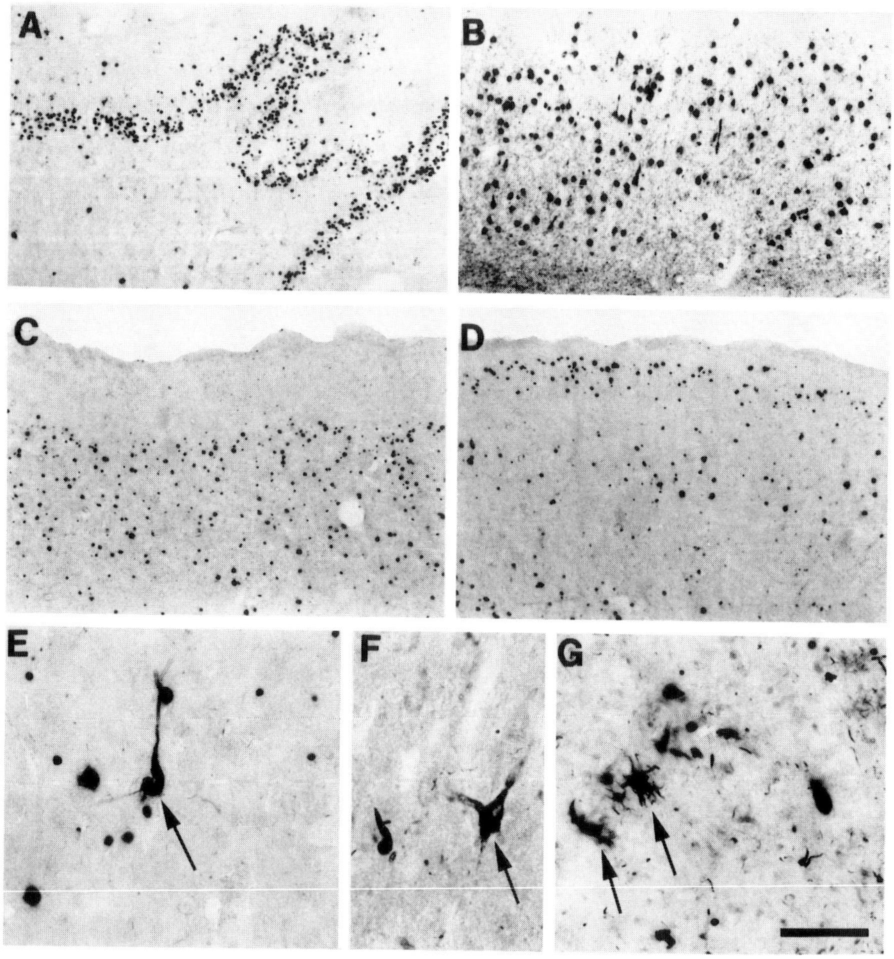


**Figure 37.** Distribution of neuropathologic changes in the hippocampus and neocortex of Pick's disease cases. There are very high densities of tau-immunoreactive Pick bodies in the granule cell layer of the dentate gyrus (A) and in the pyramidal cells of the CA1 field (B). Lower densities are found predominantly in layers II and VI of areas 20 and 9 (C, D). In some cases tau-immunoreactive NFTs (E), achromatic cells (F), and glial tangles (G) are also observed. Adapted from Buée-Scherrer *et al.* (1996b). © Springer-Verlag. Bar = 100 μm (A–D) and 30 μm (E–G).

determinants with the NFTs of AD, such as phosphorylated neurofilament epitopes identical to those found in NFTs, microtubule-associated protein tau, and ubiquitin, indicating that Pick bodies derive from altered cytoskeletal components (Ulrich *et al.*, 1979; Probst *et al.*, 1983; Munoz-Garcia and Ludwin, 1984; Perry *et al.*, 1987; Love *et al.*, 1988; Hof *et al.*, 1994c; Buée-Scherrer *et al.*, 1996b; Delacourte *et al.*, 1996). Interestingly, the biochemical profile of tau proteins in Pick disease differs from that seen in AD and other dementing illnesses. In Pick's disease cases devoid of NFTs, an unexpected tau profile characterized by a strong immunoreactivity of the 55 kDa and 64 kDa bands and very weak staining of the 69 kDa band has been reported in all areas analyzed except the basal ganglia. Such differences in the electrophoretic profile of tau proteins may be a useful tool to distinguish Pick's disease from CBD and PSP—disorders sometimes difficult to differentiate clinically and neuropathologically (Clark *et al.*, 1986; Lang *et al.*, 1994; Rinne *et al.*, 1994; Jendroska *et al.*, 1995; Buée-Scherrer *et al.*, 1996b; Delacourte *et al.*, 1996; Litvan *et al.*, 1997a).

Pick's disease cases generally contain variable densities of Pick bodies, and the hippocampal formation consistently shows higher Pick body numbers than does the neocortex (Table XIII, Fig. 37) (Hof *et al.*, 1994c). Dystrophic neurites are particularly obvious in the CA1 field of the hippocampus and in the subiculum. Many Pick's disease cases contain substantial numbers of NFTs and also frequently amyloid deposition (Hof *et al.*, 1994c) (Fig. 37). Pick's disease cases with high NFT counts tend to have fewer Pick bodies, yet no significant correlation can be observed at the regional and laminar levels between NFTs and Pick body densities, and no correlation is apparent between age and Pick body or NFT densities in these cases (Hansen *et al.*, 1988b; Hof *et al.*, 1994c). The granule cell layer of the dentate gyrus displays the highest Pick body counts of all cortical regions (Table XIII, Fig.

**Table XIII. Regional and Laminar Distribution of Pick Bodies and NFTs in Pick's Disease**[a]

| Area | Layers | Pick bodies | NFT |
|---|---|---|---|
| DG | | 490 | 1 (0) |
| CA1 | | 145 | 17 (23) |
| Subiculum | | 60 | 6 (18) |
| Entorhinal | II/III | 86 | 9 (55) |
| | V/VI | 79 | 7 (25) |
| 20 | II/III | 136 | 11 (22) |
| | V/VI | 55 | 5 (30) |
| 9 | II/III | 38 | 6 (18) |
| | V/VI | 26 | 4 (25) |
| 10 | II/III | 45 | 4 (12) |
| | V/VI | 28 | 2 (18) |
| 11p[b] | II/III | 89 | 10 (30) |
| | V/VI | 71 | 7 (45) |

[a]Data represent average counts of Pick bodies or NFTs per $mm^2$ in 16 cases of Pick's disease. Note the extremely high density in the granule cell layer of the dentate gyrus (DG) and the presence in these cases of some NFTs. Values in parentheses are averaged from 10 severe AD cases for comparison. Adapted from Hof *et al.* (1994c).
[b]Posterior orbitofrontal cortex.

37) (Hof *et al.*, 1994c). In comparison, the CA1 field of the hippocampus and layers II/III and V/VI of the entorhinal cortex exhibit lower Pick body densities, as does the subiculum, with a substantial degree of variability among cases (Table XIII, Fig. 37) (Hof *et al.*, 1994c). The inferior temporal cortex generally has more Pick bodies than the superior frontal cortex, and the posterior orbitofrontal cortex is usually severely affected (Table XIII).

In neocortical areas, the distribution of Pick bodies exhibits a strikingly different laminar pattern compared with that of NFTs in AD. Whereas in AD the NFTs are principally located in layers III and V/VI, the Pick bodies predominate in layers II and VI (Table XIII, Fig. 37) (Hof *et al.*, 1994c), suggesting that in Pick's disease other elements of the cortical circuitry are affected and that the development of the dementing syndrome involves other pathways than in AD. The massive involvement of the granule cells of the dentate gyrus indicates that the projection to the pyramidal neurons of the Ammon's horn is dramatically compromised in Pick's disease, whereas it appears mostly spared in AD. This alteration of the dentate gyrus in Pick's disease occurs in addition to changes in other components of the hippocampal circuits that appear to be affected to a comparable degree in both AD and Pick's disease. The distribution of Pick bodies in the neocortex suggests that different neuronal populations are involved as compared with AD. In particular, the small pyramidal neurons in layers II and superficial III are likely to be the most affected cell class in Pick's disease, whereas interneurons in these layers are likely to be resistant, as demonstrated by parvalbumin immunohistochemistry (Arai *et al.*, 1991). The small layer II pyramidal neurons, which are also severely affected in non-AD dementing illnesses such as PSP and Guamanian ALS/PDC (see above), are at the origin of different corticocortical projections than those of layers III and V, in that they form short-distance connections with adjacent zones of the neocortex, and the large number of Pick bodies in layer VI suggests that neurons projecting to subcortical regions, in particular the thalamus, may be severely affected in Pick's disease (Jones and Wise, 1977; Hof *et al.*, 1994c, 1995c). Moreover, the neurons in layers III and V suffer dramatic loss in Pick's disease (although they do not develop Pick bodies) and some of them appear prone to NFT formation (Arnold *et al.*, 1994; Hof *et al.*, 1994c). The development of Pick bodies in layers II and VI represents an additional alteration, which may account for only part of the typical symptomatology seen in these cases, since the severe loss in layers II and V indicates that the same corticocortical pathways affected in AD may be severely compromised in Pick's disease as well. Therefore, if the neurons prone to Pick body formation and those undergoing cell death through other mechanisms are accounted for together, Pick's disease can be considered to represent the disconnection of circuits directly affected by Pick bodies, including certain local corticocortical projections, certain corticothalamic projections, and mossy fibers from the dentate gyrus, in addition to long-association pathways similar to those affected in AD, at least in frontal and temporal cortices (Hof *et al.*, 1994c).

The circumscribed frontotemporal atrophy, the typical laminar distribution of Pick bodies, and the neuron loss in Pick's disease cases, as well as the differential lesion distribution between Pick's disease and AD, may offer pathological correlates to the neuropsychological and neuroradiological characteristics of each disorder (Wechsler *et al.*, 1982; Kamo *et al.*, 1987; Knopman *et al.*, 1989; Mendez *et al.*, 1993;

Litvan *et al.*, 1997a), and may also help to better define the molecular mechanisms responsible for the morphological changes underlying the neuropathologic diagnosis of a variety of dementing illnesses.



### 4.7. Frontal Lobe Dementia

This condition includes principally the dementing illnesses lacking distinctive histopathologic changes (Brun, 1987; Gustafson, 1987; Neary, 1990; Mann *et al.*, 1993; Neary *et al.*, 1993; Starkstein *et al.*, 1994; Hauw *et al.*, 1996; Hooten and Lyketsos, 1996), as well as some rare conditions that combine frontal lobe dementia (FLD) with motor neuron disease, Parkinson's disease, or progressive aphasias (Neary *et al.*, 1990; Snowden *et al.*, 1992; Barber *et al.*, 1995; Basun *et al.*, 1997; Foster *et al.*, 1997). The prevalence of FLD has been reported to vary, from 3% to 20% of demented patients (Neary *et al.*, 1990; Brun, 1993; Knopman, 1993; Barber *et al.*, 1995; Levy *et al.*, 1996), and positive familial history seems to be present in 50% of the cases, the pattern of inheritance suggesting the involvement of an autosomally dominant gene located on chromosome 17 (Mann and South, 1993; Basun *et al.*, 1997; Bird *et al.*, 1997; Foster *et al.*, 1997; Froelich *et al.*, 1997; Heutink *et al.*, 1997), although we know that heterogeneity in the genetic background exists in these conditions because one FLD family has been mapped to chromosome 3 (Brown *et al.*, 1995). Clinically, these diseases form part of the spectrum of dementia with lobar atrophy that includes Pick's disease, and they are characterized by insidious onset and slow progression. At early stages, there are marked personality changes with loss of social and personal awareness, disinhibition, hyperorality and stereotyped behaviors; speech disorders are also common and evolve to palilalia, echolalia, and mutism. At later stages, there is a general cognitive deterioration characterized by a massive decline of higher cortical functions (Knopman *et al.*, 1990; Brun *et al.*, 1994; Filley *et al.*, 1996; Gustafson *et al.*, 1993; Hooten and Lyketsos, 1996). Regional blood flow studies have shown frontal and temporal hypoperfusion with relative preservation of parietal and occipital blood flow (Brun, 1993; Risberg *et al.*, 1993; Starkstein *et al.*, 1994). In addition, patients who have FLD combined with motor neuron disease show a late development of fasciculations, muscular weakness, wasting, and signs of bulbar palsy, whereas those with progressive aphasia do not necessarily develop dementia. The general symptomatology is comparable to that seen in Pick's disease, and in most cases, it is not possible to clinically differentiate FLD from Pick's disease (Gustafson, 1987, 1993; Snowden *et al.*, 1992; Mendez *et al.*, 1993, 1996; Litvan *et al.*, 1997a).

The neuropathological changes in FLD consist of moderate-to-severe cortical atrophy restricted to the frontal and temporopolar regions and a widening of the ventricles (Brun, 1993; Mann and South, 1993; Neary *et al.*, 1993; Hauw *et al.*, 1996). The atrophy is symmetrical, but cases with asymmetrical gyral atrophy have also been reported, especially when progressive aphasia is present (Knopman *et al.*, 1990; Neary *et al.*, 1993; Hooten and Lyketsos, 1996). Usually, there is no gross atrophy of the hippocampus, amygdala, or striatum, and the thalamus, brainstem, and cerebellum display very mild macroscopic changes (Brun *et al.*, 1994; Hauw *et al.*, 1996). According to recent neuropathological criteria for FLD (Brun *et al.*,

1994), frontal and temporopolar cortices show a loss of pyramidal cells preferentially in layers II and III. Spongiform degeneration in layers II and III is often seen along with moderate-to-severe astrocytic gliosis in the gray and white matter (Brun, 1993; Mann and South, 1993; Brun *et al.*, 1994; Hauw *et al.*, 1996; Hooten and Lyketsos, 1996). These lesions spread frequently to the insula and anterior cingulate gyrus, whereas the posterior temporal, parietal, and occipital cortices are usually devoid of lesions. Although the involvement of the hippocampus, amygdala, and subcortical regions is mild in FLD, cases with prominent cell loss and astrogliosis in the striatum, thalamus, and limbic structures suggest the existence of neuropathological subgroups (Knopman *et al.*, 1990; Mann and South, 1993; Mann *et al.*, 1993; Neary *et al.*, 1993; Giannakopoulos *et al.*, 1995b; Cooper *et al.*, 1996; Hauw *et al.*, 1996; Hooten and Lyketsos, 1996). Furthermore, the distribution of changes in FLD cases combined with motor neuron disease or progressive aphasia differs significantly from that observed in more typical cases, as the hippocampus and subcortical structures are more severely affected (Neary, 1990; Neary *et al.*, 1993; Mann and South, 1993; Cooper *et al.*, 1996; Hooten and Lyketsos, 1996; Kinoshita *et al.*, 1997). Ultrastructural studies have also shown a substantial synapse loss (40%) in layers II and III of affected cortical regions (Liu and Brun, 1996).

Despite the heterogeneity of FLD, the key characteristics of its neuropathological diagnosis are the overall absence of distinctive cellular alterations such as Alzheimer-type changes (except mild ones compatible with age), Pick bodies, Lewy bodies, or swollen neurons. Recent studies have demonstrated, however, that in many cases of FLD, some cellular alterations are present. In particular, αB-crystallin-immunoreactive ballooned neurons have been found in the cerebral cortex of FLD, and ubiquitin-containing neurons are seen in the neocortex and hippocampus of cases with combined motor neuron disease (Cooper *et al.*, 1995; Bergmann *et al.*, 1996; Jackson and Lowe, 1996). Cases with motor neuron disease also show abnormal filamentous materials in cortical and subcortical structures that are different from Pick bodies and NFTs (Chang *et al.*, 1995). Ubiquitin-immunoreactive neurites, which probably represent degenerating dendrites, have also been demonstrated in the frontal and temporal cortices of FLD cases (Tolnay and Probst, 1995; Kinoshita *et al.*, 1997). Thus, it appears that αB-crystallin, ubiquitin, and tau may be useful markers to differentiate some of the disorders that comprise the frontotemporal degeneration spectrum, including Pick's disease (Tolnay and Probst, 1995; Cooper *et al.*, 1995, 1996; Bergmann *et al.*, 1996; Jackson and Lowe, 1996; Kinoshita *et al.*, 1997). It is not known with certainty which neuronal types are particularly vulnerable in FLD and if similar patterns of restricted cellular susceptibility exist in this disorder. In fact, immunohistochemical studies have shown that calbindin-immunoreactive neurons are affected in the frontal cortex of FLD cases with motor neuron disease, but parvalbumin-immunoreactive neurons are preserved (Ferrer *et al.*, 1993e). These results demonstrate a certain degree of cortical interneuron resistance in FLD, as is observed in AD (see above), yet at least one population of interneurons seems to be more vulnerable in FLD than in AD. Finally, pathologic tau proteins have been described in FLD cases with a biochemical profile quite similar to that seen in AD but restricted to the frontal cortex. Because there are no NFTs in FLD, it is possible that abnormal tau proteins are formed as certain subsets of pyramidal neurons degenerate as in AD, but in FLD they do not aggregate in paired helical filaments to form NFTs (Vermersch *et al.*, 1995). The tau protein gene is located near the FLD locus on chromosome 17, and mutations in tau gene have been detected in FLD families (Foster *et al.*, 1997; Dumanchin *et al.*, 1998).

The heterogeneity of FLD is reflected in our recent study of lesion distribution in 32 cases that permitted us to define four FLD subtypes with distinct neuropathological profiles (Fig. 38) (Giannakopoulos *et al.*, 1995b). In subtype A the macroscopic cerebral atrophy involves at least one neocortical area. Type A includes cases with only frontal or only temporopolar alterations. Moderate-to-severe loss of pyramidal neurons is present in layers II and III, and to a lesser extent, in V and VI. Spongiform changes are rare, there is a marked reactive astrocytosis in cortical layers as well as in white matter, and subcortical structures are spared (Giannakopoulos *et al.*, 1995b). In subtype B gross atrophy and neuron loss are present in the frontal and temporopolar cortex. The histopathological changes in the neocortex are similar to those observed in subtype A, but the striatum and substantia nigra displays mild-to-severe gliosis (Giannakopoulos *et al.*, 1995b). In subtype C moderate-to-severe frontal or temporopolar gross atrophy occurs, but widespread neuron loss and gliosis are present in the cortex as well as in the striatum and the substantia nigra, but not in the amygdala and thalamus (Giannakopoulos *et al.*, 1995b). Subtype D shows no neuron loss in the frontal and temporopolar cortex, mild neuron loss in the substantia nigra, and moderate gliosis in the striatum, substantia nigra, and thalamus (Giannakopoulos *et al.*, 1995b). One case had a combined motor neuron disease and six had a positive familial history of dementia. There were no significant differences in age at onset, duration of illness, familial history, or brain weight among the four subtypes. The clinical symptomatology was very homogeneous independent of lesion distribution, and no correlation was found between specific symptoms and involvement of cortical or subcortical areas except for the consistent presence of extrapyramidal signs in cases with neuron loss in the striatum (Giannakopoulos *et al.*, 1995b). Given the clinical similarities between FLD and Pick's disease, the question arises whether subtypes A, B, and C represent Pick variants. However, the absence of Pick bodies in FLD cases supports the possibility that FLD and Pick's disease are two etiopathogenetically distinct disorders (Neary *et al.*, 1993).

With regard to clinicopathological correlations, our study demonstrated a certain homogeneity in the clinical symptomatology in FLD patients in spite of differences in lesion distribution (Giannakopoulos *et al.*, 1995b). Although cases with severe striatal neuron loss have early extrapyramidal symptoms, the major clinical manifestations of FLD are not strictly correlated with the topography and nature of the histopathological changes (Mann *et al.*, 1993; Giannakopoulos *et al.*, 1995b; Hooten and Lyketsos, 1996). Although many investigators have tried to correlate distinctive patterns of cognitive dysfunction with the distribution of histopathological changes in FLD, clinical data show considerable overlap among the various neuropathological groups such that the clinical syndromes related to FLD cannot predict precisely those changes (Knopman *et al.*, 1990; Neary *et al.*, 1993; Giannakopoulos *et al.*, 1995b). It is thus possible that the etiopathogenetic mechanisms leading to FLD differ from those in other neurodegenerative diseases such as Pick's disease and AD in that there is neither a consistent involvement of the cytoskeleton in the development of the histopathological changes nor a depletion of identifiable neuronal populations, and it remains uncertain whether FLD represents one or several distinct disorders.

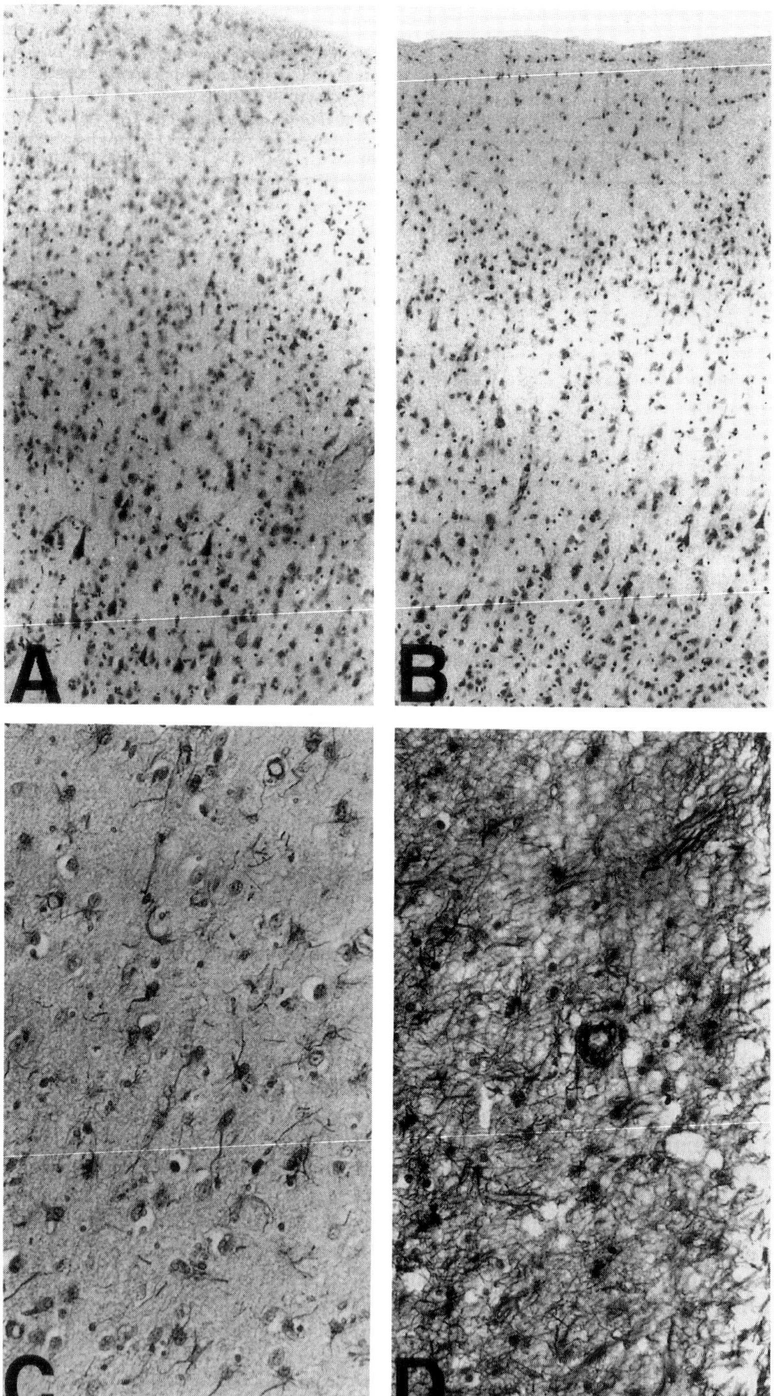

Lewy bodies are well known to occur in idiopathic Parkinson's disease, in which neurons in the substantia nigra are progressively lost, and Lewy bodies develop in some of them as well as in other brainstem nuclei (Forno, 1978). However, a substantial number of cases show Lewy bodies in hippocampus and neocortex, where these changes coexist with AD lesions (Gibb *et al.,* 1985; Bergeron and Pollanen, 1989; Braak *et al.,* 1996; Mega *et al.,* 1996). Depending on the series, it has been reported that 10–30% of elderly demented individuals have cortical and brainstem Lewy bodies, making it the most common pathologic subtype of AD, referred to as Lewy body variant of AD, or Lewy body dementia (LBD) (Perry *et al.,* 1990b; McKeith *et al.,* 1996; Mega *et al.,* 1996). Cortical Lewy bodies may also occur in the absence of marked AD pathology, a condition termed diffuse Lewy body disease (Dickson *et al.,* 1987a; Burkhardt *et al.,* 1988; Lippa *et al.,* 1994; Samuel *et al.,* 1997). Clinically, LDB is characterized by a progressive decline in cognitive function, yet prominent memory impairment may occur only in the later stage of the disease. Lewy body dementia is distinguished from AD by fluctuations in attention, cognition, and alertness; by parkinsonian motor features; and sometimes by the presence of recurrent visual hallucinations. Repeated falls, sensitivity to neuroleptics, and delusions and hallucinations are also supportive of LDB diagnosis; The classic features of parkinsonism are not characteristic of LBD patients. Patients with diffuse Lewy body disease are generally less severely demented than patients with LBD (Burkhardt *et al.,* 1988; Hansen *et al.,* 1990; Perry *et al.,* 1990b; McKeith *et al.,* 1996; Mega *et al.,* 1996; Ala *et al.,* 1997; Gangalingham *et al.,* 1997; Samuel *et al.,* 1997).

Lewy bodies are round, eosinophilic structures with a lighter central core and sometimes a halo, but in the cortex they may go unnoticed, as they are less well stained by hematoxylin-eosin and frequently lack the halo, unlike brainstem Lewy bodies (Fig. 39). However, they are labeled by antibodies to ubiquitin and nonphosphorylated neurofilament protein, but do not stain with antibodies to tau protein, which differentiate them from Pick bodies (Hansen *et al.,* 1990; Trojanowski *et al,.* 1993b). Lewy bodies are composed of 7–20 nm radiating filaments that resemble neurofilament protein and form the halo, whereas in the core of the Lewy body, these filament are present together with amorphous materials (Galloway *et al.,* 1992; Trojanowski *et al.,* 1993b). This indicates that Lewy bodies share some antigenic determinants with the NFTs of AD, and although they do not contain hyperphosphorylated tau proteins and paired helical filaments, they are composed of insoluble elements of the cytoskeleton (Galloway *et al.,* 1988, 1992; Schmidt *et al.,* 1991; Trojanowski *et al.,* 1993b). Furthermore, nonphosphorylated neurofilament protein epitopes similar to those that are dynamically incorporated into NFTs in AD (Vickers *et al.,* 1992, 1994b) are also present in cortical Lewy bodies (Wakabayashi *et al.,* 1995), demonstrating that in the cerebral cortex, Lewy bodies

←

**Figure 38.** Microscopic alterations in the neocortex of FLD cases. All panels are from subtype A cases (Giannakopoulos *et al.,* 1995b). (A, B) Frontal cortex showing a moderate neuronal loss and mild reactive gliosis (Nissl stain). (C, D) Gliosis and spongiform changes in the same region (Holzer stain). Adapted from Giannakopoulos *et al.* (1995b). Bar = 200 μm (A, B) and 100 μm (C, D).

occur in a subpopulation of pyramidal neurons that may be the same as that known to be dramatically affected in AD (Hof *et al.*, 1990a; Vickers *et al.*, 1992, 1994b). The various subunits of neurofilament protein are located in distinct regions of the Lewy body. For example, the tail domains of the middle and heavy subunits are preferentially distributed in the halo and the outer region of the Lewy body core, whereas the head domain of the middle and tail domain of the light subunits are present throughout the Lewy body (Hill *et al.*, 1991). Finally, synaptic proteins appear significantly decreased in the hippocampus and entorhinal cortex in LBD, yet synaptic alterations appear less severe than in AD (Wakabayashi *et al.*, 1994a; Samuel *et al.*, 1997).

Neuropathologically, LBD cases show comparable densities of cortical SPs, as in AD, but they have fewer NFTs. Diffuse Lewy body disease cases contain far fewer SPs and virtually no NFTs. In these conditions the degree of synaptic involvement and the number of AD-type lesions do not correlate to the severity of dementia, except for neuritic plaques and for NFTs in the entorhinal cortex, which clearly differentiates LBD from AD. There is, however, correlation between Lewy body counts and the severity of dementia (Samuel *et al.*, 1996, 1997). Topographically, the cortical atrophy in LBD is grossly comparable to that seen in AD (Mann and Snowden, 1995), suggesting that the presence of Lewy bodies in the cerebral cortex contributes to the global patterns of atrophy in a manner comparable to AD-type lesions. In the cerebral cortex, Lewy bodies are not homogeneously distributed but predominate in regions associated with the limbic system. They are present in particularly high densities in the hippocampal formation and in the anterior cingu-

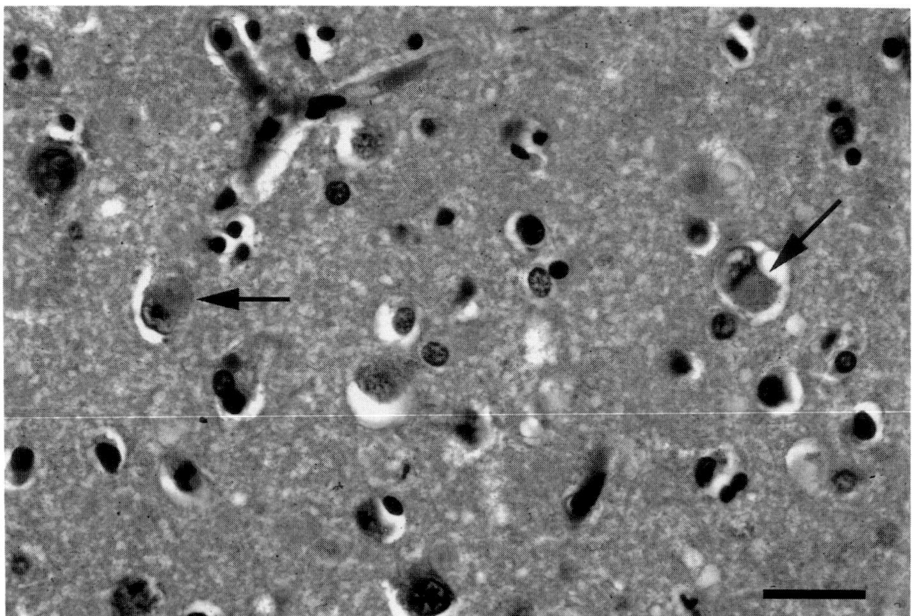

**Figure 39.** Examples of Lewy bodies in a LBD case. Two Lewy bodies are visible (arrows) in layer V of the anterior cingulate cortex. This region usually shows the highest Lewy body densities in the neocortex in LDB. Hematoxylin-eosin staining. Microphotograph courtesy of Dr. D. P. Perl. Bar = 15 μm.

late gyrus, whereas the posterior cingulate cortex and other regions of the prefrontal cortex contain lower densities of Lewy bodies (Perry *et al.,* 1990a). In the neocortex, they predominate in the frontal and temporal lobes and are preferentially located in layers IV–VI (Perry *et al.,* 1990a). In the frontal and temporal cortices, in layer II of the entorhinal cortex, and in the hippocampus, LBD cases have a neuron density comparable to that of nondemented individuals, and the percentage of remaining neurons with NFTs overlaps with the percentage obtained in AD (Lippa *et al.,* 1994, 1997), indicating that neuronal loss in the perforant path and in some corticocortical connections in LBD, is likely to be more variable and less severe than in AD. Cases with diffuse Lewy body disease may show higher densities of Lewy bodies than do nondiffuse LBD cases, although data vary among reports (Perry *et al.,* 1990a; Samuel *et al.,* 1997). The laminar and regional distribution of Lewy bodies indicates a partial overlap with the patterns observed in AD, although the devastation of cortical neurons by NFTs and neuronal losses are far greater in AD. The consistent involvement of limbic regions of the neocortex and of deep cortical layers in LBD suggests that the particular pyramidal neuron subsets that are affected differ from those vulnerable in AD. It is therefore likely that the differences in clinical symptomatology between LBD (and diffuse Lewy body disease) and AD are due in part to the involvement of other elements of the cortical circuits in LBD.

### 4.9. Other Neurodegenerative Disorders Involving Cytoskeletal Alterations

Several other neurological and dementing conditions exist that also involve some degree of cortical cytoskeletal pathology. For example, Huntington's disease involves principally the basal ganglia, but a subpopulation of neocortical projection neurons that is enriched in neurofilament protein is also affected in the course of the illness (Cudkowicz and Kowall, 1990). Amyotrophic lateral sclerosis and its variants have been mentioned above in the context of Guamanian ALS/PDC and show prominent inclusion bodies and axonal spheroids in affected upper and lower motor neurons composed of altered cytoskeletal elements, and certain cases of ALS combined with FLD or with Pick's disease demonstrate prominent ubiquitin-positive or filamentous neuronal changes (Leigh, 1994; Tolnay and Probst, 1995; Cooper *et al.,* 1995, 1996; Bergmann *et al.,* 1996; Jackson and Lowe, 1996; Kinoshita *et al.,* 1997). In addition, frontotemporal dementia associated with parkinsonism has been reported in several families with autonomous dominant inheritance of a gene associated with dementia on chromosomes 3 and 17 (Brown *et al.,* 1995; Bird *et al.,* 1997; Basun *et al.,* 1997; Foster *et al.,* 1997; Froelich *et al.,* 1997; Heutink *et al.,* 1997; Dumanchin *et al.,* 1998). In some of these cases, a pathologic profile of tau proteins has been reported with 64, 68, and 72 kDa isoforms, thus differing slightly from the tau triplet observed in AD, and numerous tangle-like lesions have been observed in neurons and glial cells in the cerebral cortex (Spillantini *et al.,* 1997).

Some conditions are relatively rare and affect principally subcortical structures with inconsistent participation of the cerebral cortex (see Jellinger, 1994b; Uitti, 1994; Takeda *et al.,* 1997, for reviews). Lesions containing paired helical filaments have been described in the hippocampus and neocortex in a variety of viral, infectious, and neurotoxic conditions, as well as in tuberous sclerosis, lipofuscinosis, and Hallervorden-Spatz disease (Wisniewski *et al.*, 1979). In Hallervorden-Spatz disease, hippocampal and neocortical NFTs can be abundant, and SPs or amyloid deposits are absent (Eidelberg *et al.*, 1987). Patients with myotonic dystrophy also show abundant cortical NFTs in the cerebral cortex, and a distinct tau protein profile has been reported in this condition, with a predominance of the tau 55 kDa isoform (Vermersch *et al.*, 1996). Widespread cortical Alzheimer-type NFTs have also been observed in the absence of SPs in non-Alzheimer, non-Pick dementia with Fahr-type calcifications (Shibayama *et al.*, 1992; Kosaka, 1994), as well as in association with verrucous dysplasias (Morán *et al.*, 1995). In these diseases it is, however, unclear whether the development of neurofibrillary changes in the affected neurons impacts on cortical function in a similar manner, as in better-characterized neurological and dementing illnesses.

Finally, it is worth mentioning the argyrophilic grains of Braak, lesions reported in isolation in the cerebral cortex of several cases as well as in association with various disorders including classic Parkinson's disease, CBD, PSP, LBD, multiple system atrophy, FLD with motor neuron disease, Huntington's disease, Pick's disease, and AD (Masliah *et al.*, 1991b; Martinez-Lage and Munoz, 1997). These lesions were first described by Braak and Braak in late-onset dementia cases showing abundant argyrophilic, small, spindle-shaped or coiled bodies in the cortical neuropil (Braak and Braak, 1987, 1989). The lesions have an unknown significance, but occur in about 11% of cases older than 65 years and have been recently shown to share antigenic determinants with the NFTs of AD, such as hyperphosphorylated tau proteins (Martinez-Lage and Munoz, 1997; Tolnay *et al.*, 1997). It is unclear whether argyrophilic grains define a separate clinical entity.

## 5. Conclusions and Future Prospects

The distribution of lesions in the cerebral cortex in a variety of neurodegenerative diseases indicates that selective neuronal vulnerability is a hallmark of all of these disorders. The selectivity is visible at the level of affected brain regions, subregions, or cortical layers, as well as at the level of specific neuron populations (Hof and Morrison, 1990, 1994, 1996; Hof *et al.*, 1990a, 1995a; De Lacoste and White, 1993; Morrison, 1993; Hof, 1997; Morrison and Hof, 1997). The notion of differential neuronal vulnerability is therefore linked to a detailed and cohesive definition of neuron typology. The definition of cortical neuron types is traditionally based on morphologic criteria such as perikaryal size, dendritic and axonal extent and arborization patterns, and to a lesser degree, synaptic target (Morrison and Hof, 1992). However, the characteristics of a cortical neuron and its categorization in a cell class depends on additional interactive criteria, such as regional and laminar location, connectivity, and neurochemical phenotype (Morrison, 1993). This approach is useful to delineate the cellular organization of neocortex as well as the cellular pathologic changes, because it takes into consideration the complex relationships between location, morphology, neurochemical phenotype, and connectivity.

In a very general sense, the most vulnerable cortical neurons group in AD is the pyramidal cells, specifically, the pyramidal cells providing long corticocortical and hippocampal projections and containing high levels of neurofilament protein. Furthermore, it appears that these neurotransmitter systems are likely to both utilize glutamate and receive extensive glutamatergic inputs as well as multiple afferents using other amino acid, monoamine, and peptide neurotransmitters. Not all corticocortical systems are equally vulnerable, however; the short projections from primary to adjacent secondary sensory areas and the local short projections between immediately adjacent cortical zones in association cortices are generally more resistant to degeneration (Hof and Morrison, 1990, 1994, 1996; Hof *et al.*, 1990a, 1995a; Morrison, 1993; Hof, 1997; Morrison and Hof, 1997). The primary visual cortex furnishes a typical example of such selective vulnerability, in that layer III neurons known to project to the secondary visual area are not affected in AD, whereas the Meynert cells and neurons in layer IVB that form long corticocortical projections from the primary visual cortex to specific visual association areas in the parieto-occipital region are clearly vulnerable (Hof and Morrison, 1990). The layer IVB neurons and the Meynert cells all show robust immunoreactivity to neurofilament protein, whereas layer III cells are less intensely stained (Hof and Morrison, 1990a; Hof and Morrison, 1995; Hof *et al.*, 1996b). Thus it is plausible that specific neurochemical and morphologic characteristics of certain pyramidal neurons may predispose them to degeneration and to NFT formation in AD (Table XIV). Conversely, interneurons containing parvalbumin, calbindin, and calretinin in association with a variety of neuropeptides are mostly resistant to degeneration in neocortical regions of AD cases (Table XV).

The relative importance of the pathologic changes observed in AD must be placed within the context of organized connections that underlie neocortical function. Integrated processing in a given sensory modality such as vision involves the

**Table XIV. Neurochemical and Morphologic Profile of Cortical Neurons Vulnerable to the Dementing Process in AD**

| Neuron types/connectivity | Neurochemical markers[a] | Vulnerability[b] |
|---|---|---|
| *Pyramidal neurons* | | |
| Long corticocortical polymodal projections | ?Glutamate receptor subunits[c] | Mostly vulnerable |
| Perforant path | High levels of nonphosphorylated neurofilament protein and microtubule-associated proteins | Vulnerable |
| Hippocampo-neocortical projections | Calbindin | Vulnerable |
| Large pyramidal neurons | Cu/Zn superoxide dismutase | Vulnerable |
| | Type II calcium/calmodulin-dependent protein kinase | Vulnerable |
| | Other protein kinases | Mostly vulnerable |
| | Apolipoprotein E | ?Mostly vulnerable |
| *Nonpyramidal neurons* | | |
| Multipolar interneurons in layers V/VI | Calbindin | Vulnerable |
| | Somatostatin | Vulnerable |
| | Substance P | Mostly vulnerable |
| Subset of bipolar neurons | Calretinin (no GluR2) | ?Mostly vulnerable |

[a]Phenotypes.
[b]Degree of vulnerability in AD.
[c]Question marks indicate insufficient data.

**Table XV. Neurochemical and Morphologic Profile of Cortical Neurons Resistant to the Dementing Process in AD[a]**

| Neuron types/connectivity | Neurochemical markers[b] | Resistance[c] |
|---|---|---|
| *Pyramidal neurons* | *Glutamate* | |
| Short corticocortical projections | ?Glutamate receptor subunits[d] | Mostly resistant |
| Small pyramidal neurons | Low levels of nonphosphorylated neurofilament protein and microtubule-associated proteins | Mostly resistant |
| Cortico-subcortical projections | ?Presenilin-1 | Mostly resistant |
| Unimodal corticocortical projections | ?Apolipoprotein J | Mostly resistant or protected |
| CA2-3 pyramidal cells | | |
| *Spiny stellate neurons* | *Glutamate* | |
| Intracortical relay cell | ?Glutamate receptor subunits | Resistant |
| *Granule cells* | *Glutamate* | |
| Dentate gyrus Mossy fibers to CA3 | Low levels of nonphosphorylated neurofilament protein | Resistant until late stages of AD |
| *Interneurons* | *GABA* | |
| Double bouquet cells | Calbindin/Somatostatin/GluR2 | Resistant |
| Basket cells | Parvalbumin/GluR2 | Resistant |
| Chandelier cells | Parvalbumin/Corticotropin-releasing factor/GluR2 | Resistant |
| Bipolar/bitufted cells | Calretinin/GluR2 | Resistant |
| Multipolar neurons | Somatostatin/NADPH/Neuropeptide Y | Mostly resistant |

[a]These tentative profiles are valid only for AD. They do not reflect the situation in other dementing illnesses.
[b]Phenotypes.
[c]Degree of resistance in AD.
[d]Question marks indicate insufficient data.

simultaneous activity of many separable visual areas that have extensive, highly ordered interconnections establishing a distributed system that subserves the integration of visual information (Felleman and Van Essen, 1991; Ungerleider, 1995). Furthermore, intellectual functions, such as cognition and language, that transcend modal specificity are presumably even more dependent upon the complex communication between neocortical regions that is provided by the corticocortical circuits. Although additional degenerative processes occur in AD that may also contribute to its symptomatology, we propose that these particular corticocortical projections degenerate, leading to a global neocortical isolation syndrome that manifests as dementia. The global loss of long corticocortical projections is likely to be the most devastating component of AD and the most directly related to dementia; the neurons providing these projections are highly specialized, and they share identifiable morphologic and neurochemical characteristics. In addition, the patterns of NFT and SP distribution in AD can also be interpreted as the reflection of a widespread breakdown of corticocortical projections (Mutrux, 1947; Pearson *et al.*, 1985; Rogers and Morrison, 1985; Duyckaerts *et al.*, 1986; Hof and Morrison, 1990; Hof *et al.*, 1990a; De Lacoste and White, 1993; Morrison, 1993; Morrison and Hof, 1997) such that both the feedforward corticocortical connections emanating from layer III and the feedback projections coming from layer V are likely to be affected.

These patterns of NFT and SP distribution are particularly evident in atypical AD cases that show a displacement of pathologic lesions on specific systems (Hof *et al.*, 1989, 1990b, 1993a, 1997c).

Although these considerations may be valid in the context of AD, comparative neuropathologic analysis of other neurodegenerative disorders suggests that many neuronal types are differentially affected by the degenerative processes specific to each disorder. Also, the distribution of these lesions in the cerebral cortex is highly heterogeneous, although different illnesses may express similar neuropathologic markers. For instance, AD and DS share certain patterns of NFT and SP distribution, whereas most dementing disorders show only rarely SPs and have very different laminar patterns of NFT distribution in the neocortex in comparison to AD. In dementia pugilistica, PSP, CBD, PEP, and Guamanian ALS/PDC the lesion distribution suggests that although all these conditions also involve corticocortical circuits to variable degrees, feedforward projections are more susceptible to degeneration than feedback projections, because NFTs predominate in layers II and III (Hof *et al.*, 1991c, 1992a,c,d, 1994a,b). For instance, the very high NFT density in the superficial third of layer III and in layer II of the temporal association neocortex suggests that in these diseases, NFT formation recruits other classes of pyramidal neurons that generally are less affected in AD cases and that do not necessarily have the same neurochemical phenotype as those vulnerable in AD (Tables XIV, XV). Interestingly, small pyramidal neurons in layers II, superficial III, and VI have been shown to be at the origin of short-distance corticocortical connections (Jones and Wise, 1977; Felleman and Van Essen, 1991; Hof *et al.*, 1995c, 1996b). It is also likely that comparable projection neurons are lost in Pick's disease cases where Pick bodies predominate in layers II and VI. In AD cases the more-severe involvement of layer V and of the lower two-thirds of layer III indicate that additional neuron populations might be differentially compromised. In addition, the presence of NFTs without significant SP formation in PSP, PEP, CBD, and Guamanian ALS/PDC agrees with the fact that NFT formation and the resultant dementia can occur in the absence of amyloid deposition, but does not negate that amyloid deposition may play an important role in the development of neurodegenerative changes in AD.

Thus, dementia can be considered as a constellation of several overlapping, yet clinically and neuropathologically distinguishable, nosological entities (Fig. 40). The differential expression of a dementing syndrome could be reflected by the neuropathologic and biochemical characteristics of each of these dementing conditions. One consistent feature among a large number of neurodegenerative diseases is the involvement of cytoskeletal elements such as neurofilament and tau proteins. Though tau proteins are a consistent marker of these conditions, each disorder emerges with an identifiable and specific profile of hyperphosphorylated tau protein isoforms (Vermersch *et al.*, 1997). Neurofilament proteins have been shown to participate in the formation of NFTs, Pick bodies, and Lewy bodies; and several animal models of neurodegeneration have demonstrated their involvement in specific lesions. For instance, transgenic mouse models have demonstrated their involvement in specific lesions. For instance, transgenic mouse models that overexpress the heavy or light subunits of neurofilament protein display a motor neuron disease similar to ALS, as do transgenic mice containing one of several mutations of superoxide dismutase gene that occur in familial ALS (Côté *et al.*, 1993; Xu *et al.*,

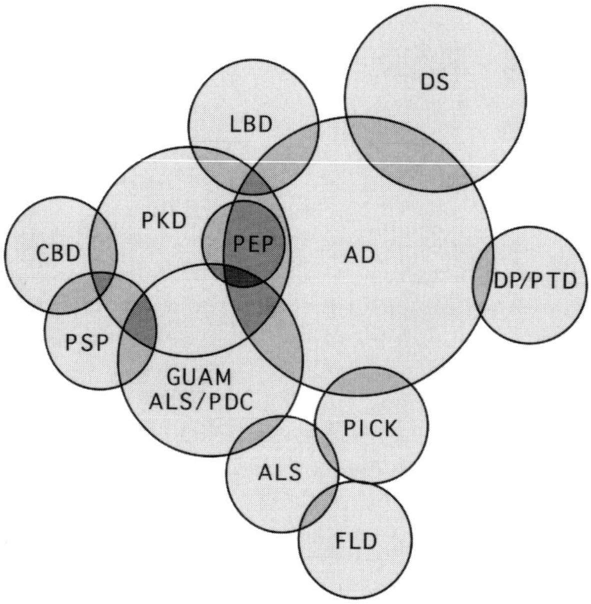


**Figure 40.** Neuropathological overlap of the major neurodegenerative disorders based on the distribution of morphologic and neurochemical alterations described in this review. Alzheimer's disease (AD) has many morphologic and biochemical features in common with diseases such as Down syndrome (DS), Lewy body dementia (LBD), and Guamanian ALS/PDC. Pick's disease has also some neuropathologic overlap with AD, as do dementia pugilistica and posttraumatic dementia (DP/PTD). Parkinson's disease (PKD), although not discussed in this review, may coexist with AD and has several features in common with Guamanian ALS/PDC, LBD, progressive supranuclear palsy (PSP), and corticobasal degeneration (CBD). Note that postencephalitic parkinsonism (PEP) is embedded within AD, PKD, and Guamanian ALS/PDC, and that CBD and PSP are related to each other but appear more distant from AD than from PKD and Guamanian ALS/PDC. Also, note the small overlap between (Caucasian) ALS, Pick's disease, and frontal lobe dementia (FLD), but the lack of overlap between Pick's disease and FLD cases with overall lack of apparent distinctive histopathologic changes. Finally, Huntington's disease (not shown), also not discussed in detail in this review, may have some neurochemical characteristics in common with AD, because neurofilament protein-containing neocortical neurons appear vulnerable in late-stage Huntington's disease. Data inferred from regional and laminar NFT distribution patterns, neuron type-specific immunohistochemistry patterns, and disease-specific tau protein isoform representation were used to design this figure.

1993; Gurney *et al.,* 1994; Lee *et al.,* 1994; Ripps *et al.,* 1995; Morrison *et al.,* 1996; Tu *et al.,* 1996; Dal Canto and Gurney, 1997). Even in the superoxide dismutase model of ALS, it is the neurons with particularly high expression of neurofilament protein that are most severely affected (Morrison *et al.,* 1996). In addition, transgenic mice that express the midsized subunit of human neurofilament protein exhibit an age-related, cell-specific variety of pathologic lesions, reminiscent of spheroids, NFTs, and Pick bodies, in the cerebral cortex as well as in some retinal ganglion neurons, further supporting a key role for neurofilament protein in neurodegeneration (Vickers *et al.,* 1994a, 1995b). Similarly, a subset of ganglion neurons in the retina of macaque monkeys with experimentally-induced glaucoma has been shown to be selectively affected, whereas other ganglion neurons as well as neuronal types containing calcium-binding proteins survive (Vickers *et al.,* 1995c). These experimental data further suggest that neurofilament protein synthesis, degradation, assembly, and transport in certain subsets of pyramidal neurons are involved in neurodegeneration across a broad spectrum of diseases (Table XIV; Fig. 40) and are critically related to disease-specific patterns of selective vulnerability. It is therefore possible that a variety of etiopathogenic factors, some of them not yet determined, can differentially affect certain elements of the cerebral circuitry, resulting in particular patterns of lesion distribution as well as specific cellular alterations. These changes may then lead to the clinical deficits typical of each disorder.

In regard to the regional specificity of *neocortical* changes in many neurodegenerative diseases, it is remarkable that those areas with the most dramatic atrophy and the highest lesion densities are the prefrontal cortex, the temporopolar cortex, and the association regions in the temporoparietal junction. This is particularly obvious in diseases such as AD, Guamanian ALS/PDC, FLD, and Pick's disease, but other dementing illnesses also show a prominent involvement of some of these regions. That these regions are uniquely well developed in human beings may account for the fact that dementing illnesses have been thus far described clinically and neuropathologically only in humans. It may be that in the human brain, these neocortical regions share certain developmental and cellular characteristics that render them more prone to the pathologic process. These cortical regions are the last to reach maturity during brain development (Kostovic, 1990; Mrzjlak *et al.*, 1990), and thus it is also conceivable that they contain neuron populations that have evolved uniquely in humans and may be those involved in the degenerative process in AD. Clearly one such subset of neurons could be a group of neurofilament protein-enriched pyramidal cells in layers III and V of these neocortical association regions, which would be consistent with the challenging hypothesis that AD is a phylogenetic disorder resulting from evolutionary changes in the hominoid genome, possibly caused by environmental factors and leading to the modification of expression of key gene products in these regions (Rapoport, 1988a,b, 1990). Although NFT-like lesions have been reported in some mammals, the fact that NFTs similar to those in humans have never been observed during brain aging in anthropoid monkeys and great apes (Dayan, 1971; Price, D. L. *et al.*, 1991; Gearing *et al.*, 1994, 1997; Martin *et al.*, 1994; Giannakopoulos *et al.*, 1997c) also supports this hypothesis.

In conclusion, it is important to keep in mind that the debilitating dementia that characterizes AD is likely to result largely from pathologic changes involving neocortical areas. Extensive age-related hippocampal and entorhinal cortex alterations can occur with only minor disruptions in daily living activities (Hof *et al.*, 1992b, 1996a; Bouras *et al.*, 1993, 1994; Giannakopoulos *et al.*, 1993; Hof and Morrison, 1996; Gómez-Isla *et al.*, 1996; Hof, 1997a; Morrison and Hof, 1997). Presumably, memory defects would show up in testing, but for the full dementia symptomatology to occur, there must be extensive and widely distributed neocortical involvement. This may be due in part to the fact that memory is highly dependent on hippocampal circuits during the early processing and acquisition stages, but may become independent of strictly hippocampal circuits once addressing and retrieval depends upon neocortical elements (Hodges *et al.*, 1992; Zola-Morgan *et al.*, 1992). This understanding may permit the very mild memory defects associated with normal aging (i.e., age-associated memory impairment, benign senescent forgetfulness) that do not prevent information retrieval (Albert, 1996; Albert and Moss, 1996) and that may relate to pathological alterations restricted to the hippocampal formation to be differentiated from the more generalized memory and cognitive defects characterizing neurodegenerative disorders such as AD, which involve critical neocortical circuits. It is necessary to explore age-related changes that occur at the subcellular level, as it appears that in normal brain aging the distribution of certain proteins, such as key neurotransmitter receptors, may be modifiable in specific circuits (Gazzaley *et al.*, 1996a,b). Such subtle changes may constitute the substrate for age-related memory impairment in the absence of substantial degeneration (Morrison and Hof, 1997). Such studies are crucial to further our knowledge of the differences that exist between normal brain aging and dementia, and to better determine the molecular and morphologic phenotypes that are most clearly linked to differential cellular vulnerability in AD and other dementing illnesses.

ACKNOWLEDGMENTS. We wish to thank all of those who have participated in our studies of brain aging and dementing illnesses in recent years. We extend special gratitude to Drs. L. Buée, A. Delacourte, A. H. Gazzaley, P. Giannakopoulos, E. A. Nimchinsky, D. P. Perl, J. C. Vickers, and B. A. Vogt for their crucial contributions and advice, and W. G. M. Janssen and A. P. Leonard for their expert technical assistance. This work was supported by NIH grants AG05138, AG06647, AG08802, and AG14382, by the Dana Foundation, and by the Brookdale Foundation.

## 6. References

Adams, C. W. M., and Bruton, C. J., 1989, The cerebral vasculature in dementia pugilistica, *J. Neurol. Neurosurg. Psychiatry* **52:**600–604.

Adams, M. M., Ungerleider, L. G., Eden, G., Bouras, C., Bertini, G., Morrison, J. H., and Hof, P. R., 1996, Neurofilament and calcium-binding proteins define visual cortical areas in human, *Soc. Neurosci. Abstr.* **22:**1059.

Akashi, T., Arima, K., Maruyama, N., Ando, S., and Inose, T., 1989, Severe cerebral atrophy in progressive supranuclear palsy: A case report, *Clin. Neuropathol.* **8:**195–199.

Ala, T. A., Yang, K. H., Sung, J. H., and Frey, W. H., II, 1997, Hallucinations and signs of parkinsonism help distinguish patients with dementia and cortical Lewy bodies from patients with Alzheimer's disease at presentation: A clinicopathological study, *J. Neurol. Neurosurg. Psychiatry* **62:**16–21.

Alajouanine, T., Delafontaine, P., and Lacan, J., 1926, Fixité du regard par hypertonie, prédominant dans le sens vertical, avec conservation des mouvements automatico-réflexes: Aspect spécial de syndrome de Parinaud par hypertonie associée à un syndrome extrapyramidal avec trouble pseudobulbaires, *Rev. Neurol.* **2:**410–418.

Albert, M. S., 1996, Cognitive and neurobiologic markers of early Alzheimer's disease, *Proc. Natl. Acad. Sci. USA* **93:**13547–13551.

Albert, M. S., and Moss, M. B., 1996, Neuropsychology of aging: Findings in humans and monkeys, in: *Handbook of Biology of Aging,* 4th Edition (E. L. Schneider, J. W. Rowe, T. E. Johnson, N. J. Holbrook, and J. H. Morrison, eds.), Academic, San Diego, pp. 217–233.

Albert, M. L., Feldman, R. G., and Willis, A. L., 1974, The "subcortical dementia" of progressive supranuclear palsy, *J. Neurol. Neurosurg. Psychiatry* **37:**121–130.

Allsop, D., Kidd, M., Lanson, M., and Tomlinson, A., 1986, Isolated senile plaque cores in Alzheimer's disease and Down's syndrome show differences in morphology, *J. Neurol. Neurosurg. Psychiatry* **49:**886–892.

Allsop, D., Haga, S., Bruton, C., Ishii, T., and Roberts, G. W., 1990, Neurofibrillary tangles in some cases of dementia pugilistica share antigens with amyloid β-protein of Alzheimer's disease, *Am. J. Pathol.* **136:**255–260.

Andressen, C., Blümcke, I., and Celio, M. R., 1993, Calcium-binding proteins: Selective markers of nerve cells, *Cell Tissue Res.* **271:**181–208.

Andrews, J. M., 1967, Neurological disease on Guam: A review of past and present investigations, *Bull. Los Angeles Neurol. Soc.* **32:**30–42.

Arai, H., Emson, P. C., Mountjoy, C. Q., Carassco, L. H., and Heizmann, C. W., 1987, Loss of parvalbumin-immunoreactive neurons from cortex in Alzheimer-type dementia, *Brain Res.* **418:**164–169.

Arai, H., Noguchi, H., Makino, Y., Kosaka, K., Heizmann, C. W., and Iizuka, R., 1991, Parvalbumin-immunoreactive neurons in the cortex of Pick's disease, *J. Neurol.* **238:**200–202.

Arima, K., Oyanagi, S., Akashi, T., and Sakata, C., 1992, Neurofibrillary tangles of progressive supranuclear palsy in the dentate fascia: An ultrastructural study of a case, *Neuropathology* **12:**51–57.

Armstrong, D. M., and Ikonomovic, M. D., 1996, AMPA-selective glutamate receptor subtype immunoreactivity in the hippocampal dentate gyrus of patients with Alzheimer's disease: Evidence for hippocampal plasticity, *Mol. Chem. Neuropathol.* **28:**59–64.

Armstrong, D. M., Benzing, W. C., Evans, J., Terry, R. D., Shields, D., and Hansen, L. A., 1989, Substance P and somatostatin coexist within neuritic plaques: Implications for the pathogenesis of Alzheimer's disease, *Neuroscience* **31:**663–671.

Armstrong, D. M., Ikonomovic, M. D., Sheffield, R., and Wenthold, R. J., 1994, AMPA-selective glutamate receptor subtype immunoreactivity in the entorhinal cortex of non-demented elderly and patients with Alzheimer's disease, *Brain Res.* **639:**207–216.

Arnold, A., Edgren, D. C., and Palladino, V. S., 1953, Amyotrophic lateral sclerosis: 50 cases observed on Guam, *J. Nerv. Ment. Dis.* **117:**135–159.

Arnold, S. E., Hyman, B. T., Flory, J., Damasio, A. R., and Van Hoesen, G. W., 1991, The topographical and neuroanatomical distribution of neurofibrillary tangles and neuritic plaques in the cerebral cortex of patients with Alzheimer's disease, *Cereb. Cortex* **1:**103–116.

Arnold, S. E., Hyman, B. T., and Van Hoesen, G. W., 1994, Neuropathologic changes in the temporal pole in Alzheimer's disease and Pick's disease, *Arch. Neurol.* **51:**145–150.

Arriagada, P. V., Growdon, J. H., Hedley-White, E. T., and Hyman, B. T., 1992a, Neurofibrillary tangles, but not senile plaques, parallel duration and severity of Alzheimer's disease, *Neurology* **42:**631–639.

Arriagada, P. V., Marzloff, K., and Hyman, B. T., 1992b, Distribution of Alzheimer-type pathologic changes in nondemented elderly individuals matches the pattern in Alzheimer's disease, *Neurology* **42:**1681–1688.

Asada, T., Yamagata, Z., Kinoshita, T., Kinoshita, A., Kariya, T., Asaka, A., and Kakuma, T., 1996, Prevalence of dementia and distribution of apoE alleles in Japanese centenarians: An almost-complete survey in Yamanashi prefecture, Japan, *J. Am. Geriat. Soc.* **44:**151–155.

Bailey-Wilson, J. E., Plato, C. C., Elston, R C., and Garruto, R. M., 1993, Potential role of an additive genetic component in the cause of amyotrophic lateral sclerosis and parkinsonism-dementia in the Western Pacific, *Am. J. Med. Genet.* **45:**68–76.

Baimbridge, K. G., Celio, M. R., and Rogers, J. H., 1992, Calcium binding proteins in the nervous system, *Trends Neurosci.* **15:**303–308.

Bálint, R., 1909, Seelenlähmung des "Schauens," optische Ataxie, räumliche Störung der Aufmerksamkeit, *Mschr. Psychiat. Neurol.* **25:**51–81.

Ball, M. J., 1977, Neuronal loss, neurofibrillary tangles, and granulovacuolar degeneration in the hippocampus with aging and dementia, *Acta Neuropathol.* **37:**111–118.

Bancher, C., Jellinger, K. A., and Zwiauer, K., 1996, Neurofibrillary tangles in the brain of a 16 month old infant, *J. Neurol. Neurosurg. Psychiatry* **54:**231.

Barbas, H., 1986, Pattern in the laminar origin of corticocortical connections, *J. Comp. Neurol.* **252:**415–422.

Barbas, H., and Blatt, G. J., 1995, Topographically specific hippocampal projections target functionally distinct prefrontal areas in the rhesus monkey, *Hippocampus* **5:**511–533.

Barber, R., Snowden, J. S., and Crauford, D., 1995, Frontotemporal dementia and Alzheimer's disease: Retrospective differentiation using information from informants, *J. Neurol. Neurosurg. Psychiatry* **59:**61–70.

Basun, H., Almkvist, O., Axelman, K., Brun, A., Campbell, T. A., Collinge, J., Forsell, C., Froelich, S., Wahlund, L. O., Wetterberg, L., and Lannfelt, L., 1997, Clinical characteristics of a chromosome 17-linked rapidly progressive familial frontotemporal dementia, *Arch. Neurol.* **54:**539–544.

Beach, T. G., and McGeer, E. G., 1992, Senile plaques, amyloid β-protein, and acetylcholinesterase fibres: Laminar distribution in Alzheimer's disease striate cortex, *Acta Neuropathol.* **83:**292–299.

Behrman, S., Carroll, J. D., Janota, I., and Matthews, W. B., 1969, Progressive supranuclear palsy: Clinicopathological study of four cases, *Brain* **92:**663–678.

Benson, D. F., Davis, R. J., and Snyder, D. B., 1988, Posterior cortical atrophy, *Arch. Neurol.* **45:**789–793.

Berg, L., McKeel, D. W., Miller, P., Baty, J., and Morris, J. C., 1993, Neuropathological indexes of Alzheimer's disease in demented and non-demented persons aged 80 years and older, *Arch. Neurol.* **50:**349–358.

Bergeron, C., and Pollanen, M., 1989, Lewy bodies in Alzheimer disease: One or two diseases? *Alzheimer Dis. Assoc. Disord.* **3:**197–204.

Bergeron, C., Pollanen, M. S., Weyer, L., and Lang, A. E., 1997, Cortical degeneration in progressive supranuclear palsy: A comparison with cortical-basal ganglionic degeneration, *J. Neuropathol. Exp. Neurol.* **56:**726–734.

Bergmann, M., Kuchelmeister, K., Schmid, K. W., Kretzschmar, H. A., and Schröder, R., 1996, Different variants of frontotemporal dementia: A neuropathological and immunohistochemical study, *Acta Neuropathol.* **92:**170–179.

Berthier, M. L., Leiguarda, R., Starkstein, S. E., Sevlever, G., and Taratuto, A. L., 1991, Alzheimer's disease in a patient with posterior cortical atrophy, *J. Neurol. Neurosurg. Psychiatry* **54:**1110–1111.

Bierer, L. M., Hof, P. R., Purohit, D. P., Carlin, L., Schmeidler, J., Davis, K. L., and Perl, D. P., 1995, Neocortical neurofibrillary tangles correlate with dementia severity in Alzheimer's disease, *Arch. Neurol.* **52:**81–88.

Bird, T. D., Wiksman, E. M., Nochlin, D., Leehey, M., Sumi, S. M., Payami, H., Poorkaj, P., Nemens, E., Rafkind, M., and Schellenberg, G. D., 1997, Chromosome 17 and hereditary dementia: Linkage studies in three non-Alzheimer families and kindreds with late-onset FAD, *Neurology* **48:**949–954.

Bliss, T. V. P., and Collingridge, G. L., 1993, A synaptic model of memory: Long-term potentiation in the hippocampus, *Nature* **361:**31–39.

Bliss, T. V. P., and Lømo, T., 1973, Long-lasting potentiation of synaptic transmission in the dentate area of the anaesthetised rabbit following stimulation of the perforant path, *J. Physiol.* **232:**331–356.

Blümcke, I., Wolf, H. K., Hof, P. R., Morrison, J. H., and Wiestler, O. D., 1994, Regional distribution of the AMPA glutamate receptor subunit GluR2(4) in human hippocampus, *Brain Res.* **682:**239–244.

Blümcke, I., Beck, H., Scheffler, B., Hof, P. R., Morrison, J. H., Wolf, H. K., Schramm, J., Elger, C. E., and Wiestler, O. D., 1996, Altered distribution of α-amino-3-hydroxy-5-methyl-4-isoxazole propionate receptor subunit GluR2(4) and *N*-methyl-D-aspartate receptor subunit NMDAR1 in the hippocampus of patients with temporal lobe epilepsy, *Acta Neuropathol.* **92:**576–587.

Bobinski, M., Wegiel, J., Tarnawski, M., Bobinski, M., Reisberg, B., De Leon, M. J., Miller, D. C., and Wisniewski, H. M., 1997, Relationships between regional neuronal loss and neurofibrillary changes in the hippocampal formation and duration and severity of Alzheimer disease, *J. Neuropathol. Exp. Neurol.* **56:**414–420.

Bolton, J. S., 1910, A contribution to the localization of cerebral function, based on the clinico-pathological study of mental disease, *Brain* **33:**26–148.

Bondareff, W., Mountjoy, C. Q., Roth, M., and Hauser, D. L., 1989, Neurofibrillary degeneration and neuronal loss in Alzheimer's disease, *Neurobiol. Aging* **10:**709–715.

Bondareff, W., Wischik, C. M., Novak, M., and Roth, M., 1991, Sequestration of tau by granulovacuolar degeneration in Alzheimer's disease, *Am. J. Pathol.* **130:**6411–6417.

Borenstein Graves, A., White, E., Koepsell, T. D., Reifler, B. V., Van Belle, G., Larson, E. B., and Raskind, M., 1990, The association between head trauma and Alzheimer's disease, *Am. J. Epidemiol.* **131:**491–501.

Bouras, C., Magistretti, P. J., and Morrison, J. H., 1986, An immunohistochemical study of six biologically active peptides in the human brain, *Hum. Neurobiol.* **5:**213–226.

Bouras, C., Hof, P. R., and Morrison, J. H., 1993, Neurofibrillary tangle densities in the hippocampal formation in a non-demented population define subgroups of patients with differential early pathologic changes, *Neurosci. Lett.* **153:**131–135.

Bouras, C., Magistretti, P. J., Morrison, J. H., and Constantinidis, J., 1987, An immunohistochemical study of pro-somatostatin-derived peptides in the human brain, *Neuroscience* **22:**781–800.

Bouras, C., Vallet, P. G., Hof, P. R., Charnay, Y., Golaz, J., and Constantinidis, J., 1990, Substance P immunoreactivity in Alzheimer disease: A study in cases presenting symmetric or asymmetric cortical atrophy, *Alzheimer Dis. Assoc. Disord.* **4:**24–34.

Bouras, C., Hof, P. R., Giannakopoulos, P., Michel, J. P., and Morrison, J. H., 1994, Regional distribution of neurofibrillary tangles and senile plaques in the cerebral cortex of elderly patients: A quantitative evaluation of a one-year autopsy population from a geriatric hospital, *Cereb. Cortex* **4:**138–150.

Braak, H., and Braak, E., 1985, On areas of transition between entorhinal allocortex and temporal isocortex in the human brain: Normal morphology and lamina-specific pathology in Alzheimer's disease, *Acta Neuropathol.* **68:**325–332.

Braak, H., and Braak, E., 1986, Ratio of pyramidal cells versus non-pyramidal cells in the human frontal isocortex and changes in ratio with aging and Alzheimer's disease, in: *Aging of the Brain and Alzheimer's Disease, Prog. Brain Res.*, Vol. 70 (D. F. Swaab, E. Fliers, M. Mirmiran, V. A. Van Gool, and F. Van Haaren, eds.), Elsevier, Amsterdam, pp. 185–212.

Braak, H., and Braak, E., 1987, Argyrophilic grains: Characteristic pathology of cerebral cortex in cases of adult onset dementia without Alzheimer's changes, *Neurosci. Lett.* **76:**124–127.

Braak, H., and Braak, E., 1988, Neuropil threads occur in dendrites of tangle-bearing nerve cells, *Neuropathol. Appl. Neurobiol.* **14:**39–44.

Braak, H., and Braak, E., 1989, Cortical and subcortical argyrophilic grains characterize a disease associated with dementia, *Neuropathol. Appl. Neurobiol.* **15:**13–26.

Braak, H., Braak, E., 1990, Neurofibrillary changes confined to the entorhinal region and an abundance of cortical amyloid in cases of presenile and senile dementia, *Acta Neuropathol.* **80:**479–486.

Braak, H., and Braak, E., 1991, Neuropathological staging of Alzheimer-related changes, *Acta Neuropathol.* **82:**239–259.

Braak, H., Braak, E., and Kalus, P., 1989, Alzheimer's disease: Areal and laminar pathology in the occipital isocortex, *Acta Neuropathol.* **77:**494–506.

Braak, H., Braak, E., Grundke-Iqbal, I., and Iqbal, K., 1986, Occurrence of neuropil threads in the senile human brain and in Alzheimer's disease: A third location of paired helical filament outside of neurofibrillary tangles and neuritic plaques, *Neurosci. Lett.* **65:**351–355.

Braak, E., Strotkamp, B., and Braak, H., (1991), Parvalbumin-immunoreactive structures in the hippocampus of the human brain, *Cell Tissue Res.* **264:**33–48.

Braak, H., Braak, E., and Bohl, J., 1992a, Retrosplenial region involvement in Alzheimer's disease, *Neurodegeneration* **1:**53–57.

Braak, H., Jellinger, K., Braak, E., and Bohl, J., 1992b, Allocortical neurofibrillary changes in progressive supranuclear palsy, *Acta Neuropathol.* **84:**478–483.

Braak, H., Braak, E., Yilmazer, D., de Vos, R. A. I., Jansen, E. N. H., and Bohl, J., 1996, Pattern of brain destruction in Parkinson's and Alzheimer's diseases, *J. Neural Trans.* **103:**455–490.

Brady, D. R., and Mufson, E. J., 1990, Amygdaloid pathology in Alzheimer's disease: Qualitative and quantitative analysis, *Dementia* **1:**5–17.

Brady, D. R., and Mufson, E. J., 1991, Alz-50 immunoreactive neuropil differentiates hippocampal complex subfields in Alzheimer's disease, *J. Comp. Neurol.* **305:**489–507.

Bramblett, G. T., Trojanowski, J. Q., and Lee, V. M. Y., 1992, Regions with abundant neurofibrillary pathology in human brain exhibit a selective reduction in levels of binding-competent tau and accumulation of abnormal tau-isoforms (A68 proteins), *Lab. Invest.* **66:**212–222.

Brandenburg, W., and Hallervorden, J., 1954, Dementia pugilistica mit anatomischem Befund, *Virchows Arch. Pathol. Anat. Physiol. Klin. Med.* **325:**680–709.

Brion, J. P., 1990, Molecular pathology of Alzheimer amyloid and neurofibrillary tangles, *Semin. Neurosci.* **2:**89–100.

Brion, J. P., and Résibois, A., 1994, A subset of calretinin-positive neurons are abnormal in Alzheimer's disease, *Acta Neuropathol.* **88:**33–43.

Brion, S., Mikol, J., and Psimaras, A., 1973, Recent finding in Pick's disease, in: *Progress in Neuropathology, Vol. 2* (H. M. Zimmerman, ed.), Grune and Stratton, New York, pp. 421–452.

Brody, H., 1955, Organization of the cerebral cortex. III. A study of aging in the human cerebral cortex, *J. Comp. Neurol.* **102:**511–556.

Brody, J. A., Hirano, A., and Scott, R. M., 1971, Recent neuropathologic observations in amyotrophic lateral sclerosis and parkinsonism-dementia of Guam, *Neurology* **21:**528–536.

Brody, J. A., Stanhope, J. M., and Kurland, L. T., 1975, Patterns of amyotrophic lateral sclerosis and parkinsonism-dementia on Guam, in: *Topics in Tropical Neurology, Contemporary Neurology Series, Vol. 12* (R. W. Hornabrook, ed.), F. A. Davis, Philadelphia, pp. 45–70.

Brown, J., Ashworth, A., Gydesen, S., Sorensen, A., Rossor, M., Hardy, J., and Collinge, J., 1995, Familial non-specific dementia maps to chromosome 3, *Hum. Mol. Genet.* **4:**1625–1628.

Brun, A., 1987, Frontal lobe degeneration of non-Alzheimer type. I. Neuropathology, *Arch. Gerontol. Geriat.* **6:**193–208.

Brun, A., 1993, Frontal lobe degeneration of non-Alzheimer type revisited, *Dementia* **4:**126–131.

Brun, A., Englund, B., Gustafson, L., Passant, U., Mann, D. M. A., Neary, D., and Snowden, J. S., 1994, Clinical and neuropathological criteria for frontotemporal dementia, *J. Neurol. Neurosurg. Psychiatry* **57:**416–418.

Buée, L., Hof, P. R., Roberts, D. D., Delacourte, A., Morrison, J. H., and Fillit, H. M., 1992, Immunohistochemical identification of thrombospondin in normal human brain and in Alzheimer's disease, *Am. J. Pathol.* **141**:783–788.

Buée, L., Ding, W., Delacourte, A., and Fillit, H., 1993, Binding of secreted human neuroblastoma proteoglycans to the Alzheimer's amyloid A4 peptide, *Brain Res.* **601**:154–163.

Buée, L., Hof, P. R., Bouras, C., Delacourte, A., Perl, D. P., Morrison, J. H., and Fillit, H. M., 1994, Pathological alterations of the cerebral microvasculature in Alzheimer's disease and related dementing disorders, *Acta Neuropathol.* **87**:469–480.

Buée, L., Pérez-Tur, J., Leveugle, B., Buée-Scherrer, V., Mufson, E. J., Loerzel, A. J., Chartier-Harlin, M. C., Perl, D. P., Delacourte, A., and Hof, P. R., 1996, Apolipoprotein E in Guamanian amyotrophic lateral sclerosis/parkinsonism-dementia complex: Genotype analysis and relationships to neuropathological changes, *Acta Neuropathol.* **91**:247–253.

Buée-Scherrer, V., Buée, L., Hof, P. R., Leveugle, B., Gilles, C., Loerel, A. J., Perl, D. P., and Delacourte, A., 1995, Neurofibrillary degeneration in amyotrophic lateral sclerosis/parkinsonism-dementia complex of Guam: Immunochemical characterization of tau proteins, *Am. J. Pathol.* **146**:924–932.

Buée-Scherrer, V., Condamines, O., Mourton-Gilles, C., Jakes, R., Goedert, M., Pau, B., and Delacourte, A., 1996a, AD2, a phosphorylation-dependent monoclonal antibody directed against tau proteins found in Alzheimer's disease, *Mol. Brain Res.* **39**:79–88.

Buée-Scherrer, V., Hof, P. R., Buée, L., Leveugle, B., Vermersch, P., Perl, D. P., Olanow, C. W., and Delacourte, A., 1996b, Hyperphosphorylated tau proteins differentiate corticobasal degeneration and Pick's disease, *Acta Neuropathol.* **91**:351–359.

Buée-Scherrer, V., Buée, L., Leveugle, B., Perl, D. P., Vermersch, P., Hof, P. R., and Delacourte, A., 1997, Pathological τ proteins in postencephalitic parkinsonism: Comparison with Alzheimer's disease and other neurodegenerative disorders, *Ann. Neurol.* **42**:356–359.

Bugiani, O., Mancardi, G. L., Brusa, A., and Ederli, A., 1979, The fine structure of subcortical neurofibrillary tangles in progressive supranuclear palsy, *Acta Neuropathol.* **45**:147–152.

Burger, P. C., and Vogel, F. S., 1973, The development of the pathologic changes of Alzheimer's disease and senile dementia in patients with Down's syndrome, *Am. J. Pathol.* **73**:457–476.

Burkhardt, C. R., Filley, C. M., Kleinschmidt-DeMasters, B. K., de la Monte, S., Norenberg, M. D., and Schneck, S. A., 1988, Diffuse Lewy body disease and progressive dementia, *Neurology* **38**:1520–1528.

Butter, C. M., Trobe, J. D., Foster, N. L., and Berent, S., 1996, Visual-spatial deficits explain visual symptoms in Alzheimer's disease, *Am. J. Ophthalmol.* **122**:97–105.

Calne, D. B., and Lees, A. J., 1988, Late progression of post-encephalitic Parkinson's syndrome, *Can. J. Neurol. Sci.* **15**:135–138.

Cambier, J., Masson, M., Dairou, R., and Henin, D., 1981, Etude anatomo-clinique d'une forme pariétale de la maladie de Pick, *Rev. Neurol.* **137**:33–38.

Campbell, M. J., and Morrison, J. H., 1989, A monoclonal antibody to neurofilament protein (SMI-32) labels a subpopulation of pyramidal neurons in the human and monkey neocortex, *J. Comp. Neurol.* **282**:191–205.

Campbell, M. J., Hof, P. R., and Morrison, J. H., 1991, A subpopulation of primate corticocortical neurons is distinguished by somatodendritic distribution of neurofilament protein, *Brain Res.* **539**:133–136.

Carlson, M. D., Penney, J. B., Jr., and Young, A. B., 1993, NMDA, AMPA, and benzodiazepine binding site changes in Alzheimer's disease visual cortex, *Neurobiol. Aging* **14**:343–352.

Carmichael, S. T., and Price, J. L., 1994, Architectonic subdivision of the orbital and medial prefrontal cortex in the macaque monkey, *J. Comp. Neurol.* **346**:366–402.

Carriedo, S. G., Yin, H. Z., and Weiss, J. H., 1996, Motor neurons are selectively vulnerable to AMPA/kainate receptor-mediated injury *in vitro*, *J. Neurosci.* **16**:4069–4079.

Castellani, R., Smith, M. A., Richey, P. L., Kalaria, R., Gambetti, P., and Perry, G., 1995, Evidence for oxidative stress in Pick disease and corticobasal degeneration, *Brain Res.* **696**:268–271.

Casson, I. R., Siegel, O., Sham, R., Campbell, E. A., Tarlau, M., DiDomenico, A., 1984, Brain damage in modern boxers, *J. Am. Med. Assoc.* **251**:2663–2667.

Celio, M. R., 1990, Calbindin D-28k and parvalbumin in the rat nervous system, *Neuroscience* **35**:375–475.

Chang, L., Cornford, M., Miller, B. L., Itabashi, H., and Mena, I., 1995, Neuronal ultrastructural abnormalities in a patient with frontotemporal dementia and motor neuron disease, *Dementia* **6**:1–8.

Chan-Palay, V., 1987, Somatostatin immunoreactive neurons in the human hippocampus and cortex shown by immunogold/silver intensification on vibratome sections: Coexistence with neuropeptide Y neurons, and effects in Alzheimer-type dementia, *J. Comp. Neurol.* **260**:201–223.

Chan-Palay, V., Zetzsche, T., and Höchli, M., 1991, Parvalbumin neurons in the hippocampus in senile dementia of the Alzheimer type, Parkinson's disease, and multi-infarct dementia, *Dementia* **2:**297–313.

Checler, F., 1995, Processing of the β-amyloid precursor protein and its regulation in Alzheimer's disease, *J. Neurochem.* **65:**1431–1444.

Chen, K. M., and Chase, T. N., 1986, Parkinsonism-dementia, in *Extrapyramidal Disorders, Handbook of Clinical Neurology,* Vol. 49 (P. J. Vinken, G. W. Bruyn, and H. L. Klawans, eds.), Elsevier, Amsterdam, pp. 167–183.

Chen, K. M., Makifuchi, T., Garruto, R. M., and Gajdusek, D. C., 1982, Parkinsonism-dementia in a Filipino migrant: A clinicopathologic case report, *Neurology* **32:**1221–1226.

Cheyette, S. R., and Cummings, J. L., 1995, Encephalitis lethargica: Lessons for contemporary neuropsychiatry, *J. Neuropsychiatry Clin. Neurosci.* **7:**125–134.

Choi, D. W., 1988, Glutamate neurotoxicity and diseases of the nervous system, *Neuron* **1:**623–634.

Choi, D. W., 1990, Cerebral hypoxia: Some new approaches and unanswered questions, *J. Neurosci.* **10:**2493–2501.

Choi, D. W., and Rothman, S. M., 1990, The role of glutamate neurotoxicity in hypoxic-ischemic neuronal death, *Annu. Rev. Neurosci.* **13:**171–182.

Clark, A. W., Manz, H. J., White, C. L., Lehmann, J., Miller, D., and Coyle, J. T., 1986, Cortical degeneration with swollen chromatolytic neurons: Its relationship to Pick's disease, *J. Neuropathol. Exp. Neurol.* **45:**268–284.

Clarke, S., 1993, Callosal connections and functional subdivision of the human occipital cortex, in: *Functional Organisation of the Human Visual Cortex* (B. Gulyas, D. Ottoson, and P. E. Roland, eds.), Pergamon, Oxford, pp. 137–150.

Clarke, S., 1994, Modular organization of human extrastriate visual cortex: Evidence from cytochrome oxidase pattern in normal and macular degeneration cases, *Eur. J. Neurosci.* **6:**725–736.

Clarke, S., 1995, Association and intrinsic connections of human extrastriate visual cortex, *Proc. R. Soc. Lond. Ser. B. (Biol. Sci.)* **257:**87–92.

Clarke, S., and Miklossy, J., 1990, Occipital cortex in man: Organization of callosal connections, related myelo- and cytoarchitecture, and putative boundaries of functional visual areas, *J. Comp. Neurol.* **298:**188–214.

Clinton, J., Ambler, M. W., and Roberts, G. W., 1991, Post-traumatic Alzheimer's disease: Preponderance of a single plaque type, *Neuropathol. Appl. Neurobiol.* **17:**69–74.

Clinton, J., Royston, M. C., Gentleman, S. M., and Roberts, G. W., 1992, Amyloid plaque: Morphology, evolution, and etiology, *Modern Pathol.* **5:**439–443.

Cochran, E. J., Fox, J. H., and Mufson, E. J., 1994, Severe panencephalic Pick's disease with Alzheimer's disease-like neuropil threads and synaptophysin immunoreactivity, *Acta Neuropathol.* **88:**479–484.

Cogan, D. G., 1985, Visual disturbances with focal progressive dementing disease, *Am. J. Ophthalmol.* **100:**68–72.

Coleman, P. D., and Flood, D. G., 1987, Neuron numbers and dendritic extent in normal aging and Alzheimer's disease, *Neurobiol. Aging* **8:**521–545.

Colon, E. J., 1972, The elderly brain: A quantitative analysis in the cerebral cortex of two cases, *Psychiatry Neurol. Neurochir.* **75:**216–270.

Colon, E. J., 1973, The cerebral cortex in presenile dementia: A quantitative analysis, *Acta Neuropathol.* **23:**281–290.

Condé, F., Lund, J. S., Jacobowitz, D. M., Baimbridge, K. G., and Lewis, D. A., 1994, Local circuit neurons immunoreactive for calretinin, calbindin D-28k, or parvalbumin in monkey prefrontal cortex: Distribution and morphology, *J. Comp. Neurol.* **341:**95–116.

Connor, J. R., Snyder, B. S., Beard, J. L., Fine, R. L., and Mufson, E. J., 1992, Regional distribution of iron and iron-regulatory proteins in the brain in aging and Alzheimer's disease, *J. Neurosci. Res.* **31:**327–335.

Conrad, C., Andreadis, A., Trojanowski, J. Q., Dickson, D. W., Kang, D., Chen, X., Wiederholt, W., Hansen, L., Masliah, E., Thal, L. J., Katzman, R., Xia, Y., Saitoh, T., 1997, Genetic evidence for the involvement of τ in progressive supranuclear palsy, *Ann. Neurol.* **41:**277–281.

Constantinidis, J., and Tissot, R., 1967, Lésions neurofibrillaires d'Alzheimer généralisées sans plaques séniles, *Schweiz. Arch. Neurol. Neurochir. Psychiat.* **100:**117–130.

Constantinidis, J., Tissot, R., and De Ajuriaguerra, J., 1970, Dystonie oculo-cervico-faciale ou paralysie progressive supranucléaire de Steele-Richardson-Olszewski: Pseudo-paralysie du regard, troubles visuo-spatiaux, pseudo-démence, altérations neuronales, *Rev. Neurol.* **122:**249–262.

Constantindis, J., Richard, J., and Tissot, R., 1974, Pick's disease: Histological and clinical correlations, *Eur. Neurol.* **11:**208–217.

Convit, A., De Leon, M. J., Tarshish, C., De Santi, S., Tsui, W., Rusinek, H., and George, A., 1997, Specific hippocampal volume reductions in individuals at risk for Alzheimer's disease, *Neurobiol. Aging* **18:**131–138.

Cooper, P. N., Jackson, M., Lennox, G., Lowe, J., and Mann, D. M. A., 1995, τ, ubiquitin, and αB-crystallin immunohistochemistry define the principal causes of degenerative frontotemporal dementia, *Arch. Neurol.* **52:**1011–1015.

Cooper, P. N., Siddons, C. A., and Mann, D. M. A., 1996, Patterns of glial cell activity in frontotemporal dementia (lobar atrophy), *Neuropathol. Appl. Neurobiol.* **22:**17–22.

Copeland, J. R. M., Davidson, I. A., Dewey, M. E., Gilmore, C., Larkin, B. A., McWilliam, C., Saunders, P. A., Scott, A., Sharma, V., and Sullivan, C., 1992, Alzheimer's disease, other dementias, depression, and pseudodementia: Prevalence, incidence, and three-year outcome in Liverpool, *Br. J. Psychiatry* **161:**230–239.

Couratier, P., Lesort, M., Sindou, P., Esclaire, F., Yardin, C., and Hugon, J., 1996, Modifications of neuronal phosphorylated τ immunoreactivity induced by NMDA toxicity, *Mol. Chem. Neuropathol.* **27:**259–264.

Corsellis, J. A. N., and Brierley, J. B., 1959, Observations on the pathology of insidious dementia following head injury, *J. Ment. Sci.* **105:**714–720.

Corsellis, J. A. N., Bruton, C. J., and Freeman-Browne, D., 1973, The aftermath of boxing, *Psychol. Med.* **3:**270–303.

Côté, F., Collard, J. F., and Julien, J. P., 1993, Progressive neuronopathy in transgenic mice expressing the human neurofilament heavy gene: A mouse model for amyotrophic lateral sclerosis, *Cell* **73:**35–46.

Cotman, C. W., and Anderson, A. J., 1995, A potential role for apoptosis in neurodegenerative and Alzheimer's diseases, *Mol. Neurobiol.* **10:**19–45.

Coyle, J. T., and Puttfarcken, P., 1993, Oxidative stress, glutamate, and neurodegenerative disorders, *Science* **262:**689–695.

Crapper McLachlan, D. R., McLachlan, C. D., Krishnan, B., Krishnan, S. S., Dalton, A. J., and Steele, J. C., 1988, Aluminum and calcium in soil and food from Guam, Palau, and Jamaica: Implications for amyotrophic lateral sclerosis and parkinsonism-dementia of Guam, *Env. Geochem. Health* **10:**45–53.

Cronin-Golomb, A., Sugiura, R., Corkin, S., and Growdon, J. H., 1993, Incomplete achromatopsia in Alzheimer's disease, *Neurobiol. Aging* **14:**471–477.

Crook, T., Bartus, R. T., Ferris, S. H., Whitehouse, P., Cohen, G. D., and Gershon, S., 1986, Age-associated memory impairment: Proposed diagnostic criteria and measures of clinical change—Report of a National Institute of Mental Health group, *Dev. Neuropsychol.* **4:**261–276.

Crystal, H. A., Horoupian, D. S., Katzman, R., and Jotkowitz, S., 1982, Biopsy-proved Alzheimer's disease presenting as a right parietal lobe syndrome, *Ann. Neurol.* **12:**186–188.

Crystal, H., Dickson, D., Fuld, P., Masur, D., Scott, R., Mehler, M., Masdeau, J., Kawas, C., Aronson, M., and Wolfson, L., 1988, Clinico-pathologic studies in dementia: Nondemented subjects with pathologically confirmed Alzheimer's disease, *Neurology* **38:**1682–1687.

Cudkowicz, M., and Kowall, N. W., 1990, Degeneration of pyramidal projection neurons in Huntington's disease, *Ann. Neurol.* **27:**200–204.

Curcio, C. A., and Drucker, D. N., 1993, Retinal ganglion cells in Alzheimer's disease and aging, *Ann. Neurol.* **33:**248–257.

Dal Canto, M. C., and Gurney, M. E., 1997, A low expressor line of transgenic mice carrying a mutant human Cu,Zn superoxide dismutase (*SOD1*) gene develops pathological changes that most closely resemble those in human amyotrophic lateral sclerosis, *Acta Neuropathol.* **93:**537–550.

Dale, G. E., Leigh, P. N., Luthert, P., Anderton, B. H., and Roberts, G. W., 1991, Neurofibrillary tangles in dementia pugilistica are ubiquitinated, *J. Neurol. Neurosurg. Psychiatry* **54:**116–118.

Daniel, S. E., de Bruin, V. M. S., and Lees, A. J., 1995, The clinical and pathological spectrum of Steele-Richardson-Olszewski syndrome (progressive supranuclear palsy): A reappraisal, *Brain* **118:**759–770.

D'Antona, R., Baron, J. C., Samson, Y., Serdaru, M., Viader, F., Agid, Y., and Cambier, J., 1985, Subcortical dementia: Frontal cortex hypometabolism detected by positron tomography in patients with progressive supranuclear palsy, *Brain* **108:**785–799.

Davies, D. C., Horwood, N., Isaacs, S. L., and Mann, D. M. A., 1992, The effect of age and Alzheimer's disease on pyramidal neuron density in the individual fields of the hippocampal formation, *Acta Neuropathol.* **83:**510–517.

Day, N. C., Williams, T. L., Ince, P. G., Kamboj, R. K., Lodge, D., and Shaw, P. J., 1995, Distribution of AMPA-selective glutamate receptor subunits in the human hippocampus and cerebellum, *Mol. Brain Res.* **31:**17–32.

Dayan, A. D., 1970a, Quantitative histological studies on the aged human brain. I. Senile plaques and neurofibrillary tangles in normal patients, *Acta Neuropathol.* **16:**85–94.

Dayan, A. D., 1970b, Quantitative histological studies on the aged human brain. II. Senile plaques and neurofibrillary tangles in senile dementia, *Acta Neuropathol.* **16:**95–102.

Dayan, A. D., 1971, Comparative neuropathology of aging: Studies on the brains of 47 species of vertebrates, *Brain* **94:**31–42.

De Boni, U., Crapper McLachlan, D. R., 1985, Controlled induction of paired helical filaments of the Alzheimer type in cultured human neurons by glutamate and aspartate, *J. Neurol. Sci.* **68:**105–118.

DeFelipe, J., Hendry, S. H. C., and Jones, E. G., 1989, Synapses of double bouquet cells in monkey cerebral cortex visualized by calbindin immunoreactivity, *Brain Res.* **503:**49–54.

DeFelipe, J., Huntley, G. W., del Río, M. R., Sola, R. G., and Morrison, J. H., 1994, Microzonal decreases in the immunostaining for non-NMDA ionotropic excitatory amino acid receptor subunits GluR2/3 and GluR5/6/7 in the human epileptogenic neocortex, *Brain Res.* **657:**150–158.

DeKosky, S. T., and Scheff, S. W., 1990, Synapse loss in frontal cortex biopsies in Alzheimer's disease: Correlation with cognitive severity, *Ann. Neurol.* **27:**457–464.

De Lacoste, M. C., and White, C. L., III, 1993, The role of cortical connectivity in Alzheimer's disease pathogenesis: A review and model system, *Neurobiol. Aging* **14:**1–16.

Delacourte, A., Défossez, A., Ceballos, I., Nicole, A., and Sinet, P. M., 1988, Preferential localization of copper zinc superoxide dismutase in the vulnerable cortical neuron in Alzheimer's disease, *Neurosci. Lett.* **92:**247–253.

Delacourte, A., Flament, S., Dibe, E. M., Hublau, P., Sablonnière, B., Hémon, B., Scherrer, V., and Défossez, A., 1990, Pathological proteins tau 64 and 69 are specifically expressed in the somatodendritic domain of the degenerating cortical neurons during Alzheimer's disease: Demonstration with a panel of antibodies against tau proteins, *Acta Neuropathol.* **80:**111–117.

Delacourte, A., Robitaille, Y., Sergeant, N., Buée, L., Hof, P. R., Wattez, A., Laroche-Cholette, A., Mathieu, J., Chagnon, P., and Gauvreau, D., 1996, Specific pathological tau protein variants characterie Pick's disease, *J. Neuropathol. Exp. Neurol.* **55:**159–168.

Delaère, P., He, Y., Fayet, G., Duyckaerts, C., and Hauw, J. J., 1993, Beta A4 deposits are constant in the brain of the oldest old: An immunohistochemical study of 20 French centenarians, *Neurobiol. Aging* **14:**191–194.

De Lima, A. D., Morrison, J. H., 1989, Ultrastructural analysis of somatostatin-immunoreactive neurons and synapses in the temporal and occipital cortex of the macaque monkey, *J. Comp. Neurol.* **283:**212–227.

Del Río, M. R., and DeFelipe, J., 1997a, Colocalization of parvalbumin and calbindin D-28k in neurons including chandelier cells of the human temporal neocortex, *J. Chem. Neuroanat.* **12:**165–173.

Del Río, M. R., and DeFelipe, J., 1997b, Synaptic connections of calretinin-immunoreactive neurons in the human neocortex, *J. Neurosci.* **17:**5143–5154.

Dewar, D., Chalmers, D. T., Graham, D. I., and McCulloch, J., 1991, Glutamate metabotropic and AMPA binding sites are reduced in Alzheimer's disease: An autoradiographic study of the hippocampus, *Brain Res.* **553:**58–64.

DeYoe, E. A., and Van Essen, D. C., 1988, Concurrent processing streams in monkey visual cortex, *Trends Neurosci.* **11:**219–225.

DeYoe, E. A., Carman, G. J., Bandettini, P., Glickman, S., Wieser, J., Cox, R., Miller, D., and Neitz, J., 1996, Mapping striate and extrastriate visual areas in human cerebral cortex, *Proc. Natl. Acad. Sci. USA* **93:**2382–2386.

Dickson, D. W., Davies, P., Mayeux, R., Crystal, H., Horoupian, D. S., Thompson, A., and Goldman, J. E., 1987a, Diffuse Lewy body disease: Neuropathological and biochemical studies of six patients, *Acta Neuropathol.* **75:**8–15.

Dickson, D. W., Ksiezak-Reding, H., Davies, P., and Yen, S. H., 1987b, A monoclonal antibody that recognizes a phosphorylated epitope on Alzheimer neurofibrillary tangles, neurofilaments, and tau proteins, immunostains granulovacuolar degeneration, *Acta Neuropathol.* **73:**254–258.

Dickson, D. W., Crystal, H. A., Mattiace, L. A., Masur, D. M., Blau, A. D., Davies, P., Yen, S. H., and

Aronson, M. K., 1991, Identification of normal and pathological aging in prospectively studied nondemented elderly humans, *Neurobiol. Aging* **13:**179–189.

Dickson, D. W., Feany, M. B., Yen, S. H., Mattyiace, L. A., and Davies, P., 1996, Cytoskeletal pathology in non-Alzheimer degenerative dementia: New lesions in diffuse Lewy body disease, Pick's disease, and corticobasal degeneration, *J. Neural Transm.* (Suppl.) **47:**31–46.

Dix, M. R., Harrison, M. J. G., and Lewis, P. D., 1971, Progressive supranuclear palsy (the Steele-Richardson-Olszewski syndrome): A report of 9 cases with particular reference to the mechanism of the oculomotor disorder, *J. Neurol. Sci.* **13:**237–256.

Droz, P., Richard, J., and Constantinidis, J., 1982, L'installation des plaques séniles et des lésions neurofibrillaires dans le cortex cérébral de l'homme âgé: Une étude statistique de quelques corrélations anatomo-histologiques, *Bol. Estud. Méd. Biol. (Mexico)* **32:**39–100.

Dumanchin, C., Camuzat, A., Campion, D., Verpillat, P., Hannequin, D., Dubois, B., Saugier-Veber, P., Martin, C., Penet, C., Charbonnier, F., Agid, Y., Frebourg, T., and Brice, A., 1998, Segregation of a missense mutation in the microtubule-associated protein tau gene with familial frontotemperal dementia and parkinsonism, *Hum. Mol. Genet.* **7:**1825–1829.

Duncan, M. W., Marini, A. M., Watters, R., Kopin, I. J., and Markey, S. P., 1992, Zinc, a neurotoxin to cultured neurons, contaminates cycad flour prepared by traditional Guamanian methods, *J. Neurosci.* **12:**1523–1537.

Duvoisin, R. C., and Yahr, M. D., 1965, Encephalitis and parkinsonism, *Arch. Neurol.* **12:**227–239.

Duyckaerts, C., Hauw, J. J., Bastenaire, F., Piette, F., Poulain, C., Rainsard, V., Javoy-Agid, F., and Berthaux, P., 1986, Laminar distribution of neocortical senile plaques in senile dementia of the Alzheimer type, *Acta Neuropathol.* **70:**249–256.

Eastwood, S. L., Burnet, P. W. J., Beckwith, J., Kerwin, R. W., and Harrison, P. J., 1994, AMPA glutamate receptors and their flip and flop mRNAs in human hippocampus, *NeuroReport* **5:**1325–1328.

Eidelberg, D., Sotrel, A., Joachim, C., Selkoe, D., Forman, A., Pendlebury, W. W., and Perl, D. P., 1987, Adult onset Hallervorden-Spatz disease with neurofibrillary pathology, *Brain* **110:**993–1013.

Einstein, G., Buranosky, R., and Crain, B. J., 1994, Dendritic pathology of granule cells in Alzheimer's disease is unrelated to neuritic plaques, *J. Neurosci.* **14:**5077–5088.

Eldridge, R., Ryan, E., Rosario, J., and Brody, J. A., 1969, Amyotrophic lateral sclerosis and parkinsonism dementia in a migrant population from Guam, *Neurology* **19:**1029–1037.

Elizan, T. S., and Casals, J., 1991, Astrogliosis in von Economo's and postencephalitic Parkinson's diseases supports probably viral etiology, *J. Neurol. Sci.* **105:**131–134.

Elizan, T. S., Hirano, A., Abrams, B. M., Need, R. L., Van Nuis, C., and Kurland, L. T., 1966, Amyotrophic lateral sclerosis and parkinsonism-dementia complex of Guam: Neurological reevaluation, *Arch. Neurol.* **14:**356–368.

Ellison, G., 1995, The *N*-methyl-D-aspartate antagonists phencyclidine, ketamine, and dizocilpine as both behavioral and anatomical models of the dementias, *Brain Res. Rev.* **20:**250–267.

Engel, P. A., Vinters, H. V., and Grunnet, M., 1992, Alzheimer's disease or plaque disease? Two cases at the frontier of a definition, *J. Geriat. Psychiatry Neurol.* **5:**200–209.

Engel, S. A., Glover, G. H., and Wandell, B. A., 1997, Retinotopic organization in human visual cortex and the spatial precision of functional MRI, *Cereb. Cortex* **7:**181–192.

Esiri, M. M., Pearson, R. C. A., and Powell, T. P. S., 1986, The cortex of the primary auditory area in Alzheimer's disease, *Brain Res.* **366:**385–387.

Eubanks, J. H., Puranam, R. S., Kleckner, N. W., Bettler, B., Heinemann, S. F., and McNamara, J. O., 1993, The gene encoding the glutamate receptor subunit GluR5 is located on human chromosome 21q21.1-22.1 in the vicinity of the gene for familial amyotrophic lateral sclerosis, *Proc. Natl. Acad. Sci. USA* **90:**178–182.

Evans, D. A., Scherr, P. A., Cook, N. A., Albert, M. S., Funkenstein, H. H., Beckett, L. A., Hebert, L. E., Wetle, T. T., Branch, L. G., Chown, M. J., Hennekens, C. H., and Taylor, J. O., 1992, The impact of Alzheimer's disease in the United States population, in: *The Oldest-Old* (R. M. Suzman, D. P. Willis, and K. G. Manton, eds.), Oxford University Press, New York, pp. 283–299.

Fallet-Bianco, C., Roudier, M., Lamour, Y., and Davous, P., 1990, Etude neuropathologique de 50 cas de démence sénile, *Rev. Neurol.* **146:**687–696.

Fama, R., Sullivan, E. V., Shear, P. K., Marsh, L., Yesavage, J. A., Tincklenberg, J. R., Lim, K. O., and Pfefferbaum, A., 1997, Selective cortical and hippocampal volume correlates of Mattis dementia rating scale in Alzheimer disease, *Arch. Neurol.* **54:**719–728.

Farooqui, A. A., and Horrocks, L. A., 1991, Excitatory amino acid receptors, neural membrane phospholipid metabolism, and neurological disorders, *Brain Res. Rev.* **16:**171–191.

Feany, M. B., and Dickson, D. W., 1995, Widespread cytoskeletal pathology characterizes corticobasal degeneration, *Am. J. Pathol.* **146:**1388–1396.

Feany, M. B., Ksiezak-Reding, H., Liu, W. K., Vincent, I., Yen, S. H. C., and Dickson, D. W., 1995, Epitope expression and phosphorylation of tau protein in corticobasal degeneration: Differentiation from progressive supranuclear palsy, *Acta Neuropathol.* **90:**37–43.

Felleman, D. J., and Van Essen, D. C., 1991, Distributed hierarchical processing in the primate cerebral cortex, *Cereb. Cortex* **1:**1–47.

Ferrer, I., and Gullotta, F., 1990, Down's syndrome and Alzheimer's disease: Dendritic spine counts in the hippocampus, *Acta Neuropathol.* **79:**680–685.

Ferrer, I., Soriano, E., Tuñon, T., Fonseca, M., and Guionnet, N., 1991, Parvalbumin immunoreactive neurons in normal human temporal neocortex and in patients with Alzheimer's disease, *J. Neurol. Sci.* **106:**135–141.

Ferrer, I., Casas, R., and Rivera, R., 1993a, Parvalbumin-immunoreactive cortical neurons in Creutzfeldt-Jakob disease, *Ann. Neurol.* **34:**864–866.

Ferrer, I., Tuñon, T., Serrano, M. T., Casas, R., Alcántara, S., Zújar, M. J., and Rivera, R. M., 1993b, Calbindin D-28k and parvalbumin immunoreactivity in the frontal cortex in patients with frontal lobe dementia of non-Alzheimer type associated with amyotrophic lateral sclerosis, *J. Neurol. Neurosurg. Psychiatry* **56:**257–261.

Ferrer, I., Tuñon, T., Soriano, E., Del Rio, A., Irairoz, I., Fonseca, M., and Guionnet, N., 1993c, Calbindin D-28k immunoreactivity in the temporal neocortex in patients with Alzheimer's disease, *Clin. Neuropathol.* **12:**53–58.

Ferrer, I., Zújar, M. J., Rivera, R., Soria, M., Vidal, A., and Casas, R., 1993d, Parvalbumin-immunoreactive dystrophic neurites and aberrant sprouts in the cerebral cortex of patients with Alzheimer's disease, *Neurosci. Lett.* **158:**163–166.

Ferrer, I., Tuñon, T., Serrano, M. T., Casas, R., Alcántara, S., Zújar, M. J., and Rivera, R. M., 1993e, Calbindin D-28k and parvalbumin immunoreactivity in the frontal cortex in patients with frontal lobe dementia of non-Alzheimer type associated with amyotrophic lateral sclerosis, *J. Neurol. Neurosurg. Psychiatry* **56:**257–261.

Filley, C. M., Kleinschmidt-De Masters, B. K., and Gross, K. F., 1996, Non-Alzheimer frontotemporal degenerative dementia: A neurobehavioral and pathologic study, *Clin. Neuropathol.* **13:**109–116.

Flament, S., Delacourte, A., and Mann, D. M. A., 1990a, Phosphorylation of tau proteins: A major event during the process of neurofibrillary degeneration: A comparative study between Alzheimer's disease and Down's syndrome, *Brain Res.* **516:**15–19.

Flament, S., Delacourte, A., Delaère, P., Duyckaerts, C., and Hauw, J. J., 1990b, Correlation between microscopical changes and tau 64 and 69 biochemical detection in senile dementia of the Alzheimer type: Tau 64 and 69 are reliable markers of the neurofibrillary degeneration, *Acta Neuropathol.* **80:**212–215.

Flament, S., Delacourte, A., Verny, M., Hauw, J. J., and Javoy-Agid, F., 1991, Abnormal tau proteins in progressive supranuclear palsy: Similarities and differences with the neurofibrillary degeneration of the Alzheimer type, *Acta Neuropathol.* **81:**591–596.

Fletcher, W. A., 1994, Ophthalmological aspects of Alzheimer's disease, *Curr. Opin. Ophthalmol.* **5:**38–44.

Fletcher, W. A., and Sharpe, J. A., 1988, Smooth pursuit dysfunction in Alzheimer's disease, *Neurology* **38:**272–277.

Fonseca, M., and Soriano, E., 1995, Calretinin-immunoreactive neurons in the normal human temporal cortex and in Alzheimer's disease, *Brain Res.* **691:**83–91.

Fonseca, M., Soriano, E., Ferrer, I., Martinez, A., and Tuñon, T., 1993, Chandelier cell axons identified by parvalbumin-immunoreactivity in the normal human temporal cortex and in Alzheimer's disease, *Neuroscience* **55:**1107–1116.

Forno, L. S., and O'Flanagan, T. J., 1973, Amyotrophic lateral sclerosis of the Guam type in a U.S. veteran, *Neurology* **23:**876–880.

Forno, L. S., Barbour, P. J., and Norville, R. L., 1978, Presenile dementia with Lewy bodies and neurofibrillary tangles, *Arch. Neurol.* **35:**818–822.

Foster, N. L., Wilhelmsen, K., Sima, A. A. F., Jones, M. Z., D'Amato, C. J., and Gilman, S., 1997, Frontotemporal dementia and parkinsonism linked to chromosome 17: A consensus conference, *Ann. Neurol.* **41:**706–715.

Fox, N. C., Warrington, E. K., Freeborough, P. A., Hartikainen, P., Kennedy, A. M., Stevens, J. M., and Rossor, M. N., 1996, Presymptomatic hippocampal atrophy in Alzheimer's disease: A longitudinal study, *Brain* **119:**2001–2007.

Francis, P. T., Sims, N. R., Procter, A. W., and Bowen, D. M., 1993, Cortical pyramidal neurone loss may cause glutamatergic hypoactivity and cognitive impairment in Alzheimer's disease: Investigative and therapeutic perspectives, *J. Neurochem.* **61**:1589–1594.

Fratiglioni, L., Grut, M., Forsell, Y., Viitanen, M., Grafstrom, M., Holmen, K., Ericsson, K., Backman, L., Ahlbom, A., and Winblad, B., 1991, Prevalence of Alzheimer's disease and other dementias in an elderly urban population: Relationship with age, sex, and education, *Neurology* **41**:1886–1892.

French, L. R., Schuman, L. M., Mortimer, J. A., Hutton, J. T., Boatman, R. A., and Christians, B., 1985, A case-control study of dementia of the Alzheimer type, *Am. J. Epidemiol.* **121**:414–421.

Freund, T. F., and Buzsáki, G., 1996, Interneurons of the hippocampus, *Hippocampus* **6**:347–470.

Froelich, S., Basun, H., Forsell, C., Lilius, L., Axelman, K., Andreadis, A., and Lannfelt, L., 1997, Mapping of a disease locus for familial rapidly progressive frontotemporal dementia to chromosome 17q12-21, *Am. J. Med. Genet. (Neuropsychiatry Genet.)* **74**:380–385.

Furey-Kurkjian, M. L., Pietrini, P., Graff-Radford, N. R., Alexander, G. E., Freo, U., Szczepanik, J., and Schapiro, M. B., 1996, Visual variants of Alzheimer disease: Distinctive neuropsychological features, *Neuropsychology* **10**:294–300.

Gabbott, P. L. A., and Bacon, S. J., 1996a, Local circuit neurons in the medial prefrontal cortex (areas 24a,b,c, and 32) in the monkey. I. Cell morphology and morphometrics, *J. Comp. Neurol.* **364**:567–608.

Gabbott, P. L. A., and Bacon, S. J., 1996b, Local circuit neurons in the medial prefrontal cortex (areas 24,a,b,c, 25, and 32) in the monkey. II. Quantitative areal and laminar distributions, *J. Comp. Neurol.* **364**:609–636.

Gajdusek, D. C., 1982, Foci of motor neuron disease in high incidence in isolated populations of East Asia and the Western Pacific, in: *Human Motor Neuron Disease, Advanced neurology,* Vol. 36 (L. P. Rowland, ed.), Raven, New York, pp. 363–393.

Gallagher, M., and Rapp, P. R., 1997, The use of animal models to study the effects of aging on cognition, *Annu. Rev. Psychol.* **48**:339–370.

Galloway, P. G., Grundke-Iqbal, I., Iqbal, K., and Perry, G., 1988, Lewy bodies contain epitopes both shared and distinct from Alzheimer neurofibrillary tangles, *J. Neuropathol. Exp. Neurol.* **47**:654–663.

Galloway, P. G., Mulvihill, P., and Perry, G., 1992, Filaments of Lewy bodies contain insoluble cytoskeletal elements, *Am. J. Pathol.* **140**:809–822.

Gangalingham, K. K., Byrne, E. J., Thornton, A., Sambrook, M. A., and Bannister, P., 1997, Motor and cognitive function in Lewy body dementia: Comparison with Alzheimer's and Parkinson's diseases, *J. Neurol. Neurosurg. Psychiatry* **62**:243–252.

García-Ladona, F. J., Palacios, J. M., Probst, A., Wieser, H. G., Mengod, G., 1994, Excitatory amino acid AMPA receptor mRNA localization in several regions of normal and neurological disease affected human brain: An *in situ* hybridization histochemistry study, *Mol. Brain Res.* **21**:75–84.

Garruto, R. M., 1991, Pacific paradigms of environmentally-induced neurological disorders: Clinical, epidemiological, and molecular perspectives, *Neurotoxicology* **12**:347–378.

Garruto, R. M., and Yase, Y., 1986, Neurodegenerative disorders of the Western Pacific: The search for mechanisms of pathogenesis, *Trends Neurosci.* **9**:368–371.

Garruto, R. M., Gajdusek, D. C., and Chen, K. M., 1980, Amyotrophic lateral sclerosis among Chamorro migrants from Guam, *Ann. Neurol.* **8**:612–619.

Garruto, R. M., Gajdusek, D. C., and Chen, K. M., 1981, Amyotrophic lateral sclerosis and parkinsonism-dementia among Filipino migrants to Guam, *Ann. Neurol.* **10**:341–350.

Garruto, R. M., Fukatsu, R., Yanagihara, R., Gajdusek, D. C., Hook, G., and Fiori, C. E., 1984, Imaging of calcium and aluminum in neurofibrillary tangle-bearing neurons in parkinsonism-dementia of Guam, *Proc. Natl. Acad. Sci. USA* **81**:1875–1879.

Gasic, G. P., and Heinemann, S., 1992, Determinants of the calcium permeation of ligand-gated cation channels, *Curr. Opin. Cell Biol.* **4**:670–677.

Gaspar, P., Duyckaerts, C., Febvret, A., Benoit, R., Beck, B., and Berger, B., 1989, Subpopulations of somatostatin 28-immunoreactive neurons display differential vulnerability in senile dementia of the Alzheimer type, *Brain Res.* **490**:1–13.

Gazzaley, A. H., Siegel, S. J., Kordower, J. H., Mufson, E. J., and Morrison, J. H., 1996a, Circuit-specific alterations of *N*-methyl-D-aspartate subunit 1 in the dentate gyrus of aged monkeys, *Proc. Natl. Acad. Sci. USA* **93**:3121–3125.

Gazzaley, A. H., Weiland, N. G., McEwen, B. S., and Morrison, J. H., 1996b, Differential regulation of NMDAR1 mRNA and protein by estradiol in the rat hippocampus, *J. Neurosci.* **16**:6830–6838.

Gazzaley, A. H., Thakker, M. M., Hof, P. R., and Morrison, J. H., 1997, Preserved number of entorhinal cortex layer II neurons in aged macaque monkeys, *Neurobiol. Aging* **18:**549–553.

Gearing, M., Rebeck, G. W., Hyman, B. T., Tigges, J., and Mirra, S. S., 1994, Neuropathology and apolipoprotein E profile of aged chimpanzees: Implications for Alzheimer's disease, *Proc. Natl. Acad. Sci. USA* **91:**9382–9386.

Gearing, M., Tigges, J., Mori, H., and Mirra, S. S., 1997, β-amyloid (Aβ) deposition in the brains of aged orangutans, *Neurobiol. Aging* **18:**139–146.

Geddes, J. F., Hughes, A. J., Lees, A. J., and Daniel, S. E., 1993, Pathological overlap in cases of parkinsonism associated with neurofibrillary tangles, *Brain* **116:**281–302.

Geddes, J. F., Vowles, G. H., Robinson, S. F. D., and Suttcliffe, J. C., 1996, Neurofibrillary tangles, but not Alzheimer-type pathology, in a young boxer, *Neuropathol. Appl. Neurobiol.* **22:**12–16.

Giaccone, G., Tagliavini, F., Linoli, G., Bouras, C., Frigerio, L., Frangione, B., and Bugiani, O., 1989, Down patients: Extracellular preamyloid deposits precede neuritic degeneration and senile plaques, *Neurosci. Lett.* **97:**232–238.

Giannakopoulos, P., Hof, P. R., Surini, M., Michel, J. P., and Bouras, C., 1993, Quantitative immunohistochemical analysis of the distribution of neurofibrillary tangles and senile plaques in the cerebral cortex of nonagenarians and centenarians, *Acta Neuropathol.* **85:**601–610.

Giannakopoulos, P., Hof, P. R., Giannakopoulos, A. S., Buée-Scherrer, V., Surini, M., Delacourte, A., and Bouras, C., 1994a, Dementia in the oldest-old: Quantitative analysis of 12 cases from a psychiatric hospital, *Dementia* **5:**348–356.

Giannakopoulos, P., Hof, P. R., Mottier, S., Michel, J. P., and Bouras, C., 1994b, Neuropathologic changes in the neocortex of 1258 cases from a geriatric hospital: Retrospective clinicopathologic evaluation of a ten year autopsy population, *Acta Neuropathol.* **87:**456–468.

Giannakopoulos, P., Hof, P. R., Vallet, P. G., Giannakopoulos, A. S., Charnay, Y., and Bouras, C., 1994c, The quantitative analysis of neuropathological changes in the cerebral cortex of centenarians, *Prog. Neuro-Psychopharmacol. Biol. Psychiatry* **19:**577–592.

Giannakopoulos, P., Hof, P. R., and Bouras, C., 1994d, Alzheimer's disease with asymmetric atrophy of the cerebral hemispheres: Morphometric analysis of four cases, *Acta Neuropathol.* **88:**440–447.

Giannakopoulos, P., Hof, P. R., and Bouras, C., 1995a, Age versus aging as a cause of dementia, *Lancet* **346:**1486.

Giannakopoulos, P., Hof, P. R., and Bouras, C., 1995b, Dementia lacking distinctive histopathology: Clinicopathological evaluation of 32 cases, *Acta Neuropathol.* **89:**346–355.

Giannakopoulos, P., Hof, P. R., Giannakopoulos, A. S., Hermann, F. R., Michel, J. P., and Bouras, C., 1995c, Regional distribution of neurofibrillary tangles and senile plaques in the cerebral cortex of very old patients, *Arch. Neurol.* **52:**1150–1159.

Giannakopoulos, P., Hof, P. R., Kövari, E., Vallet, P. G., Herrmann, F. R., and Bouras, C., 1996a, Distinct patterns of neuronal loss and Alzheimer's disease lesion distribution in elderly individuals older than 90 years, *J. Neuropathol. Exp. Neurol.* **55:**1210–1220.

Giannakopoulos, P., Hof, P. R., Savioz, A., Guimon, J., Antonarakis, S. E., and Bouras, C., 1996b, Early-onset dementias: Clinical, neuropathological, and genetic characteristics, *Acta Neuropathol.* **91:**451–465.

Giannakopoulos, P., Bouras, C., Kövari, E., Shioi, J., Tezapsidis, N., Hof, P. R., and Robakis, N. K., 1997a, Presenilin-1-immunoreactive neurons are preserved in late-onset Alzheimer's disease, *Am. J. Pathol.* **150:**429–436.

Giannakopoulos, P., Hof, P. R., Michel, J. P., Guimon, J., and Bouras, C., 1997b, Cerebral cortex pathology in aging and Alzheimer's disease: A survey of large hospital-based geriatric and psychiatric cohorts, *Brain Res. Rev.* **25:**217–245.

Giannakopoulos, P., Silhol, S., Jallageas, V., Mallet, J., Bons, N., Bouras, C., and Delaère, P., 1997c, Quantitative analysis of tau protein-immunoreactive accumulations and β-amyloid protein deposits in the cerebral cortex of the mouse lemur, *Microcebus murinus, Acta Neuropathol.* **94:**131–139.

Giannakopoulos, P., Kövari, E., Hof, P. R., French, L. E., and Bouras, C., 1998, Evidence for clusterin-mediated neuroprotection in Alzheimer's disease: An immunocytochemical study, *Acta Neuropathol.* **95:**387–394.

Gibb, W. R. G., and Lees, A. J., 1987, The progression of idiopathic Parkinson's disease is not explained by age-related changes: Clinical and pathological comparisons with post-encephalitic parkinsonian syndrome, *Acta Neuropathol.* **73:**195–201.

Gibb, W. R. G., Esiri, M. M., and Lees, A. J., 1985, Clinical and pathological features of diffuse cortical Lewy body disease (Lewy body dementia), *Brain* **110:**1131–1153.

Gibbs, C. J., Jr., and Gajdusek, D. C., 1982, An update on longer-term *in vivo* and *in vitro* studies designed to identify a virus as the cause of amyotrophic lateral sclerosis, parkinsonism-dementia, and Parkinson's disease, in: *Human Motor Neuron Diseases: Advanced Neurology,* Vol. 36 (L. P. Rowland, ed.), Raven, New York, pp. 343–353.

Ginsberg, S. D., Crino, P. B., Lee, V. M. Y., Eberwine, J. H., and Trojanowski, J. Q., 1997, Sequestration of RNA in Alzheimer's disease neurofibrillary tangles and senile plaques, *Ann. Neurol.* **41:**200–209.

Goedert, M., Jakes, R., Crowther, R. A., Cohen, P., Vanmechelen, E., Vandermeeren, M., and Cras, P., 1994, Epitope mapping of monoclonal antibodies to the paired helical filaments of Alzheimer's disease: Identification of phosphorylation sites in tau protein, *Biochem. J.* **301:**871–877.

Goedert, M., Jakes, R., Spillantini, M. G., Hasegawa, M., Smith, M. J., and Crowther, R. A., 1996, Assembly of microtubule-associated protein tau into Alzheimer-like filaments induced by sulfated glycosaminoglycans, *Nature* **383:**550–553.

Golaz, J., Bouras, C., and Hof, P. R., 1992, Motor cortex involvement in presenile dementia: Report of a case, *J. Geriat. Psychiatry Neurol.* **5:**85–92.

Gollin, P. A., Kalaria, R. N., Eikelenboom, P., Rozemuller, A., and Perry, G., 1992, $\alpha_1$-Antitrypsin and $\alpha_1$-antichymotrypsin are in the lesions of Alzheimer's disease, *NeuroReport* **3:**201–203.

Gómez-Isla, T., Price, J. L., McKeel, D. W., Jr., Morris, J. C., Growdon, J. H., and Hyman, B. T., 1996, Profound loss of layer II entorhinal cortex neurons occurs in very mild Alzheimer's disease, *J. Neurosci.* **16:**4491–4500.

Gómez-Isla, T., Hollister, R., West, H., Mui, S., Growdon, J. H., Petersen, R. C., Parisi, J. E., and Hyman, B. T., 1997, Neuronal loss correlates with but exceeds neurofibrillary tangles in Alzheimer's disease, *Ann. Neurol.* **41:**17–24.

Gómez-Ramos, P., Mufson, E. J., and Moran, M. A., 1992, Ultrastructural localization of acetylcholinesterase in neurofibrillary tangles, neuropil threads, and senile plaques in aged and Alzheimer's brain, *Brain Res.* **569:**229–237.

Gómez-Ramos, P., Bouras, C., and Morán, M. A., 1994, Ultrastructural localization of butyrylcholinesterase on neurofibrillary degeneration sites in the brains of aged and Alzheimer's disease patients, *Brain Res.* **640:**17–24.

Good, P. F., and Morrison, J. H., 1995, Morphology and kainate-receptor immunoreactivity of identified neurons within the entorhinal cortex projecting to superior temporal sulcus in the cynomolgus monkey, *J. Comp. Neurol.* **357:**25–35.

Good, P. F., Werner, P., Hsu, A., Olanow, C. W., and Perl, D. P., 1996, Evidence for neuronal oxidative damage in Alzheimer's disease, *Am. J. Pathol.* **149:**21–28.

Goodman, L., 1953, Alzheimer's disease: A clinicopathologic analysis of 23 cases with a theory on pathogenesis, *J. Nerv. Ment. Dis.* **118:**97–130.

Gottron, F., Turetsky, D., and Choi, D., 1995, SMI-32 antibody against non-phosphorylated neurofilaments identifies a subpopulation of cultured cortical neurons hypersensitive to kainate toxicity, *Neurosci. Lett.* **194:**1–4.

Graff-Radford, N. R., Bolling, J. P., Earnest, F., IV, Shuster, E. A., Caselli, R. J., and Brazis, P. W., 1993, Simultanagnosia as the initial sign of degenerative dementia, *Mayo Clin. Proc.* **68:**955–964.

Graham, D. I., Gentleman, S. M., Lynch, A., and Roberts, G. W., 1995, Distribution of β-amyloid protein in the brain following severe head injury, *Neuropathol. Appl. Neurobiol.* **21:**27–34.

Grahmann, H., and Ule, G., 1957, Beitrag zur Kenntnis der chronischen cerebralen Krankheitsbilder bei Boxern (Dementia pugilistica und traumatische Boxer-Encephalopathie), *Psychiat. Neurol.* **134:**261–283.

Greenamyre, J. T., and Young, A. B., 1989, Excitatory amino acids and Alzheimer's disease, *Neurobiol. Aging* **10:**593–602.

Greenamyre, J. T., and Maragos, W. F., 1993, Neurotransmitter receptors in Alzheimer disease, *Cerebrovasc. Brain Metab. Rev.* **5:**61–94.

Greenamyre, J. T., Penney, J. B., Young, A. B., D'Amato, C. J., Hicks, S. P., and Shoulson, I., 1985, Alterations in L-glutamate binding in Alzheimer's and Huntington's diseases, *Science* **227:**1469–1499.

Grünthal, E., 1928, Zur Hirnpathologischen Analyse der Alzheimerschen Krankheit, *Psychiat. Neurol. Wschr.* **36:**401–407.

Guiroy, D. C., Mellini, M., Miyazaki, M., Hilbich, C., Safar, J., Garruto, R. M., Yanagihara, R., Beyreuther, K., and Gajdusek, D. C., 1993, Neurofibrillary tangles of Guamanian amyotrophic lateral sclerosis, parkinsonism-dementia, and neurologically normal Guamanians contain a 4- to 4.5-kilodalton protein, which is immunoreactive to anti-amyloid β/A4-protein antibodies, *Acta Neuropathol.* **86:**265–274.

Gurney, M. E., Pu, H., Chiu, A. Y., Dal Canto, M. C., Polchow, C. Y., Alexander, D. D., Caliendo, J., Hentati, A., Kwon, Y. W., Deng, H. X., Chen, W., Zhai, P., Sufit, R. L., and Siddique, T., 1994, Motor neuron degeneration in mice that express a human Cu,Zn superoxide dismutase mutation, *Science* **264:**1772–1775.

Gustafson, L., 1987, Frontal lobe degeneration of non-Alzheimer type. II. Clinical picture and differential diagnosis, *Arch. Gerontol. Geriat.* **6:**209–223.

Gustafson, L., 1993, Clinical picture of frontal lobe degeneration of non-Alzheimer type, *Dementia* **4:**143–148.

Guterman, A., and Smith, R. W., 1987, Neurological sequelae of boxing, *Sports Med.* **4:**194–210.

Hamano, T., Yoshimura, M., Yamazaki, T., Shinkai, Y., Yanagisawa, K., Kuriyama, M., and Ihara, Y., 1997, Amyloid β-protein accumulation in the leptomeninges during aging and in Alzheimer disease, *J. Neuropathol. Exp. Neurol.* **56:**922–932.

Hanger, D. P., Brion, J. P., Gallo, J. M., Cairns, N. J., Luthert, P. J., and Anderton, B. H., 1991, Tau in Alzheimer's disease and Down's syndrome is insoluble and abnormally phosphorylated, *Biochem. J.* **275:**99–104.

Hansen, L., Salmont, D., Galasko, D., Masliah, E., Katzman, R., DeTeresa, R., Thal, L., Pay, M. M., Hofstetter, R., Klauber, M., Rice, V., Butters, N., and Alford, M., 1990, The Lewy body variant of Alzheimer's disease: A clinical and pathologic entity, *Neurology* **40:**1–8.

Hansen, L. A., DeTeresa, R., Davies, P., and Terry, R. D., 1988a, Neocortical morphometry, lesion counts, and choline acetyltransferase levels in the age spectrum of Alzheimer's disease, *Neurology* **38:**48–54.

Hansen, L. A., DeTeresa, R., Tobias, H., Alford, M., and Terry, R. D., 1988b, Neocortical morphometry and cholinergic neurochemistry in Pick's disease, *Am. J. Pathol.* **131:**507–518.

Hauw, J. J., Vignolo, P., Duyckaerts, C., Beck, H., Forette, F., Henry, J. F., Laurent, M., Piette, F., Sachet, A., and Berthaux, P., 1986, Etude neuropathologique de 12 centenaires: La fréquence de la démence sénile de type Alzheimer n'est pas particulièrement élevée dans ce groupe de personne très agées, *Rev. Neurol.* **142:**107–115.

Hauw, J. J., Verny, M., Delaère, P., Cervera, P., He, Y., and Duyckaerts, C., 1990, Constant neurofibrillary changes in the neocortex of progressive supranuclear palsy: Basic differences with Alzheimer's disease and aging, *Neurosci. Lett.* **119:**182–186.

Hauw, J. J., Duyckaerts, C., Seilhean, D., Camilleri, S., Sadzovitch, V., and Rancurel, G., 1996, The neuropathologic diagnostic criteria of frontal lobe dementia revisited: A study of ten consecutive cases, *J. Neural Transm.* (Suppl.) **47:**47–59.

Hanihara, T., Amano, N., Takahashi, T., Nagatomo, H., and Yagashita, S., 1995, Distribution of tangles and threads in the cerebral cortex in progressive supranuclear palsy, *Neuropathol. Appl. Neurobiol.* **21:**319–326.

Haug, H., Kühl, S., Mecke, E., Sass, N. L., and Wasner, K., 1984, The significance of morphometric procedures in the investigation of age changes in cytoarchitectonic structures of human brain, *J. Hirnforsch.* **4:**353–374.

Hayes, T. L., and Lewis, D. A., 1992, Nonphosphorylated neurofilament protein and calbindin immunoreactivity in layer III pyramidal neurons of human neocortex, *Cereb. Cortex* **2:**56–67.

He, Y., Ong, W. Y., Leong, S. K., and Garey, L. J., 1996, Distribution of the glutamate receptor subunit GluR1 and GABA in the human cerebral cortex: A double immunolabelling and electron microscopic study, *Exp. Brain Res.* **112:**147–157.

He, Y., Ong, W. Y., and Leong, S. K., 1997, Species differences in the distribution of the glutamate receptor subunit, GluR1, in the rat and monkey entorhinal cortex, *J. Brain Res.* **38:**27–33.

He, Y., Janssen, W. G. M., Vissavajjhala, P., and Morrison, J. H., 1998, Synaptic distribution of GluR2 in hippocampal GABAergic interneurons and pyramidal cells: A double label immunogold analysis, *Exp. Neurol.* **150:**1–13.

Heckers, S., Geula, C., and Mesulam, M. M., 1992, Acetylcholinesterase-rich pyramidal neurons in Alzheimer's disease, *Neurobiol. Aging* **13:**455–460.

Heizmann, C. W., and Braun, K., 1992, Changes in $Ca^{2+}$-binding proteins in human neurodegenerative disorders, *Trends Neurosci.* **15:**259–264.

Hendry, S. H. C., Jones, E. G., Emson, P. C., Lawson, D. E. M., Heizmann, C. W., and Streit, P., 1989, Two classes of cortical GABA neurons defined by differential calcium-binding protein immunoreactivities, *Exp. Brain Res.* **76:**467–472.

Heutink, P., Stevens, M., Rizzu, P., Bakker, E., Kros, J. M., Tibben, A., Niermeijer, M. F., van Duijn, C. M., Oostra, B. A., and van Swieten, J. C., 1997, Hereditary frontotemporal dementia is linked to chromosome 17q21-22: A genetic and clinicopathological study of three Dutch families, *Ann. Neurol.* **41:**150–159.

Higuchi, Y., Iwaki, T., and Tateishi, J., 1995, Neurodegeneration in the limbic and paralimbic system in progressive supranuclear palsy, *Neuropathol. Appl. Neurobiol.* **21:**246–254.

Hill, W. D., Lee, V. M. Y., Hurtig, H. I., Murray, J. M., and Trojanowski, J. Q., 1991, Epitopes located in spatially separate domains of each neurofilament subunit are present in Parkinson's disease Lewy bodies, *J. Comp. Neurol.* **309:**150–160.

Hinton, D. R., Sadun, A. A., Blanks, J. C., and Miller, C. A., 1986, Optic-nerve degeneration in Alzheimer's disease, *New Engl. J. Med.* **315:**485–487.

Hirano, A., 1990, Amyotrophic lateral sclerosis on Guam: In pursuit of pathology over 30 years, in: *ALS: New Advances in Toxicology and Epidemiology* (F. C. Rose and F. H. Norris, eds.), Smith-Gordon, New York, pp. 103–111.

Hirano, A., and Zimmerman, H. M., 1962, Alzheimer's neurofibrillary changes: A topographic study, *Arch. Neurol.* **7:**227–242.

Hirano, A., and Llena, J., 1986, Neuropathological features of parkinsonism-dementia complex on Guam: Reappraisal and comparative study with Alzheimer's disease and Parkinson's disease, in: *Progress in Neuropathology Vol. 6* (H. M. Zimmerman, ed.), Raven, New York, pp. 17–31.

Hirano, A., Kurland, L. T., Krooth, R. S., and Lessell, S., 1961a, Parkinsonism-dementia complex, an endemic disease on the island of Guam. I. Clinical features, *Brain* **84:**642–661.

Hirano, A., Malamud, N., and Kurland, L. T., 1961b, Parkinsonism-dementia complex, an endemic disease on the island of Guam. II. Pathological features, *Brain* **84:**662–679.

Hirano, A., Malamud, N., Elizan, T. S., and Kurland, L. T., 1966, Amyotrophic lateral sclerosis and parkinsonism-dementia complex on Guam: Further pathologic studies, *Arch. Neurol.* **15:**35–51.

Hirano, A., Arumugasamy, N., and Zimmerman, H. M., 1967, Amyotrophic lateral sclerosis: A comparison of Guam and classical cases, *Arch. Neurol.* **16:**357–363.

Hirano, A., Dembitzer, H. M., Kurland, L. T., and Zimmerman, H. M., 1968, The fine structure of some intraganglionic alterations: Neurofibrillary tangles, granulovacuolar bodies, and "rodlike" structures as seen in Guam amyotrophic lateral sclerosis and parkinsonism-dementia complex, *J. Neuropathol. Exp. Neurol.* **27:**167–182.

Höchli, M., Zetsche, T., and Chan-Palay, V., 1991, Parvalbumin-immunoreactive neurons in the normal human hippocampus, *Dementia* **2:**243–258.

Hodges, J. R., Salmon, D. P., and Butters, N., 1992, Semantic memory impairment in Alzheimer's disease: Failure of access or degraded knowledge? *Neuropsychologia* **30:**301–314.

Hof, P. R., 1997, Morphology and neurochemical characteristics of the vulnerable neurons in brain aging and Alzheimer's disease, *Eur. Neurol.* **37:**71–81.

Hof, P. R., and Morrison, J. H., 1990, Quantitative analysis of a vulnerable subset of pyramidal neurons in Alzheimer's disease. II. Primary and secondary visual cortex, *J. Comp. Neurol.* **301:**55–64.

Hof, P. R., and Bouras, C., 1991, Object recognition deficit in Alzheimer's disease: Possible disconnection of the occipito-temporal component of the visual system, *Neurosci. Lett.* **122:**53–56.

Hof, P. R., and Morrison, J. H., 1991, Neocortical neuronal subpopulations labeled by a monoclonal antibody to calbindin exhibit differential vulnerability in Alzheimer's disease, *Exp. Neurol.* **111:**293–301.

Hof, P. R., and Nimchinsky, E. A., 1992, Regional distribution of neurofilament and calcium-binding proteins in the cingulate cortex of the macaque monkey, *Cereb. Cortex* **2:**456–467.

Hof, P. R., and Morrison, J. H., 1994, The cellular basis of cortical disconnection in Alzheimer disease and related dementing conditions, in: *Alzheimer Disease* (R. D. Terry, R. Katzman, and K. L. Bick, ed.), New York: Raven, pp. 197–229.

Hof, P. R., and Morrison, J. H., 1995, Neurofilament protein defines regional patterns of cortical organization in the macaque monkey visual system: A quantitative immunohistochemical analysis, *J. Comp. Neurol.* **352:**161–186.

Hof, P. R., and Morrison, J. H., 1996, Hippocampal and neocortical involvement in normal brain aging and dementia: Morphologic and neurochemical profile of the vulnerable circuits, *J. Am. Geriat. Soc.* **44:**857–864.

Hof, P. R., Bouras, C., Constantinidis, J., and Morrison, J. H., 1989, Balint's syndrome in Alzheimer's disease: Specific disruption of the occipito-parietal visual pathway, *Brain Res.* **493:**368–375.

Hof, P. R., Cox, K., and Morrison, J. H., 1990a, Quantitative analysis of a vulnerable subset of pyramidal neurons in Alzheimer's disease. I. Superior frontal and inferior temporal cortex, *J. Comp. Neurol.* **301:**44–54.

Hof, P. R., Bouras, C., Constantinidis, J., and Morrison, J. H., 1990b, Selective disconnection of specific visual association pathways in cases of Alzheimer's disease presenting with Balint's syndrome, *J. Neuropathol. Exp. Neurol.* **49:**168–184.

Hof, P. R., Knabe, R., Bovier, P., and Bouras, C., 1991a, Neuropathological observations in a case of autism presenting with self injury behavior, *Acta Neuropathol.* **82:**321–326.

Hof, P. R., Cox, K., Young, W. G., Celio, M. R., Rogers, J., and Morrison, J. H., 1991b, Parvalbumin-immunoreactive neurons in the neocortex are resistant to degeneration in Alzheimer's disease, *J. Neuropathol. Exp. Neurol.* **50:**451–462.

Hof, P. R., Perl, D. P., Loerzel, A. J., and Morrison, J. H., 1991c, Neurofibrillary tangle distribution in the cerebral cortex of parkinsonism-dementia cases from Guam: Differences with Alzheimer's disease, *Brain Res.* **564:**306–313.

Hof, P. R., Delacourte, A., and Bouras, C., 1992a, Distribution of cortical neurofibrillary tangles in progressive supranuclear palsy: A quantitative analysis of six cases, *Acta Neuropathol.* **84:**45–51.

Hof, P. R., Bierer, L. M., Perl, D. P., Delacourte, A., Buée, L., Bouras, C., and Morrison, J. H., 1992b, Evidence for early vulnerability of the medial and inferior aspects of the temporal lobe in an 82-year-old patient with preclinical signs of dementia: Regional and laminar distribution of neurofibrillary tangles and senile plaques, *Arch. Neurol.* **49:**946–953.

Hof, P. R., Bouras, C., Buée, L., Delacourte, A., Perl, D. P., and Morrison, J. H., 1992c, Differential distribution of neurofibrillary tangles in the cerebral cortex of dementia pugilistica and Alzheimer's disease cases, *Acta Neuropathol.* **85:**23–30.

Hof, P. R., Charpiot, A., Delacourte, A., Buée, L., Purohit, D., Perl, D. P., and Bouras, C., 1992d, Distribution of neurofibrillary tangles and senile plaques in the cerebral cortex in postencephalitic parkinsonism, *Neurosci. Lett.* **139:**10–14.

Hof, P. R., Archin, N., Osmand, A. P., Dougherty, G. H., Wells, C., Bouras, C., and Morrison, J. H., 1993a, Posterior cortical atrophy in Alzheimer's disease: Analysis of a new case and reevaluation of an historical report, *Acta Neuropathol.* **86:**215–223.

Hof, P. R., Lüth, H. J., Rogers, J. H., and Celio, M. R., 1993b, Calcium-binding proteins define subpopulations of interneurons in the cingulate cortex, in: *Neurobiology of Cingulate Cortex and Limbic Thalamus* (B. A. Vogt and M. Gabriel, eds.), Birkhäuser, Boston, pp. 181–205.

Hof, P. R., Nimchinsky, E. A., Celio, M. R., Bouras, C., and Morrison, J. H., 1993c, Calretinin-immunoreactive neocortical interneurons are unaffected in Alzheimer's disease, *Neurosci. Lett.* **152:**145–149.

Hof, P. R., Perl, D. P., Loerzel, A. J., Steele, J. C., and Morrison, J. H., 1994a, Amyotrophic lateral sclerosis and parkinsonism-dementia from Guam: Differences in neurofibrillary tangle distribution and density in the hippocampal formation and neocortex, *Brain Res.* **650:**107–116.

Hof, P. R., Nimchinsky, E. A., Buée-Scherrer, V., Buée, L., Nasrallah, J., Hottinger, A. F., Purohit, D. P., Loerzel, A. J., Steele, J. C., Delacourte, A., Bouras, C., Morrison, J. H., and Perl, D. P., 1994b, Amyotrophic lateral sclerosis/parkinsonism-dementia complex of Guam: Quantitative neuropathology, immunohistochemical analysis of neuronal vulnerability, and comparison with related neurodegenerative disorders, *Acta Neuropathol.* **88:**397–404.

Hof, P. R., Bouras, C., Perl, D. P., and Morrison, J. H., 1994c, Quantitative neuropathologic analysis of Pick's disease cases: Cortical distribution of Pick bodies and coexistence with Alzheimer's disease, *Acta Neuropathol.* **87:**115–124.

Hof, P. R., Giannakopoulos, P., Vickers, J. C., Bouras, C., and Morrison, J. H., 1995a, The morphologic and neurochemical basis of dementia: Aging, hierarchical patterns of lesion distribution, and vulnerable neuronal phenotype, *Rev. Neurosci.* **6:**97–124.

Hof, P. R., Mufson, E. J., and Morrison, J. H., 1995b, The human orbitofrontal cortex: Cytoarchitecture and quantitative immunohistochemical parcellation, *J. Comp. Neurol.* **359:**48–68.

Hof, P. R., Nimchinsky, E. A., and Morrison, J. H., 1995c, Neurochemical phenotype of corticocortical connections in the macaque monkey: Quantitative analysis of a subset of neurofilament protein-immunoreactive projection neurons in frontal, parietal, temporal, and cingulate cortices, *J. Comp. Neurol.* **362:**109–133.

Hof, P. R., Bouras, C., Perl, D. P., Sparks, D. L., Mehta, N., and Morrison, J. H., 1995d, Age-related distribution of neuropathologic changes in the cerebral cortex of patients with Down's syndrome: Quantitative regional analysis and comparison with Alzheimer's disease, *Arch. Neurol.* **52:**379–391.

Hof, P. R., Giannakopoulos, P., and Bouras, C., 1996a, The neuropathological changes associated with normal brain aging, *Histol. Histopathol.* **11:**1075–1088.

Hof, P. R., Ungerleider, L. G., Webster, M. J., Gattass, R., Adams, M. M., Sailstad, C. A., and Morrison, J. H., 1996b, Neurofilament protein is differentially distributed in subpopulations of corticocortical projection neurons in the macaque monkey visual pathways, *J. Comp. Neurol.* **376:**112–127.

Hof, P. R., Vissavajjhala, P., Rosenthal, R. E., Fiskum, G., and Morrison, J. H., 1996c, Distribution of glutamate receptor subunit proteins GluR2(4), GluR5/6/7, and NMDAR1 in the canine and primate cerebral cortex: A comparative immunohistochemical analysis, *Brain Res.* **723:**77–89.

Hof, P. R., Nimchinsky, E. A., Ungerleider, L. G., and Morrison, J. H., 1997a, Morphologic and neurochemical characteristics of corticocortical projections: Emergence of circuit-specific features and relationships to degenerative changes in Alzheimer's disease, in: *Connections, Cognition, and Alzheimer's Disease* (B. T. Hyman, C. Duyckaerts, and Y. Christen, eds.), Springer-Verlag, Berlin, pp. 59–82.

Hof, P. R., Ungerleider, L. G., Adams, M. M., Webster, M. J., Gattass, R., Blumberg, D. M., and Morrison, J. H., 1997b, Callosally-projecting neurons in the macaque monkey V1/V2 border are enriched in nonphosphorylated neurofilament protein, *Vis. Neurosci.* **14:**981–987.

Hof, P. R., Vogt, B. A., Bouras, C., and Morrison, J. H., 1997c, Atypical form of Alzheimer's disease with prominent posterior cortical atrophy: A review of lesion distribution and circuit disconnection in cortical visual pathways, *Vision Res.* **37:**3609–3625.

Hooten, W. M., and Lyketsos, C. G., 1996, Frontotemporal dementia: A clinicopathological review of four postmortem studies, *J. Neuropsychiatry Clin. Neurosci.* **8:**10–19.

Howard, R. S., and Lees, A. J., 1987, Encephalitis lethargica: A report of four recent cases, *Brain* **110:**19–33.

Hubbard, B. M., Fenton, G. W., and Anderson, J. M., 1990, A quantitative histological study of early clinical and preclinical Alzheimer's disease, *Neuropathol. Appl. Neurobiol.* **16:**111–121.

Hudson, A. J., 1981, Amyotrophic lateral sclerosis and its association with dementia, parkinsonism, and other neurological disorders: A review, *Brain* **104:**217–247.

Huntley, G. W., Rogers, S. W., Moran, T., Janssen, W., Archin, N., Vickers, J. C., Cauley, K., Heinemann, S. F., and Morrison, J. H., 1993, Selective distribution of kainate receptor subunit immunoreactivity in monkey neocortex revealed by a monoclonal antibody that recognizes glutamate receptor subunits GluR5/6/7, *J. Neurosci.* **13:**2965–2981.

Huntley, G. W., Vickers, J. C., Janssen, W., Brose, N., Heinemann, S. F., and Morrison, J. H., 1994a, Distribution and synaptic localization of immunocytochemically identified NMDA receptor subunit proteins in sensory-motor and visual cortices of monkey and human, *J. Neurosci.* **14:**3603–3619.

Huntley, G. W., Vickers, J. C., and Morrison, J. H., 1994b, Cellular and synaptic localization of NMDA and non-NMDA receptor subunits in neocortex: Organizational features related to cortical circuitry, function, and disease, *Trends Neurosci.* **17:**536–543.

Huntley, G. W., Vickers, J. C., and Morrison, J. H., 1997, Quantitative localization of NMDAR1 receptor subunit immunoreactivity in inferotemporal and prefrontal association cortices of monkey and human, *Brain Res.* **749:**245–262.

Hyman, B. T., Damasio, A. R., Van Hoesen, G. W., and Barnes, C. L., 1984, Alzheimer's disease: Cell specific pathology isolates the hippocampal formation, *Science* **225:**1168–1170.

Hyman, B. T., Van Hoesen, G. W., Kromer, L. J., and Damasio, A. R., 1986, Perforant pathway changes and the memory impairment of Alzheimer's disease, *Ann. Neurol.* **20:**472–481.

Hyman, B. T., Van Hoesen, G. W., and Damasio, A. R., 1987, Alzheimer's disease: Glutamate depletion in the hippocampal perforant pathway zone, *Ann. Neurol.* **22:**37–40.

Hyman, B. T., Van Hoesen, G. W., and Damasio, A. R., 1990, Memory-related neural systems in Alzheimer's disease: An anatomic study, *Neurology* **40:**1721–1730.

Hyman, B. T., Marzloff, K., Wenninger, J. J., Dawson, T. M., Bredt, D. S., and Snyder, S. H., 1992, Relative sparing of nitric oxide synthase-containing neurons in the hippocampal formation in Alzheimer's disease, *Ann. Neurol.* **32:**818–820.

Hyman, B. T., Penney, J. B., Jr., Blackstone, C. D., and Young, A. B., 1994, Localization of non-*N*-

methyl-D-aspartate glutamate receptors in normal and Alzheimer hippocampal formation, *Ann. Neurol.* **35:**31–37.

Hyman, B. T., West, H. L., Rebeck, G. W., Lai, F., and Mann, D. M. A., 1995, Neuropathological changes in Down syndrome hippocampal formation: Effect of age and apolipoprotein E genotype, *Arch. Neurol.* **52:**373–378.

Iacopino, A. M., and Christakos, S., 1990, Specific reduction of calcium-binding protein (28-kilodalton calbindin-D) gene expression in aging and neurodegenerative diseases, *Proc. Natl. Acad. Sci. USA* **87:**4078–4082.

Iacopino, A. M., Quintero, E. M., and Miller, E. K., 1994, Calbindin-$D_{28K}$: A potential neuroprotective protein, *Neurodegeneration* **3:**1–20.

Ichimiya, Y., Emson, P. C., Mountjoy, C. Q., Lawson, D. E. M., and Heizmann, C. W., 1988, Loss of calbindin-28k immunoreactive neurones from the cortex in Alzheimer-type dementia, *Brain Res.* **475:**156–159.

Iimoto, D. S., Masliah, E., DeTeresa, R., Terry, R. D., and Saitoh, T., 1989, Aberrant casein kinase II in Alzheimer's disease, *Brain Res.* **507:**273–280.

Ikonomovic, M. D., Sheffield, R., and Armstrong, D. M., 1995, AMPA-selective glutamate receptor subtype immunoreactivity in the aged human hippocampal formation, *J. Comp. Neurol.* **359:**239–252.

Ishii, T., and Nakamura, Y., 1981, Distribution and ultrastructure of Alzheimer's neurofibrillary tangles in postencephalitic parkinsonism of Economo type, *Acta Neuropathol.* **55:**59–62.

Ishino, H., and Otsuki, S., 1976, Frequency of Alzheimer's neurofibrillary tangles in the cerebral cortex in progressive supranuclear palsy (subcortical argyrophilic dystrophy), *J. Neurol. Sci.* **28:**309–316.

Ito, H., Goto, S., Hirano, A., and Yen, S. H. C., 1991a, Immunohistochemical study of the hippocampus in parkinsonism-dementia complex on Guam, *J. Geriat. Psychiatry Neurol.* **4:**134–142.

Ito, H., Hirano, A., Yen, S. H. C., and Kato, A., 1991b, Demonstration of β amyloid protein-containing neurofibrillary tangles in parkinsonism-dementia complex on Guam, *Neuropathol. Appl. Neurobiol.* **17:**365–373.

Jamada, M., and Mehraein, P., 1968, Verteilungsmuster der senilen Veränderungen im Gehirn: Die Beteiligung des limbischen Systems bei hirnatrophischen Prozessen des Seniums und bei Morbus Alzheimer, *Arch. Psychiat. Neurol.* **211:**308–324.

Jackson, M., and Lowe, J., 1996, The new neuropathology of degenerative frontotemporal dementias, *Acta Neuropathol.* **91:**127–134.

Jacquet, M. F., Boucquey, D., Theaux, R., Smidts, M., Callewaerts, A., Brucher, J. M., and Kevers, L., 1990, L'atrophie corticale postérieure: Variante anatomo-clinique de la maladie d'Alzheimer, *Acta Neurol. Belg.* **90:**265–273.

Jagust, W. J., Davies, P., Tiller-Borcich, J. K., and Reed, B. R., 1990, Focal Alzheimer's disease, *Neurology* **40:**14–19.

Jakob, H., 1960, Zur pathologischen Anatomie der Pickschen Krankheit. I. Mitteilung. Vergleichende Untersuchungen über Ausdehnung und Schwerpunkte der Atrophie, *Arch. Psychiat. Zschr. Ges. Neurol.* **201:**269–297.

Jakob, H., 1961, Zur pathologischen Anatomie der Pickschen Krankheit. II. Mitteilung. Die feineren histologischen Veränderungen und die Frage der Lokalisation des Prozeβbeginns am Neuron, *Arch. Psychiat. Zschr. Ges. Neurol.* **202:**540–568.

Jakob, H., 1979, Die Picksche Krankheit. Eine neuropathologisch-anatomisch-klinische Studie, in: *Monographien aus dem Gesamtgebiete der Psychiatrie, Vol. 23* (H. Hippius, W. Janzarik, and C. Müller, eds.), Springer-Verlag, Berlin, pp. 1–110.

Jellinger, K., 1971, Progressive supranuclear palsy (subcortical argyrophilic dystrophy), *Acta Neuropathol.* **19:**347–352.

Jellinger, K. A., 1994a, Quantitative neuropathologic analysis of Pick's disease cases, *Acta Neuropathol.* **87:**223–224.

Jellinger, K. A., 1994b, Rare neurodegenerative disorders, in: *Neurodegenerative Diseases* (D. B. Calne, ed.), W. B. Saunders, Philadelphia, pp. 909–931.

Jendroska, K. Rossor, M. N., Mathias, C. J., and Daniel, S. E., 1995, Morphological overlap between corticobasal degeneration and Pick's disease: A clinicopathological report, *Mov. Disord.* **10:**111–114.

Jobst, K. A., Smith, A. D., Szatmari, M., Esiri, M. M., Jaskowski, A., Hindley, N., McDonald, B., and Molyneux, A. J., 1994, Rapidly progressing atrophy of medial temporal lobe in Alzheimer's disease, *Lancet* **343:**829–830.

Jones, E. G., and Wise, S. P., 1977, Size, laminar, and columnar distribution of efferent cells in the sensory-motor cortex of monkeys, *J. Comp. Neurol.* **175:**391–438.

Jordan, B. D., 1987, Neurologic aspects of boxing, *Arch. Neurol.* **44:**453–459.

Jorm, A. F., 1991, Cross-national comparisons of the occurrence of Alzheimer's and vascular dementias, *Eur. Arch. Psychiatry Clin. Neurosci.* **240:**218–222.

Kalus, P., Braak, H., Braak, E., and Bohl, J., 1989, The presubicular region in Alzheimer's disease: Topography of amyloid deposits and neurofibrillary changes, *Brain Res.* **494:**198–203.

Kamo, H., McGeer, P. L., Harrop, R., McGeer, E. G., Calne, D. B., Martin, W. R. W., and Pate, B. D., 1987, Positron emission tomography and histopathology in Pick's disease, *Neurology* **37:**439–445.

Katz, B., and Rimmer, S., 1989, Ophthalmologic manifestations of Alzheimer's disease, *Surv. Ophthalmol.* **34:**31–34.

Katzman, R., and Terry, R., 1983, Normal aging of the nervous system, in: *The Neurology of Aging* (R. Katman and R. D. Terry, eds.), F. A. Davis, Philadelphia, pp. 15–50.

Katzman, R., Terry, R., DeTeresa, R., Brown, T., Davies, P., Fuld, P., Renbing, X., and Peck, A., 1988, Clinical, pathological, and neurochemical changes in dementia: A subgroup with preserved mental status and numerous neocortical plaques, *Ann. Neurol.* **23:**138–144.

Kawamata, T., Tooyama, I., Yamada, T., Walker, D. G., and McGeer, P. L., 1993, Lactotransferrin immunohistochemistry in Alzheimer and normal human brain, *Am. J. Pathol.* **142:**1574–1585.

Kelly, M., and Kowall, N. W., 1989, Corticotropin-releasing factor immunoreactive neurons persist throughout the brain in Alzheimer's disease, *Brain Res.* **501:**392–396.

Kemper, T. L., 1984, Neuroanatomical and neuropathological changes in normal aging and in dementia, in: *Clinical Neurology of Aging* (M. L. Alberts, ed.), Oxford University Press, New York, pp. 9–52.

Kemper, T. L., 1991, Down syndrome, in: *Normal and Altered States of Function, Cerebral Cortex, Vol. 9* (A. Peters and E. G. Jones, eds.), Plenum, New York, pp. 511–526.

Kisvárday, Z. F., Gulyas, A., Beroukas, D., North, J. B., Chubb, I. W., and Somogyi, P., 1990, Synapses, axonal and dendritic patterns of GABA-immunoreactive neurons in human cerebral cortex, *Brain* **113:**793–812.

Kosaka, K., 1994, Diffuse neurofibrillary tangles with calcification: A new presenile dementia, *J. Neurol. Neurosurg. Psychiatry* **57:**594–596.

Kesslak, J. P., Yuan, D., Neeper, S., and Cotman, C. W., 1995, Vulnerability of the hippocampus to kainate excitotoxicity, *Neurosci. Lett.* **188:**117–120.

Kidd, M., 1963, Paired helical filaments in electron microscopy of Alzheimer's disease, *Nature* **197:**192–193.

Kidd, M., 1964, Alzheimer's disease: An electron microscopical study, *Brain* **87:**307–320.

Killiany, R. J., Moss, M. B., Albert, M. S., Sandor, T., Tieman, J., and Jolesz, F., 1993, Temporal lobe regions on magnetic resonance imaging identify patients with early Alzheimer's disease, *Arch. Neurol.* **50:**949–954.

Kinoshita, A., Tomimoto, H., Suenaga, T., Akiguchi, I., and Kimura, J., 1997, Ubiquitin-related cytoskeletal abnormality in frontotemporal dementia: Immunohistochemical and immuno-electron microscope studies, *Acta Neuropathol.* **94:**67–72.

Kisby, G. E., Ellison, M., and Spencer, P. S., 1992, Content of the neurotoxins cycasin (methylazoxymethanol-β-D-glucoside) and BMAA (β-*N*-methylamino-L-alanine) in cycad flour prepared by Guam Chamorros, *Neurology* **42:**1336–1340.

Kiyosawa, M., Bosley, T. M., Chawluk, J., Jamieson, D., Schatz, N. J., Savino, P. J., Sergott, R. C., Reivich, M., and Alavi, A., 1989, Alzheimer's disease with prominent visual symptoms, *Ophthalmology* **96:**1077–1086.

Knopman, D. S., Christensen, K. J., Schut, L. J., Harbaugh, R. E., Reeder, T., Ngo, T., and Frey, W., II, 1989, The spectrum of imaging and neuropsychological findings in Pick's disease, *Neurology* **39:**362–368.

Knopman, D. S., 1993, Overview of dementia lacking distinctive histopathology: Pathological designation of a progressive dementia, *Dementia* **4:**132–136.

Knopman, D. S., Mastri, A. R., Frey, W. H., Sung, J. H., and Rustan, T., 1990, Dementia lacking distinctive histologic features: A common non-Alzheimer degenerative dementia, *Neurology* **40:**251–266.

Kobayashi, K., Emson, P. C., Mountjoy, C. Q., Thornton, S. N., Lawson, D. E. M., and Mann, D. M. A., 1990, Cerebral cortical calbindin D-28k and parvalbumin neurones in Down's syndrome, *Neurosci. Lett.* **113:**17–22.

Koerner, D. R., 1952, Amyotrophic lateral sclerosis on Guam: A clinical study and review of the literature, *Ann. Intern. Med.* **37:**1204–1220.

Kostovic, I., 1990, Structural and histochemical reorganization of the human prefrontal cortex during perinatal and postnatal life, in: *The Prefrontal Cortex: Its Structure, Function, and Pathology, Prog. Brain Res., Vol. 85* (H. B. M. Uylings, C. C. Van Eden, J. P. C. de Bruin, M. A. Corner, and M. G. P. Feenstra, eds.), Elsevier, Amsterdam, pp. 223–240.

Kowall, N. W., and Beal, M. F., 1988, Cortical somatostatin, neuropeptide Y, and NADPH diaphorase neurons: Normal anatomy and alterations in Alzheimer's disease, *Ann. Neurol.* **23:**105–114.

Kowall, N. W., and Beal, M. F., 1989, Galanin-like immunoreactivity is present in human substantia innominata and in senile plaques in Alzheimer's disease, *Neurosci. Lett.* **98:**118–123.

Kowall, N. W., and Beal, M. F., 1991, Glutamate-, glutaminase-, and taurine-immunoreactive neurons develop neurofibrillary tangles in Alzheimer's disease, *Ann. Neurol.* **29:**162–167.

Ksiezak-Reding, H., Dickson, D. W., Davies, P., and Yen, S. H., 1987, Recognition of tau epitopes by anti-neurofilament antibodies that bind to Alzheimer neurofibrillary tangles, *Proc. Natl. Acad. Sci. USA* **84:**3410–3414.

Ksiezak-Reding, H., Morgan, K., Mattiace, L. A., Davies, P., Liu, W. K., Yen, S. H., Weidenheim, K., and Dickson, D. W., 1994, Ultrastructure and biochemical composition of paired helical filaments in corticobasal degeneration, *Am. J. Pathol.* **145:**1496–1508.

Kurland, L. T., 1988, Amyotrophic lateral sclerosis and Parkinson's disease complex on Guam linked to an environmental toxin, *Trends Neurosci.* **11:**51–54.

Kurylo, D. D., Corkin, S., Dolan, R. P., Rizzo, J. F., III, Parker, S. W., and Growdon, J. H., 1994, Broad-band visual capacities are not selectively impaired in Alzheimer's disease, *Neurobiol. Aging* **15:**305–311.

Kurylo, D. D., Corkin, S., Rizzo, J. F., III, and Growdon, J. H., 1996, Greater relative impairment of object recognition than visuospatial abilities in Alzheimer's disease, *Neuropsychology* **10:**74–81.

Kuskowski, M. A., Malone, S. M., Mortimer, J. A., and Dysken, M. W., 1989, Smooth pursuit eye movements in dementia of the Alzheimer type, *Alzheimer Dis. Assoc. Disord.* **3:**157–171.

Lampert, P. W., and Hardman, J. M., 1984, Morphological changes in the brains of boxers, *J. Am. Med. Assoc.* **251:**2676–2679.

Lang, A. E., Bergeron, C., Pollanen, M. S., and Ashby, P., 1994, Parietal Pick's disease mimicking cortical-basal ganglionic degeneration, *Neurology* **44:**1436–1440.

Lautenschlager, N. T., Cupples, L. A., Rao, V. S., Auerbach, S. A., Becker, R., Burke, J., Chui, H., Duara, R., Foley, E. J., Glatt, S. L., Green, R. C., Jones, R., Karlinsky, H., Kukull, W. A., Kurz, A., Larson, E. B., Martelli, K., Sadovnick, A. D., Volicer, L., Waring, S. C., Growdon, J. H., and Farrer, L. A., 1996, Risk of dementia among relatives of Alzheimer's disease patients in the MIRAGE study: What is in store for the oldest old? *Neurology* **46:**641–650.

Lee, M. K., Marszalek, J. R., and Cleveland, D. W., 1994, A mutant neurofilament subunit causes massive, selective motor neuron death: Implication for the pathogenesis of human motor neuron disease, *Neuron* **13:**975–988.

Lehéricy, S., Baulac, M., Chiras, J., Pierot, L., Martin, N., Pillon, B., Deweer, B., Dubois, B., and Marsault, C., 1994, Amygdalohippocampal MR volume measurements in the early stages of Alzheimer disease, *Am. J. Neuroradiol.* **15:**927–937.

Leigh, P. N., 1994, Pathogenic mechanisms in amyotrophic lateral sclerosis and other motor neuron disorders, in: *Neurodegenerative Diseases* (D. B. Calne, ed.), W. B. Saunders, Philadelphia, pp. 473–488.

Leininger, B. E., Gramling, S. E., Farrell, A. D., Kreutzer, J. S., and Peck, E. A., III, 1990, Neuropsychological deficits in symptomatic minor head injury patients after concussion and mild concussion, *J. Neurol. Neurosurg. Psychiatry* **53:**293–296.

Lepore, F. E., Steele, J. C., Cox, T. A., Tillson, G., Calne, D. B., Duvoisin, R. C., Lavine, L., and McDarby, J. V., 1988, Supranuclear disturbances of ocular motility in Lytico-Bodig, *Neurology* **38:**1849–1853.

Leranth, C., and Ribak, C. E., 1991, Calcium-binding proteins are concentrated in the CA2 field of the monkey hippocampus: A possible key to this region's resistance to epileptic damage, *Exp. Brain Res.* **85:**129–136.

Lessell, S., Hirano, A., Torres, J., Kurland, L. T., 1962, Parkinsonism-dementia complex: Epidemiological considerations in the Chamorros of the Mariana islands and California, *Arch. Neurol.* **7:**377–385.

Leveugle, B., Spik, G., Perl, D. P., Bouras, C., Fillit, H. M., and Hof, P. R., 1994, The iron-binding protein lactotransferrin is present in pathologic lesions in a variety of neurodegenerative disorders: A comparative immunohistochemical analysis, *Brain Res.* **650:**20–31.

Levey, A. L., Heilman, C. J., Lah, J. J., Nash, N. R., Rees, H. D., Wakai, M., Mirra, S. S., Rye, D. B., Nochlin, D., Bird, T. D., and Mufson, E. J., 1997, Presenilin-1 protein expression in familial and sporadic Alzheimer's disease, *Ann. Neurol.* **41:**742–753.

Levine, D. N., Lee, J. M., and Fisher, C. M., 1993, The visual variant of Alzheimer's disease: A clinicopathologic case study, *Neurology* **43:**305–313.

Levy, M. H., Hart, W. M., Jr., Sonstein, F. M., and Ballinger, W. E., Jr., 1995, The incredible *shrinking* brain [With comments by A. Sadun], *Surv. Ophthalmol.* **39:**315–322.

Levy, M. L., Miller, B. L., Cummings, J. L., Fairbanks, L. A., and Craig, A., 1996, Alzheimer's disease and frontotemporal dementias: Behavioral distinctions, *Arch. Neurol.* **53:**687–690.

Lewis, D. A., and Lund, J. S., 1990, Heterogeneity of chandelier neurons in monkey neocortex: Corticotropin-releasing-factor- and parvalbumin-immunoreactive populations, *J. Comp. Neurol.* **293:**599–615.

Lewis, D. A., Campbell, M. J., Terry, R. D., and Morrison, J. H., 1987, Laminar and regional distribution of neurofibrillary tangles and neuritic plaques in Alzheimer's disease: A quantitative study of visual and auditory cortices, *J. Neurosci.* **7:**1799–1808.

Lilienfeld, D. E., Perl, D. P., and Olanow, C. W., 1994, Guam neurodegeneration, in: *Neurodegenerative Diseases* (D. B. Calne, ed.), W. B. Saunders, Philadelphia, pp. 895–908.

Lippa, C. F., Hamos, J. E., Pulaski-Salo, D., Degennaro, L. G., and Drachman, D. A., 1992, Alzheimer's disease and aging: Effects on perforant pathway perikarya and synapses, *Neurobiol. Aging* **13:**405–411.

Lippa, C. F., Smith, T. W., and Swearer, J. N., 1994, Alzheimer's disease and Lewy body disease: A comparative clinicopathological study, *Ann. Neurol.* **35:**81–88.

Lippa, C. F., Pulaski-Salo, D., Dickson, D. W., and Smith, T. W., 1997, Alzheimer's disease, Lewy body disease, and aging: A comparative study of the preforant path, *J. Neurol. Sci.* **147:**161–166.

Lipton, S. A., and Rosenberg, P. A., 1995, Excitatory amino acids as a final common pathway for neurologic disorders, *New Engl. J. Med.* **330:**613–622.

Litvan, I., Hauw, J. J., Bartko, J. J., Lantos, P. L., Daniel, S. E., Horoupian, D. S., McKee, A., Dickson, D., Bancher, C., Tabaton, M., Jellinger, K., and Anderson, D. W., 1996, Validity and reliability of the preliminary NINDS neuropathologic criteria for progressive supranuclear palsy and related disorders, *J. Neuropathol. Exp. Neurol.* **55:**97–105.

Litvan, I., Agid, Y., Sastri, N., Jankovic, J., Wenning, G. K., Goetz, C. G., Verny, M., Brandel, J. P., Jellinger, K., Chaudhuri, K. R., McKee, A., Lai, E. C., Pearce, R. K. B., and Bartko, J. J., 1997a, What are the obstacles for an accurate clinical diagnosis of Pick's disease? A clinicopathological study, *Neurology* **49:**62–69.

Litvan, I., Campbell, G., Mangone, C. A., Verny, M., McKee, A., Chaudhuri, K. R., Jellinger, K., Pearce, R. K. B., and D'Olhaberriague, L., 1997b, Which clinical features differentiate progressive supranuclear palsy (Steele-Richardson-Olszewski syndrome) from related disorders? A clinicopathological study, *Brain* **120:**65–74.

Livingstone, M., and Hubel, D., 1988, Segregation of form, color, movement, and depth: Anatomy, physiology, and perception, *Science* **240:**740–749.

Liu, X., and Brun, A., 1996, Regional and laminar synaptic pathology in frontal lobe degeneration of non-Alzheimer type, *Int. J. Geriat. Psychiatry* **11:**47–55.

Love, S., Saitoh, T., Quijada, S., Cole, G. M., and Terry, R. D., 1988, Alz-50, ubiquitin, and tau immunoreactivity of neurofibrillary tangles, Pick bodies, and Lewy bodies, *J. Neuropathol. Exp. Neurol.* **47:**393–405.

Love, J., Mayer, R. J., and Landon, M., 1993, Ubiquitin in neurodegenerative diseases, *Brain Pathol.* **3:**55–65.

Lu, Y. M., Yin, H. Z., Chiang, J., and Weiss, J. H., 1996, $Ca^{2+}$-permeable AMPA/kainate and NMDA channels: High rate of $Ca^{2+}$ influx underlies potent induction of injury, *J. Neurosci.* **16:**5457–5465.

Lucassen, P. J., Chung, W. C. J., Kamphorst, W., and Swaab, D. F., 1997, DNA damage distribution in the human brain as shown by *in situ* end labeling: Area-specific differences in aging and Alzheimer disease in the absence of apoptotic morphology, *J. Neuropathol. Exp. Neurol.* **56:**887–900.

Lüers, T., and Spatz, H., 1957, Picksche Krankheit (Progressive umschriebene Großhirnatrophie), in: *Handbuch der speziellen pathologischen Anatomie und Histologie Bd XIII/1A (Nervensystem)* (W. Scholz, ed.), Springer, Berlin, pp. 614–715.

Mackenzie Ross, S. J., Graham, N., Stuart-Green, L., Prins, M., Xuereb, J., Patterson, K., and Hodges, J. R., 1996, Progressive biparietal atrophy: An atypical presentation of Alzheimer's disease, *J. Neurol. Neurosurg. Psychiatry* **61:**388–395.

Magnani, G., Bettoni, L., and Mazzuchi, A., 1982, Lesioni atrofiche bioccipitali di incerta natura all'origine di una agnosia visiva, *Riv. Neurol.* **52:**137–148.

Malamud, N., Hirano, A., and Kurland, L. T., 1961, Pathoanatomic changes in amyotrophic lateral sclerosis on Guam, *Arch. Neurol.* **5:**71–85.

Mann, D. M. A., 1988, The pathological association between Down syndrome and Alzheimer disease, *Mech. Aging Dev.* **443:**99–136.

Mann, D. M. A., and Esiri, M. M., 1989, The pattern of acquisition of plaques and tangles in the brains of patients under 50 years of age with Down's syndrome, *J. Neurol. Sci.* **89:**169–179.

Mann, D. M. A., and South, P. W., 1993, The topographic distribution of brain atrophy in frontal lobe dementia, *Acta Neuropathol.* **85:**334–340.

Mann, D. M. A., and Snowden, J. S., 1995, The topographic distribution of brain atrophy in cortical Lewy body disease: Comparison with Alzheimer's disease, *Acta Neuropathol.* **89:**178–183.

Mann, D. M. A., Yates, P. O., and Marcyniuk, B., 1985, Some morphometric observations on the cerebral cortex and hippocampus in presenile Alzheimer's disease, senile dementia of the Alzheimer type, and Down's syndrome in middle age, *J. Neurol. Sci.* **69:**139–159.

Mann, D. M. A., Yates, P. O., Marcyniuk, B., and Ravindra, C. R., 1986, The topography of plaques and tangles in Down's syndrome patients of different ages, *Neuropathol. Appl. Neurobiol.* **12:**447–457.

Mann. D. M. A., Prinja, D., Davies, C. A., Ihara, Y., Delacourte, A., Défossez, A., Mayer, R. J., and Landon, M., 1989, Immunocytochemical profile of neurofibrillary tangles in Down's syndrome patients of different ages, *J. Neurol. Sci.* **92:**247–260.

Mann, D. M. A., South, P. W., Snowden, J. S., and Neary, D., 1993, Dementia of frontal lobe type: Neuropathology and immunohistochemistry, *J. Neurol. Neurosurg. Psychiatry* **56:**605–614.

Mann, D. M. A., Iwatsubo, T., Ihara, Y., Cairns, N. J., Lantos, P. L., Bogdanovic, N., Lannfelt, L., Winblad, B., Maat-Schieman, M. L. C., and Rossor, M. N., 1996, Predominant deposition of amyloid-beta 42(43) in plaques in cases of Alzheimer's disease and hereditary cerebral hemorrhage associated with mutations in the amyloid precursor protein gene, *Am. J. Pathol.* **148:**1257–1266.

Martin, L. J., Pardo, C. A., Cork, L. C., and Price, D. L., 1994, Synaptic pathology and glial response to neuronal injury precede the formation of senile plaques and amyloid deposits in the aging cerebral cortex, *Am. J. Pathol.* **145:**1358–1381.

Martin, J. P., 1965, The globus pallidus in post-encephalitic parkinsonism, *J. Neurol. Sci.* **2:**344–365.

Martinez-Lage, P., and Munoz, D. G., 1997, Prevalence and disease association of argyrophilic grains of Braak, *J. Neuropathol. Exp. Neurol.* **56:**157–164.

Martland, H. S., 1928, Punch drunk, *J. Am. Med. Assoc.* **91:**1103–1107.

Masliah, E., Cole, G., Shimohama, S., Hansen, L., DeTeresa, R., Terry, R. D., and Saitoh, S., 1990, Differential involvement of protein kinase C isozymes in Alzheimer's disease, *J. Neurosci.* **10:**2113–2124.

Masliah, E., Cole, G. M., Hansen, L. A., Mallory, M., Albright, T., Terry, R. D., and Saitoh, T., 1991a, Protein kinase C alteration is an early biochemical marker in Alzheimer's disease, *J. Neurosci.* **11:**2759–2767.

Masliah, E., Hansen, L. A., Quijada, S., DeTeresa, R., Alford, M., Kauss, J., and Terry, R., 1991b, Late onset dementia with argyrophilic grains and subcortical tangles or atypical progressive supranuclear palsy? *Ann. Neurol.* **29:**389–396.

Masliah, E., Terry, R. D., Alford, M., DeTeresa, R., and Hansen, L. A., 1991c, Cortical and subcortical patterns of synaptophysin-like immunoreactivity in Alzheimer's disease, *Am. J. Pathol.* **138:**235–246.

Masliah, E., Iimoto, D. S., Mallory, M., Albright, T., Hansen, L., and Saitoh, T., 1992a, Casein kinase II alteration precedes tau accumulation in tangle formation, *Am. J. Pathol.* **140:**263–268.

Masliah, E., Mallory, M., Hansen, L., Alford, M., DeTeresa, R., Terry, R., Baudier, J., and Saitoh, T., 1992b, Localization of amyloid precursor protein in GAP43-immunoreactive aberrant sprouting neurites in Alzheimer's disease, *Brain Res.* **574:**312–316.

Masliah, E., Mallory, M., Hansen, L., DeTeresa, R., and Terry, R. D., 1993, Quantitative synaptic alterations in the human neocortex during normal aging, *Neurology* **43:**192–197.

Massman, P. J., Kreiter, K. T., Jankovic, J., and Doody, R. S., 1996, Neuropsychological functioning in cortical-basal ganglionic degeneration: Differentiation from Alzheimer's disease, *Neurology* **46:**720–726.

Matsuo, S. E., Shi, R. W., Billingsley, M. L., Van de Voorde, A., O'Connor, M., Trojanowski, J. Q., and Lee, V. M. Y., 1994, Biopsy-derived adult human brain tau is phosphorylated at many of the same sites as Alzheimer's disease paired helical filament tau, *Neuron* **13:**989–1002.

Mattson, M. P., Cheng, B., Davis, D., Bryant, K., Lieberburg, I., and Rydel, R. E., 1992, β-amyloid peptides destabilize calcium homeostasis and render human cortical neurons vulnerable to excitotoxicity, *J. Neurosci.* **12:**376–389.

Mawal-Dewan, M., Schmidt, M. L., Balin, B., Perl, D. P., Lee, V. M. Y., and Trojanowski, J. Q., 1996, Identification of phosphorylation sites in PHF-tau from patients with Guam amyotrophic lateral sclerosis/parkinsonism-dementia complex, *J. Neuropathol. Exp. Neurol.* **55:**1051–1059.

McCarthy, G., Spicer, M., Adrignolo, A., Luby, M., Gore, J., and Allison, T., 1995, Brain activation associated with visual motion studied by functional magnetic resonance imaging in humans, *Hum. Brain Mapp.* **2:**234–243.

McGeer, P. L., and McGeer, E. G., 1995, The inflammatory response system of brain: Implications for therapy of Alzheimer and other neurodegenerative diseases, *Brain Res. Rev.* **21:**195–218.

McGeer, P. L., Kawamata, T., and Walker, D. G., 1992, Distribution of clusterin in Alzheimer brain tissue, *Brain Res.* **579:**337–341.

McKee, A. C., Kosik, K. S., Kennedy, M. B., and Kowall, N. W., 1990, Hippocampal neurons predisposed to neurofibrillary tangle formation are enriched in type II calcium/calmodulin-dependent protein kinase, *J. Neuropathol. Exp. Neurol.* **49:**49–63.

McKeith, I. G., Galasko, D., Kosaka, K., Perry, E. K., Dickson, D. W., Hansen, L. A., Salmon, D. P., Lowe, J., Mirra, S. S., Byrne, E. J., Lennox, G., Quinn, N. P., Edwardson, J. A., Ince, P. G., Bergeron, C., Burns, A., Miller, B. L., Lovestone, S., Collerton, D., Jansen, E. N. H., Ballard, C., de Vos, R. A. I., Wilcock, G. K., Jellinger, K. A., and Perry, R. H., 1996, Consensus guidelines for the clinical and pathologic diagnosis of dementia with Lewy bodies (DLB): Report of the consortium on DLB international workshop, *Neurology* **47:**1113–1124.

McKenzie, J. E., Gentleman, S. M., Roberts, G. W., Graham, D. I., and Royston, M. C., 1994, Increased numbers of βAPP-immunoreactive neurons in the entorhinal cortex after head injury, *NeuroReport* **6:**161–164.

McKenzie, J. E., Roberts, G. W., and Royston, M. C., 1996, Comparative investigation of neurofibrillary damage in the temporal lobe in Alzheimer's disease, Down's syndrome, and dementia pugilistica, *Neurodegeneration* **5:**259–264.

Mega, M. S., Mastermann, D. L., Benson, D. F., Vinters, H. V., Tomiyasu, U., Craing, A. H., Foti, D. J., Kaufer, D., Scharre, D. W., Fairbanks, L., and Cummings, J. L., 1996, Dementia with Lewy bodies: Reliability and validity of clinical and pathologic criteria, *Neurology* **47:**1403–1409.

Meldrum, B., and Garthwaite, J., 1991, Excitatory amino acid neurotoxicity and neurodegenerative diseases, *Trends Pharmacol. Sci.* **11:**54–61.

Mendez, M. F., Mendez, M. A., Martin, R., Smyth, K. A., and Whitehouse, P. J., 1990a, Complex visual disturbances in Alzheimer's disease, *Neurology* **40:**439–443.

Mendez, M. F., Tomsak, R. L., and Remler, B., 1990b, Disorders of the visual system in Alzheimer's disease, *J. Clin. Neuro-Ophthalmol.* **10:**62–69.

Mendez, M. F., Turner, J., Gilmore, G. C., Remler, B., and Tomsak, R. L., 1990c, Bálint's syndrome in Alzheimer's disease: Visuospatial functions, *Int. J. Neurosci.* **54:**339–346.

Mendez, M. F., Selwood, A., Mastri, A. R., and Frey, W. H., II, 1993, Pick's disease versus Alzheimer's disease: A comparison of clinical characteristics, *Neurology* **43:**289–292.

Mendez, M. F., Cherrier, M., Perryman, K. M., Pachana, N., Miller, B. L., and Cummings, J. L., 1996, Frontotemporal dementia versus Alzheimer's disease: Differential cognitive features, *Neurology* **47:**1189–1194.

Mentis, M. J., Horwitz, B., Grady, C. L., Alexander, G. E., VanMeter, J. W., Maisog, J. M., Pietrini, P., Schapiro, M. B., and Rapoport, S. I., 1996, Visual cortical dysfunction in Alzheimer's disease evaluated with a temporally graded "stress test" during PET, *Am. J. Psychiatry* **153:**32–40.

Metzger, R. E., LaDu, M. J., Pan, J. B., Getz, G. S., Frail, D. E., and Falduto, M. T., 1996, Neurons of the human frontal cortex display apolipoprotein E immunoreactivity: Implications for Alzheimer's disease, *J. Neuropathol. Exp. Neurol.* **55:**372–380.

Meunier, M., Bachevalier, J., Mishkin, M., and Murray, E. A., 1993, Effects on visual recognition of combined and separate ablations of the entorhinal and perirhinal cortex in rhesus monkeys, *J. Neurosci.* **13:**5418–5432.

Miklossy, J., 1993, The geniculocalcarine pathway in man, and some putative rostral visual areas involved in visuo-spatial attention, in: *Functional Organisation of the Human Visual Cortex* (B. Gulyas, D. Ottoson, and P. E. Roland, eds.), Pergamon, Oxford, pp. 123–136.

Millspaugh, J. A., 1937, Dementia pugilistica, *US Navy Med. Bull.* **35:**297–303.

Minc-Golomb, D., Knobler, H., and Groner, Y., 1991, Gene dosage of CuZnSOD and Down's syndrome: Diminished prostaglandin synthesis in human trisomy 21, transfected cells, and transgenic mice, *EMBO J.* **10:**2119–2124.

Minoshima, S., Foster, N. L., and Kuhl, D. E., 1994, Posterior cingulate cortex in Alzheimer's disease, *Lancet* **344:**895.

Minoshima, S., Giordani, B., Berent, S., Frey, K. A., Foster, N. L., and Kuhl, D. E., 1997, Metabolic reduction in the posterior cingulate cortex in very early Alzheimer's disease, *Ann. Neurol.* **42:**85–94.

Mirra, S. S., Hart, M. N., and Terry, R. D., 1993, Making the diagnosis of Alzheimer's disease: A primer for practicing neuropathologists, *Arch. Pathol. Lab. Med.* **117:**132–144.

Mishkin, M., Ungerleider, L. G., and Macko, K. A., 1983, Object vision and spatial vision: Two cortical pathways, *Trends Neurosci.* **6:**414–417.

Mitsuyama, Y., and Seyama, S., 1981, Frequency of Alzheimer's neurofibrillary tangle in the brains of progressive supranuclear palsy, postencephalitic parkinsonism, Alzheimer's disease, senile dementia and non-demented elderly people, *Folia Psychiat. Neurol. Jpn.* **35:**189–204.

Mizutani, T., and Shimada, H., 1990, Neuropathological aspects of centenarian brains: Study of 24 autopsy cases, in: *Psychogeriatrics: Biomedical and Social Advances* (K. Hasegawa and A. Homma, eds.), Excerpta Medica, Tokyo, pp. 128–133.

Mizutani, T., and Shimada, H., 1992, Neuropathological background of twenty-seven centenarian brains, *J. Neurol. Sci.* **108:**168–177.

Montine, K. S., Kim, P. J., Olson, W. R., Markesbery, W. R., and Montine, T. J., 1997a, 4-hydroxy-2-nonenal pyrrole adducts in human neurodegenerative diseases, *J. Neuropathol. Exp. Neurol.* **56:**866–871.

Montine, K. S., Olson, S. J., Amarnath, V., Whetsell, W. O., Jr., Graham, D. G., and Montine, T. J., 1997b, Immunohistochemical detection of 4-hydroxy-2-nonenal adducts in Alzheimer's disease is associated with inheritance of APOE4, *Am. J. Pathol.* **150:**437–443.

Morán, M. A., Mufson, E. J., and Gómez-Ramos, P., 1994, Cholinesterases colocalize with sites of neurofibrillary degeneration in aged and Alzheimer's disease brains, *Acta Neuropathol.* **87:**284–292.

Morán, M. A., Probst, A., Navarro, C., and Gómez-Ramos, P., 1995, Alzheimer's disease-type neurofibrillary degeneration in verrucose dysplasias of the cerebral cortex, *Acta Neuropathol.* **90:**356–365.

Morel, F., 1943, Angiopathie topistique atteignant électivement la IV$^{e}$ couche de l'aire striée, *Mschr. Psychiat. Neurol.* **108:**248–254.

Morel, F., 1945, Les aires striée, parastriée et péristriée dans les troubles de la fonction visuelle au cours de la maladie d'Alzheimer, *Confin. Neurol.* **6:**238–242.

Morel, F., and Wildi, E., 1952, General and cellular pathochemistry of senile and presenile alterations of the brain, *Proc. First Int. Congr. Neuropathol. (Rome)* **2:**347–374.

Morel, F., and Wildi, E., 1955, Contribution à la connaissance des différentes altérations cérébrales du grand âge, *Arch. Suisses Neurol. Psychiat.* **76:**190–197.

Mori, H., Nishimura, M., Namba, Y., and Oda, M., 1994, Corticobasal degeneration: A disease with widespread appearance of abnormal tau and neurofibrillary tangles, and its relation to progressive supranuclear palsy, *Acta Neuropathol.* **88:**113–121.

Morris, R. G., Anderson, E., Lynch, G. S., and Baudry, M., 1986, Selective impairment of learning and blockade of long-term potentiation by a *N*-methyl-D-aspartate receptor antagonist, AP5, *Nature* **319:**774–776.

Morris, J. C., McKeel, J. D. W., Storandt, M., Rubin, E. H., Price, J. L., Grant, E. A., Ball, M. J., and Berg, L., 1991, Very mild Alzheimer's disease: Informant-based clinical, psychometric, and pathologic distinction from normal aging, *Neurology* **41:**469–478.

Morrison, B. M., Gordon, J. W., Ripps, M. E., and Morrison, J. H., 1996, Quantitative immunocytochemical analysis of the spinal cord in G86R superoxide dismutase transgenic mice: Neurochemical correlates of selective vulnerability, *J. Comp. Neurol.* **373:**619–631.

Morrison, J. H., 1993, Differential vulnerability, connectivity, and cell typology, *Neurobiol. Aging* **14:**51–54.

Morrison, J. H., and Hof, P. R., 1992, The organization of the cerebral cortex: From molecules to circuits, *Disc. Neurosci.* **9(2):**1–80.

Morrison, J. H., and Hof, P. R., 1997, Life and death of neurons in the aging brain, *Science* **278:**412–419.

Morrison, J. H., Rogers, J., Scherr, S., Benoit, R., and Bloom, F. E., 1985, Somatostatin immunoreactivity in neuritic plaques of Alzheimer's patients, *Nature* **314:**90–92.

Morrison, J. H., Lewis, D. A., Campbell, M. J., Huntley, G. W., Benson, D. L., and Bouras, C., 1987, A monoclonal antibody to non-phosphorylated neurofilament protein marks the vulnerable cortical neurons in Alzheimer's disease, *Brain Res.* **416:**331–336.

Morrison, J. H., Siegel, S. J., Gazzaley, A. H., Huntley, G. W., 1996a, Glutamate receptors: Emerging links between subunit proteins and specific excitatory circuits in primate hippocampus and neocortex, *Neuroscientist* **2:**272–283.

Morrison, J. H., Vissavajjhala, P., Blumberg, D. M., Hu, Y., Janssen, W. G. M., and Hof, P. R., 1996b, Calcium-binding proteins parvalbumin, calbindin, and calretinin display differential colocalization patterns with glutamate receptor subunit protein GluR2 in macaque monkey area V1, *Soc. Neurosci. Abstr.* **22:**1994.

Morrison, J. H., Hof, P. R., and Huntley, G. W., 1998, Neurochemical organization of the primate visual cortex, in: *Handbook of Chemical Neuroanatomy, Vol. 14, The Primate Nervous System, Pt. II* (F. E. Bloom, A. Björklund, and T. Hökfelt, eds.), Elsevier, Amsterdam, pp. 299–430.

Mortimer, J. A., Schuman, L. M., and French, L. R., 1981, Epidemiology of dementing illness, in: *The Epidemiology of Dementia* (J. A. Mortimer and L. M. Schuman, eds.), Oxford University Press, New York, pp. 3–23.

Mortimer, J. A., French, L. R., Hutton, J. T., and Schuman, L. M., 1985, Head injury as a risk factor for Alzheimer's disease, *Neurology* **35:**264–267.

Moss, M., and Albert, M., 1988, Alzheimer's disease and other dementing disorders, in: *Geriatric Neuropsychology* (M. Albert and M. Moss, eds.), Guilford, New York, pp. 145–177.

Moss, M. B., Killiany, R. J., Lai, Z. C., Rosene, D. L., and Herndon, J. G., 1997, Recognition memory span in rhesus monkeys of advanced age, *Neurobiol. Aging* **18:**13–19.

Motte, J., and Williams R. S., 1989, Age-related changes in the density and morphology of plaques and neurofibrillary tangles in Down syndrome brain, *Acta Neuropathol.* **77:**535–546.

Mountjoy, C. Q., Roth, M., Evans, N. J. R., and Evans, H. M., 1983, Cortical neuronal counts in normal elderly controls and demented patients, *Neurobiol. Aging* **4:**1–11.

Mrzjlak, L., Uylings, H. B. M., Van Eden, C. G., and Júdas, M., 1990, Neuronal development in human prefrontal cortex in prenatal and postnatal stages, in: *The Prefrontal Cortex: Its Structure, Function, and Pathology, Prog. Brain Res.*, Vol. 85 (H. B. M. Uylings, C. G. Van Eden, J. P. C. de Bruin, M. A. Corner, and M. G. P. Feenstra, eds.), Elsevier, Amsterdam, pp. 223–240.

Mulder, D. W., Kurland, L. T., and Iriarte, L. L. G., 1954, Neurologic diseases on the island of Guam, *U.S. Armed Forces Med. J.* **5:**1724–1739.

Münch, G., Thome, J., Foley, P., Schinzel, R., and Riederer, P., 1997, Advanced glycation end products in aging and Alzheimer's disease, *Brain Res. Rev.* **23:**134–143.

Munoz-Garcia, D., and Ludwin, S. K., 1984, Classic and generalized variants of Pick's disease: A clinicopathological, ultrastructural, and immunocytochemical comparative study, *Ann. Neurol.* **16:**467–480.

Murphy, G. M., Jr., Forno, L. S., Ellis, W. G., Nochlin, D., Levy-Lahad, E., Poorkaj, D., Bird, T. D., Jiang, Z. L., and Cordell, B., 1996, Antibodies to presenilin proteins detect neurofibrillary tangles in Alzheimer's disease, *Am. J. Pathol.* **149:**1839–1846.

Mutrux, S., 1947, Diagnostic différentiel histologique de la maladie d'Alzheimer et de la Démence sénile: Pathophobie de la zone de projection corticale, *Mschr. Psychiat. Neurol.* **113:**100–117.

Mutrux, S., 1958, Contribution à l'étude histologique différentielle de la démence sénile et de la maladie d'Alzheimer, *Psychiat. Neurol.* **136:**146–156.

Nakano, I., and Hirano, A., 1983, Neuron loss in the nucleus basalis of Meynert in parkinsonism-dementia complex of Guam, *Ann. Neurol.* **13:**87–91.

Namba, Y., Tomonaga, M., Kawasaki, H., Otomo, E., and Ikeda, K., 1991, Apolipoprotein E immunoreactivity in cerebral amyloid deposits and neurofibrillary tangles in Alzheimer's disease and Creuzfeldt-Jakob disease, *Brain Res.* **541:**163–166.

Neary, D., 1990, Non-Alzheimer's disease forms of cerebral atrophy, *J. Neurol. Neurosurg. Psychiatry* **53:**929–931.

Neary, D., and Snowden, J. S., 1987, Perceptuospatial disorder in Alzheimer's disease, *Semin. Ophthalmol.* **2:**151–158.

Neary, D., Snowden, J. S., Mann, D. M. A., Northen, B., Goulding, P. J., and McDermott, N., 1990, Frontal lobe dementia and motor neuron disease, *J. Neurol. Neurosurg. Psychiatry* **53:**23–32.

Neary, D., Snowden, J. S., and Mann, D. M. A., 1993, The clinical pathological correlates of labor atrophy, *Dementia* **4**:154–159.

Neumann, M. A., 1949, Pick's disease, *J. Neuropathol. Exp. Neurol.* **8**:255–282.

Nielson, K. A., Cummings, B. J., and Cotman, C. W., 1996, Constructional apraxia in Alzheimer's disease correlates with neuritic neuropathology in occipital cortex, *Brain Res.* **741**:284–293.

Nihei, K., McKee, A. C., and Kowall, N. W., 1993, Patterns of neuronal degeneration in the motor cortex of amyotrophic lateral sclerosis patients, *Acta Neuropathol.* **86**:55–64.

Nimchinsky, E. A., Vogt, B. A., Morrison, J.H., and Hof, P. R., 1995, Spindle neurons of the human anterior cingulate cortex, *J. Comp. Neurol.* **355**:27–37.

Nimchinsky, E. A., Hof, P. R., Young, W. G., and Morrison, J. H., 1996, Neurochemical, morphologic, and laminar characterization of cortical projection neurons in the cingulate motor areas of the macaque monkey, *J. Comp. Neurol.* **374**:136–160.

Nimchinsky, E. A., Vogt, B. A., Morrison, J. H., and Hof, P. R., 1997, Neurofilament and calcium-binding proteins in the human cingulate cortex, *J. Comp. Neurol.* **384**:597–620.

Nishiyama, E., Ohwada, J., Iwamoto, N., and Arai, H., 1993, Selective loss of calbindin D28k-immunoreactive neurons in the cortical layer II in brains of Alzheimer's disease: A morphometric study, *Neurosci. Lett.* **163**:223–226.

Nissen, M. J., Corkin, S., Buonanno, F. S., Growdon, J. H., Wray, S. H., and Bauer, J., 1985, Spatial vision in Alzheimer's disease: General findings and a case report, *Arch. Neurol.* **42**:667–671.

O'Banion, M. K., Coleman, P. D., and Callahan, L. M., 1994, Regional neuronal loss in aging and Alzheimer's disease: A brief review, *Semin. Neurosci.* **6**:307–314.

Obrenovitch, T. P., and Urenjak, J., 1997, Altered glutamatergic transmission in neurological disorders: From high extracellular glutamate to excessive synaptic efficacy, *Prog. Neurobiol.* **51**:39–87.

Ohm, T. G., 1997, Does Alzheimer's disease start early in life? *Mol. Psychiatry* **2**:21–25.

Oliver, C., and Holland, A. J., 1986, Down's syndrome and Alzheimer's disease: A review, *Psychol. Med.* **16**:307–322.

Olney, J. W., 1990, Excitotoxin-mediated neuron death in young and old age, in: *Molecular and Cellular Mechanisms of Neuronal Plasticity in Normal Aging and Alzheimer's Disease* (P. Coleman, G. Higgins, and C. Phelps, eds.), Elsevier, Amsterdam, pp. 37–51.

Olson, C. R., and Musil, S. Y., 1992a, Posterior cingulate cortex: Sensory and oculomotor properties of single neurons in behaving cat, *Cereb. Cortex* **2**:485–502.

Olson, C. R., and Musil, S. Y., 1992b, Topographic organization of cortical and subcortical projections to posterior cingulate cortex in the cat: Evidence for somatic, ocular, and complex subregions, *J. Comp. Neurol.* **323**:1–24.

Olson, C. R., Musil, S. Y., and Goldberg, M. E., 1993, Neurophysiology of posterior cingulate cortex in the behaving monkey, in: *The Neurobiology of Cingulate Cortex and Limbic Thalamus* (B. A. Vogt and M. Gabriel, eds.), Birkhäuser, Boston, pp. 366–380.

Onari, K., and Spatz, H., 1926, Anatomische Beitrage zur Lehre von der Pickschen umschriebenen Groβhirnrindenatrophie (Picksche Krankheit), *Z. Ges. Neurol. Psychiat.* **101**:470–511.

Ong, W. Y., and Garey, L. J., 1996, A light and electron microscopic study of GluR4-positive cells in human cerebral cortex, *Neurosci. Lett.* **210**:107–110.

Orban, G. A., Dupont, P., De Bruyn, B., Vogels, R., Vandenberghe, R., and Mortelmans, L., 1995, A motion area in human visual cortex, *Proc. Natl. Acad. Sci. USA* **92**:993–997.

Ottersen, O. P., Laake, J. H., Reichelt, W., Haug, F. M., and Torp, R., 1996, Ischemic disruption of glutamate homeostasis in brain: Quantitative immunocytochemical analyses, *J. Chem. Neuroanat.* **12**:1–14.

Oyanagi, K., Makifuchi, T., Ohtoh, T., Chen, K. M., Gajdusek, D. C., and Chase, T. N., 1997, Distinct pathological features of the Gallyas- and tau-positive glia in the parkinsonism-dementia complex and amyotrophic lateral sclerosis of Guam, *J. Neuropathol. Exp. Neurol.* **56**:308–316.

Pagliusi, S. R., Gerrard, P., Abdallah, M., Talabot, D., and Catsicas, S., 1994, Age-related changes in expression of AMPA-selective glutamate receptor subunits: Is calcium-permeability altered in hippocampal neurons? *Neuroscience* **61**:429–433.

Pakkenberg, B., and Gundersen, H. J. G., 1997, Neocortical neuron number in humans: Effect of sex and age, *J. Comp. Neurol.* **384**:312–320.

Pantelakis, S., 1954, Un type particulier d'angiopathie sénile du système nerveux central: l'angiopathie congophile: Topographie et fréquence, *Mschr. Psychiat. Neurol.* **128**:219–256.

Parasuraman, R., Greenwood, P. M., and Alexander, G. E., 1995, Selective impairment of spatial attention during visual search in Alzheimer's disease, *NeuroReport* **6**:1861–1864.

Paulus, W., and Selim, M., 1990, Corticonigral degeneration with neuronal achromasia and basal neurofibrillary tangles, *Acta Neuropathol.* **81:**89–94.

Pearson, R. C. A., Esiri, M. M., Hiorns, R. W., Wilcock, G. K., and Powell, T. P. S., 1985, Anatomical correlates of the distribution of the pathological changes in the neocortex in Alzheimer disease, *Proc. Natl. Acad. Sci. USA* **82:**4531–4534.

Pellegrini-Giampietro, D. E., Zukin, R. S., Bennett, M. V. L., Cho, S., and Pulsinelli, W. A., 1992, Switch in glutamate receptor subunit gene expression in CA1 subfield of hippocampus following global ischemia in rats, *Proc. Natl. Acad. Sci. USA* **89:**10499–10503.

Pellegrini-Giampietro, D. E., Pulsinelli, W. A., and Zukin, R. S., 1994, NMDA and non-NMDA receptor gene expression following global brain ischemia in rats: Effect of NMDA and non-NMDA receptor antagonists, *J. Neurochem.* **62:**1067–1073.

Perl, D. P., 1985, Relationship of aluminum to Alzheimer's disease, *Environ. Health Perspect.* **63:**149–153.

Perl, D. P., and Brody, A. R., 1980, Alzheimer's disease: X-ray spectrometric evidence of aluminum accumulation in neurofibrillary tangle-bearing neurons, *Science* **208:**297–299.

Perl, D. P., and Pendlebury, W. W., 1986, Aluminum neurotoxicity: Potential role in the pathogenesis of neurofibrillary tangle formation, *Can. J. Neurol. Sci.* **13:**441–445.

Perl, D. P., and Good, P. F., 1987, Uptake of aluminum into central nervous system along nasal-olfactory pathways, *Lancet* **1:**1028.

Perl, D. P., Gajdusek, D. C., Garruto, R. M., Yanagihara, R. T., and Gibbs, C. J., 1982, Intraneuronal aluminum accumulation in amyotrophic lateral sclerosis and parkinsonism-dementia of Guam, *Science* **217:**1053–1055.

Perry, G., Stewart, D., Friedman, R., Manetto, V., Autilio-Gambetti, L., and Gambetti, P., 1987, Filaments of Pick's bodies contain altered cytoskeletal elements, *Am. J. Pathol.* **127:**559–568.

Perry, R. H., 1986, Recent advances in neuropathology, *Br. Med. Bull.* **42:**34–41.

Perry, R. H., Irving, D., Blessed, G., Fairbairn, A., and Perry, E. K., 1990a, Senile dementia of Lewy body type: A clinically and neuropathologically distinct form of Lewy body dementia in the elderly, *J. Neurol. Sci.* **95:**119–139.

Perry, R. H., Irving, D., and Tomlinson, B. E., 1990b, Lewy body prevalence in the aging brain: Relationship to neuropsychiatric disorders, Alzheimer-type pathology, and catecholaminergic nuclei, *J. Neurol. Sci.* **100:**223–233.

Piccardo, P., Yanagihara, R., Garruto, R. M., Gibbs, C. J., Jr., and Gajdusek, D. C., 1988, Histochemical and X-ray microanalytical localization of aluminum in amyotrophic lateral sclerosis and parkinsonism-dementia of Guam, *Acta Neuropathol.* **77:**1–4.

Pick, A., 1892, Über die Beziehungen der senilen Hirnatrophie zur Aphasie, *Prag. Med. Wschr.* **17:**165–167.

Pietrini, P., Furey, M. L, Graff-Radford, N., Freo, U., Alexander, G. E., Grady, C. L., Dani, A., Mentis, M. J., and Schapiro, M. B., 1996, Preferential metabolic involvement of visual cortical areas in a subtype of Alzheimer's disease: Clinical implications, *Am. J. Psychiatry* **153:**1261–1268.

Pietrini, P., Dani, A., Furey, M. L., Alexander, G. E., Freo, U., Grady, C. L., Mentis, M. J., Mangot, M., Simon, E. W., Horwitz, B., Haxby, J. V., and Schapiro, M. B., 1997, Low glucose metabolism during brain stimulation in older Down's syndrome subjects at risk for Alzheimer's disease prior to dementia, *Am. J. Psychiatry* **154:**1063–1069.

Pike, C. J., and Cotman, C. W., 1995, Calretinin-immunoreactive neurons are resistant to β-amyloid toxicity *in vitro, Brain Res.* **671:**293–298.

Plioplys, A. V., 1987, Down's syndrome: Precocious neurofilament antigen expression, *J. Neurol. Sci.* **79:**91–100.

Pollard, H., Héron, A., Moreau, J., Ben-Ari, Y., and Khrestchatisky, M., 1993, Alterations of the GluR-B AMPA receptor subunit flip/flop expression in kainate-induced epilepsy and ischemia, *Neuroscience* **57:**545–554.

Politoff, A. L., and Monson, N., 1997, Age vs. aging in the pathogenesis of senile dementia of the Alzheimer type: Electrophysiological evidence, *Dement. Geriat. Cogn. Disord.* **8:**18–25.

Posey, W. C., 1904, Paralysis of upward movement of the eyes, *Ann. Ophthalmol.* **13:**523–529.

Potier, B., Krzywkowski, P., Lamour, Y., and Dutar, P., 1994, Loss of calbindin-immunoreactivity in CA1 hippocampal stratum radiatum and stratum lacunosum-moleculare interneurons in the aged rat, *Brain Res.* **661:**181–188.

Price, D. L., Martin, L. J., Sisodia, S. S., Wagster, M. V., Koo, E. H., Walker, L. C., Koliatsos, V. E., and

Cork, L. C., 1991, Aged non-human primates: An animal model of age-associated neurodegenerative disease, *Brain Pathol.* **1**:287–296.

Price, J. L., Davis, P. B., Morris, J. C., and White, D. L., 1991, The distribution of tangles, plaques and related immunohistochemical markers in healthy aging and Alzheimer's disease, *Neurobiol. Aging* **12**:295–312.

Probst, A., and Dufresne, J. J., 1975, Paralysie supranucléaire progressive (ou dystonie oculo-facio-cervicale), *Schweiz. Arch. Neurol. Neurochir. Psychiat.* **116**:107–134.

Probst, A., Anderton, B. H., Ulrich, J., Kohler, R., Kahn, J., and Heitz, P. U., 1983, Pick's disease: An immunocytochemical study of neuronal changes, *Acta Neuropathol.* **60**:175–182.

Probst, A., Langui, D., Lautenschlager, C., Ulrich, J., Brion, J. P., and Anderton, B. H., 1988, Progressive supranuclear palsy: Extensive neuropil threads in addition to neurofibrillary tangles: Very similar antigeneicity of subcortical neuronal pathology in progressive supranuclear palsy and Alzheimer's disease, *Acta Neuropathol.* **77**:61–68.

Quigley, B. J., Jr., and Kowall, N. W., 1991, Substance P-like immunoreactive neurons are depleted in Alzheimer's disease cerebral cortex, *Neuroscience* **41**:41–60.

Rail, D., Scholtz, C., and Swash, M., 1981, Post-encephalitic parkinsonism: Current experience, *J. Neurol. Neurosurg. Psychiatry* **44**:670–676.

Rapoport, S. I., 1988a, Brain evolution and Alzheimer's disease, *Rev. Neurol.* **144**:79–90.

Rapoport, S. I., 1988b, A phylogenic hypothesis for Alzheimer's disease, in: *Genetics and Alzheimer's Disease* (P. M. Sinet, Y. Lamour, and Y. Christen, eds.), Springer-Verlag, Berlin, pp. 62–88.

Rapoport, S. I., 1990, Topography of Alzheimer's disease: Involvement of association neocortices and connected regions: Pathological, metabolic, and cognitive correlations; relation to evolution, in: *Imaging, Cerebral Topography, and Alzheimer's Disease* (S. I. Rapoport, H. Petit, D. Leys, and Y. Christen, eds.), Springer-Verlag, Berlin, pp. 1–17.

Rapp, P. R., and Amaral, D. G., 1992, Individual differences in the cognitive and neurobiological consequences of normal aging, *Trends Neurosci.* **15**:340–345.

Rebeck, G. W., Perls, T. T., West, H. L., Sodhi, P., Lipsitz, L. A., and Hyman, B. T., 1994, Reduced apolipoprotein E4 allele frequency in the oldest old Alzheimer's patient and cognitively normal individuals, *Neurology* **44**:1513–1516.

Rebeiz, J. J., Kolodny, E. H., and Richardson, E. P., Jr., 1967, Corticodentatonigral degeneration with neuronal achromasia: A progressive disorder of late adult life, *Trans. Am. Neurol. Assoc.* **92**:23–26.

Rebeiz, J. J., Kolodny, e. H., and Richardson, E. P., Jr., 1968, Corticodentatonigral degeneration with neuronal achromasia, *Arch. Neurol.* **18**:20–33.

Reiman, E. M., Caselli, R. J., Yun, L. S., Chen, K., Bandy, D., Minoshima, S., Thibodeau, S. N., and Osborne, D., 1996, Preclinical evidence of Alzheimer's disease in persons homozygous for the ε4 allele for apolipoprotein E, *New Engl. J. Med.* **334**:752–758.

Riley, D. E., Lang, A. E., Lewis, A., Resch, L., Ashby, P., Hornykiewicz, O., and Black, S., 1990, Corticobasal ganglionic degeneration, *Neurology* **40**:1203–1212.

Rinne, J. O., Lee, M. S., Thompson, P. D., and Marsden, C. D., 1994, Corticobasal degeneration: A clinical study of 36 cases, *Brain* **117**:1183–1196.

Ripps, M. E., Huntley, G. W., Hof, P. R., Morrison, J. H., and Gordon, J. W., 1995, Transgenic mice expressing an altered murine superoxide dismutase gene provide an animal model of amyotrophic lateral sclerosis, *Proc. Natl. Acad. Sci. USA* **92**:689–693.

Risberg, J., Passant, U., Warkentin, S., and Gustafson, L., 1993, Regional cerebral blood flow in frontal lobe dementia of non-Alzheimer type, *Dementia* **4**:186–187.

Ritchie, K., and Kildea, D., 1995, Is senile dementia "age-related" or "aging-related"? Evidence from meta-analysis of dementia prevalence in the oldest old, *Lancet* **346**:931–933.

Robakis, N. K., 1994, β-amyloid and amyloid precursor protein: Chemistry, molecular biology, and neuropathology, in: *Alzheimer Disease* (R. D. Terry, R. Katzman, and K. L. Bick, eds.), Raven, New York, pp. 317–326.

Roberts, A. H., 1969, *Brain damage in boxers: A Study of the Prevalence of Traumatic Encephalopathy among Ex-Professional Boxers,* Pitman, London.

Roberts, G. W., 1988, Immunocytochemistry of neurofibrillary tangles in dementia pugilistica and Alzheimer's disease: Evidence for common genesis, *Lancet* **2**:1456–1458.

Roberts, G. W., Allsop, D., and Bruton, C., 1990a, The occult aftermath of boxing, *J. Neurol. Neurosurg. Psychiatry* **53**:373–378.

Roberts, G. W., Whitwell, H. L., Acland, P. L., and Bruton, C. J., 1990b, Dementia in a punch-drunk wife, *Lancet* **335:**918–919.

Roberts, G. W., Gentleman, S. M., Lynch, A., Graham, D. I., 1991, βA4 amyloid protein deposition in brain after head trauma, *Lancet* **338:**1422–1423.

Roberts, G. W., Gentleman, S. M., Lynch, A. Murray, L., Landon, M., and Graham, D. I., 1994, β amyloid protein deposition in the brain after severe head injury: Implications for the pathogenesis of Alzheimer's disease, *J. Neurol. Neurosurg. Psychiatry* **57:**419–425.

Rodgers-Johnson, P., Garruto, R. M., Yanagihara, R., Chen, K.-M., Gajdusek, D. C., and Gibbs, C. J., Jr., 1986, Amyotrophic lateral sclerosis and parkinsonism-dementia on Guam: A 30-year evaluation of clinical and neuropathologic trends, *Neurology* **36:**7–13.

Rogers, J., and Morrison, J. H., 1985, Quantitative morphology and regional and laminar distributions of senile plaques in Alzheimer's disease, *J. Neurosci.* **5:**2801–2808.

Rudelli, R., Strom, J. O., Welch, P. T., and Ambler, M. W., 1982, Posttraumatic premature Alzheimer's disease: Neuropathologic findings and pathogenetic considerations, *Arch. Neurol.* **39:**570–575.

Rumble, B., Retallack, R., Hilbich, C., Simms, G., Multhaup, G., Martins, R., Hockey, A., Montgomery, P., Beyreuther, K., and Masters, C. L., 1989, Amyloid A4 protein and its precursor in Down's syndrome and Alzheimer's disease, *New Engl. J. Med.* **320:**1446–1452.

Sadun, A. A., Borchert, M., DeVita, E., Hinton, D. R., and Bassi, C. J., 1987, Assessment of visual impairment in patients with Alzheimer's disease, *Am. J. Ophthalmol.* **104:**113–120.

Samuel, W., Galasko, D., Masliah, E., and Hansen, L. A., 1996, Neocortical Lewy body counts correlate with dementia in the Lewy body variant of Alzheimer's disease, *J. Neuropathol. Exp. Neurol.* **55:**44–52.

Samuel, W., Alford, M., Hofstetter, C. R., and Hansen, L., 1997, Dementia with Lewy bodies versus pure Alzheimer disease: Differences in cognition, neuropathology, cholinergic dysfunction, and synapse density, *J. Neuropathol. Exp. Neurol.* **56:**499–508.

Satoh, J., Tabira, T., Sano, M., Nakayama, H., and Tateishi, J., 1991, Parvalbumin-immunoreactive neurons in the human central nervous system are decreased in Alzheimer's disease, *Acta Neuropathol.* **81:**388–395.

Saitoh, T., Masliah, E., Jin, L. W., Cole, G. M., Wieloch, T., and Shapiro, I. P., 1991, Protein kinases and phosphorylation in neurologic disorders and cell death, *Lab. Invest.* **64:**596–616.

Sampson, V. L., Morrison, J. H., and Vickers, J. C., 1997, The cellular basis for the relative resistance of parvalbumin and calretinin immunoreactive neocortical neurons to the pathology of Alzheimer's disease, *Exp. Neurol.* **154:**295–302.

Sautière, P. E., Sindou, P., Couratier, P., Hugin, J., Wattez, A., and Delacourte, A., 1992, Tau antigenic changes induced by glutamate in rat primary culture model: A biochemical approach, *Neurosci. Lett.* **140:**206–210.

Schapiro, M. B., Ball, M. J., Grady, C. L., Haxby, J. V., Kaye, J. A., and Rapoport, S. I., 1988, Dementia in Down syndrome: Cerebral glucose utilization, neuropsychological assessment, and neuropathology, *Neurology* **38:**938–942.

Schapiro, M. B., Grady, C. L., Kumar, A., Herscovitch, P., Haxby, J. V., Moore, A. M., White, B., Friedland, R. P., and Rapoport, S. I., 1990, Regional cerebral glucose metabolism is normal in young adults with Down syndrome, *J. Cereb. Blood Flow Metab.* **10:**199–206.

Schapiro, M. B., Haxby, J. V., Grady, C. L., 1992, Nature of mental retardation and dementia in Down syndrome: Study with PET, CT, and neuropsychology, *Neurobiol. Aging* **13:**723–734.

Schenk, V. W. D., and van Mansvelt, J., 1955, The cortical degeneration in Pick's syndrome: A quantitative analysis, *Folia Psychiat. Neurol. Neurochir. Neerl.* **58:**42–62.

Schmidt, M. L., Murray, J., Lee, V. M. Y., Hill, W. D., Wertkin, A., and Trojanowski, J. Q., 1991, Epitope map of neurofilament protein domains in cortical and peripheral nervous system Lewy bodies, *Am. J. Pathol.* **139:**53–65.

Schmidt, M. L., Lee, V. M. Y., Forman, M., Chiu, T. S., and Trojanowski, J. Q., 1997, Monoclonal antibodies to a 100-kd protein reveal abundant Aβ-negative plaques throughout gray matter of Alzheimer's disease brains, *Am. J. Pathol.* **151:**69–80.

Schmidt, M. L., Huang, R., Martin, J. A., Henley, J., Mawal-Dewan, M., Hurtig, H. I., Lee, V. M. Y., and Trojanowski, J. Q., 1996, Neurofibrillary tangles in progressive supranuclear palsy contain the same tau epitopes identified in Alzheimer's disease PHFtau, *J. Neuropathol. Exp. Neurol.* **55:**534–539.

Schneider, J. A., Watts, R. L., Gearing, M., Brewer, R. P., and Mirra, S. S., 1997, Corticobasal degeneration: Neuropathological and clinical heterogeneity, *Neurology* **48:**959–969.

Schneider, J., Watts, R., Gearing, M., Brewer, R., Mirra, S., 1997, Corticobasal degeneration: neuropathologic and clinical heterogeneity, *Neurology* **48:**959–969.

Scholz, W., 1938, Studien zur Pathologie der Hirngefässe. II. Die drusige Entartung der Hirnarterien und -capillaren. (Ein Form seniler Gefässerkrankung.), *Z. Ges. Neurol. Psychiat.* **162:**694–715.

Schwab, C., Steele, J. C., and McGeer, P. L., 1996, Neurofibrillary tangles of Guam parkinson-dementia are associated with reactive microglia and complements proteins, *Brain Res.* **707:**196–205.

Schwab, C., Steele, J. C., Akiyama, H., McGeer, E. G., and McGeer, P. L., 1995, Relationship of amyloid β/A4 protein to the neurofibrillary tangles in Guamanian parkinsonism-dementia, *Acta Neuropathol.* **90:**287–298.

Senut, M. C., Roudier, M., Davous, P., Fallet-Bianco, C., and Lamour, Y., 1991, Senile dementia of the Alzheimer type: Is there a correlation between entorhinal cortex and dentate gyrus? *Acta Neuropathol.* **82:**306–315.

Sereno, M. I., Dale, A. M., Reppas, J. B., Kwong, K. K., Belliveau, J. W., Brady, T. J., Rosen, B. R., and Tootell, R. B. H., 1995, Borders of multiple visual areas in humans revealed by functional magnetic resonance imaging, *Science* **268:**889–893.

Seress, L., Gulyás, A. I., and Freund, T. F., 1992, Pyramidal neurons are immunoreactive for calbindin $D_{28k}$ in the CA1 subfield of the human hippocampus, *Neurosci. Lett.* **138:**257–260.

Sergeant, N., David, J. P., Goedert, M., Jakes, R., Vermersch, P., Buée, L., Lefranc, D., Wattez, A., and Delacourte, A., 1997, Two-dimensional characterization of paired helical filament-tau from Alzheimer's disease: Demonstration of an additional 74-kDa component and age-related biochemical modifications, *J. Neurochem.* **69:**834–844.

Shankar, S. K., Yanagihara, R., Garruto, R. M., Grundke-Iqbal, I., Kosik, K. S., and Gajdusek, D. C., 1989, Immunocytochemical characterization of neurofibrillary tangles in amyotrophic lateral sclerosis and parkinsonism-dementia of Guam, *Ann. Neurol.* **25:**146–151.

Shefer, V. F., 1972, Absolute number of neurons and thickness of the cerebral cortex during aging, senile and vascular dementia, and Pick's and Alzheimer's disease, *Zh. Nevropatol. Psikhiat. im. S.S. Korsakova* **72:**1024–1029.

Shefer, V. F., 1976, Hippocampal pathology as one of the possible factors in the pathogenesis of several dementias of old age, *Zh. Nevropatol. Psikhiat. im. S.S. Korsakova* **76:**1032–1036.

Sherriff, F. E., Bridges, L. R.,and Sivaloganathan, S., 1994, Early detection of axonal injury after human head trauma using immunohistochemistry for β-amyloid precursor protein, *Acta Neuropathol.* **87:**55–62.

Shibayam, H., Kitoh, J., Marui, Y., Kobayashi, H., Iwase, S., and Kayukawa, Y., 1983, An unusual case of Pick's disease, *Acta Neuropathol.* **59:**79–87.

Shibayama, H., Kobayashi, H., Nagakawa, M., Yamada, K., Iwata, H., Iwai, K., Takeuchi, T., Mu-Qune, X., Ishihara, R., Iwase, S., and Kitoh, J., 1992, Non-Alzheimer, non-Pick dementia with Fahr's syndrome, *Clin. Neuropathol.* **11:**2237–2250.

Shimohama, S., Homma, Y., Suenaga, T., Fujimoto, S., Taniguchi, T., Araki, W., Yamaoka, Y., Tanekawa, T., and Kimura, J., 1991, Aberrant accumulation of phospholipase C-delta in Alzheimer's brains, *Am. J. Pathol.* **139:**737–742.

Shimohama, S., Perry, G., Richey, P., Praprotnik, D., Takenawa, T., Fukami, K., Whitehouse, P. J., and Kimura, J., 1995, Characterization of the association of phospholipase C-δ with Alzheimer neurofibrillary tangles, *Brain Res.* **669:**217–224.

Shipp, S., and Zeki, S., 1989a, The organization of connections between areas V5 and V1 in macaque visual cortex, *Eur. J. Neurosci.* **1:**309–332.

Shipp, S., and Zeki, S., 1989b, The organization of connections between areas V5 and V2 in macaque visual cortex, *Eur. J. Neurosci.* **1:**333–354.

Siegel, S. J., Brose, N., Janssen, W., Gasic, G. P., Jahn, R., Heinemann, S. F., and Morrison, J. H., 1994, Regional, cellular, and ultrastructural distribution of *N*-methyl-D-aspartate receptor subunit 1 in monkey hippocampus, *Proc. Natl. Acad. Sci. USA* **91:**564–568.

Siegel, S. J., Janssen, W. G., Tullai, J. W., Rogers, S. W., Moran, T., Heinemann, S. F., and Morrison, J. H., 1995, Distribution of the excitatory amino acid receptor subunits GluR2(4) in monkey hippocampus and colocalization with subunits GluR5-7 and NMDAR1, *J. Neurosci.* **15:**2707–2719.

Silverman, S. E., Tran, D. B., Zimmermann, K. M., and Feldon, S. E., 1994, Dissociation between the detection and perception of motion in Alzheimer's disease, *Neurology* **44:**1814–1818.

Simic, G., Kostovic, I., Winblad, B., and Bodganovic, N., 1997, Volume and number of neurons of the human hippocampal formation in normal aging and Alzheimer's disease, *J. Comp. Neurol.* **379:**482–494.

Sindou, P., Lesort, M., Couratier, P., Yardin, C., Esclaire, F., and Hugon, J., 1994, Glutamate increases tau phosphorylation in primary neuronal cultures from fetal rat cerebral cortex, *Brain Res.* **646:**124–128.

Sinha, U. K., Hollen, K. M., and Miller, C. A., 1993, Abnormal neuritic architecture identified by Di-I in Pick's disease, *J. Neuropathol. Exp. Neurol.* **52:**411–418.

Sjögren, T., Sjögren, H., and Lindgren, A. G. H., 1952, Morbus Alzheimer and morbus Pick (a genetic, clinical, and patho-anatomical study), *Acta Psychiat. Neurol. Scand. (suppl.)* **82:**1–152.

Sloviter, R. S., 1989, Calcium-binding protein (calbindin-$D_{28k}$) and parvalbumin immunocytochemistry: Localization in the rat hippocampus with specific reference to the selective vulnerability of hippocampal neurons to seizure activity, *J. Comp. Neurol.* **280:**183–196.

Sloviter, R. S., Sollas, A., Barbara, N. M., and Laxer, K. D., 1991, Calcium-binding protein (calbindin-D28K) and parvalbumin immunocytochemistry in the normal and epileptic human hippocampus, *J. Comp. Neurol.* **308:**381–396.

Smith, M. A., Kutty, R. K., Richey, P. L., Yan, S. D., Stern, D., Chader, G. J., Wiggert, B., Petersen, R. B., and Perry, G., 1994a, Heme oxygenase-1 is associated with the neurofibrillary pathology of Alzheimer's disease, *Am. J. Pathol.* **145:**42–47.

Smith, M. A., Taneda, S., Richey, P. L., Miyata, S., Yan, S. D., Stern, D., Sayre, L. M., Monnier, V. M., and Perry, G., 1994b, Advanced Maillard reaction end products are associated with Alzheimer disease pathology, *Proc. Natl. Acad. Sci. USA* **91:**5710–5714.

Smith, M. A., Richey Harris, P. L., Sayre, L. M., Beckman, J. S., and Perry, G., 1997, Widespread peroxynitrite-mediated damage in Alzheimer's disease, *J. Neurosci.* **17:**2653–2657.

Snowden, J. S., Neary, D., Mann, D. M. A., Goulding, P. J., and Testa, H. J., 1992, Progressive language disorder due to lobar atrophy, *Ann. Neurol.* **31:**174–183.

Sobel, E., Louhija, J., Sulkava, R., Davanipour, Z., Kontula, K., Miettinen, H., Tikkanen, M., Kainulainen, K., and Tilvis, R., 1995, Lack of association of apolipoprotein E allele epsilon 4 with late-onset Alzheimer's disease among Finnish centenarians, *Neurology* **45:**903–907.

Solodkin, A., Veldhuizen, S. D., and Van Hoesen, G. W., 1996, Contingent vulnerability of entorhinal parvalbumin-containing neurons in Alzheimer's disease, *J. Neurosci.* **16:**3311–3321.

Sommer, B., Kohler, M., Sprengel, R., Seeburg, P. H., 1991, RNA editing in brain controls a determinant of ion flow in glutamate-gated channels, *Cell* **67:**11–19.

Sparks, D. L., and Hunsaker, J. C., III, 1992, Down's syndrome: Occurrence of ALZ-50 reactive neurons and the formation of senile plaques, *J. Neurol. Sci.* **109:**77–82.

Spencer, P. S., Nunn, P. B., Hugon, J., Ludolph, A. C., Ross, S. M., Roy, D. N., and Robertson, R. C., 1987, Guam amyotrophic lateral sclerosis-parkinsonism-dementia linked to a plant excitant neurotoxin, *Science* **237:**517–522.

Spillantini, M. G., Goedert, M., Crowther, R. A., Murrell, J. L., Farlow, M. R., and Ghetti, B., 1997, Familial multiple system tauopathy with presenile dementia: A disease with abundant neuronal and glial tau filaments, *Proc. Natl. Acad. Sci. USA* **94:**4113–4118.

Squire, L. R., and Zola-Morgan, S., 1991, The medial temporal lobe memory system, *Science* **253:**1380–1386.

Starkstein, S. E., Migliorelli, R., Tesón, A., Sabe, L., Vázquez, S., Turjanski, M., Robinson, R. G., and Leiguarda, R., 1994, Specificity of changes in cerebral blood flow in patients with frontal lobe dementia, *J. Neurol. Neurosurg. Psychiatry* **57:**790–796.

Steele, J. C., 1972, Progressive supranuclear palsy, *Brain* **95:**693–704.

Steele, J. C., Richardson, J. C., and Olszewski, J., 1964, Progressive supranuclear palsy: A heterogeneous degeneration involving brain stem, basal ganglia, and cerebellum, with vertical gaze and pseudobulbar palsy, nuchal dystonia, and dementia, *Arch. Neurol.* **10:**333–359.

Su, J. H., Cummings, B. J., and Cotman, C. W., 1992, Localization of heparan sulfate glycosminoglycan and proteoglycan core protein in aged brain and Alzheimer's disease, *Neuroscience* **51:**801–813.

Su, J. H., Anderson, A. J., Cummings, B. J., and Cotman, C. W., 1994, Immunocytochemical evidence for apoptosis in Alzheimer's disease, *NeuroReport* **5:**2529–2533.

Surbeck, B., 1961, L'angiopathie dyshorique (Morel) de l'écorce cérébrale, *Acta Neuropathol.* **1:**168–197.

Suzuki, W. A., Zola-Morgan, S., Squire, L. R., and Amaral, D. G., 1993, Lesions of the perirhinal and parahippocampal cortices in the monkey produce long-lasting impairment in the visual and tactile modalities, *J. Neurosci.* **13:**2430–2451.

Szatkowski, M., and Atwell, D., 1994, Triggering and execution of neuronal death in brain ischemia: Two phases of glutamate release by different mechanisms, *Trends Neurosci.* **17:**359–365.

Sze, C. I., Troncoso, J. C., Kawas, C., Mouton, P., Price, D. L., and Martin, L. J., 1997, Loss of the presynaptic vesicle protein synaptophysin in hippocampus correlates with cognitive decline in Alzheimer disease, *J. Neuropathol. Exp. Neurol.* **56:**933–944.

Tagliavini, F., Pilleri, G., Bouras, C., and Constantinidis, J., 1984, The basal nucleus of Meynert in patients with progressive supranuclear palsy, *Neurosci. Lett.* **44:**37–42.

Takagi, S. F., 1991, Olfactory frontal cortex and multiple olfactory processing in primates, in: *Cerebral Cortex, Vol. 9.* (A. Peters and E. G. Jones, eds.), Plenum, New York, pp. 133–152.

Takahashi, T., Amano, N., Hanihara, T., Nagatomo, H., Yagishita, S., Itoh, Y., Yamaoka, K., Toda, H., and Tanabe, T., 1996, Corticobasal degeneration: Widespread argentophilic threads and glia in addition to neurofibrillary tangles: Similarities of cytoskeletal abnormalities in corticobasal degeneration and progressive supranuclear palsy, *J. Neurol. Sci.* **138:**66–77.

Takashima, S., Becker, L. E., Armstrong, D. L., and Chan, F., 1981, Abnormal neuronal development in the visual cortex of the human fetus and infant with Down's syndrome: A quantitative and qualitative Golgi study, *Brain Res.* **221:**1–21.

Takeda, A., Arai, N., Komori, T., Kato, S., and Oda, M., 1997, Neuronal inclusions in the dentate fascia in patients with multiple system atrophy, *Neurosci. Lett.* **227:**157–160.

Talamo, B. R., Rudel, R. A., Kosik, K. S., Lee, V. M. Y., Neff, S., Adelman, L., and Kauer, J. S., 1989, Pathological changes in olfactory neurons in patients with Alzheimer's disease, *Nature* **337:**736–739.

Tellez-Nagel, I., and Wisniewski, H. M., 1973, Ultrastructure of neurofibrillary tangles in Steele-Richardson-Olszewski syndrome, *Arch. Neurol.* **29:**324–327.

Terry, R. D., and Katzman, R., 1983, Senile dementia of the Alzheimer's type: Defining a disease, in: *The Neurology of Aging* (R. Katzman and R. D. Terry, eds.), F. A. Davis, New York, pp. 51–84.

Terry, R. D., and Wisniewski, H. M., 1970, The ultrastructure of the neurofibrillary tangle and the senile plaque, in: *CIBA Foundation Symposium on Alzheimer's Disease and Related Conditions* (G. E. W. Wolstenholme and M. O'Connor, eds.), J. & A. Churchill, London, pp. 145–155.

Terry, R. D., Peck, A., DeTeresa, R., Schechter, R., and Horoupian, D. S., 1981, Some morphometric aspects of the brain in senile dementia of the Alzheimer's type, *Ann. Neurol.* **10:**184–192.

Terry, R. D., DeTeresa, R., and Hansen, L. A., 1987, Neocortical cell counts in normal human adult aging, *Ann. Neurol.* **21:**530–539.

Terry, R. D., Masliah, E., Salmon, D. P., Butters, N., DeTeresa, R., Hill, R., Hansen, L. A., and Katzman, R., 1991, Physical basis of cognitive alterations in Alzheimer's disease: Synapse loss is the major correlate of cognitive impairment, *Ann. Neurol.* **30:**572–580.

Terry, R. D., Masliah, E., and Hansen, L. A., 1994, Structural basis of the cognitive alterations in Alzheimer disease, in: *Alzheimer Disease* (R. D. Terry, R. Katzman, and K. L. Bick, eds.), Raven, New York, pp. 179–196.

Tillema, S., and Wijnberg, C. J., 1953, "Endemic" amyotrophic lateral sclerosis on Guam: Epidemiological data—A preliminary report, *Doc. Med. Geogr. Trop.* **5:**366–370.

Tissot, R., Constantinidis, J., and Richard, J., 1985, Pick's disease, in: *Neurobehavioural Disorders, Handbook of Clinical Neurology, Vol. 46* (J. A. M. Frederiks, ed.), Elsevier, Amsterdam, pp. 233–246.

Tokuda, T., Ikeda, S., Yanagisawa, N., Ihara, Y., and Glenner, G. G., 1991, Re-examination of ex-boxers' brains using immunohistochemistry with antibodies to amyloid β-protein and tau protein, *Acta Neuropathol.* **82:**280–285.

Tolnay, M., and Probst, A., 1995, Frontal lobe degeneration: Novel ubiquitin-immunoreactive neurites within frontotemporal cortex, *Neuropathol. Appl. Neurobiol.* **21:**492–497.

Tolnay, M., Spillantini, M. G., Goedert, M., Ulrich, J., Langui, D., and Probst, A., 1997, Argyrophilic grain disease: Widespread hyperphosphorylation of tau protein in limbic system, *Acta Neuropathol.* **93:**477–484.

Tomlinson, B. E., and Kitchener, D., 1972, Granulovacuolar degeneration of hippocampal pyramidal cells, *J. Pathol.* **106:**165–185.

Tomlinson, B. E., Blessed, G., and Roth, M., 1968, Observations on the brains of non-demented old people, *J. Neurol. Sci.* **7:**331–356.

Tomlinson, B. E., Blessed, G., and Roth, M., 1970, Observations on the brains of demented old people, *J. Neurol. Sci.* **11:**205–242.

Tomonaga, M., 1977, Ultrastructure of neurofibrillary tangles in progressive supranuclear palsy, *Acta Neuropathol.* **37:**177–181.

Tootell, R. B. H., and Taylor, J. B., 1995, Anatomical evidence for MT and additional cortical visual areas in humans, *Cereb. Cortex* **5:**39–55.

Tootell, R. B. H., Reppas, J. B., Kwong, K. K., Malach, R., Born, R. T., Brady, T. J., Rosen, B. R., and Belliveau, J. W., 1995, Functional analysis of human area MT and related visual cortical areas using magnetic resonance imaging, *J. Neurosci.* **15:**3215–3230.

Torres, J., Iriarte, L. L. G., and Kurland, L. T., 1957, Amyotrophic lateral sclerosis among Guamanians in California, *Calif. Med.* **86:**385–388.

Torvik, A., and Meen, D., 1966, Distribution of the brain stem lesions in postencephalitic parkinsonism, *Acta Neurol. Scand.* **42:**415–425.

Trojanowski, J. Q., Mawal-Dewan, M., Schmidt, M. L., Martin, J., and Lee, V. M. Y., 1993a, Localization of the mitogen activated protein kinase ERK2 in Alzheimer's disease neurofibrillary tangles and senile plaque neurites, *Brain Res.* **618:**333–337.

Trojanowski, J. Q., Schmidt, M. L., Shin, R. W., Bramblett, G. T., Rao, D., and Lee, V. M. Y., 1993b, Altered tau and neurofilament proteins in neurodegenerative diseases: Diagnostic implications for Alzheimer's disease and Lewy body dementias, *Brain Pathol.* **3:**45–54.

Tsubokawa, H., Oguro, K., Robinson, H. P. C., Masuzawa, T., and Kawai, N., 1995, Single glutamate channels in CA1 pyramidal neurones after transient ischaemia, *NeuroReport* **6:**527–531.

Tu, P. H., Raju, P., Robinson, K. A., Gurney, M. E., Trojanowski, J. Q., and Lee, V. M. Y., 1996, Transgenic mice carrying a human mutant superoxide dismutase transgene develop neuronal cytoskeletal pathology resembling human amyotrophic lateral sclerosis, *Proc. Natl. Acad. Sci. USA* **93:**3155–3160.

Turetsky, D. M., Canzoniero, L. M. T., Sensi, S. L., Weiss, J. H., Goldberg, M. P., and Choi, D. W., 1994, Cortical neurones exhibiting kainate-activated $Co^{2+}$ uptake are selectively vulnerable to AMPA/kainate receptor-mediated toxicity, *Neurobiol. Dis.* **1:**101–110.

Turski, L., and Turski, W. A., 1993, Towards an understanding of the role of glutamate in neurodegenerative disorders: Energy metabolism and neuropathology, *Experientia* **49:**1064–1072.

Uitti, R. J., 1994, Cerebellar degenerations, in: *Neurodegenerative Diseases* (D. B. Calne, ed.), W. B. Saunders, Philadelphia, pp. 787–806.

Ulrich, J., 1985, Alzheimer changes in non-demented patients younger than sixty-five: Possible early stages of Alzheimer's disease and senile dementia of Alzheimer type, *Ann. Neurol.* **17:**273–277.

Ulrich, J., Haugh, M., Anderton, B. H., Probst, A., Lautenschlager, C., and His, B., 1987, Alzheimer dementia and Pick's disease: Neurofibrillary tangles and Pick bodies are associated with identical phosphorylated neurofilament epitopes, *Acta Neuropathol.* **73:**240–246.

Ungerleider, L. G., 1995, Functional brain imaging studies of cortical mechanisms for memory, *Science* **270:**769–775.

Uchihara, T., Mitani, K., Mori, H., Kondo, H., Yamada, M., and Ikeda, K., 1994, Abnormal cytoskeletal pathology peculiar to corticobasal degeneration is different from that of Alzheimer's disease or progressive supranuclear palsy, *Acta Neuropathol.* **88:**379–383.

Vallet, P. G., Guntern, R., Hof, P. R., Golaz, J., Delacourte, A., Robakis, N. K., and Bouras, C., 1992, A comparative study of histological and immunohistochemical methods for neurofibrillary tangles and senile plaques in Alzheimer's disease, *Acta Neuropathol.* **83:**170–178.

Van Essen, D. C., and Maunsell, J. H. R., 1983, Hierarchical organization and functional streams in the visual cortex, *Trends Neurosci.* **6:**370–375.

Verdoorn, T. A., Burnashev, N., Monyer, H., Seeburg, P. H., and Sakmann, B., 1991, Structural determinants of ion flow through recombinant glutamate receptor channels, *Science* **252:**1715–1718.

Vermersch, P., Frigard, B., David, J. P., Fallet-Bianco, C., and Delacourte, A., 1992a, Presence of abnormally phosphorylated tau proteins in the entorhinal cortex of aged non-demented subjects, *Neurosci. Lett.* **144:**143–146.

Vermersch, P., Frigard, B., and Delacourte, A., 1992b, Mapping of neurofibrillary degeneration in Alzheimer's disease: Evaluation of heterogeneity using the quantification of abnormal tau proteins, *Acta Neuropathol.* **85:**48–54.

Vermersch, P., Delacourte, A., Javoy-Agid, F., Hauw, J. J., and Agid, Y., 1993, Dementia in Parkinson's disease: Biochemical evidence for cortical involvement using the immunodetection of abnormal tau proteins, *Ann. Neurol.* **33:**445–450.

Vermersch, P., Robitaille, Y., Bernier, L., Wattez, A., Gauvreau, D., and Delacourte, A., 1994, Biochemical mapping of neurofibrillary degeneration in a case of supranuclear palsy: Evidence for general cortical involvement, *Acta Neuropathol.* **87:**545–553.

Vermersch, P., Bordet, R., Ledoze, F., Ruchoux, M. M., Chapon, F., Thomas, P., Destée, A., Lechevallier, B., and Delacourte, A., 1995, Mise en évidence d'un profil particulier de protéines Tau pathologiques dans les cas de démences fronto-temporales, *C. R. Acad. Sci. Paris [Sciences de la Vie]* **318:**439–445.

Vermersch, P., Sergeant, N., Ruchoux, M. M., Hofmann-Radvanyi, H., Wattez, A., Petit, H., Dewailly, P., and Delacourte, A., 1996, Specific tau variants in the brains of patients with myotonic dystrophy, *Neurology* **47:**711–717.

Vermersch, P., Buée-Scherrer, V., Buée, L., David, J. P., Wattez, A., Sergeant, N., Hof, P. R., Agid, Y., Perl, D. P., Olanow, C. W., Robitaille, Y., Gauvreau, D., Petit, H., and Delacourte, A., 1997, Cortical mapping of pathological tau proteins in several neurodegenerative disorders, in: *Connections, Cognition, and Alzheimer's Disease* (B. T. Hyman, C. Duyckaerts, and Y. Christen, eds.), Springer-Verlag, Berlin, pp. 41–52.

Verny, M., Jellinger, K. A., Hauw, J. J., Bancher, C., Litvan, I., and Agid, Y., 1996, Progressive supranuclear palsy: A clinicopathological study of 21 cases, *Acta Neuropathol.* **91:**427–431.

Vickers, J. C., 1997, A cellular mechanism for the neuronal changes underlying Alzheimer's disease, *Neuroscience* **78:**629–639.

Vickers, J. C., Delacourte, A., and Morrison, J. H., 1992, Progressive transformation of the cytoskeleton associated with normal aging and Alzheimer's disease, *Brain Res.* **594:**273–278.

Vickers, J. C., Huntley, G. W., Edwards, A. M., Moran, T., Rogers, S. W., Heinemann, S. F., and Morrison, J. H., 1993, Quantitative localization of AMPA/kainate and kainate glutamate receptor subunit immunoreactivity in neurochemically identified subpopulations of neurons in the prefrontal cortex of the macaque monkey, *J. Neurosci.* **13:**2982–2992.

Vickers, J. C., Morrison, J. H., Friedrich, V. L., Jr., Elder, G. A., Perl, D. P., Katz, R. N., and Lazzarini, R. N., 1994a, Age-associated and cell-type-specific neurofibrillary pathology in transgenic mice expressing the human midsized neurofilament subunit, *J. Neurosci.* **14:**5603–5612.

Vickers, J. C., Riederer, B. M., Marugg, R. A., Buée-Scherrer, V., Buée, L., Delacourte, A., and Morrison, J. H., 1994b, Alterations in neurofilament protein immunoreactivity in human hippocampal neurons related to normal aging and Alzheimer's disease, *Neuroscience* **62:**1–13.

Vickers, J. C., Huntley, G. W., Hof, P. R., Bederson, J., DeFelipe, J., Heinemann, S. F., and Morrison, J. H., 1995a, Immunocytochemical localization of non-NMDA ionotropic excitatory amino acid receptor subunits in human neocortex, *Brain Res.* **671:**175–180.

Vickers, J. C., Lazzarini, R. A., Riederer, B. M., and Morrison, J. H., 1995b, Intraperikaryal neurofilamentous accumulations in a subset of retinal ganglion cells in aged mice that express a human neurofilament gene, *Exp. Neurol.* **136:**266–269.

Vickers, J. C., Schumer, R. A., Podos, S. M., Wang, R. F., Riederer, B. M., and Morrison, J. H., 1995c, Differential vulnerability of neurochemically identified subpopulations of retinal neurons in a monkey model of glaucoma, *Brain Res.* **680:**23–35.

Vickers, J. C., Chin, D., Edwards, A. M., Sampson, V., Harper, C., and Morrison, J. H., 1996, Dystrophic neurite formation associated with age-related β amyloid deposition in the neocortex: Clues to the genesis of neurofibrillary pathology, *Exp. Neurol.* **141:**1–11.

Victoroff, J., Ross, W., Benson, D. F., Verity, M. A., and Vinters, H. V., 1994, Posterior cortical atrophy: Neuropathologic correlations, *Arch. Neurol.* **51:**269–274.

Vieregge, P., Reinhardt, V., and Höft, B., 1991, Is progression in postencephalitic Parkinson's disease late and age-related? *J. Neurol.* **238:**299–303.

Vissavajjhala, P., Janssen, W. G. M., Hu, Y., Gazzaley, A. H., Moran, T., Hof, P. R., and Morrison, J. H., 1996, Synaptic distribution of AMPA-GluR2 subunit and its colocalization with calcium-binding proteins in rat cerebral cortex: An immunohistochemical analysis using a GluR2-specific monoclonal antibody, *Exp. Neurol.* **142:**296–312.

Vogt, B. A., Van Hoesen, G. W., and Vogt, L. J., 1990, Laminar distribution of neuron degeneration in posterior cingulate cortex in Alzheimer's disease, *Acta Neuropathol.* **80:**581–589.

Vogt, B. A., Crino, P. B., and Vogt, L. J., 1992, Reorganization of cingulate cortex in Alzheimer's disease: Neuron loss, neuritic plaques, and muscarinic receptor binding, *Cereb. Cortex* **2:**526–535.

Vogt, B. A., Vogt, L. J., Nimchinsky, E. A., and Hof, P. R., 1997, Primate cingulate cortex chemoarchitecture and its disruption in Alzheimer's disease, in: *Handbook of Chemical Neuroanatomy, Vol. 13, The Primate Nervous System. Pt. I* (F. E. Bloom, A. Björklund, and T. Hökfelt, eds.), Elsevier, Amsterdam, pp. 455–528.

Vogt Weisenhorn, D. M., Weruaga-Prieto, E., and Celio, M. R., 1996, Calretinin-immunoreactivity in organotypic cultures of the rat cerebral cortex: Effects of serum deprivation, *Exp. Brain Res.* **108:**101–112.

Vyas, S., Javoy-Agid, F., Herrero, M. T., Strada, O., Boissière, F., Hibner, U., and Agid, Y., 1997, Expression of Bcl-2 in adult human brain regions with special reference to neurodegenerative disorders, *J. Neurochem.* **69:**223–231.

Wakabayashi, K., Honer, W. G., and Masliah, E., 1994a, Synapse alterations in the hippocampal-entorhinal formation in Alzheimer's disease with and without Lewy body disease, *Brain Res.* **667**:24–32.
Wakabayashi, K., Oyanagi, K., Makifuchi, T., Ikuta, F., Homma, A., Homma, Y., Horikawa, Y., and Tokiguchi, S., 1994b, Corticobasal degeneration: Etiopathological significance of the cytoskeletal alterations, *Acta Neuropathol.* **87**:545–553.
Wakabayashi, K., Hansen, L. A., and Masliah, E., 1995, Cortical Lewy body-containing neurons are pyramidal cells: Laser confocal imaging of double-immunolabeled sections with anti-ubiquitin and SMI32, *Acta Neuropathol.* **89**:404–408.
Wakabayashi, K., Hansen, L. A., Vincent, I., Mallory, M., and Masliah, E., 1997, Neurofibrillary tangles in the dentate granule cells of patients with Alzheimer's disease, Lewy body disease, and progressive supranuclear palsy, *Acta Neuropathol.* **93**:7–12.
Walker, L. C., Kitt, C. A., Cork, L. C., Struble, R. G., Dellovado, T. L., and Price, D. L., 1988, Multiple transmitter systems contribute neurites to individual senile plaques, *J. Neuropathol. Exp. Neurol.* **47**:138–144.
Watson, J. D. G., Myers, R., Frackowiak, R. S. J., Hajnal, J. V., Woods, R. P., Mazziotta, J. C., Shipp, S., and Zeki, S., 1993, Area V5 of the human brain: Evidence from a combined study using positron emission tomography and magnetic resonance imaging, *Cereb. Cortex* **3**:79–94.
Weber, L. L., Leissring, M. A., Yang, A. J., Glabe, C. G., Cribbs, D. H., and LaFerla, F. M., 1997, Presenilin-1 immunoreactivity is localized intracellularly in Alzheimer's disease brain, but not detected in amyloid plaques, *Exp. Neurol.* **143**:37–44.
Wechsler, A. F., Verity, A., Rosenschein, S., Fried, I., and Scheibel, A. B., 1982, Pick's disease: A clinical, computed tomographic, and histologic study with Golgi impregnation observations, *Arch. Neurol.* **39**:287–290.
West, M. J., 1993, Regionally specific loss of neurons in the aging human hippocampus, *Neurobiol. Aging* **14**:287–293.
West, M. J., and Gundersen, H. J. G., 1990, Unbiased stereological estimation of the number of neurons in the human hippocampus, *J. Comp. Neurol.* **296**:1–22.
West, M. J., Coleman, P. D., Flood, D. G., and Troncoso, J. C., 1994, Differences in the pattern of hippocampal neuronal loss in normal aging and Alzheimer's disease, *Lancet* **344**:769–772.
Wilcock, G. K., and Esiri, M. M., 1982, Plaques, tangles and dementia: A quantitative study, *J. Neurol. Sci.* **56**:343–356.
Wildi, E., and Dago-Akribi, A., 1968, Altérations cérébrales chez l'homme âgé, *Bull. Schweiz. Akad. Med. Wiss.* **24**:107–132.
Wildi, E., Linder, A., and Costoulas, G., 1964, Etude statistique des altérations dégénératives cérébrales apparaissant au cours du vieillissement, *Psychiat. Neurol.* **148**:41–68.
Williams, D. B., Annegers, J. F., Kokmen, E., O'Brien, P. C., and Kurland, L. T., 1991, Brain injury and neurologic sequelae: A cohort study of dementia, parkinsonism, and amyotrophic lateral sclerosis, *Neurology* **41**:1554–1557.
Wilson, F. A. W., Ó Scalaidhe, S. P., and Goldman-Rakic, P. S., 1993, Dissociation of object and spatial processing domains in primate prefrontal cortex, *Science* **260**:1955–1958.
Wisniewski, H. M., Coblentz, J. M., and Terry, R. D., 1972, Pick's disease: A clinical and ultrastructural study, *Arch. Neurol.* **26**:97–108.
Wisniewski, K. E., Jervis, G. A., Moretz, R. C., and Wisniewski, H. M., 1979, Alzheimer neurofibrillary tangles in disease other than senile and presenile dementia, *Ann. Neurol.* **5**:288–294.
Wisniewski, K. E., Wisniewski, H. M., and Wen, G. Y., 1985, Occurrence of neuropathological changes and dementia of Alzheimer's disease in Down's syndrome, *Ann. Neurol.* **17**:272–282.
Wisniewski, H. M., Bancher, C., Barcikowska, M., Wen, G. Y., and Currie, J., 1989, Spectrum of morphological appearance of amyloid deposits in Alzheimer's disease, *Acta Neuropathol.* **78**:337–347.
Wisniewski, T., Dowjat, W. K., Permanne, B., Pahla, J., Kumar, A., Gallo, G., and Frangione, B., 1997, Presenilin-1 is associated with Alzheimer's disease amyloid, *Am. J. Pathol.* **151**:610–610.
Wolozin, B., Scicutella, A., and Davies, P., 1988, Reexpression of a developmentally regulated antigen in Down syndrome and Alzheimer disease, *Proc. Natl. Acad. Sci. USA* **85**:6202–6206.
Wong, K. Y., Allen, I. V., McQuaid, S., and McConnell, R., 1996, An immunohistochemical study of neurofibrillary tangle formation in post-encephalitic Parkinsonism, *Clin. Neuropathol.* **15**:22–25.
Xiao, J., Perry, G., Troncoso, J., and Monteiro, M. J., 1996, α-Calcium-calmodulin-dependent kinase II is associated with paired helical filaments of Alzheimer's disease, *J. Neuropathol. Exp. Neurol.* **55**:954–963.

Xu, Z., Cork, L. C., Griffin, J. W., and Cleveland, D. W., 1993, Increased expression of neurofilament subunit NF-L produces morphological alterations that resemble the pathology of human motor neuron disease, *Cell* **73**:23–33.

Yagashita, S., Itoh, Y., Amano, N., Nakano, T., and Saitoh, A., 1979, Ultrastructure of neurofibrillary tangles in progressive supranuclear palsy, *Acta Neuropathol.* **48**:27–30.

Yamaguchi, H., Nakazato, Y., Shoji, M., Okamoto, K., Ihara, Y., Morimatsu, M., and Hirai, S., 1991, Secondary deposition of beta amyloid within extracellular neurofibrillary tangles in Alzheimer-type dementia, *Am. J. Pathol.* **138**:699–705.

Yamaguchi, H., Ishiguro, K., Uchida, T., Takashima, A., Lemere, C. A., and Imahori, K., 1996, Preferential labeling of Alzheimer neurofibrillary tangles with antisera for tau protein kinase (TPK)I/glycogen synthase kinase-3β and cyclin-dependent kinase 5, a component of TPK II, *Acta Neuropathol.* **92**:232–241.

Yan, S. D., Chen, X., Schmidt, A. M., Brett, J., Godman, G., Zou, Y. S., Scott, C. W., Caputo, C., Frappier, T., Smith, M. A., Perry, G., Yen, S. H., and Stern, D., 1994, Glycated tau protein in Alzheimer disease: A mechanism for induction of oxidant stress, *Proc. Natl. Acad. Sci. USA* **91**:7787–7791.

Yang, Q., Wang, S., Karlsson, J. E., Hamberger, A., and Haglid, K. G., 1995, Phosphorylated and nonphosphorylated neurofilament proteins: Distribution in the rat hippocampus and early changes after kainic acid induced seizures, *J. Chem. Neuroanat.* **9**:217–228.

Yase, Y., 1970, Neurologic disease in the Western Pacific islands, with a report on the focus of amyotrophic lateral sclerosis found in the Kii peninsula, Japan, *Am. J. Trop. Med. Hyg.* **19**:155–166.

Yasuda, R. P., Ikonomovic, M. D., Sheffield, R., Rubin, R. T., Wolfe, B. B., and Armstrong, D. M., 1995, Reduction of AMPA-selective glutamate receptor subunits in the entorhinal cortex of patients with Alzheimer's disease pathology: A biochemical study, *Brain Res.* **678**:161–167.

Yin, H. Z., Turetsky, D., Choi, D. W., and Weiss, J. H., 1994, Cortical neurones with $Ca^{2+}$ permeable AMPA/kainate channels display distinct receptor immunoreactivity and are GABAergic, *Neurobiol. Dis.* **1**:43–49.

Yoshimura, N., 1989, Topography of Pick body distribution in Pick's disease: A contribution to understanding the relationship between Pick's and Alzheimer's diseases, *Clin. Neuropathol.* **8**:1–6.

Zeki, S. M., and Schipp, S., 1988, The functional logic of cortical connections, *Nature* **335**:311–317.

Zellweger, H., 1977, Down syndrome, in: *Congenital Malformations of the Brain and Skull, Pt. II, Handbook of Clinical Neurology, Vol. 31* (P. J. Vinken, and G. W. Bruyn, eds.), North Holland, Amsterdam, pp. 367–469.

Zhang, H., Sternberger, N. H., Rubinstein, L. J., Herman, M. M., Binder, L. I., and Sternberger, L. A., 1989, Abnormal processing of multiple proteins in Alzheimer disease, *Proc. Natl. Acad. Sci. USA* **86**:8045–8049.

Zhang, Z. X., Anderson, D. W., Lavine, L., and Mantel, N., 1990a, Patterns of acquiring parkinsonism-dementia complex on Guam: 1944 through 1985, *Arch. Neurol.* **47**:1019–1024.

Zhang, Z. X., Anderson, D. W., and Mantel, N., 1990b, Geographic patterns of parkinsonism-dementia complex on Guam: 1956 through 1985, *Arch. Neurol.* **47**:1069–1074.

Zhang, Z. X., Anderson, D. W., Mantel, N., and Román, G. C., 1995, Motor neuron disease on Guam: Temporal occurrence, 1941–1985, *Acta Neurol. Scand.* **92**:299–305.

Zhang, Z. X., Anderson, D. W., Mantel, N., and Román, G. C., 1996, Motor neuron disease on Guam: Geographic and familial occurrence, 1956–1985, *Acta Neurol. Scand.* **94**:51–59.

Zimmerman, H. M., 1945, Monthly report to the medical officer in command, *U.S. Naval Medical Officer in Command, U.S. Naval Medical Research Unit No. 2.*

Zola-Morgan, S., and Squire, L. R., 1984, Preserved learning in monkeys with medial temporal lesions: Sparing of motor and cognitive skills, *J. Neurosci.* **4**:1072–1085.

Zola-Morgan, S., Squire, L. R., Amaral, D. G., and Suzuki, W. A., 1989, Lesions of perirhinal and parahippocampal cortex that spare the amygdala and hippocampal formation produce severe memory impairment, *J. Neurosci.* **9**:4355–4370.

Zola-Morgan, S., Squire, L. R., Rempel, N. L., Clower, R. P., and Amaral, D. G., 1992, Enduring memory impairment in monkeys after ischemic damage to the hippocampus, *J. Neurosci.* **12**:2582–2596.

# 9

# The Anatomy of Dementias

GENE J. BLATT and THOMAS L. KEMPER

## 1. Introduction

In this chapter we provide an overview of the patterns of cerebral cortical and subcortical pathological changes that occur in the various dementias. When viewed in this manner, the dementias can be categorized into five major groups based on the topography of their cerebral cortical involvement. These are: (1) the preferential involvement of the temporal lobes and adjacent parietal, occipital, and frontal lobes in Alzheimer's disease (AD); (2) the predilection for frontal and parietal lobes in progressive subcortical gliosis (PSG); (3) the combined involvement of the cerebral cortex of frontal and temporal lobes in Pick's disease (PiD) and dementia of the frontal type (DFT); (4) diffuse Lewy body disease (DLBD) and Huntington's disease (HD) with prominent cortical and subcortical pathology; and (5) a group characterized predominantly by subcortical pathological changes such as corticobasal degeneration (CBD), multisystem atrophy (MSA) and progressive supranuclear palsy (PSP). Although not strictly a dementia, schizophrenia (SZP) has been included because of its prominent cerebral cortical pathology and the frequent presence of a residual state resembling dementia. First, an overview of each dementia will be presented, followed by discussion of each of the five groups. Schizophrenia will be discussed separately as a developmental disease of subcortical, limbic, and paralimbic systems.

GENE J. BLATT • Department of Anatomy and Neurobiology, Boston University School of Medicine, Boston, Massachusetts 02118. THOMAS L. KEMPER • Departments of Pathology, Anatomy and Neurobiology, and Neurology, Boston University School of Medicine, Boston, Massachusetts 02118.
*Cerebral Cortex*, Volume 14, edited by Peters and Morrison. Kluwer Academic/Plenum Publishers, New York, 1999.

## 2. Alzheimer's Disease and the Primary Involvement Pattern of the Temporal Lobe

Of the various pathological changes in AD, the distribution of neurofibrillary tangles (NFTs) and neuronal and synaptic loss have provided the closest correlations with the severity and clinical features of the disease, whereas the deposition of Aβ amyloid as diffuse plaques and the density and distribution of neuritic (senile) plaques (NPs), though prominent features of the neuropathology of AD, have provided less-close clinical correlations (Braak and Braak, 1991a; Terry *et al.*, 1991; Arriagada *et al.*, 1992a; Samuel *et al.*, 1994; Bierer *et al.*, 1995; Blennow *et al.*, 1996; Mann, 1996). Of the pathological changes that correlate strongly with the clinical features of AD, the distribution of the NFTs has been the most thoroughly documented. Although there is a strong relationship between neuronal loss in AD and the development of the neurofibrillary tangle, it dose not appear that NFT formation is the sole mechanism for the loss. The presence of high somatodendritic levels of neurofilament protein has been demonstrated in a subset of pyramidal cells in layers III and V and in layers II and IV in the entorhinal cortex that are vulnerable to the development of NFTs and neuronal loss in AD (Morrison *et al.*, 1987; Hof *et al.*, 1990; Hof *et al.*, 1990; see Hof, 1997, and Hof *et al.*, this volume, for reviews), with more than 80% of the large pyramidal cells that are immunoreactive to this protein showing cell loss in the temporal and frontal cortex (Morrison *et al.*, 1987). Recently, Sampson *et al.* (1997) have identified a similar neuronal type in layer I of the cerebral cortex that is vulnerable to NFT formation. Conversely, the calcium-binding proteins parvalbumin, calbindin, and calretinin have been identified in the nonpyramidal inhibitory neurons in the cerebral cortex that are resistant to both cell death and to the formation of NFTs in AD (Hof *et al.*, 1990, 1992; Sampson *et al.*, 1997). There does not, however, appear to be an obligate relationship between the NFT development of and cell loss, because the lateral vestibular nucleus, which has been shown to develop NFTs in AD, does not show evidence of cell loss. (Ransmayr *et al.*, 1994). Hof *et al.* (1990) have reported a similar finding in the layer III pyramidal cells in the primary visual cortex. Further, Gomez-Isla *et al.* (1997) have reported that, although NFTs and neuronal loss increase in parallel to each other in the cerebral cortex in AD, the extent of neuronal loss was seven times greater than the number of accumulated NFTs. Cell loss also occurs in AD in neurons that do not develop NFTs. For example, Lassmann (1996), from an analysis of DNA fragmentation as an index of cell death, thought that neurofibrillary pathology may not directly induce cell death, but that this change and amyloid deposition may increase the risk of cell death in response to additional minor metabolic insults. Bancher *et al.* (1997), using similar techniques, found that most of the dying cells did not bear NFTs, and further, that most were not located within areas of amyloid deposits. In this later study, and that of Troncoso *et al.* (1996), there was also evidence of a loss of glial cells, and in the study of Troncoso *et al.* (1996) of microglial cells as well, cells that do not develop NFTs.

In an early global survey of the regional cerebral cortical distribution of the various pathological changes in AD, Brun and Gustafson (1976) noted that the most marked changes were found in the temporal lobe, where they were most prominent medially, as well as in the adjoining parietal and occipital lobes and in the posterior cingulate gyrus, with less-marked degeneration found in the frontal lobes. They also noted a relative sparing of the primary sensory areas and motor cortex, a finding already noted by Mutrux (1947) and Goodman (1953). More recent global surveys of the individual pathological changes in AD have further emphasized this distribution pattern in the cerebral cortex. Arnold *et al.* (1991a), in a quantitative assessment of distribution of NFTs and NPs in 39 different cortical areas in 11 AD brains, found that NFTs were most densely distributed in the temporal and limbic lobes, with less-marked involvement of the frontal, parietal, and occipital lobes and the least involvement in the primary sensory and motor cortices. The NPs were more evenly distribution throughout the hemispheres, with less-dense distribution than that found for NFTs in the allocortical and periallocortical areas. Braak and Braak (1991a) systematically surveyed 83 brains of patients with and without known AD for evidence of staging of the progression of diffuse amyloid deposition and of neuritic pathology in AD (see Braak and Braak, this volume). The neurofibrillary changes included NFTs,neuritic (senile) plaques, and neuropil threads. Diffuse amyloid deposition was the first change noted. Because most of these deposits did not correspond to NPs and because the distribution of diffuse amyloid varied from individual to individual, it was found to be of limited use for the staging of AD. The NP itself, which failed to show a consistent pattern of occurrence, distribution, or density, was also found to be of limited use for the staging of AD. In contrast, the distribution of the neurofibrillary tangles and neuropil threads showed a well-defined, consistent pattern of distribution. With these changes, Braak and Braak noted a progression from an initial stage of involvement of the entorhinal cortex and adjacent areas 35 (their transentorhinal region) on the medial part of the temporal lobe to the adjacent limbic cortices, finally progressing to extensive involvement of the neocortex, with the most dense neurofibrillary change in the medial part of the temporal lobe. In the neocortex the most marked changes were noted throughout the temporal lobe and adjacent occipital lobe, the inferior parietal lobule, and the orbital and medial surface of the frontal lobe. The patients with extensive neocortical involvement with the neurofibrillary changes were known to have AD. Braak and Braak surmised that the transentorhinal and limbic stages represented the earlier stages of the disease. Consistent with this early medial temporal lobe involvement in AD is the striking, grossly visible atrophy of this region at an early stage in the disease (LeMay, 1986; Van Hoesen *et al.*, 1991; Killiany *et al.*, 1993; Convit *et al.*, 1996; Fox *et al.*, 1996), the prominent, early dense accumulations of NFTs in this region (Arriagada *et al.*, 1992b; Hof *et al.*, 1992); and the early loss of neurons in the entorhinal cortex at the initial clinical stages of AD (Gomez-Isla *et al.*, 1996).

The cytoarchitectonic areas that are heavily involved with NFTs and neuronal loss in the medial temporal lobe include the areas CA1; subiculum; in the prosubiculum of the hippocampal complex, the entorhinal cortex and adjacent area 35 (of Brodmann); and the amygdala; with strikingly less-densely distributed NFTs in areas CA2-4 of the hippocampus and presubiculum (Jamada and Mehraein, 1968; Kemper, 1978, 1994; Hyman *et al.*, 1994, 1986, 1990; Morys *et al.*, 1994; West *et al.*, 1994; Giannakopoulos *et al.*, 1995; Gomez-Isla *et al.*, 1996; Simic *et al.*, 1997). As pointed out by many investigators (e.g., Kemper, 1978, 1994; Van Hoesen, 1991, Hyman et al., 1984, 1986, 1990), the neurons with a predilection for developing NFTs in the hippocampus and adjacent entorhinal cortex are the ones involved with the extrinsic efferent and afferent connections of those regions. The densest accumulation of NFTs in the entorhinal cortex are in layers II and IV. The large neurons in layer II of the entorhinal cortex are the site of origin of the perforant pathway to the molecular layer of the fascia dentata and to the molecular layers of the various hippocampal fields, and are the site of convergence of neocortical projections to the entorhinal cortex. The subicular fields provide the major efferent projections back to the entorhinal cortex, projecting to a deeper layer, layer IV, which in turn projects back to the neocortex. The predominately intrinsic circuitry involving hippocampal areas CA2-4 is relatively spared. As recently pointed out by Hyman *et al.* (1990), this principle of selective involvement of extrinsic input and output neurons in the hippocampal complex can also be seen in the reciprocal interrelationship between the hippocampal formation and the amygdala, an area likewise heavily involved with NFTs and cell loss in AD (Jamada and Mehraein, 1968; Herzog and Kemper, 1980; Hyman *et al.*, 1990; Kemper, 1994; Vereecken *et al.*, 1994). All these medial temporal areas have been implicated as substrates for recent memory, and the selective involvement of key parts of this circuitry early in the natural history of AD is consistent with the early appearance of memory loss in this disease (Albert, 1996).

The neocortex, which is involved later in the course of AD, shows preferential cell loss and NFT formation in the pyramidal cells in layers III and V (Mann *et al.*, 1985; Pearson *et al.*, 1985; Braak *et al.*, 1989; Lewis *et al.*, 1987; Delaère *et al.*, 1989; Hof *et al.*, 1990; Hof and Morrison, 1990; Arnold *et al.*, 1991a; Hof *et al.*, this volume), cell layers known to give rise to corticocortical and corticosubcortical projections (Rockland and Pandya, 1979; Pandya and Yeterian, 1985, 1990; Felleman and Van Essen, 1991). These circuits, analyzed in detail by Felleman and Van Essen (1991), are complex, with feedforward and feedback projections to and from the primary sensory and motor cortices, as well as lateral projections interconnecting cortical processing areas of similar stages of of the same modality. Involvements of pyramidal cells in layers III and V would interrupt all of these interrelated cortical circuits. According to Felleman and Van Essen (1991), the long feedforward corticocortical projections arise predominantly from the pyramidal cells in layer III, but may also arise from other layers. In this regard, Morrison *et al.* (1987), Hof and Morrison (1990), and Hof *et al.* (1990) have provided evidence for the specific vulnerability of this subset of pyramidal cells in AD. Utilizing an immunostain that in primates identifies the long corticocortically projection neurons, they have shown in AD a loss of over 80% of the large neurons identified with this immunostain (Morrison *et al.*, 1987). In contrast, the subcortically projecting Betz cells in layer V of the motor cortex, also identified with this antibody, failed to develop NFTs or show cell loss in AD (Morrison *et al.*, 1987). This selective involvement of the pyramidal cells that give rise to the corticocortical circuits could result in what Hof *et al.* (1990) have called a "neocortical isolation syndrome" and provide a mechanism for dementia in AD. In this regard Mann (1996) has shown a good correlation between the extent of pyramidal cell loss and the severity of dementia in AD. Further, disruption of these corticocortical projections to the transentorhinal cortex and amygdala could further augment the memory deficit in AD.

As a consequence of neuronal loss and other pathological changes in the cerebral cortical neurons in AD, the extent of synaptic loss has shown strong correlations with the extent of dementia (Terry *et al.*, 1991; Blennow *et al.*, 1996; Samuel *et al.*, 1994; Scheff *et al.*, 1996; Masliah, this volume). Also associated with this neuronal loss is the prominent loss of myelin in the subcortical white matter and the gyral atrophy (Tomlinson *et al.*, 1970; Miller *et al.*, 1980; de la Monte, 1989), as well as loss of myelin in the corpus callosum (Janowsly *et al.*, 1996; Vermersch *et al.*, 1994). According to Janowsky *et al.* (1996), atrophy in the corpus callosum closely "tracks" the cognitive decline.

In the striatum in AD, there is a predilection for NFTs and neuronal loss to occur in the large cholinergic neurons (Oyanagi *et al.*, 1987, 1989a; Lehéricy *et al.*, 1989; Braak and Braak, 1990; Selden *et al.*, 1994), with neuronal loss comparable to that found in the nucleus basalis of Meynert (Oyanagi *et al.*, 1989a). Lehéricy *et al.* (1989). Selden *et al.* (1994) found this loss of neurons only in the ventral striatum, with the remaining neurons in this region smaller than in control brains (Selden *et al.*, 1994). Neuritic plaques appear to have a similar distribution pattern (Suenaga *et al.*, 1990). Thalamic involvement in AD is apparently common; McDuff and Sumi (1985) noted NPs and NFTs in the thalamus in 25 of 28 brains. Within the thalamus, German *et al.* (1987) noted midline nuclei and the nucleus limitans involvement, Grossi *et al.* (1989) noted medial dorsal nucleus involvement, and Braak and Braak (1991b) found limbic, midline, intralaminar, and reticular nuclei involvement. In a recent study of the medial dorsal nucleus, Paskavitz *et al.* (1996) reported a mean neuronal loss of 29% without significant change in synaptic density. They suggest that such a change may occur late in the disease. In this study the medial part of the medial dorsal nucleus was more involved than its lateral part.

In the cholinergic basal forebrain, the dopaminergic ventral mesencephalon, and the noradrenergic locus coeruleus there is marked neuronal loss in AD. In the cholinergic basal forebrain, neuronal loss has been well documented (Perry *et al.*, 1981; Whitehouse *et al.*, 1982, 1983; Arendt *et al.*, 1983, 1985, 1995a; Candy *et al.*, 1983; Coyle *et al.*, 1983; Nagai *et al.*, 1983; Pearson *et al.*, 1983; Tagliavini and Pilleri, 1983; Wilcock *et al.*, 1983, 1988; Mann *et al.*, 1984b,c; McGeer *et al.*, 1984; Rogers *et al.*, 1985; Saper *et al.*, 1985; Etienne *et al.*, 1986; Doucette and Ball, 1987; Rinne *et al.*, 1987; Allen *et al.*, 1988; Ichimiya *et al.*, 1986; Vogels *et al.*, 1990; Lehéricy *et al.*, 1993; see Kemper, this volume, for further details), with cell loss ranging from 15.5% (Vogels *et al.*, 1990) to 90% (Whitehouse *et al.*, 1982). This neuronal loss is more marked at posterior than at anterior levels of the nucleus basalis (Etienne *et al.*, 1986; Doucette and Ball, 1987; Wilcock *et al.*, 1988) and has been shown to affect ChAT- and AChE-immunostained neurons (Nagai *et al.*, 1983; McGeer *et al.*, 1984; Lehéricy *et al.*, 1993), but not those immunoreactive for neuropeptide Y (Lehéricy *et al.*, 1993). In contrast, neuronal loss in AD has not been noted in the cholinergic medial septum (Vogels *et al.*, 1990) nor of the cholinergic neurons in the brain stem (Zweig *et al.*, 1987; Jellinger, 1988; Woolf *et al.*,1989). In the dopaminergic ventral mesencephalon, neuronal loss is well documented in AD (van Domburg and ten Donkelaar, 1991; Tabaton *et al.*, 1985; Mann *et al.*, 1987; Uchihara *et al.*, 1992) and has been found to be more marked in the ventral tegmental area (VTA) than in the substantia nigra (SN) (Mann *et al.*, 1987; van Domburg and ten Donkelaar, 1991), with the VTA also showing a greater concentration of neurofibrillary tangles (van Domburg and ten Donkelaar, 1991). Van Domburg and ten Donkelaar (1991) have noted in the SN that neuronal cell loss is more marked in nonpigmented (47%) than in pigmented neurons (41%). Neuronal loss in the noradrenergic locus coeruleus (LC) has been frequently reported in AD (Forno, 1978; Tomonaga, 1979; Perry *et al.*, 1981; Tomlinson *et al.*, 1981; Bondereff *et al.*, 1982; Iversen *et al.*, 1983; Mann *et al.*, 1983, 1984a,b, 1985; Ichimiya *et al.*, 1986; Marcyniuk *et al.*, 1986; Zubenko and Moossy, 1988; Zweig *et al.*, 1988; Chan-Palay and Asan, 1989; Förstl *et al.*, 1992; German *et al.*, 1992), with the most severe loss found in the rostral part of the LC by Chan-Palay and Asan (1989) and by German *et al.* (1992), in the dorsal part by Marcyniuk *et al.* (1986), and in the midpart of the nucleus by Zweig *et al.* (1988). Neuronal loss in the serotoninergic dorsal raphe has been found in most (Curcio and Kemper, 1984; Yamamoto and Hirano, 1985; Zweig *et al.*, 1988; Aletrino *et al.*, 1992; Chen *et al.*, 1994), but not all, studies (Ichimiya *et al.*, 1986). In the serotoninergic nucleus centralis superior (median raphe nucleus), Tabaton *et al.* (1985) found evidence of neuronal loss, whereas Zweig *et al.* (1988) found none.

## 3. Progressive Subcortical Gliosis

A rare cause of dementia is progressive subcortical gliosis (PSG). Neumann (1949) and later Neumann and Cohen (1967) described the neuropathology of five cases of a unique, progressive dementia that they called progressive subcortical gliosis. Their review of the literature revealed scattered reports of similar cases (Stern, 1939; Hassin and Levitin, 1941; Stern and Reed, 1945), and subsequent additional cases were reported by Seitelberger (1968), Minagawa and Nakazato (1973), Mora *et al.* (1975), Masse *et al.* (1981), and Verity (1987). The cognitive profile in these case reports is characterized initially by changes in social behavior and motivation, and a progressive language disturbance with decreased fluency, often leading to mutism. Impaired executive function, abnormal ideation, hallucinations, delusions, and confabulation accompanied by unmanageable behavior are also commonly reported. Motor disturbances, including gait disorder with falling and difficulty swallowing, have also been noted. Memory and visuospatial skills are compromised later in the course of disease.

The most frequently reported neuropathological findings are reduced brain weight, ventricular dilatation, and gyral atrophy that usually involves at least some portion of the frontal lobes. In personally observed cases, the atrophy affected the upper parts of the frontal lobes and the superior parietal lobule (Fig. 1A–C). The unique and characteristic feature of PSG is the preserved cytoarchitecture of the cerebral cortex in association with a marked gliosis in layer I and in the deepest part of the cerebral cortex that extends into the subcortical white matter, where it is associated with loss of myelin. Other areas of involvement in individual cases include the basal ganglia, thalamus, amygdala, brainstem, and cerebellum.

## 4. Dementia of the Frontal Type

After the reports of Brun (1987), Gustafson (1987), and Neary *et al.* (1988), an increasing recognition has arisen of a newly described dementia associated with gyral atrophy of the frontal and anterior temporal lobes (Fig. 2) that lacks the diagnostic features of AD, PiD, or DLBD. Brun (1987) called it "frontal lobe degeneration of non-Alzheimer type," Neary (1997) "dementia of the frontal type" (DFT), and Knopman *et al.* (1990) "dementia lacking distinctive histological features." According to Brun and Passant (1996), it is the second most common dementia after AD. Most cases of DFT begin before age 65 and have an average clinical duration of about seven years (3–17 years), with approximately 50% of the cases having a first degree familial relative with a similar condition (Brun and Passant, 1996). Clinically, it is characterized by evidence of frontal lobe pathology such as loss of personal and social awareness, disinhibition, perseverative behavior, loss of insight, and in many cases, disturbances of expressive speech with late mutism. Memory and visual-spatial function are relatively spared (Neary *et al.*, 1988; Knopman *et al.*, 1990; Neary, 1990; Brun *et al.*, 1994; Giannakopoulos *et al.*, 1995; Brun and Passant, 1996; Gregory and Hodges, 1996).

The characteristic histological features in the affected cerebral cortex are a neuronal loss in layers II and III, a mild increase in glial cell numbers, and a spongiform change (Brun, 1987, 1993; Knopman *et al.*, 1990; Snowden *et al.*, 1992;

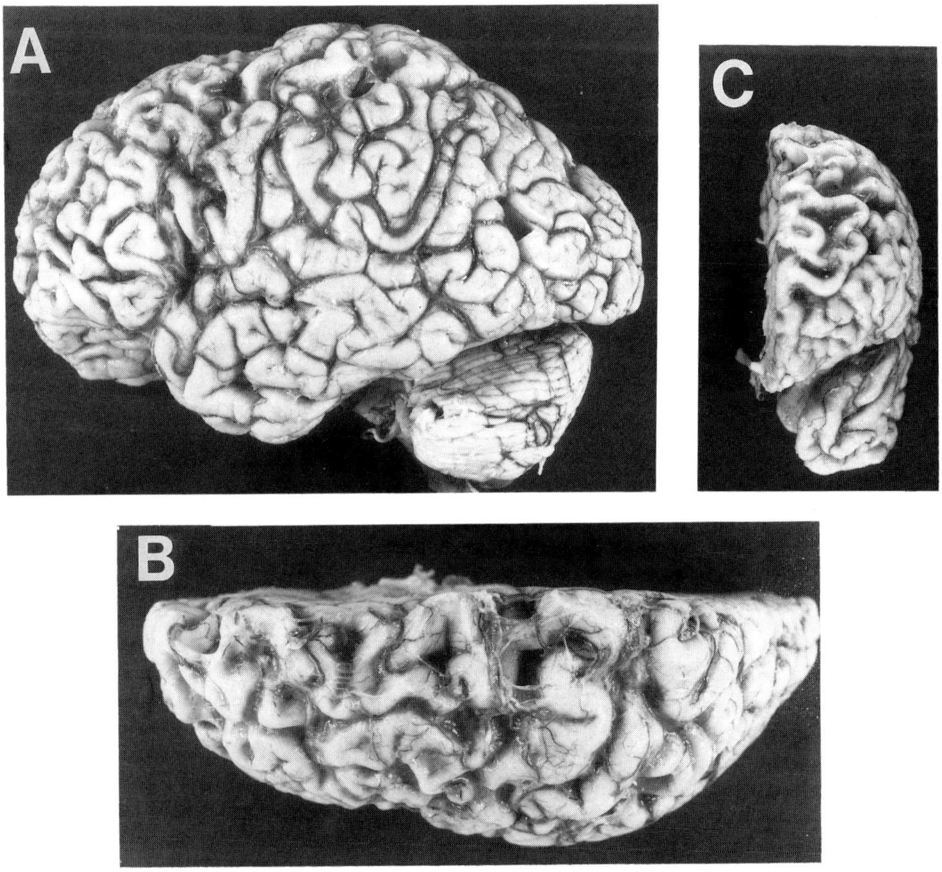


**Figure 1.** (A) Lateral view of the left cerebral hemisphere in progressive subcortical gliosis. (B) Superior view of the same hemisphere. (C) Frontal view. Note the marked gyral atrophy affecting the upper parts aof the frontal lobe and superior parietal lobule.

Mann *et al.*, 1993a; Brun *et al.*, 1994; Giannakopoulos *et al.*, 1995; Brun and Passant, 1996). The topography of the gyral atrophy is the same as that found in PiD (Mann and South, 1993), but it differs from PiD in that in PiD, there is a greater degree of gyral atrophy, more-severe neuronal loss affecting many more layers of the cerebral cortex, more-marked gliosis, lack of a prominent spongy change in the affected areas, and the presence in many PiD brains of Pick bodies and / or Pick cells (Mann *et al.*, 1993a; Brun *et al.*, 1994; Esiri *et al.*, 1997). The cerebral cortical changes in many brains are symmetrical (Brun *et al.*, 1994; Giannakopoulos *et al.*, 1995), with 13 of 34 cases in the series of Giannakopoulos *et al.* (1995) showing asymmetric atrophy.

The subcortical white matter beneath the affected cortices often shows a mild increase in astroglial cells, loss of myelinated fibers (Brun *et al.*, 1994; Giannakopoulos *et al.*, 1995) and loss of volume in the affected lobes and beyond (Mann *et al.*, 1993a). Subcortical formations are inconsistently involved, with neuronal loss of gliosis reported in the pars compacta of the substantia nigra, striatum, amygdala, hippocampus, and thalamus (Knopman *et al.*, 1990; Neary *et al.*, 1990; Giannakopoulos *et al.*, 1995; Brun *et al.*, 1994; Brun and Passant, 1996). Both Knopman *et al.* (1990) and Brun and Passant (1996) reported a lack of involvement of the cholinergic basal forebrain. It is unclear whether these various combinations of involvement of the cerebral cortex and subcortical structures represent different diseases or are different manifestations of DFT. There is, however, one consistent pattern in which the typical histological changes of amyotrophic lateral sclerosis (ALS) are combined with those of DFT (Horoupian *et al.*, 1984; Mitsuyama, 1984; Knopman *et al.*, 1990; Neary, 1990; Ferrer *et al.*, Mann *et al.*, 1993a); Brun *et al.*,

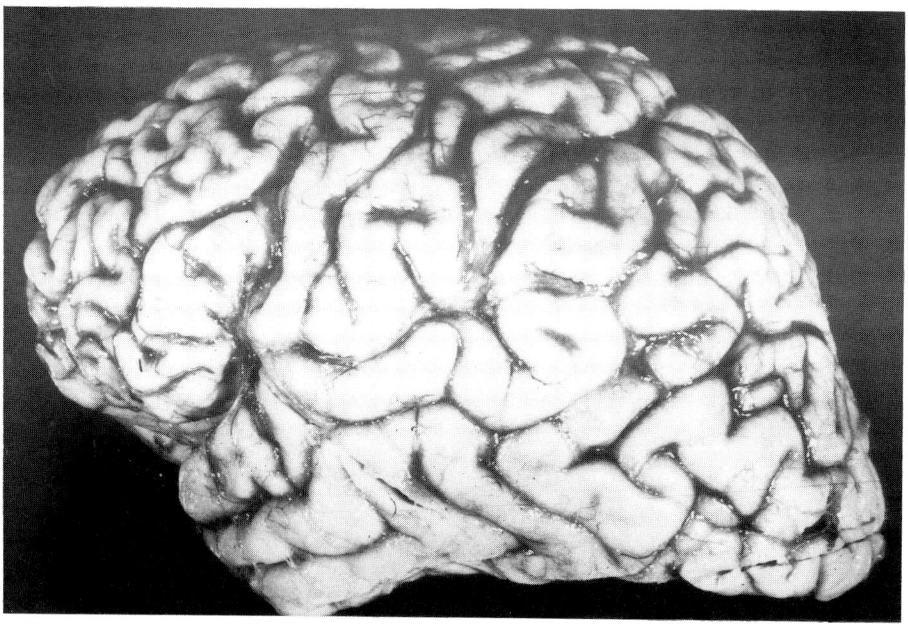

**Figure 2.** Lateral view of the left cerebral hemisphere in dementia of the frontal type. Note the marked atrophy of the gyri at tip of the frontal lobe and mild gyral atrophy of the adjacent tip of the temporal lobe.

1994: Giannakopoulos *et al.,* 1995; Brun and Passant, 1996). These usually present as a dementia, with the features of ALS occurring later (Horoupian *et al.,* 1984; Knopman *et al.,* 1990). On histological examination, the evidence of ALS includes loss of neurons in the anterior horns of the spinal cord and in the hypoglossal nucleus in combination with mild degenerative changes in the pyramidal tract. In these brains, Golgi studies of individual neurons in the affected cerebral cortical areas have shown degenerative changes in the dendrites of both pyramidal and nonpyramidal cells (Horoupian *et al.,* 1984; Ferrer *et al.,* 1991), and Ferrer *et al.* (1993) has shown marked loss of calbindin D-28k-immunoreactive neurons, a cell type that corresponds to the multipolar, bitufted, and double bouquet nonpyramidal cells in the superficial cortical layers. Neary (1990) states that patients with combined DFT and ALS also show significant loss of neurons in the pars compacta of the substantia nigra.

## 5. Pick's Disease

Pick's disease (PiD) or Pick's lobar atrophy, first described by Arnold Pick (1892) and named thirty years later, is a progressive dysfunction and atrophy of the anterior temporal lobe and parts of frontal lobe (Fig. 3) (Cummings and Duchen,

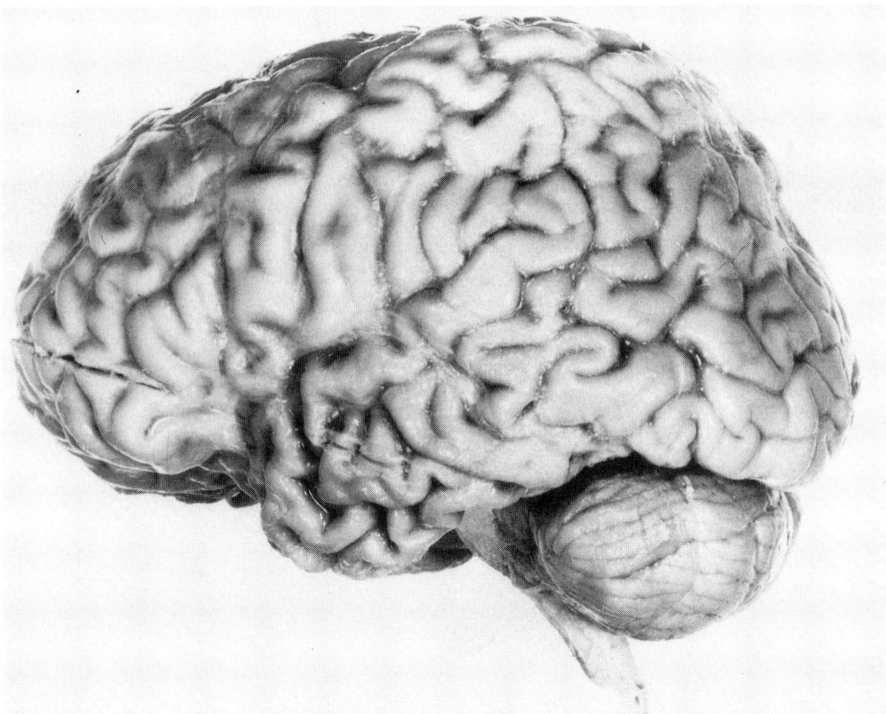

**Figure 3.** Lateral view of the left cerebral hemisphere in Pick's disease. Note the marked gyral atrophy of the tip of the temporal lobe and adjacent frontal lobe.

1981; Cummings, 1991; Cummings and Benson, 1994; Mendez and Zander, 1991; Mendez *et al.,* 1993; Hodges, 1994) resulting in dementia and ultimately death, usually within 5–10 years after onset, which is generally at 45–65 years of age (Robertson *et al.,* 1958; Cummings and Duchen, 1981; Tissot *et al.,* 1985). Arnold Pick's main interest in presenting the case of an elderly demented patient was to correlate focal brain degeneration with focally accentuated neuropsychological deficits, not to describe a new form of dementia (Forstl and Baldwin, 1994). Constantinidis *et al.* (1974) classified PiD into three types: *Group A* are patients with Pick bodies and Pick cells; *Group B* are patients with Pick cells and predominantly frontal lobe atrophy; *Group C* are patients with no Pick bodies or cells (Hulette and Crain, 1992). Specifically, PiD is characterized by alterations in personality, lack of planning, anomia, and speech disturbances (Mendez *et al.,* 1993). In cases with predominately temporal lobe involvement, progressive aphasia and selective language deficits (i.e., semantic memory loss) are observed (Graff-Radford *et al.,* 1990).

Anatomically, PiD differs from AD in that the latter has predominant atrophic changes in the hippocampus and bilateral parietal lobe (de Leon *et al.,* 1995), in contrast to the more-rostral (anterior) temporal lobe (and frontal) changes in PiD. There is severe neuronal loss in the affected areas, with layer III the most vulnerable, then layers II and V, accompanied by an increase in glia (e.g., Arnold *et al.,* 1994). PiD is characterized by severe, localized atrophy of the temporal pole (Arnold *et al.,* 1994), with sparing of the posterior two-thirds of the superior temporal gyrus (e.g., Hulette and Crain, 1992). Pick cells are found in about 50% of reported cases, mainly in the frontal lobe; Pick bodies are found in about 20% of reported cases, mainly in the temporal lobe (Tissot *et al.,* 1975). Pick bodies and other cellular alterations have also been identified in the hippocampal formation. Specifically, the granule cell layer of the dentate gyrus contains a high number of Pick bodies, and dystrophic neurites are localized to the CA1 and subiculum (Hof *et al.,* 1994). When frontal lobe involvement is present, the orbitofrontal, medial frontal, and frontal pole are most often affected. Pick bodies are often located in the posterior orbitofrontal cortex and inferior temporal cortex (Hof *et al.,* 1994), this volume). Cells containing Pick bodies are also seen in other diseases such as corticobasal degeneration, Alzheimer's disease (Dickson *et al.,* 1986), and progressive supranuclear palsy (Higuchi *et al.,* 1995). In all four Pick's patients examined by Hulette and Crain (1992), marked atrophy of the caudate nucleus was observed. In addition to neuron loss in the caudate, other subcortical nuclei, such as the amygdala, can lose most of their neurons and develop astrocytosis. Usually, there is a modest loss also of neurons in other subcortical nuclei, including the pons, putamen, globus pallidus, thalamus, pars reticulata of substantia nigra, locus coeruleus, and nucleus basalis of Meynert (Uhl *et al.,* 1983; Mizukami and Kosaka, 1989; Munoz-Garcia and Ludwin, 1984; Tagliavini and Pilleri, 1983; Yoshimura, 1989). White matter in the affected lobes has a marked pallor, and subcortical gliosis is present (Hulette and Crain, 1992).

The microscopic anatomy and biochemistry of Pick's disease is thoroughly described in this volume in the chapter by Hof *et al.* and thus will not be covered in detail here. With regard to which neurotransmitters are affected in Pick's disease, there is a paucity of information in the literature. The cortical cholinergic supply is intact (Yates *et al.,* 1980; Wood *et al.,* 1983) and ChAT activities are normal (Francis *et al.,* 1993), in marked contrast to findings in Alzheimer's disease. Serotonin receptors are reduced in cortex (Sparks and Markesbery, 1991; Francis *et al.*, 1993), but cortical somatostatin (Meier-Ruge *et al.*, 1984), GABA (Yates *et al.*, 1980; Francis *et al.*, 1993) and dopamine (Yates *et al.*, 1980; Francis *et al.*, 1993) are largely unchanged.

It is noteworthy that there are similarities and differences in cell morphology when comparing PiD with other neurologic disorders. For example, pale, enlarged cortical neurons with eccentric nuclei ("ballooned" or "achromatic" neurons) are found in both PiD and in corticobasal degeneration (CBD) (Feany and Dickson, 1996). Neuropil threads, although sometimes present in PiD, are sparse in progressive supranuclear palsy (PSP) but much more numerous in CBD, especially in the white matter (Feany and Dickson, 1996). One of the hallmarks of AD is the deposition of amyloid, a distinction that sets it apart from PiD, CBD, and PSP. Another distinction is that in PiD and PSP, there is a propensity for neurofibrillary tangles to be located in supragranular layers (i.e., higher densities in layers II and III), whereas in AD, neurofibrillary tangles are mainly found in infragranular layers (i.e., layers V and VI) (Hof *et al.*, 1994). Ultrastructurally, Pick bodies express phosphorylated neurofilament proteins, tau, ubiquitin, and tubulin (e.g., Jackson and Lowe, 1996, and Hof *et al.*, this volume). In all three disorders, however, there are numerous tau-immunoreactive neuronal and glial inclusions in cortical and subcortical regions (Feany and Dickson, 1996). Finally, the distribution of lesions and affected areas differs significantly in the three disorders. Pick's disease and CBD contain more tau-immunoreactive inclusions in the cortex versus the brainstem, whereas in PSP more inclusions are observed in the brainstem and basal ganglia (Feany and Dickson, 1996). It is now evident that no one type of neuronal inclusion can be utilized to diagnose a particular disease; rather, it is the relative proportion and distribution of each type of lesion that should be used to delineate individual disorders. Thus discrete pathological mechanisms are involved in these disorders and are not variations of the same disorder (Feany and Dickson, 1996). In addition, due to differences in distribution and proportion of these various neuronal inclusions and lesions, different populations of cortical neurons are affected, and alterations of select corticocortical and corticosubcortical projections may distinguish each type. Hof *et al.* (1994) suggest that the distribution of Pick bodies in Pick's disease leads to the conclusion that Pick's disease is a disconnection of short corticocortical pathways, corticothalamic connections, and mossy fibers from dentate granule cells. In contrast, Alzheimer's disease affects long corticocortical circuits, particularly with respect to frontal and temporal neocortical regions (Hof *et al.*, 1994).

## 6. Huntington's Disease

Huntington's disease (HD) is an autosomal-dominant disorder, characterized by chorea, rigidity, and dementia, that is associated with an excess number of a cytosine-adenine-guanine triplet repeat sequence on the Huntington gene of chromosome 4 (Huntington's Disease Collaborative Group, 1993). Normally, there are 10 to 29 repeats of this sequence, in HD there are 36 to 121 (Kremer *et al.*, 1994), with the younger patients often displaying the largest number of repeats (Craufurd and Dodge, 1993). The frequency of HD is 4–7 per 100,000, with the majority of cases showing an onset from middle to late life and an average duration of 17 years (Martin and Gusella, 1986; Lowe *et al.,* 1997).

The neuropathological hallmark of HD is atrophy of the caudate nucleus and the putamen. Based on the extent of neuronal cell loss and of gross atrophy of these nuclei, Vonsattel *et al.* (1985) have proposed a widely used staging system of severity. In grade 0 there is no evidence of neuropathology, in grade 1 the changes are evident only on microscopic examination, and in grades 2–4, there is increasingly severe neuronal loss with gross atrophy of the nuclei. In grades 3 and 4, there is also atrophy of globus pallidus. The earliest changes in the striatum are in the paraventricular part of the caudate nucleus and the dorsal part of the putamen, which then progress ventrally and laterally. Increasing severity of structural involvement of the striatum is associated with a decrease in brain weight (20% in grade 4 as compared to grade 0). Astrocyte density is only mildly increased in grade 4 brains, which show a loss of more than 95% of neurons but only a 28% increase in astrocyte density. There is a high degree of correlation between severity of striatal involvement and physical disability, with the most severe atrophic changes noted in patients with a juvenile onset (Myers *et al.,* 1988). Studies of cerebral glucose utilization (Kuhl *et al.,* 1982) have shown an early deficit in glucose uptake in the caudate and putamen that persists, with no evidence of decreased glucose uptake in other areas of the brain at any stage of the disease.

Huntington's disease affects primarily striatal projection neurons, with relative sparing of aspiney interneurons. The latter are thought to be local-circuit cells containing somatostatin, neuropeptide Y, NADPH diaphorase, and AChE. The projection neurons are GABAergic and colocalize with either enkephalin or substance P (Graveland *et al.,* 1985; Kowall *et al.,* 1987; Ferrante *et al.,* 1987). Using the rapid Golgi method to demonstrate cellular and dendritic details of individual cells, Graveland *et al.* (1985) have shown degenerative and regenerative changes in the spiney projection neurons, with relative sparing of the morphology of the aspiney neurons. Reiner *et al.* (1988), in a study of the striatal afferent plexuses of these neurons in the globus pallidus, have shown that the enkephalin-containing neurons, which project to the external segment of the globus pallidus, are affected earlier than are the axonal plexuses of the substance P-containing neurons that project to the inner segment of the globus pallidus and the pars compacta and reticulata of the substantia nigra. In advanced stages of HD, all these target areas are depleted of their respective peptides except for the pars compacta of the substantia nigra. Pars reticulata is thought to play a role in the early appearance of abnormal saccadic eye movements in HD, and the early compromise of the external division of globus pallidus, due to loss of its striatal projection, is thought to play a role in the choreiform movements. Hedreen and Folstein (1995), who noted that the earliest neurona' loss in the striatum in HD occurs in its striosomal compartment, suggest that this may account for the decreased enkephalin staining noted by Reiner *et al.* (1988). Because the striosomes normally have an inhibitory projection to the pars compacts of the substantia nigra, release of this inhibition would enhance the normally inhibitory projection to the enkephalin-producing neurons in the striatum. These investigators attribute the chorea to the increased activity of the pars compacta of the substantia nigra. They suggest that other early manifestations of HD, such as cognitive deficits and mood disorders, could be due to a similar disturbed relationship to other, nonmotor, parts of the striatum.

The filamentous material, which is present in all LBs, is thought to consist of neurofilaments that have undergone a phosphorylation, ubiquitination, proteolysis, and cross-linking. The vesicular material is thought to represent presynaptic and dense core vesicles (Nishimura *et al.*, 1994). The distribution pattern of LBs in the subcortical areas and in the brain stem closely parallels that of neurofibrillary tangles (NFTs) in AD (Hirano, 1973), and Schmidt *et al.* (1996) have noted in the amygdala of demented individuals with PD that LBs and NFTs may occur in the same neuron.

Loss of pigmented neurons in the substantia nigra pars compacta (SNpc) and the dopaminergic ventral tegmental area (VTA) has been well documented in PD, with approximately three-quarters of the SNpc neurons lost by the time of autopsy (Ma *et al.*, 1996; Halliday *et al.*, 1996), and many of the remaining cells decreased in size (Hoogendijk *et al.*, 1995; Ma *et al.*, 1996). Fearnley and Lees (1991) estimate that the symptoms of Parkinson's disease appear when 50% of cells in SNpc are lost. Within the SNpc in PD, the most pronounced cell loss occurs in the ventrolateral tier of SNpc neurons, an area relatively spared in the aged brain (van Domburg and ten Donkelaar, 1991; Fearnley and Lees, 1991; Gibb and Lees, 1991). The importance of the SNpc in the symptomatology of PD is shown by the well-documented decrease in dopamine in the striatum (Ehringer and Hornykiewicz, 1960), the parkinsonian symptoms resulting from MPTP-mediated SNpc destruction (Langston, 1995), the finding that in PD with asymmetric symptoms SNpc cell loss is greater on the contralateral side (Kempster *et al.*, 1989), and the well-known clinical response to dopamine replacement therapy. In the unpigmented neurons of the pars reticulata of the substantia nigra, Hardman *et al.* (1996) found no evidence of cell loss in PD.

Parkinson's disease cell loss in the LC is also well documented, with a neuronal loss between 21% and 94% (German *et al.*, 1992; Tomonaga, 1979; Mann *et al.*, 1983; Chui *et al.*, 1986; Chan-Palay and Asan, 1989; Zweig *et al.*, 1993; Hoogendijk *et al.*, 1995), affecting all parts of the nucleus (German *et al.*, 1992; Chan-Palay and Asan, 1989; Zweig *et al.*, 1993). In studies by Chan-Palay and Asan (1989) and Zweig *et al.* (1993), neuronal loss was most marked in the patients with dementia, and Chan-Palay and Asan (1989) also found a correlation with depression. Atrophic changes in the remaining LC cells have also been noted. In a Golgi study of individual LC neurons, Patt and Gerhard (1993) found unaffected neurons intermixed with neurons that had shortened dendrites and a spine density, and Hoogendijk *et al.* (1995) presented evidence of an increased number of small neurons.

The presence of the Lewy bodies in the dorsal motor nucleus of the vagus (DMNV) and in the nucleus basalis of Meynert (NBM) as been known since their original description by Lewy (1912). Their presence in the cholinergic basal forebrain is associated with neuronal loss (Arendt *et al.*, 1983; Candy *et al.*, 1983; Whitehouse *et al.*, 1983) amounting to 70% (Arendt *et al.*, 1983). Whitehouse *et al.* (1983) have suggested that the extent of this loss is related to the dementia in PD. Arendt *et al.* (1995) have provided evidence of compensatory reorganization of the remaining neurons in the nucleus basalis of Meynert (NBM) in PD, with a size increase in the dendritic field that is related to the extent of the cell loss. The cholinergic pedunculopontine nucleus in the brain stem also shows significant loss in PD (Jellinger, 1988), as does the dorsal motor nucleus of the vagus (Jellinger, 1990).

In addition to the frequently documented involvement of the SNpc, LC, DMNV, and NBM, the Lewy body has a wide distribution within other subcortical areas (den Hartog Jager and Bethlem, 1960; Ohama and Ikuta, 1976; Matzuk and Saper, 1985; Jellinger, 1990; Braak *et al.*, 1995), and its relationship to cell loss has been documented in some of these locations (Jellinger, 1990). Ohama and Ikuta (1976) note 20 different nuclei that are consistently involved, with seven others involved in individual cases. Of these 20 nuclei, only 12 are pigmented, indicating a lack of obligate relationship to melanin, a finding also noted by Matzuk and Saper (1985). In contrast, 19 of these 20 nuclei contain monoaminergic cell bodies. The single exception was the intermediolateral column, the source of sympathetic outflow from the spinal cord. Consistent with their predilection for monoaminergic neurons within the central nervous system, Lewy bodies also occur in peripheral autonomic ganglia (Wohlwill, 1925; Herzog, 1926; Hechst and Nussbaum, 1931; Forno and Norville, 1976), where Forno and Norville (1976) have observed them in all patients with idiopathic Parkinsonism. Also consistent with this predilection, there is no evidence of cell loss in the subthalamic nucleus (Hardman *et al.*, 1997) or in any of the thalamic nuclei (Xuereb *et al.*, 1991) in PD. Neither of these areas contain monoaminergic neurons.

More recently, attention has been drawn to the presence of LBs in the amygdala, hippocampus, and cerebral cortex in PD. Braak *et al.* (1994, 1995) have noted consistent involvement of the amygdala, with the highest LB density in the accessory cortical and central nucleus and a less-marked involvement of the periamygdaloid cortex and the medial basal nucleus. There, LBs may also be found in cell processes, where they are referred to as "Lewy neurites." These investigators suggest that the heavy involvement of the cortical amygdalar complex may be related to both the early deficiency in olfaction and to the involvement of the central nucleus with frequently occurring autonomic disturbances in PD. The central nucleus of the amygdala has projections to widespread areas of the lateral hypothalamus, brain stem, and spinal cord, many of which are in areas involved in autonomic function that are themselves sites of predilection for LB accumulation. Despite this widespread central and peripheral autonomic nervous system involvement, in a clinical study, Goetz *et al.* (1986) has noted that impairment of the autonomic nervous system function in PD may be mild.

The occurrence of Lewy bodies in the cerebral cortex, with their predilection for small- and medium-sized pyramidal neurons in layers V and VI in anterior frontal, temporal, insular, and anterior cingulate cortex (Yoshimura, 1983; Kosaka *et al.*, 1984) is well recognized. According to Braak *et al.* (1994) and de Vos *et al.* (1995), the vast majority of PD patients show this change, and according to Hughes *et al.* (1993) and Kim *et al.* (1995), all PD patients have cerebral cortical Lewy bodies. In the hippocampus in DLBD, Dickson *et al.* (1991) have described abnormal neurites that are immunoreactive to ubiquitin and occur primarily in the pyramidal layer of areas CA2 and CA3 of the hippocampus, a change that Braak *et al.* (1995) feel may be confined to area CA2. Later, Braak *et al.* (1994) found these neurites in PD brains without evidence of DLBD, and Kim *et al.* (1995) have noted that all patients with this lesion had cortical Lewy bodies. Although the origin of these neurites is unknown, Veazy *et al.* (1982) have shown that the pyramidal layer of hippocampal area CA2 is the site of a projection from the supramammillary nucleus.

There is no consensus about the relationship between neuropathological changes and dementia in PD. Vos *et al.* (1995) noted that cortical Lewy bodies are present in all PD patients, demented or not, and also noted that the CA2–3 ubiquitin-immunoreactive neurites show no correlation with the presence of dementia. Braak *et al.* (1994), who noted that many of the nondemented PD patients already had marked changes in the amygdala, felt that the neurites did not contribute to the cognitive decline. Both Bancher *et al.* (1993) and Duyckaerts *et al.* (1993) considered the subcortical changes in PD irrelevant to the dementia development. Zwieg *et al.* (1993), who found evidence for greater cell loss in the LC, VTA, and NBM, and larger numbers of Lewy bodies in the anterior cingulate gyrus in demented patients with PD, felt that the dementia could be due to the additive or synergistic effect of pathological involvement of multiple sites, a view shared by Jellinger (1990). Many investigators have emphasized the importance of Alzheimer-type pathology for dementia in PD. Bancher *et al.* (1993) found a close correlation between the pattern of superimposed cortical Alzheimer-type pathology and the degree of dementia, and de Vos *et al.* (1995) felt that the major determinant to mental decline in PD was Alzheimer-type pathology. Samuel *et al.* (1997) found a correlation of the degree of dementia with neocortical Lewy body counts and with Alzheimer-type pathology, but not with neocortical synaptic density. Vermersch *et al.* (1993), in an immunocytochemical study of abnormal tau proteins, also provided evidence of the importance of AD-type changes in the cerebral cortex for dementia in PD. In this study the lesions were most marked in the prefrontal cortex, suggesting that involvement of this area may contribute to cognitive changes. Hughes *et al.* (1993), however, found that only 9 of 31 PD patients with dementia had the pathology of AD and that the AD changes were mainly neuritic plaques. Three of these patients had numerous cortical Lewy bodies, two had possible vascular causes of the dementia, and in more than half (55%), no cause was found. Laakso *et al.* (1996), in an MRI study of hippocampal volume in PD, found that in patients without dementia the hippocampus was smaller than in control brains, and that it was even smaller in those *with* dementia.

Recently, many demented patients have been found to have Lewy bodies in the same subcortical distribution pattern as that found in PD, along with a large number of these bodies in the cerebral cortex, resulting in a disease now called diffuse Lewy body disease (DLBD). This newly described disease was first recognized by Kosaka (1984), who called it "Lewy body disease." Clinically, DLBD is characterized by progressive dementia, often with daily fluctuations in severity, agitation, hallucinations, and striking deficits in visuospatial and visuoconstructive abilities. Parkinsonian features can occur at any stage of the disease, but they may never manifest (Byrne *et al.*, 1989; Crystal *et al.*, 1990; Hansen *et al.*, 1990; Kosaka, 1993; Kalra *et al.*, 1996; McKeith *et al.*, 1996; Salmon *et al.*, 1996). Clinical features that distinguish DLBD from AD are fluctuations in cognitive function, persistent, well-formed visual hallucinations, and the motor features of Parkinsonism (McKeith *et al.*, 1996). As was noted for dementia in PD, the role of cortical and subcortical pathological changes in DLBD dementia is unclear. Lennox *et al.* (1989) stated that the dementia severity is related to the density of the cortical Lewy bodies, a finding also noted by Samuel *et al.* (1997), who found a weak correlation with neocortical neurofibrillary tangles and no correlation with dementia and senile plaque density. Pellisé *et al.* (1996), who noted abnormal ubiquitin-immunostained neurites in the same distribution as cortical Lewy bodies, suggested that they may play a significant role in the cognitive deficits in DLBD. However, Lippa *et al.* (1994), in cases of DLBD uncomplicated by AD, has noted no evidence of neuronal loss in the cerebral cortex and no evidence of gyral atrophy. Many, but not all, of these patients have the neurofibrillary tangles and the diffuse and neuritic plaques that characterize AD (Hansen *et al.,* 1990; Kosaka *et al.,* 1990; Gibb *et al.,* 1989a; Kalra *et al.,* 1996). Kosaka (1993) and Kuzuhara and Yoshimura (1993) favored the AD-type pathology and cortical and subcortical Lewy bodies as the causes of the dementia, with cortical Lewy bodies alone in some cases as the cause. McKeith *et al.* (1996), in a paper entitled "Consensus guidelines for the clinical and pathological diagnosis of dementia with Lewy bodies: Report of the consortium of DLBD international workshop," considered that the only feature essential for the pathological diagnosis of DLBD is the presence of cortical and subcortical Lewy bodies. They also commented that AD pathology is frequently present, that the relationship of DLBD to AD and to PD remains uncertain, and that AD pathology in DLBD consists primarily of diffuse or neuritic plaques, with few neocortical NFTs.

## 8. Corticobasal Degeneration

Corticobasal degeneration (CBD) was first described in the late 1960s (Rebeiz *et al.,* 1967, 1968) and is also called cortical-basal ganglionic degeneration (Riley *et al.,* 1990) or corticonigral degeneration (Lippa *et al.,* 1990). The main clinical features include asymmetrical clumsiness; arm (sometimes leg) stiffness and jerking; dystonic rigidity and akinesia of limb with rhythmic myoclonus; apraxia due to upper motor neuron involvement; and a cortical sensory disturbance that progresses to cognitive dysfunction in many patients (e.g., Ksiezak-Reding *et al.,* 1994; Lippa *et al.,* 1990; Paulus and Selim, 1990; Riley *et al.,* 1990). The average age of onset is about 60 years (Rinne *et al.,* 1994; Thompson and Marsden, 1992). Neuropathological studies show that about 20% of patients with non-idiopathic parkinson syndromes are diagnosed with CBD (Feifel *et al.,* 1994).

Imaging studies demonstrate variable cortical atrophy with asymmetric frontoparietal distribution, whereas others are generalized (e.g., Sawle *et al.,* 1991). Typically, there is asymmetrical cortical atrophy of the posterior frontal, parietal, and peri-Rolandic cortices contralateral to the most-affected limbs, and loss of pigment from substantia nigra and locus coeruleus. Histologically, CBD has four main features: (1) swollen cortical neurons with neurofilaments, (2) $tau^+$ neuropil threads, (3) $tau^+$ astrocytes, and (4) filamentous $tau^+$ protein inclusions in basal neurons and rarely, in cortical neurons as well (see Hof *et al.,* this volume, for review). Ultrastructurally, some tangles are seen in basal neurons with paired helical filaments and straight tubules similar to PSP (Wakabayashi *et al.,* 1994). The affected cortical areas show neuronal loss, microvacuolation, and astrocytosis. The key pathological feature is the appearance of swollen cortical neurons in affected areas, most frequently in layers III, V, and VI. Swollen neurons have a pale-stained, large nucleus with a surrounding area of microvacuolation. Swollen neurons are seen in cortex, fewer in subcortical areas. In some cases, swollen neurons are limited to cortex and lateral substantia nigra (Riley *et al.,* 1990). In nigral neurons there are filamentous inclusions that are basophilic and argentophilic, originally named "corticobasal inclusions" (Gibb *et al.,* 1989b), and the neurons have many similarities to tangles in PSP (Paulus and Selim, 1990). Cell loss and/or degeneration may also be seen in different cases in globus pallidus, corpus striatum, subthalamic nucleus, thalamus, red nucleus, and brain stem nuclei (e.g., Wakabayashi and Takahashi, 1996). The cerebellar dentate nucleus may be abnormal in some cases. There may be loss of myelin from corticospinal tracts, related to motor involvement; white-matter loss is also observed below the affected cortical areas (Ksiezak-Reding *et al.,* 1994). There is widespread accumulation of abnormal tau protein in cell bodies and processes in neurons and glia, similar to PSP, and some appears identical to the glial cytoplasmic inclusions seen in multiple system atrophy (MSA). These glial inclusions, although not nearly as abundant as in MSA, are found in both astrocytes (Horoupian and Chu, 1994) and oligodendrocytes (Ikeda *et al.,* 1994; Wakabayashi *et al.,* 1994).

Chemically, there is a reduction in dopamine uptake into the caudate and putamen, and also asymmetrical cerebrocortical hypometabolism agreeing with pathology shown by functional imaging studies (Riley *et al.,* 1990; Blin *et al.,* 1992; Sawle *et al.,* 1991).

## 9. Multiple System Atrophy

Multiple System Atrophy, named by Graham and Oppenheimer (1969), has been subclassified into three disorders: (1) Olivopontocerebellar atrophy (OPCA; prominent cerebellar ataxia); (2) Shy–Drager syndrome (prominent autonomic dysfunction); and (3) striatonigral degeneration (prominent parkinsonism). The initiating pathogenic mechanisms of MSA remain unknown (Quinn and Wenning, 1995). Clinically, MSA involves the degeneration of many systems and may present with combinations of extrapyramidal, weakness, pyramidal weakness, autonomic failure (Kennedy and Duchen, 1985) and cerebellar signs and symptoms such as ataxia. Atrophy with accompanying cell loss and gliosis may occur in some or many of the following structures; caudate, posterior or dorsolateral part(s) of the putamen, pale-staining substantia nigra, pars externa of globus pallidus, pontine nuclei, inferior olivary nucleus, cerebellum (i.e., Purkinje cells), middle cerebellar peduncle, and intermediolateral cell column (Papp and Lantos, 1994) of the spinal cord and Onuf's nucleus (Quinn, 1989; Quinn and Wenning, 1996). Occasionally, some patients present with frontal lobe atrophy (Robbins *et al.,* 1992) and astrocytosis in motor cortex has been observed in layer V (Fujita *et al.,* 1993). Between 7% and 20% of patients diagnosed with Parkinson's disease have MSA, with the age of onset usually 40 to 60 years. When the striatum is involved, most patients have reduced $D_2$ receptor density and a lack of response to L-dopa (Churchyard *et al.,* 1993). Patients with MSA have decreased striatal radioactivity subsequent to injection of positron-emitting ligands at dopamine receptors or uptake sites (Goldstein, 1993).

One of the main pathological features of MSA is the presence of inclusion bodies in glial cells (Papp *et al.,* 1989). In addition, neuronal cytoplasmic inclusions, neuronal nuclear inclusions, glial nuclear inclusions, and neuropil threads are seen (Kato and Nakamura, 1990; Nakazato *et al.,* 1990; Papp and Lantos, 1992). Costa *et al.* (1992) utilized the Bodian method of silver staining to reveal oligodendroglial inclusions in the cytoplasm in one case of MSA in the tegmentum pontis, putamen, substantia nigra, inferior olives, and cerebellar white matter. These inclusions were also found in locations where neuronal loss was minimal. In marked contrast, in three cases of OPCA no oligodendroglial inclusions were found, leading Costa *et al.* (1992) to suggest that this subgroup should not belong to MSA. In a later study, Costa and Duyckaerts (1993) described neuronal cytoplasmic inclusions composed of granular material, which is also argentophilic and immunoreactive for ubiquitin. The inclusions occur in neurons in hippocampus, amygdala, subiculum, fascia dentata, putamen, subthalamic nucleus, substantia nigra, pontine nuclei, inferior olives, and the reticular formation. Although histologically the neuronal and oligodendroglial inclusions are identical, ultrastructurally they have different antigenic properties (Costa and Duyckaerts, 1993). It is noteworthy that oligodentroglial inclusions are now considered to be an early and specific marker for MSA, and have not been identified in other neurological diseases or in normal brains (Costa and Duyckaerts, 1993).

## 10. Progressive Supranuclear Palsy

First described by Steele *et al.* (1964), the main clinical features of PSP are progressive supranuclear oculomotor disturbances, mainly vertical-gaze palsy (Chiu, 1995; see Grandas and Esteban, 1994 for review); symmetrical akinesia and rigidity with a parkinsonian syndrome involving disequilibrium and frequent falls (tremor is found in only 12–16% of cases; Masucci and Kurtzke, 1989); nuchal dystonia (e.g., Chui, 1995); pseudobulbar palsy; cognitive changes ("subcortical dementia") that include slow information processing, rapid forgetting, difficulty in planning and shifting conceptual sets; slowed motor execution; and problems in orienting attention. Eventually, these disturbances lead to dementia (Grafman *et al.,* 1995; Litvan, 1994a). The term "subcortical dementia" is used to distinguish it from "cortical dementia," which presents with aphasia, apraxia, and/or agnosia (see Litvan, 1994a, for review). Between 1% and 8% of patients with Parkinsonian symptoms have PSP. The age of onset is from 50 to 77 years, with a median age of 64 and a duration of six to seven years (Maher and Lees, 1986). Lantos (1994) proposes a new classification that distinguishes three types of PSP: *Type 1* or *Typical,* which conforms to original descriptions of PSP; *Type 2,* atypical histological variants of PSP; and *Type 3,* cases in which PSP is accompanied by lesions characteristic of another neurodegenerative or vascular disease. A family with autosomal-dominant PSP involving five generations is described in the literature by de Yebenes *et al.* (1995).

PSP is characterized by atrophy of midbrain and pontine tegmentum (Frasca *et al.,* 1991) and a loss of pigment from substantia nigra and locus coeruleus, all of which may play a role in the motor disability. Findings of MRI in PSP show atrophy of the midbrain and of the region around the third ventricle in slightly more than half of all cases (Savoiardo *et al.,* 1994). Usually, the cortex is grossly unremarkable. Some cases, however, show marked atrophy of frontotemporal regions, including the hippocampal formation and entorhinal cortex, whereas other cases show atrophy in area 4 in the precentral gyrus (Savoiardo *et al.,* 1994).

Neuronal loss is severe in ventromedial part of SN. This is in contrast to Parkinson's disease with Lewy bodies, which shows cell loss in the ventrolateral portion of SN. Moderate cell loss is found in various forebrain and brainstem nuclei, including basal ganglia, subthalamic nucleus, pontine and mesencephalic tegmentum, LC, periaqueductal gray, red nucleus, superior colliculi, oculomotor (CN III), trochlear nucleus (CN IV), cerebellar nuclei and inferior olive (hypertrophy in one reported case; Yagashita *et al.,* 1986), and superior cerebellar peduncle (see Lantos, 1994; Daniel *et al.,* 1995; Hauw *et al.,* 1994, for review).

Microscopically, the hallmark of PSP is abnormal neurons with tau$^+$ protein aggregates forming neurofibrillary tangles. The tangles are present mostly in the same areas where neuronal loss is observed. Collectively, tangles are seen in substantia nigra, globus pallidus, subthalamic nucleus, tegmentum of the midbrain and pons, LC, periaqueductal gray, pretectal region, nucleus basalis of Meynert, raphe nuclei, nuclei of various cranial nerves (Lantos, 1994), and in the spinal cord, most frequently in the posterior horn (Kato *et al.,* 1986). Tau$^+$ neurons were also seen in the hippocampus, in areas 23 and 24 of cingulate cortex and cortical areas 4 and 9 (Hauw *et al.,* 1990). Ultrastructurally, these tangles are composed of straight filaments 15 nm in diameter (Tellez-Nagel and Wisniewski, 1973; Cervos-Navarro and Schumacher, 1994), as well as of paired helical filaments similar to those in AD (Yagashita *et al.,* 1979; Takahashi*et al.,* 1987; Ghatak *et al.,* 1980). Astrocytes have intense immunoreactivity for tau protein (see Hof *et al.,* this volume). Hauw *et al.* (1994) suggest that a high density of neurofibrillary tangles and neuropil threads in the basal ganglia and brainstem is essential for the diagnosis of typical PSP (type 1). In an earlier study by Hauw *et al.* (1990), differences in tangle distribution in PSP are compared to those of AD. In five cases of PSP, neurofibrillary tangles were predominately located in precentral gyrus (area 4), in contrast to AD, in which association areas predominate; PSP tangles involved large pyramidal neurons and small cells, sparing the midsized pyramidal cells involved in AD; tangles predominated in layers V and VI in area 4 in PSP, as compared to layers III and V in AD; and there is a virtual absence of beta amyloid deposits in PSP (Hauw *et al.,* 1990). The heaviest involvement of area 4 could be due to retrograde degeneration of subcortical structures. The pedunculopontine nucleus and the subthalamic nucleus, also heavily affected in PSP, receive projections from layer V of area 4, as demonstrated in the monkey (Hartmann-von Monakow *et al.,* 1979; Glickstein *et al.,* 1985).

PSP is characterized by an 80–90% reduction in dopamine due to nigrostriatal dysfunction and reduced homovanilic acid in caudate and putamen (Kish *et al.,* 1985; Ruberg *et al.,* 1985), but unlike Parkinson's disease, the mesocortical and mesolimbic dopaminergic systems remain intact. The dopamine loss is more severe in the caudate than in putamen and may contribute to the high frequency of cognitive deficits in PSP (Hornykiewicz and Shannak, 1994). In contrast, the serotonin (5-HT) levels in the basal ganglia are not significantly altered in PSP patients, thus resulting in the predominance of serotoninergic inhibitory influences on motor behavior (Hornykiewicz and Shannak, 1994). In addition, there is a 28–48% loss of postsynaptic $D_2$ receptors from caudate and putamen (see Brooks, 1994, for review), and therefore, patients do not respond to L-dopa (Leenders *et al.,* 1988; Finali *et al.,* 1989; Bhatt *et al.,* 1991; Brooks *et al.,* 1990). Cell loss in nucleus basalis is not as severe as that in AD (Tagliavini *et al.,* 1984) and is not associated with a marked cholinergic deficiency in cerebral cortex (Javoy-Agid, 1994). In addition, there is up to 60% cell loss in cholinergic pedunculopontine nucleus, which received projections from the precentral gyrus, globus pallidus, and SN and projects back to SN and to oculomotor nuclei and superior colliculus (Kuypers and Lawrence, 1967; Carpenter and Sutin, 1983; Hirsch *et al.*, 1987; Jellinger *et al.*, 1988). Malessa *et al.* (1991) reports a significant reduction of ChAT-like-immunoreactivity in the pontine reticular formation in PSP patients. Cholinergic therapy (e.g., physostigmine, a cholinesterase inhibitor, RS-86) is not effective (Litvan, 1994b). Unlike in AD or PD, somatostatin levels in PSP are normal (Javoy-Agid, 1994).

## 11. Schizophrenia

Schizophrenia (SZP) afflicts 1% of the population worldwide, with one million affected in the United States. The usual age of onset is 18–30 years (Andreasen, 1995). It is now looked upon as a neurodevelopmental disorder centering on the limbic system, with arrested migration of late-generated neurons to the superficial layers (I–III) of certain cortical areas, particularly entorhinal cortex (Arnold *et al.*, 1991b,c; Jacob and Beckmann, 1994) and prefrontal cortex (Keshaven, 1994). A recent study suggests a possible role of the glutamatergic system in SZP. Keshavan *et al.* (1994) suggest that excessive synaptic pruning in prefrontal cortex in adolescence may occur in SZP, possibly involving the excitatory glutaminergic synapses on pyramidal cells and subcortical structures such as the lenticular nuclei. Along these lines, it is interesting that a reduction in synaptophysin immunostaining has been reported in prefrontal cortex in five of six schizophrenic patients (Glantz and Lewis, 1994), further evidence that increased synaptic pruning is occurring in the schizophrenic brain. Keshavan *et al.* (1994) argue that prefrontal cortex is the last cortical brain region to mature in humans, with maturation occuring through young adulthood, which coincides with the onset of SZP. They further suggest that this excessive synaptic pruning may affect the circuitry loop between prefrontal cortex and limbic and striatal structures, because synaptic pruning and disruption of glutaminergic synapses occurs in the temporal and prefrontal cortices and in lenticular nuclei.

Neuron reduction is evident in the hippocampal CA subfields (Falkai and Bogerts, 1986) and entorhinal cortex without gliosis, and the loss is most severe in the left hemisphere (Falkai *et al.*, 1988). There is a 40% reduction of both neurons and glia in the mediodorsal nucleus of the thalamus and nucleus accumbens (Pakkenberg, 1990) and significantly lower neuronal density in prefrontal, anterior cingulate, and primary motor cortices (Benes *et al.*, 1986). Parahippocampal gyrus thickness is decreased (Altschuler *et al.*, 1990), as is hippocampal volume (Bogerts *et al.*, 1990). There are alterations in GABAergic, glutamatergic, and dopaminergic intrinsic innervation of anterior cingulate cortex, suggesting additional miswiring in the schizophrenic brain (Benes *et al.*, 1995a).

Recently, the limbic system has been the focus in SZP research employing neuroimaging, neurohistological, and neuropharmacological studies. In a recent review of the neuropathology of SZP, Bachus and Kleinman (1996) provide evidence that limbic cortical regions modulate dopaminergic function in the striatum and nucleus accumbens via glutamatergic projections. Involvement of the temporal lobe in SZP has been widely reported. Enlargement of the inferior horn of the lateral ventricle has been shown in many studies, especially when MRI is combined with computerized image analysis systems such as those used by Suddath *et al.* (1989), who report an average 67% enlargement and a 20% volumetric reduction of temporal lobe gray matter in 22 chronic schizophrenic patients. Smaller temporal lobe gray matter volumes in MRI may reflect decreases in neuron size in some temporal lobe structures. Arnold *et al.* (1995) point out that, of more than 60 postmortem studies since the early 1970s, roughly half report involvement of the hippocampus in SZP. They also note that many of the psychiatric symptoms associated with SZP, such as psychosis, mood disturbances, and lack of impulse control, are common in patients with diseases affecting the hippocampal formation. Decreased neuron size in the posterior hippocampus in schizophrenics has also been reported by Benes *et al.* (1991), and changes in volume of hippocampus and parahippocampus have been demonstrated (Bogerts *et al.*, 1985, 1990; Jeste and Lohr, 1989). In 14 schizophrenic subjects, quantitative morphometric analyses by Benes *et al.* (1991) demonstrated significantly decreased hippocampal pyramidal neuron size in CA1, CA2, CA3, and CA4 subfields. A study of hippocampal pyramidal cell orientation, conducted in schizophrenic brains from the Yakolev collection, showed marked disorganization and, although not statistically significant, there appears to be a relationship between the degree of pyramidal cell disarray and the severity of behavioral impairment (Altshuler *et al.*, 1987). A study of the hippocampal formation of schizophrenics from the Yakolev collection, conducted by Jeste and Lohr (1989), demonstrated lower pyramidal cell density in the CA1–CA4 ammonic subfields, also suggesting neuropathology localized to the hippocampus. In schizophrenics, Arnold *et al.* (1995) report smaller neurons in the hippocampal formation, including the subicular and CA subfields and in entorhinal cortex. This may disrupt connections to subcortical and cortical targets and in turn might have behavioral sequelae. In their studies, decreased neuron size in hippocampal subfields also correlated with decreased microtubule-associated proteins. Immunoreactivity to MAP2 and MAP5 was decreased in these subfields, suggesting that decreased size may be reflected in structural abnormalities in these neurons (Arnold *et al.*, 1991b).

An MRI imaging study by Shenton *et al.* (1992) in 15 right-handed men with SZP and 15 controls found a 19% reduction in the volume of gray matter in the left anterior hippocampus-amygdala, a 13% reduction in the left parahippocampal gyrus, and a 15% reduction in the left superior temporal gyrus, suggesting that SZP involves localized reductions in the gray matter of the left temporal lobe. Jakob and Beckmann (1994) and Arnold *et al.* (1991b) both demonstrated alterations in the development of entorhinal cortex in SZP. Jakob and Beckmann (1994) demonstrated alterations in entorhinal cortex layer II island formations and a decrease in volume and shape of entorhinal layer II and III neurons in postmortem histological analysis of five selected cases of SZP compared with five controls. In six schizophrenic patients, Arnold *et al.* (1991b) demonstrated abnormalities in portions of the entorhinal cortex, such as aberrant invaginations of the surface, disruption of cortical layers, heterotopic displacement of neurons, and paucity of neurons in the superficial layers. Jakob and Beckmann (1994) suggest that these changes are most likely due to local migrational disturbances that occur during the second trimester of brain development, which could give rise to aberrant connections within the limbic system and related structures. Utilizing the method of NADPH-diaphorase, which labels subpopulations of neurons derived from the embryological subplate zone during cortical development, Akbarian *et al.* (1993a,b) found reduced numbers of temporal lobe neurons in layers I–III and increased numbers in the deep layers, especially in the subcortical white matter and the vestigial subplate neurons. This suggests that the later-generated neurons destined to migrate from the subplate zone become arrested in the deep layers and do not migrate correctly to the superficial layers. The entorhinal cortex is central to many limbic circuits such as entorhinal-hippocampal, entorhinal-insula, and entorhinal-orbitofrontal reciprocal projections, and because Benes *et al.* (1991) demonstrated alterations in the intrinsic hippocampal circuitry throughout the ammonic subfields, major disturbances in limbic connectivity are highly probable in SZP.

Benes (1995a) suggests that one major structure altered in SZP is the anterior cingulate gyrus. Benes and Bird (1987) have reported abnormal aggregates of neurons in anterior cingulate cortex, particularly in layer II. Benes *et al.* (1992) reported an increased density of glutamatergic vertical axons giving off increased numbers of collaterals that provide excitatory input to GABAergic interneurons in layer I and the distal dendrites of pyramidal cells in layer II. The increased outflow from these layer II pyramidal cells in turn provides increased excitation to these same GABAergic interneurons via collaterals (Benes, 1995a). Based on a hypothesis initiated by Olney and Farber (1995), Benes (1995b) postulates that either a reduced number of GABAergic cells or a reduced activity may be excitotoxically induced by the convergence of vertical axon terminals with pyramidal cell collateral axon terminals on the GABAergic interneurons. An increase in inhibitory dopaminergic projections to these GABAergic cells may occur as a compensatory reaction to the increased excitatory inputs to these interneurons (Benes, 1993, 1995a,b; see Fig. 1 in Benes, 1995a). Benes *et al.* (1993) further point out that both pyramidal cells and GABAergic interneurons may receive these dopaminergic inputs, which may be increased in SZP.

Decreased $D_2$ receptor binding has been noted in the schizophrenic brain (Carlsson, 1988), and it has been shown in the rat that medial prefrontal cortex GABAergic interneurons have $D_1$ and, in most cases, $D_2$ receptors (Vincent *et al.*, 1993, 1995). Psychotic episodes, including negative symptoms (flattened affect, apathy, poor motivation, cognitive deficits), may be in part accounted for by elevated dopamine $D_3$ receptors in the mesolimbic region and elevated 5-$HT_{1A}$ receptor in the frontal lobe, whereas positive symptoms such as hallucinations and delusions may in part be explained by altered $D_2$ receptors in temporal cortex (Joyce *et al.*, 1997).

Thalamic involvement, including cell loss and reduced tissue volume in SZP (see Bogerts, 1993, for review), was reported as early as the 1920s from analysis of brains in the Vogt collection (Funfgeld, 1925), and has been further described by Baumer (1954) and Treff and Hempel (1958). In the latter study, morphometric analysis and statistical procedures determined that there was a significant neuron reduction in the mediodorsal nucleus (MD) in eight patients as compared with eight control brains. Pakkenberg (1990) confirmed the earlier study by Treff and Hempel (1958), reporting a reduction in the number of cells in MD, but did not specify which cell populations (i.e., parvocellular vs. magnocellular) are reduced. It is not known if this reduction is due to a lower production of MD neurons during development or whether it represents an acquired loss. In contrast, an earlier study by Lesch and Bogerts (1984) found no significant differences in the volume of any of the large subnuclei of the thalamus. These investigators did, however, report a significant reduction of the periventricular gray matter that surrounds the third ventricle (Lesch and Bogerts, 1984; Bogerts, 1993). The midline thalamic nuclei have connections with many structures involved in SZP and may be an overlooked factor that has a role in this disease. Such midline thalamic nuclei include the paraventricular (which may be considered part of the epithalamus; Jones, 1985) and a group of nuclei dorsal to CM that includes the reuniens, rhomboid, and parataenial nuclei (see Bentivoglio *et al.,* 1991). The central median nucleus (CM) is usually not included in the midline nuclear group because its fibers are embedded within the internal medullary lamina. These midline and intralaminar nuclei, which receive subcortical afferents from the hypothalamus, periaqueductal gray matter, spinothalamic tract, and medullary and pontine reticular formation, are also major thalamic targets of ascending noradrenergic and serotoninergic axons from the locus coeruleus (LC) and raphe nuclei respectively, as well as cholinergic input from the midbrain (Williams *et al.,* 1995). In addition, efferents from these midline intralaminar nuclei project to the hippocampal formation (HF), amygdala, nucleus accumbens, and parts of cingulate and orbitofrontal cortex. Thus, the midline and intralaminar group are highly interconnected with many structures involved in SZP.

Although most anatomical studies in SZP focus on limbic, paralimbic, and subcortical structures and pathways, there is also evidence for the involvement of the cerebellum. A recent review of the literature by Martin and Albers (1995) suggests that the anterior vermis and corresponding fastigial nucleus lay a role in SZP via the vermian-fastigial-forebrain pathway. A decrease in the anterior vermis area was previously reported by Weinberger *et al.* (1980). These authors suggested a nonmotor function of this pathway via its cortical connections, subsequently described by Schmahmann and Pandya (1991). The connections are to parietal, temporal, and frontal lobes and to mesolimbic areas. Anatomical connections with the fastigial nucleus include many structures that are implicated in SZP, such as nucleus accumbens and cingulate and orbital gyri (Harper and Heath, 1973). It is interesting that the fastigial nucleus also has ascending projections to the limbic system, including the hippocampus and amygdala, via the dopaminergic ventral tegmental area of the midbrain (Snider *et al.,* 1976). Martin and Albers (1995) indicate that such paleocerebellar-limbic projections from the anterior vermis and fastigial nucleus may modulate sensory inputs to the hippocampus. Further, a bilateral descending hippocampal-cerebellar projection system terminates in the cerebellar vermis (Newman and Reza, 1979). Similar to mechanisms described above by Benes and co-workers, a GABA-glutamate-dopamine system may be operative in this projection system, with a Purkinje cell (GABA) projection to fastigial nucleus (possible glutamate) modulating midbrain dopaminergic neurons, thus increasing mesolimbic and mesocortical activity (Martin and Albers, 1995). Although studies suggest a cerebellar role in emotion and cognition (see Schmahmann, 1991, 1997, for review), additional studies are needed to elucidate that role in the disease process.

## 12. Discussion

### 12.1. General

The anatomy of the dementias will now be discussed and grouped into categories based on the pattern of involvement of cortical and subcortical structures (see Table I).

**Table I. Cortical and Subcortical Structures Involved in Dementias[a]**

| Disease | Neocortex | | WhM | HF | EC | Amygdala | Striatum | Thalamus | NTbs |
|---|---|---|---|---|---|---|---|---|---|
| | Lobes | Layers | | | | | | | |
| AD | L, T, P, O, F | III, V (Py) | + | CA1, Sub | III, IV | + | + | + | + |
| PSG | F, P | I, VI[b] | + | ? | ? | ± | ± | ± | ± |
| DFT | F, T | I, II, III (Py&NP) | + | O | ± | ± | ± | O | ± |
| PiD | F, T, L | II, III, V | + | CA1, Sub | + | + | + | + | + |
| HD | L, T, P, F | III, V, VI (Py) | + | CA1, Sub | II, V | + | + | + | O |
| PD/DLBD | L, F, T | V, VI, (Py)[c] | ? | CA2–3[d] | ? | + | ? | + | + |
| CBD | F, P[e] | III, V, VI[f] | + | ? | ? | + | + | + | + |
| MSA | ±F | V[g] | + | ? | + | + | ? | + | + |
| PSP | F, T | V, VI | ? | + | + | ? | + | ? | + |
| SZP | L, T, F | I, II, III | ? | + | + | O | + | + | ? |

[a] *Other structures involved: PiD:* pons, globus pallidus, SN pars reticulata. *HD:* globus pallidus, subthalamic nucleus, ventrobasal nucleus of thalamus (small cells), hypothalamus (lateral tuberal n.), n. pontis centralis caudalis; SN pars compacta and reticulata, Purkinje cells. *PD/DLBD:* Dorsal motor nucleus of the vagus, multiple brain stem monoinergic nuclei, peripheral automonic ganglia. *CBD:* Globus pallidus, subthalamic nucleus, red nucleus, "brain stem" nuclei, cerebellar nuclei. *MSA:* Widespread involvment of basal ganglia, cerebellar circuits, and central autonomic structures. *PSP:* Mesencephalic and pontine tegmentum, subthalamic nucleus, cerebellar nuclei.

[b] Gliosis without cell loss.

[c] Without cell loss.

[d] Abnormal ubiquitin-immunoreactive neurites.

[e] Peri-rolandic atrophy.

[f] Swollen neurons.

[g] Astrocytosis in area V.

As seen in Table I, all 11 dementias discussed in this chapter involve the frontal lobe (F), some with predominately cortical involvement (AD, PiD, DFT) and others with predominately subcortical involvement (HD, PD/DLBD, PSG, PSP, CBD, MSA, and possibly SZP). Within the cortex, specific cell layers are involved in each disease. Some dementias primarily involve supragranular layers (DFT, PiD, SZP), whereas others emphasize infragranular layers (PD/DLBD, CBD, PSP). Most of the dementias include limbic system structures such as hippocampus, amygdala, and entorhinal cortex, with additional involvement of other subcortical structures.

We will now review the overall patterns of organization that emerge from comparing similarities and differences in anatomical involvement in these degenerative diseases.

## 12.2. Alzheimer's Disease and Progressive Subcortical Gliosis: Head Brain/Body Brain and Architectonics

Of these dementias, the anatomical distribution of the pathological changes in Alzheimer's disease has been the most thoroughly studied, particularly its distribution of NFTs. The NFT shows a striking predilection for cytoarchitectonic areas and specific cell types, a topographical correspondence to neurons that show cell loss, and good correlations of distribution pattern with the clinical features of AD. There is early involvement of the hippocampal fields and the entorhinal cortex with NFTs, spreading at later stages to the limbic neocortical areas of the adjacent temporal lobe and eventually to the adjacent parietal, occipital, and frontal lobes. The NFTs are most densely distributed in the temporal lobe, inferior parietal lobule and adjacent occipital lobe, and the frontal lobe ventral to the inferior frontal sulcus. These neocortical areas are primarily related to the special senses, vision, hearing, and olfaction. On architectonic grounds these cortical areas show an emphasis on the granule cells, with the pediment for this architectonic trend in the paleocortex (Fig. 4 A) (Sanides, 1972; Pandya and Yeterian, 1985, 1990). Within these cortical areas, the neurons that give rise to the long corticocortical projections appear to be particularly affected, impeding the orderly flow of neocortical information and providing a substrate for the dementia. A similar vulnerability has been found in the interrelationships between the heavily involved hippocampus, entorhinal cortex, and amygdala, providing a substrate for the memory deficit, a deficit that appears early in AD. The timing of involvement of the subcortical areas, relative to the timing of NFT appearance in the cerebral cortex, is unknown. The majority of subcortical areas involved in AD, however, have in common a projection to the neuropathologically involved neocortical and mesial temporal limbic structures. When viewed in this way, AD shows a predilection for neocortical representation of the senses of the head (vision, hearing, olfaction), their related mesial temporal limbic structures (hippocampus, amygdala, and entorhinal cortex) and synaptically related subcortical structures.

An almost exact reversal of the gyral pattern of involvement in AD occurs in progressive subcortical gliosis (PSG). In PSG atrophy affects the upper parts of the frontal lobe and the superior parietal lobule. On architectonic grounds these cortical areas show an emphasis on pyramidal cells that appear to have their pediment in the hippocampus (Sanides, 1972; Pandya and Yeterian, 1985, 1990). These areas

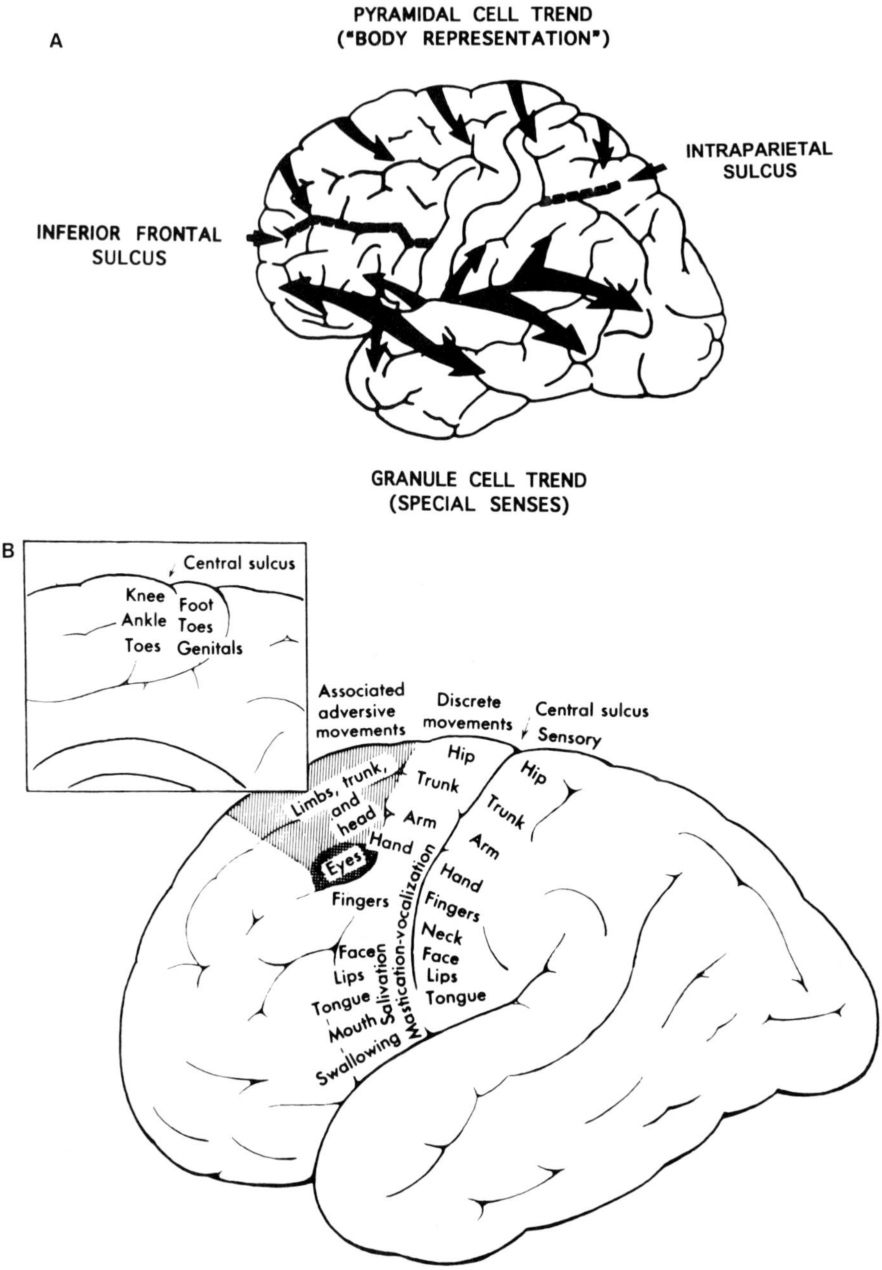


**Figure 4.** (A) A schematic drawing showing the granule and pyramidal cell architectonic trends in the human brain. Over the convexity of the brain, these two trend meet anteriorly at the level of the inferior frontal sulcus and posteriorly to the interparietal sulcus. The cerebral cortices in the granule cell trend subserve the special senses of the head, whereas those in the pyramidal cell trend subserve the cortical representation of the trunk and extremities. (Based on Sanides, 1970, and Pandya and Yeterian, 1985, 1990; Diagram from Kemper, 1994). (B) Diagram showing the homunculus of the human brain. Note that the head representation is at the level of the inferior frontal gyrus and the trunk and extremities representation is at the level of the middle and superior frontal gyri. From Haymaker and Schiller (1969) with permission.

provide cortical representation for the limbs and trunk of the body. In this sense, PSG shows an emphasis on cortical areas representing the body rather than the head. In the human brain, the pyramidal and granular architectonic trends that subserve these two different body representations meet at the inferior frontal gyrus and the interparietal sulcus (Sanides, 1972). A linc drawn between the two sulci would intersect the body representation homunculi in the pre- and postcentral gyri at approximately the neck area (Fig. 4B). This pattern of gyral atrophy is commonly observed in the brains of elderly individuals (Sandor *et al.,* 1990), but is less pronounced (Fig. 5), suggesting a possible relationship between PSG and aging. It is interesting to note that the striking feature of PSG neuropathology is the emphasis on layers I and VI and on subcortical white matter, areas with a predilection for age-

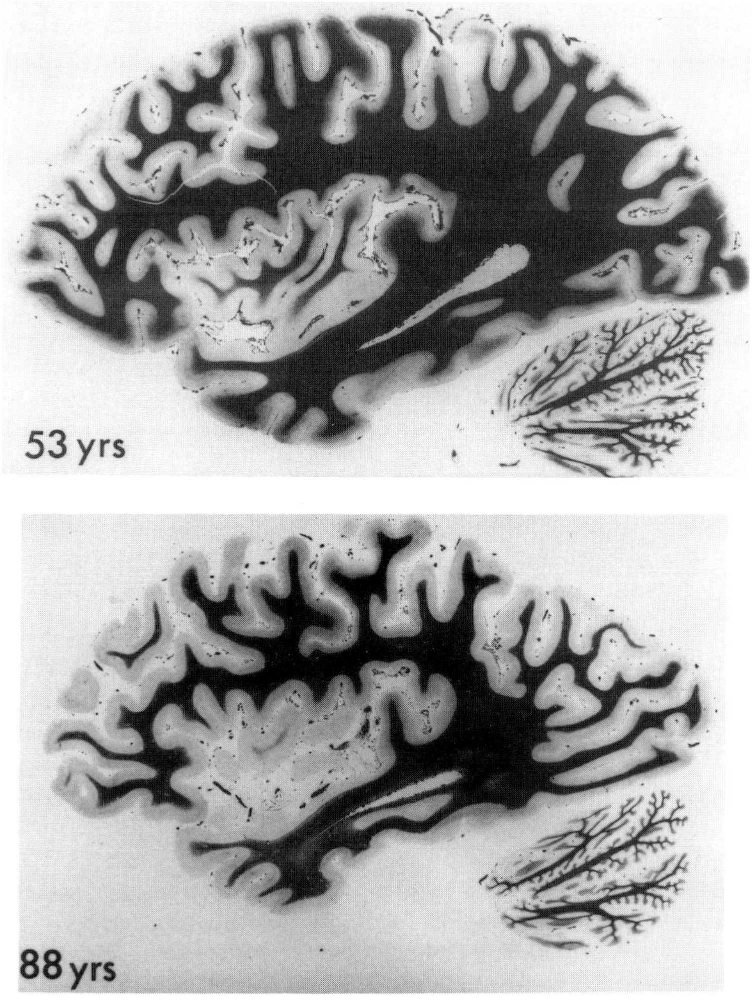


**Figure 5.** Myelin-stained parasagittal sections of the brain of mentally intact individuals ages 53 and 88 years. Age-related atrophy of the gyri in the 88 year old is confined to the upper parts of the frontal and parietal lobes and occurs in association with central white-matter loss.

related changes in nonhuman primates (see Peters, this volume). The monkeys show marked atrophy and gliosis of layer I in association with degenerative changes in dendrites, neuronal perikarya, and myelinated fibers. Similar changes in myelinated fibers are also found in the cerebral cortex and in the subcortical white matter.

## 12.3. Embryology of the Cerebral Cortex and Progressive Subcortical Gliosis

Another way to view the topography of the gliosis in PSG is that it occurs in those areas of the cerebral cortex that are the first to appear in its embryological development. In the period of early fetal growth, before the development of the cortical plate (layers II–VIa), the surface of the cerebral hemispheres is occupied by a zone of neurons embedded in a dense axonal plexus (Rickmann *et al.*, 1977; Marin-Padilla, 1978; Raedler *et al.*, 1980). This primitive (or primordial) plexiform layer becomes separated in later stages of development into two discrete layers by the migration of neurons into the primitive layer. This later migration forms cortical layers II through VIa, with the upper part of the primitive plexiform layer becoming incorporated into layer I of the cerebral cortex. The deep part becomes the lower section of layer VI (VIb), with some scattered neurons appearing in the subcortical white matter. It is precisely these two derivatives of the primitive plexiform layer that show the marked gliosis in PSG, raising the possibility that genetic or developmental predisposition factors may play a role in PSG's etiology.

## 12.4. Frontal-Type Dementia, Pick's Disease, and the Selective Vulnerability of the Frontal and Temporal Lobes

Selective gyral atrophy of the frontal and temporal lobes with predominant loss of pyramidal cells in the supragranular layers II and III occurs in both DFT and PiD. This pattern of laminar involvement is the reverse of that noted in AD wherein the cellular pathology is most prominent in the infragranular layer V. In both DFT and PiD, the atrophy, in contrast to that found in AD, is most marked anteriorly in the frontal and temporal lobes (Hof *et al.*, 1994). A relationship between DFT and PiD (and possibly also corticobasal degeneration) has been suggested by Neary (1997) and by Kertesz (1997), perhaps related to the accumulation of an abnormal tau protein (Kertesz, 1997). Further, in the brains of PiD patients, a wide variety of pathological changes has been reported (e.g., Constantinidis *et al.*, 1974). These two diseases differ in that the changes within the cerebral cortex are relatively mild in DFT, in which they are associated with a spongiform change, whereas in PiD there is a more extensive neuronal loss that is associated with gliosis and a more consistent involvement of subcortical structures. The anatomical substrate for the selective pattern of frontotemporal atrophy and neuronal loss is not known. The more superficial pyramidal cells in the cortical plate are thought to give rise to short corticocortical projections, the more deeply situated pyramidal cells in the supragranular layer to more distant corticocortical projections (Jones, 1984; Felleman and Van Essen, 1991). The tips of the frontal and temporal poles are synaptically far removed from the primary motor cortex in the frontal lobe and from the primary visual and auditory cortices in the occipital and temporal lobes, being connected to them by a long series of corticocortical projections, and they are in turn interconnected with each other (Pandya and Yeterian, 1985, 1990). This dependence of the distal association areas in the tips of the frontal and temporal lobes would render the tips particularly vulnerable to a perturbation of corticocortical circuits.

In addition to neuronal loss and other pathological changes in the supragranular pyramidal cells in PiD there is also involvement of the infragranular cell layers, the source of the cortical projection to the striatum and the amygdala (Jones, 1984; Van Hoesen, 1981). In contrast to DFT, these subcortical nuclei are involved frequently in PiD, suggesting a relationship between the involvement of the infragranular cell layer and these subcortical structures. In addition to involvement of the amygdala, there is prominent involvement of the hippocampus and entorhinal cortex in PiD, a feature also noted in Alzheimer's disease, the other dementia with a marked predilection for temporal lobe pathology.

## 12.5. Huntington's Disease and Diffuse Lewy Body Disease: Dementias with Cerebral Cortical Pathology and Prominent Subcortical Changes

In both Huntington's chorea and diffuse Lewy body disease, the cerebral cortical changes are only readily seen with special stains (DLBD) or by cell counting (Huntington's disease). Both show marked pathology in subcortical areas. In DLBD these subcortical changes are primarily in nuclei characterized by the presence of monoaminergic and cholinergic neurons; striatal involvement is characterized by the loss of dopaminergic innervation. In contrast, in HD the most conspicuous subcortical pathology is in the striatum, with marked loss of the projection neurons. In concert with this pronounced subcortical pathology, the neuronal changes in the cerebral cortex are prominent in the pyramidal cells in layers V and VI—cells of origin of cerebral cortical projections to these subcortical areas. In common with the two other dementias that show prominent involvement of the frontal and temporal lobes, namely AD and PiD, HD and DLBD both show pathological changes in the amygdala and hippocampal formation, and in Huntington's disease, in the entorhinal cortex as well.

## 12.6. Corticobasal Degeneration, Multiple System Atrophy, and Progressive Supranuclear Palsy: Dementias with Predominately Subcortical Involvement

In contrast to the other disorders discussed in this chapter, CBD, MSA, and PSP are diseases that predominantly affect subcortical structures, with cortical areas also involved. Each of these disorders has specific histological features which, in typical cases, distinguish them from one another: (1) CBD is characterized by swollen neurons in the lateral SN and in posterior frontal and/or parietal cortical layers III, V, and VI. The nigral neurons also contain filamentous inclusions, which are basophilic and argentophilic, with many similarities to the neurofibrillary tangles in PSP; (2) MSA is characterized by the presence of inclusion bodies in glial cells; this remains the most important pathological feature; (3) PSP is characterized by abnormal neurons, with tau$^+$ protein aggregates forming neurofibrillary tangles composed of either straight filaments or the paired helical filaments similar to those present in AD. It should be pointed out that these features are not unique to these diseases. It is their distribution and predominance that allows for the post-mortem diagnosis. For example, diagnosis of CBD, but not PSP, relies on the presence of significant numbers of swollen cerebral cortical and nigral neurons, slight or no involvement of the cerebellar dentate nucleus, and a paucity of tangles in brain stem nuclei. Also, some glial inclusions are found in CBD. These are identical to those in MSA but are no less numerous and if present, are usually in areas such as peri-Rolandic cortex, which is not involved in MSA. PiD may be differentiated from CBD if large numbers of Pick bodies are present.

In all three of these diseases (CBD, MSA, and PSP), there are common features in the distribution of cell loss, gliosis, and histological alterations in subcortical structures. In typical cases of PSP, MSA, and CBD, the SN, LC, caudate, putamen, globus pallidus, and subthalamic nuclei are all involved. The inferior olive and cerebellum are usually involved in MSA and PSP, but they rarely are in CBD. It is therefore clear that although similar motor structures are involved in all three diseases, at least some of the motor pathways that undergo degeneration differ in each disorder. For example, in some cases of CBD, swollen neurons are found in the lateral substantia nigra. In MSA, it is mainly the posterior or dorsolateral part(s) of the putamen that undergoes cell loss and gliosis.

Developing striatal neurons of different types migrate in clusters (Brand and Rakic, 1979) and histochemically, the striatum has a patch distribution (Graybiel and Ragsdale, 1980). These authors also described the efferent projections of the striatum as being grouped in geometrical configurations. Because the motor cortex is also involved in MSA and in PSP, it is of interest that area 4 projects bilaterally and somatotopically upon the putamen with patchlike terminations that are the most dense in the lateral regions, and extend for nearly its entire length (Kunzle, 1975). It is also of interest that in HRP labeling experiments in the monkey (Jones *et al.*, 1977; Jones and Wise, 1977), the smaller pyramidal cells in the upper portion of layer V give rise to the massive corticostriate projection. The two common features all three diseases share with regard to the cortex are involvement of the frontal lobe and of layer V. Two major targets of striatal efferents are globus pallidus and substantia nigra, which are heavily involved in all three disorders. Pallidal afferents also come from the subthalamic nucleus, likewise involved in all three. It has been known for some time that there are regional differences in the distribution and concentration of various peptides in the globus pallidus. For example, in the monkey, striatopallidal fibers immunoreactive for enkephalin are densely distributed in the lateral (external pallidal segment, whereas fibers immunoreactive for substance P are densely distributed in the medial (internal) segment (Haber and Elde, 1981; Haber and Nauta, 1983). In HD there is a reduction in both enkephalin and substance P in globus pallidus and SN (Emson *et al.*, 1980). In PSP cell loss and gliosis occur in the striatum and in both the external and internal pallidal segments. The internal pallidal neurons are known to project via ventrolateral thalamus (VL) to premotor and prefrontal areas (Alexander *et al.*, 1986), and Brooks (1994) suggests that the internal pallidal lesions probably contribute to the frontal cognitive deficits in PSP patients.

Relating PSP symptomotology to anatomy in a recent report, Destee *et al.* (1990) suggest that loss of self-activation seems to result from a damaged cortico-subcortical circuit forming a limbic loop, whereas the compulsive behavior of the obsessive type may result from the interruption of a frontal-caudal-pallidal-subthalamic circuit. Litvan (1994b) comments that specific lesions in PSP may underlie particular behavioral deficits: Basal forebrain and mediodorsal thalamic (MD) lesions may lead to memory deficits; involvement of the mesencephalic and pontine nuclei (rostral interstitial nucleus of the MLF, superior colliculus, and laterodorsal tegmental nucleus) can result in supranuclear ophthalmoplegia; and striatal and pedunculopontine nuclei lesions may contribute to the motor deficits seen in PSP patients (Litvan, 1994b).

## 12.7. Schizophrenia

Schizophrenia (SZP) is a complex disease with a large literature describing numerous case studies. In contrast, there is a paucity of information on specific neuroanatomical involvement and connectivity in SZP, and few studies attempt to link behavioral symptoms with specific anatomical circuits. It is clear that parts of the limbic system, such as anterior cingulate and entorhinal cortices and the hippocampus, are affected, as well as the vermian-fastigial-forebrain system and MD, nucleus accumbens, and other subcortical structures.

### Subcortical Circuits Involvement

Some of the symptomology of SZP may be explained by close examination of subcortical circuitry. Recently, Carlsson and Carlsson (1990) proposed a hypothesis that the cerebral cortex can control its sensory input, which is important in arousal, via dopamine and glutamatergic modulated feedback loops involving the dorsal and ventral striatum, thalamus, and mesencephalic reticular formation. Applying this model to SZP, these authors suggest that there may be a neurotransmitter imbalance of dopamine and glutamate in a feedback-regulated system controlling arousal and the processing of information signals to the cerebral cortex via the thalamus. The efferents from the dorsal striatum innervate mainly the ventrolateral thalamic nucleus (VL) which then projects to neocortex, whereas efferents from the ventral striatum project to thalamic nuclei, which in turn project to frontal and limbic cortex and the amygdala (Nauta, 1989, in Carlsson and Carlsson, 1990). These authors also suggest a modulatory role for the locus coeruleus and raphe in controlling arousal via the dorsal and ventral striatum. Presumably, MD and the midline and intralaminar groups—including the paraventricular, reuniens, rhomboid and parataenial nuclei—also play a critical role linking subcortical and cortical structures involved in SZP. These thalamic nuclei link the central striatum with limbic areas and frontal cortical areas, implicating these structures in the control of psychomotor activity. Bogerts (1993) points out that, as widely reported in the literature, thalamic lesions induce apathetic behavior, drowsiness, mental lethargy, stupor, disturbances of memory and sensory phenomena such as hyperesthesia, hyperpathia, and painful perceptions—all of the "negative" symptomology associated with SZP. The enlarged third ventricle reported in some patients may be due to atrophy of midline thalamic nuclei immediately surrounding the third ventricle.

Miller *et al.* (1989) reported that the midline and intralaminar nuclei play a role in memory and arousal. It is very likely, given the negative symtomology of SZP, that at least some of the midline intralaminar nuclei and limbic structures, such as the hippocampus and amygdala, as well as their connections, are affected in the disease pathology. The involvement of MD has been reported (see above) but no detailed analysis of the magnocellular versus parvocellular parts of MD in SZP is available. The magnocellular portion of MD received olfactory input from piriform and adjacent cortex (Price, 1990), and from mediobasal and lateral nuclei in the amygdala (Armstrong, 1990; de Olmos, 1990). Efferent projections from the magnocelluar part of MD project to medial prefrontal cortex (Giguerre and Goldman-Rakic, 1988; Ray and Price, 1993), to olfactory regions of orbitofrontal cortex (Jones, 1985), and to ventromedial cingulate cortex, inferior parietal cortex, superior temporal sulcus, and anterior insula (Giguerre and Goldman-Rakic, 1988). The parvocellular part of MD receives afferents from the globus pallidus (Alheid *et al.*, 1990), although the origin of most noncortical afferents is not clear. There are reciprocal connections with dorsolateral and dorsomedial prefrontal cortex as well as the anterior cingulate gyrus and supplementary motor area (Giguerre and Goldman-Rakic, 1988; Wiesendanger and Weisendanger, 1985). Schmahmann and Pandya (1990) also demonstrated projections from parvocellular MD to posterior parietal cortex. Damage to parvocellular MD is known to cause transient amnesia, particularly with confusion over passage of time, decreased anxiety, and obsessive thinking (Armstrong, 1990). Thus, in addition to the medial and intralaminar groups of nuclei being likely affected in SZP, parts of MD may also be involved in the disease process.

It is interesting that in schizophrenics, Pakkenberg (1990), utilizing an unbiased stereological technique, found that total numbers of glia and neurons in MD and in nucleus accumbens differed significantly from controls. This suggests possible disruption of another brain pathway: MD and the basolateral nucleus of the amygdala have major input to prefrontal cortex; parts of prefrontal cortex project to nucleus accumbens, a prominent limbic region that in turn projects in part to the ventral pallidum. Disruption along this pathway might play in role in symptomology of SZP. Pakkenberg (1990) found a normal total neuron number in the basolateral nucleus of the amygdala and ventral pallidum, which suggests that the most likely sites of disruption in connectivity within this pathway are in MD and nucleus accumbens.

Finally, it is still not clear whether the cerebellum plays a role in the disease process. Strides are being made in deciphering its possible role in cognition (Schmahmann, 1991, 1997; Schmahmann and Pandya, 1991), and it likely that GABAergic Purkinje cells projecting to the fastigial nucleus modulate midbrain dopaminergic neurons, resulting in increased mesolimbic and mesocortical activity in SZP (Martin and Albers, 1995). Much work remains in linking the limbic, subcortical, and cerebellar circuits in the disease process to definitively determine the underlying neuroanatomical substrate(s) for the residual state in SZP.

## 12.8. Conclusion

The diverse dementias reviewed here all have in common the involvement of pyramidal cells in the frontal lobes, with differing patterns of involvement of other lobes and also of subcortical areas. The tip of the frontal lobe is synaptically far removed from the primary motor cortex in the frontal lobe and from the association areas of the temporal and parietal lobes that project to it by a long series of corticocortical projections (Pandya and Yeterian, 1985, 1990). This dependence of the tip of the frontal lobe on these corticocortical projections would render it particularly vulnerable to a perturbation of corticocortical circuits. This is especially striking in human brains, because the tip appears to lack a thalamic projection (McLardy, 1950; Angevine *et al.*, 1964; Van Buren and Borke, 1972), a major source of afferent projections to the cerebral cortex. Such vulnerability is especially obvious in PiD and DFT. Another striking pattern in cortical involvement in dementias is seen in AD versus PSG. AD affects cortices involved in the special senses of the head and follows a granule-cell architectonic trend, whereas PSG involves the cortex that represents the trunk and extremities, and follows a pyramidal-cell architectonic trend. The two trends meet at the level of middle and superior frontal gyri. Further, the cortical involvement of PSG appears strikingly similar to that seen in normal aging. HD and DLBD have cortical and subcortical pathology, with the subcortical projection neurons in layers V and VI involved in both as well. A major pathology in HD is loss of projection neurons in the striatum; in DLBD it is a loss of the dopaminergic input to the striatum. Three diseases with predominent involvement of subcortical structures, CBD, MSA and PSP, all have distinguishing histological alterations, with the definitive diagnosis dependent on the distribution and predominence of these features. Finally, the description of the neuroanatomical basis of SZP is in its early stages. A number of investigators now view SZP as having a neurodevelopmental basis, so that Benes and colleagues (Benes, 1993, 1995a,b; Benes *et al.*, 1986, 1992) have emphasized a limbic involvement that includes anterior cingulate cortex, with an increased excitatory input to inhibitory interneurons and possible alterations in the dopaminergic innervation to these interneurons.

We are now at a point where research efforts need to focus on the specific neuroanatomical pathways involved in dementias, rather than simply listing involved structures. Only then will we begin to unravel the complexities that form the underlying basis of dementia symptomology.

ACKNOWLEDGMENTS. The authors would like to thank Mr. Jonathan Wisco and Drs. Ed Yeterian and Trudy Van Houten for their assistance with research articles. We would also like to thank Dr. Alan Peters for manuscript review, and Mrs. Susan Blease for assistance with photography. GJB would like to thank Faith and Aaron Blatt for their unending support. Also supported by NIA NIH 2 P01 AG00001-17A2 (TLK).

## 13. References

Akbarian, S., Bunney, W. E., and Potkin, S. G., 1993a, Altered distribution of nicotinamide-adenine dinucleotide phosphate-diaphorase cells in frontal lobe of schizophrenics implies disturbances of cortical development, *Arch. Gen. Psychiatry* **50:**169–177.

Akbarian, S., A., Vinuela, and Kim, J. J., 1993b, Distorted distribution of nicotinamide-adenine dinucleotide phosphate-diaphorase cells in temporal lobe of schizophrenics implies anomalous cortical development, *Arch. Gen. Psychiatry* **50:**178–187.

Albert, M. S., 1996, Cognitive and neurobiological markers of early Alzheimer's disease, *Proc. Natl. Acad. Sci. USA* **93:**13547–13551.

Aletrino, M. A., Vogels, O. J., Van Domburg, P. H., and ten Donkelaar, H. J., 1992, Cell loss in the nucleus raphe dorsalis in Alzheimer's disease, *Neurobiol. Aging* **13:**461–468.

Alexander, G. E., Delong, M. R., and Strick, P. L., 1986, Parallel organization of functionally segregated circuits linking basal ganglia and cortex, *Annu. Rev. Neurosci.* **9:**357–381.

Alheid, G. F., Heimer, L., and Switzer, R. C., III, 1990, Basal Ganglia, in: *The Human Nervous System.* (G. Paxinos, ed.), Academic, San Diego, pp. 483–582.

Allen, S. J., Dawbarn, D., and Wilcock, G. K., 1988, Morphometric immunohistochemical analysis of neurons in the nucleus basalis of Meynert in Alzheimer's disease, *Brain Res.* 454:275–281.

Altshuler, L. L., Conrad, A., Kovelman, J. A., and Schiebel, A., 1987, Hippocampal pyramidal cell orientation in schizophrenia, *Arch. Gen. Psychiatry* **44:**1094–1098.

Altshuler, L. L., Casanova, M. F., Goldberg, T. E., and Kleinman, I. E., 1990, The hippocampus and parahippocampus in schizophrenic, suicide, and control brains, *Arch. Gen. Psychiatry* **47:**1029–1034.

Andreasen, N. C., 1995, Symptoms, signs, and diagnosis of schizophrenia, *Lancet* **346:**477–481.

Angevine, J. B., Locke, S., and Yakovlev, P., 1964, Limbic nuclei of the thalamus and connections of the limbic cortex: Thalamocortical projections of the magnocellular medial dorsal nucleus in man, *Arch. Neurol.* **10:**165–180.

Arendt, T., Bigl, V., Arendt, A., and Tennstedt, A., 1983, Loss of neurons in the nucleus basalis of Meynert in Alzheimer's disease, paralysis agitans, and Korsakoff's disease, *Acta Neuropathol,* **61:**101–108.

Arendt, T., Bigl, V., Tennstedt, A., and Arendt, A., 1985, Neuronal loss in different parts of the nucleus basalis is related to neurite plaque formation in cortical target areas in Alzheimer's disease, *Neuroscience* **14:**1–14.

Arendt, T., Bruckner, M. K., Bigl, V., and Marcova, L., 1995a, Dendritic reorganization in the basal forebrain under degenerative conditions and its defects in Alzheimer's disease. II. Ageing, Korsakoff's disease, Parkinson's disease, and Alzheimer's disease, *J. Comp. Neurol.* **351:**189–222.

Arendt, T., Bruckner, M. K., Bigl, V., and Marcova, L., 1995b, Dendritic reorganization in the basal forebrain under degenerative conditions and its defects in Alzheimer's disease. III. The basal forebrain compared with other subcortical areas. *J. Comp. Neurol.* **351:**223–246.

Armstrong, E., 1990, Limbic thalamus: Anterior and mediodorsal nuclei, in: *The Human Nervous System* (G. Paxinos, ed.), Academic, San Diego, pp. 469–481.

Arnold, S. E., Hyman, B. T., Flory, J., Damasio, A. R., and Van Hoesen, G. W., 1991a, The topographical and neuroanatomical distribution of neurofibrillary tangles and neuritic plaques in the cerebral cortex of patients with Alzheimer's disease, *Cereb. Cortex* **1:**103–116.

Arnold, S. E., Lee, V. M., Gur, R. E., and Trojanowski, J. Q., 1991b, Abnormal expression of two microtubule-associated proteins (MAP2 and MAP5) in specific subfields of the hippocampal formation in schizophrenia, *Proc. Natl. Acad. Sci. USA* **88:**10850–10854.

Arnold, S. E., Hyman, B. T., Van Hoesen, G. W., and Damasio, A., 1991c, Some cytoarchitectual abnormalities of the entorhinal cortex in schizophrenia, *Arch. Gen. Psychiatry* **48:**625–632.

Arnold, S. E., Hyman, B. T., and Van Hoesen, G. W., 1994, Neuropathologic changes in the temporal pole in Alzheimer's and Pick's disease, *Arch. Neurol.* **51:**145–150.

Arnold, S. E., Franz, B. R., Ruben, B. A., Gur, C., Gur, R. E., Shapiro, R. M., Moberg, P. J., and Trojanowski, J. Q., 1995, Smaller neuron size in schizophrenia in hippocampal subfields that mediate cortical-hippocampal interactions, *Am. J. Psychiatry* **152:**738–748.

Arriagada, P. V., Marzloff, K., and Hyman, B. T., 1992a, Distribution of Alzheimer-type pathologic changes in nondemented elderly individuals matches the pattern of Alzheimer's disease, *Neurology* **42;**1681–1688.

Arriagada, P. V., Growden, J. H., Hedley-Whyte, T., and Hyman, B. T., 1992b, Neurofibrillary tangles, but not senile plaques, parallel duration and severity of Alzheimer's disease, *Neurology* **42:**631–639.

Bachus, S. E., and Kleinman, J. E., 1996, The neuropathology of schizophrenia, *J. Clin. Psychiatry* **57** (suppl. 11):72–83.

Bancher, C., Lassmann, H., Breitschopf, H., and Jellinger, K. A., 1997, Mechanisms of cell death in Alzheimer's disease, *J. Neural Transm. (Suppl.)* **50:**141–152.

Bancher, C., Braak, H., Fischer, P., and Jellinger, K. A., 1993, Neuropathological staging of Alzheimer lesions and intellectual status in Alzheimer's and Parkinson's disease patients, *Neurosci. Lett.* **162:**179–182.

Baumer, H., 1954, Veranderungen des Thalamus bei Schizophrenie, *J. Hirnforsch.* **1:**157–172.

Benes, F. M., Davidson, J., and Bird, E. D., 1986, Quantitative cytoarchitectural studies of the cerebral cortex of schizophrenics, *Arch. Gen. Psychiatry* **43:**31–35.

Benes, F. M., and Bird, E. D., 1987, An analysis of the arrangement of neurons in the cingulate cortex of schizophrenic patients, *Arch. Gen. Psychiatry* **44**;608–616.

Benes, F. M., Sorensen, I., and Bird, E. D., 1991, Reduced neuronal size in posterior hippocampus of schizophrenic patients, *Schizophr. Bull.* **17**:597–608.

Benes, F. M., Sorenson, I., Vincent, S. L., Bird, E. D., and Sathi, M., 1992, Increased density of glutamate-immunoreactive vertical processes in superficial laminae in cingulate cortex of schizophrenic brain, *Cereb. Cortex* **2**:502–512.

Benes, F. M., 1993, Relationship of cingulate cortex to schizophrenia and other psychiatric disorders, in: *Neurobiology of Cingulate Cortex and Limbic Thalamus* (B. A. Vogt and M. Garbiel, eds.), Birkhauser, Boston, pp. 581–605.

Benes, F. M., Vincent, S. L., and Malloy, R., 1993, Dopamine-immunoreactive axon varicosities form nonrandom contacts with GABA-immunoreactive neurons in rat medial prefrontal cortex, *Synapse* **15**:285–295.

Benes, F. M., 1995a, Altered glutamatergic and GABAergic mechanisms in the cingulate cortex of the schizophrenic brain, *Arch. Gen. Psychiatry* **52**:1015–1018.

Benes, F. M., 1995b, Is there a neuroanatomical basis for schizophrenia? *Neuroscientist* **1**:112–120.

Bentivoglio, M., Balercia, G., and Kruger, L., 1991, The specificity of the nonspecific thalamus: The midline nuclei, *Prog. Brain Res.* **87**:53–80.

Bhatt, M. H., Snow, B. J., Martin, W. R. W., Peppard, R., and Calne, D. B., 1991, Positron emission tomography in progressive supranuclear palsy, *Arch. Neurol.* **48**:354–389.

Bierer, L. M., Hof, P. R., Purohit, D. P., Carlin, L., Schmeidler, J., Davis, K. L., and Perl, D. P., 1995, Neocortical neurofibrillary tangles correlate with dementia severity in Alzheimer's disease, *Arch. Neurol.* **52**:81–88.

Blennow, K., Bogdanovic, N., Alafuzoff, I., Ekman, R., and Davidsson, P., 1996, Synaptic pathology in Alzheimer's disease: Relation to severity of dementia, but not to senile plaques, neurofibrillary tangles, or the ApoE4 allele, *J. Neural Transm. (Genet. Sect.)* **103**:603–618.

Blin, J., Vidailhet, M. J., and Pillon, B., 1992, Corticobasal degeneration: Decreased and asymmetrical glucose consumption as studied with PET, *Mov. Disord.* **7**:348–354,

Bogerts, B., 1993, Recent advances in the neuropathology of schizophrenia, *Schizophr. Bull.* **19**:431–445.

Bogerts, B., Meertz, E., and Schonfeldt-Bausch, R., 1985, Basal ganglia and limbic system pathology in schizophrenia, *Arch. Gen. Psychiatry* **42**:784–791.

Bogerts, B., Falkai, P., Haupts, M., Greve, B., Ernst, S., Tapernon-Franz, U., and Heinzmann, U., 1990, Post-mortem volume measurements of limbic system and basal ganglia structures in chronic schizophrenics: Initial results from a new brain collection, *Schizophr. Res.* **3**:295–301.

Bondereff, W., Mountjoy, C. Q., and Roth, M., 1982, Loss of neurons of origin of adrenergic projections to cerebral cortex (nucleus locus ceruleus) in senile dementia, *Neurology* **32**:164–168.

Braak, H., and Braak, E., 1990, Alzheimer's disease; Striatal amygdaloid deposits and neurofibrillary tangles, *J. Neuropathol. Exp. Neurol.* **49**:215–224.

Braak, H., and Braak, E., 1991a, Neuropathological staging of Alzheimer-related changes, *Acta Neuropathol.* **82**:239–259.

Braak, H., and Braak, E., 1991b, Alzheimer's disease affects limbic nuclei of the thalamus, *Acta Neuropathol.* **81**:261–268.

Braak, H., and Braak, E., 1992, Allocortical involvement in Huntington's disease, *Neuropathol. Appl. Neurobiol.* **18**:539–547.

Braak, H., Braak, E., and Kalus, P., 1989, Alzheimer's disease: Areal and laminar pathology in the occipital isocortex, *Acta Neuropathol.* **77**:494–506.

Braak, H., Braak, E., Yilmazer, D., de Vos, R. A., Jansen, E. N., Bohl, J., and Jellinger, K., 1994, Amygdala pathology in Parkinson's disease, *Acta Neuropathol.* **88**:493–500.

Braak, H., Braak, E., Yilmazer, D., Schultz, C., de Vos, R. A., and Jansen, E. N., 1995, Nigral and extranigral pathology in Parkinson's disease, *J. Neural Transm. (Suppl).* **46**:15–31.

Brand, S., and Rakic, P., 1979, Genesis of the primate neostriatum: [$^{3}$H]-thymidine autoradiographic analysis of the time of neuron origin in the rhesus monkey, *Neuroscience* **4**:767–778.

Brooks, D. J., 1994, PET studies in progressive supranuclear palsy, *J. Neural Transm. (Suppl.)* **42**:119–134.

Brooks, D. J., Ibañez, V., and Sawle, G. V., 1990, Differing patterns of striatal $^{18}$F-dopa uptake in Parkinson's disease, multiple system atrophy, and progressive supranuclear palsy, *Ann. Neurol.* **28**:547–555.

Brun, A., 1987, Frontal lobe degeneration of non Alzheimer type. I. Neuropathology, *Arch. Gerontol. Psychiatry* **6:**193–207.

Brun, A., 1993, Frontal lobe degeneration of non Alzheimer type revisited, *Dementia* **4:**126–131.

Brun, A., and Gustafson, L., 1976, Distribution of cerebral degeneration in Alzheimer's disease: A clinico-pathological study, *Arch. Psychiat. Nervenkr.* **233:**15–33.

Brun, A., and Passant, U., 1996, Frontal lobe degeneration of non-Alzheimer type, *Acta Neurol. Scand. (Suppl.)* **168:**28–30.

Brun, A., Englund, B., Gustafson, L., Passant, U., Mann, D. M. A., Neary, D., and Snowden, J. S., 1994, Clinical and neuropathological criteria for frontotemporal dementia, *J. Neurol. Neurosurg. Psychiatry* **57:**416–418.

Byrne, E. J., Lennox, G., Lowe, J., and Godwin-Austen, R. B., 1989, Diffuse Lewy body disease: Clinical features in 15 cases. *J. Neurol. Neurosurg. Psychiatry* **52:**709–717.

Candy, J. M., Perry, R. H., Perry, E. K., Irving, D., Blessed, G., Fairbairn, A. F., and Tomlinson, B. E., 1983, Pathological changes in the nucleus of Meynert in Alzheimer's disease and Parkinson's disease, *J. Neurol. Sci.* **59:**277–289.

Carlsson, A., 1988, The current status of the dopamine hypothesis of schizophrenia, *Neuropychopharmacology* **1:**179–186.

Carlsson, M., and Carlsson, A., 1990, Schizophrenia: A subcortical neurotransmitter imbalance syndrome? *Schizophr. Bull.* **16:**425–432.

Carpenter, M. B., and Sutin, J., 1983, *Human Neuroanatomy,* 8th Ed., Williams and Wilkens, Baltimore, Maryland.

Cervos-Navarro, J., and Schumacher, K., 1994, Neurofibrillary pathology in progressive supranuclear palsy (PSP), *J. Neural Transm. (Suppl.)* **42:**153–164.

Chan-Palay, V., and Asan, E., 1989, Alterations in catecholamine neurons of the locus coeruleus in senile dementia of the Alzheimer type and in Parkinson's disease with and without dementia and depression, *J. Comp. Neurol.* **287:**373–392.

Chen, C. P., Eastwood, S. L., McDonald, B., Hope, R. A., and Esiri, M. M., 1994, Patterns and markers of cell loss in the dorsal raphe nucleus in Alzheimer's disease, *Neuropath. Appl. Neurobiol.* **20:**503.

Chui, H. C., Mortimer, J. A., Slager, U., Zarow, C., Bondareff, W., and Webster, D. D., 1986, Pathologic correlates of dementia in Parkinson's disease, *Arch. Neurol.* **43:**991–995.

Chiu, H. F., 1995, Psychiatric aspects of progressive supranuclear palsy, *J. Gen. Hosp. Psychiatry* **17:**135–143.

Churchyard, A., Donnan, G. A., and Hughes, A., 1993, Dopa resistance in multiple system atrophy: Loss of postsynaptic D2 receptors. *Ann. Neurol.* **34:**219–226.

Clark, A. W., Parhad, I. M., Folstein, S. E., Whitehouse, P. J., Hedreen, J. C., Price, D. L., and Chase, G. A., 1983, The nucleus basalis in Huntington's disease, *Neurology* **33:**1262–1267.

Constantinidis, J., Richard, J., and Tissot, R., 1974, Pick's disease: Histological and clinical correlations, *Eur. Neurol.* **11:**208–217.

Convit, A., DeLeon, M. J., Tarshish, C., De Santi, S., Tsui, W., Rusinek, H., and George, A., 1996, Specific hippocampal volume reduction in individuals at risk for Alzheimer's disease, *Neurobiol. Aging* **18:**132–138.

Costa, C., and Duyckaerts, C., 1993, Oligodendroglial and neuronal inclusions in multiple system atrophy, *Curr. Opin. Neurol.* **6:**865–871.

Costa, C., Duyckaerts, C., Cervera, P., and Hauw, J. J., 1992, Oligodendroglial inclusions, a marker of multisystemic atrophies, *Rev. Neurol. (Paris)* **148:**274–280.

Coyle, J. T., Price, D. L., and Delong, M. R., 1983, Alzheimer's disease: A disorder of cortical cholinergic innervation, *Science* **219:**1184–1190.

Craufurd, D., and Dodge, A., 1993, Mutation size and age at onset in Huntington's disease, *J. Med. Genet.* **30:**1008–1011.

Crystal, H. A., Dickson, D. W., Lizardi, J. E., Davies, P., and Wolfson, L. I., 1990, Antemortem diagnosis of diffuse Lewy body disease, *Neurology* **40:**1523–1528.

Cudkowicz, M., and Kowall, N., 1990, Degeneration of pyramidal projection neurons in Huntington's disease cortex, *Ann. Neurol.* **27:**200–204.

Cummings, J. L. 1991, Pick's Disease, in: *Clinical Neurology* (M. Swash and J. Oxbury, eds.), Churchill-Livingstone, Edinburgh, U.K. pp. 131–139.

Cummings, J. L., and Benson, D. F., 1994, *Dementia: A Clinical Approach,* 2nd Ed., Butterworth, Boston,

Cummings, J. L., and Duchen, L. W., 1981, Kluver-Bucy syndrome in Pick's Disease: Clinical and pathological correlations, *Neurology* **31:**1415–1422.

Curcio, C. A., and Kemper, T. L., 1984, Nucleus raphe dorsalis in senile dementia: Neurofibrillary changes and neuronal cell packing density, *J. Neuropathol. Exp. Neurol.* **43:**359–368.
Daniel, S. E., de Bruin, V. M. S., and Lees, A. J., 1995, The clinical and pathological spectrum of Steele-Richardson-Olszewski syndrome (progressive supranuclear palsy): A reappraisal, *Brain* **118:**759–770.
de la Monte, S. M., 1989, Quantitation of cerebral atrophy in preclinical and end-stage Alzheimer's disease, *Ann. Neurol.* **25:**450–459.
de la Monte, S. M., Vonsattel, J.-P., and Richardson, E. P., 1988, Morphometric demonstration of atrophic changes in the cerebral cortex, white matter, and neostriatum in Huntington's disease, *J. Neuropathol. Exp. Neurol.* **47:**516–525.
de Leon, M. J., Convit, A., De Santi, S., Golomb, J., Tarshish, C., Rusinek, H., Bobinski, M., Ince, C., Miller, D. C., and Wisniewski, H. M., 1995, The hippocampus in aging and Alzheimer's disease, *Neuroimaging Clin. N. Am.* **5:**1–17.
de Olmos, J., 1990, Amygdala, in: *The Human Nervous System* (G. Paxinos, ed.), Academic, San Diego, pp. 583–755.
de Vos, R. A., Jansen, E. N., Stam, F. C., Ravid, R., and Swaab, D. F., 1995, "Lewy body disease": Clinico-pathological correlations in 18 consecutive cases of Parkinson's disease with and without dementia, *Clin. Neurol. Neurosci.* **97:**13–22.
de Yebenes, J. G., Sarasa, J. L., Daniel, S. E., and Lees, A. J., 1995, Familial progressive supranuclear palsy: Description of a pedigree and review of the literature, *Brain* **118:**1095–1103.
Delaère, P., Duyckaerts, C., Brion, J. P., Poulain, V., and Hauw, J.-J., 1989, Tau, paired helical filaments, and amyloid in the neocortex: A morphometric study of 15 cases with graded intellectual status in aging and senile dementia of Alzheimer type, *Acta Neuropathol.* **77:**645–653.
den Hartog Jager, W. A., and Bethlem, J., 1960, The distribution of Lewy bodies in the central and autonomic nervous system in idiopathic paralysis agitans, *J. Neurol. Neurosurg. Psychiatry* **23:**282–290.
Destee, A., Gray, F., Parent, M., Neuville, V., Muller, J. P., Verier, A., and Warot, P., 1990, Obsessive-compulsive behavior and progressive supranuclear palsy, *Rev. Neurol.* **146:**12–18.
Dickson, D. W., Yen, S.-H., and Houroupian, D. S., 1986, Pick-body like inclusions in the dentate fascia of the hippocampus in Alzheimer's disease, *Acta Neuropathol. (Berlin)* **71:**38–45.
Dickson, D. W., Ruan, D., Crystal, H., Mark, M. H., Davies, P., Kress, Y., and Yen, S.-H., 1991, Hippocampal degeneration differentiates diffuse Lewy body disease (DLBD) from Alzheimer's disease: Light and electron microscopic immunocytochemistry of CA2–3 neurites specific to DLBD, *Neurology* **41:**1402–1409.
Dom, R., Malfroid, M., and Baro, F., 1976, Neuropathology of Huntington's chorea: Studies of the ventrobasal complex of the thalamus, *Neurology* **26:**64–68.
Doucette, R., and Ball, M. J., 1987, Left-right symmetry of neuronal cell counts in the nucleus basalis of Meynert of control and of Alzheimer's disease brains, *Brain Res.* **422:**357–360.
Duyckaerts, C., Hauw, J.-J., Bastenaire, F., Piette, F., Poulain, C., Rainsard, V., Javoy-Agid, F., and Berthaux, P., 1986, Laminar distribution of neocortical senile plaques in senile dementia of the Alzheimer type, *Acta Neuropathol.* **70:**249–256.
Duyckaerts, C., Gaspar, P., Costa, C., Bonnet, A. M., and Hauw, J. J., 1993, Dementia in Parkinson's disease: Morphometric data, *Adv. Neurol.* **60:**447–55.
Ehringer, H., and Hornykiewicz, O., 1960, Verteilung von Noradrenalin und Dopamine (3-Hydroxy-tyramine) in gehirn des menschen und ihr verhalaten bei erkrankungen des extrapyramidales system, *Klin. Wschr.* **38:**1236–1239.
Emson, P. C., Arregui, A., Clement-Jones, V., Sandberg, B. E. B., and Rossor, M., 1980, Regional distribution of methionine-enkephalin and substance P-like immunoreactivity in normal human brain and in Huntington's disease, *Brain Res.* **199:**147–160.
Esiri, M. M., Hyman, B. T., Beyreuther, K., and Master, C. L., 1997, Ageing and dementia, in: *Greenfield's Neuropathology*, 6th Ed. (D. I. Graham and P. L. Lantos, eds.), Arnold, London, pp. 153–233.
Etienne, P., Robitaille, Y., Gauthier, S., and Nair, N. P. V., 1986, Nucleus basalis neuronal loss and neuritic plaques in advanced Alzheimer's disease, *Can. J. Physiol. Pharmacol.* **64:**318–324.
Falkai, P., and Bogerts, B., 1986, Cell loss in the hippocampus of schizophrenics, *Eur. Arch. Psychiatry Neurol. Sci.* **236:**154–161.
Falkai, P., Bogerts, B., and Rozumek, M., 1988, Limbic pathology in schizophrenia: The entorhinal region: A morphometric study, *Biol. Psychiatry* **24:**515–521.
Fama, R., Sullivan, E. V., Shear, P. K., Marsh, L., Yesavage, J. A., Tinklenberg, J. S., Lim, K. O., and

Pfefferbaum, A., 1997, Selective cortical hippocampal volume correlates of Mattis Dementia Rating Scale in Alzheimer's disease, *Arch. Neurol.* **54:**719–728.

Feany, M. B., and Dickson, D., 1996, Neurodegenerative disorders with extensive tau pathology: A comparative study and review, *Ann. Neurol.* **40:**139–148.

Fearnley, J., and Lees, A. J., 1991, Ageing and Parkinson's disease: Substantia nigra regional selectivity, *Brain* **114:**2283–2301.

Feifel, E., Brenner, M., Teiwes, R., Kucking, C. H., and Deuschl, G., 1994, Corticobasal degeneration: The significance of clinical criteria for establishing the diagnosis, *Nervenarzt* **65:**653–659.

Felleman, D. J., and Van Essen, D. C., 1991, Distributed hierachical processing in the primate cerebral cortex, *Cereb. Cortex* **1:**1–47.

Ferrante, R. J., Beal, M. F., Kowall, N. W., Richardson, E. P., and Martin, J. B., 1987, Sparing of acetylcholinesterase-containing striatal neurons in Huntington's disease, *Brain Res.* **411:**162–166.

Ferrer, I., Roig, C., Espino, A., Peiro, G., and Guiu, X. M., 1991, Dementia of frontal lobe type and motor neuron disease: A Golgi study of the frontal cortex, *J. Neurol. Neurosurg. Psychiatry* **54:**932–934.

Ferrer, I., Tuñón, T., Serrano, M. T., Casas, T., Alcántara, S., Zújar, M. J., and Rivera, R. M., 1993, Calbindin D-28k and parvalbumin immunoreactivity in the frontal cortex in patients with frontal lobe dementia of non-Alzheimer type associated with amyotrophic lateral sclerosis, *J. Neurol. Neurosurg. Psychiatry.* **56:**257–261.

Finali, G., Piccinin, G. L., Piccirilli, M., and Stefano, E., 1989, Treatment of progressive supranuclear palsy, *Clin. Ter.* **129:**339–351.

Forno, L. S., 1978, The locus coeruleus in Alzheimer's disease, *J. Neuropath. Exper. Neurol.* **37:**614–625.

Forno, L. S., 1986, The Lewy body in Parkinson's disease, *Adv. Neurol.* **45:**35–43.

Forno, L. S., 1996, Neuropathology of Parkinson's disease, *J. Neuropathol. Exp. Neurol.* **55:**259–272.

Forno, L. S., and Norville, R. L., 1976, Ultrastructure of Lewy bodies in the stellate ganglion, *Acta Neuropathol.* **34:**183–197.

Forstl, H., and Baldwin, B., 1994, Pick and focal brain atrophy, *Fortschr. Neurol. Psychiatry* **62:**345–355.

Förstl, H., Bruns, A., Luthert, P., Cairns, N., Lantos, P., and Levy, R., 1992, Clinical and neuropathological correlates of depression in Alzheimer's disease, *Psychol. Med.* **22:**877–884.

Fox, N. C., Warringtton, E. K., Freeborough, P. A., Hartikainen, P., Kennedy, A. M., Stevens, L. M., and Rossor, M. N., 1996, Presymptomatic hippocampal atrophy in Alzheimer's disease: A longitudinal MRI study, *Brain* **119:**2001–2007.

Francis, P. T., Holmes, C., Webster, M. T., Stratmann, G. C., Proctor, A. W., and Bowen, D. M., 1993, Preliminary neurochemical findings in non-Alzheimer dementia due to lobar atrophy, *Dementia* **4:**172–177.

Frasca, J., Blumbergs, P. C., Henschke, P., and Burns, R. J., 1991, A clinical and pathological study of progressive supranuclear palsy, *Clin. Exp. Neurol.* **28:**79–89.

Fujita, T., Doi, M., and Ogata, T., 1993, Cerebral cortical pathology of sporadic olivopontocerebellar atrophy, *J. Neurol. Sci.* **116:**41–46.

Funfgeld, E., 1925, Pathological anatomische Untersuchungen bei Dementia praecox mit besonderer Berücksichtigung des Thalamus opticus, *Zeitschr. Gesam. Neurol. Psychiatrie* **95:**411–463.

German, D. C., White, C. L., and Sparkman, D. R., 1987, Alzheimer's disease: Neurofibrillary tangles in nuclei that project to the cerebral cortex, *Neuroscience* **21:**305–312.

German, D. C., Manaya, K. F., White, C. L., Woodward, D. J., McIntyre, D. D., Smith, W. K., Kalaria, R. N., and Mann, D. M. A., 1992, Disease-specific patterns of locus coeruleus cell loss, *Ann. Neurol.* **32:**667–676.

Ghatak, N. R., Nochlin, D., and Hadfield, M. G., 1980, Neurofibrillary pathology in progressive supranuclear palsy, *Acta Neuropathol. (Berlin)* **52:**73–76.

Giannakopoulos, P., Hof, P. R., and Bouras, C., 1995, Dementia lacking distinctive histopathology: Clinicopathological evaluation of 32 cases, *Acta Neuropathol.* **89:**346–355.

Gibb, W. R. G., and Lees, A. J., 1991, Anatomy, pigmentation, central and dorsal subpopulations of the substantia nigra, and differential cell death in Parkinson's disease, *J. Neurol. Neurosurg. Psychiatry* **54:**388–396.

Gibb, W. R. G., Luthert, P. J., Janota, I., and Lantos, P. L., 1989a, Cortical Lewy body dementia: Clinical features and classification, *J. Neurol. Neurosurg. Psychiatry* **52:**185–192.

Gibb, W. R. G., Luthert, P. J., and Marsden, C. D., 1989b, Corticobasal degeneration, *Brain* **112:**1171–1192.

Giguerre, M., and Goldman-Rakic, P. S., 1988, Mediodorsal nucleus: Areal, laminar, and tangential distribution of afferents and efferents in the frontal lobe of rhesus monkeys, *J. Comp. Neurol.* **277**:195–213.

Glantz, L. A., and Lewis, D. A., 1994, Synaptophysin, and not rab 3A, is specifically reduced in the prefrontal cortex of schizophrenic subjects, *Soc. Neurosci. Abstr.* **20**:622.

Glickstein, M., May, J. G., III, and Mercier, B., 1985, Corticopontine projection in the macaque: The distribution of labelled cortical cells after large injections of horseradish peroxidase in the pontine nuclei, *J. Comp. Neurol.* **235**:343–359.

Goetz, C. G., Lutge, W., and Tanner, C. M., 1986, Autonomic dysfunction in Parkinson's disease, *Neurology* **36**:73–75.

Goldman-Rakic, P. S., and Selemon, L. D., Topography of corticostriatal projections in nonhuman primates and implications for functional parcellation of the neostriatum, in: *Cerebral Cortex,* Vol. 5 (E. G. Jones and A. Peters, Eds.), Plenum, New York, pp. 447–446.

Goldstein, D. S., 1993, Chemical mediators of autonomic nervous system activity. *Curr. Opin. Neurol. Neurosurg.* **6**;524–526.

Gomez-Isla, T., Hollister, R., West, H., Mui, S., Growdon, J. H., Petersen, R. C., Parisi, J. E., and Hyman, B. T., 1997, Neuronal loss correlates with, but exceeds, neurofibrillary tangles in Alzheimer's disease, *Ann. Neurol.* **41**:17–24.

Gomez-Isla, T., Price, J. L., McKeel, D. W., Jr., Morris, J. C., Growdon, J. H., and Hyman, B. T., 1996, Profound loss of layer II entorhinal cortex neurons occurs in very mild Alzheimer's disease, *J. Neurosci.* **16**:4491–4500.

Goodman, L., 1953, Alzheimer's disease; A clinicopathologic analysis of 23 cases with a theory on pathogenesis, *J. Nerv. Ment. Dis.* **118**:97–130.

Graff-Radford, N. R., Damasio, A. R., Hyman, B. T., Hart, M. N., Tranl, D., Damasio, H., Van Hoesen, G. W., and Rezai, K., 1990, Progressive aphasia in a patient with Pick's disease: A neuropsychological, radiologic, and anatomic study, *Neurology* **40**:620–626.

Grafman, J., Litvan, I., and Stark, M., 1995, Neuropsychological features of progressive supranuclear palsy, *Brain Cogn.* **28**:311–320.

Graham, J., and Oppenheimer, D., 1969, Orthostatic hypotension and nicotine sensitivity in a case of multiple system atrophy, *J. Neurol. Neurosurg. Psychiatry* **32**:28–34.

Grandas, F., and Esteban, A., 1994, Eyelid motor abnormalities in progressive supranuclear palsy, *J. Neural. Transm. (Suppl.)* **42**:33–41.

Graveland, G. A., Williams, R. S., and DiFiglia, M., 1985, Evidence for degenerative and regenerative changes in neostriatal spiney neurons in Huntington's disease, *Science* **227**:790–773.

Graybiel, A. M., and Ragsdale, C. W., Jr., 1980, Clumping of acetylcholinesterase activity in the developing striatum of the human fetus and young infant, *Proc. Natl. Acad. Sci. USA* **77**:1214–1218.

Gregory, C. A., and Hodges, J. R., 1996, Clinical features of frontal lobe dementia in comparison to Alzheimer's disease, *J. Neural. Transm. (Suppl.)* **47**:103–23.

Grossi, D., Lopez, O. L., and Martinez, A. J., 1989, Mammillary bodies in Alzheimer's disease, *Acta Neuropathol.* **80**:41–45.

Gustafson, L., 1987, Frontal lob degeneration of non-Alzheimer type. II. Clinical picture and differential diagnosis, *Arch. Gerontol. Psychiatry* **51**:353–361.

Haber, S. N., and Elde, R., 1981, Correlation between met-enkephalin and substance P immunoreactivity in the primate globus pallidus, *Neuroscience* **6**:1291–1297.

Haber, S. N., and Nauta, W. J. H., 1983, Ramifications of the globus pallidus in the rat as indicated by patterns of immunohistochemistry, *Neuroscience* **9**:243–260.

Halliday, G. M., McRitchie, D. A., Cartwright, H. R., Pamphlett, R., Hely, M. A., and Morris, J. G., 1996, Midbrain neuropathology in ideopathic Parkinson's disease and diffuse Lewy body disease, *J. Clin. Neurosci.* **3**:52–60.

Hansen, L., Salmon, D., Galasko, D., Masliah, E., Katzman, R., DeTeresa, R., Thal, L., Ray, M. M., Hofstetter, R., Klauber, M., Rice, V., Butters, N., and Alford, M., 1990, The Lewy body variant of Alzheimer's disease: A clinical and pathological entity, *Neurology* **40**:1–8.

Hardman, C. D., McRitchie, D. A., Halliday, G. M., Cartwright, H. R., and Morris, J. G., 1996, Substantia nigra pars reticulata neurons in Parkinson's disease, *Neurodegeneration* **5**:49–55.

Hardman, C. D., Halliday, G. M., McRitchie, D. A., and Morris, J. G., 1997, The subthalamic nucleus in Parkinson's disease and progressive supranuclear palsy, *J. Neuropathol. Exp. Neurol.* **56**:132–142.

Harper, J. W., and Heath, R. G., 1973, Anatomic connections of the fastigial nucleus to the rostral forebrain in the cat, *Exp. Neurol* **39**:285–292.

Hartmann-von Monakow, K., Akert, K., and Kunzle, H., 1979, Projection of precentral and premotor cortex to the red nucleus and other midbrain areas in *Macaca fascicularis, Exp. Brain Res.* 34:91–105.
Hassin, G. B., and Levitin, D., 1941, Pick's Disease: Clinicopathologic study and report of a case, *Arch. Neurol. Psychiatry* **45:**814–833.
Hauw, J. J., Verny, M., Delaere, P., Cervera, P., He, Y., and Duyckaerts, C., 1990, Constant neurofibrillary changes in the neocortex in progressive supranuclear palsy: Basic differences with Alzheimer's disease and aging, *Neurosci. Lett.* **119:**182–186.
Hauw, J. J., Daniel, S. E., Dickson, D., Horoupian, D. S., Jellinger, K., Lanto, P. L., McKee, A., Tabaton, M., and Litvan, I., 1994, Preliminary NINDS neuropathologic criteria for Steele-Richardson-Olszewski syndrome (progressive supranuclear palsy), *Neurology* **44:**2015–2019.
Haymaker, W., and Schiller, F., 1969, Cerebral landmarks: Organization of the cerebral cortex and descending and ascending pathways, in: *Bing's Local diagnosis in neurological disease, 5th Ed.* (W. Haymaker, ed.), V. V. Mosby, St. Louis, pp. 311–328.
Hechst, B., and Nussbaum, I., 1931, Beiträge zur Histopathologie der sympathischen Ganglien, *Archiv. Psychiat. Nervenkr.* **95:**556–583.
Hedreen, J. C., and Folstein, S. E., 1995, Early loss of neostriatal neurons in Huntington's disease, *J. Neuropathol. Exp. Neurol.* **54:**105–130.
Hedreen, J. C., Peyser, C. E., Folstein, S. E., and Ross, C. A., 1991, Neuronal loss in layers V and VI of cerebral cortex in Huntington's disease, *Neurosci. Lett.* **133:**257–61.
Heinsen, H., Strik, M., Bauer, M., Luther, K., Ulmer, G., Gangnus, D., Jungkunz, G., Eisenmenger, W., and Götz, M., 1994, Cortical and striatal neuronal number of Huntington's disease, *Acta Neuropathol.* **88:**320–333.
Heinsen, H., Rüb, U., Gangnus, D., Jungkunz, G., Bauer, M., Ulmer, G., Bethke, B., Schüler, M., Blöcker, F., Eisenmenger, W., Götz, M., and Strik, M., 1996, Nerve cell loss in the thalamic centromedian–parafascicular complex in patients with Huntington's disease, *Acta Neuropathol.* **91:**161–168.
Herzog, A. G., and Kemper, T. L., 1980, Amygdaloid changes in aging and dementia, *Arch. Neurol.* **37:**625–629.
Herzog, E., 1926, Beitrag zur normalen und pathologischen Histologie des Sympathicus, *Z. Neurol.* **103:**75.
Higuchi, Y., Iwaki, T., and Tateishi, J., 1995, Neurodegeneration in the limbic and paralimbic system in progressive supranuclear palsy, *Neuropathol. Appl. Neurobiol.* **21:**246–254.
Hirano, A., 1973, Progress in the pathology of motor neuron disease, in: *Progress in Neuropathology,* Vol. II (H. M. Zimmerman, ed.), Grune and Stratton, New York, pp. 181–215.
Hirsch, E. C., Graybiel, A. M., Duyckaerts, and Javoy, A. F., 1987, Neuronal loss in the pedunculopontine tegmental nucleus in Parkinson disease and in progressive supranuclear palsy, *Proc. Natl. Acad. Sci. USA* **84:**5976–5980.
Hodges, J. R., 1994, Pick's disease, in: *Dementia* (A. Burns and R. Levy, eds.), Chapman and Hall, London, pp. 739–752.
Hof, P. R., 1997, Morphology and neurochemical characteristics of the vulnerable neurons in brain aging and Alzheimer's disease, *Eur. Neurol.* **37:**71–81.
Hof, P. R., and Morrison, J. H., 1990, Quantitative analysis of a vulnerable subset of pyramidal neurons in Alzheimer's disease. II. Primary and secondary visual cortex, *J. Comp. Neurol.* **301:**55–64.
Hof, P. R., Cox, K., and Morrison, J. H., 1990, Quantitative analysis of a vunerable subset of pyramidal neurons in Alzheimer's disease. I. Superior frontal and inferior temporal cortex, *J. Comp. Neurol.* **301:**44–54.
Hof, P. R., Bierer, L. M., Perl, D. P., Delacourte, A., Buée, L., Bouras, C., and Morrison, J. H., 1992, Evidence for early vulnerability of the medial and inferior aspects of the temporal lobe in an 82-year-old patient with preclinical signs of dementia, *Arch. Neurol.* **49:**946–953.
Hof, P. R., Bouras, C., Perl, D. P., and Morrison, J. H., 1994, Quantitative neuropathologic analysis of Pick's disease cases: Cortical distribution of Pick's bodies and coexistence with Alzheimer's disease, *Acta Neuropathol.* **87:**115–124.
Hoogendijk, W. J., Pool, C. W., Troost, D., van Zwieten, E., and Swaab, D. F., 1995, Image analyzer-assisted morphometry of the locus coeruleus in Alzheimer's disease, Parkinson's disease, and amyotrophic lateral sclerosis, *Brain* **118:**131–143.
Horoupian, D., and Chu, P., 1994, Unusual case of corticobasal degeneration with tau/Gallyas-positive neuronal and glial tangles, *Acta Neuropathol. (Berlin)* **88:**592–598.
Horoupian, D. S., Thal, L., Katzman, R., Terry, R. D., Davies, P., Hirano, A., DeTeresa, R., Fuld, P. A.,

Petito, C., Blass, J., and Ellis, J. M., 1984, Dementia and motor neuron disease: Morphometric, biochemical, and Golgi study, *Ann. Neurol.* **16:**305–313.

Hornykiewicz, O., and Shannak, K., 1994, Brain monoamines in progressive supranuclear palsy: comparison with idopathic Parkinson's disease., *J. Neural Transm. (Suppl.)* **42:**219–227.

Hughes, A. J., Daniel, S. E., Blankson, S., and Lees, A. J., 1993, A clinicopathologic study of 100 cases of Parkinson's disease, *Arch. Neurol.* **50:**140–148.

Hulette, C. M., and Crain, B. J., 1992, Lobar atrophy without Pick's bodies, *Clin. Neuropathol.* **11:**151–156.

Huntington's Disease Collaborative Research Group, 1993, A novel gene containing a trinucleotide repeat that is expanded and unstable on Huntington's disease chromosome, *Cell* **7:**971–983.

Hyman, B. T., Van Hoesen, G. W., Demasio, A. R., and Barns, C. L., 1984, Alzheimer's disease: Cell-specific pathology isolates the hippocampal formation, *Science* **225:**1168–1170.

Hyman, B. T., Van Hoesen, G. W., Kromer, L. J., and Demasio, A. R., 1986, Perforant pathway changes and the memory impairment of Alzheimer's disease, *Ann. Neurol.* **20:**472–481.

Hyman, B. T., Van Hoesen, G. W., and Damasio, A. R., 1990, Memory-related neural systems in Alzheimer's disease: An anatomic study, *Neurology* **40:**1721–1730.

Ichimiya, Y., Arai, H., Kosaka, K., and Iizuka, R., 1986, Morphological and biochemical changes in the cholinergic and monoaminergic system in Alzheimer's disease, *Acta Neuropathol.* **70:**112–116.

Ikeda, K., Akiyama, H., and Haga, C., 1994, Argyrophilic thread-like structure in corticobasal degeneration and supranuclear palsy. *Neurosci. Lett.* **174:**157–159.

Iversen, L. L., Rossor, M. N., Reynolds, G. P., Hills, R., Roth, M., Mountjoy, C. Q., Foote, S. L., Morrison, J. H., and Bloom, F. E., 1983, Loss of pigmented dopamine-β-hydroxylase positive cells from locus coeruleus in senile dementia of Alzheimer's types, *Neurosci. Lett.* **39:**95–100.

Jackson, M., Gentleman, S., Lennox, G., Ward, L., Gray, T., Randell, K., Morrell, K., and Lowe, J., 1995, The cortical neuritic pathology of Huntington's disease, *Neuropathol. Appl. Neurobiol.* **21:**18–26.

Jackson, M., and Lowe, J., 1996, The new neuropathology of degenerative frontotemporal dementias, *Acta Neuropathol.* **91:**127–134.

Jacob, H., and Beckmann, H., 1994, Circumscribed malformation and nerve cell alterations in the entorhinal cortex of schizophrenics, *J. Neural. Transm.* **98:**83–106.

Jamada, M., and Mehraein, P., 1968, Verteilungsmuster der senilen Veranderungen im Gehirn: Die Beteiligung des limbischen Systems bei hirnatrophischen Prozessen des Senium und bei Morbus Alzheimer, *Arch. Psychiat. Neurol.* **211:**308–324.

Janowsky, J. S., Kaye, J. A., and Carper, R. A., 1996, Atrophy of the corpus callosum in Alzheimer's disease versus healthy aging, *J. Am. Geriat. Soc.* **44:**798–803.

Javoy-Agid, F., 1994, Cholinergic and peptidergic systems in PSP, *J. Neural. Trans. (Suppl.)* **42:**205–218.

Jellinger, K., 1988, The pedunculopontine nucleus in Parkinson's disease, progressive supranuclear palsy and Alzheimer's disease, *J. Neurol. Neurosurg. Psychiatry* **51:**540–543.

Jellinger, K., 1990, New developments in the pathology of Parkinson's disease, *Adv. Neurol.* **53:**1–16.

Jeste, D. V., Barban, L., and Parisi, J., 1984, Reduced Purkinje cell density in Huntington's disease, *Exp. Neurol.* **85:**78–86.

Jeste, D. V., and Lohr, J. B., 1989, Hippocampal pathologic findings in schizophrenia: A morphometric study, *Arch. Gen. Psychiatry* **46:**1019–1024.

Jones, E. G., 1984, Laminar distribution of cortical efferent cells, in: *Cerebral Cortex,* Vol. 1 (E. G. Jones and A. Peters, eds.), Plenum, New York, pp. 521–553.

Jones, E. G., 1985, *The Thalamus,* Plenum, New York, pp. 403–411.

Jones, E. G., and Wise, S. P., 1977, Size, laminar and columnar distribution of efferent cells in the sensory-motor cortex of monkeys, *J. Comp. Neurol.* **175:**391–438.

Jones, E. G., Coulter, J. D., Burton, H., and Porter, R., 1977, Cells of origin and terminal distribution of corticostriatal fibers arising in the sensory-motor cortex of monkeys, *J. Comp. Neurol.* **173:**53–80.

Joyce, J. N., Goldsmith, S. G., and Gurevich, E. V., 1997, Limbic circuits and monoamine receptors: Dissecting the effects of antipsychotics from disease processes, *J. Psychiat. Res.* **31:**197–217.

Kalra, S., Bergeron, C., and Lang, A. E., 1996, Lewy body disease and dementia: A review, *Arch. Int. Med.* **156:**487–493.

Kato, T., Hirano, A., Weinberg, M. N., and Jacobs, A. K., 1986, Spinal cord lesions in progressive supranuclear palsy: Some new observations, *Acta Neuropathol.* **71:**11–14.

Kato, S., and Nakamura, H., 1990, Cytoplasmic argyrophilic inclusions of pontine nuclei in patients with olivopontocerebellar atrophy: Immunochemical and ultrastructural studies, *Acta Neuropathol.* **79:**584–594.

Kato, S., Oda, M., Hayashi, H., Shimizu, T., Hayashi, M., Kawata, A., and Tanabe, H., 1995, Decrease of medullary catecholaminergic neurons in multiple system atrophy and Parkinson's disease and their preservation in amyotrophic lateral sclerosis, *J. Neurol. Sci.* **132:**216–21.

Kemper, T., 1978, Senile dementia: A focal disease in the temporal lobe, in: *Senile Dementia; A Biomedical Approach* (K. Nandy, ed.), Elsevier, New York, pp. 105–113.

Kemper, T., 1994, Neuroanatomical and neuropathological changes in normal aging and dementia, in: *Clinical Neurology of Aging,* 2nd Ed. (M. L. Albert and J. E. Knoeffel, eds.), Oxford University Press, New York.

Kempster, P. A., Gibb, W.R.G., Stern, G. M., and Lees, A. J., 1989, Asymmetry of substantia nigra neuronal loss in Parkinson's disease and its relevance to the mechanism of levodopa related motor fluctuations, *J. Neurol. Neurosurg. Psychiatry* **52:**72–76.

Kennedy, P. G., and Duchen, L. W., 1985, A quantitative study of intermediolateral cell column cells in motor neuron disease and the Shy–Drager syndrome, *J. Neurol. Neurosurg. Psychiatry* **48:**1103–1106.

Kertesz, A., 1997, Frontotemporal dementia, Pick's disease, and corticobasal degeneration. *Arch. Neurol.* **54:**1427–1429.

Keshavan, M. S., Anderson, S., and Pettegrew, J. W., 1994, Is schizophrenia due to excessive synaptic pruning in the prefrontal cortex? The Feinberg hypothesis revisited, *J. Psychiat. Res.* **28:**239–264.

Killiany, R. J., Moss, M. B., Albert, M. S., Sandor, T., Tieman, J., and Jolesz, F., 1993, Temporal lobe regions on magnetic resonance imaging identify patients with early Alzheimer's disease, *Arch. Neurol.* **50:**949–954.

Kim, H., Gearing, M., and Mirra, S. S., 1995, Ubiquitin-positive CA2–3 neurites in hippocampus coexist with cortical Lewy bodies, *Neurology* **45:**1768–1770.

Kish, S. J., Chang, L. J., and Mirchandani, L., 1985, Progressive supranuclear palsy: Relationship between extrapyramidal disturbances, dementia, and brain neurotransmitter markers, *Ann. Neurol.* **18:**530–536.

Knopman, D. S., 1993, Overview of dementia lacking distinctive histology: Pathological designation of a progressive dementia, *Dementia* **4:**132–136.

Knopman, D. S., Mastri, A. R., Frey, W., H., Sung, J. H., and Rustan, B. A., 1990, Dementia lacking distinctive histological features: A common non-Alzheimer degenerative dementia, *Neurology* **40:**251–256.

Koeppen, A. H., 1989, The nucleus pontis centralis caudalis in Huntington's disease, *J. Neurol. Sci.* **91:**129–141.

Kosaka, K., 1993, Dementia and neuropathology in Lewy body disease, *Adv. Neurol.* **60:**456–463.

Kosaka, K., Yoshimura, M., Ikeda, K., and Budka, H., 1984, Diffuse type of Lewy body disease: Progressive dementia with abundant cortical Lewy bodies and senile changes of varying degree: A new disease? *Clin. Neuropathol.* **3:**185–192.

Kosaka, K., Matsushita, M., Oyanagi, S., and Mehraein, P., 1990, A clinicopathological study of the "Lewy body diseases," *Psychiat. Neurol. Jpn.* **82:**292–311.

Kowall, N. W., Ferrante, R. J., and Martin, J. B., 1987, Patterns of cell loss in Huntington's disease, *Trends Neurosci.* **10:**24–29.

Kremer, B., Goldberg, P., Andrew, S. E., Theilmann, J., Telenius, H., Zeisler, J., Squitieri, F., Lin, B., Bassett, A., Almquist E., Bird, T. D., and Hayden, M. R., 1994, A worldwide study of the Huntington's disease mutation, *New Engl. J. Med.* **330:**1401–1406.

Ksiezak-Reding, H., Morgan, K., and Mattiace, L., 1994, Ultrastructure and biochemical composition of paired helical filaments in corticobasal degeneration, *Am. J. Pathol.* **145:**1496–1508.

Kuhl, D. E., Phelps, M. E., Markham, C. H., Metter, E. J., Riege, W. H., and Winter, J., 1982, Cerebral metabolism and atrophy in Huntington's disease determined by $^{18}$FDG and computed tomographic scan, *Ann. Neurol.* **12:**425–434.

Kunzle, H., 1975, Bilateral projections from precentral motor cortex to the putamen and other parts of the basal ganglia: An autoradiographic study in *Macaca fascicularis, Brain Res.* **88:**195–209.

Kuypers, H. G. J. M., and Lawrence, D. G., 1967, Cortical projections to the red nucleus and the brain stem in the rhesus monkey, *Brain Res.* **4:**151–188.

Kuzuhara, S., and Yoshimura, M., 1993, Clinical and neuropathological aspects of diffuse Lewy body disease in the elderly, *Adv. Neurol.* **60:**464–469.

Laakso, M. P., Partanen, K., Riekkinen, P., Lehtovirta, M., Helkala, E. L., Hallikainen, M., Hanninen, T., and Vainio, P., 1996, Hippocampal volumes in Alzheimer's disease, Parkinson's disease with and without dementia, and in vascular dementia: An MRI study, *Neurology* **46:**678–681.

Lange, H., Thörner, G., Hof, A., and Schröder, K. F., 1976, Morphometric studies of the neuropathological changes in choreatic diseases, *J. Neurol. Sci.* **28**:401–425.

Langston, J. W., 1995, MPTP as it relates to the etiology of Parkinson's disease: Neuropathological correlations, in: *Etiology of Parkinson's Disease* (J. H. Ellenberg, W. H. Koller, and J. W. Langston, eds.), Marcel Dekker, New York, pp. 3367–3399.

Lantos, P. L., 1994, The neuropathology of progressive supranuclear palsy, *J. Neural Transm. (Suppl.)* **42**:137–152.

Lassmann, H., 1996, Patterns of synaptic and nerve cell pathology in Alzheimer's disease, *Behav. Brain Res.* **78**:9–14.

Leenders, K. L., Frackowiak, R. S., and Lees, A. J., 1988, Steele-Richardson-Olszewski syndrome. Brain energy metabolism, blood flow, and fluorodopa uptake measured by positron emission tomography, *Brain* **11**:615–630.

Lehéricy, S., Hirsch, E. C., Cervera, P., Hersh, L. B., Hauw, J.-J., Ruberg, M., and Agid, Y., 1989, Selective loss of cholinergic neurons in the ventral striatum of patients with Alzheimer's disease, *Proc. Natl. Acad. Sci. USA* **86**:8580–8584.

Lehéricy, S., Hirsch, E. C., Cervera-Piérot, P., Hersh, L. B., Bakchine, S., Piette, F., Duyckaerts, C., Hauw, J.-J., Javoy-Agid, F., and Agid, Y., 1993, Heterogeneity and selectivity of the degeneration of cholinergic neurons in the basal forebrain of patients with Alzheimer's disease, *J. Comp. Neurol.* **330**:15–31.

Le May, M., Stafford, J. L., Sandor, T., Albert, M., Haykal, H., and Zamani, A., 1986, Statistical assessment of perceptual CT scan ratings in patients with Alzheimer type dementia, *J. Comput. Assis. Tomog.* **10**:802–809.

Lennox, G., Lowe, J., Landon, M., Byrne, E. J., Mayer, R. J., and Godwin-Austen, R. B., 1989, Diffuse Lewy body disease: Correlative neuropathology using anti-ubiquitin immunocytochemistry, *J. Neurol. Neurosurg. Psychiatry* **52**:1236–1247.

Lesch, A., and Bogerts, B., 1984, The diencephalon and schizophrenia: Evidence for reduced thickness of the periventricular grey matter, *Eur. Arch. Psychiatry Neurol. Sci.* **234**:212–219.

Lewis, D. A., Campbell, M. J., Terry, R. D., and Morrison, J. H., 1987, Laminar and regional distribution of neurofibrillary tangles and neuritic plaques in Alzheimer's disease: A quantitative study of visual and auditory cortices, *J. Neurosci.* **7**:1799–1808.

Lewy, F. H., 1912, Paralysis agitans. I. Pathologische Anatomie, in: *Handbuck der Neurologie,* Vol. III, part 2 (M. Lewandowsky, ed.), Springer, Berlin, pp. 920–933.

Lippa, C. F., Smith, T. W., and Fontineau, N., 1990, Corticonigral degeneration with neuronal achromasia: A clinicopathologic study of two cases, *J. Neurol. Sci.* **98**:301–310.

Lippa, C. F., Smith, T. W., and Swearer, J. M., 1994, Alzheimer's disease and Lewy body disease: A comparative clinicopathological study, *Ann. Neurol.* **35**:81–88.

Litvan, I., 1994a, Cognitive disturbances in progressive supranuclear palsy, *J. Neural Transm. (Suppl.)* **42**:69–78.

Litvan, I., 1994b, Cholinergic approaches to the treatment of progressive supranuclear palsy, *J. Neural Trans.* **42**:275–281.

Lowe, J., Lennox, G., and Leigh, P. N., 1997, Disorders of movement and system degeneration, in: *Greenfield's Neuropathology,* Vol. 2, 6th ed. (D. I. Graham and P. L. Lantos, eds.), Arnold, London, pp. 281–366.

Ma, S. Y., Rinne, J. O., Collan, Y., Roytta, M., and Rinne, U. K., 1996, A quantitative morphometrical study of neuron degeneration in the substantia nigra in Parkinson's disease, *J. Neurol Sci.* **140**:40–45.

Maher, E. R., and Lees, A. J., 1986, The clinical features and natural history of the Steele-Richardson-Olzewski syndrome (progressive supranuclear palsy), *Neurology* **36**:1005–1008.

Malessa, S., Hirsch, E. C., Cervera, P., Javoy-Agid, F., Duyckaerts, C., Hauw, J. J., and Agid, Y., 1991, Progressive supranuclear palsy: Loss of choline acetyltransferase-like immunoreactive neurons in the pontine reticular formation, *Neurology* **41**:1593–1597.

Mann, D. M. A., 1989, Subcortical afferents projection systems in Huntington's chorea, *Acta Neuropathol* **78**:551–554.

Mann, D. M. A., 1996, Pyramidal nerve cell loss in Alzheimer's disease, *Neurodegeneration* **5**:423–427.

Mann, D. M. A., and South, P. W., 1993, The topographic distribution of brain atrophy in frontal type dementia, *Acta Neuropathol.* **85**:334–340.

Mann, D. M. A., Yates, P. O., and Hawkes, J., 1983, The pathology of the human locus ceruleus, *Clin. Neuropathol.* **2**:1–7.

Mann, D. M. A., Yates, P. O., and Marcyniuk, B., 1984a, Monoaminergic neurotransmitter systems in presenile Alzheimer's disease and in senile dementia of Alzheimer type, *Clin. Neuropathol.* **3:**199–205.

Mann, D. M. A., Yates, P. O., and Marcyniuk, B., 1984b, Alzheimer's presenile dementia, senile dementia of Alzheimer type, and Down's syndrome in middle age from an age related continuum of pathological changes, *Neuropathol. Appl. Neurobiol.* **10:**185–207.

Mann, D. M. A., Yates, P. O., and Marcyniuk, B., 1984c, Changes in nerve cells of the nucleus basalis of Meynert in Alzheimer's disease and the relationship to ageing and to the accumulation of lipofuscin pigment, *Mech. Ageing Dev.* **25:**189–205.

Mann, D. M. A., Yates, P. O., and Marcyniuk, B., 1985, Correlation between senile plaque and neurofibrillary tangle counts in the cerebral cortex and neuronal counts in the cortex and subcortical structures in Alzheimer's disease, *Neurosci. Lett.* **56:**51–55.

Mann, D. M. A., Yates, P. O., and Marcyniuk, B., 1987, Dopaminergic neurotransmitter systems in Alzheimer's disease and in Down's syndrome at middle age, *J. Neurol. Neurosurg. Psychiatry* **50:**341–344.

Mann, D. M. A., Oliver, R., and Snowden, J. S., 1993b, The topographic distribution of brain atrophy in Huntington's disease and progressive supranuclear palsy, *Acta Neuropathol.* **85:**553–559.

Marcyniuk, B., Mann, D. M. A., and Yates, P. O., 1986, The topography of cell loss from locus coeruleus in Alzheimer's disease, *J. Neurol. Sci.* **76:**335–345.

Marin-Padilla, M., 1978, Dual origin of the mammalian neocortex and evolution of the cortical plate, *Anat. Embryol.* **152:**109–126.

Martin, P., and Albers, M., 1995, Cerebellum and schizophrenia: A selective review, *Schizophr. Bull.* **21:**241–250.

Martin, J. B., and Gusella, J. F., 1986, Huntington's disease pathogenesis and management, *New Engl. J. Med.* **315:**1267–1276.

Masse, G., Mikol, K. J., and Brion, S., 1981, A typical presenile dementia: Report of an anatomo-clinical case and review of the literature, *J. Neurol. Sci.* **52:**245–267.

Masucci, E. F., and Kurtzke, J. F., 1989, Tremor in progressive supranuclear palsy, *Acta Neurol. Scand.* **80:**296–300.

Matzuk, M. M., and Saper, C. B., 1985, Preservation of hypothalamic dopaminergic neurons in Parkinson's disease, *Ann. Neurol.* **18:**552–555.

Mayeux, R., Denaro, J., Hemenegildo, N., Marder, K., Tang, M.-X., Cote, L. J., and Stern, Y., 1992, A population-based investigation of Parkinson's disease with and without dementia, *Arch. Neurol.* **49:**492–497.

McDuff, T., and Sumi, S. M., 1985, Subcortical degeneration in Alzheimer's disease, *Neurology* **35:**123–126.

McGeer, P. L., McGeer, E. G., Suzuki, J., Doleman, C. E., and Nagai, T., 1984, Aging, Alzheimer's disease, and the cholinergic system of the basal forebrain, *Neurology* **34:**741–745.

McKeith, L. G., Galasko, D., Kosaka, K., Perry, E. K., Dickson, D. W., Hansen, L. A., Salmon, D. P., Lowe, J., Mirra, S. S., Byrne, E. J., Lennox, G., Quinn, N. P., Edwardson, J. A., Ince, P. G., Bergeron, C., Bruns, A., Miller, B. L., Lovestone, S., Collerton, D., Jansen, E. N., Ballard, C., de Vos, R. A., Wilcock, G. K., Jellinger, K. A., and Perry, R. H., 1996, Consensus guidelines for the clinical and pathologic diagnosis of dementia with Lewy bodies (DLB): Report of the consortium on DLB international workshop, *Neurology* **47:**1113–1124.

McLardy, T., 1950, Thalamic projection to frontal cortex in man, *J. Neurol. Neurosurg. Psychiatry* **13:**198–202.

Meier-Ruge, W., Iwangoff, P., and Reichlmeier, K., 1984, Neurochemical enzyme changes in Alzeimer's and Pick's disease, *Arch. Gerontol. Geriat.* **3:**161–165.

Mendez, M. F., and Zander, B. A., 1991, Dementia presenting with aphasia: Clinical characteristics, *J. Neurol. Neurosurg. Psychiatry* **54:**542–545.

Mendez, M. F., Selwood, A., Mastri, A. R., and Frey, W. H., 1993, Pick's disease versus Alzeimer's disease: A comparison of clinical characteristics, *Neurology* **43:**289–292.

Meyers, R. H., Vonsattel, J.-P., Stevens, T. J., Cupples, L. A., Richardson, E. P., and Bird, E. D., 1988, Clinical and neuropathological assessment of severity in Huntington's disease, *Neurology* **38:**341–347.

Miller, A. K. H., Alston, R. L., and Corsellis, J.A.N., 1980, Variations with age in the volumes of grey and white matter in the cerebral hemisphere of man: Measurements with an image analyser, *Neuropathol. Appl. Neurobiol.* **6:**119–132.

tial loss of striatal projection neurons in Huntington disease, *Proc. Natl. Acad. Sci, USA* **85:**5733–5737.

Rickmann, M., Chronwall, B. M., and Wolff, J. R., 1977, On the development of nonpyramidal neurons and axons outside the cortical plate: The early marginal zone as a pallial anlage, *Anat. Embryol.* **151:**285–307.

Riley, D., Lang, A., and Lewis, A., 1990, Cortical-basal ganglionic degeneration, *Neurology* **40:**1203–1212.

Rinne, J. O., Paljärvi, L., and Rinne, U. K., 1987, Neuronal size and density in the nucleus basalis of Meynert in Alzheimer's disease, *J. Neurol. Sci.* **79:**67–76.

Rinne, J. O., Lee, M. S., Thompson, P. D., and Marsden, C. D., 1994, Corticobasal degeneration: A clinical study of 36 cases, *Brain* **117:**1183–1196.

Robbins, T. W., James, M., and Lange, K. W., 1992, Cognitive performance in multiple system atrophy, *Brain* **1:**271–291.

Robertson, E. E., le Roux, A., and Brown, J. H., 1958, The clinical differentiation of Pick's disease, *J. Ment. Sci.* **104:**1000–1024.

Rockland, K. S., and Pandya, D. N., 1979, Laminar origins and terminations of cortical connections of the occipital lobe in the rhesus monkey, *Brain Res.* **179:**3–20.

Rogers, J. D., Brogan, D., and Mirra, S. S., 1985, The nucleus basalis of Meynert in neurological disease: A quantitative morphological study, *Ann. Neurol.* **17:**170–173.

Ruberg, M., Javoy, A. F., and Hirsch, E., 1985, Dopaminergic and cholinergic lesions in progressive supranuclear palsy, *Ann. Neurol.* **18:**523–529.

Salmon, D. P., Galasko, D., Hansen, L. A., Masliah, E., Butters, N., Thal, L. J., and Katzman, R., 1996, Neuropsychological deficits associated with diffuse Lewy body disease, *Brain Cogn.* **31:**148–165.

Sampson, V. L., Morrison, J. H., and Vickers, J. C., 1997, The cellular basis for the relative resistance of parvalbumin and calretinin immunoreactive neocortical neurons to the pathology of Alzheimer's disease, *Exp. Neurol.* **145:**295–302.

Samuel, W., Terry, R. D., DeTeresa, R., Butters, N., and Masliah, E., 1994, Clinical correlates of cortical and nucleus basalis pathology in Alzheimer dementia, *Arch. Neurol.* **51:**772–778.

Samuel, W., Alford, M., Hofstetter, C. R., and Hansen, L., 1997, Dementia with Lewy bodies versus pure Alzheimer's disease: Differences in cognition, neuropathology, cholinergic dysfunction, and synapse density, *J. Neuropathol. Exp. Neurol.* **56:**299–508.

Sandor, T., Albert, M., Stafford, J., and Kemper, T. L., 1990, Symmetrical and asymmetrical changes in brain tissue with age as measured on CT scans, *Neurobiol. Aging* **11:**21–28.

Sanides, F., 1970, Functional architecture of motor and sensory cortices in primates in the light of a new concept of neocortex evolution, in: *The Primate Brain: Advances in Primatology,* Vol. 1 (C. R. Hoback and W. Montagna, eds.), Appleton-Century Crofts, New York, pp. 137–208.

Sanides, F., 1972, Representation in the cerebral cortex and its laminar and areal lamination pattern, in: *Structure and Function of the Nervous System,* Vol. 5 (G. F. Bourne, ed.), Academic, New York, pp. 329–453.

Saper, C. B., 1985, Organization of cerebral afferent systems in the rat. II. Hypothalamocortical projections, *J. Comp. Neurol.* **237:**21–46.

Saper, C. B., German, D. C., and White, C. L., 1985, Neuronal pathology in the nucleus basalis and associated cell groups in senile dementia of the Alzheimer type: Possible role in cell loss, *Neurology* **35:**1089–1095.

Savoiardo, M., Girotti, F., Strada, L., and Cieri, E., 1994, Magnetic resonance imaging in progressive supranuclear palsy and other parkinsonian disorders, *J. Neural Transm. (Suppl)* **42:**93–110.

Sawle, G. V., Brooks, D. J., and Marsden, C. D., 1991, Corticobasal degeneration: A unique pattern of regional cortical oxygen hypometabolism and striatal fluorodopa uptake demonstrated by positron emission tomography, *Brain* **114:**541–556.

Scheff, S. W., Sparks, D. L., and Price, D. A., 1996, Quantitative assessment of synaptic density in the outer molecular layer of the hippocampal dentate gyrus in Alzheimer's disease, *Dementia* **7:**226–232.

Schmahmann, J. D., 1991, An emerging concept: The cerebellar contribution to higher function, *Arch. Neurol.* **48:**1178–1187.

Schmahmann, J. D., 1997, The cerebellum and cognition, in: *International Review of Neurobiology, Vol. 41* (J. D. Schmahmann, ed.), Academic, San Diego.

Schmahmann, J. D., and Pandya, D. N., 1990, Anatomical investigations of projections from thalamus to posterior parietal cortex in the rhesus monkey: A WGA-HRP and fluorescent tracer study, *J. Comp. Neurol.* **295:**299–326.

Schmahmann, J. D., and Pandya, D. N., 1991, An emerging concept: The cerebellar contribution to higher function, *Arch. Neurol.* **48:**1178–1187.

Schmidt, M. L., Martin, J. A., Lee, V. M., and Trojanowski, J. Q., 1996, Convergence of Lewy bodies and neurofibrillary tangles in amygdala neurons of Alzheimer's disease and Lewy body disorders, *Acta Neuropathol.* **91:**475–481.

Seitelberger, F., 1968, Neuropathology of senile dementia, *Acta Neuropathol.* **4:**109–118.

Selden, N., Geula, C., Hersh, L., and Mesulam, M. M., 1994, Human striatum: Chemoarchitecture of the caudate nucleus, putamen, and ventral striatum in health and Alzheimer's disease, *Neuroscience* **60:**621–636.

Shenton, M. E., Kikinis, R., Jolesz, F. A., Pollack, S. D., LeMay, M., Wible, C. G., Hokama, H., Martin, J., Metcalf, D., Coleman, M., and McCarley, R. W., 1992, Abnormalities of the left temporal lobe and thought disorder in schizophrenia, *N. Engl. J. Med.* **327:**604–612.

Simic, G., Kostovic, I., Winblad, B., and Bogdanovic, N., 1997, Volume and number of neurons of the human hippocampal formation in normal aging and Alzheimer's disease, *J. Comp. Neurol.* **379:**482–494.

Snider, R. S., Maiti, A., and Snider, S. R., 1976, Cerebellar pathways to ventral midbrain and nigra, *Exp. Neurol.* **53:**714–728.

Snowden, J. S., Neary, D., Mann, D.M.A., Goulding, P. J., and Testa, H. J., 1992, Progressive language disorder due to lobar atrophy, *Ann. Neurol.* **31:**174–183.

Sotrel, A., Paskevich, P., Kiely, D. K., Bird, E. D., Williams, R. S., and Meyers, R. H., 1991, Morphometric analysis of the prefrontal cortex in Huntington's disease, *Neurology* **41:**1117–1123.

Spargo, E., Everall, I. P., and Lantos, P. L., 1993, Neuronal loss in hippocampus in Huntington's disease: A comparison with HIV infection, *J. Neurol. Neurosurg. Psychiatry* **56:**487–491.

Sparks, D. L., and Markesbery, W. R., 1991, Altered serotonergic and cholinergic synaptic markers in Pick's disease, *Arch. Neurol.* **48:**796–799.

Steele, J., Rishardson, J., and Olszewski, J., 1964, Progressive supranuclear palsy: A heterogeneous degeneration involving the brain stem, basal ganglia, and cerebellum with vertical-gaze and pseudobulbar palsy, nuchal dystonia, and dementia, *Arch. Neurol.* **2:**473–486.

Stern, K., 1939, Severe dementia associated with bilateral symmetrical degeneration of the thalamus, *Brain* **62:**157–171.

Stern, K., and Reed, G. E., 1945, Presenile dementia (Alzheimer's disease): Its pathogenesis and classification, *Am. J. Psychiatry* **102:**191–197.

Suddath, R. L., Casanova, M. F., Goldberg, T. E., Daniel, D. G., Kelsoe, J. R., and Weinberger, D. R., 1989, Temporal lobe pathology in schizophrenia: A quantitative magnetic resonance imaging study, *Am. J. Psychiatry* **146:**464–472.

Suenaga, T., Hirano, A., Llena, J. F., Yen, S.-H., and Dickson, D. W., 1990, Modified Bielschowsky stain and immunohistochemical studies on striatal plaques in Alzheimer's disease, *Acta Neuropathol.* **80:**280–286.

Tabaton, M., Schenone, A., Romagnoli, P., and Mancardi, G. L., 1985, A quantitative and ultrastructural study of substantia nigra and nucleus centralis superior in Alzheimer's disease, *Acta Neuropathol.* **68:**218–223.

Tagliavini, F., and Pilleri, G., 1983, Basal nucleus of Meynert: A neuropathological study in Alzheimer's disease, simple senile dementia, Pick's disease, and Huntington's chorea, *J. Neurol. Sci.* **62:**243–260.

Tagliavini, F., Pilleri, G., Bouras, C., and Constantinidis, J., 1984, The basal nucleus of Meynert in patients with progressive supranuclear palsy, *Neurosci. Lett.* **44:**37–42.

Takahashi, H., Takeda, S., Ikuta, F., and Homma, Y., 1987, Progressive supranuclear palsy with limbic system involvement: A report of a case with ultrastructural investigation of neurofibrillary tangles in various locations, *Clin. Neuropathol.* **6:**271–276.

Tellez-Nagel, I., and Wisniewski, H. M., 1973, Ultrastructure of neurofibrillary tangles in Steele-Richardson-Olszewski syndrome, *Arch. Neurol.* **29:**324–327.

Terry, R. D., Masliah, E., Salmon, D. P., Butters, N., DeTeresa, R., Hill, R., Hansen, L. A., and Katzman, R., 1991, Physical basis of cognitive alterations in Alzheimer's disease: Synapse loss is the major correlate of cognitive impairment, *Ann. Neurol.* **30:**572–580.

Thompson, P. D., and Marsden, C. D., 1992, Corticobasal degeneration, *Ballieres Clin. Neurol.* **1:**677–686.

Tissot, R., Constantinidis, J., and Richard, J., 1975, *La Maladie de Pick,* Masson et Cie, Paris.

Tissot, R., Constantinidis, J., and Richard, J., 1985, Pick's disease, in: *Handbook of Clinical Neurology,* Vol. 2 (J.A.M. Fredericks, ed.), Elsevier, Amsterdam, pp. 233–246.

Tomlinson, B. E., Blessed, G., and Roth, M., 1970, Observations on the brain of demented old people, *J. Neurol. Sci.* **11:**205–242.

Tomlinson, B. E., Irving, D., and Blessed, G., 1981, Cells loss in the locus coeruleus in senile dementia of Alzheimer's type, *J. Neurol. Sci.* **49:**419–428.

Tomonaga, M., 1979, On the morphological changes in locus ceruleus in the senile human brain, *Jpn. J. Geriat.* **16:**545–550.

Treff, W. M., and Hempel, K. J., 1958, Die Zellidichte bei Schizophrenen and klinisch Gesunden, *J. Hirnforsch.* **4:**314–369.

Troncoso, J. C., Sukhov, R. R., Kawas, C. H., and Koliatsos, V. E., 1996, *In situ* labeling of dying cortical neurons in normal aging and in Alzheimer's disease: Correlations with senile plaques and disease progression, *J. Neuropathol. Exp. Neurol.* **55:**1134–1142.

Uchihara, T., Kondo, H., Kosaka, K., and Tsukagoshi, H., 1992, Selective loss of nigral neurons in Alzheimer's disease: A morphometric study, *Acta Neuropathol.* **83:**271–276.

Uhl, G. R., Hilt, D. C., and Hedreen, J. C., 1983, Pick's disease (lobar sclerosis): Depletion of neurons in the nucleus basalis of Meynert, *Neurology* **33:**1470–1473.

Van Buren, R., and Borke, R. C., 1972, Variations and connections of the human thalamus, Springer-Verlag, New York.

van Domburg, P. H. M. F., and ten Donkelaar, H. J., 1991, The human substantia nigra and ventral tegmental area: A neuroanatomical study with notes on aging and aging diseases, *Adv. Anat. Embryol. Cell Biol.* **121:**1–132.

Van Hoesen, G. W., 1981, The differential distribution, diversity, and sprouting of cortical projections to the amygdala in the rhesus monkey, in: *The Amygdaloid Complex* (Y. Ben-Ari, ed.), Elsevier, Amsterdam, pp. 77–90.

Van Hoesen, G. W., Hyman, B. T., and Damasio, A. R., 1991, Entorhinal cortex pathology in Alzheimer's disease, *Hippocampus* **1:**1–8.

Veazy, R. B., Amaral, D. G., and Cowan, W. M., 1982, The morphology and connections of the posterior hypothalamus in the cynomolgus monkey (*Macaca fascicularis*). II. Efferent connections, *J. Comp. Neurol.* **207:**135–156.

Vereecken, T. H., Vogels, O. J., and Nieuwenhuys, R., 1994, Neuron loss and shrinkage in the amygdala in Alzheimer's disease, *Neurobiol. Aging* **15:**45–54.

Verity, M. A., and Wechsler, A. F., 1987, Progressive subcortical gliosis of Neumann: A clinicopathologic study of two cases with review, *Arch. Gerontol. Geriat.* **6:**245–261.

Vermersch, P., Delacourte, A., Javoy-Agid, F., Hauw, J. J., and Agid, Y., 1993, Dementia in Parkinson's disease: Biochemical evidence for cortical involvement using the immunodetection of abnormal tau proteins, *Ann. Neurol.* **33:**445–450.

Vermersch, P., Scheltens, P., Barkhof, F., Steinling, M., and Leys, D., 1994, Evidence for atrophy of the corpus callosum in Alzheimer's disease, *Eur. Neurol.* **34:**83–86.

Vincent, S. L., Khan, Y., and Benes, F. M., 1993, Cellular distribution of dopamine $D_1$ and $D_2$ receptors in rat medial prefrontal cortex, *J. Neurosci.* **13:**2551–2561.

Vincent, S. L., Khan, Y., and Benes, F. M., 1995, Cellular colocalization of dopamine $D_1$ and $D_2$ receptors in rat medial prefrontal cortex, *Synapse* **17:**26–35.

Vogels, O. J. M., Broere, C. A. J., and Nieuwenhuys, R., 1990, Neuronal hypertrophy in the human supraoptic and paraventricular nucleus in aging and Alzheimer's disease, *Neurosci. Lett.* **109:**62–67.

Vonsattel, J.-P., Meyers, R. H., Stevens, T. J., Ferrante, R. J., Bird, E. D., and Richardson, E. P., 1985, Neuropathological classification of Huntington's disease. *J. Neuropathol. Exp. Neurol.* **44:**559–577.

Wakabayashi, K., and Takahashi, H., 1996, Neuropathology of corticobasal degeneration, *No. To. Shinkei.* **48:**505–512.

Wakabayashi, K., Oyanagi, K., and Makifuchi, T., 1994, Corticobasal degeneration: Etiopathological significance of the cytoskeletal alterations, *Acta Neuropathol.* **87:**545–553.

Weinberger, D. R., Kleinman, J. E., Luchins, D. J., Bigelow, L. B., and Wyatt, R. J., 1980, Cerebellar pathology in schizophrenia: A controlled postmortem study, *Am. J. Psychiatry* **137:**359–361.

West, M. J., Coleman, P. D., Flood, D. G., and Troncoso, J. C., 1994, Differences in the pattern of hippocampal neuronal loss in normal ageing and Alzheimer's disease, *Lancet* **344:**769–772.

Whitehouse, P. J., Hedreen, J. C., White, C. L., and Price, D. L., 1983, Basal forebrain neurons in the dementia of Parkinson disease, *Ann. Neurol.* **13:**243–248.

Whitehouse, P. J., Price, D. L., Struble, R. G., Clark, A. W., Coyle, J. T., and DeLong, M. R., 1982, Alzheimer's disease and senile dementia: Loss of neurons in basal forebrain, *Science* **215:**1237–1239.

Wiesendanger, R., and Weisendanger, M., 1985, The thalamic connections with medial area 6 (supplementary motor cortex) in the monkey (*Macaca fascicularis*), *Exp. Brain Res.* **59:**91–104.

Wilcock, G. K., Esiri, M. M., Bowen, D. M., and Smith, C. C. T., 1983, The nucleus basalis in Alzheimer's disease: Cell counts and cortical biochemistry, *Neuropathol. Appl. Neurobiol.* **9:**175–179.

Wilcock, G. K., Esiri, M. M., Bowen, D. M., and Hughes, A. D., 1988, The differential involvement of subcortical nuclei in senile dementia of Alzheimer's type, *J. Neurol. Neurosurg. Psychiatry* **51:**842–849.

Williams, P. L., Bannister, L. H., Berry, M. M., Collins, P., Dyson, M., Dussek, J. E., and Ferguson, M. W. J., 1995, *Gray's Anatomy,* (P. L. Williams, ed.), Churchill Livingstone, New York, pp. 1080–1093.

Wohlwill, F., 1925, Pathologisch-anatomische Befunde an excidierten sympathischen Malsganglien bei Asthma bronchiale, *Klin. Wschr.* **4:**107.

Wood, P. L., Etienne, P., and Lal, S., 1983, A post mortem comparison of the cortical cholinergic system in Alzheimer's and Pick's disease, *J. Neurol. Sci.* **62:**211–217.

Woolf, N. J., Jacobs, R. W., and Butcher, L. L., 1989, The pontomesencephalotegmental cholinergic system does not degenerate in Alzheimer's disease, *Neurosci. Lett.* **96:**277–282.

Xuereb, J. H., Perry, R. H., Candy, J. M., Perry, E. K., Marshall, E., and Bonham, J. R., 1991, Nerve cell loss in the thalamus in Alzheimer's disease and Parkinson's disease, *Brain* **114:**1363–1379.

Yagashita, S., Itoh, Y., and Amano, N., 1979, Ultrastructure of neurofibrillary tangles in progressive supranuclear palsy, *Acta Neuropathol.* **8:**27–30.

Yagashita, S., Itoh, Y., and Nakano, T., 1986, Hypertrophy of the olivary nucleus: An ultrastructural study, *Acta Neuropathol.* **69:**132–138.

Yamamoto, T., and Hirano, A., 1985, Nucleus raphe dorsalis in Alzheimer's disease: Neurofibrillary tangles and loss of large neurons, *Ann. Neurol.* **17:**573–577.

Yates, C. M., Simpson, J., Maloney, A.F.J., and Gordon, A., 1980, Neurochemical observations in a case of Pick's disease, *J. Neurol. Sci.* **48:**257–263.

Yoshimura, M., 1983, Cortical changes in the parkinsonian brain: A contribution to the delineation of "diffuse Lewy body disease," *J. Neurol.* **229:**17–32.

Yoshimura, N., 1989, Topography of Pick body distribution in Pick's disease: A contribution to understanding the relationship between Pick's and Alzheimer's diseases, *Clin. Neuropathol.* **8:**1–6.

Zubenko, G. S., and Moossy, J., 1988, Major depression in primary dementia: Clinical and neuropathological correlates, *Arch. Neurol.* **45:**1182–1196.

Zweig, R. M., Whitehouse, P. J., Casanove, M. F., Walker, L. C., Jankel, W. R., and Price, D. L., 1987, Loss of pedunculopontine neurons in progressive supranuclear palsy, *Ann. Neurol.* **22:**18–25.

Zweig, R. M., Ross, C. A., Hedreen, J. C., Steele, C., Cardillo, J. E., Whitehouse, P. J., Folstein, M. F., and Price, D. L., 1988, The neuropathology of aminergic nuclei in Alzheimer's disease, *Ann. Neurol.* **24:**233–242.

Zweig, R. M., Ross, C. A., Hedreen, J. C., Peyser, C., Cardillo, J. E., Folstein, S. E., and Price, D. L., 1992, Locus coeruleus involvement in Huntington's disease, *Arch. Neurol.* **49:**152–156.

Zweig, R. M., Cardillo, J. E., Cohen, M., Giere, S., and Hedreen, J. C., 1993, The locus coeruleus and dementia in Parkinson's disease, *Neurology* **43:**986–91.

# 10

# Age-Related Changes in Subcortical Nuclei that Project to the Cerebral Cortex

THOMAS L. KEMPER

## 1. Introduction

Subcortical afferents to the neocortex from the forebrain include projections from the claustrum, thalamus, amygdala, nucleus basalis (of Meynert), hypothalamus, mesencephalon and pons, pars compacta of the substantia nigra, ventral tegmental area, pedunculopontine nucleus, locus coeruleus, raphe dorsalis, raphe medianus (nucleus centralis superior), medial parabrachial nucleus, and scattered neurons in the tegmentum of the mesencephalon and pons (Porrino and Goldman-Rakic, 1982; Tigges *et al.*, 1982, 1983; Doty, 1983; Mesulam *et al.*, 1983a; Saper, 1985; Gaspar *et al.*, 1992). There are no known cerebral cortical projections from the medulla. Age- and disease-related neuronal changes have been documented in all of these subcortical areas with the single exception of the claustrum. In this review, the evidence for structural changes in these cortically projecting subcortical nuclei that occur in aging and in diseases affecting the cerebral cortex are examined along with their relationship to changes that occur in the cerebral cortex. The

THOMAS L. KEMPER • Departments of Pathology, Anatomy and Neurobiology, and Neurology, Boston University School of Medicine, Boston, Massachusetts 02118

*Cerebral Cortex,* Volume 14, edited by Peters and Morrison. Kluwer Academic/Plenum Publishers, New York, 1999.

majority of the studies focus on cholinergic basal forebrain, dopaminergic ventral mesencephalon, locus coeruleus, and median and dorsal raphe nuclei; these are reviewed individually. The results of studies on the other cortically projecting subcortical nuclei, for which there is little information, are incorporated in the general discussions at the end.

There are technical limitations in the interpretation of many of these studies. The validity of the techniques used in most of them has recently been called into question, and they are now being replaced with unbiased stereological methods of cell counting, which ensure uniform, systematic sampling of the areas of interest and are independent of cell size and shape (Gundersen, *et al.*, 1988; Williams and Rakic, 1988; West, 1993; Coggeshall and Lekan, 1996; Saper, 1996). Further, it is also increasingly apparent that brain tissue undergoes shrinkage or swelling during processing, and that this artifact may also affect the accuracy of cell counts, particularly when cell populations are compared at different ages (Haug, 1985; Haug and Eggers, 1991).

A major interest in nuclei that have shown age-related neuronal cell loss lies in their neurotransmitter-specific cells that project to the cerebral cortex, such as the serotonergic neurons in dorsal and median raphe, dopaminergic neurons in substantia nigra and the ventral tegmental area, and noradrenergic neurons in nucleus locus coeruleus. These cortically projecting subcortical nuclei also contain a variety of other cell types that code for different transmitters, such as the GABAergic neurons (Ford *et al.*, 1995). Many studies of these subcortical nuclei have used immunostaining to identify a specific cell type, whereas others have counted all neurons regardless of type. Few studies have focused on more than one cell type. Compromise of a cell type other than the neurotransmitter-specific cell of interest could conceivably affect the function of these nuclei and add to the deficits attributed to the neurotransmitter-specific neurons. Another problem is the need for corroborating studies showing that the nuclei are actually undergoing neuronal loss. In this regard electron microscopic and Golgi studies demonstrating degenerative changes in these nuclei would be of particular benefit. Such studies would also provide evidence of the extent of involvement of the remaining cells, a finding that might help explain the severity of the cerebral cortex deficit.

A major goal of the study of the effect of age on subcortical areas that project to the cerebral cortex is the understanding of their role in the aging human brain and in dementia. In this regard, studies of experimental animals, particularly nonhuman primates, are valuable because of the high incidence of Alzheimer's disease in humans (Evans *et al.*, 1989) and the uncertainty of whether one is dealing with the early, preclinical stages of this disease, a problem because many of the nuclei that show cell loss with age also show prominent cell loss in Alzheimer's disease. Alzheimer's disease does not appear to occur in the nonhuman primate (Peters *et al.*, 1996). A further advantage of the animal model is the ability to prepare animals with optimal methods of fixation and to be able to behaviorally assess the animals prior to sacrifice with tests that are sensitive to aging.

## 2. Subcortical Areas

### 2.1. Cholinergic Basal Forebrain

#### 2.1.1. Anatomy

Cholinergic neurons that project to cerebral cortex are found in the basal forebrain and brain stem (Gorry, 1963; Parent *et al.*, 1979; Mesulam *et al.*, 1983a,b, 1984a; Saper, 1984; Tigges and Tigges, 1985; Eckenstein and Baughman, 1987). An addition source of cholinergic innervation is an intrinsic bipolar neuron, a cortical cell type that has been identified only in the rat (Houser *et al.*, 1985; Wainer *et al.*, 1985). In both rat and rhesus monkey, Mesulam *et al.* (1983a,b) have identified, with combined choline acetyltransferase (ChAT) and acetylcholinesterase (AChE) immunostaining and with retrograde tracer techniques, four cortically projecting cholinergic areas in the forebrain and two in the brainstem, which they call Ch sectors. Ch1 corresponds to the medial septal nucleus, Ch2 to the vertical limb of the diagonal band of Broca, Ch3 to the nucleus of the horizontal limb of the diagonal band of Broca, and Ch4 to the nucleus basalis (of Meynert). In the brainstem, Mesulam *et al.* (1983b, 1984a) have also identified Ch5, which is mostly within the pedunculopontine nucleus of the pontomesencephalic reticular formation, and Ch6, which is in the periventricular area in the dorsolateral tegmental gray. The major sources of cholinergic innervation to the hippocampus are Ch1 and Ch2; to the olfactory bulb it is Ch3; to the amygdala and neocortex it is Ch4; and to the thalamus and (to a lesser extent) to the neocortex, Ch5 and Ch6 (Mesulam *et al.*, 1983b). In area Ch1 in the macaque, approximately 10% of the neurons are cholinergic; in Ch2, 70% are cholinergic; in Ch3, 1% are cholinergic; and in Ch4, at least 90% are cholinergic (Mesulam *et al.*, 1983a). Not all projection neurons in the basal forebrain are cholinergic. In the macaque, 1–12% of the basal forebrain neurons that project to the olfactory bulb are cholinergic, those that project to the hippocampus are 70–77% cholinergic, and those that project to the neocortex are 86–100% cholinergic (Mesulam *et al.*, 1983a, 1986). In the rat, 50% of the septohippocampal projecting neurons are cholinergic (Wainer *et al.*, 1985).

All areas of the neocortex are innervated by cholinergic basal forebrain neurons, with species differences in the pattern and specificity of innervation (Sillito and Murphy, 1987). In the rat, Saper (1984) found a close reciprocal correspondence between the basal forebrain projection to the cerebral cortex and the cerebral cortical projection to the basal forebrain. Further, the area innervation by individual basal forebrain neurons appears to be discrete, Bigl *et al.* (1982) noted that the vast majority of the basal forebrain neurons innervate relatively small areas of the cerebral cortex, which according to the study of Price and Stern (1983), measure less than 1–1.5 mm in diameter. In rats the highest density of cholinergic fibers is in layer V, with little regional variation (Saper, 1984; Houser *et al.*, 1985; Eckenstein and Baughman, 1987). These ChAT-positive rat cortical fibers form symmetric synapses (Wainer *et al.*, 1984; Houser *et al.*, 1985). All monkey neocortical areas also receive a cholinergic innervation with, however, marked differences from the rat in regional distribution of the fibers, the laminar pattern within the various cytoarchitectonic areas (Campbell *et al.,* 1987; Lewis, 1991; Mesulam *et al.,* 1984b), and in the topography of the cerebral cortical projection to the basal forebrain (Mesulam and Mufson, 1984). In a systematic survey of the cerebral cortex in the macaque, Mesulam *et al.* (1984b) found the highest density of cholinergic fibers in the paralimbic cortices that are immediately adjacent to the allocortices, with the lowest density in the sensory association and prefrontal cortical areas. Within the association areas, the highest density of fibers is in layers I, V, and VI, and in these lamina, and in layer IV in the primary sensory areas. Throughout the temporal lobe there is a dense band of fibers in layer I. In the frontal lobe, Lewis (1991) found the most-dense plexus from layer I to the upper part of layer III, with a distinct band in layer V. In contrast to the rat, neocortical input to Ch4 in the macaque is selective, with projections from the prepiriform cortex, orbitofrontal cortex, anterior insula, temporal pole, and entorhinal cortex. Subcortical inputs to the nucleus basalis in the macaque are from the septal nuclei, nucleus accumbens–ventral pallidum complex, and the hypothalamus (Mesulam and Mufson, 1984). These investigators therefore felt that the basal forebrain cholinergic system is in a position to act predominantly as a relay, transmitting information from limbic and paralimbic areas to the entire neocortical surface. The pattern of distribution of cholinergic fiber plexuses in human cerebral cortex is similar to that found in the monkey (Mesulam *et al.,* 1992; Mesulam and Geula, 1988, 1992). In the human cerebral cortex, Mesulam and Geula (1992) have shown an overlap of staining between AChE and ChAT positive axons; when the reaction products for the two markers are examined concurrently, they are colocalised in the same axons. As far as can be determined, the cholinergic plexus has not been examined by electron microscopy in the primate brain. Atrophy of neurons in the basal forebrain following cortical ablation has been reported in humans (Pearson *et al.,* 1983a) and in monkeys (Gorry, 1963; Pearson *et al.,* 1983a), with the most severe changes noted in monkeys after lesions of the frontal lobes.

Cholinergic input to the cerebral cortex is essential for its normal development. Following neonatal lesions in the cholinergic basal forebrain in the mouse, Hohmann *et al.* (1988) found delays in development of the cerebral cortex that were most pronounced in the superficial layers, with the extent of abnormality correlating with degree of cholinergic denervation. In a later study using this same model, Hohmann *et al.* (1991) compared layer V pyramidal cells in the somatomotor cortex on P7 and P14 ipsilateral and contralateral to the basal forebrain lesions. The ipsilateral neurons showed significant reductions in all aspects of the development of their dendritic tree as well as in cell body size on P7. This change was attenuated by P14, however, with persistent effects on cell size and apical dendritic branching.

### 2.1.2. Age-Related Changes

Numerous studies have been done on the effect of age on cell size and number in the basal forebrain in a variety of experimental animals and in humans, often with conflicting results. In a serial-section study of 7-, 15-, and 53-month-old C57Bl/NNIA mice, Hornberger *et al.* (1985) found no evidence of age-related neuronal loss of AChE-immunostained neurons the medial septum, medial preoptic area, nucleus of the diagonal band, or nucleus basalis; cells were progressively smaller, particularly in the nucleus basalis. Mesulam *et al.* (1987) also found no evidence of neuronal loss of ChAT-immunostained neurons in areas Ch1 to Ch4 of the basal forebrain in a comparison of 4- with 24–25-month-old CD-1 mice. Cells were smaller in areas Ch1, Ch2, and Ch4. In this study they measured selected subregions of the basal forebrain with automated cell-counting equipment using a gray scale. Gilad *et al.* (1987) counted the total number of AChE-immunostained neurons at three different levels of the septum in 3- and 24-month-old Wistar-Kyoto rats and 3-, 24-, and 30-month-old Brown-Norway rats. There was hypertrophy of neurons in the septum of the oldest rats of both strains, and neuronal loss in the Wistar-Kyoto rats at 24 months and in the Brown-Norway rats at 24 and 30 months. This cell loss occurred with no change in the distribution of AChE-stained terminals or of ChAT activity in the hippocampus, suggesting sprouting of the remaining terminals. Markram and Segal (1990) found no evidence of age-related neuronal loss in Nissl-stained sections of the medial septum and the vertical and horizontal limbs of the diagonal band of Broca in a comparison of 10- and 35-month-old male Hooded rats. There was, however, a reduction in the number of neurons in medial septum and vertical limb of the diagonal band of Broca that were labeled for nerve growth-factor (NGF) receptors, and a reduction of the intensity of NGF immunostaining in all three regions. Altavista *et al.* (1990), in a serial-section study of AChE immunostained neurons in three 3- and 24-month-old rats, found neuronal loss and atrophy of neurons in all areas of the cholinergic basal forebrain, most markedly in the septal nuclei. The latter showed a 66% loss of intensely stained neurons. In agreement with these age-related atrophic changes in rat basal forebrain, Aston-Jones *et al.* (1985) have shown a reduced conduction velocity from basal forebrain to cerebral cortex without change in the rate of intracortical conduction velocity. Gage *et al.* (1984) have shown, in rats with an age-related deficit in spatial learning, an improvement in this memory by the implantation in the hippocampus of fetal septal cells. Fischer *et al.* (1987) have shown improved retention of age-related loss of visual memory in rats by the continuous intracerebral infusion of nerve growth factor. In the Long-Evans rats, Koh *et al.* (1989) found that those animals with age-related spatial-memory impairments also showed atrophy and loss of staining of NGF-positive neurons in the basal forebrain and a reduction of intensely immunoreactive cells.

In the monkey, Stroessner-Johnson *et al.* (1992) compared the number and size of ChAT-immunostained neurons at five equally spaced levels throughout the medial septal nuclei of animals 10–12 and 23–25 years old. In the old monkeys there was significant cell loss only in the caudal two sections, with cell size increased throughout. In the rostral three sections, the cells were increased in size only in monkeys with behavioral impairment. Amaral (1993) made similar observations on ChAT-immunostained neurons in the medial septum of old monkeys, and found that cell loss was the most marked caudally, with the remaining neurons showing hypertrophy that was most marked rostrally. Most recently, Voytko *et al.* (1995) examined the basal forebrain at predetermined levels, counting nucleoli of neurons that were immunostaining for p75, a low-affinity nerve growth factor receptor. The number of neurons was similar at all the ages (3–33 years) and levels, but the neurons were larger in the older monkeys, with significant enlargement at the more posterior levels that project to the temporal and occipital lobes. They found significant correlations with low cell counts at intermediate levels in the older spatial-memory impaired monkeys.

In the human brain, McGeer *et al.* (1984) did a serial-section study throughout the nucleus basalis of Meynert (NBM) in 10 individuals aged 38 weeks to 95 years and found a marked age-related progressive cell loss. In this study neuronal perikarya larger than 35 μm were counted. Cell loss by the tenth decade was approximately 60% (estimated from their graph). Bigl *et al.* (1987), in a serial-section study of areas Ch1, 2, and 4 of the basal forebrain from 17 brains aged 36 to 85 years, counted neuronal nucleoli in neurons of all sizes. They found no evidence of neuronal loss up to approximately 60 years of age, with some, but not all, older individuals showing evidence of cell loss beyond this age. Analysis of the different Ch areas in the basal forebrain, as well as different parts of area Ch4, also failed to show evidence of age-related neuron loss. Lowes-Hummel *et al.* (1989), in a serial-section study using nucleolar counting of neurons more than 35 μm in diameter, counted the total number of neurons in the NBM. In 16 brains from individuals aged 35 weeks of gestation to 90 years of age, they reported an overall decrease of 23%. De Lacalle *et al.* (1991) counted neuronal nucleoli in cells of all sizes in serial sections of the nucleus basalis of Meynert of 39 brains from individuals ages 16 to 106 years. They found a linear age-related loss of neurons, with a decrease of 25.6% by age 60 and 50% by age 100. Cell size increased up to age 60, then declined. In cell-size frequency histogram, comparing brains aged 16–29 years with those aged 80 to 100 years, it could also be seen that there was a marked decrease in percentage of the smallest neurons. The decrease in cell number and size was most marked posteriorly, in an area that projects to the polar cortex of the temporal lobe; the anterior part showed no significant age-related neuronal loss. Two studies using neuronal nucleolar counts at the level of the maximum density of neurons in the NBM have shown different results. Chui *et al.* (1984) counted neurons more than 25 μm in diameter and found no evidence of age-related neuronal cell loss in 117 brains of individuals aged 25 to 87 years. Mann *et al.* (1984b) counted neurons larger than 30 μm in diameter from 20 individuals aged 40 to 90 years and found an age-related loss of 35%. Arendt *et al.* (1995a), in a study of 35 adult brains in Nissl-stained sections, counted nucleoli of neurons larger than 30 μm in the vertical limb of the diagonal band of Broca and NB and larger than 20 μm in the medial septal nuclei and found an approximately 30% loss of cells by age 80. In the Golgi stains, this cell loss was accompanied by an increase in the extent of the dendritic tree in the remaining neurons. In contrast to the frequently reported age-related neuronal cell loss in the basal forebrain in humans, Geula and Mesulam (1989) found only a modest age-related loss of AChE-stained fibers in the entorhinal cortex and in two different neocortical areas. Similarly, de Lacalle *et al.* (1994) noted that the ChAT- and AChE-stained fibers in the hippocampus and entorhinal cortex showed a trend toward a decrease in density only in brains of individuals aged 40 to 110 years.

### 2.1.3. Disease-Related Changes

In Alzheimer's disease (AD), neuronal cell loss is well documented in the basal forebrain (Perry *et al.*, 1982; Whitehouse *et al.*, 1982, 1983a; Arendt *et al.*, 1983, 1984, 1985, 1995a; Candy *et al.*, 1983; Coyle *et al.*, 1983; Nagai *et al.*, 1983; Pearson *et al.*, 1983b; Tagliavini and Pilleri, 1983; Wilcock *et al.*, 1983, 1988; Mann *et al.*, 1984b,c; McGeer *et al.*, 1984; Rogers *et al.*, 1985; Saper *et al.*, 1985; Etienne *et al.*, 1986; Doucette and Ball, 1987; Rinne *et al.*, 1987; Allen *et al.*, 1988; Ichimiya *et al.*, 1986;

Vogels *et al.*, 1990; Lehéricy *et al.*, 1993), with the percentage of cell loss ranging from 15.5% (Vogels *et al.*, 1990) to 90% (Whitehouse *et al.*, 1982). With the Golgi stain, Arendt *et al.* (1995a) have shown that many of the remaining neurons are larger and have an abnormal proliferation of dendrites throughout their dendritic field and an aberrant proliferation of perisomatic and distal dendritic processes. This change was confined to reticular neurons. Multiple factors enter into the differences in extent of reported neuronal cell loss in AD. As pointed out by several investigators (Pearson *et al.*, 1983b; Rinne *et al.*, 1987; Allen *et al.*, 1988; Vogels *et al.*, 1990), many of the neurons in the basal forebrain undergo atrophy, with a decrease in the number of large neurons and increase in number of small ones. For example, Allen *et al.* (1988), using enolase as a specific neuron marker, found a 61% decrease in large neurons and a 59% increase in small cells, with a resultant overall calculated cell loss of 29% in AD. Similarly, in Nissl-stained sections, Vogels *et al.* (1990) found a 70% decrease in neurons with a diameter larger than 37.5 μm, and an increase of approximately 30% of neurons with a diameter smaller than 12.5 μm. In this study, the total cell loss was 15.5%. Many of the studies counted only "large" neurons or neurons over a certain size, thus missing the atrophic cells (i.e., Perry *et al.*, 1982; Arendt *et al.*, 1983, 1985, 1995a; Tagliavini and Pilleri, 1983; Saper *et al.*, 1985; Etienne *et al.*, 1986; Ichimiya *et al.*, 1986; Rinne *et al.*, 1987; Wilcock *et al.*, 1988). Another variable is the site of the cell counts. Etienne *et al.* (1986), Doucette and Ball (1987), and Wilcock *et al.* (1988) all report greater cell loss at posterior than at anterior levels of the nucleus basalis. Another variable is the use of Nissl staining versus specific immunostaining for ChAT, AChE, or neuropeptide Y (NPY). For example with ChAT or AChE, Nagai *et al.* (1983), McGeer *et al.* (1984), and Lehéricy *et al.* (1993) reported an approximately 50% cell loss in the nucleus basalis in AD, whereas Lehéricy *et al.* (1993) found no evidence of loss of NPY cells in this location.

The method of cell count is another factor in the variation of reported cell loss in AD. In the single study using unbiased stereological techniques in a serial-section study throughout the nucleus basalis, Vogels *et al.* (1990) found only a 15.5% cell loss in this region and no evidence of cell loss in the medial septum. In contrast to the basal forebrain, the brain stem shows no evidence of significant loss of cholinergic neurons in AD. Woolf *et al.* (1989) found no loss of these neurons in the pontomesencephalic region, and Zweig *et al.* (1987) and Jellinger (1988) failed to find evidence of significant neuronal loss from the pedunculopontine nucleus.

Loss of cholinergic neurons in the nucleus basalis has been documented in the basal forebrain in other diseases, including idiopathic parkinsonism (Arendt *et al.*, 1983, 1995a; Candy *et al.*, 1983; Whitehouse *et al.*, 1983b) and Korsakoff's disease (Arendt *et al.*, 1983, 1995a; Tagliavini and Pilleri, 1983), but not postencephalitic parkinsonism (Arendt *et al.*, 1983), Pick's disease (Tagliavini and Pilleri, 1983), Huntington's chorea (Arendt *et al.*, 1983; Tagliavini and Pilleri, 1983), chronic alcoholism (Arendt *et al.*, 1983), or schizophrenia (Arendt *et al.*, 1983). Woolf *et al.* (1989) failed to find evidence of cell loss in the pontomesencephalic region in two cases of idiopathic parkinsonism, whereas Jellinger (1988) found a marked loss in this region in this disease as well as in progressive supranuclear palsy. In idiopathic parkinsonism, Arendt *et al.* (1995a) found that this loss was more uniformly distributed than that encountered in aging.

## 2.2. Dopaminergic Ventral Mesencephalon

### 2.2.1. Anatomy

Dopaminergic neurons are widespread in the mammalian brain, occurring the olfactory bulb, diencephalon, ventral mesencephalon, and possibly the medulla (Björklund and Lindvall, 1984). In fishes and amphibians, they are intermixed with serotonergic neurons (Oades and Halliday, 1987); in rats Descarries *et al.* (1986) report dopaminergic neurons in the serotonergic raphe nuclei that project to the striatum. In mammals the dopaminergic neurons in the ventral mesencephalon are found bilaterally in the substantia nigra pars compacta (SNpc) and various midline nuclei referred to by Tsai (1925) as the ventral tegmental area (VTA). In human studies of age-related neuronal cell loss and of cell loss in disease, the SNpc is often divided into dorsal and ventral tiers. Olszewski and Baxter (1954) refer to the ventral tier as the pars $\partial$, noting that its neurons are more uniformly distributed than in the upper tier, which they call the pars $\beta$. Gibb and Lees (1991) and Gibb (1991) note that the dorsal tier is composed of more heavily pigmented cells than the ventral cell group which they call the "ventrolateral tier." The ventrolateral cell group is not present at the rostral and caudal pole of the SNpc. At intermediate levels it is surrounded dorsally, medially, and laterally by the dorsal tier neurons (Gibb, 1991). Areas comparable in the human brain to the SNpc classification of Hassler (1937) and Braak and Braak (1986) are provided by Gibb (1991). According to Gibb's (1991) analysis of the literature, the dorsal tier predominantly projects to the caudate nucleus and the ventrolateral tier projects to the putamen. In most studies, the VTA is divided into five cell groups. In rats, cats, monkeys, and humans, Halliday and Törk (1986) recognized these cell groups as the parabrachialis pigmentosa (PBP), paranigralis (PN), central linear (CL), interfascicular (IF), and rostral linear (RL). Of these VTA nuclei, Halliday and Törk (1986) found the PBP is particularly well developed in monkeys and the PN in humans. In a Nissl-stain study of cell types in the substantia nigra (SN) and VTA in rats, cats, and monkey, Poirier *et al.* (1983) noted a pronounced increase in the total number of neurons in the SN in macaque as compared with rat and cat, with the major increase of a neuron they classified as the "compact type," a type that they feel corresponds to the dopaminergic neuron. In the SN, which included the pars compacta, reticulata, and PBP, the compact cell type accounted for 85% of the neurons in monkeys, 58% in cats, and 44% in rats. In humans, and in some nonhuman primates, these dopaminergic neurons are pigmented (Foley and Baxter, 1958; Mann and Yates, 1974; Schwyn and Fox, 1974; Bogerts, 1981; Bogerts *et al.*, 1983), with the pigment first noted in humans at 18 months of age (Foley and Baxter, 1958; Mann and Yates, 1974). Mann and Yates (1974) noted a linear increase in accumulation of melanin within individual neurons until 60 years of age, then a decline.

Details of cortically projecting cells in the dopaminergic ventral mesencephalon are available for the rhesus monkey. These projections are from the medial one-third of SNpc and, in the VTA, from PBP, PN, and IF, as well as from a group of cells lateral to the caudal part of the SNpc referred to as the A8 cell group or the "retrorubral nucleus" (Porrino and Goldman-Rakic, 1982). In rodents the dopaminergic projection from the ventral mesencephalon is confined to the prefrontal, cingulate, and perirhinal regions, with the densest projection often occurring in the deeper layers (Fallon and Loughlin, 1987; Lewis *et al.,* 1986). There is evidence in the rodent brain that layer I afferents to the cerebral cortex arise from the substantia nigra and those to the deeper layer arise from the VTA (Björklund and Lindvall, 1984; Fallon and Loughlin, 1987). In contrast, the dopaminergic innervation to the cerebral cortex in primates—including humans—is particularly dense and extensive, with the densest afferent plexus in layer I (Lewis *et al.,* 1987; Gaspar *et al.,* 1989, 1992; Berger *et al.,* 1991; van Domburg and ten Donkelaar, 1991; Williams and Goldman-Rakic, 1993). In primates all cortical areas are innervated (Lewis *et al.,* 1986, 1987), with the densest innervation in the frontal lobe, in which it shows a gradient of innervation that does not respect cytoarchitectonic borders, and which the densest innervation into the agranular frontal premotor areas, areas 6 and 8 (Williams and Goldman-Rakic, 1993). Within the cerebral cortex, the dopaminergic fibers form symmetric synapses with dendrites and their spines, which also receive another terminal forming an asymmetric synapse (Goldman-Rakic *et al.,* 1989; Smiley and Goldman-Rakic, 1993; Smiley *et al.,* 1994).

### 2.2.2. Age-Related Changes

Little is known about age-related changes in the SNpc in experimental animals. McNeill and Koek (1990) found no evidence of significant neuronal cell loss in a serial-section study of the dopaminergic cells in the mesencephalon of C57BL/6N mice 3 to 30 months of age. In this study the SN, VTA, and A8 cell group were studied as a unit. In the C57/BJ/6 mouse, Tatton *et al.* (1991) found a linear decrease in SN neurons from 8 to 102 weeks of age. In a serial-section study of the SNpc of 13 rhesus monkeys ages 5 to 32 years using the Nissl stain and unbiased stereological methods, Siddiqi *et al.* (1997) found an age-related neuronal loss of 25%. Electron microscopical studies of the remaining neurons in the older monkey by Siddiqi *et al.* (1996) showed a variety of degenerative changes in the neurons, inclusions bodies thought to be in glial cell processes, and proliferative changes in astroglial cells. In this study the most marked neuronal changes were found in a small neuron corresponding in size and configuration to the GABAergic cells in the SNpc.

Age-related neuronal loss in the SNpc has been frequently documented in the human brain and is accompanied by the accumulation of extraneuronal melanin pigment, a common finding in the SNpc of older individuals (Mann and Yates, 1974; Gibb and Lee, 1991). McGeer *et al.* (1977), in a serial-section study of the substantia nigra in 13 brains from the first to the ninth decade, found an approximately 50% loss of pigmented neurons by the ninth decade (estimated from their graph). Mann *et al.* (1984a), who studied a single level of the SNpc of 67 normative individuals ages 11 to 97 years, reported a 35% loss by age 90. In this study the cell loss was most marked after the seventh decade. Gibb and Lees (1991), also using a single H-E-stained section at the midpart of the SN from 62 brains of individuals aged 40 to 90 years, found that the loss of pigmented neurons was linear at the rate of 4.7 to 6.0% per decade. Fearnley and Lees (1991), also using a single H-E-stained section at a predetermined level of the SNpc, reported a similar finding. In 36 normative brains from individuals 21 to 91 years of age, they found a linear cell loss of pigmented neurons at a rate of 4.7% per decade. Fearnley and Lees (1991) have also analyzed the topography of the neuronal loss in this nucleus, demonstrating that it occurs primarily in the dorsal sector, and that in the ventral part the loss appeared to involve only its medial sector. With the rapid Golgi method, Cruz-Sánchez (1995) report that changes in neuronal morphology are common in aged individuals. These include contorted profiles of the cell bodies, swelling and beading of dendritic branches, and loss of dendrites and dendritic spines. Little attention has been paid to the possibility of age-related neuronal cell loss in the VTA. In a serial-section, unbiased stereological study of 14 monkeys aged 5 to 32 years, Siddiqi *et al.* (1997) found a 32% age-related neuronal loss in the PN, with no significant loss in the PBP.

#### 2.2.3. Disease-Related Changes

In Alzheimer's disease there is a similar pattern of neuronal loss to that found in aging, with the loss more marked in the VTA than in the SN (Mann *et al.*, 1987; van Domburg and ten Donkelaar, 1991), and the VTA is also the site of the maximum number of neurofibrillary tangles (van Domburg and ten Donkelaar, 1991). Van Domburg and ten Donkelaar (1991), who studied four AD brains in serial sections, counted pigmented and nonpigmented neurons using unbiased techniques of cell counting. Although cell loss was noted in the SN, the difference from the control values was not significant, whereas there was significant cell loss in the VTA in the AD brains. In this study the neuronal cell loss was greater in nonpigmented (47%) than in pigmented neurons (41%). Tabaton *et al.* (1985), Mann *et al.* (1987), and Uchihara *et al.* (1992) studied the SN and VTA at selected levels, using either cell counts (Tabaton *et al.*, 1985; Uchihara *et al.*, 1992) or nucleolar counts (Mann *et al.*, 1987). Tabaton *et al.* (1985) found a 33% loss of neurons in the SN, Mann *et al.* (1987) found a 17% loss in the SN and a 43% loss in the VTA, and Uchihara *et al.* (1992) found a loss of 21–51%, depending on location within the SN. A similar pattern of cell loss in the SN–VTA complex has been found in older individuals with Down's syndrome (Mann *et al.*, 1987), a syndrome with many features of AD.

In contrast to AD, idiopathic parkinsonism (IP), which lacks the conspicuous cerebral cortical pathology of AD, shows a different pattern of cell loss in the SN and VTA. Bogerts *et al.* (1983) did a serial-section study in IP dividing the SN into thirds, the lateral two-thirds and the medial third (which included the paranigralis) both showed a similar degree of neuronal loss, 66% and 63%, respectively. Within the SN, van Domburg and ten Donkelaar (1991), Fearnley and Lees (1991), and Gibb and Lees (1991) found the greatest cell loss in the ventrolateral tier, an area relatively spared in the aged brain. This greater neuronal loss had also been noted by Bogerts *et al.* (1983) and by Hassler (1938). In postencephalitic parkinsonism cell loss is uniformly distributed in the SN and the paranigralis (Bogerts *et al.*, 1983). In schizophrenic brains, Bogerts *et al.* (1983) found no evidence of neuronal cell loss. There was a volume reduction of the lateral SN and a neuronal-size reduction in the medial SN.

### 2.3. Locus Coeruleus

#### 2.3.1. Anatomy

Noradrenergic neurons occur in the medulla and pons, where they can be divided into three major cell groups: a locus coeruleus–subcoeruleus complex; a lateral tegmental cell system that extends from the medulla to the level of motor nucleus of the V cranial nerve that is located mainly in relationship to the lateral reticular nucleus; and a dorsal medullary cell group, located mainly in relation to the solitary track, the dorsal motor nucleus of the vagus, and the commissural nucleus (Lindvall and Bjöklund, 1983; Felten *et al.*, 1974). With the exception of the locus coeruleus, it is difficult to distinguish the location of these noradrenergic cell neurons in Nissl preparations (Felten and Sladeck, 1983). Of these cell groups, only the locus coeruleus has a cerebral cortical projection and, according to Fallon and Loughlin (1987), it is the sole source of noradrenergic projections to the cerebral cortex in rats. The LC is particularly prominent in primates (Felten and Sladeck, 1983), and in humans it is one of the most highly vascularized nuclei in the brain (Finley and Cobb, 1940). In both humans and nonhuman primates, the neurons of the LC and the subcoeruleus contain melanin pigment, which first appears at the fifth fetal month and is found even in the brain of albinos (Foley and Baxter, 1958). According to Mann and Yates (1974), who studied LC neurons from birth to age 91, melanin content within individual neurons increases until late middle-age, then declines. Virtually all pigmented cells express tyrosine hydroxylase activity (Baker *et al.*, 1989).

Locus coeruleus projections are predominantly unilateral and extensively collateralized, with 1500 neurons in the rat LC projecting to practically all parts of the nervous system (Lindvall and Bjöklund, 1983); those from the subcoeruleus and extreme dorsal, ventral, and anterior parts of the LC are the least collateralized (Felten and Sladeck, 1983). A striking example of a localized LC projection, according to Loughlin *et al.* (1986), is the projection from the LC anterior pole to the hypothalamus. In rats, multipolar, fusiform, and small oval cells have been identified in the LC, with the projection neurons shown to be multipolar and fusiform cells (Waterhouse *et al.*, 1983). Ascending subcortical projections from the LC in the monkey are located in the central tegmental tract, with a complex pattern of projections in the forebrain (Bowden *et al.*, 1978). Ascending LC subcortical projection that have been identified in the monkey include the bed nucleus of the stria terminalis; the substantia innominata; the anterior, central, and basal nucleus of the amygdala; the paraventricular, supraoptic, and dorsomedial part of the hypothalamus; and anteroventral nucleus of the thalamus (Bowden *et al.*, 1978). In the rat the projection to the neocortex, amygdala, and piriform cortex are from neurons throughout the LC, with the exception of its ventral tip (Lindvall and Bjöklund, 1983), and according to Waterhouse *et al.* (1983), the neocortical projections in the rat are from the dorsal part of the caudal LC, with the frontal projection also arising from its ventral part. In the latter study, the neurons projecting to different parts of the cerebral cortex were found to overlap each other, but were not coextensive with neurons giving rise to projections to the caudate and hippocampus. Loughlin *et al.* (1986) have localized the neocortical projections to the central part of the LC and the hippocampal projections to its dorsal part. In contrast to the rat, the cerebral cortical projection neurons of the LC in the monkey are found throughout the LC, with little topographic organization (Freedman *et al.*, 1975; Porrino and Goldman-Rakic, 1982; Doty, 1983). The cerebellar projection in the rat arise from cells throughout the LC, with the highest density of projection cells from the ventral part, and spinal cord projections from the ventral part of both LC and subcoeruleus (Loughlin *et al.*, 1986). The LC and the subcoeruleus have also been shown to be the origin of the spinal projection in cats and monkeys (Handcock and Fougerousse, 1976). These descending projections from LC and subcoeruleus give rise to an extensive innervation to regions that give rise to the parasympathetic and sympathetic outflow (Westlund and Coulter, 1980).

In the rat, layer I of the cerebral cortex contains a dense, gridlike, noradrenergic fiber plexus; layers II and III contain a sparsely innervated plexus of radially oriented fibers; layers IV and V, a moderately dense fiber plexus of obliquely oriented fibers, and layer VI contains a long, anterior-posterior orientation of fibers (Fallon and Loughlin, 1987; Papadopoulos and Parnavelas, 1991). Fallon and Loughlin (1987) note that this pattern occurs in the majority of the cortical areas, with a particularly dense innervation of layer V of the motor cortex and in layer IV of the visual cortex. In primates, the innervation of the neocortex shows a greater heterogeneity than in rats (Fallon and Loughlin, 1987). As with rodents, the innervation is widespread, but there is a greater variation in regional and laminar distribution. The highest density of innervation is found in the sensorimotor areas, with a decreasing gradient both rostrally and caudally; the least density is in the visual cortex, and in contrast to the rat, primate fibers are least numerous in layer I (Morrison *et al.*, 1982; Levitt *et al.*, 1984; Lewis *et al.*, 1986; Gaspar *et al.*, 1989; Lewis and Morrison, 1986). In human brain the highest density of fibers is in layers III and V, with variable extension into layer IV (Gaspar *et al.*, 1989). In the human brain there is no evidence of a gradient of density in the limbic cortices (Gaspar *et al.*, 1989). In rat, monkey, and human there are few NA fibers in subcortical white matter (Fallon and Loughlin, 1987; Gaspar *et al.*, 1989; Morrison *et al.*, 1982).

### 2.3.2. Age-Related Changes

Age-related neuronal loss in the LC in the ASH/TO mouse has been reported by Sturrock and Rao (1985) in a serial-section study of three mice each at ages 6, 9, 12, 15, 22, 25, and 28 months. Employing Abercombie corrections and nuclear counts, they noted a significant cell loss after 22 months of age. In a study of the C57BJ/6 mouse from 8–102 weeks of age, Tatton *et al.* (1991) noted that the LC lost a constant proportion of its neurons throughout the mouse's life span and that this loss occurred without reference to topography. In the Fisher 344 rat, Goldman and Coleman (1981) found no evidence of age-related cell loss from animals 12, 20, 27, 30, and 32 months of age in a study that used nucleolar counts. No studies of age-related neuronal loss in nonhuman primates could be found. In the human brain, age-related neuronal loss in the locus coeruelus has been documented in many studies since the classic study of Vijayashankar and Brody (1973) in which they counted nucleoli in pigmented neurons in serial sections from 24 individuals aged 14–86 years. They noted that the number of neurons was constant until the seventh decade, then declined 39% by the ninth decade. This study of Vijayashankar and Brody (1973), together with nine others (Tomlinson *et al.*, 1981; Bondareff *et al.*, 1982; Mann *et al.*, 1983; Tomonaga, 1983; Yoshinaga, 1986; German *et al.*, 1988; Lohr and Jeste, 1988; Marcyniuk *et al.*, 1989; Chan-Palay and Asan, 1989a), has been reviewed by Manaye *et al.* (1995). All showed age-related neuronal cell loss that varied from 16% (Marcyniuk *et al.*, 1989) to 57% (German *et al.*, 1988). In most of these studies, serial sections were used, with neuromelanin as the marker. The chronological pattern of cell loss is still not clear. Manaye *et al.* (1995), in their own study of 14 normative brains from 1 to 104 years of age, suggested that there was a progressive loss of cells with age, whereas Vijayashankar and Brody (1973), Tomlinson *et al.* (1981), Bondereff *et al.* (1982), and Mann *et al.* (1983) thought the decline was nonlinear. In addition to obvious technical problems such as sample size and age range, Manaye *et al.* (1995) have focused on other problems that could contribute to the data variability. One of these is the use of neuromelanin as a marker. In the human, melanin pigment appears in LC and subcoeuruleus neurons in the fifth fetal month (Foley and Baxter, 1958). In children 1–2 years old, Manaye *et al.* (1995) showed that only a few TH-positive cells evidenced neuromelanin granules, and that even in individuals in their early twenties, as many as 16–18% of TH-positive cells failed to show neuromelanin. By the fifth decade, the number of cells containing the two markers was comparable. The latter is consistent with the study by Baker *et al.* (1989) of a 53-year-old individual in which virtually all pigmented cells in the LC expressed TH activity. Further, Manaye *et al.* (1995) found, in the TH-immunostained sections of two brains of individuals ages 22 and 82 years, that an average of 20% of pigmented cells failed to express TH. Thus, using neuromelanin as a criteria in a study that includes younger individuals will underestimate the number of TH-positive cells in the LC at that age. In the study by Manaye *et al.* (1995), TH-immunostained cells were counted in serial sections throughout the LC from early childhood to 104 years of age, with an overall loss of 50% of the neurons in the LC. In contrast to the studies showing age-related neuronal cell loss in the LC, Mouton *et al.* (1994) failed to find evidence of such loss when unbiased stereological methods were used to compare the LC from six individuals aged 19–43 years with six individuals aged 65–78. An explanation for the difference in the findings of this study and the others in which cell loss is noted is difficult to determine. Mouton *et al.* (1994) report studying the main part of the LC, perhaps excluding the rostral pole, where the cell loss is reported to be the greatest, and the use of melanin as marker in the younger brains may have contributed to the difference. In the study of Chan-Palay and Asan (1989a), neuronal cell loss was significantly greater in the rostral than in the caudal part of the LC and did not occur in the subcoeruleus. Chan-Palay and Asan (1989a) attributed the location of the cell loss in the LC to that part of the nucleus that projects to the forebrain. In contrast, Marcyniuk *et al.* (1989) in a serial-section study of 24 brains from individuals aged 17 to 94, noted that the age-related cell loss occurred uniformly throughout the LC. In the studies of Zweig *et al.* (1988) and Chan-Palay and Asan (1989a), the neuronal loss was more marked in patients that were depressed.

### 2.3.3. Disease-Related Changes

Numerous studies have documented neuronal cell loss in Alzheimer's disease (AD) (Forno, 1978; Tomonaga, 1979; Perry *et al.,* 1981; Tomlinson *et al.,* 1981; Bondereff *et al.,* 1982; Iversen *et al.,* 1983; Mann *et al.,* 1983; 1984a,b, 1985; Ichimiya *et al.,* 1986; Marcyniuk *et al.,* 1986; Zubenko and Moossy, 1988; Zweig *et al.,* 1988; Chan-Palay and Asan, 1989b; Förstl *et al.,* 1992; German *et al.,* 1992), with the most severe loss found in the rostral part of the LC (Chan-Palay and Asan, 1989b; German *et al.,* 1992), where Chan-Palay and Asan (1989a) had reported the greatest age-related loss. Other studies have reported a different distribution pattern, with the greatest loss in the dorsal part of the LC (Marcyniuk *et al.,* 1986) or at the midpart of the nucleus (Zweig *et al.,* 1988). The cell loss is the most severe in early-onset AD (Forno, 1978; Zweig *et al.,* 1988). Bondareff *et al.* (1982) report that cell loss is greater in those patients with a more marked dementia, whereas Zweig *et al.* (1988) found no correlation with dementia severity. Several investigators report more-severe cell loss in AD patients with clinical signs of being depressed (Chan-Palay and Asan, 1989b; Förstl *et al.,* 1992; Zubenko and Moossy, 1988; Zweig *et al.,* 1988). Cell loss in the LC has also been reported in older individuals with Down's syndrome (Mann *et al.,* 1983; German *et al.,* 1992), in which there is the early development of an Alzheimer-like pathology. German *et al.* (1992) note that LC cell loss in Down's syndrome is more severe in the rostral pole. Pick's disease, a dementia with severe loss of cortical neurons, has also been reported to show cell loss in the LC (Arima and Akashi, 1990). Diseases without marked cerebral cortical involvement have also shown LC cell loss. These include Parkinson's disease, wherein LC cell loss is a hallmark of the disease (German *et al.,* 1992; Tomonaga, 1979; Mann *et al.,* 1983; Chui *et al.,* 1986; Chan-Palay and Asan, 1989b) and has been reported to affect all parts of the LC (German *et al.,* 1992; Chan-Palay and Asan, 1989b). Other examples are dementia pugilistica (Mann *et al.,* 1983), progressive supranuclear palsy (Tomonaga, 1979; Mann *et al.,* 1983), chronic alcoholism (Arango *et al.,* 1994), and depression of various etiologies (Zubenko and Moossy, 1988).

## 2.4. Median and Dorsal Raphe Nuclei

### 2.4.1. Anatomy

The serotonergic raphe system in the brain stem is phylogenetically old (Steinbusch and Nieuwenhuys, 1983; Stuesse *et al.,* 1995), with terminals present in practically all parts of the central nervous system (Steinbusch, 1981; Steinbusch and Nieuwenhuys, 1983; Fallon and Loughlin, 1987). The cortically projecting raphe nuclei are the nucleus centralis superior (NCS, or median raphe nucleus) and the dorsal raphe nuclei (NRD), with projections that are primarily ipsilateral (Porrino and Goldman-Rakic, 1982; Tigges *et al.,* 1982, 1983; Fallon and Loughlin, 1987; Doty, 1983; Wilson and Molliver, 1991). The NRD and NCS are located in the lower mesencephalon and upper pons, with scattered serotonergic neurons found in the adjacent reticular formation (Steinbusch, 1981; Felten and Sladek, 1983; Steinbusch and Nieuwenhuys, 1983; Fallon and Loughlin, 1987). According to Steinbusch and Nieuwenhuys (1983), these scattered reticular-formation neurons in the rat constitute the majority of the serotonergic neurons, and in the human pons, Baker *et al.* (1991) estimate that approximately half the serotonergic neurons are in the reticular formation outside the NCS. In the human pons, the largest number of serotonergic neurons are in the NCS (Baker *et al.,* 1991), and in nonhuman primates the NCS has the largest group of serotonin neurons with ascending projections (Azmitia and Gannon, 1986). Many cells within the raphe system are not serotonergic. For example, in the cat Wiklund *et al.* (1985) found that only 35% of neurons in the NCS and 75% of neurons in the NRD were serotonergic, whereas Felten and Sladek (1983) have estimated that only half of the neurons in the monkey NCS are serotonergic, with a higher proportion in the NRD; in the human brain, Baker *et al.* (1991) found that only 43% of the NCS neurons are serotonergic. In rat the majority of cortically projecting neurons in the NCS (60%) are serotonergic (O'Hearn and Molliver, 1984). Other neurotransmitters that have been identified in the NRD are dopamine, norepinephrine, GABA, CCK, leu-ENK, met-ENK, substance P, and VIP, and in the NCS, dopamine (see: Steinbusch and Nieuwenhuys, 1983, for review).

In the human brain, Baker *et al.* (1990) subdivided the dorsal raphe nucleus into a caudal subnucleus in the upper pons, ventral, ventrolateral, and dorsal subnuclei located in the mesencephalon, and an interfascicular subnucleus extending from the pons to the mesencephalon. Baker *et al.* (1991) divided the human NCS into three subnuclei: a dorsal group that interdigitates with the medial longitudinal fasciculus, a densely cellular midline group, and a less-densely cellular paramedian group.

In cerebral cortex the serotonergic axon terminals, or their varicosities, form asymmetric synapses on dendritic shafts and dendritic spines, with only about 30% to 45% of the axon beads or swellings forming synaptic junctions, suggesting that the 5-HT innervation of the cortex is largely nonjunctional (Séguéla *et al.*, 1989). Itakura *et al.* (1985) have noted 5-HT terminals on the basement membranes of capillaries, suggesting a role in cerebrovascular metabolism. Kosofsky and Molliver (1987) and Wilson and Molliver (1991) point out that there are two morphological types of serotonergic axons in the cerebral cortex, fine and beaded, present in all cortical areas. In most areas beaded axons predominate in layer I, with small numbers of them scattered throughout layers II and III, whereas fine axons predominate in layers II through VI. These investigators argue that this represents a dual innervation from the brain stem, with the beaded fibers that predominate in layer I arising from the NCS and the fine axons in layers II–VI arising from the NRD. The distribution of the two types of terminals seems to be broadly comparable to the distribution of the two types of 5-HT receptors in the cerebral cortex of *Macaca mulatta.* Lidow *et al.* (1989) found the highest concentration of 5-$HT_1$ receptors in layers I–III and the highest concentration of 5-$HT_2$ receptors in layers III and IV.

In the rat the cerebrocortical projections of the NRD are predominantly to the frontal, anterior cingulate, and piriform cortices, and of the NCS, predominantly to the hippocampus, entorhinal, piriform, and anterior cingulate cortices, with many of the neurons forming collateral projections to other areas of the brain (Fallon and Loughlin, 1987). Innervation of rat cerebral cortex is generally homogeneous throughout the different cell layers and cortical regions, with the fibers in layers I and VI running tangentially (Fallon and Loughlin, 1987). In monkey the cortical projection are more dense and laminated, and the fibers are primarily directed toward the sensory regions, with particularly dense innervation in layer IV, and with the greatest density of innervation in the primary visual and auditory cortex (Lewis *et al.*, 1986; Berger *et al.*, 1988; Wilson and Molliver, 1991). Measurements of monkey neurotransmitter levels show that 5-HT turnover is greater in occipital cortex than in other neocortical areas (Brown *et al.*, 1979). According to Wilson and Molliver (1991), the serotonergic projection from the brain stem is the most abundant ascending projection from the reticular formation in primates.

### 2.4.2. Age-Related Changes

In the rat, Lolova and Davidoff (1992), in a comparison of 3- with 28-month-old Wistar rats, found that both the NRD and the NCS showed a significant age-related decrease in size of serotonin-immunostained neurons and their processes that was more marked in the NRD, and a decreased density of these neurons that was significant only in the NCS. Van Luijtelaar *et al.* (1992) also failed to find evidence of an age-related loss of serotonin-immunostained neurons in the NRD in a serial-section study of rats aged 12, 18, and 24 months but did find aberrant serotonergic fibers in the forebrain. Kemper (1993), in a serial-section study of Nissl-stained sections of six monkeys aged 10–32 years, using the optical disector technique of Williams and Rakic (1988), reported age-related cell loss in the various subnuclei of the NRD and in the NCS that varied from 25–46%. In a similar study of the NCS in eight monkeys aged 7 to 33 years, the neuronal cell loss was 40% when the youngest three monkeys were compared with the oldest two (Kemper *et al.,* 1997). This decrease in neuron number appeared to affect two different cell populations in the NCS, one corresponding in size to the serotonergic cells in this nucleus and the other, with a greater neuron loss, corresponding to a smaller cell. Thus, the major age-related cell loss in the NCS appears to be a nonserotonergic neuron. In this study, comparisons of the changes in cell-packing density with behavioral testing showed significant correlations with overall test performance as well as with a test of memory function. In the human brain, Aletrino *et al.* (1992), also using stereological methods in a serial-section study of Nissl-stained sections of 10 brains from individuals aged 60 to 91 years, found no evidence of age-related neuronal cell loss in the NRD.

### 2.4.3. Disease-Related Changes

Neuronal loss in Alzheimer's disease has been documented in the NRD in several studies using nucleolar counts at selected levels of the nucleus. With this approach, Curcio and Kemper (1984) reported a 15% loss of large neurons without a significant effect on the number of small-to-medium neurons. Yamamoto and Hirano (1985), who counted neurons greater than 25 μm in diameter, reported a 23% loss. Zweig *et al.* (1988), using the same size criteria, reported a 10–36% loss, depending on anatomical level. In contrast, Ichimiva *et al.* (1986), who counted cells of all sizes, found no evidence of significant cell loss. Aletrino *et al.* (1992), using serial sections, stereological techniques, and no cell-size exclusions, found an average neuronal cell loss of 40%. Mann *et al.* (1984a) appear to have confused the NRD with the dorsal tegmental nucleus (DTN), as they have equated their "DTN" with supratrochlear nucleus, another name for the NRD. They report that the "DTN" did not show cell loss in Alzheimer's disease. More recently, Chen *et al.* (1994), in an abstract, reported loss of immunostained neurons in all divisions of the NRD in AD, and wrote that many of the remaining neurons in these nuclei showed neurofibrillary tangles. Ono *et al.* (1995) report neuronal loss in the NRD in patients with myotonic dystrophy only in those individuals who showed clinical hypersomnia.

In the NCS, Tabaton *et al.* (1985), with nucleolar counts at a single level, found a 37% loss, whereas Zweig *et al.* (1988) using nucleolar counts at a several levels, found no cell loss. However, these investigators did find a significant correlation between cell number in the rostral NCS levels and clinical depression.

## 3. Discussion



### 3.1. Age-Related Changes

#### 3.1.1. Summary of Changes and Their Topography

The presence and extent of age-related neuronal loss in subcortical nuclei that project to the cerebral cortex shows both a species difference and differences in the results of studies in the same nucleus in the same species. A similar conclusion was noted by Flood and Coleman (1988) in their analysis of the effect of age on cell numbers in a variety of brain areas in animals and in man. Many of these discrepancies undoubtedly reflect the inherent difficulty in conducting and interpreting the studies (Gundersen *et al.* 1988; Williams and Rakic, 1988; West, 1993). Despite such difficulties, there is evidence for age-related cell loss in many subcortical nuclei that project to the cerebral cortex. The strongest evidence is for neurons in the cholinergic basal forebrain, dopaminergic ventral mesencephalon (SNpc and VTA complex), median raphe, and locus coeruleus. Other areas that have shown evidence of cell loss are the lateral hypothalamic area, the septal nuclei, and the dorsal raphe nucleus. Although it is well documented that these nuclei project to the neocortex, it remains unclear whether neuronal loss is specifically related to the cells of origin of their neocortical projection. In the human LC, cell loss has been reported as occurring predominately in its rostral part and, to a lesser extent, in its middle part, by Chan-Palay and Asan (1989a), and by Marcyniuk *et al.* (1989) as showing no evidence of topography. The topography of the cortical projection from human LC to cerebral cortex is unknown. In monkey these LC projection neurons are found throughout the nucleus (Freedman *et al.*, 1975; Porrino and Goldman-Rakic, 1982; Doty, 1983). In a study by Loughlin *et al.* (1986) in the rat, the LC anterior pole was shown to have a strong projection to the hypothalamus, and the cerebral cortical projection was found to originate from the central part of the nucleus, midway between its dorsal and ventral parts. The relationships of age-related neuronal cell loss in the dopaminergic ventral mesencephalon to its neocortical projections is clearer: the more medial, cortically projecting part of this complex (Porrino and Goldman-Rakic, 1982) shows the most severe cell loss in humans (Fearnley and Lees, 1991). In a study by de Lacalle *et al.* (1991), age-related neuronal loss in the nucleus basalis of Meynert was most marked posteriorly, in an area that projects to the temporal pole. This area is not known to be a site of predilection for age-related changes in the cerebral cortex. Further, many neurons in these subcortical nuclei also project to others areas of the brain. This has been well documented for the LC, where individual cortically projecting neurons have been shown to project to multiple other areas of the brain (Lindvall and Björklund, 1983).

#### 3.1.2. Relationship to the Reticular Formation

The cortically projecting subcortical nuclei that show evidence of age-related neuron loss have in common a location within the reticulate core of the brain. Historically, the reticular formation has been defined as a meshlike intermixture of nerve cells and intersecting fibers within the innermost, subventricular parts of the brain (Yakovlev, 1952; Meyer, 1971). More recently, with the Golgi stain, it has been viewed as a neuronal substrate whose neurons have a characteristic dendritic configuration consisting of a few long, straight, randomly oriented intersecting dendrites with unusually long, thick, dendritic spines. In this view, the reticular formation extends as an uninterrupted formation from the spinal cord to the forebrain. In the brain stem it occupies the tegmentum, and at more rostral levels, the hypothalamus, midline, and intralaminar nuclei of the thalamus; and the basal forebrain, medial septum, and medial and central nuclei of the amygdala (Leontovich and Zhukova, 1963; Ramón-Moliner and Nauta, 1966; Scheibel and Scheibel, 1967; Ramón-Moliner, 1975). Ramón-Moliner and Nauta (1966) have labeled this cell type "isodendritic." In both human and animal brain, other parts of the reticular formation that do not project to the cerebral cortex have also shown evidence of neuron loss. In the human this age-related loss has been reported in the suprachiasmatic nucleus (Swaab *et al.,* 1985; Goudsmit *et al.,* 1990), and in the rat, in several other areas of the hypothalamus as well (Hsu and Peng, 1978; Sabel and Stein, 1981). In human amygdala, Herzog and Kemper (1980) also found that the greatest age-related neuronal loss occurred in the central and medial nuclei, nuclei that represent the most lateral extension of the reticular formation as an organized formation the forebrain. The extent to which other components of this formation also show age-related neuronal loss is unknown. The study of Rosales *et al.* (1968), showing age-related atrophy of mass intermedia of the thalamus, suggests that the atrophic change in the reticulate core of the brain may be more extensive than indicated by the available studies showing neuronal loss. Scheibel and Scheibel (1967), in their Golgi analysis of the nonspecific thalamic nuclei (which include the midline nuclei of the massa intermedia), found the neuronal architecture of this area similar to that of the reticular formation of the brain stem. Further, the only myelinated fiber system in the brain stem to show age-related loss of staining is the central tegmental tract (Yakovlev and Lecours, 1967), a major fiber system within the reticular formation of the tegmentum. The only areas investigated within the reticular formation that have not shown evidence of age-related neuronal loss are the hypothalamic paraventricular and supraoptic nuclei (Hsu and Peng, 1978; Goudsmit *et al.,* 1990; Vogels *et al.,* 1990). In a recent morphometric Golgi study of 14 different loci in the human reticular formation, Arendt *et al.* (1995a,b) found that in those areas that showed an age-related decrease in cell-packing density, the remaining isodendritic neurons showed an increase in number and extent of their dendritic arbors. This compensatory reorganization appears to be a characteristic of this specific cell type. In their Golgi study of the basal forebrain, Arendt *et al.* (1995a), recognized a multipolar neuronal type in addition to the isodendritic neurons, with the multipolar neuron failing to show these changes.

### 3.1.3. Other Areas of Cell Loss

Thus, evidence suggests that the reticular formation is a major target for age-related neuronal loss, affecting areas of the formation that project to the cerebral cortex as well as others that do not. Subcortical areas outside the reticulate core that do not project to the cerebral cortex have also shown evidence of age-related neuronal cell loss, suggesting additional areas of vulnerability. In the rat auditory system, Casey and Feldman (1982) found loss of superior olivary neurons, and Hall *et al.* (1975) found loss of Purkinje cells in the human cerebellum. The remaining studies of age-related changes in subcortical nuclei populations have found no evidence of neuronal loss. In the human brain these studies include the mammillary body (Wilkinson and Davies, 1978), the cranial nerve nuclei of the trochlear nerve (Vijayashankar and Brody, 1973) and abducens nerve (Vijayashankar and Brody, 1977), the ventral cochlear nucleus (Konigsmark and Murphy, 1970), and the inferior olive (Moatamed, 1966; Monagle and Brody, 1974). In the rat brain, Boseila *et al.* (1975) found no evidence of age-related neuronal loss in the red nucleus.

### 3.1.4. Relationship between Subcortical Changes and Changes in the Cerebral Cortex

Little is known about the structural changes in the cerebral cortex that occur as a result of neuron loss in the subcortical nuclei that project to it. In the human brain, Geula and Mesulam (1989) found only a modest age-related loss of AChE-stained fibers in the entorhinal cortex and in two different neocortical areas, and de Lacalle *et al.* (1994) noted a trend toward a decrease in density of ChAT- and AChE-stained fibers in the hippocampus and entorhinal cortex. Similarly, modern studies are also showing little age-related change in the neuronal cell layers of the cerebral cortex. Although earlier studies reported neuronal loss with age in both human (Brody, 1955, 1970; Colon, 1972; Shefer, 1972; Henderson *et al.*, 1980) and monkey (Brizzee, 1973; Brizzee *et al.*, 1975, 1980) cerebral cortex, more-recent studies in humans (Terry *et al.*, 1987; Haug, 1985; Haug and Eggers, 1991) and in the rhesus monkey (Vincent *et al.*, 1989; Peters and Sethares, 1993; Peters *et al.*, 1994) failed to find evidence of significant loss (see Peters, this volume). These latter studies on the rhesus monkey (Vincent *et al.*, 1989; Peters and Sethares, 1993; Peters *et al.*, 1994) and that of Peters *et al.* (1991) shows that in the cell layers of the cerebral cortex, there are only occasional degenerative changes in myelin, accumulation of lipofuscin pigment, and increase in microglial and oligodendroglial cells. In contrast to the paucity of changes in the cellular layers, layer I shows marked age-related changes in the rhesus monkey. These include a thinning of the layer, proliferation of glial cells process (especially in the glial-limiting membrane), thinning of the axonal fiber plexus, and degenerative changes in myelin, axons, dendrites, and neurons. These degenerative changes are of interest because layer I in the monkey has a prominent fiber plexus of cholinergic (Mesulam *et al.*, 1984b; Lewis, 1991), dopaminergic (Lewis *et al.*, 1987; Gaspar *et al.*, 1989, 1991; Berger *et al.*, 1991; van Domburg and ten Donkelaar, 1991; Williams and Goldman-Rakic, 1993), and serotonergic fibers (Kosofsky and Molliver, 1987; Lidow *et al.*, 1989; Wilson and Molliver, 1991), suggesting a relationship between the changes in layer I and those in the subcortical nuclei. Less likely is a relationship between degenerative changes in layer I and cell loss in the locus coeruleus, as the noradrenergic fiber plexus in layer I of the monkey is not well developed (Morrison *et al.*, 1982; Levitt *et al.*, 1984; Lewis *et al.*, 1986; Gaspar *et al.*, 1989; Lewis and Morrison, 1986).

Although the relationship between the age-related changes in layer I and its subcortical projections is not known, it is of interest that a relationship between these two areas is established at the earliest stages of cerebral cortical development. At this stage the prospective cerebral cortex is represented by a zone called the primitive plexiform layer, which is composed of neurons generated by the primitive ependyma embedded in an intrinsic and afferent plexus of axons. At later stages of development the primitive plexiform layer is divided into layer I and a subcortical zone, called the subplate, by the migration of neurons to the definitive cortical plate, layers II through VIa (Marin-Padilla, 1978; Raedler *et al.,* 1980). Identified axons in the primitive plexiform layer include noradrenergic, serotonergic, and cholinergic fibers (Molliver, 1982; Voigt and de Lima, 1991; Distler and Robertson, 1992). The occurrence of age-related changes in layer I—a remnant of the primitive plexiform layer—and in the reticulate core of the brain thus suggests that a major focus of these is in phylogenetically old areas of the brain. It is also interesting that the most highly and recently evolved area, the neocortex, shows little evidence of age-related change. Further, there is a greater complexity and topographical differentiation of these subcortical afferents to the cerebral cortex in monkey and human brain as compared with that of the rodent. This suggests a corresponding increase in the precision of this relationship to the various cytoarchitectonic areas in the primate brain, perhaps making some cytoarchitectonic areas more vulnerable than others to age-related changes in these afferent projections.

### 3.2. Disease-Related Changes

#### 3.2.1. Summary of Changes and Their Topography

Superimposed on age-related cell loss in reticular formation subcortical nuclei that project to the cerebral cortex is further cell loss in these nuclei in AD. This predilection for involvement in AD has been well documented since it was first pointed out by Rossor (1981) and Benton *et al.* (1982). All studies reviewed found cell loss in AD in the NB, VTA, and LC. Virtually all studies of the SN also found evidence of neuronal loss in AD. The single exception for the SN was the van Domburg and ten Donkelaar (1991) study of only four AD brains. The effect of AD on the dorsal and median raphe nuclei is less clear, with most studies showing cell loss in the dorsal raphe (Curcio and Kemper, 1984; Yamamoto and Hirano, 1985; Zweig *et al.,* 1988) and two studies (Tabaton *et al.,* 1985; Zweig *et al.,* 1988) showing cell loss in AD in the median raphe. The effect of AD on neuronal cell numbers in the cortically projecting areas in the hypothalamus and in the medial parabrachial nucleus is unknown.

This predilection in AD for involvement of reticular formation subcortical nuclei that project to the cerebral cortex is best shown by the distribution pattern of the neurofibrillary tangle (NFT), a key pathological change in AD that is associated with neuronal cell loss. With appropriate staining techniques, the NFT stands out as a distinctive feature, permitting its selective distribution pattern in these nuclei to be readily seen. Within the forebrain, the NFT is found in the midline and intralaminar nuclei of the thalamus and in the lateral, lateral tuberal, and posterior hypothalamic areas (German *et al.,* 1987; Saper and German, 1987; Grossi *et al.,* 1989; Braak and Braak, 1991a,b). In the brain stem, it is found in the mesencephalon in the substantia nigra and the VTA. In the pons NFTs have been noted in the dorsal and median raphe nuclei, pedunculopontine nucleus, medial parabrachial nucleus, and the locus coeruleus (Curcio and Kemper, 1984; Yamamoto and Hirano, 1985; Tabaton *et al.*, 1985; German *et al.*, 1987; Jellinger, 1988; Mufson *et al.*, 1988; Gibb *et al.*, 1989).

Many of these nuclei in AD brains show the same cell-loss topography as that of age-related change. As in the aged human brain, cell loss in AD is most marked in the NB at the more posterior levels (Etienne *et al.*, 1986; Doucette and Ball, 1987; Wilcock *et al.*, 1988) and is also greater in the VTA complex than in the subtantia nigra (Mann *et al.*, 1987; van Domburg and ten Donkelaar, 1991; Uchihara *et al.*, 1992). Further, as in aged human brain, the topography of neuronal loss in the LC varies from report to report, with no consistently noted pattern. Chan-Palay and Asan (1989b) and German *et al.* (1992) reported the largest age-related loss in the rostral part of the LC, Marcyniuk *et al.* (1986) in the dorsal part, and Zweig *et al.* (1988) at the midpart of the nucleus. A similar pattern in aging and in AD can be seen in hypothalamic nuclei that do not project to the cerebral cortex. The suprachiasmatic nucleus in the hypothalamus, which shows age-related cell loss, displays an even more marked loss in AD (Swaab *et al.*, 1985; Goudsmit *et al.*, 1990), and the supraoptic and paraventricular nuclei, which have shown no evidence of age-related cell loss, have also shown no cell loss in AD in these studies (Swaab *et al.*, 1985; Goudsmit *et al.*, 1990) and in a study by Mann *et al.* (1985a). Further, neuronal loss in the medial and central nuclei of the amygdala is greater in AD than in the aged brain (Herzog and Kemper, 1980).

### 3.2.2. Relationship between Subcortical Changes and Changes in the Cerebral Cortex

The neuronal loss in subcortical nuclei that project to the cerebral cortex in AD is associated with marked pathological changes in neocortex (see Braak and Braak; Ginsburg *et al.*; and Maslia and Salmon, all in this volume). The changes include dense accumulations of neuritic (senile) plaques and neurofibrillary tangles (Arnold *et al.*, 1991; Braak and Braak, 1991a) and neuronal loss that is most marked in layers III and V (Mann *et al.*, 1985b; Hof *et al.*, 1990, and this volume). Although a mechanism of the neuronal loss in these cortically projecting subcortical nuclei in AD is unknown, its topography is similar to that found in the aged brain, and its greater severity in AD is associated with florid pathological changes in the cerebral cortex. Further, in AD this subcortical cell loss in the reticular formation is also associated with cell loss and NFT formation in other brain areas that project to the cerebral cortex, such as the amygdala (Herzog and Kemper, 1980; Vereecken *et al.*, 1994) and the medial dorsal nucleus of the thalamus (Paskavitz *et al.*, 1995), as well as with areas of the brain unrelated to the cerebral cortex, such as the ventral striatum (Lehéricy *et al.*, 1989; Oyanagi *et al.*, 1991) and the retinal ganglion cells (Hinton *et al.*, 1986).

Pick's disease, another dementia with profound loss of cortical neurons, is of little help in understanding the relationship between the cerebral cortex and the subcortical nuclei, as little is known about the details of its subcortical involvement. In the NB, Tagliavini *et al.* (1983) found no cell loss, whereas Uhl *et al.* (1983) found substantial loss. In the LC some, but not all, brains have shown cell loss. In neither of these nuclei is the loss topography known. In the SN, the cell loss is greatest in the medial and central parts of the nucleus (Kosaka *et al.*, 1991). In Pick's disease, however, there are frequent pathological changes in the striatum

(Kosaka *et al.*, 1991), a major target of the nigral projection. Cell loss topography in the subcortical nuclei in Parkinson's disease, in which the predominant pathology is subcortical, is different than that found in aging and in AD, suggesting that cell loss in the latter may be related to the cerebral cortex. In PD the well-known and characteristic neuronal loss in the SN, VTA, and LC is uniformly distributed throughout these nuclei (Bogerts, *et al.*, 1983; Chan-Palay and Asan, 1989b; Fearnley and Lees, 1991; Gibb and Lees, 1991; German *et al.*, 1992). Similarly, Arendt *et al.* (1995a) noted that cell loss in the basal forebrain in PD is more uniformly distributed than that of age-related change. Another disease emphasizing subcortical pathology is progressive supranuclear palsy (PSP). In this dementing illness, there is marked cell loss and NFT formation in the SN, LC, the median and dorsal raphe nuclei, and to a lesser extent, the NB. These changes, however, occur in association with widespread, marked cell loss and NFT formation in many other areas of the brain, and nothing is known about the topography of this loss in individual nuclei (Lowe *et al.*, 1997).

## 4. Summary and Conclusions

Age-related changes in subcortical nuclei that project to the cerebral cortex appear to be confined to the reticulate core of the brain, and those in the cerebral cortex are found primarily in its molecular layer, layer I. A relationship between these two areas is established at the earliest stages of cortical development, and in the adult brain, layer I receives a heavy projection from many of the subcortical nuclei that show age-related neuronal loss. In Alzheimer's disease, in which the cerebral cortical pathology is much more marked than that found in normal aging, the topography of the cell loss in these subcortical nuclei is similar, but the extent of loss is more marked than that found in aging. This suggests that the age-related changes in the subcortical nuclei and in cerebral cortex may be directly related. If they are related, however, it is unknown which changes occur first. Another way to view the interrelationship between the age-related changes in subcortical nuclei and in the molecular layer of the cerebral cortex is that both may occur independently, and that these two phylogenetically old areas of the brain are both major targets for age-related change. A third possibility is a combination of both mechanisms.

## 5. References

Aletrino, M. A., Vogels, O. J., Van Domburg, P. H., and ten Donkelaar, H. J., 1992, Cell loss in the nucleus raphe dorsalis in Alzheimer's disease, *Neurobiol. Aging* **13:**461–468.

Allen, S. J., Dawbarn, D., and Wilcock,, G. K., 1988, Morphometric immunohistochemical analysis of neurons in the nucleus basalis of Meynert in Alzheimer's disease, *Brain Res.* **454:**275–281.

Altavista, M. C., Rossi, P., Bentivoglio, A. R., Crociani, P., and Albanese, A., 1990, Aging is associated with a diffuse impairment of forebrain cholinergic neurons, *Brain Res.* **508:**51–59.

Amaral, D. G., 1993, Morphological analysis of the brains of behaviorally characterized aged nonhuman primates, *Neurobiol. Aging* **14:**671–672.

Arango, V., Underwood, M. D., and Mann, J. J., 1994, Fewer pigmented neurons in the locus coeruleus of uncomplicated alcoholics. *Brain Res.* **650:**1–8.

Arendt, T., Bigl, V., Arendt, A., and Tennstedt, A., 1983, Loss of neurons in the nucleus basalis of Meynert in Alzheimer's disease, paralysis agitans, and Korsakoff's disease, *Acta Neuropathol.* **61:**101–108.

Arendt, T., Bigl, V., Tennstedt, A., and Arendt, A., 1984, Correlation between cortical plaque count and neuronal loss in the nucleus basalis in Alzheimer's Disease, *Neurosci. Lett.* **48:**81–85.

Arendt, T., Bigl, V., Tennstedt, A., and Arendt, A., 1985. Neuronal loss in different parts of the nucleus basalis is related to neurite plaque formation in cortical target areas in Alzheimer's disease, *Neuroscience* **14:**1–14.

Arendt, T., Bruckner, M. K., Bigl, V., and Marcova, L., 1995a, Dendritic reorganization in the basal forebrain under degenerative conditions and its defects in Alzheimer's disease. II. Ageing, Korsakoff's disease, Parkinson's disease, and Alzheimer's disease, *J. Comp. Neurol.* **351:**189–222.

Arendt, T., Bruckner, M. K., Bigl, V., and Marcova, L., 1995b, Dendritic reorganization in the basal forebrain under degenerative conditions and its defects in Alzheimer's disease. III. The basal forebrain compared with other subcortical areas, *J. Comp. Neurol.* **351:**223–246.

Arima, K., and Akashi, T., 1990, Involvement of the locus coeruleus in Pick's disease with and without Pick body formation, *Acta Neuropathol.* **79:**629–633.

Arnold, S. E., Hyman, B. T., Flory, J., Damasio, B. T., and Van Hoesen, G. W., 1991, The topographical distribution of neurofibrillary tangles and neuritic plaques in the cerebral cortex of patients with Alzheimer's disease, *Cereb. Cortex* **1:**103–116.

Aston-Jones, G., Rogers, J., Shaver, R. D., Dinan, T. G., and Moss, D. E., 1985, Age-impaired impulse flow from nucleus basalis to cortex, *Nature* **318:**462–464.

Azmitia, E. C., and Gannon, P. G., 1986, The primate serotonergic system: A review of human and animal studies and a report on *Macaca fascicularis, Adv. Neurol.* **43:**407–468.

Baker, K. B., Törk, I., Hornung, J.-P., and Halasz, P., 1989, The human locus coeruleus complex: An immunohistochemical and three dimensional reconstruction study, *Exp. Brain Res.* **77:**257–270.

Baker, K. G., Halliday, G. M., and Törk, I., 1990, Cytoarchitecture of the human dorsal raphe nucleus, *J. Comp. Neurol.* **301:**147– 161.

Baker, K. G., Halliday, G. M., Halasz, P., Hornung, J.-P., Geffen, L. B., Cotton, R. G. H., and Törk, I., 1991, Cytoarchitecture of serotonin-synthesizing neurons in the pontine tegmentum of the human brain, *Synapse* **7:**301–320.

Benton, J. S., Bowen, D. M., Allen, S. J., Haan, E. A., Davidson, A. N., Neary, D., Murphy, R. P., and Snowden, J. S., 1982, Alzheimer's disease as a disorder of isodentritic core, *Lancet* **i:**456.

Berger, B., Trottier, S., Verney, C., Gaspar, P., and Alvarez, C., 1988, Regional and laminar distribution of the dopamine and serotonin innervation in the macaque cerebral cortex: A radioautographic study, *J. Comp. Neurol.* **273:**99–119.

Berger, B., Gaspar, P., and Verney, C., 1991, Dopaminergic innervation of the cerebral cortex: Unexpected differences between rodent and primate, *Trends Neurosci.* **14:**21–27.

Bigl, V., Woolf, N. J., and Butcher, L. L., 1982, Cholinergic projection from the basal forebrain to frontal, parietal, temporal, occipital, and cingulate cortices: A combined fluorescent tracer and acetylcholinesterase analysis, *Brain Res. Bull* **8:**727–749.

Bigl, V., Arendt, T., Fischer, S., Werner, M., and Arendt, A., 1987, The cholinergic system in aging, *Gerontology* **33:**172–180.

Björklund, A., and Lindvall, O., 1984, Dopamine-containing systems in the CNS, in: *Handbook of Chemical Neuroanatomy,* Vol. 2, Pt. 1 (A. Björklund and T. Hökfelt, eds.), Elsevier, Amsterdam, pp. 55–122.

Bogerts, B., 1981, A brainstem atlas of catecholaminergic neurons in man, using melanin as a natural marker, *J. Comp. Neurol.* **197:**63–80.

Bogerts, B., Häntsch, J., and Herzer, M., 1983, A morphometric study of the dopamine-containing cell groups in the mesencephalon of normals, Parkinson patients, and schizophrenics, *Biol. Psychiatry* **18:**951–969.

Bondareff, W., Mountjoy, C. Q., and Roth, M., 1982, Loss of neurons of origin of adrenergic projections to cerebral cortex (nucleus locus coeruleus) in senile dementia, *Neurology* **32:**164–168.

Boseila, A.-W. A., Hashen, S. M., and Badawy, Y. H., 1975, Volumetric studies on the red nucleus of the rat at different ages, *Acta Anat.* **91:**175–180.

Bowden, D. M., German, D. C., and Poynter, W. D., 1978, An autoradiographic, semistereotaxic mapping of major projections from locus coeruleus and adjacent nuclei in *Macaca mulatta, J. Comp. Neurol.* **145:**257–276.

Braak, H., and Braak, E., 1986, Nuclear configuration and neuronal types of the nucleus nigra in the brain of the human adult, *Hum. Neurobiol.* **5:**71–82.

Braak, H., and Braak, E., 1991a, Neuropathological staging of Alzheimer-related changes, *Acta Neuropathol.* **82:**239–259.

Braak, H., and Braak, E., 1991b, Alzheimer's disease affects limbic nuclei of the thalamus, *Acta Neuropathol.* **81:**261–268.

Brizzee, K. R., 1973, Quantitative studies on aging changes in cerebral cortex of rhesus monkey and albino rat with notes of prolonged, low-dosage ionizing radiation in the rat, *Prog. Brain Res.* **40:**141–160.

Brizzee, K. R., Klara, P., and Johnson, J. E., 1975, Changes in microanatomy, neurocytology, and fine structure with aging, in: *Neurobiology of Aging* (J. M. Ordy and K. R. Brizzee, eds.), Plenum, New York, pp. 575–594.

Brizzee, K. R., Ordy, J. M., and Bartus, R., 1980, Localization of cellular changes within multimodal sensory regions in aged monkey brain: Possible implications for age-related cognitive loss, *Neurobiol. Aging* **1:**45–52.

Brody, H., 1955, Organization of the cerebral cortex. III. A study of aging in the human cerebral cortex, *J. Comp. Neurol.* **102:**511–556.

Brody, H., 1970, Structural changes in the aging nervous system, in: The Regulatory Role of the Nervous System in Aging: Interdisciplinary Topics in Gerontology (T. H. Blmenyhol, ed.), S. Karger, Basal, Switzerland, pp. 9–21.

Brown, R. M., Crane, A. M., and Goldman, P. S., 1979, Regional distribution of monoamines in the cerebral cortex and subcortical structures of the rhesus monkey: Concentrations and *in vivo* synthesis, *Brain Res.* **168:**133–150.

Campbell, M. J., Lewis, D. A., Foote, S. L., and Morrison, J. H., 1987, Distribution of choline acetyltransferase-, serotonin-, dopamine-β-hydroxylase-, tyrosine hydroxylase-immunoreactive fibers in monkey primary auditory cortex, *J. Comp. Neurol.* **261:**209–220.

Candy, J. M., Perry, R. H., Perry, E. K., Irving, D., Blessed, G., Fairbairn, A. F., and Tomlinson, B. E., 1983, Pathological changes in the nucleus of Meynert in Alzheimer's disease and Parkinson's disease, *J. Neurol. Sci.* **59:**277–289.

Casey, M. A., and Feldman, M. L., 1982, Aging in the rat medial nucleus of the trapezoid body. I. Light microscopy, *Neurobiol. Aging* **3:**187–195.

Chan-Palay, V., and Asan, E., 1989a, Quantitation of catecholamine neurons in the locus coeruleus in human brains of normal young and older adults and in depression, *J. Comp. Neurol.* **287:**357–372.

Chan-Palay, V., and Asan, E., 1989b, Alterations in catecholamine neurons of the locus coeruleus in senile dementia of the Alzheimer type and in Parkinson's disease with and without dementia and depression, *J. Comp. Neurol.* **287:**373–392.

Chen, C. P., Eastwood, S. L., McDonald, B., Hope, R. A., Esiri, M. M., 1994, Patterns and markers of cell loss in the dorsal raphe nucleus in Alzheimer's disease. *Neuropathol. Appl. Neurobiol.* **20:**503.

Chui, H. C., Bondereff, W., Zarow, C., and Slager, U., 1984, Stability of neuron number in the human nucleus basalis of Meynert with age, *Neurobiol. Aging* **5:**83–88.

Chui, H. C., Mortimer, J. A., Slager, U., Zarow, C., Bondereff, W., and Webster, D. D., 1986, Pathologic correlates of dementia in Parkinson's disease, *Arch. Neurol.* **43:**991–995.

Coggeshall, R. E., and Lekan, H. A., 1996, Methods for determining number of cells and synapses: A case for more uniform standards of review, *J. Comp. Neurol.* **364:**6–15.

Colon, E. J., 1972, The elderly brain: A quantitative analysis in the cerebral cortex of two cases, *Psychiat. Neurol. Neurochir.* **75:**261–270.

Coyle, J. T., Price, D. L., and DeLong, M. R., 1983, Alzheimer's disease: A disorder of cortical cholinergic innervation, *Science* **2.9:**1184–1190.

Cruz-Sánchez, F. F., Cardozo, A., and Tolosa, E., 1995, Neuronal changes in the substantia nigra with aging: A Golgi study, *J. Neuropathol. Exp. Neurol.* **54:**74–81.

Curcio, C. A., and Kemper, T. L., 1984, Nucleus raphe dorsalis in senile dementia: Neurofibrillary changes and neuronal cell packing density, *J. Neuropathol. Exp. Neurol.* **43:**359–368.

de Lacalle, S., Iraizoz, I., and Gonzalo, L. M., 1991, Differential changes in cell size and number in topographic subdivisions of human basal nucleus in normal aging, *Neuroscience* **43:**445–456.

de Lacalle, S., Lim, C., Sobreviela, T., Mufson, E. J., Hersh, L. B., and Saper, C. B., 1994, Cholinergic innervation in the human hippocampal formation, including the entorhinal cortex, *J. Comp. Neurol.* **345:**321–344.

Descarries, L., Berthelet, F., Garcia, S., and Beaudet, A., 1986, Dopaminergic projection from nucleus raphe dorsalis to neostriatum in the rat, *J. Comp. Neurol.* **249:**511–520.

Distler, P. G., and Robertson, R. T., 1992, Development of AChE-positive neuronal projections from basal forebrain to cerebral cortex in organotypic tissue slice cultures, *Dev. Brain Res.* **67:**181–196.

Doty, R. W., 1983, Nongeniculate afferents to striate cortex in macaques, *J. Comp. Neurol.* **218:**157–173.

Doucette, R., and Ball, M. J., 1987, Left–right symmetry of neuronal cell counts in the nucleus basalis of Meynert of control and of Alzheimer's disease brains, *Brain Res.* **422:**357–360.

Eckenstein, F., and Baughman, R. W., 1987, Cholinergic innervation in cerebral cortex, in: *Cerebral Cortex, Vol. 6* (E. G. Jones and A. Peters, eds.), Plenum, New York, pp. 129–160.

Etienne, P., Robitaille, Y., Gauthier, S., and Nair, N. P. V., 1986, Nucleus basalis neuronal loss and neuritic plaques in advanced Alzheimer's disease, *Can. J. Physiol. Pharmacol.* **64:**318–324.

Evans, D., Funkenstein, H. H., Albert, M. S., Scherr, P. A., Cook, N. R., Chown, M. J., Hebert, L. E., Hennekens, C. H., and Taylor, J. O., 1989, Prevalence of Alzheimer's disease in a community population of older persons higher than previously reported, *J. Am. Med. Assoc.* **262:**2551–2556.

Fallon, J. H., and Loughlin, S. E., 1987, Monoamine innervation of cerebral cortex and a theory of the role of monoamines in cerebral cortex and basal ganglia, in: *Cerebral Cortex, Vol. 6* (E. G. Jones and A. Peters, eds.), Plenum, New York, pp. 41–127.

Fearnley, J., and Lees, A. J., 1991, Ageing and Parkinson's disease: Substantia nigra regional selectivity, *Brain* **114:**2283–2301.

Felten, D. L., and Sladek, J. R., 1983, Monoamine distribution in primate brain. V. Monoaminergic nuclei: Anatomy, pathways, and local organization, *Brain Res. Bull.* **10:**171–284.

Felten, D. L., Laties, A. M., and Carpenter, M. B., 1974, Monoamine-containing cell bodies in the squirrel monkey brain, *Am. J. Anat.* **139:**153–166.

Finley, K. H., and Cobb, S., 1940, The capillary bed of the locus coeruleus, *J. Comp. Neurol.* **73:**49–58.

Fischer, W., Wictorin, K., Björklund, A., Williams, L. R., Varon, S., and Gage, F. H., 1987, Amelioration of cholinergic neuron atrophy and spacial memory impairment in aged rats by nerve growth factor, *Nature* **329:**65–68.

Flood, D. G., and Coleman, P. D., 1988, Neuron numbers and sizes in aging brain: Comparison of human, monkey, and rodent data, *Neurobiol. Aging* **9:**453–463.

Foley, J. M., and Baxter, D., 1958, On the nature of pigment granules in the cells of the locus coeruleus and substantia nigra, *J. Neuropathol. Exp. Neurol.* **17:**586–598.

Ford, B., Holmes, C. J., Mainville, L., and Jones, B. E., 1995, GABAergic neurons in the rat pontomesencephalic tegmentum: Codistribution with cholinergic and other tegmental neurons projecting to the posterior hypothalamus, *J. Comp. Neurol.* **363:**177–196.

Forno, L. S., 1978, The locus caeruleus in Alzheimer's disease. *J. Neuropathol. Exp. Neurol.* **37:**614.

Förstl, H., Bruns, A., Luthert, P., Cairns, N., Lantos, P., and Levy, R., 1992, Clinical and neuropathological correlates of depression in Alzheimer's disease, *Psychol. Med.* **22:**877–884.

Freedman, R., Foote, S. L., and Bloom, F. E., 1975, Histochemical characterization of neocortical projection of the nucleus locus coeruleus in the squirrel monkey, *J. Comp. Neurol.* **164:**209–232.

Gage, F. H., Björklund, A., Stenevi, U., Dunnett, S. B., and Kelly, P. A., 1984, Intrahippocampal septal grafts ameliorate learning impairments in aged rats, *Science* **225:**533–535.

Gaspar, P., Berger, B., Febvret, A., Vigny, A., and Henry, J. P., 1989, Catecholamine innervation of the human cerebral cortex as revealed by comparative immunohistochemistry of tyrosine hydroxylase and dopamine-beta-hydroxylase, *J. Comp. Neurol.* **279:**249–271.

Gaspar, P., Stepniewska, I., and Kaas, J. H., 1992, Topography and collateralization of dopaminergic projections to motor and lateral prefontal cortex in owl monkeys, *J. Comp. Neurol.* **325:**1–21.

German, D. C., White, C. L., and Sparkman, D. R., 1987, Alzheimer's disease: Neurofibrillary tangles in nuclei that project to the cerebral cortex, *Neuroscience* **21:**305–312.

German, D. C., Walker, B. S., Manaya, K., Smith, W. K., Woodward, D. J., and North, A. J., 1988, The human locus coeruleus: Computer reconstruction of cellular distribution, *J. Neurosci.* **8:**1776–1788.

German, D. C., Manaya, K. F., White, C. L., Woodward, D. J., McIntire, D. D., Smith, W. K., Kalaria, R. N., and Mann, D. M. A., 1992, Disease-specific patterns of locus coeruleus cell loss, *Ann. Neurol.* **32:**667–676.

Geula, C., and Mesulam, M.-M., 1989, Cortical cholinergic fibers in aging and Alzheimer's disease: A morphometric study, *Neuroscience* **33:**469–481.

Gibb, W. R. G., 1991, Neuropathology of the substantia nigra, *Eur. Neurol. (suppl. 1)* **31:**48–59.

Gibb, W. R. G., and Lees, A. J., 1991, Anatomy, pigmentation, central and dorsal subpopulations of the substantia nigra, and differential cell death in Parkinson's disease, *J. Neurol. Neurosurg. Psychiatry* **54:**388–396.

Gibb, W. R. G., Mountjoy, C. Q., Mann, D. M. A., and Lees, A.-J., 1989, The substantia nigra and ventral tegmental area in Alzheimer's disease and Down's syndrome, *J. Neurol. Neurosurg. Psychiatry* **52:**193–200.

Gilad, G. M., Rabey, J. M., Tizabi, Y., and Gilad, V. H., 1987, Age-dependent loss and compensatory changes of septohippocampal cholinergic neurons in two rat strains differing in longevity and response to stress, *Brain Res.* **436:**311–322.

Goldman, G., and Coleman, P. D., 1981, Neuronal numbers in locus coeruleus do not change with age in Fisher 344 rat, *Neurobiol. Aging* **2:**33–36.

Goldman-Rakic, P. S., Leranth, C., Williams, M. S., Mons, N., and Geffard, M., 1989, Dopamine synaptic complex with pyramidal neurons in primate cerebral cortex, *Proc. Natl. Acad. Sci. USA* **86:**9015–9019.

Gorry, J. D., 1963, Studies on the comparative anatomy of the ganglion basale of Meynert, *Acta Anat.* **55:**1–104.

Goudsmit, E., Hofman, M. A., Fliers, E., and Swaab, D. F., 1990, The supraoptic and paraventricular nuclei of the human hypothalamus in relation to sex, age, and Alzheimer's disease, *Neurobiol. Aging* **11:**529–536.

Grossi, D., Lopez, O. L., and Martinez, A. J., 1989, Mammillary bodies in Alzheimer's disease, *Acta Neuropathol. Scand.* **80:**41–45.

Gundersen, H. J. G., Bendsten, T. F., Korbo, L., Marcussen, N., Møller, A., Nielsen, K., Nyengaard, J. R., Pakkenberg, B., Sørensen, F. B., Vesterby, A., and Weset, M. J., 1988, Some new, simple, and efficient stereological methods and their use in pathological research and diagnosis, *Acta. Pathol. Microbiol. Immunol. Scand.* **96:**379–384.

Hall, T. C., Miller, A. K. H., and Corsellis, J. A. N., 1975, Variation in human Purkinje cell population according to age and sex, *Neuropathol. Appl. Neurobiol.* **1:**267–292.

Halliday, G. M., and Törk, I., 1986, Comparative anatomy of the ventromedial mesencephalic tegmentum in the rat, cat, monkey, and human, *J. Comp. Neurol.* **252:**423–445.

Handcock, M. B., and Fougerousse, C. L., 1976, Spinal projections from the nucleus locus coeruleus and nucleus subcoeruleus in the cat and monkey as demonstrated by the retrograde transport of horseradish peroxidase, *Brain. Res. Bull.* **1:**229–234.

Hassler, R., 1937, Zur Normalanatomie der Substantia nigra. Versucheiner architektonischen Gliederung, *J. Psychol. Neurol.* **48:**1–55.

Hassler, R., 1938, Zur Pathologie der Paralysis agitans und des postenzephalitischen Parkinsonismus, *J. Psychiat. Neurol.* **48:**387–476.

Haug, H., 1985, Are neurons of the human cerebral cortex really lost during aging? A morphometric examination, in: *Senile Dementia of the Alzheimer Type* (J. Traber and W. H., eds.), Springer-Verlag, Berlin, pp. 150–1163.

Haug, H., and Eggers, R., 1991, Morphometry of the human cortex cerebri and corpus striatum during aging, *Neurobiol. Aging* **12:**336–338.

Henderson, G., Tomlinson, B. E., and Gibson, P. H., 1980, Cell counts in the human cerebral cortex in normal adults throughout life using an image analysis computer, *J. Neurol. Sci.* **46:**113–136.

Herzog, A. G., and Kemper, T. L., 1980, Amygdaloid changes in aging and senile dementia, *Arch. Neurol.* **37:**625–629.

Hinton, D. R., Sadum, A. A., Blanks, J. C., and Miller, C. A., 1986, Optic nerve degeneration in Alzheimer's disease, *New Engl. J. Med.* **315:**485–487.

Hof, P. R., Bouras, C., Constantinidis, J., and Morrison, J. H., 1990, Selective disconnection of specific visual association pathways in cases of Alzheimer's disease presenting as Balint's syndrome, *J. Neuropathol. Exp. Neurol.* **49:**168–184.

Hohmann, C. F., Brooks, A. R., and Coyle, J. T., 1988, Neonatal lesions of the basal forebrain cholinergic neurons result in abnormal cortical development, *Brain Res.* **470:**253–264.

Hohmann, C. F., Kwiterovich, K. K., Oster-Granite, M. L., and Coyle, J. T., 1991, Newborn basal forebrain lesions disrupt cortical cytodifferentiation as visualized by rapid Golgi staining, *Cereb. Cortex* **1:**143–157.

Hornberger, J. C., Buell, S. J., Flood, D. G., McNeill, T. H., and Coleman, P. D., 1985, Stability of numbers but not size of mouse forebrain cholinergic neurons to 53 months, *Neurobiol. Aging* **6:**269–275.

Houser, C. R., Crawford, G. D., Salvaterra, P. M., and Vaughn, J. E., 1985, Immunocytochemical localization of choline acetyltransferase in rat cerebral cortex: A study of cholinergic neurons and synapses, *J. Comp. Neurol.* **234:**17–34.

Hsu, H. K., and Peng, M. T., 1978, Hypothalamic neuron number of old female rats, *Gerontology* **24:**434–440.

Ichimiya, Y., Arai, H., Kosaka, K., and Iizuka, R., 1986, Morphological and biochemical changes in the cholinergic and monoaminergic system in Alzheimer's disease, *Acta Neuropathol.* **70:**112–116.

Itakura, T., Yakate, H., Kimura, H., Kamei, I., Nakakita, K., Naka, Y., Nakai, K., Imai, H., and Komai, N., 1985, 5-hydroxytryptamine innervation of vessels in the rat cerebral cortex: Immunohistochemical finding and hydrogen clearance studies, *J. Neurosurg.* **62:**42–47.

Iversen, L. L., Rossor, M. N., Reynolds, G. P., Hills, R., Roth, M., Mountjoy, C. Q., Foote, S. L., Morrison, J. H., and Bloom, F. E., 1983, Loss of pigmented dopamine-β-hydroxylase positive cells from locus coeruleus in senile dementia of Alzheimer's type, *Neurosci. Lett.* **39:**95–100.

Jellinger, K., 1988, The pedunculopontine nucleus in Parkinson's disease, progressive supranuclear palsy, and Alzheimer's disease, *J. Neurol. Neurosurg. Psychiatry* **51:**540–543.

Kemper, T. L., 1993, The relationship of cerebral cortical changes to nuclei in the brainstem, *Neurobiol. Aging* **14:**659–660.

Kemper, T. L., Moss, M. B., Rosene, D. L., and Killiany, R. J., 1997, Age-related neuronal loss in the nucleus centralis superior of the rhesus monkey, *Acta Neuropathol.* **94:**124–130.

Koh, S., Chang, P., Collier, T. J., and Loy, R., 1989, Loss of NGF receptor immunoreactivity in basal forebrain neurons of aged rats: Correlation with spatial memory impairment, *Brain Res.* **498:**397–404.

Konigsmark, B., and Murphy, E., 1970, Neuronal populations in the human brain, *Nature* **228:**1335–1336.

Kosaka, K., Ikeda, K., Kobayashi, K., and Mehraein, P., 1991, Striatopallidonigral degeneration in Pick's disease: A clinicopathological study of 41 cases, *J. Neurol.* **238:**151–160.

Kosofsky, B. E., and Molliver, M., 1987, The serotoninergic innervation of cerebral cortex: Different classes of axon terminals arise from dorsal and median raphe nuclei, *Synapse* **1:**153–168.

Lehéricy, S., Hirsch, E. C., Cervera, P., Hersh, L. B., Hauw, J.-J., Ruberg, M., and Agid, Y., 1989, Selective loss of cholinergic neurons in the ventral striatum of patients with Alzheimer's disease, *Proc. Natl. Acad. Sci. USA* **86:**8580–8584.

Lehéricy, S., Hirsch, E. C., Cervera-Piérot, P., Hersh, L. B., Bakchine, S., Piette, F., Duyckaerts, C., Hauw, J.-J., Javoy-Agid, F., and Agid, Y., 1993, Heterogeneity and selectivity of the degeneration of cholinergic neurons in the basal forebrain of patients with Alzheimer's disease, *J. Comp. Neurol.* **330:**15–31.

Leontovich, T. A., and Zhukova, G. P., 1963, The specificity of the neuronal structure and topography of the reticular formation in the brain and spinal cord of carnivora, *J. Comp. Neurol.* **121:**347–379.

Levitt, P., Rakic, P., and Goldman-Rakic, P., 1984, Regional-specific distribution of catecholamine afferents in primate cerebral cortex: A fluorescence histochemical analysis, *J. Comp. Neurol.* **227:**23–36.

Lewis, D. A., 1991, Distribution of choline acetyltransferase immunoreactive axons in monkey frontal cortex, *Neuroscience* **40:**363–374.

Lewis, D. A., and Morrison, J. H., 1986, Noradrenergic innervation of monkey prefontal cortex: A dopamine-β-hydroxylase immunohistochemical study, *J. Comp. Neurol.* **282:**317–330.

Lewis, D. A., Campbell, M. J., Foote, S. L., and Morrison, J. H., 1986, The monoaminergic innervation of primate neocortex, *Hum. Neurobiol.* **5:**181–188.

Lewis, D. A., Campbell, M. J., Foote, S. L., Goldstein, M., and Morrison, J. H., 1987, The distribution of tyrosine hydroxylase-immunoreactive fibers in primate neocortex is widespread but regionally specific, *J. Neurosci.* **7:**279–290.

Lidow, M. S., Goldman-Rakic, P. S., Gallager, D. W., and Rakic, P., 1989, Quantitative autoradiographic mapping of serotonin 5-HT1 and 5-HT2 receptors and uptake sites in the neocortex of the rhesus monkey, *J. Comp. Neurol.* **280:**27–42.

Lindvall, O., and Björklund, A., 1983, Dopamine- and norepinephrine-containing neuron systems: Their anatomy in the rat brain, in: *Chemical Neuroanatomy* (G. Emson, ed.), Raven, New York, pp. 229–255.

Lolova, I., and Davidoff, M., 1992, Age-related changes in serotonin-immunoreactive neurons in the rat nucleus raphe dorsalis and nucleus centralis superior: A light microscope study, *Mech. Aging Dev.* **62:**279–289.

Lohr, J. B., and Jeste, D. V., 1988, Locus coeruleus morphometry in aging and schizophrenia, *Acta Psychia. Scand.* **77**:689–697.

Loughlin, S. E., Foote, S. L., and Bloom, F. E., 1986, Efferent projections of nucleus locus coeruleus: Topographic organization of cells of origin demonstrated by three-dimensional reconstruction, *Neuroscience* **2**:291–306.

Lowe, J., Lennox, G., and Leigh, P. N., 1997, Disorders of movement and system degenerations, in: *Greenfield's Neuropathology,* Vol. 2 (D. I. Graham and P. L. Lantos, eds.), Oxford University Press, New York, pp. 281–366.

Lowes-Hummel, P., Gertz, H.-J., Ferszt, R., and Cervos-Navarro, J., 1989, The basal nucleus of Meynert revised: The nerve cell number decreases with age, *Arch. Gerontol. Geriat.* **8**:21–27.

Manaye, K. F., McIntire, D. D., Mann, D. M. A., and German, D. C., 1995, Locus coeruleus cell loss in the aging human brain: A non-random process, *J. Comp. Neurol.* **358**:79–87.

Mann, D. M. A., and Yates, P. O., 1974, Lipoprotein pigments: Their relationship to ageing in the human nervous system. II. The melanin content of pigmented nerve cells, *Brain* **97**:489–498.

Mann, D. M. A., and Yates, P. O., 1983, Serotonin nerve cells in Alzheimer's disease, *J. Neurol. Neurosurg. Psychiatry* **46**:96.

Mann, D. M. A., Yates, P. O., and Hawkes, J., 1983, The pathology of the human locus coeruleus, *Clin. Neuropathol.* **2**:1–7.

Mann, D. M. A., Yates, P. O., and Marcyniuk, B., 1984a, Monoaminergic neurotransmitter systems in presenile Alzheimer's disease and in senile dementia of Alzheimer types, *Clin. Neuropathol.* **3**:199–205.

Mann, D. M. A., Yates, P. O., and Marcyniuk, B., 1984b, Alzheimer's presenile dementia, senile dementia of Alzheimer type, and Down's syndrome in middle age from an age related continuum of pathological changes. *Neuropathol. Appl. Neurobiol.* **10**:185–207.

Mann, D. M. A., Yates, P. O., and Marcyniuk, B., 1984c, Changes in nerve cells of the nucleus basalis of Meynert in Alzheimer's disease and the relationship to ageing and to the accumulation of lipofuscin pigment, *Mech. Ageing Dev.* **25**:189–205.

Mann, D. M. A., Yates, P. O., and Marcyniuk, B., 1985a, Changes in Alzheimer's disease in the magnocellular neurons in the supraoptic and paraventricular nuclei of the hypothalamus and their relationship to the noradrenergic deficit, *Clin. Neuropathol.* **4**:127–134.

Mann, D. M. A., Yates, P. O., and Marcyniuk, B., 1985b, Correlation between senile plaque and neurofibrillary tangle counts in the cerebral cortex and neuronal counts in the cortex and subcortical structures in Alzheimer's disease, *Neuroscience Lett.* **56**:51–55.

Mann, D. M. A., Yates, P. O., and Marcyniuk, B., 1987, Dopaminergic neurotransmitter systems in Alzheimer's disease and in Down's syndrome at middle age, *J. Neurol. Neurosurg. Psychiatry* **50**:341–344.

Marcyniuk, B., Mann, D. M. A., and Yates, P. O., 1986, The topography of cell loss from locus coeruleus in Alzheimer's disease, *J. Neurol. Sci.* **76**:335–345.

Marcyniuk, B., Mann, D. M. A., and Yates, P. O., 1989, The topography of nerve cell loss from locus coeruleus in elderly persons, *Neurobiol. Aging* **10**:5–9.

Marín-Padilla, M., 1978, Dual origin of the mammalian neocortex and evolution of the cortical plate, *Anat. Embryol.* **152**:109–126.

Markram, H., and Segal, M., 1990, Regional changes in NGF receptor immunohistochemical labeling in the septum of the aged rat, *Neurobiol. Aging* **11**:481–484.

McGeer, P. L., McGeer, E. G., and Suzuki, J. S., 1977, Aging and extrapyramidal function, *Arch. Neurol.* **34**:33–35.

McGeer, P. L., McGeer, E. G., Suzuki, J., Doleman, C. E., and Nagai, T., 1984, Aging, Alzheimer's disease, and the cholinergic system of the basal forebrain, *Neurology* **34**:741–745.

McNeill, T. H., and Koek, L. L., 1990, Differential effects of advancing age on neurotransmitter cell loss in the substantia nigra and striatum of C57BL/6N mice, *Brain Res.* **521**:107–117.

Mesulam, M.-M., and Mufson, E. J., 1984, Neural inputs into the nucleus basalis of the substantia innominata (Ch4) in the rhesus monkey, *Brain* **107**:253–274.

Mesulam, M.-M., and Geula, C., 1988, Nucleus basalis (Ch4) and cortical cholinergic innervation in the human brain: Observations based on the distribution of acetylcholinesterase and choline acetyltransferase, *J. Comp. Neurol.* **275**:216–240.

Mesulam, M.-M., and Geula, C., 1992, Overlap between acetylcholinesterase-rich and choline acetyltransferase-positive (cholinergic) axons in human cerebral cortex, *Brain Res.* **577**:112–120.

Mesulam, M.-M., Mufson, E. J., Levey, A. I., and Wainer, B. H., 1983a, Cholinergic innervation of cortex by the basal forebrain: Cytochemistry and cortical connections of the septal area, diagonal band, nucleus basalis (substantia innominata), and hypothalamus in the rhesus monkey, *J. Comp. Neurol.* **214:**170–197.

Mesulam, M.-M., Mufson, E. J., Wainer, B. H., and Levey, A. I., 1983b, Central cholinergic pathway in the rat: An overview based on an alternative nomenclature (Ch1-Ch6), *Neuroscience* **10:**1185–1983.

Mesulam, M.-M., Mufson, E. J., Levey, A. I., and Wainer, B. H., 1984a, Atlas of cholinergic neurons in the forebrain and upper brainstem of the macaque based on monoclonal choline acetyltransferase immunohistochemistry and acetylcholinesterase histochemistry, *Neuroscience* **12:**669–686.

Mesulam, M.-M., Rosen, A. D., and Mufson, E. J., 1984b, Regional variations in cortical cholinergic innervations: Chemoarchitectonics of acetylcholinesterase-containing fibers in the macaque brain, *Brain Res.* **311:**245–258.

Mesulam, M.-M., Mufson, E. J., and Wainer, B. H., 1986, Three-dimensional representation and cortical projection topography of the nucleus basalis (Ch4) in the macaque: Concurrent demonstration of choline acetyltransferase and retrograde transport with a stabilized tetramethylbenzidine method for horseradish peroxidase, *Brain Res.* **367:**301–308.

Mesulam, M.-M., Mufson, E. J., and Rogers, J., 1987, Age-related shrinkage of cortically projecting cholinergic neurons: A selective effect, *Ann. Neurol.* **22:**31–36.

Mesulam, M.-M., Hersh, L. B., Mash, D. C., and Geula, C., 1992, Differential cholinergic innervation within functional subdivisions of the human cerebral cortex: A choline acetyltransferase study, *J. Comp. Neurol.* **318:**316–328.

Meyer, A., 1971, *Historical Aspects of Cerebral Anatomy,* Oxford University Press, New York.

Moatamed, F., 1966, Cell frequencies in the human inferior olivary nuclear complex, *J. Comp. Neurol.* **128:**109–116.

Molliver, M. E., 1982, Role of monoamines in the development of the neocortex, *Neurosci. Res. Bull. Prog.* **20:**492–507.

Monagle, R. D., and Brody, H., 1974, The effects of age upon the main nucleus of the inferior olive in human, *J. Comp. Neurol.* **155:**61–66.

Morrison, J. H., Foote, S. L., O'Connor, D., and Bloom, F. E., 1982, Laminar, tangential, and regional organization of the noradrenergic innervation of monkey cortex: Dopamine-β-hydroxylase immunohistochemistry, *Brain Res. Bull.* **9:**309–319.

Mouton, P. R., Pakkenberg, B., Jørgen, H., Gundersen, G., and Price, D. L., 1994, Absolute number and size of pigmented locus coeruleus neurons in young and aged individuals, *J. Chem. Neuroanat.* **7:**185–190.

Mufson, E. J., Mash, D. C., and Hersh, L. B., 1988, Neurofibrillary tangles in cholinergic pedunculopontine neurons in Alzheimer's disease, *Ann. Neurol.* **24:**623–629.

Nagai, T., McGeer, P. L., Peng, J. H., McGeer, E. G., and Doleman, C. E., 1983, Choline acetyltransferase immunohistochemistry in brain of Alzheimer's disease patients and controls, *Neuroscience* **36:**195–199.

Oades, R. D., and Halliday, G. M., 1987, Ventral tegmental (A10) system: Neurobiology. I. Anatomy and connectivity, *Brain Res. Rev.* **12:**117–165.

O'Hearn, E., and Molliver, M. E., 1984, Organization of the raphe cortical projections in rat: A quantitative retrograde study, *Brain Res. Bull.* **13:**709–726.

Olszewski, J., and Baxter, D., 1954, *Cytoarchitecture of the Human Brain Stem,* S. Karger, Basal, Switzerland, and New York.

Ono, S., Kanda, F., Takahashi, K., Fukuoka, Y., Jinnai, K., Kurisaki, H., Mitake, S., Inagaki, T., Nagao, K., 1995, Neuronal cell loss in the dorsal raphe nucleus and the superior central nucleus in myotonic dystrophy: a clinicopathological correlation. *Acta Neuropathol.* **89:**122–125.

Oyanagi, K., Takahashi, H., Wakabayashi, K., and Ikuta, F., 1991, Large neurons in the neostriatum in Alzheimer's disease and progressive supranuclear palsy: A topographic, histologic, and ultrastructural investigation, *Brain Res.* **544:**221–226.

Papadopoulos, G. C., and Parnavelas, J. G., 1991, Monoamine systems in the cerebral cortex: Evidence for anatomical specificity, *Prog. Neurobiol.* **36:**195–200.

Parent, A., Gravel, S., and Oliver, A., 1979, The extrapyramidal and limbic system: Relationship at the globus pallidus level: A comparative histochemical study in the rat, cat, and monkey, in: *Advances in Neurology,* Vol. 24 (L. J. Poirier, T. L. Sourkes, and P. J. Bédard, eds.), New York, Raven, pp. 1–11.

Paskavitz, J. F., Lippa, C. F., Hamos, J. E., Pulaski-Salo, D., and Drachman, D. A., 1995, Role of the dorsomedial nucleus of the thalamus in Alzheimer's disease, *J. Geriat. Psychiatry Neurol.* **8:**32–37.

Pearson, R. C. A., Gatter, K. C., and Powell, T. P. S., 1983a, Retrograde cell degeneration in the basal nucleus in the monkey and man, *Brain Res.* **261**:321–326.

Pearson, R. C. A., Sonfroniew, M. V., Cuello, A. C., Powell, T. P. S., Eckentein, F., Esiri, M. M., and Wilcock, G. K., 1983b, Persistence of cholinergic neurons in the basal nucleus in a brain with senile dementia of the Alzheimer type demonstrated by immunohistochemical staining for choline acetyltransferase, *Brain Res.* **289**:375–379.

Perry, E. K., Tomlinson, B. E., Blessed, G., Perry, R. H., Cross, A. J., and Crow, T. J., 1981, Neuropathological and biochemical observations of the noradrenergic system in Alzheimer's disease, *J. Neurol. Sci.* **51**:279–287.

Perry, R. H., Candy, J. M., Perry, E. K., Irving, D., Blessed, G., Fairbairn, A. F., and Tomlinson, B. E., 1982, Extensive loss of choline acetyltransferase activity is not reflected by neuronal loss in the nucleus of Meynert in Alzheimer's disease, *Neurosci. Lett.* **33**:311–315.

Peters, A., and Sethares, C., 1993, Aging and the Meynert cells in rhesus monkey primary visual cortex, *Anat. Rec.* **236**:721–729.

Peters, A., Josephson, K., and Vincent, S. L., 1991, Effects of aging in the neuroglial cells and pericytes within area 17 of the rhesus monkey cerebral cortex, *Anat. Rec.* **229**:384–398.

Peters, A., Leahu, D., Moss, M. B., and McNally, K., 1994, The effect of aging on area 46 of the frontal cortex of the aging monkey, *Cereb. Cortex* **6**:621–635.

Peters, A., Rosene, D. L., Moss, M. B., Kemper, T. L., Abraham, C. R., Tigges, J., and Albert, M., 1996, Neurobiological bases of age-related cognitive decline in the rhesus monkey, *J. Neuropathol. Exp. Neurol.* **55**:861–874.

Poirier, L. J., Giguere, M., and Marchand, R., 1983, Comparative morphology of the substantia nigra and ventral tegmental area in the monkey, cat, and rat, *Brain Res. Bull.* **11**:371–397.

Porrino, L. J., and Goldman-Rakic, P. S., 1982, Brainstem innervation of prefrontal and anterior cingulate cortex in rhesus monkey revealed by retrograde transport of HRP, *J. Comp. Neurol.* **205**:63–76.

Price, J. L., and Stern, R., 1983, Individual cells in the nucleus basalis-diagonal band complex have restricted axonal projections to the cerebral cortex in the rat, *Brain Res.* **269**:352–356.

Raedler, E., Raedler, A., and Feldhaus, S., 1980, Dynamical aspects of neocortical histogenesis in the rat, *Anat. Embryol.* **158**:253–269.

Ramón-Moliner, E., 1975, Specialized and generalized dendritic patterns, in: *Perspectives in Neurobiology: Golgi Centennial Symposium* (M. Santini, ed.), Raven, New York, pp. 87–100.

Ramón-Moliner, E., and Nauta, W. J. H., 1966, The isodendritic core of the brain stem, *J. Comp. Neurol.* **126**:311–335.

Rinne, J. O., Paljärvi, L., and Rinne, U. K., 1987, Neuronal size and density in the nucleus basalis of Meynert in Alzheimer's disease, *J. Neurol. Sci.* **79**:67–76.

Rogers, J. D., Brogan, D., and Mirra, S. S., 1985, The nucleus basalis of Meynert in neurological disease: A quantitative morphological study, *Ann. Neurol.* **17**:173–170.

Rosales, R. K., LeMay, M. J., and Yakovlev, P. I., 1968, Anatomical and pneumoencephalographic study of the massa intermedia with regard to its development and its absence in relation to age and sex, *Trans. Am. Neurol. Assoc.* **68**:300.

Rossor, M. N., 1981, Parkinson's disease and Alzheimer's disease as disorders of the isodendritic core, *Br. Med. J.* **283**:1588–1590.

Sabel, B. A., and Stein, D. G., 1981, Extensive loss of subcortical neurons in the aging rat brain, *Exp. Neurol.* **73**:507–516.

Saper, C. B., 1984, Organization of cerebral cortical afferent systems in the rat. I. Magnocellular basal nucleus, *J. Comp. Neurol.* **222**:313–342.

Saper, C. B., 1985, Organization of cerebral afferent systems in the rat. II. Hypothalamocortical projections, *J. Comp. Neurol.* **237**:21–46.

Saper, C. B., 1996, Any way you cut it: A new journal policy for the use of unbiased counting methods, *J. Comp. Neurol.* **364**:5.

Saper, C. B., and German, D. C., 1987, Hypothalamic pathology in Alzheimer's disease, *Neurosci. Lett.* **74**:364–370.

Saper, C. B., German, D. C., and White, C. L., 1985, Neuronal pathology in the nucleus basalis and associated cell groups in senile dementia of the Alzheimer type: Possible role in cell loss, *Neurology* **35**:1089–1095.

Scheibel, M. E., and Scheibel, A. B., 1967, Structural organization of nonspecific thalamic nuclei and their projection toward cortex, *Brain Res.* **6**:60–94.

Schwyn, R. C., and Fox, C. A., 1974, The primate substantia nigra: A Golgi and electron microscopic study, *J. Hirnforsch.* **15:**95–126.

Séguéla, P., Watkins, K. C., and Descarries, L., 1989, Ultrastructural relationships of serotonin axon terminals in the cerebral cortex of the adult rat, *J. Comp. Neurol.* **289:**129–142.

Shefer, V. F., 1972, Absolute numbers of neurons and thickness of the cerebral cortex during aging, senile and vascular dementia, and Pick's and Alzheimer's disease, *Zh. Nevropatol. Psikhiatr. i.m. Korsakova* **72:**1024–1029.

Sillito, A. M., and Murphy, P. C., 1987, The cholinergic modulation of cortical function, in: *Cerebral Cortex, Vol. 6* (E. G. Jones and A. Peters, eds.), Plenum, New York, pp. 161–185.

Siddiqi, Z. A., Peters, A., Kemper, T. L., and Feldman, M. L., 1996, Age-related changes in the rhesus monkey: Morphology of neurons in substantia nigra-pars compacta. *Soc. Neurosci. Abstr.* **22:**1565.

Siddiqi, Z. A., Kemper, T. L., Peters, A., and Rosene, D. L., 1997, Age-related changes in the dopaminergic brainstem nuclei in the rhesus monkey. *Soc. Neurosci. Abstr.* **23:**2002.

Smiley, J. F., and Goldman-Rakic, P. S., 1993, Heterogeneous targets of dopamine synapses in monkey prefrontal cortex demonstrated by serial section electron microscopy: A laminar analysis using the silver-enhanced diaminobenzidine sulfide (SEDS) immunolabelling McNeill technique, *Cereb. Cortex* **7:**223–238.

Smiley, J. F., Levey, A. I., Ciliax, B. J., and Goldman-Rakic, P. S., 1994, D1 dopamine receptor immunoreactivity in human and monkey cerebral cortex: Predominant and extrasynaptic localization in dendritic spines, *Proc. Natl. Acad. Sci. USA* **91:**5720–5724.

Steinbusch, H. W. M., 1981, Distribution of serotonin-immunoreactivity in the central nervous system of the rat: Cell bodies and terminals, *Neuroscience* **6:**557–618.

Steinbusch, H. W. M., and Neiuwenhuys, R., 1983, The raphe nuclei of the rat brainstem: A cytoarchitectonic and immunohistochemical study, in: *Chemical Neuroanatomy* (P. C. Emson, ed.), Raven, New York, pp. 131–207.

Stroessner-Johnson, H. M., Rapp, P. R., and Amaral, D. G., 1992, Cholinergic cell loss and hypertrophy in the medial septal nucleus of behaviorally characterized aged rhesus monkey, *J. Neurosci.* **12:**1936–1944.

Stuesse, S. L., Stuesse, D. C., and Cruce, W. L., 1995, Raphe nuclei in three cartilaginous fishes, *Hydrolagus colliei, Heterodontus francisci,* and *Squalus acanthias, J. Comp. Neurol* **358:**414–427.

Sturrock, R. R., and Rao, K. A., 1985, A quantitative histological study of neuronal loss from the locus coeruleus of ageing mice, *Neuropathol. Appl. Neurobiol.* **11:**55–60.

Swaab, D. F., Fliers, E., and Partiman, T. S., 1985, The suprachiasmatic nucleus of the human brain in relation to age, sex, and senile dementia, *Brain Res.* **342,**37–44.

Tabaton, M., Schenone, A., Romagnoli, P., and Mancardi, G. L., 1985, A quantitative and ultrastructural study of substantia nigra and nucleus centralis superior in Alzheimer's disease, *Acta Neuropathol.* **68:**218–223.

Tagliavini, F., and Pilleri, G., 1983, Basal nucleus of Meynert: A neuropathological study in Alzheimer's disease, simple senile dementia, Pick's disease, and Huntington's chorea, *J. Neurol. Sci.* **62:**243–260.

Tatton, W. G., Greenwood, C. E., Verrier, M. C., Holland, D. P., Kwan, M. M., and Biddle, F. E., 1991, Different rates of age-related loss for four murine monoaminergic neuronal populations, *Neurobiol. Aging* **12:**543–556.

Terry, R. D., DeTeresa, R., and Hansen, L. A., 1987, Neocortical cell counts in normal adult aging, *Ann. Neurol.* **21:**530–539.

Tigges, J., and Tigges, M., 1985, Subcortical sources of direct projections to visual cortex, in: *Cerebral Cortex, Vol. 3* (A. Peters and E. G. Jones, eds.) Plenum, New York, pp. 351–378.

Tigges, J., Tigges, M., Cross, N. A., McBride, R. L., Letbetter, W. D., and Anschel, S., 1982, Subcortical structures projecting to visual cortical areas in squirrel monkey, *J. Comp. Neurol.* **209:**29–40.

Tigges, J., Walker, L. C., and Tigges, M., 1983, Subcortical projections to the occipital and parietal lobes of the chimpanzee brain, *J. Comp. Neurol.* **220:**106–115.

Tomlinson, B. E., Irving, D., and Blessed, G., 1981, Cells loss in the locus coeruleus in senile dementia of Alzheimer's types, *J. Neurol. Sci.* **49:**419–428.

Tomonaga, M., 1979, On the morphological changes in locus ceruleus in the senile human brain, *Jpn. J. Geriat.* **16:**545–550.

Tomonaga, M., 1983, Neuropathology of the locus coeruleus: A semi-quantitative study, *J. Neurol.* **230:**231–240.

Tsai, C., 1925, The optic tracts and centres of the opossum, *Didelphis virginiana*, *J. Comp. Neurol.* **39:**173–215.

Uchihara, T., Kondo, H., Kosaka, K., and Tsukagoshi, H., 1992, Selective loss of nigral neurons in Alzheimer's disease: A morphometric study, *Acta Neuropathol.* **83:**271–276.

Uhl, G. R., Hilt, D. C., Hedreen, J. C., Whitehouse, P. J., and Price, D. L., 1983, Pick's disease (lobar sclerosis): Depletion of neurons in the nucleus basalis of Meynert, *Neurology* **33:**1470–1473.

van Domburg, P. H. M. F., and ten Donkelaar, H. J., 1991, The human substantia nigra and ventral tegmental area: A neuroanatomical study with notes on aging and aging diseases, *Adv. Anat. Embryol. Cell Biol.* **121:**1–132.

van Luijtelaar, M. G., Tonnaer, J. A., and Steinbusch, H. W., 1992, Aging of the serotonergic system in the rat forebrain: An immunocytochemical and neurochemical study, *Neurobiol. Aging* **13:**201–215.

Vereecken, T. H., Vogels, O. J., and Nieuwenhuys, R., 1994, Neuronal loss and shrinkage in the amygdala in Alzheimer's disease, *Neurobiol. Aging* **15:**45–54.

Vijayashankar, N., and Brody, H., 1973, The neuronal population of the nuclei of the trochlear nerve and the locus coeruleus in the human, *Anat. Rec.* **172:**421–472.

Vijayashankar, N., and Brody, H., 1977, A study of aging in the human abducens nucleus, *J. Comp. Neurol.* **173:**433–438.

Vincent, A. E., Peters, A., and Tigges, J., 1989, Effects of aging on the neurons within area 17 of rhesus monkey cerebral cortex, *Anat. Rec.* **223:**329–341.

Vogels, O. J. M., Broere, C. A. J., and Nieuwenhuys, R., 1990, Neuronal hypertrophy in the human supraoptic and paraventricular nucleus in aging and Alzheimer's disease, *Neurosci. Lett.* **109:**62–67.

Voigt, T., and de Lima, A. D., 1991, Serotonergic innervation of the ferret cerebral cortex. II. Postnatal development, *J. Comp. Neurol.* **314:**415–428.

Voytko, M. L., Sukhov, R. R., Walker, L. C., Brekler, S. J., Price, D. L., and Koliatsos, V. E., 1995, Neuronal number and size are preserved in the nucleus basalis of aged rhesus monkeys, *Dementia* **6:**131–141.

Wainer, B. H., Bolam, J. P., Freund, T. F., Henderson, Z., Totterdell, S., and Smith, A. D., 1984, Cholinergic synapses in the rat brain: A correlated light and electron microscopic immunohistochemical study employing a monoclonal antibody against choline acetyltransferase, *Brain Res.* **308:**69–76.

Wainer, B. H., Levey, A. I., Rye, D. B., Mesulam, M.-M., and Mufson, E. J., 1985, Cholinergic and noncholinergic septohippocampal pathways, *Neurosci. Lett.* **54:**45–52.

Waterhouse, B. D., Lin, C.-S., Burne, R. D., and Woodward, D., 1983, The distribution of neocortical projection neurons in the locus coeruleus, *J. Comp. Neurol.* **217:**418–431.

West, M. J., 1993, New stereological methods for counting neurons, *Neurobiol. Aging* **14:**275–285.

Westlund, K. N., and Coulter, J. D., 1980, Descending projections of the locus coeruleus and subcoeruleus/medial parabrachial nuclei in monkey: Axonal transport studies and dopamine-β-hydroxylase immunohistochemistry, *Brain Res. Rev.* **2:**235–264.

Whitehouse, P. J., Hedreen, J. C., Jones, B. E., and Price, D. L., 1983a, A computer analysis of neuron size in the nucleus basalis of Meynert in patients with Alzheimer's disease, *Ann. Neurol.* **14:**149–150.

Whitehouse, P. J., Hedreen, J. C., White, C. L., and Price, D. L., 1983b, Basal forebrain neurons in the dementia of Parkinson disease, *Ann. Neurol.* **13:**243–248.

Whitehouse, P. J., Price, D. L., Struble, R. G., Clark, A. W., Coyle, J. T., and DeLong, M. R., 1982, Alzheimer's disease and senile dementia: Loss of neurons in basal forebrain, *Science* **215:**1237–1239.

Wilcock, G. K., Esiri, M. M., Bowen, D. M., and Smith, C. C. T., 1983, The nucleus basalis in Alzheimer's disease: Cell counts and cortical biochemistry, *Neuropathol. Appl. Neurobiol.* **9:**175–179.

Wilcock, G. K., Esiri, M. M., Bowen, D. M., and Hughes, A. D., 1988, The differential involvement of subcortical nuclei in senile dementia of Alzheimer's type, *J. Neurol. Neurosurg. Psychiatry* **51:**842–849.

Wiklund, L., Léger, L., Persson, M., and Mancardi, G. L., 1985, Monoaminergic cell distribution in the cat brain stem: A fluorescence histochemical study with quantitation of indolaminergic and locus coeruleus cell groups, *J. Comp. Neurol.* **203:**613–647.

Wilkinson, A., and Davies, I., 1978, The influence of age and dementia on the neuron population of the mammillary bodies, *Age and Aging* **7:**151–160.

Williams, S. M., and Goldman-Rakic, P. S., 1993, Characterization of the dopaminergic innervation of the primate frontal cortex using a dopamine-specific antibody, *Cereb. Cortex* **3:**199–222.

Williams, R. W., and Rakic, P., 1988, Three-dimensional counting: An accurate and direct method to estimate numbers of cells in sectioned material, *J. Comp. Neurol.* **278:**344–352.

Wilson, M. A., and Molliver, M. E., 1991, The organization of serotonergic projections to cerebral cortex in primates: Regional distribution of axon terminals, *Neuroscience* **44:**537–553.

Woolf, N. J., Jacobs, R. W., and Butcher, L. L., 1989, The pontomesencephalotegmental cholinergic system does not degenerate in Alzheimer's disease, *Neurosci. Lett.* **96:**277–282.

Yakovlev, P. I., 1952, Patterns of neuronal assemblies and the anatomical substratum of seizures, *Epilepsia* **1:**51–68.

Yakovlev, P. I., and Lecours, A. R., 1967, The myelogenetic cycles of regional maturation of the brain, in: *Regional Development of the Brain in Early Life* (A. Minkowski, ed.), Blackwell, London, pp. 3–70.

Yamamoto, T., and Hirano, A., 1985, Nucleus raphe dorsalis in Alzheimer's disease: Neurofibrillary tangles and loss of large neurons, *Ann. Neurol.* **17:**573–577.

Yoshinaga, R., 1986, Morphometric study of the human locus coeruleus: The changes with ageing and degenerative neurological diseases, *Fukuoka Igaku Zasshi* **77:**293–308.

Zubenko, G. S., and Moossy, J., 1988, Major depression in primary dementia: Clinical and neuropathological correlates, *Arch. Neurol.* **45:**1182–1196.

Zweig, R. M., Whitehouse, P. J., Casanove, M. F., Walker, L. C., Jankel, W. R., and Price, D. L., 1987, Loss of pedunculopontine neurons in progressive supranuclear palsy, *Ann. Neurol* **22:**18–25.

Zweig, R. M., Ross, C. A., Hedreen, J. C., Steele, C., Cardillo, J. E., Whitehouse, P. J., Folstein, M. F., and Price, D. L., 1988, The neuropathy of aminergic nuclei in Alzheimer's disease, *Ann. Neurol.* **24:** 233–242.

# 11

# Ultrastructural Changes in Dementing Illnesses

IMAHARU NAKANO and ASAO HIRANO

## 1. Introduction

Dementia is an organic mental disorder characterized by a general loss of intellectual abilities involving impairment of memory, judgment, and abstract thinking, as well as changes in personality. Because conditions that can cause dementia frequently exhibit characteristic pathological profiles in the brain that can only be adequately detected by electron microscopy (EM), EM has made and is expected to continue making conspicuous contributions to our understanding of the neuropathologies of such conditions. Although there are a lot of dementia-causing illnesses (Tomlinson, 1992), in this review we confine ourselves to typical dementing diseases in which ultrastructural brain changes, both intraneuronal and interstitial, have been intensively investigated.

## 2. Alzheimer's Disease

Alzheimer's disease (AD) or dementia of the Alzheimer type, is a condition involving progressive dementia with personality changes in middle or old age.

---

IMAHARU NAKANO • Department of Neurology, Jichi Medical School, Tochigi-ken, 329-0498 Japan. ASAO HIRANO • Division of Neuropathology, Department of Pathology, Montefiore Medical Center, Bronx, New York 10467.

*Cerebral Cortex,* Volume 14, edited by Peters and Morrison. Kluwer Academic/Plenum Publishers, New York, 1999.

Although the substantial neuropathological changes responsible for the dementia in this condition are likely to be the loss of cerebral cortical neurons, including those in the hippocampus, the conspicuous morphological features in AD brains are, among others, Alzheimer's neurofibrillary tangles (NFTs) and senile plaques (SPs).

### 2.1. Alzheimer's Neurofibrillary Tangles

NFTs are abnormal neuronal intracytoplasmic argentophilic inclusions that develop due to the thickening and increasing tortuosity of fibrils, the pattern differing according to the site in the brain and the type of neurons affected; in general, they are flare-shaped in hippocampal pyramidal neurons and globose-shaped in neurons of the deep nuclei (Fig. 1A). NFTs can be widespread in the central nervous system and have several preferential regions, such as the Ammon's horn, parahippocampal gyrus, amygdala, and substantia innominata (Hirano and Zimmerman, 1961).

The major proteinaceous component of NFTs is the microtubule-associated protein tau (Grundke-Iqbal *et al.,* 1986; Nukina and Ihara, 1986), which is abnormally phosphorylated (Lee *et al.,* 1991). NFTs are also ubiquitinated (Mori *et al.,* 1987).

The ultrastructure of NFTs was first described by Terry (1963) and Kidd (1964, 1963); NFTs are composed of bundles of abnormal filaments, longitudinal sections of which show regular constrictions at about 80-nm intervals (Hirano *et al.,* 1968; Kidd, 1964, 1963) with some variation in the intervals (Hirano, 1981) (Fig. 1B). The filaments are approximately 10 nm at the narrowest point and about 20 nm at the widest point (Hirano *et al.,* 1968). Filament cross sections show circular profiles of about 15-nm diameter, with frequent central dots (Hirano *et al.,* 1968), and oblique sections show archiform features (Hirano *et al.,* 1968; Kidd, 1963, 1964). The fundamental configuration of the filaments is still controversial; although the paired helical filament (PHF) model first postulated by Kidd (1963, 1964) is the most prevalent, being supported by the results with the quick-freeze, deep-etch, and replica method (Ohtsubo *et al.,* 1990a), the twisted-ribbon model has also been proposed (Pollanen *et al.,* 1995).

In addition to PHFs, NFTs are occasionally composed of straight filaments approximately 15-nm diameter (Hirano *et al.,* 1968) (Fig. 1C). A recent study indicated a transition between constricted and straight filaments (Miyakawa *et al.,* 1994).

Cholinergic neurons in the nucleus basalis of Meynert (nbM), which have been the target of extensive studies on AD and other dementing conditions (Nakano and Hirano, 1984a, 1983; Whitehouse *et al.,* 1982, 1981), are one of the preferential locations of NFTs. NFTs here are globose; electron microscopy of neurons containing such NFTs reveals aggregations of irregularly arranged PHF bundles, among which are often large lipofuscin granules with a prominent mosaic pattern (Fig. 1A). These characteristic lipofuscin granules are observed in neurons in the nbM in neuropsychiatric (Averback, 1981a,b) as well as nonneurological (Nakano and Hirano, 1984b; Morimura *et al.,* 1985) conditions.

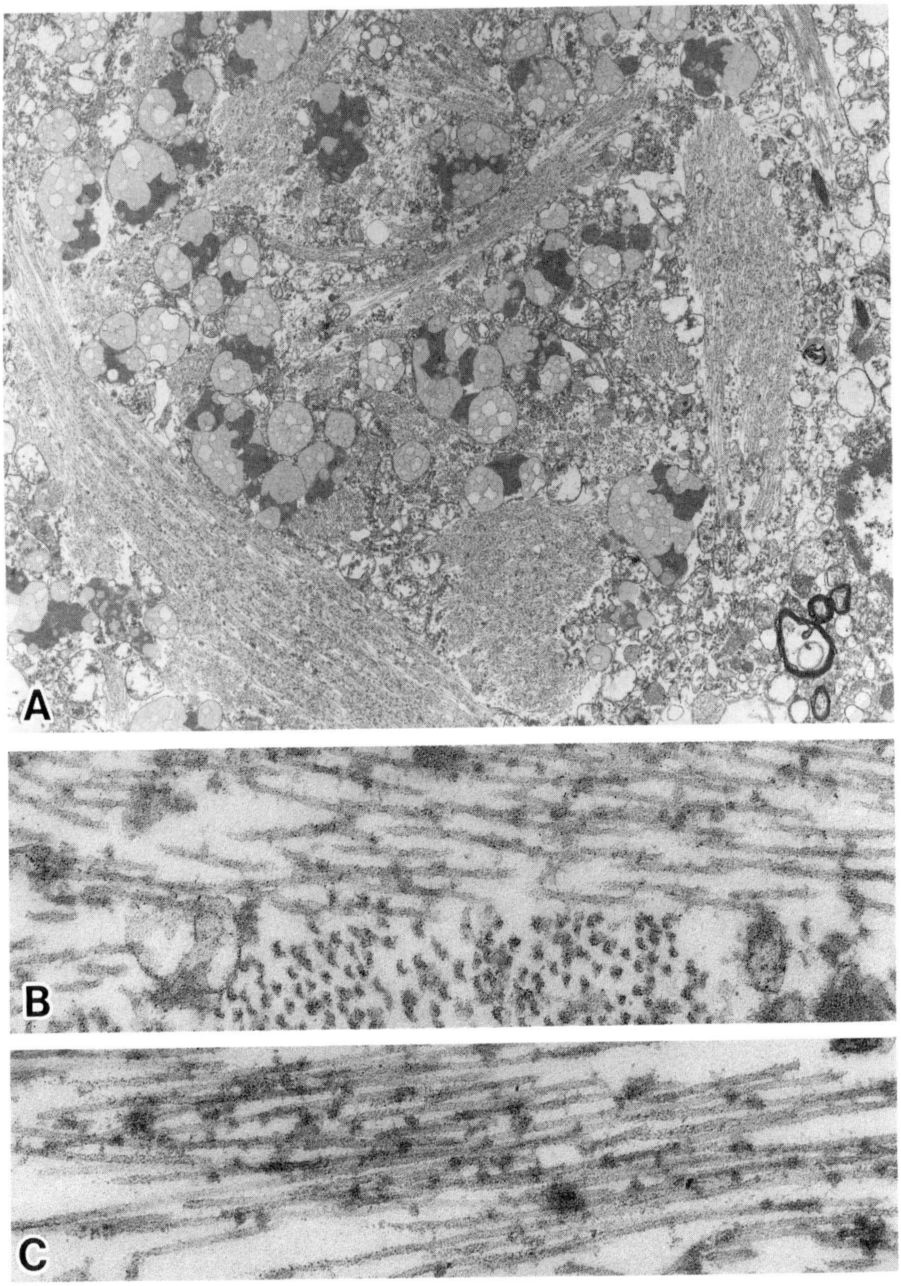


**Figure 1.** Alzheimer's neurofibrillary tangle. (A) A globose form of neurofibrillary tangle in a neuron in the nucleus of the diagonal band of Broca. X6,000. (B) Higher magnification of A shows about 80-nm regular constrictions on tangentially viewed filaments and characteristic curvilinear profiles of slightly tilted cross sections of the filaments. X80,000. (C) Almost-straight tubules composing a tangle in a pyramidal neuron of the Ammon's horn. X80,000.

The silver impregnation-staining method for the AD cerebral cortex reveals numerous argentophilic, curly threadlike profiles ("neuropil threads" or "curly fibers"). They are tau- and ubiquitin-positive (Iwatsubo *et al.*, 1992; Ihara, 1988) and consist of PHFs (Perry *et al.*, 1991; Ohtsubo *et al.*, 1990b; Yamaguchi *et al.*, 1990).

## 2.2. Ghost Tangles

When neurons containing NFTs die, the filaments are expelled into the extracellular space, resulting in ghost tangles (GTs) (Fig. 2). The filaments comprising GTs are usually straight with vague margins (Yamaguchi *et al.*, 1991a,b; Nakano and Hirano, 1984b; Okamoto *et al.*, 1983) (Fig. 2B), although they occasionally exhibit traces of regular constrictions of PHFs (Probst *et al.*, 1982). The morphological alterations may correspond to immunohistochemical (Bondareff *et al.*, 1990; Schmidt *et al.*, 1988) and chemical (Endoh *et al.*, 1993) changes that take place in

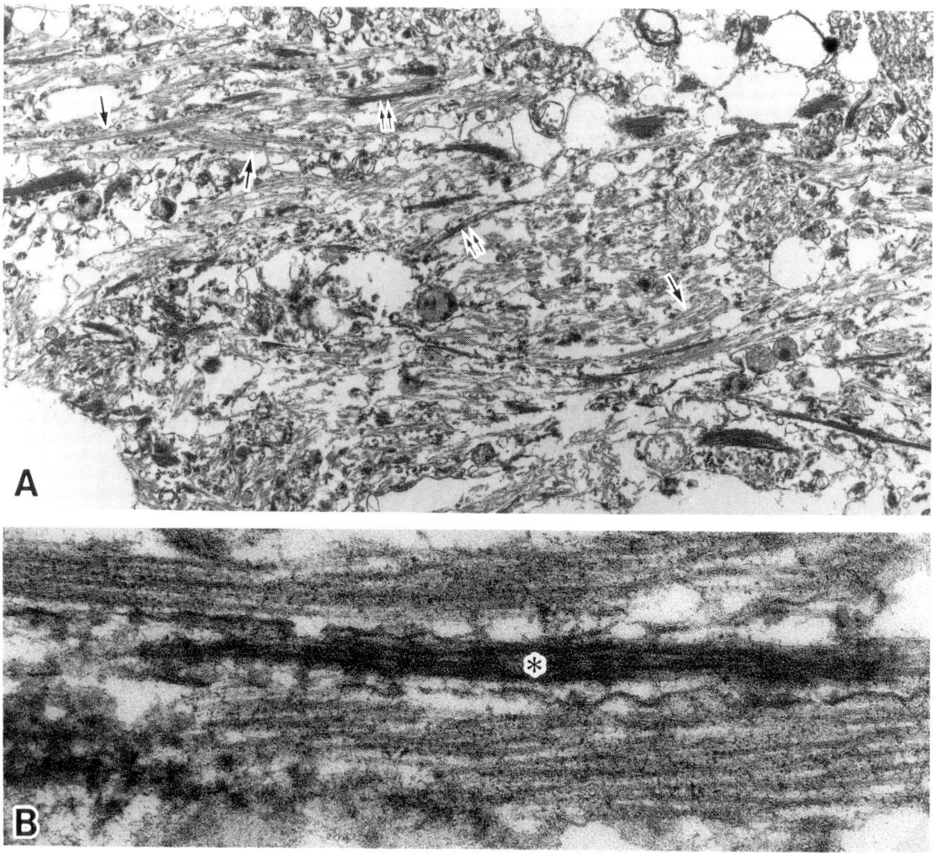


**Figure 2.** Ghost tangle. (A) A ghost tangle in the nucleus basalis of Meynert. Aggregates of tangle filaments (arrows) are invaded by much-denser glial filaments (double arrows). X8,000. (B) Filaments composing the ghost tangle are straight, with blurred margins. An invading astrocytic process with a dense bundle of glial filaments (*) can be seen. X90,000.

GTs. Many astrocytic processes without a basal lamina invade GTs, dividing them into small bundles of abnormal filaments (Yamaguchi *et al.*, 1991a,b; Probst *et al.*, 1982; Okamoto *et al.*, 1983) (Fig. 2A, B).

An EM study of GTs in a 25-year-duration case of juvenile Alzheimer disease (Nakano *et al.*, 1992a) revealed that the astrocytic processes invading GTs were quite few, resulting in the formation of large bundles, and that astrocytic processes invading or bordering GTs often had a fairly well-developed basal lamina, with a lamina lucida and densa (Fig. 3) (Nakano *et al.*, 1996). The scarcity of GT-invading astrocytic processes indicates that the processes that had once invaded the GTs had since withdrawn. The basal lamina formation associated with GTs implies that astrocytes recognize GTs as a nonneural tissue, because one of the functions of the basal lamina is to separate certain parenchymal tissue from connective tissue, thus serving as a boundary between two different tissue elements (Furthmayr, 1988).

### 2.3. Astrocytic Paired Helical Filaments

Although the NFT was for a long time considered to be a specific marker of neurons in the central nervous system, recent EM investigation has demonstrated the presence of NFTs in neurons of the peripheral nervous system, such as the upper cervical ganglia of an elderly patient (Kawasaki *et al.*, 1987), the dorsal root ganglia in patients with progressive supranuclear palsy (Nishimura *et al.*, 1993), and the adrenal pheochromocytoma cells, a sort of paraganglion cell (Izumiyama *et al.*, 1990). Still, these cells are neurons or homologues of neurons.

In the above-mentioned case of juvenile Alzheimer's disease (Nakano *et al.*, 1992a), we found abnormal filaments free in the astrocytic cytoplasm that were

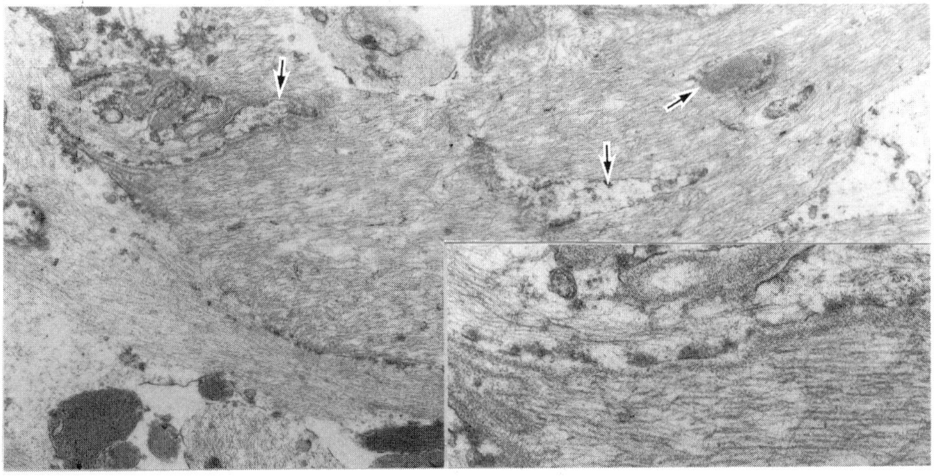

**Figure 3.** A large ghost tangle bundle in a case of juvenile AD is bordered or invaded by astrocytic processes (arrows), which are partially accompanied by a basal lamina. X15,000. Insert: Higher magnification of one of the astrocytic processes clearly shows a basal lamina with a lamina densa and a lamina lucida, and well-developed hemidesmosome-like profiles. X45,000.

ultrastructurally indistinguishable from neuronal PHFs (Fig. 4). Almost-straight tubules of approximately 15-nm diameter were often observed too, especially in the astrocytes with a hydropic cytoplasm (Fig. 5). Such cells were identified as astrocytes on the basis of basal lamina, aggregrates of glial filaments, and gap junctions with adjacent cells. Immunoelectron microscopy demonstrated that such astrocytic abnormal filaments were tau- and ubiquitin-positive (Fig. 6). Thus, these filaments are also indistinguishable from neuronal paired helical filaments (PHFs) or straight tubules in terms of their antigenicity.

Similar reports of astrocytic PHFs in AD (Yamazaki *et al.,* 1995; Ikeda *et al.,* 1992a,b) followed ours; Ikeda *et al.* (1992a,b) indicated that, in addition to the possibility of their formation by glial cells, such astrocytic PHFs are ghost tangle filaments engulfed by astrocytes, whereas our observations (Yamazaki *et al.,* 1995; Nakano *et al.,* 1992a) strongly support *de novo* formation of PHFs by astrocytes. Moreover, abnormal filaments exhibiting positive tau immunoreactivity, but with a size and constriction interval different from those of neuronal PHFs, were found in the muscle cell vacuoles of inclusion body myositis (Askanas *et al.,* 1994).

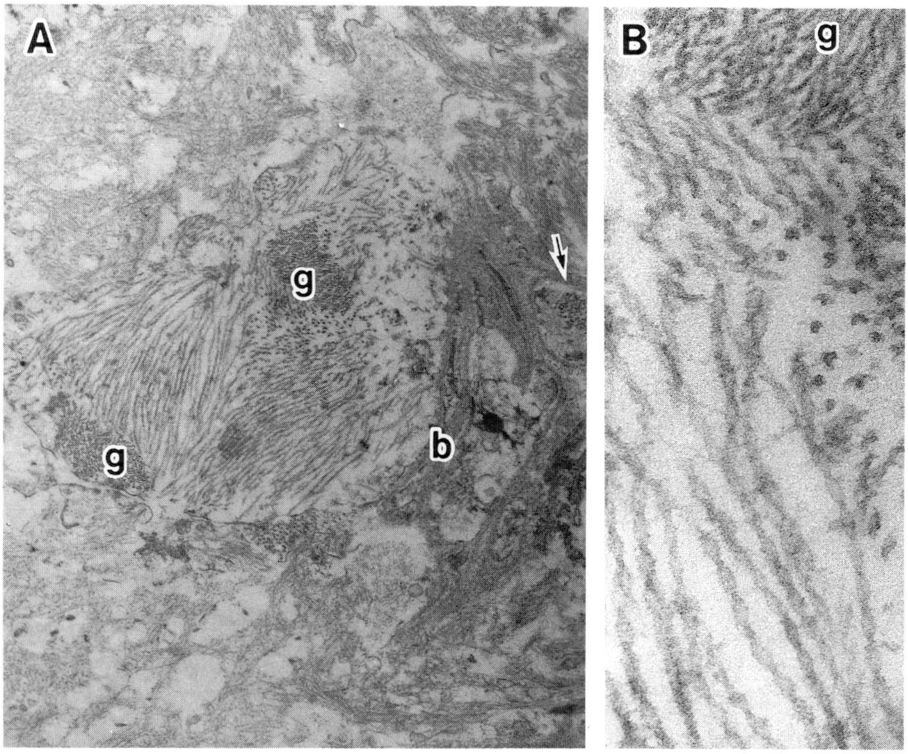


**Figure 4.** Astrocytic tangle in a case of juvenile AD. (A) An astrocytic process, identified as such from a basal lamina (b) and glial filaments (g), contains abnormal, larger filaments. Another astrocytic process carries abnormal filaments (arrow). X20,000. (B) Higher magnification of A shows paired helical filaments with regular constrictions and characteristic crescentic cross sections adjacent to glial filaments (g). X100,000.

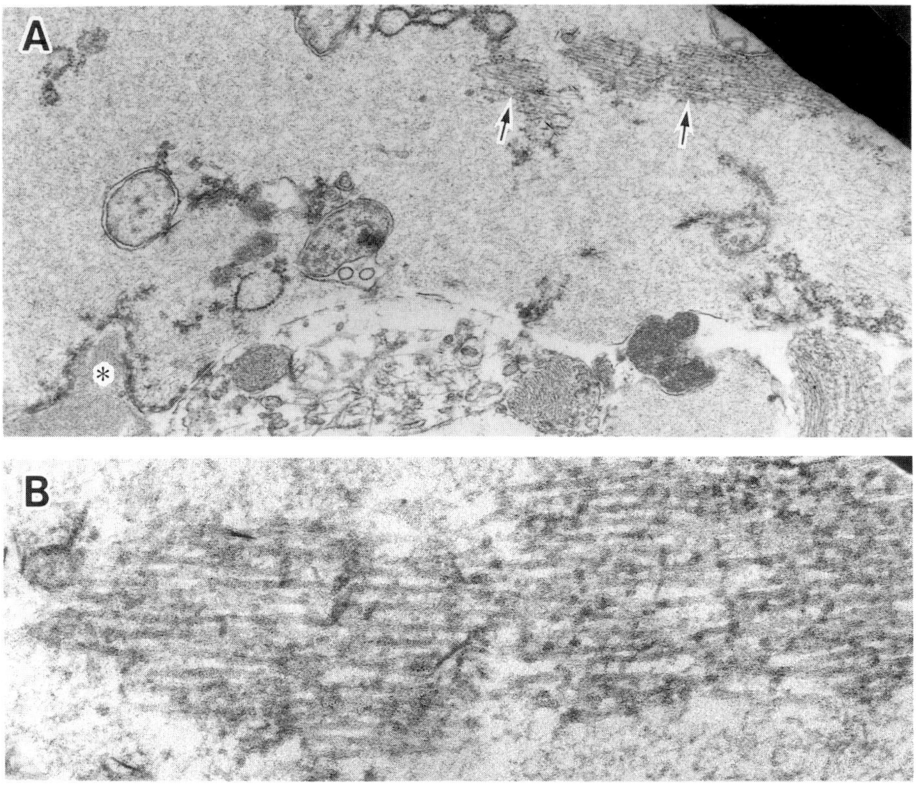


**Figure 5.** Astrocytic tangle in a case of juvenile AD. (A) A hydropic astrocyte, which has a basal lamina (*), contains two bundles of abnormal filaments (arrows). X24,000. (B) Higher magnification of the bundle discloses straight tubules or only slightly constricted filaments. X100,000.

## 2.4. Anchorage Densities of Perivascular Astrocytes

The perivascular astrocytes in devastated brain areas show unique anchorage densities along the basal lamina-lined cell membranes that face the perivascular space (Fig. 7A, B). Although the anchorage densities were first described in sudanophilic leukodystrophy and spinocerebellar degeneration (Nakano *et al.*, 1992b), they were also observed in Pick disease (Yamazaki *et al.*, 1994), old cerebral infarcts (Fig. 7C), and adrenoleukodystrophy (unpublished data), as well as in AD (Wegiel and Wisniewski, 1994). The anchorage densities are 200–300 nm away from the cell membrane and run parallel with it, usually receiving filaments from deeply located glial filaments. In turn the densities send isolated 13–16-nm filaments obliquely or almost perpendicularly to unusually prominent hemidesmosome-like profiles on the inner leaflet of the cell membrane.

## 2.5. Senile Plaques

Senile plaques (SPs), which are associated with aging and, particularly, with AD, are abnormal, roughly spherical argentophilic structures mainly located in the cerebral cortex. Typical SPs consist of a β-protein amyloid core (Selkoe, 1991; Glenner and Wong, 1984), a pale intermediate zone, and an outermost zone of argentophilic granular material.

On EM, the amyloid core was found to consist of approximately 10-nm filaments (Terry and Wisniewski, 1970; Terry *et al.,* 1964) intermingled in the center but forming radiating bundles at the periphery. The argentophilic structures surrounding the amyloid core are distended dystrophic neurites that contain dense bodies, tubulovesicular structures, PHFs, Hirado bodies, accumulations of neurofilaments, microtubules, mitochondria, and synaptic vesicles (Wisniewski and Terry, 1973; Hirano *et al.,* 1972; Terry and Wisniewski, 1970).

SPs resembling those in humans are found in aged nonhuman primates (Nakano *et al.,* 1993; Wisniewski and Terry, 1973) (Fig. 8), and dogs (Ishihara *et al.,* 1991; Wisniewski *et al.,* 1970), but typical PHFs are not seen in SPs in these animals.

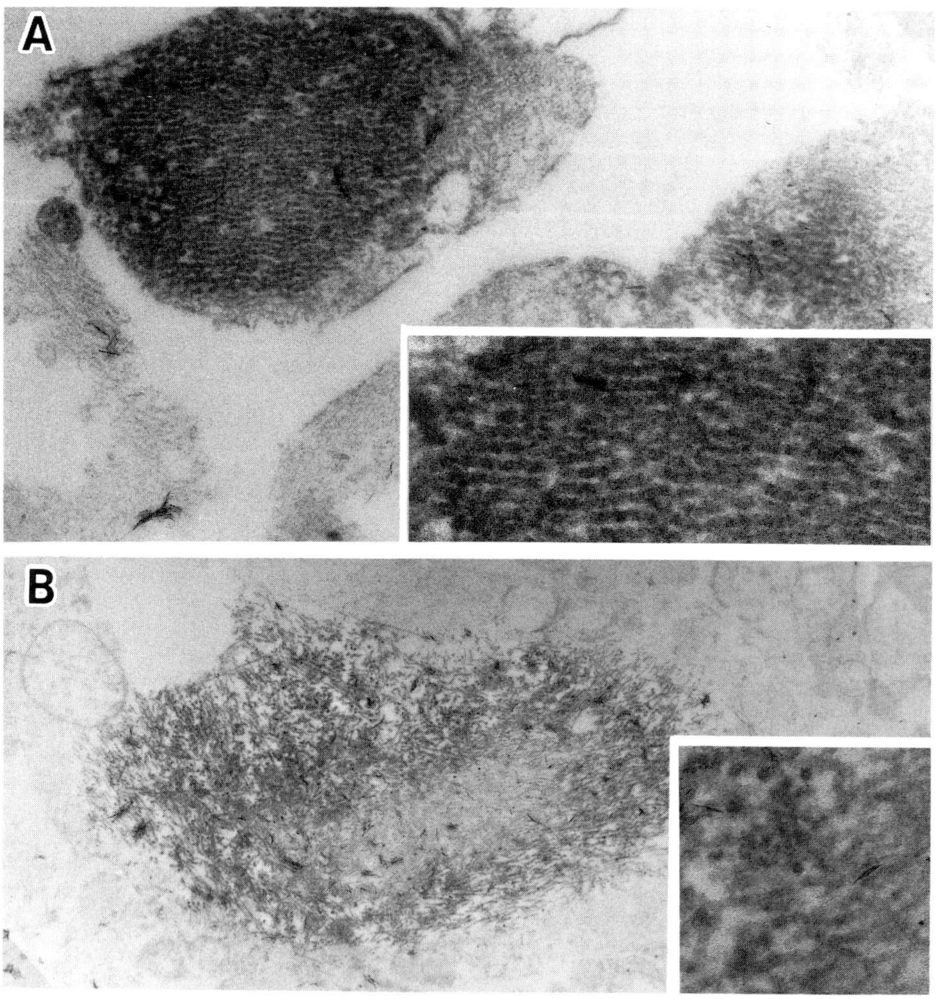


**Figure 6.** Immunoelectron microscopy of astrocytic tangles are positive for anti-tau protein (A) and anti-ubiquitin (B) antibodies, respectively. (A) X20,000, inset: X90,000 (B) X16,000, inset: X80,000.

The time sequence of SP formation has not been yet determined. SPs used to be classified as primitive plaques (aggregations of altered neuronal processes without obvious amyloid), classic plaques (typical plaques with an amyloid core and halo), and compact (burned-out) plaques (amyloid core alone) (Fig. 9). Wisniewski and Terry (1973) regarded the neuronal changes (primitive plaques) to be

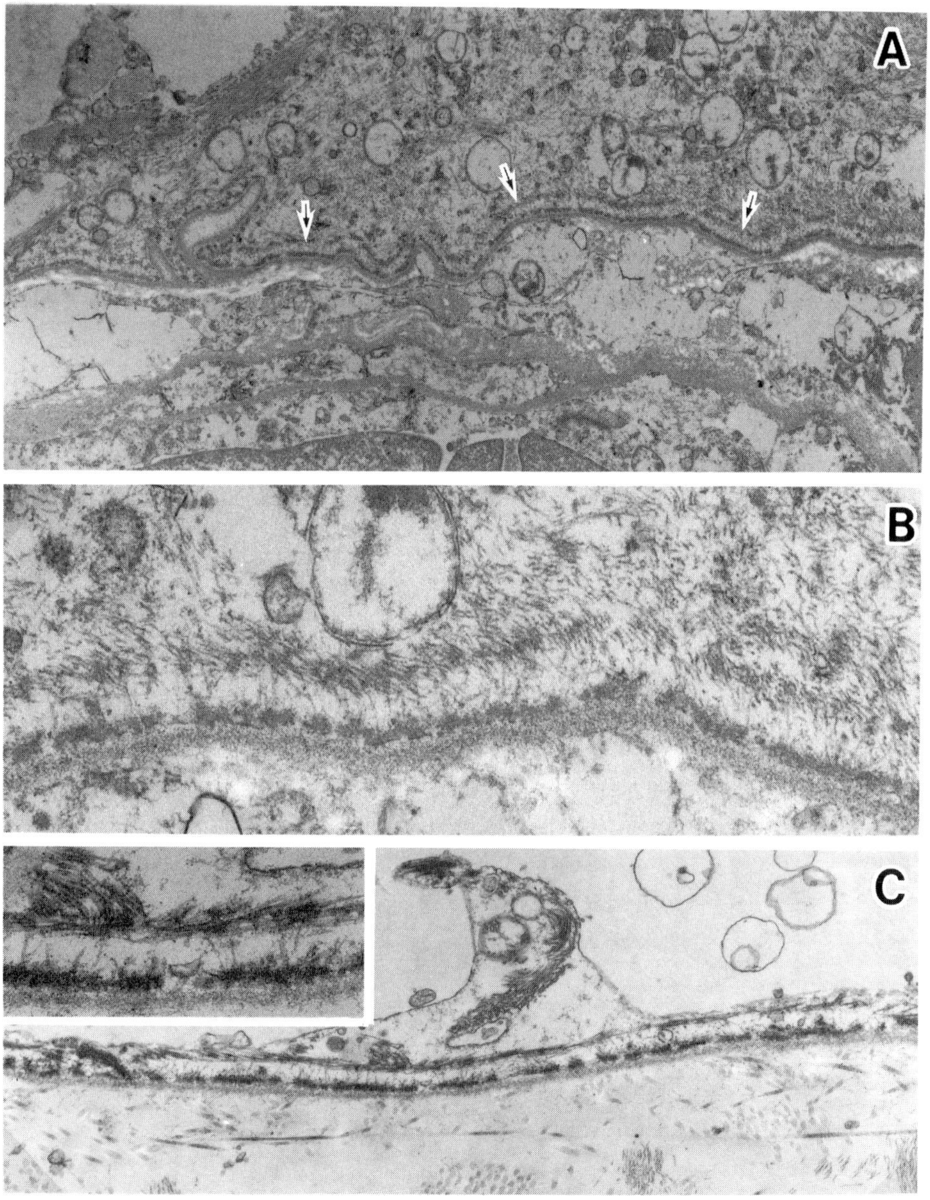


**Figure 7.** Astrocytic anchorage densities. (A) A perivascular astrocyte with well-developed hemidesmosome-like structures in juvenile AD has an anchorage density (arrows). X10,000. (B) Higher magnification of (A) clearly shows that glial filaments merge into the anchorage density. X45,000. (C) An anchorage density in a perivascular astrocyte in an old infarct. X20,000. Inset: X45,000.

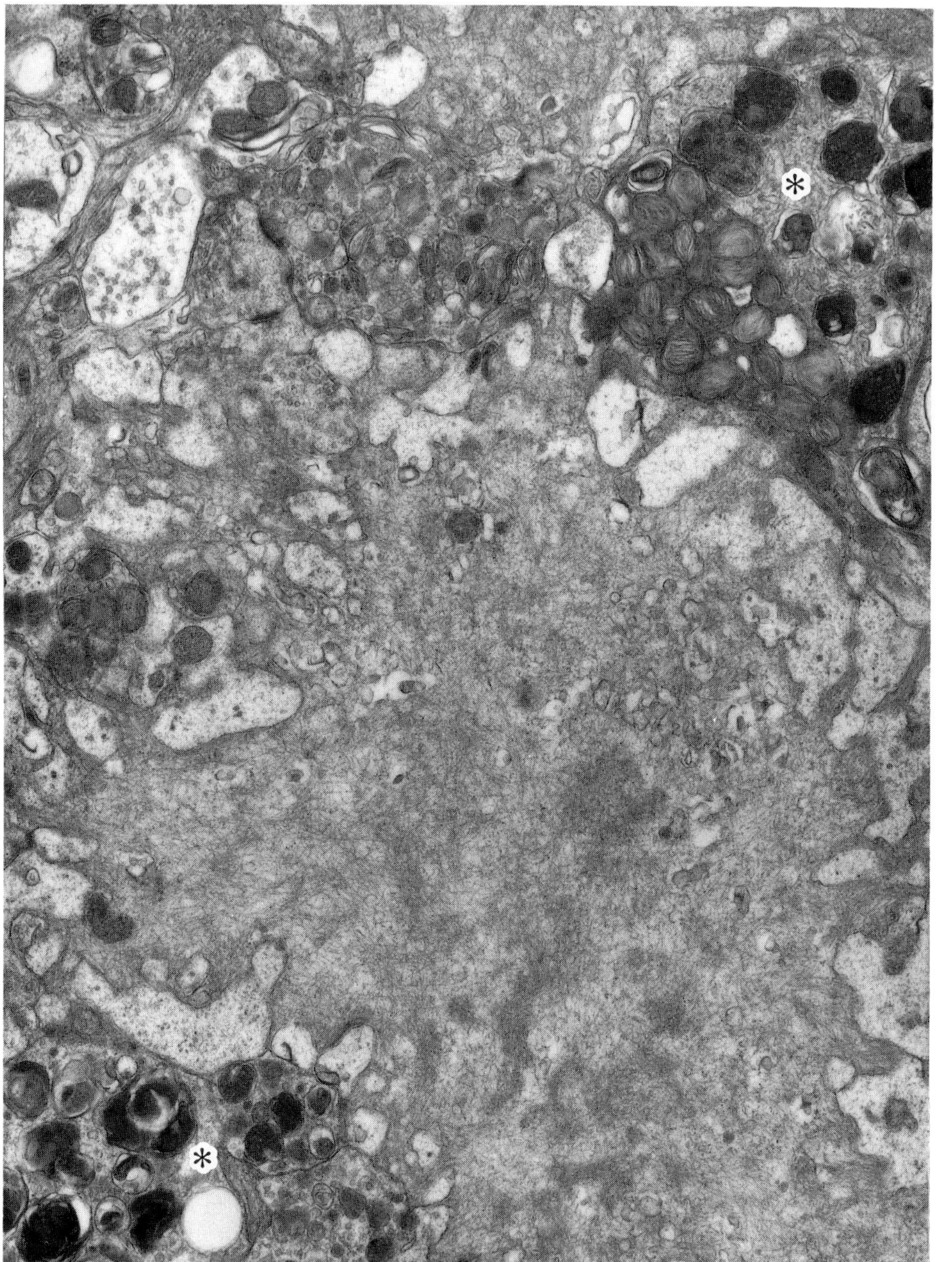


**Figure 8.** A senile plaque of an aged monkey.. An amyloid core is surrounded by light astrocytic processes and dystrophic neurites (*). X20,000.

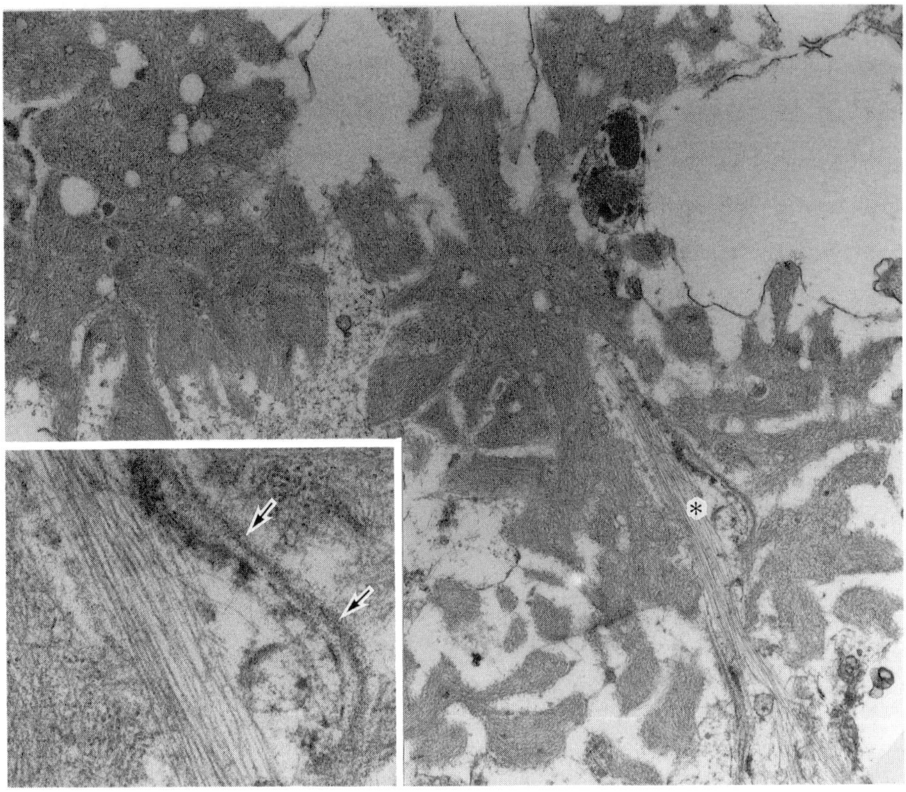

**Figure 9.** A burned-out plaque in juvenile AD invaded by an astrocytic process with glial filaments (*). X20,000. Inset: Higher magnification shows a basal lamina (arrows) with a lamina densa and a lamina lucida on the astrocyte. X60,000.

primary, and the amyloid precipitation to be secondary. This notion was essentially supported by a recent finding in the brains of aged nonhuman primates (Martin *et al.*, 1994). On the other hand, some investigators consider newly described diffuse-type plaques as the initial stage of SPs (Ikeda *et al.*, 1990; Yamaguchi *et al.*, 1989). There is no evidence, however, that diffuse plaques develop into more mature plaques.

## 2.6. Hirano Bodies (Eosinophilic Rodlike Structures)

The pyramidal cell layers of the Ammon's horn in AD frequently contain a large number of Hirano bodies, which are bright eosinophilic rodlike or round structures observed in contact with neurons or in the neuropil. Hirano bodies were first described in the Ammon's horn of patients with amyotrophic lateral sclerosis (ALS) or Parkinsonism-dementia complex on Guam (Hirano, 1965), and then observed in various conditions, including Pick disease (Schochet *et al.*, 1968) and Creutzfeldt-Jakob disease (Hirano, 1994). Although the Ammon's horn of normal aged people also contains Hirano bodies, they are more predominant in AD (Schmidt *et al.*, 1989; Gibson and Tomlinson, 1977).

Since the first demonstration of immunoreactivity of Hirano bodies toward actin (Goldman, 1983), epitopes of high-molecular-weight microtubule-associated protein (Peterson *et al.*, 1988), tau protein (Galloway *et al.*, 1987), and the C-terminal domain of β-amyloid precursor protein (Munoz *et al.*, 1993), but not ubiquitin (Manetto *et al.*, 1989), have been detected in the inclusion bodies.

EM demonstrated that Hirano bodies were present in synaptic terminals or dystrophic neurites within SPs in the Ammon's horn. Hirano bodies consist of highly organized crystalloid arrays displaying a lattice-like, or "herringbone," configuration (Fig. 10) (Hirano, 1965). Three-dimensional analysis of the inclusion bodies indicated that they consisted of interlaced arrays of 6–10-nm parallel filaments (Schochet and McCormick, 1972), with later support from a study on Hirano-body development in brindled mice (Peterson *et al.*, 1986) and by results obtained with the quick-freeze, deep-etch, and replica method (Fig. 11) (Izumiyama *et al.*, 1991), whereas O'Brien *et al.* (1980) maintained that Hirano bodies were stacked sheets of membrane-bound ribosomal particles. On the other hand, Mori *et al.* (1986), based on digital processing of electron micrographs, proposed that Hirano bodies are composed of helical strands.

Although light microscopically Hirano bodies were found to be confined to the pyramidal cell layer of the Ammon's horn, an EM study revealed them in neurons in various regions of the central nervous system, peripheral nerves, astrocytomas, muscle fibers, and an inflammatory cell of a leptomeningeal vessel (Hirano, 1994, 1992).

## 2.7. Granulovacuolar Degeneration

Hippocampal pyramidal neurons in AD frequently contain granulovacuolar degeneration (GVD), or Simchowicz bodies. The inclusions are 3–4-μm round vacuoles containing a dense core, which is basophilic on hematoxylin and eosin and argentophilic on silver-impregnation staining. Besides AD, GVDs are observed

**Figure 10.** Hirano body. X160,000.

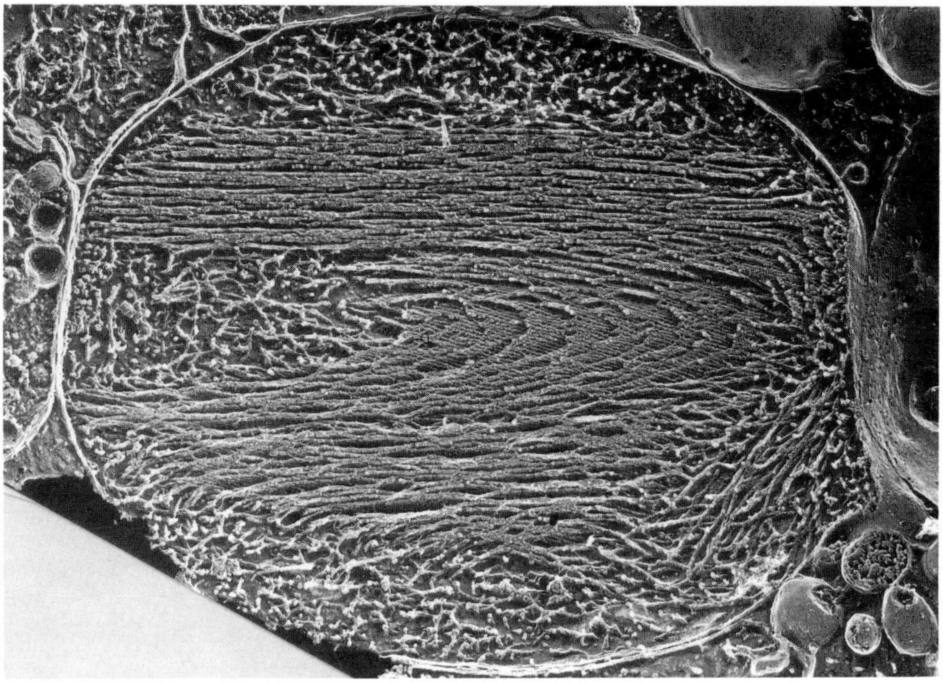

**Figure 11.** A Hirano body with the quick-freeze, deep-etch, and replica method. X38,000. From Izumiyama, N., *et al.* (1991).

in aged brains, Pick's disease, and the Parkinsonism-dementia complex of Guam (Hirano *et al.,* 1968), among others. They are confined to the soma of the hippocampal pyramidal neurons, with some exceptions (Steele *et al.,* 1964).

In EM, GVDs appear as cytoplasmic membrane-bound inclusions containing a dense core of granular material (Oyanagi and Ikuta, 1974; Hirano *et al.,* 1968). Although the origin of GVDs is unknown, they were considered to be lysosomal in nature (Terry and Wisniewski, 1972). Ultrastructural observation of undulated membranes lining GVD-limiting membranes as well as flat cisterns enclosing electron dense cytoplasmic material led Oyanagi and Ikuta (1974) to speculate that GVDs may be a product of autophagocytosis. Recently, Okamoto *et al.* (1991a) strengthened this notion by clearly showing flat cisterns demarcating an electron-dense cytoplasmic mass (Fig. 12).

## 3. Pick's Disease

Pick's disease, which is characterized by prominent neurobehavioral alterations including personality changes, was first described on clinical and macroscopic pathological bases alone. The typical gross pathology of the brain comprises well-circumscribed, often asymmetric, cortical atrophy (lobar atrophy). The major histological findings are marked neuron loss in the involved cerebral cortex associ-

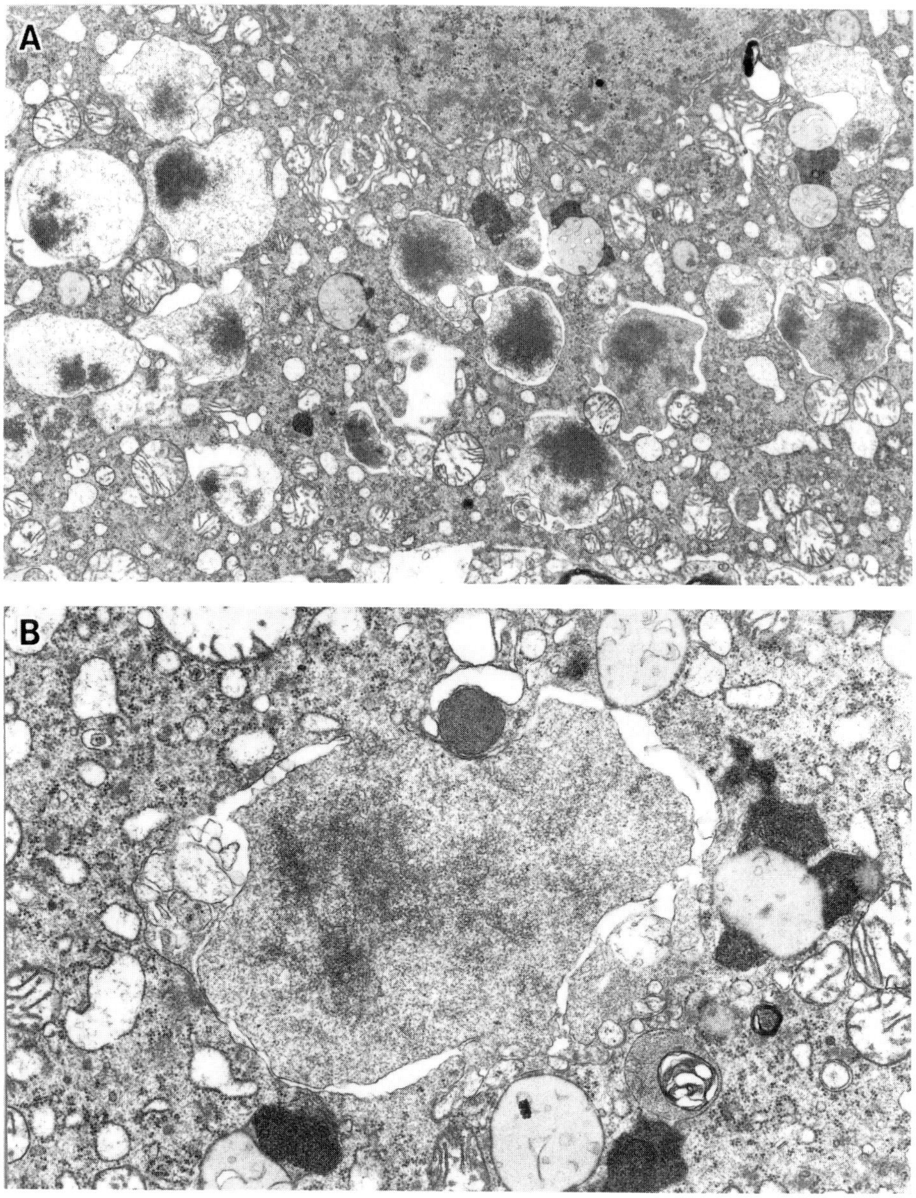


**Figure 12.** Granulovacuolar degeneration (GVD). (A) Many GVDs can be seen in a neuron in the hippocampus. X12,800. (B) An area of a GVD at early stage incompletely surrounded by flat cisterns contains a fine granular material. X24,200. From Okamoto, K., *et al.* (1991), Reexamination of granulovascular degeneration, *Acta Neuropathol.* **82:**340 (Fig. 4). © Springer-Verlag.

ated with prominent gliosis, Pick bodies in the remaining neurons, and ballooned achromatic neurons.

Pick bodies are round or ellipsoidal argentophilic inclusions in the neuronal cytoplasm in Pick's disease. They are most often found in the pyramidal neurons in the hippocampus and granule cells of the dentate gyrus.

Ultrastructurally, Pick bodies are composed of randomly arranged abnormal filaments. Oyanagi *et al.* (1979) reported that the filaments were approximately 15-nm tubular structures (Fig. 13A, C), frequently with a geometrical arrangement. On the other hand, Pick bodies were found to contain twisted tubules with constrictions at approximately 160-nm intervals (Fig. 13D), in addition to the 15-nm

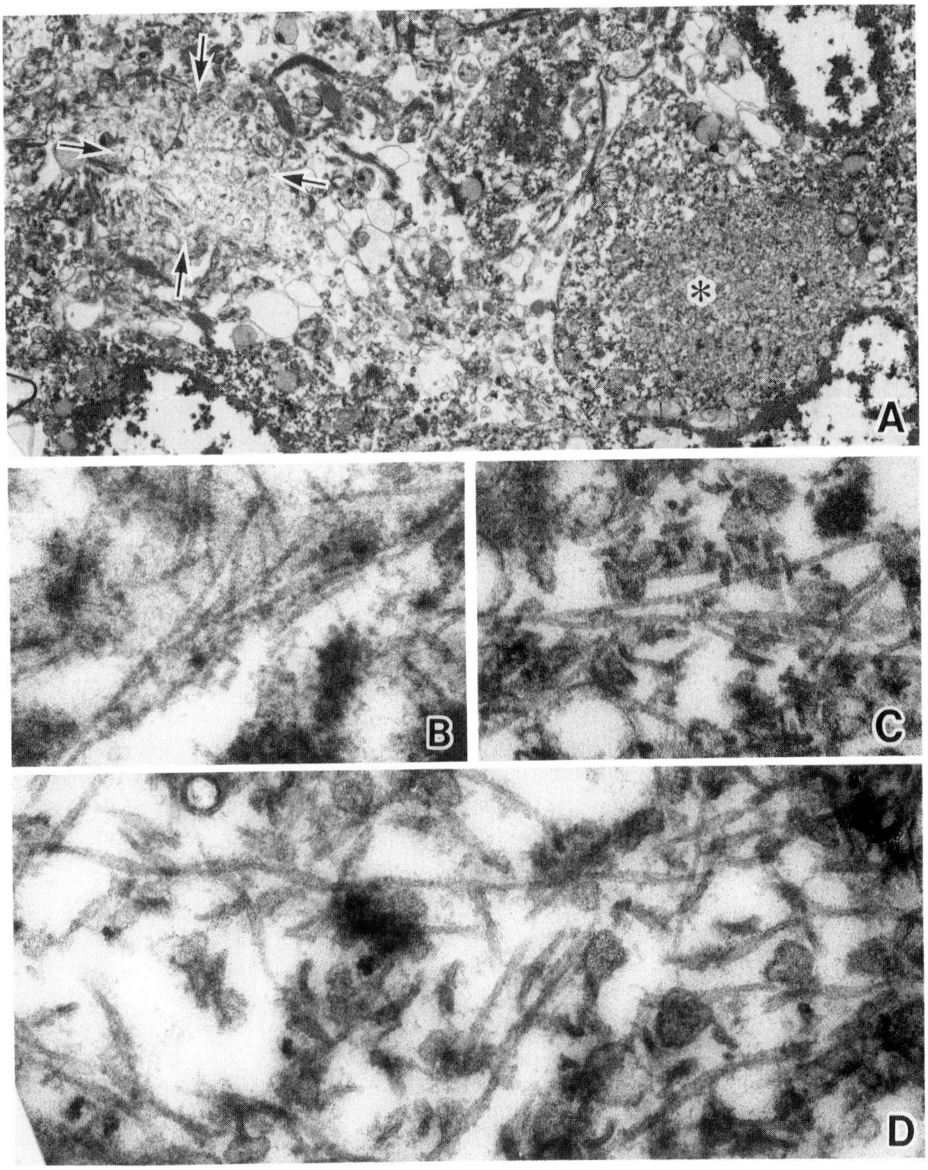


**Figure 13.** Pick body (PB). (A) An intraneuronal PB (*) and an extraneuronal one (arrows) can be seen in Pick's disease. X4,600. Higher magnification shows that straight filaments in the extracellular PB (B) have blurred margins in comparison with those in the intracellular PB (C). B, C: X85,700. (D) Twisted tubules with constrictions of about 160-nm intervals in a PB. X98,000. Courtesy of Ms. Yoko Izumiyama, Tokyo Institute of Psychiatry, from Izumiyama *et al.* (1994), Extracellular or ghost Pick bodies and their lack of tau immunoreactivity: A histological, immunohistochemical and electron microscopic study, *Acta Neuropathol.* **87:**277 (Fig. 3). © Springer-Verlag.

straight tubules (Kato and Nakamura, 1990; Munoz-Garcia and Ludwin, 1984). The ballooned neurons seen in Pick's disease were reported to contain Pick body-type filaments in addition to neurofilaments (Murayama *et al.*, 1990).

Recently, Izumiyama *et al.* (1994) reported the ultrastructure and immunohistochemistry of extracellular of "ghost" Pick bodies. They are composed of straight tubules intermingled with thin or thick glial bundles; the tubules with a diameter of approximately 13-nm are a little smaller than their intracellular counterparts (Fig. 13A, B). Tau protein epitopes, which are clearly detectable in intracellular Pick bodies (Murayama *et al.*, 1990; Kato and Nakamura, 1990), are absent in ghost Pick bodies, just as they are in the ghost tangles (Bondareff *et al.*, 1990; Schmidt *et al.*, 1988).

The tau protein-associated changes in this condition are not confined to the Pick bodies; Yamazaki *et al.* (1994) found tau-positive glial inclusions in a long-duration case of Pick's disease. On EM, they comprised variously sized bundles of approximately 15-nm-thick straight tubules that were free in the astrocytic cytoplasm (Fig. 14). The astrocytic inclusions in Pick's disease were confirmed by Ishizu *et al.* (1995), who additionally reported tubular/filamentous inclusions in oligodendrocytes in this condition.

## 4. Progressive Supranuclear Palsy and Corticobasal Degeneration

### 4.1. Progressive Supranuclear Palsy

This condition, first described by Steele *et al.* (1964), is characterized clinically by progressive extrapyramidal features, supranuclear ophthalmic palsy, and dementia; and pathologically by degeneration of various regions of the brain, such as the pallidum, subthalamic nucleus, substantia nigra and dentate nucleus, as well as widespread neurofibrillary tangles. The NFTs in this condition are mainly distributed in the above-mentioned deep structures, but some progressive supranuclear palsy (PSP) cases also show NFTs in the limbic system (Takahashi *et al.*, 1987). In addition, NFTs have been described in the dorsal root ganglia (Nishimura *et al.*, 1993), and even in astroglia (glial fibrillary tangle) (Nishimura *et al.*, 1992).

In PSP the NFTs are usually composed of intermingled bundles of approximately 15-nm straight tubules (Fig. 15) (Kato *et al.*, 1992; Powell *et al.*, 1974; Roy *et al.*, 1974; Tellez-Nagel and Wisniewski, 1973). More than a few cases of PSP, however, have NFTs containing PHFs as elements (Amano *et al.*, 1989; Yagishita *et al.*, 1979; Tomonaga, 1977). In the dorsal root ganglia, NFTs also consist of PHFs (Nishimura *et al.*, 1993), but astrocytic inclusions are composed of only straight tubules indistinguishable from their neuronal counterparts (Nishimura *et al.*, 1992).

### 4.2. Corticobasal Degeneration

Corticobasal degeneration (CDB), first described by Rebeiz *et al.* (1968) and intensively reappraised by Gibb *et al.* (1989), is a rare condition characterized by

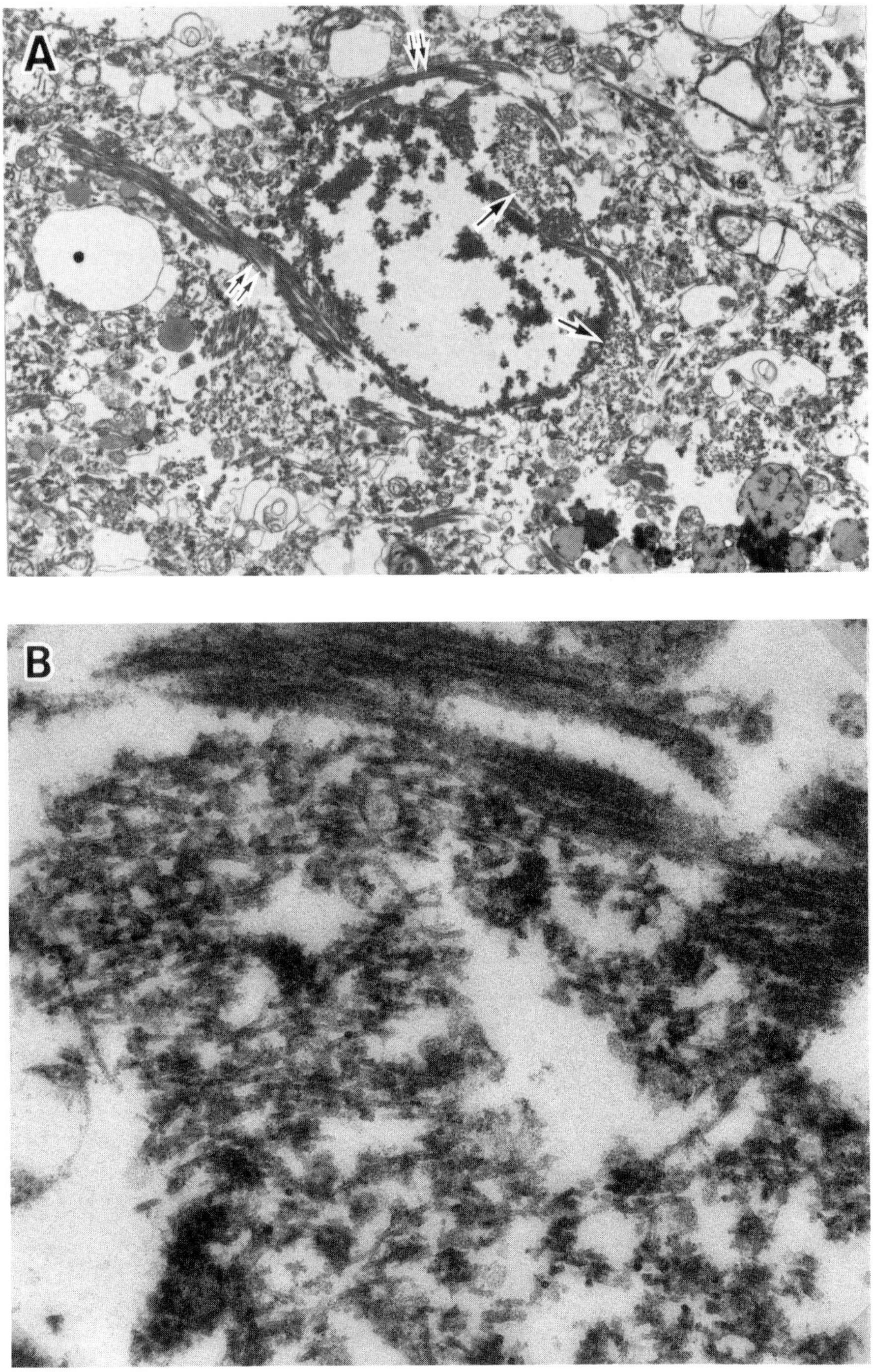


**Figure 14.** Straight astrocytic tubules in Pick disease. (A) An astrocyte with glial filaments (double arrows) has juxtanuclear aggregates of abnormal filaments (arrows). X6,000. (B) Higher magnification of A shows the abnormal filaments are straight tubules indistinguishable from those in Pick bodies. X60,000.

asymmetric involvement of the cerebral cortex associated with neuronal achromasia and by extrapyramidal structures such as the substantia nigra and pallidum. Although CBD was first thought to be a unique entity independent of PSP because of the few NFTs, the recent observation of several CBD cases with frequent NFTs and a tau protein abnormality similar to that seen in PSP (Nukina *et al.*, 1992; Paulus and Selim, 1990) triggered a controversy about the similarity of the two conditions (Mori *et al.*, 1994; Wakabayashi *et al.*, 1994).

On EM, NFTs in CBD were found to be composed of straight tubules or PHFs (Mori *et al.*, 1994; Wakabayashi *et al.*, 1994), as seen in PSP. Cortical achromatic neurons contain approximately 10-nm neurofilaments as well as approximately 20-nm scattered filaments coated with a granular material (Wakabayashi *et al.*, 1994). Tau-positive glial tangles in CBD are located in oligodendroglia and composed of approximately 15-nm straight tubules (Wakabayashi *et al.*, 1994).

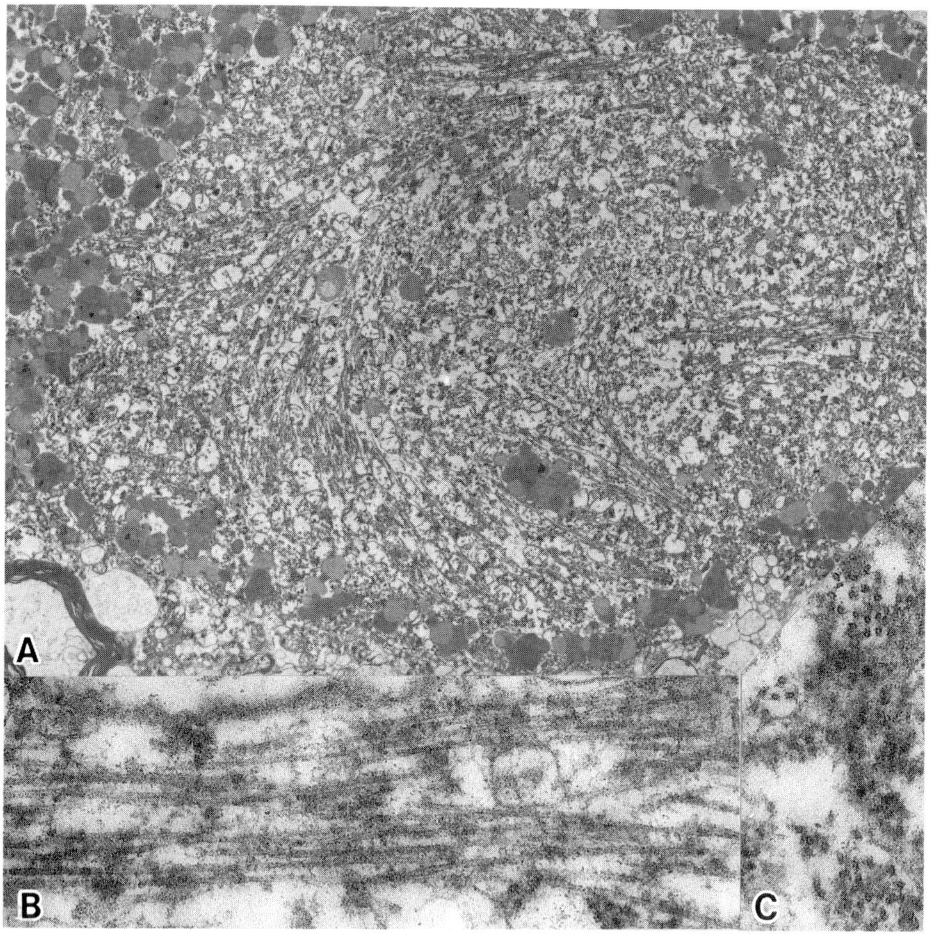


**Figure 15.** Neurofibrillary tangle in PSP. (A) A hypothalamic neuron in PSP contains a neurofibrillary tangle. X6,000. (B, C) Higher magnification of A shows the tangle is composed of approximately 15-nm straight tubules. (B) Tangential view. (C) Cross section. B, C: X90,000.

In 1961 Okazaki *et al.* reported two cases of dementia with flexion contractures but no parkinsonism. The brains exhibited numerous Lewy bodies in various regions, including the cerebral cortex. After about 15 years of silence, this condition attracted attention again in Japan under the new name of diffuse Lewy body disease (DLBD) (Kosaka *et al.,* 1988; Yoshimura, 1983; Kosaka *et al.,* 1980; Kosaka, 1978; Kosaka *et al.,* 1976).

Clinically, DLBD is characterized by progressive dementia and/or parkinsonism, and pathologically by numerous Lewy bodies distributed in various areas, including the cerebral cortex (cortical Lewy bodies). In addition, the cerebral cortex in DLBD frequently has changes seen in Alzheimer's disease, especially senile plaques (Dickson *et al.,* 1989; Kosaka *et al.,* 1988), thus leading some investigators to propose that this condition is a subtype of Alzheimer's disease (Hansen *et al.,* 1993), although other researchers disagree (Kosaka, 1995; Perry *et al.,* 1990). Kosaka *et al.* (1988) suggested that the dementia can be attributed to coexistent Alzheimer changes. Recent investigators, however, indicated a relation of cortical Lewy body density with the dementia seen in this condition (Samuel *et al.,* 1996; Lennox *et al.,* 1989a).

Cortical Lewy bodies are vaguely circumscribed, being slightly eosinophilic inclusions that usually have a quite vague core, causing difficulty in their identification in H&E sections. Lewy bodies, the cortical or brain-stem type, are strongly immunopositive for an anti-ubiquitin antibody (Kuzuhara *et al.,* 1988), so this staining method is the most sensitive for detecting cortical Lewy bodies (Lennox *et al.,* 1989b; Kuzuhara *et al.,* 1988).

Although ultrastructural studies of cortical Lewy bodies have been limited (Dickson *et al.,* 1989; Kuzuhara *et al.,* 1988; Kosaka, 1978; Kosaka *et al.,* 1976), their fine structures essentially resemble those of Lewy bodies in the brain stem (Duffy and Tennyson, 1965). The inclusions are composed of crisscrossing filaments intermingled with a fine, dense, granular material (Fig. 16). The filaments vary in size from approximately 15 to 20 nm, and the granulofilamentous elements show a tendency to aggregate in the center. Unlike brain-stem Lewy bodies, the central aggregations are much more vague, corresponding to the light microscopical appearance, and no radial arrangement of peripherally located filaments is evident. The structures that are ubiquitinated are said to be the peripheral filaments (Kuzuhara *et al.,* 1988).

## 6. Creutzfeldt–Jakob Disease

Creutzfeldt–Jakob disease (CJD) is a relentlessly progressive, fatal neurological disease involving dementia, ataxia, myoclonus, and cortical blindness. CJD belongs to a more generic disease group, the spongiform encephalopathies (prion diseases), which include a variety of neurodegenerative diseases: CJD, Gerstmann-Sträussler-Scheinker disease, kuru, and fatal familial insomnia (Medori *et al.,* 1992) in humans; scrapie in sheep; mink encephalopathy; bovine spongiform encephalopathy; and chronic wasting disease in deer and elk (Williams and Young, 1982;

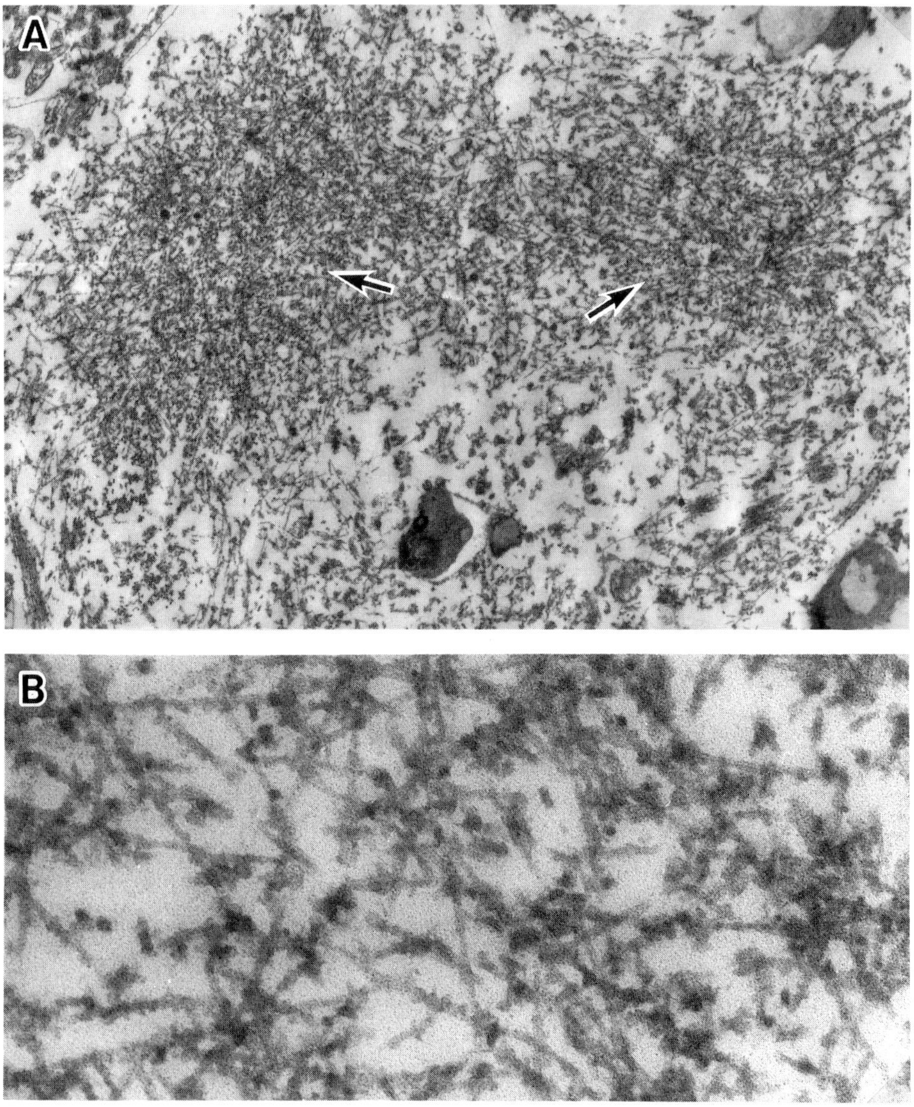


**Figure 16.** Cortical Lewy body. (A) A cortical Lewy body with two cores (arrows). X12,000. (B) Higher magnification of A shows the Lewy bodies are composed of abnormal 15–20-nm filaments frequently decorated by a granular material. X90,000. Courtesy of Dr. Saburo Yagishita, Kanagawa Rehabilitation Center.

Prusiner, 1995). All these conditions are considered attributable to prions, proteinaceous infectious particles seemingly devoid of nucleic acid genomes (Prusiner, 1995, 1982). Prions are apparently composed of a single protein designated as $PrP^{Sc}$, an infectious isoform of a normal prion protein ($PrP^{C}$), which in humans is encoded by the PrP gene located on chromosome 20 (Liao *et al.,* 1986).

The histological triad in the gray matter of CJD is neuronal loss, spongiform change, and astrocytosis. Electron microscopic studies of the sponginess in the CJD brain (Hirano *et al.,* 1972; Bignami and Forno, 1970; Gonatas *et al.,* 1965), and in experimental kuru, CJD, and scrapie (Lampert *et al.,* 1972), have shown that it results predominantly from the empty swelling of neuronal processes, especially synaptic terminals (Fig. 17). But these studies as well as later ultrastructural studies on CJD brains (Chou *et al.*, 1980) and experimental prion diseases (Baker *et al.*, 1990; Liberski *et al.*, 1990; Manuelidis, 1985) failed to reveal convincing virus particles or pathogens (Jeffrey *et al.*, 1995), although some profiles were suggested to be

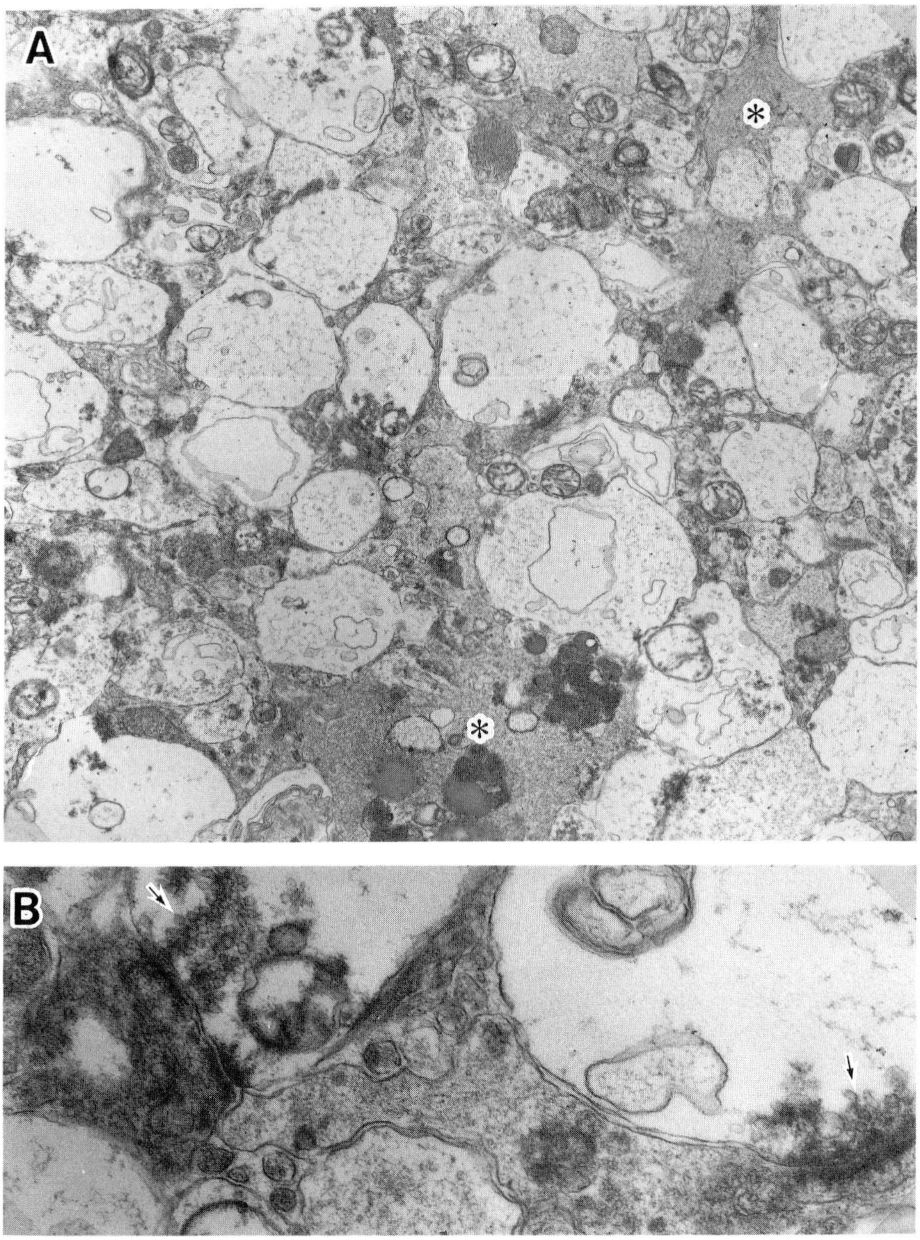


**Figure 17.** Spongy change in Creutzfeldt-Jakob disease. (A) Many vacuoles can be seen in the cerebral cortex in Creutzfeldt-Jakob disease. Processes of reactive astrocytes (*) separate the vacuoles. X10,000. (B) Higher magnification of A reveals that the majority of the vacuoles are enlarged presynaptic terminals with synaptic vesicles (arrows). X40,000.

significant in relation to the pathogenic mechanism (Liberski *et al.*, 1992; Chou *et al.*, 1980).

Kuru and some prion disease involving a variant PrP, including Gerstmann-Sträussler-Scheinker disease, show amyloid plaques consisting of prion protein (kuru plaques) (Fig. 18). Unlike senile plaques, the association of kuru plaques with dystrophic neurites is generally inconspicuous, although they can have a radiating crown of abnormal neuronal processes (Lantos, 1992). A new variant of sporadic CJD with a suspected relation to bovine spongiform encephalopathy ("mad cow disease") also exhibits many plaques (Will *et al.*, 1996).

## 7. Amyotrophic Lateral Sclerosis with Dementia

A small number of patients with amyotrophic lateral sclerosis (ALS) are affected by neuropsychiatric symptoms including dementia (Hudson, 1981). The condition is named ALS with dementia, and has been intensively investigated in Japan (Nakano, 1993; Yoshida *et al.*, 1992; Mitsuyama, 1984; Yuasa, 1970). Meyer (1929) seems to have been the first to have made an autopsy report on this condition. After several similar cases were reported, Van Reeth *et al.* (1961) published a case report on ALS with dementia in which Bunina bodies were clearly presented one year before Bunina's publication on the inclusion body named after her.

Some cases ALS with dementia show changes of the cerebral cortex that can adequately explain the mental changes (Yoshida *et al.*, 1992; Horoupian *et al.*, 1984). However, despite the overt behavioral abnormality and dementia, the cerebral neocortical alterations in this condition such as the mild sponginess of the neuropil or minimal loss of superficial cortical neurons are—as causes—frequently subtle and unconvincing. Recently, overt and constant loss of neurons associated with gliosis was noticed in the surface layer of the medial cortex in the temporal pole, and at the border between CA1 and subiculum at the level of the pes hippocampi (Nakano, 1993). In addition, the amygdala and gyrus ambients showed frequent neuronal loss with gliosis, giving rise to the possibility that such temporal lobe lesions play a role in dementia in this condition.

Another new finding in ALS is ubiquitinated intracytoplasmic inclusions in the extramotor cortical neurons (Okamoto *et al.*, 1996; Okamoto *et al.*, 1992; Wightman *et al.*, 1992; Okamoto *et al.*, 1991b). The inclusions, usually ovoid or crescentic, are mainly observed in the granule cells in the dentate gyrus, and cortical neurons mainly in the parahippocampal gyrus and frontal lobe.

Electron microscopic observation of the inclusions revealed that they do not have a limiting membrane and thus are loose aggregations of 15–17-nm filamentous profiles attached by electron-dense granules (Fig. 19). A small number of mitochondria and/or lipofuscin granules are frequently entrapped in the inclusions. Immuno-EM with antibodies against ubiquitin demonstrated that the antibodies decorate with the filamentous and granular structures of the inclusions (Fig. 20). Okamoto *et al.* (1996) surmised that the granular material may be altered ribosomes.

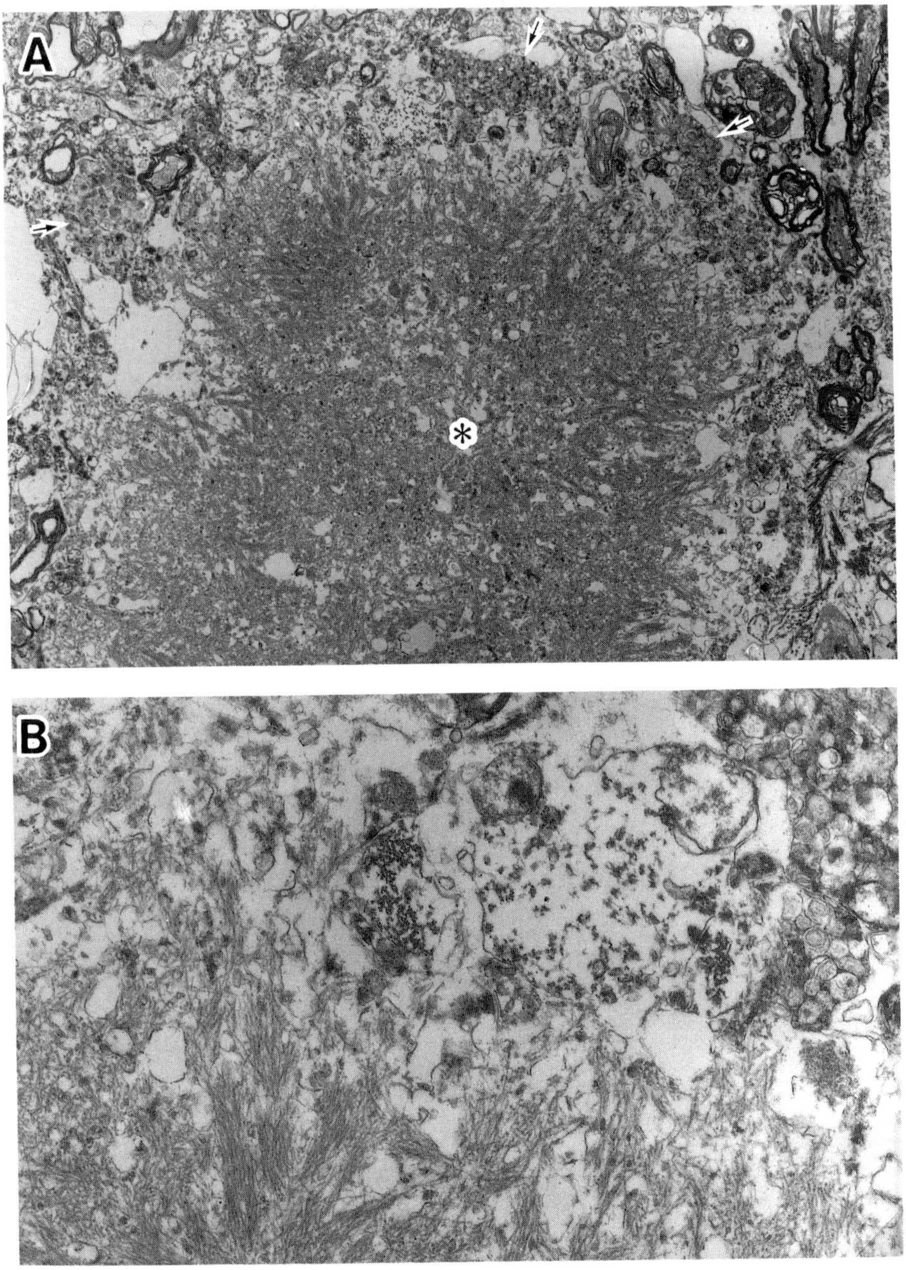


**Figure 18.** Kuru plaque. (A) A kuru plaque (*) associated with occasional dystrophic neurites (arrows) in the cerebral cortex in a long-survival case of Creutzfeldt-Jakob disease. X5,000. (B) Higher magnification of A clearly shows radiating bundles of amyloid filaments and adjacent dystrophic neurites. X20,000.

## 8. AIDS–Dementia Complex

AIDS–Dementia complex (ADC) (Price and Brew, 1988; Navia *et al.*, 1986a,b), more recently referred to as HIV-1-associated cognitive/motor complex (A Work ing Group of the American Academy of Neurology AIDS Task Force, 1991), is a clinical neuropsychiatric disorder that is a frequent and usually later consequence of infection by human immunodeficiency virus type 1 (HIV-1). It is important to know that ADC is always caused by direct HIV-1 brain infection; there seems to be more than one set of neuropathological processes that can bring it about (Budka, 1991; Price *et al.*, 1991).

The brains of ADC patients frequently contain multinucleated giant cells (MGCs), which are macrophage-related (Koenig *et al.*, 1986) and are now considered to be a marker for HIV-1 infection. Although MGCs are generally thought to arise through cell fusion, occasional nuclear bridges or extensions can be seen (Mizusawa *et al.*, 1988), suggesting amitotic cell division. *In situ* hybridization and immunocytochemical methods have demonstrated the virus in MGCs, macrophages (Wiley *et al.*, 1986), and microglial cells (Price *et al.*, 1988; Vazeux *et al.*, 1987). On EM, virus particles from 70 to 160 nm in diameter with an eccentric,

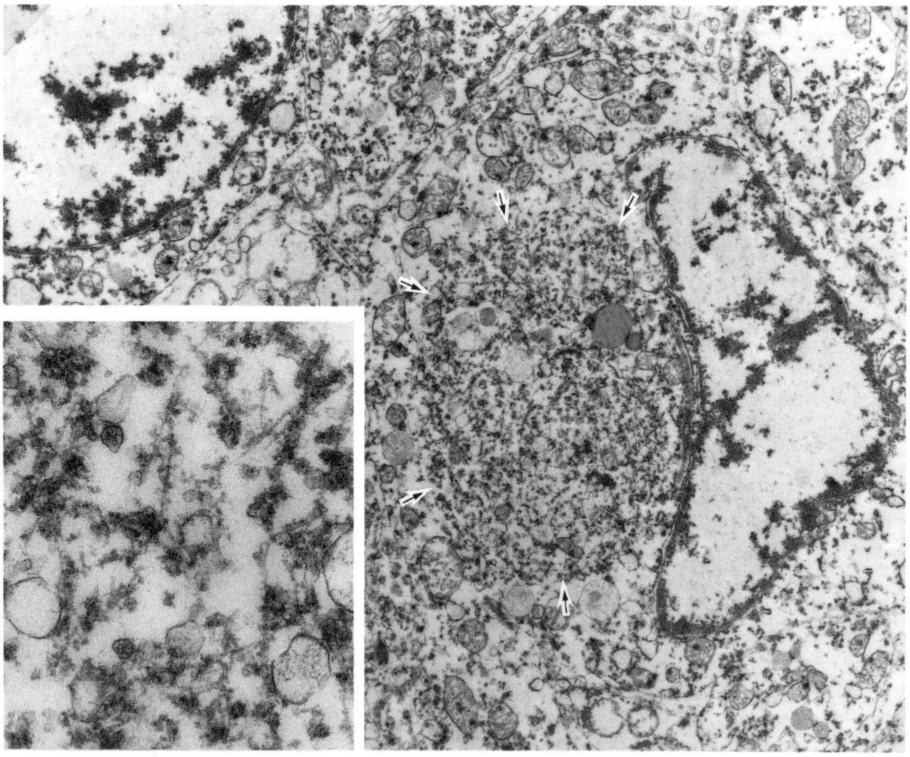

**Figure 19.** A round inclusion (arrows) in a granule cell in the hippocampal dentate gyrus in ALS with dementia, X8,000. Inset: The inclusion consists of approximately 15-nm filaments with associated granules. X40,000.

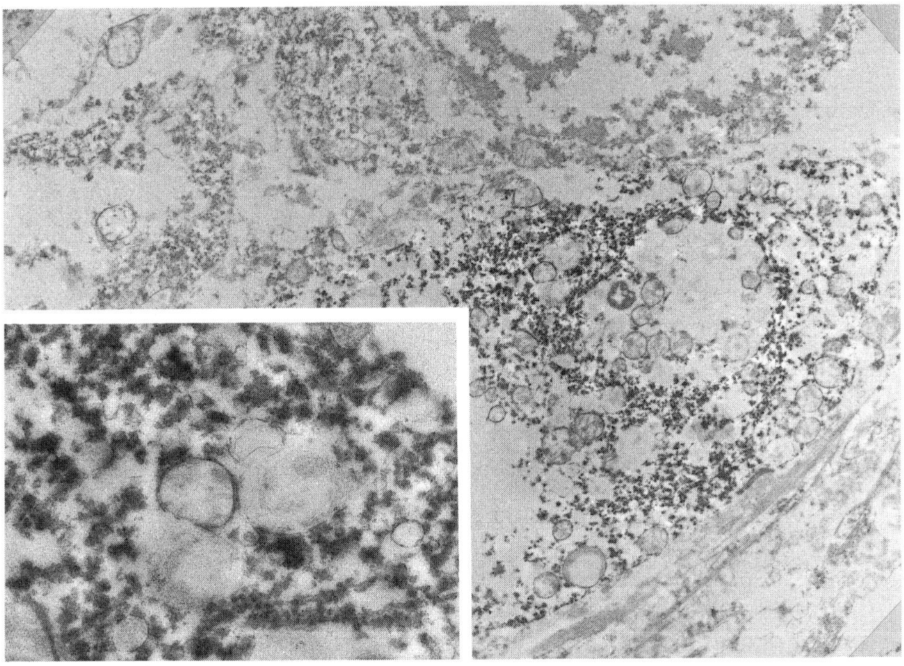

**Figure 20.** Inclusions in the dentate granule cells in ALS with dementia are ubiquitinated. X6,000. Inset: X40,000.

dense nucleoid and a conical or tapered, less-dense core shell can be found free in the cytoplasm or in the cisterns of MGCs and mononucleated macrophage-like cells in the brain (Fig. 21) (Budka, 1991; Meyenhofer *et al.*, 1987); a few extracellular particles can be seen in the proximity of particle-containing cells. With regard to neuroectodermal cells, HIV-1 particles (Mirra and del Rio, 1989; Epstein *et al.*, 1985) and DNA (Takahashi *et al.*, 1996) have been reported in astrocytes. Although an ultrastructural study also revealed the presence of HIV-1 particles in oligodendrocytes (Gyorkey *et al.*, 1987), the identity of these cells was challenged by others (Budka and Lassmann, 1988; Sharer and Prineas, 1988). There is no evidence for neuronal infection by HIV-1, thus leading most investigators to favor the idea that the only cells in the brain that can be readily demonstrated to contain HIV-1 are macrophages, microglial cells, and MGCs (Sharer, 1992). The neuronal loss in the cerebral cortex seen in ADC (Everall *et al.*, 1993) without overt neuronal infection by HIV-1 may suggest an indirect neuronal injury via infected macrophages and microglial cells (Epstein and Gendelman, 1993; Lipton, 1991).

## 9. Conclusion

As shown in this review, the electron microscope is always helpful for neuropathological investigation of the brain and spinal cord affected with dementing illness.

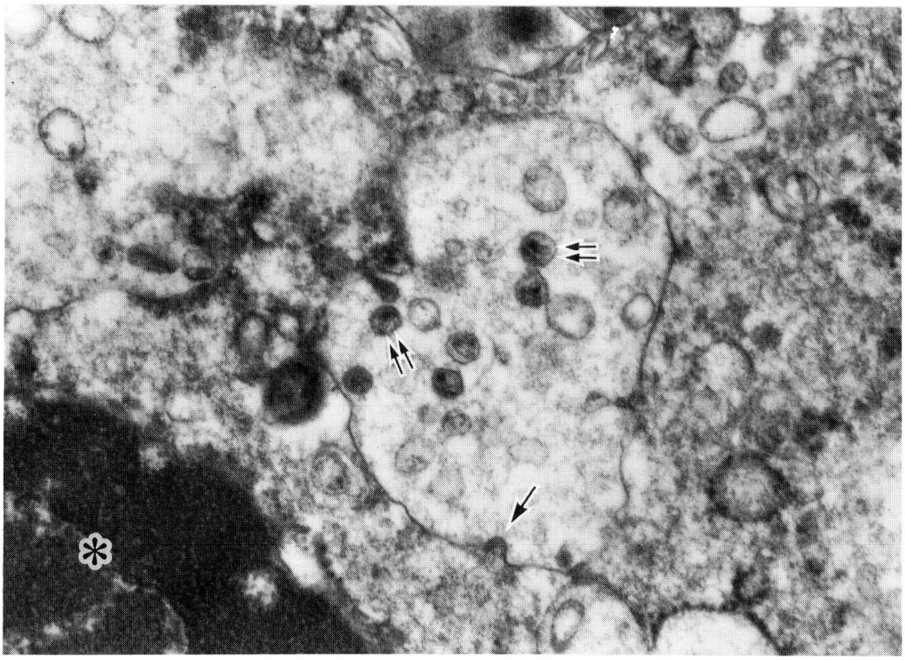

**Figure 21.** HIV virions (double arrows) in cisterns of a macrophage. Budding can be also seen (arrow). Nucleus (*), X21,450. Courtesy of Dr. Hidehiro Mizusawa, Tokyo Medical and Dental University.

ACKNOWLEDGMENTS. This study was supported in part by the Research Committee of CNS Degenerative Diseases, Ministry of Health and Welfare, Japan.

## 10. References

Amano, N., Iwabuchi, K., Yokoi, S., Yagishita, S., Itoh, Y., Saitoh, A., Nagatomo, H., and Matsushita, M., 1989, The reappraisal study of the ultrastructure of Alzheimer's neurofibrillary tangles in three cases of progressive supranuclear palsy, *Brain Nerve (Tokyo)* **41:**35–44.

Askanas, V., Engel, W. K., Bilak, M., Alvarez, R. B., and Selkoe, D. J., 1994, Twisted tubulofilaments of inclusion body myositis muscle resemble paired helical filaments of Alzheimer brain and contain hyperphosphorylated tau, *Am. J. Pathol.* **144:**177–187.

Averback, P., 1981a, Lesions of the nucleus ansae peduncularis in neuropsychiatric disease, *Arch. Neurol.* **38:**230–235.

Averback, P., 1981b, Structural lesions of the brain in young schizophrenics, *Can. J. Neurol. Sci.* **8:**73–76.

A Working Group of the American Academy of Neurology AIDS Task Force, 1991, Nomenclature and research case definitions for neurologic manifestations of human immunodeficiency virus-type 1 (HIV-1) infection, *Neurology* **41:**778–785.

Baker, H. F., Duchen, L. W., Jacobs, J. M., and Ridley, R. M., 1990, Spongiform encephalopathy transmitted experimentally from Creutzfeldt-Jakob and familial Gerstmann-Sträussler-Scheinker diseases, *Brain* **113:**1891–1909.

Bignami, A., and Forno, L. S., 1970, Status spongiosus in Jakob-Creutzfeldt disease: Electron microscopic study of a cortical biopsy, *Brain* **93:**89–94.

Bondareff, W., Wischik, C. M., Novak, M., Amos, W. B., Klug, A., and Roth, M., 1990, Molecular analysis of neurofibrillary degeneration in Alzheimer's disease: An immunohistochemical study, *Am. J. Pathol.* **137**:711–723.

Budka, H., 1991, Neuropathology of human immunodeficiency virus infection, *Brain Pathol.* **1**:163–175.

Budka, H., and Lassmann, H., 1988, Human immunodeficiency virus in glial cells? (letter), *J. Infect. Dis.* **157**:203.

Chou, S. M., Payne, W. N., Gibbs, C. J., Jr., and Gajdusek, D. C., 1980, Transmission and scanning electron microscopy of spongiform change in Creutzfeldt-Jakob disease, *Brain* **103**:885–904.

Dickson, D. W., Crystal, H., Mattiace, L. A., Kress, Y., Schwagerl, A., Ksiezak-Reding, H., Davies, P., and Yen, S.-H. C., 1989, Diffuse Lewy body disease: Light and electron microscopic immunocytochemistry of senile plaques, *Acta Neuropathol.* **78**:572–584.

Duffy, P. E., and Tennyson, V. M., 1965, Phase and electron microscopic observations of Lewy bodies and melanin granules in the substantia nigra and locus ceruleus in Parkinson's disease, *J. Neuropathol. Exp. Neurol.* **24**:398–414.

Endoh, R., Ogawara, M., Iwatsubo, T., Nakano, I., and Mori, H., 1993, Lack of the carboxyl terminal sequence of tau in ghost tangles of Alzheimer's disease, *Brain Res.* **601**:164–172.

Epstein, L. G., and Gendelman, H. E., 1993, Human immunodeficiency virus type 1 infection of the nervous system: Pathogenetic mechanisms, *Ann. Neurol.* **33**:429–436.

Epstein, L. G., Sharer, L. R., Cho, E.-S., Meyenhofer, M. F., Navia, B. A., and Price, R. W., 1985, HTLV-III/LAV-like retrovirus particles in the brains of patients with AIDS encephalopathy, *AIDS Res.* **1**:447–454.

Everall, I., Luthert, P., and Lantos, P., 1993, A review of neuronal damage in human immunodeficiency virus infection: Its assessment, possible mechanism, and relationship to dementia, *J. Neuropathol. Exp. Neurol.* **52**:561–566.

Furthmayr, H., 1988, Basement membranes, in: *The Molecular and Cellular Biology of Wound Repair* (R. A. F. Clark and P. M. Henson, eds.), Plenum, New York, pp. 525–558.

Galloway, P. G., Perry, G., Kosik, K. G., and Gambetti, P., 1987, Hirano bodies contain tau protein, *Brain Res.* **403**:337–340.

Gibb, W. R., Luthert, P. J., and Marsden, C. D., 1989, Corticobasal degeneration, *Brain* **112**:1171–1192.

Gibson, P. H., and Tomlinson, B. E., 1977, Numbers of Hirano bodies in the hippocampus of normal and demented people with Alzheimer's disease, *J. Neurol. Sci.* **33**:199–206.

Glenner, G. G., and Wong, C. W., 1984, Alzheimer's disease: Initial report of the purification and characterization of a novel cerebrovascular amyloid protein, *Biochem. Biophys. Res. Comm.* **120**:885–890.

Goldman, J. E., 1983, The association of actin with Hirano bodies, *J. Neuropathol. Exp. Neurol.* **42**:146–152.

Gonatas, N. K., Terry, R. D., and Weiss, M., 1965, Electron microscopic study in two cases of Jakob-Creutzfeldt disease, *J. Neuropathol. Exp. Neurol.* **24**:575–598.

Grundke-Iqbal, I., Iqbal, K., Quinlan, M., Tung, Y.-C., Zaidi, M. S., and Wisniewski, H. M., 1986, Microtubule-associated protein tau: A component of Alzheimer paired helical filaments, *J. Biol. Chem.* **261**:6084–6089.

Gyorkey, F., Melnick, J. L., and Gyorkey, P., 1987, Human immunodeficiency virus in brain biopsies of patients with AIDS and progressive encephalopathy, *J. Infect. Dis.* **155**:870–876.

Hansen, L. A., Masliah, E., Galasko, D., and Terry, R. D., 1993, Plaque-only Alzheimer disease is usually the Lewy body variant, and vice versa, *J. Neuropathol. Exp. Neurol.* **52**:648–654.

Hirano, A., 1965, Pathology of amyotrophic lateral sclerosis, in: *Slow, Latent, and Temperate Virus Infections* (D. C. Gajdusek, C. J. Gibbs, Jr., and M. Alpers, eds.), NINDB Monograph No. 2, NIH, Washington, pp. 23–37.

Hirano, A., 1981, *A Guide to Neuropathology*, Igaku-Shoin, New York.

Hirano, A., 1992, *A Guide to Neuropathology*, 3rd Ed., Igaku-Shoin, Tokyo.

Hirano, A., 1994, Hirano bodies and related neuronal inclusions, *Neuropathol. Appl. Neurobiol.* **20**:3–11.

Hirano, A., and Zimmerman, H. M., 1961, Alzheimer's neurofibrillary changes: A topographic study, *Arch. Neurol.* **5**:401–415.

Hirano, A., Dembitzer, H. M., Kurland, L. T., and Zimmerman, H. M., 1968, The fine structure of some intraganglionic alterations: Neurofibrillary tangles, granulovacuolar bodies, and "rod-like" structures seen in Guam amyotrophic lateral sclerosis and parkinsonism-dementia complex, *J. Neuropathol. Exp. Neurol.* **27**:176–182.

Hirano, A., Ghatak, N. R., Johnson, A. B., Partnow, M. J., and Gomori, A. J., 1972, Argentophilic plaques in Creutzfeldt-Jakob disease, *Arch. Neurol.* **26**:530–542.

Horoupian, D. S., Thal, L., Katzman, R., Terry, R. D., Davies, P., Hirano, A., DeTeresa, R., Fuld, P. A., Petito, C., Blass, J., and Ellis, J. M., 1984, Dementia and motor neuron disease: Morphometric, biochemical, and Golgi studies, *Ann. Neurol.* **16:**305–313.

Hudson, A. J., 1981, Amyotrophic lateral sclerosis and its association with dementia, parkinsonism, and other neurological disorders, *Brain* **104:**217–247.

Ihara, Y., 1988, Massive somatodendritic sprouting of cortical neurons in Alzheimer's disease, *Brain Res.* **459:**138–144.

Ikeda, S., Yanagisawa, N., Allsop, D., and Glenner, G. G., 1990, Early senile plaques in Alzheimer's disease demonstrated by histochemistry, immunocytochemistry, and electron microscopy, *Hum. Neuropathol.* **21:**1221–1226.

Ikeda, K., Akiyama, H., Haga, C., and Haga, S., 1992a, Evidence that neurofibrillary tangles undergo glial modification, *Acta Neuropathol.* **85:**101–104.

Ikeda, K., Haga, C., Akiyama, H., Kase, K., and Iritani, S., 1992b, Coexistence of paired helical filaments and glial filaments in astrocytic processes within ghost tangles, *Neurosci. Lett.* **148:**126–128.

Ishihara, T., Gondo, T., Takahashi, M., Uchino, F., Ikeda, S., Allsop, D., and Imai, K., 1991, Immunohistochemical and immunoelectron microscopical characterization of cerebrovascular and senile plaque amyloid in aged dogs' brains, *Brain Res.* **548:**196–205.

Ishizu, K., Kuroda, S., Nishinaka, T., Yoshiki, S., and Namba, M., 1995, Glial tangles in Pick's disease, *Neuropathology* **15:**163–174.

Iwatsubo, T., Hasegawa, M., Esaki, Y., and Ihara, Y., 1992, Lack of ubiquitin immunoreactivities at both ends of neuropil threads: Possible bidirectional growth of neuropil threads, *Am. J. Pathol.* **140:**277–282.

Izumiyama, N., Asami, E., Itoh, Y., and Ohtsubo, K., 1990, Alzheimer's neurofibrillary tangles and paired helical filaments in the pheochromocytoma cells of the adrenal medulla: Electron microscopic and immunoelectron microscopic observations, *Acta Neuropathol.* **81:**213–216.

Izumiyama, N., Ohtsubo, K., Tachikawa, T., and Nakamura, H., 1991, Elucidation of three-dimensional ultrastructure of Hirano bodies by the quick-freeze, deep-etch, and replica method, *Acta Neuropathol.* **81:**248–254.

Izumiyama, Y., Ikeda, K., and Oyanagi, S., 1994. Extracellular or ghost Pick bodies and their lack of tau immunoreactivity: A histological, immunohistochemical and electron microscopic study, *Acta Neuropathol.* **87:**277–283.

Jeffrey, M., Goodbrand, I. A., and Goodsir, C. M., 1995, Pathology of the transmissible spongiform encephalopathies with special emphasis on ultrastructure, *Micron* **26:**277–298.

Kato, S., and Nakamura, H., 1990, Presence of two different fibril subtypes in the Pick body: An immunoelectron microscopic study, *Acta Neuropathol.* **81:**125–129.

Kato, S., Hirano, A., and Llena, J. F., 1992, Immunohistochemical, ultrastructural, and immunoelectron microscopic studies of spinal cord neurofibrillary tangles in progressive supranuclear palsy, *Neuropathol. Appl. Neurobiol.* **18:**531–538.

Kawasaki, H., Murayama, S., Tomonaga, M., Izumiyama, N., and Shimada, H., 1987, Neurofibrillary tangles in human upper cervical ganglia: Morphological study with immunohistochemistry and electron microscopy, *Acta Neuropathol.* **75:**156–159.

Kidd, M., 1963, Paired helical filaments in electron microscopy of Alzheimer's disease, *Nature* **197:**192–193.

Kidd, M., 1964, Alzheimer's disease: An electron microscopical study, *Brain* **87:**307–320.

Koenig, S., Gendelman, H. E., Orenstein, J. M., Dal Canto, M. C., Pezeshkpour, G. H., Yungbluth, M., Janotta, F., Aksamit, A., Martin, M. A., and Fauchi, A. S., 1986, Detection of AIDS virus in macrophages in brain tissue from AIDS patients with encephalopathy, *Science* **233:**1089–1093.

Kosaka, K., 1978, Lewy bodies in cerebral cortex: Report of three cases, *Acta Neuropathol.* **42:**127–134.

Kosaka, K., 1995, Diffuse Lewy body disease, *Clin. Neurol.* **35:**1455–1456.

Kosaka, K., Oyanagi, S., Matsushita, M., and Hori, A., 1976, Presenile dementia with Alzheimer-, Pick-, and Lewy-body changes, *Acta Neuropathol.* **36:**221–233.

Kosaka, K., Matsushita, M., Oyanagi, S., and Mehraein, P., 1980, A clinicopathological study of "Lewy body disease." *Psychiat. Neurol. Jpn.*, **82:**292–311.

Kosaka, K., Tsuchiya, K., and Yoshimura, M., 1988, Lewy body disease with and without dementia: A clinicopathological study of 35 cases, *Clin. Neurol.* **7:**299–305.

Kuzuhara, S., Mori, H., Izumiyama, N., Yoshimura, M., and Ihara, Y., 1988, Lewy bodies are ubiquitinated: A light and electron microscopic immunocytochemical study, *Acta Neuropathol.* **75:**345–353.

Lampert, P. W., Gajdusek, D. C., and Gibbs, C. J., Jr., 1972, Subacute spongiform virus encephalopathies: Scrapie, kuru, and Creutzfeldt-Jakob disease: A review, *Am. J. Pathol.* **68:**626–652.

Lantos, P. L., 1992, From slow virus to prion: A review of transmissible spongiform encephalopathy, *Histopathology* **20:**1–11.

Lee, V. M.-Y., Balin, B. J., Otvos, L., Jr., and Trojanowski, J. Q., 1991, A68: A major subunit of paired helical filaments and derivatized forms of normal tau, *Science* **251:**675–678.

Lennox, G., Lowe, J., Landon, M., Byrne, E. J., Mayer, R. J., and Godwin-Austen, R. B., 1989a, Diffuse Lewy body disease: Correlative neuropathology using anti-ubiquitin immunocytochemistry, *J. Neurol. Neurosurg. Psychiatry* **52:**1236–1247.

Lennox, G., Lowe, J., Morrell, K., Landon, M., and Mayer, R. J., 1989b, Anti-ubiquitin immunocytochemistry is more sensitive than conventional techniques in the detection of diffuse Lewy body disease, *J. Neurol. Neurosurg. Psychiatry* **52:**67–71.

Liao, Y.-C. J., Lebo, R. B., Clawson, G. A., and Smuckler, E. A., 1986, Human prion protein cDNA: Molecular cloning, chromosomal mapping, and biological implications, *Science* **233:**364–367.

Liberski, P. P., Yanagihara, R., Asher, D. M., Gibbs, C. J., Jr., and Gajdusek, D. C., 1990, Reevaluation of the ultrastructural pathology of experimental Creutzfeldt-Jakob disease, *Brain* **113:**121–137.

Liberski, P. P., Budka, H., Sluga, E., Barcikowska, M., and Kwiecinski, H., 1992, Tubulovesicular structures in Creutzfeldt-Jakob disease, *Acta Neuropathol.* **84:**238–243.

Lipton, S. A., 1991, HIV-related neurotoxicity, *Brain Pathol.* **1:**193–199.

Manetto, V., Adbul-Karim, F. W., Perry, G., Tabaton, M., Autilio Gambetti, L., and Gambetti, P., 1989, Selective presence of ubiquitin in intracellular inclusions, *Am. J. Pathol.* **134:**505–513.

Manuelidis, E. E., 1985, Creutzfeldt-Jakob disease, *J. Neuropathol. Exp. Neurol.* **44:**1–17.

Martin, L. J., Pardo, C. A., Cork, L. C., and Price, D. L., 1994, Synaptic pathology and glial responses to neuronal injury precede the formation of senile plaques and amyloid deposits in the aging cerebral cortex, *Am. J. Pathol.* **145:**1358–1381.

Medori, R., Tritschler, H. J., LeBlanc, A., Villare, F., Manetto, V., Chen, H. Y., Xue, R., Leal, S., Montagna, P., Cortelli, P., Tinuper, P., Avoni, P., Moghi, M., Baruzzi, A., Hauw, J. J., Ott, J., Lugaresi, E., Autilio-Gambetti, L., and Gambetti, P., 1992, Fatal familial insomnia, a prion disease with a mutation at codon 178 of the prion protein gene, *N. Engl. J. Med.* **326:**444–449.

Meyenhofer, M. F., Epstein, L. G., Cho, E.-S., and Sharer, L. R., 1987, Ultrastructural morphology and intracellular production of human immunodeficiency virus (HIV) in brain, *J. Neuropathol. Exp. Neurol.* **46:**474–484.

Meyer, A., 1929, Über eine der amyotrophischen Lateralsklerose nahestehende Erkrankung mit psychischen Störungen: Zugleich ein Beitrag zur Frage der spastischen Psendosklerose (A. Jakob, ed.), *Zschr. Gesamte Neurol. Psychiat.* **121:**107–128.

Mirra, S. S., and de Rio, C., 1989, The fine structure of acquired immunodeficiency syndrome encephalopathy, *Arch. Pathol. Lab. Med.* **113:**858–865.

Mitsuyama, Y., 1984, Presenile dementia with motor neuron disease in Japan: Clinico-pathological review of 26 cases, *J. Neurol. Neurosurg. Psychiatry* **47:**953–959.

Miyakawa, T., Katsuragi, S., Yamashita, K., and Ohuchi, K., 1994, Morphological investigation of neurofibrillary tangles in Alzheimer's disease, *Jpn. J. Psychiat. Neurol.* **48:**43–47.

Mizusawa, H., Hirano, A., and Llena, J. F., 1988, Nuclear bridges within multinucleated giant cells in subacute encephalitis of acquired immunodeficiency syndrome (AIDS), *Acta Neuropathol.* **76:**166–169.

Mori, H., Tomonaga, M., Baba, N., and Kanaya, K., 1986, The structure analysis of Hirano bodies by digital processing on electron micrographs, *Acta Neuropathol.* **71:**32–37.

Mori, H., Kondo, J., and Ihara, Y., 1987, Ubiquitin is a component of paired helical filaments in Alzheimer's disease, *Science* **235:**1641–1644.

Mori, H., Nishimura, M., Namba, Y., and Oda, M., 1994, Corticobasal degeneration: A disease with widespread appearance of abnormal tau and neurofibrillary tangles, and its relation to progressive supranuclear palsy, *Acta Neuropathol.* **88:**113–121.

Morimura, Y., Hirano, A., and Llena, J. F., 1985, Electronmicroscopic observation of the nucleus basalis of Meynert in human autopsy cases, *Acta Neuropathol.* **68:**130–137.

Munoz, D. G., Wang, D., and Greenberg, B. D., 1993, Hirano bodies accumulate C-terminal sequence of β-amyloid precursor protein (β-APP) epitopes, *J. Neuropathol. Exp. Neurol.* **52:**14–21.

Munoz-Garcia, D., and Ludwin, S. K., 1984, Classic and generalized variants of Pick's disease: A clinicopathological, ultrastructural, and immunocytochemical comparative study, *Ann. Neurol.* **16:**467–480.

Murayama, S., Mori, H., Ihara, Y., and Tomonaga, M., 1990, Immunocytochemical and ultrastructural studies of Pick's disease, *Ann. Neurol.* **27:**394–405.

Nakano, I., 1993, Temporal lobe lesions in amyotrophic lateral sclerosis with or without dementia: a neuropathological study, *Neuropathology* **13:**215–227.

Nakano, I., and Hirano, A., 1983, Neuron loss in the nucleus basalis of Meynert in parkinsonism-dementia complex of Guam, *Ann. Neurol.* **13:**87–91.

Nakano, I., and Hirano, A., 1984a, Parkinson's disease: Neuron loss in the nucleus basalis without concomitant Alzheimer's disease, *Ann. Neurol.* **15:**415–418.

Nakano, I., and Hirano, A., 1984b, Nucleus basalis of Meynert: Electron-microscopic observation of the nucleus basalis of Meynert, *Neurol. Med. (Tokyo)* **20:**264–267.

Nakano, I., Iwatsubo, T., Otsuka, N., Kamei, M., Matsumura, K., and Mannen, T., 1992a, Paired helical filaments in astrocytes: Electron microscopy and immunohistochemistry in a case of atypical Alzheimer's disease, *Acta Neuropathol.* **83:**228–232.

Nakano, I., Kato, S., Yazawa, I., and Hirano, A., 1992b, Anchorage densities associated with hemidesmosome-like structures in perivascular reactive astrocytes, *Acta Neuropathol.* **84:**85–88.

Nakano, I., Hayashi, M., Fujisawa, K., and Mori, H., 1993, Electron microscopic observation of senile plaques (SPs) in aged non-human primates, *J. Neuropathol. Exp. Neurol. (Abstr.)* **52:**264.

Nakano, I., Iwatsubo, T., Otsuka, N., Kamei, M., Matsumura, K., and Mannen, T., 1996, Some unusual responses of astrocytes to ghost tangles in a long duration case of juvenile Alzheimer's disease: An electron microscopic study, *J. Neurol. Sci.* **136:**41–46.

Navia, B. A., Jordan, B. D., and Price, R. W., 1986a, The AIDS dementia complex. I. Clinical features, *Ann. Neurol.* **19:**517–524.

Navia, B. A., Cho, E.-S., Petito, C. K., and Price, R. W., 1986b, The AIDS-dementia complex. II. Neuropathology, *Ann. Neurol.* **19:**525–535.

Nishimura, M., Namba, Y., Ikeda, K., and Oda, M., 1992, Glial fibrillary tangles with straight tubules in the brains of patients with progressive supranuclear palsy, *Neurosci. Lett.* **143:**35–38.

Nishimura, M., Namba, Y., Ikeda, K., Akiguchi, I., and Oda, M., 1993, Neurofibrillary tangles in the neurons of spinal dorsal root ganglia of patients with progressive supranuclear palsy, *Acta Neuropathol.* **85:**453–457.

Nukina, N., and Ihara, Y., 1986, One of the antigenic determinants of paired helical filaments is related to tau protein, *J. Biochem.* **99:**1541–1544.

Nukina, N., Quan, Y., Nakano, I., and Otomo, E., 1992, Widespread tau abnormality is a case of corticobasal degeneration, *Clin. Neurol. (Tokyo)* **32:**1093–1101.

O'Brien, L., Shelley, K., Towfighi, J., and McPherson, A., 1980, Crystalline ribosomes are present in brains from senile humans, *Proc. Natl. Acad. Sci. USA* **77:**2260–2264.

Ohtsubo, K., Izumiyama, N., Shimada, H., Tachikawa, T., and Nakamura, H., 1990a, Three-dimensional structure of Alzheimer's neurofibrillary tangles of the aged human brain revealed by the quick-freeze, deep-etch, and replica method, *Acta Neuropathol.* **79:**480–485.

Ohtsubo, K., Izumiyama, N., Kuzuhara, S., Mori, H., and Shimada, H., 1990b, Curly fibers are tau-positive strands in the pre- and post-synaptic neurites, consisting of paired helical filaments: Observations by the freeze-etch and replica method, *Acta Neuropathol.* **81:**111–115.

Okamoto, K., Hirano, A., Yamaguchi, H., and Hirai, S., 1983, The fine structure of eosinophilic stages of Alzheimer's neurofibrillary tangles, *J. Clin. Electron Microsc.* **16:**77–82.

Okamoto, K., Hirai, S., Iizuka, T., Yanagisawa, T., and Watanabe, M., 1991a, Reexamination of granulovacuolar degeneration, *Acta Neuropathol.* **82:**340–345.

Okamoto, K., Hirai, S., Yamazaki, T., Sun, X. Y., and Nakazato, Y., 1991b, New ubiquitin-positive intraneuronal inclusions in the extra-motor cortices in patients with amyotrophic lateral sclerosis, *Neurosci. Lett.* **129:**233–236.

Okamoto, K., Murakami, N., Kusaka, H., Yoshida, M., Hashizume, Y., Nakazato, Y., Matsubara, E., and Hirai, S., 1992, Ubiquitin-positive intraneuronal inclusions in the extramotor cortices of presenile dementia with motor neuron disease, *J. Neurol.* **239:**426–430.

Okamoto, K., Hirai, S., Amari, M., and Sakurai, A., 1996, Electron micrograph of ubiquitin-positive intraneuronal inclusions in the extra-motor cortices in patients with amyotrophic lateral sclerosis, *Neuropathology* **16:**112–116.

Okazaki, H., Lipkin, L. E., and Aronson, S. M., 1961, Diffuse intracytoplasmic ganglionic inclusions (Lewy type) associated with progressive dementia and quadriparesis in flexion, *J. Neuropathol. Exp. Neurol.* **24:**398–414.

Oyanagi, S., and Ikuta, F., 1974, An ultrastructural observation on granulovacuolar degenerations in senile dementia, with special reference to the process of their formation, *Brain Nerve (Tokyo)* **26:**783–788.

Oyanagi, S., Tanaka, M., Omori, T., Matsushita, M., and Ishii, T., 1979, Regular arrangements of tubular structures in Pick bodies formed in an autopsied case of Pick's disease, *Shinkei Kenkyu no Shinpo* **23:**441–451.

Paulus, W., and Selim, M., 1990, Corticonigral degeneration with neuronal achromasia and basal neurofibrillary tangles, *Acta Neuropathol.* **81:**89–94.

Perry, G., Kawai, M., Tabaton, M., Onorato, M., Mulvilhill, P., Richey, P., Mornadi, A., Connolly, J. A., and Gambetti, P., 1991, Neuropil treads of Alzheimer's disease show a marked alteration of the normal cytoskeleton, *J. Neurosci.* **11:**1748–1755.

Perry, R. H., Irving, D., Blessed, G., Fairbairn, A., and Perry, E. K., 1990, Senile dementia of Lewy body type: A clinically and neuropathologically distinct form of Lewy body dementia in the elderly, *J. Neurol. Sci.* **95:**119–139.

Peterson, C., Suzuki, K., Kress, Y., and Goldman, J. E., 1986, Abnormalities of dendritic actin organization in the brindled mouse, *Brain Res.* **382:**205–212.

Peterson, C., Kress, Y., Vallee, R., and Goldman, J. E., 1988, High molecular weight microtubule-associated proteins bind to actin lattices, *Acta Neuropathol.* **77:**168–174.

Pollanen, M. S., Markiewicz, P., Goh, M. C., and Bergeron, C., 1995, Alzheimer paired helical filaments: A comparison with the twisted ribbon model, *Acta Neuropathol.* **90:**194–197.

Powell, H. C., London, G. W., and Lampert, R. W., 1974, Neurofibrillary tangles in progressive supranuclear palsy: Electron microscopic observations, *J. Neuropathol. Exp. Neurol.* **33:**98–106.

Price, R. W., and Brew, B. J., 1988, The AIDS dementia complex, *J. Infect. Dis.* **158:**1079–1083.

Price, R. W., Brew, B., Sidtis, J., Rosenblum, M., Scheck, A. C., and Cleary, P., 1988, The brain in AIDS: Central nervous system HIV-1 infection and AIDS dementia complex, *Science* **239:**586–592.

Price, R. W., Sidtis, J. J., and Brew, B. J., 1991, AIDS dementia complex and HIV-1 infection: A view from the clinic, *Brain Pathol.* **1:**155–162.

Probst, A., Ulrich, J., and Heitz, Ph. U., 1982, Senile dementia of Alzheimer type: Astroglial reaction to extracellular neurofibrillary tangles in the hippocampus, *Acta Neuropathol.* **57:**75–79.

Prusiner, S. B., 1982, Novel proteinaceous infectious particles cause scrapie, *Science* **216:**136–144.

Prusiner, S. B., 1995, The prion diseases, *Scientific American* January: 30–37.

Rebeiz, J. J., Kolodny, E. H., and Richardson, E. P., 1968, Corticodentatonigral degeneration with neuronal achromasia, *Arch. Neurol.* **18:**20–33.

Roy, S., Datta, C. K., Hirano, A., Ghatak, N. R., and Zimmerman, H. M., 1974, Electron microscopic study of neurofibrillary tangles in Steele-Richardson-Olszewski syndrome, *Acta Neuropathol.* **29:**175–179.

Samuel, W., Galasko, D., Masliah, E., and Hansen, L. A., 1996, Neocortical Lewy body counts correlate with dementia in the Lewy body variant of Alzheimer's disease, *J. Neuropathol. Exp. Neurol.* **55:**44–52.

Schmidt, M. L., Gur, R. E., Gur, R. C., and Trojanowski, J. Q., 1988, Intraneuronal and extracellular neurofibrillary tangles exhibit mutually exclusive cytoskeletal antigens, *Ann. Neurol.* **23:**184–189.

Schmidt, M. L., Lee, V. M.-Y., and Trojanowski, J. Q., 1989, Analysis of epitopes shared by Hirano bodies and neurofilament proteins in normal and Alzheimer's disease hippocampus, *Lab. Invest.* **60:**513–522.

Schochet, S. S., Jr., and McCormick, W. F., 1972, Ultrastructure of Hirano bodies, *Acta Neuropathol.* **21:**50–60.

Schochet, S. S., Jr., Lampert, R. W., and Lindenberg, R., 1968, Fine structure of the Pick and Hirano bodies in a case of Pick's disease, *Acta Neuropathol.* **11:**330–337.

Selkoe, D. J., 1991, The molecular pathology of Alzheimer's disease, *Neuron* **6:**487–498.

Sharer, L. R., 1992, Pathology of HIV-1 infection of the central nervous system: A review, *J. Neuropathol. Exp. Neurol.* **51:**3–11.

Sharer, L. R., and Prineas, J. W., 1988, Human immunodeficiency virus in glial cells, continued, (letter), *J. Infect. Dis.* **157:**204.

Steele, J. C., Richardson, J. C., and Olszewski, J., 1964, Progressive supranuclear palsy: A heterogeneous degeneration involving the brain stem, basal ganglia, and cerebellum with vertical gaze and pseudobulbar palsy, nuchal dystonia, and dementia, *Arch. Neurol.* **10:**333–359.

Takahashi, H., Takeda, S., Ikuta, F., and Homma, Y., 1987, Progressive supranuclear palsy with limbic system involvement: Report of a case with ultrastructural investigation of neurofibrillary tangles in various locations, *Clin. Neuropathol.* **6:**271–276.

Takahashi, K., Wesselingh, S. L., Griffin, D. E., McArthur, J. C., Johnson, R. T., and Glass, J. D., 1996, Localization of HIV-1 in human brain using polymerase chain reaction/*in situ* hybridization and immunohistochemistry, *Ann. Neurol.* **39:**705–711.

Tellez-Nagel, I., and Wisniewski, H. M., 1973, Ultrastructure of neurofibrillary tangles in Steele-Richardson-Olszewski syndrome, *Arch. Neurol.* **29:**324–327.

Terry, R. D., 1963, The fine structure of neurofibrillary tangles in Alzheimer's disease, *J. Neuropathol. Exp. Neurol.* **22:**629–642.

Terry, R. D., and Wisniewski, H. M., 1970, The ultrastructure of the neurofibrillary tangle and the senile plaque, in: *Alzheimer's Disease and Related Conditions* (G. E. W. Wolstenholme and M. O'Connor, eds.), J. and A. Churchill, London, pp. 145–168.

Terry, R. D., and Wisniewski, H. M., 1972, Ultrastructure of senile dementia and of experimental analogs, in: *Aging and The Brain* (C. M. Gaintz, eds.), Plenum, New York, 1972, pp. 89–116.

Terry, R. D., Gonatas, N. K., and Weiss, M., 1964, Ultrastructural studies in Alzheimer's presenile dementia, *Am. J. Pathol.* **44:**269–297.

Tomlinson, B. E., 1992, Ageing and the dementias, in: *Greenfield's Neuropathology*, 5th Ed. (J. H. Adams and L. W. Duchen, eds.), Oxford University Press, New York, pp. 1284–1410.

Tomonaga, M., 1977, Ultrastructure of neurofibrillary tangles in progressive supranuclear palsy, *Acta Neuropathol.* **37:**177–181.

Van Reeth, P. Ch., Périer, O., Coërs, C., and van Bogaert, L., 1961, Démence de Pick associée à une sclérose latérale amyotrophique atypique (Etude anatomoclinique), *Acta Neurol. Psychiat. Belg.* **61:**309–325.

Vazeux, R., Brousse, N., Jarry, A., Henin, D., Marche, C., Vedrenne, C., Mikol, J., Wolff, M., Michon, C., Rozenbaum, W., Bureau, J.-F., Montagnier, L., and Brahic, M., 1987, AIDS subacute encephalitis: Identification of HIV-infected cells, *Am. J. Pathol.* **126:**403–410.

Wakabayashi, K., Oyanagi, K., Makifuchi, T., Ikuta, F., Homma, A., Homma, Y., Horikawa, Y., and Tokiguchi, S., 1994, Corticobasal degeneration: Etiopathological significance of the cytoskeletal alterations, *Acta Neuropathol.* **87:**545–553.

Wegiel, J., and Wisniewski, H. M., 1994, Rosenthal fibers, eosinophilic inclusions, and anchorage densities with desmosome-like structures in astrocytes in Alzheimer's disease, *Acta Neuropathol.* **87:**355–361.

Whitehouse, P. J., Price, D. L., Clark, A. W., Coyle, J. T., and DeLong, M. R., 1981, Alzheimer's disease: Evidence for selective loss of cholinergic neurons in the nucleus basalis, *Ann. Neurol.* **10:**122–126.

Whitehouse, P. J., Price, D. L., Struble, R. G., Clark, A. W., Coyle, J. T., and DeLong, M. R., 1982, Alzheimer's disease and senile dementia: Loss of neurons in the basal forebrain, *Science* **215:**1237–1239.

Wightman, G., Anderson, V. E. R., Martin, J., Swash, M., Anderton, B. H., Neary, D., Mann, D., Luthert, P., and Leigh, P. N., 1992, Hippocampal and neocortical ubiquitin-immunoreactive inclusions in amyotrophic lateral sclerosis, *Neurosci. Lett.* **139:**269–274.

Wiley, C. A., Schrier, R. D., Nelson, J. A., Lampert, P. W., and Oldstone, M. B. A., 1986, Cellular localization of human immunodeficiency virus infection within the brains of acquired immune deficiency syndrome patients, *Proc. Natl. Acad. Sci. USA* **83:**7089–7093.

Will, R. G., Ironside, J. W., Zeidler, M., Cousens, S. N., Estibeiro, K., Alperovitch, A., Poser, S., Pocchiari, M., Hofman, A., and Smith, P. G., 1996, A new variant of Creutzfeldt-Jakob disease in the UK, *Lancet* **347:**921–925.

Williams, E. S., and Young, S., 1982, Spongiform encephalopathy of Rocky Mountain elk, *J. Wildl. Dis.* **18:**465–471.

Wisniewski, H. M., and Terry, R. D., 1973, Re-examination of the pathogenesis of the senile plaque, in: *Progress in Neuropathology*, Vol. 2 (H. M. Zimmerman, ed.), Grune and Stratton, New York, pp. 1–26.

Wisniewski, H. M., Johnson, A. B., Raine, C. S., Kay, W. J., and Terry, R. D., 1970, Senile plaques and cerebral amyloidosis in aged dogs, *Lab. Invest.* **23:**287–296.

Yagishita, S., Itoh, Y., Amano, N., Nakano, T., and Saitoh, A., 1979, Ultrastructure of neurofibrillary tangles in progressive supranuclear palsy, *Acta Neuropathol.* **48:**27–30.

Yamaguchi, H., Nakazato, Y., Hirai, S., Shoji, M., and Harigaya, Y., 1989, Electron micrograph of diffuse plaques: Initial stage of senile plaque formation in the Alzheimer brain, *Am. J. Pathol.* **135:**593–597.

Yamagauchi, H., Nakazato, Y., Hirai, S., Shoji, M., and Ihara, Y., 1990, Ultrastructure of the neuropil threads in the Alzheimer brain: Their dendritic origin and accumulation in the senile plaques, *Acta Neuropathol.* **80:**368–374.

Yamaguchi, H., Nakazato, Y., Kawarabayashi, T., Ishiguro, K., Ihara, Y., Morimatsu, M., and Hirai, S., 1991a, Extracellular neurofibrillary tangles associated with degenerating neurites and neuropil threads in Alzheimer-type dementia, *Acta Neuropathol.* **81:**603–609.

Yamaguchi, H., Nakazato, Y., Shoji, M., Okamoto, K., Ihara, Y., Morimatsu, M., and Hirai, S., 1991b, Secondary deposition of beta amyloid within extracellular neurofibrillary tangles in Alzheimer-type dementia, *Am. J. Pathol.* **138:**699–705.

Yamazaki, M., Nakano, I., Imazu, O., Kaieda, R., and Terashi, A., 1994, Astrocytic straight tubules in the brain of a patient with Pick's disease, *Acta Neuropathol.* **88:**587–591.

Yamazaki, M., Nakano, I., Imazu, O., and Terashi, A., 1995, Paired helical filaments and straight tubules in astrocytes: An electron microscopic study in dementia of the Alzheimer type, *Acta Neuropathol.* **90:**31–36.

Yoshida, M., Murakami, N., Hashizume, Y., and Takahashi, A., 1992, A clinicopathological study of 13 cases of motor neuron disease with dementia, *Clin. Neurol.* **32:**1193–1202.

Yoshimura, M., 1983, Cortical changes in the parkinsonian brain: A contribution to the delineation of "diffuse Lewy body disease," *J. Neurol.* **229:**17–32.

Yuasa, R., 1970, Amyotrophic lateral sclerosis with dementia, *Clin. Neurol.* **10:**569–577.

# 12

# Locating Genetic Modifiers for Inherited Neurodegenerative Diseases

LINDSAY A. FARRER

## 1. Introduction

Advances in molecular and statistical genetics have heralded discoveries of the basis of many disorders, including those affecting the central nervous system. The principal strategy, often referred to as "reverse genetics," does not require any knowledge of the biochemistry or pathogenesis of a disease. This approach, however, presumes the existence of a major disease susceptibility gene whose various forms, called alleles, are faithfully transmitted from one generation to the next in a predictable Mendelian fashion. The first step entails localizing the gene to a chromosome using one of several methodologies, most frequently genetic linkage analysis. Then other strategies, including positional cloning and physical mapping, are employed to isolate (i.e., clone) the gene. Next the gene is sequenced to deduce the predicted primary protein structure that it encodes and is evaluated for mutations associated with the disease. Finally, experiments are pursued to delineate the

LINDSAY A. FARRER • Departments of Medicine (Genetics Program) and Neurology, Boston University School of Medicine, and Department of Epidemiology and Biostatistics, Boston University School of Public Health, Boston, Massachusetts 02118.

*Cerebral Cortex,* Volume 14, edited by Peters and Morrison. Kluwer Academic/Plenum Publishers, New York, 1999.

expression of the gene product in various tissues from individuals with and without disease.

For some disorders, such as cystic fibrosis and several forms of cancer, deciphering the genetic abnormality has led to the development of promising therapies (e.g., supplying a deficient product, targeted gene therapy). In these situations, knowledge of the primary genetic defect provided sufficient understanding about pathogenesis. However, there are a growing number of genetic disorders whose primary defect has been identified but is a weak predictor of the natural history. Not surprisingly, CNS diseases are well represented among this group. These conditions will not manifest without a defect in the primary gene, but when or how it develops is at best only modestly correlated with the mutation, suggesting that other factors govern expression of the disease gene. These factors, called modifiers, may be genetic or environmental.

This chapter seeks to sort out the evidence for the existence of genetic modifiers for inherited neurodegenerative diseases. Several strategies for localizing modifier genes are discussed, and each strategy combines novel analytical approaches with the powerful tools of molecular genetics. Elucidation of such modifiers may provide a better understanding of disease pathogenesis. Because defects in the primary genes for this group of disorders are not easy to target directly, effective therapies or prevention may be effected by regulating expression of modifier genes (i.e., regulating the regulator). Although genetic modifiers have been implicated, but not proven, for a variety of Mendelian diseases, few attempts have been made to locate such modifiers for several reasons, including inability to demonstrate the existence of discrete major modifier loci, lack of analytic strategies for permitting individuals without the primary defect to transmit the modifying effect, and need for a large number of families with multiple affected members.

## 2. Conceptualization of Genetic Modifiers

Phenotypic traits, especially disease traits, are traditionally considered to depend solely on one gene (Mendelian inheritance) or many genes (multifactorial or polygenic inheritance). In fact, the expression of genetic traits due to the interactions of two loci are common. In many such cases, the penetrance or expressivity of the primary gene is influenced by the second gene, called a “modifier gene” when expressivity is influenced. When penetrance is suppressed altogether, the word “epistasis” is used. The best-analyzed example of a modifying gene in humans is offered by the ABO blood-group system. Occurrence of the ABH antigens in saliva depends on the secretor gene. The “Bombay” phenotype is the result of homozygosity for the rare *h* allele of the *H-h* system, which suppresses expression of ABO and secretor substances (see Fig. 1).

## 3. Wilson’s Disease

Wilson’s disease (WD) is an autosomal recessive copper metabolism disorder that results in the accumulation of copper in tissues, principally the liver, basal

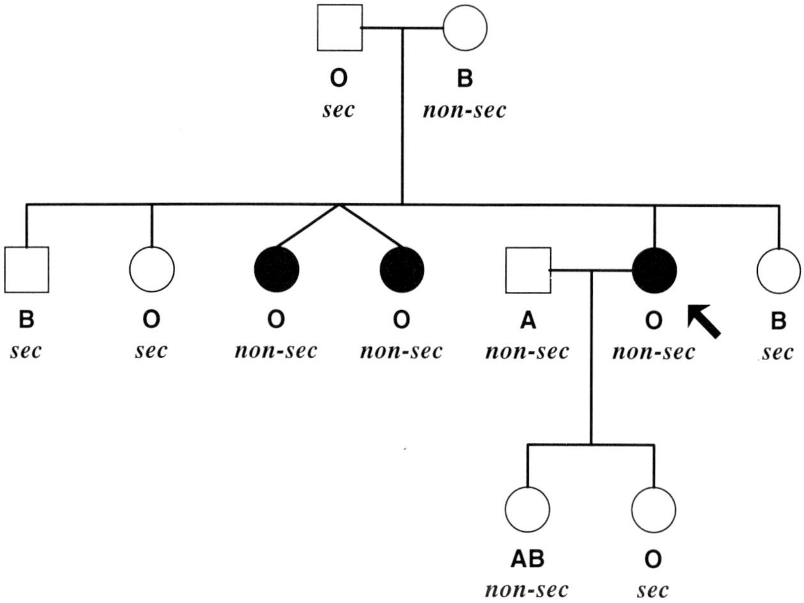


**Figure 1.** Pedigree showing suppression of the ABO blood group and secretor genes ("Bombay" phenotype). ABO blood group and secretor phenotypes are shown below pedigree symbols. In the ABO system, alleles A and B are codominant to each other and dominant to allele *O*. Thus, persons with ABO type A may have genotypes *A/A* or *A/O*, and type B persons may have genotypes *B/B* or *B/O*. In the secretor system, secretor is dominant to nonsecretor. Thus, secretors may be heterozygous for the nonsecretor allele. The proband (arrow) was ascertained because initial studies excluded her as being the biological mother of both children; the true mother of these children must have transmitted the ABO *B* and *sec* alleles to the first and second children, respectively. This peculiar inheritance pattern, originally reported by Bhende *et al.* (1952), was later demonstrated to be due to the suppression of ABO and secretor substances by a genetically independent system, *H-h*. Apparently, persons having the rare genotype *h/h* do not express the B or secretor genes (Levine *et al.*, 1955).

ganglia, and kidneys. Although WD is a relatively rare disorder (gene frequency = 0.0056 worldwide) (Scheinberg and Sternlieb, 1984), it is the most frequent cause of chronic liver disease in children (Danks, 1989). The incidence is probably underestimated, because many cases are undetected due to inadequate diagnosis. Patients usually present with liver disease or extrapyramidal syndrome in the first or second decade of life, but some biochemically normal patients manifest psychiatric symptoms in middle age (Sass-Kortsak and Bearn, 1978; Dening, 1985). Left untreated, WD is lethal, but if detected in time the disease is reversible or preventable by oral ingestion of copper-chelating agents such as penicillamine (Scheinberg and Sternlieb, 1975), zinc (Hill *et al.*, 1986), and trientine (Scheinberg *et al.*, 1987). However, many patients cannot tolerate the side effects of penicillamine or do not respond well to zinc therapy. The diagnosis depends on the demonstration of low serum copper and ceruloplasmin levels, increased urinary copper excretion, decreased incorporation of radiocopper into ceruloplasmin, and evidence for copper deposition in the cornea (Kayser-Fleischer ring) and liver. Affected individuals have severely reduced levels of ceruloplasmin, but the disease is not caused by a molecular defect in ceruloplasmin (Danks, 1989).

One of the puzzling aspects of WD is the extensive variability in liver pathology, neurological symptoms, and age at onset (Walshe, 1976; Passwell *et al.*, 1977), sometimes even within a family (Arima and Sano, 1968; Scheinberg and Sternlieb, 1984; Bonné-Tamir *et al.*, 1990). Presentation of WD in childhood is usually accompanied by hepatic disease, whereas neurological symptoms predominate when onset occurs in adolescence or young adulthood. However, there are many exceptions to this pattern, including patients in their sixth decade who present with liver disease without neurological features or Kayser-Fleischer rings (Danks, 1989).

To assess the extent of and identify possible factors contributing to this variability, we obtained the literature, hospital neurological clinics and records, and clinical and family history data on persons affected with WD living in Israel between 1958 and 1984 (L. Farrer and B. Bonné-Tamir, unpublished). From a population of 51 families, representing a diversity of Middle Eastern, North African, and European backgrounds, blood samples were collected from affected individuals in 21 families, their parents, sibs, and other relatives, for quantitative determinations of plasma copper and ceruloplasmin, liver tests, and DNA analysis. Although the majority of patients had the hepatic form of the disease, both hepatic and neurologic cases were found among all ethnic groups. In fact, affected sibs in several inbred families who most likely inherited two copies of the same mutant allele had different symptoms.

We also investigated the effect of sex on age at onset and type of clinical symptoms in these patients using analysis of variance and multiple regression techniques. Clinical type was treated as a quantitative trait (1 = hepatic; 2 = hepatic + neurologic; 3 = neurologic), and analyses adjusted for ethnic background. Sex had no apparent influence on age at onset ($n = 87$, $p > 0.5$). However, there was a significant tendency for affected females to have neurologic symptoms and for affected males to have liver disease, after adjusting for age at onset ($n = 80$, $p[\text{sex}] = 0.04$, $p[\text{onset}] = 0.0001$). This finding is remarkable, because the opposite association (i.e., females more likely to have liver disease) was observed in a series of British and Taiwanese patients (Strickland and Leu, 1975). We did not find any significant association between sex and primary clinical form among 53 Sardinian (Giagheddu *et al.*, 1985) and 49 Turkish (Aksoy and Erdem, 1972) literature cases ($p[\text{sex}] = 0.36$ and 0.94, respectively). These findings are consistent with the suggestion of genetic heterogeneity for the major locus of WD (Cox *et al.*, 1972; Strickland *et al.*, 1973).

The cause of WD, a disorder recognized since the beginning of this century (Wilson, 1912), has been recently identified as a defect in a gene that codes for a copper ATPase (Bull *et al.*, 1993; Petrukhin *et al.*, 1993; Tanzi *et al.*, 1993; Yamaguchi *et al.*, 1993). This gene, dubbed ATP7B, was cloned from the candidate region previously defined by linkage-mapping studies (Frydman *et al.*, 1985; Bowcock *et al.*, 1988; Yuzbasiyan-Gurkan *et al.*,. 1988; Stewart *et al.*, 1993; White *et al.*, 1993; Bowcock *et al.*, 1994). The product of ATP7B is a copper-binding P-type ATPase protein homologous to the Menkes disease gene (ATP7A). Mutation analysis has so far uncovered more than 25 pathogenic mutations affecting different domains of the putative protein (Tanzi *et al.*, 1993; Figus *et al.*, 1995; Houwen *et al.*, 1995; Thomas *et al.*, 1995a,b; Waldenström *et al.*, 1996; Kalinsky *et al.*, 1998). Most of these mutations are very rare and appear to be private (i.e., occurring in families descending from a common ancestor), although several have been observed in diverse ethnic groups (Figus *et al.*, 1995; Thomas *et al.*, 1995a; Waldenström *et al.*, 1996; Kalinsky *et al.*,

1998). The ATP7A and ATP7B gene products are 56% identical, but many of the domains have between 65% and 89% identity. Expression studies revealed that ATP7B is expressed primarily in the liver, kidney, and placenta (Bull *et al.*, 1993; Petrukhin *et al.*, 1993; Yamaguchi *et al.*, 1993). Preliminary studies have not revealed a strong correlation between the type or location of mutation and clinical outcome (Thomas *et al.*, 1995a,b; Kalinsky *et al.*, 1998); however, any conclusions would be speculative because most patients are compound heterozygotes (i.e., have two different mutations).

Linkage studies representing populations from North America, Europe, North Africa, the Middle East, and East Asia unequivocally support the existence of a single locus (Yuzbasiyan-Gurkan *et al.*, 1988; Figus *et al.*, 1989; Chuang *et al.*, 1991; Farrer *et al.*, 1991; Scheffer *et al.*, 1992). Thus, there is no evidence that multiple WD loci are responsible for phenotypic heterogeneity. Moreover, although different mutations at the WD locus may account for some of the clinical variability (Thomas *et al.*, 1995a), allelic heterogeneity would not account for intrafamilial variation, particularly among patients in consanguineous families who inherited two mutant alleles that are identical by descent (Arima and Sano, 1968; Bonné-Tamir *et al.*, 1990). Studies of clinical patterns in affected individuals, including those homozygous for the same ATP7B mutation, and copper ceruloplasmin levels in heterozygotes and normal homozygotes identified by linkage, suggest that phenotypic variation in WD may be due in part to an interaction of WD locus with other genetic and nongenetic modifiers (Farrer *et al.*, 1988; Bonné-Tamir *et al.*, 1990; Kalinsky *et al.*, 1998).

## 4. Huntington's Disease

Huntington's disease (HD), a dominantly inherited neurodegenerative disorder caused by the expansion of the trinucleotide repeat CAG, located at one end of the HD gene on chromosome 4p16 (HD Collaborative Research Group, 1993). The primary symptoms of chorea and dementia progress inexorably until death. Onset is usually between 30 and 50 years of age, although symptoms may appear at any time, from infancy until the ninth decade of life (Conneally, 1984). The tendency for age at onset to be more homogeneous within families (Myers *et al.*, 1982; Pericak-Vance *et al.*, 1983; Farrer *et al.*, 1984) and similarity of onset among monozygous twins (Hayden, 1981; Sudarsky *et al.*, 1983) supports a role for genetic modifiers, including those influencing longevity in normal persons (Farrer *et al.*, 1984).

To seek additional evidence for the existence of genetic modifiers of age at onset, one can evaluate the heritability of age at onset in HD families. Heritability estimates ($h^2$) are derived for a specified relative set by computing the intraclass correlation, $r$, and dividing it by the degree of relationship. Table I shows $h^2$ values for various relative sets obtained from data in the National HD Roster at Indiana University using the FCOR program (Bailey-Wilson and Elston, 1994). These estimates, which are consistently higher than 100% and show similar heritability in parents and sibs, favor the hypothesis of a major dominant gene. It is now apparent that about half of the total variation in age at onset is explained by the HD mutation

**Table I. Estimates of Heritability for Onset Age of Huntington's Disease**

| Relative set | Number of pairs | Kinship coefficient | $r$ | $h^2$ |
|---|---|---|---|---|
| Parent | 2341 | 0.5 | 0.64 | 128 |
| Sib | 2532 | 0.5 | 0.64 | 128 |
| Half-sib | 40 | 0.25 | 0.80 | 320 |
| Grandparent | 648 | 0.25 | 0.48 | 192 |
| Avuncular | 2164 | 0.25 | 0.43 | 172 |
| First cousin | 1127 | 0.125 | 0.47 | 376 |
| Great grandparent | 81 | 0.125 | 0.02 | 16 |
| Great-avuncular | 313 | 0.125 | 0.29 | 232 |
| Half-avuncular | 33 | 0.125 | 0.49 | 392 |

itself. The length of the CAG repeat tract is inversely related to age at onset, the correlation ranging from −69 to −75 (Duyao *et al.*, 1993; Snell *et al.*, 1993; Andrew *et al.*, 1993). Notably, because CAG repeat length is only modestly correlated with age at onset, and even less so for the range of CAG repeat lengths accounting for most affected persons, there are likely other modifiers.

Evidence of specific modifiers is emerging. In one study, we demonstrated that the father's age at the time of birth of an HD gene carrier is negatively associated with the eventual age at onset in his offspring (Farrer *et al.*, 1992). Although the exact mechanism underlying the paternal age effect is unknown, it us unlikely to be due to Mendelian or mitochondrial inheritance or related to an environmental factor because the effect is uniquely paternal, not maternal, in origin. One possible mechanism involves genomic imprinting, but this is very speculative. To evaluate the hypothesis that age at onset is influenced by the normal HD allele, we compared transmission patterns of genetically linked markers at the D4S10 locus in the normal parent with age at onset in the affected offspring (Farrer *et al.*, 1993). Analysis of information from 21 sibships in 14 kindreds showed a significant tendency for sibs who have similar onset ages to share the same D4S10 allele from the normal parent. Affected sibs who inherited different D4S10 alleles from the normal parent tended to have more-variable ages at onset. These findings suggest that the expression of HD is modulated by the normal HD allele or a closely linked locus. The normal copy of the HD gene has been implicated more directly by studies showing a correlation of the length of the CAG repeat associated with the normal HD allele and age at onset (Snell *et al.*, 1993; Myers *et al.*, personal communication).

## 5. Machado–Joseph Disease

Machado–Joseph disease (MJD) is an autosomal-dominant, progressive, multisystem degenerative disease that affects primarily the motor system. The clinical manifestations include variable combinations of cerebellar ataxia, progressive external ophthalmoplegia, pyramidal signs, dystonia, rigidity, amyotrophy, and peripheral neuropathy (Rosenberg *et al.*, 1976a,b; Romanul *et al.*, 1977; Fowler *et al.*, 1977; Dawson, 1977; Coutinho and Sequeiros, 1981).

MJD typically strikes its victims in middle-adult life. The mean age at onset is 38 years, but symptoms may manifest at any time between the ages of 1 and 73 (Barbeau *et al.*, 1984). Three subtypes of MJD are recognized on the basis of clinical presentation and age at onset (Coutinho and Andrade, 1978). Cerebellar ataxia, external ophthalmoplegia, and pyramidal signs (Type 2) may be associated with prominent extrapyramidal signs (Type 1—Joseph disease) or with significant peripheral amyotrophy (Type 3—Machado disease). The frequency distribution of the three phenotypes has been reported to be approximately 15%, 45%, and 40%, respectively (Barbeau *et al.*, 1984; Sequeiros and Cotinho, 1993). Greater variability of phenotype and age at onset among than within them led to the suggestion of environmental or genetic modifiers (Woods and Schaumburg, 1972; Bharucha *et al.*, 1986; Rosenberg, 1992; Sudarsky *et al.*, 1992; Sequeiros, 1993). Subsequent identification of an unstable and expanded trinucleotide sequence as the primary defect in MJD provided a partial explanation for the variability in age at onset. Several studies have demonstrated that age at onset is inversely related to CAG repeat length, with correlations ranging from −0.67 to −0.87 (Kawaguchi *et al.*, 1994; Maciel *et al.*, 1995; Maruyama *et al.*, 1995; Ranum *et al.*, 1995; Takiyama *et al.*, 1995). It is evident, however, that CAG repeat length accounts for only 45–60% of the variation in age at onset. In fact, the range of onset age for any specific CAG repeat length is large enough to preclude its prediction when counseling presymptomatic carriers (DeStefano *et al.*, 1995).

To determine if familial factors independent of CAG repeat length modulate onset age of MJD, DeStefano *et al.* (1996) analyzed onset ages and CAG repeat lengths of 136 MJD patients from 23 kindreds of Portuguese descent using two different approaches. In the first approach, onset ages from affected individuals within a sibship were treated as repeated measures, with CAG repeat length on the expanded allele included as a covariate. Using this model the estimated correlation among sibs was 0.43, indicating that an environmental or genetic factor common to sibs influences onset age. In the second approach, residuals from the regression of age at onset on CAG repeat number were computed and correlations among the residuals were calculated for various types of relative pairs. The sib pair correlation obtained by this method was the same as that obtained using the repeated measures model. Positive correlations were also observed for avuncular ($r = 0.22$) and first-cousin pairs ($r = 0.28$), supporting the hypothesis that a genetic factor is influencing age at onset. For parent–offspring pairs there was a negative correlation of $r = -0.24$, but this is likely an artifact of ascertainment bias associated with collection of parent–offspring pairs for a late-onset disease.

Further support for a genetic basis for onset-age variability was derived from *commingling analysis,* the purpose of which is to determine whether there is evidence for multiple distributions of a phenotype in a large group of individuals. If the data suggest that the phenotype is the result of a mixture of two or three distributions, a possible conclusion is that these distributions are determined by different genotypes. Commingling analysis of onset ages adjusted for CAG repeat length identi-

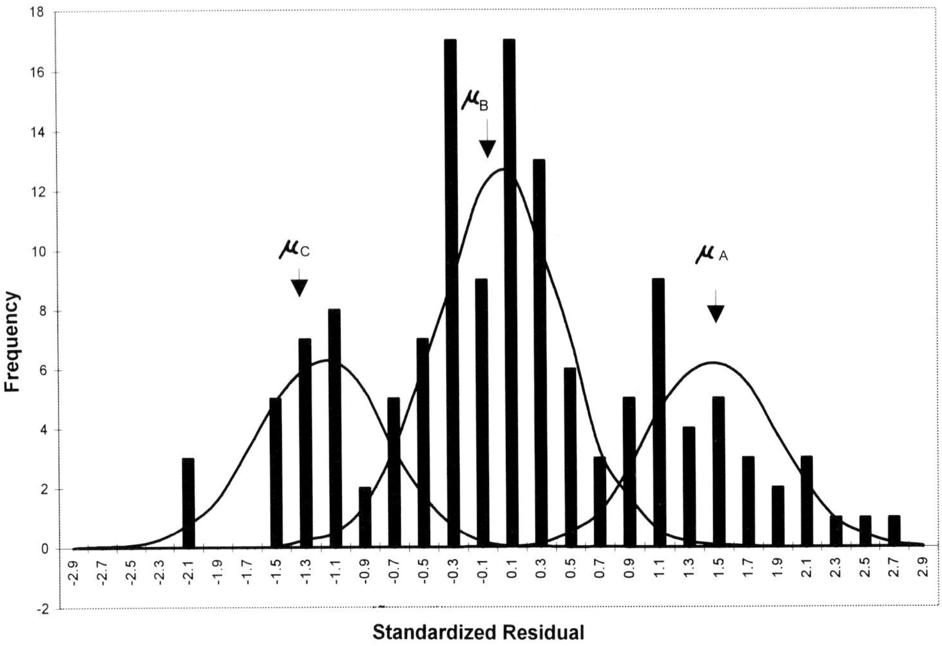


**Figure 2.** Histogram of standardized residuals from regression of onset age of Machado-Joseph disease on CAG repeat length. Three distributions identified by commingling analysis corresponding to populations A, B, and C with means $\mu_A$, $\mu_B$, and $\mu_C$, respectively, are indicated. The units of the x-axis are SDs. The SD of the residuals was 8.7 years. Reprinted from the *American Journal of Human Genetics* **59:**124 with permission from the University of Chicago Press. © 1996 by the American Society of Human Genetics. All rights reserved.

fied three distributions in this population of affected individuals, which is consistent with the segregation of a major gene influencing age at onset (DeStefano *et al.*, 1996; see Fig. 2).

## 6. Approaches for Identifying Genetic Modifiers

### 6.1. Genetic Case-Control Paradigms

The case-control method is one of the most widely used and rigorous study designs to show a cause-and-effect relationship between a treatment or risk factor and a disease. The genetic case-control method is a specialized paradigm of the general case-control design and was initially proposed by Kidd (1982, 1988) for studying phenotypic variation in complex genetic disorders. Unlike the typical case-control study in which subjects are divided according to disease phenotype, subjects in a genetic case-control study are first classified according to disease genotype. Comparisons of risk factors can then be made between individuals with the same disease genotype but different disease outcomes (Fig. 3). Another design is comparison of individuals with the same outcome but who have different genotypes (Fig. 4). As an example of the latter, it is possible to genotype individuals in the popula-

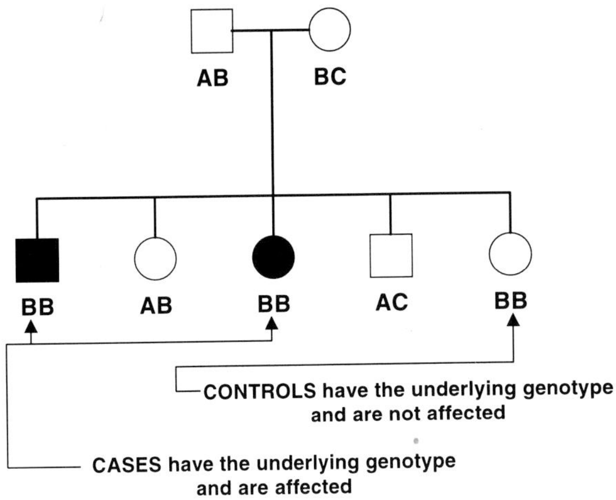


**Figure 3.** Genetic case-control paradigm.

tion for Tay-Sachs disease by screening with a biochemical assay for the gene product or else directly with a gene probe. One could then design a study to discover whether, in a group of normal individuals, heterozygotes ("cases") for Tay-Sachs have protection against particular infectious agents as compared with homozygotes ("controls"). It is noteworthy that many investigations comparing inheritance of gene polymorphisms with disease incidence (e.g., Ordovas *et al.,* 1986; Blum *et al.,* 1990; Comings *et al.,* 1991) are not true genetic case-control studies because genotypes predisposing toward disease are not distinguishable. The case-control method, however, can be applied to a group of genetic diseases for which the basic defect is unknown if the chromosomal location of the primary susceptibility gene *is* known. In this approach genotypes for unaffected relatives can be inferred by

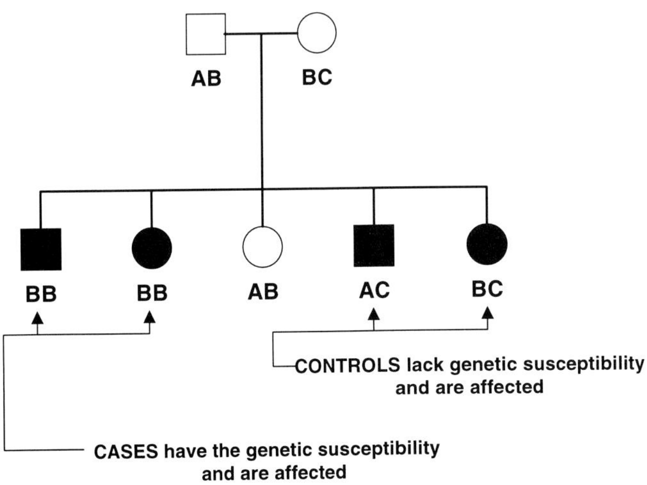


**Figure 4.** Genetic case-control paradigm.

analysis of linked markers. The specificity and sensitivity of these determinations are dependent upon the heterozygosity level and distance between the marker and the disease loci.

### Example: Modifiers for Phenotypic Variation in Wilson Disease

Genetic case-control paradigms are particularly suited for identifying factors, both biological (including genetic) and environmental, that interact with the WD gene, ATP7B. First, sibs of WD patients—who share a genetic background but are known to be the same or different at the WD locus—are a far superior control group to a group of unrelated obligate heterozygotes and normal subjects from the general population, because in the former group one can trace the segregation of particular genetic variants and assess their influence on a clinical or biochemical outcome. A second advantage of this approach is that the potential sample size of WD carriers can be greatly increased by screening ATP7B mutation or analysis of linked markers among normal sibs of WD patients. Third, characterization of ATP7B, identification of many of its mutations, and preliminary data on its function provide a rationale to test a limited number of candidate modifier loci. Screening of a limited number of potential etiologic genes minimizes the laboratory effort and the confounding effects of multiple comparisons.

**Table II. Cloned Human Copper Metabolism Genes**

| Gene symbol | Name | Functional role | Chromosomal location |
|---|---|---|---|
| | | Genes involved in copper transport | |
| ALB | albumin | transports intestinal copper | 4q11–q13 |
| ATP7A | Menkes disease gene | copper-transporting ATPase; heavy metal binding | Xq13.3 |
| CP | ceruloplasmin | ferroxidase; copper transport | 3q23–q25 |
| | | Genes directly involved in other facets of copper metabolism | |
| COX4 | cytochrome c oxidase subunit IV | electron transport chain | 16q24.1 |
| COX8 | cytochrome c oxidase subunit VIII | electron transport chain | 11q12–q13 |
| DBH | dopamine β-hydroxylase | catecholamine production | 9q34.3 |
| LOX | lysyl oxidase | cross-linking of collagen and elastin | 5q23.3–31.2 |
| TYR | tyrosinase | melanin production | 11q14–q21 |
| | | Genes indirectly involved in copper metabolism | |
| MT1 | metallothionein 1 | heavy metal detoxification | 16q13 |
| MT2A | metallothionein 2 | heavy metal detoxification | 16q13 |
| SOD1 | superoxide dismutase 1 | free radical detoxification | 21q22.1 |
| SOD2 | superoxide dismutase 2 | free radical detoxification | 6q21 |
| TF | transferrin | iron transport | 3q21 |
| TFRC | transferrin receptor | iron transport | 3q26.2–qter |

One can postulate the influence of several specific genes that modify expression of ATP7B based on genetic information and on clinical and biochemical aspects of the disorder. Cloned genes that have a role in copper metabolism are grouped in Table II according to known involvement in copper transport, demonstrated role in other facets of copper metabolism, and suggested role in copper metabolism. For example, the discovery that the WD gene (ATP7B) and the Menkes disease gene (ATP7A) have sequence and functional homology (Tanzi *et al.*, 1993) and our own finding that sex is associated with the clinical form of the disease in some populations (see section 3) provide two compelling arguments to test the recently cloned gene for Menkes disease, an X-linked disorder of copper transport (Chelly *et al.*, 1992; Mercer *et al.*, 1992; Vulpe *et al.*, 1992). These studies have shown that the Menkes and WD genes encodes a 1,500-amino-acid protein with features of a P-type copper-transporting ATPase. It is noteworthy that a much lower level of the ATP7B transcript is present in brain that is the site of high ATP7A expression (Tanzi *et al.*, 1993). This fact, added to strong homology between the two proteins and the observation of sex differences in expression in some populations, supports our hypothesis that ATP7A is a modifier of the WD gene. Ceruloplasmin is perhaps the next most logical candidate because of its deficiency in all WD patients. It has been shown that heredity influences ceruloplasmin levels in normal individuals (Cox, 1966). Thus, the familial variation of ceruloplasmin in WD heterozygotes and normal variation that is not attributable to ATP7B (Farrer *et al.*, 1988; Bonné-Tamir *et al.*, 1990) may be caused by variation at the ceruloplasmin (CP) locus.

There are several analytical approaches for evaluating the relationship between specific clinical measures of WD in affected persons and polymorphisms within candidate modifier loci. In one approach, one can test for homogeneity of allele-frequency distributions for polymorphisms at the modifier loci among groups of patients according to a clinical characteristic, such as primary clinical form (hepatic or neurologic) or age at onset (childhood, adolescence, or adulthood). Allele-frequency differences can be assessed using $\chi^2$ tests, taking into account variation of marker allele frequencies among racial or ethnic populations. Obviously, allele differences in small groups have to be relatively large to be detected. For example, to demonstrate an allelic association between a locus with 2 alleles and a bivariate clinical phenotype that is evenly distributed among 20 patients with 80% power ($\beta$) at a significance level ($\alpha$) of 0.05, the frequency of the associated allele must be 60% higher in one group than in the other. An allelic frequency difference of 30% could be detected under the same criteria for $\alpha$ and $\beta$ in an evenly divided sample of 50 patients.

To incorporate information on the primary mutation into analysis of the influence of potential modifiers, patients can be grouped based on the domain of the protein affected by the mutation. However, mutational differences probably don't account for a substantial portion of the clinical outcomes or variations because there appears to be one common mutation among all Caucasian families (Tanzi *et al.*, 1993; Thomas *et al.*, 1995a). Consideration of mutation type would probably not be profitable also because a substantial portion of families have rare or private mutations that cannot be easily identified (Figus *et al.*, 1995; Thomas *et al.*, 1995a; Waldenström *et al.*, 1996; Kalinsky *et al.*, 1998). Nonetheless, possible associations between the types of ATP7B mutations, as well as demographic factors, can be evaluated using generalized estimating equations (GEE), which allow for incorporation of a correlation structure that accounts for potential similarity among members of the same family, that is, a subset of those having the same mutation (Liang and Zeger, 1986). Use of the GEE approach to account for family data enabled DeStefano *et al.* (1998) to detect associations between the type of mutation and the clinical outcomes among individuals with Waardenburg syndrome, a disorder characterized by a variable constellation of sensorineural hearing loss, dystopia canthorum, and pigmentary disturbances of the eyes and skin.

In contrast to the methods above, which examine allelic associations, another approach entails evaluation of cosegregation of alleles for candidate modifier loci with the clinical phenotype in families where multiple forms of the clinical variant are present. To test the null hypothesis of random association between primary clinical form and modifier locus alleles inherited from each parent, one can apply an extension of the sib-pair analysis method (Penrose, 1935; Haseman and Elston, 1972; Lange, 1986) that we developed for this purpose (Farrer *et al.,* 1993). In this approach, one affected member of each subship is randomly selected as the proband and used to set the "phase" or pattern between clinical group and marker genotype for that sibship. This person is then discarded from the analysis. The remaining affected sibs will be classified as concordant (i.e., two cosegregating events, one per parent), single discordant (i.e., one cosegregating and one "recombination" event) or double discordant (i.e., two recombination events). The numbers of concordant and discordant cases are compared to the expected 1:1 ratio by a $\chi^2$ test with one degree of freedom. To detect a small genetic influence, say a concordance ratio of 3:2, approximately 100 meioses (excluding probands) would need to be scored. However, only 36 and 16 meioses, respectively, would be needed to detect concordance ratios of 2:1 and 3:1, indicating a gene that has a moderate genetic influence.

### 6.2. Genetic Linkage Analysis

In traditional genetic linkage analysis, one compares the segregation of a disease (and by inference the mutations underlying it) with allelic forms of genetic markers in families. A genetic marker is an inherited characteristic whose forms have appreciable frequency in the population. There are several classes of DNA markers, including those defined by single-base changes (restriction fragment-length polymorphisms, or RFLPs) or by variation in the number of elements in a motif or short tandem-repeat sequences (variable number of tandem repeats, or VNTFs). The latter group of markers are more useful for linkage studies because of their greater abundance, multiallelic nature (thereby ensuring high heterozygosity levels in the population), and relative efficiency of detection with the aid of polymerase chain reaction (PCR) methods. Markers that appear to cosegregate nonrandomly with the disease are said to be linked.

Various strategies can be used to assess linkage. When the mode of inheritance of the disease is known and multigeneration data are available, the lod score method, which entails tracking the transmission of disease and marker alleles from parents to children, is generally employed (Morton, 1955). With the aid of a computer program, one calculates the probability of observing the allelic segregation pattern in a pedigree for a given recombination frequency (Θ) and compares that with the probability given independent assortment of alleles (i.e., Θ = 0.5). The logarithm of odds, or lod score, is derived from the numerical quotient of these probabilities. The Θ value corresponding to the highest lod score is taken as the best estimate of recombination frequency (and hence distance) between the loci. A lod score of 3 (corresponding to odds of $10^3$ to 1) for a given Θ is the minimum value that establishes linkage. A lod score of −2 (odds of 0.01 to 1) means that linkage between the two loci can be excluded up to a distance defined by the given Θ value. Because lod scores at a specified Θ value can be added across pedigrees, the investigator can make decisions about testing more families until a resolution of linkage or no linkage can be made, if the total lod score is between −2 and 3.

The lod score method may have limited utility in searches for modifier genes for neurodegenerative diseases, because the mode of inheritance of the modifier is unknown and inference of a modifier locus genotype is impossible in persons lacking the disease. Thus, other nonparametric approaches should be considered. One such approach, the sib-pair method, is predicated on the principle that siblings who are concordant for a trait should share marker alleles more frequently than random expectation if the genes for the trait and the marker are linked (Penrose, 1935). Conversely, if the loci are linked, sibs discordant for the trait should be more dissimilar for the marker alleles than expected by chance. There have been many extensions of the method (Haseman and Elston, 1972; Green and Woodrow, 1977; Suarez, 1978; Lange, 1986), including a generalized algorithm that considers identity by state relationships in extended kindreds (Weeks and Lange, 1988). These methods are particularly useful for studies of traits in which the persons available are of a single generation or scattered among several branches of a kindred, which would be the case when assessing the informative persons for linkage studies of modifier genes. Unlike the lod method, however, these methods do not give information about genetic distance, are relatively sensitive to assumptions about marker allele frequencies, and require a very large number of families (at least 100 to 200, depending on the number of genes contributing to the expression of the trait).

### 6.2.1. Example: Modifiers for Onset Age of Machado–Joseph Disease

**6.2.1a. Basic Strategy.** Arguably, the most efficacious strategy for identifying loci that modify onset age in Machado-Joseph disease is a multifaceted approach of linkage analysis in kindreds and affected sib pairs. Age at onset is a quantitative trait, which is known to be influenced by the number of CAG repeats in the expanded disease allele, among other factors. Because the expansion is unstable during transmission, related individuals may have different number of repeats. Therefore, any analysis to identify modifiers must account for the CAG repeat length in each affected individual. The residuals obtained from regression of onset age on CAG repeat number reflect the variability among individuals for onset age that remains after accounting for CAG length. In one approach lod score analysis can be applied using extended kindreds to identify quantitative trait loci (QTL), which contribute to the variability of these residuals. The parametric lod score approach requires specification of the genetic model and may fail to reveal a true linkage if the distribution of the quantitative trait for each genotype of the QTL is grossly misspecified. For this reason, alternative approaches should be considered. The sib-pair design examines the estimated proportion of identical-by-descent alleles shared by a sib pair at a marker locus relative to the squared difference between the sibs for the quantitative trait, with no assumptions made about the genetic model for the trait locus. As with the lod score approach, the trait of interest is the residuals obtained from regression of onset age on CAG repeat length. Residuals can be computed only for MJD-affected individuals for whom a DNA sample has been obtained; onset age and CAG repeat length must be known for the phenotype of interest to be available. Given this restriction on the phenotype and the fact that only a select group of kindreds can independently demonstrate linkage, a sib-pair approach may make more efficient use of resources by targeting nuclear families with multiple affected offspring for genotyping. A complete genome scan entailing the genotyping of approximately 300 evenly spaced and highly polymorphic markers, and a replicate scan that targets regions showing evidence for linkage in the initial scan, will minimize false positive results and facilitate testing of oligogenic models.

A potentially major caveat of a linkage-based approach for identifying age-at-onset genes for MJD concerns the determination of that age, information usually provided by the patient and/or close relatives and defined as the age when a disturbance in gait or balance first appeared. This recollection of the initiation of symptoms may not be accurate, however, and recognition of early signs of the disease may be biased by previously diagnosed family members. Our experience with analysis of onset age for neurological diseases indicates a bias toward onset ages ending in 0 or 5 (Farrer *et al.*, 1993). The use of multiple informants to establish age at onset addresses some of these issues and provides a reasonably accurate measurement of this characteristic.

**6.2.1b. Lod Score Analysis.** A lod score or parametric approach to linkage analysis requires specification of the genetic model. Although a genetic model has not been elucidated for a locus modifying age at onset, the commingling analysis of DeStefano *et al.* (1996), which detected the presence of three distributions in onset age after accounting for CAG repeat length, provides reasonable initial estimates for mean and variance parameters to describe this quantitative trait locus. More-precise estimates could be obtained from formal genetic modeling studies such as segregation analysis, but this may not be practical because of the sample size and family ascertainment requirements for these experiments.

How well can lod score analysis work? We carried our simulation studies using the program SIMLINK, version 4.1 (Ploughman and Boehnke, 1989), to estimate the power of a lod score analysis to detect linkage between a marker locus and a gene that modifies onset age of MJD in six MJD kindreds comprised of multiple affected members in at least two different sibships. These kindreds were selected from more than 25 pedigrees that were informative for localization studies of the MJD gene (Sequeiros *et al.*, 1994; Twist *et al.*, 1995). The phenotype of interest in this analysis is onset age adjusted for CAG repeat, which was obtained by regression of onset age on CAG repeat length. Therefore phenotypic information was available only on MJD individuals with known onset age and CAG repeat length; all other individuals were coded as unknown; all unaffected persons are unknown with respect to the phenotype at the modifier locus. Specification of the genetic model for a quantitative trait involves defining the normal distribution according to the phenotypic mean and standard deviation for each genotype at the trait locus and also the allelic frequencies at this locus. Phenotypic means of 29, 40, and 52, corresponding to onset ages for individuals with genotypes *D/D, D/d,* and *d/d,* respectively, were used in the simulations based on the results of commingling analyses (Fig. 2). The standard deviation for each genotype was assumed to be 3.8. The frequency of the *D,* or "early-onset," allele was estimated to be 0.5, the least powerful model for lod score analysis. Frequencies of 0.75 and 0.25 for the *D* allele were assumed in separate simulations. In the simulations the marker was assumed to be a six-allele system, with a polymorphism information content (PIC) of 0.796. Most of the markers one would probably use to screen for linkage will have PIC values $> 0.8$. A range of values for the recombination fraction between the marker locus and the modifier locus were assumed. Two hundred replicate pedigrees were simulated to minimize the standard error of the simulation results.

Table III shows that this collection of families can together yield a maximum lod score of 7.14 for linkage between a genetic marker and a locus modifying age at onset. Using highly polymorphic markers, it would be possible to establish linkage to a gene modifying onset age up to a distance of 20 cM, provided age at onset is determined by the same gene in all of these families. Of importance, none of these families is sufficient for establishing linkage to a modifier locus independently, despite the fact the linkage to the disease could be demonstrated in each of them. Power to detect linkage would increase by the addition of only a few affected individuals. Furthermore, the simulation results reflect a frequency of 0.5 for each allele at the modifying locus, the least powerful case for linkage detection. If frequencies of 0.75 and 0.25 are assumed for the alleles promoting early and late onset, respectively, the maximum lod score summed over all families increases to 8.43, and the maximum for a single family increases to 2.7, close to the minimum threshold of 3.

Figure 5 illustrates the detection of linkage between a modifier gene for age at

**Table III. Expected Mean and Maximum Lod Scores for a Quantitative Trait Locus Modifying Onset Age of Machado-Joseph Disease**

| | Mean Lod Score at θ | | | |
|---|---|---|---|---|
| Pedigree | 0.0 | 0.05 | 0.1 | Maximum |
| B15 | .42 | .37 | .33 | 1.02 |
| NA1 | .69 | .47 | .36 | 1.81 |
| B16 | .50 | .43 | .32 | 2.09 |
| P1 | .22 | .20 | .15 | .73 |
| P8 | .22 | .16 | .14 | .73 |
| P14 | .19 | .17 | .14 | .75 |
| Total | 2.24 | 1.80 | 1.44 | 7.14 |

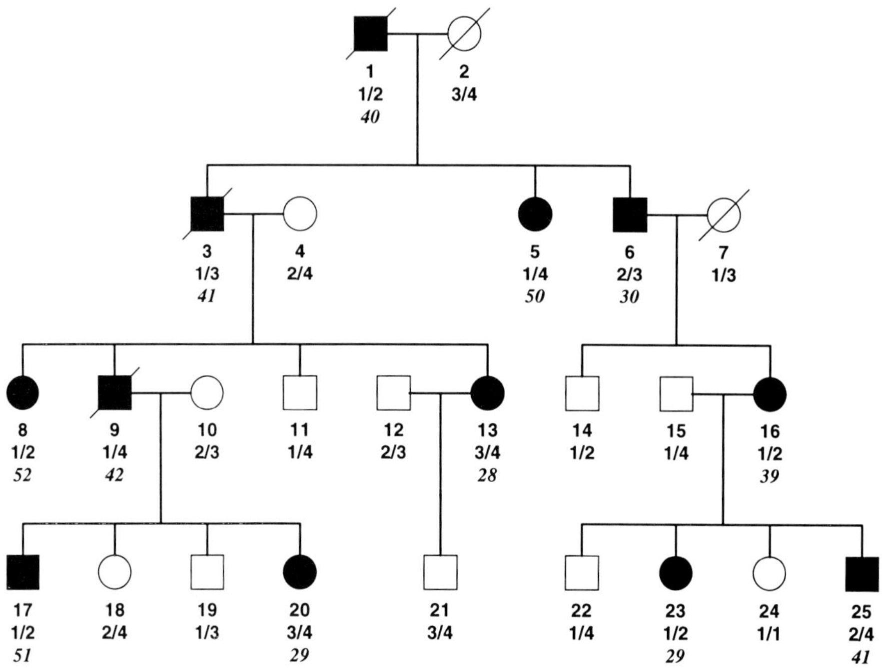


**Figure 5.** Pedigree demonstrating linkage between a locus determining onset age of Machado-Joseph disease (with alleles $D$ = early onset and $d$ = late onset) and a genetic marker. Affected individuals are denoted by shaded symbols. A slash indicates the person is deceased. Numbers below each symbol indicate position in the pedigree. Marker phenotypes are represented by two integers separated by a slash. Ages at onset shown in italics. See text for interpretation, including inference of genotypes for age at onset modifier locus. Adapted from Kant (1996).

onset and a marker locus using the lod score approach. A modifier genotype of $D/D$ can be inferred for individuals number 6, 13, 20 and 23, all of whom manifested symptoms at a very young age, whereas individuals 5, 8 and 17 are assumed to have the $d/d$ genotype, given their very late onset. Based on this information, parents 1, 3 and 9 can be inferred to be heterozygous ($D/d$) for the modifier. Inspection of these data indicates that affected individuals possessing the combination of marker allele "1," and by inference, modifier allele $d$ (which originated in great-grand-parent number 1) had onset in middle-to-late life, whereas those not inheriting this combination had a relatively early onset. Maximum lod scores for this pedigree range from 2.4 to 3.3, depending on the assumed frequency of the modifier locus alleles.

To test this approach directly, lod score analysis was conducted on eight MJD kindreds using the MLINK program in FASTLINK (Cottingham *et al.*, 1993). Genotype data for 25 RFLP and 14 dinucleotide repeat markers located on various chromosomes were available on members of these kindreds from prior studies to determine the location of the primary MJD gene defect (Twist *et al.*, 1995). The genetic model used was as described for the simulations. In separate analyses, allelic frequencies of 0.5, 0.1, and 0.9 for the "early-onset" allele at the modifier locus were assumed to assess the sensitivity of lod score results to the genetic model. The majority of markers gave inconclusive results with lod scores between −1.5 and 1.

**Table IV. Lod Scores for Tests of Linkage between a Quantitative Trait Locus Modifying Age at Onset of Machado-Joseph Disease and Several DNA Markers**

| | | Marker | | | | | | |
|---|---|---|---|---|---|---|---|---|
| θ | *f*(*D*) | SST | MS140 | D12S58 | PLA2 | HMHC | D2S99 | D14S76 |
| 0.00 | .5 | −1.95 | −1.33 | −2.70 | −1.64 | −3.53 | −1.55 | −4.76 |
| | .1 | −2.32 | −2.67 | −4.54 | −3.33 | −4.13 | −2.50 | −4.93 |
| | .9 | −2.79 | −2.61 | −3.66 | −1.27 | −3.21 | −3.40 | −4.30 |
| 0.05 | .5 | −0.89 | −0.96 | −0.75 | −0.94 | −1.64 | −1.06 | −1.77 |
| | .1 | −1.09 | −1.79 | −1.92 | −1.67 | −1.99 | −1.61 | −2.08 |
| | .9 | −1.25 | −1.66 | −1.69 | −0.61 | −1.65 | −2.12 | −1.35 |
| 0.10 | .5 | −0.55 | −0.69 | −0.27 | −0.60 | −1.04 | −0.78 | −1.04 |
| | .1 | −0.63 | −1.23 | −1.12 | −0.94 | −1.24 | −1.15 | −1.20 |
| | .9 | −0.76 | −1.14 | −0.92 | −0.27 | −1.04 | −1.32 | −0.58 |
| 0.15 | .5 | −0.35 | −0.49 | −0.05 | −0.40 | −0.67 | −0.58 | −0.63 |
| | .1 | −0.37 | −0.85 | −0.66 | −0.52 | −0.78 | −0.85 | −0.73 |
| | .9 | −0.47 | −0.79 | −0.48 | −0.09 | −0.69 | −0.79 | −0.20 |

*Note:* Lod scores were calculated at several recombination frequencies (θ) summed over 8 kindreds for which genotype data were available from previous studies. The frequency of the early onset allele, *f*(*D*), was varied in separate analyses. Data from Kant (1996).

Table IV shows a selection of markers that yielded lod scores less than −2 and thus demonstrates several locations excluded in this set of families. The results were dependent, however, on the allele frequencies at the modifier locus. A properly designed genome scan using closely and evenly spaced markers would have greater power to detect as well as exclude linkage. The markers used in Table IV have a substantially lower PIC than the highly polymorphic tetranucleotide repeat markers one would utilize in a genome screen (Murray *et al.*, 1994). Also, the families included in this experiment were chosen because genotype data were available from previous studies, not because they were optimal for a modifier gene search.

### 6.2.2. Affected-Sib-Pair Approach

One can also use affected-sib-pair (ASP) methods to determine if there is a relationship between the number of identical-by-descent (IBD) alleles shared by a pair of sibs and the squared difference in age at onset adjusted for CAG repeat length for the pair. Two alleles are IBD if they are copies of the same ancestral gene. Although the strongest ASP methods require approximately twice the sample size of lod score methods (Demenais and Amos, 1989), they do not require any assumption about the mode of inheritance of a particular trait. Affected-sib-pair analysis can be carried out using one or more different approaches and computer programs. For example, the algorithms for quantitative traits (Haseman and Elston, 1972; Amos *et al.*, 1989) have been implemented in the SIBPAL program of the SAGE package. In this approach, the proportion of IBD alleles is estimated for each marker locus. The square of the difference in the quantitative trait for the sib pair is then regressed on the estimated proportion of IBD alleles for each marker locus. Given that linkage exists between the quantitative trait and the marker locus, the expectation is that higher proportions of shared IBD alleles would be associated with smaller squared differences in the quantitative trait. A negative regression coefficient is, therefore, indicative of linkage. The SIBPAL program accounts for the dependency among sib pairs that exists when multiple pairs from a single nuclear family are incorporated by calculating the effective sample size (Wilson and Elston, 1993). The effective sample size is used in significance testing of the regression coefficient.

### 6.2.3. Locating Modifiers Using Discordant Sib Pairs

Most studies examining the power of sib-pair linkage analysis have focused on bivariate qualitative traits. If the trait has high heritability, and highly polymorphic, closely spaced markers are used, the power can approach 100% in samples as small as 100 sib pairs (Risch, 1990). Few studies have been published about power for quantitative trait loci. Several studies suggest that at least several hundred (Demenais and Amos, 1989) or if the heritability is less than 50%, several thousand (Blackwelder and Elston, 1982; Fulker and Cardon, 1994) sib pairs would be needed to map quantitative trait loci. Notably, most of these results were simulated for markers with low polymorphism information content relative to markers commonly used today. More recently, Risch and Zhang (1995) proposed a strategy of pre-selecting sib pairs on the basis of extreme phenotype values. The power of this discordant sib pair (DSP) approach increases with sibling recurrence risk, and equals that using affected sib pairs when the sibling recurrence risk equals 50% (Risch and Zhang, 1995). Thus, for example, DSPs would appear to be very efficient for mapping modifier loci responsible for the major clinical features of WD, which tend to cluster in families. However, although this method can reduce the amount of genotyping 10–40-fold over conventional designs, it is not practical for rare disorders in which one cannot afford to exclude families or, in the case of modifier-linkage studies, exclude affected individuals lacking the trait of interest.

In studies of common familial diseases, DSPs may be valuable for locating modifiers for symptoms with a high recurrence risk among siblings. Rogus and Krolewski (1996) showed by simulation that given high penetrance, low phenocopy rate, and high sibling recurrence risk (e.g., >50%), the DSP approach can be quite powerful for detecting linkage to qualitative traits, even in the presence of a significant environmental contribution. In the case of neurodegenerative conditions, the DSP approach could prove rewarding for mapping genes that determine presence or absence of specific symptoms (e.g., aphasia, agraphia, parkinsonian movements) or differential response to various drug therapies among siblings with Alzheimer disease.

## 6.3. Special Case: Primary Mutation and Its Modifier in One Genetic System

The approaches discussed in sections 6.2.1–6.2.4 for locating genetic modifier loci assume that the genes governing disease susceptibility and expression are distinct. Specifically, in the case of autosomal dominant conditions, these methods track the marker and disease alleles transmitted by the affected parent only; they do not allow for the possibility that how or when the disease is expressed is influenced by the normal allele. For example, Baker *et al.* (1991) showed that age at death of affected members in a kindred with hereditary Gertsmann-Sträussler syndrome, which is caused by a dominantly transmitted mutation in the prion protein gene (Collinge *et al.,* 1989), is modulated by the form of the normal gene transmitted by the unaffected parent. Linkage studies implicated a role for the normal HD allele in the determination of age at onset, and this was subsequently confirmed and attributed to variability in the length of the CAG repeat tract on the normal chromosome among affected individuals (see section 4).

Detection of normal allele effect on disease expression may require several analytical approaches if variability in the protein is not evident in the general population. Without evidence for such variation, it would be uncertain whether an intragenic polymorphism would actually be biologically meaningful, that is, a modifier. To illustrate this point, consider the case of MJD. In addition to the variable number of CAG repeats, other features of the MJD gene have been characterized, including two polymorphic single-base substitutions that occur within and near the CAG tract (Kawaguchi *et al.,* 1994). One of these polymorphisms (CGG/GGG) has been implicated in the stability of the CAG repeat (Igarashi *et al.,* 1996). There are also several other polymorphisms within MJD, which we have identified (Gaspar *et al.,* 1996). In addition to these known characteristics of the MJD gene, it is possible that genes or sequences near MJD on the normal chromosome may affect the onset age. Therefore one could test at least three hypotheses regarding the normal allele and onset age, namely: (1) the size of the CAG repeat tract on the normal chromosome is associated with onset age, (2) an association exists between specific alleles at other polymorphic sites within MJD and onset age, and (3) an unknown polymorphism within MJD or a closely linked gene on the normal MJD chromosome is associated with differences in onset age between siblings. Each hypothesis addresses a different question, and therefore, different methods of analysis are appropriate for testing each hypothesis.

### 6.3.1. Size of CAG Repeat and the Normal Chromosome

Several studies have found no relationship between onset age of MJD and the number of CAG repeats in the normal chromosome (Maciel *et al.,* 1995; Ranum *et al.,* 1995; Takiyama *et al.,* 1995). More recently, Durr *et al.* (1996) reported a negative correlation between age of onset and number of CAG repeats. These studies all evaluated the association between size of the normal chromosome and onset age by simple and multiple regression techniques in which all observations were treated as independent, and existing relationships among individuals were ignored. An alternative method entails evaluating the influence of the size of the normal allele on onset age, using a repeated-measures model in a population of sibships with multiple affected individuals. Treating the onset age of affected individuals within a sibship as repeated measures would allow for estimation of a covariance parameter, which can account for similarity among sibs.

### 6.3.2. Association between Onset Age and Alleles at Polymorphic Sites within Machado–Joseph Disease

Polymorphic sites within the MJD gene can be studied using lod score and sib-pair analysis as outlined in sections 6.2.1–6.2.4. In this situation of limited family material, however, the genetic information can be better exploited by the transmission/disequilibrium test (TDT) (Spielman *et al.,* 1993). The TDT compares the number of times that parents heterozygous for a marker transmit an associated marker allele to an affected offspring with the number of times they transmit the alternate allele. This test does not require multiple affected family members; all affected individuals with a DNA sample available on the unaffected parent can be included. Using the TDT, linkage will be detected only if linkage disequilibrium in the population is present. Such disequilibrium is expected under the hypothesis that the polymorphic site directly influences onset age. Because TDT is applicable to qualitative traits only, it is necessary to dichotomize individuals into early-onset and late-onset groups. By selecting individuals from the most extreme (i.e., 20%) portion of the age-at-onset distribution, a modifier gene explaining as little as 5% of the variation in onset age can be detected at the 0.0001 $\alpha$ level with fewer than 300 observations (Allison, 1997). In this special case of modifier linkage in which one is specifically interested in testing association with alleles on the normal chromosome, transmissions from the unaffected parent only should be included in the TDT statistic.

#### 6.3.3. Association between Transmission Patterns of MJD-Linked Loci from the Unaffected Parent and Onset Age in Sibs

Following the approach of Farrer *et al.* (1993) in onset age studies of HD, sibships with $\geq 3$ affected individuals are required for this analysis. In the HD study, sibs were classified as early- and late-onset based on the deviation from the within-sibship mean onset age. At the time of that study, the primary defect had not yet been identified, which, as it turns out, accounts for a significant portion of the interfamilial variation in onset age. Therefore, it was necessary to use a within-sibship rather than population-based threshold for distinguishing early- and late-onset individuals. Because the MJD gene defect has already been defined as an expanded tract of CAG, sibs should be characterized as early- or late-onset based on their residual value obtained by regressing onset age on CAG repeat length in the disease allele. The resultant regression line derived from this procedure is a logical population-based threshold: individuals with positive residuals will be classified as late-onset and those with negative residuals will be classified as early-onset. Using available DNA markers, including the CAG repeat length, it can be determined which normal allele (arbitrarily named "haplotype 1" or "haplotype 2") was inherited by each sibling. To test the null hypothesis of random association between age at onset (early or late) and normal MJD allele, a proband can be designated at random from each sibship. This proband is used to set "phase" (e.g., if the proband was late-onset, the normal haplotype inherited by this proband is designated the "late-onset" haplotype in this family). The remaining sibs are then classified as concordant or discordant, and the proportion of concordant sibs over all families is compared with the expected value of 0.5 under the null hypothesis using a $\chi^2$ test.

## 7. Conclusions

Before embarking on an ambitious project to collect and analyze clinical and genotype information on families with a phenotypically variable genetic disorder, an investigator should be cognizant of the limitations. The approaches described in this chapter are a prerequisite for efforts to clone genes that regulate other genes in neurodegenerative diseases. Obviously, such efforts would be severely compromised or perhaps doomed if no modifier gene exists or if parameters of the genetic model for the modifier are poorly estimated. Disentangling the influences of a major gene and its modifiers may not be straightforward. The genetic epidemiological approaches described herein require prior empiric or modeling studies to demonstrate the familial nature of the clinical trait or marker and the likely existence of an underlying major gene for the characteristics of interest. Although several examples of such preliminary studies have been mentioned (e.g., Table 1; Farrer *et al.*, 1984, 1988, 1992; Bonné-Tamir *et al.*, 1990; DeStefano *et al.*, 1996, 1998), more-detailed descriptions of methods to assess familial aggregation should be consulted (Khoury *et al.*, 1993; Farrer and Cupples, 1998).

Studies of the primary genes for neurodegenerative diseases and their protein products in persons and transgenic animals with and without the underlying mutations will undoubtedly increase our understanding of the cascade of events leading to neuron loss and the aberrant neurochemistry associated with these diseases. However, these experiments will not likely provide an explanation for the variable expression and age at onset. Identification of genes that govern age at onset would provide the rationale for experiments that elucidate epistatic interaction leading to onset in either early or late life. This understanding may give critical insight into the development of new drugs or other therapeutic modalities for delaying or even preventing the onset of symptoms. It is entirely possible that relatively few of these genetic switches may be responsible for a large number of neurodegenerative diseases with delayed onset, particularly those caused by a trinucleotide repeat expansion, for example, HD, MJD, DRPLA, spinocerebellar ataxia types 1 and 2, myotonic dystrophy, spinal bulbar muscular atrophy, and Friedreich's ataxia (Ashley and Warren, 1995), or those following a complex pattern of inheritance, such as Alzheimer's disease (Farrer, 1997).

ACKNOWLEDGMENTS. I thank Dr. Anita DeStefano for performing the linkage simulation studies and providing helpful comments on the manuscript. This work was supported in part by grant NS31687 from the National Institutes of Health.

## 8. References

Aksoy, S., and Erdem, S., 1972, *New Istanbul Contr. Clin. Sci.* **2:**92–97.

Allison, D. B., 1997, Transmission-disequilibrium tests for quantitative traits, *Am. J. Hum. Genet.* **60:**676–690.

Amos, C. L., Elston, R. C., Wilson, A. F., and Bailey-Wilson, J. E., 1989, A more powerful test of linkage for quantitative traits, *Genet. Epidemiol.* **6:**435–449.

Andrew, S. E., Goldberg, Y. P., Kremer, B., Telenius, H., Theilmann, J., Adam, S., Starr, E., Squitieri, F., Lin, B., Kalchman, M. A., Graham, R. K., and Hayden, M. R., 1993, The relationship between trinucleotide (CAG) repeat length and clinical features of Huntington's disease, *Nature Genet.* **4:**398–403.

Arima, M., and Sano, I., 1968, Genetic studies of Wilson's disease in Japan, in: *Birth Defects: Original Article Series, Vol. 4, No. 2, Wilson Disease* (D. Bergsma, ed.), Alan Liss, New York, pp. 54–59.

Ashley, C. T., Jr., and Warren, S. T., 1995, Trinucleotide repeat expansion and human disease, *Annu. Rev. Genet.* **29:**703–728.

Bailey-Wilson, J. E., and Elston, R. C., 1994, *Statistical Analysis for Genetic Epidemiology (SAGE)*, LSU Medical Center, New Orleans, LA.

Baker, H. F., Poulter, M., Crow, T. J., Frith, C. D., Lofthouse, R., Ridley, R. M., 1991, Amino acid polymorphism in human prion protein and age at death in inherited prion disease, *Lancet* **337:**1286.

Barbeau, A., Roy, M., Cunha, L., de Vincente, A. N., Rosenberg, R. N., Nyhan, W. L., MacLeod, P. M., Chazot, G., Langston, L. B., Dawson, D. M., Coutinho, P., 1984, The natural history of Machado-Joseph disease, *Can. J. Neurol. Sci.* **11:**510–525.

Bharucha, N. E., Bharucha, E. P., and Bhabha, S. K., 1986, Machado-Joseph disease in India, *Arch. Neurol.* **43:**142–144.

Bhende, Y. M., Deshpande, C. K., Bhatia, H. M., Sanger, R., Race, R. R., Morgan, W. T. J., and Watkins, W. M., 1952, A "new" blood-group character related to the ABO system, *Lancet* **1:**903–904.

Blackwelder, W. C., and Elston, R. C., 1982, Power and robustness of sib-pair linkage tests and extension to larger sibships, *Comm. Stat. Theory Methods* **11:**449.

Blum, K., Noble, E. P., Sheridan, P. J., Montgomery, A., Ritchie, T., Jagadeeswaran, P., Nogami, H., Briggs, A. H., and Cohn, J. B., 1990, Allelic association of human dopamine $D_2$ receptor gene in alcoholism, *J. Am. Med. Assoc.* **263:**2055–2060.

Bonné-Tamir, B., Brydman, M., Agger, M., Bekker, R., Bowcock, A. M., Hebert, J. M., Cavalli-Sforza, L. L., and Farrer, L. A., 1990, Wilson's disease in Israel: A genetic epidemiological study, *Ann. Hum. Genet.* **54:**155–168.

Bowcock, A. M., Farrer, L. A., Hebert, J. M., Agger, M., Buys, C. H. M. C., Scheffer, H., Sternlieb, I., Scheinberg, I. H., Frydman, M., Chajek-Saul, T., Bonne-Tamir, B., and Cavalli-Sforza, L. L., 1988, Eight closely linked loci place the Wilson's disease locus within 13q14-q21, *Am. J. Hum. Genet.* **43:**664–674.

Bowcock, A. M., Tomfohrde, J., Weissenbach, J., Bonne-Tamir, B., St. George-Hyslop, P., Giagheddu, M., Cavalli-Sforza, L. L., and Farrer, L. A., 1994, Refining the position of Wilson disease by linkage disequilibrium with polymorphic microsatellites, *Am. J. Hum. Genet.* **54:**79–87.

Bull, P. C., Thomas, G. R., Rommens, J. M., Forbes, J. R., and Cox, D. W., 1993, The Wilson disease gene is a putative copper transporting P-type ATPase similar to the Menkes gene, *Nature Genet.* **5:**327–337.

Chelly, J., Tumer, Z., Tonnesen, T., Petterson, A., Ishikawa-Brush, Y., Tommerup, N., Horn, N., and Monaco, A. P., 1992, Isolation of a candidate gene for Menkes disease that encodes a potential heavy metal binding protein, *Nature Genet.* **3:**14–19.

Chuang, L.-M., Tai, T.-Y., Wang, T.-R., Chang, Y.-C., Chen, K.-H., Lin, R.-S., and Lin, B. J., 1991, Esterase D and retinoblastoma gene loci are tightly linked to Wilson's disease in Chinese pedigrees from Taiwan, *Hum. Genet.* **87:**465–468.

Collinge, J., Harding, A. E., Owen, F., Poulter, M., Lofthouse, R., Boughey, A. M., Shah, T., and Crow, T. J., 1989, Diagnosis of Gertsmann-Strässler syndrome in familial dementia with prion protein gene analysis, *Lancet* **2:**15–17.

Comings, D. E., Comings, B. G., Muhleman, D., Dietz, G., Shahbahrami, B., Tast, D., Knell, E., Kocsis, P., Baumgarten, R., Kovacs, B. W., Levy, D. L., Smith, M., Borison, R. L., Evans, D. D., Klein, D. N., MacMurray, J., Tosk, J. M., Sverd, J., Gysin, R., and Flanagan, S. D., 1991, The dopamine $D_2$ receptor locus as a modifying gene in neuropsychiatric disorders, *J. Am. Med. Assoc.* **266:**1793–1800.

Conneally, P. M., 1994, Huntington's disease: Genetics and epidemiology, *Am. J. Hum. Genet.* **35:**506–526.

Cottingham, R. W., Jr., Idury, R. M., and Schaffer, A. A., 1993, Faster sequential linkage computations, *Am. J. Hum. Genet.* **53:**252–263.

Coutinho, P., and Andrade, C., 1978, Autosomal dominant system degeneration in Portuguese families of the Azores Islands, *Neurology* **28:**703–708.

Coutinho, P., and Sequeiros, J., 1981, Aspects cliniques et pathologiques de la maladie de Machado-Joseph, *J. Genet. Hum.* **29:**203–209.

Cox, D. W., 1966, Factors influencing serum ceruloplasmin level in normal individuals, *J. Lab. Clin. Med.* **68:**893–904.

Cox, D. W., Fraser, F. C., and Sass-Kortsak, A., 1972, A genetic study of Wilson disease: Evidence for heterogeneity, *Am. J. Hum. Genet.* **24:**646–666.

Danks, D. M., 1989, Disorders of copper transport, in: *Metabolic Basis of Inherited Disease* (A. L. Beaudet, W. S. Sly, and D. Valle, eds.), McGraw-Hill, New York, pp. 1411–1431.

Dawson, D. M., 1977, Ataxias in families from the Azores, *New Engl. J. Med.* **296:**1529–1530.

Demenais, F. M., and Amos, C. I., 1989, Power of the sib-pair and lod-score methods for linkage analysis of quantitative traits, in: *Multipoint Mapping and Linkage Based Upon Affected Pedigree Members* (R. C. Elston, J. W. MacCluer, S. E. Hodge, and M. A. Spence, eds.), Alan R. Liss, New York, pp. 201–206.

Dening, T. R., 1985, Psychiatric aspect of Wilson's disease, *Br. J. Psychiatry* **147:**677–682.

DeStefano, A. L., Cupples, L. A., Arnos, K. S., Asher, J. H., Baldwin, C. T., Carey, M., da Silva, E. O., Friedman, T. B., Greenberg, J., Landa, B., Milunsky, A., Morell, R., Nance, W. E., Pandya, A., Ramesar, R., Read, A., Tassebejhi, M., Wilcox, E. R., and Farrer, L. A., 1998, Correlation between Waardenburg syndrome phenotype and genotype in a population of individuals with identified PAX3 mutations, *Hum. Genet.* **102:**499–506.

DeStefano, A. L., Farrer, L. A., Maciel, P., Gaspar, C., Rouleau, G. A., Coutinho, P., and Sequeiros, J., 1995, Gender equality in Machado-Joseph disease, *Nature Genet.* **1:**118–119.

DeStefano, A. L., Cupples, L. A., Maciel, P., Gaspar, C., Radvany, J., Dawson, D. M., Sudarsky, L., Corwin, L., Coutinho, P., MacLeod, P., Sequeiros, J., Rouleau, G. A., and Farrer, L. A., 1996, A familial factor independent of CAG repeat length influences age at onset of Machado-Joseph disease, *Am. J. Hum. Genet.* **58:**119–127.

Durr, A., Stevanin, G., Cancel, G., Duyckaerts, C., Abbas, N., Didierjean, O., Chneiweiss, H., Benomar, A., Lyon-Caen, O., Julien, J., Serdaru, M., Penet, C., Agid, Y., and Brice, A., 1996, Spinocerebellar ataxia 3 and Machado-Joseph disease: Clinical, molecular, and neuropathological features, *Ann. Neurol.* **39:**490–499.

Duyao, M., Ambrose, C., Myers, R. H., Novelletto, A., Persichetti, F., Frontali, M., Folstein, S., Ross, C., Franz, M., Abbott, M., Gray, J., Conneally, P., Young, A., Penny, J., Hollingsworth, Z., Shoulson, I., Lazzarini, A., Falek, A., Koroshetz, W., Sax, D., Bird, E., Vonsattel, J., Bonilla, E., Alvir, J., Bickham Conde, J., Cha, J.-H., Dure, L., Gomez, F., Ramos, M., Sanchez-Ramos, J., Snodgrass, S., de Young, M., Wexler, N., Moscowitz, C., Penchaszadeh, G., MacFarlane, H., Anderson, M., Jenkins, B., Srinidhi, J., Barnes, G., Gusella, J., and MacDonald, M., 1993, Trinucleotide repeat length instability and age at onset in Huntington's disease, *Nature Genet.* **4:**387–392.

Farrer, L. A., 1997, Genetics and the dementia patient, *Neurologist* **3:**13–30.

Farrer, L. A., and Cupples, L. A., 1998, Determining the genetic component of a disease, in: *Approaches to Gene Mapping Studies to Complex Human Diseases* (J. L. Haines and M. A. Pericak-Vance, eds.), Wiley, New York, pp. 93–130.

Farrer, L. A., Conneally, P. M., and Yu, P. L., 1984, The natural history of Huntington's disease: Possible role of aging genes, *Am. J. Med. Genet.* **18:**115–123.

Farrer, L. A., Bonné-Tamir, B., Frydman, M., Magazanik, A., Kidd, K. K., Bowcock, A. M., and Cavalli-Sforza, L. L., 1988, Predicting genotypes at loci for autosomal recessive disorders using linked genetic markers: Application to Wilson's disease, *Hum. Genet.* **79:**109–117.

Farrer, L. A., Bowcock, A. M., Hebert, J. M., Bonné-Tamir, B., Sternlieb, I., Giagheddu, M., St. George-Hyslop, P., Frydman, M., Lößner, J., Demelia, L., Carcassi, C., Lee, R., Bekker, R., Bale, A. E., Donis-Keller, H., Scheinberg, I. H., and Cavalli-Sforza, L. L., 1991, Predictive testing for Wilson disease using tightly linked and flanking DNA markers, *Neurology* **41:**992–999.

Farrer, L. A., Cupples, L. A., Kiely, D. K., Conneally, P. M., and Myers, R. H., 1992, Inverse relationship between age at onset and paternal age suggests involvement of genetic imprinting, *Am. J. Hum. Genet.* **50:**528–535.

Farrer, L. A., Cupples, L. A., Wiater, P., Conneally, P. M., Gusella, J. F., and Myers, R. H., 1993, The normal Huntington disease (HD) allele, or a closely linked gene, influences age at onset of HD, *Am. J. Hum. Genet.* **53:**125–130.

Figus, A., Lampis, R., and DeVoto, M., *et al.,* 1989, Carrier detection and early diagnosis of Wilson's disease by restriction fragment length polymorphism analysis, *J. Med. Genet.* **26:**78–82.

Figus, A., Angius, A., Loudianos, G., Bertini, C., Dessi, V., Loi, A., Deiana, M., Lovicu, M., Olla, N., Sole, G., De Virgiliis, S., Lilliu, F., Giulia Farci, A. M., Nurchi, A., Giacchino, R., Barabino, A., Marazzi, M., Zancan, L., Greggio, N. A., Marcellini, M., Solinas, A., Deplano, A., Barbare, C., Devoto, M., Ozsoylu, S., Kocak, N., Akar, N., Karayalcin, C., Mokini, V., Cullufi, P., Balestrieri, A., Cao, A., and Pirastu, M., 1995, Molecular pathology and haplotype analysis of Wilson disease in Mediterranean populations, *Am. J. Hum. Genet.* **57:**1318–1324.

Fowler, H. L., DeMagalhaes, J., and Rogers, F. M., 1977, Azorean disease of the nervous system (letter), *New Eng. J. Med.* **297:**729.

Frydman, M., Bonne-Tamir, B., Farrer, L. A., Conneally, P. M., Magazanik, A., Ashbel, S., and Goldwitch,

Z., 1985, Assignment of the gene for Wilson disease to chromosome 13: Linkage to the esterase-D locus, *Proc. Natl. Acad. Sci. USA* **82:**1819–1821.

Fulker, D. W., and Cardon, L. R., 1994, A sib-pair approach to interval mapping of quantitative trait loci, *Am. J. Hum. Genet.* **54:**1092–1103.

Gaspar, C., Lopes-Cendes, I., DeStefano, A., Maciel, P., Silveira, I., Coutinho, P., MacLeod, P., Sequeiros, J., Farrer, L. A., and Rouleau, G. A., 1996, Linkage disequilibrium analysis of MJD1 closely linked markers in Machado-Joseph disease patients with different ethnic origins, *Hum. Genet.* **98:**620–624.

Giagheddu, M., Demelia, L., Puggioni, G., Nurchi, A. M., Contu, L., Pirari, G., Deplano, A., and Rachele, M. G., 1985, Epidemiologic study of hepatolenticular degeneration (Wilson Disease) in Sardinia (1902–1983), *Acta Neurol. Scand.* **72:**43–55.

Green, J. R., and Woodrow, J. C., 1977, Sibling method for detecting HLA-linked genes in disease, *Tissue Antigens* **9:**31–35.

Haseman, J. K., and Elston, R. C., 1972, The investigation of linkage between a quantitative trait and a marker locus, *Behav. Genet.* **2:**3–19.

Hayden, M. R., 1981, *Huntington's Chorea,* Springer-Verlag, New York.

Hill, G. M., Brewer, G. J., Juni, J. E., Prasad, A. S., and Dick, R. D., 1986, Treatment of Wilson's disease with zinc. II. Validation of oral $^{64}$copper with copper balance, *Am. J. Med. Sci.* **262:**344–349.

Houwen, R. H., Juyn, J., Hoogenrad, T. U., Ploos van Amstel, J. K., and Berger, R., 1995, H714Q mutation in Wilson disease is associated with late, neurological presentation, *J. Med. Genet.* **32:**480–482.

The Huntington's Disease Collaborative Group, 1993, A novel gene containing a trinucleotide repeat that is expanded and unstable on Huntington's disease chromosomes, *Cell* **72:**971–983.

Igarashi, S., Takiyama, Y., Cancel, G., Rogaeva, E. A., Sasaki, H., Wakisaka, A., Zhou, Y.-X., Takano, H., Endo, K., Sanpei, K., Oyake, M., Tanaka, H., Stevanin, G., Abbas, N., Dürr, A., Rogaev, E. I., Sherrington, R., Tsuda, T., Ikeda, M., Cassa, E., Nishizawa, M., Benomar, A., Julien, J., Weissenbach, J., Wang, G.-X., Agid, Y., St. George-Hyslop, P. H., Brice, A., and Tsuji, S., 1996, Intergenerational instability of the CAG repeat of the gene for Machado-Joseph disease (MJD1) is affected by the genotype of the normal chromosome: Implications for the molecular mechanisms of the stability of the CAG repeat, *Hum. Mol. Genet.* **5:**923–932.

Kalinsky, H., Funes, A., Maron, H., Zeldin, A., Pel-Or, Y., Korostishevsky, M., Gershoni-Baruch, R., Farrer, L. A., and Bonné-Tamir, B., 1998, Novel ATP7B mutations causing Wilson disease in several Israeli ethnic groups, *Hum. Mut.* **11:**145–151.

Kant, A. J., 1996, Using lod score analysis in the search for a gene modifying age at onset of Machado-Joseph disease, M.A. thesis, Boston University School of Medicine.

Kawaguchi, Y., Okamoto, T., Taniwaki, M., Aizawa, M., Inoue, M., Katayama, S., Kawakami, H., Nakamura, S., Nishimura, M., Akiguchi, I., Kimura, J., Narumiya, S., and Kakizuka, A., 1994, CAG expansion in a novel gene for Machado-Joseph disease at chromosome 14q32.1, *Nature Genet.* **8:**221–228.

Khoury, M. J., Beaty, T. H., and Cohen, B. H., 1993, *Fundamentals of Genetic Epidemiology,* Oxford University Press, New York.

Kidd, K. K., 1982, Genetic linkage markers in the study of psychiatric disorders, in: *Biological Markers in Psychiatry and Neurology* (E. Usdin, and I. Hanin, eds.), Pergamon, New York, pp. 459–466.

Kidd, K. K., 1988, Genetic research on affective disorders, in: *Proceedings of the Indo–US Symposium on Affective Disorders, Bangalore, India* (S. M. Channabasavanna and S. M. Shah, eds.), *National Institute of Mental Health and Neuroscience,* Bangalore, India, pp. 79–91.

Lange, K., 1986, The affected sib-pair method using identity by state relations, *Am. J. Hum. Genet.* **39:**148–150.

Levine, P., Robinson, E., Celano, M., Briggs, O., and Falkinburg, L., 1955, Gene interaction resulting in suppression of blood group substance B, *Blood* **10:**1100–1108.

Liang, K., and Zeger, S. L., 1986, Longitudinal data analysis using generalized linear models, *Biometrika* **17:**13–22.

Maciel, P., Gaspar, C., DeStefano, A. L., Silveira, I., Coutinho, P., Radvany, J., Dawson, D. M., Sudarsky, L., Guimaraes, J., Loureio, L., Lopes-Cendes, I., Rooke, K., Rosenberg, R., MacLeod, P., Farrer, L. A., Sequeiros, J., and Rouleau, G. A., 1995, Correlation between CAG repeat length and clinical features in Machado-Joseph disease, *Am. J. Hum. Genet.* **57:**54–61.

Maruyama, H., Nakamura, S., Matsuyama, Z., Sakai, T., Doyu, M., Sobue, G., Seto, M., Tsujihata, M., Oh-i, T., Nishio, T., Sunohara, N., Takahashi, R., Hayashi, M., Nishino, I., Ohtake, T., Oda, T., Nishimura, M., Saida, T., Matsumoto, H., Baba, M., Kawaguchi, Y., Kakizuka, A., and Kawakami, H.,

1995, Molecular features of the CAG repeats and clinical manifestations of Machado-Joseph disease, *Hum. Mol. Genet.* **4:**807–812.

Mercer, J. F. B., Livingston, J., Hall, B., Paynter, J. A., Begy, C., Chandrasekharappa, S., Lockhart, P., Grimes, A., Bhave, M., Siemieniak, D., and Glover, T. W., 1992, Isolation of a partial candidate gene for Menkes disease by positional cloning, *Nature Genet.* **3:**20–25.

Morton, N. E., 1955, Sequential tests for the detection of linkage, *Am. J. Hum. Genet.* **7:**277–318.

Murray, J. C., Buetow, K. H., Weber, J. L., Ludwigsen, S., Scherpbier-Heddema, T., Manion, F., Quillen, J., Sheffield, V. C., Sunden, S., Duyk, G. M., Weissenbach, J., Gyapay, G., Dib, C., Morrisette, J., Lathrop, G. M., Vignal, A., White, R., Matsunami, N., Gerken, S., Melis, R., Albertsen, H., Plaetke, R., Odelberg, S., Ward, D., Dausset, J., Cohen, D., and Cann, H., 1994. A comprehensive human linkage map with centimorgan density, *Science* **265:**2049–2054.

Myers, R. H., Madden, J. J., Teague, J. L., and Falek, A., 1982, Factors related to age at onset of Huntington's disease, *Am. J. Hum. Genet.* **34:**481–488.

Ordovas, J. M., Schaefer, E. J., Salem, D., Ward, R. H., Glueck, C. J., Vergani, C., Wilson, P. W. F., and Karathanasis, S. K., 1986, Apolipoprotein A–I gene polymorphism associated with premature coronary artery disease and familial hypoalphalipoproteinemia, *New Engl. J. Med.* **314:**671–677.

Passwell, J., Adam, A., Garfinkel, D., Streiffler, M., and Cohen, B. E., 1977, Heterogeneity of Wilson disease in Israel, *Isr. J. med. Sci.* **13:**15–19.

Penrose, L. S., 1935, The detection of autosomal linkage in data which consists of pairs of brothers and sisters of unspecified parentage, *Ann. Eugen.* **6:**133–138.

Pericak-Vance, M. A., Elston, R. C., Conneally, P. M., and Dawson, D. V., 1983, Age-of-onset heterogeneity in Huntington's disease families, *Am. J. Med. Genet.* **14:**49–59.

Petrukhin, K., Fischer, S. G., Pirastu, M., Tanzi, R. E., Chernov, I., Devoto, M., Brzustowicz, L. M., Cayanis, E., Vitale, E., Russo, J. J., Masteoane, D., Boukhgalter, B., Wasco, W., Figus, A. L., Loudianos, J., Cao, A., Sternlieb, I., Evgrafov, O., Parano, E., Pavone, L., Warburton, D., Ott, J., Penchaszadeh, G. K., Scheinberg, I. H., and Gilliam, T. C., 1993, Mapping, cloning, and genetic characterization of the region containing the Wilson disease gene, *Nature Genet.* **5:**338–343.

Ploughman, L. M., and Boehnke, M., 1989, Estimating the power of a proposed linkage study for a complex genetic trait, *Am. J. Hum. Genet.* **44:**543–551.

Ranum, L. P. W., Lundgren, J. K., Schut, L. J., Ahrens, M. J., Perlman, S., Aita, J., Bird, T. A., Gomez, C., and Orr, H. T., 1995, Spinocerebellar ataxia type I and Machado-Joseph disease: Incidence of CAG repeat expansion among adult-onset ataxia patients from 311 families with dominant, recessive, or sporadic ataxia, *Am. J. Hum. Genet.* **57:**603–608.

Risch, N., 1990, Linkage strategies for genetically complex traits. III. The effect of marker polymorphism on analysis of affected relative pairs, *Am. J. Hum. Genet.* **46:**242–253.

Risch, N., Zhang, H., 1995, Extreme discordant sib pairs for mapping quantitative trait loci in humans, *Science* **268:**1584–1589.

Rogus, J. J., and Krolewski, A. S., 1996, Using discordant sib pairs to map loci for qualitative traits with high sibling recurrence risk, *Am. J. Hum. Genet.* **59:**1376–1381.

Romanul, F. C. A., Fowler, H. L., Radvany, J., Feldman, R. G., and Feingold, M., 1977, Azorean disease of the nervous system, *New Engl. J. Med.* **26:**1505–1508.

Rosenberg, R. N., 1992, Machado-Joseph disease: An autosomal dominant motor system degeneration, *Mov. Discord.* **7:**193–203.

Rosenberg, R. N., Nyhan,m W. L., and Bay, C., 1976a, Autosomal dominant striationigral degeneration: A clinical, pathological, and biochemical study of a new genetic disorder, *Trans. Am. Neurol. Assoc.* **101:**78–80.

Rosenberg, R. N., Nyhan, W. L., Bay, C., and Shore, P., 1976b, Autosomal dominant striatonigral degeneration, *Neurology* **26:**703–714.

Sass-Kortsak, A., and Bearn, A. G., 1978, Hereditary disorders of copper metabolism: Wilson's disease (hepatolenticular degeneration) and Menke's disease (kinky-hair or steely-hair syndrome), in: *The Metabolic Basis of Inherited Disease, 4th Ed.* (J. B. Stanbury, J. B. Wyngaarden, and D. S. Fredrickson, eds.), McGraw-Hill, New York, pp. 1098–1126.

Scheffer, H., Houwen, R. H., te Meerman, G. J., Loessner, J., and Buys, C. H., 1992, Minimum recombinant analysis in Wilson disease: Identification of crossovers as a means to localize the Wilson disease locus, *Cytogenet. Cell Genet.*, in press.

Scheinberg, H., and Sternlieb, I., 1975, Pregnancy in penicillamine-treated patients with Wilson's disease, *New Engl. J. Med.* **293:**1300.

Scheinberg, I. H., and Sternlieb, I., 1984, *Wilson's Disease,* W. B. Saunders, Philadelphia, PA.

Scheinberg, H., Jaffe, M., and Sternlieb, I., 1987, The use of trientine in preventing the effects of interrupting penicillamine therapy in Wilson's disease, *New Engl. J. Med.* **317:**209–213.

Sequeiros, J., 1993, Machado-Joseph disease: Epidemiology, genetics, and genetic epidemiology, in: *Handbook of Cerebellar Diseases* (R. Lechtenberg, ed.), Marcel Dekker, New York, pp. 345–351.

Sequeiros, J., and Coutinho, P., 1993, Epidemiology and clinical aspects of Machado-Joseph disease, in: *Advances in Neurology* (A. Harding, T. Deufel, and S. Chamberlain, eds.), Raven, New York, pp. 139–153.

Sequeiros, J., Silveira, I., Maciel, P., Coutinho, P., Manaia, A., Gaspar, C., Burlet, P., Loureiro, L., Guimaraes, J., Tanaka, H., Takiyama, Y., Sakamoto, H., Nishizawa, M., Nomura, Y., Segawa, M., Tsuji, S., Melki, J., Munnich, A., 1994, Genetic linkage studies of Machado-Joseph disease with chromosome 14q STRPs in 16 Portuguese–Azorean kindreds, *Genomics* **21:**645–648.

Snell, R. G., MacMillan, J. C., Cheadle, J. P., Fenton, I., Lazarou, L. P., Davies, P., MacDonald, M. E., Gusella, J. F., Harper, P. S., and Shaw, D. J., 1993, Relationship between trinucleotide repeat expansion and phenotypic variation in Huntington's disease, *Nature Genet.* **4:**393–397.

Spielman, R. S., McGinnis, R. E., and Ewens, W. J., 1993, Transmission test for linkage disequilibrium: The insulin gene region and insulin-dependent diabetes mellitus (IDDM), *Am. J. Hum. Genet.* **52:**506–516.

Stewart, E. A., White, A., Tomfohrde, J., Osborne-Lawrence, S., Prestridge, L., Bonne-Tamir, B., Schienberg, I. H., St. George-Hyslop, P., Giagheddu, M., Kim, J.-W., Seo, J. K., Lo, H.-Y., Ivanova-Smolenskaya, I. A., Limborska, S. A., Cavalli-Sforza, L. L., Farrer, L. A., and Bowcock, A. M., 1993, Polymorphic microsatellites and Wilson disease, *Am. J. Hum. Genet.* **53:**864–873.

Strickland, G. T., and Leu, M.-L., 1975, Wilson's disease: Clinical and laboratory manifestations in 40 patients, *Medicine* **54:**113–137.

Strickland, G. T., Frommer, D., Leu, M.-L., Pollard, R., Sherlock, S., and Cumings, J. N., 1973, Wilson disease in the United Kingdom and Taiwan, *Q. J. Med.* **42:**619–638.

Suarez, B. K., 1978, The affected sib pair IBD distribution for HLA-linked disease susceptibility genes, *Tissue Antigens* **12:**87–93.

Sudarsky, L., Myers, R. H., and Walshe, T. M., 1983, Huntington's disease in monozygotic twins reared apart, *J. Med. Genet.* **20:**408–411.

Sudarsky, L., Corwin, L., and Dawson, D. M., 1992, Machado-Joseph disease in New England: Clinical description and distinction from the olivopontocerebellar atrophies. *Mov. Disord.* **7:**204–208.

Takiyama, Y., Igarashi, S., Rogaeva, E. A., Endo, K., Rogaev, E. I., Tanaka, H., Sherrington, R., Sanpei, K., Liang, Y., Saito, M., Tsuda, T., Takano, H., Ikeda, M., Lin, C., Chi, H., Kennedy, J. L., Lang, A. E., Wherrett, J. R., Segawa, M., Nomura, Y., Yuasa, T., Weissenbach, J., Yoshida, M., Nishizawa, M., Kidd, K. K., Tsuji, S., and St. George-Hyslop, P. H., 1995, Evidence for intergenerational instability in the CAG repeats in the MJD1 gene and for conserved haplotypes at flanking markers amongst Japanese and Caucasian subjects with Machado-Joseph disease, *Hum. Mol. Genet.* **4:**1137–1146.

Tanzi, R. E., Petrukhin, K., Chernov, I., Pellequer, J. L., Wasco, W., Ross, B., Tomano, D. M., Parano, E., Pavone, L., Brzustowicz, L. M., Devoto, M., Peppercorn, J., Bush, A. I., Sternlieb, I., Pirastu, M., Gusella, J. F., Evgrafov, O., Penchaszadeh, G. K., Honig, B., Edelman, I. S., Soares, M. B., Scheinberg, I. H., and Gilliam, T. C., 1993, The Wilson disease gene is a copper transporting ATPase with homology to the Menkes disease gene, *Nature Genet.* **5:**344–350.

Thomas, G. R., Forbes, J. R., Roberts, E. A., Walshe, J. M., and Cox, D. W., 1995a, The Wilson disease gene: Spectrum of mutations and their consequences, *Nature Genet.* **9:**210–217.

Thomas, G. R., Jensson, Ó., Gudmundsson, G., Thorsteinsson, L., and Cox, D. W., 1995, Wilson disease in Iceland: A clinical and genetic study, *Am. J. Hum. Genet.* **56:**1140–1146.

Twist, E. C., Causaubon, L. K., Ruttledge, M., Rao, V. S., MacLeod, P. M., Radvany, J., Zhao, Z., Rosenberg, R. N., Farrer, L. A., and Rouleau, G. A., 1995, Machado-Joseph disease maps to the same region of chromosome 14 as the spinocerebellar ataxia type 3 locus, *J. Med. Genet.* **32:**25–31.

Vulpe, C., Levinson, B., Whitney, S., Packman, S., and Gitschier, J., 1992, Isolation of a candidate gene for Menkes disease and evidence that it encodes a copper-transporting ATPase, *Nature Genet.* **3:**7–13.

Waldenström, E., Lagerkvist, A., Dahlman, T., Westermark, K., and Landegren, U., 1996, Efficient detection of mutations in Wilson disease by manifold sequencing, *Genomics* **37:**303–309.

Walshe, J. M., 1976, Wilson's disease (hepatolenticular degeneration), in: *Handbook of Clinical Neurology: Metabolic and Deficiency Diseases of the Nervous System,* Vol. 27 (P. J. Vinken and G. S. Bruyn, eds.), North Holland, Amsterdam, pp. 379–414.

Weeks, D. E., and Lange, K., 1988, The affected-pedigree-member method of linkage analysis, *Am. J. Hum. Genet.* **42:**315–326.

White, A., Tomfohrde, J., Stewart, E., Barnes, R., Le Paslier, D., Weissenbach, Cavalli-Sforza, L. L., Farrer, L. A., and Bowcock, A. M., 1993, A 4.5 megabase yeast artificial chromosome contig from human chromosome 13q14.3 ordering 9 polymorphic microsatellites (22 STSs) tightly linked to the Wilson disease locus, *Proc. Natl. Acad. Sci. USA* **90:**10105–10109.

Wilson, A. F., and Elston, R. C., 1993, Statistical validity of the Haseman–Elston sub-pair test in small samples, *Genet. Epidemiol.* **10:**593–598.

Wilson, S. A. K., 1912, Progressive lenticular degeneration: A familial nervous disease associated with cirrhosis of the liver, *Brain* **34:**295–509.

Woods, B. T., and Schaumburg, H. H., 1972, Nigro-spino-dentatal degeneration with nuclear ophthalmoplegia, *J. Neurol. Sci.* **17:**149–166.

Yamaguchi, Y., Heiny, M. E., and Gitlin, J. D., 1993, Isolation and characterization of a human liver cDNA as a candidate gene for Wilson disease, *Biochem. Biophys. Res. Comm.* **197:**271–277.

Yuzbasiyan-Gurkan, V., Brewer, G. J., Boerwinkle, E., and Venta, P. J., 1988, Linkage of Wilson's disease gene to chromosome 13 in North-American pedigrees, *Am. J. Hum. Genet.* **42:**825–829.

# 13

# Early Features of Alzheimer's Disease

MARILYN S. ALBERT and MARK B. MOSS

## 1. Introduction

Alzheimer's disease (AD), the most common dementing disorder, ultimately produces severe impairments that affect all cognitive domains (e.g., memory, executive function, language, spatial ability, and attention). Early in the course of disease, however, the pattern of deficits is more circumscribed. Memory changes are generally the earliest cognitive change seen in AD. Changes in executive function ability also show significant declines early in the course of AD. Neuropathologic studies in autopsy-confirmed AD patients who had a range of severity during life provide strong evidence for neurobiologic changes responsible for the pattern of cognitive deficits seen early in the disease. Together with recent neuroimaging studies, they suggest that a brain network with multiple nodes is selectively vulnerable in AD and that the cognitive deficits are related to the involvement of this brain network. We will review the early cognitive changes associated with AD and their neurobiologic correlates.

---

MARILYN S. ALBERT • Departments of Psychiatry and Neurology, Massachusetts General Hospital, Harvard Medical School, Boston, Massachusetts 02118. MARK B. MOSS • Department of Anatomy and Neurobiology, Boston University School of Medicine, Boston, Massachusetts 02118.

*Cerebral Cortex,* Volume 14, edited by Peters and Morrison. Kluwer Academic/Plenum Publishers, New York, 1999.

## 2. Memory Changes in Early Alzheimer's Disease

Difficulty with the acquisition of new information is generally the first and most salient symptom to emerge in patients with AD. When clinical neuropsychological tests are used to evaluate memory in AD patients, it is clear that recall and recognition performance are impaired in both the verbal and nonverbal domains (Wilson *et al.,* 1983; Storandt and Hill, 1989). Experimental studies have examined AD patients to determine whether the manner in which information is lost over brief delays is in any way unique to this patient group. The results of these studies suggest that a comparison of immediate- and delayed-recall performance may be a useful diagnostic measure for identifying patients with AD.

The first such study was conducted by Moss *et al.* (1986). They compared patients with AD to three other groups: amnesic patients who had alcoholic Korsakoff's syndrome (KS), dementing patients with Huntington's disease (HD), and normal controls (NC). All of the subjects were administered the Delayed Recognition Span Test (DRST). This task employs disks on which are placed a variety of stimuli (words, colors, faces, patterns, and so on). During the recognition portion of the task, the disks are placed on a board one at a time (there are 16 disks in all). As each disk is added, the board is hidden from view. The subject is then asked to point to the disk that was added during the delay interval. To do this the subject must keep track of an increasingly long series of disks. The disks are then added one at a time until the subject makes an error. This yields a delayed recognition span for each of the stimuli sets. All of the patient groups were impaired in their recognition performance with respect to controls, but there is overlap among the patient groups. There was no significant difference among the three patient groups in their ability to recognize new spatial, color, pattern, or facial stimuli; patients with HD performed significantly better than the other two groups when verbal stimuli were used. In the verbal condition of the DRST (words on the disks), a recall paradigm was added. At 15 seconds and at 2 minutes after completion of the last verbal recognition trial, subjects were asked to recall the words that had been on the disks. In this condition AD patients differ considerably from the other patients. They recall significantly fewer words over this brief delay interval (2 min) than either HD or KS patients. Although all three patient groups were equally impaired relative to normal controls at the 15-second interval, patients with AD recalled significantly fewer words than either the HD or the KS groups at the 2-minute interval; in fact, only the AD group performed significantly worse at the longer, as compared with the shorter, interval (Fig. 1). It is notable that by the end of the 2-minute interval, 11 of the 12 AD patients could recall fewer than 3 of the 16 words presented repeatedly during recognition testing. Of these 11 patients, 7 were unable to recall any of the 16 words at the longer interval. Furthermore, whereas the KS, HD, and normal control subjects lost an average of 10–15% of the verbal information between the 15-second and the 2-minute delay intervals, patients with AD lost an average of 75% of the material. This pattern of recall performance demonstrated for the first time that patients with AD lose more information over a brief delay than other patients with amnesic or dementing disorders.

A similar pattern of results has since been reported by numerous other investigators (e.g., Butters *et al.,* 1988; Hart *et al.,* 1988; Welsh *et al.,* 1992). The findings of

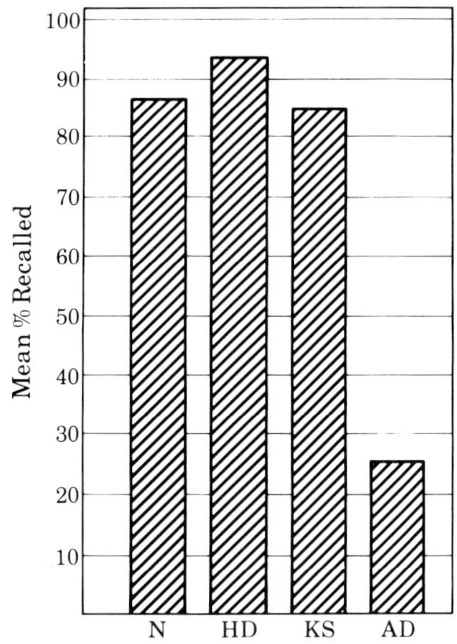


**Figure 1.** The difference between immediate and delayed recall on the verbal recall portion of the Delayed Recognition Span Test. The groups compared are normal controls (N), patients with Alzheimer's disease (AD), patients with Huntington's disease (HD), and patients with alcoholic Korsakoff's syndrome (KS).

Hart *et al.* (1988) are particularly notable. They administered a continuous recognition task to AD patients and controls, and equated both groups of subjects for retention 90 seconds after the task was completed. They then retested the subjects at 10 minutes, at 2 hours, and at 48 hours after completion of the task. The AD patients showed a greater loss of information than the controls between the 90-second interval and the 10-minute interval, but not between the 10-minute and the 2-hour or 48-hour intervals, suggesting that intervals of 10 minutes or less may be optimal for differentiating AD patients from other patient groups and from controls.

Since these findings were first reported, additional patient groups have been compared to AD patients on tasks of this nature. They likewise recalled more information after a delay than patients with AD. Milberg and Albert (1989) compared the performance of AD patients with that of progressive supranuclear palsy (PSO) patients. The two groups were equated for overall level of impairment on the basis of the Mattis Dementia Rating Scale (Mattis, 1976) and were equivalent in years of education. There was no difference between the two patient groups on most of the tasks administered (e.g., Vocabulary, Digit Span Forward, Similarities, Block Design). There was, however, a striking difference between the groups on both of the memory tasks. The AD patients were significantly impaired in comparison to the PSP patients on tests of both verbal and nonverbal memory. A comparison of patients with AD and patients with frontotemporal dementia (FTD) (Moss and Albert, 1988) also demonstrates the severe recall deficits of the AD patients. Here again, patients with AD and patients with FTD, equated for overall level of cognitive impairment, were administered the Delayed Recognition Span Test described earlier. As in the previous study, total recall on the 15-second and the 2-minute delay intervals (i.e., the "savings score") differentiated the two groups. The retention of the FTD patients over the 2-minute delay interval approaches normality, whereas the AD patients lose a substantial amount of information. In general, these findings suggest that the nature and severity of the AD patients' memory disturbance in relation to delays spanning the first 10 minutes after encoding is likely to be the result of a unique pattern of neuropathological and/or neurochemical dysfunction.

It is important to note that the alterations in memory associated with early AD are substantially different from those associated with age-related memory changes. Although there are significant declines in delayed-recall performance as individuals get older, much accumulated data indicates that these differences pertain to the fact that it takes older individuals longer to learn new information, but once learned, it is retained well over numerous delay intervals (e.g., Wickelgren, 1975; Craik, 1977). For example, if one compares the difference between recall scores on immediate and delayed recall, there are no statistically significant age-related differences (Peterson *et al.,* 1992). Thus, if one allows healthy older subjects to learn the material well (i.e., to the point where few errors are made), they do not forget what they have learned any more rapidly than the young. However, if healthy older subjects are not given the opportunity to learn material to the same level of proficiency as younger individuals, less information, on average, will be retained by the older persons after a delay.

## 3. Executive Function Changes in Early Alzheimer's Disease

In addition to memory problems, mildly impaired AD patients are substantially impaired with a set of abilities collectively referred to as *executive functions.* This conclusion derives from a study (Lafleche and Albert, 1995) that examined the performance of AD patients on a variety of tasks assessing a broad range of executive abilities: concurrent manipulation of information (e.g., set-shifting, sequencing, and monitoring), cue-directed attention (e.g., the ability to use cues to direct attention), and concept formation (e.g., abstraction). Only very mildly impaired AD patients were included, in order to determine whether an impairment in executive function precedes or coexists with significant deficits in spatial and language function.

Executive functions were assessed by seven tests. Because many of the tasks were complex, they often reflected more than one aspect of executive function. It nevertheless appeared reasonable to describe them in terms of three broad areas of executive function. Three of the tests primarily evaluated the concurrent manipulation of information, one assessed cue-directed attention, and three evaluated concept formation.

The tests that primarily evaluated the concurrent manipulation of information included the Trail Making (Reitan, 1958), Controlled Word Association (Benton and Hamsher, 1976), and Self-Ordering tests (Petrides and Milner, 1982).

It should be noted that both Self-Ordering and Cued Reaction Time were included in the test battery because they have been shown to differentiate patients with frontal lobe lesions from patients with limited temporal lobe lesions, that is, lesions that do not extend posteriorly beyond the hippocampus (Petrides and Milner, 1982; Alivisatos and Milner, 1989). Patients with frontal excisions are impaired on both tasks, whereas patients with limited temporal lobe lesions can perform both tasks as well as can controls. It was therefore hypothesized that impairments on these tasks were unlikely to be the result of the medial temporal lobe dysfunction that occurs in AD, but rather, implicated frontal system dysfunction. Four additional cognitive domains were examined to determine whether there is a differential impairment of cognitive function in mildly impaired patients with Alzheimer's disease. These included standard tests of memory, language, visuospatial function, and simple attention. The tests of simple attention were the Trails A (Reitan, 1958) and the Digit Span Forward of the WAIS-R (Wechsler, 1981). The Boston Naming Test (Kaplan *et al.,* 1983) assessed confrontation naming, the CERAD Word List Learning Test (Morris *et al.,* 1989) evaluated verbal memory, and the CERAD Figure Copying Test (Morris *et al.,* 1989) assessed spatial ability.

The results (see Table I) indicate that mildly impaired AD patients differ significantly from controls on a number of executive function tasks. Their performance, however, is not impaired on all of the tasks uniformly. Most tasks revealing a significant deficit were those requiring concurrent manipulation of information (set-shifting, self-monitoring, or sequencing). These tasks included Controlled Word Association (Word Fluency—FAS), Trails B and Self-Ordering. Performance on the Hukok Test, which assesses the ability to perceive and synthesize patterns of

**Table I. Performance of Mild AD Patients and Controls on Executive Function Tasks**

| Test | Group | M | SD | *t* | *p* |
|---|---|---|---|---|---|
| Trail making (part B) | AD[a] | 194.40 | 82.97 | −2.55 | 0.01 |
| | NC[b] | 131.55 | 72.26 | | |
| Verbal fluency (FAS) | AD | 34.20 | 13.96 | 4.46 | 0.002 |
| | NC | 51.90 | 10.99 | | |
| Self-ordering | AD | 16.40 | 5.04 | −3.53 | 0.001 |
| | NC | 11.10 | 4.42 | | |
| Hukok | AD | 7.90 | 2.85 | 4.89 | 0.0001 |
| | NC | 11.95 | 2.37 | | |
| Similarities | AD | 16.20 | 6.88 | ns | |
| | NC | 19.65 | 2.62 | | |
| Gorham proverbs | AD | 8.78 | 6.91 | ns | |
| | NC | 10.38 | 5.33 | | |
| Cued reaction time | AD | 0.9 | 1.1 | ns[c] | |
| | NC | 0.9 | 1.0 | | |

[a]Alzheimer's disease patients
[b]normal controls
[c]statistically not significant

relationships in a sequence of visual stimuli of increasing complexity, also differed significantly between patients and controls. By contrast, performance on tasks assessing cue-directed attention and verbal concept formation were not significantly impaired in either group of patients.

It therefore seems most likely that the underlying disability of the mild AD patients on executive function tasks is the result of a primary difficulty with the concurrent manipulation of information. For example, AD patients perform well the Trails A test, a test that requires sustained attention and simple sequencing, but their performance is impaired on the Trails B, which requires tracking two overlearned sequences simultaneously (i.e., numbers and letters) and switching rapidly from one sequence to the other.

This is consistent with the findings of Sahakian *et al.* (1990) in mildly impaired patients, using a task that requires shifting between stimulus dimensions, and of Baddeley *et al.*, (1986), who observed that mild-to-moderately impaired AD patients had difficulty coordinating two concurrent tasks. These results are also consistent with the findings of investigators reporting executive function deficits in AD patients with a broader range of severity (Becker, 1988; Morris and Baddeley, 1988; Lafleche *et al.*, 1990; Nestor *et al.*, 1991). Deficits in the central executive system (Morris and Baddeley, 1988) attributed to difficulties with "working memory," also agree with these findings. Moreover, these results suggest that executive function deficits precede language and spatial impairment in most patients, consistent with the findings of Grady *et al.* (1988), who reported that deficits in memory and executive function preceded impairments in language and visuospatial functions. The results are also consistent with those of Welsh *et al.* (1992), who reported that memory deficits preceded language and spatial deficits.

## 4. Changes in Brain Structure and Function in Early Alzheimer's Disease that May Underly the Memory Deficit

The most likely explanation for the memory abnormalities that characterize early-stage AD pertains to the hippocampal damage seen in these patients (Ball, 1977; Hyman *et al.*, 1985). In the hippocampal formation, neuronal loss and abnormal formations within the cells (e.g., neurofibrillary tangles and neuritic plaques) are seen primarily in the entorhinal cortex and subiculum, the primary pathways for conveying information into and out of the hippocampus (see Rosene, this volume). It has been suggested that abnormalities in these regions produce a functional isolation of the hippocampus (Hyman *et al.*, 1985; 1986), indicating that neuropathological damage to medial temporal lobe structures may be responsible for the marked memory impairment evident in the early stages of AD. These results were first observed in patients with end-stage AD; however, they have recently been extended to patients with very mild AD (Gomez-Isla, *et al.*, 1996). Most striking is the fact that the entorhinal cortex has neuronal loss of approximately 60% and 40% in layers II and IV, respectively, of the entorhinal cortex. Because this region is known to be critically important for the acquisition and retention of new information (Zola-Morgan, *et al.*, 1994), abnormalities here are likely to be responsible for the severe anterograde memory loss evident early in the course of AD.

MRI studies focusing on mildly impaired AD patients are entirely consistent with the neuropathological data cited above. They have uniformly reported significant and striking differences between MRI measures of the medial temporal lobe in AD patients and controls. Measures of the hippocampal formation, the parahippocampal gyrus, the amygdalo-hippocampal complex, and the temporal horn of the lateral ventricles have demonstrated significant differences between mildly impaired AD patients and controls across a wide range of studies, using a variety of techniques (Seab *et al.*, 1988; Kesslak *et al.*, 1991; Jack *et al.*, 1992; Convit *et al.*, 1993; Killiany *et al.*, 1993; Ikeda *et al.*, 1994; Lehericy *et al.*, 1994; Laasko *et al.*, 1995). These studies are consistent with computerized tomography (CT) studies showing suprasellar cistern/temporal horn abnormalities in AD (e.g., Sandor *et al.*, 1988, 1992; George *et al.*, 1990). A recent study (Killiany *et al.*, submitted) indicates that in order to identify AD in the prodromal phase (i.e., have evidence of recent declines in memory but status still "questionable"), very specific medial temporal lobe measures are needed. In this study MRI measures were obtained when the subjects were questionable and the measures that predicted subsequent "conversion" to AD were examined. Two sets of MRI measures were employed. The first set pertained to regions of interest (ROIs) used in previous studies with mild AD patients and controls. These ROIs were as follows and represented the total volume of each region: the hippocampal formation, the amygdala, the basal forebrain, the anterior cingulate gyrus, the head of the caudate nucleus, the temporal horn, the lateral ventricles, and the third ventricle. In the previous study, a discriminant function analysis, in which measurements of the hippocampal formation and the temporal horn were combined, differentiated 100% of the AD patients from controls (e.g., Killiany *et al.*, 1993).

For the second set of MRI measures, two ROIs were examined: the entorhinal cortex, representative of a limbic structure process (as described above), and the superior temporal sulcus region (Gomez-Isla *et al.*, 1997), representative of high-order association cortex, which is believed to be affected later in the disease process (Braak and Braak, 1993). As in previous studies, the first set of ROIs (hippocampus, amygdala, basal forebrain, anterior cingulate gyrus, head of the caudate nucleus, temporal horn, lateral ventricles, third ventricle) revealed significant differences between the mildly impaired AD patients and normal controls. Somewhat surprisingly, however, none of these measures was able to differentiate the controls from questionable subjects who went on to develop AD during the longitudinal follow-up period (*converters*'). The second set of ROIs (entorhinal cortex and superior temporal sulcus banks) readily distinguished controls from patients with mild AD. Importantly, these measures were also able to demonstrate substantial differences between controls and individuals who at the time of the MRI did not have sufficient clinical difficulties to warrant a diagnosis of dementia, but subsequently converted over the follow-up period to a diagnosis of AD. The converters had a 30% smaller entorhinal cortex volume in comparison to the controls—highly statistically significant. The AD patients demonstrated a 50% loss in comparison to the controls. The measurement of the banks of the superior temporal gyrus in the converters showed a less substantial volume loss (about 20%) in comparison to controls, but still highly significant. The AD patients demonstrated a 36% loss in comparison to the controls. Using these two MRI measures, a discriminant function analysis (adjusted for age and gender) demonstrated an overall accuracy of 93%. These

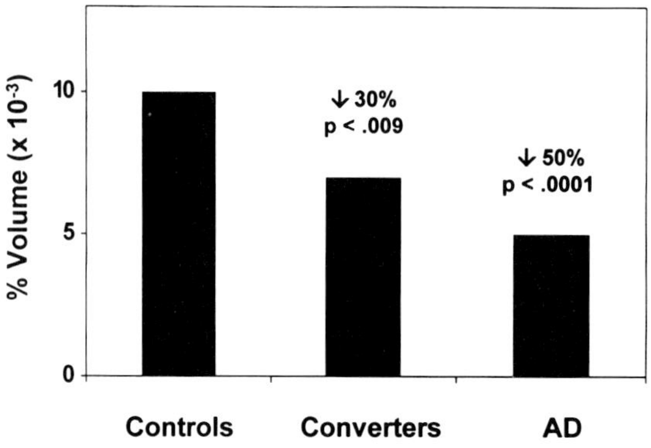


**Figure 2.** A comparison of the percentage volume of the entorhinal cortex measured on MRI scans in *controls,* in subjects who later progressed to the point where they met criteria for probable AD (*converters*), and in patients with mild AD.

findings suggest that measures of the entorhinal cortex and the banks of the superior temporal sulcus are likely to be highly useful in differentiating cases of AD in prodromal phase from controls unlikely to develop dementia.

It is particularly interesting to note that the degree of loss in the measure of the entorhinal cortex between controls and questionable subjects was comparable to that seen in neuropathological studies of this region in similar subjects. In a study reporting the neuropathological examination of 20 cases with varying degrees of dementia at the time of death, stereological cell counts in questionable cases in the entorhinal cortex demonstrated profound loss of neurons, with a 60% and 40% loss from layers II and IV, respectively, and an average loss in the entire region of 32% (Gomez-Isla *et al.,* 1996). By contrast, in severe AD patients layer II lost over 90% of its neurons, and overall loss was more than 60%. These figures compare closely with an overall volume loss of 30% in questionable cases and 50% in individuals with severe AD.

In the same cases, the banks of superior temporal gyrus showed a 10% loss that was not significant in the questionable cases, but an approximately 50% loss in established moderate and severe AD (Gomez-Isla *et al.,* 1997). The comparable figures from the current structural MRI study are 20% and 36% for volume loss based on MRI scans in questionable and established AD. These data support the hypothesis that the present MRI findings are likely to be a reflection of underlying AD neuropathology, and that the measure of entorhinal cortex will be especially useful in preclinical prediction of AD.

Recent functional neuroimaging data supports the conclusions based on neuropathological and structural neuroimaging data. In addition, the functional data suggest that a distributed brain network may be involved in early AD.

Functional imaging techniques, such as single photon emission computed tomography (SPECT) and positron-emission tomography (PET), have demonstrated regional abnormalities in brain perfusion in patients with Ad (see Jagust, 1996, for review). Temporoparietal abnormalities have emerged as the most consistent functional alteration seen in mild-to-moderately impaired AD patients. Among studies that have compared AD patients at different stages of disease, the temporal region appears to be the most sensitive and specific in distinguishing mildly impaired AD patients from controls, consistent with the structural findings described above.

As with the studies of prodromal AD patients, functional neuroimaging techniques also identify selected brain regions that are affected in the earliest stages of AD. A recently completed study using SPECT successfully identified over 80% of the questionable subjects who would later progress to the point where they met criteria for AD (Johnson *et al.*, in press).

Interestingly, the SPECT findings suggested that four brain regions were selectively involved in these early cases: hippocampus, posterior cingulate, anterior thalamus, and anterior cingulate. As described above, the hippocampal formation has been implicated for some time in the early stages of AD. The other brain regions significantly associated with the development of AD in this study have, however, received relatively less attention. Nevertheless, a review of the literature suggests that they may also be involved in the very early stages of the disease.

For example, decreased perfusion in the posterior cingulate has been reported in two recent PET studies related to the development of AD (Reiman *et al.*, 1996; Minoshima *et al.*, 1995), and cingulate activation is frequently found in activation studies of memory (Ungerleider, 1995). In addition, three of the four brain regions important in the present study for discriminating converters from controls are thought to be important for memory function in general. Recent memory studies in rodents indicate that hippocampus, posterior cingulate, and anterior thalamus comprise a memory system critical for learning relationships among cues (e.g., spatial, temporal, and so on) (Sutherland and Rudy, 1989, 1991) and neuroanatomical studies in monkeys demonstrate strong interconnections between these three regions (Pandya *et al.*, 1981; Vogt *et al.*, 1987; Van Hoesen, 1991). There is also increasing neuropathological evidence that all three sites are affected early in the course of AD (Braak and Braak, 1993). The SPECT findings thus suggest that there are alterations in a distributed brain network with multiple nodes that may underly the earliest stages of AD. It is important to note that the structural brain changes seen in early AD differ considerably from what is seen with respect to age-related change. Data in monkeys and humans indicate that age-related neuronal loss is highly selective within the hippocampal formation. For example, the subiculum shows a significant age-related loss in humans and a similar trend in monkeys; however, the CA1, CA2 and CA3 fields show no evidence of age-related neuronal loss (Amaral, 1993; Rosene, 1993; West *et al.*, 1994; see West, this volume), nor does the entorhinal cortex (Gomez-Isla, *et al.*, 1996). These data suggest that a comparison of activation patterns in the hippocampal subfields (e.g., with newer functional imaging MR techniques) may differentiate patients in the very earliest stages of Alzheimer's disease from those with age-related changes in memory.

It should also be noted that blood for genetic analysis was available on the subjects ($n = 130$) and was examined for apoE genotype because the apoE-4 allele has been identified as a risk factor for AD (Saunders *et al.*, 1993). The addition of the apoE genotype did not add to the discrimination between controls and converters in the analyses of structural MRI and SPECT described above. The negative findings regarding apoE status as a predictor of conversion to AD may derive from the fact that we have very few subjects who are homozygous for apoE-4. Recent reports suggest that the maximum effect of apoE status is seen among individuals who are homozygous for apoE-4. Moreover, given the brief follow-up period for the cohort (i.e., approximately two years), we would expect any apoE-4 effect to be revealed by a decreased age of onset. Thus, among a group that has few individuals with an apoE-4/4 genotype, there would be a low likelihood of finding a relationship between apoE status and the development of AD over a brief follow-up period.

## 5. Changes in Brain Structure in Early Alzheimer's Disease Related to Executive Function Deficits

The development of cognitive symptoms in addition to memory deficits in AD patients (e.g., cognitive inflexibility, language dysfunction, and so on) is likely due to the progression of neuropathological change to cortical regions known to underly these functions or to a dysfunction of subcortical regions that project to cortical areas.

With respect to the executive function changes seen in the early stages of AD, two explanations appear likely. The first pertains to pathological changes in subcortical structures, such as the basal forebrain, that modulate cortical function, (Whitehouse *et al.*, 1982). This is consistent with findings that the basal forebrain receives afferent projections from numerous subcortical structures and projects to numerous cortical and subcortical regions (Mesulam *et al.*, 1983; Arendt *et al.*, 1985) and can therefore serve as a source of integrated information to the cortex. The second likely source for early AD executive function deficits pertains to the loss of neocortical synapses (Pearson *et al.*, 1985; Hamos *et al.*, 1989; DeKosky and Scheff, 1990; Terry *et al.*, 1991; Masliah *et al.*, 1993) and long corticocortical projection systems (Lewis *et al.*, 1987; Morrison *et al.*, 1986; Scheff *et al.*, 1993) seen in AD. The partial degeneration of an intracortical projection system early in the course of disease could produce difficulties in tasks that require the rapid and simultaneous integration of multiple types of information. It is therefore of interest to note recent reports indicating that decreased synaptic density correlates more highly with measures of cognitive function in AD patients than do other neuropathological markers such as plaques and tangles (DeKosky and Scheff, 1990; Terry *et al.*, 1991). This is also consistent with the hypothesis that abnormalities in executive function can result from damage that is not anatomically based in the frontal lobes (Stuss and Gow, 1992). It appears unlikely that executive function abnormalities in early AD can be attributed to dysfunction of the frontal lobes per se, as there is little neuropathological evidence to suggest that the frontal lobes have extensive pathology early in the course of the disease (Brun and Gustafson, 1976). For example, PET scan data generally demonstrate that frontal declines in glucose metabolism are a late phenomenon in Alzheimer's disease (Chase *et al.*, 1984; Foster *et al.*, 1984; Cutler *et al.*, 1985; Duara *et al.*, 1986; Haxby *et al.*, 1988).

## 6. Summary

In the early stage of Alzheimer's disease, most patients demonstrate a dramatic decline in the ability to recall information over brief delay intervals, and show impairments in tasks that require them to shift a predominant set. There is considerable evidence to indicate that the memory difficulty results from neuronal loss (and the deposition of neuritic plaques and tangles) in the entorhinal cortex and in adjacent regions of the medial temporal lobe. Mild AD patients also have distinct executive function deficits, though the neuropathology that underlies these deficits is less clear. Understanding these brain–behavior relationships will shed light on the disease and may lead to improved methods for early diagnosis.

## 7. References

Alivisatos, B., and Milner, B., 1989, Effects of frontal or temporal lobectomy on the use of advance information in a choice reaction time task, *Neuropsychologia* **27:**495–503.

Amaral, D., 1993, Morphological analyses of the brain of behaviorally characterized aged nonhuman primates, *Neurobiol. Aging* **14:**671–672.

Arendt, T., Bigl, V., Tennsted, A., and Arendt, A., 1985, Neuronal loss in different parts of the nucleus basalis is related to neuritic plaque formation in cortical target areas in Alzheimer's disease, *Neuroscience* **14:**1–14.

Baddeley, A., Logie, R., Bressi, S., Della Sala, S., and Spinnler, H., 1986, Dementia and working memory, *Q. J. Exp. Psychol.* **38:**603–618.

Ball, M. H., 1997, Neuronal loss, neurofibrillay tangles, and granulovacuolar degeneration in the hippocampus with aging and dementia, *Acta Neuropathol.* **37:**11–118.

Becker, J. T., 1998, Working memory and secondary memory deficits in Alzheimer's disease, *J. Clin. Exp. Neuropsychol.* **10:**739–753.

Benton, A. L., and Hamsher, K., 1976, *Multilingual Aphasia Examination,* University of Iowa Press.

Braak, H., and Braak, E., 1993, Alzheimer neuropathology and limbic circuits, in: *Neurobiology of Cingulate Cortex and Limbic Thalamus: A Comprehensive Handbook,* Birkhauser, Boston, pp. 606–626.

Brun, A., and Gustafson, L., 1976, Distribution of cerebral degeneration in Alzheimer's disease: A clinico-pathological study, *Arch. Psychiat. Nervenk.* **223:**15–33.

Butters, N., Salmon, D., Heindel, W., and Granholm, E., 1988, Episodic, semantic and procedural memory: Some comparisons of Alzheimer and Huntington disease patients, in: *Aging and the Brain* (R. D. Terry, ed.), Raven, New York, pp. 63–87.

Chase, T. N., Foster, N. L., Fedio, P., Brooks, R., Mansi, L., and DiChiro, G., 1984, Regional cortical dysfunction in Alzheimer's disease as determined by positron emission tomography, *Ann. Neurol.* **15:**S170–S174.

Convit, A., *et al.,* 1993, Hippocampal atrophy in early Alzheimer's disease: Anatomic specificity and validation, *Psychiatry Q.* **64:**371–387.

Craik, F. I. M., 1977, Age differences in human memory, in: *Handbook of the Psychology of Aging* (J. E. Birren and K. W. Schaie, eds.) Van Nostrand Reinhold, New York.

Cutler, N., *et al.,* 1985, Clinical history, brain metabolism, and neuropsychological function in Alzheimer's disease, *Ann. Neurol.* **18:**298–309.

DeKosky, S. T., and Scheff, S. W., 1990, Synapse loss in frontal cortex biopsies in Alzheimer's disease: Correlation with cognitive severity, *Ann. Neurol.* **27:**457–464.

Duara, R., *et al.,* 1986, Positron emission tomography in Alzheimer disease, *Neurology* **36:**879–887.

Foster, N. L., *et al.,* 1984, Cortical abnormalities in Alzheimer's disease, *Ann. Neurol.* **16:**649–654.

George, A. E., *et al.,* 1990, CT diagnostic features of Alzheimer disease: Importance of the choroidal/hippocampal fissure complex, *Am. J. Neuroradiol.* **11:**101–7.

Gomez-Isla, T., Price, J., McKeel, D., Morris, J., Growdon, J., and Hyman, B., 1996, Profound loss of layer II entorhinal cortex neurons occurs in very mild Alzheimer's disease, *J. Neurosci.* **16:**4491–4500.

Gomez-Isla, T., *et al.,* 1997, Neuronal loss correlates with but exceeds neurofibrillary tangles in Alzheimer's disease, *Ann. Neurol.* **41:**17–24.

Grady, C. L., *et al.,* 1988, Longitudinal study of the early neuropsychological and cerebral metabolic changes in dementia of the Alzheimer type, *J. Clin. Exp. Neuropsychol.* **10:**576–596.

Hamos, J. E., De Gennaro, L. J., Drachman, D. A., 1989, Synaptic loss in Alzheimer's disease and other dementias, *Neurology* **39:**355–361.

Hart, R. P., Kwentus, J. A., Harkins, S. W., and Taylor, J. R., 1988, Rate of forgetting in mild Alzheimer's type dementia, *Brain Cogn.* **7**:31–38.
Haxby, J. V., *et al.*, 1988, Heterogenous anterior–posterior metabolic patterns in dementia of the Alzheimer type, *Neurology* **38**:1853–1863.
Hyman, B. T., van Hoesen, G. W., Kromer, C., Damasio, A. R., 1985, Alzheimer's disease: Cell specific pathology isolates the hippocampal formation, *Science* **225**:1168–1170.
Hyman, B., van Hoesen, G., Kromer, L., and Damasio, A., 1986, *Ann. Neurol.* **20**:472–481.
Ikeda, M., *et al.*, 1994, MRI-based quantitative assessment of the hippocampal region in very mild to moderate Alzheimer's disease, *Neuroradiology* **36**:7–10.
Jack, C., Petersen, R., O'Brien, P., and Tangalos, E., 1992, MR-based hippocampal volumetry in the diagnosis of Alzheimer's disease, *Neurology* **42**:183–188.
Jagust, W., 1996, Functional imaging patterns in Alzheimer's disease, *Ann. N.Y. Acad. Sci.* **777**:30–36.
Johnson, K., Jones, K., Holman, B. L., Becker, J. A., Spiers, P., Satlin, A., and Albert, M. S., Preclinical prediction of Alzheimer's disease using SPECT, *Neurology*, in press.
Kaplan, E., Goodglass, H., and Weintraub, S., 1983, *Boston Naming Test*, Lea and Febiger, Philadelphia, PA.
Kesslak, J., Nalcioglu, O., and Cotman, C., 1991, Quantification of magnetic resonance scans for hippocampal and parahippocampal atrophy in Alzheimer's disease, *Neurology* **41**:51–54.
Killiany, R. J., Moss, M. B., Albert, M. S., Sandor, T., Tieman, J., and Jolesz, F., 1993, Temporal lobe regions on magnetic resonance imaging identify patients with early Alzheimer's disease, *Arch. Neurol.* **50**:949–954.
Killiany, R. J., Gomez-Isla, T., Moss, M., Kikinis, R., Jolesz, F., Sandor, T., Hyman, B. T., and Albert, M. S., Preclinical prediction of Alzheimer's disease using structural MRI measures of the entorhinal cortex, submitted.
Laakso, M., *et al.*, 1995, The interuncal distance in Alzheimer disease and age-associated memory impairment, *Am. J. Neuroradiol.* **16**:727–734.
Lafleche, G. C., Stuss, D. T., Nelson, R. F., and Picton, T. W., 1990, Memory scanning and structured learning in Alzheimer's disease and Parkinson's disease, *Can. J. Aging* **9**:120–134.
Lafleche, G., and Albert, M., 1995, Executive function deficits in mild Alzheimer's disease, *Neuropsychology* **9**:313–320.
Lehericy, S., et al., 1994, Amygdalohippocampal MR volume measurements in the early stages of Alzheimer's disease, *Am. J. Neuroradiol.* **15**:929–937.
Lewis, D. A., Campbell, M. J., Terry, R. D., and Morrison, J. H., 1987, Laminar and regional distributions of neurofibrillary tangles and neuritic plaques in Alzheimer's disease: A quantitative study of visual and auditory cortices, *J. Neurosci.* **7**:1799–1808.
Masliah, E., Mallory, B. S., Hanson, L., DeTeresa, R., and Terry, R. D., 1993, Quantitative synaptic alterations in the human cortex during normal aging, *Neurology* **43**:192–197.
Mattis, S., 1976, Mental status examination for organic mental syndrome in the elderly patient, in: *Geriatric Psychiatry* (L. Bellak and T. B. Karasu, eds.), Grune and Stratton, New York, pp. 71–121.
Mesulam, M. M., Mufson, E. J., Levey, A. J., and Wainer, B. H., 1983, Cholinergic innervation of cortex by the basal forebrain: Cytochemistry and cortical connections of the septal area, diagonal band nuclei, nucleus basalis (substantia innominata), and hypothalamus in the rhesus monkey, *J. Comp. Neurol.* **214**:170–197.
Milberg, W., and Albert, M., 1989, Cognitive differences between patients with PSP and Alzheimer's Disease, *J. Clin. Exp. Neuropsychol.* **11**:605–614.
Minoshima, S., Frey, K., Koeppe, R., Foster, N., and Kuhl, D. A., 1995, Diagnostic approach to Alzheimer's disease using three-dimensional stereotactic surface projections of Fluorine-18-FDG PET, *J. Nucl. Med.* **36**:1238–1248.
Morris, R. G., and Baddeley, A. D., 1988, Primary and working memory functioning in Alzheimer-type dementia, *J. Clin. Exp. Neurol.* **10**:276–279.
Morris, J. C., *et al.*, 1989, The consortium to establish a registry for Alzheimer's disease (CERAD). I. Clinical and neuropsychological assessment of Alzheimer's disease, *Neurology* **39**:1159–1165.
Morrison, J. H., *et al.*, 1986, The laminar and regional distribution of neocortical somatostatin and neuritic plaques: Implications for Alzheimer's disease as a global neocortical disconnection syndrome, in: *Biological Substrates of Alzheimer's disease* (A. Scheibel and E. Weschler, eds.) Academic, New York, pp. 115–131.

Moss, M. B., and Albert, M. S., 1988, Alzheimer's disease and other dementing disorders, in: *Geriatric Neuropsychology* (M. S. Albert and M. B. Moss, eds.), Guilford, New York, pp. 145–178.

Moss, M. B., Albert, M. S., Butters, N., and Payne, M., 1986, Differential patterns of memory loss among patients with Alzheimer's disease, Huntington's disease, and alcoholic Korsakoff's syndrome, *Arch. Neurol.* **43:**239–246.

Nestor, P. G., Parasuraman, R., and Haxby, J. V., 1991, Speed of information processing and attention in early Alzheimer's dementia, *Dev. Neuropsychol.* **7:**243–236.

Pandya, D., van Hoesen, G., and Mesulam, M.-M., 1981, Efferent connections of the cingulate gyrus in the rhesus monkey, *Exp. Brain Res.* **42:**319–330.

Pearson, R. C., Esiri, M. M., Hiorns, R. W., Wilcock, G. K., and Powell, T. P., 1985, Anatomical correlates of the distribution of the pathological changes in the neocortex in Alzheimer's disease, *Proc. Natl. Acad. Sci. USA* **82:**4531–4534.

Peters, A., Leahu, D., Moss, M., and McNally, K., 1994, The effects of aging on area 46 of the frontal cortex of the rhesus monkey, *Cereb. Cortex* **6:**621–635.

Peterson, R., Smith, G., Kokmen, E., Ivnik, R., and Tangalos, E., 1992, Memory function in normal aging, *Neurology* **42:**396–401.

Petrides, M., and Milner, B., 1982, Deficits in subject-ordered tasks after frontal and temporal lobe lesions in man, *Neuropsychologia* **20:**249–262.

Reiman, E. M., *et al.*, 1996, Preclinical evidence of a genetic risk factor for Alzheimer's disease in apoliprotein E type 4 homozygotes using positron emission tomography, *New Engl. J. Med.* **334:**752–758.

Reitan, R. M., 1958, Validity of the Trail Making Test as an indicator of organic brain damage, *Percept. Motor Skills* **8:**271–276.

Rosene, D., 1993, Comparing age-related changes in the basal forebrain and hippocampus of the rhesus monkey, *Neurobiol. Aging* **14:**669–670.

Sahakian, B., Downes, J., Eagger, S., Evenden, J., Levy, R., Philpot, M., Roberts, A., and Robbins, T., 1990, Sparing of attentional relative to mnemonic function in a subgroup of patients with dementia of the Alzheimer type, *Neuropsychologia* **28:**1197–1213.

Sandor, T., Albert, M., Stafford, J., and Harpley, S., 1988, Use of computerized CT analysis to discriminate between Alzheimer patients and normal control subjects, *Am. J. Neuroradiol.* **9:**1181–1187.

Sandor, T., Jolesz, F., Tieman, J., Kikinis, R., Jones, K., and Albert, M., 1992, Comparative analysis of CT and MRI scans in Alzheimer patients and controls, *Arch. Neurol.* **49:**381–384.

Saunders, A. M., *et al.*, 1993, Association of apolipoprotein E allele E4 with late-onset familial and sporadic Alzheimer's disease, *Neurology* **43:**1467–1472.

Scheff, S. W., and Price, D. A., 1993, Synapse loss in the temporal lobe in Alzheimer's disease, *Ann. Neurol.* **33:**190–199.

Seab, J. P., Jagust, W. J., Wong, S. T., Roos, M. S., Reed, B. R., and Budinger, T. F., 1988, Quantitative NMR measurements of hippocampal atrophy in Alzheimer's disease, *Magn. Reson. Med.* **8:**200–208.

Storandt, M., and Hill, R. D., 1989, Very mild senile dementia of the Alzheimer type. II. Psychometric test performance, *Arch. Neurol.* **46:**383–386.

Stuss, D. T., and Gow, C. A., 1992, Frontal dysfunction after traumatic brain injury, *Neuropsychiatry, Neuropsychol. Behav. Neurol.* **5:**272–282.

Sutherland, R., and Rudy, J., 1989, Configurational association theory: The role of the hippocampal formation in learning, memory, and amnesia, *Psychobiology* **17:**129–144.

Sutherland, R., and Rudy, J., 1991, Exceptions to the rule of space, *Hippocampus* **1:**250–252.

Terry, R. D., *et al.*, 1991, Physical basis of cognitive alterations in Alzheimer's disease: Synapse loss is the major correlate of cognitive impairment, *Ann. Neurol.* **30:**572–580.

Ungerleider, L., 1995, Functional brain imaging studies of cortical mechanisms for memory, *Science* **270:**769–775.

Van Hoesen, G., 1991, Hippocampal cortical and subcortical neural systems: Their origins and targets in the monkey, in: *Learning and Memory: Discussions in Neuroscience*, Vol. 6 (L. Squire, M. Mishkin, and S. Shimamura, eds.) Elsevier, Amsterdam, pp. 20–28.

Vogt, B., Pandya, D., and Rosene, D., 1987, Cingulate cortex of the rhesus monkey: Cytoarchitecture and thalamic afferents, *J. Comp. Neurol.* **262:**256–270.

Wechsler, D., 1981, *The Wechsler Adult Intelligence Scale—Revised*, Psychological Corporation, New York.

Welsh, K., Butters, N., Hughes, J., Mohs, R., and Heyman, A., 1992, Detection and staging in Alzheimer's disease: Use of the neuro-psychological measures developed for the Consortium to Establish a Registry for Alzheimer's Disease, *Arch. Neurol.* **49:**448–452.

West, M., Coleman, P., Flood, D., and Troncoso, J., 1994, Differences in the pattern of hippocampal neuronal loss in normal aging and Alzheimer's disease, *Lancet* **344:**769–772.

Whitehouse, P. J., Price, D. L., Struble, R. G., Clark, A. C., Coyle, J. T., and DeLong, M. R., 1982, Alzheimer's disease and senile dementia: Loss of neurons in the basal forebrain, *Science* **215:**1237–1239.

Wickelgren, W., 1975, Age and storage dynamics in continuous recognition memory, *Dev. Psychol.* **11:** 165–169.

Wilson, R., Bacon, L., Fox, P., and Kaszniak, A., 1983, Primary memory and secondary memory in dementia of the Alzheimer type, *J. Clin. Neuropsychol.* **5:**337–344.

Zola-Morgan, S., Squire, L., and Ramos, S., 1994, Severity of memory impairment in monkeys as a function of locus, and extent of damage within the medial temporal lobe memory system, *Hippocampus* **4:**483–495.

# 14

# Temporal Sequence of Alzheimer's Disease-Related Pathology

HEIKO BRAAK and EVA BRAAK

## 1. Introduction

Neuropathologic hallmarks of Alzheimer's disease (AD) include the progressive deposition of virtually insoluble proteinaceous material in both extracellular and intraneuronal locations. The extracellular deposits consist mainly of Aβ-amyloid-protein (Joachim and Selkoe, 1989; Beyreuther and Masters, 1991; Selkoe, 1991, 1993, 1994); abnormally phosphorylated tau protein (PHF-tau, paired helical filament tau) dominates among the intraneuronal changes (Goedert *et al.*, 1991, 1992; Iqbal and Grundke-Iqbal, 1991; Schmidt *et al.*, 1991; Dickson *et al.*, 1992; Goedert, 1993; Iqbal *et al.*, 1993, 1994; Price and Sisodia, 1994; Trojanowski *et al.*, 1995). A variety of other substances accompany both the Aβ-amyloid deposits and the abnormal tau protein.

Clinically, the insidious onset of AD is characterized by a subtle decline of memory functions in a state of clear consciousness. As time passes, emotional disturbances, personality changes, deterioration of language functions, impairment in visuospatial tasks, and eventually dysfunction of the motor system are gradually added to the initial symptoms. The speed of intellectual decay and the duration of

HEIKO BRAAK and EVA BRAAK • Department of Anatomy, J. W. Goethe University, D-60590 Frankfurt, Germany.

*Cerebral Cortex,* Volume 14, edited by Peters and Morrison. Kluwer Academic/Plenum Publishers, New York, 1999.

the disease vary (Reisberg, 1988; Reisberg *et al.*, 1992; Franssen *et al.*, 1993; Kazee *et al.*, 1993; Berg and Morris, 1994; Corey-Bloom *et al.*, 1994).

Progression of the clinical symptoms reflects a similarly gradual development of the disease-related brain lesions. The pattern of destruction evolves more or less symmetrically (Grünthal, 1930; Moossy *et al.*, 1988; Arnold *et al.*, 1991). The process starts in specific predilection sites, then spreads in a predictable sequence into other brain areas (Kemper, 1978; Hyman *et al.*, 1984, 1990; van Hoesen and Hyman, 1990; Braak and Braak, 1991, 1994, 1995; Hyman and Gomez-Isla, 1994). Many neuronal types, cortical areas, and subcortical nuclei are damaged; others remain unscathed. Changes in the neuronal cytoskeleton are tolerated for a long time, but eventually the affected neurons die. In the initial stages of the disease, there is no significant loss of nerve cells, glial cells, or other cellular components of the brain. Conspicuous cell loss is not a predominant feature even in fully developed AD (Regeur *et al.*, 1994). Whether mild neuronal loss is insignificant or deleterious to brain function depends on the distribution of the lesions. A modest nonrandom, symmetrical loss of nerve cells in both hemispheres probably impairs function, whereas a numerically comparable but random destruction is likely to escape recognition. Therefore, the nonrandom character is an important feature of AD-related brain lesions.

## 2. Anatomical Considerations

Alzheimer's disease is mainly a disorder of the cerebral cortex, the chief controlling entity of the brain. Most of the brain functions that distinguish man from other mammals depend on the undisturbed maturation and structural integrity of the cerebral cortex. In the course of AD a limited number of subcortical nuclei gradually become involved, most of which are closely related to the cerebral cortex. The complex and characteristic lesional pattern that slowly develops is much more easily understood and interpreted if the major connections of the structures involved are briefly reviewed (depicted in simplified diagrams, Figs. 1–3).

The telencephalic cortex consists of two basic types of gray matter—neocortex and allocortex. The more or less uniformly built neocortex predominates, making up about 95% of the total surface area. Apart from a few regional variations, it shows a six-layered organization. The heterogeneously composed allocortex is small in comparison. Allocortical areas range from simply organized trilaminate fields to highly differentiated fields with more than six layers (Brodmann, 1910; Vogt and Vogt, 1919; Rose, 1935; Braak, 1980, 1984; Rosene and van Hoesen, 1987; Duvernoy, 1988; Amaral and Insausti, 1990; Zilles, 1990; Braak *et al.*, 1996b). The mammalian brain typically contains two allocortical cores, the olfactory bulb and the hippocampus. A stepwise increase in structural differentiation is observed if lines are traced from each of these cores through neighboring areas into the refined neocortex (Sanides, 1962, 1969; Pandya and Yeterian, 1985, 1990). In the human brain, the olfactory bulb and related structures are markedly reduced in comparison to corresponding areas of macrosmatic mammals. The hippocampal formation and regions related to it, however, are well developed (Stephan, 1975). Transitional zones exist between the mature neocortical and allocortical territories. A belt of

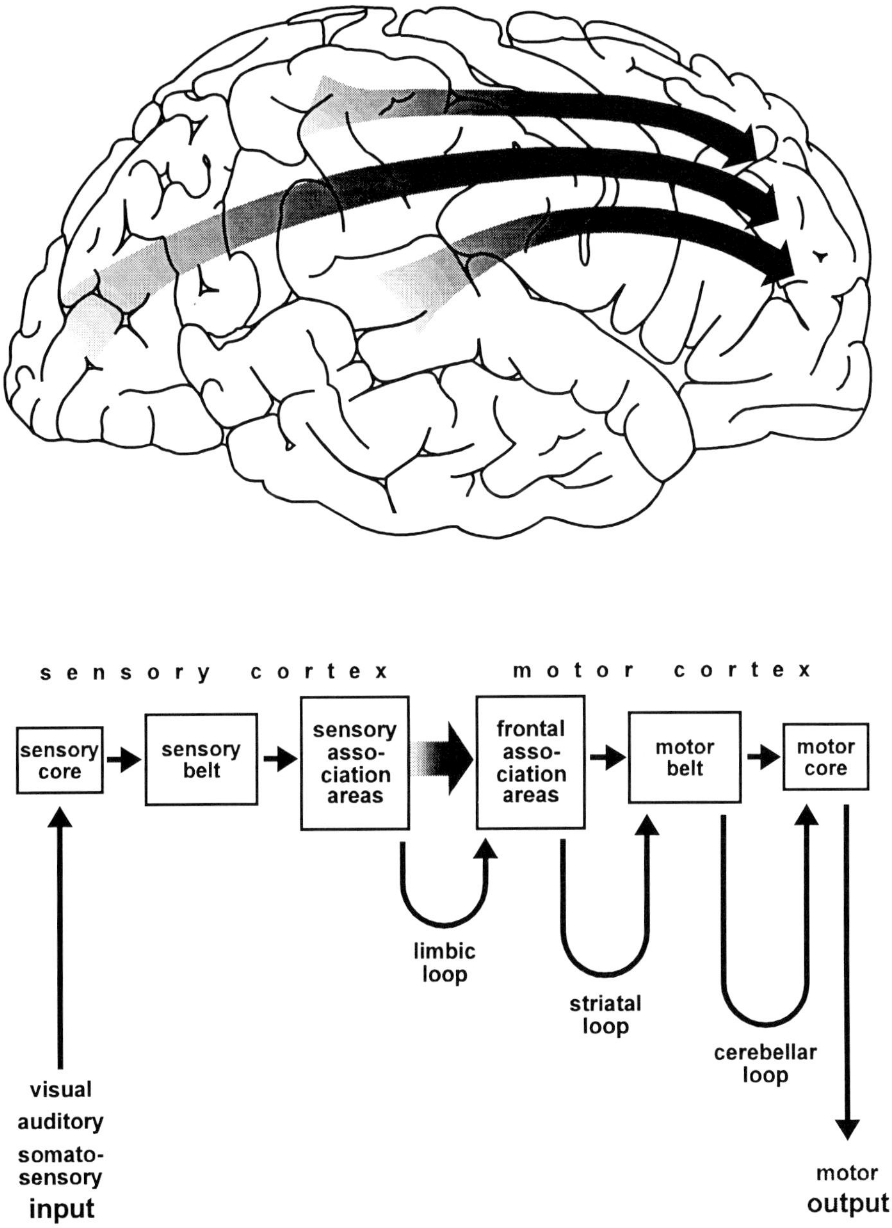


**Figure 1.** Somatosensory, visual, and auditory information proceeds through the respective core and belt fields of the neocortex to a variety of association areas and is then transported via long corticocortical pathways to the prefrontal association areas. Tracts generated from this highest organizational level of the brain guide the data back via the frontal belt areas to the frontal core, the primary motor area; striatal and cerebellar loops provide the major routes for this transport from prefrontal cortex to primary motor field. From Braak *et al.* (1996b) with permission.

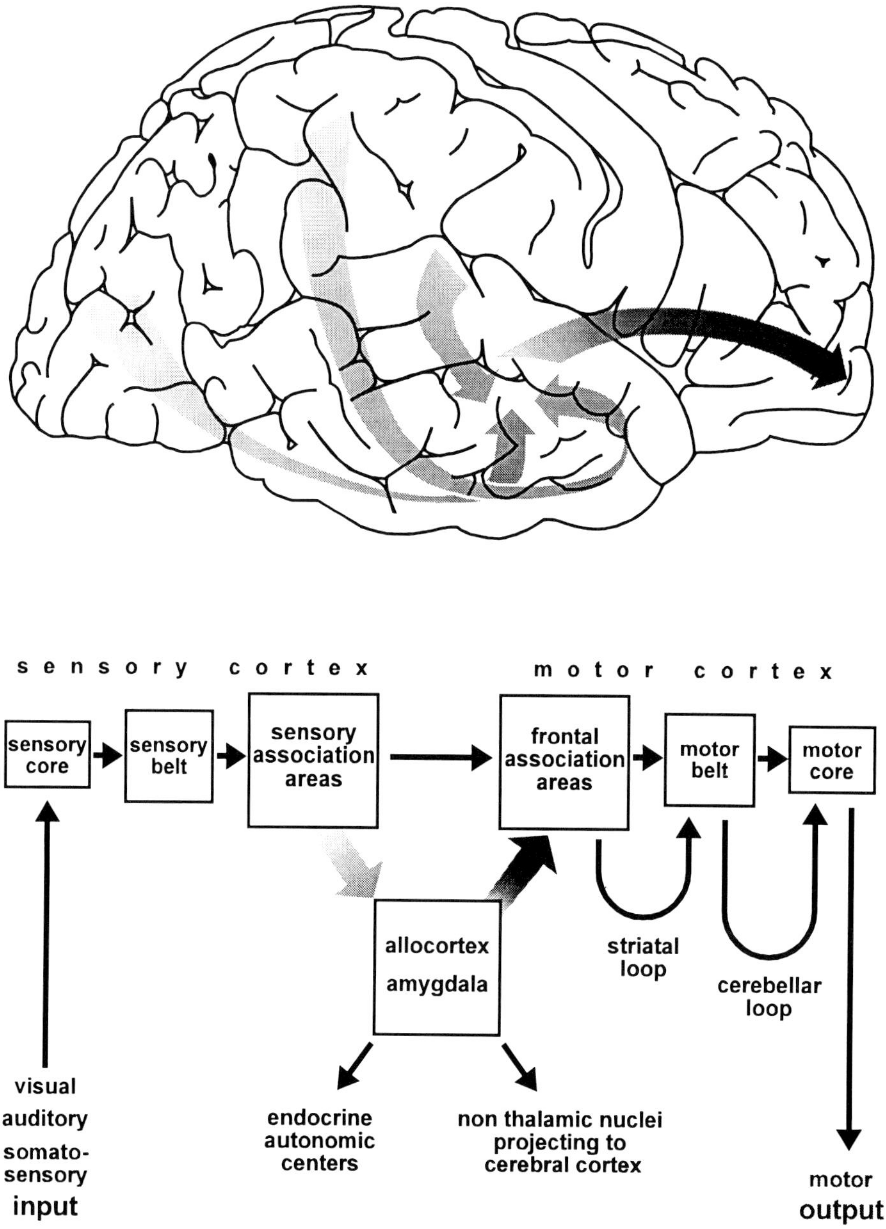


**Figure 2.** Part of the data stream from the sensory association areas to the prefrontal cortex branches off, eventually converging upon the entorhinal region and the amygdala, establishing the afferent leg of the limbic loop. In human brain the stream of neocortical information thus provides the most important input to the limbic system. Projections from the entorhinal region, the amygdala, and the hippocampal formation contribute to the efferent leg of the limbic loop, which reaches the prefrontal cortex. The limbic system thus exerts an important influence on the prefrontal cortex. From Braak *et al.* (1996b) with permission.

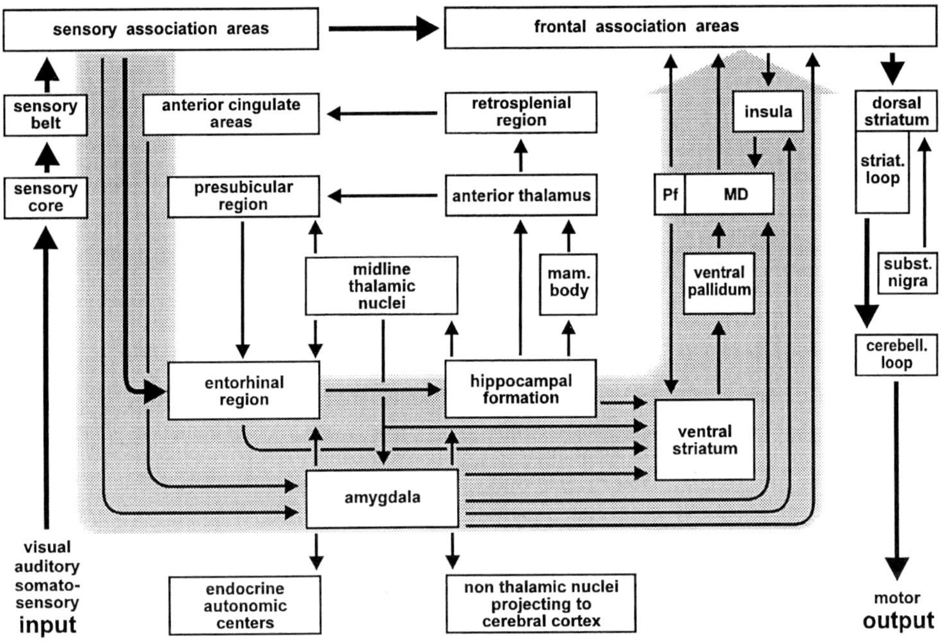


**Figure 3.** Limbic loop in greater detail. The gray arrow emphasizes the loop's strategic position between neocortical sensory association areas and prefrontal cortex. The hippocampal formation, entorhinal region, and amygdala are densely interconnected; together the three represent the highest organizational level of the limbic system. (cerebell.loop, cerebellar loop; mam.body, mamillary body; MD, mediodorsal thalamic nucleus, Pf, parafascicular nucleus, striat. loop, striatal loop, subst. nigra, substantia nigra). From Braak *et al.* (1996b) with permission.

periallocortical areas accompanies the allocortex proper; pro-neocortical fields are allied to the mature neocortex. Together, these two transition zones are occasionally referred to as the *mesocortex*. Closely related to the allocortex is the subcortical nuclear complex of the amygdala (Amaral *et al.*, 1992).

The parietal, occipital, and temporal neocortices are each comprised of a core field (primary somatosensory, visual, and auditory areas), a belt region of parasensory areas, and an extensive territory of related higher-order processing ("association") areas outside the core and the belt. In the normal human brain, the primary sensory areas give rise to short feedforward projections into layer IV of the belt areas, which in turn project to adjoining association areas (Rockland and Pandya, 1979). In this manner the data progresses sequentially from the core through the parasensory belt to a variety of association areas, from which the data are conveyed via long corticocortical pathways to the frontal association areas (prefrontal cortex). The remarkable extensiveness of the prefrontal cortex is a hallmark of the human brain; tracts generated in this highest organizational level guide the data into the frontal core field. Feedback projections terminating in layer I provide a minor pathway; the striatal and cerebellar loops represent the major routes for this data transfer (Fig. 2). Portions of the basal ganglia, many nuclei of the lower brain stem, and the cerebellum thus participate in the regulation of cortical output (Alexander *et al.*, 1990; Alheid *et al.*, 1990; Braak *et al.*, 1996a,b).

In addition to this system of short and long projections from the neocortical sensory association areas to the prefrontal cortex, there is an orderly sequence of corticocortical connections that eventually converges upon the entorhinal region (port of entry: transentorhinal region) and the amygdala (port of entry: lateral nucleus) (Felleman and van Essen, 1991; Amaral *et al.*, 1992). These connections establish the afferent leg of the limbic loop and are a comparatively recent development phylogenetically. They can be considered to result from the marked, relative enlargement of the neocortex that occurred late in the development of higher primates. Many allocortical structures that process neocortical data are highly refined, and are also a late development phylogenetically and ontogenetically.

The transentorhinal and entorhinal regions extend over both the ambient gyrus and the anterior portions of the parahippocampal gyrus (Ikeda *et al.*, 1989; Witter *et al.*, 1989; Witter, 1993; Heinsen *et al.*, 1994; Insausti *et al.*, 1995; Solodkin and van Hoesen, 1996). Small wartlike elevations with shallow grooves on their free surfaces indicate the position of the entorhinal region and make its approximate borders visible to the unaided eye (Klingler, 1948). No cellular layer of the entorhinal cortex corresponds to any layer of the neocortex, so to avoid confusion, the terminology of Rose (1935) is used throughout this text (Fig. 4a,d,e). The periallocortical transentorhinal region is located along the lateral circumference of the entorhinal cortex. This region is mostly hidden in the depths of the rhinal sulcus (Fig. 4b,c), mediating between entorhinal allocortex on the one hand and perirhinal pro-neocortex of the temporal lobe on the other. It is most extensive in the human brain and decreases markedly in size going down the primate scale (Braak and Braak, 1992).

Projections from the hippocampal formation, the entorhinal region, and the amygdala contribute to the efferent leg of the limbic loop; this loop exerts an important influence upon the prefrontal cortex. Some of the hippocampal, entorhinal, and amygdalar efferents terminate in the ventral striatum (i.e., the limbic subdivision of the putamen and the accumbens nucleus) (Nauta, 1979; Heimer *et al.*, 1982, 1991; Alexander *et al.*, 1990; Alheid *et al.*, 1990; Pandya and Yeterian, 1990). This input is supplemented by projections originating in the midline nuclei of the thalamus. The data are then transferred via the ventral pallidum and the magnocellular portion of the mediodorsal thalamic nucleus to the prefrontal cortex, particularly its orbitomedial portions. The amygdala generates supplementary projections to the magnocellular portion of the mediodorsal thalamic nucleus and the orbitomedial prefrontal cortex (Amaral *et al.*, 1992).

The predominance of input from and output to the neocortex is thus an essential feature of the limbic loop in the human brain. In lower mammals (i.e., small rodents), the major input to entorhinal cortex and amygdala is provided by the olfactory bulb and related areas. It is thus questionable whether experimental investigations in small rodents lead to significant understanding of AD-related brain destruction.

With phylogenetic advance and upon reaching the primate brain, the olfactory input loses considerably in importance. Developing an upright gait had the far-reaching consequence that the upper limbs—freed from their usual tasks—turned into flexible, nonspecialized organs that greatly fostered further development. In a similar way, the reduction of olfactory abilities freed the entorhinal cortex and thus rendered development of new components possible. The human entorhinal region now serves predominantly as a link between the neocortex and the limbic system.

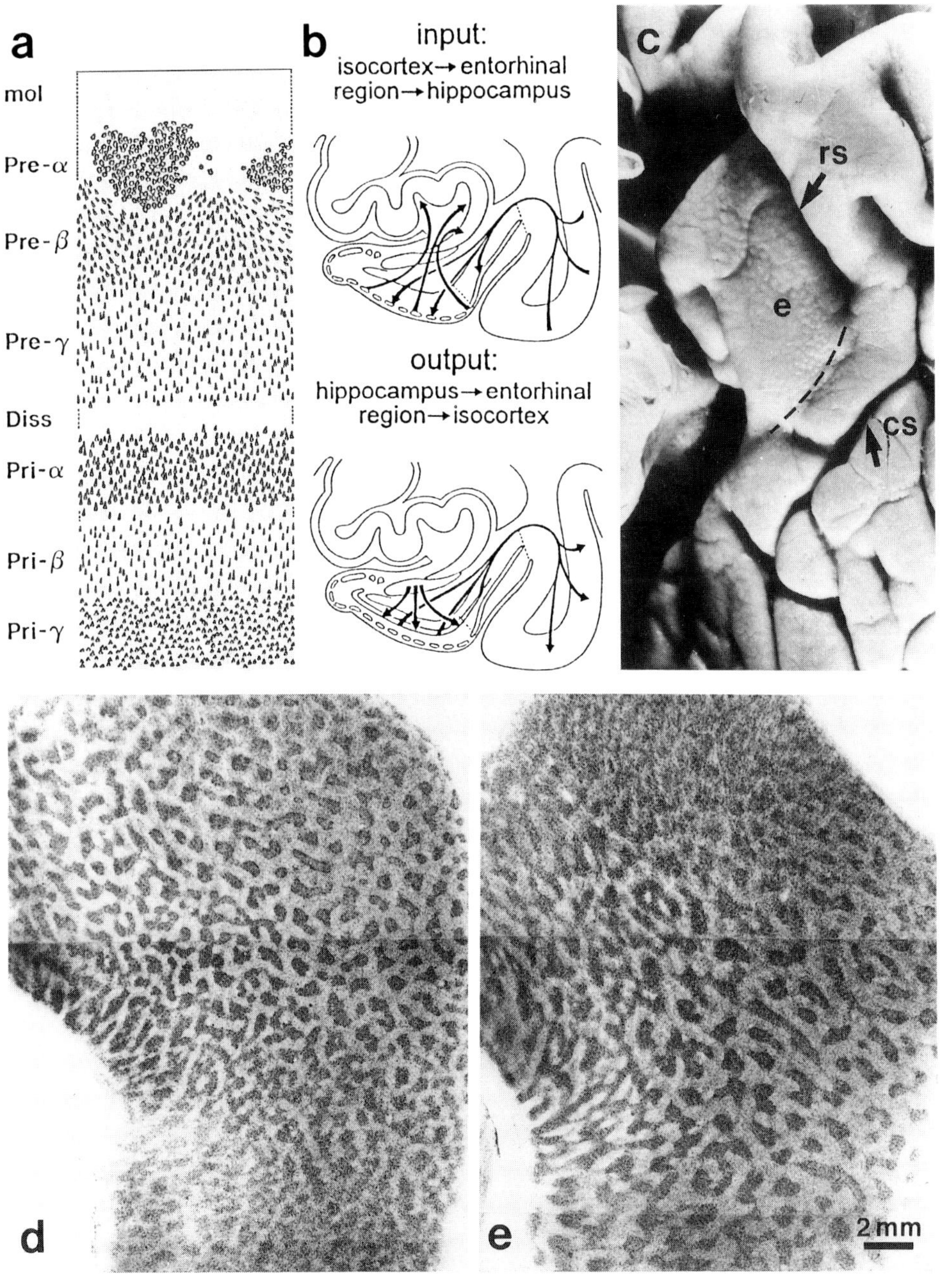


**Figure 4.** (a) Schematic drawing of the lamination pattern of the entorhinal cortex (allocortex). (mol, molecular layer; pre-α, pre-β, pre-γ: layers of the outer main stratum; Diss, lamina dissecans; pri-α, pri-β, pri-γ: layers of the inner main stratum). Note the cellular islands in layer pre-α. (b) Diagram of major entorhinal connections. Layer pre-α receives chiefly neocortical data along with additional input from limbic circuits, then sends it to the hippocampus. A dense feedback projection from the hippocampus terminates in deep-layer pri-α, from which the data are transferred back to the neocortex. The entorhinal region thus serves predominantly as an interface between neocortex and hippocampus. (c) The entorhinal region, *e*, spreads over anteromedial portions of the temporal lobe. The surface exhibits small wartlike elevations. (cs, collateral sulcus; rs, rhinal sulcus) (*a, b, c:* From Braak and Braak, 1992, with permission.) (d, e) Tangential sections through layer pre-α of two individuals. The total area occupied by the entorhinal region and the size and shape of individual cell islands differ among individuals. From Braak and Braak (1995) with permission.

The thalamic midline nuclei (medial portions of the central medial nucleus, the reuniens nucleus, and the thalamic paraventricular nucleus) receive subicular projections and are connected bidirectionally with the anterior cingulate cortex and the basolateral amygdala. Further input derives from the magnocellular nuclei of the basal forebrain, the hypothalamic tuberomamillary nucleus, and the dopaminergic nuclei of the ventral tegmentum. Their output heads into the CA1 sector, the subiculum, the presubicular and entorhinal regions, and the amygdala, establishing a short thalamo-allocortical circuit ("reuniens circuit," Fig. 5A) (Jones, 1985; Braak *et al.,* 1996b). In addition, the midline nuclei generate projections to the ventral striatum. Other limbic circuits originate in the mamillary nuclei and the anterior thalamus, both of which receive substantial input from the subiculum through the fornix. A portion of the anterior thalamic nuclei generate feedback projections to the subiculum and the entorhinal cortex. Two circuits emerge from the anterior thalamus. The first is initiated by short projections via the presubicular region to the entorhinal cortex (Fig. 5b). The second begins with projections to the retrosplenial region (Fig. 5c). This crescent-shaped portion of the cingulate pro-neocortex, located behind the splenium of the corpus callosum, receives an abundance of visual data and is bidirectionally connected to the anterogenual region and adjoining cingulate areas. The anterior cingulate pro-neocortex, in turn, receives afferents from and projects to both the thalamic midline nuclei and the basolateral amygdala (Vogt, 1985; Braak and Braak, 1993).

Finally, it should be noted that the amygdala not only markedly contributes to the efferent leg of the limbic loop but also sends projections to all nonthalamic nuclei, which diffusely project upon the cerebral cortex and other components of the central nervous system, that is, the cholinergic nuclei of the basal forebrain, the histaminergic tuberomamillary nucleus of the hypothalamus, the dopaminergic nuclei of the ventral tegmentum, the serotonergic anterior raphe nuclei, and the noradrenergic locus coeruleus (Saper, 1990; Mesulam *et al.,* 1992). In addition, the amygdala is the chief organ involved in processing viscerosensory information, and is the central guiding entity for important nuclei regulating viscero-motor functions (Fig. 6).

The components of the limbic loop thus receive a broad range of afferents that function in the integration of olfactory, somatosensory, auditory, and in particular, visual exteroceptive data with interoceptive stimuli from a variety of nuclei that process viscerosensory data. They generate projections that not only influence the prefrontal cortex (stimulation of the somato-motor system) but also guide a variety of nuclei that regulate endocrine and viscero-motor functions (Fig. 6). The components of the limbic loop are thus important in regulating not only affect-related movement, emotional stability, learning abilities, and memory functions, but in regulating endocrine and autonomic functions as well (Amaral, 1987; Hyman *et al.,* 1990; Damasio and Damasio, 1991; Squire and Zola-Morgan, 1991; Squire, 1992; Zola-Morgan and Squire, 1993).

## 3. Extracellular Amyloid Deposits

The fluid in the extracellular space of the brain contains many components, among them soluble proteins of unknown function that probably result from the

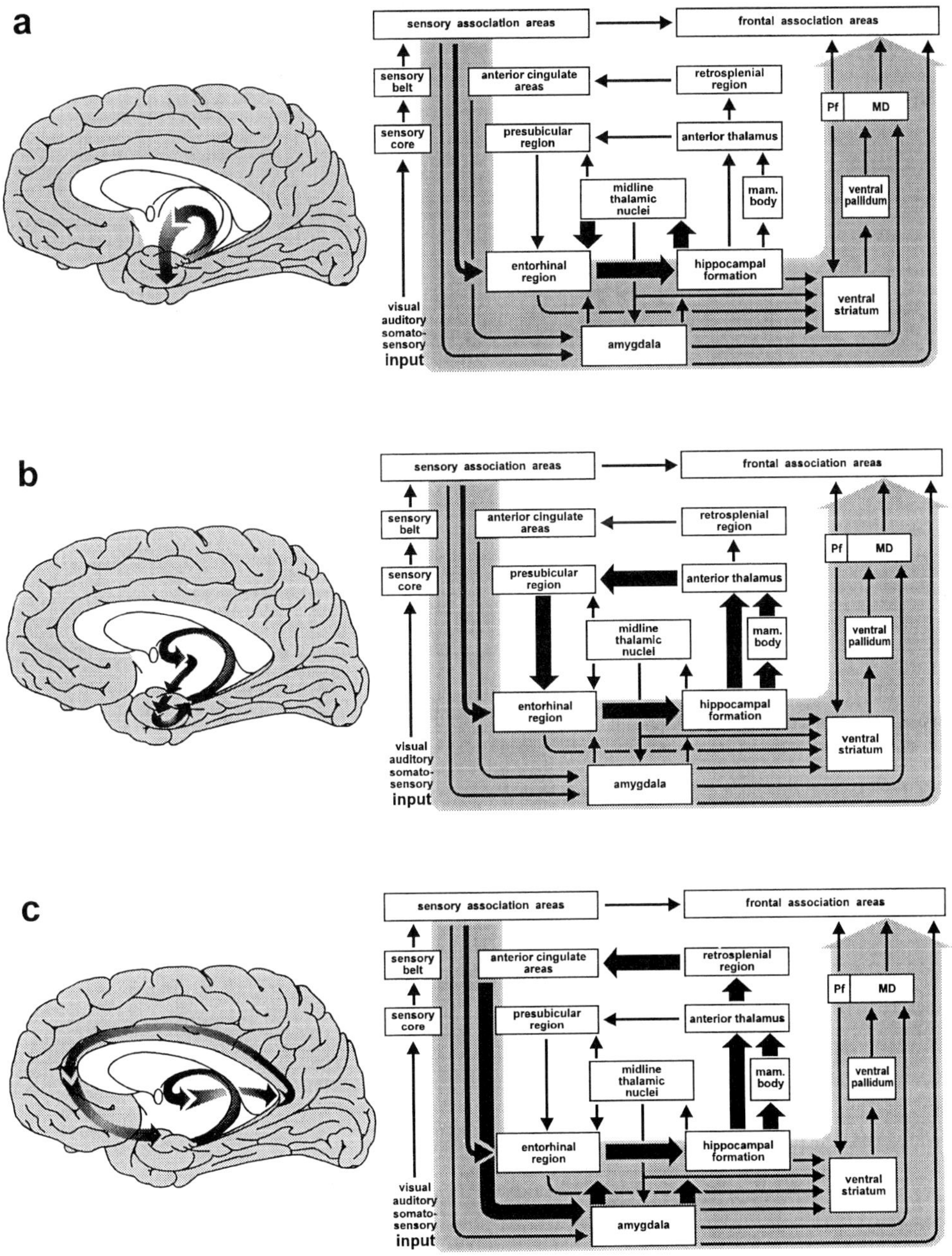


**Figure 5.** Schematic representation of important limbic circuits. (a) "Reuniens circuit": Afferent and efferent projections of the midline thalamic nuclei establish short thalamo-allocortical circuits. (b, c) This circuit is supplemented by others, including the anterior thalamus. A relatively short circuit is formed by projections to the presubicular region, *b,* and a more extended one includes the retrosplenial and anterogenual regions of cingulate gyrus and amygdala (c). (mam., body mamillary body; MD, mediodorsal thalamic nucleus; Pf, parafascicular nucleus. From Braak *et al.* (1996a) with permission.

processing of a normal constituent of the neuronal membrane, the amyloid precursor protein (APP). Under certain conditions and for poorly understood reasons abnormal processing occurs, resulting in the production of Aβ-protein, a hydrophobic self-aggregating peptide that eventually precipitates to plaque-like deposits (Selkoe, 1991, 1993, 1994; Haass, 1996). Whether all neuronal types in the central nervous system, and whether dendritic, somatic, or axonal portions of the neuronal membrane, are capable of producing normal APP-derived material and/or the Aβ-protein to the same degree is unknown. Cortical deposits of amyloid protein appear gradually at specific predilection sites. Initial deposits generally occur in the basal portions of the temporal neocortex, mostly in the form of diffuse plaques in

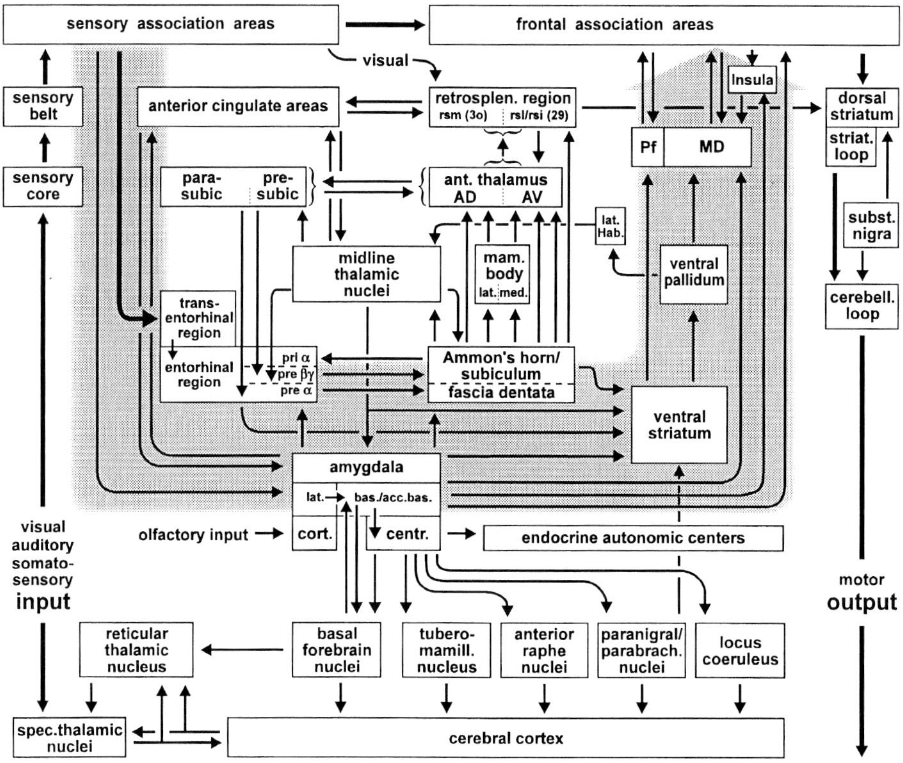


**Figure 6.** Schematic drawing of the limbic loop and related structures in greater detail. The amygdala integrates exteroceptive sensory data with interoceptive stimuli from autonomic centers. Many amygdalar efferents terminate in the nuclei regulating endocrine and autonomic functions. In addition, the amygdala generates efferent connections to all nonthalamic nuclei that project upon the cerebral cortex in a nonspecific manner. (ant.thalamus AD/AV, anterodorsal and anteroventral nuclei of the anterior thalamus; cerebell.loop, cerebellar loop; lat.Hab., lateral habenula; lat.bas./acc.bas.cort.centr., lateral, basal, accessory basal, cortical, and central amygdalar nuclei; mam.body lat./med., lateral and medial nuclei of the mamillary body; MD, mediodorsal thalamic nucleus; paranigral/parabrach., paranigral and pigmented parabrachial nuclei; parasubic presubic, parasubiculum, presubiculum; Pf, parafascicular nucleus; pri-α, pre-β, pre-γ, pre-α, layers of the entorhinal region; retrosplen.region rsm (30) rsl/rsi (29), medial retrosplenial area and lateral and intermediate retrosplenial areas; spec. thalamic nuclei, specific projection nuclei of the thalamus; striat. loop, striatal loop; subst. nigra, substantia nigra; tuberomamill., tuberomamillary nucleus). From Braak *et al.* (1996b) with permission.

layers III and V (Yamaguchi *et al.*, 1988; Wisniewski *et al.*, 1989; Ulrich *et al.*, 1990; Delaère *et al.*, 1991; Braak and Braak, 1991).



### 3.1. Transiently Developed Forms of Amyloid Deposits

Inconspicuous but extensive cloudlike amyloid deposits with blurred boundaries are formed transiently in some of the deep cortical layers and can best be observed in layer VIb of the temporal pro-neocortex and in the entorhinal layer pri-γ (Fig. 7). These formations appear at Stage A, are best-developed at Stage B, vanishing from the tissue at Stage C (see section 3.3). Faintly tinged amyloid protein strands widely infiltrate the tissue and tend to merge into each other, often generating droplike expansions extending into the white matter, eventually condensing into tightly packed granules (white matter plaques) (Braak *et al.*, 1989). There is no relationship between the depth of amyloid protein infiltration into white matter and the duration of the disease.

No amyloid deposits are found in deeper portions of white matter. Even fiber tracts running through amyloid-laden portions of gray matter (fornix, mamillothalamic tract, anterior commissure) only occasionally show a few precipitations. The perforant path, in contrast, is transiently decorated with dense amyloid deposits (Fig. 7g). Projection cells in the external entorhinal cellular layers give rise to the perforant path, which terminates in the hippocampal formation (Mizutani and Kasahara, 1995; Cabalka *et al.*, 1992; Hyman *et al.*, 1988, 1986; van Hoesen 1982). These cells are particularly prone to develop neurofibrillary changes of the Alzheimer type. The tract gives off terminal ramifications to all places that show amyloid deposits in diseased brain. In this location amyloid precipitations develop close to axon terminals of tangle-bearing nerve cells and their dendritic targets. Perforant path fibers contact only a part of the dendritic tree of CA1 projection neurons. Dendritic segments outside of the pathway initially remain free of deposits. It may be speculated, therefore, that axon terminals of tangle-bearing nerve cells produce the amyloid.

The parvocellular layer of the presubiculum displays uniform amyloid deposits even at advanced stages of AD (Kalus *et al.*, 1989). Similarly, uniform bandlike formations develop often—though not inevitably—in the CA1 pyramidal cell layers and in entorhinal layers pre-β/γ (Fig. 7d). Layer pre-α remains free of amyloid deposits. Such bandlike deposits have not been reported in cases of fully developed AD. They are transiently formed and represent an intermediate type of precipitation between the inconspicuous cloudlike formations and the permanent spherical deposits.

### 3.2. Mature Forms of Amyloid Deposits

Sharply outlined globular amyloid deposits of variable diameters represent the mature, more stable forms. Dotlike initiation sites probably occur, leading to self-aggregation of the material (Jarrett and Landsbury, 1993). In contrast to the transient forms, globular plaques rarely fuse; they appear as diffuse deposits (Probst *et al.*, 1991a,b, 1989, 1987) or as cored plaques (which can easily be mistaken for neuritic plaques). Unusually numerous glial cells or distended argyrophilic neurites

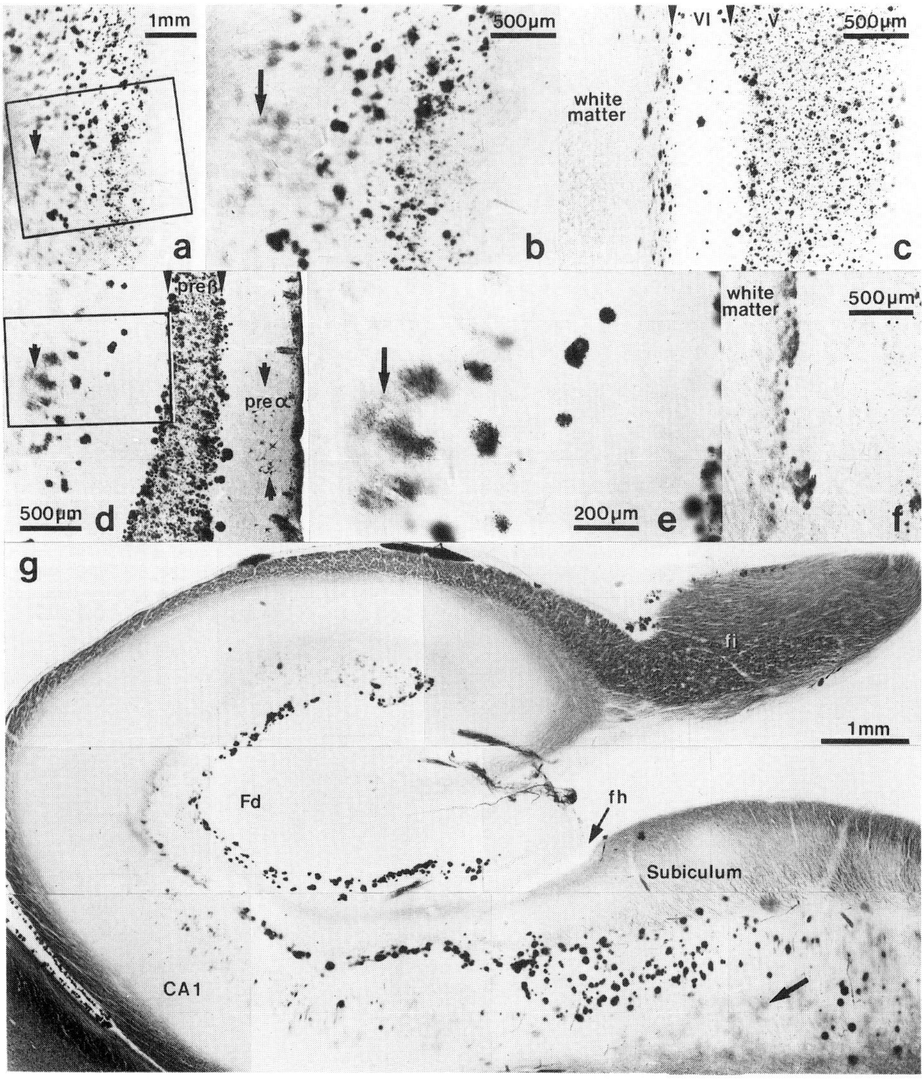


**Figure 7.** Micrographs of spherical, sharply outlined, and cloudlike amyloid deposits with blurred boundaries. (a,b) Amyloid deposits with ill-defined boundaries in the deep layers of the entorhinal region. Framed area of *a* is shown in *b* at higher magnification. White matter is to the left. (c) Layer VI of the neocortex is almost spared of amyloid deposits (anterior cingulate neocortex). (d,e) Bandlike amyloid deposits fill layers pre-β and pre-γ of the entorhinal region. Layer pre-α usually remains devoid of deposits. Arrows point to an island of layer pre-α with ghost tangles. (e) framed area of *d* at higher magnification; arrow points to the amyloid deposits with blurred boundaries of layer pri-γ. (f) White-matter amyloid deposits beneath the subiculum. (g) The perforant path is transiently decorated with amyloid deposits as it runs through the molecular layers of both the fascia dentata (Fd) and the CA1 sector (CA1). Cloudlike amyloid deposits are present between the border of the subiculum and the white matter (arrow). (fi, hippocampal fimbria; fh, hippocampal fissure).

are not found within the reaches of mere amyloid deposits, and there is only a chance contact with the capillaries of the brain tissue.

Unknown local changes in the extracellular milieu may initiate precipitation of the abnormal protein. Extraneuronal remnants of neurofibrillary tangles occasionally serve as nucleation sites for amyloid deposits (tangle-associated amyloid deposits, or TAADs). Amyloid precipitations are also frequently seen in close connection with the abluminal basement membrane of cortical vessels (amyloid angiopathy), a change that generally begins in middle portions of the molecular layer of the cerebral cortex and then extends into both the cortical gray and the leptomeninges. The predilection sites of amyloid angiopathy differ from those of spherical deposits. Individual cases may even show a severe degree of amyloid angiopathy in the absence of cortical amyloid deposits and vice versa. Amyloid angiopathy thus develops independently of spherical amyloid deposits (Vinters, 1992, 1987).

Spherical deposits predominate in neocortical layers III and V. Tiny dark dots are a specific feature of layer IVcβ in the striate area (Fig. 15c) (Braak *et al.*, 1989). Subpial portions of layer I contain confluent plaques, whereas layers II and VI are frequently spared or contain only a few deposits. The number of deposits increases until a certain level of density is reached. Even in the amyloid-laden cortices at end-stage C, a notable amount of the gray matter is still devoid of such precipitations. Given that the material is continually produced over years and slowly degraded, one would expect it to eventually occur throughout the gray matter. Because there are no signs of active elimination of the insoluble deposits from the cortex, unknown factors may exist that lead to the final cessation of amyloid protein production (Hyman *et al.*, 1993; Braak *et al.*, 1996a).

An inverse relationship is observed between the degree of myelination and the density of amyloid deposition. Sparsely myelinated cortical layers and areas display denser deposits than those rich in myelin. Densely myelinated layers, such as layers IV and Vb of the neocortex (harboring a wealth of myelinated pyramidal cell axon collaterals constituting the outer and inner lines of Baillarger) and the particularly myelin-rich molecular layer of the allocortex, remain free of deposits or show only few. Cortical oligodendrocytes may thus play a role in regulating the normal APP cleavage or in preventing Aβ-protein production. Areas and layers formed of pyramidal cells rich in lipofuscin granules also tend to show denser accumulations of globular amyloid deposits than those with sparsely pigmented neurons. The growth of amyloid deposits is probably not a passive process, but is mediated by many yet-unknown processes.

High densities of tangle-bearing neurons (up to Stage IV, Fig. 8) occasionally develop even though no parts of these cells are in contact with amyloid deposits (Braak and Braak, 1991 Goedert, 1993), contradicting the concept that specific forms of insoluble amyloid deposits exert a noxious influence upon sensitive neighboring nerve cells and thereby induce intraneuronal pathology. In general, amyloid deposits do not precede the development of neurofibrillary tangles and neuropil threads.

### 3.3. Evolutionary Stages of Aβ-Amyloid Deposition

Three stages in the gradual evolution of cortical amyloid deposits can be distinguished (Braak and Braak, 1991). Typically, the initial patches develop in the

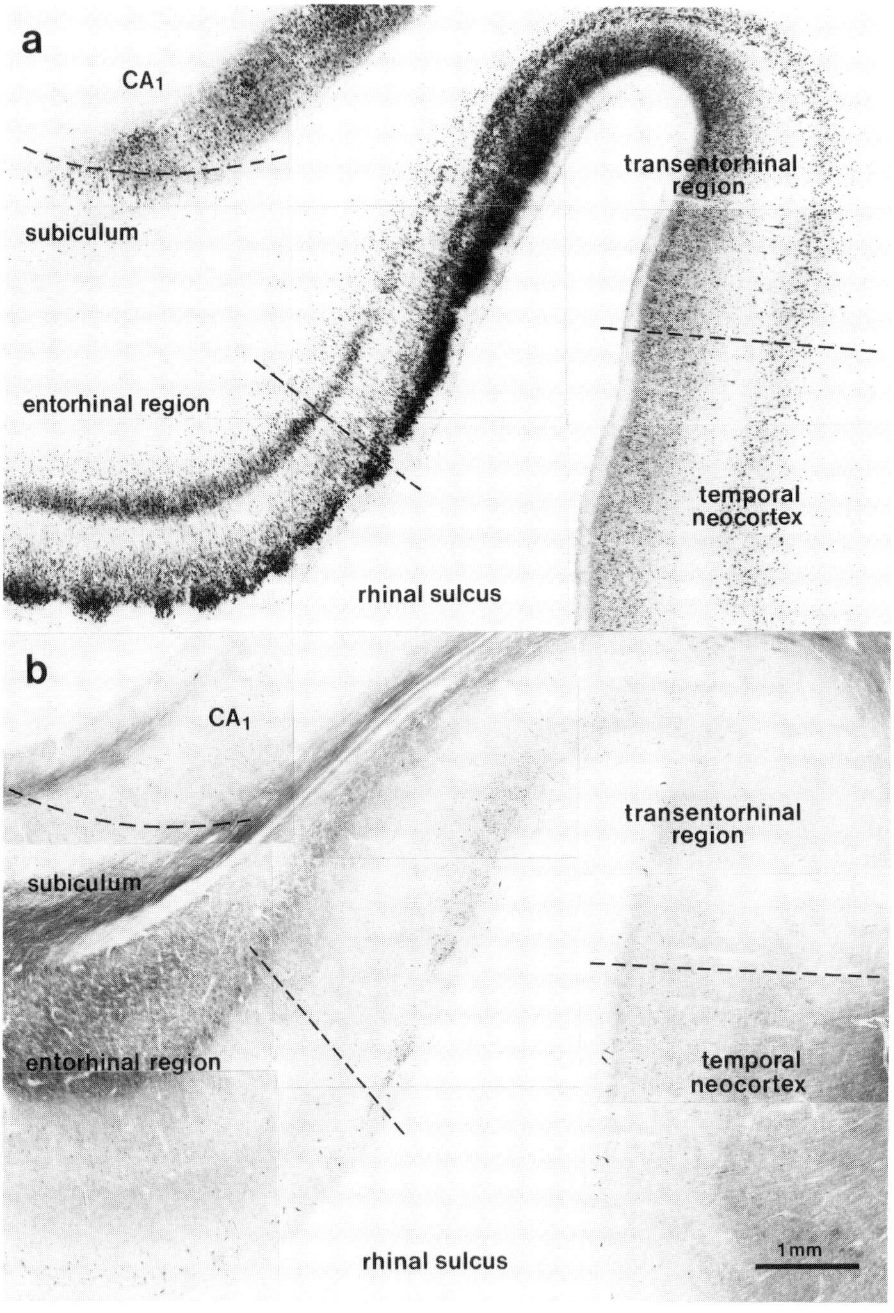


**Figure 8.** Micrographs of two consecutive sections of the entorhinal and transentorhinal regions stained for (a) neurofibrillary changes (Gallyas, 1971) and (b) amyloid deposits (Campbell *et al.*, 1987). 100 μm; case with Stages IV/0: numerous neurofibrillary tangles and neuropil threads (Stage IV) are present in the absence of Aβ-amyloid (Stage 0).

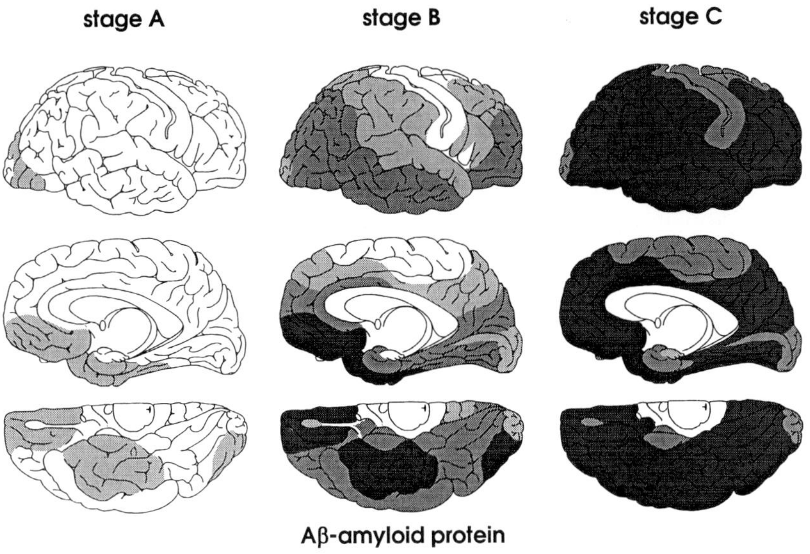


**Figure 9.** Distribution pattern of amyloid deposits in the course of AD. Stage A: Initial deposits are found in the basal neocortex. Stage B: Amyloid deposits occur in virtually all neocortical association areas. The hippocampal formation is only mildly involved. The perforant path is decorated with spherical deposits. Stage C: In the end stage, the deposits extend throughout the entire cerebral cortex. Increasing density of shading indicates increasing density of amyloid deposits. From Braak and Braak (1991) with permission.

basal neocortex, most often in poorly myelinated temporal areas such as the perirhinal or the ectorhinal fields (Stage A; Fig. 9). In the adult, layer III and V pyramidal cells of these areas are rich in lipofuscin pigment. Thereafter, bandlike deposits develop in the parvopyramidal layer of the presubiculum and in entorhinal layers pre-$\beta/\gamma$ (Fig. 7). Spherical deposits in the basal temporal neocortex rapidly increase in number and spread into the adjoining neocortical association areas, sparing for a while the belt regions and primary fields. Then the perforant path becomes decorated with beadlike amyloid deposits as it pierces the subiculum and travels through the molecular layers of CA1 and through the fascia dentata of the hippocampal formation (Stage B, Fig. 7). The end stage exhibits deposits in virtually all cortical areas, including the densely myelinated primary areas of the neocortex (Stage C).

## 4. Intraneuronal Neurofibrillary Changes

The most outstanding neuropathologic feature of AD is the progressive deposition of abnormally phosphorylated tau protein in a few susceptible types of nerve cells. During this process, tau protein gradually loses its normal capacity to stabilize the microtubules of the neuronal cytoskeleton (PHF-tau, paired helical filament tau).

## 4.1. Neurofibrillary Tangles, Neuropil Threads, and Neuritic Plaques

Three forms in which the abnormal material occurs are neurofibrillary tangles (NFTs), neuropil threads (NTs), and argyrophilic dystrophic neurites in neuritic plaques (NPs). The NFTs develop within the somata of nerve cells, whereas the NTs are found in distal portions of the dendrites (Braak *et al.,* 1986, 1988; Braak and Braak, 1988; Yamaguchi *et al.,* 1990; McKee *et al.,* 1991; Markesbery *et al.,* 1993). In general, the first brain changes seen are NTs and NFTs; NPs develop later (Braak and Braak, 1994, 1991). These changes take a long time to develop. Both NFTs and NPs are modified markedly as they appear, grow to maturity, and eventually disappear from the tissue.

The distribution pattern of NPs differs from that of NFTs, NTs, and amyloid deposits. Neuritic plaques consist of spherical aggregations of distended neurites containing the PHF-tau protein, dystrophic non-argyrophilic neurites, changed astrocytes, and microglial cells. Amyloid is generally present in the form of a compact core and peripheral infiltrations. The distribution of NPs is patchy and occurs at much lower densities than that of amyloid deposits. The factors inducing NPs and the circumstances accompanying their disappearance from the tissue are unknown. A few cortical areas remain virtually free of NPs; others (ectorhinal area, striate area) exhibit high NP densities. The cortex in the depths of the sulci generally exhibits a higher NP density than that covering the crests of the gyri (Simchowicz, 1911; Grünthal, 1930; von Braunmühl, 1957; Braak *et al.,* 1989).

Only a small number of the many CNS neuronal types are prone to developing NFTs and NTs. Susceptible types include glutamatergic, GABAergic (γ-aminobutyric acid), dopaminergic, noradrenergic, adrenergic, serotonergic, and cholinergic nerve cells. In the cerebral cortex, NFT-bearing nerve cells are all included in the class of pyramidal (projection) neurons (Feldman, 1984). Cells furnishing long ipsilateral corticocortical connections are particularly prone to accumulating the material (Morrison *et al.,* 1987; Hof and Morrison, 1990a,b). Pyramidal cells are sturdy constituents of the cortex and may survive for a long time despite marked cytoskeletal alterations and changes in their cellular processes (Braak and Braak, 1985b, 1997a; Braak *et al.,* 1995). Tangle-bearing neurons eventually die, however, for reasons yet undiscovered (Coleman and Flood, 1987; Callahan and Coleman, 1995; Lassmann *et al.,* 1995). After deterioration of the parent cell, the NFT is converted into a less densely twisted, less argyrophilic extraneuronal "ghost" tangle (Alzheimer, 1907; Bancher *et al.,* 1989; Tomlinson, 1989; Probst *et al.,* 1991a; Yamaguchi *et al.,* 1991; Braak *et al.,* 1994). The lipofuscin granules of tangle-bearing and dying cortical projection cells then become extraneuronal, and specific staining of these aggregates can be used to indicate the former position of the lost nerve cells. In CA2 and less frequently in other parts of the hippocampal formation, a portion of the ghost tangles becomes decorated with clusters of coarse argyrophilic neurites (tangle-associated neuritic clusters; TANCs) (Munoz and Wang, 1992). Even ghost tangles and extraneuronal pigment deposits are—in the long run—removed by macrophages.

Individual AD cases exhibit variable numbers of ghost tangles. Should longer remissions occur, or if the progress of the illness were to be stopped by some event, the exclusive presence of ghost tangles would indicate the previously occurring

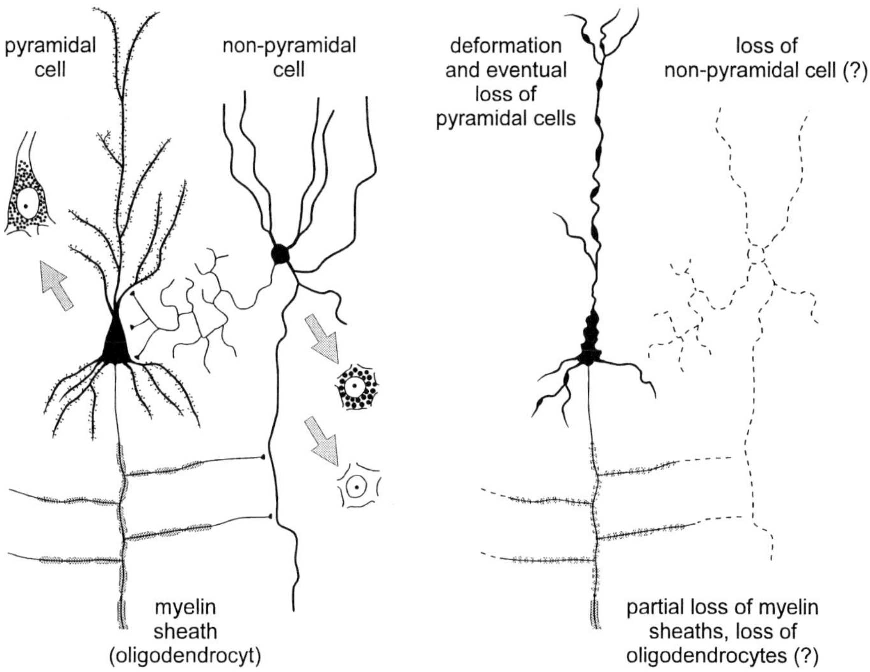


**Figure 10.** Left side: Cortical functional units consist of projection neurons (pyramidal cells), related oligodendrocytes, and local circuit neurons (nonpyramidal cells). Pyramidal cells generally contain small, evenly distributed lipofuscin granules; nonpyramidal cells are either free of pigment or filled with coarse granules (arrows). In general, the pyramidal cells have spiny dendrites and a myelinated axon that enters the white matter. Most nonpyramidal cells, by contrast, have smooth dendrites and a short, locally branching axon. Functional maturity is achieved only after myelination of the pyramidal cell axon and establishment of bidirectional synaptic contacts between the two neuronal types. Right side: AD-related destruction possibly includes premature dysfunction of specific oligodendrocytes, induction of intraneuronal cytoskeletal changes, appearance of neurofibrillary tangles and neuropil threads, and partial loss of varicose dendrites. A loss of local circuit neurons may occur secondarily to the disappearance of projection cells. From Braak and Braak (1997a) with permission.

pathologic process. Therefore, it is important to note that ghost tangles alone have not been observed in the material at our disposal; fresh NFTs and NTs are present in every case. Once the destructive process has begun, it relentlessly and immutably progresses; there is no spontaneous remission in AD. The first NFTs and NTs are very significant, as they mark the beginning of the disease.

It is now time to recall that functional maturity of specific neuronal types (e.g., cortical projection cells) is achieved only after myelination of the axon and establishment of the bidirectional synaptic contacts to local circuit neurons (Fig. 10). The possible changes in late-maturing cortical oligodendroglial cells will be discussed later. Little is known about AD-related changes in cortical nonpyramidal cells (Unger and Lange, 1992; Brion and Résibois, 1994). Many types of local circuit neurons are particularly vulnerable to disconnection of their afferents, to hypoxia, or to other noxious influences, and they tend to disappear rapidly from the tissue (Sloper *et al.*, 1980; Fairén *et al.*, 1984; Braak and Braak, 1985b; Fonseca *et al.*, 1993). Cortical local circuit neurons resist the development of AD-related neurofibrillary changes, although loss of such cells may occur secondarily to the disappearance of cortical projection cells (Braak and Braak, 1986; Lewis *et al.*, 1987; Hof *et al.*, 1991, 1993; Hof and Morrison, 1991).

### 4.2. Evolutionary Stages of Neurofibrillary Tangles and Neuropil Threads

The spread of intraneuronal changes is by no means uniform. In particular, NFTs and NTs exhibit involvement of specific architectonic units (Brun and Gustafson, 1976; Kemper, 1978; Hyman *et al.*, 1984, 1990; Braak and Braak, 1985a; van Hoesen and Hyman, 1990; Arnold *et al.*, 1991; Fewster *et al.*, 1991; Price *et al.*, 1991; van Hoesen *et al.*, 1991; Arriagada *et al.*, 1992a,b; van Hoesen and Solodkin, 1993; Hyman and Gómez-Isla, 1994). The first lesions develop in the transentorhinal region, from which the changes extend in a predictable pattern into both hippocampus and neocortex. This sequential development in the distribution pattern of NFTs and NTs is subject to little variation, thus providing a basis for distinguishing the stages in the evolution of such changes (Fig. 14) (Braak and Braak, 1991). It iscurrently impossible for the successive development of the destructive process to be directly observed. The concept of stages is necessarily derived from cross-sectional data and is essentially artificial, because the disease-related changes develop continually rather than in well-defined steps. Thus, a rough, six-stage categorization is recommended for practical reasons only.

#### 4.2.1. Pre-Tangle Phase

The evolution of cytoskeletal changes is best examined in the early stages of AD, that is, in the transentorhinal region of comparatively young individuals. Under these conditions it can be studied in the absence of amyloid deposits, vascular changes, or other pathologic lesions. Specific antibodies reacting with the abnormally phosphorylated tau protein (AT8) (Braak *et al.*, 1994) display the altered pyramidal cells with all their neuronal processes; the image resembles a successful Golgi impregnation. Initially, the immunoreactive neurons exhibit no obvious destruction of their neurites (Fig. 11a). During certain phases of cell division, AT8-positive tau proteins transiently appear. Mature nerve cells have lost their capacity to undergo mitosis, but it may be speculated that the specific neuronal types prone to developing AD-related changes first lose control of their cell cycle; changes in the neuronal cytoskeleton would then appear to be an abortive initiation of the process of cell division (Lee *et al.*, 1993; Preuss *et al.*, 1995; Vincent *et al.*, 1996).

**Figure 11.** a,b Micrographs of pre-α pyramidal cells in the transentorhinal region immunostained with the antibody AT8 directed against abnormally phosphorylated tau protein. (a) Note the Golgi-like staining of the axon (ax) and the dendrites (pre-tangle phase). (b) Eventually destruction becomes visible within the distal portions of the dendrites, which follow a tortuous course and exhibit varicosities and short appendages. These are shown at higher magnification in c, d, and e. Lower portion of the panel: Schematic drawing summarizing AT8 immunostaining with corresponding Gallyas silver staining. The progression of pathologic alterations of the neuronal cytoskeleton is shown from the pre-tangle phase (group 1) to the formation of a "late" ghost tangle (group 5). From Braak *et al.* (1994) with permission.

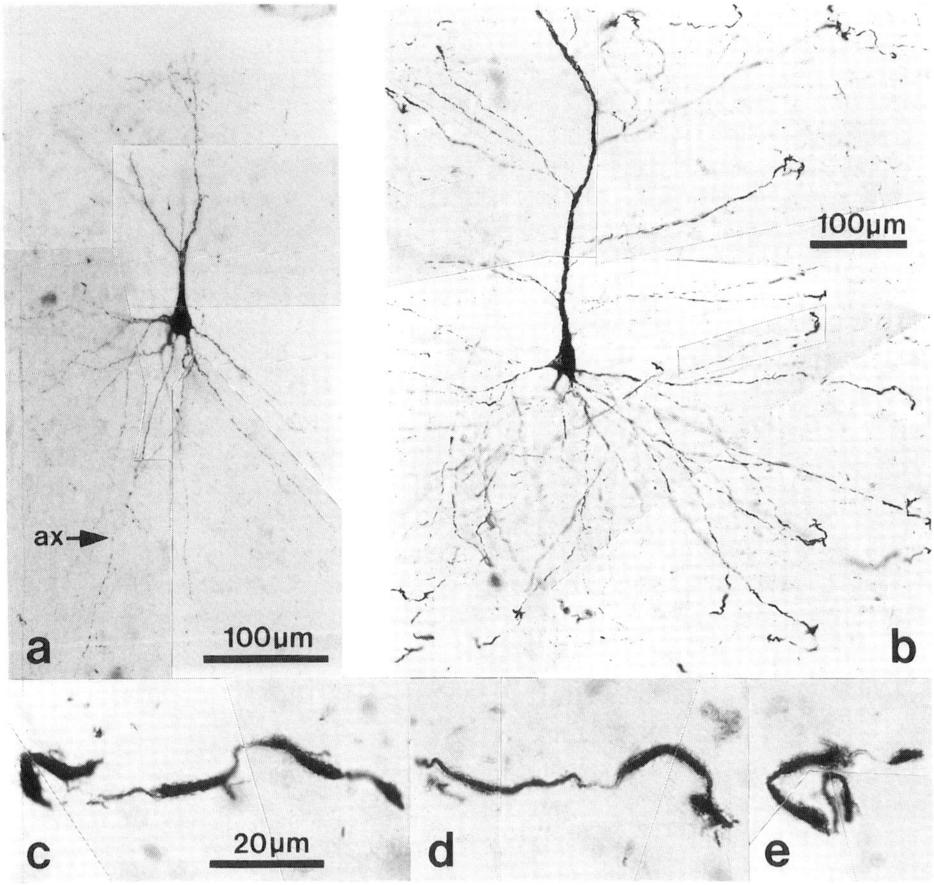
ax
100µm
a
100µm
b
c
20µm
d
e

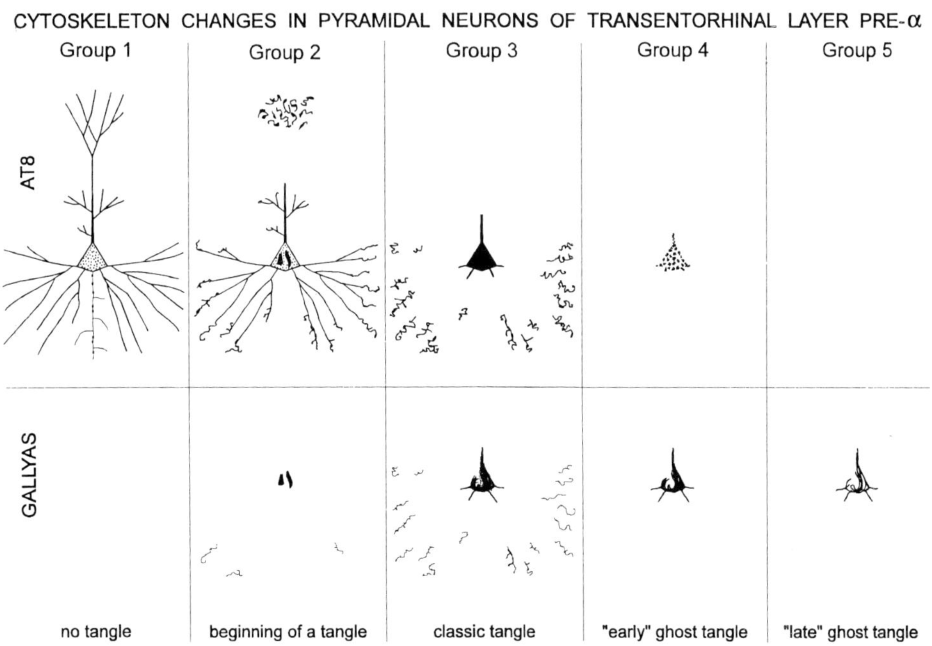
CYTOSKELETON CHANGES IN PYRAMIDAL NEURONS OF TRANSENTORHINAL LAYER PRE-α
Group 1
Group 2
Group 3
Group 4
Group 5
AT8
GALLYAS
no tangle
beginning of a tangle
classic tangle
"early" ghost tangle
"late" ghost tangle

It is important to note that the form of abnormal tau protein initially contained in the AT8-immunoreactive cells is soluble and is distributed evenly throughout the cell body and neuronal processes; aggregation to solid, argyrophilic NFTs or NTs has not yet taken place. It is unknown just how long such cells are able to maintain the seemingly unaltered structure of their neurites, but sooner or later the destruction becomes visible, with the most conspicuous changes seen in distal portions of the dendrites, which follow a curved and tortuous course and exhibit many varicosities and short appendages (Fig. 11b–e). Next, the altered portions of the dendrites probably lose their connection to the proximal stem. At this level of destruction, slender NTs appear within the changed dendrites, and shortly thereafter NFT formation begins in the soma (Fig. 11, lower portion). During this process, in addition to the initially soluble form of the abnormal tau, more and more insoluble aggregations appear (Braak *et al.,* 1994).

The first traces of abnormal material are generally associated with somatic deposits of lipofuscin or neuromelanin; central portions of larger NFTs thus show a dense feltwork around the pigment granules. Possibly the pigment provides initiation sites and/or supports cross-linking reactions. Other neuronal inclusions, for example, Lewy bodies, Lewy neurites, Hirano bodies, and granulovacuolar bodies, have no close association with the initial NFTs. It is not yet known just which local changes within the dendritic tips are responsible for the formation of NTs.

### 4.2.2. Transentorhinal Stages I and II

The transentorhinal region is the first cortical territory to be involved in AD. In Stage I only a few NFTs and NTs are encountered, whereas in Stage II the transentorhinal region is increasingly affected, and there is mild involvement of the entorhinal region proper and of sector CA1 (Figs. 12a–f, 14). The initial neurofibrillary changes develop preferentially in the absence of amyloid deposits (Fig. 20, second row).

Stage II pathology slightly hampers the transmission of neocortical information—via the entorhinal region—to the hippocampal formation. However, transentorhinal Stages I and II show no obvious impairment of intellectual capacities. As in many other neurodegenerative diseases, the pathologic process underlying AD can be identified over extended periods of time before the onset of clinical symptoms, with brain destruction slowly progressing until it can no longer be compensated for by individual reserve capacities. Stages I and II thus represent the clinically silent period of AD.

### 4.2.3. Transiently Developed Forms of Neurofibrillary Changes

Transient forms of NFTs and NTs develop in specific pyramidal cells of the hippocampal CA1 sector. Changed cells are found only in a segment of CA1 located between the prosubiculum and the opposing extremity abutting upon CA2. The change is confined to portions of the apical dendrite and its major side branches in CA1 projection cells. Within the stratum lacunosum-moleculare, these dendrites display conspicuous spindle-shaped dilations filled with argyrophilic and AT8-immunoreactive material (Fig. 13). The change appears at Stage II and is best developed at Stage III. The varicose dendritic segments vanish from the tissue at Stages IV–V, leaving no extracellular remnants that—like ghost tangles—could be de-

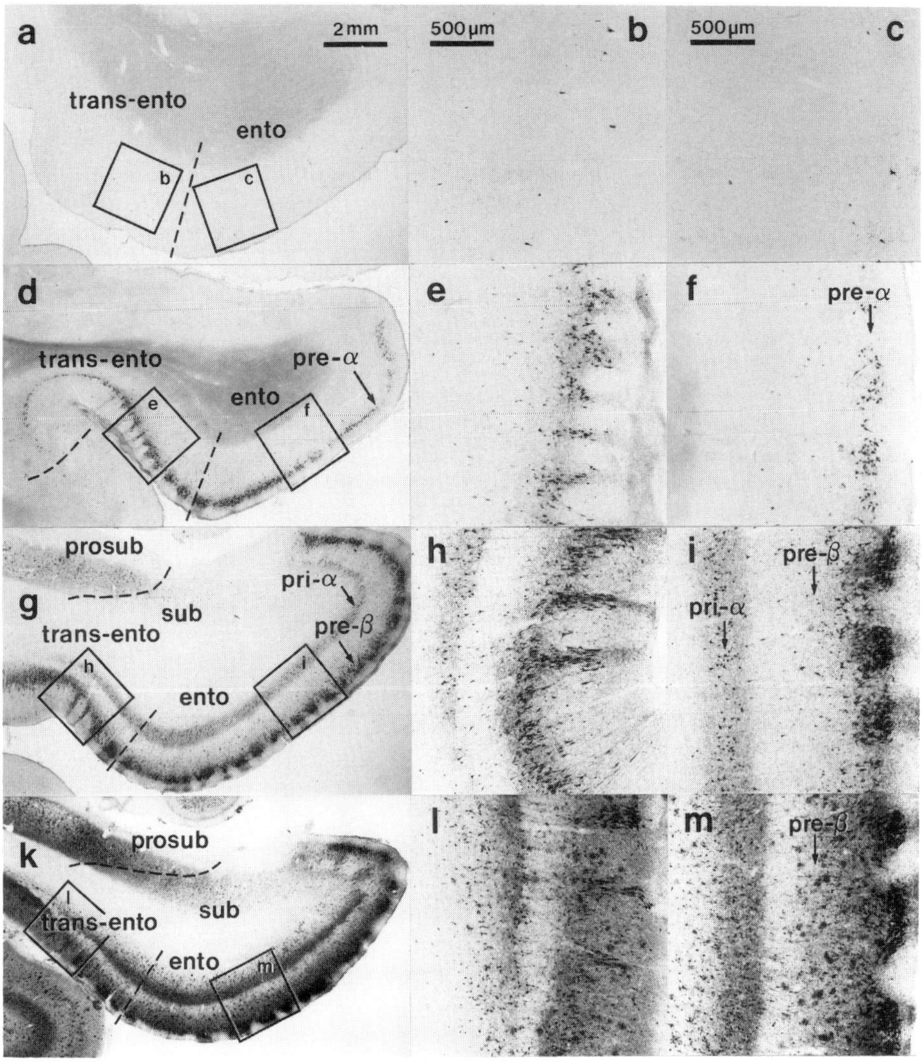


**Figure 12.** Gradual progress of destruction of the entorhinal and transentorhinal regions in the course of Alzheimer's disease. Insets provide an overview, with framed areas indicating the location of micrographs taken at higher magnification. On the left, portions of the transentorhinal region (trans-ento) are shown with the pial surface oriented downward. These are supplemented by portions of the entorhinal cortex (ento) with the pial surface directed to the right-hand border. At higher magnification the framed areas are shown: the entorhinal region (middle column) and the transentorhinal region (right column). Gallyas silver impregnation for neurofibrillary changes; 100 µm.

(a,b,c) Stage I: Very few neurofibrillary tangles and neuropil threads are encountered in the transentorhinal region.

(d,e,f) Stage II: Neurofibrillary tangles are virtually restricted to layer pre-α; they are numerous in the transentorhinal region and less densely placed in the entorhinal region.

(g,h,i) Stage III: Abundant neurofibrillary tangles mark the pre-α cellular islands in both the transentorhinal and entorhinal regions. Note the dense staining between cellular islands and pial surface caused by neuropil threads. Further neurofibrillary tangles mark the deep layer pri-α.

(k,l,m) Stage IV: Both regions show a considerable increase in the density of neurofibrillary changes. At this stage many new neurofibrillary tangles are formed in layer pre-β, and numerous neuritic plaques appear in pre-β and pri-β (pro-sub, prosubiculum; sub, subiculum).

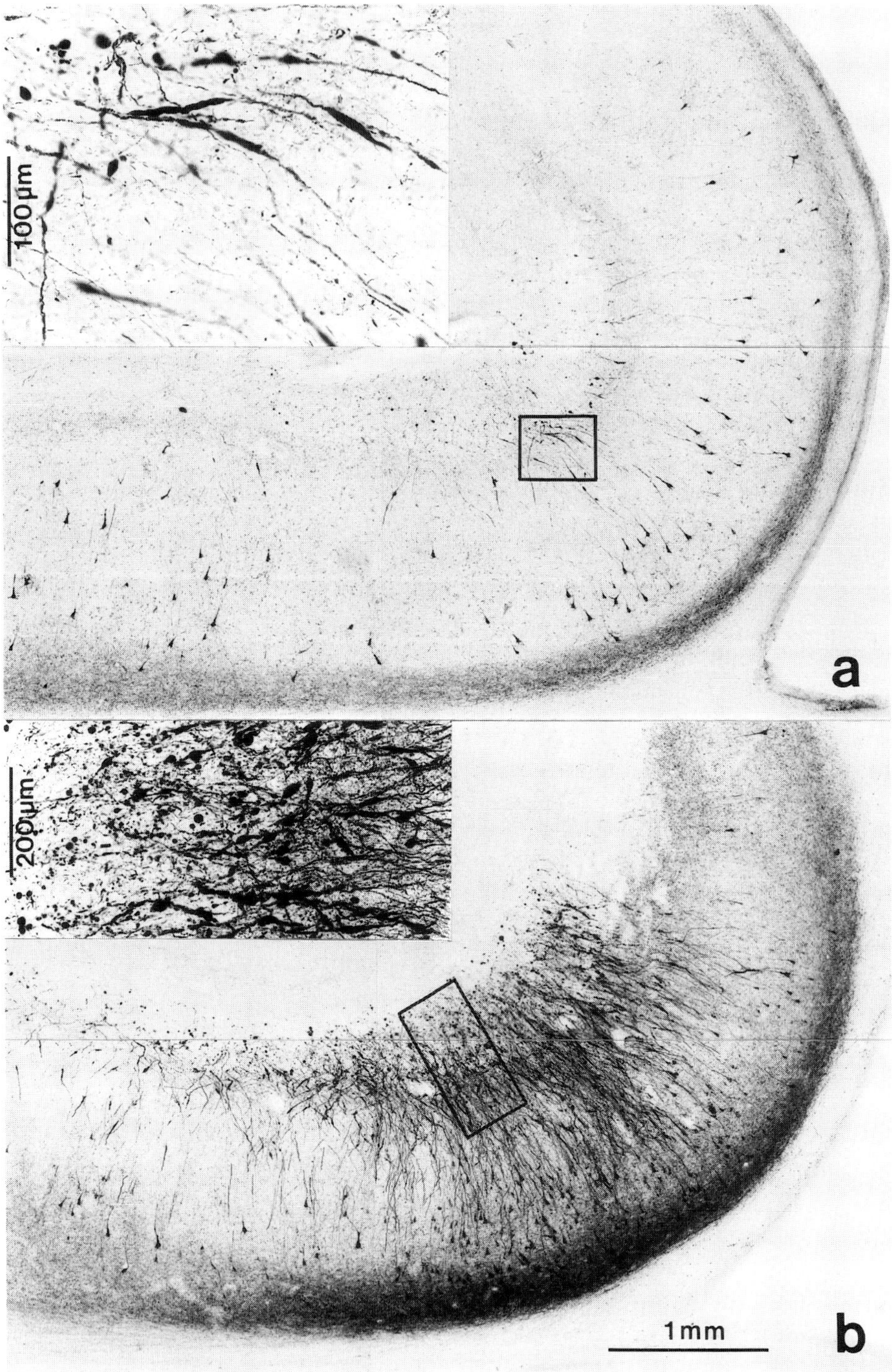


**Figure 13.** The apical dendrites of CA1 pyramidal cells transiently develop spindle-shaped dilations (insets correspond to the framed areas at higher magnification). Immunostaining with the antibody AT8, 80 µm thick. (a) The change appears in Stage II and is best developed in Stage III (b).

tected at light microscopic level. Apparently, the material is more easily degraded than are extraneuronal NFTs.

#### 4.2.4. Limbic Stages III and IV

Cases at limbic stages III or IV already reveal severe destruction of the cortex, but it is confined to a few allocortical regions and adjoining areas (Fig. 14). Due to the lack of general neocortical destruction, the limbic stages fail to meet the frequently used criteria for neuropathologic diagnosis of AD (Tierney *et al.,* 1988; Mirra *et al.,* 1991; Gearing *et al.,* 1995).

Stage III is characterized by severe destruction of the entorhinal layer pre-α and by additional involvement of the deep layer pri-α (Fig. 12g–i). A considerable proportion of ghost tangles is seen in these layers at this stage. Mild changes also occur in the hippocampus, the anterodorsal thalamic nucleus, the basolateral amygdaloid complex, and the magnocellular nuclei of the basal forebrain. Furthermore, some NFTs are seen in the thalamic reuniens and the hypothalamic tuberomamillary nucleus (Braak and Braak, 1993).

At Stage IV the changes begin in layer pre-β. Sector CA1 is infested with NFTs, but the subiculum is still only mildly affected (Fig. 12k–m). At this stage the destructive process expands markedly from the entorhinal region into the amygdala, the hippocampal formation, and in particular, the adjoining association areas of the temporal neocortex. Neurofibrillary tangles and neuropil threads develop in the large cholinergic striatal interneurons and in lipofuscin- or melanin-laden nerve cells of the oral raphe nuclei and locus coeruleus.

The clinical protocols of many individuals at Stage III or IV note an impair-

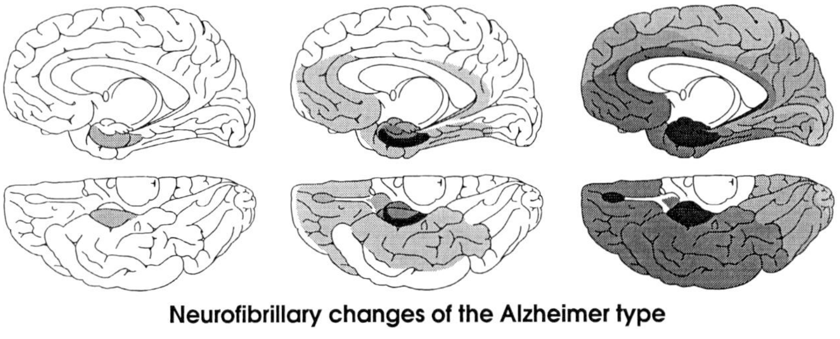


**Figure 14.** Distribution pattern of neurofibrillary changes (NFTs/NTs) in the course of AD. Six stages can be distinguished. Stages I and II show alterations that are virtually confined to a single layer of the transentorhinal and entorhinal regions. Stages III and IV show a severely affected entorhinal territory and additional involvement of many related limbic structures. Stages V and VI are marked by a devastating destruction of the neocortex. Increasing density of shading indicates increasing density of NFTs and NTs. From Braak and Braak (1991) with permission.

ment of cognitive functions and the presence of subtle personality changes, but in other individuals the appearance of symptoms is still obscured by individual reserve capacities. At first glance it seems strange that the destruction of just two entorhinal cellular layers could cause functional disturbances. However, the specific lesions bilaterally hamper both the feedforward and the feedback projections necessary for normal function of the neocortex/hippocampus interrelation. In addition, the local pathologic process interrupts the limbic loop at multiple sites (Figs. 6, 14). Because of the common presence of initial clinical symptoms and characteristic brain lesions, Stages III to IV represent the morphologic counterpart of incipient AD (Jellinger *et al.,* 1991; Braak *et al.,* 1993; Bancher *et al.,* 1993, 1996; Duyckaerts *et al.,* 1994, 1995).

#### 4.2.5. Neocortical Stages V and VI

The leading feature of Stage V is severe destruction of neocortical association areas. Only the primary motor field, the primary sensory areas, and their belt regions are uninvolved (or mildly affected). The entorhinal layers pre-α, pri-α, pre-β, the prosubiculum, and adjoining portions of the CA1 sector are infested with NFTs, NTs, and ghost tangles. Dense feltworks of NTs occur in the subiculum, parasubiculum, and retrosplenial areas. Stage V also shows a severe involvement of the amygdala, basal portions of the claustrum, and the reticular nucleus of the thalamus. Melanin-laden nerve cells of the substantia nigra begin to develop NFTs.

In Stage VI the entorhinal layer pre-α is generally denuded of nerve cells; ghost tangles represent the only remnants. A feature facilitating the distinction of Stage VI from Stage V is the appearance of globose NFTs in granule cells of the fascia dentata. At Stage VI the primary sensory areas of the neocortex exhibit a wealth of NPs and small NFTs in layers II and III, and the striate area shows a remarkably dense network of NTs in layer V (Fig. 15d,e). The density of NFTs and NTs increases stepwise as one proceeds from the primary visual field via the parastriate field into the occipital association areas (Morel, 1945; Mutrux, 1947; Rogers and Morrison, 1985; Lewis *et al.,* 1987; Braak *et al.,* 1989). The boundary between the parastriate area and the peristriate region—not easily detected in the normal human brain—appears with outstanding clarity in end-stage case AD (Fig. 16). Only the primary motor cortex still exhibits a low intensity of neurofibrillary change. The clinical protocols of Stage V or VI cases generally note the presence of severe dementia (Fig. 14).

### 4.3. Stages I to VI: Development Repeats Cortical Myelination Progression in Reverse Order

The order and pattern of NFT- and NT-related destruction bear a striking resemblance to the inverse sequence of cortical myelination in early development (Figs. 17, 18). In general, the poorly myelinated cortical areas develop NFTs and NTs earlier in the course of AD and at higher density than do the richly myelinated fields (Braak and Braak 1996).

The human cortex exhibits a late and prolonged postnatal myelination, which

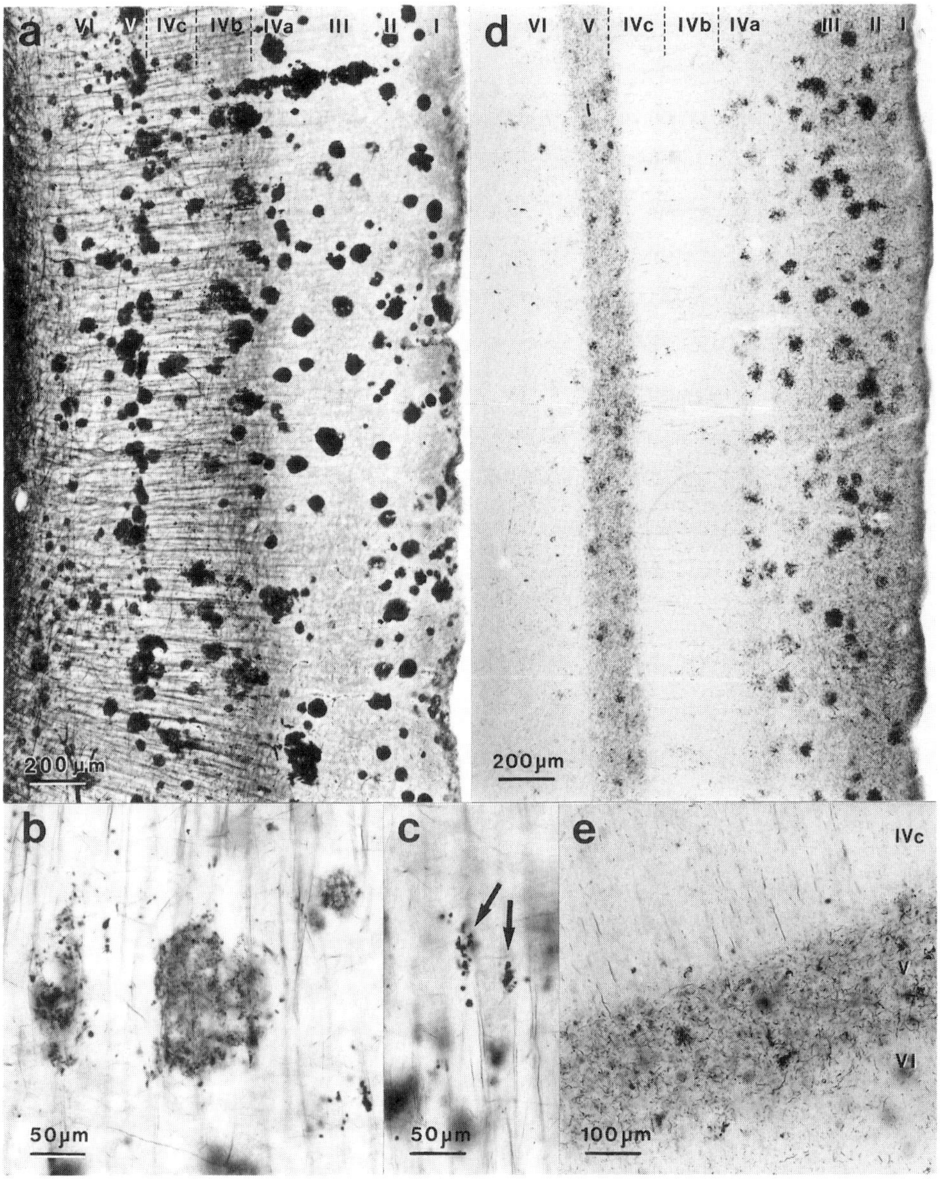


**Figure 15.** Striate area. Note that the laminar distribution of amyloid deposits (a) differs considerably from that of neurofibrillary changes (d). Layers are indicated at the upper margin. (b, c) Weakly stained amyloid plaques and intensely argyrophilic dots (arrows) in layer IVc. (e) Dense line of neuropil threads in layer V. Note the vertical neuropil threads extending from this line into portions of layer IVc. From Braak *et al.* (1989) with permission.

may last well into adulthood (Vogt and Vogt, 1919; Flechsig, 1920; Yakovlev and Lecours, 1967). The first traces of myelin appear in the primary motor area of the neocortex, then in the sensory core areas. The myelination process next spreads via the border fields into related association areas. The result is a remarkably dense myelination of the neocortical core fields.

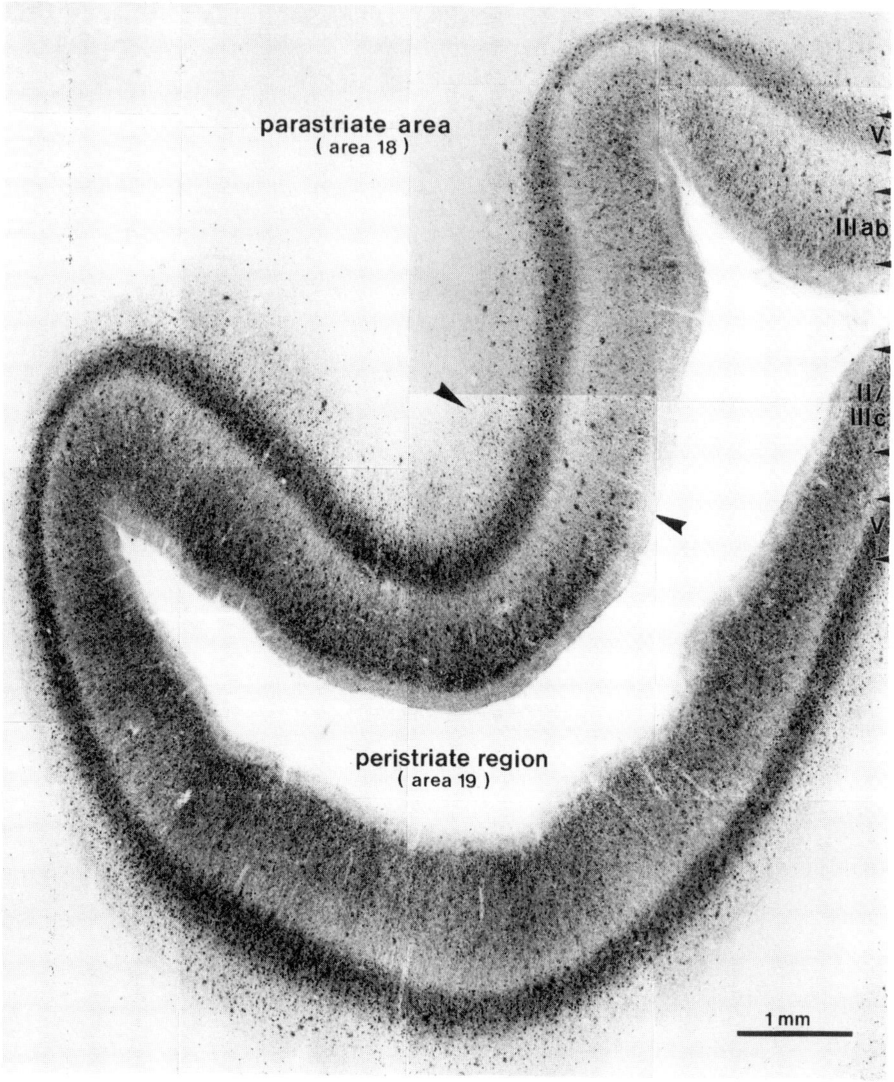


**Figure 16.** Parastriate area and peristriate region. The parastriate area shows numerous NPs in sublayer IIIab. Sublayer IIIc is almost devoid of neurofibrillary changes, as is layer IV. Layer V contains a medium number of NFTs and NTs. The peristriate region, in comparison, is dominated by a wealth of NFTs and NTs from layer II down to layer IIIc; note the sharply drawn lower border of layer IIIc. Layer IV is smaller and less pallid than in the parastriate field. Layer V is rich in NFTs and NTs. The boundary between the parastriate area and the peristriate region is easily discernable using these criteria (arrowheads).

With increasing distance from the core fields, there is a graduated decrease in the average density of myelin. The neocortical association areas near the transentorhinal region (and the region itself) are particularly poorly myelinated (Vogt and Vogt, 1919; Hopf, 1954); here is the focus of the very first neurofibrillary changes. The changes proceed into the neocortex, successively reaching association areas of the third, the second, and then the first temporal gyrus. Involvement of the auditory core and other core areas is observed only in the final stages of the disease (Figs. 17, 18).

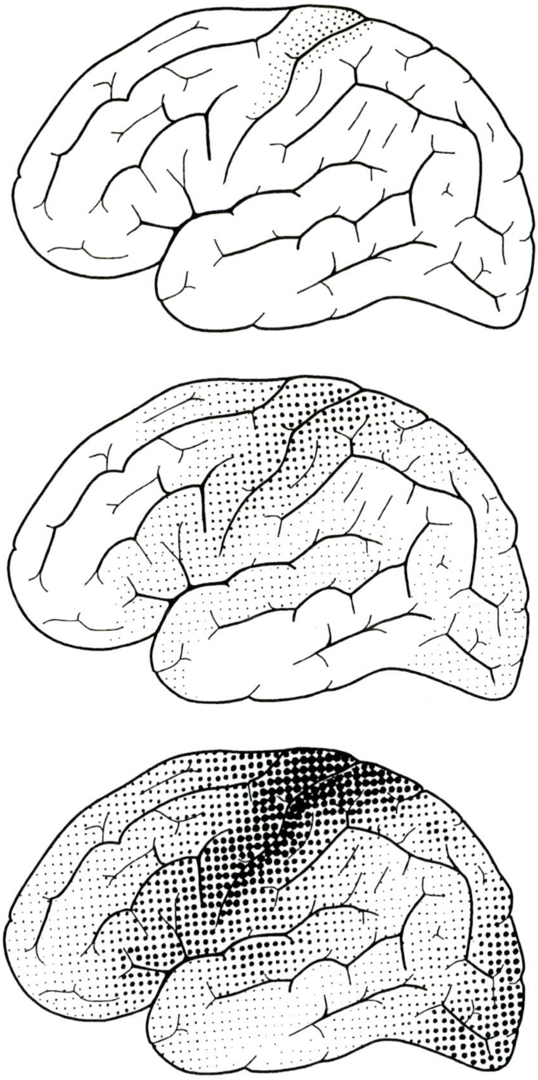

**Figure 17.** Cortical myelination from early development to adulthood. The first traces of myelin appear in the primary motor area (upper part), then in the sensory core areas. Myelination next spreads, via border fields, into related association areas (middle part). In the brain of the human adult, core fields show remarkably dense myelination (lower part). Slightly modified from Vogt and Vogt (1919).

Regressive changes in the brain often repeat the process of maturation, but in reverse order (Bachevalier and Mishkin, 1992; Reisberg *et al.,* 1992). Premature dysfunction of cortical oligodendrocytes could explain the similarities between cortical myelination and the development of AD-related changes. Factors released by oligodendrocytes exert influence upon related nerve cells and suppress disordered outgrowth of the neuronal processes (Vaughan, 1984; Schwab, 1990; Hardy and Reynolds, 1993; Cadelli *et al.,* 1992; Kapfhammer and Schwab, 1994). Nerve cell stability increases with myelination of the axons; neuritic instability observed in old age may thus result from decreasing influence of oligodendrocytes. One possible scenario is that a lack of oligodendrocytic factors results in alterations of the neuronal cytoskeleton and eventually, in the induction of NFTs and NTs (Braak and Braak, 1997a).

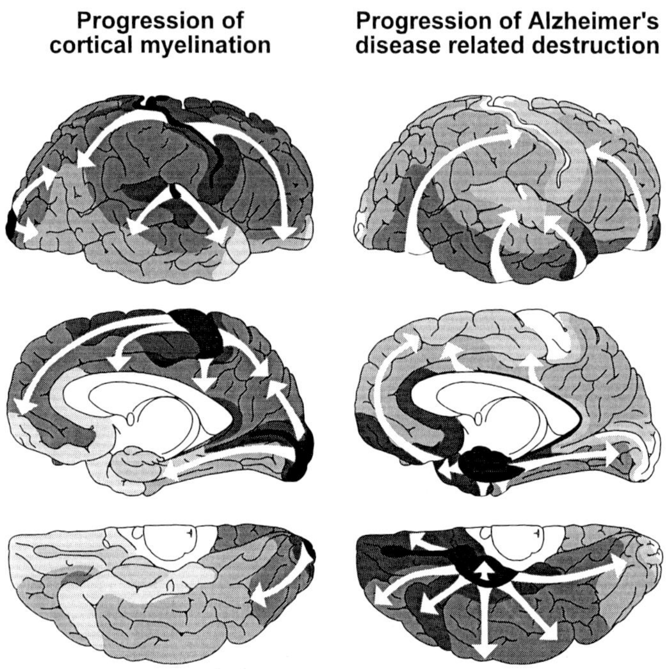


**Figure 18.** Left side: Cortical myelination begins in neocortical core fields and progresses via belt areas to related association areas (indicated by different shading of the boxes). Right side: AD-related cortical destruction begins in the transentorhinal region, from which the changes extend into adjoining areas, eventually reaching the neocortical core fields. Note that the sequence of destruction inversely recapitulates that of cortical myelination. From Braak and Braak (1996) with permission.

## 5. Age, Amyloid Deposits, and Neurofibrillary Changes

The relationships among age, amyloid-deposition, and neurofibrillary changes can be studied—using a staging procedure—through the assessment of very large numbers of nonselected autopsy cases (Figs. 19–21) (Harrell *et al.*, 1993). Despite of their frequent occurrence, neither amyloid deposits nor neurofibrillary changes inevitably accompany senescence (Fig. 21, right side) (Dickson *et al.*, 1991; Braak *et al.*, 1995; Peters *et al.*, 1996). No single feature permits the definition of a specific form of NFTs and NTs exclusively related to age, and there is no continuum uniting age-changes with AD-related pathology. All that can be said is that cases free of amyloid deposits or neurofibrillary changes decrease with advancing age (Figs. 19, 20, first row).

A small number of cases develop initial amyloid deposits and/or the neurofibrillary changes at a surprisingly young age (Figs. 19, 20, second rows) (Sugihara *et al.*, 1995) and in otherwise healthy brain. Apparently, advanced age is no prerequisite for the evolution of these changes.

The age distributions from amyloid Stage A to Stage C and from neurofibrillary stages I to VI show a marked shift toward higher age categories (Figs. 19, 20; third and fourth rows), corroborating the assumption that early stages occur predominantly in young individuals, whereas more-advanced stages gradually appear with age (Ohm *et al.*, 1995; Braak and Braak, 1997c). The mean of the stages of

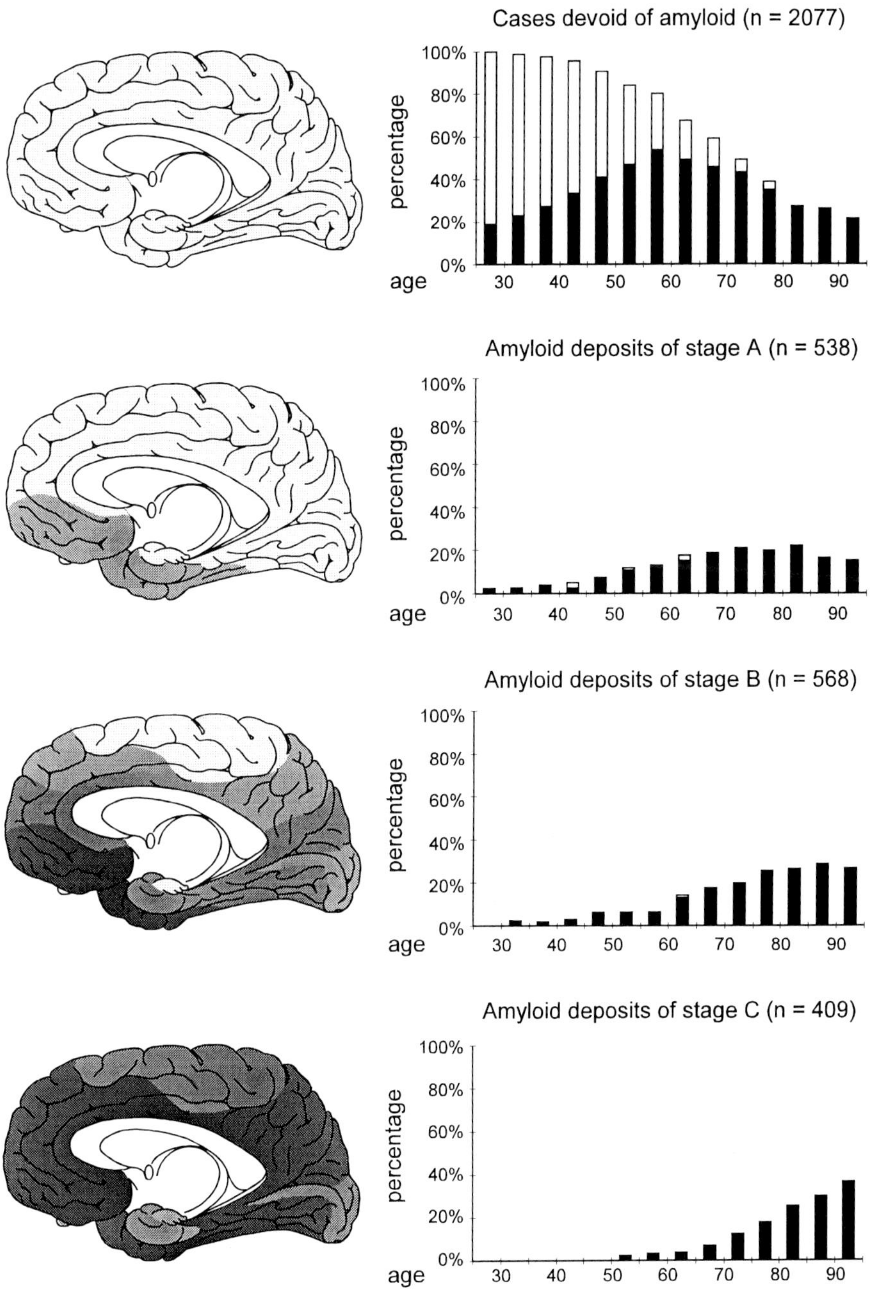


**Figures 19 and 20.** Development of AD-related changes in a total number of 3592 nonselected autopsy cases. The evolution of amyloid deposits is shown in Figure 19 and that of NFTs and NTs in Figure 20. The first row displays the frequency of cases devoid of changes in relation to the total number of cases in the various age categories. The second, third, and fourth rows are similarly designed, and show the evolution of AD-related changes. The blank areas of the bars in Figure 19 present a subgroup devoid of neurofibrillary changes. Black areas indicate a subgroup expressing NFTs/NTs, regardless of stage. Figure 20 is similarly designed, with blank bars marking a subgroup devoid of amyloid deposits, and black areas referring to a subgroup with amyloid deposits.

**Development of Neurofibrillary Changes (n = 3592)**

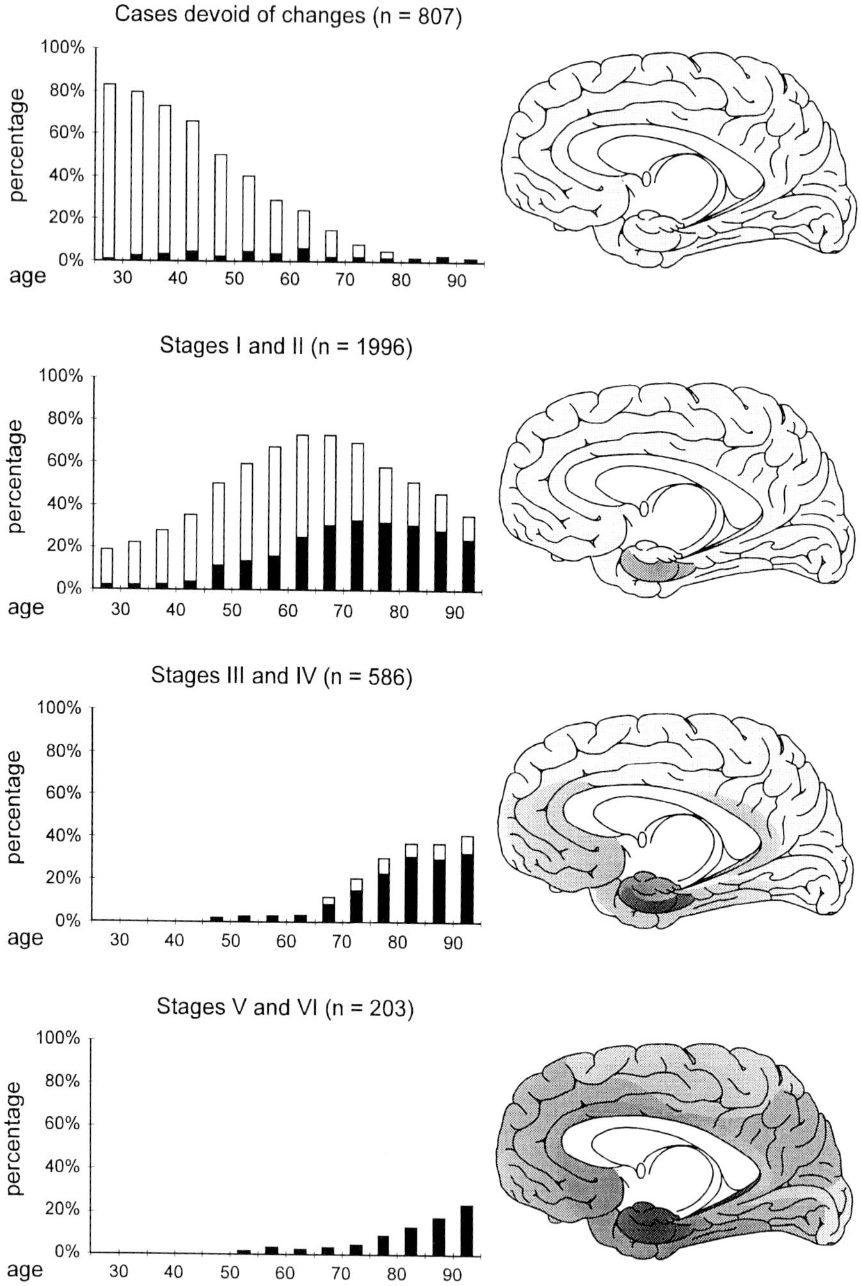

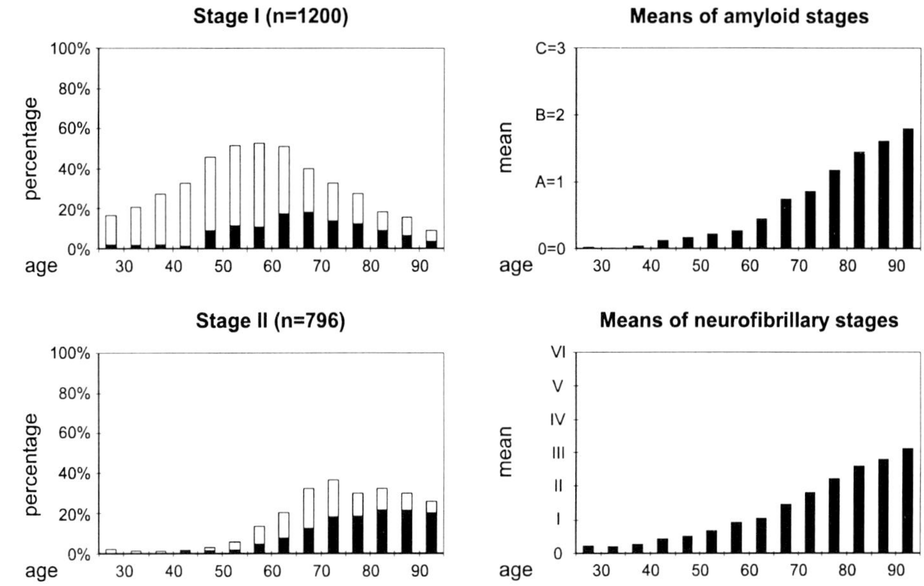


**Figure 21.** Left: Relative frequency of cases showing Stages I and II in the development of neurofibrillary changes. Initial neurofibrillary changes develop in the absence of amyloid deposits. Some individuals develop Stage I pathology early in life. Old age is not a prerequisite for the evolution of neurofibrillary changes, although there is a shift toward higher age categories with progress from Stage I to Stage II. Right: Arithmetic means of stages of both amyloid deposition (A=1, B=2, C=3) and neurofibrillary changes increases with age. Age is a risk factor for Alzheimer's disease.

both the amyloid deposition and neurofibrillary pathology likewise increases with age; age is a risk factor for AD (Fig. 21).

As noted previously, individuals with neurofibrillary Stage I or Stage II pathology show no clinical symptoms. Not all Stage I cases live long enough to arrive at end stages V–VI. Therefore, data on the appearance of Stage I are of interest because they provide information about the actual spread of the disease. To find the first neurofibrillary changes in a voluminous organ such as the human brain might seem an almost hopeless undertaking; however, this task can be coped with if the predilection sites are known (Braak and Braak, 1997b,c). In higher age categories, the relative frequency of stage I/II cases increases, reaches a culmination point, then decreases (Fig. 21, left side). The neocortical stages V and VI, in comparison, continue to increase with increasing age (Fig. 20, fourth row).

Acknowledgments. This study was supported by the Deutsche Forschungsgemeinschaft and the Bundesministerium für Bildung, Wissenschaft, Forschung und Technologie. The skillful assistance of Ms. Ingrid Szasz (drawings) is gratefully acknowledged.

## 6. References

Alexander, G. E., Crutcher, M. D., and DeLong, M. R., 1990, Basal ganglia-thalamocortical circuits: Parallel substrates for motor, oculomotor, "prefrontal," and "limbic" functions, *Prog. Brain Res.* **85:**119–146.

Alheid, G. F., Heimer, L., and Switzer, R. C., 1990, Basal ganglia, in: *The Human Nervous System* (G. Paxinos, ed.), Academic, New York, pp. 483–582.

Alzheimer, A., 1907, Über eine eigenartige Erkrankung der Hirnrinde, *Centralbl. Nervenheilk. Psychiat.*, Leipzig, **30:**177–179.

Amaral, D. G., 1987, Memory: Anatomical organization of candidate brain regions, in: *Handbook of Physiology: The Nervous System. V. Higher Functions of the Nervous System,* (J. M. Brookhart and V. B. Mountcastle, eds.), Bethesda: *Am. Physiol. Soc.,* **5:**211–294.

Amaral, D. G., and Insausti, R., 1990, Hippocampal formation, in: *The Human Nervous System,* (G. Paxinos, ed.), Academic, San Diego, pp. 711–755.

Amaral, D. G., Price, J. L., Pitkänen, A., and Carmichael, S. T., 1992, Anatomical organization of the primate amygdaloid complex, in: *The Amygdala: Neurobiological Aspects of Emotion, Memory, and Mental Dysfunction* (J. P. Aggleton, ed.), Wiley-Liss, New York, pp. 1–66.

Arnold, S. E., Hyman, B. T., Flory, J., Damasio, A. R., and Van Hoesen, G. W., 1991, The topographical and neuroanatomical distribution of neurofibrillary tangles and neuritic plaques in the cerebral cortex of patients with Alzheimer's disease, *Cereb. Cortex* **1:**103–116.

Arriagada, P. V., Growdon, J. H., Hedley-Whyte, E. T., and Hyman, B. T., 1992a, Neurofibrillary tangles, but not senile plaques, parallel duration and severity of Alzheimer's disease, *Neurology* **42:**631–639.

Arriagada, P. V., Marzloff, L., and Hyman, B. T., 1992b, Distribution of Alzheimer-type pathologic changes in nondemented elderly individuals matches the pattern in Alzheimer's disease, *Neurology* **42:**1681–1688.

Bachevalier, J., and Mishkin, M., 1992, Ontogenetic development and decline of memory functions in nonhuman primates, in: *Neurodevelopment, Aging, and Cognition* (I. Kostovic, S. Knezevic, H. M. Wisniewski, and G. J. Spilich, eds.), Birkhäuser, Boston, MA, pp. 37–59.

Bancher, C., Brunner, C., Lassmann, H., Budka, H., Jellinger, K., Wiche, G., Seitelberger, F., Grundke-Iqbal, I., and Wisniewski, H. M., 1989, Accumulation of abnormally phosphorylated $\tau$ precedes the formation of neurofibrillary tangles in Alzheimer's disease, *Brain Res.* **477:**90–99.

Bancher, C., Braak, H., Fischer, P., and Jellinger, K. A., 1993, Neuropathological staging of Alzheimer lesions and intellectual status in Alzheimer's and Parkinson's disease, *Neurosci. Lett.* **162:**179–182.

Bancher, C., Jellinger, K., Lassmann, H., and Fischer, P., 1996, Correlations between mental state and quantitative neuropathology in the Vienna longitudinal study on dementia, *Eur. Arch. Psychiatry Clin. Neurosci.* **246:**137–146.

Berg, L., and Morris, J. C., 1994, Diagnosis, in: *Alzheimer Disease* (R. D. Terry, R. Katzman, and K. L. Bick, ed.), Raven, New York, pp. 9–25.

Beyreuther, K., and Masters, C. L., 1991, Amyloid precursor protein (APP) and βA4 amyloid in the etiology of Alzheimer's disease: Precursor product relationships in the derangement of neuronal function, *Brain Pathol.* **1:**241–252.

Braak, E., Braak, H., and Mandelkow, E. M., 1994, A sequence of cytoskeleton changes related to the formation of neurofibrillary tangles and neuropil threads, *Acta Neuropathol.* **87:**554–567.

Braak, H., 1980, *Architectonics of the Human Telencephalic Cortex,* Springer-Verlag, Berlin, pp. 1–147.

Braak, H., 1984, Architectonics as seen by lipofuscin stains, in: *Cerebral Cortex, Vol. 1* (A. Peters and E. G. Jones, eds.) Plenum, New York, pp. 59–104.

Braak, H., and Braak, E., 1985a, On areas of transition between entorhinal allocortex and temporal isocortex in the human brain: Normal morphology and lamina-specific pathology in Alzheimer's disease, *Acta Neuropathol.* **68:**325–332.

Braak, H., and Braak, E., 1985b, Golgi preparations as a tool in neuropathology, with particular reference to investigations of the human telencephalic cortex, *Prog. Neurobiol.* **25:**93–139.

Braak, H., and Braak, E., 1986, Ratio of pyramidal cells versus non-pyramidal cells in the human frontal isocortex and changes in ratio with ageing and Alzheimer's disease, *Prog. Brain Res.* **70:**185–212.

Braak, H., and Braak, E., 1988, Neuropil threads occur in dendrites of tangle-bearing nerve cells, *Neuropathol. Appl. Neurobiol.* **14:**39–44.

Braak, H., and Braak, E., 1991, Neuropathological staging of Alzheimer-related changes, *Acta Neuropathol.* **82:**239–259.

Braak, H., and Braak, E., 1992, The human entorhinal cortex: Normal morphology and lamina-specific pathology in various diseases, *Neurosci. Res.* **15:**6–31.

Braak, H., and Braak, E., 1993, Alzheimer neuropathology and limbic circuits, in: *Neurobiology of Cingulate Cortex and Limbic Thalamus: A Comprehensive Handbook* (B. A. Vogt, and M. Gabriel, eds.), Birkhäuser, Boston, MA, pp. 606–626.

Braak, H., and Braak, E., 1994, Pathology of Alzheimer's disease, in: *Neurodegenerative Diseases* (D. B. Calne, ed.), W. B. Saunders, Philadelphia, PA, pp. 585–613.

Braak, H., and Braak, E., 1995, Staging of Alzheimer's disease-related neurofibrillary changes, *Neurobiol. Aging* **16:**271–284.

Braak, H., and Braak, E., 1996, Development of Alzheimer-related neurofibrillary changes in the neocortex inversely recapitulates cortical myelogenesis, *Acta Neuropathol.* **92:**197–201.

Braak, H., and Braak, E., 1997a, Cortical destruction in Alzheimer's disease, in: *Cell Death in Diseases of the Nervous System* (V. E. Koliatsos and Ratan, R., eds.), Humana, Totowa, New Jersey.

Braak, H., and Braak, E., 1997b, Time course and pattern of cortical destruction in Alzheimer's disease, in: *Proceedings of the Fifth International Conference on Alzheimer's Disease and Related Disorders, Osaka* (K. Iqbal, B. Winblad, and H. Wisniewski, eds.), Wiley, Chichester, U.K.

Braak, H., and Braak, E., 1997c, Frequency of stages of Alzheimer-related lesions in different age categories, *Neurobiol. Aging* **18:**351–357.

Braak, H., Braak, E., Grundke-Iqbal, I., and Iqbal, K., 1986, Occurrence of neuropil threads in the senile human brain and in Alzheimer's disease: A third location of paired helical filaments outside of neurofibrillary tangles and neuritic plaques, *Neurosci. Lett.* **65:**351–355.

Braak, H., Braak, E., Ohm, T. G., and Bohl, J., 1988, Silver impregnation of Alzheimer's neurofibrillary changes counterstained for basophilic material and lipofuscin pigment, *Stain Technol.* **63:**197–200.

Braak, H., Braak, E., and Kalus, P., 1989, Alzheimer's disease: Areal and laminar pathology in the occipital isocortex, *Acta Neuropathol.* **77:**494–506.

Braak, H., Duyckaerts, C., Braak, E., and Piette, F., 1993, Neuropathological staging of Alzheimer-related changes correlates with psychometrically assessed intellectual status, in: *Third International Conference on Alzheimer's Disease and Related Disorders* (B. Corian, K. Iqbal, M. Nicolini, B. Winblad, H. Wisniewski, and P. Zatta, eds.), Wiley, Chichester, U.K. pp. 131–137.

Braak, H., Braak, E., Yilmazer, D., Schultz, C., and Bohl, J., 1995, Age-related changes of the human cerebral cortex, in: *Neuropathologic Diagnostic Criteria for Brain Banking* (F. F. Cruz-Sánchez, R. Rávid, and M. L. Cuzner, eds.), IOS, Amsterdam, pp. 14–19.

Braak, H., Braak, E., Yilmazer, D., de Vos, R. A. I., Jansen, E. N. H., and Bohl, J., 1996a, Pattern of brain destruction in Parkinson's and Alzheimer's diseases, *J. Neural. Transm.* **103:**455–490.

Braak, H., Braak, E., Yilmazer, D., and Bohl, J., 1996b, Functional anatomy of human hippocampal formation and related structures, *J. Child Neurol.* **11:**265–275.

Braak, H., Braak, E., Bohl, J., and Reintjes, R., 1996c, Age, neurofibrillary changes, Aβ-amyloid, and the onset of Alzheimer's disease, *Neurosci. Lett.* **210:**87–90.

Brion, J. P., and Resibois, A., 1994, A subset of calretinin-positive neurons are abnormal in Alzheimer's disease, *Acta Neuropathol.* **88:**33–43.

Brodmann, K., 1910, Feinere Anatomie des Groβhirns, in: *Handbuch der Neurologie* (M. Lewandowsky, ed.), Springer, Berlin **1:**206–307.

Brun, A., and Gustafson, L., 1976, Distribution of cerebral degeneration in Alzheimer's disease, *Arch. Psychiat. Nervenkr.* **223:**15–33.

Cabalka, L. M., Hyman, B. T., Goodlett, C. R., Ritchie, T. C., and van Hoesen, G. W., 1992, Alteration in the pattern of nerve terminal protein immunoreactivity in the perforant pathway in Alzheimer's disease and in rats after entorhinal lesions, *Neurobiol. Aging* **13:**283–291.

Cadelli, D. S., Bandtlow, C. E., and Schwab, M. E., 1992, Oligodendrocyte- and myelin-associated inhibitors of neurite outgrowth: Their involvement in the lack of CNS regeneration, *Exp. Neurol.* **115:**189–192.

Callahan, L. M., and Coleman, P. D., 1995, Neurons bearing neurofibrillary tangles are responsible for selected synaptic deficits in Alzheimer's disease, *Neurobiol. Aging* **16:**311–314.

Campbell, S. K., Switzer, R. C., and Martin, T. L., 1987, Alzheimer's plaques and tangles: A controlled and enhanced silver-staining method, *Soc. Neurosci. Abstr.* **13:**678.

Coleman, P. D., and Flood, D. G., 1987, Neuron numbers and dendritic extent in normal aging and Alzheimer's disease, *Neurobiol. Aging* **8:**521–545.

Corey-Bloom, J., Galasko, D., and Thal, L. J., 1994, Clinical features and natural history of Alzheimer's disease, in: *Neurodegenerative Diseases* (D. B. Calne, ed.), W. B. Saunders, Philadelphia, PA, pp. 631–645.

Damasio, A. R., and Damasio, H., 1991, Disorders of higher brain function, in: *Comprehensive Neurology* (R. N. Rosenberg, ed.), Raven, New York, pp. 639–657.

Delaère, P., Duyckaerts, C., He, Y., Piette, F., and Hauw, J. J., 1991, Subtypes and differential laminar

distributions of βA4 deposits in Alzheimer's disease: Relationship with intellectual status of 26 cases, *Acta Neuropathol.* **81:**328–335.
Dickson, D. W., Crystal, H. A., Mattiace, L. A., Masur, D. M., Blau, A. D., Davies, P., Yen, S. H., and Aronson, M. K., 1991, Identification of normal and pathological aging in prospectively studied nondemented elderly humans, *Neurobiol. Aging* **13:**179–189.
Dickson, D. W., Ksiezak-Reding, H., Liu, W. K., Davies, P., Crowe, A., and Yen, S. H. C., 1992, Immunocytochemistry of neurofibrillary tangles with antibodies to subregions of tau protein: Identification of hidden and cleaved tau epitopes and a new phosphorylation site, *Acta Neuropathol.* **84:**596–605.
Duvernoy, H. M., 1988, *The Human Hippocampus*, Bergmann, Munich.
Duyckaerts, C., He, Y., Seilhean, D., Delaère, P., Piette, F., Braak, H., and Hauw, J. J., 1994, Diagnosis and staging of Alzheimer's disease in a prospective study involving aged individuals, *Neurobiol. Aging (Suppl. 1)* **15:**140–141.
Duyckaerts, C., Delaère, P., He, Y., Camilleri, S., Braak, H., Piette, F., and Hauw, J. J., 1995, The relative merits of tau and amyloid markers in the neuropathology of Alzheimer's disease, in: *Treating Alzheimer's and Other Dementias* (M. Bergener and S. I. Finkel, eds.), Springer-Verlag, New York, pp. 81–89.
Fairén, A., DeFelipe, J., and Regidor, J., 1984, Nonpyramidal neurons: General account, in: *Cerebral Cortex., Vol. 1* (A. Peters and E. G. Jones, eds.), Plenum, New York, pp. 201–253.
Feldman, M. L., 1984, Morphology of the neocortical pyramidal neuron, in: *Cerebral Cortex, Vol. 1* (A. Peters and E. G. Jones, eds.), Plenum, New York, pp. 123–200.
Felleman, D. J., and van Essen, D. C., 1991, Distributed hierarchical processing in the primate cerebral cortex, *Cereb. Cortex* **1:**1–47.
Fewster, P. H., Griffin-Brooks, S., MacGregor, J., Ojalvo-Rose, E., and Ball, M. J., 1991, A topographical pathway by which histopathological lesions disseminate through the brain of patients with Alzheimer's disease, *Dementia* **2:**121–132.
Flechsig, P., 1920, *Anatomie des menschlichen Gehirns und Rückenmarks auf myelogenetischer Grundlage*, Thieme, Leipzig.
Fonseca, M., Soriano, E., Ferrer, I., Martinez, A., and Tuñon, T., 1993, Chandelier cell axons identified by parvalbumin-immunoreactivity in the normal human temporal cortex and in Alzheimer's disease, *Neuroscience* **55:**1107–1116.
Franssen, E. H., Kluger, A., Torossian, C. L., and Reisberg, B., 1993, The neurologic syndrome of severe Alzheimer's disease, *Arch. Neurol.* **50:**1029–1039.
Gallyas, F., 1971, Silver staining of Alzheimer's neurofibrillary changes by means of physical development, *Acta Morphol. Acad. Sci. Hungary* **19:**1–8.
Gearing, M., Mirra, S. S., Hedreen, J. C., Sumi, S. M., Hansen, L. A., and Heyman, A., 1995, The consortium to establish a registry for Alzheimer's disease (CERAD). Part X. Neuropathology confirmation of the clinical diagnosis of Alzheimer's disease, *Neurology* **45:**461–466.
Goedert, M., 1993, Tau protein and the neurofibrillary pathology of Alzheimer's disease, *Trends Neurosci.* **16:**460–465.
Goedert, M., Spillantini, M. G., and Crowther, R. A., 1991, Tau proteins and neurofibrillary degeneration, *Brain Pathol.* **1:**279–286.
Goedert, M., Spillantini, M. G., Cairns, N. J., and Crowther, R. A., 1992, Tau proteins of Alzheimer paired helical filaments: Abnormal phosphorylation of all six brain isoforms, *Neuron* **8:**159–168.
Grünthal, E., 1930, Die pathologische Anatomie der senilen Demenz und der Alzheimerschen Krankheit, in: *Handbuch der Geisteskrankheiten: Die Anatomie der Psychosen, Vol. 11* (O. Blumke, ed.), Springer-Verlag, Berlin, pp. 638–672.
Haass, C., 1996, The molecular significance of amyloid β-peptide for Alzheimer's disease, *Eur. Arch. Psychiatry Clin. Neurosci.* **246:**118–123.
Hardy, R., and Reynolds, R., 1993, Neuron-oligodendroglial interactions during central nervous system development, *J. Neurosci. Res.* **36:**121–126.
Harrell, L. E., Callaway, R., and Powers, R., 1993, Autopsy in dementing illness: Who participates? *Alzheimer Dis. Assoc. Disord.* **7:**80–87.
Heimer, L., Switzer, R. C., and Van Hoesen, G. W., 1982, Ventral striatum and ventral pallidum: Components of the motor system? *Trends Neurosci.* **5:**83–87.
Heimer, L., de Olmos, J., Alheid, G. F., and Zaborszky, L., 1991, "Perestroika" in the basal forebrain: Opening the border between neurology and psychiatry, *Prog. Brain Res.* **87:**109–165.

Heinsen, H., Henn, R., Eisenmenger, W., Götz, M., Bohl, J., Bethke, B., Kockemann, U., and Püschel, K., 1994, Quantitative investigations on the human entorhinal area: Left–right asymmetry and age-related changes, *Anat. Embryol.* **190:**181–194.

Hof, P. R., and Morrison, J. H., 1990a, Quantitative analysis of a vulnerable subset of pyramidal neurons in Alzheimer's disease. I. Superior frontal and inferior temporal cortex, *J. Comp. Neurol.* **301:**44–54.

Hof, P. R., and Morrison, J. H., 1990b, Quantitative analysis of a vulnerable subset of pyramidal neurons in Alzheimer's disease. II. Primary and secondary visual cortex, *J. Comp. Neurol.* **301:**55–64.

Hof, P. R., and Morrison, J. H., 1991, Neocortical neuronal subpopulations labeled by a monoclonal antibody to calbindin exhibit differential vulnerability in Alzheimer's disease, *Exp. Neurol.* **111:**293–301.

Hof, P. R., Cox, K., Young, W. G., Celio, M. R., Rogers, J., and Morrison, J. H., 1991, Parvalbumin-immunoreactive neurons in the neocortex are resistant to degeneration in Alzheimer's disease, *J. Neuropathol. Exp. Neurol.* **50:**451–462.

Hof, P. R., Nimchinsky, E. A., Celio, M. R., Bouras, C., and Morrison, J. H., 1993, Calretinin-immunoreactive neocortical interneurons are unaffected in Alzheimer's disease, *Neurosci. Lett.* **152:**145–149.

Hopf, A., 1954, Die Myeloarchitektonik des Isocortex temporalis beim Menschen, *J. Hirnforsch.* **1:**208–279.

Hyman, B. T., and Gomez-Isla, T., 1994, Alzheimer's disease is a laminar, regional, and neural system specific disease, not a global brain disease, *Neurobiol. Aging* **15:**353–354.

Hyman, B. T., van Hoesen, G. W., Damasio, A. R., and Barnes, C. L., 1984, Alzheimer's disease: Cell-specific pathology isolates the hippocampal formation, *Science* **225:**1168–1170.

Hyman, B. T., van Hoesen, G. W., Kromer, L. J., and Damasio, A. R., 1986, Perforant pathway changes and the memory impairment of Alzheimer's disease, *Ann. Neurol.* **20:**472–481.

Hyman, B. T., Kromer, L. J., and van Hoesen, G. W., 1988, A direct demonstration of the perforant pathway terminal zone in Alzheimer's disease using the monoclonal antibody Alz-50, *Brain Res.* **450:**392–397.

Hyman, B. T., van Hoesen, G. W., and Damasio, A. R., 1990, Memory-related neural systems in Alzheimer's disease: An anatomic study, *Neurology* **40:**1721–1730.

Hyman, B. T., Marzloff, K., and Arriagada, P. V., 1993, The lack of accumulation of senile plaques or amyloid burden in Alzheimer's disease suggests a dynamic balance between amyloid deposition and resolution, *J. Neuropathol. Exp. Neurol.* **52:**594–600.

Ikeda, J., Mori, K., Oka, S., and Watanabe, Y., 1989, A columnar arrangement of dendritic processes of entorhinal cortex neurons revealed by a monoclonal antibody, *Brain Res.* **505:**176–179.

Insausti, R., Tuñon, T., Sobreviela, T., Insausti, A. M., and Gonzalo, L. M., 1995, The human entorhinal cortex: A cytoarchitectonic analysis, *J. Comp. Neurol.* **355:**171–198.

Iqbal, K., and Grundke-Iqbal, I., 1991, Alzheimer's disease: From cytoskeletal protein pathology to neuronal degeneration, in: *Alzheimer's Disease: Basic Mechanisms, Diagnosis, and Therapeutic Strategies* (K. Iqbal, D. R. C. McLachlan, B. Winblad, and H. M. Wisniewski, eds.), Wiley, Chichester, U.K., pp. 173–180.

Iqbal, K., Braak, H., Braak, E., and Grundke-Iqbal, I., 1993, Silver labeling of Alzheimer neurofibrillary changes and brain β-amyloid, *J. Histotechnol.* **16:**335–342.

Iqbal, K., Alonso, A. C., Gong, C. X., Khatoon, S., Singh, T. J., and Grundke-Iqbal, I., 1994, Mechanism of neurofibrillary degeneration in Alzheimer's disease, *Mol. Neurobiol.* **9:**119–123.

Jarrett, J., and Lansbury, P., 1993, Seeding "one-dimensional crystallization" of amyloid: A pathogenic mechanism in Alzheimer's disease and scrapie? *Cell* **73:**1055–1058.

Jellinger, K., Braak, H., Braak, E., and Fischer, P., 1991, Alzheimer lesions in the entorhinal region and isocortex in Parkinson's and Alzheimer's diseases, *Ann. N.Y. Acad. Sci.* **640:**203–209.

Joachim, C. L., and Selkoe, D. J., 1989, Minireview: Amyloid protein in Alzheimer's disease, *J. Gerontol. (Biol. Sci.)* **44:**77–82.

Jones, E. G., 1985, *The Thalamus*, Plenum, New York.

Kalus, P., Braak, H., Braak, E., and Bohl, J., 1989, The presubicular region in Alzheimer's disease: Topography of amyloid deposits and neurofibrillary changes, *Brain Res.* **494:**198–203.

Kapfhammer, J. P., and Schwab, M. E., 1994, Inverse patterns of myelination and GAP-43 expression in the adult CNS: Neurite growth inhibitors as regulators of neuronal plasticity, *J. Comp. Neurol.* **340:**194–206.

Kazee, A. M., Eskin, T. A., Lapham, L. W., Gabriel, K. R., McDaniel, K. D., and Hamill, R. W., 1993, Clinicopathologic correlates in Alzheimer's disease: Assessment of clinical and pathologic diagnostic criteria, *Alzheimer. Dis. Assoc. Disord.* **7:**152–164.

Kemper, T. L., 1978, Senile dementia: A focal disease in the temporal lobe, in: *Senile dementia: A Biomedical Approach* (E. Nandy, ed.), Elsevier, Amsterdam, pp. 105–113.

Klingler, J., 1948, Die makroskopische Anatomie der Ammonsformation, *Denkschr. Schweiz. Naturforsch. Ges.* **78(1)**:1–80.

Lassmann, H., Bancher, C., Breitschopf, H., Wegiel, J., Bobinski, M., Jellinger, K., and Wisniewski, H. M., 1995, Cell death in Alzheimer's disease evaluated by DNA fragmentation *in situ, Acta Neuropathol.* **89**:35–41.

Lee, S., Christakos, S., and Small, M. B., 1993, Apoptosis and signal transduction: Clues to a molecular mechanism, *Curr. Opin. Cell Biol.* **5**:286–291.

Lewis, D. A., Campbell, M. J., Terry, R. D., and Morrison, J. H., 1987, Laminar and regional distributions of neurofibrillary tangles and neuritic plaques in Alzheimer's disease: A quantitative study of visual and auditory cortices, *J. Neurosci.* **7**:1799–1808.

Markesbery, W. R., Wang, H. Z., Kowall, N. W., Kosik, K. S., and McKee, A. C., 1993, Morphometric image analysis of neuropil threads in Alzheimer's disease, *Neurobiol. Aging* **14**:303–307.

McKee, A. C., Kosik, K. S., and Kowall, N. W., 1991, Neuritic pathology and dementia in Alzheimer's disease, *Ann. Neurol.* **30**:156–165.

Mesulam, M. M., Hersh, L. B., Mash, D. C., and Geula, C., 1992, Differential cholinergic innervation within functional subdivisions of the human cerebral cortex: A choline acetyltransferase study, *J. Comp. Neurol.* **318**:316–328.

Mirra, S. S., Heyman, A., McKeel, D., Sumi, S. M., Crain, B. J., Brownlee, L. M., Vogel, F. S., Hughes, J. P., van Belle, G., and Berg, L., 1991, The consortium to establish a registry for Alzheimer's disease (CERAD). II. Standardization of the neuropathologic assessment of Alzheimer's disease, *Neurology* **41**:479–486.

Mizutani, T., and Kasahara, M., 1995, Degeneration of the intrahippocampal routes of the perforant and alvear pathways in senile dementia of Alzheimer type, *Neurosci. Lett.* **184**:141–144.

Moossy, J., Zubenko, G. S., Martinez, A. J., and Rao, G. R., 1988, Bilateral symmetry of morphological lesions in Alzheimer's disease, *Arch. Neurol.* **45**:251–254.

Morel, F., 1945, Les aires striée et parastriée et péristriée dans le troubles de la fonction visuelle au cours de la maladie d'Alzheimer, *Confin. Neurol.* **6**:238–242.

Morrison, J. H., Lewis, D. A., Campbell, M. J., Huntley, G. W., Benson, D. L., and Bouras, C., 1987, A monoclonal antibody to non-phosphorylated neurofilament protein marks the vulnerable cortical neurons in Alzheimer's disease, *Brain Res.* **416**:331–336.

Munoz, D. G., and Wang, D., 1992, Tangle-associated neuritic clusters: A new lesion in Alzheimer's disease and aging suggests that aggregates of dystrophic neurites are not necessarily associated with β/A4, *Am. J. Pathol.* **140**:1167–1178.

Mutrux, S., 1947, Diagnostic différentiel histologique de la maladie d'Alzheimer et de la démence sénile: Pathophobie de la zone de projection corticale, *Monatsschr. Psychiat. Neurol.* **113**:100–117.

Nauta, H. J. W., 1979, A proposed conceptual reorganization of the basal ganglia and telencephalon, *Neuroscience* **4**:1875–1881.

Ohm, T. G., Müller, H., Braak, H., and Bohl, J., 1995, Close-meshed prevalence rates of different stages as a tool to uncover the rate of Alzheimer's disease-related neurofibrillary changes, *Neuroscience* **64**:209–217.

Pandya, D. N., and Yeterian, E. H., 1985, Architecture and connections of cortical association areas, in: *Cerebral Cortex, Vol. 4* (A. Peters and E. G. Jones, eds.), Plenum, New York, pp. 3–61.

Pandya, D. N., and Yeterian, E. H., 1990, Prefrontal cortex in relation to other cortical areas in rhesus monkey: Architecture and connections, *Prog. Brain Res.* **85**:63–94.

Peters, A., Rosene, D. L., Moss, M. B., Kemper, T. L., Abraham, C. R., Tigges, J., and Albert, M. S., 1996, Neurobiological bases of age-related cognitive decline in the rhesus monkey, *J. Neuropathol. Exp. Neurol.* **55**:861–874.

Preuss, U., Döring, F., Illenberger, S., and Mandelkow, E. M., 1995, Cell cycle-dependent phosphorylation and microtubule binding of tau protein stably transfected into Chinese hamster ovary cells, *Mol. Biol. Cell* **6**:1397–1410.

Price, D. L., and Sisodia, S. S., 1994, Cellular and molecular biology of Alzheimer's disease and animal models, *Annu. Rev. Med.* **45**:435–446.

Price, J. L., Davis, P. B., Morris, J. C., and White, D. L., 1991, The distribution of tangles, plaques and related immunohistochemical markers in healthy aging and Alzheimer's disease, *Neurobiol. Aging* **12**:295–312.

Probst, A., Brunnschweiler, H., Lautenschlager, C., and Ulrich, J., 1987, A special type of senile plaque, possibly an initial stage, *Acta Neuropathol.* **74:**133–141.

Probst, A., Anderton, B. H., Brion, J. P., and Ulrich, J., 1989, Senile plaque neurites fail to demonstrate anti-paired helical filament and anti-microtubule-associated protein-tau immunoreactive proteins in the absence of of neurofibrillary tangles in the neocortex, *Acta Neuropathol.* **77:**430–436.

Probst, A., Langui, D., and Ulrich, J., 1991a, Alzheimer's disease: A description of the structural lesions, *Brain Pathol.* **1:**229–239.

Probst, A., Langui, D., Ipsen, S., Robakis, N., and Ulrich, J., 1991b, Deposition of β/A4 protein along neuronal plasma membranes in diffuse senile plaques, *Acta Neuropathol.* **83:**21–29.

Regeur, L., Jensen, G. B., Pakkenberg, H., Evans, S. M., and Pakkenberg, B., 1994, No global neocortical nerver cell loss in brains from patients with senile dementia of Alzheimer's type, *Neurobiol. Aging* **15:**347–352.

Reisberg, B., 1988, Functional assessment staging (FAST), *Psychopharmacol. Bull.* **24:**653–659.

Reisberg, B., Pattschull-Furlan, A., Franssen, E., Sclan, S. G., Kluger, A., Dingcong, L., and Ferris, S. H., 1992, Dementia of the Alzheimer type recapitulates ontogeny inversely on specific ordinal and temporal parameters, in: *Neurodevelopment, Aging, and Cognition* (I. Kostovic, S. Knezevic, H. M. Wisniewski, and G. J. Spillich, eds.), Birkhäuser, Boston, MA, pp. 345–369.

Rockland, K. S., and Pandya, D. N., 1979, Laminar origins and terminations of cortical connections of the occipital lobe in the rhesus monkey, *Brain Res.* **179:**3–20.

Rogers, J., and Morrison, J. H., 1985, Quantitative morphology and regional and laminar distribution of senile plaques in Alzheimer's disease, *J. Neurosci.* **5:**2801–2808.

Rose, M., 1935, Cytoarchitektonik und Myeloarchitektonik der Großhirnrinde, in: *Handbuch der Neurologie,* Vol. 1 (O. Bumke and O. Foerster, eds.), Springer-Verlag, Berlin, pp. 588–778.

Rosene, D. L., and van Hoesen, G. W., 1987, The hippocampal formation of the primate brain: A review of some comparative aspects of cytoarchitecture and connections, in: *Cerebral Cortex, Vol. 6* (E. G. Jones and A. Peters, eds.), Plenum, New York, pp. 345–456.

Sanides, F., 1962, Die Architektonik des menschlichen Stirnhirns, in: *Monographien aus dem Gesamtgebiete der Neurologie und Psychiatrie, Vol. 98* (M. Müller, H. Spatz, and P. Vogel, eds.) Springer-Verlag, Berlin.

Sanides, F., 1969, Comparative architectonics of the neocortex of mammals and their evolutionary interpretation, *Ann. N.Y. Acad. Sci.* **167:**404–423.

Saper, C. P., 1990, Cholinergic system, in: *The Human Nervous System* (G. Paxinos, ed.), Academic, New York, pp. 1095–1113.

Schmidt, M. L., Lee, V. M. Y., and Trojanowski, J. Q., 1991, Comparative epitope analysis of neuronal cytoskeletal proteins in Alzheimer's disease senile plaque neurites and neuropil threads, *Lab. Invest.* **64:**352–357.

Schwab, M. E., 1990, Myelin-associated inhibitors of neurite growth and regeneration in the CNS, *Trends Neurosci.* **13:**452–456.

Selkoe, D. J., 1991, The molecular pathology of Alzheimer's disease, *Neuron* **6:**487–498.

Selkoe, D. J., 1993, Physiological production of the β-amyloid protein and the mechanism of Alzheimer's disease, *Trends Neurosci.* **16:**403–409.

Selkoe, D. J., 1994, Alzheimer's disease: A central role for amyloid, *J. Neuropathol. Exp. Neurol.* **53:**438–447.

Simchowicz, T., 1911, Histologische Studien über die senile Demenz, in: *Histologische und histopathologische Arbeiten über die Großhirnrinde mit besonderer Berücksichtigung der pathologischen Anatomie der Geisteskrankheiten, Vol. 4* (F. Nissl and A. Alzheimer, eds.), Fischer, Jena, Germany, pp. 267–444.

Sloper, J. J., Johnson, P., and Powell, T. P. S., 1980, Selective degeneration of interneurons in the motor cortex of infant monkeys following controlled hypoxia: A possible cause of epilepsy, *Brain Res.* **198:**204–209.

Solodkin, A., and van Hoesen, G. W., 1996, Entorhinal cortex modules of the human brain, *J. Comp. Neurol.* **365:**610–627.

Squire, L. R., 1992, Memory and the hippocampus: A synthesis from findings with rats, monkeys, and humans, *Psychol. Rev.* **99:**195–231.

Squire, L. R., and Zola-Morgan, S., 1991, The medial temporal lobe memory system, *Science* **253:**1380–1386.

Stephan, H., 1975, Allocortex, in: *Handbuch der Mikroskopischen Anatomie des Menschen, Vol. 4, Part 9,* (W. Bargmann, ed.), Springer-Verlag, Berlin, pp. 1–998.

Sugihara, S., Ogawa, A., Nakazato, Y., and Yamaguchi, H., 1995, Cerebral β amyloid deposition in patients with malignant neoplasms: Its prevalence with aging, and effects of radiation therapy on vascular amyloid, *Acta Neuropathol.* **90:**135–141.

Tierney, M. C., Fisher, R. H., Lewis, A. J., Zorzitto, M. L., Snow, W. G., Reid, D. W., and Nieuwstraten, P., 1988, The NINCDS-ADRDA work group criteria for the clinical diagnosis of probable Alzheimer's disease: A clinicopathologic study of 57 cases, *Neurology* **38:**359–364.

Tomlinson, B. E., 1989, The neuropathology of Alzheimer's disease: Issues in need of resolution, *Neuropathol. Appl. Neurobiol.* **15:**491–512.

Trojanowski, J. Q., Shin, R. W., Schmidt, M. L., and Lee, V. M. Y., 1995, Relationship between plaques, tangles, and dystrophic processes in Alzheimer's disease, *Neurobiol. Aging* **16:**335–345.

Ulrich, J., Meier-Ruge, W., Probst, A., Meier, E., and Ipsen, S., 1990, Senile plaques: Staining for acetylcholinesterase and A4 protein: A comparative study in the hippocampus and entorhinal cortex, *Acta Neuropathol.* **80:**624–628.

Unger, J. W., and Lange, W., 1992, NADPH-diaphorase-positive cell populations in the human amygdala and temporal cortex: Neuroanatomy, peptidergic characteristics, and aspects of aging and Alzheimer's disease, *Acta Neuropathol.* **83:**636–646.

van Hoesen, G. W., 1982, The primate parahippocampal gyrus: New observations regarding its cortical connections in the monkey, *Trends Neurosci.* **5:**345–350.

van Hoesen, G. W., and Hyman, B. T., 1990, Hippocampal formation: Anatomy and the patterns of pathology in Alzheimer's disease, *Prog. Brain Res.* **83:**445–457.

van Hoesen, G. W., and Solodkin, A., 1993, Some modular features of temporal cortex in humans as revealed by pathological changes in Alzheimer's disease, *Cereb. Cortex* **3:**465–475.

van Hoesen, G. W., Hyman, B. T., and Damasio, A. R., 1991, Commentary: Entorhinal cortex pathology in Alzheimer's disease, *Hippocampus* **1:**1–8.

Vaughan, D. W., 1984, The structure of neuroglial cells, in: *Cerebral Cortex, Vol. 2* (E. G. Jones and A. Peters, eds.), Plenum, New York, pp. 285–329.

Vincent, I., Rosado, M., and Davies, P., 1996, Mitotic mechanisms in Alzheimer disease? *J. Cell Biol.* **132:**413–425.

Vinters, H. V., 1987, Cerebral amyloid angiopathy: A critical review, *Stroke* **18:**311–324.

Vinters, H. V., 1992, Cerebral amyloid angiopathy and Alzheimer's disease: Two entities or one? *J. Neurol. Sci.* **112:**1–3.

Vogt, B. A., 1985, Cingulate cortex, in: *Cerebral Cortex, Vol. 4* (A. Peters and E. G. Jones, eds.), Plenum, New York, pp. 89–149.

Vogt, C., and Vogt, O., 1919, Allgemeinere Ergebnisse unserer Hirnforschung, *J. Psychol. Neurol.* **25:**279–462.

von Braunmühl, A., 1957, Alterserkrankungen des Zentralnervensystems: Senile Involution, Senile Demenz, Alzheimersche Krankheit, in: *Handbuch der speziellen pathologischen Anatomie und Histologie.* Vol. 13/1A, (O. Lubarsch, F. Henke, and R. Rössle, eds.), Springer-Verlag, Berlin, pp. 337–539.

Wisniewski, H. M., Bancher, C., Barcikowska, M., Wen, G. Y., and Currie, J., 1989, Spectrum of morphological appearance of amyloid deposits in Alzheimer's disease, *Acta Neuropathol.* **78:**337–347.

Witter, M. P., 1993, Organization of the entorhinal-hippocampal system: A review of current anatomical data, *Hippocampus (Suppl.)* **3:**33–44.

Witter, M. P., van Hoesen, G. W., and Amaral, D. G. 1989, Topographical organization of the entorhinal projection to the dentate gyrus of the monkey, *J. Neurosci.* **9:**216–228.

Yakovlev, P. I., and Lecours, A. R., 1967, The myelogenetic cycles of regional maturation of the brain, in: *Regional Development of the Brain in Early Life* (A. Minowski, ed.), Blackwell, Oxford, pp. 3–70.

Yamaguchi, H., Hirai, S., Morimatsu, M., Shoji, M., and Ihara, Y., 1988, A variety of cerebral amyloid deposits in the brains of the Alzheimer-type dementia demonstrated by β-protein immunostaining, *Acta Neuropathol.* **76:**541–549.

Yamaguchi, H., Nakazato, Y., Shoji, M., Ihara, Y., and Hirai, S., 1990, Ultrastructure of neuropil threads in the Alzheimer brain: Their dendritic origin and accumulation in the senile plaques, *Acta Neuropathol.* **80:**368–374.

Yamaguchi, H., Nakazato, Y., Kawarabayashi, T., Ishiguro, K., Ihara, Y., Morimatsu, M., Hirai, S., 1991, Extracellular neurofibrillary tangles associated with degenerating neurites and neuropil threads in Alzheimer-type dementia, *Acta Neuropathol.* **81:**603–609.

Zilles, K., 1990, Cortex, in: *The Human Nervous System,* (G. Paxinos, ed.), Academic, New York, pp. 757–802.

Zola-Morgan, S., and Squire, L. R., 1993, Neuroanatomy of memory, *Annu. Rev. Neurosci.* **16:**547–563.

# 15

# Neuropathological Correlates of Dementia in Alzheimer's Disease

ELIEZER MASLIAH and DAVID P. SALMON

## 1. Introduction

### 1.1. Synapse Formation and Remodeling as the Microanatomical Correlate of Cognition

Memory and learning are fundamental components of cognition. During the course of Alzheimer's disease (AD) there is a progressive deterioration of these mental functions, resulting in the loss of intellectual and social abilities of the individual (Khachaturian, 1985; Katzman *et al.*, 1988; Terry *et al.*, 1991). Therefore, understanding the basic mechanisms involved in learning and memory is of fundamental importance for understanding the pathogenesis of dementia in AD. Memory formation following learning is presumed to result from modification in the efficacy of neural circuitry, either through strengthening of preexisting synapses, or formation of new contacts (Stewart and Rusakov, 1995). Recent studies postulate that the memory process includes a short-term and a long-term component

ELIEZER MASLIAH • Departments of Neurosciences and Pathology, University of California, San Diego, La Jolla, California 92093-0624. DAVID P. SALMON • Department of Neurosciences, University of California, San Diego, La Jolla, California 92093-0948.

*Cerebral Cortex*, Volume 14, edited by Peters and Morrison. Kluwer Academic/Plenum Publishers, New York, 1999.

(Addersen and Trommald, 1995), with the former mediated by the limbic system and the latter involving changes in the neocortex (Andersen and Trommald, 1995).

Studies in the chick following passive-avoidance learning have shown that initial acquisition of memory involves population changes in the fine spatial organization of synaptic vesicles and active zones in synapses, whereas longer-term changes are more prominent and primarily involve an increase in the densities of synapses and dendritic spines (Stewart and Rusakov, 1995). The short-term synaptic changes are as dynamic as the molecular changes, which have been considered the substrate of short-term memory formation (Stewart and Rusakov, 1995); thus, structural synaptic changes remain a probable mechanism for long-term memory. A possible long-lasting consequence of such spatial learning is a change in size, shape, or number of excitatory synapses (Andersen and Trommald, 1995; Van Reempts *et al.,* 1992). For example, within the limbic system, spatial learning may induce new excitatory synapses in a small subset of hippocampal CA1 neurons, because a higher spine density without alteration in dendritic length or branching has been established (Andersen and Trommald, 1995). Irrespective of spine density, the dendritic synapses are regularly spaced, suggesting the operation of an intersynaptic dispersing force. With unchanged dendritic length, the higher spine density reflects an increased number of excitatory synapses per neuron associated with spatial learning (Andersen and Trommald, 1995).

Because increased synaptic density appears to be associated with long-term learning and memory, synaptic loss might result in cognitive deficits. Recent studies have shown that impaired spatial learning in aged rats is associated with synaptic loss within the fronto-parietal cortex. In contrast, unimpaired aged rats showed synaptic densities similar to young controls (Chen *et al.,* 1995a). Furthermore, synapse loss in the impaired aged rats was ameliorated by chronic infusion of nerve growth factor (NGF) into the lateral ventricles (Chen *et al.,* 1995a). Remarkably, the reversal in cognitive performance in the water maze was correlated with synaptic sprouting in the neocortex. Taken together, these studies suggest that learning and memory are associated with structural modifications both at the pre- and postsynaptic sites, indicating that the integrity of the synaptic apparatus is a basic prerequisite to maintaining intact cognitive functioning. Therefore, depending on the neural circuitry affected, damage to the pre- and/or postsynaptic sites might result in various patterns of cognitive and motor dysfunction (Masliah *et al.,* 1993a).

### 1.2. Pre- and Postsynaptic Damage as the Basis for Dementia

During progression of neurodegeneration, damage might occur at the pre- (axonal) or postsynaptic (dendritic) site (Masliah *et al.,* 1993a, 1996a) (Table I). Although in most neurodegenerative disorders both ends are injured, it is possible that the severity and characteristics of damage might indicate where the insult begins. For example, recent studies have shown that in AD there is widespread damage to presynaptic terminals and axons (Honer *et al.,* 1992a; Lassmann *et al.,* 1992; Masliah *et al.,* 1991a,b, 1993b, 1989; DeKosky and Scheff, 1990; Hamos *et al.,*

**Table I. Structural Alterations in the Synapto-Dendritic Apparatus in Neurodegenerative Disease**

| | Presynaptic | Postsynaptic |
|---|---|---|
| Alzheimer's disease | Mild-to-severe primary loss | Secondary atrophy |
| HIV encephalitis | Mild secondary loss | Severe damage; vacuolization |
| Lewy body variant | Mild-to-moderate primary loss | Moderate vacuolization |
| Prion disease | Moderate loss | Severe vacuolization; spine loss |
| Pick's disease | Severe loss | Secondary atrophy |

1989; Davies *et al.,* 1987) whereas Creutzfeldt-Jacob disease (CJD) and human immunodeficiency virus (HIV) encephalitis show extensive damage to the postsynaptic site (Masliah *et al.,* 1992a, 1996a; Gray, 1986). In contrast, the atrophic dendritic alterations in AD appear to be secondary to the severe synaptic damage, whereas in HIV encephalitis synaptic alterations are less prominent and appear to be concomitant to the extensive dendritic alterations. In this context, selectivity for synaptic or dendritic damage might occur as a result of direct effect of specific neurotoxins, receptor specificity, or failure of trophic factors to maintain the synaptic structure (Fig. 1). Moreover, synaptic dysfunction might be associated with accumulation of abnormally processed synaptic protein fragments [i.e., amyloid-β protein (Aβ), non-Aβ component of AD amyloid (NAC)], or abnormal axoplasmic flow, or it may be secondary to neuronal cell death (Masliah *et al.,* 1996c) (Fig. 1). Further development of experimental transgenic and knockout murine animal models will help to elucidate the molecular bases for selective synapto-dendritic damage in neurodegenerative disorders (Mucke *et al.,* 1995a). Our chapter reviews the most recent approaches to understanding the neuropathological

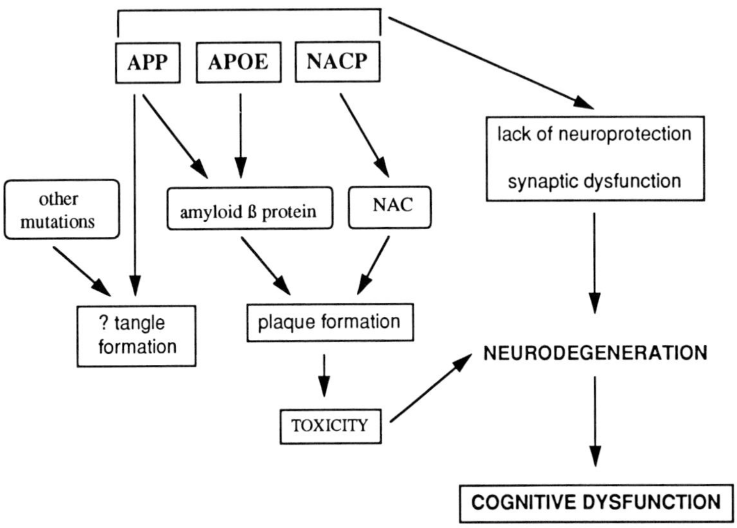


**Figure 1.** Mechanisms of neurodegeneration in Alzheimer's disease.

basis for the cognitive alterations in AD, as well as the evidence supporting a role for axosynaptic damage and tangle and plaque formation in the pathogenesis of AD.

### 1.3. Approaches to Understanding the Structural Basis of Alzheimer's Disease Dementia

Two basic approaches have been developed to understand the structural basis of cognitive alterations in AD. The first consists of correlating the severity of the dementia, as assessed neuropsychologically in the patients before death, with the various neurochemical and neuropathological alterations determined in the postmortem brains. The second consists of correlating cognitive performance with neuroimaging studies. Advantages of the clinicopathological approach are: (1) data are derived directly from affected patients for whom the diagnosis of the neurodegenerative condition is anatomically confirmed, (2) this more complete view of the disease allows one to evaluate the progression of the clinico-neuropathological alterations, and (3) because data implicitly address some of the possible mechanisms, the formation can be compared with that of validated animal models. The main disadvantage is that long delays between premortem psychometric testing and postmortem examination might interfere with the validity of the results. Furthermore, significant correlations must be interpreted with caution, because they might reflect parallel phenomena rather than a cause–effect relationship. Some studies have attempted to circumvent these problems by correlating cognitive functioning with neuropathological and neurochemical markers obtained from biopsy material. This approach has been shown to more closely reflect the correlation between cognition and neuropathological and neurochemical alterations (DeKosky *et al.*, 1992; Davies *et al.*, 1987); however, due to problems of logistics, risk, and cost, this approach is rarely used.

A number of statistical issues must also be considered in studies that examine the relationship between the degree of dementia and the nature and severity of neuropathological alterations. In correlational studies where several linear regressions have been performed, it is necessary to correct for multiple comparisons, taking into consideration the size of the sample and the number of variables included (Milliken and Johnson, 1984). Multivariate analysis, factor analysis, and the Bonferoni formula are some approaches that are often used to correct for these factors (Milliken and Johnson, 1984). *Multivariate analysis* (stepwise regression) allows a determination of the degree to which a number of independent parameters add to the strength of the correlation (Milliken and Johnson, 1984), and *factor analysis* allows a large group of variables to be reduced to a few primary factors that can be correlated with neuropathologic features (Milliken and Johnson, 1984). It has also been recommended that when linear regression analysis is performed between cognitive and neuropathological parameters in AD, control cases should not be included because their lack of lesions would artificially skew the statistical analysis (Terry *et al.*, 1991). However, in the case of linear regression with other neuroanatomical parameters that indicate neuronal integrity (e.g., neuronal and synaptic density) in AD cases, it might be useful to compare the correlation coefficient both with and without control cases. Recent studies of disease progression have shown that the biological significance of a correlative analysis might be enhanced by including cases representing the whole spectrum of the disease (Masliah *et al.*, 1994b, 1995a), that is, cases ranging from mild to severe AD assessed neuropathologically and cases with early-to-late AD in terms of clinical staging (Masliah *et al.*, 1994b, 1995a).

Additional procedures for correlating neuroanatomical parameters with cognition in dementia patients include the assessment of brain atrophy by magnetic resonance imaging (MRI) and assessment of neuronal dysfunction by single photon emission computerized tomography (SPECT) and positron-emission tomography (PET). These studies have shown significant correlations between severity of the overall cerebral atrophy (by neuroimaging) and scores on the Mini-Mental State Examination (MMSE) (Davis *et al.*, 1995). Other studies have shown that hippocampus volume correlates with specific memory measures: memory quotient and paired associate learning from the Wechsler Memory Scale (WMS), and intrusion errors and recognition discrimination from the Selective Reminding Test (Deweer *et al.*, 1995). By contrast, except for intrusions, no correlations were found between memory variables and the volume of amygdala (Deweer *et al.*, 1995). Taken together, these neuroimaging studies suggest that the volume of the hippocampus is related to quantitative and qualitative aspects of memory performance in patients with AD (Deweer *et al.*, 1995).

Single photon emission computerized tomography has been used to measure regional brain uptake of technetium-99m hexamethylpropyleneamine oxine in AD. Reductions were found in all regions, most markedly in temporal and posterior parietal cortex (McKeith *et al.*, 1993). Significant correlations were found in AD between memory impairment and decreased temporal uptake bilaterally, and between duration of illness and reduced uptake in most brain regions (McKeith *et al.*, 1993). Furthermore, SPECT has been shown to provide useful information in dementia, particularly the differentiation of AD, frontotemporal dementia, and CJD. However, it does not distinguish Parkinson's disease (PD) from AD (Read *et al.*, 1995). These approaches have the advantage of allowing a direct correlation between neuroimaging and neuropsychological assessment performed shortly after imaging, and they allow a correlation of the volume of structures, such as hippocampus and frontal cortex, with tests of cognitive performance. It is still not possible, however, to obtain live imaging at the cellular level that will yield more-detailed information about neuronal, synaptic, plaque, or tangle density. Finally, a more recent approach consists of correlating cognitive performance with neuropathological markers in experimental animal models of neurodegeneration (Heyser *et al.*, 1996; Masliah *et al.*, 1996d; Moechars *et al.*, 1996; Perry *et al.*, 1995; Zheng *et al.*, 1995). The main advantage of this approach is that it allows more-direct testing of the molecular mechanisms involved. Furthermore, animal models are amenable to manipulations in which the deficient or mutant molecule can be reintroduced and its effects on synaptic plasticity and behavior reassessed. Of course, the main disadvantage is that most of these models have been developed in rodents, and therefore any analogies or comparisons with neurodegenerative disorders in humans must be made with caution. Nonetheless, the models hold future promise for the development of novel therapeutic strategies.

## 2. Neuropathological Basis for Cognitive Decline in Alzheimer's Disease

### 2.1. Characteristics of the Neurodegenerative Process

The neurodegenerative process in AD is characterized by the progressive and irreversible deafferentation of the limbic system, association neocortex, and basal forebrain (Masliah *et al.*, 1993b; Hof *et al.*, 1990; Palmer and Gershon, 1990; Wilcock *et al.*, 1988; Hyman *et al.*, 1984; Perry *et al.*, 1977), accompanied by neuritic plaques (NPs), neurofibrillary tangles (NFTs) and neuropil threads (NTs) (see Terry *et al.*, 1994, for review). This neurodegenerative process is followed by reactive astrogliosis (Dickson *et al.*, 1988) and microglial cell proliferation (Masliah *et al.*, 1991c; Rogers *et al.*, 1988;). Although plaque and tangle formation are necessary to establish the diagnosis (Mirra *et al.*, 1993), the main factors that determines AD specificity are the anatomical patterns of disconnection and neuronal loss (Masliah *et al.*, 1993a, 1993b). The resulting clinicopathological characteristics are dependent on the specific neuronal populations affected. Other neuropathological features that have been observed in AD include amyloid angiopathy, the presence of Hirano bodies, and granulovacuolar degeneration in the hippocampus (Hirano, 1994). Table II summarizes some of the main neuropathological parameters used for correlational analysis with dementia. The neuritic plaques in AD contain Aβ, which is derived from amyloid-β precursor protein (APP) (Selkoe, 1993; Selkoe *et al.*, 1989). Embedded in the amyloid plaque core are dystrophic neurites, reactive astrocytes, and microglia (Masliah *et al.*, 1993c, 1991c; Dickson *et al.*, 1988; Terry and Wisniewski, 1970; Terry *et al.*, 1964), which are believed to be involved in the process of amyloid formation (Frackowiak *et al.*, 1992; Wisniewski *et al.*, 1991, 1989).

**Table II. Neuropathological Markers of Alzheimer's Disease**

| | | |
|---|---|---|
| Markers of neurodegeneration | | |
| | Neuronal cell counts | |
| | Synaptic counts | |
| | Glial cell counts | |
| | | astroglia |
| | | microglia |
| Markers of neurofibrillary pathology | | |
| | Tangle counts | |
| | Neuropil thread density | |
| | Neuritic plaque density | |
| Markers of amyloid deposition | | |
| | Area covered by Aβ-reactive material | |
| | Total plaque counts | |
| | Amyloid angiopathy | |
| Others | | |
| | Hirano bodies | |
| | Granulovacuolar degeneration | |

The neuritic population in the plaque is a mixed one, composed of spherical neurites containing synaptic proteins and APP (Type I), and fusiform neurites containing cytoskeletal proteins and paired helical filaments (PHF) (Type II) (Wang and Munoz, 1995; Masliah *et al.,* 1994b, 1993c; Dickson *et al.,* 1988). The precise mechanisms by which neuritic plaques are formed and their relationship to the overall neurodegenerative process in AD is not yet clear. However, recent studies indicate that aberrantly processed APP derivatives could be involved in the pathophysiological process that leads to neurodegeneration and plaque formation in AD (Mattson *et al.,* 1993a; Selkoe, 1993; Golde *et al.,* 1992; Seubert *et al.,* 1992; Sisodia *et al.,* 1990).

## 2.2. Clinico-Neuropathological Correlation in Alzheimer's Disease and Related Disorders

Clinicopathological studies in neurodegeneration have permitted assessment of the accuracy of neurological and neuropsychometric tests in predicting neuropathological diagnosis (Table III). Studies at the Alzheimer's Disease Research Center (ADRC) at the University of California at San Diego have shown that out of 137 patients clinically diagnosed with probable or possible AD, 123 (90%) had AD neuropathology; this included 29 with AD accompanied by Lewy bodies and 14 with AD and one or more infarcts (Galasko *et al.,* 1994). Cases of vascular and mixed dementia (AD and infarcts) had lower rates of agreement with pathological findings. Possible AD cases were more likely than probable AD cases to show pathological features other than AD. Clinicians predicted the presence or absence of AD pathological findings significantly better than chance. In patients with AD pathological lesions, male gender and older age at onset were significantly associated with shorter duration from onset to death (Galasko *et al.,* 1994). Overall, other studies have shown similar results; clinicians accurately predicted AD pathological findings or their absence in most cases (~85%) (Table 3). Attributing other dementias to AD, misdiagnosing patients with AD and Lewy bodies, and misjudging the vascular contribution to dementia were the major areas of inaccuracy (Jellinger *et al.,* 1990; Molsa *et al.,* 1985; Victoroff *et al.,* 1995; Wade *et al.,* 1987).

**Table III. Percentage Agreement between Clinical and Pathological Diagnosis in Different Series**

| Group | n | % agreement | | | | |
|---|---|---|---|---|---|---|
| | | AD | AD + LBD | MID[a] | Mixed | Other |
| Molsa *et al.,* 1985 | 58 | 82 | – | 42 | 20 | 100 |
| Wade *et al.,* 1987 | 65 | 85 | – | 25 | 31 | 100 |
| Jellinger *et al.,* 1990 | 445 | 86 | 58 | 67 | 54 | 50 |
| Galasko *et al.,* 1994 | 170 | 86 | 75 | 60 | 32 | 67 |

[a]multi-infarct dementia.

## 2.3. Correlations with the Progression of Neurodegenerative Alterations

Recent studies by Braak and Braak (1991), in conjunction with others (Masliah *et al.*, 1994b; Berg *et al.*, 1993; Arriagada *et al.*, 1992a,b; Mann *et al.*, 1988; Hyman *et al.*, 1986), suggest that neurodegeneration in AD begins in the entorhinal cortex and spreads to the hippocampus, temporal cortex, frontal parietal region, and then to the subcortical nuclei. As a result, the first circuit to be disconnected is the perforant pathway, which associates the entorhinal cortex with the dentate gyrus of the hippocampus (Hyman *et al.*, 1987, 1986; Cotman and Nadler, 1978). As the disease spreads, neurodegeneration occurs in the intrahippocampal circuitry, resulting in disconnection between the dentate granular cells and CA3, CA1, and subiculum (Table IV, Fig. 2). All of these major limbic system circuitries are crucial for learning and memory. Degeneration next spreads to the neocortex, where major association fibers projecting from layer III neurons become disconnected (Hof and Morrison, 1991, 1994; Masliah *et al.*, 1993b) (Table IV, Fig. 2). These circuitries appear to play a major role in learning and higher cognitive functions, that is, in language, orientation, judgment, problem solving, and visuospatial and constructional abilities (Eccles, 1984, 1981) (Table IV, Fig. 2). Finally, subcortical nuclei, including the nucleus basalis of Meynert, are affected, resulting in loss of cholinergic input to the neocortex (Perry *et al.*, 1991a). This circuitry has also been implicated recently in memory and attention (Voytko *et al.*, 1994) (Table IV, Fig. 2).

To better understand the biological significance of the Braak classification, it is important to define its correlation with the clinical stages of AD. For this purpose we correlated neuropathological indicators of AD with cognitive performance in a group of cases covering the whole spectrum of the disease, including cases with incipient (or early) AD (Masliah, 1995b) (Table V). Linear regression analysis showed a positive correlation between the Blessed Information-Memory-Concentration score and the Clinical Dementia Rating score (CDR) (Morris *et al.*, 1988), and a very strong correlation with both Braak stage ($r = 0.9$, $p < 0.001$, $n = 54$) and

**Table IV. Neural Circuitries and Neuropsychometric Tests**

| Test | | Brain region | Neural circuitry |
|---|---|---|---|
| DRS | Memory | Mediotemporal | Perforant pathway/septo-hippocampal; NbM[a] → temporal |
| | Construction | Temporal/parietal | Association |
| | Conceptualization | Frontal | Association |
| | Initiation | Frontal | Association |
| | Attention | Frontal | Association |
| MMSE | Memory | Medio temporal | Perforant pathway/septo-hippocampal |
| | Orientation | Frontotemporal | Association |
| | Language | Frontotemporal, parietal | Association |
| Blessed | Memory | Mediotemporal | Perforant pathway/septo-hippocampal |
| | Orientation | Frontotemporal | Association |
| | Attention | Frontal | Association |

[a]Nucleus basalis of Meynert.

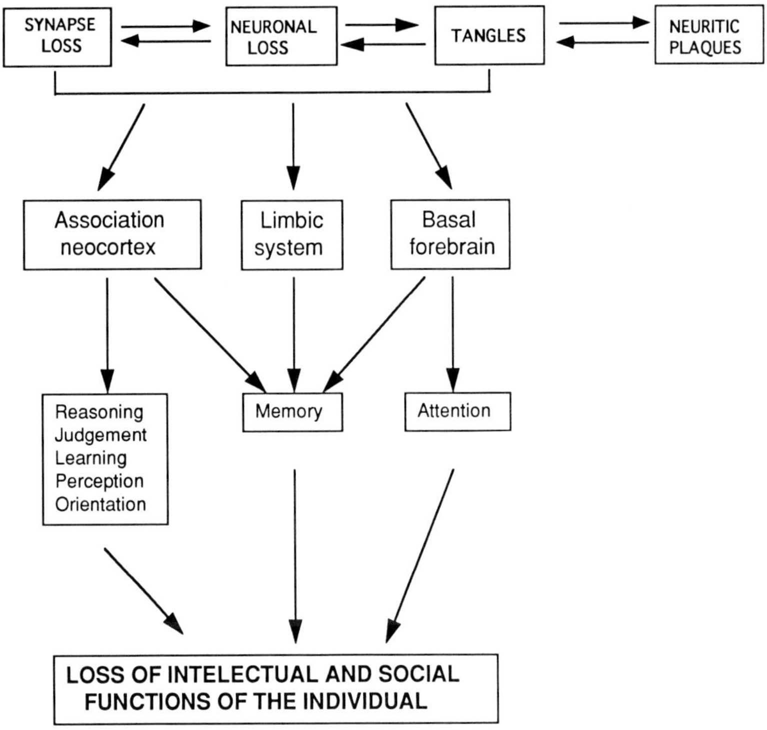


**Figure 2.** The pathological basis of Alzheimer's disease.

disease duration ($r = 0.808$, $p < 0.001$, $n = 49$) (Masliah, 1995b). Further supporting the concept that the Braak staging system reflects disease progression, the system correlated strongly in this study with Blessed, MMSE, and Mattis Dementia Rating Scale (DRS) scores (Table VI), consistent with a recent report by Bancher *et al.* (1993). In fact, the correlation coefficient between cognition and Braak stage was stronger than comparisons between cognition and other neuropathological parameters, including brain weight, plaques, and tangles (Table VI). Overall neocortical plaque density displayed a rapid increase as the disease progressed, reaching a plateau in early stages, whereas neocortical tangles were absent in the early stages (Masliah, 1995b).

Controversy still exists as to how the neurodegenerative process in AD begins. One hypothesis proposes amyloid deposition as the trigger leading to degeneration first in the entorhinal cortex and then in the other brain regions (Cummings and Cotman, 1995; Selkoe, 1989). Other hypotheses propose that neurodegeneration begins with injury to synaptic terminals in the molecular layer projecting field of the dentate gyrus for the perforant pathway, which in turn results in retrograde damage to neuronal cell bodies in the entorhinal cortex (Heinanen *et al.*, 1995; Masliah, 1995c; Masliah *et al.*, 1994a; Martin *et al.*, 1994). Factors that might be involved in damaging synaptic terminals are neurotoxins and lack of sustained synaptotrophic APP effects (Mucke *et al.*, 1994). It is also possible that synaptic damage is secondary to neuronal dysfunction; a recent study found that tangle-bearing neurons showed significant reduction in message levels for growth-associ-

**Table V. Clinicopathological Characteristics**

| | | | Plaques | | | Tangles | | | | |
|---|---|---|---|---|---|---|---|---|---|---|
| | Age | Duration | MF | IP | ST | MF | IP | ST | Braak | Blessed (IMC) |
| Controls | 80.0 ± 1.5<br>(20) | – | 0.06 ± 0.06<br>(18) | 1.1 ± 0.7<br>(14) | 0.4 ± 0.4<br>(16) | 0.0 ± 0.0<br>(18) | 0.0 ± 0.0<br>(14) | 0.0 ± 0.0<br>(16) | N/A<br>(16) | 3.8 ± 2.4<br>(6) |
| AD | 80.2 ± 0.4<br>(199) | 8.6 ± 0.4<br>(114) | 38.7 ± 1.0<br>(173) | 39.5 ± 0.9<br>(171) | 36.1 ± 1.0<br>(174) | 2.5 ± 0.2<br>(174) | 3.9 ± 0.3<br>(172) | 5.7 ± 0.4<br>(175) | 5.7 ± 0.1<br>(75) | 25.3 ± 0.7<br>(107) |
| P< | – | – | 0.001 | 0.001 | 0.001 | 0.001 | 0.001 | 0.001 | – | 0.001 |

Abbreviations: MF, midfrontal; IP, inferior parietal; ST, superior temporal.

ated protein 43 (GAP-43) and synaptophysin as compared with adjacent nontangle-bearing neurons in the parahippocampal cortex of AD patients (Callahan and Coleman, 1995). However, synaptic loss is usually more widespread and severe than neuronal loss, and in studies of AD in various stages, it appears that synaptic loss in the outer molecular layer of the hippocampal dentate gyrus appears before neurofibrillary tangle formation in the entorhinal cortex (Masliah *et al.,* 1994a). Nonetheless, a recent study has shown that in very mild AD cases, there is substantial loss of neurons in layer II of the entorhinal cortex (Gómez-Isla *et al.,* 1996). Neuronal number in AD was inversely proportional to NFT formation and to neuritic plaques, but was not significantly related to either diffuse or total plaques, supporting a contention that neuronal loss in layer II distinguishes very mild AD from nondemented aging (Gómez-Isla *et al.,* 1996). Therefore, an important goal is to determine why the neurons in layer II entorhinal cortex are the most vulnerable to degeneration, and why the process spreads to other selectively vulnerable areas, as described by Braak, Morrison, and others (Hof and Morrison, 1994; Braak and Braak, 1991).

## 2.4. Correlations with Neuropathological Markers

Recent studies have shown that in addition to the traditionally described lesions (plaques and tangles) found in the AD brain (Braak and Braak, 1991; Dickson *et al.,* 1988; Yamaguchi *et al.,* 1988; Terry and Wisniewski, 1970; Terry *et al.,* 1964; Alzheimer, 1907), the disease is characterized by neuronal loss (Hof *et al.,* 1990;

**Table VI. Correlations between Clinical and Neuropathological Alterations in Alzheimer's Disease**

| | IMC | | | MMSE | | | DRS | | |
|---|---|---|---|---|---|---|---|---|---|
| | r | n | p | r | n | p | r | n | p |
| Epidemiological | | | | | | | | | |
| Age | −0.17 | (128) | 0.04 | −0.10 | (109) | 0.27 | 0.31 | (51) | 0.02 |
| Duration | 0.36 | (108) | 0.0001 | −0.37 | (109) | 0.0001 | −0.38 | (54) | 0.005 |
| Education | −0.01 | (89) | 0.90 | 0.09 | (91) | 0.39 | −0.15 | (53) | 0.28 |
| Neuropathological | | | | | | | | | |
| Brain weight | −0.33 | (125) | 0.0001 | 0.26 | (106) | 0.0066 | 0.16 | (50) | 0.24 |
| Braak | 0.75 | (48) | 0.0001 | −0.70 | (48) | 0.0001 | 0.75 | (29) | 0.0001 |
| MF plaques | 0.14 | (125) | 0.116 | −0.13 | (108) | 0.15 | −0.15 | (50) | 0.29 |
| IP plaques | 0.25 | (124) | 0.005 | −0.27 | (108) | 0.004 | −0.40 | (51) | 0.0038 |
| ST plaques | 0.22 | (123) | 0.01 | −0.17 | (106) | 0.07 | −0.26 | (51) | 0.05 |
| HP plaques | 0.08 | (121) | 0.32 | −0.06 | (107) | 0.52 | −0.02 | (50) | 0.83 |
| MF tangles | 0.35 | (125) | 0.0001 | −0.32 | (108) | 0.0007 | −0.36 | (50) | 0.009 |
| IP tangles | 0.34 | (124) | 0.0001 | −0.28 | (108) | 0.003 | −0.41 | (51) | 0.002 |
| ST tangles | 0.37 | (123) | 0.0001 | −0.32 | (106) | 0.0006 | −0.24 | (51) | 0.08 |
| HP tangles | 0.20 | (121) | 0.02 | −0.18 | (107) | 0.05 | −0.01 | (50) | 0.98 |
| MF amyloid | 0.01 | (111) | 0.90 | −0.05 | (108) | 0.59 | −0.28 | (50) | 0.04 |
| IP amyloid | −0.04 | (111) | 0.66 | −0.02 | (108) | 0.80 | −0.12 | (51) | 0.38 |
| ST amyloid | −0.08 | (106) | 0.36 | −0.11 | (105) | 0.23 | −0.01 | (51) | 0.91 |
| HP amyloid | 0.02 | (113) | 0.78 | −0.01 | (104) | 0.98 | −0.01 | (47) | 0.90 |

Abbreviations: MF, midfrontal; IP, inferior parietal; ST, superior temporal; HP, hippocampus.

Terry *et al.,* 1981), disruption of the neuritic cytoskeleton with altered corticocortical connectivity (Masliah *et al.,* 1993b; Hof *et al.,* 1990; Morrison *et al.,* 1987), and extensive synapse loss (Liu *et al.,* 1996; Honer *et al.,* 1992a; Lassmann *et al.,* 1992; Lippa *et al.,* 1992; Masliah *et al.,* 1989, 1991a,d; DeKosky and Scheff, 1990; Hamos *et al.,* 1989; Davies *et al.,* 1987). It has been hypothesized that AD dementia could be caused by either the presence of these specific lesions alone or by the synergistic effect of some or all of the lesions (Samuel *et al.,* 1994a; Terry *et al.,* 1991; DeKosky and Scheff, 1990) (Fig. 2). Other groups have shown that neuronal loss in specific areas of the neocortex and subcortical regions is correlated with clinical alterations seen in AD (Neary *et al.,* 1986); depending on severity, the neuronal loss varies from 20–40% (Fig. 3) (Terry *et al.,* 1981). A recent review of our material shows that loss of large pyramidal neurons in midfrontal, inferior parietal, and superior temporal cortex correlates with tests of global cognition, including the Blessed and MMSE; however, no significant correlations were observed with the DRS (Table VII). Of the three regions, the one that showed the strongest correlate with cognition was the inferior parietal (Table VII), a region that plays a role in reading, writing, and speaking integration (Orrison, 1995) and in attention and visuospatial constructional process as well. Furthermore, neuronal loss within this region also correlates with neocortical tangle and plaque formation (Table VIII). In contrast, small-neuron and glial cell counts in the three regions did not significantly correlate with tests of cognition (Table VII). The lack of significant astroglial correlations (Table 8) is consistent with the notion that astroglial reactions occur later in the disease in response to ongoing synaptic terminal and neuron damage (Liu *et al.,* 1996).

Other studies have shown that neuronal loss is related not only to cognitive changes but also to psychiatric alterations in AD patients (Forstl *et al.,* 1993). For example, delusional misidentifications were associated with lower counts in hippocampal CA1. Delusions and hallucinations were observed in patients with less-severe cell loss in the parahippocampal gyrus and with lower cell counts in the dorsal raphe nucleus (Forstl *et al.,* 1994). In contrast, loss of neurons was higher in the nucleus basalis of Meynert of depressed than nondepressed AD patients, there were no differences in the neuronal counts in the substantia nigra (Forstl *et al.,* 1994). A decrease in locus coeruleus neurons has been associated with depressed mood in patients with AD (Bondareff *et al.,* 1987) and depression contributed significantly to the variance of neuronal counts in the locus coeruleus even when covariance for gender, age at onset, cognitive impairment, and cortical AD pathology was accounted for (Forstl *et al.,* 1992).

Correlations with neuronal loss do not completely explain all of the clinical alterations observed in AD, however. An alternative hypothesis is that AD dementia is also associated with disruption of neuritic substructure and loss of synaptic contacts in specific neocortical and subcortical areas (Masliah *et al.,* 1991b; McKee *et al.,* 1991) (Figs. 1 and 2). Studies of lesion progression in AD have shown that synapse loss appears first in the molecular layer of the hippocampus dentate gyrus and is correlated with abnormal expression of APP in the entorhinal cortex (Masliah *et al.,* 1994a). Damage to this circuit correlates with the early symptoms of memory loss characteristic of the disorder (Hyman *et al.,* 1984, 1986) (Fig. 2). Cases displaying clinical and neuropathological features of early or incipient AD showed a significant 20% loss of synapses in the outer molecular layer of the hippocampal dentate gyrus (Masliah *et al.,* 1994a), a finding recently confirmed by Heinonen *et*

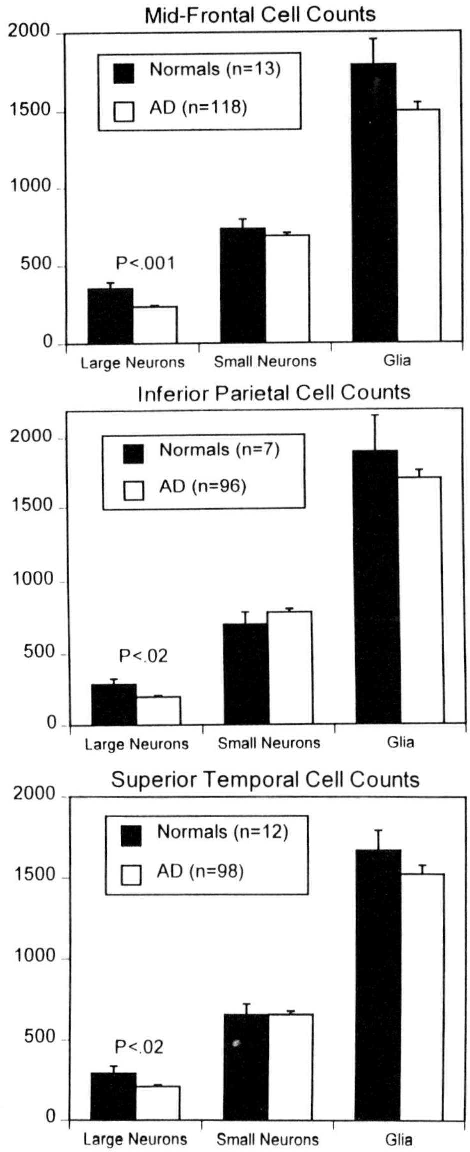


**Figure 3.** Neuronal alterations in Alzheimer's disease.

al. (1995). No significant loss of synapses was observed in neocortical regions or in entorhinal cortex in early AD (Masliah *et al.,* 1994a). Cases displaying more severe AD showed a progressively more significant loss of synapses in the neocortex, as well as in the outer molecular layer of the hippocampal dentate gyrus (Masliah *et al.,* 1994a). This was reflected by a strong negative correlation between the loss of synapses in the limbic system (Samuel *et al.,* 1994b) and frontal cortex (Masliah, 1995b; Masliah *et al.,* 1994a; Terry *et al.,* 1991) and the severity of global dementia ($r = -0.6$, $p = 0.0001$, $n = 49$) (Table VII). Further supporting these findings, measurements by electron microscopy and immunocytochemistry have both shown very strong correlations between synaptic numbers in the frontal cortex and tests of global cognition in AD (Blennow *et al.*, 1996; Samuel *et al.*, 1994a; Lassmann *et al.*, 1992; DeKosky and Scheff, 1990; Weiler *et al.*, 1990). Furthermore, the concomitant atrophy and loss of dendritic spines in AD is also correlated with the severity of dementia (Anderson, 1995). More recently, correlative studies between tests of cognition and immunochemical quantification of various synaptic proteins have confirmed this view (Blennow *et al.*, 1996; Zhan *et al.*, 1993; Lassmann *et al.*, 1992). A study by Dickson *et al.* (1995) showed that the levels of another synaptic-like marker, EP10, in all cortical areas and in hippocampus were correlated ($p < 0.05$–0.01) with the Blessed score, but not with the Fuld Object Memory Examination (FOME). The authors concluded that their univariate analysis suggests that synaptic markers may not be the best structural correlate of dementia, whereas markers indicative of cytoskeletal changes (e.g., NFTs and PHF protein accumulation) may be better correlates of dementia in the elderly (Dickson *et al.*, 1995).

Ultrastructural analyses have shown a highly significant AD-related decline in synaptic numbers in laminae III and V in both superior and middle temporal gyrus (Scheff and Price, 1993) (Table IX). There were no correlations between synaptic density and synaptic apposition length or density of senile plaques (Scheff and Price, 1993). In contrast, the entorhinal cortex showed no change in synaptic density between controls and AD cases in either lamina III or V (Scheff *et al.*, 1993). This preservation of synapses may be related to a plastic response that is greater in the entorhinal area than in other areas of the cortex (Scheff *et al.*, 1993). In more advanced AD cases, as well as in cases of the Lewy body variant of AD (LBV), there is an approximate 30–50% synapse loss in frontal, parietal, and temporal cortex (Masliah *et al.*, 1989, 1991a,b, 1993d; Scheff and Price, 1993; Lassmann *et al.*, 1992; Scheff *et al.*, 1990; Davies *et al.*, 1987) (Table 9). Although midfrontal cortex synapse density was the strongest correlate of AD dementia severity in LBV, followed by NFTs (Blennow *et al.*, 1996; Samuel *et al.*, 1994a; Arriagada *et al.*, 1992a; Lassmann *et al.*, 1992; Terry *et al.*, 1991; DeKosky and Scheff, 1990; Weiler *et al.*, 1990), there was

**Table VII. Correlations between Clinical and Neurodegenerative Alterations in Alzheimer's Disease**

| | | IMC | | | MMSE | | | DRS | | |
|---|---|---|---|---|---|---|---|---|---|---|
| | | r | n | p | r | n | p | r | n | p |
| Neuronal cell counts | MF | −0.32 | (36) | 0.05 | 0.40 | (36) | 0.01 | 0.06 | (21) | 0.77 |
| | IP | −0.566 | (29) | 0.001 | 0.50 | (29) | 0.005 | 0.04 | (21) | 0.86 |
| | ST | −0.41 | (29) | 0.02 | 0.45 | (29) | 0.014 | 0.27 | (14) | 0.33 |
| Small cell counts | MF | −0.05 | (36) | 0.74 | 0.003 | (36) | 0.98 | −0.31 | (21) | 0.17 |
| | IP | −0.21 | (29) | 0.26 | 0.25 | (29) | 0.18 | −0.13 | (15) | 0.64 |
| | ST | −0.12 | (29) | 0.53 | 0.22 | (29) | 0.24 | 0.10 | (14) | 0.73 |
| Glial cell counts | MF | 0.15 | (36) | 0.36 | −0.15 | (36) | 0.35 | −0.41 | (21) | 0.06 |
| | IP | 0.14 | (29) | 0.45 | −0.11 | (29) | 0.55 | −0.38 | (15) | 0.15 |
| | ST | 0.19 | (29) | 0.32 | 0.13 | (29) | 0.48 | 0.33 | (14) | 0.24 |
| Synaptophysin-IR presynaptic terminals | MF | −0.574 | (49) | 0.0001 | 0.54 | (49) | 0.0001 | 0.44 | (30) | 0.01 |

Abbreviations: MF, midfrontal; IP, inferior parietal; ST, superior temporal; IR, immunoreactivity.

**Table VIII. Correlations between Markers of Neurodegeneration and Neuropathology in Alzheimer's Disease**

| | Braak | | | Synaptophysin | | | Tangles | | | | | | | | | Plaques | | | | | | | | |
|---|---|---|---|---|---|---|---|---|---|---|---|---|---|---|---|---|---|---|---|---|---|---|---|---|
| | | | | | | | MF | | | IP | | | ST | | | MF | | | IP | | | ST | | |
| | r | n | p | r | n | p | r | n | p | r | n | p | r | n | p | r | n | p | r | n | p | r | n | p |
| Neuronal cell counts | | | | | | | | | | | | | | | | | | | | | | | | |
| MF | −0.17 | (36) | 0.30 | −0.07 | (36) | 0.64 | −0.37 | (36) | 0.02 | −0.25 | (36) | 0.12 | 0.40 | (35) | 0.01 | −0.24 | (36) | 0.15 | −0.38 | (36) | 0.02 | −0.28 | (35) | 0.09 |
| IP | −0.52 | (29) | 0.003 | 0.11 | (29) | 0.56 | −0.55 | (29) | 0.001 | −0.49 | (29) | 0.006 | 0.32 | (28) | 0.09 | −0.54 | (29) | 0.002 | −0.39 | (29) | 0.03 | −0.35 | (28) | 0.06 |
| ST | −0.41 | (29) | 0.02 | −0.22 | (29) | 0.24 | −0.33 | (29) | 0.08 | −0.28 | (29) | 0.13 | 0.44 | (28) | 0.01 | −.053 | (29) | 0.003 | −0.51 | (29) | 0.004 | −0.44 | (28) | 0.01 |
| Small cell counts | | | | | | | | | | | | | | | | | | | | | | | | |
| MF | 0.21 | (36) | 0.20 | −0.25 | (36) | 0.13 | 0.12 | (36) | 0.47 | 0.06 | (36) | 0.72 | 0.01 | (35) | 0.91 | 0.14 | (36) | 0.39 | −0.06 | (36) | 0.69 | 0.03 | (35) | 0.84 |
| IP | −0.16 | (29) | 0.38 | −0.25 | (29) | 0.18 | −0.39 | (29) | 0.03 | −0.35 | (29) | 0.05 | 0.21 | (28) | 0.26 | −0.13 | (29) | 0.49 | −0.17 | (29) | 0.36 | −0.24 | (28) | 0.21 |
| ST | 0.02 | (29) | 0.89 | −0.36 | (29) | 0.05 | −0.17 | (29) | 0.36 | 0.03 | (29) | 0.87 | 0.11 | (28) | 0.55 | −0.10 | (29) | 0.60 | 0.11 | (29) | 0.53 | −0.05 | (28) | 0.76 |
| Glial cell counts | | | | | | | | | | | | | | | | | | | | | | | | |
| MF | 0.03 | (36) | 0.84 | −0.12 | (36) | 0.47 | 0.45 | (36) | 0.005 | 0.20 | (36) | 0.22 | 0.25 | (35) | 0.14 | 0.12 | (36) | 0.47 | 0.02 | (36) | 0.90 | 0.10 | (35) | 0.54 |
| IP | 0.20 | (29) | 0.28 | 0.01 | (29) | 0.95 | 0.33 | (29) | 0.08 | 0.30 | (29) | 0.10 | 0.44 | (28) | 0.01 | 0.31 | (29) | 0.09 | 0.09 | (29) | 0.60 | 0.02 | (28) | 0.90 |
| ST | −0.12 | (29) | 0.51 | 0.37 | (29) | 0.04 | −0.05 | (29) | 0.75 | −0.09 | (29) | 0.62 | 0.07 | (28) | 0.08 | 0.01 | (29) | 0.92 | 0.05 | (29) | 0.70 | 0.14 | (28) | 0.46 |
| Synaptophysin-IR presynaptic terminals | | | | | | | | | | | | | | | | | | | | | | | | |
| MF | −0.48 | (48) | 0.0005 | – | – | – | −0.26 | (49) | 0.05 | −0.30 | (49) | 0.05 | 0.23 | (48) | 0.10 | −0.26 | (49) | 0.06 | −0.37 | (49) | 0.007 | −0.23 | (48) | 0.10 |

Abbreviations: MF, midfrontal; IP, inferior parietal; ST, superior temporal; IR, immunoreactivity.

**Table IX. Synaptic Pathology in Alzheimer's Disease**

| Cortical region | % Loss |
|---|---|
| Frontal | 15–60 |
| Temporal | 10–30 |
| Parietal | 10–30 |
| Insular | 10 |
| Cingulate | 10 |
| Motor | none |
| Entorhinal | variable |
| Hippocampus | |
| C3-4 | 15 |
| C1-2 | 25 |
| Subiculum | 50 |
| OML | 10–30 |
| Subcortical regions | % Loss |
| Nucleus basalis | 30 |
| Basal ganglia | none |
| Amygdala | none |
| SN/LC | variable |

only a weak correlation between midfrontal synapse loss and Blessed score (Masliah *et al.*, 1996b; Samuel *et al.*, 1996). In the nucleus basalis of AD patients, NFTs were the strongest correlate of dementia severity, followed by synapse density (Samuel *et al.*, 1994a). Stepwise regression analyses showed midfrontal synapse density to be the strongest predictor of performance on cognitive tests emphasizing higher cortical functions, but NFTs in the nucleus basalis were the strongest predictor of performance on memory-oriented tests (Samuel *et al.*, 1994a).

Further evidence of a central role for synaptic damage in the pathogenesis of AD is provided by studies showing that the dystrophic neurites of plaques contain synaptic vesicles, synaptic proteins, and neurotransmitters (Masliah *et al.*, 1991a, 1994b; Masliah and Terry, 1993; Lassmann *et al.*, 1992; Weiler *et al.*, 1990; Armstrong *et al.*, 1989). In addition, ultrastructural studies have shown that in AD the synapses are swollen and contain abnormal accumulation of cytoskeletal proteins, vesicles, and lysosomes (Masliah *et al.*, 1993c, 1991a; Gonatas and Gambetti, 1970; Gonatas *et al.*, 1967). In a recent paper, Jones and Harris (1995) review in detail several studies describing the characteristics of synaptic pathology in AD, and the alterations of perforated synapses in neurodegenerative disorders.

As a result of synaptic damage, there is concomitant loss of neurotransmitter markers such as choline acetyltransferase (ChAT) (cholinergic terminals) (Perry *et al.*, 1991a; Davies, 1979), somatostatin (peptidergic terminals) (Armstrong *et al.*, 1989), and aspartate uptake (glutaminergic terminals) (Cowburn *et al.*, 1988). Some studies have shown that ChAT levels correlated with the Blessed score in Brodman areas 9, 40, and 22, and with somatostatin levels in area 9 only; ChAT, but not somatostatin, correlated with plaques (area 22) and NFTs (areas 40 and 22) (Dournaud *et al.*, 1995). Decreases in ChAT and somatostatin were correlated in areas 9 and 22. These results indicate that the somatostatinergic deficit in AD is more regionally restricted than the cholinergic deficit. The correlation between somatostatin and ChAT as observed in areas 9 and 22 may indicate that somatostatin- and acetylcholine-containing elements in frontal and temporal lobes are relevant to the cognitive decline observed in AD (Dournaud *et al.*, 1995). Further analysis of our Alzheimer's Disease Research Center at UCSD (ADRC) cohort showed significant correlations between tests of global cognition and ChAT levels in inferior parietal ($-0.31$, $p = 0.012$, $n = 65$) and midfrontal ($-0.25$, $p = 0.03$, $n = 68$) regions, but not with ChAT levels in superior temporal cortex or hippocampus.

Another important neurotransmitter deficit that might be centrally responsible for cognitive alterations in AD (Masliah *et al.*, 1996e) is the significant loss of glutaminergic markers in the midtemporal, inferofrontal, and cingulate cortices and in the hippocampal region (Scott *et al.*, 1995; Cowburn *et al.*, 1988; Cross *et al.*, 1986). Compared with controls, AD cases displayed a 34% decrease in the levels of D-[$^3$H]-aspartate binding, a 30% decrease in L-[$^3$H]-aspartate binding, and a 48% loss of synaptophysin immunoreactivity. Increased levels of brain spectrin degradation products correlated with decreased levels of D-[$^3$H]-aspartate binding ($r = -0.587$, $p = 0.01$) and L-[$^3$H]-aspartate binding ($r = -0.570$, $p = 0.02$), and with decreased synaptophysin immunoreactivity ($r = -0.55$, $p = 0.02$) (Masliah *et al.*, 1996e). Levels of D-[$^3$H]-aspartate were correlated with levels of L-[$^3$H]-aspartate binding ($r = 0.728$, $p = 0.001$). Levels of L-[$^3$H]-aspartate uptake directly correlated with synaptophysin immunoreactivity ($r = 0.488$, $p = 0.04$, $n = 16$) and inversely correlated with the Blessed score ($r = 0.495$, $p = 0.04$, $n = 16$) (Masliah *et al.*, 1996e). These results suggest that decreased glutamate transporter activity is associated with increased excitotoxicity and neurodegeneration, and support the possibility that abnormal functioning of this system might be involved in the pathogenesis of the synaptic damage in AD (Masliah *et al.*, 1996e).

## 2.5. Correlations with Neurofibrillary Cytoskeletal Alterations

Recent studies have shown that in addition to synaptic pathology, the severity of dementia in AD patients was positively related to the number of NFTs in neocortex, but not to plaque numbers (Arriagada *et al.*, 1992a); that NFTs accumulate in a consistent pattern reflecting hierarchic vulnerability of individual cytoarchitectural fields; that NFTs appear in the entorhinal cortex, CA1/subiculum field of the hippocampal formation, and the amygdala early in the disease process; and that plaque number was also related to a hierarchic vulnerability of certain brain areas to Aβ accumulation, but the Aβ distribution pattern was different from that of NFTs (Arriagada *et al.*, 1992a). Other studies have confirmed this view and have shown that of the various neuropathological markers of AD, only NFTs showed a significant association with CDR score, and only for neocortical regions (Bierer *et al.*, 1995). In particular, NFT density in superior temporal cortex had the strongest correlation with dementia (Bierer *et al.*, 1995). Moreover, NFT density in areas 9, 40, and 22 correlated with cognitive decline and with plaque number in areas 22 and 17/18 (Dournaud *et al.*, 1995).

These data support the notion that neocortical neuronal degeneration as indicated by NFT formation is a critical determinant of the clinical progression of AD, and that medial temporal lobe structures may represent the initial site of NFT formation (Samuel *et al.*, 1994a,b; Hansen *et al.*, 1991; Braak and Braak, 1991). Although plaque density correlates with age at death, there is no correlation between plaque counts and dementia severity. These results further suggest that the clinical presentation of dementia may be closely related to neurodegeneration in neocortical regions within the temporal lobe (Arriagada *et al.*, 1992a). Furthermore, a study by Berg *et al.* (1993) showed that densities of neocortical NFTs were related to degree of dementia, whereas densities of neocortical SPs were not. The authors concluded that: (1) neocortical senile plaque densities differentiate very old subjects with AD from nondemented controls, but there is a need for more postmortem studies of older persons who are free of dementia; and (2) among the microscopic lesions studied, densities of neocortical NFTs related most closely to degree and duration of dementia (Berg *et al.*, 1993). Although these studies have shown that neuropil threads and neurofibrillary pathology could contribute to the dementia in AD (Samuel *et al.*, 1994b; Masliah *et al.*, 1992b; Delaère *et al.*, 1989), it is important to remember that a subgroup of AD cases shows little or no fibrillary pathology yet displays very significant clinical alterations (Terry *et al.*, 1987). In a subset of these cases, the most likely explanation for dementia is the concomitant presence of Lewy bodies; that is, a significant proportion of the plaque-only AD cases turn out to be LBV (Hansen *et al.*, 1993). In the rest of the cases, although no actual tangles are found, further analysis with Alz50 and other antibodies that recognize tau revealed extensive abnormal immunoreactivity in the pyramidal neurons in hippocampus, entorhinal cortex, and neocortex, indicating cytoskeletal anomalies associated with altered axoplasmic flow (Masliah *et al.*, 1994a; Hansen *et al.*, 1991).

We also found high correspondence between premortem dementia and accumulation of NFTs, with concurrent loss of synapse density in several brain regions (Samuel *et al.*, 1994a,b) (Table VI). To better understand the contribution of the pathology within hippocampus and the memory deficits in AD, we examined these same clinicopathological relationships in the context of seven subregions of the hippocampal formation (Samuel *et al.*, 1994b) (Fig. 4) and found that NFT counts

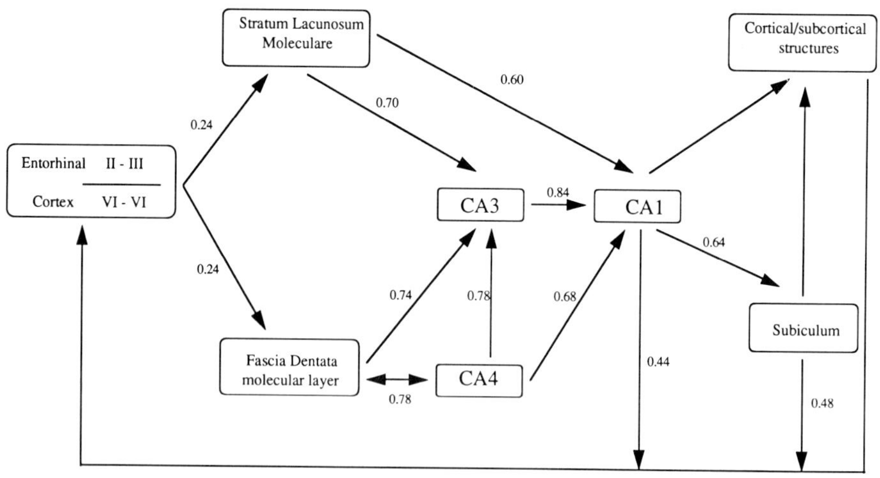


**Figure 4.** Major interconnective pathways within the human hippocampal formation.

within the CA1, subiculum, and CA4 regions were the most strongly correlated with degree of dementia, and that NFTs in these subregions appeared significantly clustered on factor analysis (Samuel *et al.*, 1994b). Synapse loss was most strongly correlated with dementia severity when it occurred in the molecular layers of dentate fasciculus and stratum lacunosum, CA2/3, and CA4; synapse loss in these subregions appeared significantly clustered on factor analysis. In general, these results were compatible with a two-component model of hippocampal connectivity and function in the context of AD. The first component consists of subregions preceding CA1 in a hypothesized input-processing sequence intrinsic to the hippocampus that summates neuronal excitation and that influences cognition primarily through synapse density. The second component consists of an "output module," mainly CA1 and subiculum, that receives the processed signal, passes it on to extrahippocampal cortical and subcortical targets, and affects cognition primarily by NFT accumulation in output neurons. A "net pathology" score combining standardized *z*-scores for synapse density and NFTs was significantly correlated with all three mental status measures (Blessed, MMSE, DRS) in all hippocampal subregions except the entorhinal cortex, and stepwise regressions on these data found net pathology in CA4 to be the most independent significant predictor of premortem dementia (Samuel *et al.*, 1994b).

Neurofibrillary tangle density closely correlated with neuronal loss and glial cell proliferation in our analyses, supporting the notion that overall NFT formation reflects the extent of neurodegeneration in AD (Table VIII). This is further supported by studies with platelet-derived growth factor B (Masliah *et al.*, 1995b). Linear regression analysis of synapse, neuronal, and tangle counts showed that the strongest correlations were of midfrontal synapse counts with both inferior parietal region tangles and Braak stage (Table VIII), indicating that during the neurodegenerative process in AD there is extensive loss of axons in association cortices, which are believed to be involved in higher cognitive functioning. This view is further supported by studies using the monoclonal antibody against phosphorylated neurofilaments (SMI312) that have corroborated extensive damage to association axons (Masliah *et al.*, 1993b; Hof *et al.*, 1992).

In addition to the characteristic NFT formation in the neuronal cell body, cytoskeletal pathology in AD is also characterized by the formation of neuropil threads (Perry *et al.*, 1991b; Yamaguchi *et al.*, 1990; Braak and Braak, 1988; Braak *et al.*, 1986). Golgi impregnation and immunoelectron microscopic studies have shown that these structures probably correspond to PHF accumulation in the terminal dendritic field of tangle-bearing neurons (Braak and Braak, 1988). Previous immunohistochemical and ultrastructural studies have suggested a strong correlation between dendritic thread counts (tau meshwork) and dementia severity (Delaère *et al.*, 1989). In contrast, other studies have shown that thread counts do not significantly correlate with Blessed scores (Masliah *et al.*, 1992b). Furthermore, the major histopathologic correlate of degree of synaptic pathology is the presence of anti-tau-immunolabeled dystrophic axons, not dendritic thread counts. A study by Delaère *et al.* (1989) differs from ours (Masliah *et al.*, 1992b) in that their work included cases with few or no plaques and tangles. In addition, whereas our study was done in the frontal cortex, their semiquantitative evaluation of anti-tau immunoreactivity was done in the temporal cortex, a region characteristically more severely affected by neurofibrillary pathology than the frontal and parietal cortices, where high plaque densities often overshadow tangle formation (Terry *et al.*, 1981). Recently, a detailed paper by McKee *et al.* (1991) showed a strong correlation between the cognitive test scores of demented and nondemented cases and the degree of neuritic pathology (a measure of combined dendritic and axonal alterations). In our cases we also observed greater axonal pathology (Masliah *et al.*, 1992b) in severely demented cases than in less cognitively impaired AD cases.

The cytoskeletal pathology in dendrites is associated with synapse clustering rather than with synapse loss or axonal pathology. Furthermore, presynaptic boutons were apposed to some threads. In this regard, a previous immunoelectron microscopic study showed that nerve terminals were forming synapses on dendritic processes containing PHF (Yamaguchi *et al.*, 1990) and that presynaptic terminals contain PHF as well as other abnormal filaments (Ellisman *et al.*, 1987; Gonatas *et al.*, 1967). These findings could be explained in three different ways: (1) modifications of the cytoskeletal proteins may accompany sprouting and synaptogenesis; (2) these synapses remain on the processes of the tangle-bearing neurons, and when neuronal degeneration is complete they will eventually be lost; and/or (3) these synapses originally formed on the threads of non-tangle-bearing neurons or pretangles. Supporting the latter possibility, previous studies (Masliah *et al.*, 1992b) showed that some of the threads originated in anti-tau-immunoreactive, non-tangle-bearing neurons. Moreover, diverse denervated rat studies have shown that during the process of sprouting, there is an increased synthesis of cytoskeletal proteins (Caceres *et al.*, 1988; Busciglio *et al.*, 1987; Caceres and Dotti, 1985). Caceres *et al.* (1988) showed an increase in microtubule-associated protein in the hippocampal molecular layer during the first days after entorhinal cortex lesion. In addition, these authors also found a significant modification in tau during reinnervation, suggesting that sprouting terminals may regulate an increase in dendritic cytoskeletal proteins by a transsynaptic message system. Taken together, these data may imply that the subpopulation of threads with synapses represents sprouting dendrites (Ihara, 1988).

The positive correlation between plaque counts and mean density of threads (Masliah *et al.*, 1992b) might indicate that increased accumulation of tau proteins in threads is associated with increased number of plaques. Dendritic threads and dystrophic axons in plaques share the same complement of tau antigenic determinants, but they differ in that threads rarely contain anti-neurofilament immunoreactivity (Schmidt *et al.*, 1991, 1990). Electron microscopic studies of neuropil threads inside and outside the plaques have shown that even though they do not differ ultrastructurally, the threads inside plaques tend to have a wider diameter, a view supported by laser confocal microscopic analysis (Masliah *et al.*, 1992b). The increase in thread size associated with plaques could be partially responsible for the increase in tau immunoreactivity, but the mechanisms by which plaque formation might induce this are unknown.

Thread counts were correlated with the decrease in size of presynaptic terminals in layers II and III of the frontal cortex, possibly indicating that damaged synapses are partially collapsed in the amyloid-free regions as a result of the loss of vesicles and other organelles in the synaptic apparatus. This is supported by our recent immunoelectron microscopic study (Yamaguchi *et al.*, 1990). Earlier morphometric electron microscopic study by Davies and co-workers (Davies *et al.*, 1987) described an increase in the synaptic disk diameter in layer V of frontal and temporal cortex, but not in layers II and III. Furthermore, we also found a significant increase in the size of the synaptophysin-immunolabeled terminals in the regions where anti-tau-immunoreactive dystrophic axons occurred, indicating that dilatation of presynaptic terminals and synapses is associated with the neuritic changes scattered in neuropil, as well as those associated with plaques. This view is supported by our previous observations showing that increased size of presynaptic boutons is associated with plaque formation (Masliah *et al.,* 1990) even though these areas show an accentuation of the overall synapse loss (Masliah *et al.,* 1990). Studies of these dilated and dystrophic plaque axons with histochemical techniques and with antibodies against GAP-43 have suggested that some of these neuritic components might represent an aberrant sprouting response in areas of synapse loss (Six *et al.,* 1992; Masliah *et al.,* 1991a,d; Sima *et al.,* 1986).

In conclusion, these results suggest that synaptic damage may precede the formation of dendritic threads and tangles, and that thread formation does not necessarily induce the synaptic pathology, but NFT formation might be an important contributing factor.

### 2.6. Association with Lewy Bodies

About 50–60% of the senile dementia patients show at autopsy neuropathological changes characteristic of AD (Lee *et al.,* 1992). The rest show combined forms of AD with cerebrovascular pathology (10%), cerebrovascular alterations alone (10%), and Lewy body disease (20%) (Lee *et al.,* 1992; Crystal *et al.,* 1990; Hansen *et al.,* 1990; Kosaka, 1990; Perry *et al.,* 1990; Lennox *et al.,* 1989; Gibb and Lees, 1988). Less than 1% show other, more-rare neuropathological conditions. Lewy body disease is also known by other names, including diffuse Lewy body disease (DLBD), which is defined by some as a primary dementia wherein cortical Lewy bodies are found, recognizing that most cases show concomitant AD pathology (Crystal *et al.,* 1990; Dickson *et al.,* 1987, 1989; Lennox *et al.,* 1989; Burkhardt *et al.,* 1988; Yoshimura, 1983). Others prefer to use the term "LBV" when combined AD and PD (Parkinson's disease) pathology is found, restricting the diagnosis of DLBD to those cases with abundant cortical and subcortical Lewy bodies without the neuropathological alterations of AD (Hansen *et al.,* 1993, 1990, 1989). Other terms used include senile dementia of the Lewy body type, combined AD and PD, and Lewy body dementia. In addition, a quarter of the patients who show the primary clinical picture of PD will develop dementia, which neuropathologically will typically show idiopathic PD in association with AD changes (Vermersch *et al.,* 1993; Ditter and Mirra, 1987). A recent consensus meeting at Newcastle in the United Kingdom (McKeith *et al.,* 1996) adopted a definition for probable dementia with Lewy bodies that includes recurrent visual hallucinations; spontaneous parkinsonism motor features; progressive cognitive decline that impairs occupational and social function—often with prominent deficits in fronto-subcortical skills and visuospatial abilities, and with two of the core features of fluctuating cognitive decline with pronounced variation in attention and arousal.

The following factors could be involved in the dementia associated (Masliah and Terry, 1994) with Lewy body disease: (1) synaptic, axonal, and neuronal damage and loss in association cortex, limbic structures, and neostriatal system (Masliah and Terry, 1994; Wakabayashi *et al.*, 1994; Masliah *et al.*, 1993d,b); (2) neurofibrillary and cytoskeletal changes associated with formation of neuritic plaques, NFTs, neuropil threads, and Lewy bodies (Samuel *et al.*, 1996; Hansen *et al.*, 1993, 1991; Dickson *et al.*, 1991; Hansen *et al.*, 1990); and (3) extent of amyloid burden (Lippa *et al.*, 1994). All of these factors and their topographical distribution are to some extent associated with the severity and characteristics of the dementia. For example, dementia could present in patients with cortical and subcortical Lewy bodies, variable amounts of amyloid, and few or no NFTs (Hansen *et al.*, 1993); or, dementia could show only Lewy bodies, with plaques or tangles (Samuel *et al.*, 1996; Hansen *et al.*, 1993). The common denominators appear to be the presence of Lewy bodies and damage and loss of presynaptic terminals and neurons in specific circuits. Although LBV and AD groups had similar cognitive test scores on their last evaluation prior to death, lower neocortical NFT and plaque counts and lower modified Braak stages (Bancher *et al.*, 1993) were observed in LBV (Samuel *et al.*, 1996). Neocortical NFT counts correlated with impaired neuropsychological test performance in AD but not in LBV. Plaque counts did not correlate with mental status in either group. In contrast, Lewy body concentrations in four neocortical areas correlated significantly with dementia severity in LBV. The association of AD lesions in the neocortex with dementia in LBV was comparatively weaker than that observed for Lewy body concentrations (Samuel *et al.*, 1996). Frontal cortex synaptic counts correlated strongly with dementia in the AD group but only weakly in the LBV group (Wakabayashi *et al.*, 1994). We quantified by microdensitometer the immunoreactivity with monoclonal antibodies against synaptic vesicle proteins [synaptotagmin, synaptobrevin (SP 10/11), SNAP25 (SP 12/14) (Honer *et al.*, 1992b), and synaptophysin-like molecules (EP10, SP6, SP15)] (Masliah *et al.*, 1994b; Honer *et al.*, 1992a) in the molecular layers of dentate gyrus, CA4, CA3, CA1, subiculum, and entorhinal cortex in AD, LBV, and DLBD. A significant decrease in SP6 immunoreactivity was observed in almost all regions in AD (28.4–70.1%, mean 41.3%), LBV (19.0–42.5%, mean 26.8%), and DLBD (19.9–31.7%, mean 27.1%) as compared with controls. In addition, SP6 immunoreactivity in the outer molecular layer of the dentate gyrus correlated strongly with tangle count in the entorhinal cortex ($r = -0.70$, $p < 0.002$), suggesting loss of perforant pathway projections. Although the decrease in SP12 and SP15 immunoreactivity was less pronounced, the mean values were decreased in demented subjects. Furthermore, SP12 and SP15 labeled a large number of neuritic plaques, indicating that: (1) in the hippocampal/entorhinal formation, the decrease of synapse protein immunoreactivity in AD is more severe than that in LBV or DLBD; (2) SP6 immunoreactivity in the molecular layer and CA subregions was positively correlated with MMSE scores of LBV patients ($r = 0.55$ to $0.62$, $p < 0.05$); and (3) synaptic markers detect a subset of dystrophic neurites in the plaques (Wakabayashi *et al.*, 1994). Compared with AD, the LBVs had 39% more GAP-43-positive plaques in the frontal cortex and 53% more in the hippocampus (Masliah *et al.*, 1993d). These neuritic alterations were accompanied by an accumulation of amyloid precursor protein and phosphorylated neurofilaments. Synapse loss was the same in LBV and AD. These results suggest more-extensive aberrant neuronal sprouting in LBV than in AD.

The findings also suggest that neocortical Lewy bodies combined with entorhinal NFTs or subcortical PD-type pathology may equalize the degree of dementia seen in LBV with that encountered in classical AD (Masliah *et al.*, 1993c). In addition to synaptic damage, Lewy body formation, amyloid deposition, and mild NFT formation (Masliah *et al.,* 1993d), LBV is also accompanied by disruption of the axonal fibers in the CA2–CA3 region (Dickson *et al.,* 1991). These alterations suggest abnormal axoplasmic flow of neurons projecting from CA4/CA3 (Reyes *et al.,* 1993). More-recent studies have shown that in addition to these alterations, the pyramidal neurons in the neocortex of LBV cases display significant fragmentation of their cytoskeleton (Smith *et al.,* 1995). In conjunction with the other neuropathological markers, these cytoskeletal alterations in the hippocampus and neocortex might contribute to memory impairment in LBV (Reyes *et al.,* 1993).

Because the accumulation of Lewy bodies correlates with cognitive decline, it is possible that Lewy body formation alters synaptic function. Previous studies have shown that neocortical and subcortical Lewy bodies immunoreact with antibodies against small and large synaptic vesicle proteins (Wakabayashi *et al.,* 1992, 1994; Nishimura *et al.,* 1994). Ultrastructurally, some elements suggestive of synaptic vesicles have been observed to be associated with Lewy bodies (Nishimura *et al.,* 1994). To better understand whether only *specific* synaptic vesicle proteins are trapped in the Lewy bodies or whether a wider variety of the proteins are associated with the Lewy bodies, double-immunolabeling studies were performed in the superior temporal LBV cortex using a polyclonal antibody against ubiquitin (which recognize Lewy bodies) and the previously mentioned monoclonal antibodies against synaptic vesicle proteins and synaptophysin-like molecules (Wakabayashi *et al.,* 1994). Laser confocal imaging of the double-labeled sections revealed that only SP15 immunolabeled the Lewy bodies. Furthermore, Lewy body counts correlated with synaptic counts in the neocortex of LBV cases ($r = 0628$, $p < 0.02$, $n = 15$). These results indicate that probably only the components necessary for assembly of integral synaptic vesicle proteins are altered by Lewy body formation, and therefore the cytoskeletal alterations accompanying their formation might specifically affect the integration of synaptophysin-like proteins in the synaptic vesicles, supporting the contention that Lewy body formation might alter synaptic structure and function and thus contribute to dementia through this mechanism.

## 2.7. Correlations with Amyloid Deposition

The original studies by Blessed and Tomlinson suggested that amyloid deposition and plaque formation might be the major correlates of cognitive alteration in AD (Blessed *et al.,* 1968), but more detailed studies in which control and LBV cases were not included in the linear regression analysis did not support this view (Terry *et al.,* 1991). Furthermore, studies employing semiquantitative assessment of amyloid angiopathy have shown that this parameter is not correlated with tests of cognition (Table VI), thus arguing that Aβ deposition in senile plaques is a secondary event because plaque numbers do not correlate with cognition as strongly as do NFTs (Nagy *et al.,* 1995; Bierer *et al.,* 1995; Berg *et al.,* 1993; Arriagada *et al.,* 1992a). However, in most studies in which amyloid was poorly correlated with cognition, plaque counts reflected the total number of diverse plaque types. More-recent studies suggest that *diffuse* plaques (which lack dystrophic neurites) might be an age-related phenomena, whereas *neuritic* plaques are more indicative of AD (Davis *et al.,* 1995). Furthermore, diffuse plaques are not associated with accentuation of synaptic loss in AD as are neuritic plaques (Masliah *et al.,* 1990), indicating that neuritic plaque numbers might be more closely related to the pathogenic process in AD and might therefore be correlated with neurodegeneration and cognitive decline (McKee *et al.,* 1991). Supporting this possibility, previous studies (McKee *et al.,* 1991) as well as a review of many cases from our cohort have found that neuritic plaque counts within the neocortex are more closely correlated with tests of global cognition than are total plaque counts (Table VI). The region showing the strongest correlation between neuritic plaques and cognition was the inferior parietal cortex (Table VI). This probably reflects the progression of the neuropathogenic process in AD, because plaque formation likely occurs later in this region than in temporal and frontal cortex and the density of lesions in more-advanced cases is lower, making a saturation effect less likely. In agreement with this possibility, linear regression analysis has shown a strong correlation between Braak stage and inferior parietal plaques ($r = 0.712$, $p = 0.0001$, $n = 48$); also, neuronal loss in the inferior parietal region and synapse loss in the midfrontal region were mildly correlated with plaque counts in the inferior parietal (Table VIII). Furthermore, Nagy *et al.* (1995) found that only the percentage of cortical area occupied by amyloid in the parietal lobe correlated with the severity of cognitive deficit in the CERAD-diagnosed AD cases. In this study, the density of *all* types of plaques showed no correlation with the extent of cognitive deficits, but the densities of neuritic plaques did correlate. Both amyloid load and tangle densities were positively correlated with disease duration. All these correlations were reduced or absent in a subgroup of cases fulfilling diagnostic criteria for AD (Nagy *et al.,* 1995). A more recent study based on computer image analysis found a strong correlation between Aβ load in entorhinal cortex and cognition as measured on various scales (e.g., $r = -0.93$ versus the Blessed IMC) (Cummings and Cotman, 1995), suggesting that the size of the cortical area affected by Aβ deposition is an important factor in the clinical manifestation of dementia (Cummings and Cotman, 1995).

Other studies (Nagy *et al.,* 1995) have shown that amyloid load and NFT densities were highly significantly—but selectively—correlated, with the most striking correlations being those between temporal lobe NFT density and frontal and parietal lobe amyloid load, and between temporal lobe NFT density and frontal and parietal lobe NFT densities. Correlations involving plaque density as a measure of amyloid load were almost always less significant than those involving the percentage of cortical area occupied by amyloid, suggesting that the latter measures amyloid load more accurately (Nagy *et al.,* 1995). However, the highest correlations for NFT densities were with densities of neuritic plaques. Overall, this study highlights the relevance of neuritic changes (revealed by NFTs and neuritic plaques) and the lower impact of amyloid on the dementia of AD (Nagy *et al.,* 1995). Further adding to the controversy surrounding the role of plaque formation in dementia, a recent study showed that high loads of diffuse plaques in the cortex of elderly individuals might be associated with very mild cognitive alterations, and concluded that Aβ deposition is an early change in AD (Morris *et al.,* 1996). However, as emphasized in previous sections, diffuse amyloid load is a variable process during aging and is not sensitive or specific enough to really predict AD-type dementia in the same way as can both neuronal loss in the layer II of the entorhinal cortex (Gomez-Isla *et al.,* 1996) and synapse loss in the outer molecular layer (Masliah *et al.,* 1994a). Taken together, these results suggest that altered functioning of vulnerable neurons might result in synaptic loss, which in turn is associated with cognitive dysfunction and amyloid deposition. Although Aβ deposition may or may not be the initial triggering factor, the eventual accumulation of the deposits throughout the brain may contribute to the overall degenerative process (Fig. 1).

## 3. Mechanisms of Synaptic Pathology in Alzheimer's Disease

### 3.1. Amyloid-β Precursor Protein and Synaptic Damage

Possible mechanisms involved in the pathogenesis of synaptic damage in AD could be related to either abnormal function of synaptic proteins or direct toxic effects at the presynaptic site (Zhong *et al.*, 1994; Masliah and Terry, 1993) (Fig. 5). Alternatively, abnormal functioning of cytoskeletal proteins might result in abnormal axonal flow, which in turn might lead to synaptic damage (Suzuki and Terry, 1967). Recent studies have shown that APP, which is believed to be centrally involved in AD (Selkoe, 1989), might play an important role as a synaptic regulator (Mucke *et al.*, 1994; Roch *et al.*, 1994; Small *et al.*, 1994; Alvarez *et al.*, 1992; Askanas et al., 1992; Schubert *et al.*, 1991). Moreover, APP metabolism appears to be abnormal in AD (Zhong *et al.*, 1994; Sisodia *et al.*, 1990). Furthermore, recent studies in transgenic mice have shown that overexpression of mutated human APP is accompanied by synaptic damage, neurodegeneration, and plaque formation (Games *et al.*, 1995).

Abnormal APP functioning could lead to neurodegeneration in AD by: (1) direct toxicity of elevated levels of aggregated Aβ (Selkoe, 1993; Golde *et al.*, 1992);

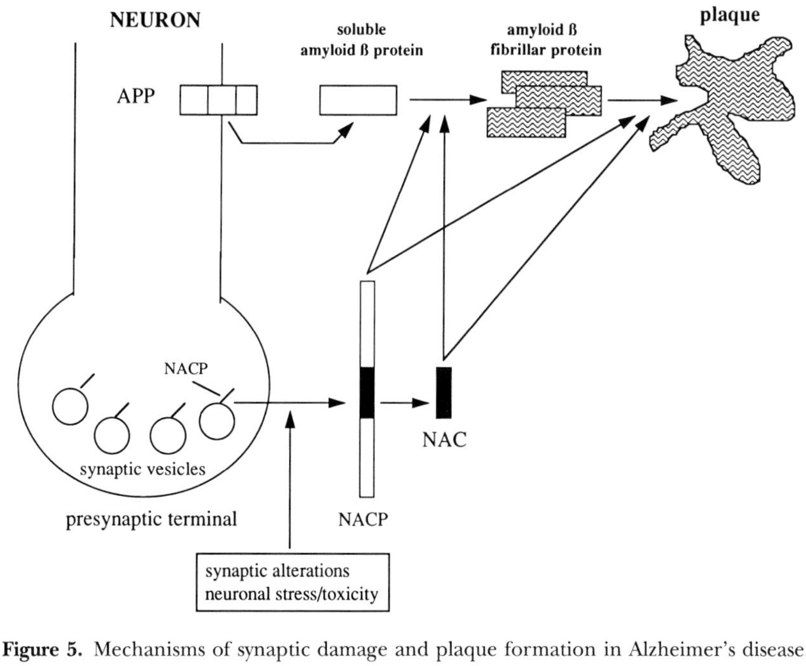


**Figure 5.** Mechanisms of synaptic damage and plaque formation in Alzheimer's disease.

(2) disruption of synaptic function (Masliah, 1995c); (3) deficient neuroprotective activity against excitotoxicity (Mucke *et al.,* 1995b; Mattson *et al.,* 1993a,b); or (4) any combination of the three. The direct toxic role of fibrillar Aβ alone *in vivo* is controversial. In addition, the amyloid load in the brain of AD patients in general does not correlate with the cognitive deficits and with the extent of the neurodegenerative process (Arriagada *et al.,* 1992a,b; Terry *et al.,* 1991; DeKosky and Scheff, 1990), with the exception of a recent study by Cummings and Cotman (1995). Therefore, although Aβ might play a limited role in promoting neuronal dysfunction, an additional process, such as excitotoxicity, might be required to satisfactorily explain AD pathology. It has been postulated that secreted APP (sAPP) might prevent the toxicity associated with activation of glutamate receptors by stabilizing intracellular calcium levels (Mattson *et al.,* 1993b,c,d). Furthermore, neuroprotection might also be achieved by modulating the levels of glutamate receptors and/or activity of glutamate/aspartate uptake system. With respect to the latter, studies have shown that potentially neurotoxic neurotransmitters like glutamate are cleared from the synaptic cleft by high-affinity $Na^+$-dependent uptake carriers located in both neurons and glia (Kanai *et al.,* 1993, 1994; Greenamyre and Porter, 1994; Balcar and Li, 1992) (Fig. 6). It is now clear that these, as well another not fully characterized glutamate transporters, are present in specific cellular populations and regions of the brain and are necessary for clearing excitotoxic neurotransmitters (Kanai *et al.,* 1995; Shashidharan *et al.,* 1994; Manfras *et al.,* 1994). Thus, deficient glutamate transporter activity might result in neurodegeneration by causing accumulation of excitotoxins at the synaptic site. Support ing a role for decreased glutamate transporter activity in neurodegeneration, recent studies have shown that: (1) in AD the high-affinity glutamate/aspartate up

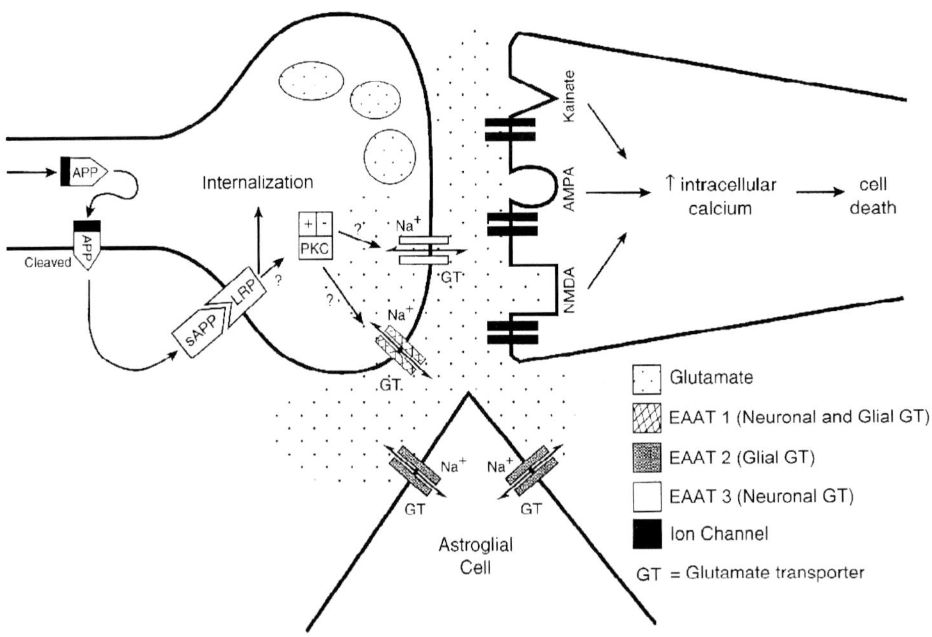


**Figure 6.** Proposed role for APP in regulating function of glutamate transporters.

take system is 40–50% decreased in neocortex (Masliah *et al.*, 1996c; Scott *et al.*, 1995; Cowburn *et al.*, 1990, 1988), (2) in amyotrophic lateral sclerosis there is selective loss of the glial glutamate transporter (GLT-1) (Rothstein *et al.*, 1995), and (3) chronic inhibition of glutamate transporter activity with L(-)-threo-3-hydroxy aspartate in experimental models results in neurodegeneration (Rothstein *et al.*, 1993). In addition, we have recently shown that glutamate transporter deficits occur early in the progression of AD and are correlated with the extent of the neurodegenerative process as assessed by levels of brain spectrin degradation products (Masliah *et al.*, 1996e). These results suggest that abnormal functioning of APP might be associated with decreased activity of glutamate transporters in AD, which in turn leads to increased excitotoxicity and neurodegeneration (Fig. 6).

### 3.2. Apolipoprotein E and Synaptic Damage

Apolipoprotein E (apoE) is a 34-kDa protein that plays an important role in cholesterol transport, uptake, and redistribution (Paik *et al.*, 1985; Mahley and Innerarity, 1983). Recent studies have shown that approximately 64% of AD cases are associated with the presence of allele ε4 of apoE (Saunders *et al.*, 1993), suggesting that abnormally functioning apoE might lead to neurodegeneration in AD (see Roses *et al.*, 1996, for review). The precise mechanisms by which abnormal functioning of apoE might lead to AD are not yet fully understood. Although some studies support a role for apoE in amyloid clearance (Strittmatter *et al.*, 1993; Schmechel *et al.*, 1993), others suggest a role in protection against excessive tau phosphorylation (Strittmatter *et al.*, 1994), intracellular calcium influx (Hartmann *et al.*, 1994), and synaptic plasticity (Poirier, 1994) (Fig. 5).

Supporting a role for apoE in synaptic plasticity, recent *in vitro* studies have shown that apoE complexed with other lipids promotes neuritic outgrowth (Holtzman *et al.*, 1995; Nathan *et al.*, 1994; Ignatius *et al.*, 1987). *In vivo*, apoE mediates delivery of cholesterol to synaptic membranes during the process of synaptic plasticity and regeneration after injury (Masliah *et al.*, 1995c; Poirier *et al.*, 1993). Supporting a role of apoE in regeneration, previous studies in dorsal root ganglion cells and in an immortalized CNS-derived neuronal cell line have shown that lipid-complexed human apoE3 increases neuritic outgrowth, whereas apoE4 decreases it (Holtzman *et al.*, 1995; Nathan *et al.*, 1994). It has been suggested that the differential effects of apoE isoforms might be related to the interactions of apoE with its receptor at multiple subcellular compartments (Holtzman *et al.*, 1995). Furthermore, kinetic studies with apoE have shown that E3 is metabolized differently from E4, supporting the possibility that the E4 isoform might represent a dysfunctional molecule (Gregg *et al.*, 1986). This is of particular interest for understanding the pathogenesis of AD, because the presence of E4 is associated with a higher risk for this neurodegenerative disorder (Saunders *et al.*, 1993).

Furthermore, studies in apoE-deficient mice have shown that this molecule plays an important role in the maintenance of the synapto-dendritic complex during aging (Masliah, 1995a,d; Masliah *et al.*, 1996d). The neurodegenerative process in the neocortex and limbic system of aged apoE-deficient mice is accompanied by behavioral alterations (Gordon *et al.*, 1995) that are reversed by infusion of apoE (Masliah *et al.*, 1996d). In addition, cholinergic deficits are also more extensive in

AD with apoE $\epsilon4$ (Poirier *et al.*, 1995). Examination of apoE immunoreactivity using Western blot analysis of brain homogenates showed an average 20% decrease in both AD and LBV cases as compared with controls; this correlated with the extent of the synaptic damage (Masliah *et al.*, 1996b), and is consistent with previous studies showing a decrease in apoE in the cortex and hippocampus in AD patients with apoE $\epsilon4$ (Bertrand *et al.*, 1995). ApoE mRNA was more abundant in AD as compared with controls, but only in apoE $\epsilon3$ cases (Yamada *et al.*, 1995). Thus, these studies suggest that the presence of a malfunctioning apoE in AD might play a role in synaptic dysfunction and neurodegeneration.

### 3.3. Non-Aβ Component Precursor and Presynaptic Damage

The novel non-Aβ component precursor (NACP), also called "synuclein" and "phosphoneuroprotein 14" (Ueda *et al.*, 1993; Nakajo *et al.*, 1993; Maroteaux and Scheller, 1991), is a 19-kD (Iwai *et al.*, 1994; Ueda *et al.*, 1993) presynaptic protein loosely associated with synaptic vesicles (Iwai *et al.*, 1994) that is encoded by chromosome 4 (Chen *et al.*, 1995b; Campion *et al.*, 1995). The non-Aβ component (NAC) is a 35-amino acid peptide derived from the larger precursor protein NACP (Ueda *et al.*, 1993). Secondary-structure analysis has shown that NAC has a strong tendency to form β-pleated structures (Iwai *et al.*, 1995), and immunocytochemical analysis reveals that antibodies raised against synthetic peptides corresponding to the NAC sequence recognize amyloid fibrils in the plaques (Masliah *et al.*, 1996c; Ueda *et al.*, 1993) (Fig. 5). Furthermore, NAC immunoreactivity is associated mainly with plaques observed in AD, but not those seen in normal aging (Masliah *et al.*, 1996c). Whereas NACP is associated with dystrophic neurites in plaques, NAC is associated with dense amyloid deposits in the plaque core (Masliah *et al.*, 1996c). Early AD cases showed onefold higher levels of NACP immunoreactivity than did moderate and severe AD cases (Iwai *et al.*, 1996), and these levels correlated with tangle counts and Blessed scores, but not with plaque counts, suggesting that the abnormal accumulation of NACP during the early stages of AD might play an important role in the mechanisms of neurodegeneration and synaptic damage. Genetic association and linkage studies using a polymorphic dinucleotide repeat sequence in the NACP gene showed that one of the NACP polymorphisms (NACP allele 2) has a significant association with healthy elderly individuals who have apoE risk, indicating a possible protective function for the allele (Xia *et al.*, 1996). Nevertheless, screening of AD families failed to establish linkage between NACP and AD (Campion *et al.*, 1995; Brookes and St. Clair, 1994).

Recent studies have suggested that abnormal functioning of NACP might be involved in the pathogenesis of AD by promoting amyloid deposition and interfering with synaptic function (Masliah *et al.*, 1996c; Iwai *et al.*, 1994) (Fig. 5). As to the first possibility, the high propensity of NAC for amyloid fibrils suggests that the co-deposition of NAC and Aβ may represent an early event in plaque formation. In this regard, NAC might differ from other plaque constituents such as acetylcholinesterase, immunoglobulin, and heparan sulfate proteoglycan, whose inclusion might be a secondary phenomenon (Brookes and St. Clair, 1994). Furthermore, NACP binds Aβ and promotes aggregation and fibrillar formation (Yoshimoto *et al.*, 1995). This binding ability is specifically dependent on the 35-amino acid NAC sequence within NACP (Yoshimoto *et al.,* 1995). Transglutaminase catalyses the formation of the covalent NAC polymers as well as aggregating with Aβ (Jensen *et al.,* 1995). The transglutaminase-reactive amino acid residues in NAC were identified as Gln79 and Lys80, the latter localized in a consensus motif conserved in the NACP/synuclein gene family (Jensen *et al.,* 1995). Therefore, transglutaminase might participate in amyloid formation and NACP/synuclein modification.

In conclusion, analysis of molecules associated with amyloid in the core of senile plaques has revealed that many of them, including APP, apoE and NACP, are important functional molecules of presynaptic terminals. Abnormal functioning of these synapse-related proteins with amyloidogenic potential might play a central role in the pathogenesis of AD (Fig. 5). These molecules look promising for the development of novel treatments.

ACKNOWLEDGMENTS. This work was supported by NIH Grants AG05131 and AG10689, by funding from the Alzheimer Disease and Related Disorders Association and the Ruth K. Broad Biomedical Research Foundation, and by NIH Grant AG12963. The authors would like to thank Mrs. Margaret Mallory and Mr. Richard DeTeresa for their invaluable help.

NOTE ADDED IN PROOF. Recent neuropathological and genetic studies have linked NACP/αsynuclein to the pathogenesis of Parkinson's and Lewy body disease (see Spillantini, M. G., *et al.,* 1997, αsynuclein in Lewy bodies, *Nature* **388:**839–840).

## 4. References

Alvarez, J., Moreno, R. D., Llanos, O., Inestrosa, N. C., Brandan, E., Colby, T., and Esch, F. S., 1992, Axonal sprouting induced in the sciatic nerve by the amyloid precursor protein (APP) and other antiproteases, *Neurosci. Lett.* **144:**130–34.

Alzheimer, A., 1907, Uber eine eigenartige Erkrankung der Hirnrinde, *Alg. Zschr. Psychiatr.* **64:**146–148.

Andersen, P., and Trommald, M., 1995, Possible strategies for finding the substrate for learning-induced changes in the hippocampal cortex, *J. Neurobiol.* **26:**396–402.

Anderson, B., 1995, Dendritic correlates of dementia severity in Alzheimer's disease, *Med. Sci. Res.* **23:**597–599.

Armstrong, D. M., Benzing, W. C., Evans, J., Terry, R. D., Shields, D., and Hansen, L. A., 1989, Substance P and somatostatin coexist within neuritic plaques: Implication for the pathogenesis of Alzheimer's disease, *Neuroscience* **31:**663–671.

Arriagada, P. V., Growdon, J. H., Hedley-Whyte, E. T., and Hyman, B. T., 1992a, Neurofibrillary tangles, but not senile plaques, parallel duration and severity of Alzheimer's disease, *Neurology* **42:**631–639.

Arriagada, P. V., Marzloff, K., and Hyman, B. T., 1992b, Distribution of Alzheimer-type pathologic changes in nondemented elderly individuals matches the pattern in Alzheimer's disease, *Neurology* **42:**1681–1688.

Askanas, V., Engel, W. K., and Alvarez, R. B., 1992, Strong immunoreactivity of β-amyloid precursor protein, including the β-amyloid protein sequence, at human neuromuscular junctions, *Neurosci. Lett.* **143:**96–100.

Balcar, V. J., and Li, Y., 1992, Heterogeneity of high affinity uptake of L-glutamate and L-aspartate in the mammalian central nervous system, *Life Sci.* **51:**1467–1478.

Bancher, C., Braak, H., Fischer, P., and Jellinger, K. A., 1993, Neuropathological staging of Alzheimer lesions and intellectual status in Alzheimer's and Parkinson's disease patients, *Neurosci. Lett.* **162:**179–182.

Berg, L., McKeel, D. W., Jr., Miller, J. P., Baty, J., and Morris, J. C., 1993, Neuropathological indexes of

Alzheimer's disease in demented and nondemented persons aged 80 years and older, *Arch. Neurol.* **50:**349–358.

Bertrand, P., Poirier, J., Oda, T., Finch, G. E., and Pasinetti, G. M., 1995, Association of apolipoprotein E genotype with brain levels of apolipoprotein E and apolipoprotein J (clusterin) in Alzheimer's disease, *Mol. Brain Res.* **33:**174–178.

Bierer, L. M., Hof, P. R., Purohit, D. P., Carlin, L., Schmeidler, J., Davis, K. L., and Perl, D. P., 1995, Neocortical neurofibrillary tangles correlate with dementia severity in Alzheimer's disease, *Arch. Neurol.* **52:**81–88.

Blennow, K., Bogdanovic, N., Alafuzoff, I., Ekman, R., and Davidsson, P., 1996, Synaptic pathology in Alzheimer's disease: Relation to severity of dementia, but not to senile plaques, neurofibrillary tangles, or the apoE4 allele, *J. Neural Transm.* **103:**603–618.

Blessed, G., Tomlinson, B. E., and Roth, M., 1968, The association between quantitative measures of dementia and senile change in the cerebral grey matter of elderly subjects, *Br. J. Psychiatry* **114:**797–811.

Bondareff, W., Mountjoy, C. Q., Roth, M., Rossor, M. N., Iversen, L. L., Reynolds, G. P., and Hauser, D. L., 1987, Neuronal degeneration in locus ceruleus and cortical correlates of Alzheimer disease, *Alzheimer Dis. Assoc. Disord.* **1:**256–262.

Braak, H., and Braak, E., 1988, Neuropil threads occur in dendrites of tangle-bearing nerve cells, *Neuropathol. Appl. Neurobiol.* **14:**39–44.

Braak, H., and Braak, E., 1991, Neuropathological stageing of Alzheimer-related changes, *Acta Neuropathol.* **82:**239–259.

Braak, H., Braak, E., Grundke-Iqbal, I., and Iqbal, K., 1986, Occurence of neuropil threads in the senile brain and in Alzheimer's disease: A third location of paired helical filaments outside the neurofibrillary tangles and neuritic plaques, *Neurosci. Lett.* **65:**351–355.

Brookes, A. J., and St. Clair, D., 1994, Synuclein proteins and Alzheimer's disease, *Trends Neurosci.* **17:**404–405.

Burkhardt, C. R., Filley, C. M., Kleinschmidt-DeMasters, B. K., de la Monte, S., Norenberg, M. D., and Schneck, S. A., 1988, Diffuse Lewy body disease and progressive dementia, *Neurology* **38:**1520–1528.

Busciglio, J., Ferreira, A., Steward, O., and Caceres, A., 1987, An immunocytochemical and biochemical study of the microtubule-associated protein tau during post-lesion afferent reorganization in the hippocampus of adult rats, *Brain Res.* **419:**244–252.

Caceres, A., and Dotti, C., 1985, Immunocytochemical localization of tubulin and the high molecular weight microtubule-associated protein 2 in Purkinje cell dendrites deprived of climbing fibers, *Neuroscience* **16:**133–150.

Caceres, A., Busciglio, J., Ferreira, A., and Steward, O., 1988, An immunocytochemical and biochemical study of the microtubule-associated protein MAP-2 during post-lesion dendritic remodeling in the central nervous system of adult rats, *Mol. Brain Res.* **3:**233–246.

Callahan, L. M., and Coleman, P. D., 1995, Neurons bearing neurofibrillary tangles are responsible for selected synaptic deficits in Alzheimer's disease, *Neurobiol. Aging* **16:**311–314.

Campion, D., Martin, C., Heilig, R., Charbonnier, F., Moreau, V., Flaman, J. M., Petit, J. L., Hannequin, D., Brice, A., and Frebourg, T., 1995, The NACP/synuclein gene: Chromosomal assignment and screening for alterations in Alzheimer disease, *Genomics* **26:**254–257.

Chen, K. S., Masliah, E., Mallory, M., and Gage, F. H., 1995a, Synaptic loss in cognitively impaired rats is ameliorated by chronic NGF infusion, *Neurosci.* **68:**19–27.

Chen, X., de Silva, R. H. A., Pettenati, M. J., Rao, P. N., St. George-Hyslop, P., Roses, A. D., Xia, Y., Horsburgh, K., Ueda, K., and Saitoh, T., 1995b, The human NACP/alpha-synuclein gene: Chromosome assignment to 4q21.2-q22 and Taql RFLP analysis, *Genomics* **26:**425–427.

Cotman, C. W., and Nadler, J. V., 1978, Reactive synaptogenesis in the hippocampus, in: *Neuronal Plasticity* (C. W. Cotman, ed.), Raven, New York, pp. 227–271.

Cowburn, R., Hardy, J., Roberts, P., and Briggs, R., 1988, Presynaptic and postsynaptic glutamatergic function in Alzheimer's disease, *Neurosci. Lett.* **86:**109–113.

Cowburn, R. F., Hardy, J. A., and Roberts, P. J., 1990, Glutamatergic neurotransmission in Alzheimer's disease, *Biochem. Soc. Trans.* **18:**390–392.

Cross, A. J., Skan, W. J., and Slater, P., 1986, The association of [$^3$H]d-aspartate binding and high-affinity glutamate uptake in the human brain, *Neurosci. Lett.* **63:**121–124.

Crystal, H. A., Dickson, D. W., Lizardi, J. E., Davies, P., and Wolfson, L. I., 1990, Antemortem diagnosis of diffuse Lewy body disease, *Neurology* **40:**1523–1528.

Cummings, B. J., and Cotman, C. W., 1995, Image analysis of beta-amyloid load in Alzheimer's disease and relation to dementia severity, *Lancet* **346:**1524–1528.

Davies, C. A., Mann, D. M. A., Sumpter, P. Q., and Yates, P. O., 1987, A quantitative morphometric analysis of the neuronal and synaptic content of the frontal and temporal cortex in patients with Alzheimer's disease, *J. Neurol. Sci.* **78:**151–164.

Davies, P., 1979, Neurotransmitter-related enzymes in senile-dementia of the Alzheimer type, *Brain Res.* **171:**319–327.

Davis, P. C., Gearing, M., Gray, L., Mirra, S. S., Morris, J. C., Edland, S. D., Lin, T., and Heyman, A., 1995, The CERAD experience. Pt. VIII: Neuroimaging–neuropathology correlates of temporal lobe changes in Alzheimer's disease, *Neurology* **45:**178–179.

DeKosky, S. T., and Scheff, S. W., 1990, Synapse loss in frontal cortex biopsies in Alzheimer's disease: Correlation with cognitive severity, *Ann. Neurol.* **27:**457–464.

DeKosky, S. T., Harbaugh, R. E., Schmitt, F. A., Bakay, R. A., Chui, H. C., Knopman, D. S., Reeder, T. M., Shetter, A. G., Senter, H. J., and Markesbery, W. R., 1992, Cortical biopsy in Alzheimer's disease: Diagnostic accuracy and neurochemical, neuropathological, and cognitive correlations, Intraventricular Bethanecol Study Group, *Ann. Neurol.* **32:**625–632.

Delaère, P., Duyckaerts, C., Brion, J. P., Poulain, V., and Hauw, J. J., 1989, Tau, paired helical filaments, and amyloid in the neocortex: A morphometric study of 15 cases with graded intellectual status in aging and senile dementia of Alzheimer type, *Acta Neuropathol.* **77:**645–653.

Deweer, B., Lehericy, S., Pillon, B., Baulac, M., Chiras, J., Marsault, C., Agid, Y., and Dubois, B., 1995, Memory disorders in probable Alzheimer's disease: The role of hippocampal atrophy as shown with MRI, *J. Neurol. Neurosurg. Psychiatry* **58:**590–597.

Dickson, D. W., Davies, P., Mayeux, R., Crystal, H., Horoupian, D. S., Thompson, A., and Goldman, J. E., 1987, Diffuse Lewy body disease: Neuropathological and biochemical studies of six patients, *Acta Neuropathol.* **75:**8–15.

Dickson, D. W., Farlo, J., Davies, P., Crystal, H., Fuld, P., and Yen, S. C., 1988, Alzheimer disease: A double immunohistochemical study of senile plaques, *Am. J. Pathol.* **132:**86–101.

Dickson, D. W., Crystal, H., Mattiace, L. A., Kress, Y., Schwagerl, A., Ksiezak-Reding, H., Davies, P., and Yen, S.-H. C., 1989, Diffuse Lewy body disease: Light and electron microscopic immunocytochemistry of senile plaques, *Acta Neuropathol.* **78:**572–584.

Dickson, D. W., Ruan, D., Crystal, H., Mark, M. H., Davies, P., Kress, Y., and Yen, S.-H., 1991, Hippocampal degeneration differentiates diffuse Lewy body disease (DLBD) from Alzheimer's disease: Light and electron microscopic immunocytochemistry of CA2-3 neurites specific to DLBD, *Neurology* **41:**1402–1409.

Dickson, D. W., Crystal, H. A., Bevona, C., Honer, W., Vincent, I., and Davies, P., 1995, Correlations of synaptic and pathological markers with cognition of the elderly, *Neurobiol. Aging* **16:**285–304.

Ditter, S. M., and Mirra, S. S., 1987, Neuropathologic and clinical features of Parkinson's disease in Alzheimer's disease patients, *Neurology* **37:**754–760.

Dournaud, P., Deleare, P., Hauw, J. J., and Epelbaum, J., 1995, Differential correlation between neurochemical deficits, neuropathology, and cognitive status in Alzheimer's disease, *Neurobiol. Aging* **16:**817–823.

Eccles, J. C., 1981, The modular operation of the cerebral neocortex considered as the material basis of mental events, *Neuroscience* **6:**1839–1856.

Eccles, J. C., 1984, The cerebral neocortex: A theory of its operation, in: *Cerebral Cortex,* Vol. 2 (E. G. Jones and A. Peters, eds.), Plenum, New York, pp. 1–38.

Ellisman, M., Ranganathan, R., Deerinck, T., Young, S., Terry, R. D., and Mirra, S., 1987, Neuronal fibrillar cytoskeleton and endomembrane system organization in Alzheimer disease, in: *Alterations in the Neuronal Cytoskeleton in Alzheimer Disease* (G. Perry, ed.), Plenum, New York, pp. 61–73.

Forstl, H., Burns, A., Luthert, P., Cairns, N., Lantos, P., and Levy, R., 1992, Clinical and neuropathological correlates of depression in Alzheimer's disease, *Psychol. Med.* **22:**877–884.

Forstl, H., Burns, A., Levy, R., Cairns, N., Luthert, P., and Lantos, P., 1993, Neuropathological correlates of behavioural disturbance in confirmed Alzheimer's disease, *Br. J. Psychiatry* **163:**364–368.

Forstl, H., Burns, A., Levy, R., and Cairns, N., 1994, Neuropathological correlates of psychotic phenomena in confirmed Alzheimer's disease, *Br. J. Psychiatry* **165:**53–59.

Frackowiak, J., Wisniewski, H. M., Wegiel, J., Merz, G. S., Iqbal, K., and Wang, K. C., 1992, Ultrastructure of the microglia that phagocytose amyloid and the microglia that produce β-amyloid fibrils, *Acta Neuropathol.* **84:**225–233.

Galasko, D., Hansen, L. A., Katzman, R., Wiederholt, W., Masliah, E., Terry, R., Hill, L. R., Lessin, P., and Thal, L. J., 1994, Clinical-neuropathological correlations in Alzheimer's disease and related dementias, *Arch. Neurol.* **51:**888–895.

Games, D., Adams, D., Alessandrini, R., Barbour, R., Berthelette, P., Blackwell, C., Carr, T., Clemes, J., Donaldson, T., Gillespie, F., Guido, T., Hagopian, S., Johnson-Wood, K., Khan, K., Lee, M., Leibowitz, P., Lieberburg, I., Little, S., Masliah, E., McConlogue, L., Montoya-Zavala, M., Mucke, L., Paganini, L., Penniman, E., Power, M., Schenk, D., Seubert, P., Snyder, B., Soriano, F., Tan, H., Vitale, J., Wadsworth, S., Wolozin, B., and Zhao, J., 1995, Alzheimer-type neuropathology in transgenic mice overexpressing V717F β-amyloid precursor protein, *Nature* **373:**523–527.

Gibb, W. R. G., and Lees, A. J., 1988, The relevance of the Lewy body to the pathogenesis of idiopathic Parkinson's disease, *J. Neurol. Neurosurg. Psychiatry* **51:**745–752.

Golde, T. E., Estus, S., Younkin, L. H., Selkoe, D. J., and Younkin, S. G., 1992, Processing of the amyloid protein precursor to potentially amyloidogenic derivatives, *Science* **255:**728–730.

Gómez-Isla, T., Price, J. L., McKeel, D. W., Jr., Morris, J. C., Growdon, J. H., and Hyman, B. T., 1996, Profound loss of layer II entorhinal cortex neurons occurs in very mild Alzheimer's disease, *J. Neurosci.* **16:**4491–4500.

Gonatas, N. K., and Gambetti, P., 1970, *The Pathology of the Synapse in Alzheimer's Disease: Ciba Foundation Symposium on Alzheimer's Disease and Related Conditions,* J & A Churchill, New York.

Gonatas, N. K., Anderson, W. W., and Evangelista, I., 1967, The contribution of altered synapses in the senile plaque: An electron microscopic study in Alzheimer's disease, *J. Neuropathol. Exp. Neurol.* **26:**25–39.

Gordon, I., Grauer, E., Genis, I., Sehayek, E., and Michaelson, D. M., 1995, Memory deficits and cholinergic impairments in apolipoprotein E-deficient mice, *Neurosci. Lett.* **199:**1–4.

Gray, E. G., 1986, Spongiform encephalopathy: A neurocytologist's viewpoint with a note on Alzheimer's disease, *Neuropathol. Appl. Neurobiol.* **12:**149–172.

Greenamyre, J. T., and Porter, R. H. P., 1994, Anatomy and physiology of glutamate in the CNS, *Neurology* (Suppl) **44:**S7–S13.

Gregg, R. E., Zech, L. A., Schaefer, E. J., Stark, D., Wilson, D., and Brewer, H. B., Jr., 1986, Abnormal *in vivo* metabolism of apolipoprotein $E_4$ in humans, *J. Clin. Invest.* **78:**815–821.

Hamos, J. E., DeGennaro, L. J., and Drachman, D. A., 1989, Synaptic loss in Alzheimer's disease and other dementias, *Neurology* **39:**355–361.

Hansen, L. A., Masliah, E., and Terry, R. D., 1989, A neuropathologic subset of Alzheimer's disease with concomitant Lewy body disease and spongiform change, *Acta Neuropathol.* **78:**194–201.

Hansen, L., Salmon, D., Galasko, D., Masliah, E., Katzman, R., DeTeresa, R., Thal, L., Pay, M. M., Hofstetter, R., and Klauber, M., 1990, The Lewy body variant of Alzheimer's disease: A clinical and pathologic entity, *Neurology* **40:**1–7.

Hansen, L. A., Masliah, E., Quijada-Fawcett, S., and Rexin, D., 1991, Entorhinal neurofibrillary tangles in Alzheimer disease with Lewy bodies, *Neurosci. Lett.* **129:**269–272.

Hansen, L. A., Masliah, E., Galasko, D., and Terry, R. D., 1993, Plaque only Alzheimer disease is usually the Lewy body variant, and vice versa, *J. Neuropathol. Exp. Neurol.* **52:**648–654.

Hartmann, H., Eckert, A., and Muller, W. E., 1994, Apolipoprotein E and cholesterol affect neuronal calcium signalling: The possible relationship to beta-amyloid neurotoxicity, *Biochem. Biophys. Res. Comm.* **200:**1185–1192.

Heinanen, O., Soininen, H., Sorvari, H., Kosunene, O., Paljarvi, L., Koivisto, E., and Riekkinen, P. J., 1995, Loss of synaptophysin-like immunoreactivity in the hippocampal formation is an early phenomenon in Alzheimer's disease, *Neuroscience* **64:**375–384.

Heyser, C. J., Masliah, E., Samimi, A., Campbell, I. L., and Gold, L. H., 1996, Progressive cognitive decline paralleled by inflammatory neurodegeneration in interleukin-6 transgenic mice, unpublished.

Hirano, A., 1994, Hirano bodies and related neuronal inclusions, *Neuropathol. Appl. Neurobiol.* **20:**3–11.

Hof, P. R., and Morrison, J. H., 1991, Neocortical neuronal subpopulations labeled by a monoclonal antibody to calbindin exhibit differential vulnerability in Alzheimer's disease, *Exp. Neurol.* **111:**293–301.

Hof, P. R., and Morrison, J. H., 1994, The cellular basis of cortical disconnection in Alzheimer disease and related dementing conditions, in: *Alzheimer Disease* (R. D. Terry, R. Katzman, and K. L. Bick, eds.), Raven, New York, pp. 197–230.

Hof, P. R., Cox, K., and Morrison, J. H., 1990, Quantitative analysis of a vulnerable subset of pyramidal neurons in Alzheimer's disease. I. Superior frontal and inferior temporal cortex, *J. Comp. Neurol.* **301**:44–54.

Hof, P. R., Bierer, L. M., Perl, D. P., Delacourte, A., Buee, L., Bouras, C., and Morrison, J. H., 1992, Evidence for early vulnerability of the medial and inferior aspects of the temporal lobe in an 82-year-old patient with preclinical signs of dementia, *Arch. Neurol.* **49**:946–953.

Holtzman, D. M., Pitas, R. E., Kilbridge, J., Nathan, B., Mahley, R. W., Bu, G., and Schwartz, A. L., 1995, Low density lipoprotein receptor-related protein mediates apolipoprotein E-dependent neurite outgrowth in a central nervous system-derived neuronal cell line, *Proc. Natl. Acad. Sci. USA* **92**:9480–9484.

Honer, W. G., Dickson, D. W., Gleeson, J., and Davies, P., 1992a, Regional synaptic pathology in Alzheimer's disease, *Neurobiol. Aging* **13**:375–382.

Honer, W. G., Kaufmann, C. A., and Davies, P., 1992b, Characterization of a synaptic antigen of interest in neuropsychiatric illness, *Biol. Psychiatry* **31**:147–158.

Hyman, B. T., VanHoesen, G. W., Damasio, A. R., and Barnes, C. L., 1984, Alzheimer's disease: Cell-specific pathology isolates the hippocampal formation, *Science* **225**:1168–1170.

Hyman, B. T., Van Hoesen, G. W., Kromer, L. J., and Damasio, A. R., 1986, Perforant pathway changes in the memory impairment of Alzheimer's disease, *Ann. Neurol.* **20**:472–481.

Hyman, B. T., Kromer, L. J., and Van Hoesen, G. W., 1987, Reinnervation of the hippocampal perforant pathway zone in Alzheimer's disease, *Ann. Neurol.* **21**:259–267.

Ignatius, M. J., Shooter, E. M., Pitas, R. E., and Mahley, R. W., 1987, Lipoprotein uptake by neuronal growth cones *in vitro*, *Science* **236**:959–962.

Ihara, Y., 1988, Massive somatodendritic sprouting of cortical neurons in Alzheimer's disease, *Brain Res.* **459**:138–144.

Iwai, A., Masliah, E., Yoshimoto, M., De Silva, R., Ge, N., Kittel, A., and Saitoh, T., 1994, The precursor protein of non-Aβ component of Alzheimer's disease amyloid (NACP) is a presynaptic protein of the central nervous system, *Neuron* **14**:467–475.

Iwai, A., Yoshimoto, M., Masliah, E., and Saitoh, T., 1995, Non-Aβ component of Alzheimer's disease amyloid (NAC) is amyloidogenic, *Biochemistry* **34**:10139–10145.

Iwai, A., Masliah, E., Sundsmo, M. P., DeTeresa, R., Mallory, M., Salmon, D. P., and Saitoh, T., 1996, The synaptic protein NACP is abnormally expressed during the progression of Alzheimer's disease, *Brain Res.*, in press.

Jellinger, K., Danielczyk, W., Fischer, P., and Gabriel, E., 1990, Clinicopathological analysis of dementia disorders in the elderly, *J. Neurol. Sci.* **95**:239–258.

Jensen, P. H., Sorensen, E. S., Petersen, T. E., Gliemann, J., and Rasmussen, L. K., 1995, Residues in the synuclein consensus motif of the alpha-synuclein fragment, NAC, participate in transglutaminase-catalysed cross-linking to Alzheimer-disease amyloid beta A4 peptide, *Biochem. J.* **310**:91–94.

Jones, D. G., and Harris, R. J., 1995, An analysis of contemporary morphological concepts of synaptic remodelling in the CNS: Perforated synapses revisited, *Rev. Neurosci.* **6**:177–219.

Kanai, Y., Smith, C. P., and Hediger, M. A., 1993, The elusive transporters with a high affinity for glutamate, *Trends Neurosci.* **16**:359–365.

Kanai, Y., Stelzner, M., Nubberg, S., Khawaja, S., Hebert, S. C., Smith, C. P., and Hediger, M. A., 1994, The neuronal and epithelial human high affinity glutamate transporter, *J. Biol. Chem.* **269**:20599–20606.

Kanai, Y., Nussberger, S., Romero, M. F., Boron, W. F., Hebert, S. C., and Hediger, M. A., 1995, Electrogenic properties of the epithelial and neuronal high affinity glutamate transporter, *J. Biol. Chem.* **270**:16561–16568.

Katzman, R., Terry, R., DeTeresa, R., Brown, T., Davies, P., Fuld, P., Renbing, X., and Peck, A., 1988, Clinical, pathological, and neurochemical changes in dementia: A subgroup with preserved mental status and numerous neocortical plaques, *Ann. Neurol.* **23**:138–144.

Khachaturian, Z. S., 1985, Diagnosis of Alzheimer's disease, *Arch. Neurol.* **42**:1097–1105.

Kosaka, K., 1990, Diffuse Lewy body disease in Japan, *J. Neurol.* **237**:197–204.

Lassmann, H., Weiler, R., Fischer, P., Bancher, C., Jellinger, K., Floor, E., Danielczyk, W., Seitelberger, F., and Winkler, H., 1992, Synaptic pathology in Alzheimer's disease: Immunological data for markers of synaptic and large dense-core vesicles, *Neuroscience* **46**:1–8.

Lee, E. H. Y., Hung, H. C., Lu, K. T., Chen, W. H., and Chen, H. Y., 1992, Protein synthesis in the hippocampus associated with memory facilitation by corticotropin-releasing factor in rats, *Peptides* **13**:927–937.

Lennox, G., Lowe, J., Landon, M., Byrne, J., Mayer, J., and Godwin-Austen, R. B., 1989, Diffuse Lewy body disease: Correlative neuropathology using anti-ubiquitin immunocytochemistry, *J. Neurol. Neurosurg. Psychiatry* **52:**1236–1247.

Lippa, C. F., Hamos, J. E., Pulaski-Salo, D., DeGennaro, L. J., and Drachman, D. A., 1992, Alzheimer's disease and aging: Effects on perforant pathway perikarya and synapses, *Neurobiol. Aging* **13:**405–411.

Lippa, C. F., Smith, T. W., and Swearer, J. M., 1994, Alzheimer's disease and Lewy body disease: A comparative clinicopathological study, *Ann. Neurol.* **35:**81–88.

Liu, X., Erikson, C., and Brun, A., 1996, Cortical synaptic changes and gliosis in normal aging, Alzheimer's disease, and frontal lobe degeneration, *Dementia* **7:**128–134.

Mahley, R. W., and Innerarity, T. L., 1983, Lipoprotein receptors and cholesterol homeostasis, *Biochem. Biophys. Acta* **737:**197–222.

Manfras, B. J., Rudert, W. A., Trucco, M., and Boehm, B. O., 1994, Cloning and characterization of a glutamate transporter cDNA from human brain and pancreas, *Biochem. Biophys. Acta* **1195:**185–188.

Mann, D. M. A., Marcyniuk, B., Yates, P. O., Neary, D., and Snowden, J. S., 1988, The progression of the pathological changes of Alzheimer's disease in the frontal and temporal neocortex examined both at biopsy and autopsy, *Neuropathol. Appl. Neurobiol.* **14:**177–195.

Maroteaux, L., and Scheller, R. H., 1991, The rat brain synucleins: Family of proteins transiently associated with neuronal membrane, *Mol. Brain Res.* **11:**335–343.

Martin, L. J., Pardo, C. A., Cork, L. C., and Price, D. L., 1994, Synaptic pathology and glial responses to neuronal injury precede the formation of senile plaques and amyloid deposits in the aging cerebral cortex, *Am. J. Pathol.* **145:**1358–1381.

Masliah, E., 1995a, Mechanisms of synaptic dysfunction in Alzheimer's disease, *Histol. Histopathol.* **10:**509–519.

Masliah, E., 1995b, The natural evolution of the neurodegenerative alterations in Alzheimer's disease, *Neurobiol. Aging* **16:**280–282.

Masliah, E., and Terry, R., 1993, The role of synaptic proteins in the pathogenesis of disorders of the central nervous system, *Brain Pathol.* **3:**77–85.

Masliah, E., and Terry, R., 1994, The role of synaptic pathology in the mechanisms of dementia in Alzheimer's disease, *Clin. Neurosci.* **1:**192–198.

Masliah, E., Terry, R. D., DeTeresa, R. M., and Hansen, L. A., 1989, Immunohistochemical quantification of the synapse-related protein synaptophysin in Alzheimer disease, *Neurosci. Lett.* **103:**234–239.

Masliah, E., Terry, R. D., Mallory, M., Alford, M., and Hansen, L., 1990, Diffuse plaques do not accentuate synapse loss in Alzheimer disease, *Am. J. Pathol.* **137:**1293–1297.

Masliah, E., Hansen, L., Albright, T., Mallory, M., and Terry, R. D., 1991a, Immunoelectron microscopic study of synaptic pathology in Alzheimer disease, *Acta Neuropathol.* **81:**428–433.

Masliah, E., Terry, R. D., Alford, M., DeTeresa, R. M., and Hansen, L. A., 1991b, Cortical and subcortical patterns of synaptophysin-like immunoreactivity in Alzheimer disease, *Am. J. Pathol.* **138:**235–246.

Masliah, E., Mallory, M., Hansen, L., Alford, M., Albright, T., Terry, R., Shapiro, P., Sundsmo, M., and Saitoh, T., 1991c, Immunoreactivity of CD45, a protein phosphotyrosine phosphatase, in Alzheimer disease, *Acta Neuropathol.* **83:**12–20.

Masliah, E., Mallory, M., Hansen, L., Alford, M., Albright, T., DeTeresa, R., Terry, R. D., Baudier, J., and Saitoh, T., 1991d, Patterns of aberrant sprouting in Alzheimer disease, *Neuron* **6:**729–739.

Masliah, E., Ge, N., Morey, M., DeTeresa, R., Terry, R. D., and Wiley, C. A., 1992a, Cortical dendritic pathology in human immunodeficiency virus encephalitis, *Lab. Invest.* **66:**285–291.

Masliah, E., Ellisman, M., Carragher, B., Mallory, M., Young, S., Hansen, L., DeTeresa, R., and Terry, R. D., 1992b, Three-dimensional analysis of the relationship between synaptic pathology and neuropil threads in Alzheimer disease, *J. Neuropathol. Exp. Neurol.* **51:**404–414.

Masliah, E., Mallory, M., Deerinck, T., DeTeresa, R., Lamont, S., Miller, A., Terry, R. D., Carragher, B., and Ellisman, M., 1993a, Re-evaluation of the structural organization of neuritic plaques in Alzheimer's disease, *J. Neuropathol. Exp. Neurol.* **52:**619–632.

Masliah, E., Mallory, M., DeTeresa, R., Alford, M., and Hansen, L., 1993b, Differing patterns of aberrant neuronal sprouting in Alzheimer's disease with and without Lewy bodies, *Brain Res.* **617:**258–266.

Masliah, E., Mallory, M., Hansen, L., Alford, M., DeTeresa, R., and Terry, R., 1993c, An antibody against phosphorylated neurofilaments identifies a subset of damaged association axons in Alzheimer's disease, *Am. J. Pathol.* **142:**871–882.

Masliah, E., Miller, A., and Terry, R. D., 1993d, The synaptic organization of the neocortex in Alzheimer's disease, *Medical Hypotheses* **41:**334–340.

Masliah, E., Mallory, M., Hansen, L., DeTeresa, R., Alford, M., and Terry, R., 1994b, Synaptic and neuritic alterations during the progression of Alzheimer's disease, *Neurosci. Lett.* **174:**67–72.

Masliah, E., Honer, W. G., Mallory, M., Voigt, M., Kushner, P., and Terry, R. D., 1994a, Topographical distribution of synaptic-associated proteins in the neuritic plaques of Alzheimer disease hippocampus, *Acta Neuropathol.* **87:**135–142.

Masliah, E., Mallory, M., Ge, N., Alford, M., Veinbergs, I., and Roses, A. D., 1995a, Neurodegeneration in the CNS of apoE-deficient mice, *Exp. Neurol.* **136:**107–122.

Masliah, E., Mallory, M., Alford, M., Hansen, L., and Saitoh, T., 1995b, PDGF-BB is associated with (neuritic) neurofibrillary alterations in Alzheimer's disease, *Neurobiol. Aging* **16:**549–556.

Masliah, E., Mallory, M., Alford, M., Ge, N., and Mucke, L., 1995c, Abnormal synaptic regeneration in hAPP695 transgenic and apoE knockout mice, in: *Research Advances in Alzheimer's Disease and Related Disorders* (K. Iqbal, J. A. Mortimer, B. Winblad, and H. M. Wisniewski, eds.), Wiley, New York, pp. 405–414.

Masliah, E., Mallory, M., Alford, M., Veinbergs, I., and Roses, A. D., 1995d, ApoE role in maintaining the integrity of the aging central nervous system, in: *Apolipoprotein E and Alzheimer's Disease* (Y. Christen and A. D. Roses, eds.), Springer-Verlag, New York, pp. 59–73.

Masliah, E., Ge, N., Achim, C. L., DeTeresa, R., and Wiley, C. A., 1996a, Patterns of neurodegeneration in HIV encephalitis, *NeuroAIDS* **1:**161–173.

Masliah, E., Mallory, M., Veinbergs, I., Miller, A., and Samuel, W., 1996b, Alterations in apolipoprotein expression during aging and neurodegeneration, *Prog. Neurobiol.,* in press.

Masliah, E., Iwai, A., Mallory, M., Ueda, K., and Saitoh, T., 1996c, Altered presynaptic protein NACP is associated with plaque formation and neurodegeneration in Alzheimer's disease, *Am. J. Pathol.* **148:**201–210.

Masliah, E., Samuel, W., Veinbergs, I., Mallory, M., Mante, M., and Saitoh, T., 1997, Neurodegeneration and cognitive impairment in apoE-deficient mice is ameliorated by infusion of recombinant apoE, *Brain Res.* **751:**307–314.

Masliah, E., Alford, M., Salmon, D., DeTeresa, R., Mallory, M., and Hansen, L.,1996e, Deficient glutamate transporter activity in Alzheimer's disease is associated with accumulation of brain spectrin degradation products and neurodegeneration, *Ann. Neurol.,* in press.

Mattson, M. P., Cheng, B., Culwell, A. R., Esch, F. S., Lieberburg, I., and Rydel, R. E., 1993a, Evidence for excitoprotective and intraneuronal calcium-regulating roles for secreted forms of the β-amyloid precursor protein, *Neuron* **10:**243–254.

Mattson, M. P., Tomaselli, K. J., and Rydel, R. E., 1993b, Calcium-destabilizing and neurodegenerative effects of aggregated β-amyloid peptide are attenuated by basic FGF, *Brain Res.* **621:**35–49.

Mattson, M. P., Cheng, B., and Smith-Swintosky, V. L., 1993c, Mechanisms of neurotrophic factor protection against calcium- and free radical-mediated excitotoxic injury: Implications for treating neurodegenerative disorders, *Exp. Neurol.* **124:**89–95.

McKee, A. C., Kosik, K. S., and Kowall, N. W., 1991, Neuritic pathology and dementia in Alzheimer's disease, *Ann. Neurol.* **30:**156–165.

McKeith, I. G., Bartholemew, P. H., Irvine, E. M., Cook, J., Adams, R., and Simpson, A. E., 1993, Single photon emission computerised tomography in elderly patients with Alzheimer's disease and multi-infarct dementia: Regional uptake of technetium-labelled HMPAO related to clinical measurements, *Br. J. Psychiatry* **163:**597–603.

McKeith, I. G., Galasko, D., Kosaka, K., Perry, E. K., Dickson, D. W., Hansen, L. A., Salmon, D. P., Lowe, J., Mirra, S. S., Byrne, E. J., Quinn, N. P., Edwardson, J. A., Ince, P. G., Bergeron, C., Burns, A., Miller, B. L., Lovestone, S., Collerton, D., Jansen, E. N. H., de Vos, R. A. I., Wilcock, G. K., Jellinger, K. A., and Perry, R. H., 1996, Clinical and pathological diagnosis of dementia with Lewy bodies (DLB): Report of the CDLB International Workshop, *Neurology* **47:**1113–1124.

Milliken, G. A., and Johnson, D. E., 1984, *Analysis of Messy Data,* Lifetime Learning Publications, Belmont.

Mirra, S. S., Hart, M. N., and Terry, R. D., 1993, Making the diagnosis of Alzheimer's disease: A primer for practicing pathologists, *Arch. Pathol. Lab. Med.* **117:**132–144.

Moechars, D., Lorent, D., De Strooper, B., Dewachter, I., and Van Leuven, F., 1996, Expression in brain amyloid precursor protein mutated in the alpha-secretase site causes disturbed behavior, neuronal degeneration, and premature death in transgenic mice, *EMBO J.* **15:**1265–1274.

Molsa, P. K., Paljarvi, L., Rinne, U. K., and Sako, E., 1985, Validity of clinical diagnosis in dementia: A prospective clinco-pathological study, *J. Neurol. Neurosurg. Psychiatry* **38:**1085–1090.

Morris, J. C., McKeel, D. W., Jr., Fulling, K., Torack, R. M., and Berg, L., 1988, Validation of clinical diagnostic criteria for Alzheimer's disease, *Ann. Neurol.* **24:**17–22.

Morris, J. C., Storandt, M., McKeel, D. W., Jr., Rubin, E. H., Price, J. L., Grant, E. A., and Berg, L., 1996, Cerebral amyloid deposition and diffuse plaques in "normal" aging: Evidence for presymptomatic and very mild Alzheimer's disease, *Neurology* **46:**707–719.

Morrison, J. H., Lewis, D. A., and Campbell, M. J., 1987, Distribution of neurofibrillary tangles and nonphosphorylated neurofilament protein-imunoreactive neurons in cerebral cortex: Implications for loss of corticocortical circuits in Alzheimer's disease, in: *Molecular Neuropathology of Aging: Banbury Report,* Vol. 27 (P. Davies and C. E. Finch, eds.), Cold Springs Harbor Laboratory, New York, pp. 109–124.

Mucke, L., Masliah, E., Johnson, W. B., Ruppe, M. D., Rockenstein, E. M., Forss-Petter, S., Pietropaolo, M., Mallory, M., and Abraham, C. R., 1994, Synaptotrophic effects of human amyloid β protein precursors in the cortex of transgenic mice, *Brain Res.* **666:**151–167.

Mucke, L., Masliah, E., and Campbell, I. L., 1995a, Transgenic models to assess the neuropathogenic potential of HIV-1 proteins and cytokines, *Curr. Topics Microbiol. Immunol.* **202:**187–205.

Mucke, L., Abraham, C. R., Ruppe, M. D., Rockenstein, E. M., Toggas, S. M., Alford, M., and Masliah, E., 1995b, Protection against HIV-1 gp120-induced brain damage by neuronal overexpression of human amyloid precursor protein (hAPP), *J. Exp. Med.* **181:**1551–1556.

Nagy, Z., Esiri, M. M., Jobst, K. A., Morris, J. H., King, E. M., McDonald, B., Litchfield, S., Smith, A., Barneston, L., and Smith, A. D., 1995, Relative roles of plaques and tangles in the dementia of Alzheimer's disease: Correlations using three sets of neuropathological criteria, *Dementia* **6:**21–31.

Nakajo, S., Tsukada, K., Omata, K., Nakamura, Y., and Nakaya, K., 1993, A new brain-specific 14-kDa protein is a phosphoprotein: Its complete amino acid sequence and evidence for phosphorylation, *Eur. J. Biochem.* **217:**1057–1063.

Nathan, B. P., Bellosta, S., Sanan, D. A., Weisgraber, K. H., Mahley, R. W., and Pitas, R. E., 1994, Differential effects of apolipoproteins E3 and E4 on neuronal growth *in vitro, Science* **264:**850–852.

Neary, D., Snowden, J. S., Mann, D. M. A., Bowen, D. M., Sims, N. R., Northern, B., Yates, P. O., and Davison, A. N., 1986, Alzheimer's disease: A correlative study, *J. Neurol. Neurosurg. Psychiatry* **49:**229–237.

Nishimura, M., Tomimoto, H., Suenaga, T., Nakamura, S., Namba, Y., Ikeda, K., Akiguchi, I., and Kimura, J., 1994, Synaptophysin and chromogranin A immunoreactivities of Lewy bodies in Parkinson's disease brains, *Brain Res.* **634:**339–344.

Orrison, W. W., Jr., 1995, *Atlas of Brain Function,* Thieme Medical Publishers, New York.

Paik, Y.-K., Chang, D. J., Reardon, C. A., Davies, G. E., Mahley, R. W., and Taylor, J. M., 1985, Nucleotide sequence and structure of the human apolipoprotein E gene, *Proc. Natl. Acad. Sci USA* **82:**3445–3449.

Palmer, A. M., and Gershon, S., 1990, Is the neuronal basis of Alzheimer's disease cholinergic or glutamatergic? *FASEB J.* **2745:**2752.

Perry, E. K., Perry, R. H., Blessed, G., and Tomlinson, B. E., 1977, Neurotransmitter enzyme abnormalities in senile dementia: CAT and GAD activities in necropsy tissue, *J. Neurol. Sci.* **34:**247–265.

Perry, E. K., McKeith, I., and Thompson, P., 1991a, Topography, extent, and clinical relevance of neurochemical deficits in dementia of Lewy body type, Parkinson's disease, and Alzheimer's disease, *Ann. N.Y. Acad. Sci.* **640:**197–202.

Perry, G., Kawai, M., and Tabaton, M., 1991b, Neuropil threads of Alzheimer's disease show a marked alteration of the normal cytoskeleton, *J. Neurosci.* **11:**1748–1755.

Perry, R. H., Irving, D., Blessed, G., Fairbairn, A., and Perry, E. K., 1990, Senile dementia of Lewy body type: A clinically and neuropathologically distinct form of Lewy body dementia in the elderly, *J. Neurol. Sci.* **95:**119–139.

Perry, T. A., Torres, E., Czech, C., Beyreuther, K., Richards, S.-J., and Dunnett, S. B., 1995, Cognitive and motor function in transgenic mice carrying excess copies of the 695 and 751 amino acid isoforms of the amyloid precursor protein gene, *Alzheimer Res.* **1:**5–14.

Poirier, J., 1994, Apolipoprotein E in animal models of CNS injury and in Alzheimer's disease, *Trends Neurosci.* **17:**525–530.

Poirier, J., Baccichet, A., Dea, D., and Gauthier, S., 1993, Cholesterol synthesis and lipoprotein reuptake during synaptic remodeling in hippocampus in adult rats, *Neuroscience* **55:**81–90.

Poirier, J., Delisle, M.-C., Quirion, R., Aubert, I., Farlow, M., Lahiri, D., Hui, S., Bertrand, P., Nalbantoglu, J., Gilfix, B. M., and Gauthier, S., 1995, Apolipoprotein E4 allele as a predictor of cholinergic deficits and treatment outcome in Alzheimer disease, *Proc. Natl. Acad. Sci. USA* **92:**12260–12264.

Read, S. L., Miller, B. L., Mena, I., Kim, R., Itabashi, H., and Darby, A., 1995, SPECT in dementia: Clinical and pathological correlation, *J. Am. Geriat. Soc.* **43:**1243–1247.

Reyes, E., Gamboa, A., and Masliah, E., 1993, Atypical diffuse Lewy body disease with neuritic abnormalities, *Clin. Neuropathol.* **12:**330–334.

Roch, J.-M., Masliah, E., Roch-Levecq, A.-C., Sundsmo, M. P., Otero, D. A. C., Veinbergs, I., and Saitoh, T., 1994, Increase of synaptic density and memory retention by a peptide representing the trophic domain of the amyloid β/A4 protein precursor, *Proc. Natl. Acad. Sci. USA* **91:**7650–7654.

Rogers, J., Luber-Narod, J., Styren, S. D., and Civin, W. H., 1988, Expression of immune system-associated antigens by cells of the human central nervous system: Relationship to the pathology of Alzheimer's disease, *Neurobiol. Aging* **9:**339–349.

Roses, A. D., Einstein, G., Gilbert, J., Goedert, M., Han, S.-H., Huang, D., Hulette, C., Masliah, E., Pericak-Vance, M. A., Saunders, A. M., Schmechel, D. E., Strittmatter, W. J., Weisgraber, K. H., and Xi, P.-T., 1996, Morphological, biochemical, and genetic support for an apolipoprotein E effect on microtubular metabolism, *Ann. N.Y. Acad. Sci.* **777:**147–157.

Rothstein, J. D., Jin, L., Dykes-Hoberg, M., and Kuncl, R. W., 1993, Chronic inhibition of glutamate uptake produces a model of slow neurotoxicity, *Proc. Natl. Acad. Sci. USA* **90:**6591–6595.

Rothstein, J. D., Van Kammen, M., Levey, A. I., Martin, L. J., and Kuncl, R. W., 1995, Selective loss of glial glutamate transporter GLT-1 in amyotrophic lateral sclerosis, *Ann. Neurol.* **38:**73–84.

Samuel, W., Terry, R. D., DeTeresa, R., Butters, N., and Masliah, E., 1994a, Clinical correlates of cortical and nucleus basalis pathology in Alzheimer dementia, *Arch. Neurol.* **51:**772–778.

Samuel, W., Masliah, E., and Terry, R., 1994b, Hippocampal connectivity and Alzheimer's dementia: Effects of pathology in a two-component model, *Neurology* **44:**2081–2088.

Samuel, W., Hansen, L. A., Galasko, D., Masliah, E., and Katzman, R., 1996, Neocortical Lewy body density correlates with dementia in the Lewy body variant of Alzheimer's disease, *J. Neuropathol. Exp. Neurol.* **55:**44–52.

Saunders, A. M., Strittmatter, W. J., Schmechel, D., St. George-Hyslop, P. H., Pericak-Vance, M. A., Joo, S. H., Rosi, B. L., Gusella, J. F., Crapper-MacLachlan, D. R., Alberts, M. J., Hulette, C., Crain, B., Goldgaber, D., and Roses, A. D., 1993, Association of apolipoprotein E allele E4 with late-onset familial and sporadic Alzheimer's disease, *Neurology* **43:**1467–1472.

Scheff, S. W., and Price, D. A., 1993, Synapse loss in the temporal lobe in Alzheimer's disease, *Ann. Neurol.* **33:**190–199.

Scheff, S. W., DeKosky, S. T., and Price, D. A., 1990, Quantitative assessment of cortical synaptic density in Alzheimer's disease, *Neurobiol. Aging* **11:**29–37.

Scheff, S. W., Sparks, D. L., and Price, D. A., 1993, Quantitative assessment of synaptic density in the entorhinal cortex in Alzheimer's disease, *Ann. Neurol.* **34:**356–361.

Schmechel, D. E., Saunders, A. M., Strittmatter, W. J., Crain, B. J., Hulette, C. M., Joo, S. H., Pericak-Vance, M. A., Goldgaber, D., and Roses, A. D., 1993, Increased amyloid beta-peptide deposition in cerebral cortex as a consequence of apolipoprotein E genotype in late-onset Alzheimer disease, *Proc. Natl. Acad. Sci. USA* **90:**9649–9653.

Schmidt, M. L., Lee, V. M.-Y., and Trojanowski, J. Q., 1990, Relative abundance of tau and neurofilament epitopes in hippocampal neurofibrillary tangles, *Am. J. Pathol.* **136:**1069–1075.

Schmidt, M. L., Lee, V. M., and Trojanowski, J. Q., 1991, Comparative epitope analysis of neuronal cytoskeletal proteins in Alzheimer's disease, senile plaque neurites, and neuropil threads, *Lab. Invest.* **64:**352–357.

Schubert, W., Prior, R., Weidemann, A., Dircksen, H., Multhaup, G., Masters, C. L., and Beyreuther, K., 1991, Localization of Alzheimer beta A4 amyloid precursor protein at central and peripheral synaptic sites, *Brain Res.* **563:**184–194.

Scott, H. L., Tannenberg, A. E. G., and Dodd, P. R., 1995, Variant forms of neuronal glutamate transporter sites in Alzheimer's disease cerebral cortex, *J. Neurochem.* **64:**2193–2202.

Selkoe, D. J., 1989, Amyloid β protein precursor and the pathogenesis of Alzheimer's disease, *Cell* **58:**611–612.

Selkoe, D. J., 1993, Physiological production of the β-amyloid protein and the mechanisms of Alzheimer's disease, *Trends Neurosci.* **16:**403–409.

Selkoe, D. J., Podlisny, M. B., Joachim, C. L., Vickers, E. A., Lee, G., Fritz, L. C., and Oltersdorf, T., 1988, Beta-amyloid precursor protein of Alzheimer disease occurs as 110- to 135-kilodalton membrane-associated proteins in neural and nonneural tissue, *Proc. Natl. Acad. Sci. USA* **85:**7341–7345.

Seubert, P., Vigo-Pelfrey, C., Esch, F., Lee, M., Dovey, H., Davis, D., Sinha, S., Schlossmacher, M., Whaley, J., Swindlehurst, C., McCormack, R., Wolfert, R., Selkoe, D., Lieberburg, I., and Schenk, D., 1992, Isolation and quantification of soluble Alzheimer's β-peptide from biological fluids, *Nature* **359:**325–327.

Shashidharan, P., Huntley, G. W., Meyer, T., Morrison, J. H., and Plaitakis, A., 1994, Neuron-specific human glutamate transporter: Molecular cloning, characterization, and expression in human brain, *Brain Res.* **662:**245–250.

Sima, A. A. F., Clark, A. W., Sternberger, N. A., and Sternberger, L. A., 1986, Lewy body dementia without Alzheimer changes, *Can. J. Neurol. Sci.* **13:**490–497.

Sisodia, S. S., Koo, E. H., Beyreuther, K., Unterbeck, A., and Price, D. L., 1990, Evidence that β-amyloid protein in Alzheimer's disease is not derived by normal processing, *Science* **248:**492–494.

Six, J., Lubke, U., Lenders, M.-B., Vandermeeren, M., Mercken, M., Villanova, M., Van de Voorde, A., Gheuenst, J., Martin, J.-J., and Cras, P., 1992, Neurite sprouting and cytoskeletal pathology in Alzheimer's disease: A comparative study with monoclonal antibodies to growth-associated protein B-50 (GAP43) and paired helical filaments, *Neurodegeneration* **1:**247–255.

Small, D. H., Nurcombe, V., Reed, G., Clarris, H., Moir, R., Beyreuther, K., and Masters, C. L., 1994, A heparin-binding domain in the amyloid protein precursor of Alzheimer's disease is involved in the regulation of neurite outgrowth, *J. Neurosci.* **14:**2117–2127.

Smith, M. C., Mallory, M., Hansen, L. A., and Masliah, E., 1995, Fragmentation of the neuronal cytoskeleton in the Lewy body variant of Alzheimer's disease, *NeuroReport* **6:**673–676.

Stewart, M. G., and Rusakov, D. A., 1995, Morphological changes associated with stages of memory formation in the chick following passive avoidance training, *Behav. Brain Res.* **66:**21–28.

Strittmatter, W. J., Saunders, A. M., Schmechel, D., Pericak-Vance, M., Enghild, J., Salvesen, G. S., and Roses, A. D., 1993, Apolipoprotein E: High-avidity binding to β-amyloid and increased frequency of type 4 allele in late-onset familial Alzheimer disease, *Proc. Natl. Acad. Sci. USA* **90:**1977–1981.

Strittmatter, W. J., Weisgraber, K. H., Goedert, M., Saunders, A. M., Huang, D., Corder, E. H., Dong, L. M., Jakes, R., Alberts, M. J., and Gilbert, J. R., 1994, Hypothesis: Microtubule instability and paired helical filament formation in the Alzheimer disease brain are related to apolipoprotein E genotype, *Exp. Neurol.* **125:**163–171.

Suzuki, K., and Terry, R. D., 1967, Fine structural localization of acid phosphatase in senile plaques in Alzheimer's presenile dementia, *Acta Neuropathol.* **8:**276–284.

Terry, R. D., and Wisniewski, H. M., 1970, The ultrastructure of the neurofibrillary tangle and the senile plaque, in: *Ciba Foundation Symposium on Alzheimer's Disease and Related Conditions* (G. E. W. Wolstenholme and M. O'Connor, eds.), J & A Churchill, London, pp. 145–168.

Terry, R. D., Gonatas, N. K., and Weiss, M., 1964, Ultrastructural studies in Alzheimer's presenile dementia, *Am. J. Pathol.* **44:**269–297.

Terry, R. D., Peck, A., DeTeresa, R., Schechter, R., and Horoupian, D. S., 1981, Some morphometric aspects of the brain in senile dementia of the Alzheimer type, *Ann. Neurol.* **10:**184–192.

Terry, R. D., Hansen, L. A., DeTeresa, R., Davies, P., Tobias, H., and Katzman, R., 1987, Senile dementia of the Alzheimer type without neocortical neurofibrillary tangles, *J. Neuropathol. Exp. Neurol.* **46:**262–268.

Terry, R. D., Masliah, E., Salmon, D. P., Butters, N., DeTeresa, R., Hill, R., Hansen, L. A., and Katzman, R., 1991, Physical basis of cognitive alterations in Alzheimer disease: Synapse loss is the major correlate of cognitive impairment, *Ann. Neurol.* **30:**572–580.

Terry, R. D., Hansen, L., and Masliah, E., 1994, Structural alterations in Alzheimer disease, in: *Alzheimer Disease* (R. D. Terry and R. Katzman, eds.), Raven, New York, pp. 179–196.

Ueda, K., Fukushima, H., Masliah, E., Xia, Y., Iwai, A., Otero, D., Kondo, J., Ihara, Y., and Saitoh, T., 1993, Molecular cloning of a novel amyloid component in Alzheimer's disease, *Proc. Natl. Acad. Sci. USA* **90:**11282–11286.

Van Reempts, J., Dikova, M., Werbrouck, L., Clincke, G., and Borgers, M., 1992, Synaptic plasticity in rat hippocampus associated with learning, *Behav. Brain Res.* **51:**179–183.

Vermersch, P., Delacourte, A., Javoy-Agid, F., Hauw, J.-J., and Agid, Y., 1993, Dementia in Parkinson's disease: Biochemical evidence for cortical involvement using the immunodetection of abnormal tau proteins, *Ann. Neurol.* **33:**445–450.

Victoroff, J., Mack, W. J., Lyness, S. A., and Chui, H. C., 1995, Multicenter clinicopathological correlation in dementia, *Am. J. Psychiatry* **152:**1476–1484.

Voytko, M. L., Olton, D. S., Richardson, R. T., Gorman, L. K., Tobin, J. R., and Price, D. L., 1994, Basal forebrain lesions in monkeys disrupt attention, but not learning and memory, *J. Neurosci.* **14:**167–186.

Wade, J. P. H., Mirsen, T. R., Hachinski, V. C., Fisman, M., Lau, C., and Merskey, H., 1987, The clinical diagnosis of Alzheimer's disease, *Arch. Neurol.* **44:**24–29.

Wakabayashi, K., Takahashi, H., Obata, K., and Ikuta, F., 1992, Immunocytochemical localization of synaptic vesicle-specific protein in Lewy body-containing neurons in Parkinson's disease, *Neurosci. Lett.* **138:**237–240.

Wakabayashi, K., Honer, W. G., and Masliah, E., 1994, Synapse alterations in the hippocampal-entorhinal formation in Alzheimer's disease with and without Lewy body disease, *Brain Res.* **667:**24–32.

Wang, D., and Munoz, D. G., 1995, Qualitative and quantitative differences in senile plaque dystrophic neurites of Alzheimer's disease and normal aged brain, *J. Neuropathol. Exp. Neurol.* **54:**548–556.

Weiler, R., Lassmann, H., Fischer, P., Jellinger, K., and Winkler, H., 1990, A high ratio of chromogranin A to synaptin/synaptophysin is a common feature of brains in Alzheimer and Pick disease, *FEBS Lett.* **263:**337–339.

Wilcock, G. K., Esiri, M. M., Bowen, D. M., and Hughes, A. O., 1988, The differential involvement of subcortical nuclei in senile dementia of Alzheimer's type, *J. Neurol. Neurosurg. Psychiatry* **51:**842–849.

Wisniewski, H. M., Wegiel, J., Wang, K. C., Kujawa, M., and Lach, B., 1989, Ultrastructural studies of the cells forming amyloid fibers in classical plaques, *Can. J. Neurol. Sci.* **16:**535–542.

Wisniewski, H. M., Barcikowska, M., and Kida, E., 1991, Phagocytosis of β–A4 amyloid fibrils of the neuritic neocortical plaques, *Acta Neuropathol.* **81:**588–590.

Xia, Y., de Silva, H. A. R., Rosi, B. L., Yamaoka, L. H., Rimmler, J. B., Pericak-Vance, M. A., Roses, A. D., Chen, X., Masliah, E., DeTeresa, R., Iwai, A., Sundsmo, M., Thomas, R. G., Hofstetter, C. R., Gregory, E., Hansen, L. A., Katzman, R., Thal, L. J., and Saitoh, T., 1996, Genetic studies in Alzheimer's disease with an NACP/alpha-synuclein polymorphism, *Ann. Neurol.* **40:**207–215.

Yamada, T., Kondo, A., Takamatsu, J., Tateishi, J., and Goto, I., 1995, Apolipoprotein E mRNA in the brains of patients with Alzheimer's disease, *J. Neurol. Sci.* **129:**56–61.

Yamaguchi, H., Hirai, S., Morimatso, M., Shoji, M., and Ihara, Y., 1988, A variety of cerebral amyloid deposits in the brains of Alzheimer-type dementia demonstrated by β-protein immunostaining, *Acta Neuropathol.* **76:**541–549.

Yamaguchi, N., Nakazato, Y., Shoji, M., Ihara, Y., and Hirai, S., 1990, Ultrastructure of the neuropil threads in the Alzheimer brain: Their dendritic origin and accumulation in the senile plaques, *Acta Neuropathol.* **80:**368–374.

Yoshimoto, M., Iwai, A., Kang, D., Otero, D. A. C., Xia, Y., and Saitoh, T., 1995, NACP, the precursor protein of non-amyloid β/A4 protein (Aβ) component of Alzheimer disease amyloid, binds Aβ and stimulates Aβ aggregation, *Proc. Natl. Acad. Sci. USA* **92:**9141–9145.

Yoshimura, M., 1983, Cortical changes in the parkinsonian brain: A contribution to the delineation of "diffuse Lewy body disease," *J. Neurol.* **229:**17–32.

Zhan, S. S., Beyreuther, K., and Schmitt, H. P., 1993, Quantitative assessment of the synaptophysin immunoreactivity of the cortical neuropil in various neurodegenerative disorders with dementia, *Dementia* **4:**66–74.

Zheng, H., Jiang, M., Trumbauer, M. E., Sirinathsinghji, D. J. S., Hopkins, R., Smith, D. W., Heavens, R. P., Dawson, G. R., Boyce, S., Conner, M. W., Stevens, K. A., Slunt, H. H., Sisodia, S. S., Chen, H. Y., and Van der Ploeg, L. H. T., 1995, β-amyloid precursor protein-deficient mice show reactive gliosis and decreased locomotor activity, *Cell* **81:**525–531.

Zhong, Z., Quon, D., Higgins, L. S., Higaki, J., and Cordell, B., 1994, Increased amyloid production from aberrant β-amyloid precursor proteins, *J. Biol. Chem.* **269:**12179–12184.

# 16

# Multifocal Cortical Neurodegeneration in Alzheimer's Disease

BRENT A. VOGT, ALEX MARTIN, KENT E. VRANA, JOHN R. ABSHER, LESLIE J. VOGT, and PATRICK R. HOF

## 1. Introduction

Cortical atrophy is well known in Alzheimer's disease (AD), however, there are different interpretations of its location and extent. Neurodegeneration and atrophy are relatively consistent in medial temporal areas, including hippocampal, entorhinal, and perirhinal cortices, and they follow a uniform pattern. This is not the only region of atrophy, however, since it occurs in prefrontal, parietotemporal, occipital, and cingulate cortices. Furthermore, although gross focal atrophy is not always present, focal neurodegeneration may occur with focal changes in glucose metabolism. The concept of multifocal cortical degeneration leads to the view of multiple structural and functional subsystem disruption in AD.

---

BRENT A. VOGT, KENT E. VRANA, JOHN R. ABSHER, and LESLIE J. VOGT • Cingulum NeuroSciences Institute, Winston-Salem, North Carolina 27101; and Departments of Physiology, Pharmacology, and Neurology, Wake Forest University School of Medicine, Winston-Salem, North Carolina 27157-1083. ALEX MARTIN • National Institute of Mental Health, Bethesda, Maryland 20892-1366. PATRICK R. HOF • Kastor Neurobiology of Aging Laboratories, Fishberg Research Center for Neurobiology and Department of Geriatrics and Adult Development, and Department of Opthalmology, Mount Sinai School of Medicine, New York, New York, 10029.

*Cerebral Cortex,* Volume 14, edited by Peters and Morrison. Kluwer Academic/Plenum Publishers, New York, 1999.

Although the argument will be made that multifocal cortical neurodegeneration accounts for clinical subgroups, genetic and neuropathological observations suggest that differences among subgroups might reflect unique mechanisms of neurodegeneration and subtypes of the disease. To demonstrate subtypes, it must be shown that two or more symptomatologies are related to a corresponding number of unique etiologies (Jorm, 1985). Although the mechanisms of neurodegeneration in AD are not known, posterior cortical atrophy with Bálint syndrome-like presentation is an example of a likely subtype that has unique clinical presentations; multifocal atrophies in medial temporal, occipitoparietal, and posterior cingulate cortices; and impairments in visuomotor processing subsystems that differ from other AD cases. In another example, an exon 9 deletion in the presenilin 1 gene leads to large senile plaques (SP) in the cerebral cortex, degeneration of corticospinal projection neurons, and spastic paraparesis (Crook *et al.*, 1998). These and other examples of multifocal cortical atrophy with disruption of functional subsystems support the hypothesis that AD is comprised of subtypes.

A systematic methodology is not yet available whereby neuropathological analyses can assess subtypes at the molecular and cellular levels and relate such changes to alterations in brain metabolism and clinical outcomes. There may be genetic and neuropathological subtypes associated with different patterns of neurodegeneration *within* AD, yet this does not presume that links have been established to all clinical endpoints and are subtypes *of* AD. The subtype hypothesis is stated this way to provide a rationale for analyzing different pathological endpoints and provide a context for future clinical studies. Hence, the term *subtypes* in this chapter refers to a preliminary view that there are neuropathological subtypes within AD.

Selection of either the subgroup or subtype hypothesis has a profound impact on study design and statistical analysis. Multiple biological subtypes suggest different patterns of transmitter system reorganization and that subgroups of patients will respond differently to drug therapies. Furthermore, this is a framework for hypotheses about brain–behavior relationships that are impaired in AD and lead to the goal of defining clinical subtypes based on specific patterns of neocortical neurodegeneration. Because neocortical neurodegeneration is not uniform in AD, and laminar patterns of neurodegeneration have a significant impact on transmitter systems, the pathological state of the cerebral cortex remains important whether or not neuropathological subtypes have yet been correlated with unique clinical endpoints. This chapter seeks to draw the line between subgroups and subtypes at the biological level. It is our view that subtypes cannot be defined solely on the basis of clinical and neuropsychological assessments. However, genetic, glucose metabolism, and multifocal neurodegeneration support the subtypes hypothesis. This chapter considers the theoretical requirements for biological subtypes and the logic, format, and issues raised are summarized here for each section:

*Neuropsychological subgroups.* Neuropsychological heterogeneity in AD is considered in terms of patterns of glucose hypometabolism, interpretive problems for such information, and some of the problems relating specific neuropsychological alterations to brain pathology.

*Genetic factors: subgroups or subtypes?* Although the genetic transmission of AD and genetic risk factors could all converge on a single brain pathology and some *in vitro* studies attempt to demonstrate links between these mechanisms, we explore the hypothesis that impairments in multiple gene systems argue for multiple mechanisms of neurodegeneration. Therefore, arguments supporting bio logical subtypes include genetic defects associated with specific clinical end points.

*Models of heterogeneity, postmortem measures, and Aβ neurotoxicity.* A theoretical discussion considers the two essential models used to assess AD heterogeneity: stochastic subgroups and subtypes. One hypothesis is that subtypes can be defined by the density of senile plaques (SP) and neurofibrillary tangles (NFT). However, immunohistochemistry for long amyloid-β peptide (Aβ42) appears to provide a link to neurotoxicity subserving focal cortical atrophies and neurodegeneration.

*Laminar patterns in neocortical neurochemistry.* Demonstration of laminar and regional vulnerabilities of cortical neurons in the pathogenesis of AD is a crucial argument for unique etiologies. The profound chemical and connectional differences among neurons in superficial and deep neocortical layers is provided including intermediate neurofilament proteins, calretinin, nerve growth factor, and nitric oxide synthase-expressing neurons. These phenotypes may be differentially impacted by neurotoxic processes in AD.

*Laminar patterns of neuron vulnerability in frontal/anterior cingulate dementias.* Laminar patterns of neurodegeneration are considered in frontal lobe dementias because different etiologies lead to neurodegeneration in different layers, cingulate cortex is always involved, and there are AD cases with focal frontotemporal atrophy.

*Multifocal atrophies and subsystem disruption in Alzheimer's disease.* Case studies are presented with postmortem assessments for frontotemporal atrophy, parietotemporal atrophy, and posterior cortical atrophy in AD. These cases demonstrate pathologically unique impairments, and because posterior cingulate cortex (PCC) is profoundly impacted in all of them, this region provides an important platform for assessing multifocal degeneration and the subtype hypothesis.

*Layer-selective neurodegeneration in posterior cingulate cortex.* A rationale is provided for studying PCC and for why laminar patterns of neurodegeneration suggest multiple etiologies. A multivariate model of neurons in four layers of PCC is used to define subtypes within AD, and the characteristics of each potential subtype are considered. Finally, the possible role of apolipoprotein E (ApoE) genotype is considered in the context of multivariate statistics, and it is concluded that few direct relationships exist between genotype and laminar patterns of degeneration.

*Neocortical neurochemistry and the subtypes hypothesis.* Efforts have been made to subtype AD based on neurotransmitter systems. Earlier receptor binding findings are reanalyzed in terms of a multivariate model of three systems to determine which alterations in binding best describe subtypes independently characterized with laminar patterns of neurodegeneration in PCC.

*Models of multifocal cortical neurodegeneration and the subtypes hypothesis.* Although the clinical, pathological, and mechanistic bases of AD subtypes are not established, frontotemporal atrophy and other multifocal cortical atrophies provide evidence for subtypes with unique subsystem impairments and clinical outcomes. Multivariate models of neurodegeneration and associated genetic mutations should lead to a more complete accounting of the subtypes of AD and provide a statistical framework for assessing multiple mechanisms of neurodegeneration.

## 2. Neuropsychological Subgroups

The assessment of cognition has played an unusually prominent role in investigations of AD because of the lack of a noninvasive biological marker that would support definitive diagnosis. One of the central findings that emerged from neuropsychological investigations was the discovery, or actually the rediscovery and clarification of the fact that clinically, AD is a heterogeneous disorder. Specifically, this claim refers to the fact that not all patients present with the same set of cognitive symptoms. Moreover, the differences between the types of patients under consideration here are not simply quantitative, and therefore not related to differences in the severity, disease duration, or rate of progression in any simple or straight forward manner. Rather, the claim of subgroups refers to the fact that AD patients can present with qualitatively distinct patterns of impaired and preserved cognitive abilities and skills. Thus, for example, one patient may have relatively severe visuospatial deficits concurrent with intact object naming and word-finding skills, whereas another patient may present with the opposite symptom profile. Groups of patients like these define qualitatively distinct subgroups, even though they all may have other symptoms in common, such as difficulty learning and remembering recently presented information. Moreover, and of critical importance for the present discussion, these different profiles are associated with correspondingly different profiles of cortical dysfunction as measured by positron emission tomography (PET). Thus, in the above example, the first patient would have more severe hypometabolism of the posterior parietal cortex (especially in the right hemisphere) relative to other cortical regions, whereas the second patient would have most-severe hypometabolism, limited to the left temporal lobe (see Martin, 1987, 1990, for reviews). These brain-imaging data provide a strong argument against the suggestion that individual patient profiles represent preexisting patterns of strengths and weaknesses that become exaggerated by a global pathological process.

How many subgroups are there? If there were a biological marker that identified subjects prior to onset of overt, clinically relevant symptoms, the answer to this question would be obtained relatively easily. Once identified, serial evaluation of at-risk subjects would soon yield information on the following questions. First, what is the earliest symptom? Is an episodic or declarative memory problem always the first symptom to appear? If so, is this impairment global, affecting both verbal and nonverbal processes—in which case it would indicate bilateral pathology of the medial temporal lobes—or can the problem be material-specific, thus suggesting lateralized pathology? If other problems can precede impaired episodic memory, what are they? Can the initial symptom be a word-finding problem, suggesting focal involvement of the left temporal cortex, or a visuospatial problem that suggests focal parietal lobe disease? Second, regardless of the nature of the initial symptom, how does the dementia progress? Does progression always proceed in a fixed manner, or do different symptoms emerge in a seemingly random fashion?

The lack of a biological marker complicates our ability to answer these questions because the available data are tempered by two important constraints. The first constraint is defined by the type of symptoms that motivate people to present themselves for evaluation. Some cognitive deficits, such as language deficits and verbal memory problems, have considerably greater impact on activities of daily living than other symptoms, such as nonverbal object-recognition and memory problems. As a result, focal involvement of the lateral and medial aspects of the left temporal lobe may be seen early in the course of the disease more often than focal involvement of the right temporal lobe. Yet it would be a mistake to conclude that this asymmetry indicates greater vulnerability to AD of the left than right temporal lobe. Thus, the true incidence of the types of cognitive problems present during the earliest stages of the disease may be distorted, and the likelihood of identifying the full range of preclinical deficits is reduced by this symptom presentation bias.

The second constraint is dictated not by the patients but by the clinician. Current diagnostic criteria for probable AD demand that the patient be impaired in at least two domains of cognition. There is no question that this is a valid diagnostic criterion in that it increases the probability of a correct diagnosis and greatly reduces false positives. Nevertheless, it should also be apparent that this criterion eliminates patients who initially present with a single deficit. Thus, the absence of a diagnostic marker has led to both patient-based and clinician-based biases that, in turn, may constrain and potentially distort answers to questions about the range of cognitive deficits associated with the initial presentation of AD, its progression, and the number of qualitatively distinguishable subgroups.

With these caveats in mind, it can be confidently stated that AD typically is associated with dysfunction in primarily four cognitive domains: (1) Deficits in episodic memory, which are assessed by tests of learning and remembering verbal and nonverbal material and are associated with bilateral involvement of the medial temporal lobe, particularly the hippocampus and parahippocampal, perirhinal and entorhinal cortices; (2) Deficits in semantic memory, which are assessed by tests of object naming and word knowledge and are associated with involvement of the left temporal cortex, and also deficits in nonverbal object recognition as assessed by tests of recognition and pattern matching, which are associated with involvement of the right temporal cortex; (3) Deficits in visuospatial abilities and skills as assessed by tests of spatial attention, object localization, orientation, and constructional ability and are associated with involvement of the parietal lobes; (4) Deficits in regulating and controlling thought and behavior and in manipulating the products of memory, as evidenced by personality changes and social/occupational dysfunction and assessed by measures of problem solving, conceptualization, and working memory. This last constellation of symptoms, which are often associated with frontal lobe involvement, can emerge early in the course of the disease, as suggested by Royall *et al.* (1994) and by the case presented below that displays frontal lobe involvement and early disruption of working memory and problem solving associated with paranoid behaviors. Consistent with this framework, PET has shown that cerebral hypometabolism occurs in the medial temporal region, the posterior region of the temporal and parietal lobes, the PCC and, to a lesser extent, the prefrontal cortex (Haxby *et al.,* 1988). Because glucose hypometabolism in individuals with very early AD is greatest in PCC (Minoshima *et al.,* 1997), this region may play a greater role in memory and visuospatial deficits than was previously appreciated.

Within this general framework, single cases present with contrasting patterns of impaired and preserved cognitive difficulties that may focus on any single domain, especially during the early stages of the disease. Typically, and perhaps as a result of diagnostic criteria, patients have difficulty learning and remembering recently presented material. In addition to memory impairment, the best-documented subgroups consist of patients with deficits in semantic memory and word-finding ability, concurrent with normal or near-normal visuospatial skills, and patients with the opposite profile of impaired and preserved abilities (Becker *et al.*, 1988; Fisher *et al.*, 1996; Haxby *et al.*, 1985; Martin *et al.*, 1986). Disease progression in these relatively focal cases is characterized by two factors: continued deterioration with the affected cognitive domain and, by implication, within its functional brain substrate (Martin *et al.*, 1987) and spreading or widening of symptoms to other, previously intact domains (Becker *et al.*, 1988). The fact that the disease continues to progress within a single cognitive domain suggests that the progression of pathology is not random, but rather carves a course that appears to parallel the functional organization of the affected system (Martin *et al.*, 1987).

As noted above, these patients are also characterized by contrasting patterns of cerebral hypometabolism in the expected regions, based on their most prominent cognitive deficit (Foster *et al.*, 1983; Haxby *et al.*, 1985; Martin *et al.*, 1986). Specifically, left temporal lobe hypometabolism in patients with object-naming deficits and other semantic memory problems, and parietal lobe hypometabolism, often more pronounced on the right than the left, in patients with visuospatial deficits. As we noted previously (Martin *et al.*, 1987, 1990), these brain-imaging data are consistent with the functional neuroanatomy of object and spatial vision, with object perception and knowledge mediated largely by the ventral, occipitotemporal processing stream, and with spatial perception and spatial cognition mediated by the dorsal, occipitoparietal processing stream (Ungerleider and Mishkin, 1982; see Ungerleider and Haxby, 1994, for review). Thus, these visual processing systems seem to be at greater risk than other functional systems for developing AD neuropathology, and can be affected in relative isolation from each other in individual cases.

In addition, it appears that even within one of these subgroups, patients can differ with regard to important characteristics. For example, as noted above, many of the patients with prominent visuospatial difficulties have little difficulty in visually recognizing objects. They perform at near normal levels on tests of object naming and face recognition, while being profoundly impaired on tests that require spatial analyses, such as route finding, judging the orientation of a line, or copying abstract designs (Becker *et al.*, 1988; Fisher *et al.*, 1996; Martin *et al.*, 1986). It is this pattern of deficits that is associated with severe posterior parietal hypometabolism relative to other neocortical regions. The fact that these patients can name line drawings of objects indicates that early visual processing stages are intact, and indeed, in these patients, PET reveals relative sparing of the occipital lobes. In contrast, other patients show a more profound visual disorder that affects basic visual processes and thus results in both spatial impairments and object recognition deficits (Levine *et al.*, 1993; Furey-Kurkjian *et al.*, 1996) (Fig. 1). Studies of these visually impaired cases have raised two important questions about variability of pathology associated with AD. First, is the hippocampus and adjacent cortex invariably involved during the early states of the disease? Second, are primary sensory and motor areas always better preserved than parasensory and limbic association cortices?

At autopsy the medial temporal lobe structures contain SP and NFT as documented in many studies (Hirano and Zimmerman, 1962; Van Hoesen *et al.*, 1991;

Braak and Braak, 1991; Hof *et al.,* 1992; Gómez-Isla *et al.,* 1996). In addition, significant memory problems commonly occur in typical cases, as well as in the above-described cases with relatively focal language and visuospatial impairments. There have also been a number of reports of autopsy-documented cases of AD that initially presented with an isolated amnesic syndrome that lasted several years before progressing into a global dementia (Neary *et al.,* 1986). These findings suggest that the medial temporal region may be the only region involved early in the disease, but leaves open the question of whether the disease can initially begin outside of this region. Reports from the visually impaired patients and their families indicate that the visual problems were the earliest and sole symptom experienced by these patients, interfering with daily activities long before the onset of clinically apparent memory difficulties (Crystal *et al.,* 1992; Levine *et al.,* 1993; Pietrini *et al.,* 1996). This suggests that the hippocampus and adjacent structures were not the initial site of AD-associated pathology in these cases. Consistent with these clinical reports, the psychometric data have shown that the visual impairments were markedly more severe than the memory problems in these patients. Nevertheless, it must be noted that when formally tested, memory functions have not been normal, even for verbal material presented auditorily (Furey-Kurkjian *et al.,* 1996). Since PCC may be associated with very early impairments (Minoshima *et al.,* 1997), it remains to be documented that episodic memory functions and associated medial temporal

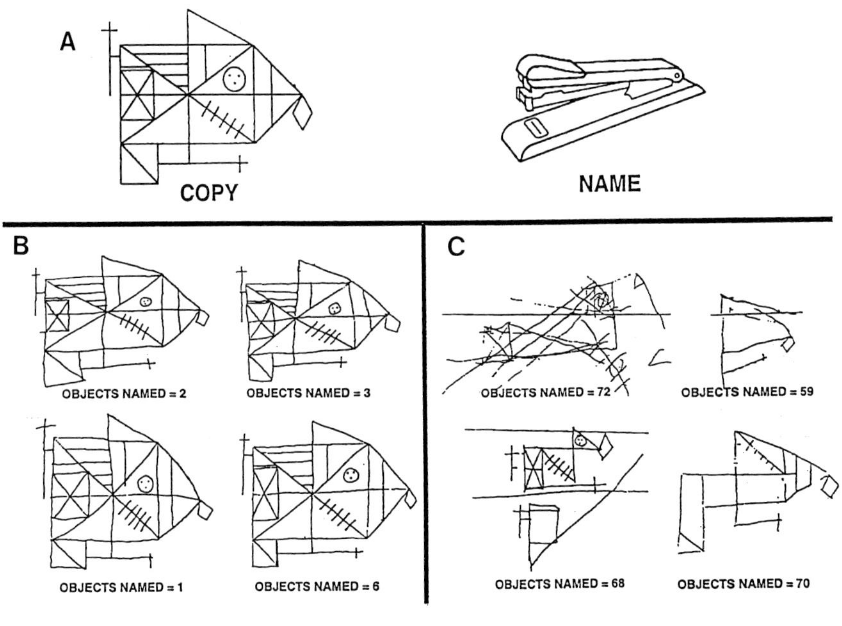


**Figure 1.** Examples of visuospatial/constructional skill and object-naming ability in AD patients forming qualitatively distinct subgroups. (A) On the left is a complex, meaningless figure commonly used to test visuospatial and constructional skill (Osterrieth, 1944). On the right is an example of 1 of 85 objects used to test object-naming ability (Kaplan *et al.,* 1976). (B) Four AD patients with intact visuoconstructional skills and severely impaired object-naming ability. (C) Four patients with the opposite cognitive profile. Relative to other cortical regions, PET revealed substantial left temporal lobe glucose hypometabolism for the patients in panel B and substantial right parietal glucose hypometabolism for the patients in panel C. AD was verified at autopsy in at least one patient in each subgroup.

lobe structures can be truly unaffected in the early stages of the disease in some cases or whether disruption at other sites accounts for impaired memory.

The visually impaired patients also show that, contrary to common belief, AD can affect a primary sensory area. In fact, the data suggest that primary visual cortex can be the most impaired region and perhaps the site of initial pathology. Autopsy studies have verified that the visually impaired patients have SP and NFT in the occipital lobe, including calcarine cortex, as discussed in detail below. PET studies have revealed that the most severe hypometabolism is in the dorsal visual stream, extending from the calcarine region to posterior parietal cortex (Pietrini *et al.,* 1996). Such cases raise another question about the regional variation of AD that has potential etiological significance: why, in at least some cases, does AD affect primary visual, but apparently not primary auditory, tactile, or motor cortices?

The PCC has been implicated in every postmortem study that evaluated this region in terms of neurodegeneration, alterations in the cholinergic, serotonergic and noradrenergic systems, and PET studies (see below for detailed discussion and references). Furthermore, cases of posterior cortical atrophy in AD have a more severe neurodegeneration and tau immunoreactivity in PCC than they do in posterior parietal cortex (Hof *et al.,* 1997; Vogt *et al.,* 1997). In spite of these early and profound changes, the involvement of PCC in AD is poorly appreciated, and this region is not considered in routine neuropathological assessments of the disease. One reason for this oversight is the lack of neuropsychological tests to elucidate impairments of cingulate cortex function, including those that test the specific contribution of this region to visuospatial function, reviewed by Olson *et al.* (1993). It is also possible that early neurodegeneration in PCC does not always involve deposition of classical lesions but undergoes neurodegeneration via other mechanisms that go undetected by routine neuropathological assessments.

The conundrum of current AD investigations is that having observed subgroups with joint clinical and cerebrally localized impairments of glucose metabolism, postmortem assessments simply verify the presence of a single disease based on established criteria. There are no systematic methodologies for assessing the extent to which subgroups represent unique subtypes. The extent to which subgroups based on neuropsychological testing represent unique subtypes cannot be determined from testing and PET studies alone, and it is not known how many different mechanisms of neurodegeneration there are in AD. Thus, although defining clinically unique subtypes of AD largely remains a goal, the evidence for multifocal degeneration is clear and biological subtypes can be assessed and should dovetail with clinical analyses based on more refined neuropsychological testing.

## 3. Genetic Factors: Subgroups or Subtypes?

Five genetic loci cosegregate with familial and sporadic forms of AD (Hardy, 1996; Yankner, 1995, for reviews) and inherited and somatic mutations of mitochondrial DNA (mtDNA) occur in AD (Shofffner *et al.,* 1993). These genes encode fundamentally different proteins that subserve disparate functions, and although it is possible that these genes converge on a single end-stage pathology, it is more likely that they do so via distinct pathways early in the disease and in distinct regional and laminar patterns. Our premise, therefore, is that different genetic variables contribute to distinct neuropathologies such as those associated with presenilin-1 (PS) and amyloid precursor protein (APP) genes. The strongest evidence for the concept of genetically linked subtypes is spastic paraparesis, degeneration of the corticospinal tract, and large SP without cores following deletion of exon 9 from the PS1 gene (Crook *et al.,* 1998). It may also be true that mtDNA mutations lead to selective vulnerability in regions with high levels of oxidative metabolism such as the retrosplenial areas of PCC (Matsunami *et al.,* 1989). Indeed, this region experiences the first glucose hypometabolism early in AD (Minoshina *et al.,* 1997). Furthermore, although homozygous inheritance of the ε4 allele of the ApoE gene is a risk factor that is not itself responsible for specific patterns of cortical degeneration, the differential interaction of the ε2 and ε4 alleles with frank genetic missense mutations in other genes suggests synergistic genetic subtypes and a potential for neuropathological and clinical subtypes associated with each.

### 3.1. Frank Genetic Subtypes: Autosomal Dominant Mutations

There are four classes of heritable genetic mutations associated with AD: mutations in the APP, in one of the two PS genes, mtDNA variants, and the recent identification of a new inherited genetic risk for late-onset AD on chromosome 12 (Pericak-Vance *et al.,* 1997). The APP is a large membrane-spanning protein of unknown function. It serves as the precursor for the 40–43-amino acid amyloid β-peptide (Aβ), which itself is a major component of SP. The APP gene is on chromosome 21 and has been implicated in AD through genetic analyses of rare familial forms of early-onset AD and Down's syndrome. There are five documented autosomal dominant forms of inherited, early-onset AD that are associated with mutations in the APP gene (see Hardy, 1997, for review). Although the number of families affected by APP mutations is small, the nature of the mutations provides insight into the disease process and has led to animal models of AD (Games *et al.,* 1995; LaFerla *et al.,* 1995). All of the mutations cluster near the site encoding the Aβ processed peptide and it is thought that miscleavage of APP results in either an increased production of Aβ and/or formation of a less soluble form of the processed peptide.

Although there is still controversy over the role of Aβ in neurodegeneration, evidence is accumulating for it as a potent neurotoxin in primary CNS culture and clonal cell line systems. Aβ42 and Aβ25-35 induce apoptosis in hippocampal neurons (Forloni *et al.,* 1993; Loo *et al.,* 1993). Although Aβ40 can induce apoptosis in cultured rat cortical neurons (Estus *et al.,* 1997), it is the aggregation state of this peptide that is particularly important. The toxic intermediates of Aβ may be free radicals. Aβ25-35 induces lipid peroxidation (Kruman *et al.,* 1997) and increases hydrogen peroxide and lipid peroxides (Behl *et al.,* 1994), which themselves induce oxidative damage of mitochondrial DNA (Bozner *et al.,* 1997). Finally, as discussed below for cases of focal cortical atrophy, Aβ42 accumulates in those layers and areas that experience the highest level of neurodegeneration. In a case of frontotemporal atrophy, Aβ42 is highest in layers II-IIIab of prefrontal cortex with most severe neuron losses. The progressive involvement of visual cortical areas in posterior cortical atrophy also suggests that increasing neurodegeneration is associated with a graded increase in the levels of Aβ (Hof *et al.*, 1997).

Missense mutations in the PS genes cosegregate with early-onset AD (Clark *et al.*, 1995; Levy-Lahad *et al.*, 1995; Hutton *et al.*, 1996). The PSs are transmembrane proteins thought to contain seven or eight membrane-spanning domains (Li and Greenwald, 1996; Doan *et al.*, 1996). The two related genes in this family reside on chromosome 14 (PS1) and chromosome 1 (PS2). While the functions of these proteins remain obscure, their mutation is associated with autosomal dominant forms of early-onset AD. There are now more than 40 genetic mutations that, with the exception of a single case of a deletion leading to altered PS1 mRNA splicing, convert a single amino acid to another amino acid. Mutations in the PSs lead to increased insoluble Aβ production (Scheuner *et al.*, 1996; Borchelt *et al.*, 1996; Citron et al., 1997). Moreover, Guo *et al.* (1997) demonstrated that PS1 mutations increase Aβ-induced apoptosis, and Maury *et al.* (1997) reported that PS mutations themselves can produce amyloidogenic peptides.

### 3.2. Different Clinical and Neuropathological Outcomes: PS and APP Mutations

Comparison of clinical outcomes associated with autosomal dominant mutations in the APP and PS genes suggests that they are different and may have differences in underlying pathology. The APP mutations (Lantos *et al.*, 1992; Mann *et al.*, 1992; Cairns *et al.*, 1993; Mullan *et al.*, 1993) are associated with earlier onset than are PS mutations ($<50$ vs. $>50$), more rapid progression, early progressive aphasia, and more frequent myoclonus, seizures, and psychosis (Haltia *et al.*, 1994; Lampe *et al.*, 1994; Kennedy *et al.*, 1995; Fox *et al.*, 1997; Harvey *et al.*, 1998). The brains of individuals with APP mutations are said to be "entirely typical," although they have elevated amyloid burden (above citations). In contrast, two cases with PS mutations were "unusual microscopically and both showed symmetric cortical atrophy which was particularly severe in the frontal and temporal lobes" (Harvey *et al.*, 1998, p. 46). Other cases have been reported with moderate spongiform changes (Haltia *et al.*, 1994) and high levels of Aβ42 (Haltia *et al.*, 1994; Lampe *et al.*, 1994; Gómez-Isla *et al.*, 1997).

The most profound finding of a relationship between an autosomal dominant mutation and unique patterns of clinical and neuropathological outcomes has been the consequences of deletion of exon 9 in the PS1 gene (Crook *et al.*, 1998). This mutation results in early spastic paraparesis associated with large SP without cores throughout the cerebral cortex and degeneration of the corticospinal tract. Although motor cortex is not generally impacted in AD, these cases show that even this region is at risk in some patients. Although some of the studies discussed above do not employ a large number of cases, all of the findings together suggest that correlative autosomal dominant mutations are associated with different patterns of Aβ deposition, neurodegeneration, and clinical outcomes. As these findings are solidified, this may provide important confirmation of the subtypes hypothesis of AD.

### 3.3. Mitochondrial DNA Mutations

Variants in mtDNA are maternally transmitted and also accumulate with age, particularly in the eighth decade (Corral-Debrinski *et al.*, 1992). Among other things, these genes encode transfer RNA in the mitochondria. The somatic mutations could result from oxygen free-radical damage. Although the functional consequences of such mutations are unknown, it is possible that disruption of mitochondrial function and passage of similar mutations to successive generations of mitochondria within the neuron may lead to significantly impaired oxidative phosphorylation and possibly cell death as a consequence of overproduction of oxygen free radicals. Because reductions in PCC neurons between the ages of 55 and 95 do not occur in neurologically intact individuals (Vogt *et al.*, 1998), it is likely that a combination of age-related and inherited mutations (Shoffner *et al.*, 1993) are responsible for neurodegenerative processes in AD.

### 3.4. Genetic Risk Factor: The ApoE Gene

Unlike the APP and PS genes, the ApoE gene on chromosome 19 does not act as the locus for the autosomal dominant mutations that lead to AD. Rather, the three naturally occurring alleles for this gene modulate the *likelihood* of being afflicted with the disease. The ϵ4 allele that is rare in the general population is a risk factor for AD, while the ϵ2 allele may be protective and is found much less frequently in AD than in the general population (Corder *et al.*, 1994). Biochemical explanations for the functional role of the ApoE gene in the etiology of AD include a differential interaction with the tau protein in NFT and with Aβ.

There are a number of different ways to consider the risk liability of ApoE genotype. It has been suggested that the homozygous ϵ4 individual has an eightfold greater risk of AD (Roses, 1995). Another statistical treatment suggests that this same genotype increases the risk by 10 years, with a single allele contributing one-half of this value. Another estimate is that the ApoE genetic locus accounts for approximately 50% of the attributable risk (Nalbantoglu *et al.*, 1994). In contrast, a study of 310 families by Blacker *et al.* (1997) reported the greatest risk for ϵ4 homozygotes between ages 60 and 65, suggesting greatest risk in early-onset cases and no role for transmission of a single ϵ4 allele. Attempts to directly link ApoE genotype and AD neuropathology have shown only an increase in SP in ϵ4 homozygotes, and the ϵ4 risk liability is not limited to AD but includes critical coronary heart disease, hypertension, and epilepsy, as discussed below. The crucial question in relation to AD is, what other genes are responsible for specific (i.e., focal) neocortical pathology?

The essential argument for genetic subtypes and preliminary indications of their neuropathological impact arise from the interaction of different ApoE alleles on other genetic systems. While homozygous expression of the ϵ4 allele increases the risk for AD, it exerts differential modulatory effects on other genes. The ϵ4 allele decreases the age at onset for early-onset AD associated with APP mutations (Houlden *et al.*, 1993), whereas the ϵ4 allele may not have the same effect on the early-onset disease associated with presenilin mutations (Van Broeckhoven *et al.*,

1994). The same gene, therefore, exerts different effects depending on the genetic environment. Conversely, the ϵ2 allele may be protective against AD and so this allele is underrepresented in AD populations (Talbot *et al.*, 1994; Corder *et al.*, 1994). There are a number of plausible molecular explanations for this finding based on binding of the tau protein in NFT and interactions with Aβ. However, this protective effect is absent in early-onset AD (Maestre *et al.*, 1995) and in black populations (van Duijn *et al.*, 1994). Thus, expression of the ApoE alleles influences the onset and progression of AD via their interactions with the products of other genes. For example, the ϵ4 allele is associated with higher densities of SP in AD (Schmechel *et al.*, 1993; Rebeck *et al.*, 1993), and these SP have greater amounts of Aβ40 which is not fibrillogenic (Gearing *et al.*, 1996). However, PS1 mutations elevate the longer form of Aβ (Borchelt *et al.*, 1996; Scheuner *et al.*, 1996), suggesting that homozygous expression of the ApoE ϵ4 allele and a PS1 mutation will increase SP densities still further.

Studies of PCC suggest that, although homozygous expression of the ApoE ϵ4 allele is associated with greater numbers of SP, the ϵ4 risk is associated only with the most severe neurodegeneration and does not account for most laminar patterns in degeneration (Vogt *et al.*, 1998). Because the ApoE risk is associated with altered processing of the APP throughout the body, including cortical deposition of SP in nondemented individuals with critical coronary heart disease and hypertension (Sparks *et al.*, 1996), and in epileptogenic cortex (Gouras *et al.*, 1997), the significance of the ApoE risk may be its influence on lipid metabolism. The issue raised by genomic studies of the ApoE is to what extent other genes may be responsible for neocortical pathology including disease subtypes and how the functions of these unknown genes are altered in ϵ4 homozygotes. For example, Jordan *et al.* (1998) provide evidence that ApoE isoform modulates Aβ-induced toxicity in cultured hippocampal pyramidal neurons. To the extent that homozygous expression of the ϵ4 allele is associated with Aβ release *in vivo,* such as in the most severe type of neurodegeneration in PCC (Vogt *et al.*, 1998), ApoE genotype may be relevant to multifocal neuron losses in one group of cases.

## 4. Models of Heterogeneity, Postmortem Measures, and Aβ Neurotoxicity

The demonstration of subgroups with impairments in different cognitive domains and associated focal glucose hypometabolism does not, in itself, prove that there are subtypes of AD. Multiple frank genetic mutations supports the expectation of biological subtypes because they themselves could lead to different disease outcomes. Although evidence for multiple etiologies must arise from direct studies of the brain, neuropathological models of AD emphasize SP and/or NFT and usually employ a single disease model with differences among cases ascribed to disease progression. Difficulties with this approach are associated with the relevance of SP and NFT densities to neocortical neurodegeneration and the assumption that AD is uniform. Efforts have been made to subgroup AD neuropathology beyond progression and comorbidities with other dementias. Armstrong *et al.* (1996) used multivariate methods to assess AD, and reported evidence for five subtypes based on SP and NFT densities in many parts of the CNS. This is an important approach and derives from a number of earlier efforts by the same group (Armstrong and Wood, 1994; Armstrong and Myers, 1992). The conclusion that subtypes were identified is a matter for debate, however, until the influence of disease progression is defined for each and unique mechanisms are shown for each subtype.

There are two general models to account for AD heterogeneity: stochastic subgroups and subtypes. The stochastic-subgroups hypothesis postulates a single disease etiology and heterogeneity resulting from random involvement of different cortical regions. One consequence of the probabilistic expression of AD pathology in the cerebral cortex is subgroups of cases with cortical lobar specializations in glucose hypometabolism, multifocal degeneration, and different patterns of clinical symptomatology. The two-group statistical analysis to assess the uniform-disease hypothesis (i.e., all AD cases are essentially the same when compared with controls) is not sufficient to analyze disease heterogeneity embodied by the stochastic-subgroups model. For this purpose multivariate analyses are used, such as principal components or cluster analysis. If different manifestations of the disease can be characterized by a multivariate model, this implies that the subgrouping may be valid, although it says nothing about disease etiology.

The *subtypes hypothesis* postulates that under the broad clinical and pathological diagnosis of definite AD, there are subtypes differentiated by multiple etiologies. Once a subgroup has been identified with neuropathological criteria that require a unique mechanism for neurodegeneration, the null hypothesis is tested by statistically comparing the subgroups with principal components analysis (PCA). The PCA seeks to identify those features that best distinguish among the postulated subtypes. The difference in approach for the subtypes model is that one begins with groups that have an *a priori* rationale for expecting a unique etiology for each subtype. The multivariate model seeks to identify the smallest number of variables that define a neuropathological subtype, and the obvious candidates for such a model are the SP and NFT markers used for diagnosis. Because there is some overlap between layers with greatest losses of large neurons (Mountjoy *et al.,* 1983) and layers with most NFT, that is, layers III and V (Pearson *et al.,* 1985), it is generally thought that NFT reflect total neurodegeneration. Even in these layers, however, a loss of eight neurons can occur for each NFT (Vogt *et al.,* 1998). Thus, neuron densities need to be considered as an alternative measure to NFT. In the following paragraphs, we consider aspects of neurodegeneration in the AD brain, the extent to which each is related to neurodegeneration and clinical outcomes, and which variables are best entered into multivariate statistical analyses.

The relevance of SP to disease progression and neurodegeneration is a matter of controversy, and the following concerns are raised about SP density as a measure of multifocal degeneration, subtypes, and disease progression. First, there is almost no relationship between neuropsychological measures of cortical function and the density of SP (Neary *et al.,* 1986; Nagy *et al.,* 1995). Second, high densities of SP have been observed in nondemented individuals with critical coronary artery disease and hypertension (Sparks *et al.,* 1990, 1996). Third, SP densities do not correlate with neuron densities (Terry *et al.,* 1981; Mann *et al.,* 1985; Vogt *et al.,* 1990). Fourth, SP do not parallel the duration and severity of AD (Arriagada *et al.,* 1992). Two studies that combined neocortical biopsy and autopsy analysis failed to show an increase in SP after onset of the dementia (Mann *et al.*, 1988; Bennett *et al.*, 1993). Fifth, the topographical distribution of SP is homogeneous throughout neocortex (Arnold *et al.*, 1991) and is probably not related to focal cognitive impairments. Thus, although SP density provides a diagnostic standard, it is unlikely they are related to subtypes or that they account for disease progression. This is not to say, however, that Aβ is not engaged early in the disease and does not lead to neuron losses depending upon the local genetic and metabolic environment. Indeed, one of the main conclusions of studies of focal cortical atrophies is that areas and layers with greatest neurodegeneration are associated with highest levels of Aβ42 (see section 7, Multifocal Atrophies).

Although neocortical NFT can correlate with dementia severity in early stages of the disease (Bierer *et al.*, 1995), NFT do not provide a strong measure of cortical neurodegeneration. The following concerns are noteworthy. First, there is only a weak relationship between NFT density and neuropsychological test performance (Neary *et al.*, 1986). Second, biopsy and postmortem assessments identified some AD cases in which NFT densities actually *decreased* in number rather than increased (Mann *et al.*, 1988; Bennett *et al.*, 1993). Third, although there is a high and negative correlation between NFT and neuron densities in entorhinal cortex (Mountjoy *et al.*, 1983), this relationship does not hold in other neocortical areas, where a NFT is not deposited in most neurons (Vogt *et al.*, 1990). Fourth, although it has been reported that NFT parallel the duration and severity of AD (Arriagada *et al.*, 1992), the best correlation with duration of the disease was for NFT averaged over 13 areas, whereas the correlations of NFT with individual areas were not significant. Fifth, in posterior cortical atrophy in AD, neuron densities in PCC are not necessarily related to NFT densities or laminar distributions (Hof *et al.*, 1997; Vogt *et al.*, 1997). Finally, NFT may be a final common pathway for a number of different mechanisms of neuron death, including oxidative stress (Blass *et al.*, 1990) and glutamate neurotoxicity (De Boni and McLachlan, 1985). These observations suggest that, while NFT are valuable for differential diagnosis, they may not always be useful for assessing multifocal neurodegeneration and subtypes because they are poorly related to cell loss, the course of the disease, dementia severity, and a single mechanism of cell death.

Although NFT may not accurately estimate total neurodegeneration, they have been correlated with different apraxias (Giannakopoulos *et al.*, 1998). Ideomotor and dressing apraxias are associated with NFT in anterior cingulate cortex, whereas constructional apraxia is related to NFT in superior parietal, posterior cingulate, and occipital cortices. This study also reported that SP distributions do not correlate to the different apraxias. Thus, NFT may be valuable for defining the neocortical substrates of some clinical deficits.

Neurochemical studies by Bondareff *et al.* (1987) argued that high and low levels of neurodegeneration in the locus coeruleus are associated with different levels of neocortical noradrenaline and choline acetyltransferase and that individuals with high losses of neurons and low levels of noradrenaline had a higher risk for depression and other behavioral problems. Our observations of β-adrenoceptor binding (Vogt *et al.*, 1991) verify that there may be differential impairment of the noradrenergic inputs to the cerebral cortex, and fundamentally different responses of cortical neurons to these changes. In other words, the receptor changes do not fit a single-etiology model. Evidence discussed in detail below suggests that alterations in the noradrenergic system may represent, in part, unique disease mechanisms and AD subtypes. It will be necessary, however, to provide a broader context of neocortical neurodegeneration within which these subtypes can be specified.

Terry *et al.* (1991) reported that synaptic density was highly correlated with dementia severity. They also indicated that SP and NFT added little to multivariate, stepwise regression analyses. Although it is possible that terminal fields of cortical neuron projections recognized by synaptic markers can be used to assess subtypes, there is little regional or laminar specificity associated with massive losses of cortical neuron projections, and laminar patterns of synaptic losses have not been identified. A strong correlation between neuropsychological and neuropathological measures of AD in an analysis by Neary *et al.* (1986) was with large cortical neurons. Although neurons and other measures were only sampled in layers III and V, it is important that the clinical course of the dementia is related to neurodegeneration. The substantial loss of cortical synapses, therefore, may be due to cortical neuron degeneration, and relationships between neuron and synapse degeneration will require further analysis.

## 5. Laminar Patterns in Neocortical Neurochemistry

It is generally accepted that large neocortical neurons, mainly in layers IIIc and Va, preferentially degenerate in AD (Mountjoy *et al.*, 1983; Hansen *et al.*, 1988) and form NFT (Pearson *et al.*, 1985; Giannakopoulos *et al.*, 1993). Furthermore, differences among cases are thought to result from progressive loss of neurons over the clinical course of the disease, with increases in NFT related to disease progression (Arriagada *et al.*, 1992). However, changes in the laminar patterns in neurodegeneration in PCC, as described below, require different mechanisms to explain neurodegeneration throughout all layers, versus cases with no degeneration in PCC in relation to age-matched controls. If differences among cases were simply quantitative in terms of SP and/or NFT densities, different mechanisms would be unlikely. However, because the chemoarchitecture, connections, and neurochemistry differ for each neocortical layer, qualitatively different laminar patterns in neurodegeneration may require fundamentally different mechanisms.

A summary of some of the major connections and unique chemoarchitecture of layers II–V in PCC area 23 provides a context for interpreting laminar patterns of neurodegeneration in AD. One of the major outputs of layer III–V pyramidal neurons in area 23 is to other cortical regions, including anterior cingulate, inferior parietal, prefrontal, superior temporal, and posterior parahippocampal cortices (Baleydier and Mauguière, 1980; Pandya *et al.*, 1981; Vogt and Pandya, 1987; Van Hoesen *et al.*, 1993). No extrinsic projections are known for neurons in layers II and IV. Neurons in these layers likely have primarily intrinsic connections as they do in human area 19 (Shkol'nik-Yarros, 1971). It is possible that differences in cytoskeletal organization are associated with the different transport needs of medium-large extrinsically projecting pyramids in layers III and V versus small-medium intrinsic projection pyramids in layers II and IV. Cytoskeletal differences, therefore, could provide clues as to the selective vulnerability of neurons in different layers to AD pathogenesis.

Even within layers III and V, there are phenotypically unique populations of pyramidal neurons. Although neurons in both layers express intermediate neurofilament proteins (i.e., SMI32 immunoreactivity; Hof *et al.,* 1995; Nimchinsky *et al.,* 1997; Vogt *et al.,* 1997), these neurofilaments are located mainly in sublayers IIIc and Va, as shown in Figure 2. A small percentage of area 23 projection neurons in these layers (about 20%) is SMI32-immunoreactive in double-labeling studies (Hof *et al.,* 1995), but most neurons in layers II, IIIab, IV, and Vb are not SMI32-immunoreactive. One of the surprising features of neurodegeneration in PCC in AD is the preservation of the largest neurons in layer IIIc, even in cases with such severe neurodegeneration that laminar architecture is almost completely lost (Vogt *et al.,* 1998). The preserved neurons may be phenotypically unique and express these intermediate neurofilaments. Since neurons are lost in both layers IV and V in another neuropathological subtype of AD (STIV-V), this also suggests sparing of neurons with intermediate neurofilaments, because almost no neurons in layer IV are SMI32-immunoreactive. It is possible that a dependence of pyramidal neurons on transport mechanisms requiring intermediate neurofilaments protects them from neurodegeneration associated with hyperphosphorylation of tau or other impairments of cytoskeletal function engaged by AD.

Neurons containing calcium-binding proteins appear to be resistant to neurodegenerative processes in AD (Hof *et al.,* 1993b), and differences in their laminar distribution could provide neuroprotective mechanisms for cortical interneurons. The three major calcium-binding proteins expressed by cortical neurons are in different subpopulations in PCC (Hof *et al.,* 1992; Nimchinsky *et al.,* 1997). Whereas calbindin and parvalbumin tend to be in interneurons spread evenly throughout the cortical layers, calretinin is largely expressed by neurons in layers II and IIIab, as shown in Figure 2. We report that neurons in layers II and IIIab are last to form NFT in PCC (i.e., during years 11+; Vogt, 1998). The calretinin-containing neurons in layers II and/or IIIab could provide a neuroprotective function over many years

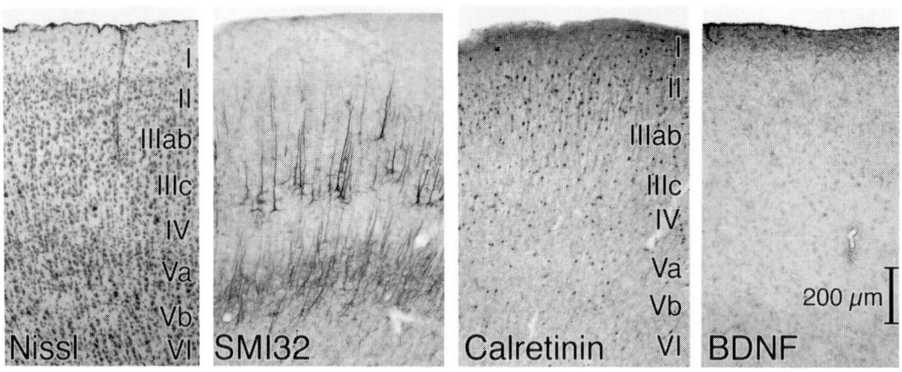


**Figure 2.** Lamination pattern in monkey area 23a (Nissl) and three neurochemical markers. SMI32 is an antibody to intermediate neurofilaments, and it reacts primarily with neurons in layers IIIc and Va. Calretinin is a calcium-binding protein that is primarily expressed by interneurons in layers I-II, lesser in layer IIIab, and least throughout deeper layers. Brain-derived neurotrophic factor (BDNF) is expressed by neurons throughout the full depth of the cortex. The laminar distributions of these and other markers of chemically unique neurons suggest that laminar patterns of neurodegeneration in area 23 of AD require unique mechanisms of cell death.

of disease expression. In late-stage AD, however, impaired production of calretinin by these neurons could lead to hyperexcitation of pyramidal neurons in these layers and formation of NFT both in area 23 as well as in areas that receive inputs from these layers, such as areas 29 and 30 in the callosal sulcus (Vogt and Pandya, 1987).

Multilayer (vs. layer-specific) neurodegenerative processes in AD may be associated with altered production of nerve growth factor (NGF) or nitric oxide. Growth factors have long been suspected to play a role in degeneration of basal forebrain neurons in AD, and reductions in the density of NGF receptor-expressing neurons in the basal forebrain have been reported (Mufson *et al.,* 1989; Loy *et al.,* 1990). NGF plays a supportive role for neocortical neurons, including those in PCC, and NGF and its receptors p75 and trk are produced by neurons in most layers of rat cerebral cortex (Pitts and Miller, 1995). NGF- and brain-derived neurotrophic factor-immunoreactive neurons occur throughout monkey PCC (Fig. 2), and although it is not known how NGF levels change in PCC in AD, altered NGF production could be associated with cell loss. Alternatively, nitric oxide mediates glutamate neurotoxicity (Dawson *et al.,* 1991b), and neurons expressing nitric oxide synthase are in all layers (Dawson *et al.,* 1991b). As suggested in the former study, these neurons are ideally situated to mediate widely distributed and nonspecific neurodegenerative events. Excitatory stimulation of these neurons by glutamate could lead to neurodegeneration in most cortical layers. Thus, layer-specific differences in the expression of intermediate neurofilament proteins by pyramids or calretinin by interneurons could provide for layer-selective vulnerabilities of neurons, while impaired growth factor synthesis or nitric oxide metabolism could subserve neuron death in all layers.

## 6. Laminar Patterns of Neuron Vulnerability in Frontal/Anterior Cingulate Dementias

The selective vulnerability of neocortical neurons in different layers provides one basis for assessing different mechanisms, as discussed above. We begin here with the frontal lobe dementias for three reasons. First, they have a number of unique laminar patterns and markers of neurodegeneration, which suggest mechanistic differences for cell death. Second, the frontal lobe dementias always involve cingulate cortex. Their consideration, therefore, provides a context for assessing this region in AD. Third, there are cases of AD with frontotemporal atrophy and behavioral deficits that raise questions about localization of multifocal patterns of neurodegeneration and the possibility of disease subtypes. Codegeneration of anterior cingulate cortex and PCC in AD cases with focal frontotemporal atrophy suggests that dense and reciprocal connections between frontal cortex and the cingulate gyrus are relevant to the biological bases of AD. Finally, impairments in executive control function occur early in AD (Royall *et al.,* 1994) and suggest that impaired frontal lobe function plays a pivotal rather than ancillary role in AD pathology.

Many frontal lobe dementias are associated with greatest neuron losses in layers II and III and little or none in layers V or VI, while Lewy body dementia has little or no change in total neuron densities and formation of Lewy bodies in neurons in layers V and VI. In all instances of frontal lobe dementia there is coinvolvement of anterior cingulate cortex. In fact, Lewy body dementia has greater involvement of anterior cingulate cortex than prefrontal cortex and this emphasizes the need to assess contributions of cingulate cortex to frontal lobe dementias, including cases of frontal lobe atrophy in AD.

Pick's disease is characterized by a massive frontotemporal atrophy and spherical inclusions called Pick bodies in cortical, basal ganglia, and some brainstem structures (Jakob, 1979; Tissot *et al.*, 1985; Brion *et al.*, 1991). In the cerebral cortex, Pick bodies are prominent in the granule cell layer of the dentate gyrus and in the prefrontal and temporal neocortex, whereas the parietal and occipital regions are generally less affected or spared (Yoshimura, 1989; Brion *et al.*, 1991; Hof *et al.*, 1994). The distribution of Pick bodies and ballooned neurons in neocortex preferentially involves layers II-III and VI (Hof *et al.*, 1994). Clinically, Pick's disease is associated with mood disturbances, irritability contrasting with incoherent joviality, early loss of social awareness, and bulimia and verbal stereotypia not usually seen in AD (Jakob, 1979; Tissot *et al.*, 1985; Brion *et al.*, 1991; Mendez *et al.*, 1993).

Other frontal lobe dementias, frequently classified clinically as atypical Pick's disease, are not associated with Pick bodies, SP, NFT, or Lewy bodies. In these cases there is moderate-to-severe atrophy confined to the frontal and temporopolar regions and widening of the ventricular system (Brun, 1993; Mann *et al.*, 1993; Neary *et al.*, 1993; Giannakopoulos *et al.*, 1995; Bergmann *et al.*, 1996). According to new neuropathological criteria for frontal lobe dementia (Brun *et al.*, 1994), frontal and temporopolar cortices show pyramidal cell loss mostly in layers II and III, whereas layers V-VI are less affected. Spongiform degeneration of layers II and III is often seen, along with moderate-to-severe astrocytic gliosis in the gray and white matter (Brun, 1993; Mann and South, 1993; Mann *et al.*, 1993; Giannakopoulos *et al.*, 1995; Jackson and Lowe, 1996). The lesions frequently include the insula and cingulate cortex, while the posterior temporal, parietal, and occipital cortices are generally spared. In some cases there is limited atrophy in frontal cortex associated with neuron losses in layers I-III only and significant involvement of anterior cingulate, anterior insular, and anterior temporal pole cortex (Gustafson *et al.*, 1990). Clinically, these cases form part of a spectrum of lobar atrophy and are characterized by insidious onset and slow progression. In the early stages, there are marked personality changes with loss of social and personal awareness, disinhibition, impulsivity, hyperorality, and stereotyped behavior (Brun, 1993; Gustafson, 1993). Memory and spatial orientation are relatively spared. Despite neuropathological groups in frontal lobe dementia, no clear clinicopathological correlations can be established, indicating that the neuropathological profile of frontal lobe dementia is variable, the only consistent feature being the absence of dysmorphic subcellular structures such as the Pick bodies and NFT.

Progressive supranuclear palsy and corticobasal degeneration, both characterized by progressive impairment in motor function and dementia, exhibit a preferential distribution of neurofibrillary degenerative changes in the hippocampal formation and a particular involvement of the primary motor cortex, with severe atrophy of the precentral gyrus and high numbers of dystrophic neurites, NFT, and ballooned neurons (Feany and Dickson, 1995; Feany *et al.*, 1995; Hauw *et al.*, 1990; Hof *et al.*, 1992b). The anterior cingulate, prefrontal, parietal, and temporal neocortical regions are less severely affected, and the occipital cortex is mostly spared in these cases (Hauw *et al.*, 1990; Hof *et al.*, 1992b).

Lewy body dementia exhibits yet another pattern of pathology in the frontal, temporal, and anterior cingulate cortices. Lewy body inclusions can be widely dispersed in the neocortex, however, they arise primarily in the hippocampal formation and anterior cingulate cortex (Kosaka *et al.*, 1984; Sima *et al.*, 1986; Lippa *et al.*, 1994; Braak *et al.*, 1996). In spite of unchanged neuron densities in layers V and VI of frontal cortex (Lippa *et al.*, 1994), there is a high density of Lewy bodies in these layers, and it can be almost three times higher in anterior cingulate cortex than in prefrontal cortex. Although there are diffuse SP in temporal cortex in Lewy body dementia, there are no NFT and little or no neurodegeneration. Finally, the density of Lewy bodies in most neocortical areas is related to dementia severity (Lennox *et al.*, 1989). These cases initially present with a fluctuating confusional state, visual hallucinations, behavioral disturbances, and mild extrapyramidal features that can be distinguished from parkinsonian impairments (Perry *et al.*, 1990).

These observations demonstrate that differences in clinical symptomatology are related to the involvement of neuronal subsets of anterior cingulate and frontal cortices. This differential vulnerability can be characterized in terms of laminar patterns of neurodegeneration in the frontal/cingulate dementias and markers of cell pathology. Therefore, differences in laminar patterns of neurodegeneration in AD might represent subtypes, based on the likelihood of different underlying mechanisms of cell death. Furthermore, strong and reciprocal connections between prefrontal and cingulate cortices imply that cases of AD with focal frontotemporal atrophy will significantly involve cingulate cortex.

## 7. Multifocal Atrophies and Subsystem Disruption in Alzheimer's Disease

### 7.1. Frontotemporal Atrophy and Case FG

There is a subgroup of AD cases with frontal lobe glucose hypometabolism and a high degree of behavioral deficits typical of frontal lobe damage. Grady *et al.* (1990) observed more inappropriate behaviors in cases with greatest prefrontal glucose hypometabolism. Mann *et al.* (1992b) reported that two groups with equivalent parietal glucose hypometabolism and impaired neuropsychological measures of parietal function could be distinguished in terms of frontal lobe glucose metabolism, measures of executive functions, and rate of disease progression. Cases with most rapid progression and impaired executive functions had greatest glucose hypometabolism in prefrontal cortex. Finally, Royall *et al.* (1994) observed early impairment of executive control functions in AD that could be attributed to disrupted frontal lobe functions.

Although large neuron degeneration has been reported in mid-prefrontal cortex, there is a wide range of degeneration, with many cases having a normal density of neurons (Terry *et al.*, 1981). Brun and Englund (1981) report only modest losses of neurons in prefrontal cortex. Furthermore, there is heterogeneity in NFT densities, which are significantly different for early- versus late-onset cases (Hansen *et al.*, 1988). In the context of disease heterogeneity, frontal cortex is a region that is differentially impacted by AD and in cases where frontotemporal atrophy is particularly striking and associated with behavioral deficits that suggest a clinicopathological subtype.

Our case FG is an example of AD with focal frontotemporal degeneration. Case FG was an intelligent, right-handed male who was an electrical engineer. At the age of 85, FG was driving the family on a long trip, during which he became profoundly paranoid of truck drivers to the point that he would follow them to verify that they were not endangering the family. Upon arrival at their destination, he immediately returned to the car to continue driving to another destination. Within two years he was evaluated neurologically and said to have increased problems with spatial disorientation, recent memory, and forgetfulness. His immediate recall was fairly good, but he had a problem remembering current events. No active hallucinations or delusions were noted at this time. The CT scans showed mild to moderate cortical atrophy and ventricular enlargement, findings that were confirmed one year later, as shown in Figure 3. At 89 FG suffered from mood instability and physical aggression toward his wife. In the past two years he showed many menacing behaviors such as threatening to hit his wife with a metal cane or kitchen chair. He was overresponsive to and agitated by TV violence to the point that the TV was removed from his room. He was frequently concerned about "people being around and coming into the house." On one visit, a psychiatrist observed that FG was alert, and semi-cooperative, and well groomed, although he was wearing a leather jacket, shirt, and sweater all buttoned and tucked into his pants. There was mild psychomotor agitation, and he attempted to leave the examination and resisted being redirected. There was some loss of facial expression, but speech was at a normal rate, volume, and tone. Mood was euthymic with an appropriate, bright, and stable affect. His thought was severely disorganized, with most responses being incoherent and nonsensical. He had visual and auditory hallucinations. His praxis was poor, and he could not name colors, misnamed a pen, but was able to name a watch and knew his full name. Neuropsychological tests showed a global deterioration in all cognitive domains, including attention, memory, registration, and recall, as well as language, naming, praxis/executive function, and perceptual disturbances. There was a decline in speech to the point where complete sentences were not always used, and FG frequently spoke with adjectives until his death at age 90.

Postmortem assessment showed profound frontotemporal atrophy (Fig. 3B). There were no Lewy bodies, Pick bodies, or ballooned neurons in prefrontal or any other neocortical region assessed, suggesting that this was not a case of frontal lobe dementia. The ApoE genotype was homozygous for the $\epsilon 4$ allele and this is an infrequent finding in frontal lobe dementia (Pickering-Brown *et al.*, 1995). There was a high density of SP throughout all neocortical regions, including prefrontal, inferior parietal, and PCC. Throughout neocortex there were a limited number of layer V NFT and in the hippocampal formation and entorhinal/perirhinal cortices there was a pattern of NFT typical of AD. Heavy $A\beta 42$ deposition in all cortical regions, including prominent buildup in prefrontal cortex, underscored the AD pathogenesis of neocortical neurodegeneration.

Tau-ir NFT were particularly high around the fundus of mid-rostrocaudal levels of the cingulate sulcus (Fig. 3; area 24d). This region had a high density of NFT in layer V and numerous neuritic plaques in layers I–III. This region contains one of the cingulate motor areas as previously reviewed (Vogt *et al.*, 1997) and it is interesting in this context because FG suffered from dysexecutive syndrome and disinhibi-

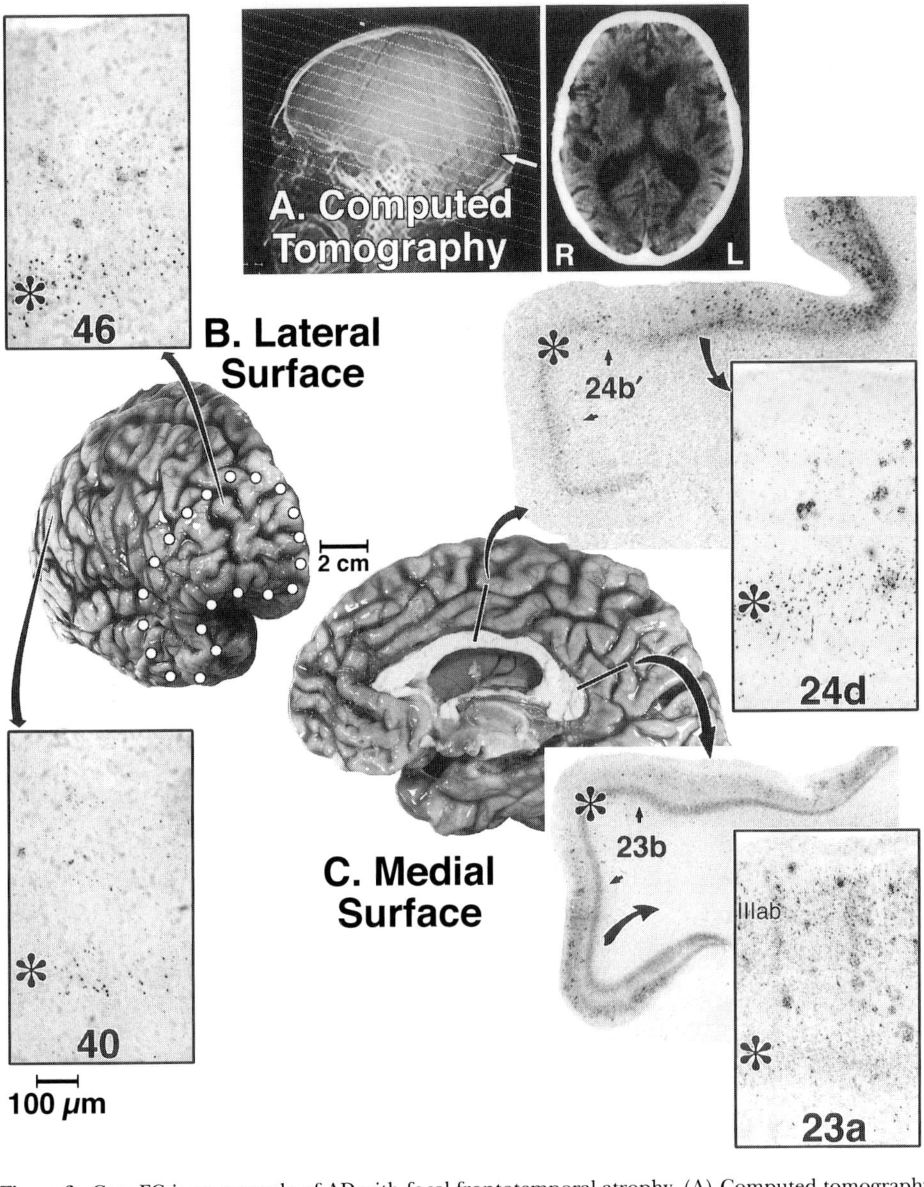


**Figure 3.** Case FG is an example of AD with focal frontotemporal atrophy. (A) Computed tomography two years before death shows mild cerebral atrophy and ventricular swelling. Postmortem assessment demonstrated clear frontotemporal atrophy (B, outlined with dots) and moderate dilation of sulci on the superior and medial surface (C). Insets are tau-immunoreactive NFT and neuritic plaques in areas 46, 40, 24d, and 23a. The greatest number of NFT were in layer V (asterisks). Area 23a had the highest density of neuropil threads of any neocortical area analyzed and an equal density of NFT in layers IIIab and Va. The highest level of neurofibrillary degeneration so far observed in this case is in posterior cingulate area 23. Additionally, there was marked involvement of area 24d around the fundus of the cingulate sulcus, suggesting an involvement of a cingulate motor area in this case.

tion. Also of interest in terms of the cingulate gyrus is that area 23a had equal densities of NFT in layers V and II-IIIab, and area 23 had the highest density of neuropil threads in neocortex (Fig. 3).

The density of thioflavine S-stained, mature, and burned-out SP was about the same in all neocortices, as shown in Figure 4. Immunohistochemistry for antibodies selective for Aβ40 and Aβ42 showed that the clumps of thioflavine S-stained, mature, and burned-out SP were similar to those immunoreactive for Aβ40 (Fig. 5). The very substantial neurodegeneration in this case was in layers II and III, as shown for PCC in Figure 5C and 5G and there was a striking match between this cell loss and uniform deposition of Aβ42 in layer III (Fig. 5B,F). There was more Aβ42 in area 40, while it was at highest levels in prefrontal area 46 (Fig. 5D). Notice in area 46 that the highest levels of Aβ42 were in layers II-IIIab, as indicated with the bracket (Fig. 5D). These layers are the site of greatest neurodegeneration in area 46. Neuron densities are best analyzed in perikaryal drawings that do not include glial and endothelial cells, and Figure 6 is a series of perikaryal drawings from

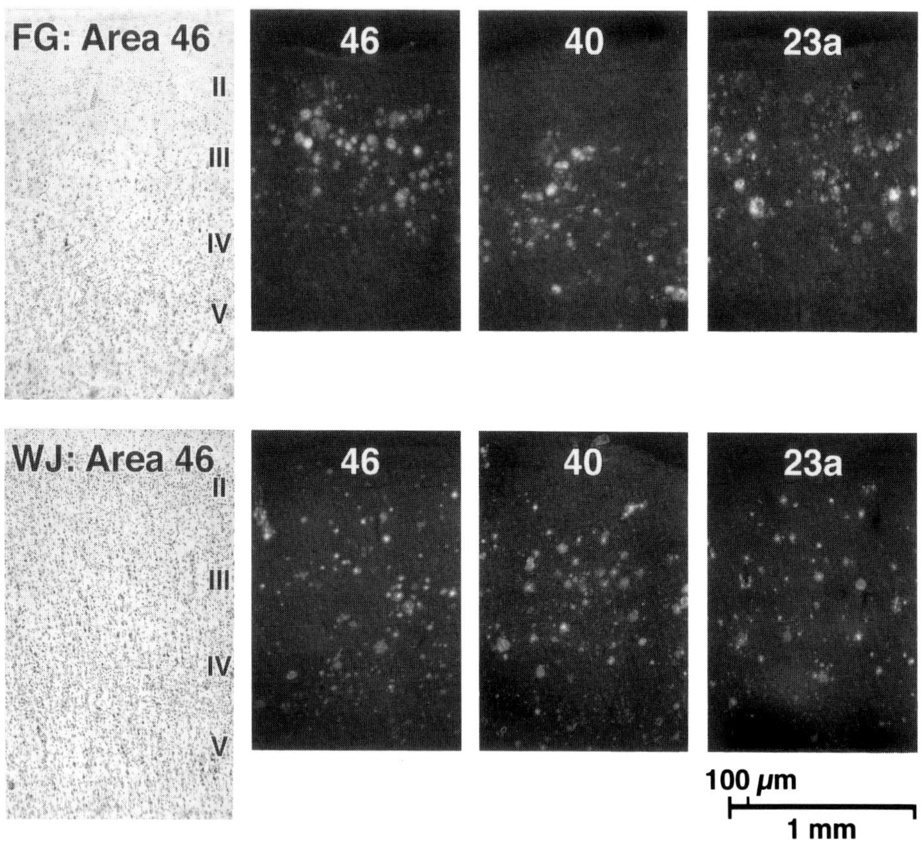


**Figure 4.** Representative densities of thioflavine S-stained SP in areas 46, 40, and 23a for cases FG and WJ. The densities in SP do not differentiate between these cases, although they are extreme examples of focal cortical atrophy, and FG was an ϵ4 homozygote and WJ was an ϵ3 homozygote.

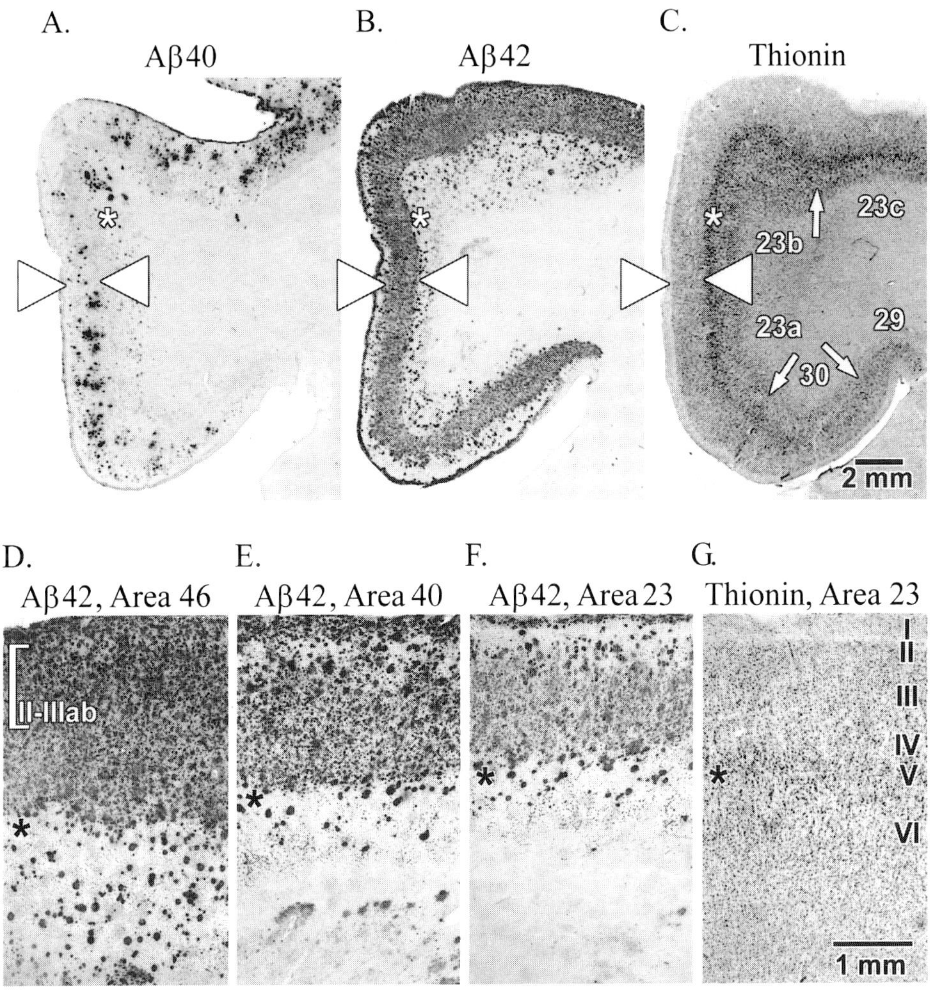


**Figure 5.** Aβ40 and Aβ42 immunohistochemistry in PCC (A–C, F, G) and prefrontal area 46 (D), and parietal area 40 (E). Aβ40-ir mature and burned out plaques form clumps in layer III of all areas and parallels the appearance of thioflavine S-stained SP. Aβ42 has a uniform distribution in all areas and is associated mainly with diffuse and immature plaques in layer III and mature plaques in layers I-II and V-VI. Although many neurons are lost throughout cortex, the most dense deposits of Aβ42 are in layers II-IIIab of area 46, where gross focal atrophy (Fig. 3) and neurodegeneration (Fig. 6) occur. There is a clear association in PCC of neuron degeneration in layers II-III (C, G) and maximal deposits of Aβ42 (B, F). Large white arrows emphasize layer III in A–C.

case FG. Most neuron losses in areas 46 and 23 were in superficial II-IIIab and deep layers V and VI. In area 40, neurodegeneration was significant in layers IV and V, however, layers II-III had much less degeneration, and there was no overt cortical shrinkage. The striking association between focal prefrontal atrophy, the highest levels of Aβ42, and a match between highest Aβ42 and neurodegeneration in layers II-IIIab suggests that Aβ42 neurotoxicity may be a key factor in multifocal cortical atrophies. Not all cell loss, however, was related to Aβ42 early on. In area 23, layers II-IIIab also had a high number of NFT. Layer IV had severe neuron losses in all areas, but only moderate Aβ42 deposition, while neurons in layer IIIc were relatively free of NFT and degeneration. There were neuron losses in deeper layers with only scattered Aβ42-ir SP, but this could have been due to neurofibrillary changes or the actions of Aβ42 on the apical dendrites of pyramidals neurons.

Overall, it is unlikely that a single mechanism of neurodegeneration was responsible for these complex patterns of neurodegeneration and disease markers. The severe neurodegeneration in prefrontotemporal and cingulate areas likely impaired the regulation of executive motor functions. Although the extensive de-

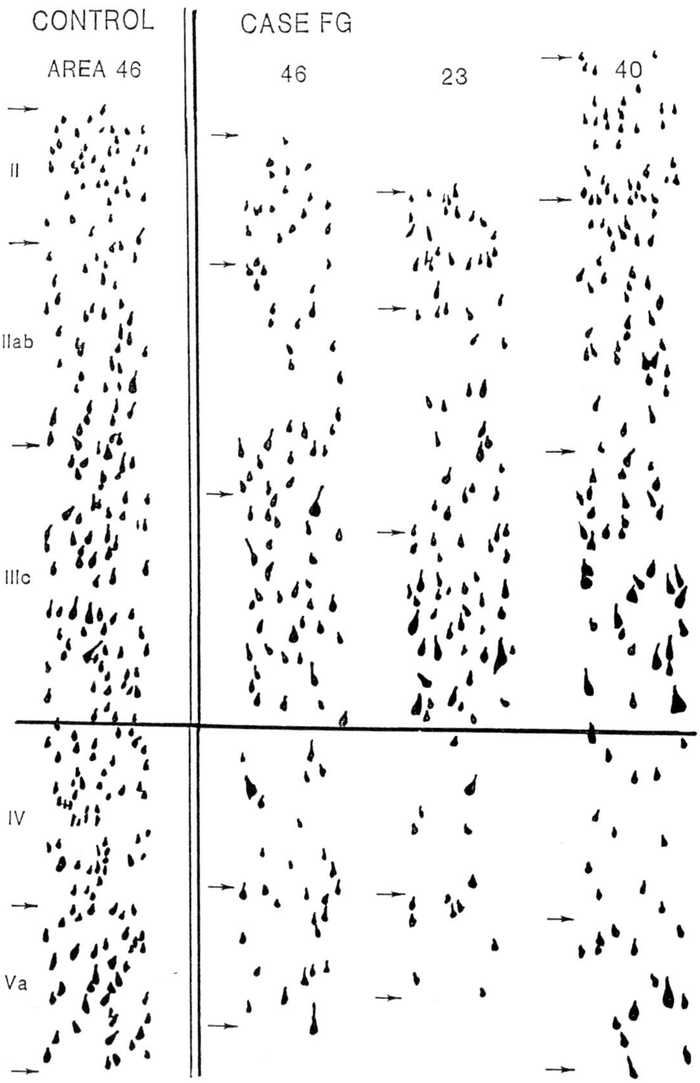


**Figure 6.** Drawings of neuronal perikarya through areas 46, 23, and 40 for case FG and a control are shown on the left to guide the eye in terms of neuron densities for each layer. Although these areas are atrophied and show substantial neurodegeneration, particularly in layers IIIab, IV, and V, parietal area 40 is less impacted, as expected from the focal atrophy shown in Figure 3.

generation in these regions is only partially accounted for by NFT density, there are some close matches between the deposition of Aβ42 and neurodegeneration that may lead to an understanding of the mechanisms of focal cortical neurodegeneration and subtypes of the disease.



### 7.2. Parietotemporal Atrophy and Case WJ

As discussed previously, the "typical" case of AD is characterized by bilateral hypometabolism of the medial temporal region, and posterior temporal and parietal cortices, with associated deficits in learning, memory, language abilities, and visuospatial skills, respectively. Some patients, however, may have deficits limited to word-finding and other language abilities and show relatively focal hypometabolism of the left temporal lobe, or to visuospatial abilities, with hypometabolism limited to parietal cortex. Case WJ had early language deficits and severe left temporal pole atrophy that later advanced to a visuospatial deficit and a swath of parietotemporal atrophy at postmortem. Although this case had more extensive cortical involvement than the earlier case, it provides insight into neuronal substrates of cortical function.

The first sign of a cognitive problem in case WJ was impairment in word finding. Magnetic resonance imaging subsequently showed evidence of profound temporal atrophy in the left hemisphere, and the most severe atrophy at postmortem was in the same region (Fig. 7). Because the language impairment began years before the memory and visuospatial impairments, the first sites of disrupted neuronal function may have been in superior temporal and temporal pole cortices rather than in hippocampal, entorhinal, and parietal cortices. Case WJ was a right-handed male with a Ph.D. in biochemistry. He was very athletic, enjoyed frequent outdoor trips, and did not use alcohol or smoke. When he was about 67 his wife noted that he had word-finding problems, substitution of the wrong words, and spelling and pronunciation difficulties. Her experiences in special education and her husband's high level of education made these changes more easily identified than might have been the case in other families. It is also interesting that his mother had possible AD, and WJ's wife reported that his dementia did not appear to follow a similar course to his mother's. Two years later he was taking a cross-country camping trip alone and got lost because he took wrong turns and could not read maps correctly. By age 74 he was speaking with pronouns only, no nouns, and he could not write. He would get lost going to a local gasoline station. A neurological assessment at that time was unremarkable beyond the dementing illness, and CT showed mild cortical atrophy in addition to severe atrophy in the left temporal pole. Within one year MR studies confirmed the earlier CT observations and it was noted that there were bilateral areas of white matter high signal intensities, particularly underlying parietal cortex. Profound left-hemisphere atrophy and bilateral leukoaraiosis can be seen in Figure 7. A 7-month treatment with Cognex did not alter the progression of WJ's language and visuospatial deficits. There was no evidence of behavioral impairments characteristic of case FG, nor oculomotor deficits encountered in posterior cortical atrophy in AD, as discussed below. To the time of his death at age 77, WJ had a relatively restricted cognitive deficit.

At postmortem the brain weighed 1320 gm and had severe left-hemisphere atrophy, particularly in parietotemporal cortex (Fig. 7). There was a relative preser-

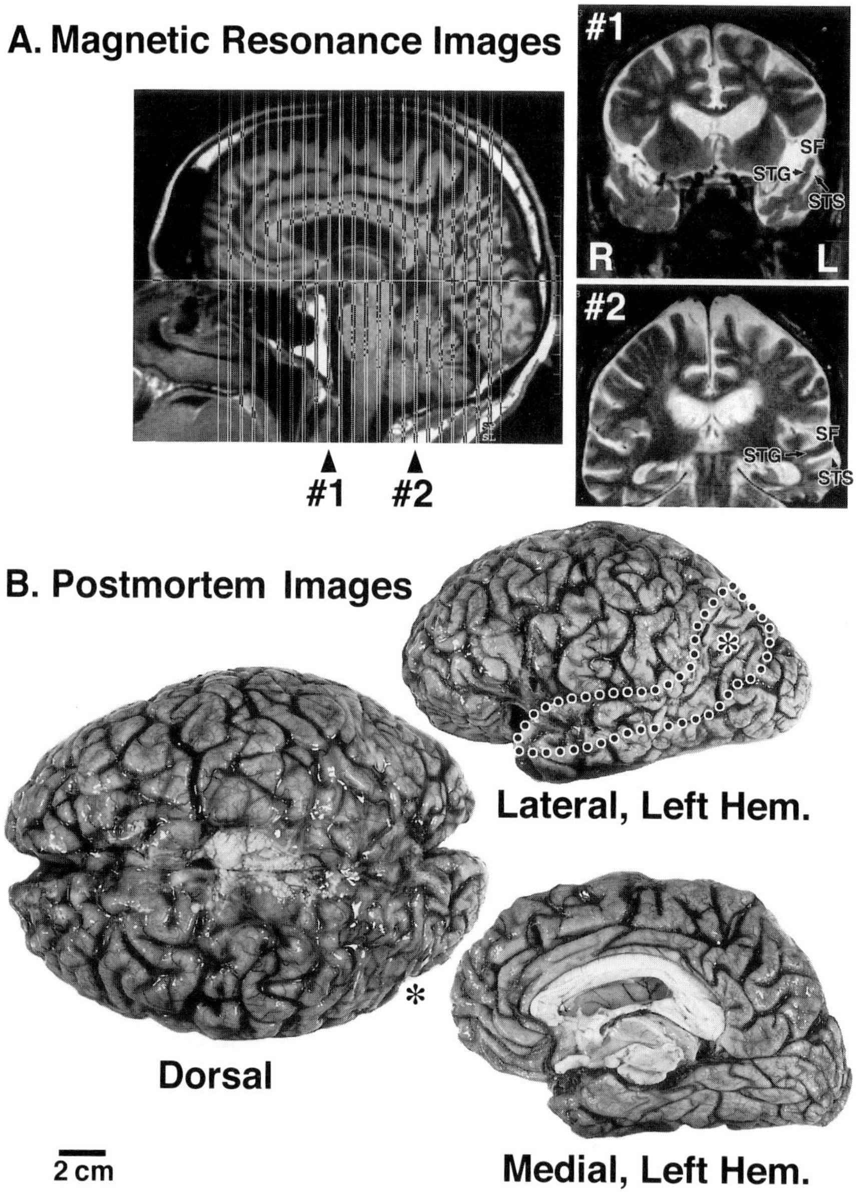


**Figure 7.** Case WJ is an example of focal parietotemporal atrophy in AD. (A) Magnetic resonance images taken 3 years before death. Section 1 shows the severe left-temporal pole atrophy, and section 2 shows the severe superior temporal gyrus atrophy. (B) The postmortem images confirm the region of focal atrophy (outlined with circles). The asterisk on the left lateral surface is on posterior parietal cortex. The dorsal view can be used to compare the left and right hemispheres. The medial hemisphere has significant atrophy as seen by widening of vertically oriented sulci along the superior surface, including the paracentral and parieto-occipital sulci and marginal ramus of the cingulate sulcus. The left medial surface was reoriented for ease of comparison of regions with atrophy.

vation of most of prefrontal cortex bilaterally except for area 10 of the frontal pole. Atrophy in the temporal pole was profound and localized to the left hemisphere (asterisk in Fig. 7). The medial surface had a moderate degree of atrophy, with widening of a number of the vertically oriented sulci including the marginal ramus of the cingulate sulcus. Histological analyses showed that the high-signal intensity in the white matter were areas of watery astrocytosis associated with severe axon degeneration. The density of thioflavine S-stained SP was typical for advanced stages of AD and relatively uniform throughout neocortex (Fig. 4). Subcortical structures appeared normal and the locus coeruleus and raphe nuclei had normal densities of neurons, although some reduction in pigmentation, and occasional NFT. Neocortical neurodegeneration was severe throughout the left hemisphere, including areas 10, 40, and 23. There was almost no laminar architecture, and neurons mainly in layer IIIc were preserved, and provided landmarks for assessing neuron densities.

This case documents an early language deficit in AD that is associated with profound and focal atrophy in left superior temporal and temporal pole cortices. Tyrrell *et al.* (1990) report significant glucose hypometabolism in left anterior temporal lobe associated with progressive aphasia in cases with focal cortical atrophy. Their study supports the view that rostral, left-temporal atrophy contributed to the language deficit in case WJ. The disease in WJ appears to have spread from this region into parietal cortex within two years of onset of language impairment, contributing the visuospatial deficit and producing an enlarged area of atrophy at the time of death. This case also shows involvement of PCC to a similar extent of lateral neocortical areas, and raises the question regarding the extent to which focal lobar damage and disruption of functional subsystems in AD represent etiologically unique subtypes.

### 7.3. Posterior Cortical Atrophy

Although AD patients who first experience visuospatial deficits tend to have an earlier onset of the disease and appear to be less frequent than the typical pattern of early memory impairment (Albert *et al.,* 1990), they are of particular interest because early visuospatial deficits are often associated with posterior cortical atrophy in AD (Benson *et al.,* 1988). This contrasts with the common pattern of AD pathology that tends to have less severely affected visual occipital areas in comparison to frontal, parietal, and temporal cortices (Arnold *et al.,* 1991). In addition, the significant involvement of PCC in these cases (Hof *et al.,* 1993a, 1997; Vogt *et al.,* 1997) raises the possibility that they also represent a unique subtype.

The visual impairment is not diffuse, but rather affects specific aspects of visual function, and it is important to discuss the selective involvement of particular components of the visual circuitry in the neurodegenerative process in AD (Hof *et al.,* this volume). Several clinical and neuropathological studies have demonstrated the occurrence of AD cases with atypical neuro-ophthalmological presentation (Cogan, 1985; Morel, 1945; Fletcher, 1994; Benson *et al.,* 1988). Most patients with posterior cortical atrophy initially display a complex visual syndrome characterized by paralysis of visual fixation associated with an optic ataxia and a disturbance of visual attention known as Bálint's syndrome (Bálint, 1909; Grünthal, 1928; Benson *et al.*, 1988; Fletcher, 1994). Deficits in motion perception and target tracing are usual, although not always consistent (Neary and Snowden, 1987; Fletcher and Sharpe, 1988; Kiyosawa *et al.*, 1989; Kurylo *et al.*, 1994; Butter *et al.*, 1996).

The posterior cortical atrophy in this syndrome is shown in Figure 8A. This distribution of gross atrophy has been demonstrated with CT and MRI, and at autopsy, and it predominates in the parietotemporo-occipital junction and in occipital and posterior cingulate cortices (Nissen *et al.*, 1985; Benson *et al.*, 1988; Berthier *et al.*, 1991; Hof and Bouras, 1991; Hof *et al.*, 1989, 1990a, 1993a; Mendez *et al.*, 1990a,b,c; Graff-Radford *et al.*, 1993; Levine *et al.*, 1993; Levy *et al.*, 1995; Pietrini *et al.*, 1996). The clinical symptomatology of AD patients with posterior cortical atrophy suggests involvement of cortical pathways linking the primary visual regions to the posterior parietal, cingulate, and visual association cortices in the early stages of the dementia (Hof and Bouras, 1991; Hof *et al.*, 1989, 1990a, 1993a, 1997, and this volume). There is a gradient of NFT densities from area V1 to the visual association regions in the parietal cortex, as well as in the PCC in these cases, all of whom suffered from visuomotor impairment resembling Bálint's syndrome. The same gradient is also apparent with Aβ immunohistochemistry.

In cases with visuomotor impairment, the number of NFT in layers II-III and V-VI is almost twice as high in area V2 as in V1, and further increases in areas V3/V4, 7b/7m, and 23. Also, prefrontal areas 9, 46, and 45 have far fewer NFT than the occipital (except V1), parietal, and posterior cingulate regions, indicating a shift of NFT distribution onto the occipital and parietal cortices. In these cases V1 and V2 contain comparable NFT densities, and in some cases up to three times as high NFT densities in layers II-III as in prefrontal cortex. The most striking differences in NFT densities are in areas V3/V4, MT, 7b/7m, and 23, where they can be as high as 8.5-fold in layers II-III and 5.4 fold in layers V-VI. Figure 8B shows the severe neurofibrillary changes in areas 29, 30, and 23a of PCC and the relatively smaller involvement of medial parietal area 7. In these cases, the inferior temporal cortex contains higher NFT counts than prefrontal areas, yet the densities in areas TE and TEO are lower than those in the posterior cingulate and inferior parietal cortices, particularly area MT. Finally, SP distribution also has a progressive density increase from area V1 to areas V2, V3/V4, MT, 7b/7m, and 23. Area V2 contains about 30% more SP than area V1. Layers I–III of areas MT and 23 show the highest SP densities.

These patterns of NFT may correspond to functional systems in human occipital cortex (Watson *et al.*, 1993; McCarthy *et al.*, 1995; Sereno *et al.*, 1995; Tootell *et al.*, 1995; DeYoe *et al.*, 1996). In particular, the region corresponding to area MT consistently contains more NFT than cortex of the occipital fields. The region reported by Orban *et al.* (1995) involved in detecting kinetic contours may correspond to the inferotemporal and medial parietal areas 7b/7m that are severely affected in posterior cortical atrophy. The consistent and severe involvement of PCC is particularly relevant in light of recent observations that this area is involved in visuospatial functions. Neurons in monkey area 23 respond to visual stimuli of large textured patterns and the orbital position of the eye (Olson *et al.*, 1993), while lesions in PCC impair spatial memory (Murray *et al.*, 1989). These data suggest that cortical networks linking the parietal, cingulate, and occipital regions are involved in AD and are selectively affected with posterior cortical atrophy (Hof *et al.*, 1993a, 1997).

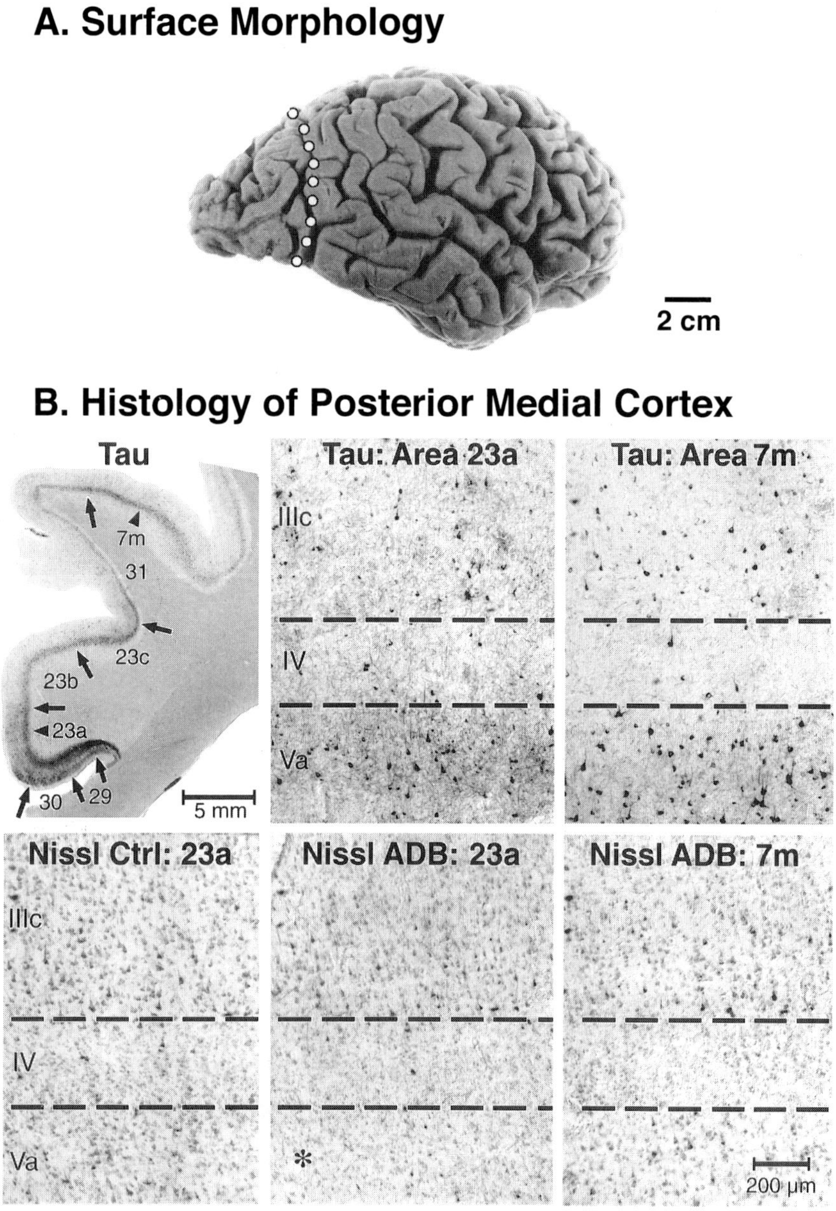


**Figure 8.** Two cases of AD with posterior cortical atrophy. (A) Surface morphology of a case reported by Hof *et al.* (1993a). Dots emphasize the profound atrophy in the occipitoparietal region in this case. (B) Histology from another case of posterior cortical atrophy (ADB) in tau-ir and Nissl-stained tissues. Tau-ir neuropil threads and neuritic plaques were more dense in areas 29 and 30 of posterior cingulate cortex than in medial parietal area 7m. The density of NFT in layer Va was approximately equal in areas 23a and 7m. Neuron degeneration, however, was significantly different in these areas, with area 23a having severe degeneration in layer Va (asterisk) that did not occur in area 7m (Ctrl = age-matched control case). Thus, the density of tau-ir NFT do not accurately reflect total neurodegeneration, and the AD pathogenic process could have started in PCC rather than parietal cortex because of the longer standing (more severe) neurodegeneration.

These atypical AD cases serve as excellent examples of clinicopathological correlations between SP and NFT distributions and clinical symptoms that are likely to involve disruption of subsets of corticocortical connections. They are consistent with the observation of Victoroff *et al.* (1994) that posterior cortical atrophy is clinically homogeneous but displays neuropathologic heterogeneity. The disturbances in motion perception and target tracing frequently observed in AD (Neary and Snowden, 1987; Sadun *et al.*, 1987; Fletcher and Sharpe, 1988; Katz and Rimmer, 1989; Kiyosawa *et al.*, 1989; Butter *et al.*, 1996) are more severe in posterior cortical atrophy and may be related to the very high densities of lesions in areas MT, 7b/7m, and 23. Although additional studies of such cases using functional imaging will provide key information that is missing between the neurologic symptoms and the affected cortical circuits, current observations suggest close relationships between functional subsystems, connectivity patterns, and pathogenic mechanisms. The association of impaired cortical subsystems, gradients in NFT and Aβ peptides, and clinical outcomes suggests that posterior cortical atrophy may be a subtype of AD.

## 8. Layer-Selective Neurodegeneration in Posterior Cingulate Cortex

Cases of AD with multifocal lobar degeneration suggest unique mechanisms of neurodegeneration to the extent that specific subsystems are involved and that each subsystem requires unique mechanisms to produce isolated patterns of cell death. Another conclusion from this survey is that PCC is profoundly impacted in the multifocal disease process(es). Neurofibrillary degeneration is often greater in PCC than it is in lateral neocortical areas, i.e., those areas that are usually invoked as a rationale for impaired functions. Glucose hypometabolism in very early AD is greater in PCC than in lateral neocortex and the parahippocampal gyrus (Minoshima *et al.*, 1997), and it may be related to disorientation in time and place (Hirono *et al.*, 1998). Thus, an early and profound impact of AD in PCC is suggested by multifocal cortical atrophies and glucose metabolism. Furthermore, PCC is engaged in learning and memory (Valenstein *et al.*, 1987) and in visuospatial functions (Murray *et al.*, 1989; Olson *et al.*, 1993). Hence, functional impairments often attributed to parietal and medial temporal cortices may commonly result from very early disruption of PCC.

Another approach to neuropathological heterogeneity, multifocal degeneration, and possible subtypes within AD is assessment of a consistently sampled cortical area in a large group of cases. Such an analysis does not begin with parahippocampal areas because this region appears to undergo a stereotyped pattern of neurodegeneration that is closely linked to NFT formation. In contrast, although the posterior cingulate region has been implicated in many studies of AD, the range of variability in each data set, including neurochemical, ligand binding, NFT, and neuron density, is extraordinary. Does such variability simply reflect a common theme, such as large neuron degeneration, or does it represent qualitatively different processes, such as large, medium, and/or small neuron degeneration early in the disease according to unique subtype mechanisms? There are five reasons for assessing PCC in this context and for approaching AD from a cingulocentric perspective. First, PCC has a well-documented involvement in AD, including neurodegeneration (Brun and Englund, 1981; Mountjoy *et al.*, 1983; Vogt, *et al.*, 1990), NFT and SP formation (Mountjoy *et al.*, 1983; Vogt *et al.*, 1990; Braak and Braak, 1993), and has significant reductions in choline acetyltransferase activity (Rossor *et al.*, 1982; Procter *et al.*, 1988). Second, postmortem samples can be more consistently sampled from PCC than from any other neocortical region, because the splenium of the corpus callosum provides a uniform landmark, standardizing tissue collection from multiple institutions. Third, there is a full range of cytoarchitectural differentiation from periallocortical areas 26–30 to neocortical area 23. These areas are differentially impacted in AD and provide insight into disease progression (Vogt *et al.*, 1990; Braak and Braak, 1993). Fourth, there are strong connections between PCC and mid-prefrontal, posterior parietal, superior temporal, and parahippocampal cortices. This ensures that cingulate cortex is linked directly to those regions that have most frequently been implicated in the disease. Fifth, PET studies show that PCC may be one of the earliest cortical regions to be involved in AD, including individuals with the earliest memory-loss symptoms (Minoshima *et al.*, 1997).

Neurodegeneration is a better measure of cognitive decline than SP or NFT densities (Neary *et al.*, 1986; Mann *et al.*, 1988; Bennett *et al.*, 1993), and the highest degree of variability in neuron degeneration is in PCC (0–80% losses; Brun and Englund, 1981). Although this latter study interpreted changes in terms of a graded disease progression in a single-etiology model, a subsequent study of PCC failed to support this view (Vogt *et al.*, 1990). Neurodegeneration does not follow a single or bilaminar pattern, and NFT and SP are a poor reflection of total neurodegeneration. Rather, there are qualitatively different laminar patterns in neurodegeneration in PCC and they may represent subtypes.

A consideration of the subtype hypothesis employed principal components analysis (PCA) and pattern-recognition methods to assess the extent to which laminar patterns of neurodegeneration represent statistically unique subgroups (Vogt *et al.*, 1998). Five subgroups were identified based on neurons in each layer in 160-μm-wide traverses of area 23a; neurons in each layer were not corrected for atrophy for reasons discussed in the article cited. Because each subgroup represented a different pattern of neuron loss that may require different pathogenic yet unknown events, they were referred to as subtypes (ST) within AD as follows: STO, no neurodegeneration; STIIIab, losses mainly in layer IIIab; STIV-V, losses mainly in layers IV-V; STII-V, losses of about 60% in most layers; STSevere, losses of more than 80% in most layers with almost no evidence of laminar architecture. The latter two subtypes are not different only in terms of total neurodegeneration, because they differ in age of disease onset (STSevere = 63 ± 3.4; STII-V = 74 ± 1.4; mean ± SEM), proportion of ApoE ε4 homozygotes (STSevere = 42%; STII-V = 20%), SP (STSevere = 40 ± 3.9/1.9 mm$^2$; STII-V = 27 ± 1.8), and NFT (STSevere = 47 ± 9/mm$^2$; STII-V = 20 ± 4.5). A subset of cases from the above study (i.e., STIIIab, STIV-V, and STII-V) are used here to demonstrate unique laminar patterns of neurodegeneration, assess the hypothesis that they represent independent subtypes, and evaluate the ApoE risk factor in relation to neocortical neuron losses and the rationale for subtypes rather than subgroups.

The PCA was used to determine a principal components score, or eigenvector, for cases using neuron counts in layers IIIab, IIIc, IV, and Va. A three-dimensional plot, or eigenvector projection, of these values in Figure 9 shows that three subgroups are formed as predicted from previous studies. Neurons for layers II, Vb, and VI have been explored to assess their contributions to improving subgrouping of cases, however, they did not improve the analysis, and layer VI has a wide range of neuron densities because it is near the apex of the ventral bank of the cingulate gyrus (i.e., densities are overwhelmingly affected by curvature). The individuals in Figure 9 are coded according to our previous *a priori* assessment of neurodegeneration by determining layers with greatest proportionate loss of neurons in comparison to controls. The fact that there is no overlap among the members of each subgroup suggests that the multilayer model of neurons defines three subgroups. To the extent that each laminar pattern reflects unique mechanisms of cell death, this model defines subtypes within AD.

Subtypes defined by neurons in layers III–Va are independent entities and do not merge with long durations of the disease. If this did occur, of course, there would be substantial overlap of cases, particularly at the perimeter of each subgroup in the eigenvector projections, but this is not so. Direct demonstration of subtype independence is provided in Figure 10, which shows perikaryal drawings for individual cases with different disease durations and from each subtype. Differences among the three subtypes can be visualized in this figure, and it can be seen that the loss of neurons in any layer is essentially the same for short- and long-duration cases, i.e., 2–3 and 10 years, respectively. Finally, the range of disease durations is similar (STIIIab, 3–15; STIV-V, 2–10; STII-V, 1–14 years), the age at disease onset is not significantly different (STIIIab, 69 ± 4; STIV-V, 75 ± 2; STII-V, 74 ± 2 years), and SP densities, verifying their inclusion under the diagnostic AD umbrella (STIIIab, 32 ± 2; STIV-V, 22 ± 5; STII-V, 27 ± 2).

Because PCC is significantly impacted in AD, it must be considered to what extent these impairments are reflected throughout neocortex. Perikaryal strips were analyzed through the right hemisphere of case WJ (presented in detail earlier). This hemisphere had no focal atrophy, and Figure 11 shows samples from a number of neocortical areas. Although neuron losses occur in each layer, they are most profound in layers IV and Va, and there is a more limited loss in layer IIIab. This same distribution of neurodegeneration occurs throughout neocortex in this

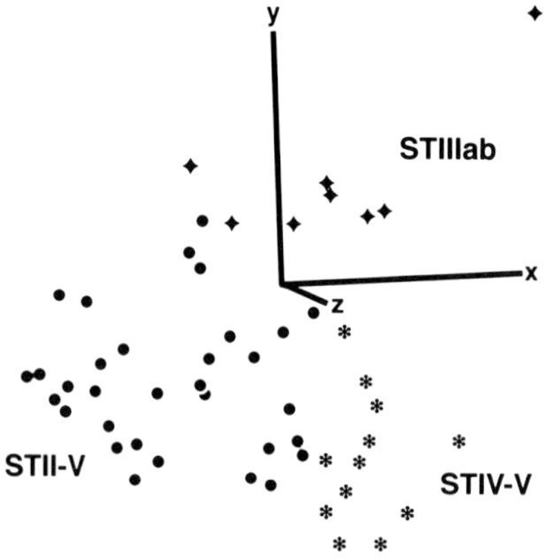


**Figure 9.** Neurons in layers IIIab, IIIc, IV, and Va were included in a PCA for this eigenvector projection. Eigenvalues for each of the three AD subgroups segregate into different sectors of the graph, suggesting that a model including these layers can define statistically unique subgroups.

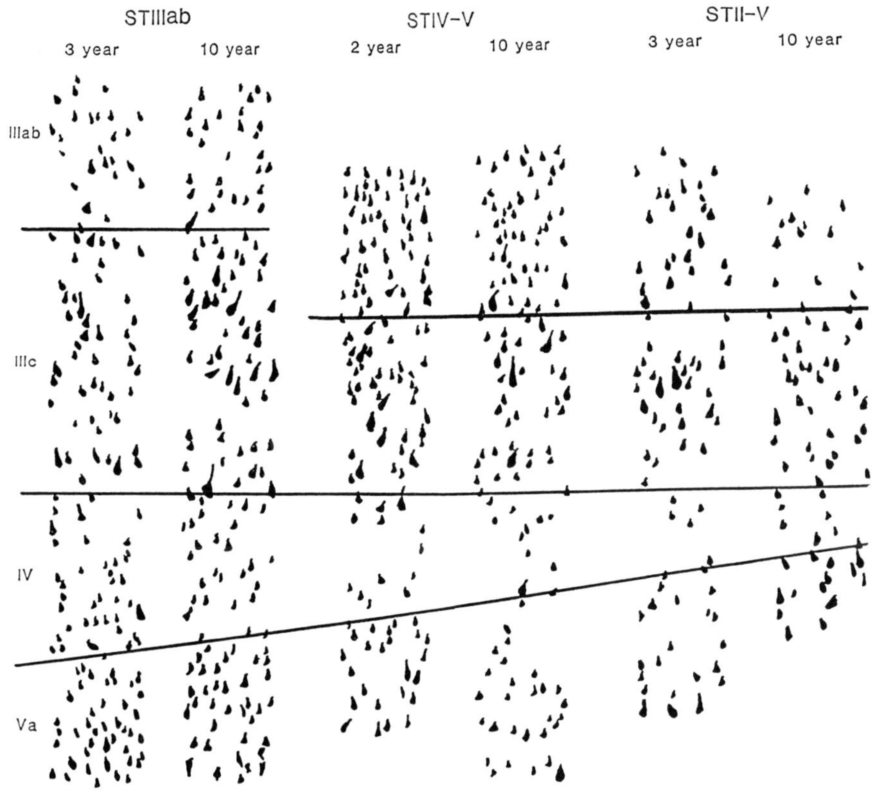


**Figure 10.** Neurodegeneration in area 23a of AD is not uniform, although it is relatively stable over the course of the clinical expression of the disease as shown by perikaryal strips from six cases. In STIIIab the greatest neurodegeneration is in layer IIIab for 3- and 10-year durations. In STIV-V, greatest degeneration is in the deep cortical layers, whereas in STII-V, degeneration is significant throughout the cortex, with no laminar preference. There is no means of accounting for these patterns and progressions with a single-disease state that has a single progression.

hemisphere, verifying that disruptions in PCC can reflect alterations throughout the neocortex.

Because homozygous expression of the ApoE $\epsilon4$ allele is associated with elevated densities of SP (Schmechel *et al.,* 1993; Sparks *et al.,* 1996) and severe neurodegeneration in PCC (Vogt *et al.,* 1998), it might be considered whether a direct link can be established between ApoE genotype and neurodegeneration. Subtypes based on laminar patterns of PCC neurodegeneration are ideal for direct tests of the relevance of the $\epsilon4$ allele to neocortical neuron losses. First, although there are significant differences in the laminar distribution of neurodegeneration in the three subtypes discussed above, the proportion of cases homozygous for the $\epsilon4$ allele is about the same: STIIIab, 29%; STIV-V, 15%; STII-V, 20%. Second, there were enough $\epsilon3$ (n = 9) and $\epsilon4$ (n = 6) homozygotes in STII-V to compare total neurons in layers IIIab–Va statistically. The differences between these two groups ($\epsilon4/4 = 97 \pm 9$; $\epsilon3/3 = 117 \pm 7$) were not significant. Third, coding the cases in these three groups according to ApoE genotype (3/3, 3/4, 4/4) and visual inspection of the eigenvector projection for neurons in layers III–Va for each shows that there is no subgrouping of cases according to ApoE genotype. Finally, an ApoE risk factor ($\epsilon2/\epsilon3 = 2$, $\epsilon3/\epsilon3 = 3$, $\epsilon2/\epsilon4 = 4$, $\epsilon3/\epsilon4 = 5$, $\epsilon4/\epsilon4 = 6$) in the neuron model does not identify new subgroups, nor does it alter subgroups defined with neurons alone. Thus, ApoE genotype is not related to laminar specificities in neurodegeneration. Observations by Blacker *et al.* (1997) that the homozygous ε4 risk is expressed primarily at ages 60–65 and lack of a role for heterozygous expression of the ε4 allele suggest a restricted risk. Although there is an association between STSevere, with its early onset, high density of SP, and high proportion of ε4 homozygotes, the ApoE ε4 risk is not relevant to other subtypes with late-onset disease.

In conclusion, each subtype of AD based on a unique laminar pattern of neurodegeneration progresses independently of the others. The different laminar patterns of neuron losses may require unique mechanisms for neurodegeneration. It is unlikely that any single protein, transmitter, or large-neuron model of AD etiopathology will account for the full breadth of AD heterogeneity.

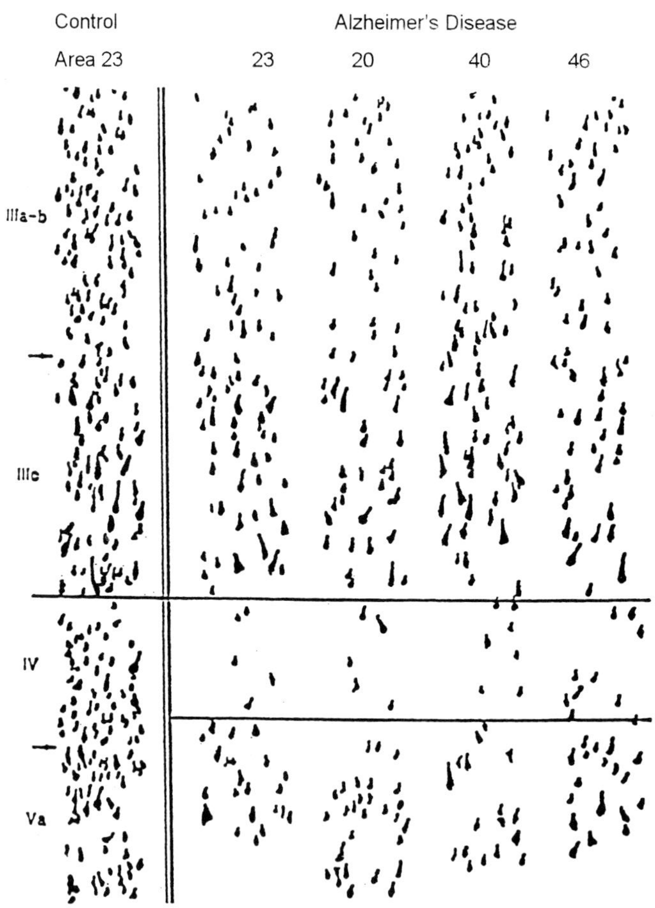


**Figure 11.** Perikaryal strips were drawn from the least atrophied right hemisphere of case WJ. They show that alterations in PCC area 23 can serve as a reference to a widespread neurodegeneration. Thus area 23a shows loss of neurons mainly in layers IV and V, and a more restricted loss in layer IIIab. This same pattern is reflected throughout lateral neocortical areas.

Although transmitter systems provide a direct and logical means of modifying AD symptomatology, there has been little success with this approach. Among the reasons for failure are that such interventions do not alter the course of neurodegenerative events, treatments are often started too late in the dementing process, and there may be substantial differences in the responses of transmitter systems to multiple etiologies. If the latter scenario proves correct, determination of disease subtype may be as important to therapeutic intervention as is assessing the stage of disease progression. In this context the issue is not how much a single system is impacted, but rather to what extent multiple systems are involved and the underlying cortical pathology associated with each. This is an ideal problem for multivariate analysis.

Neurodegenerative events in brainstem transmitter systems have been demonstrated, and in some instances they are associated with behavioral impairments. These observations provide evidence for subtypes and suggest new perspectives for defining subtypes based on cortical reorganization. Neurons in the locus coeruleus degenerate in AD, and there is a commensurate loss of neocortical norepinephrine (Mann *et al.*, 1980; Cross *et al.*, 1981; Rossor *et al.*, 1984). Bondareff *et al.* (1987) reassessed losses of norepinephrinergic function in AD and proposed that there may be two subtypes based on neurodegeneration in the locus coeruleus and reductions of neocortical norepinephrine and choline acetyltransferase activity. The most severe reductions in the latter compounds were associated with the most profound losses of neurons in the locus coeruleus, high densities of cortical SP and NFT, and early age at disease onset. In contrast, late age at disease onset was associated with less severe loss of neurons in the locus coeruleus and higher levels of cortical norepinephrine. Other studies emphasize the extensive losses of cortical choline acetyltransferase activity (Rossor *et al.*, 1982), and it has been suggested that early-onset cases have greater losses than do late-onset cases (Bird *et al.*, 1983). Finally, Procter *et al.* (1995) report that cases with significant reductions of paroxetine binding to the serotonin transporter are associated with prominent behavioral symptoms, including depression.

Although there is little doubt that the impact of AD on noradrenergic, cholinergic, and serotonergic systems is heterogeneous, the extent to which these differences reflect different mechanisms is controversial. Because neurochemically unique inputs to neocortex are altered in a bimodal fashion and these changes have an impact on cortical transmitter receptor systems, it can be postulated that subgroups based on subcortical transmitter system alterations may provide important clues for different mechanisms of the disease. As such, neurotransmitter system alterations are important for understanding neocortical reorganization that should be part of characterizing AD subtypes. Studies of AD neocortex have identified alterations in most transmitter systems, including changes in receptor density, G-protein coupling, and enzyme stimulation. Since AD has multisystem defects, an approach is needed that quantitatively characterizes more than one system. Furthermore, because neurons often degenerate according to a laminar instead of a diffuse pattern, transmitter receptors are best analyzed autoradiographically in sections of neocortex. In this manner an exact relationship can be drawn between layers with and without neurodegeneration and changes in ligand binding in the same layers.

The same multivariate statistics and pattern-recognition approach used above to assess laminar patterns of neurodegeneration can be applied to transmitter systems. Analysis of laminar patterns in ligand binding in layers III–Va can be used to study direct corollaries of neurodegeneration. Here we reassess earlier findings and use them to evaluate subtypes of AD with this new approach. The receptor binding protocols of most interest in this context are those for postsynaptic receptors expressed by neocortical neurons, including β-adrenoceptors, $GABA_A$ receptors (Vogt *et al.,* 1991), and muscarinic receptors (Vogt *et al.,* 1992). The rationale for studying each system can be briefly summarized. Kalaria *et al.* (1989) reported that there was a significant elevation in β-adrenoceptors in prefrontal cortex and suggested that this is a consequence of dennervation supersensitivity. Although we were able to confirm that β-adrenoceptor binding increased in PCC in AD, it did not occur in all cases, and when laminar patterns of neurodegeneration were considered, it was remarkably selective for subgroups of cases. Interestingly, there was no laminar specificity in the elevation, i.e., it occurred significantly in all layers when there was an increase. This was not the case for $GABA_A$ binding. Levels of GABA and $GABA_A$ binding are reduced in AD (Rossor *et al.,* 1984; Reisine *et al.,* 1978; Chu *et al.,* 1987). Our studies showed a reduction in muscimol binding to $GABA_A$ receptors, a loss that was layer-specific, and one that occurred only in those layers that had the greatest loss of neurons.

Studies of muscarinic receptors in AD neocortex have produced a wide range of findings because binding is assessed outside the context of neurodegeneration. Most studies incorrectly presume that neurodegeneration is uniform throughout neocortex and that a reduction in binding is consistent with cholinergic denervation. Because changes in $M_2$ binding (i.e., $^3$H-oxotremorine-M in the presence of 50–100 nM pirenzepine; OXO-M/PZ) in AD are linearly and inversely related to neurons in layer V, alterations in binding are likely due to cortical neurons and not cholinergic afferents (Vogt *et al.,* 1992).

Recent studies of laminar patterns in neurodegeneration in PCC in a large sample of cases (Vogt *et al.,* 1998) show that there are a few cases with no neurodegeneration (subtype 0, or ST0) and cases with neurodegeneration mainly in layer IIIa-b (STIIIab), and show that earlier cases with selective losses of neurons in layers IV or V-VI statistically merge to form STIV-V. The autoradiography data previously published for each ligand has been recalculated to accommodate these laminar patterns. No ligand binding data are available for STII-V or STSevere. Ligand binding (muscimol, OXO-M/PZ, cyanopindolol) for 8 control and 12 AD cases was segregated by subtype. Although binding in all AD cases has some statistically significant differences from control values, there is at least one subgroup for each ligand that has either no or little change in binding as compared with controls, and there is at least one subgroup that has very significant differences from the control values. Alterations in binding associated with subtypes provide independent validation for subtypes according to laminar patterns of neurodegeneration. In addition, reductions in $GABA_A$ binding were associated primarily with the layers of greatest neurodegeneration, elevations in β-adrenoceptor binding in all cases were due entirely to binding in the STIV-V, and elevations in OXO-M/PZ binding were greatest in STIIIab and STIV-V.

Because changes in no single transmitter system can be used to characterize the cortical subtypes, alterations in many systems can be assessed with PCA in a multivariate model. Subtypes based on neurodegeneration use neurons in layers IIIab, IIIc, IV, and Va; these layers are included here in a multivariate model of ligand binding. Because the major differences in muscimol and OXO-M/PZ binding between STIIIab and STIV-V lie in layer IIIc, binding in this layer plays a particularly important role in discriminating among subtypes. Figure 12A shows three-dimensional plots of the eigenvectors for binding of these three ligands in layers III–Va. The control and STO values cosegregate because they are not distinguished by neuron densities or ligand binding in the midcortical layers, and there are too few cases in STO to argue for statistical significance. The control and STO cases are separated from STIIIab and STIV-V; however, there is overlap between cases in STIIIab and STIV-V, indicated in Fig. 12A with arrows. Because the differences between STIIIab and STIV-V are not great in layer V in group statistics, this layer was removed from the analysis and the eigenvector projection shown in Figure 12B. This change provides for dissociation of the cases in these two classes. Finally, because the differences between these subgroups in cyanopindolol binding was not layer-specific, this ligand was removed entirely from the analysis and the results shown in Figure 12C. The STIIIab and STIV-V subtypes are maximally dissociated under these conditions, and even STO shows some dissociation from control values, although a frank separation of these groups does not occur with three cases.

A model that incorporates ligand binding for postsynaptic receptors in the midcortical layers uncovers three AD subtypes that are defined with an independent sample of AD cases using laminar patterns of neuron loss. The multivariate

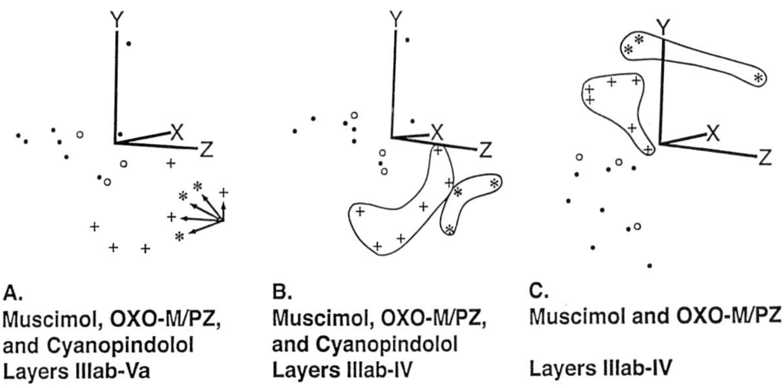


**Figure 12.** Exploring subgroups of AD cases with PCA and eigenvector projections. Receptor binding was entered for each case. (A) Specific binding for three ligands in layers IIIab, IIIc, IV, and Va. Although cases for STO are expected to intermingle with control values because they have the same neuron densities in each layer, it was surprising that some overlap occurred between cases in STIIIab and STIV-V. Removal of binding for each ligand in layer Va resulted in a segregation of cases for these latter two subtypes. Because elevations in cyanopindolol binding occurred in all layers, this variable was removed from the analysis of specific layers to see if there was any improvement in the subtype segregation. (C) The receptor binding that best describes these three AD subtypes is that of muscimol and OXO-M/PZ in layers IIIab, IIIc, and IV. (● = control; ○ = STO; * = STIIIab; + = STIV–V).

model demonstrates which layers and ligands provide the greatest definition of subtypes based on ligand binding in multiple transmitter systems. Two issues are raised by these observations. First, Bondareff *et al.* (1987) suggested that one group of AD cases had a substantial reduction in neurons in the locus coeruleus, losses of norepinephrine and choline acetyltransferase activities. We propose that this may be true for STIV-V. It may also be true for the subtypes with most severe neurodegeneration, because these have the earliest age of disease onset (Vogt *et al.*, 1998). Second, because layer IIIc was particularly useful in discriminating between STIIIab and STIV-V, this layer should be considered in more detail in terms of its contribution to disease progression as well as to neuropathological subtypes.

## 10. Models of Multifocal Cortical Neurodegeneration and the Subtypes Hypothesis

Alzheimer's disease is not a global cortical disease, although atrophy in its late stages can attain a profound level. To say that it is global implies that all areas are equally involved, with a similar pattern of large neuron losses. Corollaries of the global viewpoint emphasize changes that occur equally in all cortical areas, such as reductions in choline acetyltransferase activity for the cholinergic hypothesis, and elevations in SP according to the ApoE genotype hypothesis. These observations do not account for multifocal cortical neurodegeneration, however, and they fail to explain in a significant way the primary etiologies of cortical cell death.

Assessing each cortical area in terms of a multifocal model suggests that different areas undergo different structural and temporal patterns of neuron loss, Aβ and NFT deposition, and glucose hypometabolism, such as those in entorhinal and posterior cingulate cortices. What appears to be a single swath of neurodegeneration on the medial surface extending from entorhinal to cingulate cortex in fact represents at least two independent foci of cell death. Entorhinal cortex has early NFT formation, particularly in layers II and V, and no glucose hypometabolism, while PCC has variable NFT formation, early glucose hypometabolism, and late-stage formation of NFT in layers II-IIIab. Most important, entorhinal cortex is part of medial temporal areas that undergo a consistent pattern of neuron loss and mark the stages of disease progression. In contrast, PCC has an extremely varied involvement with multiple laminar patterns of neurodegeneration that are not linked directly to disease progression. Although the global model and the null hypothesis that all differences in AD cortex can be accounted for by disease progression is popular, the multifocal model and subtypes hypothesis are gaining support.

To demonstrate subtypes there must be qualitative rather than quantitative differences among changes in brain structure and function, and unique etiologies of cell death must be associated with each subtype. What is the evidence for subtypes of AD? First, early observations show that the APP and PS mutations are associated with different clinical and neuropathological endpoints; and most important, spastic paraparesis. Second, neurodegeneration does not proceed in a single and uniform manner in neocortex, but rather according to different laminar patterns. Third, disruption of different functional subsystems in neocortex may require different mechanisms to trigger neurodegeneration such as that in frontotemporal or posterior cortical atrophies. Fourth, age at disease onset stands as an important divide where the rate of progression, clinical impairments, and extent of neurodegeneration differ qualitatively and may not be explained by a single etiology and progression.

The distance between genetic mutations and clinical and neuropathological endpoints is great and it can only be bridged by understanding a complicated series of molecular events that lead from impaired genomic DNA and environmental risk factors to multifocal patterns of neurodegeneration in the final Alzheimer's syndrome. Although all of the facts have yet to be provided and a full accounting is not complete, there is building evidence that *a priori* hypotheses for AD research should begin with multifocal and subtypes hypotheses and appropriate statistical strategies. This review proposes an incremental approach to the subtypes problem, in which AD is characterized by genetic, neurochemical, and multifocal neurodegeneration in multivariate models. Eventually, such models will be constrained by clinical observations, and the subtypes of Alzheimer's disease will be understood in their entirety.

ACKNOWLEDGMENTS. We thank the individuals and institutions that provided case materials for these studies, including Drs. Francine M. Benes and Stephen L. Vincent at the Harvard Brain Tissue Resource Center, McLean Hospital, Belmont, Massachusetts (PHS grant #MH/NS31862); Dr. Gary W. Van Hoesen and Loren Spence at the Department of Anatomy, University of Iowa, Iowa City, Iowa; Dr. Constantin Bouras at the Department of Psychiatry, University of Geneva, Geneva, Switzerland; and Drs. Venkat R. Challa and Christopher C. Colenda at Wake Forest University School of Medicine, Winston-Salem, North Carolina. This research was supported by PHS grants AG05138, DA06634, the Brookdale Foundation, and the Wake Forest University School of Medicine DNA/Brain Resource Center. Finally, we thank Dr. Michael W. Miller for providing monkey PCC immunoreacted with an antibody to brain-derived neurotrophic factor, and Linda Arcure for her substantial assistance in preparing the illustrations.

## 11. References

Albert, M. S., Duffy, F. H., and McAnulty, G. B., 1990, Electrophysiologic comparisons between two groups of patients with Alzheimer's disease, *Arch. Neurol.* **47:**857–863.

Armstrong, R. A., and Myers, D., 1992, Principal components analysis of Alzheimer's disease based on neuropathological data: A study of 79 patients, *Neurosci. Res. Comm.* **11:**1–9.

Armstrong, R. A., and Wood, L., 1994, The identification of pathological subtypes of Alzheimer's disease using cluster analysis, *Acta Neuropathol.* **88:**60–66.

Armstrong, R. A., Wood, L., Myers, D., and Smith, C. U. M., 1996, The use of multivariate methods in the identification of subtypes of Alzheimer's disease: A comparison of principal components and cluster analysis, *Dementia* **7:**215–220.

Arnold, S. E., Hyman, B. T., Flory, J., Damasio, A. R., and Van Hoesen, G. W., 1991, The topographical distribution of neurofibrillary tangles and neuritic plaques in the cerebral cortex of patients with Alzheimer's disease, *Cereb. Cortex* **1:**103–116.

Arriagada, P. V., Growdon, J. H., Hedley-Whyte, E. T., and Hyman, B. T., 1992, Neurofibrillary tangles, but not senile plaques, parallel duration and severity of Alzheimer's disease, *Neurology* **42:**631–639.

Baleydier, C., and Mauguière, F., 1980, The duality of the cingulate gyrus in monkey: Neuroanatomical study and functional hypotheses, *Brain* **103:**525–554.

Bálint, R., 1909, Seelenlähmung des "Schauens," optische Ataxie, räumliche Störung der Aufmerksamkeit, *Mschr. Psychiatr. Neurol.* **25:**51–81.

Becker, J. T., Huff, F. J., Nebes, R. D., Holland, A., and Boller, F., 1988, Neuropsychological function in Alzheimer's disease: Pattern of impairment and rates of progression, *Arch. Neurol.* **45:**263–268.

Behl, C., Davis, J. B., Lesley, R., and Schubert, D., 1994, Hydrogen peroxide mediates amyloid β protein toxicity, *Cell* **77:**817–827.

Bennett, D. A., Cochran, E. J., Saper, C. B., Leverenz, J. B., Gilley, D. W., and Wilson, R. S., 1993, Pathological changes in frontal cortex from biopsy to autopsy in Alzheimer's disease, *Neurobiol. Aging* **14:**589–596.

Benson, D. F., Davis, R. J., and Snyder, D. B., 1988, Posterior cortical atrophy, *Arch. Neurol.* **45:**789–793.

Bergmann, M., Kuchelmeister, K., Schmid, K. W., Kretzschmar, H. A., and Schröder, R., 1996, Different variants of frontotemporal dementia: A neuropathological and immunohistochemical study, *Acta Neuropathol.* **92:**170–179.

Berthier, M. L., Leiguarda, R., Starkstein, S. E., Sevlever, G., and Taratuto, A. L., 1991, Alzheimer's disease in a patient with posterior cortical atrophy, *J. Neurol. Neurosurg. Psychiatry* **54:**1110–1111.

Bierer, L. M., Hof, P. R., Purohit, D. P., Carlin, L., Schmeidler, J., Davis, K. L., Perl, D. P., 1995, Neocortical neurofibrillary tangles correlate with dementia severity in Alzheimer's disease. *Arch. Neurol.* **52:**81–88.

Bird, T. D., Stranaham, S., Sumi, S. M., and Radkind, M., 1983, Alzheimer's disease: Choline acetyltransferase activity in brain tissue from clinical and pathological subtypes, *Ann. Neurol.* **14:**284–293.

Blacker, D., Haines, J. L., Rodes, L., Terwedow, R. H., Harrell, L. E., Perry, R. T., Bassett, S. S., Meyers, D., Albert, M. S., and Tanzi, R., 1997, ApoE-4 and age at onset of Alzheimer's disease: The NIMH genetics initiative, *Neurology* **48:**139–147.

Blass, J. P., Baker, A. C., Ko, L.-W., and Black, R. S., 1990, Induction of Alzheimer antigens by an uncoupler of oxidative phosphorylation, *Arch. Neurol.* **47:**864–869.

Bondareff, W., Mountjoy, C. Q., and Roth, M., 1982, Loss of neurons of origin of the adrenergic projection to cerebral cortex (nucleus locus ceruleus) in senile dementia, *Neurology* **32:**164–168.

Bondareff, W., Mountjoy, C. Q., Roth, M., Rossor, M. N., Iversen, L. L., and Reynolds, G. P., 1987, Age and histopathologic heterogeneity in Alzheimer's disease, *Arch. Gen. Psychiatry* **44:**412–417.

Borchelt, D. R., Thinakaran, G., Eckman, C. B., Lee, M. K., Ratovitsky, T., Yager, D., Slunt, H. H., Wang, R., Seeger, M., Levey, A. I., Gandy, S. E., Copeland, N. G., Jenkins, N. A., Price, D. L., Younkin, S. G., and Sisodia, S. S., 1996, Familial Alzheimer's disease-linked presenilin 1 variants elevate Aβ1-42/1-40 ratio *in vitro* and *in vivo, Neuron* **17:**1005–1013.

Bozner, P., Grishko, V., LeDoux, S. P., Wilson, G. L., Chyan, Y.-C., and Pappolla, M. A., 1997, The amyloid β protein induces oxidative damage of mitochondrial DNA, *J. Neuropathol. Exp. Neurol.* **56:**1356–1362.

Braak, H., and Braak, E., 1991, Neuropathological stageing of Alzheimer-related changes, *Acta Neuropathol.* **82:**239–259.

Braak, H., and Braak, E., 1993, Alzheimer neuropathology and limbic circuits, in: *Neurobiology of Cingulate Cortex and Limbic Thalamus* (B. A. Vogt and M. Gabriel, eds.), Birkhäuser, Boston, pp. 606–626.

Braak, H., Braak, E., Yilazer, de Vos, R. A. I., Jansen, E. N. H., and Bohl, J., 1996, Pattern of brain destruction in Parkinson's and Alzheimer's diseases, *J. Neural. Transm.* **103:**455–490.

Brion, S., Plas, J., and Jeanneau, A., 1991, La maladie de Pick: Point de vue anatomo-clinique. *Rev. Neurol.* **147:**693–704.

Brun, A., 1993, Frontal lobe degeneration of non-Alzheimer type revisited, *Dementia* **4:**126–131.

Brun, A., and Englund, E., 1981, Regional pattern of degeneration in Alzheimer's disease: Neuronal loss and histopathological grading, *Histopathology* **5:**549–564.

Brun, A., Englund, B., Gustafson, L., Passant, U., Mann, D. M. A., Neary, D., and Snowden, J. S., 1994, Consensus statement: Clinical and neuropathological criteria for frontotemporal dementia, *J. Neurol. Neurosurg. Psychiatry* **57:**416–418.

Butter, C. M., Trobe, J. D., Foster, N. L., and Berent, S., 1996, Visual-spatial deficits explain visual symptoms in Alzheimer's disease, *Am. J. Ophthalmol.* **122:**97–105.

Cairns, N. J., Chadwick, A., Lantos, P. L., Levy, R., and Rossor, M. N., 1993, βA4 protein deposition in familial Alzheimer's disease with the mutation in codon 717 of the βA4 amyloid precursor protein gene and sporadic Alzheimer's disease, *Neurosci. Lett.* **149:**137–147.

Campbell, M. J., and Morrison, J. H., 1989, Monoclonal antibody to neurofilament protein (SMI32) labels a subpopulation of pyramidal neurons in the human and monkey neocortex, *J. Comp. Neurol.* **282:**191–205.

Campbell, M. J., Hof, P. R., and Morrison, J. H., 1991, A subpopulation of primate corticocortical neurons is distinguished by somatodendritic distribution of neurofilament protein, *Brain Res.* **539:**133–136.

Chu, D. C. M., Penney, J. B., and Young, A. B., 1987, Cortical $GABA_B$ and $GABA_A$ receptors in Alzheimer's disease: A quantitative autoradiographic study, *Neurology* **37:**1454–1459.

Citron, M., Westaway, D., Xia, W., Carlson, G., Diehl, T., Lee, M., Davis, A., Motter, R., Perry, B., Yao, H., Stromme, R., Rommens, J., Schenk, D., St. George Hyslop, P., and Selkoe, D. J., 1997, Mutant presenilins of Alzheimer's disease increase production of 42-residue amyloid beta-protein in both transfected cells and transgenic mice, *Nature Med.* **3:**67–72.

Clark, R. F., Hutton, M., Fuldner, R. A., and Goate, A., 1995, The structure of the presenilin 1 (S182) gene and identification of six novel mutations in early onset Alzheimer's disease families, *Nature Genet.* **11:**219–222.

Cogan, D. G., 1985, Visual disturbances with focal progressive dementing disease, *Am. J. Ophthalmol.* **100:**68–72.

Corder, E. H., Saunders, A. M., Strittmatter, W. J., Schmechel, D. E., Gaskell, P. C., Small, G. W., Roses, A. D., Haiones, J. L., and Pericak-Vance, M. A., 1993, Gene dose of apolipoprotein E type 4 allele and the risk of Alzheimer's disease in late onset families, *Science* **261:**921–923.

Corder, E. H., Saunders, A. M., Strittmatter, W. J., Schmechel, D. E., Gaskell, P. C., Conneally, K. E., Small, G. W., Roses, A. D., and Paricak-Vance, M. A., 1994, Apolipoprotein E type 2 allele decreases the risk of late-onset Alzheimer's disease, *Nature Genet.* **7:**180–184.

Corral-Debrinski, M., Horton, T., Lott, M. T., Shoffner, J. M., Beal, M. F., and Wallace, D. G., 1992, Mitochondrial DNA deletions in human brain: Regional variability and increase with advanced age, *Nature Genet.* **2:**324–329.

Cronin-Golomb, A., Suguira, R., Corkin, S., and Growdon, J. H., 1993, Incomplete achromatopsia in Alzheimer's disease, *Neurobiol. Aging* **14:**471–477.

Crook, R., Verkkoniemi, A., Perez-Tur, J., et al., 1998, A variant of Alzheimer's disease with spastic paraparesis and unusual plaques due to deletion of exon 9 of presenilin 1, *Nature Med.* **4:**452–455.

Cross, A. J., Crow, T. J., Perry, E. K., Blessed, G., and Tomlinson, B. E., 1981, Reduced dopamine-β-hydroxylase activity in Alzheimer's disease, *Br. Med. J.* **282:**93–94.

Crystal, M. A., Horoupian, D. S., Katzman, R., and Jotkowitz, S., 1982, Biopsy-proved Alzheimer's disease presenting as a right parietal lobe syndrome, *Ann. Neurol.* **12:**186–188.

Dawson, V. L., Bredt, D. S., Fotuhi, D. S., Hwang, P. M., and Snyder, S. H., 1991a, Nitric oxide synthase and neuronal NADPH-diaphorase are identical in brain and peripheral tissues, *Proc. Natl. Acad. Sci. USA* **88:**7797–7801.

Dawson, V. L., Dawson, T. M., London, E. D., Bredt, D. S., and Snyder, S. H., 1991b, Nitric oxide mediates glutamate neurotoxicity in primary cortical cultures, *Proc. Natl. Acad. Sci. USA* **88:**6368–6371.

De Boni, U., and McLachlan, D. R. C., 1985, Controlled induction of paired helical filaments of the Alzheimer type in cultured human neurons by glutamate and aspartate, *J. Neurol. Sci.* **68:**105–118.

Doan, A., Thinakaran, G., Borchelt, D. R., Slunt, H. H., Ratovitsky, T., Selkoe, D. J., Seeger, M., Gandy, S. E., Price, D. L., and Sisodia, S. S., 1996, Protein topology of presenilin 1, *Neuron* **17:**1023–1030.

DeYoe, E. A., Carman, G. J., Bandettini, P., Glickman, S., Wieser, J., Cox, R., Miller, D., and Neitz, J., 1996, Mapping striate and extrastriate visual areas in human cerebral cortex, *Proc. Natl. Acad. Sci. USA* **93:**2382–2386.

Estus, S., Tucker, M., van Rooyan, C., Wright, S., Brigham, E. F., Wogulis, M., and Rydel, R. E., 1997, Aggregated amyloid-β protein induces cortical neuronal apoptosis and concomitant apoptotic pattern of gene induction, *J. Neurosci.* **17:**7736–7745.

Feany, M. B., and Dickson, D. W., 1995, Widespread cytoskeletal pathology characterizes corticobasal degeneration, *Am. J. Pathol.* **146:**1388–1396.

Feany, M. B., Ksiezak-Reding, H., Liu, W. K., Vincent, I., Yen, S. H. C., and Dickson, D. W., 1995, Epitope expression and phosphorylation of tau protein in corticobasal degeneration: Differentiation from progressive supranuclear palsy, *Acta Neuropathol.* **90:**37–43.

Fisher, N. J., Rourke, B. P., Bieliauskas, L., Giordani, B., Berent, S., and Foster, N., 1996, Neuropsychological subgroups of patients with Alzheimer's disease, *J. Clin. Exp. Neuropsychol.* **18:**349–370.

Fletcher, W. A., 1994, Ophthalmological aspects of Alzheimer's disease, *Curr. Opin. Ophthalmol.* **5:**38–44.

Fletcher, W. A., Sharpe, J. A., 1988, Smooth pursuit dysfunction in Alzheimer's disease, *Neurology* **38:**272–277.

Forloni, G., Chiesa, R., Smiroldo, S., Verga, L., Salmona, M., Tagliavini, F., Angeretti, N., 1993, Apoptosis mediated neurotoxicity induced by chronic application of β amyloid fragment 25-35, *NeuroReport* **4:**523–526.

Foster, N. L., Chase, T. N., Fedio, P., Petronsas, N. J., Brooks, R. A., and DiChiro, G., 1983, Alzheimer's disease: Focal cortical changes shown by positron emission tomography, *Neurology* **33:**961–965.

Fox, N. C., Kennedy, A. M., and Harvey, R. J., *et al.,* 1997, Clinicopathological features of familial Alzheimer's disease associated with the M139V mutation in the presenilin 1 gene, *Brain* **120:**491–501.

Freed, D. M., Corkin, S., Growdon, J. H., and Nissen, M. J., 1989, Selective attention in Alzheimer's disease: Characterizing cognitive subgroups of patients, *Neuropsychologia* **27:**325–339.

Furey-Kurkjian, M. L., Pietrini, P., Graff-Radford, N. R., Alexander, G. E., Freo, U., Szczepanik, J., and Schapiro, M. B., 1996, Visual variant of Alzheimer's disease: Distinctive neuropsychological features, *Neuropsychology* **10:**294–300.

Games, D., Adams, D., Alessandrini, R., Blackwell, C., Clemens, J., Hagopian, S., Lieberburg, I., McConiogue, L., Mucke, L., Schenk, D., Wadsworth, S., Wolozin, B., and Zhao, J., 1995, Alzheimer-type neuropathology in transgenic mice overexpressing V717F β-amyloid precursor protein, *Nature* **373:**523–527.

Gearing, M., Mori, H., and Mirra, S. S., 1996, Aβ-peptide length and apolipoprotein E genotype in Alzheimer's disease, *Ann. Neurol.* **39:**395–399.

Giannakopoulos, P., Hof, P. R., Surini, M., Michel, J. P., and Bouras, C., 1993, Quantitative immunohistochemical analysis of the distribution of neurofibrillary tangles and senile plaques in the cerebral cortex of nonagenarians and centagenarians, *Acta Neuropathol.* **85:**602–610.

Giannakopoulos, P., Hof, P. R., and Bouras, C., 1995, Dementia lacking distinctive histopathology: Clinicopathological evaluation of 32 cases, *Acta Neuropathol.* **89:**346–355.

Giannakopoulos, P., Duc, M., Gold, G., Hof, P. R., Michel, J.-P., Bouras, C., 1998, Pathologic correlates of apraxia in Alzheimer disease, *Arch. Neurol.* **55:**689–695.

Gómez-Isla, T., Wasco, W., Pettingell, W. P., and Gurubhagavatula, S., *et al.,* 1997, A novel presenilin-1 mutation: Increased β-amyloid and neurofibrillary changes, *Ann. Neurol.* **41:**809–813.

Gómez-Isla, T., Price, J. L., McKeel, D. W., Jr., Morris, J. C., Growdon, J. H., and Hyman, B. T., 1996, Profound loss of layer II entorhinal cortex neurons occurs in very mild Alzheimer's disease, *J. Neurosci.* **16:**4491–4500.

Gouras, G. K., Relkin, N. R., Sweeney, D., Munoz, D. G., Mackenzie, I. R., and Gandy, S., 1997, Increased apolipoprotein E ϵ4 in epilepsy with senile plaques, *Ann. Neurol.* **41:**402–404.

Grady, C. l., Haxby, J. V., Shapiro, M. B., Kumar, A., Ball, M. J., Heston, L., and Rapoport, S. I., 1990, Subgroups in dementia of the Alzheimer type identified using positron emission tomography, *J. Neuropsychiat.* **2:**373–384.

Graff-Radford, N. R., Bolling, J. P., Earnest, F., IV, Shuster, E. A., Caselli, R. J., and Brazis, P. W., 1993, Simultanagnosia as the initial sign of degenerative dementia, *Mayo Clin. Proc.* **68:**955–964.

Grünthal, E., 1928, Zür Hirnpathologischen Analyse der Alzheimerschen Krankheit, *Psychiat.-Neurol. Wschr.* **36:**401–407.

Guo, Q., Sopher, B. L., Furukawa, K., Pham, D. G., Robinson, N., Martin, G. M., and Mattson, M. P., 1997, Alzheimer's presenilin mutation sensitizes neurons to apoptosis induced by trophic factor withdrawal and amyloid β-peptide: Involvement of calcium and oxyradicals, *J. Neurosci.* **17:**4212–4222.

Gustafson, L., 1993, Clinical picture of frontal lobe degeneration of non-Alzheimer type, *Dementia* **4:**143–148.

Gustafson, L., Brun, A., and Risberg, J., 1990, Frontal lobe degeneration of non-Alzheimer type, *Adv. Neurol.* **51:**65–71.

Haltia, M., Viitanen, M., Sulkava, R., and Ala-Hurula, V., *et al.,* 1994, Chromosome 14-encoded Alzheimer's disease: Genetic and clinicopathological description, *Ann. Neurol.* **36:**362–367.

Halliday, G. M., McCann, H. L., Pamphlett, R., Brooks, W. S., Creasey, H., McCusker, E., Cotton, R. G. H.,

Broe, G. E., and Harper, C. G., 1992, Brain stem serotonin-synthesizing neurons in Alzheimer's disease: A clinicopathological correlation, *Acta Neuropathol.* **84:**638–650.

Hansen, L. A., DeTeresa, R., Davies, P., and Terry, R. D., 1988, Neocortical morphometry, lesion counts, and choline acetyltransferase levels in the age spectrum of Alzheimer's disease, *Neurology* **38:**48–54.

Hardy, J., 1996, New insights into the genetics of Alzheimer's disease, *Ann. Med.* **28:**255–258.

Hardy, J., 1997, Amyloid, the presenilins, and Alzheimer's disease, *Trends Neurosci.* **20:**154–159.

Harvey, R. J., Ellison, D., Hardy, J., Hutton, M., Roques, P. K., Collinge, J., Fox, N. C., and Rossor, M. N., 1998, Chromosome 14 familial Alzheimer's disease: The clinical and neuropathological characteristics of a family with a leucine to serine substitution at codon 250 of the presenilin 1 gene, *J. Neurol. Neurosurg. Psychiatry* **64:**44–49.

Hauw, J. J., Verny, M., Delaère, P., Cervera, P., He, Y., and Duyckaerts, C., 1990, Constant neurofibrillary changes in the neocortex of progressive supranuclear palsy: Basic differences with Alzheimer's disease and aging, *Neurosci. Lett.* **119:**182–186.

Haxby, J. V., Grady, C. L., and Koss, E., 1988, Heterogeneous anterior-posterior metabolic patterns in dementia of the Alzheimer type, *Neurology* **38:**1053–1063.

Haxby, J. V., Duara, R., Grady, C. L., Culter, N. R., and Rapoport, S. I., 1985, Relations between neuropsychological and cerebral metabolic asymmetries in early Alzheimer's disease, *J. Cereb. Blood Flow Metab.* **5:**193–200.

Hirano, A., and Zimmerman, H. M., 1962, Alzheimer's neurofibrillary changes: A topographic study, *Arch. Neurol.* **7:**227–242.

Hirono, N., Mori, E., Ishii, K., Ikejiri, Y., Imamura, T., Shimomura, T., Hashimoto, M., Yamashita, H., and Sasaki, M., 1998, Hypofunction in the posterior cingulate gyrus correlates with disorientation for time and place in Alzheimer's disease, *J. Neurol. Neurosurg. Psychiatry* **64:**552–554.

Hof, P. R., and Bouras, C., 1991, Object recognition deficit in Alzheimer's disease: Possible disconnection of the occipito-temporal component of the visual system, *Neurosci. Lett.* **122:**53–56.

Hof, P. R., and Morrison, J. H., 1991, Neocortical neuronal subpopulations labeled by a monoclonal antibody to calbindin exhibit differential vulnerability in Alzheimer's disease, *Exp. Neurol.* **111:**293–301.

Hof, P. R., and Nimchinsky, E. A., 1992, Regional distribution of neurofilament and calcium-binding proteins in the cingulate cortex of the macaque monkey, *Cereb. Cortex* **2:**456–467.

Hof, P. R., Bouras, C., Constantinidis, J., and Morrison, J. H., 1989, Bálint's syndrome in Alzheimer's disease: Specific disruption of the occipitoparietal visual pathway, *Brain Res.* **493:**368–375.

Hof, P. R., Bouras, C., Constantinidis, J., and Morrison, J. H., 1990a, Selective disconnection of specific visual association pathways in cases of Alzheimer's disease presenting with Bálint's syndrome, *J. Neuropathol. Exp. Neurol.* **49:**168–184.

Hof, P. R., Cox, K., and Morrison, J. H., 1990b, Quantitative analysis of a vulnerable subset of pyramidal neurons in Alzheimer's disease. I. Superior frontal and inferior temporal cortex, *J. Comp. Neurol.* **301:**44–54.

Hof, P. R., Delacourte, A., and Bouras, C., 1992c, Distribution of cortical neurofibrillary tangles in progressive supranuclear palsy: A quantitative analysis of six cases, *Acta Neuropathol.* **84:**45–51.

Hof, P. R., Bierer, L. M., Perl, D. P., Delacourte, A., Buée, L., Bouras, C., and Morrison, J. H., 1992a, Evidence for early vulnerability of the medial and inferior aspects of the temporal lobe in an 82-year-old patient with preclinical signs of dementia, *Arch. Neurol.* **49:**946–953.

Hof, P. R., Bouras, C., Buée, L., Delacourte, A., Perl, D. P., and Morrison, J. H., 1992b, Differential distribution of neurofibrillary tangles in the cerebral cortex of dementia pugilistica and Alzheimer's disease cases, *Acta Neuropathol.* **85:**23–30.

Hof, P. R., Archin, N., Osmand, A. P., Dougherty, A. P., Wells, C., Bouras, C., and Morrison, J. H., 1993a, Posterior cortical atrophy in Alzheimer's disease: Analysis of a new case and re-evaluation of a historical report, *Acta Neuropathol.* **86:**215–223.

Hof, P. R., Nimchinsky, E. A., Celio, M. R., Bouras, C., and Morrison, J. H., 1993b, Calretinin-immunoreactive neocortical interneurons are unaffected in Alzheimer's disease, *Neurosci. Lett.* **152:**145–149.

Hof, P. R., Bouras, C., Perl, D. P., and Morrison, J. H., 1994, Quantitative neuropathologic analysis of Pick's disease cases: Cortical distribution of Pick bodies and coexistence with Alzheimer's disease, *Acta Neuropathol.* **87:**115–124.

Hof, P. R., Nimchinsky, E. A., and Morrison, J. H., 1995, Neurochemical phenotype of corticocortical connections in the macaque monkey: Quantitative analysis of a subset of neurofilament protein-immunoreactive projection neurons in frontal, parietal, temporal, and cingulate cortices, *J. Comp. Neurol.* **362:**109–133.

Hof, P. R., Vogt, B. A., Bouras, C., and Morrison, J. H., 1997, Patterns of corticocortical disconnection in atypical forms of Alzheimer's disease with prominent posterior cortical atrophy, *Vision Res.* **37:**3609–3625.

Houlden, H., Collinge, J., and Kennedy, A., *et al.*, 1993, ApoE genotype and Alzheimer's disease, *Lancet* **342:**737–738.

Hutton, M., Busfield, F., Wragg, M., Crook, R., Perez-Tur, J., Wright, K., Lenden, C., Martinez, A., Houlden, H., Roberts, G., Roques, P., Rossor, M., Venter, J. C., Adams, M. D., Phillips, C. A., Fulder, R. A., Hardy, J., and Goate, A., 1996, Complete analysis of the presenilin 1 gene in early onset Alzheimer's disease, *NeuroReport* **7:**801–805.

Hyman, B. T., Marzloff, K., and Arriagada, P. V., 1993, The lack of accumulation of senile plaques or amyloid burden in Alzheimer's disease suggests a dynamic balance between amyloid deposition and resolution, *J. Neuropathol. Exp. Neurol.* **52:**594–600.

Hyman, B. T., West, H. L., Gomez-Isla, T., and Mui, S., 1995, Quantitative neuropathology in Alzheimer's disease: Neuronal loss in high-order association cortex parallels dementia, in: *Research Advances in Alzheimer's Disease and Related Disorders* (K. Iqbal, J. A. Mortimer, and H. M. Wisniewski, eds.), Wiley, New York, pp. 453–460.

Jackson, M., Lowe, J., 1996, The new neuropathology of degenerative frontotemporal dementias, *Acta Neuropathol.* **91:**127–134.

Jakob, H., 1979, Die Picksche Krankheit: Eine neuropathologisch-anatomisch-klinische Studie, in: *Monographien aus der Gesamtgebiet der Psychiatrie* (H. Hippius, W. Jazarik, and C. Müller, eds.), Bd 23, Springer-Verlag, Berlin, pp. 1–110.

Jordan, J., Galindo, M. F., Miller, R. J., Reardon, C. A., Getz, G. S., and La Du., M. J., 1998, Isoform-specific effect of apolipoprotein E on cell survival and β-amyloid-induced toxicity in rat hippocampal pyramidal neuronal cultures, *J. Neurosci.* **18:**195–204.

Jorm, A. F., 1985, Subtypes of Alzheimer's dementia: A conceptual analysis and critical review, *Psychol. Med.* **15:**543–553.

Kalaria, R. N., Andorn, A. C., Tabaton, M., Whitehouse, P. J., Harik, S. L., and Unnerstall, J. R., 1989, Adrenergic receptors in aging and Alzheimer's disease: Increased $\beta_2$-receptors in prefrontal cortex and hippocampus, *J. Neurochem.* **53:**1772–1781.

Kaplan, E., Goodglass, H., and Weintraub, S., 1976, *Boston Naming Test* (experimental edition), Boston, Boston Veterans Administration Medical Center.

Katz, B., and Rimmer, S., 1989, Ophthalmologic manifestations of Alzheimer's disease, *Surv. Ophthalmol.* **34:**31–34.

Kennedy, A. M., Newman, S. K., Frackowiak, R. S. J., Cunningham, V. J., Roques, P., Stevens, J., Neary, D., Bruton, C. J., Warrington, E. K., and Rossor, M. N., 1995, Chromosome 14 linked familial Alzheimer's disease: A clinico-pathological study of a single pedigree, *Brain* **118:**185–205.

Kiyosawa, M., Bosley, T. M., Chawluk, J., Jamieson, D., Schatz, N. J., Savino, P. J., Sergott, R. C., Reivich, M., and Alavi, A., 1989, Alzheimer's disease with prominent visual symptoms, *Ophthalmology* **96:**1077–1086.

Kosaka, K., Yoshimura, M., Ikeda, K., and Budka, H., 1984, Diffuse type of Lewy body disease: Progressive dementia with abundant cortical Lewy bodies and senile changes of varying degree: A new disease? *Clin. Neuropathol.* **3:**185–192.

Kruman, I., Bruce-Keller, A. J., Bredesen, D., Waeg, G., and Mattson, M. P., 1997, Evidence that 4-hydroxynonenal mediates oxidative stress-induced neuronal apoptosis, *J. Neurosci.* **17:**5089–5100.

Kurylo, D. D., Corkin, S., Dolan, R. P., Rizzo, J. F., III, Parker, S. W., and Growdon, J. H., 1994, Broad-band visual capacities are not selectively impaired in Alzheimer's disease, *Neurobiol. Aging* **15:**305–311.

LaFerla, F. M., Tinkle, B. T., Bieberich, C. J., Haudenschild, C. C., and Jay, G., 1995, The Alzheimer's Aβ peptide induces neurodegeneration and apoptotic cell death in transgenic mice, *Nature Genet.* **9:**21–29.

Lampe, T. H., Bird, T. D., Nochin, D., Nemens, E., Risse, S. C., Sumi, S. M., Koerler, R., Leaird, B., Wier, M., and Raskind, M. A., 1994, Phenotype of chromosome 14-linked familial Alzheimer's disease in a large kindred, *Ann Neurol.* **36:**368–378.

Lantos, P. L., Luthert, P. J., Hanger, D., Anderton, B. H., Mullan, M., and Rossor, M., 1992, Familial

Alzheimer's disease with the amyloid precursor protein position 717 mutation and sporadic Alzheimer's disease have the same cytoskeletal pathology, *Neurosci. Lett.* **137:**221–224.

Lennox, G., Lowe, J., Byrne, E. J., Landon, M., Mayer, R. J., and Godwin-Austen, R. B., 1989, Diffuse Lewy body disease, *Lancet:*323–324.

Levine, D. N., Lee, J. M., and Fisher, C. M., 1993, The visual variant of Alzheimer's disease: A clinicopathologic case study. *Neurology* **43:**305–313.

Levy, M. H., Hart, W. M., Jr., Sonstein, F. M., and Ballinger, W. E., Jr., 1995, The incredible shrinking brain [with comments by A. Sadun], *Surv. Ophthalmol.* **39:**315;322.

Levy-Lahad, E., Wasco, W., Poorkaj, P., Romano, D. M., Oshima, J., Crowley, A. C., Guenette, S. Y., Galas, D., Wijsman, E. M., and Tanzi, R. E., 1995, Candidate gene for the chromosome 1 familial Alzheimer's disease locus, *Science* **269:**973–977.

Li, X., Greenwald, I., 1996, Membrane topology of the *C. elegans* SEL-12 presenilin, *Neuron* **17:**1015–1021.

Lippa, C. F., Smith, T. W., and Swearer, J. M., 1994, Alzheimer's disease and Lewy body disease: A comparative clinicopathological study, *Ann. Neurol.* **35:**81–88.

Loo, D. T., Copani, A., Pike, C. J., Edward, E. R., Walencewicz, A. J., and Cotman, C. W., 1993, Apoptosis is induced by β-amyloid in cultured central nervous system neurons, *Proc. Natl. Acad. Sci. USA* **90:**7951–7955.

Loy, R., Heyer, D., Clagett-Dame, M., and DiStefano, P. S., 1990, Localization of NGF receptors in normal and Alzheimer's basal forebrain with monoclonal antibodies against the truncated form of the receptor, *J. Neurosci. Res.* **27:**651–664.

Maestre, G., Ottamen, R., and Stern, Y., *et al.*, 1995, Apolipoprotein E and Alzheimer's disease: Ethnic variation in genetic risk, *Ann. Neurol.* **37:**254–259.

Mann, D. M. A., Yates, P. O., and Marcyniuk, B., 1985, Some morphometric observations on the cerebral cortex in presenile Alzheimer's disease and Down's syndrome in middle age, *J. Neurol. Sci.* **69:**139–159.

Mann, D. M. A., Marcyniuk, B., Yates, P. O., Neary, D., and Snowden, J. S., 1988, The progression of the pathological changes of Alzheimer's disease in frontal and temporal neocortex examined both at biopsy and at autopsy, *Neuropathol. Appl. Neurobiol.* **14:**177–195.

Mann, D. M. A., Jones, D., Snowden, J. S., Neary, D., and Hardy, J., 1992a, Pathological changes in the brain of a patient with familial Alzheimer's disease having a missense mutation at codon 717 in the amyloid precursor protein, *Neurosci. Lett.* **137:**225–228.

Mann, D. M. A., and South, P. W., 1993, The topographic distribution of brain atrophy in frontal lobe dementia, *Acta Neuropathol.* **85:**334–340.

Mann, D. M. A., South, P. W., Snowdon, J. S., and Neary, D., 1993, Dementia of frontal lobe type: Neuropathology and immunohistochemistry, *J. Neurol. Neurosurg. Psychiatry* **56:**605–614.

Mann, U., Lincoln, J., Yates, P. O., Stamp, J. E., and Toper, S., 1980, Changes in the monoamine containing neurons of the human CNS in senile dementia, *Br. J. Psychiatry* **136:**533–541.

Mann, U., Mohr, E., Gearing, M., and Chase, T. N., 1992b, Heterogeneity in Alzheimer's disease: Progression rate segregated by distinct neuropsychological and cerebral metabolic profiles, *J. Neurol. Neurosurg. Psychiatry* **55:**956–959.

Martin, A., 1987, Representation of semantic and spatial knowledge in Alzheimer's patients: Implications for models of preserved learning and amnesia, *J. Clin. Exp. Neuropsychol.* **9:**191–224.

Martin, A., 1990, Neuropsychology of Alzheimer's disease: The case for subgroups, in: *Modular Deficits in Alzheimer-type Dementia* (Schwartz, M. F., ed.), MIT Press, Cambridge, MA, pp. 145–175.

Martin, A., Brouwers, P., Lalonde, F., Cox, C., Teleska, P., Fedio, P., Foster, N. L., and Chase, T. N., 1986, Toward a behavioral typology of Alzheimer's patients, *J. Clin. Exp. Neuropsychol.* **8:**594–610.

Matsunami, K. I., Kawashima, T., and Satake, H., 1989, Mode of [$^{14}$C] 2-deoxy-D-glucose uptake into retrosplenial cortex and other memory-related structures of the monkey during a delayed response, *Brain Res. Bull.* **22:**829–838.

Maury, C. P. J., Nurmiaho-Lassila, E.-L., and Liljestrom, M., 1997, Alzheimer's disease-associated presenilins 1 and 2: Accelerated amyloid fibril formation of mutant 410 Cys→Tyr and 141 Asn→Ile peptides, *Biochem. Biophys. Res. Comm.* **235:**249–252.

McCarthy, G., Spicer, M., Adrignolo, A., Luby, M., Gore, J., and Allison, T., 1995, Brain activation associated with visual motion studied by functional magnetic resonance imaging in humans, *Hum. Brain Mapp.* **2:**234–243.

McKhann, G., Drachman, D., and Folstein, M., 1984, Clinical diagnosis of Alzheimer's disease, *Neurology* **34:**939–944.

Mendez, M. F., Mendez, M. A., Martin, R., Smyth, K. A., and Whitehouse, P. J., 1990a, Complex visual disturbances in Alzheimer's disease, *Neurology* **40:**439–443.

Mendez, M. F., Tomsak, R. L., and Remler, B., 1990b, Disorders of the visual system in Alzheimer's disease, *J. Clin. Neuro-ophthalmol.* **10:**62–69.

Mendez, M. F., Turner, J., Gilmore, G. C., Remler, B., and Tomsak, R. L., 1990c, Bálint's syndrome in Alzheimer's disease: Visuospatial functions, *Int. J. Neurosci.* **54:**339–346.

Mendez, M. F., Selwood, A., Mastri, A. R., and Frey, W. H., II, 1993, Pick's disease versus Alzheimer's disease: A comparison of clinical characteristics, *Neurology* **43:**289–292.

Minoshima, S., Giordani, B., Berent, S., Frey, K. A., Foster, N. L., and Kuhl, D. E., 1997, Metabolic reduction in the posterior cingulate cortex in very early Alzheimer's disease, *Ann. Neurol.* **42:**85–94.

Mirra, S. S., Heyman, A., McKeel, J., Sumi, S. M., Crain, B. J., Brownlee, L. M., Hughes, J. P., Vogel, F. S., van Belle, G., and Berg, L., 1991, The consortium to establish a registry for Alzheimer's disease (CERAD), *Neurology* **41:**479–486.

Morel, F., 1945, Les aires striée, parastriée et péristriée dans les troubles de la fonction visuelle au cours de la maladie d'Alzheimer, *Confin. Neurol.* **6:**238–242.

Mountjoy, C. Q., Roth, M., Evans, N. J. R., and Evans, H. M., 1983, Cortical neuronal counts in normal elderly controls and demented patients, *Neurobiol. Aging* **4:**1–11.

Mufson, E. J., Bothwell, M., Hersh, L. B., and Kordower, J. H., 1989, Nerve growth factor receptor immunoreactive profiles in the normal aged human basal forebrain: Colocalization with cholinergic neurons, *J. Comp. Neurol.* **285:**196–217.

Mullan, M., Tsuji, S., and Miki, T., *et al.*, 1993, Clinical comparison of Alzheimer's disease in pedigrees with the codon 717 Val to Ile mutation in the amyloid precursor protein gene, *Neurobiol. Aging* **14:**407–419.

Murray, E. A., Davidson, M., Gaffan, D., Olton, D. S., and Suomi, S., 1989, Effects of fornix transection and cingulate cortical ablation on spatial memory in rhesus monkeys, *Exp. Brain Res.* **74:**173–186.

Nagy, Z., Esiri, M. M., Jobst, K. A., Morris, J. H., King, E., McDonal, B., Litchfield, S., Smith, A., Barnetson, L., and Smith, A. D., 1995, Relative roles of plaques and tangles in the dementia of Alzheimer's disease: Correlations using three sets of neuropathological criteria, *Dementia* **6:**21–31.

Nalbantoglu, J., Gilfix, B. M., Bertrand, P., Robitaille, Y., Gauthier, S., Rosenblatt, D. S., and Poirier, J., 1994, Predictive value of apolipoprotein E genotyping in Alzheimer's disease: Results of an autopsy series and an analysis of several combined studies, *Ann. Neurol.* **36:**889–895.

Neary, D., and Snowden, J. S., 1987, Perceptuospatial disorder in Alzheimer's disease, *Semin. Ophthalmol.* **2:**151–158.

Neary, D., Snowden, J. S., Mann, D. M. A., Bowen, D. M., Sims, N. R., Northen, B., Yates, P. O., and Davison, A. N., 1986, Alzheimer's disease: A correlative study, *J. Neurol. Neurosurg. Psychiatry* **49:**229–237.

Neary, D., Snowden, J. S., and Mann, D. M. A., 1993, The clinical pathological correlates of lobar atrophy, *Dementia* **4:**154–159.

Nielson, K. A., Cummings, B. J., and Cotman, C. W., 1996, Constructional apraxia in Alzheimer's disease correlates with neuritic neuropathology in occipital cortex, *Brain Res.* **741:**284–293.

Nimchinsky, E. A., Hof, P. R., Young, W. G., and Morrison, J. H., 1996, Neurochemical, morphologic, and laminar characterization of cortical projection neurons in the cingulate motor areas of the macaque monkey, *J. Comp. Neurol.* **374:**136–160.

Nimchinsky, E. A., Vogt, B. A., Morrison, J. H., and Hof, P. R., 1997, Neurofilament and calcium-binding proteins in the human cingulate cortex, *J. Comp. Neurol.,* in press.

Nissen, M. J., Corkin, S., Buonanno, F. S., Growdon, J. H., Wray, S. H., and Bauer, J., 1985, Spatial vision in Alzheimer's disease: General findings and a case report, *Arch. Neurol.* **42:**667–671.

Olson, C. R., Musil, S. Y., and Goldberg, M. E., 1993, Posterior cingulate cortex and visuospatial cognition: Properties of single neurons in the behaving monkey, in: *Neurobiology of Cingulate Cortex and Limbic Thalamus* (B. A. Vogt and M. Gabriel, eds.), Birkhäuser, Boston, pp. 366–380.

Orban, G. A., Saunders, R. C., Vandenbussche, E., 1995, Lesions of the superior temporal cortical motion areas impair speed discrimination in the macaque monkey. *Eur. J. Neurosci.* **7:**2261–2276.

Osterrieth, P., 1944, Le test de copie d'une figure complexe, *Arch. Psychol.* **30;**206–356.

Pandya, D. N., Van Hoesen, G. W., and Mesulam, M.-M., 1981, Efferent connections of the cingulate gyrus in the rhesus monkey, *Brain Res.* **42:**319–330.

Pearson, R. C. A., Esiri, M. M., Hiorns, R. W., Wilcock, G. K., and Powell, T. P. S., 1985, Anatomical correlates of the distribution of the pathological changes in the neocortex in Alzheimer disease, *Proc. Natl. Acad. Sci. USA* **82:**4531–4534.

Pericak-Vance, M. A., Bass, M. P., Yamaoka, L. H., Scott, W. K., Terdow, H. A., Menold, M. M., Conneally, P. M., Small, G. W., Vance, J. M., Saunders, A. M., Roses, A. D., and Haines, J. L., 1997, Complete genomic screen in late-onset familial Alzheimer disease: Evidence for a new locus on chromosome 12, *J. Am. Med. Assoc.* **278:**1237–1241.

Perry, R. H., Irving, D., Blessed, G., Fairbairn, A., and Perry, E. K., 1990, Senile dementia of Lewy body type: A clinically and neuropathologically distinct form of Lewy body dementia in the elderly, *J. Neurol. Sci.* **95:**119–139.

Pickering-Brown, S. M., Siddons, M., Mann, D. M. A., Owen, F., Neary, D., and Snowden, J. S., 1995, Apolipoprotein E allelic frequencies in patients with lobar atrophy, *Neurosci. Lett.* **188:**205–207.

Pietrini, P., Furey, M. L., Graff-Radford, N., Freo, U., Alexander, G. E., Grady, C. L., Dani, A., Mentis, M. J., and Shapiro, M. B., 1996, Preferential metabolic involvement of visual cortical areas in a subtype of Alzheimer's disease: Clinical implications, *Am. J. Psychiatry* **153:**1261–1268.

Pitts, A. F., and Miller, M. W., 1995, Expression of nerve growth factor, p75, and trk in the somatosensory and motor cortices of mature rats: Evidence for local trophic support circuits, *Somatosensory Motor Res.* **12:**329–342.

Procter, A. W., Lowe, S. L., Palmer, A. M., Francis, P. T., Esiri, M. M., Stratmann, G. C., Najlerahim, A., Patel, A. J., Hunt, A., and Bowen, D. M., 1988, Topographical distribution of neurochemical changes in Alzheimer's disease, *J. Neurol. Sci.* **84:**125–140.

Procter, A. W., Francis, P. T., Chen, P., Chessell, I. P., Dijk, S., Clarke, N. A., Webster, M.-T., and Bowen, D. M., 1995, The neurochemical pathology of Alzheimer's disease, in: *Neurobiology of Alzheimer's Disease* (D. Dawbarn and S. J. Allen, eds.), BIOS Scientific, London.

Rebeck, G. W., Reiter, J. S., Strickland, D. K., and Hyman, B. T., 1993, Apolipoprotein E in sporadic Alzheimer's disease: Allelic variation and receptor interactions, *Neuron* **11:**575–580.

Reisine, T. D., Yamamura, H. I., Bird, E. D., Spokes, E., and Enna, S. J., 1978, Pre- and postsynaptic neurochemical alterations in Alzheimer's disease, *Brain Res.* **159:**477–481.

Roses, A. D., 1995, Apolipoprotein E genotyping in the differential diagnosis, not prediction, of Alzheimer's disease, *Ann. Neurol.* **38:**6–14.

Rossor, M. N., Garrett, N. J., Johnson, A. L., Mountjoy, C. Q., Roth, R., and Iversen, L. L., 1982, A post-mortem study of cholinergic and GABA systems in senile dementia, *Brain* **105:**313–330.

Rossor, M. N., Iversen, L. L., Reynolds, G. P., Mountjoy, C. Q., and Roth, M., 1984, Neurochemical characteristics of early and late onset types of Alzheimer's disease, *Br. Med. J.* **288:**961–964.

Royall, D. R., Mahurin, R. K., and Cornell, J., 1994, Bedside assessment of frontal degeneration: Distinguishing Alzheimer's disease from non-Alzheimer's cortical dementia, *Exp. Aging Res.* **20:**95–103.

Sadun, A. A., Borchert, M., DeVita, E., Hinton, D. R., and Bassi, C. J., 1987, Assessment of visual impairment in patients with Alzheimer's disease, *Am. J. Ophthalmol.* **104:**113–120.

Scheuner, D., Eckman, C., Jensen, M., Song, X., Citron, M., Suzuki, N., Bird, T. D., Hardy, J., Hutton, M., Larson, E., Levy-Lahad, E., Peskind, E., Tanzi, R., Wasco, W., Lannfelt, L., Selko, D., and Younkin, S., 1996, Secreted amyloid β-protein similar to that in senile plaques of Alzheimer's disease is increased *in vivo* by the presinilin 1 and 2 and APP mutations linked to familial Alzheimer's disease, *Nature Med.* **2:**864–870.

Schmechel, D. E., Saunders, A. M., Strittmater, W. J., Crain, B. J., Joo, S. H., Pericak-Vance, M. A., Goldgaber, D., and Roses, A. D., 1993, Increased amyloid β-peptide deposition in cerebral cortex as a consequence of apolipoprotein E genotype in late-onset Alzheimer disease, *Proc. Natl. Acad. Sci. USA* **90:**9649–9653.

Sereno, M. I., Dale, A. M., Reppas, J. B., Kwong, K. K., Belliveau, J. W., Brady, T. J., Rosen, B. R., and Tootell, R. B. H., 1995, Borders of multiple visual areas in humans revealed by functional magnetic resonance imaging, *Science* **268:**889–893.

Shkol'nik-Yarros, E. G., 1971, *Neurons and Interneuronal Connections of the Central Visual System,* Plenum, New York, pp. 94–101.

Shoffner, J. M., Brown, M. D., Torroni, A., Lott, M. T., Mirra, S. S., Beal, M. F., Watts, R. L., Hansen, L. A., Fayad, M., and Wallace, D. C., 1993, Mitochondrial DNA variants observed in Alzheimer disease and Parkinson disease patients, *Genomics* **17:**171–184.

Sima, A. A. F., Clark, A. W., Sternberger, N. A., and Sternberger, L. A., 1986, Lewy body dementia without Alzheimer changes, *Can. J. Neurol. Sci.* **13:**490–497.

Sparks, D. L., Hunsaker, J. C., III, Scheff, S. W., Kryscio, R. J., Hensen, J. L., and Markesbery, W. R., 1990,

Cortical senile plaques in coronary artery disease, aging, and Alzheimer's disease, *Neurobiol. Aging* **11:**601–607.

Sparks, D. L., Scheff, S. W., Liu, H., Landers, T., Danner, F., Coyne, C. M., and Hunsker, J. C., III, 1996, Increased density of senile plaques, but not neurofibrillary tangles, in non-demented individuals with the apolipoprotein E4 allele: Comparison to confirmed Alzheimer's disease patients, *J. Neurol. Sci.* **138:**97–104.

Talbot, C., Lenden, C., and Craddock, N., *et al.*, 1994, Protection against Alzheimer's disease with apoE ϵ2, *Lancet* **343:**1432–1433.

Terry, R. D., Peck, A., DeTeresa, R., Schechter, R., and Horoupian, D. S., 1981, Some morphometric aspects of the brain in senile dementia of the Alzheimer type, *Ann. Neurol.* **10:**184–192.

Terry, R. D., Masliah, E., Salmon, D. P., Butters, N., DeTeresa, R., Hill, R., Hansen, L. A., and Katzman, R., 1991, Physical basis of cognitive alterations in Alzheimer's disease: Synapse loss is the major correlate of cognitive impairment, *Ann. Neurol.* **30:**572–580.

Tissot, R., Constantinidis, J., and Richard, J., 1985, Pick's disease, *Handbk. Clin. Neurol.* **46:**233–246.

Tomlinson, B. E., Irving, D., and Blessed, G., 1981, Cell loss in the locus ceruleus in senile dementia of Alzheimer type, *J. Neurol. Sci.* **49:**419–428.

Tootell, R. B. H., Reppas, J. B., Kwong, K. K., Malach, R., Born, R. T., Brady, T. J., Rosen, B. R., and Belliveau, J. W., 1995, Functional analysis of human area MT and related visual cortical areas using magnetic resonance imaging, *J. Neurosci.* **15:**3215–3230.

Tootell, R. B. H., and Taylor, J. B., 1995, Anatomical evidence for MT and additional cortical visual areas in humans, *Cereb. Cortex* **5:**39–55.

Tyrrell, P. J., Warrington, E. K., Frackowiak, R. S. J., and Rossor, M. N., 1990, Heterogeneity in progressive aphasia due to focal cortical atrophy, *Brain* **113:**1321–1336.

Ungerleider, L. G., and Mishkin, M., 1982, Two cortical visual systems, in: *Analysis of Visual Behavior* (D. J. Ingle, M. A. Goodale, and R. J. W. Mansfield, eds.), MIT Press, Cambridge, MA, pp. 549–586.

Underleider, L. G., and Haxby, J. V., 1994, "What" and "where" in the human brain, *Curr. Opin. Neurobiol.* **4:**157–165.

Valenstein, E., Bowers, D., Verfaellie, M., Heilman, K. M., Day, A., and Watson, R. T., 1987, Retrosplenial amnesia, *Brain* **110:**1631–1646.

Van Broeckhoven, C., Backhovens, H., Cruts, M., Martin, J. J., Crook, R., Houlden, H., and Hardy, J., 1994, ApoE genotype does not modulate age of onset in families with chromosome 14 encoded Alzheimer's disease, *Neurosci. Lett.* **169:**179–180.

van Duijn, C. M., de Knijff, P., Cruts, M., Wehnert, A., Havekes, L. M., Hofman, A., and Van Broeckhoven, C., 1994, Apolipoprotein E4 allele in a population-based study of early-onset Alzheimer's disease. *Nature Genet.* **7:**74–78.

Van Hoesen, G. W., Hyman, B. T., and Damasio, A. R., 1991, Entorhinal cortex pathology in Alzheimer's disease, *Hippocampus* **1:**1–8.

Van Hoesen, G. W., Morecraft, R. J., and Vogt, B. A., 1993, Connections of the monkey cingulate cortex, in: *Neurobiology of Cingulate Cortex and Limbic Thalamus* (B. A. Vogt and M. Gabriel, eds.), Birkhäuser, Boston, pp. 249–284.

Victoroff, J., Ross, W., Benson, D. F., Verity, M. A., and Vinters, H. V., 1994, Posterior cortical atrophy: Neuropathologic correlations, *Arch. Neurol.* **51:**269–274.

Vogt, B. A., and Pandya, D. N., 1987, Cingulate cortex of the rhesus monkey. II. Cortical afferents, *J. Comp. Neurol.* **262:**271–289.

Vogt, B. A., Van Hoesen, G. W., Vogt, L. J., 1990, Laminar distribution of neuron degeneration in posterior cingulate cortex in Alzheimer's disease, *Acta Neuropathol.* **80:**581–589.

Vogt, B. A., Crino, P. B., and Volicer, L., 1991, Laminar alterations in gamma-aminobutyric $acid_A$, muscarinic and beta adrenoceptors, and neuron degeneration in cingulate cortex in Alzheimer's disease, *J. Neurochem.* **57:**282–290.

Vogt, B. A., Crino, P. B., and Vogt, L. J., 1992, Reorganization of cingulate cortex in Alzheimer's disease: Neuron loss, neuritic plaques, and muscarinic receptor binding, *Cereb. Cortex* **2:**526–535.

Vogt, B. A., Vogt, L. J., Nimchinsky, E. A., and Hof, P. R., 1997, Primate cingulate cortex chemoarchitecture and its disruption in Alzheimer's disease, in: *Handbook of Chemical Neuroanatomy, Vol. 13: The Primate Nervous System, Part 1* (F. E. Bloom, A. Björklund, and T. Hökfelt, eds.), Elsevier, Amsterdam, pp. 455–528.

Vogt, B. A., Vogt, L. J., Vrana, K. E., Gioia, L., Meadows, R. S., Challa, V. R., Hof, P. R., and Van Hoesen,

G. W., 1998, Multivariate analysis of laminar patterns of neurodegeneration in posterior cingulate cortex in Alzheimer's disease, *Exp. Neurol.*, in press.

Watson, J. D. G., Myers, R., Frackowiak, R. S. J., Hajnal, J. V., Woods, R. P., Mazziotta, J. C., Shipp, S., and Zeki, S., 1993, Area V5 of the human brain: Evidence from a combined study using positron emission tomography and magnetic resonance imaging, *Cereb. Cortex* **3**:79–94.

Wilcock, G. K., Esiri, M. M., Bowen, D. M., and Smith, C. C. T., 1982, Alzheimer's disease: Correlation of cortical choline acetyltransferase activity with the severity of dementia and histological abnormalities, *J. Neurol. Sci.* **57**:407–417.

Yankner, B. A., 1995, Mechanisms of neuronal degeneration in Alzheimer's disease, *Neuron* **16**:921–932.

Yoshimura, N., 1989, Topography of Pick body distribution in Pick's disease: A contribution to understanding the relationship between Pick's and Alzheimer's diseases, *Clin. Neuropathol.* **8**:1–6.

17

# Molecular Pathology of Alzheimer's Disease and Related Disorders

STEPHEN D. GINSBERG, M. LUISE SCHMIDT, PETER B. CRINO, JAMES H. EBERWINE, VIRGINIA M.-Y. LEE, and JOHN Q. TROJANOWSKI

## 1. Introduction

Alzheimer's disease (AD) is characterized clinically by memory deficits and progressive cognitive decline leading to incapacitating dementia and death. Gross examination of the AD brain indicates that the cerebral cortex is most severely affected, with notable atrophy of the medial temporal lobe, temporal pole (including the subcortical amygdala), midfrontal, and orbitofrontal cortex. Pathologically, the disease is characterized by the deposition of filamentous material in

STEPHEN D. GINSBERG, M. LUISE SCHMIDT, VIRGINIA M.-Y. LEE, and JOHN Q. TROJANOWSKI • The Center for Neurodegenerative Disease Research, Department of Pathology and Laboratory Medicine, University of Pennsylvania School of Medicine, Philadelphia, Pennsylvania 19104-4283. PETER B. CRINO • Department of Neurology, University of Pennsylvania School of Medicine, Philadelphia, Pennsylvania 19104. JAMES H. EBERWINE • Departments of Pharmacology and Psychiatry, University of Pennsylvania School of Medicine, Philadelphia, Pennsylvania 19104.

*Cerebral Cortex,* Volume 14, edited by Peters and Morrison. Kluwer Academic/Plenum Publishers, New York, 1999.

intracellular and extracellular compartments in the form of neurofibrillary tangles (NFTs), neuropil threads (NTs), and senile plaques (SPs), predominantly within the hippocampal formation (entorhinal cortex, Ammon's horn—CA1, CA2, CA3, and CA4—dentate gyrus, and subicular complex) and neocortex (the secondary association and polymodal cortex). Additionally, subsets of neurons appear selectively vulnerable to neurodegeneration within the AD brain, including CA1 pyramidal neurons of the hippocampal formation and layers II, III, and V corticocortical projection neurons of secondary association cortex, leading to regionally specific synaptic and connectivity-based degeneration (see Hof *et al.,* this volume). A diagnosis of AD is made postmortem on aged patients who display progressive cognitive decline and the presence of abundant NFTs and SPs with dystrophic processes—the so-called *neuritic plaque*—upon microscopic evaluation of hippocampus and neocortex.

The dementia associated with unsuccessful aging has been documented for centuries (Berchtold and Cotman, 1998), but Alzheimer (1907) was the first to describe the clinicopathological entity that bears his name. Recent advances in neuroscience, including cellular and molecular approaches to evaluating neurodegenerative diseases, have yielded insight into the underlying structural components of both tangle and plaque lesions. Genetic studies of the familial forms of Alzheimer's disease (FAD), along with the more prevalent sporadic form of AD, have demonstrated that what we designate "AD" is a heterogeneous disorder with differing causes (Fig. 1) but similar phenotype and neuropathological hallmarks (see Vogt *et al.,* this volume). Therefore, distinct factors may activate degenerative pathways that share NFTs, NTs, and SPs as final common pathological features.

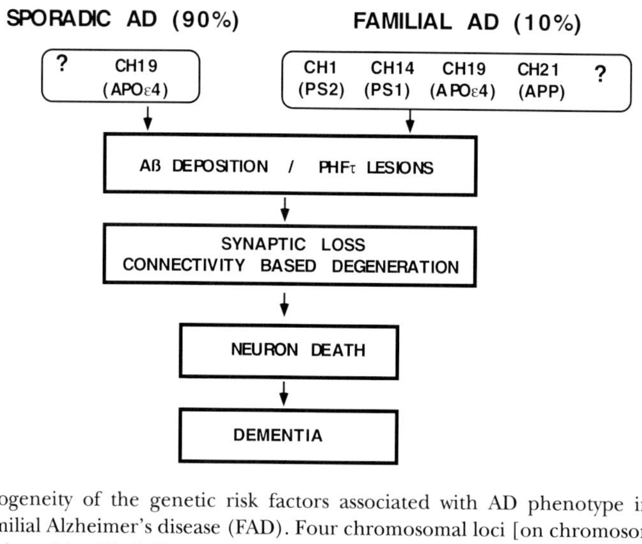


**Figure 1.** Heterogeneity of the genetic risk factors associated with AD phenotype in patients with sporadic and familial Alzheimer's disease (FAD). Four chromosomal loci [on chromosomes (CH) 1, 14, 19, and 21] have been identified. Homozygosity for the ε4 allele of the apolipoprotein E (apoE) gene on CH 19 is a risk factor for sporadic AD and some FAD cases, whereas mutations in the PS1 (CH 14, the majority of early-onset FAD), PS2 (CH 1, Volga German pedigree), and APP (CH 21, very few cases) genes are associated with the development of FAD. Convergence of pathogenic mechanisms for the forms of AD is hypothesized because characteristic accumulation of neurofibrillary lesions (PHFt) and deposition of amyloid-β protein (Aβ) is observed, and these lesions may be indicative of a final common degenerative pathway to synaptic and circuit dissolution, frank neuron loss, and resultant dementia.

The intracellular formation of NFTs in somatodendritic compartments and NTs within distal dendrites and axons is believed to be one of the key abnormalities in the AD brain and may result from the progressive neuronal degeneration observed with hippocampus and neocortex of affected individuals, which leads to subsequent synaptic loss and ultimately to loss of functional circuits within the telencephalon. Neuritic plaques, that is, extracellular deposits of amyloid-β protein (Aβ) that also contain dystrophic neurites, are believed to be repositories of dying cellular processes and reactive glial cells. The Aβ-containing amyloid fibrils in these lesions may be toxic to neurons and their processes in the local milieu. Tangle and plaque lesions are not restricted to the AD brain; they are observed in lower numbers in nondemented senescent humans. Additionally, a small number of central nervous system disorders are characterized by tangle formation and neuron loss, suggesting that a pathway leading to cell death involves the formation of NFTs independent of Aβ deposition.

## 2. Genetics of Familial Alzheimer's Disease and Sporadic Alzheimer's Disease

### 2.1. Familial Alzheimer's Disease Caused by Mutations in the Amyloid-β Precursor Protein and Presenilin Genes

Recent advances in the genetics of FAD and sporadic AD have demonstrated the heterogeneous nature of this disorder (Fig. 1). For example, five autosomal dominant mutations in the amyloid-β precursor protein (APP) gene on chromosome 21 account for a very small percentage of FAD kindreds, but provide genetic evidence for the abnormal function of APP processing and Aβ deposition in the pathogenesis of AD (Selkoe, 1996; Hendriks *et al.*, 1992; Chartier-Harlin *et al.*, 1991; Goate *et al.*, 1991; Murrell *et al.*, 1991). Autosomal dominant mutations in genes encoding two recently cloned polypeptides known as *presenilin-1* (PS1), or S182, on chromosome 14, and *presenilin-2* (PS2), or STM2, on chromosome 1 account for most FAD kindreds (Levy-Lahad *et al.*, 1995; Rogaev *et al.*, 1995; Sherrington *et al.*, 1995). To date, approximately 50 mutations have been identified within the PS1 gene, and two mutations have been detected in the PS2 gene (Cruts and Van Broeckhoven, 1998; Haas, 1997; Lendon *et al.*, 1997; Levy-Lahad *et al.*, 1995; Rogaev *et al.*, 1995; Sherrington *et al.*, 1995). Not all known FAD kindreds have been linked to PS1, PS2, and APP mutations; therefore other gene mutations linked to FAD are likely to be identified.

The mutations within presenilin genes in FAD have generated much interest in the functional characterization of these proteins, with the hope that analyses will lead to insight into the pathogenesis of FAD, and ultimately, of sporadic AD, which share similar clinical and pathophysiological features. PS1 and PS2 are transmembrane proteins of approximately 50 kiloDaltons (kDa) that have 67% sequence homology and are localized primarily to the endoplasmic reticulum and Golgi apparatus by immunocytochemistry (Cook *et al.*, 1996; Kovacs *et al.*, 1996; Lee *et al.*, 1996; Suzuki *et al.*, 1996; Rogaev *et al.*, 1995; Sherrington *et al.*, 1995). The mRNA and protein of PS1 and PS2 are distributed throughout mouse, rat, monkey, and human tissues, including brain, with notable enrichment in the hippocampus, entorhinal cortex, and cerebellum (Lah *et al.*, 1997; Kovacs *et al.*, 1996; Lee *et al.*, 1996; Suzuki *et al.*, 1996). The presenilins belong to a distinct gene family with little homology to known mammalian genes. However, they are similar to *Caenorhabditis elegans* sel-12 (~50% homology) and spe-4 (~25% homology) genes and protein products (Levitan and Greenwald, 1995; L'Hernault and Arduengo, 1992). Sel-12 interacts with lin-12 of the LIN-1/Notch superfamily of receptors, which functions to regulate cell fate in *C. elegans* (Levitan and Greenwald, 1995). The human PS1 and PS2 proteins can substitute functionally in *C. elegans* cell-fate development for the wild-type sel-12 protein, and PS1/PS2 FAD mutants have lower activity in cell-fate recovery (Levitan *et al.*, 1996). Notch and the presenilins are also coexpressed in neural progenitor cells within the ventricular zone of mouse and rat embryos (Berezovska *et al.*, 1997). Although the function of the presenilins in mammalian systems—especially their role in the development and/or progression of FAD—remains unknown, the sequence and functional homology to sel-12 suggests a potential role in mammalian cell fate/differentiation. Yeast two hybrid system analysis has demonstrated that PS1 interacts with catenins (notably β-catenin) and related members of the Armadillo gene family in *Drosophila,* molecules that play a role in cell–cell adherens junctions, and are also involved in the Wnt signaling pathway in dividing cells (Kosik, 1998; Murayama *et al.*, 1998; Zhou *et al.*, 1998). Furthermore, knockout of the PS1 gene in mouse causes severe skeletal deformations, CNS hemorrhage, and subsequent death within hours of birth, indicating that PS1 expression is essential for normal mammalian development of skeletal tissues and neurogenesis (Shen *et al.*, 1997; Wong *et al.*, 1997). Transgenic mouse lines expressing either the wild-type human PS1 protein or mutant PS1 can rescue the PS1 knockout mouse from the embryonic lethal phenotype, suggesting that the mutations in PS1 cause a loss of function during development (Davis *et al.*, 1998; Qian *et al.*, 1998). Cell culture and transgenic mouse studies have also detected an interaction between mutations in the PS1 and PS2 genes observed in FAD kindreds and increased production of the amyloidogenic fragment Aβ1-42 (Mehta *et al.*, 1998; Citron *et al.*, 1997; Hardy, 1997; Tomita *et al.*, 1997; Borchelt *et al.*, 1996; Duff *et al.*, 1996). Specifically, an increased ratio of Aβ1-40/Aβ1-42 caused by defective presenilins may act as a seed or nidus for fibrillar Aβ deposition, as Aβ1-42 rapidly forms fibrils *in vitro* and is the principal component of SP lesions (Dickson, 1997; Iwatsubo *et al.*, 1994; Schmidt *et al.*, 1994; Jarrett *et al.*, 1993; Roher *et al.*, 1993a). Although our current understanding of the mechanisms leading to AD pathogenesis is rudimentary, the aforementioned genetic studies have begun to elucidate genomic alterations that have functional sequelae, and they ultimately may lead to further characterization of antemortem genetic markers in individuals with a predisposed risk factor for dementia, as well as to palliative gene therapies to correct genetic mutations that cause neurodegeneration.

### 2.2. ApoE as a Risk Factor for Alzheimer's Disease

In addition to the APP and presenilin mutations in FAD, the risk of developing sporadic AD (approximately 90% for affected patients) has been linked to the apolipoprotein E (apoE) gene on chromosome 19. Three major alleles have been identified—apoE ε2, ε3, and ε4—with the ε3 allele being the most common. Individuals homozygous for the apoE ε4 allele are more likely to develop sporadic AD or FAD at an earlier age of onset as compared with individuals with one (heterozygous) or no copies of the ε4 allele (Blacker *et al.*, 1997; Corder *et al.*, 1993; Saunders *et al.*, 1993; Strittmatter *et al.*, 1993a). In contrast, the presence of the ε2 allele may decrease the risk for AD, but it is associated with a higher risk of hemorrhage due to cerebral amyloid angiopathy (Nicoll *et al.*, 1997; Corder *et al.*, 1994). Apolipoprotein E is a 34 kDa glycoprotein synthesized principally in the liver, and it functions in lipid transport—notably cholesterol—and metabolic pathways (Mahley, 1988). It binds to soluble forms of the Aβ peptide *in vitro,* and isoform-specific interactions have been demonstrated between apoE and amyloid fibril formation (Haas *et al.*, 1997; Näslund *et al.*, 1995; Sanan *et al.*, 1994; Wisniewski *et al.*, 1994; Strittmatter *et al.*, 1993b). Apolipoprotein E also binds the microtubule-associated protein (MAP) tau in its nonphosphorylated state *in vitro* in an isoform-specific manner, with the ε3 allele binding with a higher affinity to tau than does the ε4 allele (Strittmatter *et al.*, 1994). Furthermore, apoE increases neurite outgrowth *in vitro* (Puttfarcken *et al.*, 1997; Nathan *et al.*, 1994) and may play a role in neuronal viability. The brain is rich in apoE mRNA and protein (Corder *et al.*, 1994; Strittmatter *et al.*, 1994; Elshourbagy *et al.*, 1985), with astrocytes being the primary cell type for protein localization. In addition to being observed within the normal brain, apoE immunoreactivity has also been localized to NFT and SP lesions in AD and other neurodegenerative disorders (Bao *et al.*, 1996; Näslund *et al.*, 1995; Rebeck *et al.*, 1993; Namba *et al.*, 1991). However, the function of apoE in the normal brain (as well as its role in the pathogenesis of AD) is unknown. For example, apoE knockout mice have altered long-term potentiation and learning responses and lack of apoE reduces amyloid deposition in double transgenic (apoE and APP constructs) mice (Raber *et al.*, 1998; Veinbergs *et al.*, 1998; Bales *et al.*, 1997). One hypothesis is that apoE—notably the ε4 allele—may act as a pathological chaperone to increase deposition or block reuptake of soluble Aβ peptides and/or hyperphosphorylated tau.

## 3. Neurofibrillary Tangles: Pathophysiological Alterations in Hippocampus and Neocortex

### 3.1. Tau Structure and Function

Tau is an abundant, low molecular weight, microtubule-associated protein (MAP) found in the CNS and PNS, and is localized primarily to axonal compartments in normal brain (Couchie *et al.*, 1992; Binder *et al.*, 1985). Abnormally phosphorylated tau is observed within neurofibrillary lesions (Fig. 2) of the AD brain and other tangle-bearing disorders (Table 1). The paired helical filaments (PHFs) and straight filaments (SFs) of these lesions have somatodendritic, distal dendritic, and axonal distribution. Neurofibrillary tangles and neuropil threads are observed primarily in the cerebral cortex, with the hippocampal formation, parahippocampal gyrus, temporal association areas, and parietal association areas being particularly involved (Fig. 2). In contrast, the primary motor and sensory cortices

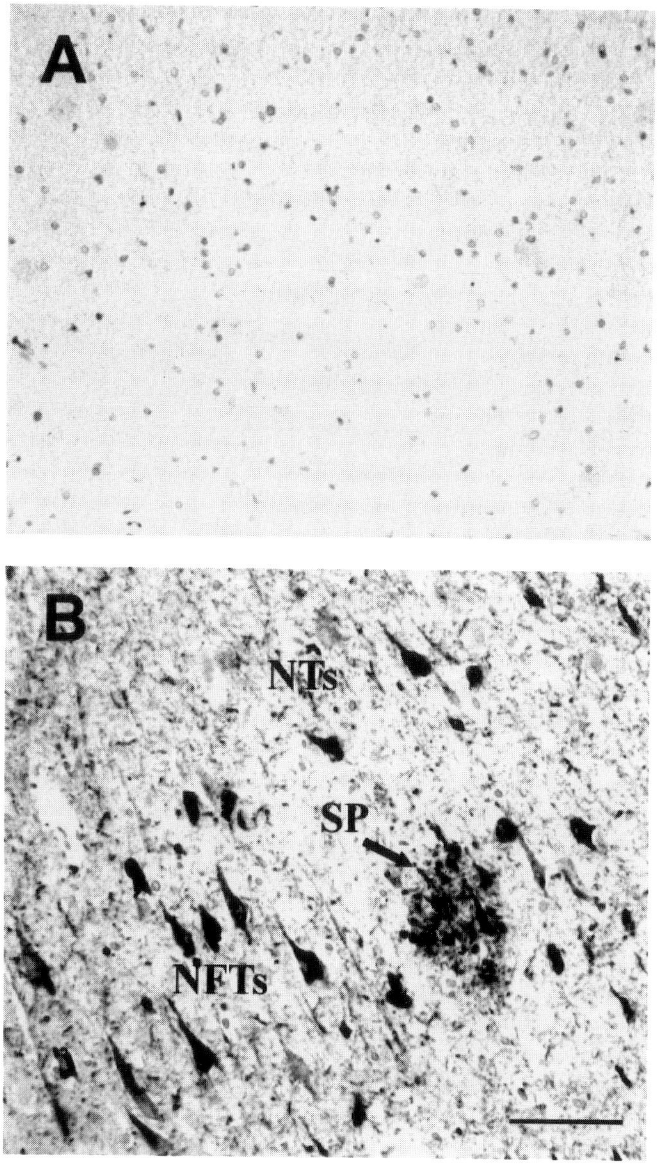


**Figure 2.** Neurofibrillary pathology in the hippocampal formation of AD brain. Tissue sections were immunostained with antibodies directed against hyperphosphorylated tau (PHFtau; antibody PHF1) and counterstained lightly with hematoxylin. (A) Tissue section from an age-matched normal control. Note the absence of immunoreactive neurofibrillary lesions. (B) Representative distribution of tau-positive neurofibrillary lesions within the CA1-subiculum region of a patient with AD. Note the presence of NFTs, NTs, and a dystrophic neurite-containing neuritic plaque (SP; arrow). Bar = 100 μm.

and the occipital lobe are relatively spared, but rare AD cases exhibit visual deficits associated with Bálint's syndrome and display numerous NFTs throughout occipital and posterior parietal association areas (Hof *et al.,* 1990). Neurofibrillary tangles associated with AD are also observed with varying degrees of severity in several subcortical regions, notably the amygdala (Braak and Braak, 1994). Cloning of the

**Table I. Neurodegenerative Diseases that have Neurofibrillary Pathology**

| |
|---|
| Alzheimer's disease |
| Down's syndrome |
| Lewy body variant of Alzheimer's disease |
| Dementia pugilistica |
| Inclusion body myositis |
| Sporadic dementia with tangles |
| Presenile familial dementia |
| Argyrophilic-grain dementia |
| Guamanian ALS/PDC |
| Pick's disease |
| Progressive supranuclear palsy |
| Multiple system atrophy |
| Multiple system tauopathy with presenile dementia |
| Chromosome 17-linked frontotemporal dementia and Parkinsonism (FTDP-17) |
| Corticobasal degeneration |
| Niemann-Pick disease type C |
| Hallervorden-Spatz disease |
| Subacute sclerosing panencephalitis |
| Postencephalitic parkinsonism |
| Diffuse neurofibrillary tangle disease with calcification |
| Gerstmann–Sträussler–Scheinker syndrome |
| Creutzfeldt–Jakob disease |
| Prion protein cerebral amyloid angiopathy |

tau gene has demonstrated that multiple tau isoforms are produced from a single gene on chromosome 17 through alternative mRNA splicing (Goedert *et al.*, 1989a, 1988; Lee *et al.*, 1988). Six isoforms of tau protein are found in the adult human brain. All of the isoforms share three or four tandem repeats of 31 or 32 amino acids in the carboxy-terminus of the molecule, and these tandem-repeat regions contain microtubule (MT)-binding domains (Goode and Feinstein, 1994; Butner and Kirschner, 1991; Goedert *et al.*, 1989a,b). The six isoforms differ from each other by the presence or absence of 29- or 58-amino acid inserts in the amino-terminal region (Fig. 3). In fetal brain, tau is developmentally regulated in that only the shortest isoform, with three tandem repeats and no inserts, is expressed (Goedert *et al.*, 1997, 1989a).

As a neuronal cytoskeletal constituent and member of the low molecular-weight MAPs, tau functions to stabilize MTs in the polymerized configuration (Hirokawa *et al.*, 1988). Tau also promotes polymerization of tubulin monomers into MTs (Goode and Feinstein, 1994). Phosphorylation negatively regulates tau in its ability to bind to MTs (Drechsel *et al.*, 1992), and occurs at many proline-directed serine/threonine sites, although phosphorylation at non-proline-directed sites has been observed (Goedert *et al.*, 1997; Morishima-Kawashima *et al.*, 1995; Trojanowski and Lee, 1994). The phosphorylation sites have been mapped for tau protein isoforms by mass spectrometry and phosphorylation dependent/independent antibodies. Earlier, fetal tau was thought to be phosphorylated at 12 serine/threonine sites (serine residues 198, 199, 202, 235, 262, 396, 400, and 404; threonine residues 181, 205, 217, and 231, according to the longest adult tau nomenclature) (Goedert *et al.*, 1989b), and the adult forms were thought to be phosphorylated on at least 9 serine/threonine sites (serine residues 199, 202, 235, 262, 396, and 404; threonine residues 181, 205, and 231) (Trojanowski and Lee, 1994; Hasegawa *et al.*, 1993; Poulter *et al.*, 1993; Watanabe *et al.*, 1993; Kanemaru *et al.*, 1992). However, studies of biopsy-derived human brain tau indicate that reassessment of the normal sites of tau phosphorylation in human brain is needed, because normal tau undergoes rapid dephosphorylation postmortem or after surgical excision (Matsuo *et al.*, 1994; see below). Tau is phosphorylated by several kinases *in vitro,* including MAP kinase, glycogen synthase kinase-3, cyclin-dependent kinase-5, cAMP-dependent protein kinase, $Ca^{2+}$/calmodulin-dependent protein kinase-2 (CaM-KII), and MARK, a novel family of kinases that phosphorylate several MAPs (Drewes *et al.*, 1997; Litersky *et al.*, 1996; Sperber *et al.*, 1995; Hanger *et al.*, 1992; Mandelkow *et al.*, 1992).

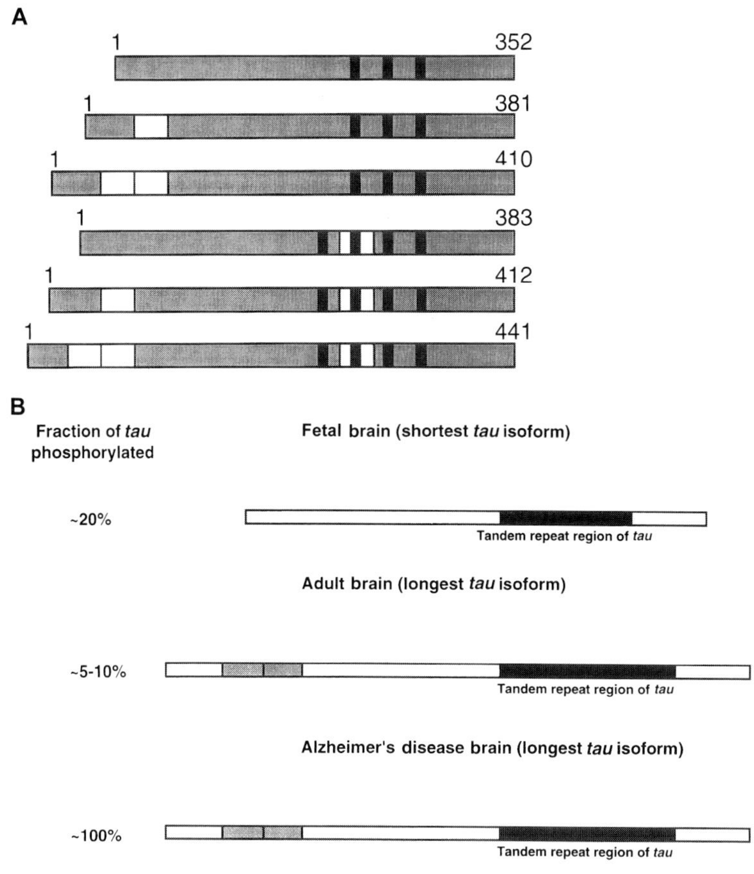


**Figure 3.** Schematic representations of fetal, adult, and AD tau protein. (A) Illustration of the six normal adult human brain tau isoforms. The numbers represent amino acid residues. The 352-amino acid isoform is expressed in fetal brain, and all six tau isoforms are expressed in adult brain. The regions of high homology are shown in gray, and the insert regions that distinguish the tau isoforms are depicted in white. The three- or four-tandem repeat regions in the carboxyl-third of the molecule (microtubule-binding regions) are shown as black bars. (B) Representation of the phosphorylation state of fetal, normal adult, and AD tau. The approximate fraction of tau molecules that are phosphorylated is shown on the left for fetal tau and for the longest adult and AD tau isoforms, indicating the hyperphosphorylation of tau proteins in AD brain.

Conversely, tau is dephosphorylated *in vitro* by protein phosphatase 2A (PP2A) and by protein phosphatase 2B (PP2B or calcineurin) (Sontag *et al.*, 1996, 1995; Goedert *et al.*, 1995a, 1992; Matsuo *et al.*, 1994; Mawal-Dewan *et al.*, 1994).

## 3.2. Paired Helical Filament Tau Structure and Localization to Neurofibrillary Tangles and Dystrophic Processes in Alzheimer's Disease

The major structural feature of neurofibrillary lesions found within the telencephalon of AD patients is the paired helical filament (PHF), which consist of two filaments twisted into a helix-like configuration (Fig. 4). Electron microscopy has revealed that the width of a PHF alternates between approximately 8 and 20 nm,

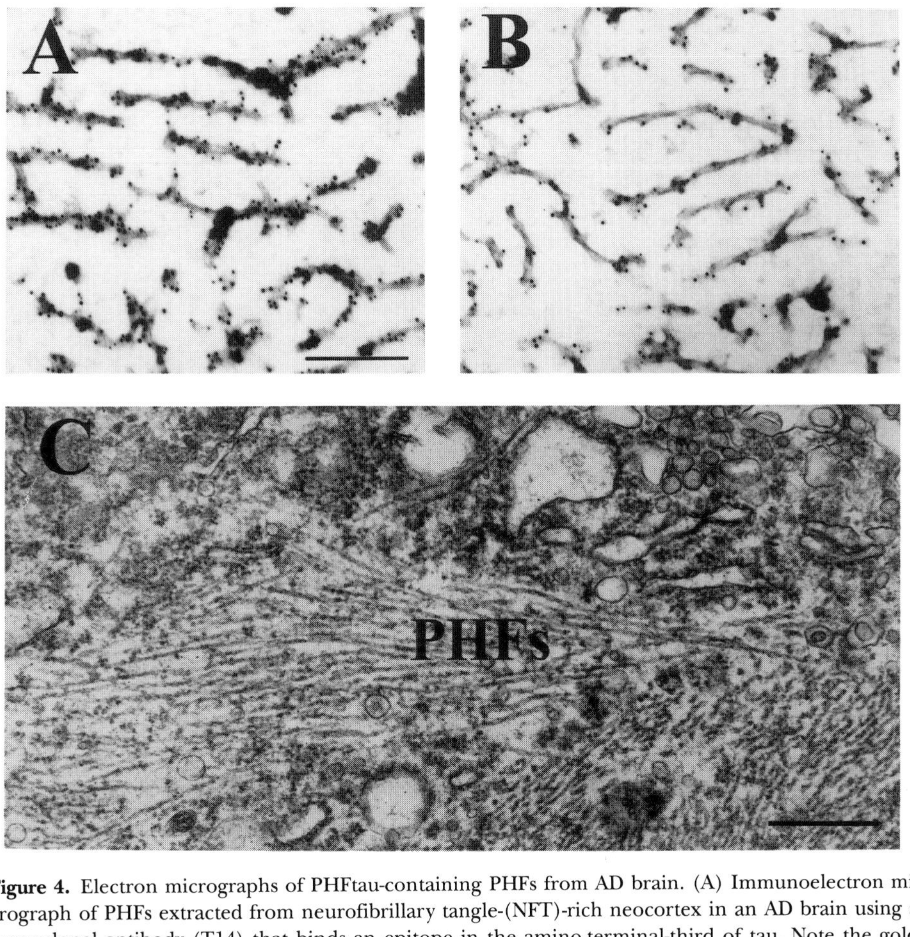


**Figure 4.** Electron micrographs of PHFtau-containing PHFs from AD brain. (A) Immunoelectron micrograph of PHFs extracted from neurofibrillary tangle-(NFT)-rich neocortex in an AD brain using a monoclonal antibody (T14) that binds an epitope in the amino-terminal-third of tau. Note the gold particles (black puncta) decorating the PHFs. (B) A preparation similar to A, immunolabeled with a monoclonal antibody (T46) that recognizes the carboxyl-terminal of tau. (C) Electron micrograph of the cytoplasm of an NFT-bearing neuron from AD neocortex. Note the abundance of PHFs throughout the cytoplasm and the marginization of remaining organelles. Bar = 200 nm (A,B); 100 nm (C).

with a periodicity of 80 nm (Crowther and Wischik, 1985). Paired helical filaments can be observed as dispersed filaments or in tangles—elements that differ in their reaction to denaturing agents. Specifically, PHFs in dispersed filaments are soluble in guanidine or sodium dodecyl sulfate, whereas PHFs in tangle fragments are far less soluble. The major species of both dispersed filaments and tangle fragments is the PHF, although a minor species (approximately 10%) is the straight filament (SF). Several techniques, including protein chemistry, molecular cloning, and immunoelectron microscopy, have been employed to demonstrate that hyperphosphorylated tau, known as *PHFtau,* is the building block of neurofibrillary lesions in the AD brain (Lee *et al.*, 1991; Goedert *et al.*, 1988; Kondo *et al.*, 1988; Kosik *et al.*, 1988, 1986; Wischik *et al.*, 1988; Crowther and Wischik, 1985). The PHFtau is composed of all six adult tau isoforms and is abnormally phosphorylated as well as hyperphosphorylated. "Abnormal phosphorylation" indicates that more amino acids in the PHFtau molecule are phosphorylated than in normal tau; "hyperphosphorylation" indicates that for each phosphorylated serine/threonine residue, a greater proportion of PHFtau molecules are phosphorylated than in normal adult tau (Fig. 3). Tau extracted from PHFs is observed on polyacrylamide gels as three major bands of approximately 60, 64, and 68 kDa, along with a variable background smear caused by partially degraded and cross-linked PHFs (Lee *et al.*, 1991; Greenberg and Davies, 1990). Phosphorylation sites in the PHFtau molecule have been identified via methods similar to those that identify normal fetal and adult tau, for example, mass spectrometry and phosphorylation-dependent anti-tau antibodies. PHFtau is phosphorylated on at least 21 serine/threonine sites (serine residues 198, 199, 202, 208, 210, 214, 231, 235, 262, 396, 400, 404, 409, 412, 413, and 422; threonine residues 181, 205, 212, 217, and 403) (Goedert *et al.*, 1995b, 1994, 1993; Morishima-Kawashima *et al.*, 1995; Seubert *et al.*, 1995; Matsuo *et al.*, 1994; Otvos *et al.*, 1994; Hasegawa *et al.*, 1993, 1992; Lee *et al.*, 1991). Several earlier studies have demonstrated that the phosphorylation state of PHFtau partially recapitulated the phosphorylation state of fetal rat and human tau, but not adult tau (Morishima-Kawashima *et al.*, 1995; Mawal-Dewan *et al.*, 1994; Bramblett *et al.*, 1993; Goedert *et al.*, 1993; Watanabe *et al.*, 1993; Kanemaru *et al.*, 1992). For example, Ser-202 and Ser-306 are phosphorylated in both PHFtau and fetal tau, but not in any of the six adult forms of tau isolated from postmortem tissue of normal adult cerebral cortex (Bramblett *et al.*, 1993; Goedert *et al.*, 1993). However, subsequent studies using rapidly isolated tau proteins from fresh cortical biopsy tissue in human cortex (Matsuo *et al.*, 1994) and fresh tissue from nonhuman primate and rodent cortex (Seubert *et al.*, 1995; Garver *et al.*, 1994; Mawal-Dewan *et al.*, 1994) suggest that adult tau is phosphorylated at nearly all of the same sites as PHFtau, although to a significantly lesser extent. Therefore, previously identified differences in the phosphorylation state of normal CNS tau and PHFtau from postmortem brain samples may reflect a more rapid dephosphorylation of normal CNS tau relative to PHFtau at equivalent postmortem intervals.

The aberrant hyperphosphorylation state of PHFtau has functional sequelae in the reduction or inability of PHFtau to bind to MTs. The loss of MT-binding presumably causes depolymerization of the MTs and subsequent disruption of axonal transport, and ultimately, neuronal degeneration (Fig. 5) The diminished MT-binding properties of PHFtau are regulated by the phosphorylation state, as dephosphorylation of PHFtau restores the ability to bind MTs (Iqbal *et al.*, 1994;

Bramblett *et al.*, 1993). Additionally, filament formation may be induced by interactions of PHFtau released from MTs with other molecules, such as RNA or proteoglycans (Arrasate *et al.*, 1997; Hasegawa *et al.*, 1997; Goedert *et al.*, 1996; Kampers *et al.*, 1996; Perez *et al.*, 1996). Therefore, these studies suggest that the accumulation of PHFs within neurofibrillary lesions of the AD brain may be initiated by aberrant activation of protein kinases and/or inactivation of protein phosphatases, with tau being the substrate for both, but with other factors contributing to PHF formation (Fig. 5). The formal characterization of this hypothetical cascade of events, primarily within hippocampal and neocortical circuits, will be critical to elucidate the underlying mechanisms of degeneration that cause the pervasive dementia and debilitation associated with AD.

### 3.3. Neurofibrillary Pathology in Other Tangle-Bearing Diseases

The NFTs, NTs, and dystrophic neurites observed principally within telencephalic regions of the AD brain are also hallmark lesions of other CNS disorders (Table 1). For example, both neurofibrillary and Aβ lesions are observed in patients with Down's syndrome (Flament *et al.*, 1990; Mann *et al.*, 1986), dementia pugilistica (Roberts *et al.*, 1990; Roberts, 1988), and inclusion-body myositis (Askanas *et al.*, 1994, 1992; Mendell *et al.*, 1991). NFTs and SPs, along with Lewy bodies, are also pathological hallmarks of the Lewy body variant of AD (LBVAD) (Trojanowski *et al.*, 1998a, 1998b; Cummings, 1995; Hansen *et al.*, 1990). Interestingly, α-synuclein has been demonstrated recently to be a major protein constitu-

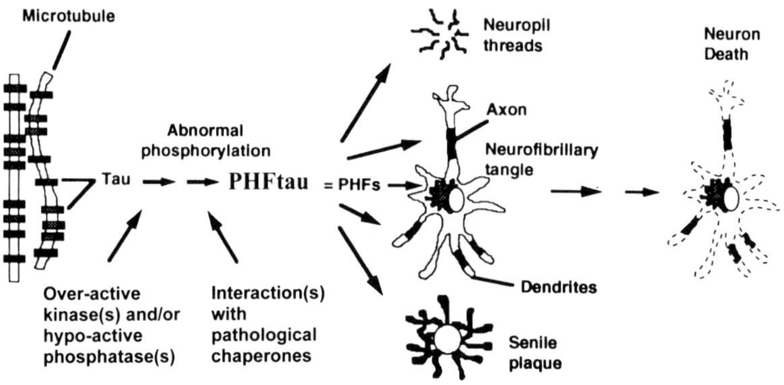


**Figure 5.** A hypothetical mechanism for the conversion of normal brain tau (rectangles overlying two MTs on the left) into PHFtau. PHFtau is generated in neuronal perikarya and their processes as a consequence of overactivity or inappropriate activation of kinase(s) and/or hypoactivity or loss of phosphatase(s) that regulate the phosphorylation state of CNS tau. Filament formation may be augmented when the excess pool of unbound tau interacts with potential pathological chaperones such as RNA and heparan sulfate proteoglycan. PHFtau then accumulates as filamentous aggregates in neuronal processes (i.e., NTs; top), in neuronal perikarya as NFTs (middle), and in dystrophic neurites associated with neuritic senile plaques (bottom). PHFtau loses the ability to bind MTs, which could lead to MT depolymerization, axonal transport disruption, and neuronal dysfunction and/or degeneration in AD. Accumulation of PHFs in neurons and dendrites could exacerbate this process by sequestering normal tau and by physically blocking transport of proteins and organelles in neuronal perikarya, axons, and dendrites. Death of the neuron (right) is the culmination of this intracellular cascade of events.

ent of Lewy body filaments in Lewy body-containing diseases (including FAD) and glial cytoplasmic inclusions seen in multiple system atrophy (Lippa *et al.*, 1999, 1998; Baba *et al.*, 1998; Takeda, *et al.*, 1998a; Tu *et al.*, 1998; Spillantini *et al.*, 1997b). Neurofibrillary pathology in the absence, or low abundance, of SPs is observed in patients with sporadic, familial presenile, and argyrophilic grain dementias (Reed *et al.*, 1997; Bancher and Jellinger, 1994; Sumi *et al.*, 1992; Itagaki *et al.*, 1989a; Braak and Braak, 1987); amyotrophic lateral sclerosis/parkinsonism–dementia complex (ALS/PDC) of Guam (Mawal-Dewan *et al.*, 1996; Buee-Scherrer *et al.*, 1995; Hirano *et al.*, 1966); Pick's disease (Murayama *et al.*, 1990; Perry *et al.*, 1987); progressive supranuclear palsy (Schmidt *et al.*, 1996; Flament *et al.*, 1991; Bancher *et al.*, 1987); multiple system atrophy including glial NFT inclusions (Papp and Lantos, 1992; Papp *et al.*, 1989); multiple system tauopathy with presenile dementia including neuronal and glial NFT inclusions (Spillantini *et al.*, 1997a); chromosome 17-linked frontotemporal dementia and parkinsonism (FTDP-17) (Reed *et al.*, 1998; Spillantini *et al.*, 1998a; Bird *et al.*, 1997; Foster *et al.*, 1997); corticobasal degeneration (Ksiezak-Reding *et al.*, 1994; Mori *et al.*, 1994; Wakabayashi *et al.*, 1994; Paulus and Selim, 1990); Niemann-Pick disease type C (Auer *et al.*, 1995; Love *et al.*, 1995; Suzuki *et al.*, 1995); Hallervorden-Spatz disease (Eidelberg *et al.*, 1987); subacute sclerosing panencephalitis (McQuaid *et al.*, 1994, 1993; Mandybur *et al.*, 1977); postencephalitic parkinsonism (Buée-Scherrer *et al.*, 1997; Wong *et al.*, 1996; Geddes *et al.*, 1993); and diffuse neurofibrillary tangles with calcification (Kosaka, 1994). In addition, neurofibrillary lesions are observed along with prion protein amyloid deposits in Gerstmann-Sträussler-Scheinker syndrome (Tagliavini *et al.*, 1993; Ghetti *et al.*, 1989), Creutzfeldt-Jakob disease (Hsiao *et al.*, 1992), and prion protein cerebral amyloid angiopathy (Ghetti *et al.*, 1996). Similar to PHFs in the AD brain, hyperphosphorylated PHFtau is the structural component of these neurofibrillary lesions. The PHFtau extracted from the brains of Down's syndrome, ALS/PDC of Guam, Niemann-Pick type C, and Gerstmann–Sträussler–Scheinker syndrome displays three major bands similar in electrophoretic mobility to those observed in AD (Auer *et al.*, 1995; Buée-Scherrer *et al.*, 1995; Tagliavini *et al.*, 1993; Flament *et al.*, 1990). Immunocytochemistry using phosphate-dependent anti-tau antibodies has demonstrated that similar epitopes exist within the NFTs in AD brain and in several other diseases, including Down's syndrome, ALS/PDC of Guam, and Niemann-Pick type C (Mawal-Dewan *et al.*, 1996; Auer *et al.*, 1995; Buée-Scherrer *et al.*, 1995). In contrast, only the 60 kDa and 64 kDa species are observed in PHFtau extracted from Pick's disease, and only the 64 kDa and 68 kDa species are observed in PHFtau from patients with progressive supranuclear palsy and corticobasal degeneration (Buée-Scherrer *et al.*, 1996; Delacourte *et al.*, 1996; Ksiezak-Reding *et al.*, 1994; Flament *et al.*, 1991). Additionally, the morphology of PHFs differs in progressive supranuclear palsy, corticobasal degeneration, and postencephalitic parkinsonism as compared with AD, in that SFs are the major species of filament (Ksiezak-Reding *et al.*, 1994; Geddes *et al.*, 1993; Takahashi *et al.*, 1989). However, neurofibrillary lesions from AD brain and progressive supranuclear palsy contain the same hyperphosphorylated epitopes of PHFtau as those assessed by phosphorylation-dependent anti-tau antibodies (Schmidt *et al.*, 1996; Bancher *et al.*, 1987). Therefore, these studies indicate some heterogeneity in the structure and configuration of PHFtau-containing lesions, yet there is overlap in the hyperphosphorylated tau epitopes that constitute the building blocks of NFTs and NTs.

Recent genetic analyses of kindreds with FTDP-17 has demonstrated multiple mutations within the tau gene on chromosome 17, including missense mutations in exons 9, 10, 12, and 13, as well as point mutations in the 5′ splice donor site of exon 10 (Fig. 6); (Clark *et al.*, 1998; Hutton *et al.*, 1998; Poorkaj *et al.*, 1998; Spillantini *et al.*, 1998b; D'Souza *et al.*, 1999; Iijima *et al*, 1999; Rizzv *et al.*, 1999). FTDP-17 kindreds display wide phenotypic variety that may reflect the function (i.e., dysfunction, and/or toxic gain of function) of individual tau mutations, as mutations in exons 9, 10, and 12 lie within the MT-binding domains of tau, whereas mutations in exon 13 affect the carboxy terminus of the MT-binding region (Clark *et al.*, 1998; Hong *et al.*, 1998; Hutton*et al.*, 1998; Poorkaj *et al.*, 1998; Spillantini *et al.*, 1998b). Further understanding of the "tauopathic" process(es) that occur in FTDP-17 may provide critical information underlying the mechanism(s) of disease in sporadic AD and related NFT-bearing diseases.

## 4. Senile Plaques: Pathophysiological Alterations in Hippocampus and Neocortex

### 4.1. Amyloid Deposition and Alzheimer's Disease

Amyloid deposition is a potentially pathological process whereby straight fibrils of approximately 5–10 nm in diameter accumulate within extracellular compartments of the CNS, vasculature, and peripheral tissues (Cohen and Connors, 1987). Amyloid fibrils are highly insoluble and readily observed by Congo red birefringence or thioflavine S (TS) fluorescence due to a preferred β-pleated sheet conformation. Many proteins, including Aβ, are capable of forming amyloid fibrils *in vivo* (Cohen and Connors, 1987). The extracellular accumulation of abundant SPs within the hippocampus and neocortex (frontal, temporal, and parietal lobes) of elderly patients is an absolute requirement for a postmortem diagnosis of AD. Although several proteins are associated with SPs in the AD brain, including NACP (nonamyloid component of plaques; a 35 amino acid proteolytic fragment of the α-synuclein protein that is found in Lewy bodies) (Takeda *et al.*, 1998a; Irizarry *et al.*, 1996; Masliah *et al.*, 1996; Iwai *et al.*, 1995), α1-anti-chymotrypsin, chromogranin A, serum amyloid P, heparan sulfate proteoglycan, several complement proteins (Clq, C3c, C5b-9), and apoE (Selkoe, 1996, 1997; Pasinetti, 1996; Schmidt *et al.*, 1994), the major protein purified from extracellular plaques (SPs and amyloid around the cerebrovasculature) is the ~4 kDa Aβ protein (Masters *et al.*, 1985; Glenner and Wong, 1984).

Although SPs are a distinguishing pathological hallmark of AD, heterogeneity exists in their morphology and composition. Specifically, SPs are categorized as neuritic plaques or diffuse plaques (Fig. 7). *Neuritic plaques* have a dense Aβ-containing core and fibrillar corona, as well as an abundance of PHFtau-containing dystrophic neurites dispersed throughout the corona and core regions. They are associated with inflammatory activation; microglial cells permeate the cytological boundaries of the fibrillar corona, and reactive astrocytes are observed around the periphery (Ginsberg *et al.*, 1997; Itagaki *et al.*, 1989b; Dickson *et al.*, 1988; McGeer *et al.*, 1987). Neuritic plaques are the classical SPs observed within the telencephalon of the AD brain, and they are readily observed by Congo red or TS

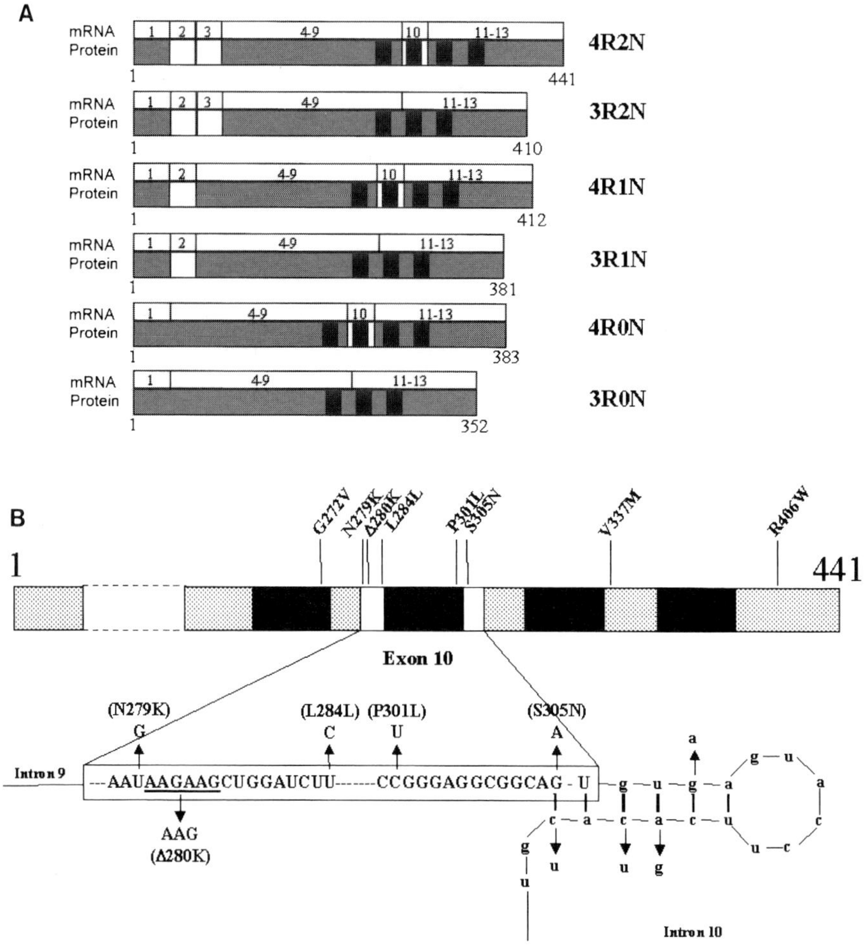


**Figure 6.** Tau mutations in FTDP-17. (A) The six alternatively spliced brain tau isoforms (see Figure 3A) are depicted to illustrate the corresponding exons in the tau mRNA. The top mRNA panel for each isoform represents the exons that comprise specific regions of the tau protein. Black boxes represent MT-binding domains and white boxes represent amino-terminal inserts. The inclusion of exons 2 and 3 results in tau isoforms with two (2N), one (1N), or no (0N) amino-terminal inserts, and the inclusion of exon 10, which encodes the second MT-binding repeat, results in tau isoforms with four (4R) or three (3R) MT-binding repeats. (B) Tau mutations in FTDP-17 are illustrated in a diagram of the longest brain isoform of tau (411 amino acid residues, 4R2N-four MT-binding repeats [black boxes] and 2 amino-terminal inserts [hatched line white box]). This schematic depicts G272V (missense mutation glycine to valine at residue 272) in the first MT-binding repeat in exon 9, N279K, A280K (AAG deletion abolishing exon 10 inclusion), L284L, P301L, and S305N within exon 10, V337M in the inter-repeat region between MT-binding repeats 3 and 4 in exon 12, and R406W in the carboxy-terminal region of exon 13. The proposed RNA stem-loop structure at the splice site of exon 10 is also shown to illustrate intronic point mutations (two C to U mutations and two G to A mutations).

staining because the amyloid fibrils lie in a β-pleated sheet conformation. *Diffuse plaques,* as their name suggests, lack a definitive amyloid core, contain few or no dystrophic neurites, do not produce an inflammatory response, nor are they visualized as readily as neuritic SPs by Congo red or TS. Rather, these lesions are identified best by immunohistochemistry using antibodies directed against the Aβ peptide (Li *et al.,* 1994; Schmidt *et al.,* 1994, 1991; Masliah *et al.,* 1990). Furthermore, the amyloidogenic fragment Aβ1-42 is deposited early and in great abun-



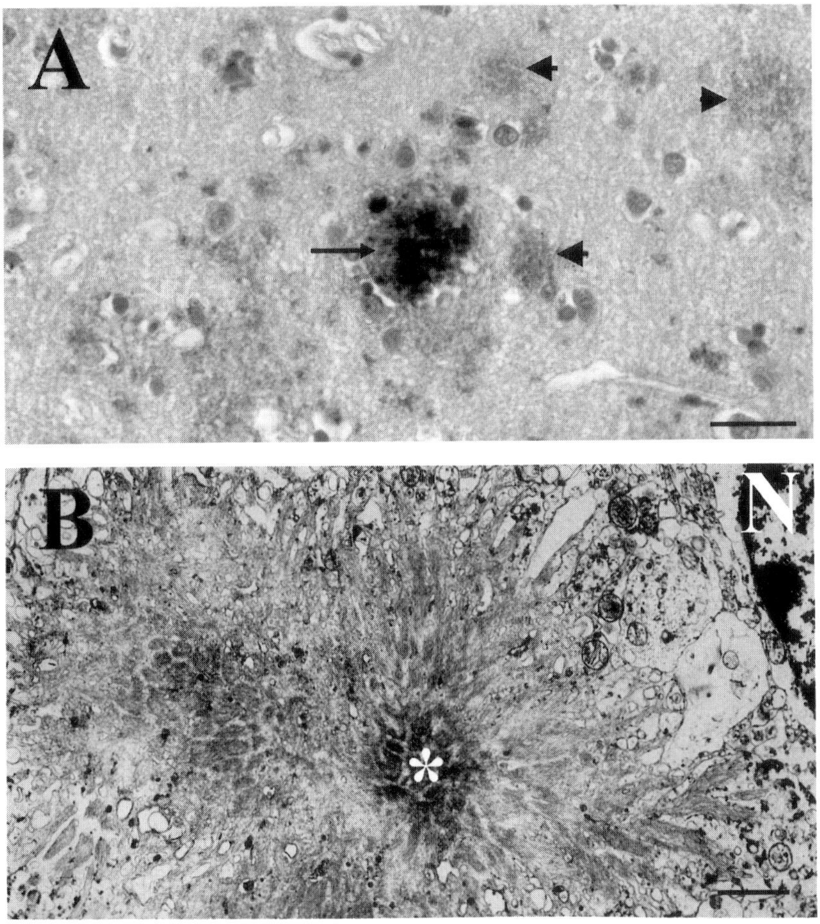


**Figure 7.** Plaque morphology in the AD brain. (A) Double-labeled immunohistochemical image of temporal neocortex from an AD patient delineating an intensely labeled dystrophic neurite-containing neuritic plaque (arrow) from pale diffuse plaques (arrowheads) that do not contain dystrophic processes. The tissue section is immunostained with antibodies directed against amyloid β-protein (labeling both diffuse and neuritic plaques) and amyloid β-precursor protein (black puncta observed within the neuritic plaque), and is lightly counterstained with hematoxylin. Barr = 100 μm. (B) Electron micrograph illustrating amyloid fibrils within a SP from temporal neocortex in AD. The fibrils orient in a radiating array, creating a "star" type morphology (asterisk); N, nucleus. Bar = 100 nm.

dance within neuritic and diffuse plaques, but other modified and unmodified species of Aβ (e.g., isomerized, racemized, pyroglutamated) are present as well; vascular amyloid also consists of both Aβ1-40 and Aβ1-42 species (Iwatsubo *et al.*, 1996, 1994; Saido *et al.*, 1995; Suzuki *et al.*, 1994; Gowing *et al.*, 1994; Roher *et al.*, 1993,b). Diffuse plaques have been termed "pre-amyloid plaques" because they are thought to represent early-stage precursors that ultimately mature into neuritic plaques. Alternatively, these lesions may have little or no pathological sequelae.

### 4.2. Amyloid Deposition and Other Neurological Disorders

In addition to AD, several other neurological diseases involve cerebral amyloid deposition, although not all of the deposits are due to Aβ accumulation (Table 2). For example, Down's syndrome (trisomy 21) patients invariably develop AD pathology by their fourth decade of life, and SPs have been identified in some DS patients by early adolescence (Leverenz and Raskind, 1998; Lemere *et al.*, 1996; Mann *et al.*, 1986; Wisniewski *et al.*, 1985). The coincidence of AD pathology in Down's syndrome and the localization of the APP gene to chromosome 21 was the initial impetus for genetic analyses of FAD on the chromosome 21 locus, with a very small percentage of FAD kindreds actually having a mutation in the APP gene (Chartier-Harlin *et al.*, 1991; Goate *et al.*, 1991; Murrell *et al.*, 1991). An accumulation of Aβ is also observed in the brains of individuals subjected to repeated head trauma, such as dementia pugilistica patients and individuals with acute head trauma due to motor vehicle accidents (Roberts *et al.*, 1990; Roberts, 1988). Intracellular accumulation of soluble Aβ(1-40) has also been observed in the brains of pediatric patients with mucopolysaccharidosis (Hurler's syndrome or MPS I and Sanfilippo's syndrome or MPS III), autosomal recessive syndromes in which specific lysosomal hydrolase deficiencies lead to the accumulation of heparan sulfates (Ginsberg *et al.*,

**Table II. Neurodegenerative Disorders that Display Amyloid Plaques**

| |
|---|
| Amyloid β-protein (Aβ)-containing plaques |
| Alzheimer's disease |
| Down's syndrome |
| Dementia pugilistica |
| Acute head trauma |
| Cerebral amyloid angiopathy |
| Hereditary cerebral hemorrhage with amyloidosis-Dutch type |
| Cystatin C-containing plaques |
| Hereditary cerebral hemorrhage with amyloidosis-Icelandic type |
| Prion protein (PrP)-containing plaques |
| Gerstmann–Sträussler–Scheinker syndrome |
| Fatal familial insomnia |
| Kuru |
| Creutzfeldt–Jakob disease |

1999b). Cerebral amyloid angiopathy (Vinters, 1987) and hereditary cerebral hemorrhage with amyloidosis-Dutch type (HCHWA-D)—a rare autosomal dominant disorder that leads to stroke (Van Broeckhoven *et al.*, 1990)—are also characterized pathologically by Aβ accumulation in the cerebrovasculature and by diffuse plaques within gray matter. Genetic analyses have determined that a single missense cytosine-to-guanine mutation occurs in HCHWA-D, resulting in the substitution of glutamine for glutamic acid at codon 693 of APP within the Aβ domain (Levy *et al.*, 1990). The Icelandic variant, termed "HCHWA-I", is similar clinically to HCHWA-D because amyloid angiopathy is observed in cerebral vessel walls, but diffuse plaques are not found in gray matter, and amyloidosis results from a variant of cystatin C, a cysteine protease inhibitor, not from Aβ (Cohen *et al.*, 1983). Point mutations in the prion protein (PrP) gene also lead to amyloid plaques consisting of PrP in the brains of individuals with Gerstmann–Sträussler–Scheinker syndrome, fatal familial insomnia, kuru, and familial forms of Creutzfeldt-Jakob disease (CJD) (Prusiner, 1998). As with AD, familial forms of CJD are rare, and the majority of cases are sporadic (of unknown transmission) as well as a few documented iatrogenic (caused by infection) cases (Prusiner, 1998).

### 4.3. Genetics and Proteolytic Processing of Amyloid-β Precursor Protein and Amyloid-β Protein

The Aβ protein is derived from one or more of the APPs, and a single copy of the APP gene is located on chromosome 21 in humans (Goldgaber *et al.*, 1987; Kang *et al.*, 1987; Robakis *et al.*, 1987; Tanzi *et al.*, 1987). In addition to APP, there are two APP-like proteins (APLPs) known as APLP1 and APLP2, that are highly homologous to APP but lack the Aβ domain (Slunt *et al.*, 1994; Wasco *et al.*, 1993, 1992). The APP gene is conserved across species; homologues of APP have been identified on chromosome 16 in mouse and in the genome of *Drosophila* and *C. elegans* (Daigle and Li, 1993; Luo *et al.*, 1992; Lovett *et al.*, 1987). Furthermore, subtle behavioral deficits caused by a deletion in the APP homologue in *Drosophila* are ameliorated by addition of the human APP gene, suggesting a conservation of APP function across phylogeny (Luo *et al.*, 1992).

The human APP gene encodes 19 exons over approximately 400 kilobases, and multiple APP isoforms are derived by alternative mRNA splicing (Yoshikai *et al.*, 1990; Lemaire *et al.*, 1989). Three major isoforms of APP that contain the Aβ sequence on exons 16 and 17 are designated $APP_{695}$, $APP_{751}$, and $APP_{770}$ (Golde *et al.*, 1990; Ponte *et al.*, 1988; Kang *et al.*, 1987). APP isoforms are delineated by the number of amino acids encoded (e.g., the $APP_{695}$ isoform contains 695 amino acids). The $APP_{751}$ isoform includes a 56-amino acid insert encoded by exon 7 that shares high sequence homology to the Kunitz protease inhibitors (KPI), and $APP_{770}$ contains the KPI insert along with a 19-amino acid insert encoded by exon 8 that is homologous to the MRC-OX-2 antigen found on thymocytes (Kitaguchi *et al.*, 1988; Ponte *et al.*, 1988; Tanzi *et al.*, 1988). Amyloid-β precursor protein mRNA is expressed in a wide variety of tissues, with the highest expression in brain and kidney (Golde *et al.*, 1990; Manning *et al.*, 1988; Neve *et al.*, 1988; Tanzi *et al.*, 1988). A tissue-specific distribution of APP isoforms is observed, with $APP_{695}$ expressed principally in brain, specifically within neurons (Manning *et al.*, 1988; Bahmanyar *et al.,* 1987; Goedert, 1987), whereas the KPI-containing $APP_{751}$ and $APP_{770}$ are expressed in glial cells, endothelial cells, and peripheral tissues (Golde *et al.,* 1990; Neve *et al.,* 1988; Tanzi *et al.,* 1988).

Full-length APP has the primary structure of a transmembrane glycoprotein with a single membrane-spanning domain, similar to cell-surface receptors of the epidermal growth factor superfamily (Kang *et al.,* 1987). Amyloid-β precursor protein, migrating at approximately 110–135 kDa by gel electrophoresis, undergoes significant posttranslational modification, including N-linked glycosylation, O-linked glycosylation, and sulfation (Oltersdorf *et al.,* 1990; Weidemann *et al.,* 1989). The full-length, membrane-associated APP is processed into several secreted derivatives, including Aβ (Fig. 8). The 43 amino acid Aβ peptide is localized internally within APP; therefore, two cleavage sites are necessary to release Aβ from the parent APP molecule. Two pathways and three proteases have been proposed to process the proteolytic cleavage(s) of Aβ (Fig. 8). One pathway, the α-secretory, or non-amyloidogenic, pathway has an amino-terminal cleavage (by an undetermined protease or proteases, termed the "α-secretase") within the Aβ peptide itself (at amino acid Aβ16) and a carboxy-terminal cleavage within the presumed transmembrane domain (by an undetermined protease or proteases, termed the "γ-secretase") to release the carboxy-terminal fragment (CTF) of the Aβ peptide, termed the P3 fragment (Golde and Younkin, 1995; Zhong *et al.,* 1994; Esch *et al.,* 1990; Sisodia *et al.,* 1990). Because the Aβ peptide is cleaved by the α-secretase, amyloid deposition

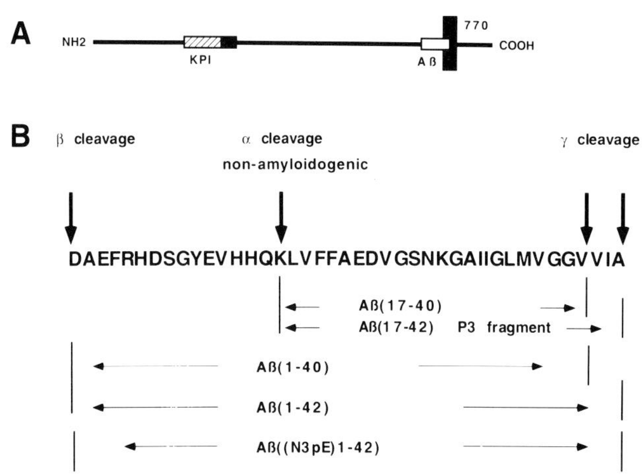


**Figure 8.** Representation of amyloid β-precursor protein (APP) and generation of amyloid-β protein (Aβ) peptide fragments. (A) Schematic diagram of the largest alternatively spliced APP protein containing 770 residues. Two alternatively spliced exons of 56 amino acids (hatched box) containing a Kunitz protease inhibitor (KPI) domain and 19 amino acids (gray box) homologous to the MRC-OX-2 antigen are depicted. The Aβ fragment (white box) is located at the carboxyl-terminal region of the APP molecule and is partially embedded within a single membrane-spanning domain (vertical dark gray box). Because the Aβ peptide is embedded within the APP molecule, two cleavages are necessary to release the fragment. (B) Processing of the APP protein into the various Aβ peptides. The amino acid sequence for Aβ protein is depicted along with the α-, β-, and γ-cleavage sites to generate the P3 fragment, Aβ1-40, and the amyloidogenic Aβ1-42 peptides. The amino-terminal conversion of glutamate to pyroglutamate at the third amino acid residue also is illustrated for Aβ1-42 to exemplify one of several amino terminal modifications of Aβ.

is precluded. The second pathway, the β-secretory, or amyloidogenic pathway, cleaves the amino-terminal APP after the methionine-preceding amino acid Aβ1 (i.e., aspartate) to produce a secreted form of Aβ peptide that may make amyloid fibrils, depending on where in the carboxy-terminal Aβ sequence the γ-secretase proteolytic cleavage occurs (Fig. 8) (Seubert *et al.*, 1995, 1993; Golde *et al.*, 1992). Specifically, the γ-secretase cleavage within the putative transmembrane domain that produces Aβ1-42 forms amyloid fibrils *in vitro,* whereas γ-secretase cleavage that produces Aβ1-40 is less amyloidogenic (Jarrett *et al.*, 1993; Roher *et al.*, 1993a). Furthermore, immunohistochemical analyses of AD brain have demonstrated that Aβ1-42 is the principal component of SP lesions, although Aβ1-40 is also detected in SPs, and the P3 fragment is observed in diffuse plaques (Dickson, 1997; Higgins *et al.*, 1996; Iwatsubo *et al.*, 1996, 1995, 1994; Schmidt *et al.*, 1994; Roher *et al.*, 1993a,b). One hypothesis is that the amyloidogenic Aβ1-42 peptide acts as a nidus for SP deposition, with subsequent accumulation of the non-amyloidogenic APP derivatives (e.g., soluble Aβ1-40, P3, and other truncated CTFs) as the SPs mature into neuritic plaques. In addition to carboxy-terminal variations in Aβ, species with amino-terminal modifications (e.g., intramolecular conversions, isomerization, racemization, and proteolysis) have been detected (Gowing *et al.*, 1994; Roher *et al.*, 1993a; Mori *et al.*, 1992). Further, the conversion of the third amino-terminal residue from glutamate to pyroglutamate [AβN3(pE)] via intramolecular dehydration has been shown to generate a modified form of Aβ that is a major deposited species in SPs within the brains of AD and Down's syndrome patients (Saido *et al.*, 1996, 1995). The change from glutamate to pyroglutamate at position Aβ3 may add to the insolubility of Aβ deposits, as aminopeptidases can only digest free amino-terminal residues. Pyroglutamate-specific aminopeptidases (McDonald and Barrett, 1986), therefore, may be important regulators of Aβ clearance, and dysfunction of these proteases may lead to SP formation. In addition to the secretion of Aβ through α- and β-secretory pathways, Aβ peptides are generated within neurons, residing primarily within the endoplasmic reticulum/intermediate compartments, where they may exert toxic effects from intracellular stores (Skovronsky *et al.*, 1998; Chyung *et al.*, 1997; Cook *et al.*, 1997; Hartmann *et al.*, 1997; Turner *et al.*,1996; Wertkin *et al.*, 1993).

### 4.4. Function of Amyloid-β Protein

Although the genetic and proteolytic processing of APP and Aβ have been well characterized, the functions(s) of APP and Aβ, and their role in the pathogenesis of SPs and AD are not well understood. Mutations in the APP gene have been observed in FAD (a few kindred), HCHWA-D, and extra copies of the APP gene in trisomy 21 lead to subsequent plaque pathology in the CNS and cerebrovasculature. Less is known about the biological significance of Aβ. Studies have demonstrated neurotoxicity *in vitro* when Aβ is administered in a fibrillar conformation, suggesting that fibrillar Aβ in SPs is deleterious to neurons in the local milieu (Yankner, 1996; Pike *et al.*, 1993; Busciglio *et al.*, 1992). *In vivo* assessment of Aβ toxicity has proven to be more difficult, as microinjection of Aβ peptides into rat and monkey cerebral cortex have yielded both positive and negative observations (Frautschy *et al.*, 1998, 1991; Podlinsky *et al.*, 1993; Kowall *et al.*, 1991).

However, co-injection of Aβ and heparan sulfate proteoglycan have caused cytotoxicity (Snow *et al.*, 1994), suggesting that heparan sulfate proteoglycan may act as a pathological chaperone of Aβ, similar to its enhancement of PHF formation of tau proteins *in vitro* (Goedert *et al.*, 1996; Perez *et al.*, 1996). Accumulation of Aβ immunoreactivity has also been observed in association with SPs and synaptic degeneration within the cerebral cortex of aged rhesus monkeys, suggesting that Aβ deposition is a constituent of the cascade of events associated with senescence (Martin *et al.*, 1994). Recently, several proteins with phosphotyrosine interaction/phosphotyrosine binding (PID/PTB) domains, including X11 and Fe65 and their homologues, have been identified by yeast two hybrid screens to bind the YENPTY motif within the cytoplasmic domain of the APP molecule, and to regulate levels of soluble Aβ/APP processing pathways (Borg *et al.*, 1998, 1996; Sastre *et al.*, 1998; Zambrano *et al.*, 1997).

Because SPs are abundant in AD brain and requisite for the diagnosis of AD, Aβ deposition has received much attention in terms of a possible mechanism for the pathogenesis of AD. However, several lines of evidence argue against secreted Aβ and the resultant extracellular SPs being solely responsible for AD and the associated dementia. First, SPs, like NFTs, are observed in the brains of cognitively normal aged individuals, as well as in AD and other disorders (Table 2) (Dickson *et al.*, 1991; Katzman *et al.*, 1998; see also Hof *et al.*, this volume). Second, clinicopathological data suggest that SP density, or "plaque burden," does not correlate well with the severity of dementia, demonstrating a disparity between the pathological lesion and cognitive deficits. In contrast, neocortical and hippocampal NFTs and neuronal cell loss are good predictors of the duration and severity of dementia (Bobinski *et al.*, 1997; Arriagada *et al.*, 1992; Morris *et al.*, 1991; Terry *et al.*, 1991). Third, injection of Aβ into the brains of animals does not induce overt neuron loss (Podlinsky *et al.*, 1993; Frautschy *et al.*, 1991; Kowall *et al.*, 1991), and there is little or no significant neuron loss in transgenic mice that develop abundant SPs (Calhoun *et al.*, 1998; Irizarry *et al.*, 1997a,b; Games *et al.*, 1995). Interestingly, traumatic brain injury induces significantly more CA3 hippocampal pyramidal neuron loss in transgenic mice overexpressing mutant APP (PDAPP mice) than wild-type mice, and this neuronal loss is linked to a surge in Aβ levels on the side ipsilateral to the injury (Smith *et al.*, 1998).

## 5. Other Lesions and Cytopathology: Pathology in Hippocampus and Neocortex

### 5.1. Hirano Bodies

These rodlike structures were first described by Hirano *et al.* (1966) in the brains of patients with ALS/PDC of Guam. Hirano bodies are composed of filaments approximately 6–10 nm in length, arrayed in a parallel sheetlike conformation, and localized principally to and around pyramidal neurons and adjacent neuropil in the CA1 region of the hippocampal formation (see chapter by Nakano and Hirano, this volume). These inclusions are observed in the aged normal brain, and they are abundant in the brains of AD patients (Gibson and Tomlinson, 1977). Immunohistochemical analyses have found that the building blocks of Hirano bodies are likely to be microfilaments, as these inclusions cross-react with antibodies directed against actin and actin-binding proteins (Schmidt *et al.*, 1989; Galloway *et al.*, 1987). The relationship between Hirano body formation and neurodegeneration is unclear. The abundance of Hirano bodies within regions that are highly vulnerable to neurodegeneration in AD suggests the possibility that the aberrant cross-linking of microfilaments such as actin in the form of Hirano bodies may interact with neurofilaments, MTs, and MAPs, to be a permissive factor in NFT pathogenesis.

### 5.2. AMY Plaques

Recently, several monoclonal antibodies generated against immunogens isolated from extracts of PHFtau within AD brain did not recognize PHFtau or NFTs. Rather, they recognize a novel 100 kDa protein and abundant plaque-like lesions (AMY plaques) (Schmidt *et al.*, 1997). The AMY plaques are not visualized by stains for amyloid and they do not colocalize with Aβ deposits. However, AMY plaques and SPs are found within the same cortical regions. The AMY plaques are observed readily within the brains of AD and elderly Down's syndrome patients, and are rare within the brains of age-matched elderly controls or patients with other neurodegenerative or neuropsychiatric disorders (Schmidt *et al.*, 1997). The identification of the novel 100 kDa protein and the morphological characterization of AMY plaques suggest that lesions other than NFTs, NTs, and SPs exist in AD brain and may play a role in the pathogenesis of AD.

### 5.3. Granulovacuolar Degeneration

Granulovacuolar degeneration (Simchowicz, 1911), occurs principally within hippocampal pyramidal cells (notably in CA1 and the subiculum) and refers to the appearance of one or more membrane-bound cytoplasmic vacuoles containing a small argyrophilic core. Granulovacuolar degeneration is an age-related phenomenon readily observed in the hippocampus of successfully aged nondemented individuals, and more frequently, within the hippocampal formation of AD patients (Mann, 1978; Ball and Lo, 1977; Tomlinson and Kitchener, 1972). In AD the pyramidal cells undergoing granulovacuolar degeneration may also display neurofibrillary pathology (Kahn *et al.*, 1985; Grünthal, 1930). Hippocampal neurons with abundant granulovacuoles are also observed in other neurodegenerative disorders, including Down's syndrome, Pick's disease, and ALS/PDC of Guam (Esiri *et al.*, 1997). By immunohistochemistry, the granulated core region cross-reacts with antibodies directed against neurofilament proteins and tau, suggesting that the vacuolar cisternae are sites of cytoskeletal elements (Dickson *et al.*, 1987; Kahn *et al.*, 1985). The relationship of granulovacuolar degeneration to neurodegenerative processes is unknown. The vacuolation of CA1 and subicular pyramidal neurons may reflect age-related phagocytic or endosomal/lysosomal alterations that are accelerated in AD, because these cells represent populations of neurons that are particularly vulnerable to neurodegeneration.

### 5.4. Nonneuronal Cells and Inflammatory Changes in Alzheimer's Disease

In addition to alterations in neuronal populations, the recruitment and activation of nonneuronal cells (e.g., microglial and astroglial cells) is readily observed in AD, notably in association with SPs (Ginsberg *et al.*, 1997; Itagaki *et al.*, 1989b; Dickson *et al.*, 1988; McGeer *et al.*, 1987). Microglia, that is, cells in the CNS that are of macrophage lineage (Hickey and Kimura, 1988), are involved in phagocytosis of cellular debris, including Aβ fibrils (Paresce *et al.*, 1996; Martin *et al.*, 1994; Wisniewski *et al.*, 1991; Perlmutter *et al.*, 1990). For example, microglial cells express receptors for advanced glycation end products (RAGE), further supporting the phagocytic role of these cells (Christie *et al.*, 1996; El Khoury *et al.*, 1996; Paresce *et al.*, 1996; Yan *et al.*, 1996). Macrophage-derived cells in AD brain also display characteristics of activated microglia, including expression of major histocompatibility (MHC) antigens (Tooyama *et al.*, 1990; McGeer *et al.*, 1987), as well as production of cytokines that cause local inflammation. Specifically, microglia in the cerebral cortex of AD patients display immunoreactivity to cytokines, including the interleukins (IL-1β, IL-6), transforming growth factor (TGFβ), tumor necrosis factor (TNFα), and other neurotoxic factors (Dickson, 1997; Giulian *et al.*, 1995; Dickson *et al.*, 1993). Components of the complement cascade, mediators of inflammatory responses and immunity in the periphery, are expressed by microglial cells and are also localized to neuritic plaques (Eikelenboom and Veerhuis, 1996; Pasinetti, 1996; Walker and McGeer, 1992; Ishii and Haga, 1984). Consequently, anti-inflammatory compounds have been hypothesized to have a therapeutic benefit in the treatment or delayed progression of AD and are being tested in multicenter clinical trials (Rogers *et al.*, 1993; McGeer and Rogers, 1992). Microglia also may secrete Aβ peptides and thereby contribute to the accumulation of amyloid fibrils in SPs (Martin *et al.*, 1994; Francowiak *et al.*, 1992; Wisniewski *et al.*, 1989). Therefore, microglial cells may play divergent roles in the AD brain, including the phagocytosis of cellular debris in association with SPs, secretion of Aβ, release of cytokines, and complement-associated inflammatory factors.

Astrogliosis is also a prominent feature of the cellular pathology associated with AD. Glial activation, as assessed by abundant glial fibrillary acidic protein (GFAP) immunoreactivity, is observed in regions of profound cell loss and NFT accumulation, including the CA1 subfield of the hippocampus and neuritic plaques throughout hippocampus and neocortex (Arnold *et al.*, 1996). In contrast to microglial cells, astrocytes are not observed within the cytological confines of SPs; rather, GFAP-immunoreactive astrocytic processes tend to ramify around the periphery of neuritic plaques (Ginsberg *et al.*, 1997; Itagaki *et al.*, 1989b). The function(s) of astrocytes within the AD brain remain unknown, although they may be involved in the clearance of cellular debris as well as the secreting of apoE and pro-inflammatory substances such as cytokines (Dickson, 1997; Pasinetti, 1996).

## 6. Analysis of Vulnerable Cell Types and Pathological Lesions in Diseased Brain

### 6.1. Cell Death in Alzheimer's Disease

In addition to NFT, NT, and SP formation in the AD brain, regionally specific neuron loss, quantified by several methodologies, including unbiased stereological approaches, is well documented in hippocampus and neocortex, as well as in subcortical nuclei with widespread cortical projections (Simic *et al.*, 1997; Gomez-Isla *et al.*, 1996; West *et al.*, 1994; Coleman and Flood, 1987). Neuronal loss in AD is abundant in layer II/III stellate cells of entorhinal cortex, hippocampal and subicular pyramidal cells, and corticocortical pyramidal neurons of layers II/III and V of secondary association and polymodal neocortex. In contrast, neocortical and hippocampal nonpyramidal neurons do not appear as vulnerable, nor is cell loss particularly severe in pyramidal layers of the primary motor, sensory, and visual cortices. In addition to extensive neuronal loss, alterations in dendroarchitecture have been observed in AD. Both aberrant dendritic sprouting and dendritic regression have been reported within the cerebral cortex of AD patients (Coleman and Flood, 1987; Scheibel and Tomiyasu, 1978). The sequela of regionally specific neuronal degeneration and presumed damage caused by NFTs, NTs, and SPs is the actual loss of functional circuits related to the dying cells, termed *connectivity-based degeneration* (see Hof *et al.*, this volume). An example of connectivity-based degeneration in AD is hippocampal disconnection (Hyman *et al.*, 1990, 1984). Neuronal degeneration and pathological lesions are abundant throughout the hippocampal formation, notably in entorhinal cortex (particularly layer II stellate cell islands that project to the dentate gyrus, and layer IV cells that receive a major projection back from the subicular complex) (Gomez-Isla *et al.*, 1996; West *et al.*, 1994). Dentate granule cells do not show degenerative morphology; however, SPs abound in the molecular layer of the dentate gyrus and in the hilus/CA4 regions where the majority of perforant path axons terminate, suggesting the association of SPs with axonal and synaptic degeneration. Within Ammon's horn, degeneration of CA1 pyramidal cells and NFTs is a hallmark of AD-related pathology; CA3 and CA2 pyramidal neurons appear less sensitive to neurodegeneration. Subicular neurons also degenerate, and abundant NFTs are observed throughout the subicular complex and adjacent parahippocampal gyrus. Therefore afferent, efferent, and intrinsic hippocampal circuitry is compromised by degeneration of the neurons that supply these connections. Because the entorhinal-hippocampal-subicular circuit is critical for memory processing (Squire, 1986), the disconnection of the hippocampal formation contributes to the pervasive memory dysfunction of AD.

Although progressive neuronal loss of vulnerable cell populations is a hallmark of AD, the mechanism(s) by which these neurons die remains unknown. Cell death can be categorized into two broad classes: active cell death (*apoptosis*), which requires the activation of cell-death genes and the synthesis of new proteins but does not initiate an inflammatory cascade, and passive cell death (*necrosis*), which involves cell membrane degeneration, ionic imbalances, inflammatory cell recruitment, and eventually cytolysis (Martin*et al.*, 1998; Oppenheim, 1991; Wyllie, 1981).

The two cell-death pathways can be delineated by morphological criteria. Necrosis manifests as nuclear pyknosis and cytoplasmic swelling with associated vacuolar cisternae and may include the presence of inclusion bodies such as NFTs; furthermore, when dead-cell cytoplasm is scavenged by phagocytic cells, the insoluble inclusions remain as "tombstones" (e.g., ghost tangles) of the perikaryon. In contrast, apoptosis causes the condensation of nuclear chromatin into characteristic apoptotic bodies, and membrane blebbing is observed on the cell surface of plasma membranes. The DNA strand breaks (with intact histones) that occur by apoptotic cell death can be visualized by agarose gel electrophoresis as a distinct laddering pattern, whereas DNA strand breaks and histone degeneration in necrotic cell death lead to a DNA smear.

Cell degeneration and death associated with AD have been traditionally considered necrotic, with long-term progressive onset and the presence of dystrophic processes along with intracellular and extracellular inclusions. Additionally, studies have demonstrated the loss of neurotransmitters, second-messenger molecules, and cytoskeletal elements (O'Neill *et al.*, 1992; Bowen *et al.*, 1983) found in intracellular stores that may be released upon membrane disintegration. For example, the MAP protein tau is elevated in the CSF of AD patients (Trojanowski *et al.*, 1996a,b), suggesting that this intracellular protein is released into extracellular spaces following the degeneration/death of neurons that contain NFTs or give rise to NTs. Defects in the endosomal-lysosomal system, including the synthesis and release of acid hydrolases such as cathepsins, have also been identified in AD, notably in vulnerable neuronal populations (Cataldo *et al.*, 1997, 1996; Nixon *et al.*, 1992). A growing body of evidence also suggests that AD is associated with aberrant oxidative stress (i.e., lipid peroxidation, protein nitrosylation, increased protein carbonyls, and DNA strand breakage) and these toxic products of reactive oxygen and reactive nitrogen species ultimately lead to necrosis and neurodegeneration (Gabbita *et al.*, 1998; Hensley *et al.*, 1998; Hermon *et al.*, 1998; Marcus *et al.*, 1998; Markesbery and Lovell, 1998; Montine *et al.*, 1998, 1997; Pratico *et al.*, 1998; Stadtman, 1992; Smith *et al.*, 1991). Advanced glycation end products (AGE), another marker for necrotic cell death, are caused by the reaction of glucose molecules with senescent proteins to produce Amadori products and Maillard reaction products (Yin, 1995; Horiuchi and Araki, 1994) and are increased in the AD brain, and it is hypothesized that glycated Aβ may act as a nidus for SP formation (Takeda *et al.*, 1998b; Smith *et al.*, 1994; Vitek *et al.*, 1994).

Recent studies using techniques that identify DNA fragmentation *in situ*, such as the TdT-mediated deoxyuridine triphosphate nick-end labeling (TUNEL) and *in situ* tailing (IST) methods, have identified populations of neurons that may undergo apoptotic-like changes in the AD brain (Anderson *et al.*, 1996; Lassmann *et al.*, 1995; Smale *et al.*, 1995), suggesting that both active and passive cell-death programs may occur. The reliance on DNA fragmentation as a marker for apoptotic cell death must be interpreted carefully, because DNA strand breakage also occurs during necrosis (Stadelmann *et al.*, 1998; Gold *et al.*, 1994). Additionally, cellular resolution is critical for delineating neurons from glial cells that may also undergo programmed cell death as part of their turnover. Optimally, several methodological criteria (e.g., TUNEL/IST labeling, agarose gel electrophoresis, electron microscopy) and clinicopathological criteria (e.g., agonal state of the patient, postmortem delay) must be employed to determine unequivocally whether apoptotic programs of death occur in AD and senescent brain.

### 6.2. Elucidating Pathologic Mechanisms: RNA Species Assessment in Alzheimer's Disease Lesions

One of the major technical difficulties in analyzing neuronal degeneration in postmortem samples of the AD brain is the regional and cellular selectivity of the neuropathology, because certain populations of cells degenerate, or harbor NFTs, whereas adjacent cells do not display overt cytological alterations. A goal is to characterize single elements—both normal and pathological—for protein, mRNA, and DNA content in order to identify quantitative differences in expression that may provide insight into mechanisms, and ultimately treatments, for neurodegenerative diseases.

Although the polypeptide composition of vulnerable neurons, NFTs, NTs, and SPs has been characterized extensively within the AD brain, few data exist on the nonproteinaceous components of these lesions. We have determined that NFTs, NTs, and SPs in the AD brain contain cytoplasmic RNA species by acridine orange (AO) histofluorescence (Ginsberg *et al.,* 1998a, 1997). Acridine orange is a fluorescent dye that intercalates selectively into nucleic acids (Schümmelfeder, 1958; von Bertalanffy and Bickis, 1956) and has been used previously to detect RNA and DNA in normal and developing brain (Topaloglu and Sarnat, 1989; Mai *et al.,* 1984; Schmued *et al.,* 1982). Acridine orange histofluorescence detects the presence of RNA species in NFTs, NTs, and SPs, as pretreatment with RNase, but not DNase, abolishes staining (Fig. 9). By double-labeling with AO and TS, the majority of NFTs within the entorhinal cortex and CA1 subfield of the hippocampus are found to contain RNA, whereas a subpopulation of SPs show AO labeling (Fig. 9). Quantitative analyses demonstrate that approximately 80% of entorhinal and CA1 NFTs and approximately 55% of SPs contain cytoplasmic RNA species (Ginsberg *et al.,* 1997a). Furthermore, double-labeling studies of AO and astroglial/microglial markers indicate that astrocytic processes do not colocalize to AO-labeled SPs, whereas microglial cells appear to be a minor contributor to the RNA signal in SPs (Fig. 10). RNA sequestration to NFTs has also been observed in the brains of patients with ALS/PDC of Guam, corticobasal degeneration, progressive supranuclear palsy, and Pick's disease (Pick bodies also sequester RNAs) (Ginsberg *et al.,* 1998a). In contrast, other pathological lesions in neurodegenerative disorders that do not contain appreciable levels of RNA as assessed by AO histofluorescence include Lewy bodies, glial cytoplasmic inclusions, and Hirano bodies (Ginsberg *et al.,* 1998a).

Acridine orange labeling of NFTs suggests that cytoplasmic RNA species become trapped in the intracellular, protease-insensitive PHFtau aggregates within the somatodendritic domain of tangle-bearing neurons. The functional sequelae of RNA interactions with NFTs are unknown, but we speculate that the sequestration of mRNAs may impede translation and ultimately accelerate neuronal degeneration. Similarly, RNA observed in NTs may also be sequestered by the PHFtau aggregates present in the dystrophic processes containing NTs. Because polyribosomes and multiple mRNAs have been identified within dendritic compartments (Martone *et al.*, 1996; Miyashiro *et al.*, 1994; Garner *et al.*, 1988; Davis *et al.*, 1987; Steward and Levy, 1987), and dendrites can synthesize proteins locally (Crino *et al.*, 1998; Crino and Eberwine, 1996), the accumulation of mRNAs within NTs could impair their translation. Additionally, RNAs facilitate the aggregation of recombinant tau protein into PHFs *in vitro* (Kampers *et al.*, 1996; Hasegawa *et al.*, 1997; Friedhoff *et*

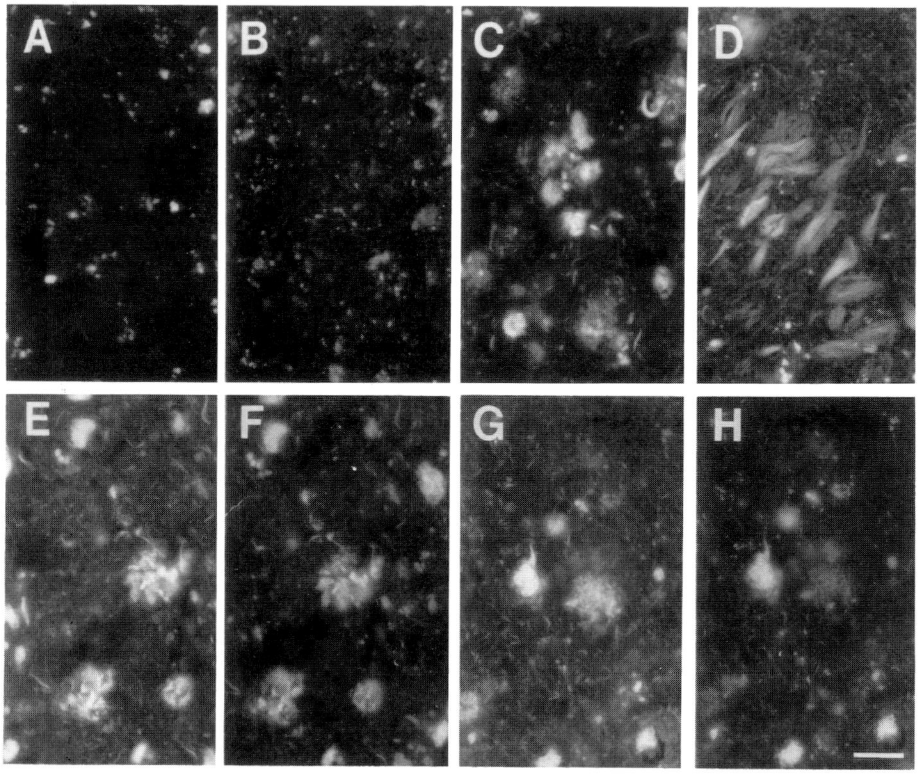


**Figure 9.** AO histochemistry localizing cytoplasmic RNA species within NFTs and SPs of patients with AD. Used with permission from Lippincott Raven Press. (A) Untreated entorhinal section demonstrating the presence of autofluorescent lipofuscin granules and the lack of autofluorescent NFTs and SP. (B) Entorhinal section pretreated with RNase prior to AO histochemistry. Note the paucity of NFT and SP labeling, suggesting that AO identifies cytoplasmic RNA species within these profiles. (C) Entorhinal section pretreated with DNase prior to AO histochemistry. AO-labeled SPs and NTs are clearly delineated, further suggesting the presence of RNA, but not DNA, within these neuropathological markers of AD. (D) Hippocampal section pretreated with proteinase K prior to AO histochemistry. NFTs are identified, indicating that fluorescence labeling is not due to nonspecific protein binding. (E) Thioflavine-S (TS) labeling of a tissue section from the CA1-subicular region double-stained with TS and AO. Note the presence of TS-positive SPs and NTs. (F) Same tissue section as E, viewed with a rhodamine filter to demonstrate labeling with AO. Note that most profiles colocalize TS and AO. The TS staining intensity is greater, however, as both E an F were processed using the same exposure and development times. (G) Similar to E, TS labeling of a tissue section from layers II/III entorhinal cortex double-stained with TS and AO. An intensely labeled NFT and SP are each identified using a fluorescein isothiocyanate (FITC) filter. (H) Same tissue section as G, using a rhodamine filter, demonstrating colocalization of AO and TS to the NFT, but not the SP, within this field. Bar = 50 μm.

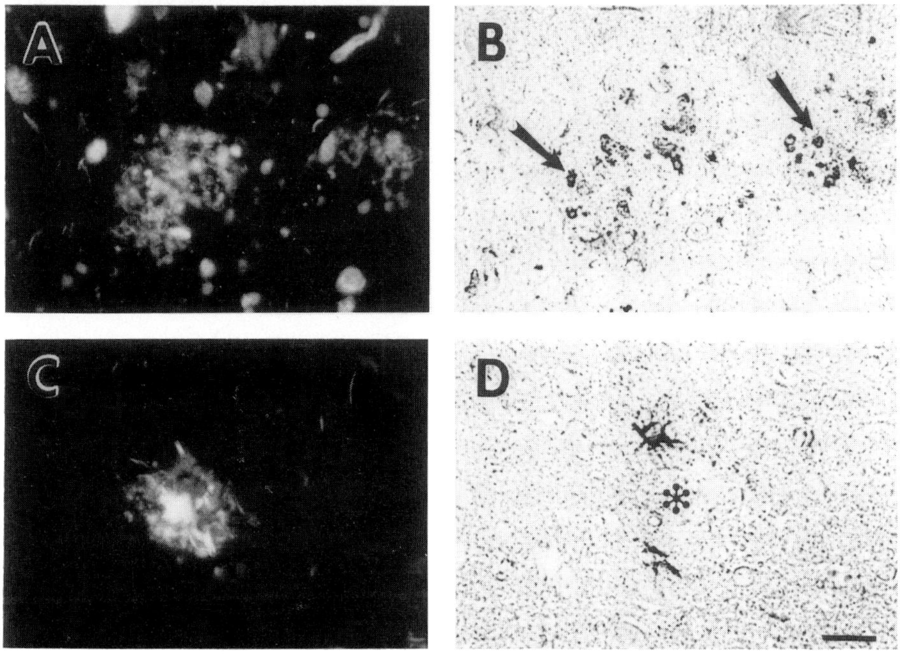


**Figure 10.** Double-labeled preparations indicating association of astrocytes, microglial cells, and AO-labeled SPs in AD entorhinal cortex. Used with permission from Lippincott Raven Press. (A) Epifluorescence photomicrograph of AO-labeled SPs within layers II/III of entorhinal cortex. (B) Brightfield photomicrograph of the same section as A, illustrating the presence of CD68-immunoreactive microglial cells within the cytological confines (arrows) of the AO-labeled SPs. (C) Similar to A, this epifluorescence photomicrograph identifies an AO-labeled entorhinal cortex SP. (D) Brightfield photomicrograph of the same section as B, demonstrating the apposition, but not colocalization, of GFAP-immunoreactive astrocytes and AO-labeled SP (asterisk). Bar = 25 μm.

*al.*, 1998a,b), and may act as pathological chaperones under the appropriate conditions *in vivo*. In contrast to NTs and NFTs, SPs, are extracellular lesions. Therefore, it remains unknown how RNA species that are localized to cytosolic compartments accumulate in extracellular deposits. Because RNA species are found in neuritic plaques (Ginsberg *et al.*, 1999a), we hypothesize that RNA observed in SPs originates from degenerating neurites with ruptured plasma membranes that adhere to protease-insensitive Aβ fibrils; diffuse plaques are devoid of AO histofluorescence (Ginsberg *et al.*, 1998a). Furthermore, the presence of cytoplasmic RNA species within SPs also provides indirect evidence that amyloid plaques contain the cellular remnants of degenerating neurons and reactive glial cells.

## 6.3. mRNA Species in Single Profiles Isolated from the Alzheimer's Disease Hippocampus

Acridine orange allows the identification of cytoplasmic RNAs within NFTs, NTs, and SPs, but the individual subtypes (rRNA, tRNA, and mRNA) cannot be ascertained by this method. However, molecular biological techniques such as reverse transcriptase-polymerase chain reaction (RT-PCR) and single-cell RNA amplification can be utilized to characterize mRNA transcripts within lesions of the AD brain. For example, studies by Eberwine and colleagues (1995) allow the identification and quantitative analysis of mRNA species in immunohistochemically identified single cells in fixed tissue sections of human brain and primary cell cultures (Crino *et al.*, 1998, 1996; Crino and Eberwine, 1996; Ginsberg *et al.*, 1998b, 1999a). Compartmental assessment of mRNAs can also be evaluated by the amplified antisense RNA (aRNA) technique, because mRNAs localized to the somal cytoplasm can be differentiated from mRNAs present in dendrites and growth cones (Crino *et al.*, 1998; Crino and Eberwine, 1996; Eberwine *et al.*, 1995; Miyashiro *et al.*, 1994). The quantitation of known mRNAs by single-cell analysis provides a paradigm by which anatomically defined structures can be characterized and potentially differentiated from neighboring cells. This type of experimental design is optimal for analyzing gene expression in AD brain (Chow *et al.*, 1998; Ginsberg *et al.*, 1999a), as alterations within selectively vulnerable cells may be observed relative to the surrounding neurons, neuropil, glia, and blood vessels.

The ability to analyze mRNA species from individual cells is an approach we have adapted to assess the mRNA content of single neurons, SPs, and NFTs within the AD brain (Ginsberg *et al.*, 1998b, 1999a). Specifically, the array of mRNAs, or the "expression profile," amplified from individual CA1 pyramidal neurons within the hippocampus of age-matched normal controls is used as a normative database to probe for potential differences in the mRNA content of SPs and NFTs within the CA1 pyramidal cell layer of the hippocampus in AD patients (Fig. 11). Detection of differences in gene expression between normal neurons and pathological lesions will further our knowledge on the nonproteinaceous structure of these elements and also provide specific mRNAs to target for functional characterization, including mRNA turnover, gene translation efficiency, and interactions with RNA binding proteins. Moreover, mRNAs that display changes in the AD brain may correlate with protein defects that can be exploited by animal models of neurodegeneration, ultimately leading to therapeutic interventions.

The technique of aRNA amplification allows linear amplification of cDNAs isolated from a variety of sources, including immunohistochemically identified neurons, NFTs, and SPs, with the potential for adding exponential amplification steps (e.g., PCR) (Eberwine *et al.*, 1992; VanGelder *et al.*, 1990). In the first step of the aRNA procedure, a 66 base pair oligonucleotide primer consisting of 24 TTPs and a T7 RNA polymerase promoter sequence (oligo-dT[24]T7) is annealed to each mRNA that has a polyadenylated poly [$A^+$] tail. This step ("*in situ* transcription") is performed on the entire immunolabeled tissue section. The following procedures are then conducted on individual cells selected for the expression of a specific marker by immunohistochemistry and then aspirated from tissue sections using a micropipette connected to a micromanipulator, allowing for single-cell specificity. When single-stranded cDNAs containing the oligo-dT(24)T7 RNA polymerase promoter from individual immunohistochemically identified structures are converted into double-stranded DNA, the promoter region becomes functional and can then be recognized by the T7 RNA polymerase, to synthesize RNA molecules from the double-stranded cDNA template (Fig. 12). Due to the orientation of the T7 RNA polymerase promoter sequence, the RNA that is synthesized from the double-stranded cDNA template is antisense to the poly [$A^+$] mRNA in the tissue and can be used as a probe to detect mRNA species. Although specific mRNAs in individual cells may be of very low abundance, the RNA technique is advantageous because of a linear amplification (as opposed to the exponential amplification of PCR-based methods) of the original message population. Each round of aRNA amplification results in an approximately 1000-fold amplification, and two rounds

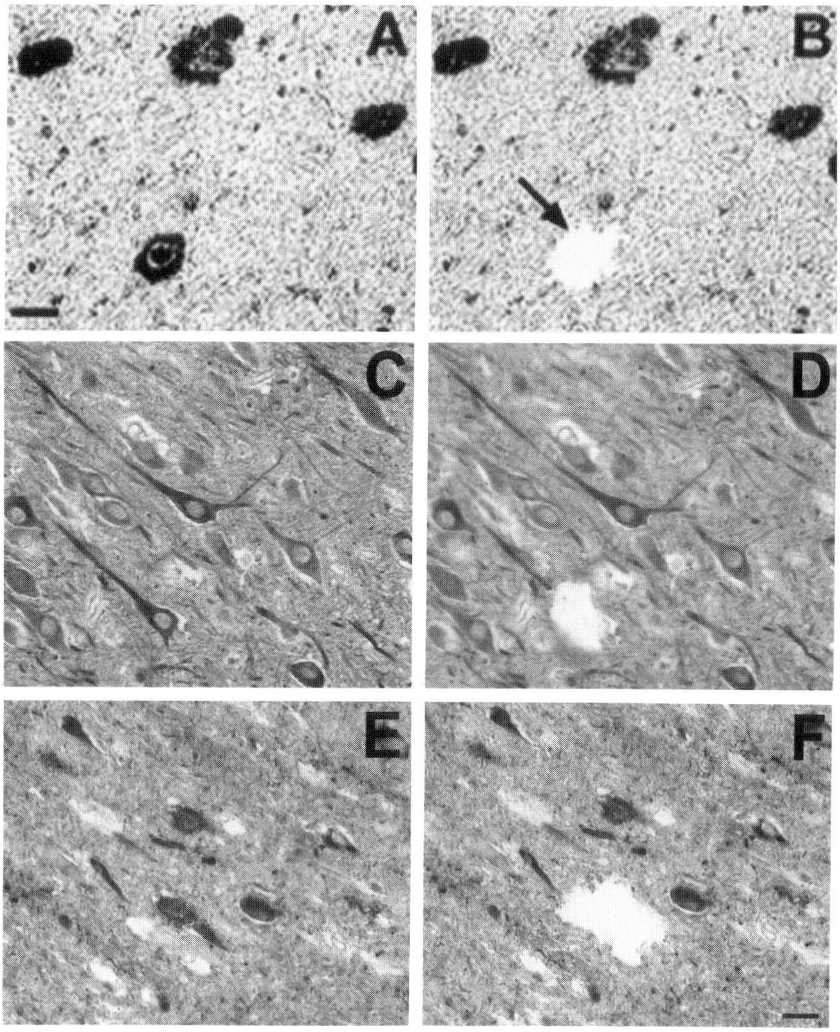


**Figure 11.** Single-cell aspiration of SPs, NFTs, and CA1 pyramidal neurons. (A,B) Tissue section from AD hippocampal formation, immunostained with an antibody directed against Aβ. The same field is shown in A prior to aspiration with a micropipette and in B following aspiration (arrow). (C,D) Tissue section from the hippocampal formation of an age-matched normal control immunostained with an antibody directed against epitopes of the medium and heavy neurofilament subunits. CA1 pyramidal neurons located in stratum pyramidale are visualized prior to (C) and following (D) aspiration with a micropipette (arrow, E,F). Tissue section from the hippocampal formation of an AD patient immunostained with an antibody directed against PHFtau. The same field is shown in (E) prior to aspiration with a micropipette and (F) following aspiration (arrow). Bar = 25 μm (A,B); 30 μm (C–F).

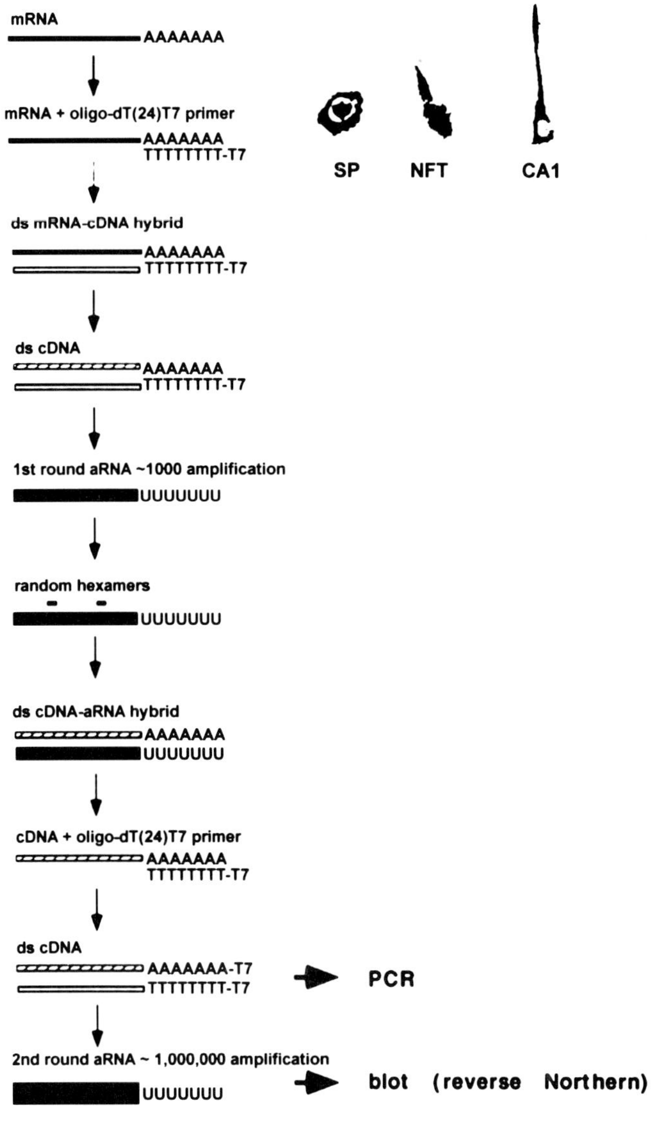


**Figure 12.** Methodology for the linear amplification of mRNAs from single cells. An oligo-dT(24)T7 primer is annealed to poly ($A^+$) mRNAs on the section (*in situ* transcription). A double-stranded (ds) mRNA-cDNA hybrid is formed using dNTPs and reverse transcriptase. The hybrid is converted into ds cDNA, forming a functional T7 RNA polymerase promoter. Following the removal of tertiary structures and blunt ending (not shown), the first round of aRNA synthesis occurs with the addition of T7 RNA polymerase and NTPs. A second round of aRNA amplification begins by annealing random hexamers to the newly formed aRNA, and synthesizing a cDNA strand. The oligo-dT(24)T7 primer is then reintroduced which binds to the poly ($A^+$) sequence on the newly synthesized cDNA strand, and a ds cDNA template is formed. From the second round amplification, the amplified ds cDNA (prior to the aRNA amplification step) can be used as template for PCR (first arrowhead). Candidate mRNAs can be identified via reverse Northern blotting (second arrowhead) by probing with radiolabeled second round aRNA products to known cDNA clones or expressed sequence tagged cDNAs (ESTs).

yield an amplification of approximately $1 \times 10^6$ from the original mRNA in the desired cell (Fig. 12). Candidate mRNAs are identified by radiolabeling the second-round aRNA products with $^{32}$P-CTP or fluorescently tagged NTPs and hybridization to known cDNA clones (e.g., via reverse Northern blotting or large-scale gene microarrays) that have been heat-denatured and adhered to nitrocellulose membranes in a slot blot apparatus. Thus, this method allows for studies of molecular neuropathology at the single-cell level in tissue sections.

Recently, we have identified neurons and pathological lesions in paraffin-embedded tissue sections of the hippocampal formation of AD patients and age-matched normal controls by immunohistochemistry prior to aRNA amplification. For example, normal neurons and neuropil are identified by antibodies directed against epitopes of the medium and heavy neurofilament subunits (e.g., RMdO20) (Lee *et al.,* 1987); NFTs, and NTs are visualized by antibodies directed against PHFtau (e.g., PHF1) (Greenberg *et al.,* 1992); and neuritic SPs are visualized by PHF1 and by antibodies directed against Aβ (e.g., 2332) (Schmidt *et al.,* 1994) (Fig. 11). Provided that the immunohistochemical procedures are carried out using standard RNase-free precautions, aRNA amplification can be successfully employed to visualize the expression profiles of SPs, NFTs, and CA1 pyramidal cells (Ginsberg *et al.,* 1998b, 1999a).

Emerging data sets indicate that CA1 pyramidal neurons contain a complement of genes that encode cytoskeletal elements, glutamate receptors, protein kinases/phosphatases, and neurotrophins/neurotrophin receptors (Fig. 13). Similar single-cell analyses have been performed on SPs aspirated from the CA1–subicular region in AD patients (Ginsberg *et al.,* 1999a), and the data suggest that SPs contain multiple mRNA species, notably genes enriched in dendritic compartments, such as cAMP response element-binding protein (CREB) (Crino *et al.,* 1998; Crino and Eberwine, 1996). Although the expression profile for SPs differs from normal CA1 neurons in terms the relative amounts of several genes, including the neurotrophin receptors trkB and trkC, and the apoptotic cell-death gene bax (Fig. 14), the overall pattern of gene expression is similar, suggesting that components of SPs that contain mRNA are of neuronal origin. Furthermore, these elements are likely to contain dendritic processes due to the quantitative expression of cDNA clones known to be dendrite-enriched (Crino *et al.,* 1998; Crino and Eberwine, 1996; Martone *et al.,* 1996; Mayford *et al.,* 1996; Burgin *et al.,* 1990). Thus, evidence accumulates for the hypothesis that amyloid plaques contain the remnants of dying neurons. Noneuronal messages (e.g., glial fibrillary acidic protein mRNA that is present in astrocytes, and Il-1 and Il-6 mRNAs that are enriched in microglial cells) are not abundant in SPs or in CA1 pyramidal neurons (Ginsberg *et al.,* 1999a). Additionally, recent single-cell aRNA analysis of NFT-bearing neurons in CA1 hippocampus of AD brains using reverse Northern blots of (>100 cDNAs) and (>15,000 expressed sequence tagged cDNAs; ESTs) on large scale microarrays indicates significant differences in classes of mRNAs including protein phosphatases, cytoskeletal elements, glutamate receptors, and dopamine receptors (Ginsberg *et al.,* 1998b).

The detection of an individual mRNA species in a plaque or a tangle provides valuable information on a single gene and has relevance toward the regulation and synthesis of related protein products. The advantage of the aforementioned aRNA protocol is that multiple mRNAs can be characterized from a single structure and

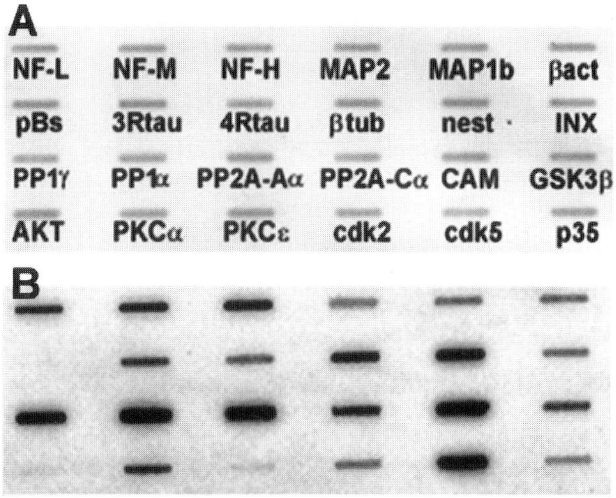


**Figure 13.** Representative reverse Northern blot demonstrating an mRNA expression profile of a single CA1 pyramidal neuron from normal control hippocampus. (A) Bromphenol blue (0.1%) is added to each linearized cDNA to ensure equal loading of samples on an array (24 lanes are depicted). (B) The same array is shown following hybridization with radiolabeled aRNA from a single CA1 neuron. The varying signal intensities demonstrate the relative abundance of specific mRNAs analyzed. Low, medium, and high molecular weight neurofilament subunits (NF-L, NF-M, and NF-H), microtubule-associated proteins (MAP2 and MAP1b), β-actin (βact), bluescript vector (pBs), 3- and 4-repeat tau (3Rtau, 4Rtau), β-tubulin (βtub), nestin (nest), internexin (INX), protein phosphatase 1 (PP1γ, PP1α), protein phosphatase 2 (PP2A-Aα, PP2A-Cα), CaM Kinase II (CAM), glycogen synthase kinase 3β (GSK-3β), protein kinase B (AKT), protein kinase C (PKCα, PKCε), cyclin-dependent kinase (cdk2, cdk5, and brain-specific activator p35), focal adhesion kinase (FAK), stress-activated protein kinase (SAPK3, SAPK4), jnk kinase (JNK1), and erk kinase (ERK1, ERK2).

compared by both qualitative and quantitative criteria to other identified structures. Moreover, the interpretation of mRNA content within SPs is relevant to how SPs are perceived. Specifically, *in vitro* data has elucidated neurotoxic properties of secreted Aβ (Yankner, 1996), and by inference, SPs are considered to be toxic entities. Notwithstanding potential toxic effects of Aβ and SPs in the neocortex and hippocampal formation in AD and other neurodegenerative disorders, the present aRNA analyses suggest that SPs are the repositories for dying neurons, and may shed insight into the mechanisms of RNA sequestration, fibril accumulation, and ultimately the relationship(s) of SPs to somatodendritic and axonal degeneration within the AD brain. For example, increasing evidence indicates that Aβ is generated inside the neurons (Skovronsky *et al.*, 1998; Turner *et al.*, 1996; Wertkin *et al.*, 1993), where it may have intracellular toxic effects—killing from inside, rather than outside, a neuron. Use of the aRNA method may clarify how this occurs.

## 7. Conclusions

Alzheimer's disease is essentially a pervasive, fatal disorder associated with senescence. Population demographics of the United States indicate that senior citizens are one of the fastest growing segments of our society; thus management of

AD will become increasingly important in terms of health care allocation and quality-of-life issues. For example, an important public health issue is now being raised as a result of longitudinal studies that indicate estrogen replacement therapy may reduce the risk of AD (Baldereschi *et al.*, 1998; Kawas *et al.*, 1997). Understanding the structural abnormalities in hippocampus and neocortex in AD and other neurodegenerative disorders is critical for elucidating neurodegenerative mechanisms and for developing new therapies. Presently, we have only a partial understanding of the mechanisms that lead to the accumulation of two hallmark lesions of AD, NFTs, and SPs. Multidisciplinary approaches have led to the characterization of the protein building blocks of NFTs, SPs, and the genes from whence they derive. Specific antibodies have led to the characterization of sites for the hyperphosphorylation of the PHFtau molecule and to the assessment of amyloidogenic fragments of the Aβ peptide, but despite these basic advances, we know little about the mechanisms underlying filament or fibril assembly or about causal or secondary

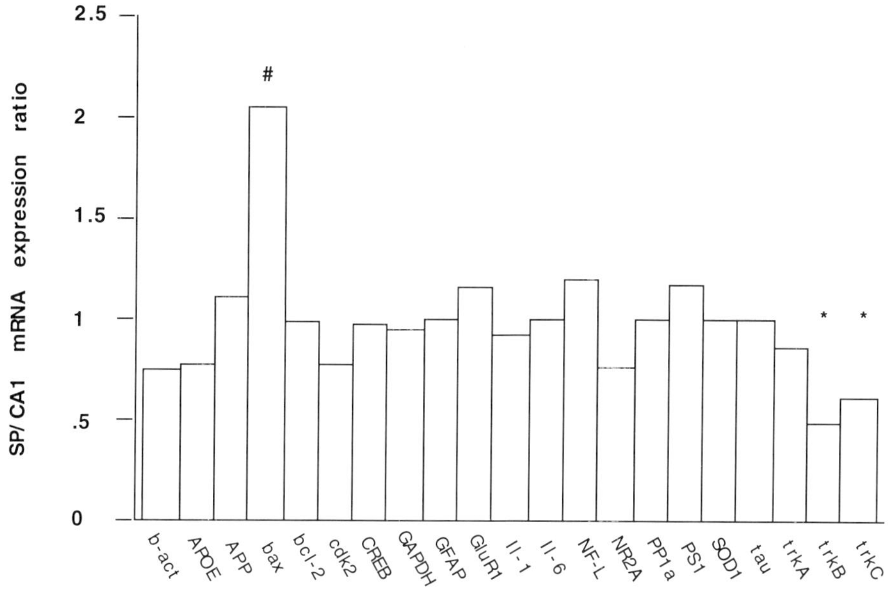


**Figure 14.** Histogram illustrating the relative ratios of selected mRNAs amplified from single SPs from AD brain and single CA1 pyramidal cells from age-matched control brain. A total of 15 SPs from five AD brains and 15 CA1 neurons from four control brains were analyzed. A ratio of 1 indicates a similar level of mRNA expression; a higher ratio indicates a greater amount, and a lower ratio a lesser amount, of message in a single SP relative to a single CA1 neuron. The majority of mRNAs analyzed indicate similar expression levels in SPs and CA1 neurons. For example, glial fibrillary acidic protein (GFAP) and cyclic AMP response element binding protein (CREB) expression, expressed at near-background and high levels, respectively, in CA1 neurons (see Fig. 13) are expressed at similar levels of intensity in SPs. In contrast, bax mRNA levels are significantly up-regulated in SPs relative to CA1 pyramidal neurons, (#), and trkB and trkC expression is significantly down-regulated in SPs relative to CA1 neurons (*). (b-act, β-actin; apoE, apolipoprotein E; APP, amyloid precursor protein; bax, proapoptotic cell-death gene bax; bcl-2, antiapoptotic cell-death gene bcl-2; cdk2, cyclin-dependent kinase-2; GAPDH, glyceraldehyde 3′-phosphate dehydrogenase; GluR1, glutamate receptor subunit 1; Il-1, interleukin 1; Il-6, interleukin 6; NF-L, low molecular-weight neurofilament subunit; NR2A, *N*-methyl-D-aspartate receptor 2A; PP1a, protein phosphatase 1α subunit; PS1, presenilin-1; SOD1, superoxide dismutase 1; tau, low molecular-weight microtubule-associated protein tau; trkA–C, high-affinity tyrosine kinase receptors, A, B, and C.)

relationships leading to the development of AD. Furthermore, the selective vulnerability of specific neuronal populations that undergo subsequent connectivity-based degeneration makes experimental assessment difficult due to the multiplicity of cell types within the mammalian CNS. Thus the aRNA technique may be of great value in assessing genes that vary in the specific cell types that are vulnerable to degeneration in AD.

The next level of understanding the pathogenesis of AD lies in recapitulating the mechanism(s) of neuronal degeneration in suitable *in vivo* and *in vitro* models. For example, cell culture manipulations of human-derived neurons, transgenic animal development, and gene-targeting technologies are likely to be the substrates for investigation into the mechanisms of NFT and SP formation and their relationships to AD pathogenesis. Ultimately, however, these analyses must be brought into the context of human cortical neurobiology and neuropathology. Multicenter AD clinical trials and subsequent pathological assessments are the critical means by which outcomes are currently determined. The goal of treatment or at least delaying the onset/progression of AD depends upon continuous extraction and interpretation of data gathered in a variety of venues, including postmortem pathological analyses, culture dishes, animal models, RNA/DNA blots, and gene chip technologies.

ACKNOWLEDGMENTS. This work was supported by grants from the National Institute on Aging, of the National Institutes of Health, the Dana Foundation, and a Pioneer award from the Alzheimer's Association. We thank Dr. T. E. Golde for helpful discussions; Dr. M. Hong for the FTDP-17 tau illustrations; Mr. T-S Chiu for graphics; and our colleagues in the Center for Neurodegenerative Disease Research, Departments of Pathology and Laboratory Medicine, Neurology, Psychiatry, and the Penn Alzheimer Center for their assistance; and the families of the patients who made this research possible.

## 8. References

Alzheimer, A., 1907, Uber eine eigenartige Erkrangkung der Hirnrinde, *Allg. Z. Psychiatr.,* **64:**146–148.

Anderson, A. J., Su, J. H., and Cotman, C. W., 1996, DNA damage and apoptosis in Alzheimer's disease: Colocalization with c-jun immunoreactivity, relationship to brain area, and effect of postmortem delay, *J. Neurosci.* **16:**1710–1719.

Arnold, S. E., Franz, B. R., Trojanowski, J. Q., Moberg, P. J., and Gur, R. E. 1996, Glial fibrillary acidic protein-immunoreactive astrocytosis in elderly patients with schizophrenia and dementia, *Acta Neuropathol.* **91:**269–277.

Arriagada, P. V., Growdon, J. H., Hedley-Whyte, E. T., and Hyman, B. T., 1992, Neurofibrillary tangles, but not senile plaques, parallel duration and severity of Alzheimer's disease, *Neurology* **42:**631–639.

Askanas, V., Engel, W. K., and Alvarez, R. B., 1992, Light and electron microscopic localization of β-amyloid protein in muscle biopsies of patients with inclusion-body myositis, *Am. J. Pathol.* **141:**31–36.

Askanas, V., Engel, W. K., Bilak, M., Alvarez, R. B., and Selkoe, D. J., 1994, Twisted tubulofilaments of inclusion body myositis muscle resemble paired helical filaments of Alzheimer brain and contain hyperphosphorylated tau, *Am. J. Pathol.* **144:**177–187.

Auer, I. A., Schmidt, M. L., Lee, V.M.-Y., Curry, B., Suzuki, K., Shin, R.-W., Pentchev, P. G., Carstea, E. D., and Trojanowski, J. Q., 1995, Paired helical filament tau (PHFtau) in Niemann-Pick type C disease is similar to PHFtau in Alzheimer's disease, *Acta Neuropathol.* **90:**547–551.

Baba, M., Nakajo, S., Tu, P. H., Tomita, T., Nakaya, K., Lee, V. M., Trojanowski, J. Q., and Iwatsubo, T., 1998, Aggregation of α-synuclein in Lewy bodies of sporadic Parkinson's disease and dementia with Lewy bodies, *Am. J. Pathol.* **152:**879–884.

Bahmanyar, S., Higgins, G. A., Goldgaber, D., Lewis, D. A., Morrison, J. H., Wilson, M. C. Shankar, S. K., and Gajdusek, D. C., 1987, Localization of amyloid β-protein mRNA in brains from patients with Alzheimer's disease, *Science* **237:**77–88.

Baldereschi, M., Di Carlo, A., Lepore, V., Bracco, L., Maggi, S., Grigoletto, F., Scarlato, G., and Amaducci, L., 1998, Estrogen-replacement therapy and Alzheimer's disease in the Italian Longitudinal Study on Aging, *Neurology* **50:**996–1002.

Bales, K. R., Verina, T., Dodel, R. C., Du, Y., Altstiel, L., Bender, M., Hyslop, P., Johnstone, E. M., Little, S. P., Cummins, D. J., Piccardo, P., Ghetti, B., and Paul, S. M., 1997, Lack of apolipoprotein E dramatically reduces amyloid β-peptide deposition, *Nature Genet.* **17:**263–264.

Ball, M. J., and Lo, P., 1977, Granulovacuolar degeneration in the aging brain and in dementia, *J. Neuropathol. Exp. Neurol.* **36:**474–487.

Bancher, C., and Jellinger, K. A., 1994, Neurofibrillary tangle predominant form of senile dementia of Alzheimer type: A rare subtype in very old subjects, *Acta Neuropathol.* **88:**565–570.

Bancher, C., Lassmann, H., Budka, H., Grundke-Iqbal, I., Iqbal, K., Wiche, G., Seitelberger, F., and Wisniewski, H. M., 1987, Neurofibrillary tangles in Alzheimer's disease and progressive supranuclear palsy: Antigenic similarities and differences: Microtubule-associated protein tau antigenicity is prominent in all types of tangles, *Acta Neuropathol.* **74:**39–46.

Bao, F., Arai, H., Matsushita, S., Higuchi, S., and Sasaki, H., 1996, Expression of apolipoprotein E in normal and diverse neurodegenerative disease brain, *NeuroReport* **7:**1733–1739.

Berchtold, N. C., and Cotman, C. W., 1998, Evolution in the conceptualization of dementia and Alzheimer's disease: Greco-Roman period to the 1960s, *Neurobiol. Aging* **19:**173–189.

Berezovska, O., Xia, M., Page, K., Wasco, W., Tanzi, R. E., and Hyman, B. T., 1997, Developmental regulation of presenilin mRNA expression parallels Notch expression, *J. Neuropathol. Exp. Neurol.* **56:**40–44.

Binder, L. I., Frankfurter, A., and Rebhun, L. I., 1985, The distribution of tau in the mammalian central nervous system, *J. Cell Biol.* **101:**1371–1378.

Bird, T. D., Wijsman, E. M., Nochlin, D., Leehey, M., Sumi, S. M., Payami, H., Poorkaj, P., Nemens, E., Rafkind, M., and Schellenberg, G. D., 1997, Chromosome 17 and hereditary dementia: linkage studies in three non-Alzheimer families and kindreds with late-onset FAD, *Neurology* **48:**949–954.

Blacker, D., Haines, J. L., Rodes, L., Terwedow, H., Go, R. C., Harrell, L. E., Perry, R. T., Bassett, S. S., Chase, G., Meyers, D., Albert, M. S., and Tanzi, R., 1997, ApoE-4 and age at onset of Alzheimer's disease: The NIMH genetics initiative, *Neurology* **48:**139–147.

Bobinski, M., Wegiel, J., Tarnawski, M., Bobinski, M., Reisberg, B., De Leon, M. J., Miller, D. C., and Wisniewski, H. M., 1997, Relationships between regional neuronal loss and neurofibrillary changes in the hippocampal formation and duration and severity of Alzheimer disease, *J. Neuropathol. Exp. Neurol.* **56:**414–420.

Borchelt, D. R., Thinakaran, G., Eckman, C. B., Lee, M. K., Davenport, F., Ratovitsky, T., Prada, C.-M., Kim, G., Seekins, S., Yager, D., Slunt, H. H., Wang, R., Seeger, M., Levey, A. I., Gandy, S. E., Copeland, N. G., Jenkins, N. A., Price, D. L., Younkin, S. G., and Sisodia, S. S., 1996, Familial Alzheimer's disease-linked presenilin 1 variants elevate Aβ1-42/40 ratio *in vitro* and *in vivo*, *Neuron* **17:**1005–1013.

Borg, J. P., Ooi, J., Levy, E., and Margolis, B., 1996, The phosphotyrosine interaction domains of X11 and FE65 bind to distinct sites on the YENPTY motif of amyloid precursor protein, *Mol. Cell. Biol.* **16:**6229–6241.

Borg, J. P., Yang, Y., De Taddeo-Borg, M., Margolis, B., and Turner, R. S., 1998, The X11α protein slows cellular amyloid precursor protein processing and reduces Aβ40 and Aβ42 secretion, *J. Biol. Chem.* **273:**14761–14766.

Bowen, D. M., Allen, S. J., Benton, J. S., Goodhart, M. J., Haan, E. A., Palmer, A. N., Sims, N. R., Smith, C.C.T., Spillane, J. A., Esiri, M. M., Neary, D., Snowdon, J. S., Wilcock, G. K., and Davison, A. N., 1983, Biochemical assessment of serotonergic and cholinergic dysfunction and cerebral atrophy in Alzheimer's disease, *J. Neurochem.* **41:**266–272.

Braak, H., and Braak, E., 1987, Argyrophilic grains: Characteristic pathology of cerebral cortex in cases of adult onset dementia without Alzheimer changes, *Neurosci. Lett.* **76:**124–127.

Braak, H., and Braak, E., 1994, Pathology of Alzheimer's disease, in: *Neurodegenerative Diseases* (D. B. Calne, Ed.), W. B. Saunders, Philadelphia, pp. 585–613.

Bramblett, G. T., Goedert, M., Jakes, R., Merrick, S. E., Trojanowski, J. Q., and Lee, V.M.-Y., 1993, Abnormal tau phosphorylation at Ser 396 in Alzheimer's disease recapitulates development and contributes to reduced microtubule binding, *Neuron* **10:**1089–1099.

Bueé-Scherrer, V., Buée, L., Hof, P. R., Leveugle, B., Gilles, C., Loerzel, A. J., Perl, D. P., and Delacourte, A., 1995, Neurofibrillary degeneration in amyotrophic lateral sclerosis/parkinsonism-dementia complex of Guam: Immunochemical characterization of tau proteins, *Am. J. Pathol.* **68:**924–932.

Buée-Scherrer, V., Hof, P. R., Buée, L., Leveugle, B., Vermersch, P., Perl, D. P., Olanow, C. W., and Delacourte, A., 1996, Hyperphosphorylated tau proteins differentiate corticobasal degeneration and Pick's disease, *Acta Neuropathol.* **91:**351–359.

Buée-Scherrer, V., Buée, L., Leveugle, B., Perl, D. P., Vermersch, P., Hof, P. R., and Delacourte, A., 1997, Pathological tau proteins in postencephalitic parkinsonism: Comparison with Alzheimer's disease and other neurodegenerative disorders, *Ann. Neurol.* **42;**356–359.

Burgin, K. E., Waxham, M. N., Rickling, S., Westgate, S. A., Mobley, W. C., and Kelly, P. T., 1990, *In situ* hybridization histochemistry of $Ca^{2+}$/calmodulin-dependent protein kinase in developing rat brain, *J. Neurosci.* **10:**1788–1798.

Busciglio, J., Lorenzo, A., and Yankner, B. A., 1992, Methodological variables in the assessment of β amyloid neurotoxicity, *Neurobiol. Aging* **13:**609–612.

Butner, K. A., and Kirschner, M. W., 1991, Tau protein binds to microtubules through a flexible array of distributed weak sites, *J. Cell. Biol.* **115:**717–730.

Calhoun, M. E., Wiederhold, K.-H., Abramowski, D., Phinney, A. L., Probst, A., Sturchler-Pierrat, C., Staufenbiel, M., Sommer, B., and Jucker, M., 1998, Neuron loss in APP transgenic mice, *Nature* **395;**755–756.

Cataldo, A. M., Barnett, J. L., Picroni, C., and Nixon, R. A., 1997, Increased neuronal endocytosis and protease delivery to early endosomes in sporadic Alzheimer's disease: neuropathologic evidence for a mechanism of increased β-amyloidogenesis, *J. Neurosci.* **17:**6142–6151.

Cataldo, A. M., Hamilton, D. J., Barnett, J. L., Paskevich, P. A., and Nixon, R. A., 1996, Properties of the endosomal-lysosomal system in the human central nervous system: Disturbances mark most neurons in populations at risk to degenerate in Alzheimer's disease, *J. Neurosci.* **16:**186–199.

Chartier-Harlin, M.-C., Crawford, F., Houlden, H., Warren A., Hughes, D., Fidani, L., Goate, A., Rossor, M., Roques, P., Hardy, J., and Mullan, M., 1991, Early-onset Alzheimer's disease caused by mutation at codon 717 of the β-amyloid precursor protein gene, *Nature* **353:**844–846.

Chow, N., Cox, C., Callahan, L. M., Weimer, J. M., Guo, L., and Coleman, P. D., 1998, Expression profiles of multiple genes in single neurons of Alzheimer's disease, *Proc. Natl. Acad. Sci. USA* **95:**9620–9625.

Christie, R. H., Freeman, M., and Hyman, B. T., 1996, Expression of the macrophage scavenger receptor, a multifunctional lipoprotein receptor, in microglia associated with senile plaques in Alzheimer's disease, *Am. J. Pathol.* **148:**399–403.

Chyung, A.S.C., Greenberg, B. D., Cook, D. G., Doms, R. W., and Lee, V.M.-Y., 1997, Novel β-secretase cleavage of β-amyloid precursor protein in the endoplasmic reticulum/intermediate compartment of NT2N cells, *J. Cell. Biol.* **138**(3):671–80.

Citron, M., Westaway, D., Xia, W., Carlson, G.T.D., Levesque, G., Johnson-Wood, K., Lee, M. K., Seubert, P., Davis, A., Kholodenko, D., Motter, R., Sherrington, R., Perry, B., Yao, H., Strome, R., Lieberburg, I., Rommens, J., Kim, S., Schenk, D., Fraser, P., St. George-Hyslop, P., and Selkoe, D. J., 1997, Mutant presenilins of Alzheimer's disease increase production of 42-residue amyloid-β protein in both transfected cells and transgenic mice, *Nature Med.* **3:**67–72.

Clark, L. N., Poorkaj, P., Wszolek, Z., Geschwind, D. H., Nasreddine, Z. S., Miller, B., Li, D., Payami, H., Awert, F., Markopoulou, K., Andreadis, A., D'Souza, I., Lee, V.M.-Y., Reed, L., Trojanowski, J. Q., Zhukareva, V., Bird, T., Schellenberg, G., and Wilhelmsen, K. C., 1998, Pathogenic implications of mutations in the tau gene in pallido-ponto-nigral degeneration and related neurodegenerative disorders linked to chromosome 17, *Proc. Natl. Acad. Sci. USA* **95:**13103–13107.

Cohen, A. S., and Connors, L. H., 1987, The pathogenesis and biochemistry of amyloidosis, *J. Pathol.* **151:**1–10.

Cohen, D. H., Feiner, H., Jensson, O., and Frangione, B., 1983, Amyloid fibril in hereditary cerebral hemorrhage with amyloidosis (HCHWA) is related to the gastroentero-pancreatic neuroendocrine protein, gamma, trace, *J. Exp. Med.* **158:**623–628.

Coleman, P. D., and Flood, D. G., 1987, Neuron numbers and dendritic extent in normal aging and Alzheimer's disease, *Neurobiol. Aging* **8:**521–545.

Cook, D. G., Sung, J. C., Golde, T. E., Felsenstein, K. M., Wojczyk, B. S., Tanzi, R. E., Trojanowski, J. Q., Lee, V.M.-Y., and Doms, R. W., 1996, Expression and analysis of presenilin 1 in a human neuronal system: Localization in cell bodies and dendrites, *Proc. Natl. Acad. Sci. USA* **93:**9223–9228.

Cook, D. G., Forman, M. Sung, J. C., Leight, S., Kolson, D. L., Iwatsubo, T., Lee, V.M.-Y., and Doms, R. W., 1997, Alzheimer's Aβ(1-42) is generated in the endoplasmic reticulum/intermediate compartment of NT2N cells, *Nature Med.* **3:**1021–1023.

Corder, E. H., Saunders, A. M., Strittmatter, W. J., Schmechel, D. E., Gaskell, P. C., Small, G. W., Roses, A. D., Haines, J. L., and Pericak-Vance, M. A., 1993, Gene dose of apolipoprotein E type 4 allele and the risk of Alzheimer's disease in late onset families, *Science* **261:**921–923.

Corder, E. H., Saunders, A. M., Risch, N. J., Strittmatter, W. J., Schmechel, D. E., Gaskell, P. C., Jr., Rimmler, J. B., Locke, P. A., Conneally, P. M., Schmader, K. E., Small, G. W., Roses, A. D., Haines, J. L., and Pericak-Vance, M. A., 1994, Protective effect of apolipoprotein E type 2 allele for late onset Alzheimer disease, *Nature Genet.* **7:**180–184.

Couchie, D., Mavilia, C., Georgieff, I. S., Liem, R. K., Shelanski, M. L., and Nunez, J., 1992, Primary structure of high molecular weight tau present in the peripheral nervous system, *Proc. Natl. Acad. Sci. USA* **89:**4378–4381.

Crino, P. B., and Eberwine, J., 1996, Molecular characterization of the dendritic growth cone: Regulated mRNA transport and local protein synthesis, *Neuron* **17:**1173–1187.

Crino, P. B., Dichter, M. A., Trojanowski, J. Q., and Eberwine, J. H., 1996, Embryonic markers in tuberous sclerosis: single cell molecular pathology, *Proc. Natl. Acad. Sci. USA* **93:**14152–14157.

Crino, P. B., Khodakhah, K., Becker, K., Ginsberg, S. D., Hemby, S., and Eberwine, J. H., 1998, Presence and phosphorylation of transcription factors in dendrites *Proc. Natl. Acad. Sci. USA* **95:**2313–2318.

Crowther, R. A., and Wischik, C. M., 1985, Image reconstruction of the Alzheimer paired helical filament, *EMBO J.* **4:**3661–3665.

Cruts, M., and Van Broeckhoven, C., 1998, Presenilin mutations in Alzheimer's disease, *Hum. Mut.* **11:**183–190.

Cummings, J. L., 1995, Lewy body disease with dementia: pathophysiology and treatment, *Brain Cogn.* **28:**266–280.

Daigle, I., and Li, C., 1993, apl-1, A *Caenorhabditis elegans* gene encoding a protein related to the human β-amyloid protein precursor *Proc. Natl. Acad. Sci. USA* **90:**12045–12049.

Davis, J. A., Naruse, S., Chen, H., Eckman, C., Younkin, S., Price, D. L., Borchelt, D. R., Sisodia, S. S., and Wong, P. C., 1998, An Alzheimer's disease-linked PS1 variant rescues the developmental abnormalities of PS1-deficient embryos, *Neuron* **20:**603–609.

Davis, L., Banker, G. A., and Steward, O., 1987, Selective transport of RNA in hippocampal neurons in culture, *Nature* **330:**477–480.

Delacourte, A., Robitaille, Y., Sergeant, N., Buée, L., Hof, P. R., Wattez, A., Laroche-Cholette, A., Mathieu, J., Chagnon, P., and Gauvreau, D., 1996, Specific pathological tau protein variants characterize Pick's disease, *J. Neuropathol. Exp. Neurol.* **55;**159–168.

Dickson, D. W., 1997, The pathogenesis of senile plaques, *J. Neuropathol. Exp. Neurol.* **56:**321–339.

Dickson, D. W., Ksiezak-Reding, H., Davies, P., and Yen, S. H., 1987, A monoclonal antibody that recognizes a phosphorylated epitope in Alzheimer neurofibrillary tangles, neurofilaments, and tau proteins immunostains granulovacuolar degeneration, *Acta Neuropathol.* **73:**254–258.

Dickson, D. W., Farlo, J., Davies, P., Crystal, H., Fuld, P., and Yen, S. H., 1988, Alzheimer's disease: A double-labeling immunohistochemical study of senile plaques, *Am. J. Pathol.* **132:**86–101.

Dickson, D. W., Crystal, H. A., Mattiace, L. A., Masur, D. M., Blau, A. D., Davies, P., Yen, S. H., and Aronson, M. K., 1991, Identification of normal and pathological aging in prospectively studied nondemented elderly humans, *Neurobiol. Aging* **13:**179–189.

Dickson, D. W., Lee, S. C., Mattiace, L. A., Yen, S.-H.C., and Brosnan, C. F., 1993, Microglia and cytokines in neurological disease, with special reference to AIDS and Alzheimer's disease, *Glia* **7:**75–83.

Drecshel, D. N., Hyman, A. A., Cobb, M. H., and Kirschner, M., 1992, Modulation of the dynamic instability of tubulin assembly by the microtubule associated protein tau, *Mol. Biol. Cell* **3:**1147–1154.

Drewes, G., Ebneth, A., Preuss, U., Mandelkow, E.-M., and Mandelkow, E., 1997, MARK, a novel family of protein kinases that phosphorylate microtubule-associated proteins and trigger microtubule disruption, *Cell* **89:**297–308.

D'Souza, I., Poorkaj, P., Hong, M., Nochlin, D., Lee., V.M.-Y., Bird, T. D., and Schellenberg, G. D., 1999,

Missense and silent tau gene mutations cause fronto temporal dementia with parkinsonism—chromosome-17 type by affecting alternative splicing, *Proc. Natl. Acad. Sci. USA,* in press.

Duff, K., Eckman, C., Zehr, C., Yu, X., Prada, C.-M., Perez-tur. J., Hutton, M., Buée, L., Harigaya, Y., Yager, D., Morgan, D., Gordon, M. N., Holcomb, L., Refolo, L., Zenk, B., Hardy, J., and Younkin, S., 1996, Increased Aβ42 (43) in brains of mice expressing mutant presenilin 1, *Nature* **383:**710–713.

Eberwine, J., Yeh, H., Miyashiro, K., Cao, Y., Nair, S., Finnell, R., Zettel, M., and Coleman, P., 1992, Analysis of gene expression in single live neurons, *Proc. Natl. Acad. Sci. USA* **89:**3010–3014.

Eberwine, J., Crino, P., and Dichter, M., 1995, Single-cell mRNA amplification: Implications for basic and clinical neuroscience, *Neuroscientist* **1:**200–211.

Eidelberg, D., Sotrel, A., Joachim, C., Selkoe, D., Forman, A., Pendlebury, W. W., and Perl, D. P., 1987, Adult onset Hallervorden-Spatz disease with neurofibrillary pathology: A discrete clinicopathological entity, *Brain* **110:**993–1013.

Eikelenboom, P., and Veerhuis, R., 1996, The role of complement and activated microglia in the pathogenesis of Alzheimer's disease, *Neurobiol. Aging* **17:**673–680.

El Khoury, J., Hickman, S. E., Thomas, C. A., Cao, L., Silverstein, S. C., and Loike, J. D., 1996, Scavenger receptor-mediated adhesion of microglia to β-amyloid fibrils, *Nature* **382:**716–719.

Elshourbagy, N. A., Liao, W. S., Mahley, R. W., and Taylor, J. M., 1985, Apolipoprotein E mRNA is abundant in the brain and adrenals, as well as in the liver, and is present in other peripheral tissues of rats and marmosets, *Proc. Natl. Acad. Sci. USA* **82:**203–207.

Esch, F. S., Keim, P. S., Beattie, E. C., Blacher, R. W., Culwell, A. R., Oltersdorf, T., McClure, D., and Ward, P. J., 1990, Cleavage of amyloid-β peptide during constitutive processing of its precursor, *Science* **248:**1122–1124.

Esiri, M. M., Hyman, B. T., Beyreuther, K., and Masters, C. L., 1997, Aging and dementia, in: *Greenfield's Neuropathology,* 6th Ed., Vol. 2 (D. I. Graham and P. L. Lantos, Eds.), Arnold, London, pp. 153–233.

Flament, S., Delacourte, A., and Mann, D. M., 1990, Phosphorylation of tau proteins: A major event during the process of neurofibrillary degeneration: A comparative study between Alzheimer's disease and Down's syndrome, *Brain Res.* **516:**15–19.

Flament, S., Delacourte, A., Verny, M., Hauw, J. J., and Javoy-Agid, F., 1991, Abnormal tau proteins in progressive supranuclear palsy: Similarities and differences with the neurofibrillary degeneration of the Alzheimer type, *Acta Neuropathol.* **81:**591–596.

Foster, N. L., Wilhelmsen, K., Sima, A. A., Jones, M. Z., D'Amato, C. J., and Gilman, S., 1997, Frontotemporal dementia and parkinsonism linked to chromosome 17: a consensus conference. Conference participants, *Ann. Neurol.* **41:**706–715.

Francowiak, J., Wisniewski, H. M., Wegiel, J., Merz, G. S., Iqbal, K., and Wang, K. C., 1992, Ultrastructure of the microglia that phagocytose amyloid and the microglia that produce β-amyloid fibrils, *Acta Neuropathol.* **84:**225–233.

Frautschy, S. A., Baird, A., and Cole, G. M., 1991, Effects of injected Alzheimer β-amyloid cores in rat brain, *Proc. Natl. Acad Sci. USA* **88:**8362–8366.

Frautschy, S. A., Horn, D. L., Sigel, J. J., Harris-White, M. E., Mendoza, J. J., Yang, F., Saido, T. C., and Cole, G. M., 1998, Protease inhibitor coinfusion with amyloid-β protein results in enhanced deposition and toxicity in rat brain, *J. Neurosci.* **18:**8311–8321.

Friedhoff, P., Schneider, A., Mandelkow, E. M., and Mandelkow, E., 1998a, Rapid assembly of Alzheimer-like paired helical filaments from microtubule-associated protein tau monitored by fluorescence in solution, *Biochemistry* **37:**10223–10230.

Friedhoff, P., von Bergen, M., Mandelkow, E. M., Davies, P., and Mandelkow, E., 1998b, A nucleated assembly mechanism of Alzheimer paired helical filaments, *Proc. Natl. Acad. Sci. USA* **95**(26):15712–15717.

Gabbita, S. P., Lovell, M. A., and Markesbery, W. R., 1998, Increased nuclear DNA oxidation in the brain in Alzheimer's disease, *J. Neurochem.* **71:**2034–2040.

Galloway, P. G., Perry, G., and Gambetti, P., 1987, Hirano body filaments contain actin and actin associated proteins, *J. Neuropathol. Exp. Neurol.* **46:**185–199.

Games, D., Adams, D., Alessandrini, R., Barbour, R., Berthelette, P., Blackwell, C., Carr, T., Clemens, J., Donaldson, T., Gillespie, F., Guido, T., Hagoplan, S., Johnson-Wood, K., Khan, K., Lee, M., Leibowitz, P., Lieberburg, I., Little, S., Masliah, E., McConlogue, L., Montoya-Zavala, M., Mucke, L., Paganini, L., Penniman, E., Power, M., Schenk, D., Seubert, P., Snyder, B., Soriano, F., Tan, H., Vitale, J., Wadsworth, S., Wolozin, B., and Zhao, J., 1995, Alzheimer-type neuropathology in transgenic mice overexpressing V717F β-amyloid precursor protein, *Nature* **373:**523–527.

Garner, C. C., Tucker, R. P., and Matus, A., 1988, Selective localization of messenger RNA for cytoskeletal protein MAP2 in dendrites, *Nature* **336:**674–677.

Garver, T. D., Harris, K. A., Lehman, R.A.W., Lee, V.M.-Y., Trojanowski, J. Q., and Billingsley, M. L., 1994, Tau phosphorylation in human, primate, and rat brain: Evidence that a pool of tau is highly phosphorylated *in vivo* and is rapidly dephosphorylated *in vitro*, *J. Neurochem.* **63:**2279–2287.

Geddes, J. F., Hughes, A. J., Lees, A. J., and Daniel, S. E., 1993, Pathological overlap in cases of parkinsonism associated with neurofibrillary tangles: A study of recent cases of postencephalitic parkinsonism and comparison with progressive supranuclear palsy and guamanian parkinsonism-dementia complex, *Brain*, **116:**281–302.

Ghetti, B., Tagliavini, F., Masters, C. L., Beyreuther, K., Giaccone, G., Verga, L., Farlow, M. R., Conneally, P. M., Dlouhy, S. R., Azzarelli, B., and Bugiani, O., 1989, Gerstmann-Straussler-Scheinker disease. II. Neurofibrillary tangles and plaques with PrP-amyloid coexist in an affected family, *Neurology* **39:**1453–1461.

Ghetti, B., Piccardo, P., Spillantini, M. G., Ichimiya, Y., Porro, M., Perini, F., Kitamoto, T., Tateishi, J., Seiler, C., Frangione, B., Bugiani, O., Giaccone, G., Prelli, F., Goedert, M., Dlouhy, S. R., and Tagliavini, F., 1996, Vascular variant of prion protein cerebral amyloidosis with tau-positive neurofibrillary tangles: The phenotype of the stop codon 145 mutation in PRNP, *Proc. Natl. Acad. Sci. USA* **93:**744–748.

Gibson, P., and Tomlinson, B. E., 1977, The numbers of Hirano bodies in the hippocampus of normal and demented subjects with Alzheimer's disease, *J. Neurol. Sci.* **33:**199–206.

Ginsberg, S. D., Crino, P. B., Lee, V.M.-Y., Eberwine, J. H., and Trojanowski, J. Q., 1997, Sequestration of RNA in Alzheimer's disease neurofibrillary tangles and senile plaques, *Ann. Neurol.* **41:**200–209.

Ginsberg, S. D., Galvin, J. E., Chiu, T.-S., Lee, V.M.-Y., Masliah, E., and Trojanowski, J. Q., 1998a, RNA sequestration to pathological lesions of neurodegenerative disorders, *Acta Neuropathol.* **96:**487–494.

Ginsberg, S. D., Hemby, S. E., Lee, V.M.-Y., Eberwine,. J. H., and Trojanowski, J. Q., 1998b, Single cell mRNA analysis of neurofibrillary tangle bearing CA1 neurons from Alzheimer's disease brain, *Soc. Neurosci. Abstr.* **24:**962.

Ginsberg, S. D., Crino, P. B., Hemby, S. E., Weingarten, J. A., Lee, V.M.-Y., Eberwine, J. H., and Trojanowski, J. Q., 1999a, Predominance of neuronal mRNAs in individual Alzheimer's disease senile plaques, *Ann. Neurol.*, **45**(2):174–81.

Ginsberg, S. D., Galvin, J. E., Lee, V.M.-Y., Rorke, L. B., Dickson, D. W., Wolfe, J. H., Jones, M. Z., and Trojanowski, J. Q., 1999b, Accumulation of intracellular amyloid-β peptide (Aβ 1-40) in mucopolysaccharidosis brains, *J. Neuropathol. Exp. Neurol.*, in press.

Giulian, D., Haverkamp, L. J., Li, J., Karshin, W. L., Yu, J., Tom, D., Li, X., and Kirkpatrick, J. B., 1995, Senile plaques stimulate microglia to release a neurotoxin found in Alzheimer brain, *Neurochem. Int.* **27:**119–137.

Glenner, G. G., and Wong, C. W., 1984, Alzheimer's disease: Initial report of the purification and characterization of a novel cerebrovascular amyloid protein, *Biochem. Biophys. Res. Commun.* **120:**885–890.

Goate, A., Chartier-Harlin, M. C., Mullan, M., Brown, J., Crawford, F., Fidani, L., Giuffra, L., Haynes, A., Irving, N., James, L., Mant, R., Newton, P., Rooke, K., Roques, P., Talbot, C., Pericak-Vance, M., Roses, A., Williamson, R., Rossor, M., Owen, M., and Hardy, J., 1991, Segregation of a missense mutation in the amyloid precursor protein gene with familial Alzheimer's disease, *Nature* **349:**704–706.

Goedert, M., 1987, Neuronal localization of amyloid-β protein precursor mRNA in normal human brain and in Alzheimer's disease, *EMBO J.* **6:**3627–3632.

Goedert, M., Wischik, C. M., Crowther, R. A., Walker, J. E., and Klug, A., 1988, Cloning and sequencing of the cDNA encoding a core protein of the paired helical filament of Alzheimer disease: Identification as the microtubule-associated protein tau, *Proc. Natl. Acad. Sci. USA* **85:**4051–4055.

Goedert, M., Spillantini, M. G., Potier, M. C., Ulrich, J., and Crowther, R. A., 1989a, Cloning and sequencing of the cDNA encoding an isoform of microtubule-associated protein tau containing four tandem repeats: differential expression of tau protein mRNAs in human brain, *EMBO J.* **8:**393–399.

Goedert, M., Spillantini, M. G., Jakes, R., Rutherford, D., and Crowther, R. A., 1989b, Multiple isoforms of human microtubule-associated protein tau: Sequences and localization in neurofibrillary tangles of Alzheimer's disease, *Neuron* **3:**519–526.

Goedert, M., Cohen, E. S., Jakes, R., and Cohen, P., 1992, p42 map kinase phosphorylation sites in microtubule-associated protein tau are dephosphorylated by protein phosphatase 2A1, *FEBS Lett.* **312:**95–99.

Goedert, M., Jakes, R., Crowther, R. A., Six, J., Luebke, U., Vandermeeren, M., Cras, P., Trojanowski, J. Q., and Lee, V.M.-Y., 1993, The abnormal phosphorylation of tau protein at serine 202 in Alzheimer's disease recapitulates phosphorylation during development, *Proc. Natl. Acad. Sci. USA* **90:**5066–5070.

Goedert, M., Jakes, R., Crowther, R. A., Cohen, P., Vanmechelen, R., Vandermeeren, M., and Cras, P., 1994, Epitope mapping of monoclonal antibodies to the paired helical filaments of Alzheimer's disease: Identification of phosphorylation sites in tau protein, *Biochem. J.* **301:**871–877.

Goedert, M., Jakes, R., Qi, Z., Wang, J. H., and Cohen, P., 1995a, Protein phosphatase 2A is the major enzyme in brain that dephosphorylates tau protein phosphorylated by proline-directed protein kinases or cyclic AMP-dependent protein kinase, *J. Neurochem.* **65;**2804–2807.

Goedert, M., Jakes, R., and Vanmechelen, E., 1995b, Monoclonal antibody AT8 recognizes tau protein phosphorylated at both serine 202 and threonine 205, *Neurosci. Lett.* **189:**167–169.

Goedert, M., Jakes, R., Spillantini, M. G., Hasegawa, M., Smith, M. J., and Crowther, R. A., 1996, Assembly of microtubule-associated protein tau into Alzheimer-like filaments induced by sulfated glycosaminoglycans, *Nature* **383:**550–553.

Goedert, M., Trojanowski, J. Q., and Lee, M.-Y., 1997, The neurofibrillary pathology of Alzheimer's disease, in: *The Molecular and Genetic Basis of Neurological Disease,* 2nd Ed. (R. N. Rosenberg, S. B. Prusiner, S., DiMauro, and R. L. Barchi, eds.), Butterworth-Heinemann, Boston, pp. 613–627.

Gold, R., Schmied, M., Giergerich, G., Breitschopf, H., Hartung, H. P., Toyka, K., and Lassmann, H., 1994, Differentiation between cellular apoptosis and necrosis by combined use of in situ tailing and nick translation techniques, *Lab. Invest.* **71:**219–225.

Golde, T. E., and Younkin, S. G., 1995, Processing of the amyloid-β protein precursor, in: *Pathobiology of Alzheimer's disease* (A. M. Goate and F. Ashall, eds.), Academic, London, pp. 115–143.

Golde, T. E., Estus, S., Usiak, M., Younkin, L. H., and Younkin, S. G., 1990, Expression of β-amyloid protein precursor mRNAs: Recognition of a novel alternatively spliced form and quantitation in Alzheimer's disease using PCR, *Neuron* **4:**253–267.

Golde, T. E., Estus, S., Younkin, L. H., Selkoe, D. J., and Younkin, S. G., 1992, Processing of the amyloid-β protein precursor to potentially amyloidogenic derivatives, *Science* **255:**728–730.

Goldgaber, D., Lerman, M. I., McBride, O. W., Saffiotti, U., and Gajdusek, D. C., 1987, Characterization and chromosomal localization of a cDNA encoding brain amyloid of Alzheimer's disease, *Science* **235:**877–880.

Gómez-Isla, T., Price, J. L., McKeel, D. W., Jr., Morris, J. C., Growdon, J. H., and Hyman, B. T., 1996, Profound loss of layer II entorhinal cortex neurons occurs in very mild Alzheimer's disease, *J. Neurosci.* **16:**4491–4500.

Goode, B. L., and Feinstein, S. C., 1994, Identification of a novel microtubule binding and assembly domain in the developmentally regulated inter-repeat region of tau, *J. Cell Biol.* **124:**769–782.

Gowing, E., Roher, A. E., Woods, A. S., Cotter, R. J., Chaney, M., Little, S. P., and Ball, M. J., 1994, Chemical characterization of Aβ 17-42 peptide, a component of diffuse amyloid deposits of Alzheimer disease, *J. Biol. Chem.* **269:**10987–10990.

Greenberg, S. G., and Davies, P., 1990, A preparation of Alzheimer paired helical filaments that displays distinct tau proteins by polyacrylamide gel electrophoresis, *Proc. Natl. Acad. Sci. USA* **87:**5827–5831.

Greenberg, S. C., Davies, P., Scheim, J. D., and Binder, L. I., 1992, Hydrofluoric acid-treated PHF proteins display the same biochemical properties as normal tau, *J. Biol. Chem.* **267:**564–569.

Grünthal, E., 1930, Die pathologische Anatomie der senilen Demenz und der Alzheimerschen Krankheit, in: *Handbuch der Geisteskrankheiten. Die Anatomie der Psychosen,* Vol. 11 (O. Bumke, ed.), Springer-Verlag, Berlin, pp. 638–672.

Haas, C., 1997, Presenilins: Genes for life and death, *Neuron* **18:**687–690.

Haas, C., Cazorla, P., Miguel, C. D., Valdivieso, F., and Vazquez, J., 1997, Apolipoprotein E forms stable complexes with recombinant Alzheimer's disease β-amyloid precursor protein, *Biochem. J.* **325:**169–175.

Hanger, D. P., Hughes, K., Woodgett, J. R., Brion, J.-P., and Anderton, B. H., 1992, Glycogen synthase kinase-3 induces Alzheimer's disease-like phosphorylation of tau: Generation of paired helical filament epitopes and neuronal localization of the kinase, *Neurosci. Lett.* **147:**58–62.

Hansen, L. A., Salmon, D., Galasko, D., Masliah, E., Katzman, R., DeTeresa, R., Thal, L., Pay, M. M., Hofstetter, R., Klauber, M., Rice, V., Butters, N., and Alford, M., 1990, The Lewy body variant of Alzheimer's disease: a clinical and pathologic entity, *Neurology* **40:**1–8.

Hardy, J., 1997, The Alzheimer family of diseases: Many etiologies, one pathogenesis? *Proc. Natl. Acad. Sci. USA* **94:**2095–2097.

Hartmann, T., Bieger, S. C., Bruhl, B., Tienari, P. J., Ida, N., Allsop, D., Roberts, G. W., Masters, C. L., Dotti, C. G., Unsicker, K., and Beyreuther, K., 1997, Distinct sites of intracellular production for Alzheimer's disease Aβ40/42 amyloid peptides, *Nature Med.* **3:**1016–1020.

Hasegawa, M., Crowther, R. A., Jakes, R., and Goedert, M., 1997, Alzheimer-like changes in microtubule-associated protein tau induced by sulfated glycosaminoglycans. Inhibition of microtubule binding, stimulation of phosphorylation, and filament assembly depend of the degree of sulfation, *J. Biol. Chem.* **273:**33118–33124.

Hasegawa, M., Morishima-Kawashima, M., Takio, K., Suzuki, M., Titani, K., and Ihara, Y., 1992, Protein sequence and mass spectrometric analyses of tau in the Alzheimer's disease brain, *J. Biol. Chem.* **267:**17047–17054.

Hasegawa, M., Watanabe, A., Takio, K., Suzuki, M., Arai, T., and Ihara, Y., 1993, Characterization of two distinct monoclonal antibodies to paired helical filaments: Further evidence for fetal-type phosphorylation of the tau in paired helical filaments, *J. Neurochem.* **60:**2068–2077.

Hendriks, L., van Duijn, C. M., Cras, P., Cruts, M., Van Hul, W., van Harskamp, F., Warren, A., McInnis, M. G., Antonarakis, S. E., Martin, J. J., Hofman, A., and Van Broeckhoven, C., 1992, Presenile dementia and cerebral haemorrhage linked to a mutation at codon 692 of the β-amyloid precursor protein gene, *Nature Genet.* **1:**218–221.

Hensley, K., Maidt, M. L., Yu, Z., Sang, H., Markesbery, W. R., and Floyd, R. A., 1998, Electrochemical analysis of protein nitrotyrosine and dityrosine in the Alzheimer brain indicates region-specific accumulation, *J. Neurosci.* **18:**8126–8132.

Hermon, M., Cairns, N., Egly, J.-M., Fery, A., Labudova, O., and Lubec, G., 1998, Expression of DNA excision-repair-cross-complementing proteins p80 and p89 in brain of patients with Down syndrome and Alzheimer's disease, *Neurosci. Lett.* **251:**45–48.

Hickey, W. F., and Kimura, H., 1988, Perivascular microglial cells of the CNS are bone marrow-derived and present antigen *in vivo, Science* **239:**290–292.

Higgins, L. S., Murphy, G. M., Jr., Forno, L. S., Catalano, R., and Cordell, B., 1996, P3 β-amyloid peptide has a unique and potentially pathogenic immunohistochemical profile in Alzheimer's disease brain, *Am. J. Pathol.,* **149:**585–596.

Hirano, A., Malamud, N., Elizan, T. S., and Kurland, L. T., 1966, Amyotrophic lateral sclerosis and Parkinsonism-dementia complex on Guam, *Arch. Neurol.* **15:**35–51.

Hirokawa, N., Shiomura, Y., and Okabe, S., 1988, Tau proteins: The molecular structure and mode of binding on microtubules, *J. Cell Biol.* **107:**1449–1459.

Hof, P. R., Bouras, C., Constantinidis, J., and Morrison, J. H., 1990, Balint's syndrome in Alzheimer's disease: Specific disruption of the occipitoparietal visual pathway, *Brain Res.* **493:**368–375.

Hong, M., Zhukareva, V., Vogelsberg-Ragaglia, V., Wszolek, Z., Reed, L., Miller, B. I., Geschwind, D. H., Bird, T. D., McKeel, D., Goate, A., Morris, J. C., Wilhelmsen, K. C., Schellenberg, G. D., Trojanowski, J. Q., and Lee, V.M.-Y., 1998, Mutation-specific functional impairments in distinct tau isoforms of hereditary FTDP-17, *Science* **282**(5395):1914–1917.

Horiuchi, S., and Araki, N., 1994, Advanced glycation end products of the Maillard reaction and their relation to aging, *Gerontology* **2:**10–15.

Hsiao, K., Dlouhy, S. R., Farlow, M. R., Cass, C., Da Costa, M., Conneally, P. M., Hodes, M. E., Ghetti, B., and Prusiner, S. B., 1992, Mutant prion proteins in Gerstmann-Straussler-Scheinker disease with neurofibrillary tangles, *Nature Genet.* **1:**68–71.

Hutton, M., Lendon, C. L., Rizzu, P., Baker, M., Froelich, S., Houlden, H., Pickering-Brown, S., Chakraverty, S., Isaacs, A., Grover, A., Hackett, J., Adamson, J., Lincoln, S., Dickson, D., Davies, P., Petersen, R. C., Stevens, M., de Graaff, E., Wauters, E., van Baren, J., Hillebrand, M., Joosse, M., Kwon, J. M., Nowotny, P., Che, L. K., Norton, J., Morris, J. C., Reed, L. A., Trojanowski, J., Basun, H., Lannfelt, L., Neystat, M., Fahn, S., Dark, F., Tannenberg, T., Dodd, P. R., Andreadis, A., Snowden, J., Craufurd, D., Neary, D., Owen, F., Oostra, B. A., Hardy, J., Goate, A., van Swieten, H., Mann, D., Lynch, T., and Heutink, P., 1998, Association of missense and 5′-splice-site mutations in tau with the inherited dementia FTDP-17, *Nature* **393:**702–705.

Hyman, B. T., Van Hoesen, G. W., Damasio, A. R., and Barnes, C. L., 1984, Alzheimer's disease: cell-specific pathology isolates the hippocampal formation, *Science* **225:**1168–1170.

Hyman, B. T., Van Hoesen, G. W., and Damasio, A. R., 1990, Memory-related neural systems in Alzheimer's disease: An anatomic study, *Neurology* **40:**1721–1730.

Iijima, M., Ishino, H., Tabira, T., Trojanowski, J. Q., Lee, V.M.-Y., Schmidt, M. L., Matsumoto, T., Yamashita, Y., Yoshioka, S., and Ishino, H., 1999, A familial presenile dementia with abundant tau positive neurofibrillary tangles and glial inclusions but no senile plaques, *NeuroReport* **10:**497–501.

Iqbal, K., Zaidi, T., Bancher, C., and Grundke-Iqbal, I., 1994, Alzheimer paired helical filaments: Restoration of the biological activity by dephosphorylation, *FEBS Lett.* **349:**104–108.

Irizarry, M. C., Kim, T. W., McNamara, M., Tanzi, R. E., George, J. M., Clayton, D. F., and Hyman, B. T., 1996, Characterization of the precursor protein of the non-Aβ component of senile plaques (NACP) in the human central nervous system, *J. Neuropathol. Exp. Neurol.* **55:**889–895.

Irizarry, M. C., McNamara, M., Fedorchak, K., Hsiao, K., and Hyman, B. T., 1997a, APPsw transgenic mice develop age-related Aβ deposits and neuropil abnormalities, but no neuronal loss in CA1, *J. Neuropathol. Exp. Neurol.* **56:**965–973.

Irizarry, M. C., Soriano, F., McNamara, M., Page, K. J., Schenk, D., Games, D., and Hyman, B. T., 1997b, Aβ deposition is associated with neuropil changes, but not with overt neuronal loss in the human amyloid precursor protein V717F (PDAPP) transgenic mouse, *J. Neurosci.* **17:**7053–7059.

Ishii, T., and Haga, S., 1984, Immuno-electron-microscopic localization of complement in amyloid fibrils of senile plaques, *Acta Neuropathol.* **63:**296–300.

Itagaki, S., McGeer, P. L., Akiyama, H., Beattie, B. L., Walker, D. G., Moore, G. R., and McGeer, E. G., 1989a, A case of adult-onset dementia with argyrophilic grains, *Ann. Neurol.* **26:**685–689.

Itagaki, S., McGeer, P. L., Akiyama, H., Zhu, S., and Selkoe, D., 1989b, Relationship of microglia and astrocytes to amyloid deposits of Alzheimer disease, *J. Neuroimmunol.* **24:**173–182.

Iwai, A., Yoshimoto, M., Masliah, E., and Saitoh, T., 1995, Non-Aβ component of Alzheimer's disease amyloid (NAC) is amyloidogenic, *Biochemistry* **34:**10139–10145.

Iwatsubo, T., Okada, A., Suzuki, N., Mizusawa, H., Nukina, N., and Ihara, Y., 1994, Visualization of Aβ42(43) and Aβ40 in senile plaques with end-specific Aβ monoclonals: Evidence that an initially deposited species is Aβ42(43), *Neuron* **13:**45–53.

Iwatsubo, T., Mann, D.M.A., Odaka, A., Suzuki, N., and Ihara, Y., 1995, Amyloid-β protein (Aβ) deposition: Aβ42(43) precedes Aβ40 in Down's syndrome, *Ann. Neurol.* **37:**294–299.

Iwatsubo, T., Saido, T. C., Mann, D.M.A., Lee, V.M.-Y., and Trojanowski, J. Q., 1996, Full-length amyloid-β[1-42(43)] and amino-terminally modified and truncated Aβ42(43) deposit in diffuse plaques, *Am. J. Pathol.* **149:**1823–1830.

Jarrett, J. T., Berger, E. P., and Lansbury, P. T., Jr., 1993, The carboxy terminus of the β amyloid protein is critical for the seeding of amyloid formation: Implications for the pathogenesis of Alzheimer's disease, *Biochemistry* **32:**4694–4697.

Kahn, J., Anderton, B. H., Probst, A., Ulrich, J., and Esiri, M. M., 1985, Immunohistological study of granulovacuolar degeneration using monoclonal antibodies to neurofilaments, *J. Neurol. Neurosurg. Psychiatry* **48:**924–926.

Kampers, T., Friedhoff, P., Biernat, J., Mandelkow, E. M., and Mandelkow, E., 1996, RNA stimulates aggregation of microtubule-associated protein tau into Alzheimer-like paired helical filaments, *FEBS Lett.* **399:**344–349.

Kanemaru, K., Takio, K., Miura, R., Titani, K., and Ihara, Y., 1992, Fetal-type phosphorylation of the tau in paired helical filaments, *J. Neurochem.* **58:**1667–1675.

Kang, J., Lemiare, H.-G., Unterbeck, A., Salbaum, J. M., Masters, C. L., Grzeschik, K.-H., Multhaup, G., Beyreuther, K., and Müller-Hill, B., 1987, The precursor of Alzheimer's disease amyloid A4 protein resembles a cell-surface receptor, *Nature* **325:**733–736.

Katzman, R., Terry, R., DeTeresa, R., Brown, T., Davies, P., Fuld, P., Renbing, X., and Peck, A., 1988, Clinical, pathological, and neurochemical changes in dementia: A subgroup with preserved mental status and numerous neocortical plaques, *Ann. Neurol.* **23:**138–144.

Kawas, C., Resnick, S., Morrison, A., Brookmeyer, R., Corrada, M., Zonderman, A., Bacal, C., Lingle, D. D., and Metter, E., 1997, A prospective study of estrogen replacement therapy and the risk of developing Alzheime 's disease: the Baltimore Longitudinal Study of Aging, *Neurology* **48:**1517–1521.

Kitaguchi, N., Takahashi, Y., Tokushima, Y., Shiojiri, S., and Ito, H., 1988, Novel precursor of Alzheimer's disease amyloid-β protein shows protease inhibitory activity, *Nature* **331:**530–532.

Kondo, J., Honda, T., Mori, H., Hamada, Y., Miura, R., Ogawara, M., and Ihara, Y., 1988, The carboxyl third of tau is tightly bound to paired helical filaments, *Neuron* **1:**827–834.

Kosaka, K., 1994, Diffuse neurofibrillary tangles with calcification: A new presenile dementia, *J. Neurol. Neurosurg. Psychiatry* **57:**594–596.

Kosik, K. S., 1998, Presenilin interactions and Alzheimer's disease, *Science* **279:**463–465.

Kosik, K. S., Joachim, C. L., and Selkoe, D. J., 1986, Microtubule-associated protein tau is a major antigenic component of paired helical filaments in Alzheimer's disease, *Proc. Natl. Acad. Sci. USA* **83:**4044–4048.

Kosik, K. S., Orecchio, L. D., Binder, L., Trojanowski, J. Q., Lee, V.M.-Y., and Lee, G., 1988, Epitopes that span the tau molecule are shared with paired helical filaments, *Neuron* **1:**817–825.

Kovacs, D. M., Fausett, H. J., Page, K. J., Kim, T.-W., Moir, R. D., Merriam, D. E., Hollister, R. D., Hallmark, O. G., Mancini, R., Felsenstein, K. M., Hyman, B. T., Tanzi, R. E., and Wasco, W., 1996, Alzheimer-associated presenilins 1 and 2: Neuronal expression in brain and localization to intracellular membranes in mammalian cells, *Nature Med.* **2:**224–229.

Kowall, N. W., Beal, M. F., Busciglio, J., Duffy, L. K., and Yankner, B. A., 1991, An *in vivo* model for the neurodegenerative effects of β-amyloid and protection by substance P, *Proc. Natl. Acad. Sci. USA* **88:**7247–7251.

Ksiezak-Reding, H., Morgan, K., Mattiace, L. A., Davies, P., Liu, W. K., Yen, S. H., Weidenheim, K., and Dickson, D. W., 1994, Ultrastructure and biochemical composition of paired helical filaments in corticobasal degeneration, *Am. J. Pathol.* **145:**1496–1508.

L'Hernault, S. W., and Arduengo, P. M., 1992, Mutation of a putative sperm membrane protein in *Caenorhabditis elegans* prevents sperm differentiation but not its associated meiotic divisions, *J. Cell Biol.* **119:**55 –68.

Lah, J. J., Heilman, C. J., Nash, N. R., Rees, H. D., Yi, H., Counts, S. E., and Levey, A. I., 1997, Light and electron microscopic localization of presenilin-l in primate brain, *J. Neurosci.* **17:**1971–1980.

Lassmann, H., Bancher, C., Breitschopf, H., Wegiel, J., Bobinski, M., Jellinger, K., and Wisniewski, H. M., 1995, Cell death in Alzheimer's disease evaluated by DNA fragmentation *in situ, Acta Neuropathol.* **89:**35–41.

Lee, G., Cowan, N., and Kirschner, M., 1988, The primary structure and heterogeneity of tau protein from mouse brain, *Science* **239:**285–288.

Lee, M. K., Slunt, H. H., Martin, L. J., Thinakaran, G., Kim, G., Gandy, S. E., Seeger, M., Koo, E., Price, D. L., and Sisodia, S. S., 1996, Expression of presenilin 1 and 2 (PS1 and PS2) in human murine tissues, *J. Neurosci.* **16:**7513–7525.

Lee, V.M.-Y., Carden, M. J., Schlaepfer, W. W., and Trojanowski, J. Q., 1987, Monoclonal antibodies distinguish several differentially phosphorylated states of the two largest rat neurofilament subunits (NF-H and NF-M) and demonstrate their existence in the normal nervous system of adult rats, *J. Neurosci.* **7:**3474–3488.

Lee, V.M.-Y., Balin, B. J., Otvos, L., Jr., and Trojanowski, J. Q., 1991, A68: A major subunit of paired helical filaments and derivatized forms of normal tau, *Science* **251:**675–678.

Lemaire, H. G., Salbaum, J. M., Multhaup, G., Kang, J., Bayney, R. M., Unterbeck, A., Beyreuther, K., and Muller-Hill, B., 1989, The PreA4(695) precursor protein of Alzheimer's disease A4 amyloid is encoded by 16 exons, *Nucl. Acids Res.* **17:**517–522.

Lemere, C. A., Blusztajn, J. K., Yamaguchi, H., Wisniewski, T., Saido, T. C., and Selkoe, D. J., 1996, Sequence of deposition of heterogeneous amyloid-β peptides and apoE in Down syndrome: implications for initial events in amyloid plaque formation, *Neurobiol. Dis.* **3:**16–32.

Lendon, C. L., Ashall, F., and Goate, A. M., 1997, Exploring the etiology of Alzheimer disease using molecular genetics, *JAMA* **277:**825–831.

Leverenz, J. B., and Raskind, M. A., 1998, Early amyloid deposition in the medial temporal lobe of young Down syndrome patients: a regional quantitative analysis, *Exp. Neurol.* **150:**296–304.

Levitan, D., and Greenwald, I., 1995, Facilitation of lin-12-mediated signaling by sel-12, a *Caenorhabditis elegans* S182 Alzheimer's disease gene, *Nature* **377:**351–354.

Levitan, D., Doyle, T. G., Brousseau, D., Lee, M. K., Thinakaran, G., Slunt, H. H., Sisodia, S. S., and Greenwald, I., 1996, Assessment of normal and mutant human presenilin function in *Caenorhabditis elegans, Proc. Natl. Acad. Sci. USA* **93:**14940–14944.

Levy, E., Carman, M. D., Fernandez-Madrid, I. J., Power, M. D., Lieberburg, I., van Duinen, S. G., Bots, G. T., Luyendijk, W., and Frangione, B., 1990, Mutation of the Alzheimer's disease amyloid-β gene in hereditary cerebral hemorrhage, Dutch type, *Science* **248:**1124–1126.

Levy-Lahad, E., Wasco, W., Poorkaj, P., Romano, D. M., Oshima, J., Pettingell, W. H., Yu, C.-E., Jondro, P. D., Schmidt, S. D., Wang, K., Crowley, A. C., Fu, Y.-H., Guenette, S. Y., Galas, D., Nemens, E., Wijsman, E. M., Bird, T. D., Schellenberg, G. D., and Tanzi, R. E., 1995, Candidate gene for the chromosome 1 familial Alzheimer's disease locus, *Science* **269:**973–977.

Li, Y.-T., Woodruff-Pak, D. S., and Trojanowski, J. Q., 1994, Amyloid plaques in cerebellar cortex and the integrity of Purkinje cell dendrites, *Neurobiol. Aging* **15:**1–9.

Lippa, C. F., Fujiwara, H., Mann, D.M.A., Giasson, B., Baba, M., Schmidt, M. L., Nee, L. E., O'Connell, B., Pollen, D. A., St. George-Hyslop, P., Gheti, B., Nochlin, D., Bird, T. D., Cairns, N. J., Lee, V.M.-Y., Iwatsubo, T., and Trojanowski, J. Q., 1998, Lewy bodies contain altered α-synuclein in brains of many familial Alzheimer's disease patients with mutations in presenilin and amyloid precursor protein genes, *Am. J. Pathol.* **153:**1365–1370.

Lippa, C. F., Schmidt, M. L., Lee, V.M.-Y., and Trojanowski, J. Q., 1999, Antibodies to α-synuclein detect Lewy bodies in many Down's syndrome brains with Alzheimer's disease, *Ann. Neurol.* **45**(3):353–357.

Litersky, J. M., Johnson, G. V., Jakes, R., Goedert, M., Lee, M., and Seubert, P., 1996, Tau protein is phosphorylated by cyclic AMP-dependent protein kinase and calcium/calmodulin-dependent protein kinase II within its microtubule-binding domains at Ser-262 and Ser-356, *Biochem. J.* **316:**655–660.

Love, S., Bridges, L. R., and Case, C. P., 1995, Neurofibrillary tangles in Niemann-Pick disease type C, *Brain* **118:**119–129.

Lovett, M., Goldgaber, D., Ashley, P., Cox, D. R., Gajdusek, D. C., and Epstein, C. J., 1987, The mouse homolog of the human amyloid-β protein (AD-AP) gene is located on the distal end of mouse chromosome 16: Further extension of the homology between human chromosome 21 and mouse chromosome 16, *Biochem. Biophys. Res. Comm.* **144:**1069–1075.

Luo, L., Tully, T., and White, K., 1992, Human amyloid-β precursor protein ameliorates behavioral deficit of flies deleted for Appl gene, *Neuron* **9:**595–605.

Mahley, R. W., 1988, Apolipoprotein E: Cholesterol transport protein with expanding role in cell biology, *Science* **240:**622–630.

Mai, J. K., Schmidt-Kastner, R., and Tefett, H.-B., 1984, use of acridine orange for histologic analysis of the central nervous system, *J. Histochem. Cytochem.* **32:**97–104.

Mandelkow, E. M., Drewes, G., Biernat, J., Gustke, N., Van Lint, J., Vandenheede, J. R., and Mandelkow, E., 1992, Glycogen synthase-3 and the Alzheimer-like state of microtubule-associated protein tau, *FEBS Lett.* **314:**315–321.

Mandybur, T. I., Nagpaul, A. S., Pappas, Z., and Niklowitz, W. J., 1977, Alzheimer neurofibrillary change in subacute sclerosing panencephalitis, *Ann. Neurol.* **1:**103–107.

Mann, D.M.A., 1978, Granulovacuolar degeneration in pyramidal cells of the hippocampus, *Acta Neuropathol.* **42:**149–151.

Mann, D. M., Yates, P. O., Marcyniuk, B., and Ravindra, C. R., 1986, The topography of plaques and tangles in Down's syndrome patients of different ages, *Neuropathol. App. Neurobiol.* **12:**447–457.

Manning, R. W., Reid, C. M., Lampe, R. A., and Davis, L. G., 1988, Identification in rodents and other species of an mRNA homologous to the human β-amyloid precursor, *Brain Res.* **427:**293–297.

Marcus, D. L., Thomas, C., Rodriguez, C., Simberkoff, K., Tsai, J. S., Strafaci, J. A., and Freedman, M. L., 1998, Increased peroxidation and reduced antioxidant enzyme activity in Alzheimer's disease, *Exp. Neurol.* **150:**40–44.

Markesbery, W. R., and Lovell, M. A., 1998, Four-hydroxynonenal, a product of lipid peroxidation, is increased in the brain in Alzheimer's disease, *Neurobiol. Aging* **19:**33–36.

Martin, L. J., Portera-Caillau, C., Ginsberg, S. D., and Al-Abdulla, N. A., 1998, Animal models and degenerative disorders of the human brain, *Lab. Anim. Sci.* **27:**2–9.

Martin, L. J., Pardo, C. A., Cork, L. C., and Price, D. L., 1994, Synaptic pathology and glial responses to neuronal injury precede the formation of senile plaques and amyloid deposits in the aging cerebral cortex, *Am. J. Pathol.* **145:**1358–1381.

Martone, M. E., Pollock, J. A., Jones, Y. Z., and Ellisman, M. H., 1996, Ultrastructural localization of dendritic messenger RNA in adult rat hippocampus, *J. Neurosci.* **16:**7437–7446.

Masliah, E., Iwai, A., Mallory, M., Ueda, K., and Saitoh, T., 1996, Altered presynaptic protein NACP is associated with plaque formation and neurodegeneration in Alzheimer's disease, *Am. J. Pathol.* **148:**201–210.

Masliah, E., Terry, R. D., Mallory, M., Alford, M., and Hansen, L. A. 1990, Diffuse plaques do not accentuate synapse loss in Alzheimer's disease, *Am. J. Pathol.* **137:**1293–1297.

Masters, C. L., Simms, G., Weinman, N. A., Multhaup, G., McDonald, B. L., and Beyreuther, K., 1985, Amyloid plaque core protein in Alzheimer disease and Down syndrome, *Proc. Natl. Acad. Sci. USA* **82:**4245–4249.

Matsuo, E. S., Shin, R.-W., Billingsley, M. L., Van de Voorde, A., O'Connor, M., Trojanowski, J. Q., and Lee, V.M.-Y., 1994, Biopsy-derived adult human brain tau is phosphorylated at many of the same sites as Alzheimer's disease paired helical filament tau, *Neuron* **13:**989–1002.

Mawal-Dewan, M., Henley, J., Van de Voorde, A., Trojanowski, J. Q., and Lee, V.M.-Y., 1994, The phosphorylation state of tau in the developing rat brain is regulated by phosphoprotein phosphatases, *J. Biol. Chem.* **269:**30981–30987.

Mawal-Dewan, M., Schmidt, M. L., Balin, B., Perl, D. P., Lee, V.M.-Y., and Trojanowski, J. Q., 1996, Identification of phosphorylation sites in PHF-tau from patients with Guam amyotrophic lateral sclerosis/parkinsonism dementia complex, *J. Neuropathol. Exp. Neurol.* **55:**1051–1059.

Mayford, M., Baranes, D., Podsypanina, K., and Kandel, E. R., 1996, The 3′-untranslated region of CaMKII α is a cis-acting signal for the localization and translation of mRNA in dendrites, *Proc. Natl. Acad. Sci. USA* **93:**13250–13255.

McDonald, J. K., and Barrett, A. J., 1986, *Pyroglutamyl Peptidase I., Pyroglutamyl Peptidase II.*, Academic, London.

McGeer, P. L., and Rogers, J., 1992, Anti-inflammatory agents as a therapeutic approach to Alzheimer's disease, *Neurology* **42:**447–449.

McGeer, P. L., Itagaki, S., Tago, H., and McGeer, E. G., 1987, Reactive microglia in patients with senile dementia of the Alzheimer types are positive for the histocompatibility glycoprotein HLA-DR, *Neurosci. Lett.* **79:**195–200.

McQuaid, S., Kirk, J., Zhou, A. L., and Allen, I. V., 1993, Measles virus infection of cells in perivascular infiltrates in the brain in subacute sclerosing panencephalitis: Confirmation by non-radioactive *in situ* hybridization, immunocytochemistry, and electron microscopy, *Acta Neuropathol.* **85:**154–158.

McQuaid, S., Allen, I. V., McMahon, J., and Kirk, J., 1994, Association of measles virus with neurofibrillary tangles in subacute sclerosing panencephalitis: A combined *in situ* hybridization and immunocytochemical investigation, *Neuropathol. Appl. Neurobiol.* **20:**103–110.

Mehta, N. D., Refolo, L. M., Eckman, C., Sanders, S., Yager, D., Perez-Tur, J., Younkin, S., Duff, K., Hardy, J., and Hutton, M., 1998, Increased Aβ42(43) from cell lines expressing presenilin 1 mutations, *Ann. Neurol.* **43:**256–258.

Mendell, J. R., Sahenk, Z., Gales, T., and Paul, L., 1991, Amyloid filaments in inclusion body myositis: Novel findings provide insight into nature of filaments, *Arch. Neurol.* **48:**1229–1234.

Miyashiro, K., Dichter, M., and Eberwine, J., 1994, On the nature and differential distribution of mRNAs in hippocampal neurites: Implications for neuronal functioning, *Proc. Natl. Acad. Sci. USA* **91:** 10800–10804.

Montine, K. S., Kim, P. J., Olson, S. J., Markesbery, W. R., and Montine, T. J., 1997, 4-Hydroxy-2-nonenal pyrole adducts in human neurodegenerative disease, *J. Neuropathol. Exp. Neurol.* **56:**866–871.

Montine, T. J., Markesbery, W. R., Morrow, J. D., and Roberts, L. J., Jr., 1998, Cerebrospinal fluid F2-isoprostane levels are increased in Alzheimer's disease, *Ann. Neurol.* **44:**410–413.

Mori, H., Takio, K., Ogawara, M., and Selkoe, D. J., 1992, Mass spectrometry of purified amyloid-β protein in Alzheimer's disease, *J. Biol. Chem.* **267:**17082–17086.

Mori, H., Nishimura, M., Namba, Y., and Oda, M., 1994, Corticobasal degeneration: A disease with widespread appearance of abnormal tau and neurofibrillary tangles, and its relation to progressive supranuclear palsy, *Acta Neuropathol.* **88:**113–121.

Morishima-Kawashima, M., Hasegawa, M., Takio, K., Suzuki, M., Yoshida, H., Titani, K., and Ihara, Y., 1995, Proline-directed and non-proline-directed phosphorylation of PHF-tau, *J. Biol. Chem.* **270:**823–829.

Morris, J. C., McKeel, D. W., Jr., Storandt, M., Rubin, E. H., Price, J. L., Grant, E. A., Ball, M. J., and Berg, L., 1991, Very mild Alzheimer's disease: Informant-based clinical, psychometric, and pathologic distinction from normal aging, *Neurology* **41:**469–478.

Murayama, M., Tanaka, S., Palacino, J., Murayama, O., Honda, T., Sun, X., Yasutake, K., Nihonmatsu, N., Wolozin, B., and Takashima, A., 1998, Direct association of presenilin-1 with β-catenin, *FEBS Lett.* **433:**73–77.

Murayama, S., Mori, H., Ihara, Y., and Tomonaga, M., 1990, Immunocytochemical and ultrastructural studies of Pick's disease, *Ann. Neurol.* **27:**394–405.

Murrell, J., Farlow, M., Ghetti, B., and Benson, M. D., 1991, A mutation in the amyloid-β precursor protein associated with hereditary Alzheimer's disease, *Science* **254:**97–99.

Namba, Y., Tomonaga, M., Kawasaki, H., Otomo, E., and Ikeda, K., 1991, Apolipoprotein E immunoreactivity in cerebral amyloid deposits and neurofibrillary tangles in Alzheimer's disease and kuru plaque amyloid in Creutzfeldt-Jakob disease, *Brain Res.* **541**:163–166.

Näslund, J., Thyberg, J., Tjernberg, L. O., Wernstedt, C., Karlström, A. R., Bogdanovic, N., Gandy, S. E., Lanfelt, L., Terenius, L., and Nordstedt, C., 1995, Characterization of stable complexes involving apolipoprotein E and the amyloid-β peptide in Alzheimer's disease brain, *Neuron* **15**:219–228.

Nathan, B. P., Bellosta, S., Sanan, D. A., Weisgraber, K. H., Mahley, R. W., and Pitas, R. E., 1994, Differential effects of apolipoproteins E3 and E4 on neuronal growth *in vitro, Science* **264**:850–852.

Neve, R. L., Finch, E. A., and Dawes, L. R., 1988, Expression of the Alzheimer amyloid-β precursor gene transcripts in the human brain, *Neuron* **1**:669–677.

Nicoll, J.A.R., Burnett, C., Love, S., Graham, D. I., Dewar, D., Ironside, J. W., Stewart, J., and Vinters, H. V., 1997, High frequency of apolipoprotein E e2 allele in hemorrhage due to cerebral amyloid angiopathy, *Ann. Neurol.* **41**:716–721.

Nixon, R. A., Cataldo, A. M., Paskevich, P. A., Hamilton, D. J., Wheelock, T. R., and Kanaley-Andrews, L., 1992, The lysosomal system in neurons: Involvement at multiple stages of Alzheimer's disease pathogenesis, *Ann. N.Y. Acad. Sci.* **674**:65–88.

O'Neill, C. J., Winblad, W. B., and Cowburn, R. F., 1992, Neurotransmitter, receptor, and signal transduction disturbances in Alzheimer's disease, *Acta. Neurol. Scand. (Suppl.)* **139**:59–62.

Oltersdorf, T., Ward, P. J., Henriksson, T., Beattie, E. C., Neve, R., Lieberburg, I., and Fritz, L. C., 1990, The Alzheimer amyloid-β precursor protein: Identification of a stable intermediate in the biosynthetic/degradative pathway, *J. Biol. Chem.* **265**;4492–4497.

Oppenheim, R. W., 1991, Cell death during development of the nervous system, *Annu. Rev. Neurosci.* **14**:453–501.

Otvos, L., Jr., Feiner, L., Lang, E., Szendrei, G. I., Goedert, M., and Lee, V. M., 1994, Monoclonal antibody PHF-1 recognizes tau protein phosphorylated at serine residues 396 and 404, *J. Neurosci. Res.* **39**:669–673.

Papp, M. I., and Lantos, P. L., 1992, Accumulation of tubular structures in oligodendroglial and neuronal cells as the basic alteration in multiple system atrophy, *J. Neurol. Sci.* **107**:172–182.

Papp, M. I., Kahn, J. E., and Lantos, P. L., 1989, Glial cytoplasmic inclusions in the CNS of patients with multiple system atrophy (striatonigral degeneration, olivopontocerebellar atrophy and Shy-Drager syndrome), *J. Neurol. Sci.* **94**:79–100.

Paresce, D. M., Ghosh, R. N., and Maxfield, F. R., 1996, Microglial cells internalize aggregates of the Alzheimer's disease amyloid-β protein via a scavenger receptor, *Neuron* **17**:553–565.

Pasinetti, G. M., 1996, Inflammatory mechanisms in neurodegeneration and Alzheimer's disease: The role of the complement system, *Neurobiol. Aging* **17**:707–716.

Paulus, W., and Selim, M., 1990, Corticonigral degeneration with neuronal achromasia and basal neurofibrillary tangles, *Acta Neuropathol.* **81**:89–94.

Perez, M., Valpuesta, J. M., Medina, M., Montejo de Garcini, E., and Avila, J., 1996, Polymerization of tau into filaments in the presence of heparin: The minimal sequence required for tau–tau interaction, *J. Neurochem.* **67**:1183–1190.

Perlmutter, L. S., Barron, E., and Chui, H. C., 1990. Morphologic association between microglia and senile plaque amyloid in Alzheimer's disease, *Neurosci. Lett.* **119**:32–36.

Perry, G., Stewart, D., Friedman, R., Manetto, V., Autilio-Gambetti, L., and Gambetti, P., 1987, Filaments of Pick's bodies contain altered cytoskeletal elements, *Am. J. Pathol.* **127**:559–568.

Pike, C. J., Burdick, S., Walencewicz, A. J., Glabe, C. G., and Cotman, C. W., 1993, Neurodegeneration induced by β-amyloid peptides *in vitro:* The role of peptide assembly state, *J. Neurosci.* **13**:1676–1687.

Podlinsky, M. B., Stephenson, D. T., Frosch, M. P., Tolan, D. R., Lieberburg, I., Clemens, J. A., and Selkoe, D. J., 1993, Microinjection of synthetic amyloid-β protein in monkey cerebral cortex fails to produce acute neurotoxicity, *Am. J. Pathol.* **142**:17–24.

Ponte, P., Gonzalez-DeWhitt, P., Schilling, J., Miller, J., Hsu, D., Greenberg, B., Davis, K., Wallace, W., Lieberburg, I., and Fuller, F., 1988, A new A4 amyloid mRNA contains a domain homologous to serine proteinase inhibitors, *Nature* **331**:525–527.

Poorkaj, P., Bird, T. D., Wijsman, E., Nemens, E., Garruto, R. M., Anderson, L., Andreadis, A., Wiederholt, W. C., Raskind, M., and Schellenberg, G. D., 1998, Tau is a candidate gene for chromosome 17 frontotemporal dementia, *Ann. Neurol.* **43**:815–825.

Poulter, L., Barratt, D., Scott, C. W., and Caputo, C. B., 1993, Locations and immunoreactivities of phosphorylation sites on bovine and porcine tau proteins and a PHF-tau fragment, *J. Biol. Chem.*, **268:**9636–9644.

Pratico, D., Lee, V.M.-Y., Trojanowski, J. Q., Rokach, J., and Fitzgerald, G. A., 1998, Increased F2-isoprostanes in Alzheimer's disease: evidence for enhanced lipid peroxidation *in vivo, FASEB J.* **12**(15):1777–1783.

Prusiner, S. B., 1998, Prions, *Proc. Natl. Acad. Sci. USA* **95:**13363–13383.

Qian, S., Jiang, P., Guan, X. M., Singh, G., Trumbauer, M. E., Yu, H., Chen, H. Y., Van de Ploeg, L. H., and Zheng, H., 1998, Mutant human presenilin 1 protects presenilin 1 null mouse against embryonic lethality and elevates Aβ1-42/43 expression, *Neuron.* **20:**611–617.

Raber, J., Wong, D., Buttini, M., Orth, M., Bellosta, S., Pitas, R. E., Mahley, R. W., and Mucke, L., 1998, Isoform-specific effects of human apolipoprotein E on brain function revealed in ApoE knockout mice: increased susceptibility of females, *Proc. Natl. Acad. Sci. USA* **95:**10914–10919.

Rebeck, G. W., Reiter, J. S., Strickland, D. K., and Hyman, B. T., 1993, Apolipoprotein E in sporadic Alzheimer's disease: Allelic variation and receptor interactions, *Neuron* **11:**575–580.

Reed, L. A., Grabowski, T. J., Schmidt, M. L., Morris, J. C., Goate, A., Solodkin, A., Van Hoesen, G. W., Schelper, R. L., Talbot, C. J., Wragg, M. A., and Trojanowski, J. Q., 1997, Autosomal dominant dementia with widespread neurofibrillary tangles, *Ann. Neurol.* **42:**564–572.

Reed, L. A., Schmidt, M. L., Wszolek, Z. K., Balin, B. J., Soontornniyomkij, V., Lee, V. M., Trojanowski, J. Q., and Schelper, R. L., 1998, The neuropathology of a chromosome 17-linked autosomal dominant parkinsonism and dementia ("pallido-ponto-nigral degeneration"), *J. Neuropathol. Exp. Neurol.* **57:**588–601.

Rizzu, P., Van Swieten J. C., Joosse, M., Hasegawa, M., Stevens, M., Tibben, A., Niermeijer, M. F., Hillebrand, M., Ravid, R., Oostra, B. A., Goedert, M., van Duijn, C. M., and Heutink, P., 1999, High prevalence of mutations in the microtubule-associated protein tau in a population study of frontotemporal dementia in the netherlands, *Am. J. Hum. Genet.* **64**(2):414–421.

Robakis, N. K., Ramakrishna, M., Wolfe, G., and Wisniewski, H. M., 1987, Molecular cloning and characterization of a cDNA encoding the cerebrovascular and the neuritic plaque amyloid peptides, *Proc. Natl. Acad. Sci. USA* **84:**4190–4194.

Roberts, G. W., 1988, Immunocytochemistry of neurofibrillary tangles in dementia pugilistica and Alzheimer's disease: Evidence for common genesis, *Lancet* **2:**1456–1458.

Roberts, G. W., Allsop, D., and Bruton, C., 1990, The occult aftermath of boxing, *J. Neurol. Neurosurg. Psychiatry* **53:**373–378.

Rogaev, E. I., Sherrington, R., Rogaeva, E. A., Levesque, G., Ikeda, M., Liang, Y., Chi, H., Lin, C., Holman, K., Tsuda, T., Mar, L., Sorbi, S., Nacmlas, B., Placentini, S., Amaducci, L., Chumakov, I., Cohen, D., Lannfelt, L., Fraser, P. E., Rommens, J. M., and St. George-Hyslop, P. H., 1995, Familial Alzheimer's disease in kindreds with missense mutations in a gene on chromosome 1 related to the Alzheimer's disease type 3 gene, *Nature* **376:** 775–778.

Rogers, J., Kirby, L. C., Hempleman, S. R., Berry, D. L., McGeer, P. L., Kaszniak, A. W., Zalinski J., Cofield, M., Mansukhani, L., Wilson, P., and Kogan, F., 1993, Clinical trial of indomethacin in Alzheimer's disease *Neurology* **43:**1609–1611.

Roher, A. E., Lowenson, J. D., Clarke, S., Wolkow, C., Wang, R., Cotter, R. J., Reardon, I. M., Zürcher-Neely, H. A., Heinrikson, R. L., Ball, M. J., and Greenberg, B. D., 1993a, Structural alterations in the peptide backbone of β-amyloid core protein may account for its deposition and stability in Alzheimer's disease, *J. Biol. Chem.* **268:**3072–3083.

Roher, A. E., Palmer, K. C., Yurewicz, E. C., Ball, M. J., and Greenberg, B. D., 1993b, Morphological and biochemical analyses of amyloid plaque core proteins purified from Alzheimer disease brain tissue, *J. Neurochem.* **61:**1916–1926.

Saido, T. C., Iwatsubo, T., Mann, D. M., Shimada, H., Ihara, Y., and Kawashima, S., 1995, Dominant and differential deposition of distinct β-amyloid peptide species, A beta N3(pE), in senile plaques, *Neuron* **14:**457–466.

Saido, T. C., Yamao-Harigaya, W., Iwatsubo, T., and Kawashima, S. 1996, Amino- and carboxyl-terminal heterogeneity of β-amyloid peptides deposited in human brain, *Neurosci. Lett.* **215:**173–176.

Sanan, D. A., Weisgraber, K., Russell, S. J., Mahley, R. W., Huang, D., Saunders, A., Schmechel, D., Wisniewski, T., Frangione, B., Roses, A. D., and Strittmatter, W. J., 1994, Apolipoprotein E associates with Aβ amyloid peptide to form novel monofibrils: Isoform apoE4 associates more efficiently than apoE3, *J. Clin. Invest.* **94:**860–869.

Sastre, M., Turner, R. S., and Levy, E., 1998, X11 interaction with amyloid-β precursor protein modulates its cellular stabilization and reduces amyloid β-protein secretion, *J. Biol. Chem.* **273:**22351–22357.

Saunders, A. M., Strittmatter, W. J., Schmechel, D., St. George-Hyslop, P. H., Pericak-Vance, M. A., Joo, S. H., Rosi, B. L., Gusella, J. F., Crapper-MacLachlan, D. R., Alberts, M. J., Hulette, C., Crain, B., Goldgaber, D., and Roses, A. D., 1993, Association of apolipoprotein E allele e4 with late-onset familial and sporadic Alzheimer's disease, *Neurology* **43:**1467–1472.

Scheibel, A. B., and Tomiyasu, U., 1978, Dendritic sprouting in Alzheimer's presenile dementia, *Exp. Neurol.* **60:**1–8.

Schmidt, M. L., Lee, M.V.-Y., and Trojanowski, J. Q., 1989, Analysis of epitopes shared by Hirano bodies and neurofilament proteins in normal and Alzheimer's disease hippocampus, *Lab. Invest.* **60:**513–522.

Schmidt, M. L., Lee, V.M.-Y., and Trojanowski, J. Q., 1991, Comparative epitope analysis of neuronal cytoskeletal proteins in Alzheimer's disease senile plaque neurites and neuropil threads, *Lab. Invest.* **64;**352–357.

Schmidt, M. L., DiDario, A. G., Otvos, L., Jr., Hoshi, M., Kant, J. A., Lee, V.M.-Y., and Trojanowski, J. Q., 1994, Plaque-associated neuronal proteins: A recurrent motif in neuritic amyloid deposits throughout diverse cortical areas of the Alzheimer's disease brain, *Exp. Neurol.* **130:**311–322.

Schmidt, M. L., Huang, R., Martin, J. A., Henley, J., Mawal-Dewan, M., Hurtig, H. I., Lee, V.M.-Y., and Trojanowski, J. Q., 1996, Neurofibrillary tangles in progressive supranuclear palsy contain the same tau epitopes identified in Alzheimer's disease PHFtau, *J. Neuropathol. Exp. Neurol.* **55:**534–539.

Schmidt, M. L., Lee, V.M.-Y., Formam, M., Chiu, T.-S., and Trojanowski, J. Q., 1997, Monoclonal antibodies to a novel 100 kDa protein reveal abundant Aβ-negative plaques throughout grey matter of Alzheimer's disease brains, *Am. J. Pathol.* **151:**69–80.

Schmued, L. C., Swanson, L. W., and Sawchenko, P. E., 1982, Some fluorescent counterstains for neuroanatomical studies, *J. Histochem. Cytochem.* **30:**123–128.

Schümmelfeder, N., 1958, Histochemical significance of the polychromatic fluorescence induced in tissue stained with acridine orange, *J. Histochem. Cytochem.* **6;**392–393.

Selkoe, D. J., 1996, Amyloid-β protein and the genetics of Alzheimer's disease, *J. Biol. Chem.* **271:**18295–18298.

Selkoe, D. J., 1997, Alzheimer's disease: genotypes, phenotypes, and treatments, *Science* **275:**630–631.

Seubert, P., Oltersdorf, T., Lee, M. G., Barbour, R., Blomquist, C., Davis, D. L., Bryant, K., Fritz, L. C., Galasko, D., Thal, L. J., Lieberburg, I., and Schenk, D. B., 1993, Secretion of β-amyloid precursor protein cleaved at the amino terminus of the β-amyloid peptide, *Nature* **361:**260–263.

Seubert, P., Mawal-Dewan, M., Barbour, R., Jakes, R., Goedert, M., Johnson, G.V.W., Litersky, J. M., Schenk, D., Lieberburg, I., Trojanowski, J. Q., and Lee, V.M.-Y., 1995, Detection of phosphorylated Ser 262 in fetal tau, adult tau, and paired helical filament tau, *J. Biol. Chem.* **270:**18917–18922.

Shen, J., Bronson, R. T., Chen, D. F., Xia, W., Selkoe, D. J., and Tonegawa, S., 1997, Skeletal and CNS defects in presenilin-1 deficient mice, *Cell* **89:**629–639.

Sherrington, R., Rogaev, E. I., Liang, Y., Rogaeva, E. A., Levesque, G., Ikeda, M., Chi, H., Lin, C., Li, G., Holman, K., Tsuda, T., Mar, L., Foncin, J.-F., Brunl, A. C., Montesi, M. P., Sorbi, S., Rainero, I., Pinessi, L., Nee, L., Chumakov, I., Pollen, D., Brookes, A., Sanseau, P., Pollinsky, R. J., Wasco, W., Da Silva, H.A.R., Haines, J. L., Pericak-Vance, M. A., Tanzi, R. E., Roses, A. D., Fraser, P. E., Rommens, J. M., and St. George-Hyslop, P. H., 1995, Cloning of a gene bearing missense mutations in early onset amilial Alzheimer's disease, *Nature* **375:**754–760.

Simchowicz, T., 1911, Histologische Studien über der senile Demenz, in: *Histologische und histopathologische Arbeiten über die Grosshirnrinde mit besonderer Berücksichtigung der pathologischen Anatomie der Geisteskrankheiten, Vol. 4* (F. Nissl and A. Alzheimer, eds.), Fischer, Jena, Germany, pp. 267–444.

Simic, G., Kostovic, I., Winblad, B., and Bogdanovic, N., 1997, Volume and number of neurons in the human hippocampal formation in normal aging and Alzheimer's disease, *J. Comp. Neurol.* **379:**482–494.

Sisodia, S. S., Koo, E. H., Beyreuther, K., Unterbeck, A., and Price, D. L., 1990, Evidence that β-amyloid protein in Alzheimer's disease is not derived by normal processing, *Science* **248:**492–495.

Skovronsky, D. M., Doms, R. W., and Lee, V.M.-Y., 1998, Detection of a novel intraneuronal pool of insoluble amyloid-β protein that accumulates with time in culture, *J. Cell Biol.* **141:**1031–1039.

Slunt, H. H., Thinakaran, G., Von Koch, C., Lo, A.C.Y., Tanzi, R. E., and Sisodia, S. S., 1994, Expression of a ubiquitous, cross-reactive homologue of the mouse β-amyloid precursor protein (APP), *J. Biol. Chem.* **269;**2637–2644.

Smale, G., Nichols, N. R., Brady, D. R., Finch, C. E., and Horton, W. E., Jr., 1995, Evidence for apoptotic cell death in Alzheimer's disease, *Exp. Neurol.* **133:**225–230.

Smith, C. D., Carney, J. M., Starke-Reed, P. E., Oliver, C. N., Stadtman, E. R., Floyd, R. A., and Markesbery, W. R., 1991, Excess brain protein oxidation and enzyme dysfunction in normal aging and in Alzheimer disease, *Proc. Natl. Acad. Sci. USA* **88:**10540–10543.

Smith, D. H., Nakamura, M., McIntosh, T. K., Wang, J., Rodriguez, A., Chen, X.-H., Raghupathi, R., Saatman, K. E., Clemens, J., Schmidt, M. L., Lee, V.M.-Y., and Trojanowski, J. Q., 1998, Brain trauma induces massive hippocampal neuron death linked to a surge in β-amyloid levels in mice overexpressing mutant amyloid precursor protein, *Am. J. Pathol.* **153:**1005–1010.

Smith, M. A., Taneda, S., Richey, P. L., Miyata, S., Yan, S.-D., Stern, D., Sayre, I. M., Monnier, V. M., and Perry, G., 1994, Advanced Maillard reaction products are associated with Alzheimer disease pathology, *Proc. Natl. Acad. Sci. USA* **91:**5710–5714.

Snow, A. D., Sekiguchi, R., Nochlin, D., Fraser, P., Kimata, K., Mizutani, A., Arai, M., Schreier, W. A., and Morgan, D. G., 1994, An important role of heparan sulfate proteoglycan (perlecan) in a model system for the deposition and persistence of fibrillar Aβ-amyloid in rat brain, *Neuron* **12:**219–234.

Sontag, E., Nunbhakdi-Craig, V., Bloom, G. S., and Mumby, M. C., 1995, A novel pool of protein phosphatase 2A is associated with microtubules and is regulated during the cell cycle. *J. Cell Biol.* **128:**1131–1144.

Sontag, E., Nunbhakdi-Craig, V., Lee, G., Bloom, G. S., and Mumby, M. C., 1996, Regulation of the phosphorylation state and microtubule-binding activity of tau by protein phosphatase 2A, *Neuron* **17:**1201–1207.

Sperber, B. R., Leight, S., Goedert, M., and Lee, V.M.-Y., 1995, Glycogen synthase kinase-3β phosphorylates tau protein at multiple sites in intact cells, *Neurosci. Lett.* **197:**149–153.

Spillantini, M. G., Bird, T. D., and Ghetti, B., 1998a, Frontotemporal dementia and Parkinsonism linked to chromosome 17: a new group of tauopathies, *Brain Pathol.* **8:**387–402.

Spillantini, M. G., Goedert, M., Crowther, R. A., Murrell, J. R., Farlow, M. R., and Ghetti, B., 1997a, Familial multiple system tauopathy with presenile dementia: A disease with abundant neuronal and glial tau filaments, *Proc. Natl. Acad. Sci. USA* **94:**4113–4118.

Spillantini, M. G., Murrell, J. R., Goedert, M., Farlow, M. R., Klug, A., and Ghetti, B., 1998b, Mutation in the tau gene in familial multiple system tauopathy with presenile dementia, *Proc. Natl. Acad. Sci. USA* **95:**7737–7741.

Spillantini, M. G., Schmidt, M. L., Lee, V.M.-Y., Trojanowski, J. Q., Jakes, R., and Goedert, M., 1997b, α-synuclein in Lewy bodies, *Nature* **388:**839–840.

Squire, L. R., 1986, Mechanisms of memory, *Science* **232:**1612–1619.

Stadelmann, C., Bruck, W., Bancher, C., Jellinger, K., and Lassmann, H., 1998, Alzheimer disease: DNA fragmentation indicates increased neuronal vulnerability, but not apoptosis, *J. Neuropathol. Exp. Neurol.* **57:**456–464.

Stadtman, E. R., 1992, Protein oxidation and aging, *Science* **257:**1220–1224.

Steward, O., and Levy, W. B., 1987, Preferential localization of polyribosomes under the base of dendritic spines in granule cells of the dentate gyrus, *J. Neurosci.* **7:**284–291.

Strittmatter, W. J., Saunders, A. M., Schmechel, D., Pericak-Vance, M., Enghild, J., Salvesen, G. S., and Roses, A. D., 1993a, Apolipoprotein E: High-avidity binding to β-amyloid and increased frequency of type 4 allele in late-onset familial Alzheimer disease, *Proc. Natl. Acad. Sci. USA* **90:**1977–1981.

Strittmatter, W. J., Weisgraber, K. H., Huang, D. Y., Dong, L.-M., Salvesen, G. S., Pericak-Vance, M., Schmechel, D., Saunders, A. M., Goldgaber, D., and Roses, A. D., 1993b, Binding of human apolipoprotein E to synthetic amyloid-β peptide: Isoform-specific effects and implications for late-onset Alzheimer disease, *Proc. Natl. Acad. Sci. USA* **90;**8098–8102.

Strittmatter, W. J., Saunders, A. M., Goedert, M., Weisgraber, K. H., Dong, L.-M., Jakes, R., Huang, D. Y., Pericak-Vance, M., Schmechel, D., and Roses, A. D., 1994, Isoform-specific interactions of apolipoprotein E with microtubule-associated protein tau: Implications for Alzheimer's disease, *Proc. Natl. Acad. Sci. USA* **91:**11183–11186.

Sumi, S. M., Bird, T. D., Nochlin, D., and Raskind, M. A., 1992, Familial presenile dementia with psychosis associated with cortical neurofibrillary tangles and degeneration of the amygdala, *Neurology* **42;**120–127.

Suzuki, K., Parker, C. C., Pentchev, P. G., Katz, D., Ghetti, B., D'Agostino, A. N., and Carstea, E. D., 1995, Neurofibrillary tangles in Niemann-Pick disease type C, *Acta Neuropathol.* **89:**227–238.

Suzuki, N., Iwatsubo, T., Odaka, A., Ishibashi, Y., Kitada, C., and Ihara, Y., 1994, High tissue content of soluble Aβ 1-40 is linked to cerebral amyloid angiopathy. *Am. J. Pathol.* **145:**452–460.

Suzuki, T., Nishiyama, K., Murayama, S., Yamamoto, A., Sato, S., Kanazawa, I., and Sakaki, Y., 1996, Regional and cellular presenilin 1 gene expression in human and rat tissues, *Biochem. Biophys. Res. Comm.* **219:**708–713.

Tagliavini, F., Giaccone, G., Prelli, F., Verga, L., Porro, M., Trojanowski, J. Q., Farlow, M. R., Frangione, B., Ghetti, B., and Bugiani, O., 1993, A68 is a component of paired helical filaments of Gerstmann-Straussler-Scheinker disease, Indiana kindred, *Brain Res.* **616:**325–329.

Takahashi, H., Oyanagi, K., Takeda, S., Hinokuma, K., and Ikuta, F., 1989, Occurrence of 15-nm-wide straight tubules in neocortical neurons in progressive supranuclear palsy, *Acta Neuropathol.* **79:**233–239.

Takeda, A., Mallory, M., Sundsmo, M., Honer, W., Hansen, L., and Masliah, E., 1998a, Abnormal accumulation of NACP/α-synuclein in neurodegenerative disorders, *Am. J. Pathol.* **152:**367–372.

Takeda, A., Yasuda, T., Miyata, T., Goto, Y., Wakai, M., Watanabe, M., Yasuda, Y., Horie, K., Inagaki, T., Doyu, M., Maeda, K., and Sobue, G., 1998b, Advanced glycation end products co-localized with astrocytes and microglial cells in Alzheimer's disease brain, *Acta Neuropathol.* **95:**555–558.

Tanzi, R. E., Gusella, J. F., Watkins, P. C., Bruns, G.A.P., St. George-Hyslop, P., Van Keuren, M. L., Patterson, D., Pagan, S., Kurnit, D. M., and Neve, R. L., 1987, Amyloid-β protein gene: cDNA, mRNA distribution, and genetic linkage near the Alzheimer locus, *Science* **235:**880–884.

Tanzi, R. E., McClatchey, A. I., Lamperti, E. D., Villa-Komaroff, L., Gusella, J. F., and Neve, R. L., 1988, Protease inhibitor domain encoded by an amyloid-β protein precursor mRNA associated with Alzheimer's disease, *Nature* **331:**528–530.

Terry, R., Masliah, E., Salmon, D. P., Butters, N., DeTeresa, R., Hill, R., Hansen, L. A., and Katzman, R., 1991, Physical basis of cognitive alterations in Alzheimer's disease: Synapse loss is the major correlate of cognitive impairment, *Ann. Neurol.* **30:**572–580.

Tomita, T., Maruyama, K., Saido, T. C., Kume, H., Shinozaki, K., Tokuhiro, S., Capell, A., Walter, J., Grünberg, J., Haas, C., Iwatsubo, T., and Obata, K., 1997, The presenilin 2 mutation (N1411) linked to familial Alzheimer disease (Volga German families) increases the secretion of amyloid-β protein ending at the 42nd (or 43rd) residue, *Proc. Natl. Acad. Sci. USA* **94:**2025–2030.

Tomlinson, B. E., and Kitchener, D., 1972, Granulovacuolar degeneration of hippocampal pyramidal cells, *J. Pathol.* **106:**165–185.

Tooyama, I., Kimura, H., Akiyama, H., and McGeer, P. L., 1990, Reactive microglia express class I and class II major histocompatibility complex antigens in Alzheimer's disease, *Brain Res.* **523:**273–280.

Topaloglu, H., and Sarnat, H. B., 1989, Acridine orange-RNA fluorescence maturing neurons in the perinatal rat brain, *Anat. Rec.* **224:**88–93.

Trojanowski, J. Q., and Lee, V.M.-Y., 1994, Paired helical filament tau in Alzheimer's disease: The kinase connection, *Am. J. Pathol.* **144:**449–453.

Trojanowski, J. Q., Clark, C. M., Arai, H., and Lee, V.M.-Y., 1996a, Elevated levels of tau in cerebrospinal fluid: Implications for the antemortem diagnosis of Alzheimer's disease, *Alzheimer Dis. Rev.* **1:**77–83.

Trojanowski, J. Q., Clark, C. M., Schmidt, M. L., Arai, H., Zhu, C., Mawal-Dewan, M., and Lee, V.M.-Y., 1996b, Implications of tau-rich neurofibrillary lesions for the pathobiology and diagnosis of Alzheimer's disease, in: *Current Neurology,* Vol. 16 (S. H. Appell, ed.), Houghton-Mifflin, Boston, pp. 93–113.

Trojanowski, J. Q., Galvin, J. E., Schmidt, M. L., Tu, P.-H., Iwatsubo, T., and Lee, M.-Y., 1998a, Mechanisms of neuron death in neurodegenerative diseases of the elderly. Role of the Lewy body, in: *Handbook of the Aging Brain,* (E. Wang and S. H. Snyder, eds.), Academic Press, San Diego, pp. 143–152.

Trojanowski, J. Q., Goedert, M., Iwatsubo, T., and Lee, M.-Y., 1998b, Fatal attractions: abnormal protein aggregation and neuron death in Parkinson's disease and Lewy body dementia, *Cell Death Diff.* **5:**832–837.

Tu, P.-H., Galvin, J. E., Baba, M., Giasson, B., Tomita, T., Leight, S., Nakajo, S., Iwatsubo, T., Trojanowski, J. Q., and Lee, V.M.-Y., 1998, Glial cytoplasmic inclusions in white matter oligodendrocytes of multiple system atrophy brains contain insoluble α-synuclein, *Ann. Neurol.* **44:**415–422.

Turner, R. S., Suzuki, N., Chyung, A.S.C., Younkin, S. G., and Lee, V.M.-Y., 1996, Aβ40 and Aβ42 are generated intracellularly in cultured human neurons and their secretion increases with maturation, *J. Biol. Chem.* **271:**8966–8970.

Van Broeckhoven, C., Haan, J., Bakker, E., Hardy, J. A., Van Hul, W., Wehnert, A., Vegter-Van der Vlis, M., and Roos, R. A., 1990, Amyloid-β protein precursor gene and hereditary cerebral hemorrhage with amyloidosis (Dutch), *Science* **248:**1120–1122.

VanGelder, R., von Zastrow, M., Yool, A., Dement, W., Barchas, J., and Eberwine, J., 1990, Amplified RNA (aRNA) synthesized from limited quantities of heterogeneous cDNA, *Proc. Natl. Acad. Sci. USA* **87:**1663–1667.

Veinbergs, I., Jung, M. W., Young, S. J., Van Uden, E., Groves, P. M., and Masliah, E., 1998, Altered long-term potentiation in the hippocampus of apolipoprotein E-deficient mice, *Neurosci. Lett.* **249:**71–74.

Vinters, H. V., 1987, Cerebral amyloid angiopathy: A critical review, *Stroke* **18:**311–324.

Vitek, M. P., Bhattacharya, K., Glendening, J. M., Stopa, E., Vlassara, H., Bucala, R., Manogue, K., and Cerami, A., 1994, Advanced glycation end products contributes to amyloidosis in Alzheimer disease, *Proc. Natl. Acad. Sci USA* **91:**4766–4770.

von Bertalanffy, L., and Bickis, I., 1956, Identification of cytoplasmic basophilia (ribonucleic acid) by fluorescence microscopy, *J. Histochem. Cytochem.* **4:**481–493.

Wakabayashi, K., Oyanagi, K., Makifuchi, T., Ikuta, F., Homma, A., Homma, Y., Horikawa, Y., and Tokiguchi, S., 1994, Corticobasal degeneration: Etiopathological significance of the cytoskeletal alterations, *Acta Neuropathol.* **87:**545–553.

Walker, D. G., and McGeer, P. L., 1992, Complement gene expression in human brain: Comparison between normal and Alzheimer disease cases, *Mol. Brain Res.* **14:**109–116.

Wasco, W., Bupp, K., Magendantz, M., Gusella, J. F., Tanzi, R. E., and Solomon, F., 1992, Identification of a mouse brain cDNA that encodes a protein related to the Alzheimer disease-associated amyloid-β protein precursor, *Proc. Natl. Acad. Sci. USA* **89:**10758–10762.

Wasco, W., Gurubhagavatula, S. D., Paradis, M., Romano, D. M., Sisodia, S. S., Hyman, B. T., Neve, R. L., and Tanzi, R. E., 1993, Isolation and characterization of APLP2 encoding a homologue of the Alzheimer's associated amyloid-β protein precursor, *Nature Genet.* **5:**95–100.

Watanabe, A., Hasegawa, M., Suzuki, M., Takio, K., Morishima-Kawashima, M., Titani, K., Arai, T., Kosik, K. S., and Ihara, Y., 1993, *In vivo* phosphorylation sites in fetal and adult rat tau, *J. Biol. Chem.* **268:**25712–25717.

Weidemann, A., Konig, G., Bunke, D., Fischer, P., Salbaum, J. M., Masters, C. L., and Beyreuther, K., 1989, Identification, biogenesis, and localization of precursors of Alzheimer's disease A4 amyloid protein, *Cell* **57:**115–126.

Wertkin, A. M., Turner, R. S., Pleasure, S. J., Golde, T. E., Younkin, S. G., Trojanowski, J. Q., and Lee, V.M.-Y., 1993, Human neurons derived from a teratocarcinoma cell line express solely the 695-amino acid amyloid-β precursor protein and produce intracellular β-amyloid or A4 peptides, *Proc. Natl. Acad. Sci. USA* **90:**9513–9517.

West, M. J., Coleman, P. D., Flood, D. G., and Troncoso, J. C., 1994, Differences in the pattern of hippocampal neuronal loss in normal aging and Alzheimer's disease, *Lancet* **344:**769–772.

Wischik, C. M., Novak, M., Thogersen, H. C., Edwards, P. C., Runswick, M. J., Jakes, R., Walker, J., Milstein, C., Roth, M., and Klug, A., 1988, Isolation of a fragment of tau derived from the core of the paired helical filament of Alzheimer's disease, *Proc. Natl. Acad. Sci. USA* **85:**4506–4510.

Wisniewski, K. E., Dalton, A. J., Crapper McLachlan, D. R., Wen, G. Y., and Wisniewski, H. M., 1985, Alzheimer's disease in Down's syndrome: Clinicopathologic studies, *Neurology* **35:**957–961.

Wisniewski, H. M., Wegiel, J., Wang, K. C., Kujawa, M., and Lach, B., 1989, Ultrastructural studies of the cells forming amyloid fibers in classical plaques, *Can. J. Neurol. Sci.* **16:**535–542.

Wisniewski, H. M., Barcikowska, M., and Kida, E., 1991, Phagocytosis of β/A4 amyloid fibrils of the neuritic neocortical plaques, *Acta Neuropathol.* **81:**585–590.

Wisniewski, T., Castano, E. M., Golabek, A., Vogel, T., and Frangione, B., 1994, Acceleration of Alzheimer's fibril formation by apolipoprotein E *in vitro, Am. J. Pathol.* **145:**1030–1035.

Wong, K. T., Allen, I. V., McQuaid, S., and McConnell, R., 1996, An immunohistochemical study of neurofibrillary tangle formation in post-encephalitic Parkinsonism, *Clin. Neuropathol.* **15:**22–25.

Wong, P. C., Zheng, H., Chen, H., Becher, M. W., Sirinathsinghji, D. J., Trumbauer, M. E., Chen, H. Y., Price, D. L., Van der Ploeg, L. H., and Sisodia, S. S., 1997, Presenilin 1 is required for Notchl and DII1 expression in the paraxial mesoderm, *Nature* **387:**288–292.

Wyllie, A. H., 1981, Cell death: A new classification separating apoptosis from necrosis, in: *Cell Death in Biology and Pathology* (I. D. Bowen and R. A. Lockshin, eds.), Chapman, and Hall, London, pp. 9–34.

Yan, S. D., Chen, X., Fu, J., Chen, M., Zhu, H., Roher, A., Slattery, T., Zhao, L., Nagashima, M., Morser, J., Migheli, A., Nawroth, P., Stern, D., and Schmidt, A. M., 1996, RAGE and amyloid-β peptide neurotoxicity in Alzheimer's disease, *Nature* **382:**685–691.

Yankner, B. A., 1996, Mechanisms of neuronal degeneration in Alzheimer's diseases, *Neuron* **16:**921–932.

Yin, D., 1995, Studies on age pigments evolving into a new theory of biological aging, *Gerontology* **2:**159–172.

Yoshikai, S., Sasaki, H., Dohura, K., Furuya, H., and Sakaki, Y., 1990, Genomic organization of the human amyloid-β protein precursor gene, *Gene* **87:**257–263.

Zambrano, N., Buxbaum, J. D., Minopoli, G., Fiore, F., De Candia, P., De Renzis, S., Faraonio, R., Sabo, S., Cheetham, J., Sudol, M., and Russo, T., 1997, Interaction of the phosphotyrosine interaction/phosphotyrosine binding-related domains of Fe65 with wild-type and mutant Alzheimer's β-amyloid precursor proteins, *J. Biol. Chem.* **272:**6399–6405.

Zhong, Z., Higaki, J., Murakami, K., Wang, Y., Catalano, R., Quon, D., and Cordell, B., 1994, Secretion of β-amyloid precursor protein involves multiple cleavage sites, *J. Biol. Chem.* **269:**627–632.

Zhou, J., Liyanage, U., Medina, M., Ho, C., Simmons, A. D., Lovett, M., and Kosik, K. S., 1998, Presenilin 1 interaction in the brain with a novel member of the Armadillo family, *NeuroReport* **8:**2085–2090.

# 18

# Glutamate Receptors and Excitotoxic Mechanisms in Alzheimer's Disease

JAMES W. GEDDES

## 1. Introduction

Large pyramidal neurons, the neurons most vulnerable in Alzheimer's disease (AD), use glutamate as a neurotransmitter and receive glutamatergic inputs (Fonnum, 1984; Hof and Morrison, 1994). Loss of these neurons would be expected to result in decreased glutamate levels, loss of glutamate receptors, and loss of presynaptic glutamate transport sites. Of potentially greater interest is the possibility that glutamate may play a causal role in AD neuropathology. Glutamate and its receptors play critical roles in long-term potentiation—a cellular model of learning and memory (Asztely and Gustafsson, 1996; Riedel and Reymann, 1996)—and in spatial memory (Morris *et al.*, 1986), thus glutamate receptor dysfunction could contribute to the memory deficits evident in AD. Overstimulation of glutamate receptors can result in neuron death through excitotoxic mechanisms (Choi, 1992). Moreover, glutamate can cause cytoskeletal disruption resembling that associated with neurofibrillary pathology in AD (De Boni and McLachlan, 1985; Mattson, 1990; Couratier *et al.*, 1996). In this chapter the evidence of disrupted glutamatergic transmission in Alzheimer's disease is reviewed, and the possibility that it may contribute to, or result from, the disease process is discussed.

---

JAMES W. GEDDES • Sanders-Brown Center on Aging and Department of Anatomy and Neurobiology, University of Kentucky, Lexington, Kentucky 40536-0230.

*Cerebral Cortex,* Volume 14, edited by Peters and Morrison. Kluwer Academic/Plenum Publishers, New York, 1999.

## 2. Components of Glutamatergic Pathways

### 2.1. The Ionotropic Receptors

Glutamate receptors are coupled to ionophores and to G proteins. The three ionotropic receptors (Table I) are named after selective the agonists *N*-methyl-D-aspartic acid (NMDA), α-amino-3-hydroxy-5-methyl-4-isoxazole-propionate (AMPA), and kainic acid (KA) (Monaghan *et al.,* 1989). The AMPA receptors were formerly referred to as "quisqualate receptors", but quisqualate also activates metabotropic glutamate receptors, and AMPA has greater specificity than quisqualate for this site (Monaghan *et al.,* 1989). During the past decade, the subunits that assemble into ionotropic glutamate receptors have been identified (Nakanishi, 1992; Hollmann and Heinemann, 1993; Westbrook, 1994). The G-protein-coupled (metabotropic) receptors (Table II) are divided into three groups based on sequence similarity (Nicoletti *et al.,* 1996).

#### 2.1.1. The AMPA Receptors

The AMPA receptors mediate fast synaptic transmission and are permeable to $Na^+$, $K^+$, and in some cases, $Ca^{2+}$. These receptors desensitize rapidly in response to agonist activation, and recover slowly. The AMPA receptors are formed from four different subunits, GluR1–GluR4 (GluRA-GluRD), arranged in a pentameric assembly (see Fletcher and Lodge, 1996, for review). One subunit (GluR2) confers low $Ca^{2+}$ permeability. If GluR2 is absent, AMPA receptors are permeable to $Ca^{2+}$ (e.g., the AMPA channels in cultured Purkinje neurons). The unique properties of GluR2 are the result of a single amino acid substitution (Sommer *et al.,* 1991; Seeburg, 1996). In GluR2 a specific arginine (R) residue confers low $Ca^{2+}$ permeability. In GluR1, GluR3, and GluR4, glutamine (Q) is present at the same position. Interestingly, the mRNA of GluR2 encodes glutamine, but post-transcriptional editing removes an amino group from a specific adenine base, altering the codon for glutamine to arginine (Sommer *et al.,* 1991; Kohler *et al.,* 1993; Seeburg *et al.,* 1998). In addition to this post-transcriptional editing, each GluR has inserts that can differ in orientation (*flip* and *flop*), generated by alternative splicing of the primary transcript (Sommer *et al.,* 1990). Flip and flop differ in densensitization rate and response to drugs modifying densensitization. Post-transcriptional modifications include phosphorylation of GluR1 by $Ca^{2+}$/calmodulin II kinase (Soderling, 1996), and by protein kinase-A and protein kinase-C (Roche *et al.,* 1996).

**Table I. Ionotropic Glutamate Receptors**

| Receptor/agonist | Properties | Function |
|---|---|---|
| AMPA | Activation of $Na^+$, $K^+$ channels | Fast excitation |
| KA | Activation of $Na^+$, $K^+$ channels | Unknown |
| NMDA | Activation of $Na^+$, $K^+$, and $Ca^{2+}$ channels | Slow excitation |

**Table II. Metabotropic Glutamate Receptors[a]**

| Group | mGlu receptors | Transduction pathways | Putative role in neurodegeneration | Mechanisms |
|---|---|---|---|---|
| Group I | mGlu1<br>mGlu5 | ↑PI<br>↓$I_K$<br>PLD? | Enhancement of neuronal toxicity | ↑NMDA currents<br>↑Glu release<br>Depolarization and VOCC opening |
| Group II | mGlu2<br>mGlu3 | ↓cAMP<br>↓$I_{Ca}$ (L,N.P-VOCC)<br>↑cAMP ($Gi^{\beta\gamma?}$) | Neuroprotection against excitotoxicity or apoptosis | ↓cAMP<br>↓$I_{Ca}$<br>↓Glu release |
| Group III | mglu4<br>muGlu6<br>mGlu7<br>mGlu8 | ↓cAMP<br>↓$I_{Ca}$ (N,P-VOCC) | Neuroprotection? | ↓Glu release |

[a]Adapted from Nicoletti *et al.* (1996).
[b]Abbreviations: Glu, glutamate; PI, polyphosphoinositide; PLD, phospholipase D; VOCC, voltage-operated $Ca^{2+}$ channel.

Phosphorylation of GluR1 plays an important role in long-term potentiation (Barria *et al.,* 1997).

The highest densities of [$^3$H]AMPA binding sites are in hippocampus, particularly CA1 (Cotman and Monaghan, 1987). In neocortex, [$^3$H]AMPA binding sites are greatest in outer cortical layers (I–III). The distribution of [$^3$H]AMPA binding sites is very similar to that of NMDA receptors. Antibodies specific to GluR1, GluR2/3 and GluR4 label the pyramidal and nonpyramidal neuron somata, dendrites, and axon initial segments (Vickers *et al.,* 1993; Huntley *et al.,* 1994a; Vickers *et al.,* 1995). These antibodies can also label astrocytes in primate neocortex (Vickers *et al.,* 1993).

### 2.1.2. The Kainate Receptors

Kainic acid interacts with GluR1-4 with low affinity, but there is also a family of high-affinity KA subunits, designated GluR5-7, KA1, and KA2 (see Nakanishi, 1992; Seeburg, 1993; Westbrook, 1994, for reviews). These have limited homology with other cloned receptors and are 50 times less abundant than AMPA receptors. Similar to AMPA receptors, KA receptors are permeable to $Na^+$ and $K^+$ upon activation, and also undergo rapid desensitization. The GluR5 and GluR6 subunits can make homomeric channels responsive to agonists. In contrast, GluR7, KA1, and KA2 can form functional channels when combined with GluR5 or GluR6, but not by themselves. These subunits are therefore designated as kainate binding proteins (Henley, 1994). The GluR5 and GluR6 subunits can undergo RNA editing similar to GluR2. However, GluR2 exists almost entirely in the edited form, whereas both edited and unedited forms of GluR5 and GluR6 are found in the brain. There are also several alternatively spliced forms of both GluR5 and GluR6 (see Fletcher and Lodge, 1996, for review).

Although kainate receptors have been identified by ligand binding studies and *in situ* hybridization, it has been difficult to identify receptor channels with kainate-type properties using patch clamp recordings (reviewed in Lerma, 1998). Kainic acid can activate both AMPA and kainate receptors, although with different potency. Previously available antagonists did not have sufficient specificity to distinguish between KA and AMPA receptors. Recently, a selective AMPA antagonist has been identified, 2,3-benzodiazepine (GYKI53665) (Paternain *et al.,* 1995). This has enabled investigation of the functional role of KA receptors in synaptic transmission (Lerma, 1998). Presynaptic KA receptors can influence glutamate release (Chittajallu *et al.,* 1996). Kainate receptors have also been found in astrocytes and O2A progenitor cells, and there is a high concentration of KA receptors on Bergmann glia in the chick cerebellum. In the adult CNS, there is a high density of KA receptors in CA3 of the hippocampal formation. Kainate receptors may also play a role in neuronal development. The precise role of the KA receptor, however, remains to be determined.

The [$^3$H]KA binding sites are predominantly localized in telencephalon, with a distribution largely complementary to NMDA receptors (Cotman and Monaghan, 1987). The highest density is in CA3 stratum lucidum, the terminal zone of the mossy fiber pathway. There is also a high density of [$^3$H]KA bindings sites in the inner molecular layer of the dentate gyrus. In cerebral cortex, KA sites have higher densities in layers V and VI than in layers I–III. In monkey prefrontal cortex, approximately 70% of the neurons immunoreactive for GluR2/3 are also immunoreactive for GluR5/6/7 (Vickers *et al.,* 1993; Huntley *et al.,* 1994a). All neurons immunoreactive for GluR5/6/7 also immunostain with antibodies against GluR2/3 (see Huntley *et al.,* 1994a, for review).

### 2.1.3. The NMDA Receptors

Application of an agonist (glutamate or NMDA) under polarized conditions does not activate the NMDA receptor-associated ion channel, which is normally blocked by $Mg^{2+}$. The $Mg^{2+}$ block is removed by depolarization, and the activated channel is permeable to $Na^+$, $K^+$, and $Ca^{2+}$. The NMDA receptor-associated ion channel activates slowly, but remains activated for several hundred milliseconds in the presence of the agonist. Glycine is required as a co-agonist and potentiates the binding of glutamate to the receptor. The sensitivity of the site for glycine is very high ($K_D$ of 0.2 μM), suggesting it might always be occupied *in vivo.* However, recent evidence suggests that glycine does act in a regulatory capacity (Parsons *et al.,* 1998).

Functional NMDA receptors can be reconstituted from two subunit types, NR1 and NR2 (NR2A–NR2D) (see Sucher *et al.,* 1996, for review). The NMDAR1 subunit shows some sequence homology with AMPA subunits, with four putative transmembrane segments. The NR2A, -2B, -2C, and -2D subunits share only ~20% homology with NR1. Each NMDA receptor contains at least one NR1 subunit, and one or more NR2s in a probable pentameric complex, although the precise stoichiometry is unknown. The different NR2 subunits influence the strength of the $Mg^{2+}$ block, glycine sensitivity, and kinetics. The 2A subunit is present in forebrain and cerebellum, whereas cerebellum alone contains the highest level of 2C but no 2B. There are at least eight splice variants of NR1, and these are important in receptor regulation and localization, which are both mediated via phosphorylation (Ehlers *et al.,* 1995). The NR1 and NR2 subunits contain an asparagine (N) at the site equivalent to the Q/R site of AMPA receptors. This mediates the high $Ca^{2+}$ permeability and $Mg^{2+}$ channel blocker.

The NMDA receptor represents a binding complex. Ligand binding has identified three key sites: the *transmitter recognition site,* with which agonists (L-glutamate, NMDA) and competitive antagonists (CPP) interact; a *modulatory site,* which binds glycine; and a *phencyclidine binding site* located within the ion channel labeled with TCP or MK-801. Agonist-preferring and antagonist-preferring forms of the NMDA receptor complex have also been described based on pharmacology (Monaghan *et al.,* 1988), and there is pharmacological heterogeneity of channel blockers for NMDA receptors in different brain regions. The pharmacologic differences in binding to NMDA receptor subtypes corresponds to alterations in the NR2 subunit composition (Monaghan and Larsen, 1997).

In addition to the NMDAR1 and R2 subunits, an NMDA receptor-like complex has been purified from brain synaptic membranes. The complex includes a glutamate binding protein, a glycine/TCP binding protein, and other proteins of unknown function (Kumar *et al.,* 1991, 1994, 1995; Aistrup *et al.,* 1996; Babcock *et al.,* 1996). This complex has ligand binding characteristics similar to that of the NMDA receptor, although its functional significance remains to be determined.

The NMDA receptors are located throughout the brain, but predominate within telencephalon. The highest densities of NMDA-sensitive [$^3$H]glutamate binding sites are in strata oriens and radiatum of hippocampal areas CA1 (Cotman and Monaghan, 1987). Within neocortex, the highest levels are in layers I/III. Electrophysiology studies demonstrate an NMDA receptor involvement in synaptic transmission in layers II/III (cortical integration layers) and V/VI (cortical output layers). Transmission between pyramidal neurons in these layers involves both NMDA and non-NMDA receptors. In contrast, transmission between pyramidal cells and interneurons involves primarily non-NMDA receptors. In layer IV (termination of thalamocortical afferents) and layer I, excitatory synaptic transmission involves non-NMDA receptors (see Kaczmarek *et al.,* 1997, for review). In monkey prefrontal cortex, neurons immunpositive for NMDAR1 also contain GluR2/3 and GluR5/6/7 subunits (Huntley *et al.,* 1994a).

## 2.2. The Metabotropic Glutamate Receptors

Eight members of metabotropic glutamate receptor (mGluR) family have been identified and cloned (see Westbrook, 1994; Miller *et al.,* 1995, for reviews). These may be classified into three groups according to their transduction mechanisms, pharmacology, and sequence homology (Table II). As noted in the table, Group I receptors enhance neuronal toxicity induced by NMDA or other ionotropic GluR agonists, whereas Group II and III receptors are neuroprotective (Nicoletti *et al.,* 1996).

## 2.3. The Transporters

The primary mechanism for removal of extracellular glutamate is by $Na^+$-dependent, high-affinity transport. Five glutamate transporters have been cloned

(Table III). Glial transporters GLT1 and EAAC1 are responsible for the majority of glutamate transport and the maintenance of low extracellular glutamate levels (Rothstein *et al.*, 1996; Tanaka *et al.*, 1997). Neuronal transporter EAAC1 is not as effective in lowering extracellular glutamate levels, though loss of EAAC1 does result in neurotoxicity (Rothstein *et al.*, 1996).

## 3. Alzheimer's Disease Neuropathology

The neuropathology of AD is described in detail elsewhere in this volume (see Braak and Braak; Hof *et al.*) and is briefly summarized below. Neuropathologic abnormalities present in the brain of individuals with AD include neuron loss, neurofibrillary tangles, neuropil threads, and senile plaques. Neuron loss in AD primarily involves association regions of neocortex, hippocampus, and entorhinal cortex rather than primary sensory and motor areas (Hirano and Zimmerman, 1962; Brun and Englund, 1981; Terry *et al.*, 1981; Mountjoy *et al.*, 1983; Mann, 1985; Hof *et al.*, 1995). Pyramidal neurons are particularly vulnerable, being decreased 23–46% in neocortical layers III and V (Mann, 1985; Pearson *et al.*, 1985; Lewis *et al.*, 1987; Braak and Braak, 1991). In contrast, spiny stellate cells, small pyramidal cells, and inhibitory interneurons in other cortical lamina are largely resistant. Synapse density is decreased 29–42% in lamina III and V in vulnerable neocortical regions, although this may be partially compensated for by an increase in synapse size (Scheff *et al.*, 1990, 1993). A similar decline is observed in the intensity of synaptophysin immunohistochemical staining (Masliah *et al.*, 1989; Terry *et al.*, 1991). Within the hippocampal formation, pyramidal neurons in CA1 and subiculum show a predilection for neurofibrillary pathology, whereas CA3 pyramidal neurons and dentate gyrus granule neurons are largely resistant (Ball, 1977). In the entorhinal cortex, stellate cells in layers II and IV are the most vulnerable. The entorhinal cortex and transentorhinal cortex are affected very early in the disease process. Tangles in these regions are also common in nondemented elderly individuals, but not to the same extent as in Alzheimer's disease (Hyman *et al.*, 1984; Braak and Braak, 1991). In contrast to the increase in dendritic extent observed during

**Table III. $Na^+$-Dependent Glutamate Transporters**

| Rat transporter | Human transporter | Location | References |
|---|---|---|---|
| GLT1 | EAAT2 | Astrocytes, some neurons | (Pines *et al.*, 1992; Arriza *et al.*, 1994; Rothstein *et al.*, 1994) |
| GLAST | EAAT1 | Astrocytes | (Storck *et al.*, 1992; Tanaka 1993; Arriza *et al.*, 1994; Rothstein *et al.*, 1994) |
| EAAC1 | EAAT3 | Neurons | (Kanai and Hediger 1992; Rothstein *et al.*, 1994) |
| | EAAT4 | Cerebellum | (Fairman *et al.*, 1995) |
| | EAAT5 | Retina | (Arriza *et al.*, 1997) |

normal aging, there is dendritic regression in vulnerable regions in AD (Flood and Coleman, 1986; Coleman and Flood, 1987).



## 4. Alterations in Glutamatergic Pathways in Alzheimer's Disease

### 4.1. The NMDA Receptors

#### 4.1.1. Autoradiographic Studies

In 1985 Greenamyre and colleagues reported a massive decrease in [$^3$H]glutamate binding in cerebral cortex of AD patients as compared with nondemented elderly individuals. Specifically, they observed a 57% decline in "low-affinity quisqualate sites," which were subsequently shown to correspond largely to NMDA-sensitive sites (Greenamyre *et al.,* 1987). In contrast, the high-affinity quisqualate sites (AMPA receptors) were unchanged, as was cortical binding to GABA, and muscarinic receptors, suggesting a selective loss of NMDA receptors in AD. This result simulated considerable interest, as the selective decline in glutamate binding sites could conceivably account for the memory impairment and dementia observed in AD.

In contrast to loss of NMDA binding sites in cerebral cortex observed by Greenamyre and colleagues (1985), studies by our group did not reveal a marked loss of [$^3$H]glutamate binding to NMDA-sensitive sites in the hippocampal formation in AD (Geddes *et al.,* 1986). The NMDA receptor density was slightly lower in AD as compared with nondemented individuals; however, the changes were not significant. One of the seven AD cases examined did display a marked decrease in binding density in CA1, but also had a marked neuronal loss in this region. In a subsequent study, Greenamyre *et al.* (1987) also examined the hippocampal formation, where a large decline in NMDA-sensitive [$^3$H]glutamate binding was observed, reaching almost 90% in the stratum moleculare in CA1. Quisqualate-sensitive [$^3$H]glutamate binding sites were also markedly decreased (45–69%) in the hippocampal formation, whereas muscarinic receptor binding was unchanged.

The cause of the discrepancy between our findings and those of Greenamyre and colleagues (1985) is uncertain. It may reflect differences in the severity in the cases studied by the two groups, or possibly reflect different methods. Greenamyre and colleagues (1985) performed their binding studies in the presence of $Cl^-$ and $Ca^{2+}$ ions, and they defined the NMDA sites as L-[$^3$H]glutamate binding not displaced by 2.5 μM quisqualate. Subsequent analysis indicated that at least 50% of the glutamate binding under these conditions is not displaceable by NMDA (Young *et al.,* 1988). Using autoradiographic conditions established by Monaghan for rat brain (Monaghan and Cotman, 1985), our binding sites were performed in the absence of $Cl^-$ and $Ca^{2+}$ ions, and defined NMDA binding as [$^3$H]glutamate binding not displaced by 200 μM NMDA. At the time of these studies, [$^3$H]glutamate was the best available ligand for the NMDA receptor, although its affinity is low as compared with ligands used for autoradiographic studies of other receptors, such as muscarinic and $GABA_A$ sites. Limitations of autoradiographic studies are discussed in a previous review by Greenamyre and Maragos (1993).

Subsequent studies utilized [$^3$H]1-[1-(2-thienyl)-cyclohexyl]piperidine ([$^3$H]TCP), which binds to the ion channel sites of the NMDA receptor. The distribution of [$^3$H]TCP binding was very similar to that observed for NMDA-sensitive [$^3$H]glutamate binding. We observed an average 40% decline in [$^3$H]TCP binding in CA1 *stratum pyramidale/radiatum,* but no significant decline in dentate gyrus molecular layer or in *stratum pyramidale/lucidum* of CA3. Some AD patients had a significant loss of binding in CA1, whereas others had little loss. Results from Maragos *et al.* (1987) were very similar. In a comparison of 6 control, 3 non-AD dementia, and 4 AD cases, the binding of 20 nM [$^3$H]TCP was reduced in the AD group by 30–55%, although the changes were not significant. After extending the study with an additional 14 cases and using 40 nM TCP, pooled results from the two experiments revealed significant decreases in [$^3$H]TCP binding site density in CA1, subiculum, and CA3, but no significant decline in dentate gyrus.

In addition to the studies discussed above, two other groups used autoradiography to examine NMDA receptors in AD. Jansen and colleagues (1990) observed an average reduction of 46% in binding to several sites on the NMDA receptor (glutamate, phencyclidine, and glycine) in CA1 of hippocampus and entorhinal area. Represa *et al.* (1988) observed a 40% reduction in binding to the NMDA-sensitive site in the CA1 region, and significant losses in CA3 and the dentate gyrus molecular layer.

In a further attempt to resolve the discrepant results obtained with NMDA-sensitive [$^3$H]glutamate binding in AD, the Cotman (Ulas *et al.,* 1992) and the Young (Penney *et al.,* 1990) laboratories examined the NMDA receptor in autoradiographic studies involving a larger number of cases. In a postmortem study of 10 AD, 9 control, and 6 demented non-AD cases Penney *et al.* (1990) observed a 50% loss of NMDA-sensitive [$^3$H]glutamate binding in stratum pyramidale of CA1, and 40% loss of [$^3$H]TCP binding. They observed a mild loss of binding to the NMDA receptor complex in CA3 (approximately 20%) and no significant loss of NMDA receptors in dentate gyrus or subiculum. Penney and colleagues (1990) also reanalyzed the data from their previous study (Greenamyre *et al.,* 1987) in light of the observation that under the earlier binding conditions, approximately 50% of the quisqualate-insensitive glutamate binding was not displaceable by NMDA. The reanalysis confirmed a marked (70%) loss of NMDA-sensitive [$^3$H]glutamate binding in CA1 and dentate gyrus. However, loss of "other" sites ([$^3$H]glutamate binding displaced by 1 mM glutamate but not by 100 μM NMDA) was also evident. The large loss of [$^3$H]glutamate binding sites in the 1987 Greenamyre study, as compared with subsequent studies, may therefore be the result of variations in disease severity.

Ulas *et al.* (1992) also reexamined NMDA binding sites in AD using the ligands [$^3$H]glutamate, [$^3$H]MK-801, [$^3$H]CPP, and [$^3$H]glycine. In the hippocampal formation obtained from 10 AD and 8 nondemented individuals, the distribution of binding sites for each ligand were similar. In CA1 there was a 34% decline in [$^3$H]MK-801 binding (noncompetitive NMDA receptor antagonist; binds to ion channel), a 35% decrease in NMDA-sensitive [$^3$H]glutamate binding, and no significant changes in the binding of [$^3$H]CPP (NMDA receptor antagonist) or [$^3$H]glycine (NMDA receptor co-agonist). The results suggest a selective loss of binding to "agonist-preferring" versus "antagonist-preferring" binding sites (see Monaghan *et al.,* 1988). The results also demonstrated considerable intersubject variability in

NMDA binding, with some AD patients showing markedly reduced binding and others showing no change or even showing an increased. The variation could not be accounted for by the extent of neuron loss or neurofibrillary/neuritic plaque pathology. Glycine stimulated [$^3$H]glutamate binding to a similar extent in control and AD cases (Ulas *et al.*, 1992), in contrast to a decrease in glycine-stimulated [$^3$H]glutamate binding in cortical membranes from AD patients (Procter *et al.*, 1989a,b; Steele *et al.*, 1989). Preservation of the glycine modulatory site of the NMDA receptor in AD was also observed by Palmer and Burns (1994). They noted that the modulatory redox and polyamine sites were also intact.

Despite the earlier discrepant findings, recent studies generally agree on the status of hippocampal NMDA receptors in AD. There is an approximately 40% loss in binding site densities in the CA1 region of the hippocampal formation, observed using ligands against both the glutamate binding site and the ion channel (Represa *et al.*, 1988; Jansen *et al.*, 1990; Penney *et al.*, 1990; Ulas *et al.*, 1992). Although this is similar in magnitude to the neuron loss in CA1, there does not appear to be a correlation between neuron loss and decrease in binding site density. Most studies observe no loss of binding in the dentate gyrus molecular layer and little or no loss in CA3.

#### 4.1.2. Membrane Binding Studies

In addition to the autoradiographic studies, NMDA receptors have been examined in synaptosomal and membrane preparations in a series of studies by Cowburn and colleagues (1988a,b,c; 1990) and by Proctor and colleagues (1989a,b) and Steele and colleagues (1989). These studies revealed no evidence for a change in the affinity or density of NMDA-dependent L-[$^3$H]glutamate binding sites in the cortex, hippocampus, or caudate of AD patients as compared with control subjects. Similar results have been obtained by other investigators. Mouradian (1988) examined [$^3$H]MK-801 binding in several cortical regions and found no consistent differences in binding densities or affinity constants. Using [$^3$H]TCP, Simpson (1988) noted a loss of binding in frontal cortex, but not in hippocampus. More recently, Scheuer and colleagues (1996) observed a 20% decrease in [$^3$H]MK-801 binding in frontal, but not parietal, cortex.

Although Proctor and colleagues (1989a,b) and Steele and colleagues (1989) observed no decrease in the density of NMDA sites in washed membrane preparations, they did observe a decreased ability of added glycine to stimulate [$^3$H]glutamate or [$^3$H]MK-801 binding, suggesting impaired coupling of the agonist binding site and ion channel to the glycine recognition site.

### 4.2. The Kainic Acid Receptors

The first study of [$^3$H]KA binding in Alzheimer's disease revealed an expansion of binding sites in the inner molecular layer of the dentate gyrus (Geddes *et al.*, 1985), indicative of sprouting in response to pathology in entorhinal cortex. This was accompanied by intensification of acetylcholineterase histochemical staining in the outer molecular layer, similar to the sprouting response observed in rat brain following entorhinal cortex lesions. Sprouting has been confirmed with additional markers (Geddes *et al.*, 1990b) and by other investigators (Hyman *et al.*, 1987a). Jansen (1990) commented that [$^3$H]KA binding in the dentate gyrus molecular layer was more prominent in some AD sections than in control brains.

Further examination of [$^3$H]KA binding in the hippocampal formation indicated no significant loss of binding in CA1, and moderate decrease in binding to CA3 stratum lucidum in one study (Geddes *et al.*, 1986), but not in another (Geddes *et al.*, 1992). Similarly, Jansen (1990) did not detect a reduction in [$^3$H]KA binding sites in any part of the hippocampus. Penney *et al.* (1990) observed a significant (approximately 30%) decline in [$^3$H]KA binding in CA1 and prosubiculum, but not in other hippocampal regions. Dewar (1991) detected a modest decline in [$^3$H]KA binding in subiculum, but not in other hippocampal regions. In contrast to the relative sparing of [$^3$H]KA binding in the hippocampal formation observed in the above studies, Represa and colleagues (1988) reported a 53% loss of [$^3$H]KA binding in CA3 and the dentate gyrus molecular layer, with almost undetectable binding in some cases. However, as illustrated by Jansen (1990), [$^3$H]KA binding appears normal in this region even in AD cases exhibiting a marked loss of other ligands. A membrane binding study found no change in the receptor density of affinity of [$^3$H]KA binding sites in hippocampus or neocortex of AD cases as compared with nondemented controls (Cowburn *et al.*, 1989).

The relative sparing of [$^3$H]KA binding sites, even in regions with marked neuronal loss such as CA1, is intriguing. As discussed above, kainic acid binding sites may be present in glia (see also Gallo and Russell, 1995). Under the autoradiographic conditions used, [$^3$H]glutamate binding to implanted astrocytes was not displaceable by kainic acid (Bridges *et al.*, 1987). However, in a human temporal lobe epilepsy case, intense [$^3$H]kainic acid binding was observed in the region of a glioblastoma (Geddes *et al.*, 1990a). It appears that [$^3$H]KA may also bind to the GLT1 $Na^+$-dependent glutamate transporter that is predominantly localized to glia.

### 4.3. The AMPA Receptors

As noted above, Greenamyre and colleagues (1987) observed a 45–69% loss of quisqualate-sensitive [$^3$H]glutamate binding in presence of $Cl^-$ and $Ca^{2+}$ in several regions of the hippocampal formation (CA1, dentate gyrus molecular layer, and parahippocampal cortex) in AD. Jansen (1990), the first to use the more specific [$^3$H]AMPA ligand, also found a significant (38%) loss of binding in CA1. A similar loss binding (36%) was evident in CA3, although this was not statistically significant. The decrease in binding was due to a change in receptor density, not affinity. We did not observe a significant loss of [$^3$H]AMPA binding in CA1 or elsewhere in the hippocampal formation (Geddes *et al.*, 1992). Although the average [$^3$H]AMPA binding in CA1 in AD cases was 30% less than in controls, there was considerable overlap in individual values. A significant increase in [$^3$H]AMPA binding was evident in the infragranular layer of the hippocampus (Geddes *et al.*, 1992). This was most pronounced in AD cases that also exhibited increased [$^3$H]KA binding in the dentate gyrus molecular layer. The increase in [$^3$H]AMPA and [$^3$H]KA binding may reflect sprouting as a compensatory response to the neuronal degeneration. Alternatively, the presence of both AMPA and KA binding sites on glia (Patneau *et al.*, 1994; Seifert and Steinhauser, 1995) suggests that the increased binding site density observed in AD may reflect the increased glial density in vulnerable regions (Vijayan *et al.,* 1991).



### 4.4. The Metabotropic Receptors

Only a single study has examined metabotropic glutamate receptors in AD. Dewar and colleagues (1990) used [$^3$H]glutamate as the ligand (in the presence of 2.5 mM $CaCl_2$), and blocked binding to ionotropic glutamate receptors using 100 μM NMDA and 100 μM AMPA. Nonspecific binding was determined using 5 μM quisqualate. Specific [$^3$H]glutamate binding was displaced by mGluR agonists, including trans-ACPD (70%), ibotenate (72%), and L-AP4 (30%). In AD patients, [$^3$H]glutamate metabotropic binding was reduced in subiculum and CA1 (regions with the greatest binding density in controls), but not in other regions of the hippocampal formation. The same study observed a similar pattern of changes in [$^3$H]AMPA binding sites.

Although the alterations in mGluRs in AD are not well characterized, there is evidence that mGluRs may influence the pathogenesis of AD. Activation of mGluR subtypes 1α/β and 5α accelerate processing of amyloid precursor protein (APP) into non-amyloidogenic APPs (Jolly-Tornetta *et al.,* 1998; Lee *et al.,* 1996). The apoptosis and excitotoxicity induced by β-amyloid is inhibited by agonists of Group II and Group III mGluRs (Copani *et al.,* 1995).

### 4.5. *In Situ* Hybridization Studies of Glutamate Receptors

The first *in situ* hybridization study of GluR1 mRNA in AD indicated increased expression in all hippocampal subregions of AD relative to controls (Harrison *et al.,* 1990). A subsequent study noted a similar distribution of GluR1 mRNA in the hippocampal formation of control and AD cases, but did not report any changes in expression (Potier *et al.,* 1992). Pellegrini-Giampietro *et al.* (1994) observed marked intersubject variability in GluR1 mRNA expression, similar to previous autoradiographic studies (Geddes *et al.,* 1992). Some AD cases had a marked loss of GluR1 mRNA levels in CA1, CA3, and dentate gyrus; others had levels comparable to those observed in controls. The loss of GluR1 mRNA was apparently not due to cell loss. Their study also revealed a negative correlation between tissue storage duration and GluR1 mRNA levels (Potier *et al.,* 1992). In AD there is an increase in unedited GluR2 mRNA (from less than 0.1% to 1%) in prefrontal cortex. This would result in increased calcium influx through AMPA receptors and thereby play a causal role in the neurodegeneration. However, the increase in unedited GluR2 mRNA may be secondary to gliosis (Akbarian *et al.,* 1995).

There is decreased expression of NMDAR1 mRNA in Layers II–VI of entorhinal cortex in AD cases, although only the decrease in layer II (34%) is statistically significant ($P < 0.05$) (Ulas and Cotman, 1997). Reduced levels of NMDAR1 mRNA are also found in perirhinal cortex, but not in CA1, hilus, or subiculum. As in autoradiographic binding studies (Ulas and Cotman, 1997), there was substantial intersubject variation in levels of NMDAR1 mRNA in both the AD and the control group.

### 4.6. Immunohistochemical Studies of Glutamate Receptors

Only the AMPA subunits have been examined in AD using immunohistochemistry. Using antibodies against GluR1, GluR2/3, and GluR4, Hyman and colleagues (1994) observed almost identical patterns of immunostaining in tissues from control and AD patients. The only difference found was more intense immunoreactivity in the inner versus the outer layer in 3 of 15 AD patients, whereas staining was homogenous across this layer in the control and the remaining AD specimens. Armstrong and colleagues (Ikonomovic *et al.,* 1995) observed a variable loss of GluR1 and GluR2/3 immunostaining in vulnerable hippocampal regions (e.g., CA1), which appeared to correspond to the extent of neurofibrillary pathology and cell loss. The intensity of immunostaining in remaining neurons was comparable to, or greater than, that observed in controls. An increased intensity of immunostaining was also evident in less vulnerable regions, particularly in the supragranular and polymorphic layer of the dentate gyrus (Ikonomovic *et al.,* 1995; Armstrong and Ikonomovic, 1996). This is similar to results obtained with [$^3$H]AMPA binding (Geddes *et al.,* 1992). Armstrong and colleagues (1994) also observed a marked loss of GluR1 and GluR2/3 immunoreactivity in entorhinal cortex of AD patients, particularly in vulnerable layers II and III. The loss of GluR2 precedes neurofibrillary tangle formation in this region (Ikonomovic *et al.,* 1997).

### 4.7. Glutamate Levels

There are conflicting results regarding levels of glutamate in postmortem hippocampal or neocortical tissue obtained from AD patients. Procter *et al.* (1988) reported increased glutamate levels in temporal and parietal cortex. Smith and colleagues (1983) observed no altered levels in biopsy samples of temporal neocortex, nor did they find evidence of decreased glutamate release as compared with controls. Similarly, Tarbit and colleagues (1980) found no significant changes in postmortem hippocampal tissue. However, studies by Ellison *et al.* (1986), Sasaki *et al.* (1986), Hyman *et al.* (1987), and Lowe and Bowen (1990) report decreased glutamate levels in neocortical areas in AD. The results of Hyman and colleagues (1987b) are particularly striking. They measured the glutamate concentration in the dentate gyrus molecular layer (the terminal zone of the entorhinal projection to the hippocampal formation) and observed an 80% decrease in glutamate concentration in AD.

### 4.8. Glutamate Transport

Several studies have demonstrated a reduction in $Na^+$-dependent [$^3$H]D-aspartate binding or uptake in the neocortex and hippocampus in AD. The D-aspartate (D-asp) is taken up by the $Na^+$-dependent glutamate transporters, but not by the vesicular transporters. Hardy and colleagues (1987) observed a 60% decrease in synaptosomal [$^3$H]D-asp uptake in cortical and hippocampal, but not subcortical, regions in AD patients. A 40–60% decrease in $Na^+$-dependent [$^3$H]D-asp binding to extensively washed and lysed membrane preparations has also been observed in hippocampus (Cowburn *et al.*, 1988b), in various cortical regions (Cross *et al.*, 1987; Procter *et al.*, 1987; Cowburn *et al.*, 1988a; Chalmers *et al.*, 1990), and in caudate nucleus (Simpson *et al.*, 1988). Scott *et al.* (1995) observed an overall 44% decreased [$^3$H]D-asp binding site density in midtemporal, frontal, cingulate, and hippocampal cortices in AD, and a reduction in $B_{max}$ in these same regions, as well as in less vulnerable brain areas (occipital and motor cortices). In contrast, Rothstein *et al.* (1992) observed no loss of high-affinity, $Na^+$-dependent [$^3$H]L-glutamate transport in synaptosomes from spinal cord, motor cortex, sensory cortex, visual cortex, striatum, and hippocampus of AD cases. Masliah and colleagues (1996) found that in midfrontal cortex from AD cases as compared with controls, [$^3$H]D-asp binding was decreased by 34%, [$^3$H]L-asp binding was decreased by 30%, and synaptophysin immunoreactivity (indicative of synaptic density) was decreased by 48%.

The loss of [$^3$H]D-asp binding in Alzheimer's disease has been interpreted as a loss of presynaptic glutamatergic terminals. Although this is consistent with the loss of synapses observed in several cortical regions in AD, it is now clear that $Na^+$-dependent D-aspartate binding is not a marker of presynaptic terminals. Unilateral decortication or striatal quinolinic acid lesions did not reduce striatal D-asp binding in an autoradiographic assay (Greenamyre *et al.*, 1990). Moreover, knockout of the predominantly glial GLT1 transporter abolishes 94.2% of glutamate uptake in a crude synaptosomal preparation (Tanaka *et al.*, 1997). The loss of [$^3$H]D-asp binding in AD is therefore likely due to decreased binding to glial uptake sites. Deafferentation of hippocampus (fimbria-fornix lesions) or striatum (corticoststriatal lesions) results in decreased expression of glial glutamate transporters (GLT1 and GLAST) and decreased D-aspartate binding in the target areas (Ginsberg *et al.*, 1995, 1996; Levy *et al.*, 1995). However, the loss of glutamate transporters may be transient (Ginsberg *et al.*, 1996). Amygdala kindling results in decreased expression of GLAST, increased EAAC1, and no change in GLT1 expression in piriform cortex and amygdala (Miller *et al.*, 1997). Thus, both decreased and increased neuronal activity could influence expression of glutamate transporters and [$^3$H]D-aspartate binding in Alzheimer's disease. In AD frontal cortex, there is a selective loss of GLT1 (EAAT2) (Li *et al.*, 1997).

## 5. Glutamate Receptors and Excitotoxic Insult

The term "excitotoxic" was coined by John Olney (1969), and the toxic effects of glutamate were first described by Lucas and Newhouse (1957). *Excitotoxicity* refers to the ability of glutamate (or related excitatory amino acids) to induce the death of neurons through overexcitation. Choi and colleagues (1987) identified two components of glutamate neuronal toxicity. The first is neuronal swelling, resulting from elevations in intracellular $Na^+$. The second is delayed neuronal degeneration, resulting from a sustained elevation in intracellular $Ca^{2+}$. *In vitro,* the latter component of excitotoxic insult can be prevented by NMDA receptor antagonists. There is controversy, however, concerning the efficacy of NMDA antagonists against excitotoxic insults, such as ischemia, *in vivo. In vitro* small subsets of neurons with $Ca^{2+}$-permeable AMPA/KA receptors are exquisitely sensitive to AMPA and KA toxicity

(Lu *et al.*, 1995). These $Ca^{2+}$-permeable AMPA/KA receptors are most abundant at early stages of neuronal development (Pellegrini-Giampietro *et al.*, 1992). It is therefore unclear if $Ca^{2+}$-permeable AMPA receptors might contribute to neurodegeneration in the adult CNS.

Although most studies of excitotoxic insult have focused on the NMDA receptor, excessive stimulation of AMPA and KA receptors can also induce neuron death, although not as rapidly as via NMDA receptors (Koh and Choi, 1991; Choi, 1992; Chen *et al.*, 1995; Simonian *et al.*, 1996; Portera-Cailliau *et al.*, 1997). Although excitotoxic neuron death is typically thought of as necrotic, there is increasing evidence that glutamate and glutamate receptor agonists can trigger both apoptotic and necrotic neuron death (Ankarcrona *et al.*, 1995; Chen *et al.*, 1995; Ferrer *et al.*, 1995; Portera-Cailliau *et al.*, 1995; Simonian *et al.*, 1996; Lesort *et al.*, 1997; Portera-Cailliau *et al.*, 1997). Some of the neuron death following cerebral ischemia is apoptotic (Choi, 1996), and blockade of NMDA receptors may unmask apoptotic mechanisms (Gwag *et al.*, 1995; Pang and Geddes, 1997).

The mechanism of neuron death in Alzheimer's disease is unclear. Three groups have detected increased DNA fragmentation in both neurons and glia in AD, and interpreted this as evidence of apoptosis. (Su *et al.*, 1994; Lassmann *et al.*, 1995; Smale *et al.*, 1995). However, the terminal deoxynucleotide transferase-mediated dUPT-biotin nick-end labeling (TUNEL) technique can also label necrotic cell death (Pang and Geddes, 1997). Morphologically, it can be difficult to distinguish between apoptotic and necrotic death (see Choi, 1996).

## 6. Relationship between Glutamatergic Pathways and Alzheimer's Disease Pathology?

The 30–40% loss in NMDA and AMPA receptor density in CA1 of the hippocampal formation is approximately the same magnitude as neuronal loss (West *et al.*, 1994; but see also Simic *et al.*, 1997), suggesting that receptor deficits could result from neuropathology. However, studies to date have not found a strong correlation between the extent of neuronal loss or tangle density and the receptor density (Geddes *et al.*, 1992; Ulas *et al.*, 1992). In most studies CA1 is the only region in which decreased AMPA and NMDA receptor density is observed. Autoradiographic, *in situ* hybridization, and immunhistochemical studies indicate marked individual heterogeneity in AMPA and NMDA receptors in CA1; some individuals display little or no loss, whereas others exhibit massive deficits (Geddes *et al.*, 1992; Ulas *et al.*, 1992; Pellegrini-Giampietro *et al.*, 1994; Ikonomovic *et al.*, 1995). The preservation of glutamate receptors in the wake of neuropathology may reflect receptor up-regulation in remaining neurons, or a compensatory sprouting response (Geddes *et al.*, 1985). Studies by Scheff and colleagues (Scheff *et al.*, 1990; Scheff and Price, 1993) reveal that decreased synaptic density in vulnerable brain regions is compensated for by a corresponding increase in synapse size. A similar increase in glutamate receptor density could account for the maintenance of glutamate receptors in neocortical regions despite the neuron and synapse loss. Confounding this interpretation is the presence of AMPA and KA receptors on glial cells (Gallo and Russell, 1995). The sparing—and in some cases increase—of

[$^3$H]KA and [$^3$H]AMPA binding sites in AD may reflect glial proliferation in response to the neuronal degeneration.

The approximate 40% loss in $Na^+$-dependent [$^3$H]D-asp binding in neocortical regions is similar to the magnitude of synapse loss. With the cloning of the glutamate transporters, it has become clear that $Na^+$-dependent glutamate transporter sites are predominantly located on glia, not on the presynaptic nerve terminal. The mechanisms underlying the decrease in transport sites are uncertain. A possible explanation is down-regulation of glial transporters in response to decreased glutamatergic input, although an animal study demonstrates that the down-regulation of glutamate transporters following lesions is transient (Ginsberg *et al.*, 1996). Alternatively, β-amyloid deposits, associated with senile plaques, may contribute to the reduction in glutamate uptake (Harris *et al.*, 1996). There is a selective loss of the GLT1 (EAAT2) transporter in amyotrophic lateral sclerosis (Rothstein *et al.*, 1995), and also decreased levels in frontal cortex in AD (Li *et al.*, 1997). Decreased glutamate transport could conceivably play a major role in the neurodegeneration characteristic of AD.

Although excitotoxic mechanisms have long been implicated in the pathogenesis of AD, it has been difficult to reconcile the acute nature of most excitotoxic insults (e.g., ischemia) with the slow, progressive decline characteristic of AD. A possible explanation is that an age-related decline in energy metabolism, leading to a "slow" or "indirect" excitotoxic death of vulnerable neurons, may to contribute to late-onset neurodegenerative disorders, including AD (Albin and Greenamyre, 1992; Beal, 1992). The indirect excitotoxic hypothesis proposes that impairment of energy metabolism results in the loss of ATP, inability to maintain ion pumps ($Na^+/K^+$ ATPase), membrane depolarization, and removal of the voltage-dependent $Mg^{2+}$ block of the NMDA receptor. This series of events is proposed to result in low or ambient levels of glutamate becoming neurotoxic. The indirect excitotoxic hypothesis was first postulated by Novelli *et al.* (1988) who reported that several mechanisms of energy impairment (anoxia, hypoglycemia, inhibition of oxidative phosphorylation) markedly exacerbate the vulnerability of primary neuronal cultures to the toxic action of glutamate and NMDA.

In support of the indirect excitotoxic hypothesis, there is clear evidence of impaired energy metabolism in AD (Chandrasekaran *et al.*, 1997). However, it is uncertain if this plays a causal role (Davis *et al.*, 1997) or whether it results from the neuropathology (Chandrasekaran *et al.*, 1997). Moreover, most experimental support for the indirect excitotoxic hypothesis is based on observations shortly after acute metabolic impairment (Novelli *et al.*, 1988; Zeevalk and Nicklas, 1992). There is also evidence for NMDA-independent toxicity resulting from metabolic impairment. In primary cultures of rat hippocampal neurons, the metabolic poison 3-nitropropionic acid (3NP), an irreversible inhibitor of succinate dehydrogenase, cause both rapid and delayed neuron death (Pang and Geddes, 1997). The rapid neuron death is necrotic, is exacerbated by elevated levels of extracellular glutamate, and is blocked by NMDA receptor antagonists. The NMDA antagonists do not block the delayed, apoptotic neuronal death induced by 3NP. The NMDA receptor blockade may delay, but not prevent, neuronal death resulting from metabolic impairment. The slow or indirect excitotoxic hypothesis is compelling, and suggests that a chronic, mild impairment in energy metabolism could result in enhanced glutamate toxicity.

A component of the senile plaque, β-amyloid, can enhance glutamate toxicity in primary cultures of both rodent and human (Koh *et al.*, 1990) cortical neurons. The enhanced toxicity may be mediated by impairment of astrocytic glutamate uptake, glucose uptake, or energy metabolism, or via free radical mechanisms (Hensley *et al.*, 1994; Kaneko *et al.*, 1995; Harris *et al.*, 1996; Mark *et al.*, 1997; Parpura-Gill *et al.*, 1997). *In vivo* there is little evidence for β-amyloid toxicity or β-amyloid-mediated enhancement of glutamate toxicity. More recently, it was demonstrated that mutations in presenilin-1 associated with early onset familial AD, also enhance the vulnerability of hippocampal neurons to glutamate toxicity (Guo *et al.*, 1999).

Could excitotoxic mechanisms contribute to neurofibrillary pathology in AD? Glutamate was able to induce the formation of paired helical filaments (present in neurofibrillary tangles) in cultured human spinal cord neurons (De Boni and McLachlan, 1985). However, these results have not been replicated. Mattson (1990) observed that administration of glutamate to primary cultures of rat hippocampal neurons resulted in increased Alz50 immunoreactivity. Alz50 is a monoclonal antibody against the N-terminal region of human tau, and which immunostains neurofibrillary tangles in immunohistochemical studies (Hyman *et al.*, 1988). Subsequently, glutamate was shown to increase tau phosphorylation in primary neurons (Sindou *et al.*, 1994). Excessive tau phosphorylation is evident in paired helical filaments. Other studies, however, suggest that glutamate administration results in tau dephosphorylation (Davis *et al.*, 1995). *In vivo* intrahippocampal injection of the excitotoxin quinolinate or the metabolic inhibitor 3NP results in rapid tau dephosphorylation (Pang *et al.*, 1996). Excitotoxic insult, metabolic impairment, and ischemia (focal, global, and postmortem) result in tau dephosphorylation, but also in a relative sparing of tau as compared with MAP2 (Geddes *et al.*, 1994; Schwab *et al.*, 1994; Pang *et al.*, 1996). This resembles the presence of tau, but loss of MAP2, in neurofibrillary tangles (McKee *et al.*, 1989). Thus, excitotoxic insult results in cytoskeletal disruption that resembles some, but not all, aspects of neurofibrillary pathology in AD.

## 7. Summary and Conclusions

Neurofibrillary pathology and neuron loss in AD primarily affect large pyramidal neurons in the neocortex and hippocampal formation, which use glutamate as a neurotransmitter and receive glutamatergic input. This would be expected to result in marked losses of glutamate receptors and possibly glutamate transporters. Surprisingly, the only consistent finding of glutamate receptor deficits in AD is a decrease in NMDA and AMPA receptors in CA1 of the hippocampal formation. Even this loss of receptors in CA1 is not evident in all AD cases. There is little if any loss of glutamate receptors in vulnerable neocortical regions, despite the observation that neocortical tangles provide one of the best neuropathologic correlates of cognitive decline (see Geddes *et al.*, 1997). Unbiased quantitative estimates of neuron density do not demonstrate neuron loss in the neocortex in AD (Regeur *et al.*, 1994) and there are conflicting reports regarding neuron loss in the hippocampus (West *et al.*, 1994; Simic *et al.*, 1997). The alterations in glutamate receptor density may therefore be consistent with the magnitude of neuron loss in AD. Alternatively, the relative sparing of glutamate receptors may result from receptor up-regulation, axonal sprouting, or increases in synapse size. The presence of KA and AMPA receptors on glia confounds interpretation of receptor changes in AD and may account for the greater loss of NMDA versus AMPA and KA receptors in hippocampal area CA1.

In contrast to ionotropic glutamate receptors, a decline in $Na^+$-dependent [$^3$H]D-aspartate binding to glutamate transport sites is consistently observed in vulnerable neocortical regions in AD. This is consistent with decreased levels of GLT1 (EAAT2) in AD frontal cortex (Li *et al.*, 1997). The decrease in transport sites may represent down-regulation in response to deafferentation, or result from β-amyloid toxicity.

Evidence that glutamatergic systems may play a causal role in the pathogenesis of AD remains circumstantial. Many of the vulnerable neurons possess moderate-to-high densities of NMDA receptors, although the density of NMDA receptors in less-vulnerable regions (e.g., dentate gyrus) can exceed that of vulnerable neurons (e.g., entorhinal cortex). The hypothesis that an age-related decline in energy metabolism may exacerbate glutamate toxicity and contribute to neurodegeneration in late-onset neurodegenerative disorders is compelling. Experimental support for this hypothesis, however, is largely derived from acute metabolic impairment, and it is unclear whether the metabolic impairment observed in AD is a cause of, or a result of, neuropathology. Both metabolic impairment and excitotoxic insult results in cytoskeletal disruption that resembles some, but not all, aspects of neurofibrillary pathology.

## 8. References

Aistrup, G. L., Szentirmay, M., Kumar, K. N., Babcock, K. K., Schowen, R. L., and Michaelis, E. K., 1996, Ion channel properties of a protein complex with characteristics of a glutamate/N-methyl-D-aspartate receptor, *FEBS Lett.* **394:**141–148.

Akbarian, S., Smith, M. A., and Jones, E. G., 1995, Editing for an AMPA receptor subunit RNA in prefrontal cortex and striatum in Alzheimer's disease, Huntington's disease, and schizophrenia, *Brain Res.* **699:**297–304.

Albin, R. L., and Greenamyre, J. T., 1992, Alternative excitotoxic hypotheses, *Neurology* **42:**733–738.

Ankarcrona, M., Dypbukt, J. M., Bonfoco, E., Zhivotovsky, B., Orrenius, S., Lipton, S. A., and Nicotera, P., 1995, Glutamate-induced neuronal death: A succession of necrosis or apoptosis depending on mitochondrial function, *Neuron* **15:**961–973.

Armstrong, D. M., and Ikonomovic, M. D., 1996, AMPA-selective glutamate receptor subtype immunoreactivity in the hippocampal dentate gyrus of patients with Alzheimer disease: Evidence for hippocampal plasticity, *Mol. Chem. Neuropathol.* **28:**59–64.

Armstrong, D. M., Ikonomovic, M. D., Sheffield, R., and Wenthold, R. J., 1994, AMPA-selective glutamate receptor subtype immunoreactivity in the entorhinal cortex of non-demented elderly and patients with Alzheimer's disease, *Brain Res.* **639:**207–216.

Arriza, J. L., Fairman, W. A., Wadiche, J. I., Murdoch, G. H., Kavanaugh, M. P., and Amara, S. G., 1994, Functional comparisons of three glutamate transporter subtypes cloned from human motor cortex, *J. Neurosci.* **14:**5559–5569.

Arriza, J. L., Eliasof, S., Kavanaugh, M. P., and Amara, S. G., 1997, Excitatory amino acid transporter 5, a retinal glutamate transporter coupled to a chloride conductance, *Proc. Natl. Acad. Sci. USA* **94:**4155–4160.

Asztely, F., and Gustafsson, B., 1996, Ionotropic glutamate receptors: Their possible role in the expression of hippocampal synaptic plasticity, *Mol. Neurobiol.* **12:**1–11.

Babcock, K. K., Chen, X., Eggeman, K. T., Kumar, K. N., Decedue, C. J., and Michaelis, E. K., 1996, A synaptic membrane glycine-, glutamate- and thienylcyclohexylpiperidine-binding protein: Isolation and immunochemical characterization, *Neurochem. Int.* **29:**507–519.

Ball, M. J., 1977, Neuronal loss, neurofibrillary tangles, and granulovacuolar degeneration in the hippocampus with ageing and dementia: A quantitative study, *Acta Neuropathol. (Berlin)* **37:**111–118.

Barria, A., Muller, D., Derkach, V., Griffith, L. C., and Soderling, T. R., 1997, Regulatory phosphorylation of AMPA-type glutamate receptors by CaM-KII during long term potentiation, *Science* **276:** 2042–2045.

Beal, M. F., 1992, Does impairment of energy metabolism result in excitotoxic neuronal death in neurodegenerative illnesses? *Ann. Neurol.* **31:**119–130.

Beal, M. F., Brouillet, E., Jenkins, B. G., Ferrante, R. J., Kowall, N. W., Miller, J. M., Storey, E., Srivastava, R., Rosen, B. R., and Hyman, B. T., 1993, Neurochemical and histologic characterization of striatal excitotoxic lesions produced by the mitochondrial toxin 3-nitropropionic acid, *J. Neurosci.* **13:**4181–4192.

Braak, H., and Braak, E., 1991, Neuropathological staging of Alzheimer-related changes, *Acta Neuropathol. (Berlin)* **82:**239–259.

Bridges, R. J., Kesslak, J. P., Nieto-Sampedro, M., Broderick, J. T., Yu, J., and Cotman, C. W., 1987, A L-[3H]glutamate binding site on glia: An autoradiographic study on implanted astrocytes, *Brain Res.* **415:**163–168.

Brouillet, E., Jenkins, B. G., Hyman, B. T., Ferrante, R. J., Kowall, N. W., Srivastava, R., Roy, D. S., Rosen, B. R., and Beal, M. F., 1993, Age-dependent vulnerability of the striatum to the mitochondrial toxin 3-nitropropionic acid, *J. Neurochem.* **60:**356–359.

Brun, A., and Englund, E., 1981, Regional pattern of degeneration in Alzheimer's disease: Neuronal loss and histopathological grading, *Histopathology* **5:**549–564.

Chalmers, D. T., Dewar, D., Graham, D. I., Brooks, D. N., and McCulloch, J., 1990, Differential alterations of cortical glutamatergic binding sites in senile dementia of the Alzheimer type, *Proc. Natl. Acad. Sci. USA* **87:**1352–1356.

Chandrasekaran, K., Hatanpaa, K., Rapoport, S. I., and Brady, D. R., 1997, Decreased expression of nuclear and mitochondrial DNA-encoded genes of oxidative phosphorylation in association neocortex in Alzheimer disease, *Mol. Brain Res.* **44:**99–104.

Chen, Q., Harris, C., Brown, C. S., Howe, A., Surmeier, D. J., and Reiner, A., 1995, Glutamate-mediated excitotoxic death of cultured striatal neurons is mediated by non-NMDA receptors, *Exp. Neurol.* **136:**212–224.

Chittajallu, R., Vignes, M., Dev, K. K., Barnes, J. M., Collingridge, G. L., and Henley, J. M., 1996, Regulation of glutamate release by presynaptic kainate receptors in the hippocampus, *Nature* **379:**78–81.

Choi, D. W., 1987, Ionic dependence of glutamate neurotoxicity, *J. Neurosci.* **7:**369–379.

Choi, D. W., 1992, Excitotoxic cell death, *J. Neurobiol.* **23:**1261–1276.

Choi, D. W., 1996, Ischemia-induced neuronal apoptosis, *Curr. Opin. Neurobiol.* **6:**667–672.

Coleman, P. D., and Flood, D. G., 1987, Neuron numbers and dendritic extent in normal aging and Alzheimer's disease, *Neurobiol. Aging* **8:**521–545.

Copani, A., Bruno, V., Battaglia, G., Leanza, G., Pellitteri, R., Russo, A., Stanzani, S., and Nicoletti, F., 1995, Activation of metabotropic glutamate receptors protects cultured neurons against apoptosis induced by beta-amyloid peptide, *Mol. Pharmacol.* **47:**890–897.

Cotman, C. W., and Monaghan, D. T., 1987, Anatomical organization of excitatory amino acid receptors and their pathways, *Trends Neurosci.* **10:**273–280.

Couratier, P., Lesort, M., Sindou, P., Esclaire, F., Yardin, C., and Hugon, J., 1996, Modifications of neuronal phosphorylated tau immunoreactivity induced by NMDA toxicity, *Mol. Chem. Neuropathol.* **27:**259–273.

Cowburn, R., Hardy, J., Roberts, P., and Briggs, R., 1988a, Presynaptic and postsynaptic glutamatergic function in Alzheimer's disease, *Neurosci. Lett.* **86:**109–13.

Cowburn, R., Hardy, J., Roberts, P., and Briggs, R., 1988b, Regional distribution of pre- and postsynaptic glutamatergic function in Alzheimer's disease, *Brain Res.* **452:**403–407.

Cowburn, R. F., Hardy, J. A., and Roberts, P. J., 1988c, Characterisation of $Na(^{+})$-independent L-[$^{3}$H]glutamate binding sites in human temporal cortex, *J. Neurochem.* **50:**1872–1878.

Cowburn, R. F., Hardy, J. A., Briggs, R. S., and Roberts, P. J., 1989, Characterisation, density, and distribution of kainate receptors in normal and Alzheimer's diseased human brain, *J. Neurochem.* **52:**140–147.

Cowburn, R. F., Hardy, J. A., and Roberts, P. J., 1990, Glutamatergic neurotransmission in Alzheimer's disease, *Biochem. Soc. Trans.* **18:**390–392.

Cross, A. J., Slater, P., Simpson, M., Royston, C., Deakin, J. F., Perry, R. H., and Perry, E. K., 1987, Sodium dependent D-[3H]aspartate binding in cerebral cortex in patients with Alzheimer's and Parkinson's diseases, *Neurosci. Lett.* **79:**213–217.

Davis, D. R., Brion, J. P., Couck, A. M., Gallo, J. M., Hanger, D. P., Ladhani, K., Lewis, C., Miller, C. C., Rupniak, T., Smith, C., *et al.,* 1995, The phosphorylation state of the microtubule-associated protein tau as affected by glutamate, colchicine, and beta-amyloid in primary rat cortical neuronal cultures, *Biochem. J.* **309:**941–949.

Davis, R. E., Miller, S., Herrnstadt, C., Ghosh, S. S., Fahy, E., Shinobu, L. A., Galasko, D., Thal, L. J., Beal, M. F., Howell, N., and Parker, W. D., Jr., 1997, Mutations in mitochondrial cytochrome c oxidase genes segregate with late-onset Alzheimer disease, *Proc. Natl. Acad. Sci. USA* **94:**4526–4531.

De Boni, U., and McLachlan, D. R. C., 1985, Controlled induction of paired helical filaments of the Alzheimer type in cultured human neurons, by glutamate and aspartate, *J. Neurol. Sci.* **68:**105–118.

Dewar, D., Chalmers, D. T., Shand, A., Graham, D. I., and McCulloch, J., 1990, Selective reduction of quisqualate (AMPA) receptors in Alzheimer cerebellum, *Ann. Neurol.* **28:**805–810.

Dewar, D., Chalmers, D. T., Graham, D. I., and McCulloch, J., 1991, Glutamate metabotropic and AMPA binding sites are reduced in Alzheimer's disease: An autoradiographic study of the hippocampus, *Brain Res.* **553:**58–64.

Ehlers, M. D., Tingley, W. G., and Huganir, R. L., 1995, Regulated subcellular distribution of NR1 subunit of the NMDA receptor, *Science* **269:**1734–1737.

Ellison, D. W., Beal, F. M., Mazurek, M. F., and Bird, E. D., 1986, A postmortem study of amino acid neurotransmitters in Alzheimer's disease, *Ann. Neurol.* **20:**616–621.

Fairman, W. A., Vandenberg, R. J., Arriza, J. L., Kavanaugh, M. P., and Amara, S. G., 1995, An excitatory amino-acid transporter with properties of a ligand-gated chloride channel, *Nature* **375:**599–603.

Ferrer, I., Martin, F., Serrano, T., Reiriz, J., Perez-Navarro, E., Alberch, J., Macaya, A., and Planas, A. M., 1995, Both apoptosis and necrosis occur following intrastriatal administration of excitotoxins, *Acta Neuropathol. (Berlin)* **90:**504–510.

Fletcher, E. J., and Lodge, D., 1996, New developments in the molecular pharmacology of alpha-amino-3-hydroxy-5-methyl-4-isoxazole proprionate and kainate receptors, *Pharmacol. Ther.* **70:**65–89.

Flood, D. G., and Coleman, P. D., 1986, Failed compensatory dendritic growth as a pathophysiological process in Alzheimer's disease, *Can. J. Neurol. Sci. (Suppl. 4)* **13:**475–479.

Fonnum, F., 1984, Glutamate: A neurotransmitter in mammalian brain, *J. Neurochem.* **42:**1–11.

Gallo, V., and Russell, J. T., 1995, Excitatory amino acid receptors in glia: Different subtypes for distinct functions? *J. Neurosci. Res.* **42:**1–8.

Geddes, J. W., Monaghan, D. T., Cotman, C. W., Lott, I. T., Kim, R. C., and Chui, H. C., 1985, Plasticity of hippocampal circuitry in Alzheimer's disease, *Science* **230:**1179–1181.

Geddes, J. W., Chang-Chui, H., Cooper, S. M., Lott, I. T., and Cotman, C. W., 1986, Density and distribution of NMDA receptors in the human hippocampus in Alzheimer's disease, *Brain Res.* **399:**156–161.

Geddes, J. W., Cahan, L. D., Cooper, S. M., Kim, R. C., Choi, B. H., and Cotman, C. W., 1990a, Altered distribution of excitatory amino acid receptors in temporal lobe epilepsy, *Exp. Neurol.* **108:**214–220.

Geddes, J. W., Wong, J., Choi, B. H., Kim, R. C., Cotman, C. W., and Miller, F. D., 1990b, Increased expression of the embryonic form of a developmentally regulated mRNA in Alzheimer's disease, *Neurosci. Lett.* **109:**54–61.

Geddes, J. W., Ulas, J., Brunner, L. C., Choe, W., and Cotman, C. W., 1992, Hippocampal excitatory amino acid receptors in elderly normal individuals and those with Alzheimer's disease: non-*N*-methyl-D-aspartate receptors, *Neuroscience* **50:**23–34.

Geddes, J. W., Schwab, C., Wilson, L., Craddock, S., and Pettegrew, L. C., 1994, Alterations in tau immunostaining in the rat hippocampal formation following transient cerebral ischemia, *J. Cereb. Blood Flow Metab.* **14:**554–564.

Geddes, J. W., Tekirian, T. L., Soultanian, N. S., Ashford, J. W., Davis, D. G., and Markesbery, W. R., 1997, Comparison of neuropathologic criteria for the diagnosis of Alzheimer's disease, *Neurobiol. Aging,* in press.

Ginsberg, S. D., Martin, L. J., and Rothstein, J. D., 1995, Regional deafferentation down-regulates subtypes of glutamate transporter proteins, *J. Neurochem.* **65:**2800–2803.
Ginsberg, S. D., Rothstein, J. D., Price, D. L., and Martin, L. J., 1996, Fimbria-fornix transections selectively down-regulate subtypes of glutamate transporter and glutamate receptor proteins in septum and hippocampus, *J. Neurochem.* **67:**1208–1216.
Greenamyre, J. T., and Maragos, W. F., 1993, Neurotransmitter receptors in Alzheimer disease, *J. Cereb. Brain Metab. Rev.* **5:**61–94.
Greenamyre, J. T., Penney, J. B., Young, A. B., D'Amato, C. J., Hicks, S. P., and Shoulson, I., 1985, Alterations in L-glutamate binding in Alzheimer's and Huntington's diseases, *Science* **227:**1496–1499.
Greenamyre, J. T., Penney, J. B., D'Amato, C. J., and Young, A. B., 1987, Dementia of the Alzheimer's type: Changes in hippocampal L-[3H]glutamate binding, *J. Neurochem.* **48:**543–551.
Greenamyre, J. T., Higgins, D. S., and Young, A. B., 1990, Sodium-dependent D-aspartate 'binding' is not a measure of presynaptic neuronal uptake sites in an autoradiographic assay, *Brain Res.* **511:**310–318.
Guo, Q., Fu, W., Sopher, B. L., Miller, M. W., Ware, C. B., Martin, G. M., and Mattson, M. P., 1999, Increased vulnerability of hippocampal neurons to excitotoxic necrosis in presenilin-1 mutant knock-in mice, *Nature Med.* **5:**101–106.
Gwag, B. J., Lobner, D., Koh, J. Y., Wie, M. B., and Choi, D. W., 1995, Blockade of glutamate receptors unmasks neuronal apoptosis after oxygen-glucose deprivation *in vitro, Neuroscience* **68:**615–619.
Hardy, J. A., Cowburn, R. F., Barton, A., Reynolds, G., Lofdahl, E., O'Carroll, A.-M., Wester, P., and Winblad, B., 1987, Region-specific loss of glutamate innervation in Alzheimer's disease, *Neurosci. Lett.* **73:**77–80.
Harris, M. E., Wang, Y., Pedigo, N. W., Jr., Hensley, K., Butterfield, D. A., and Carney, J. M., 1996, Amyloid beta peptide (25-35) inhibits Na(+)-dependent glutamate uptake in rat hippocampal astrocyte cultures, *J. Neurochem.* **67:**277–286.
Harrison, P. J., Barton, A. J., Najlerahim, A., and Pearson, R. C. A., 1990, Distribution of a kainate/AMPA receptor mRNA in normal and Alzheimer brain, *NeuroReport* **1:**149–152.
Henley, J. M., 1994, Kainate-binding proteins: Phylogeny, structures, and possible functions, *Trends Pharmacol. Sci.* **15:**182–190.
Hensley, K., Carney, J. M., Mattson, M. P., Aksenova, M., Harris, M., Wu, J. F., Floyd, R. A., and Butterfield, D. A., 1994, A model for beta-amyloid aggregation and neurotoxicity based on free radical generation by the peptide: Relevance to Alzheimer disease, *Proc. Natl. Acad. Sci. USA* **91:**3270–3274.
Hirano, A., and Zimmerman, H. M., 1962, Alzheimer's neurofibrillary changes: A topographic study, *Arch. Neurol.* **7:**227–242.
Hof, P. R., and Morrison, J. H., 1994, The cellular basis of cortical disconnection in Alzheimer's disease and related dementing conditions, in: *Alzheimer's Disease* (R. D. Terry, R. Katzman, and K. L. Bick, eds.), Raven, New York, pp. 197–229.
Hof, P. R., Giannakopoulos, P., Vickers, J. C., Bouras, C., and Morrison, J. H., 1995, The morphologic and neurochemical basis of dementia: Aging, hierarchical patterns of lesion distribution, and vulnerable neuronal phenotype, *Rev. Neurosci.* **6:**97–124.
Hollmann, M., and Heinemann, S. F., 1993, Cloned glutamate receptors, *Annu. Rev. Neurosci.* **17:**31–108.
Huntley, G. W., Vickers, J. C., Janssen, W., Brose, N., Heinemann, S. F., and Morrison, J. H., 1994a, Distribution and synaptic localization of immunocytochemically identified NMDA receptor subunit proteins in sensory-motor and visual cortices of monkey and human, *J. Neurosci.* **14:**3603–3619.
Huntley, G. W., Vickers, J. C., and Morrison, J. H., 1994b, Cellular and synaptic localization of NMDA and non-NMDA receptor subunits in neocortex: Organizational features related to cortical circuitry, function, and disease, *Trends Neurosci.* **17:**536–543.
Hyman, B. T., Van Hoesen, G. W., Damasio, A. R., and Barnes, C. L., 1984, Alzheimer's disease: Cell-specific pathology isolates the hippocampal formation, *Science* **225:**1168–1170.
Hyman, B. T., Kromer, L. J., and Van Hoesen, G. W., 1987a, Reinnervation of the hippocampal perforant pathway zone in Alzheimer's disease, *Ann. Neurol.* **21:**259–267.
Hyman, B. T., Van Hoesen, G. W., and Damasio, A. R., 1987b, Alzheimer's disease: Glutamate depletion in the hippocampal perforant pathway zone, *Ann. Neurol.* **22:**37–40.
Hyman, B. T., Van Hoesen, G. H., Wolozin, B. L., Davies, P., Kromer, L. J., and Damasio, A. R., 1988, Alz-50 antibody recognizes Alzheimer-related neuronal changes, *Ann. Neurol.* **23:**371–379.
Hyman, B. T., Penney, J. B., Jr., Blackstone, C. D., and Young, A. B., 1994, Localization of non-*N*-

methyl-D-aspartate glutamate receptors in normal and Alzheimer hippocampal formation, *Ann. Neurol.* **35:**31–37.

Ikonomovic, M. D., Mizukami, K., Davies, P., Hamilton, R., Sheffield, R., and Armstrong, D. M., 1997, The loss of GluR2(3) immunoreactivity precedes neurofibrillary tangle formation in the entorhinal cortex and hippocampus of Alzheimer brains, *J. Neuropathol. Exp. Neurol.* **56:**1018–1027.

Ikonomovic, M. D., Sheffield, R., and Armstrong, D. M., 1995, AMPA-selective glutamate receptor subtype immunoreactivity in the hippocampal formation of patients with Alzheimer's disease, *Hippocampus* **5:**469–486.

Jansen, K. L., Faull, R. L., Dragunow, M., and Synek, B. L., 1990, Alzheimer's disease: Changes in hippocampal *N*-methyl-D-aspartate, quisqualate, neurotensin, adenosine, benzodiazepine, serotonin, and opioid receptors: An autoradiographic study, *Neuroscience* **39:**613–627.

Jolly-Tornetta, C., Gao, Z. Y., Lee, V. M., and Wolf, B. A., 1998, Regulation of amyloid precursor protein secretion by glutamate receptors in human Ntera 2 neurons, *J. Biol. Chem.* **273:**14015–14021.

Kaczmarek, L., Kossut, M., and Skangiel-Kramska, J., 1997, Glutamate receptors in cortical plasticity: Molecular and cellular biology, *Physiol. Rev.* **77:**217–255.

Kanai, Y., and Hediger, M. A., 1992, Primary structure and functional characterization of a high-affinity glutamate transporter, *Nature* **360:**467–471.

Kaneko, I., Yamada, N., Sakuraba, Y., Kamenosono, M., and Tutumi, S., 1995, Suppression of mitochondrial succinate dehydrogenase, a primary target of beta-amyloid, and its derivative racemized at Ser residue, *J. Neurochem.* **65:**2585–2593.

Koh, J. Y., and Choi, D. W., 1991, Selective blockade of non-NMDA receptors does not block rapidly triggered glutamate-induced neuronal death, *Brain Res.* **548:**318–321.

Koh, J. Y., Yang, L. L., and Cotman, C. W., 1990, Beta-amyloid protein increases the vulnerability of cultured cortical neurons to excitotoxic damage, *Brain Res.* **533:**315–320.

Kohler, M., Burnashev, N., Sakmann, B., and Seeburg, P. H., 1993, Determinant of Ca2+ permeability in both TM1 and TM2 of high affinity kainate receptors channels: Diversity by RNA editing, *Neuron* **10:**491–500.

Kumar, K. N., Tilakaratne, N., Johnson, P. S., Allen, A. E., and Michaelis, E. K., 1991, Cloning of cDNA for the glutamate-binding subunit of an NMDA receptor complex, *Nature* **354:**70–73.

Kumar, K. N., Babcock, K. K., Johnson, P. S., Chen, X., Eggeman, K. T., and Michaelis, E. K., 1994, Purification and pharmacological and immunochemical characterization of synaptic membrane proteins with ligand-binding properties of *N*-methyl-D-aspartate receptors, *J. Biol. Chem.* **269:**27384–27393.

Kumar, K. N., Babcock, K. K., Johnson, P. S., Chen, X., Ahmad, M., and Michaelis, E. K., 1995, Cloning of the cDNA for a brain glycine-, glutamate-, and thienylcyclohexylpiperidine-binding protein, *Biochem. Biophys. Res. Commun.* **216:**390–398.

Lassmann, H., Bancher, C., Breitschopf, H., Wegiel, J., Bobinski, M., Jellinger, K., and Wisniewski, H. M., 1995, Cell death in Alzheimer's disease evaluated by DNA fragmentation *in situ, Acta Neuropathol. (Berlin)* **89:**35–41.

Lee, R. K., Jimenez, J., Cox, A. J., and Wurtman, R. J., 1996, Metabotropic glutamate receptors regulate APP processing in hippocampal neurons and cortical astrocytes derived from fetal rats, *Ann. N.Y. Acad. Sci.* **777:**338–343.

Lee, R. K., Wurtman, R. J., Cox, A. J., and Nitsch, R. M., 1995, Amyloid precursor protein processing is stimulated by metabotropic glutamate receptors, *Proc. Natl. Acad. Sci. USA* **92:**8083–8087.

Lerma, J., 1998, Kainate receptors: An interplay between excitatory and inhibitory synapses, *FEBS Lett.* **430:**100–104.

Lesort, M., Esclaire, F., Yardin, C., and Hugon, J., 1997, NMDA induces apoptosis and necrosis in neuronal cultures: Increased APP immunoreactivity is linked to apoptotic cells, *Neurosci. Lett.* **221:**213–216.

Levy, L. M., Lehre, K. P., Walaas, S. I., Storm-Mathisen, J., and Danbolt, N. C., 1995, Down-regulation of glial glutamate transporters after glutamatergic denervation in the rat brain, *Eur. J. Neurosci.* **7:**2036–2041.

Lewis, D. A., Campbell, M. J., Terry, R. D., and Morrison, J. H., 1987, Laminar and regional distributions of neurofibrillary tangles and neurite plaques in Alzheimer's disease: A quantitative study of visual and auditory cortices, *J. Neurosci.* **7:**1799–1809.

Li, S., Mallory, M., Alford, M., Tanaka, S., and Masliah, E., 1997, Glutamate transporter alterations in Alzheimer disease are possibly associated with abnormal APP expression, *J. Neuropathol. Exp. Neurol.* **56:**901–911.

Lowe, S. L., Bowen, D. M., Francis, P. T., and Neary, D., 1990, Ante mortem cerebral amino acid concentrations indicate selective degeneration of glutamate-enriched neurons in Alzheimer's disease, *Neuroscience* **38:**571–577.

Lu, Y. M., Yin, H. Z., and Weiss, J. H., 1995, Ca2+permeable AMPA/kainate channels permit rapid injurious Ca2+ entry, *NeuroReport* **6:**1089–1092.

Lucas, D. R., and Newhouse, J. P., 1957, The toxic effect of sodium L-glutamate on the inner layers of the retina, *AMA Arch. Opthalmol.* **58:**193–201.

Mann, D. M., 1985, The neuropathology of Alzheimer's disease: A review with pathogenetic, aetiological, and therapeutic considerations, *Mech. Aging Dev.* **31:**213–255.

Maragos, W. F., Chu, D. C., Young, A. B., D'Amato, C. J., and Penney, J. J., 1987, Loss of hippocampal [3H]TCP binding in Alzheimer's disease, *Neurosci. Lett.* **74:**371–376.

Mark, R. J., Pang, Z., Geddes, J. W., Uchida, K., and Mattson, M. P., 1997, Amyloid beta-peptide impairs glucose transport in hippocampal and cortical neurons: Involvement of membrane lipid peroxidation, *J. Neurosci.* **17:**1046–1054.

Masliah, E., Alford, M., DeTeresa, R., Mallory, M., and Hansen, L., 1996, Deficient glutamate transport is associated with neurodegeneration in Alzheimer's disease, *Ann. Neurol.* **40:**759–766.

Masliah, E., Terry, R. D., DeTeresa, R. M., and Hansen, L. A., 1989, Immunohistochemical quantification of the synapse-related protein synaptophysin in Alzheimer disease, *Neurosci. Lett.* **103:**234–239.

Mattson, M. P., 1990, Antigenic changes similar to those seen in neurofibrillary tangles are elicited by glutamate and calcium influx in cultured hippocampal neurons, *Neuron* **4:**105–117.

McKee, A. C., Kowall, N. W., and Kosik, K. S., 1989, Microtubular reorganization and dendritic growth response in Alzheimer's disease, *Ann. Neurol.* **26:**652–659.

Miller, S., Kesslak, J. P., Romano, C., and Cotman, C. W., 1995, Roles of metabotropic glutamate receptors in brain plasticity and pathology, *Ann. N.Y. Acad. Sci.* **757:**460–474.

Miller, H. P., Levey, A. I., Rothstein, J. D., Tzingounis, A. V., and Conn, P. J., 1997, Alterations in glutamate transporter protein levels in kindling-induced epilepsy, *J. Neurochem.* **68:**1564–1570.

Monaghan, D. T., and Cotman, C. W., 1985, Distribution of *N*-methyl-D-aspartate-sensitive L-[3H]glutamate-binding sites in rat brain, *J. Neurosci.* **5:**2909–2919.

Monaghan, D. T., Olverman, H. J., Nguyen, L., Watkins, J. C., and Cotman, C. W., 1988, Two classes of *N*-methyl-D-aspartate recognition sites: Differential distribution and differential regulation by glycine, *Proc. Natl. Acad. Sci. USA* **85:**9836–9840.

Monaghan, D. T., Bridges, R. J., and Cotman, C. W., 1989, The excitatory amino acid receptors: Their classes, pharmacology, and distinct properties in the function of the central nervous system, *Annu. Rev. Pharmacol. Toxicol.* **29:**365–402.

Monaghan, D. T., and Larsen, H., 1997, NR1 and NR2 subunit contributions to *N*-methyl-s-aspartate receptor channel blocker pharmacology, *J. Pharmacol. Exp. Ther.* **280:**614–620.

Morris, R. G. M., Anderson, E., and Lynch, G. S., 1986, Selective impairment of learning and blockade of long-term potentiation by an *N*-methyl-D-aspartate receptor antagonist, AP5, *Nature* **319:**774–776.

Mountjoy, C. Q., Roth, M., Evans, N. J., and Evans, H. M., 1983, Cortical neuronal counts in normal elderly controls and demented patients, *Neurobiol. Aging* **4:**1–11.

Mouradian, M. M., Contreras, P. C., Monahan, J. B., and Chase, T. N., 1988, [3H]MK-801 binding in Alzheimer's disease, *Neurosci. Lett.* **93:**225–230.

Nakanishi, S., 1992, Molecular diversity of glutamate receptors and implications for brain function, *Science* **258:**597–603.

Nicoletti, F., Bruon, V., Copani, A., Casabona, G., and Knopfel, T., 1996, Metabotropic glutamate receptors: A new target for the therapy of neurodegenerative disorders, *Trends Neurosci.* **19:**267–271.

Novelli, A., Reilly, J. A., Lysko, P. G., and Henneberry, R. C., 1988, Glutamate becomes neurotoxic via the *N*-methyl-D-aspartate receptor when intracellular energy levels are reduced, *Brain Res.* **451:**205–212.

Olney, J. W., 1969, Brain lesions, obesity, and other disturbances in mice treated with monosodium glutamate, *Science* **16ⁱ:**719–721.

Palmer, A. M., and Burns, M. A., 1994, Preservation of redox, polyamine, and glycine modulatory domains of the *N*-methyl-D-aspartate receptor in Alzheimer's disease, *J. Neurochem.* **62:**187–196.

Pang, Z., and Geddes, J. W., 1997, Mechanisms of cell death induced by the mitochondrial toxin 3-nitropropionic acid: Acute excitotoxic necrosis and delayed apoptosis, *J. Neurosci.* **17:**3064–3073.

Pang, Z., Umberger, G. H., and Geddes, J. W., 1996, Neuronal loss and cytoskeletal disruption following intrahippocampal administration of the metabolic inhibitor malonate: Lack of protection by MK-801, *J. Neurochem.* **66:**474–484.

Parpura-Gill, A., Beitz, D., and Uemura, E., 1977, The inhibitory effects of beta-amyloid on glutamate and glucose uptakes by cultured astrocytes, *Brain Res.* **754:**65–71.

Parsons, C. G., Danysz, W., Hesselink, M., Hartmann, S., Lorenz, B., Wollenburg, C., and Quack, G., 1998, Modulation of NMDA receptors by glycine—introduction to some basic aspects and recent developments, *Amino Acids* **14:**207–216.

Paternain, A. V., Morales, M., and Lerma, J., 1995, Selective antagonism of AMPA receptors unmasks kainate receptor-mediated responses in hippocampal neurons, *Neuron* **14:**185–189.

Patneau, D. K., Wright, P. W., Winters, C., Mayer, M. L., and Gallo, V., 1994, Glial cells of the oligodendrocyte lineage express both kainate- and AMPA-preferring subtypes of glutamate receptor, *Neuron* **12:**357–371.

Pearson, R. C. A., Esiri, M. M., Hiorns, R. W., Wilcock, G. K., and Powell, T. P. S., 1985, Anatomical correlates of the distribution of the pathological changes in the neocortex in Alzheimer's disease, *Proc. Natl. Acad. Sci. USA* **82:**4531–4534.

Pellegrini-Giampietro, D. E., Bennett, M. V., and Zukin, R. S., 1992, Are Ca(2+)-permeable kainate/AMPA receptors more abundant in immature brain? *Neurosci. Lett.* **144:**65–69.

Pellegrini-Giampietro, D. E., Bennett, M. V., and Zukin, R. S., 1994, AMPA/kainate receptor gene expression in normal and Alzheimer's disease hippocampus, *Neuroscience* **61:**41–49.

Penney, J. B., Maragos, W. F., Greenamyre, J. T., Debowey, D. L., Hollingsworth, Z., and Young, A. B., 1990, Excitatory amino acid binding sites in the hippocampal region of Alzheimer's disease and other dementias, *J. Neurol. Neurosurg. Psychiatry* **53:**314–320.

Pines, G., Danbolt, N. C., Bjoras, M., Zhang, Y., Bendahan, A., Eide, L., Koepsell, H., Storm-Mathisen, J., Seeberg, E., and Kanner, B. I., 1992, Cloning and expression of a rat brain L-glutamate transporter, *Nature* **360:**464–467.

Portera-Cailliau, C., Hedreen, J. C., Price, D. L., and Koliatsos, V. E., 1995, Evidence for apoptotic cell death in Huntington disease and excitotoxic animal models, *J. Neurosci.* **15:**3775–3787.

Portera-Cailliau, C., Price, D. L., and Martin, L. J., 1997, Non-NMDA and NMDA receptor-mediated excitotoxic neuronal deaths in adult brain are morphologically distinct: Further evidence for an apoptosis-necrosis continuum, *J. Comp. Neurol.* **378:**88–104.

Potier, M. C., Spillantini, M. G., and Carter, N. P., 1992, The human glutamate receptor cDNA GluR1: Cloning, sequencing, expression, and localization to chromosome 5, *DNA Seq.* **2:**211–218.

Procter, A. W., Palmer, A. M., Bowen, D. M., Murphy, E., and Neary, D., 1987, Glutamatergic denervation in Alzheimer's disease: A cautionary note, *J. Neurol. Neurosurg. Psychiatry* **50:**825.

Procter, A. W., Lowe, S. L., Palmer, A. M., Francis, P. T., Esiri, M. M., Stratmann, G. C., Najlerahim, A., Patel, A. J., Hunt, A., and Bowen, D. M., 1988, Topographical distribution of neurochemical changes in Alzheimer's disease, *J. Neurol. Sci.* **84:**125–40.

Procter, A. W., Stirling, J. M., Stratmann, G. C., Cross, A. J., and Bowen, D. M., 1989a, Loss of glycine-dependent radioligand binding to the *N*-methyl-D-aspartate-phencyclidine receptor complex in patients with Alzheimer's disease, *Neurosci. Lett.* **101:**62–66.

Procter, A. W., Wong, E. H., Stratmann, G. C., Lowe, S. L., and Bowen, D. M., 1989b, Reduced glycine stimulation of [3H]MK-801 binding in Alzheimer's disease, *J. Neurochem.* **53:**698–704.

Regeur, L., Jensen, G. B., Pakkenberg, H., Evans, S. M., and Pakkenberg, B., 1994, No global neocortical nerve cell loss in brains from patients with senile dementia of Alzheimer's type, *Neurobiol. Aging* **15:**347–352.

Represa, A., Duyckaerts, C., Tremblay, E., Hauw, J. J., and Ben-Ari, Y., 1988, Is senile dementia of the Alzheimer type associated with hippocampal plasticity? *Brain Res.* **457:**355–359.

Riedel, G., and Reymann, K. G., 1996, Metabotropic glutamate receptors in hippocampal long-term potentiation and learning and memory, *Acta Physiol. Scand.* **157:**1–19.

Roche, K. W., O'Brien, R. J., Mammen, A. L., Bernhardt, J., and Huganir, R. L., 1996, Characterization of multiple phosphorylation sites on the AMPA receptor GluR1 subunit, *Neuron* **16:**1179–1188.

Rothstein, J. D., Martin, L. J., and Kuncl, R. W., 1992, Decreased glutamate transport by the brain and spinal cord in amyotrophic lateral sclerosis, *N. Eng. J. Med.* **326:**1464–1468.

Rothstein, J. D., Martin, L., Levey, A. I., Dykes-Hoberg, M., Jin, L., Wu, D., Nash, N., and Kuncl, R. W., 1994, Localization of neuronal and glial glutamate transporters, *Neuron* **13:**713–725.

Rothstein, J. D., Van Kammen, M., Levey, A. I., Martin, L. J., and Kuncl, R. W., 1995, Selective loss of glial glutamate transporter GLT-1 in amyotrophic lateral sclerosis, *Ann. Neurol.* **38:**73–84.

Rothstein, J. D., Dykes-Hoberg, M., Pardo, C. A., Bristol, L. A., Jin, L., Kuncl, R. W., Kanai, Y., Hediger,

M. A., Wang, Y., Schielke, J. P., and Welty, D. F., 1996, Knockout of glutamate transporters reveals a major role for astroglial transport in excitotoxicity and clearance of glutamate, *Neuron* **16:**675–686.

Sasaki, H., Muramoto, O., Kanazawa, I., Arai, H., Kosaka, K., and Iizuka, R., 1986, Regional distribution of amino acid transmitters in post-mortem brains of presenile and senile dementia of Alzheimer type, *Ann. Neurol.* **19:**263–269.

Scheff, S. W., and Price, D. A., 1993, Synapse loss in the temporal lobe in Alzheimer's disease, *Ann. Neurol.* **33:**190–199.

Scheff, S. W., DeKosky, S. T., and Price, D. A., 1990, Quantitative assessment of cortical synaptic density in Alzheimer's disease, *Neurobiol. Aging* **11:**29–37.

Scheff, S. W., Sparks, L., and Price, D. A., 1993, Quantitative assessment of synaptic density in the entorhinal cortex in Alzheimer's disease, *Ann. Neurol.* **34:**356–361.

Scheuer, K., Maras, A., Gattaz, W. F., Cairns, N., Förstl, H., and Müler, W. E., 1996, Cortical NMDA receptor properties and membrane fluidity are altered in Alzheimer's disease, *Dementia* **7:**210–214.

Schwab, C., Bondada, V., Sparks, D. L., Cahan, L. D., and Geddes, J. W., 1994, Postmortem changes in the levels and localization of microtubule-associated proteins (tau, MAP2 and MAP1B) in the rat and human hippocampus, *Hippocampus* **4:**210–225.

Scott, H. L., Tannenberg, A. E., and Dodd, P. R., 1995, Variant forms of neuronal glutamate transporter sites in Alzheimer's disease cerebral cortex, *J. Neurochem.* **64:**2193–2202.

Seeburg, P. H., 1993, The molecular biology of mammalian glutamate receptor channels, *Trends Neurosci.* **16:**359–365.

Seeburg, P. H., 1996, The role of RNA editing in controlling glutamate receptor channel properties, *J. Neurochem.* **66:**1–5.

Seeburg, P. H., Higuchi, M., and Sprengel, R., 1998, RNA editing of brain glutamate receptor channels: mechanism and physiology, *Brain Res. Rev.* **26:**217–229.

Seifert, G., and Steinhauser, C., 1995, Glial cells in the mouse hippocampus express AMPA receptors with an intermediate Ca2+ permeability, *Eur. J. Neurosci.* **7:**1872–1881.

Simic, G., Kostovic, I., Winblad, B., and Bogdanovic, N., 1997, Volume and number of neurons of the human hippocampal formation in normal aging and Alzheimer's disease, *J. Comp. Neurol.* **379:**482–494.

Simonian, N. A., Getz, R. L., Leveque, J. C., Konradi, C., and Coyle, J. T., 1996, Kainic acid induces apoptosis in neurons, *Neuroscience* **75:**1047–1055.

Simpson, M. D., Royston, M. C., Deakin, J. F., Cross, A. J., Mann, D. M., and Slater, P., 1988, Regional changes in [3H]D-aspartate and [3H]TCP binding sites in Alzheimer's disease brains, *Brain Res.* **462:**76–82.

Sindou, P., Lesort, M., Couratier, P., Yardin, C., Esclaire, F., and Hugon, J., 1994, Glutamate increases tau phosphorylation in primary neuronal cultures from fetal rat cerebral cortex, *Brain Res.* **646:**124–128.

Smale, G., Nichols, N. R., Brady, D. R., Finch, C. E., and Horton, W. E., Jr., 1995, Evidence for apoptotic cell death in Alzheimer's disease, *Exp. Neurol.* **133:**225–230.

Smith, C. C. T., Bowen, D. M., Neary, D., and Davison, A. N., 1983, Amino acid release from biopsy samples of temporal neocortex from patients with Alzheimer's disease, *Brain Res.* **264:**138–141.

Soderling, T. R., 1996, Modulation of glutamate receptors by calcium/calmodulin-dependent protein kinase II, *Neurochem. Int.* **28:**359–361.

Sommer, B., Keinanen, K., Verdoorn, T. A., Wisden, W., Burnashev, N., Herb, A., Kohler, M., Takagi, T., Sakmann, B., and Seeburg, P. H., 1990, Flip and flop: A cell-specific functional switch in glutamate-operated channels of the CNS, *Science* **249:**1580–1585.

Sommer, B., Kohler, M., Sprengel, R., and Seeburg, P. H., 1991, RNA editing in brain controls a determinant of ion flow in glutamate-gated channels, *Cell* **67:**11–19.

Steele, J. E., Palmer, A. M., Stratmann, G. C., and Bowen, D. M., 1989, The *N*-methyl-D-aspartate receptor complex in Alzheimer's disease: Reduced regulation by glycine but not zinc, *Brain Res.* **500:**369–373.

Storck, T., Schulte, S., Hofmann, K., and Stoffel, W., 1992, Structure, expression, and functional analysis of a Na(+)-dependent glutamate/aspartate transporter from rat brain, *Proc. Natl. Acad. Sci. USA* **89:**10955–10959.

Su, J. H., Anderson, A. J., Cummings, B. J., and Cotman, C. W., 1994, Immunohistochemical evidence for apoptosis in Alzheimer's disease, *NeuroReport* **5:**2529–2533.

Sucher, N. J., Awobuluyi, M., Choi, Y.-B., and Lipton, S. A., 1996, NMDA receptors: From genes to channels, *Trends Pharmacol. Sci.* **17:**348–355.

Tanaka, K., 1993, Cloning and expression of a glutamate transporter from mouse brain, *Neurosci. Lett.* **159:**183–186.

Tanaka, K., Watase, K., Manabe, T., Yamada, K., Watanabe, M., Takahashi, K., Iwama, H., Nishikawa, T., Ichihara, N., Kikuchi, T., Okuyama, S., Kawashima, N., Hori, S., Takimoto, M., and Wada, K., 1997, Epilepsy and exacerbation of brain injury in mice lacking the glutamate transporter GLT-1, *Science* **276:**1699–1702.

Tarbit, I., Perry, E. K., Perry, R. H., Blessed, G., and Tomlinson, B. E., 1980, Hippocampal free amino acids in Alzheimer's disease, *J. Neurochem.* **35:**1246–1249.

Terry, R. D., Peck, A., DeTeresa, R., Schechter, R., and Horoupian, D. S., 1981, Some morphometric aspects of the brain in senile dementia of the Alzheimer type, *Ann. Neurol.* **10:**184–192.

Terry, R. D., Masliah, E., Salmon, D. P., Butters, N., DeTeresa, R., Hill, R., Hansen, L. A., and Katzman, R., 1991, Physical basis of cognitive alterations in Alzheimer's disease: Synapse loss is the major correlate of cognitive impairment, *Ann. Neurol.* **30:**572–580.

Ulas, J., and Cotman, C. W., 1997, NMDAR1 gene expression in Alzheimer's disease, *Neuroscience,* in press.

Ulas, J., Brunner, L. C., Geddes, J. W., Choe, W., and Cotman, C. W., 1992, *N*-methyl-D-aspartate receptor complex in the hippocampus of elderly normal individuals and those with Alzheimer's disease, *Neuroscience* **49:**45–61.

Vickers, J. C., Huntley, G. W., Edwards, A. M., Moran, T., Rogers, S. W., Heinemann, S. F., and Morrison, J. H., 1993, Quantitative localization of AMPA/kainate and kainate glutamate receptor subunit immunoreactivity in neurochemically identified subpopulations of neurons in the prefrontal cortex of the macaque monkey, *J. Neurosci.* **13:**2982–2992.

Vickers, J. C., Huntley, G. W., Hof, P. R., Bederson, J., DeFelipe, J., and Morrison, J. H., 1995, Immunocytochemical localization of non-NMDA ionotropic excitatory amino acid receptor subunits in human neocortex, *Brain Res.* **671:**175–180.

Vijayan, V. K., Geddes, J. W., Anderson, K. J., Chang-Chui, H., Ellis, W. G., and Cotman, C. W., 1991, Astrocyte hypertrophy in the Alzheimer's disease hippocampal formation, *Exp. Neurol.* **112:**72–78.

West, M. J., Coleman, P. D., Flood, D. G., and Troncoso, J. C., 1994, Differences in the pattern of hippocampal neuronal loss in normal ageing and Alzheimer's disease, *Lancet* **344:**769–772.

Westbrook, G. L., 1994, Glutamate receptor update, *Curr. Opin. Neurobiol.* **4:**337–346.

Young, A. B., Greenamyre, J. T., Hollingsworth, Z., Albin, R., D'Amato, C., Shoulson, I., and Penney, J. B., 1988, NMDA receptor losses in putamen from patients with Huntington's disease, *Science* **241:**981–983.

Zeevalk, G. D., and Nicklas, W. J., 1992, Evidence that the loss of the voltage-dependent Mg2+ block at the *N*-methyl-D-aspartate receptor underlies receptor activation during inhibition of neuronal metabolism, *J. Neurochem.* **59:**1211–1220.

# 19

# Nerve Growth Factor Systems in Alzheimer's Disease

E. J. MUFSON and J. H. KORDOWER

## 1. Introduction

In many neurological disorders, specific neuronal populations are selectively vulnerable. This is especially true for patients with Alzheimer's disease (AD). A consistent and early pathological feature of AD is the degeneration of cholinergic basal forebrain neurons (CBF) (Bartus *et al.*, 1982; Whitehouse *et al.*, 1983; Mufson *et al.*, 1989a), a cell population that projects widely to the entire cortical mantle (Mesulam *et al.*, 1983). Although there are numerous neuropathological and neurochemical deficits in the AD brain, cortical cholinergic dysfunction resulting from atrophy and degeneration of cholinergic basal forebrain (CBF) neurons is the pathologic feature that correlates best with the cognitive impairment (Bierer *et al.*, 1995). During the last fifteen years it has become increasingly clear that neurotrophic molecules play a key role in the survival and maintenance of some adult neuronal populations, including the cholinergic neurons of the basal forebrain (see Hefti *et al.*, 1989, for review). The viability and phenotypic expression of CBF

E. J. MUFSON and J. H. KORDOWER • Research Center for Brain Repair, Department of Neurological Sciences, Rush Alzheimer's Disease Center, Rush Presbyterian-St. Luke's Medical Center, Chicago, Illinois 60612.

*Cerebral Cortex,* Volume 14, edited by Peters and Morrison. Kluwer Academic/Plenum Publishers, New York, 1999.

neurons are potently modified by nerve growth factor (NGF), the prototype neurotrophic molecule first characterized by Levi-Montalcini and coworkers (see Levi-Montalcini and Angelleti, 1968; Thoenen and Barde, 1980). Nerve growth factor is just one member of a superfamily of neurotrophic factors that includes brain-derived neurotrophic factor (BDNF), neurotrophin-3 (NT-3), neurotrophin-4 (NT-4), neutotrophin-5 (NT-5), and neurotrophin-6 (NT-6) (Maisonpierre *et al.*, 1990; Bothwell, 1991; Ibanez *et al.*, 1992; Ip *et al.*, 1992; Jing *et al.*, 1992; Barbacid, 1994; Koster *et al.*, 1994). As few as six regions of amino acid non-identity confer specificity for each neurotrophin (Ip *et al.*, 1993). A potential role for NGF as a trophic molecule for CBF neurons was initially inferred from retrograde transport studies by Schwaab and colleagues (Schwaab *et al.*, 1979; Seiler and Schwaab, 1984). In these experiments, retrogradely labeled perikarya were located within the septal/diagonal band complex or the nucleus basalis magnocellularis following injection of $^{125}$I-labeled NGF into cortex and hippocampus, respectively. Since then, a plethora of empirical evidence has confirmed NGF's role in CBF function. Autoradiographic studies revealed that neurons that bind radio-labeled NGF co-distribute with CBF neurons (Richardson *et al.*, 1986). Nerve growth factor is found in highest concentrations within target zones of CBF perikarya, such as the cerebral cortex, hippocampus, and olfactory bulb (Whittemore and Seiger, 1987). Also, NGF increases choline acetyltransferase (ChAT) levels in cultured septal cholinergic perikarya (Hefti *et al.*, 1985; Martinez *et al.*, 1985). Intraventricular injections of NGF in neonatal rodents increases ChAT activity in the basal forebrain, hippocampus, cortex, and striatum (Gnahn *et al.*, 1983; Mobley *et al.*, 1985). Importantly, NGF may also enhance the function of these neurons and rescue them from various brain insults (see Hefti *et al.*, 1989 for review). For example, infusion of NGF, transplantation of NGF-containing mouse salivary gland, or grafting of cells that have been genetically modified to secrete NGF rescue septal cholinergic cells that undergo degeneration following transection of the fimbria-fornix system (e.g., Hefti *et al.*, 1986; Williams *et al.*, 1986; Springer *et al.*, 1987; Emerich *et al.*, 1994; Kordower *et al.*, 1994a; Armstrong *et al.*, 1987). Some of these approaches attenuate cognitive impairments resulting from septohippocampal axotomy (Will and Hefti, 1985; Fisher *et al.*, 1987; Springer *et al.*, 1987). Interestingly, aged rats exhibiting cognitive impairments have degenerative changes of CBF perikarya and alterations in NGF that may be analogous to those observed in animals with experimental damage to the septohippocampal pathway (see Collier and Coleman, 1991; Scott and Crutcher, 1994; Scott *et al.*, 1994, for reviews). In this regard, aged rats with, but not without, memory impairments show exaggerated neuron loss and atrophy of CBF/p75$^{NTR}$ neurons. These animals also display reduced levels of cortical cholinergic markers (Biegon *et al.*, 1986; Fischer *et al.*, 1987; Koh and Loy, 1988; Koh *et al.*, 1989; Markram and Segal, 1990) relative to aged rats with normal cognition. Recently, Cooper and coworkers (1994) demonstrated that retrograde transport of radio-labeled NGF from hippocampus declines in aged rats. This study further revealed that CBF neurons in the aged rats exhibit diminished trkA receptor gene expression, but retain stable p75$^{NTR}$ mRNA expression. These findings suggest an association between age-related changes in the retrograde transport of NGF, expression of its signal-transducing high-affinity trkA receptor, and morphological alterations of CBF neurons.

## 2. Nerve Growth Factor Receptor Subtypes

Over the last few years, evidence has accumulated demonstrating that NGF recognizes at least two classes of cell-surface receptors (Meakin and Shooter, 1991): a fast-dissociating, low-affinity NGF receptor and a slow-dissociating, high-affinity NGF receptor (Fig. 1) (see also Peterson and Gage, this volume). The low-affinity receptor is a 75-kDa transmembrane glycoprotein (see Bothwell, 1991, for review). Because this receptor binds to virtually all members of the NGF family (Rodriguez-Tebar *et al.*, 1990; Bothwell, 1991; Lindsay *et al.*, 1994), it has been termed the "p75 neurotrophin receptor" ($p75^{NTR}$) (Chao and Hempstead, 1995; Chao, 1995). In the classic retrograde model of NGF–$p75^{NTR}$ interactions, it was postulated that this complex is internalized and retrogradely transported to CBF consumer neurons (Fig. 2), where signal transduction occurs. Although observations employing modern molecular biologic procedures indicate that the physiological actions of NGF require complexing to its high-affinity receptor (Fig. 2), several important functions for the low-affinity $p75^{NTR}$ have been suggested, including presentation or recruitment of NGF to the trkA receptor (Jing *et al.*, 1992), signaling through a G-protein-mediated mechanism (Feinstein and Larhammar, 1990), and a relationship with molecules capable of signaling or providing substrates for the high-affinity receptor, trkA (Parada *et al.*, 1992).

The trkA receptor is a 140-kDa transmembrane glycoprotein with a cytoplasmic protein kinase domain (Hempstead *et al.*, 1991; see Bothwell, 1991; Lindsay *et al.*, 1994, for reviews). This receptor is essential for NGF signal transduction and mediates its biological effects (Chao, 1992). It was at first controversial whether coexpression of both the low-affinity $p75^{NTR}$ and the high-affinity trkA receptor were necessary for the biological activity of NGF (see Jing *et al.*, 1992; Parada *et al.*, 1992). It is now clear that the trkA receptor alone is sufficient to mediate cellular responses to NGF (Weskamp and Reichardt, 1991; Ibanez *et al.*, 1992). Like most kinase growth factor receptors, trkA signals through receptor oligomerization (Heldin, 1995). Activation of the tyrosine kinase activity of trkA via NGF binding leads to receptor trans- and autotyrosine phosphatidylation (PY) (Fig. 1) and to PY of second messengers, including phosphatidylinositol-3 kinase (PI-3; Soltoff *et al.*, 1992). The PI-3 kinase messenger is involved in protein trafficking and endocytosis of ligand-receptor complexes (see Kaplan and Stephens, 1994). Because microinjection of NGF into cells does not induce a biological response (Heumann *et al.*, 1984), it is assumed that binding to receptors located on the cell surface mediates these effects.

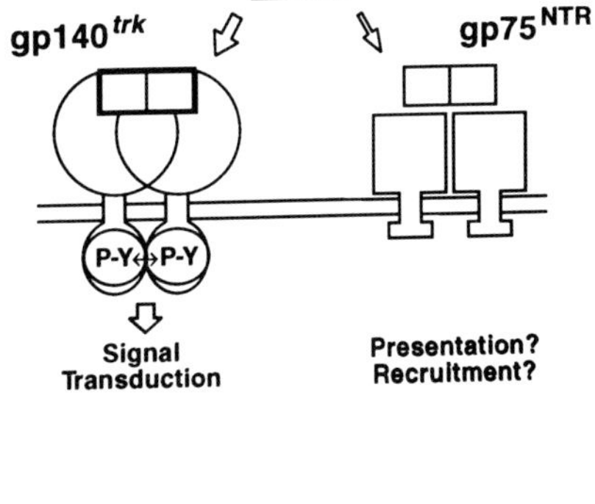


**Figure 1.** Diagrammatic representation of the proposed molecular structure of trkA and $p75^{NTR}$. NGF is represented as a dimeric molecule. P-Y indicates tyrosine-phosphorylated residues in the catalytic kinase domain (square boxes) of the $gp^{140trk}$ receptor. Adapted from Jing *et al.* (1992).

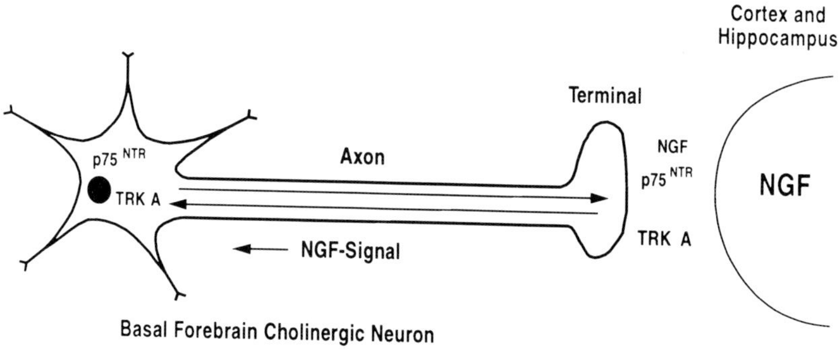


**Figure 2.** Diagrammatic representation of the proposed interaction between a cholinergic basal forebrain (CBF) neuron and its NGF-producing target site. In the classic retrograde transport model, $p75^{NTR}$ is produced within a CBF neuron and is anterogradely transported to its target site, where it binds with NGF and is then retrogradely transported to CBF consumer neurons. This model must now include the trkA receptor, which is also produced by CBF neurons and anterogradely transported to its target site, where it interacts with NGF and possibly with $p75^{NTR}$. For clarity, this relationship is shown outside of the neurons' target site. A receptor trophin complex is then retrogradely transported to CBF neurons. Accumulating evidence indicates trkA receptor participation in the signaling processes initiated by NGF.

## 3. Nerve Growth Factor in Neurological Disease

Almost twenty years have passed since it was first hypothesized that degenerative events in human neurologic disorders result from reductions in trophic support (Appel, 1981). This hypothesis was particularly favored in AD, where reductions in cortical NGF were postulated to underlie the massive degeneration of CBF neurons (Hefti and Weiner, 1983; Hefti *et al.*, 1989). The initial observations that cortical NGF mRNA was unchanged in AD (Goedert *et al.*, 1986) tempered enthusiasm for this hypothesis. However, the current explosion of information concerning the relative importance of the low-affinity $p75^{NTR}$ and high-affinity trkA receptors in NGF signal transduction has reinvigorated interest in the role NGF may play in CBF neuronal degeneration in AD. In this chapter we review the data investigating the expression of NGF and its low- and high-affinity receptors in AD relative to normal aged subjects. The findings gained from these studies have led us to postulate that an alteration in the production and utilization of the trkA receptor is a major pathological event underlying CBF degeneration in AD.

## 4. Nerve Growth Factor Protein Levels in Normal Aged Human Brain

To better understand disease-related changes in NGF, it is imperative to develop knowledge of its levels in normal aged human brain. Together with our coworkers at the University of Cincinnati and the University of Alabama, we determined NGF expression within cortical production sites using a specific bioassay for this neurotrophin (Crutcher *et al.*, 1993; Scott *et al.*, 1995). As seen in Fig. 3, aged controls displayed high levels of NGF-like immunoreactivity within the hippocampal formation (1100 pg/gm tissue), amygdala (690 pg/gm), temporal cortex (450 pg/gm), frontal cortex (330 pg/gm), parietal cortex (330 pg/gm), occipital cortex (270 pg/gm), putamen (230 pg/gm), and cerebellum (60 pg/gm). Based upon normal aged control samples and AD hippocampal and cortical samples, we found that postmortem delay did not affect this assay, and samples from the hippocampus revealed no gender difference in NGF levels in either aged control or AD individuals. Levels of NGF were high in the caudate nucleus, medial thalamus, and choroid plexus as well. These findings are similar to the distribution of NGF-like immunoreactivity found in rat and monkey brain (see Scott and Crutcher, 1994, for review). Interestingly, NGF immunoreactivity was also high (1100 pg/gm tissue) in the nucleus basalis of Meynert. Because there is little NGF mRNA within the nucleus basalis, this likely reflects large amounts of NGF which have been retrogradely transported from the target sites.

## 5. Nerve Growth Factor-Like Immunoreactivity in Hippocampus

It has recently been possible to assess the distribution of NGF within its hippocampal production sites in tissue from adult monkeys and normal aged humans

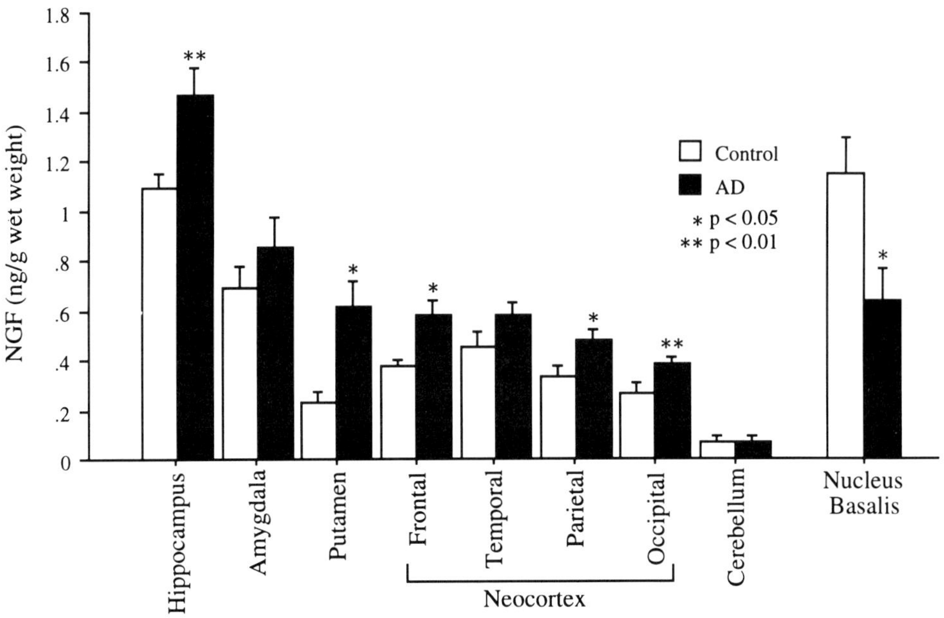


**Figure 3.** NGF levels in AD across various regions of the human brain. Increases in NGF-like immunoreactivity relative to the control group are pronounced in neo- and limbic cortices, areas traditionally known to be affected in AD. Note the 50% decline in NGF within the nucleus basalis. From Scott *et al.*, "Nerve Growth Factor in Alzheimer's disease: Increased levels throughout the brain coupled with declines in nucleus basalis," *The Journal of Neuroscience* **15(9)**:6215 (1995), with permission.

(Mufson *et al.*, 1994) using a polyclonal antibody raised against NGF (Conner *et al.*, 1992; Conner and Varon, 1992). Immunohistochemistry revealed a diffuse staining pattern within the neuropil of the hippocampus and no evidence of immunoreactive cell bodies or fibers (Fig. 4). Nerve growth factor immunoreactivity was the most intense within the hilus (CA4) of the dentate gyrus and extended to the CA3 and CA2 regions of the hippocampus (Fig. 4). Within CA2 the pattern of staining was trilaminar, with the densest labeling in stratum radiatum, intermediate labeling in stratum oriens, and the lightest labeling in stratum pyramidal. At no level of the hippocampus was NGF-like immunoreactivity seen within CA1 or adjacent subicular and entorhinal cortices. Although we did not observe NGF-immunoreactive perikarya, Conner and Varon (1992) have shown NGF-reactive neurons within the rat cortex, hippocampus, olfactory bulb, amygdala, striatum, thalamus, hypothalamus, and brainstem, as well as within non-cholinergic neurons of the basal forebrain in rats pretreated with the axoplasmic transport blocker colchicine. Thus, the inability to visualize NGF-immunoreactive somata in these regions is likely due to our inability to employ colchicine in primates to block axoplasmic flow.

## 6. Target-Derived Nerve Growth Factor within Aged Cholinergic Basal Forebrain Neurons

We also employed the polyclonal NGF antibody (Conner and Varon, 1992) to immunohistochemically visualize retrogradely transported NGF within basal forebrain cholinergic neurons (Mufson *et al.*, 1994, 1995). A continuum of NGF-immunoreactive neurons was seen within the basal forebrain subfields, including the medial septum (Ch1), vertical limb of the diagonal band (Ch2; Fig. 5), and the nucleus basalis (Ch4; Fig. 6). The NGF reaction product most often appeared as a granular or punctate particulate, localized primarily within the cell somata and within primary dendritic processes (Fig. 6). The intensity of staining ranged from a few granules to reactions so intense that they appeared to fill the entire cell body (Fig. 6A, B). We also found a few scattered NGF-immunopositive fibers displaying beaded-varicosities that coursed through the subfields of the basal forebrain (Fig. 6). Dual immunohistochemical staining revealed that the NGF reaction product was exclusively found within $p75^{NTR}$-immunoreactive basal forebrain neurons (Mufson *et al.*, 1994). These findings support the classic model of NGF retrograde transport from its cortical production sites to CBF consumer neurons. However, sophisticated molecular biological assays have recently provided new evidence that in addition to target-derived NGF signaling of the CBF, these neurons may also

**Figure 4.** Photomicrographs of NGF immunoreactivity within monkey hippocampus. (A) Hippocampus immunoreacted with anti-NGF antibodies preadsorbed to purified β-NGF results in a complete absence of hippocampal staining. (B) NGF immunoreactivity within the neuropil of the hilar (CA4) and CA3–CA2 hippocampal subfields. (C) High-power photomicrograph of CA2 (arrows in B). Note the trilaminar appearance of the reaction product, which is heavy in stratum radiatum (RA), moderate in stratum oriens (OR), and light in the stratum pyramidal (PY). Bars - 200 μm (A, B); 40 μ (C). From Mufson *et al.* (1994) with permission.

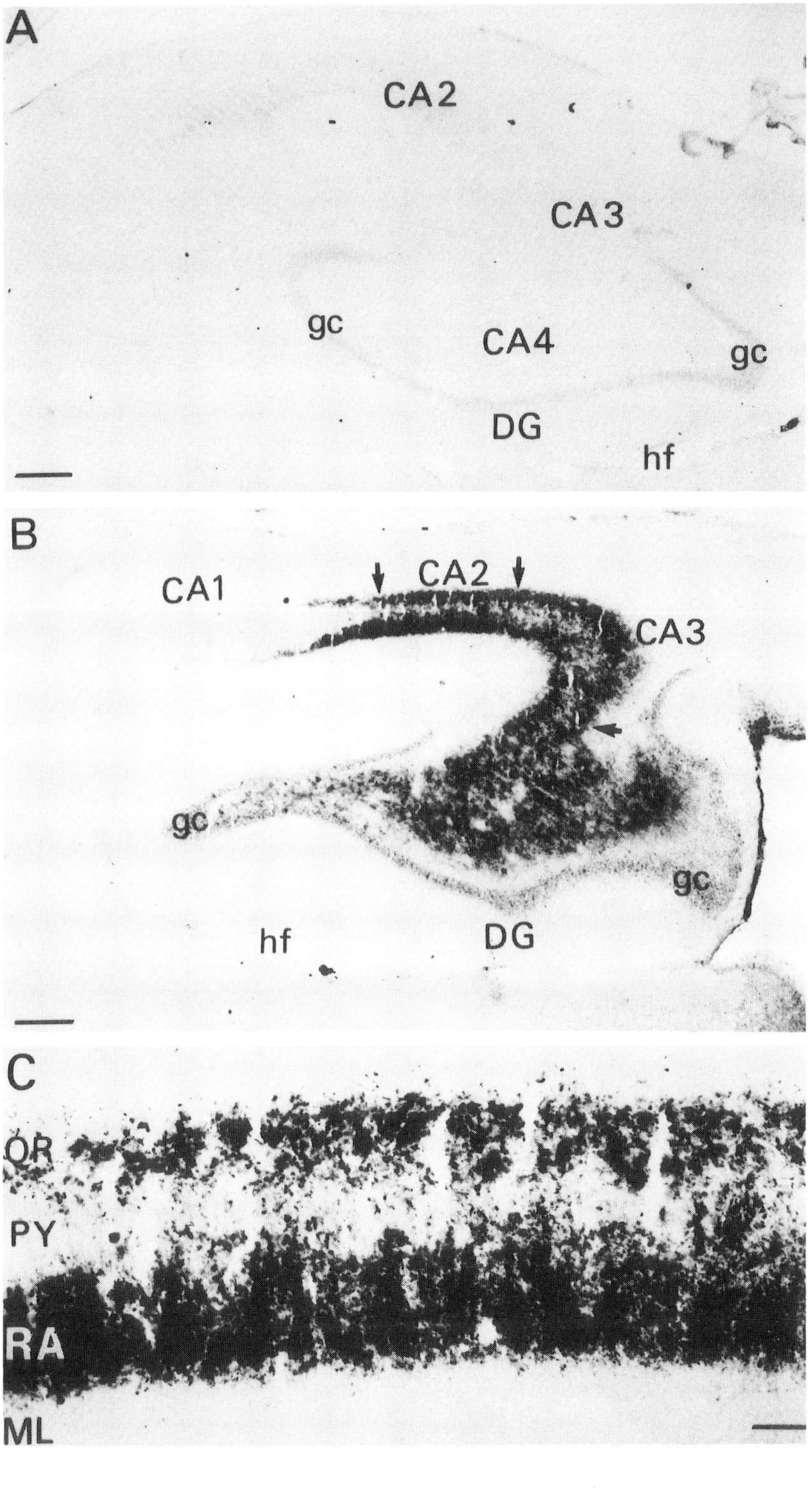
A
CA2
CA3
gc
CA4
gc
DG
hf
B
CA1
CA2
CA3
gc
gc
hf
DG
C
OR
PY
RA
ML

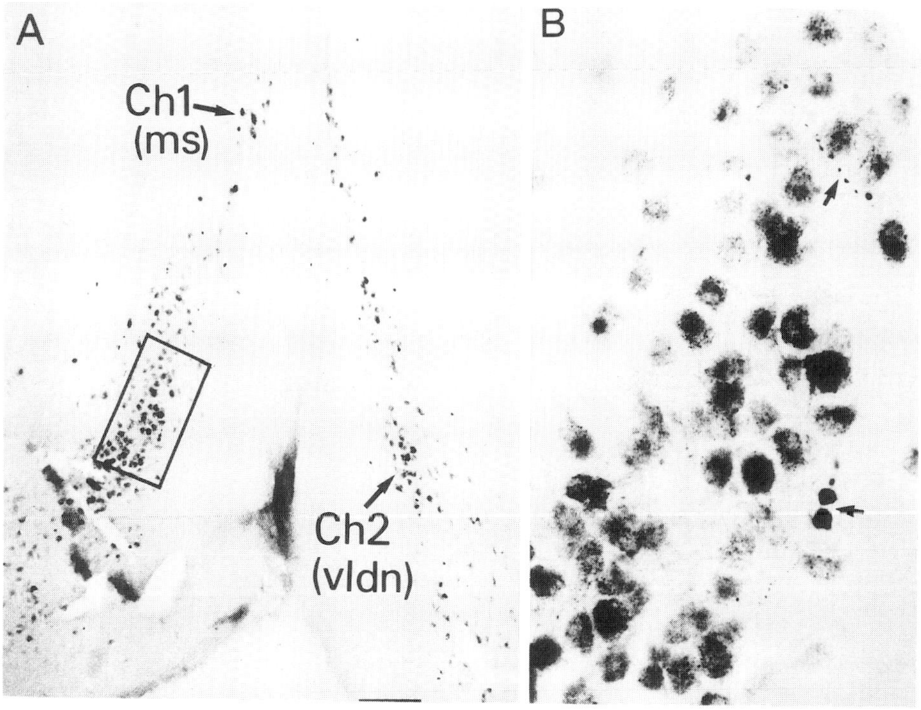


**Figure 5.** (A) Low-power photomicrograph of NGF-like-immunoreactive (ir) neurons within the medial septum (Ch1) and vertical limb nucleus of the diagonal band (Ch2) in monkey. (B) Higher magnification of NGF-ir neurons in the boxed area of A. Arrows indicate the location of NGF-like-ir fibers within Ch2. Bars = 100 μm (A); 20 μm (B).

be modulated by NGF produced locally in these perikarya (Saporito and Carswell, 1995), suggesting an autoregulatory function for this neurotrophin.

## 7. Nerve Growth Factor Receptors and Cholinergic Basal Forebrain Neurons

Immunohistochemical and *in situ* hybridization studies in rats (Yan and Johnson, 1988; Batchelor *et al.,* 1989; Steninger *et al.,* 1993; Gibbs and Pfaff, 1994; Sobreviela *et al.,* 1995; Holtzman *et al.,* 1995), monkeys (Kordower *et al.,* 1988), and humans (Hefti and Mash, 1989; Kordower *et al.,* 1989; Mufson *et al.,* 1996, 1989a, 1997; Higgins and Mufson, 1989) have demonstrated that neurons within the CBF contain the p75$^{NTR}$ (Figs. 7–9) and the trkA receptor (Figs. 10, 11). It is well established that these CBF neurons provide the major cholinergic innervation to the cortex and hippocampus (Fig. 12) (Mesulam *et al.,* 1983). Co-distribution and colocalization studies (Kordower *et al.,* 1988, 1989, 1994b; Mufson *et al.,* 1989a, 1996; Sobreviela *et al.,* 1995) have shown a high degree of overlap between CBF neurons expressing p75$^{NTR}$, the specific cholinergic marker, choline acetyltransferase (ChAT), and the acetylcholine-degrading enzyme, acetylcholinesterase

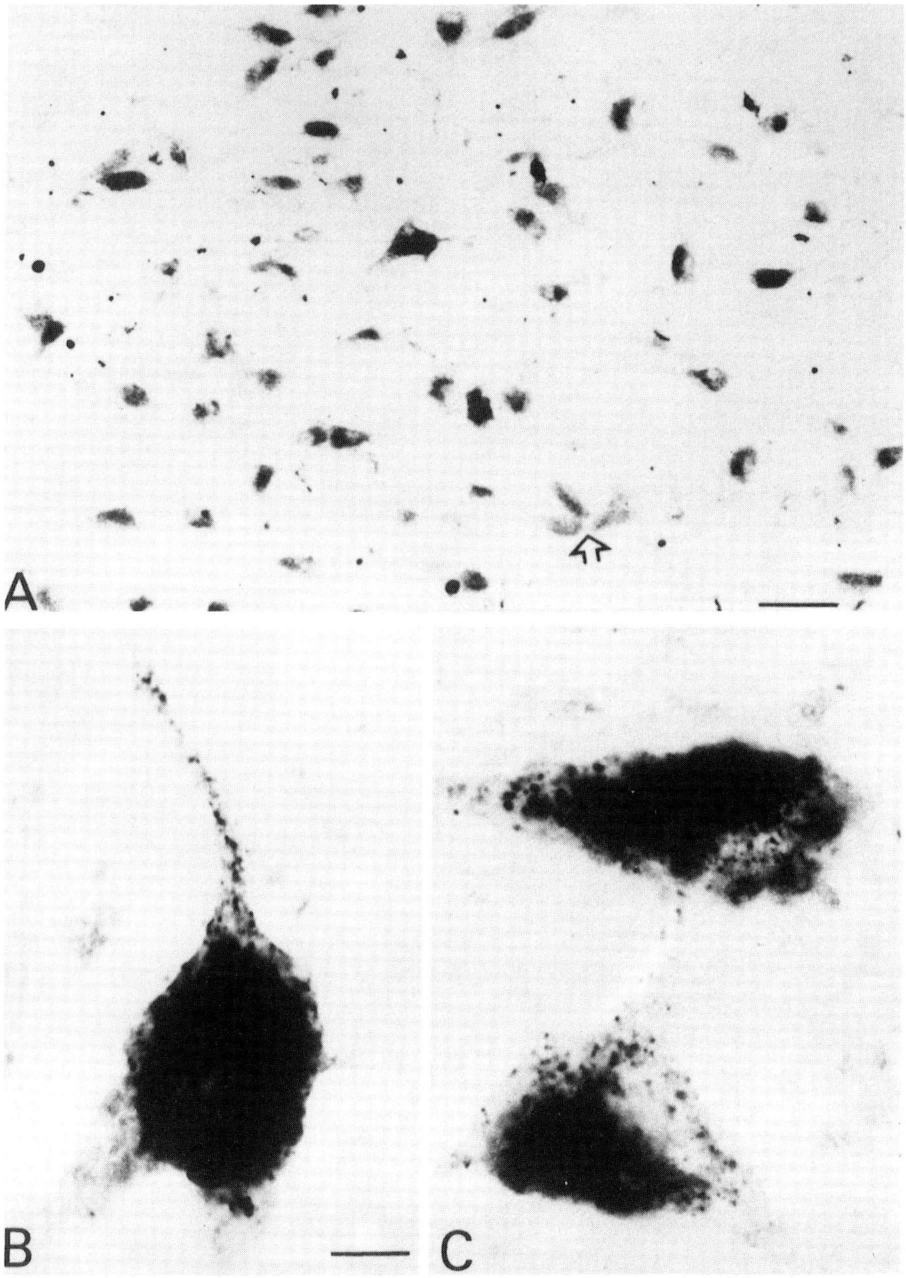


**Figure 6.** NGF immunostaining within the anterolateral division of the human nucleus basalis (Ch4al). (A) Numerous NGF-containing neurons exhibiting different levels of immunoreactivity. Open arrow indicates a cluster of lightly immunopositive neurons. (B, C) Higher magnification micrographs of NGF-immunostained nucleus basalis neurons. Bars = 100 μm (A); 50 μm (B, C). From Mufson *et al.* (1994) with permission.

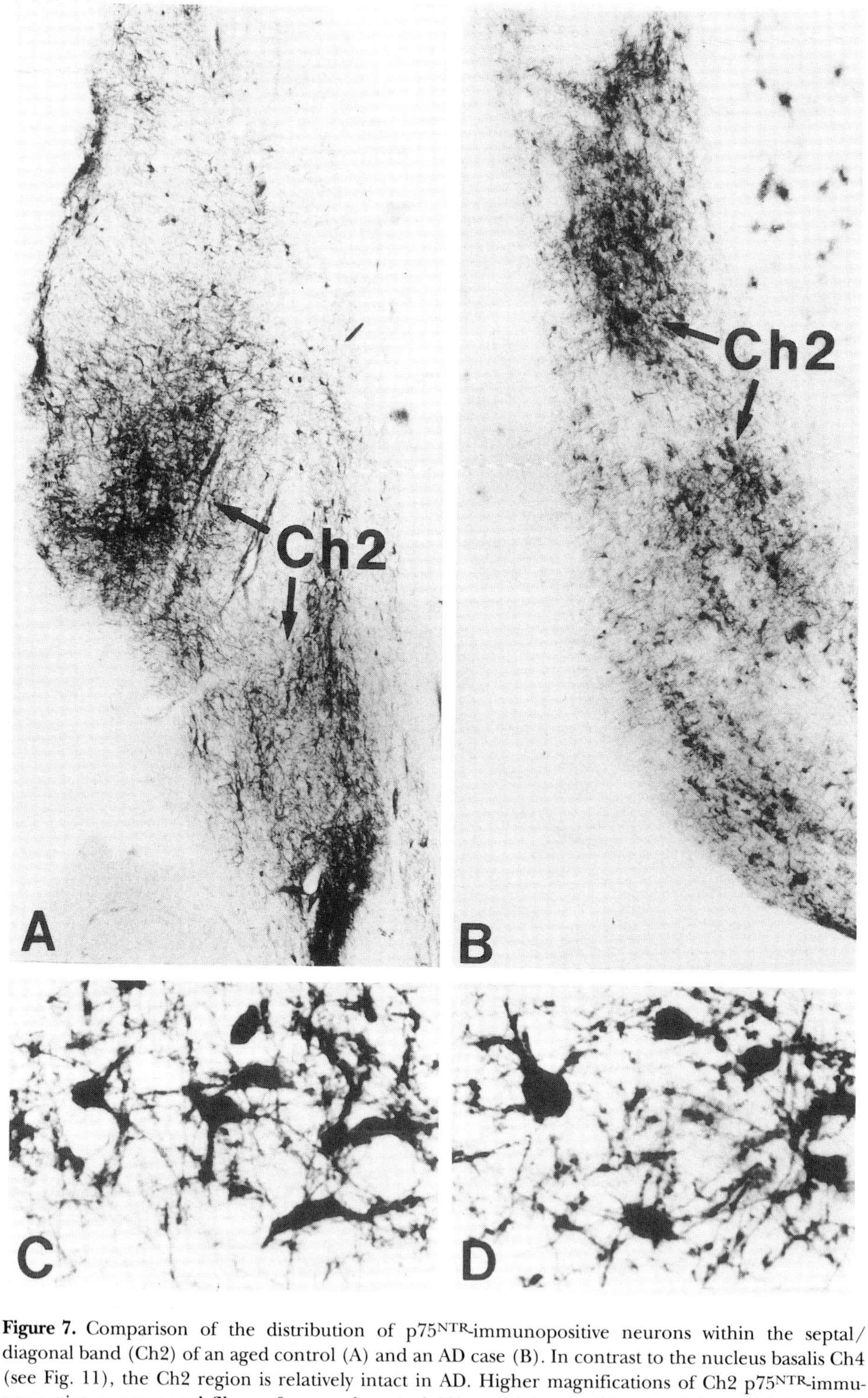


**Figure 7.** Comparison of the distribution of p75$^{NTR}$-immunopositive neurons within the septal/diagonal band (Ch2) of an aged control (A) and an AD case (B). In contrast to the nucleus basalis Ch4 (see Fig. 11), the Ch2 region is relatively intact in AD. Higher magnifications of Ch2 p75$^{NTR}$-immunoreactive neurons and fibers of an aged control (C) and AD case (D), × 180. From Mufson *et al.* (1989b) with permission.

(AChE) (Fig. 13). Furthermore, these studies indicate that the p75$^{NTR}$ antibody is an excellent marker for CBF neurons in nonhuman primate (Kordower *et al.*, 1988) and human (Mufson *et al.*, 1989a) brain. The following is a brief overview detailing the positive profile distribution of p75$^{NTR}$ and trkA receptors in the primate basal forebrain.

## 8. Low-Affinity p75$^{NTR}$ and High-Affinity trkA Receptor within the Basal Forebrain

In both monkeys and humans, a continuum of neurons located throughout the basal forebrain express p75$^{NTR}$, trkA and NGF immunoreactivity. This region extends from the medial septum rostrally to the caudal aspect of the globus pallidus, a distance in humans of approximately 1.4 cm (Mufson *et al.*, 1989a, 1994, 1996) (Figs. 5–11). In primates (< 10%) p75$^{NTR}$-positive neurons are seen within the medial septal (Ch1) region (Fig. 7). These neurons are relatively small (12–20 μm by 20–40 μm) and mainly ovoid. Ventrally, p75$^{NTR}$- and trkA-containing neurons extend into the vertical limb nucleus of the diagonal band (Ch2) (Fig. 7C). Combined p75$^{NTR}$-Nissl preparations reveal that 80–90% of these neurons are p75$^{NTR}$-immunopositive. An occasional p75$^{NTR}$- and trkA-containing fusiform neuron is seen along the ventral surface of the brain, probably corresponding to the horizontal limb of the diagonal band (Ch3) (Mesulam *et al.*, 1983).

Caudally, p75$^{NTR}$-immunoreactive neurons are located in the bed nucleus of the stria terminalis, beneath the globus pallidus, and embedded within the fibers of the internal capsule (Fig. 8A). At this level, a large oval mass of receptor-positive neurons located ventral to the anterior commissure can be seen, which corresponds to the anteromedial subfield of the nucleus basalis (Ch4am) (Fig. 8A). The predominant cell type consists of large (24 by 80 μm) multipolar neurons with proximal dendrites that can be followed for up to 320 μm from the perikarya in a single plane. Smaller oval and fusiform receptor-containing neurons are also intermixed in this area. Nissl-counterstained sections demonstrate that virtually all (> 95%) of Ch4 neurons express p75$^{NTR}$. At the level where the anterior commissure emerges from the temporal lobe, the anteromedial subdivision aggregates with a group of p75$^{NTR}$- and trkA-containing neurons that corresponds to Ch4al of the nucleus basalis (Fig. 8B). The Ch4am and Ch4al subfields are demarcated by a vascular structure or a central rarification in density of p75$^{NTR}$- or ChAT-containing profiles (Fig. 13). At this and other more caudal levels, p75$^{NTR}$- and trkA-containing neurons of similar morphology to those seen in the nucleus basalis extend into

→

**Figure 8.** p75$^{NTR}$ immunocytochemistry in normal aged brain showing the location within the nucleus basalis of positive neurons corresponding to Ch4am, Ch4al, and Ch3; medial to the left, dorsal toward the top. (A) p75$^{NTR}$ staining in the rostral aspect of Ch4am. Curved arrows indicate p75$^{NTR}$-containing neurons embedded within fibers coursing through the olfactory tubercle. Single black arrow points to p75$^{NTR}$-positive fibers coursing within the external capsule; open arrow indicates a cluster of p75$^{NTR}$-containing subputaminal neurons; double black arrow shows p75$^{NTR}$-immunoreactive neurons in the bed nucleus of the stria terminalis, × 10. (B) p75$^{NTR}$ immunoreactivity within the nucleus basalis corresponding to Ch4am, Ch4al, and Ch3. Open arrows indicate subputaminal p75$^{NTR}$-positive neurons, single arrow and arrowhead point to p75$^{NTR}$-immunoreactive neurons adjacent to the anterior commissure, and double arrow indicates interstitial p75$^{NTR}$-containing neurons within the internal medullary lamina, × 10. From Mufson *et al.* (1989a) with permission.

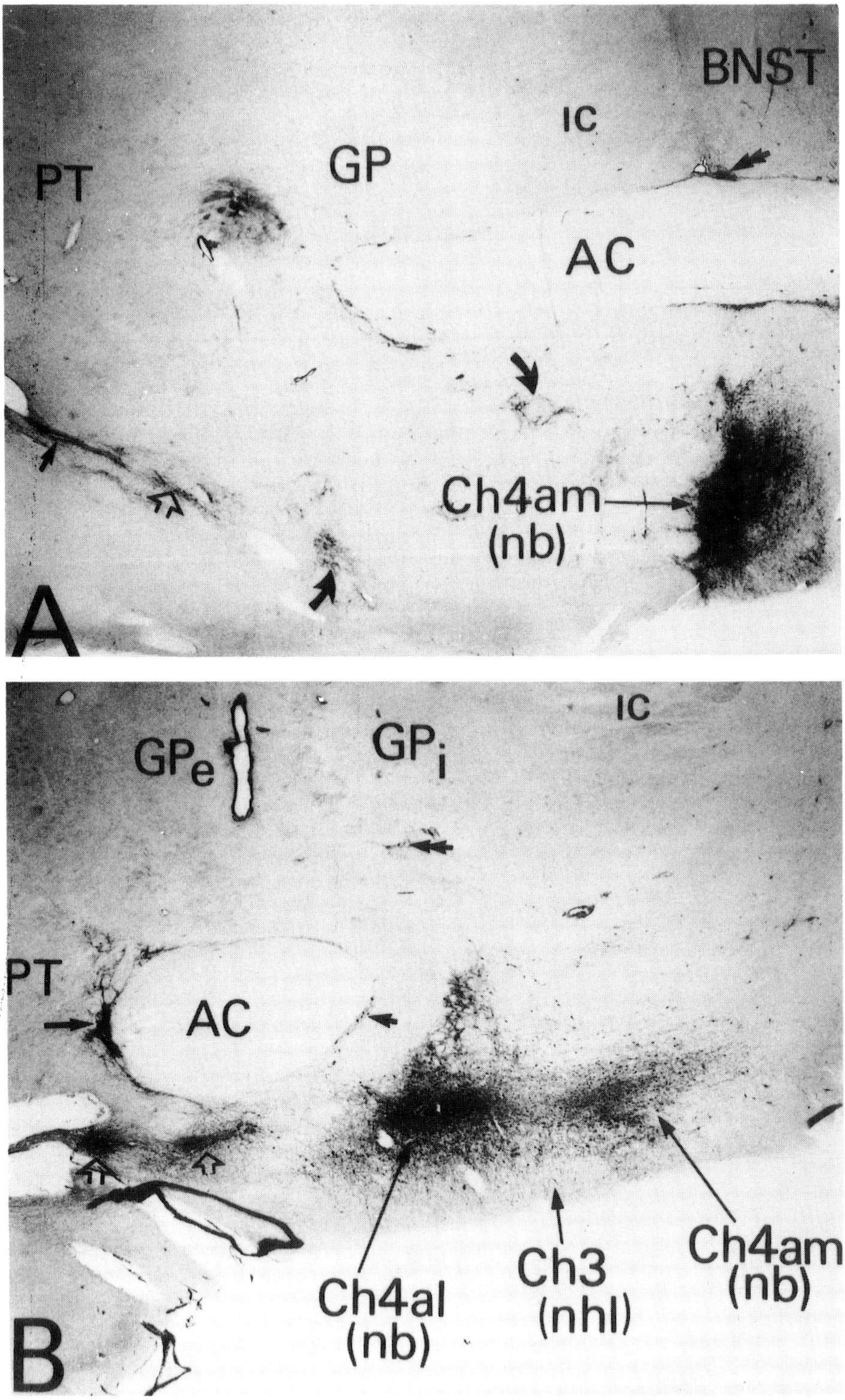
BNST
IC
GP
PT
AC
Ch4am
(nb)
A
IC
GPe
GPi
PT
AC
Ch4al
(nb)
Ch3
(nhl)
Ch4am
(nb)
B

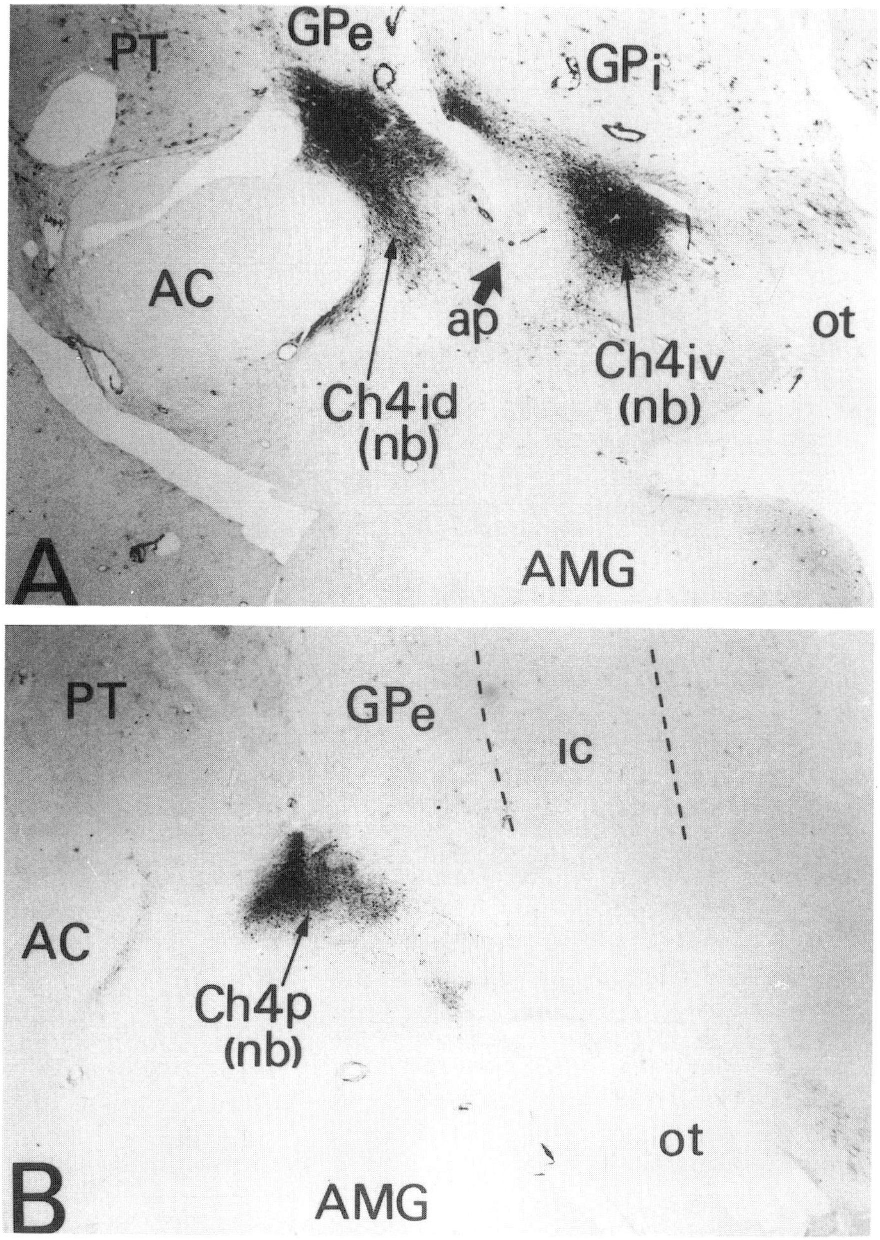


**Figure 9.** NGF receptor immunostaining in the brain of an aged normal individual showing the intermediate (Ch4i) and posterior subfields (Chp) of the nucleus basalis. (A) $p75^{NTR}$ immunoreactivity within the intermediodorsal (Ch4id) and (B) intermedioventral (Ch4iv) subsectors of the nucleus basalis. Dashed lines indicate position of internal capsule, ×10. From Mufson *et al.* (1989a) with permission.

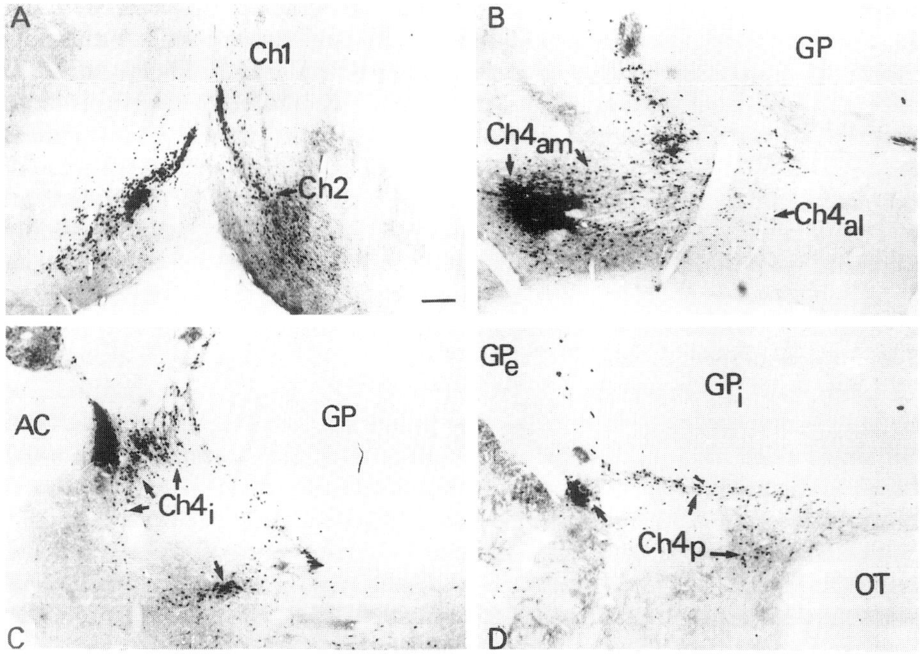


**Figure 10.** Low-power photomicrographs showing the distribution of trk-immunoreactive neurons in the subfields of monkey basal forebrain (A) Ch1/Ch2 (septal/diagonal band) complex. (B) Anterior (Ch4am-al) nucleus basalis. (C) Intermediate (Ch4i) subdivision of nucleus basalis. (D) Posterior (Ch4p) subdivision of nucleus basalis. AC, anterior commissure; GP, globus pallidus; GPe, external division of globus pallidus; GPi, internal division of globus pallidus; OT, optic tract. Bars = 1000 $\mu$m. From Kordower *et al.* (1994) with permission.

the medullary laminae of the globus pallidus (Fig. 9). These positive neurons, as well as those embedded within the anterior commissure and the internal capsule, correspond to the interstitial components of Ch4 (Mesulam *et al.*, 1983; Mufson *et al.*, 1989a). At the level of the mammillary bodies, the remaining $p75^{NTR}$ and trkA receptor-containing neurons are located ventral to the globus pallidus and lateral to the temporal limb of the anterior commissure, forming an intermediate division. This subdivision is defined by the ansa peduncularis, which divides it into dorsal (Ch4id) and ventral (Ch4iv) subfields (Figs. 9A, 10, 11). Posterior to the ansa peduncularis at the level of the diencephalon-mesencephalon interface, the remaining $p75^{NTR}$ and trkA receptor-containing neurons from a compact posterior (Ch4p) subgroup nested below the putamen and globus pallidus, and above the lateral half of the optic tract (Figs. 9–11). These cells continue through to the level of the lateral geniculate nucleus.

## 9. Colocalization of $p75^{NTR}$ and trk within the Primate Basal Forebrain

We used a double chromagen immunocytochemical method to determine the extent to which trk-containing neurons in the primate CBF colocalized with the p75NTR. In these studies tissue from adult and aged monkeys was stained with an antibody that reacts with all trks, but primarily recognizes the trkA receptor (Steininger *et al.*, 1993; Kordower *et al.*, 1994b; Chen *et al.*, 1996). All monkeys displayed an extensive, but incomplete, colocalization between trk- and p75NTR-immunoreactive neurons within the basal forebrain (Fig. 14). In addition to dual-labeled perikarya, many basal forebrain neurons only expressed trk immunoreactivity (Fig. 14). Conversely, a few basal forebrain neurons expressed only p75NTR (Fig. 14). Colocalization experiments revealed that 68–73% of all basal forebrain neurons (septal/diagonal band complex and nucleus basalis) colocalized both trk and p75NTR. In contrast, 23–28% of basal forebrain neurons expressed only the trk receptor, whereas approximately 4% of the basal forebrain neurons expressed only the p75NTR. Although there was some individual variability, a similar degree of trk/p75NTR colocalization was observed in all subdivisions of the basal forebrain. Separate quantitation of double-labeled neurons was performed for hippocampal projecting septal/diagonal band (Ch1/Ch2) neurons and the cortical and amygdaloid projecting neurons of the nucleus basalis (Ch4) subfields. Between 42 and 75% of the neurons within the septal/diagonal band complex colocalized the trk and p75NTR, whereas 66–76% of neurons within the nucleus basalis subfield also displayed both trk and p75NTR immunoreactivity. This nonhuman primate study

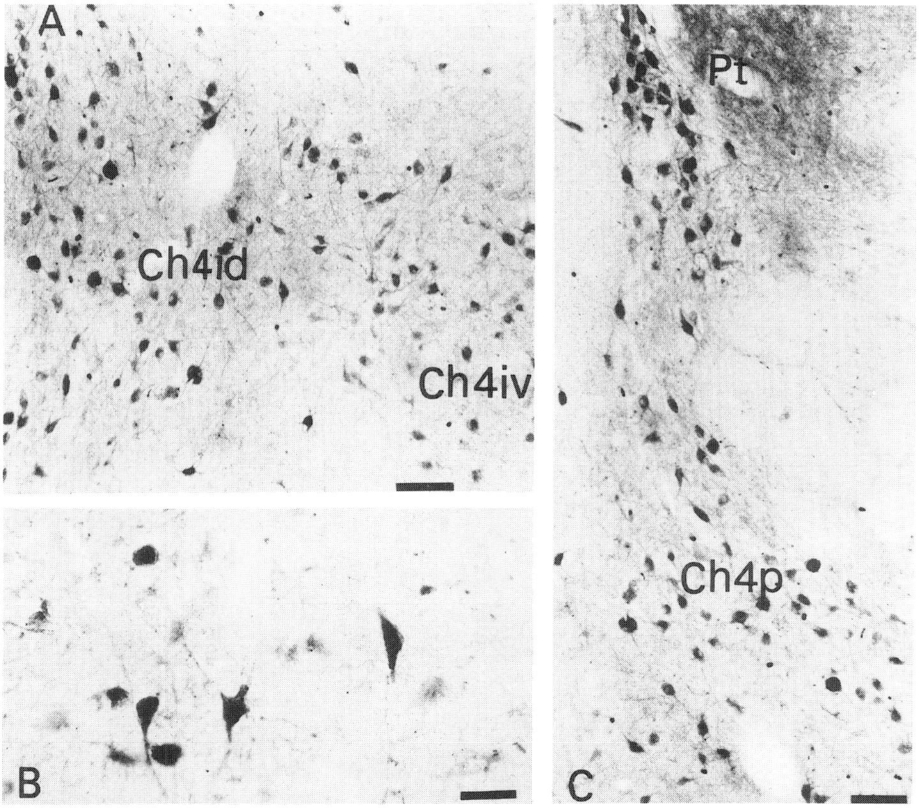


**Figure 11.** Photomicrographs of trkA-immunoreactive neurons within (A) intermediate dorsal (Ch4id) and ventral (Ch4iv) and (B) posterior (Ch4p) portions of the nucleus basalis, as well as the medial septal nucleus (C) in an aged control. Pt, putamen. Bars = 50 μm. From Mufson *et al.* (1997) with permission.

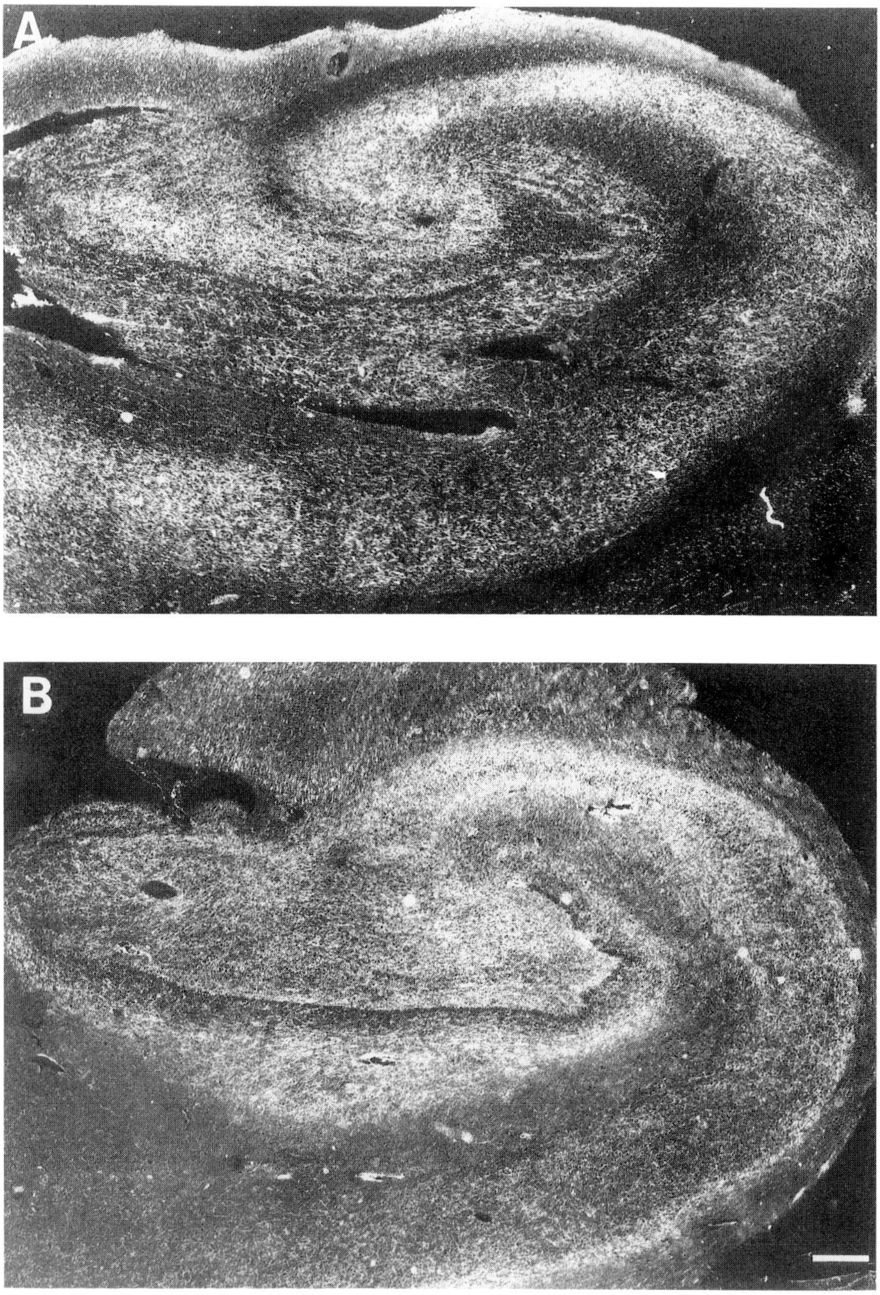


**Figure 12.** Cholinergic and p75$^{NTR}$ innervation of the human hippocampal formation demonstrated by immunohistochemistry. (A) Photomicrograph illustrating the general appearance of ChAT staining in a 66 year old. (B) Photomicograph illustrating distribution of p75$^{NTR}$-immunoreactive fibers within the brain of a 47 year old. Note the close similarity of the innervation patterns, consistent with the finding of a colocalization of ChAT and p75$^{NTR}$ within basal forebrain cholinergic neurons. Bar = 0.1 cm. From deLacalle *et al.* (1994) with permission.

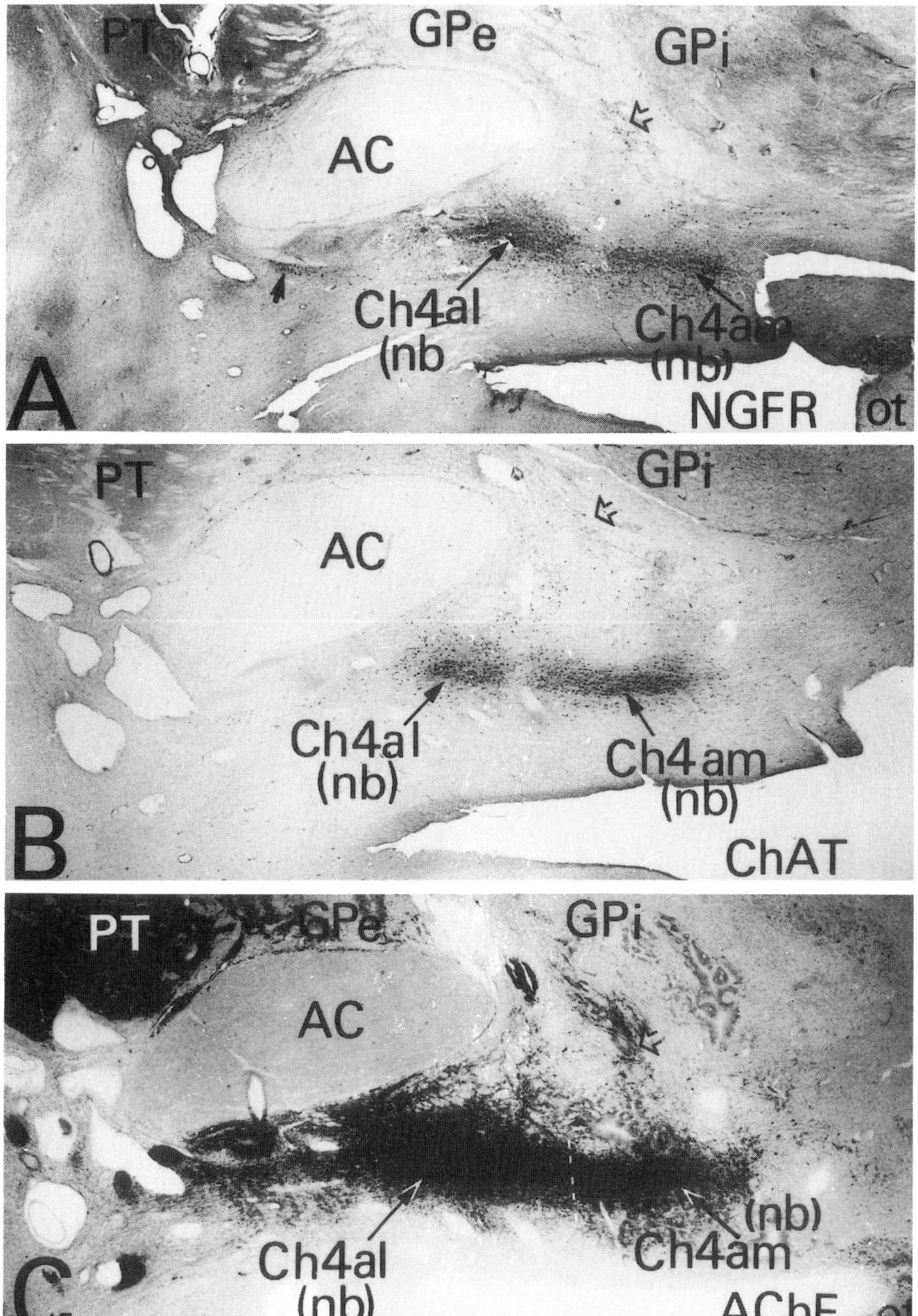


**Figure 13.** Adjacent sections histochemically reacted for $p75^{NTR}$ (A), ChAT (B), and AChE (C) at the level of emergence of the anterior commissure showing the concordance of all three markers within the anteromedial and anterolateral subsectors of the human nucleus basalis (Ch4) from a 77-year-old normal female patient. The dotted line in C indicates the approximate border between Ch4am and Ch4al, which is obscured by the intense AChE neuropil staining, ×10. From Mufson *et al.* (1989b) with permission.

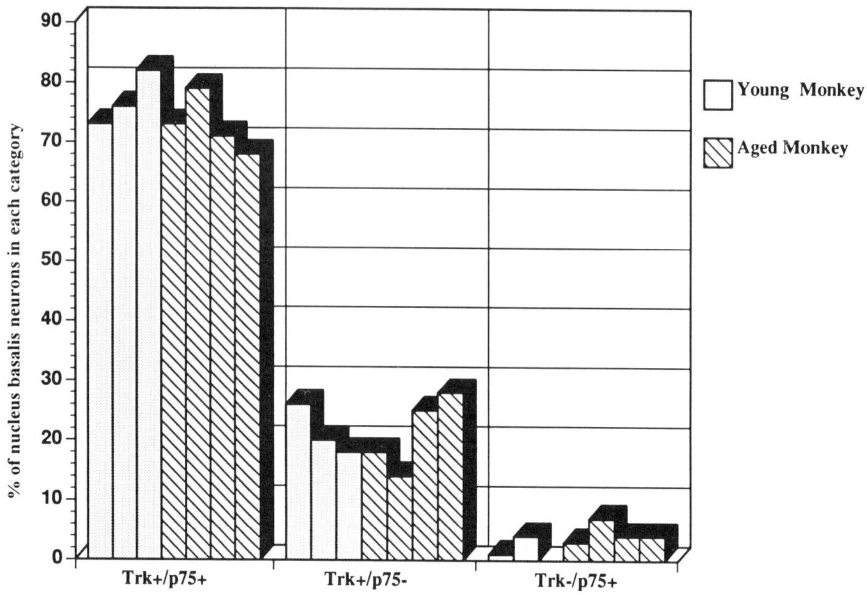


**Figure 14.** Histogram showing the percentage of nucleus basalis neurons colocalizing trk and p75$^{NTR}$, or trk and p75$^{NTR}$ alone, in young and aged monkey. Note the lack of age-related change in the number of neurons containing each marker (see text for details).

also revealed that there were no age-related changes in the degree of colocalization (Fig. 14).

## 10. Alterations in the Nerve Growth Factor System in Alzheimer's Disease

### 10.1. Cytoskeletal Abnormalities within p75$^{NTR}$-Containing Cholinergic Basal Forebrain Neurons

In AD the CBF neurons display extensive neurofibrillary degeneration. To determine the extent to which p75$^{NTR}$-containing CBF neurons exhibit cytoskeletal lesions, we stained basal forebrain sections from AD and aged controls with thioflavin-S or an antibody against paired helical filament (PHF). A few senile plaques (SPs) were seen scattered in the AD basal forebrain (Fig. 15B). Of particular interest is the observation derived from our fluorescence microscopic analysis, which revealed a large number of p75$^{NTR}$-immunoreactive neurons invested with differing degrees of neurofibrillary material (Fig. 15C). Some neurons were thioflavin-S-negative or only minimally fluorescent for this amyloid marker. Others were more heavily invested, suggesting a greater involvement in the disease process (Fig. 15D, E). In addition, many neurofibrillary tangle-bearing profiles observed in the basal forebrain were ghosts; that is, neurons appeared to be largely devoid of cytoplasm and organelles consisting of neurofibrillary remnants (Fig. 15D). This pattern of amyloid expression is suggestive of a continuum of degenerative events.

The neurofibrillary tangles observed within the CBF neurons were globose and fluoresced as a dense core mass (Fig. 15C), in contrast to the more wispy appearance of the neurofibrillary tangles observed within the cortex (Fig. 15A). These observations potentially represent differences in the pattern of amyloid aggregation between the two cell types.

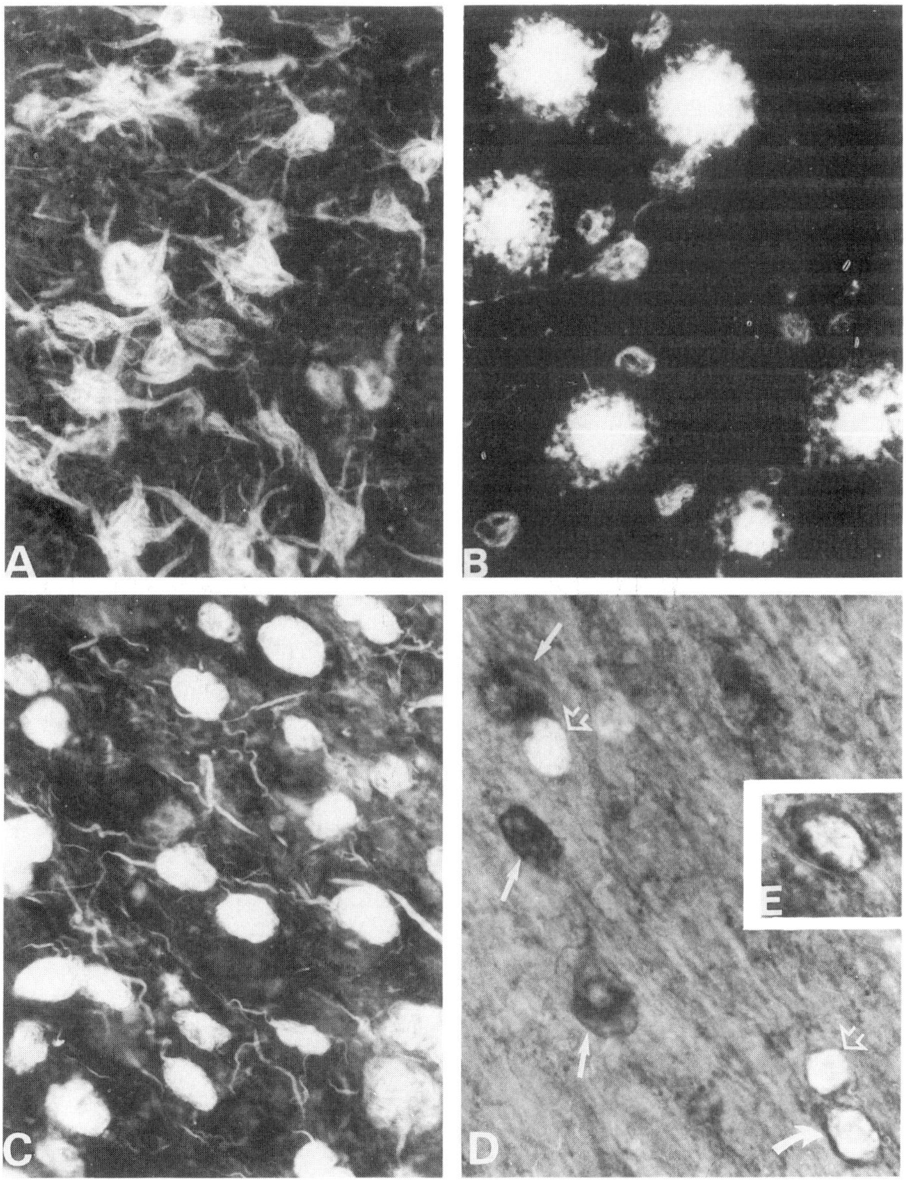


**Figure 15.** Thioflavin-S stained sections showing neurofibrillary tangles (NFTs) and senile plaques (SPs) in AD. (A) NFTs within layer II of entorhinal cortex. (B) SPs within superior temporal cortex. (C) NFTs within the nucleus basalis (Ch4am). (D) Ch4am $p75^{NTR}$-immunoreactive neurons counterstained with thioflavin S. Solid white arrows indicate $p75^{NTR}$-containing neurons, open arrows indicate ghost tangles, and curved arrow indicates a tangle-bearing $p75^{NTR}$ neuron. (E) Another example of a tangle-bearing neuron containing $p75^{NTR}$. (A–E, × 380). From Mufson *et al.* (1989b) with permission.

### 10.2. Loss of $p75^{NTR}$ Protein- and mRNA-Containing Cholinergic Basal Forebrain Neurons

To determine the effect of AD on $p75^{NTR}$-immunoreactive neurons, we used immunohistochemical and *in situ* hybridization protocols (Kordower *et al.,* 1989; Mufson *et al.,* 1989b; Higgins and Mufson, 1989). Aged human control cases averaged 190,000 $p75^{NTR}$-immunopositive neurons within the CBF. There was an apparent 1:1 relationship between the number of viable CBF neurons and the number of perikarya expressing the message for $p75^{NTR}$. There was a striking reduction in the number of CBF neurons expressing the protein or mRNA for $p75^{NTR}$ within the subfields of the nucleus basalis (Figs. 7, 16) as compared with the septal/diagonal band complex in AD (Fig. 7). Significant reductions in the number of $p75^{NTR}$-immunostained neurons were observed in the anteromedial (35.1%), anterolateral (76.4%), intermediate (62.1%), and posterior (76.5%) subdivisions of the nucleus basalis. Interestingly, the remaining CBF neurons continue to express both the protein (Figs. 7, 16, 17) and mRNA for $p75^{NTR}$ (Fig. 18), and select AD cases exhibit an increase in $p75^{NTR}$ mRNA expression (Fig. 18D).

These findings indicate a regionally specific loss of $p75^{NTR}$-containing neurons within the basal forebrain in AD. In these cases the neurons projecting to the temporal cortex were the most severely affected. In contrast, hippocampal projecting basal forebrain neurons were affected to a lesser degree (see also Vogels *et al.,* 1990), suggesting that efforts to rescue and restore telencephalic cholinergic neurotransmission in AD with growth factors such as NGF might be best focused upon atrophied hippocampal projecting CBF neurons in the diseased brain. On the other hand, neuron replacement strategies might be best directed toward the target areas of the anterolateral, intermediate, and posterior subfields of the nucleus basalis, because these neurons are lost in greatest numbers in AD.

## 11. Altered Transport of Target-Derived Nerve Growth Factor

Currently, little is known about the ability of CBF-containing neurons to retrogradely transport NGF in AD. We combined NGF immunohistochemistry with quantitative optical densitometry to evaluate whether retrogradely transported NGF is altered within CBF consumer neurons in AD (Mufson *et al.,* 1994, 1995). In aged control brains, there was virtually a complete overlap in the number and distribution of neurons immunoreactive for NGF and $p75^{NTR}$ (Fig. 19A, B), illustrating that almost all nucleus basalis neurons retrogradely transport NGF. In contrast, there was a dissociation in the distribution of NGF- and $p75^{NTR}$-containing neurons within the nucleus basalis in AD. Although reduced in absolute number relative to aged-matched controls, $p75^{NTR}$-immunoreactive perikarya were observed in the AD brain (Fig. 19C). However, adjacent sections revealed that many of these residual neurons were devoid of NGF staining (Fig. 19D). Further, many CBF perikarya appeared shrunken with blunted dendrites or displayed a ballooned appearance (Fig. 19D). Of those neurons containing detectable NGF immunoreactivity, optical density measurements revealed a significant 32% reduction in staining intensity relative to age-matched controls ($p < 0.001$) (Fig. 20). By contrast, vir-

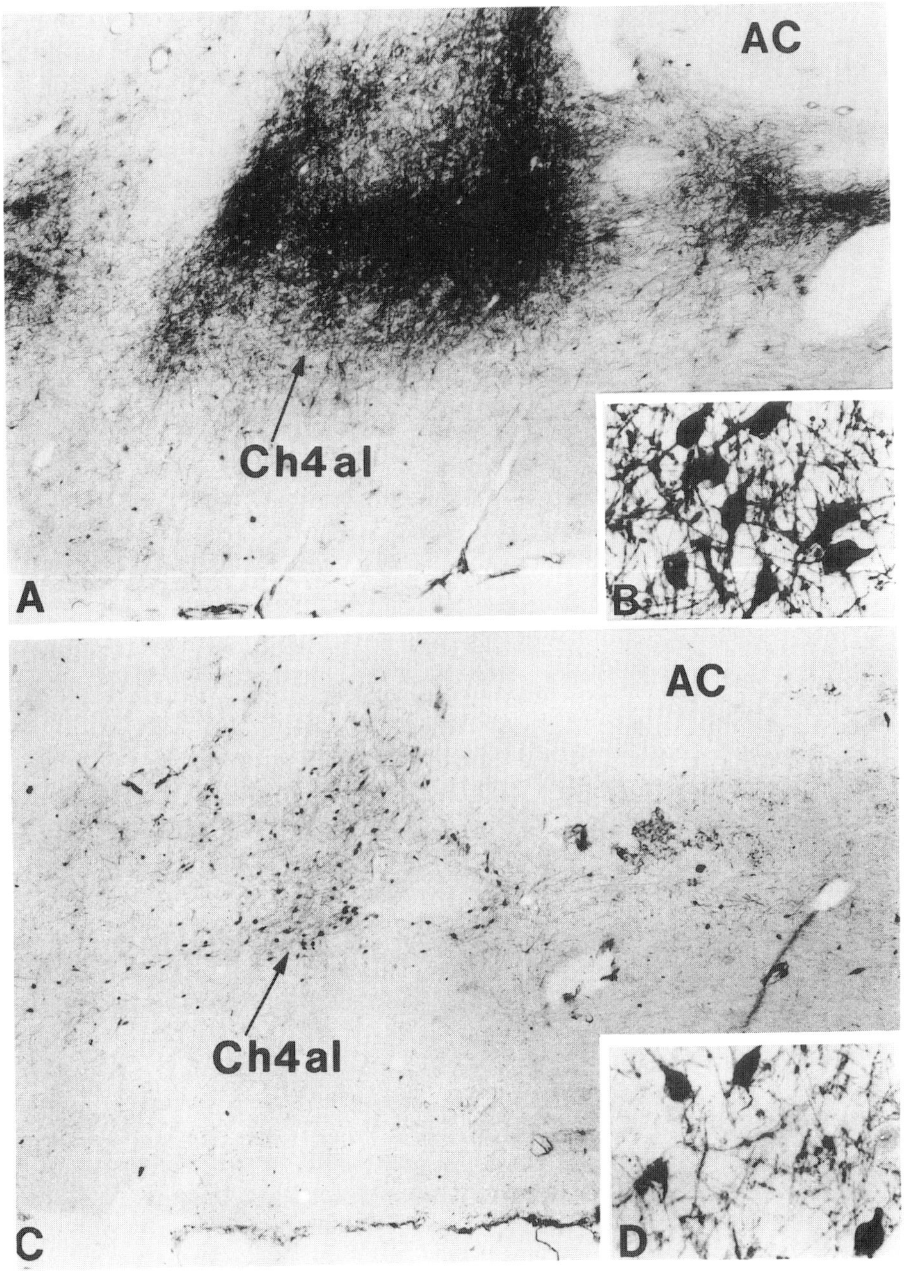


**Figure 16.** Comparison of the loss of nucleus basalis Ch4al $p75^{NTR}$-immunoreactive neurons in aged control (A, B) and in AD (C, D). Higher magnifications (insets) show the dystrophic appearance of Ch4al neurons and the decrease in $p75^{NTR}$-immunoreactive neuropil staining. (A, C, ×19; B and D, ×171). AC, anterior commissure. From Mufson *et al.* (1989b) with permission.

tually all remaining CBF neurons were intensely $p75^{NTR}$-immunoreactive in AD (Fig. 19C), demonstrating the specificity of this finding.

In AD the CBF neurons displayed a significant reduction of NGF immunoreactivity, and many CBF neurons failed to exhibit any NGF reactivity at all, suggesting that CBF neurons in AD are receiving inadequate trophic support. Although the exact mechanism leading to a diminished NGF expression within CBF neurons is unknown, previous studies have indicated that this defect does not occur at the level of NGF synthesis within the cortex (Crutcher *et al.,* 1993; Scott *et al.,* 1995).

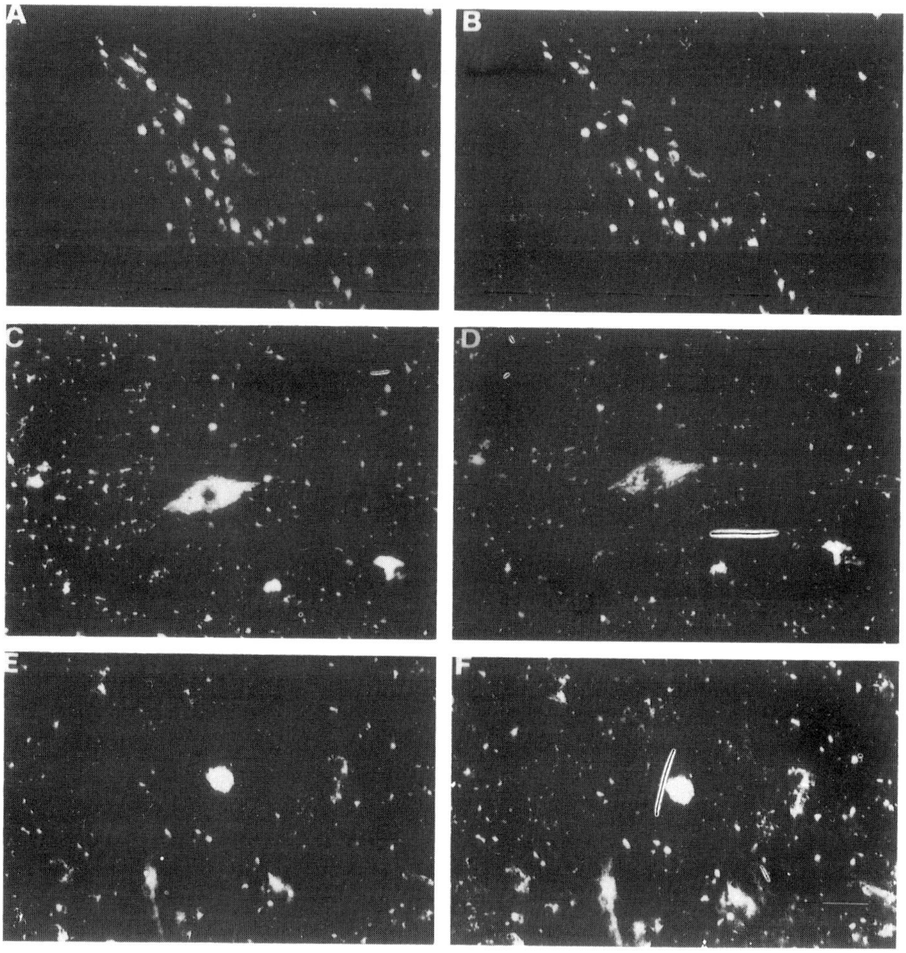


**Figure 17.** Double-labeled profiles of basal forebrain neurons from an AD case stained for both ChAT (left panel) and $p75^{NTR}$ (right panel). (A) With Texas Red, nucleus basalis neurons stained for ChAT (B) With FITC, the virtually identical population stained for $p75^{NTR}$. (C, D) More higher powered views demonstrate that healthy-appearing neurons within the nucleus basalis dual stained for ChAT (C) and $p75^{NTR}$ (D). Similarly, the neuron in E and F, in a state of degeneration, still colocalized ChAT (E) and $p75^{NTR}$ (F). Bars = 250 μm (A, B) and 50 μm (C, D). Reprinted from *Neurobiol. Aging* **10,** Kordower *et al.,* Nerve growth factor receptor and choline acetyltransferase remain colocalized in the nucleus basalis (Ch4) of Alzheimer's patients (1989) with kind permission from Elsevier Science Ltd., the Boulevard, Langford Lane, Kidlington OX5 1GB, UK.

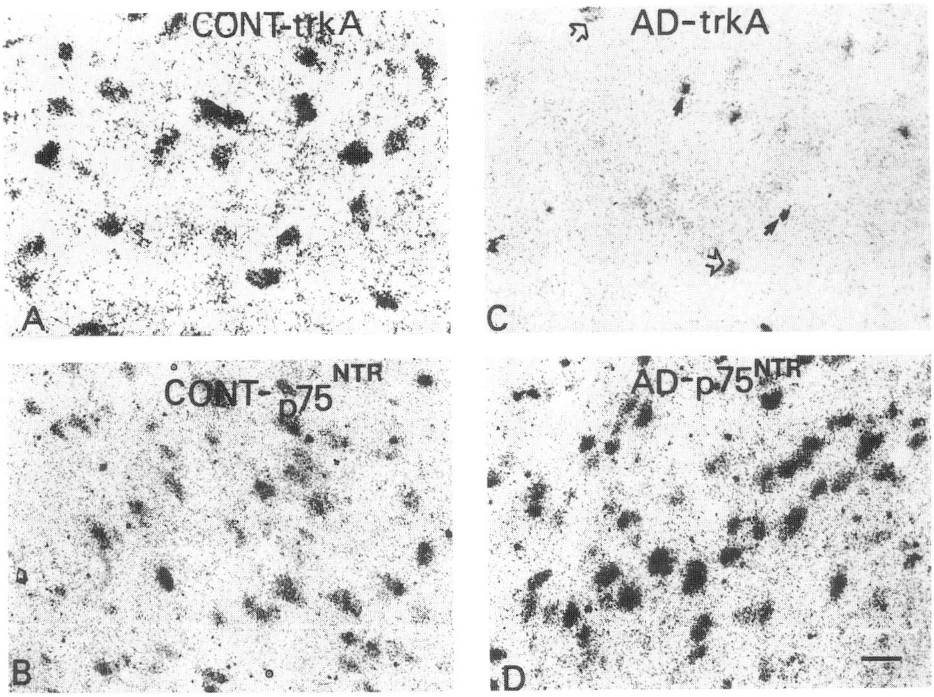


**Figure 18.** Photomicrographs of trkA- and $p75^{NTR}$ mRNA-positive neuronal hybridization within the nucleus basalis. (A) TrkA mRNA-hybridized neurons within the nucleus basalis of an aged control. (C) Reduction in trkA mRNA expression in remaining neurons in an AD patient. Black arrows indicate neurons exhibiting heavy, open arrows light, trkA mRNA hybridization. (B) $p75^{NTR}$ mRNA-hybridized neurons within the nucleus basalis of an aged control patient. (D) An example of more-intense $p75^{NTR}$ hybridization within nucleus basalis neurons in an AD patient. Note the approximate 1:1 relationship between trkA (A) and $p75^{NTR}$ (C) mRNA positive cells in the aged controls. Bar = 70 μm. From Mufson *et al.* (1996) with permission.

The fact that NGF is synthesized at normal levels and NGF protein is increased in AD cortex further supports our findings. It appears that NGF accumulates in the AD cortex because it is ineffectively transported to the basal forebrain cholinergic consumer neurons. In the following section of this chapter we provide evidence generated from studies of the high-affinity receptor indicating that the reduction of NGF transported by CBF neurons in AD is the result of a defect in the expression of the trkA receptor.

## 12. Increased Nerve Growth Factor Levels in Neo- and Limbic Cortex

Because we observed a defect in retrogradely transported NGF within CBF neurons, we investigated whether NGF protein levels within the cortical production sites were altered in AD using a sensitive two-site Elisa assay (Crutcher *et al.*, 1993; Scott *et al.*, 1995). Evaluation of NGF levels revealed that protein levels were signifi-

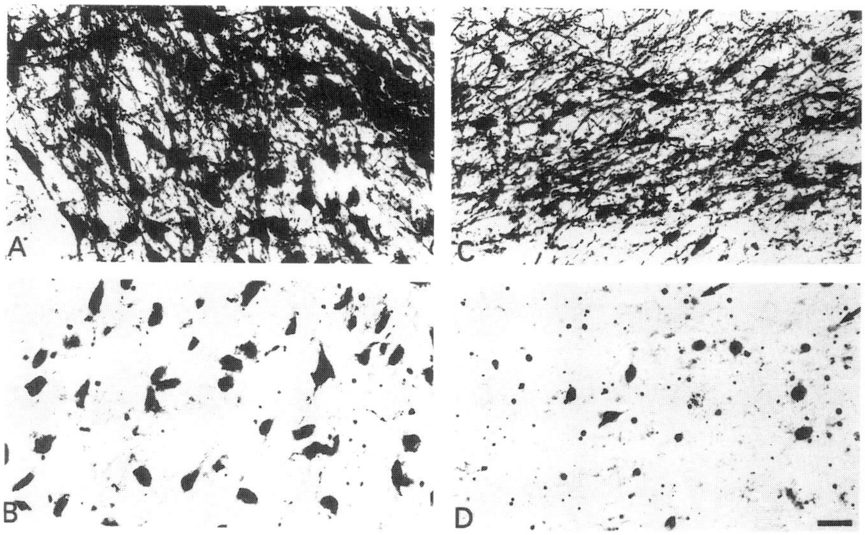


**Figure 19.** Photomicrographs of sections immunostained for p75$^{NTR}$ and NGF within the nucleus basalis from normal aged control (A, B) and an AD case (B, D). Note the dissociation in the expression of p75$^{NTR}$- and NGF-immunoreactive neurons within the nucleus basalis in AD. Panel B shows the paucity of intensely NFG-immunoreactive neurons as compared with the extensive p75$^{NTR}$ neuronal staining (A). The black arrows in B indicate examples of lightly NGF-immunostained neurons. Open arrows in B indicate NGF-immunostained neurons that exhibited an oval or ballooned appearance. Bar = 50 μm. From Mufson *et al.* (1995), with permission.

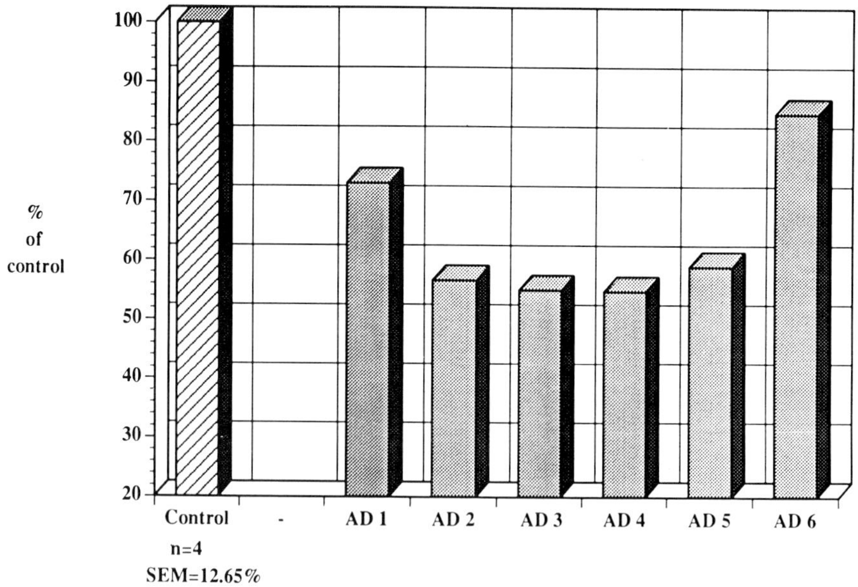


**Figure 20.** Histogram showing the difference in optical density measurements of NGF-immunostained nucleus basalis neurons between normal aged controls and AD cases (see text for details). From Mufson *et al.* (1995) with permission.

cantly increased in AD as compared with aged controls in the amygdala, hippocampus, frontal cortex, superior temporal cortex, inferior parietal lobule, and occipital cortex (Fig. 3). These findings were similar to those previously reported by Crutcher *et al.* (1993) in which AD-related increases in frontal and occipital poles were observed. In contrast, tissue samples of the nucleus basalis of Meynert displayed a 50% AD-related NGF reduction, a decline in the CBF reminiscent of the reduction in NGF-like immunoreactivity we reported within CBF neurons in AD (Fig. 21) (Mufson *et al.,* 1995). The accumulation of NGF seen within the AD cortex was not found in patients with Parkinson's disease (Fig. 21), suggesting an AD-specific response.

These findings indicate that there are increases in neocortical NGF levels coupled with decreases in the retrograde transport of this neurotrophin by CBF consumer neurons in AD. The accumulation of NGF within the cortex indicates that NGF protein is abundantly available in AD brain. The mechanism underlying the defect in retrogradely transported NGF by CBF neurons remains an area of great interest. Because $p75^{NTR}$ protein and mRNA are stable (Goedert *et al.,* 1986) or increased (Emfors *et al.,* 1990) within CBF neurons in AD, it is unlikely that a defect in the production of this ubiquitous receptor is the primary factor responsible for the inability of CBF perikarya to normally retrogradely transport available cortical NGF. These observations, in conjunction with our findings of an increase in NGF levels in AD cortex (Crutcher *et al.,* 1993; Scott *et al.,* 1995), led us to hypothesize that a defect in the production and/or utilization of the high-affinity trkA receptor is the major pathological alteration in the NGF system that underlies CBF degeneration in AD. The following sections outline the data underlying this hypothesis.

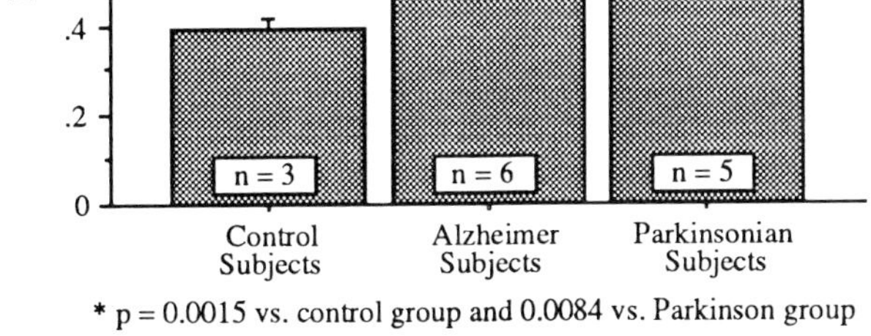


**Figure 21.** NGF levels in the superior frontal gyrus (mear ± SEM). Increased levels were seen in this region in AD but not Parkinson's disease when compared with age-matched control samples. From Scott *et al.,* "Nerve Growth Factor in Alzheimer's disease: Increased levels throughout the brain coupled with declines in nucleus basalis," *The Journal of Neuroscience* **15(9)**:6218 (1995) with permission.

## 13. trkA within Cholinergic Basal Forebrain Neurons

Because the high-affinity trkA receptor, which is synthesized in CBF neurons (Gibbs and Pfaff, 1994; Holtzman *et al.*, 1995; Shelton *et al.*, 1995; LeSauteur *et al.*, 1996; Mufson *et al.*, 1996), is essential for its signal transduction, a defect in this system would deprive NGF consumer neurons of required trophic support, which could then result in atrophy or cell death. Therefore, we combined *in situ* hybridization, antibody immunocytochemistry, quantitative optical densitometry, and reverse transcription polymerase chain reaction (RT-PCR) technology to evaluate whether trkA mRNA and protein are altered within CBF neurons in AD.

### 13.1. Alterations in trkA Gene and Protein Expression in Nucleus Basalis

Tissue from aged control brains either hybridized for trkA gene expression or immunostained for trkA protein using the mAb 5C3 antibody, which recognizes the extracellular domain of the human trkA receptor (LeSauteur *et al.*, 1996), revealed numerous intensely labeled neurons scattered throughout all subfields of the cholinergic basal forebrain (Figs. 18, 22). In the aged brain, adjacent sections immunostained for $p75^{NTR}$ revealed an approximate 1:1 relationship between the number of viable nucleus basalis cholinergic neurons and the number of neurons expressing detectable trkA mRNA (Fig. 18). Both qualitative and quantitative analyses indicated that trkA mRNA and protein-immunoreactive neurons within the nucleus basalis were reduced in AD relative to aged controls (Figs. 22, 23). As seen with $p75^{NTR}$ staining, many of the remaining immunoreactive neurons within the nucleus basalis were either shrunken with blunted dendrites or displayed a swollen, globose appearance (Figs. 1C, D; 2B, C; 3B–D; 6D). In the diseased brain there were many fewer trkA mRNA- and protein-containing neurons within the nucleus basalis (Fig. 22), indicative of degeneration of this neuronal population in AD. Interestingly, many remaining CBF neurons failed to express the message or the protein for trkA (Figs. 18, 22). Of those remaining neurons exhibiting detectable trkA mRNA and protein, optical density measurements revealed significant reductions in hybridization intensity (66%) and immunoreactivity (35%) relative to aged controls (Fig. 23). The reduced immunoreactivity for trkA receptors was not restricted to atrophied neurons; diminished immunoreactivity could be seen in many of the remaining large neurons in the nucleus basalis of AD patients (Fig. 22).

### 13.2. Reverse Transcription Polymerase Chain Reaction and Western Blot Analysis of trkA

Further support for the extensive reduction in trkA within the nucleus basalis was generated from studies using RT-PCR and biochemical assays of trkA levels in aged control and AD brain. The RT-PCR study revealed that trkA mRNA levels were reduced in the nucleus basalis in AD relative to aged controls (Fig. 24), an effect specific for trkA, as gene expression for the housekeeping gene G3PDH is similar in AD and aged controls (Fig. 24). Detergent-solubilized protein extracts from nucleus basalis, caudate, and frontal cortex of aged normal and AD brains were resolved by SDS-PAGE and analyzed by Western blotting with mAb 5C3. Reduced trkA density in AD was seen in the nucleus basalis (~ 65% of the intensity of control; $p < 0.02$) and frontal cortex (~ 52% of the intensity of control; $p < 0.02$), but not in the striatum (~ 97% of the intensity of control; $p < 0.05$) (Fig. 25).

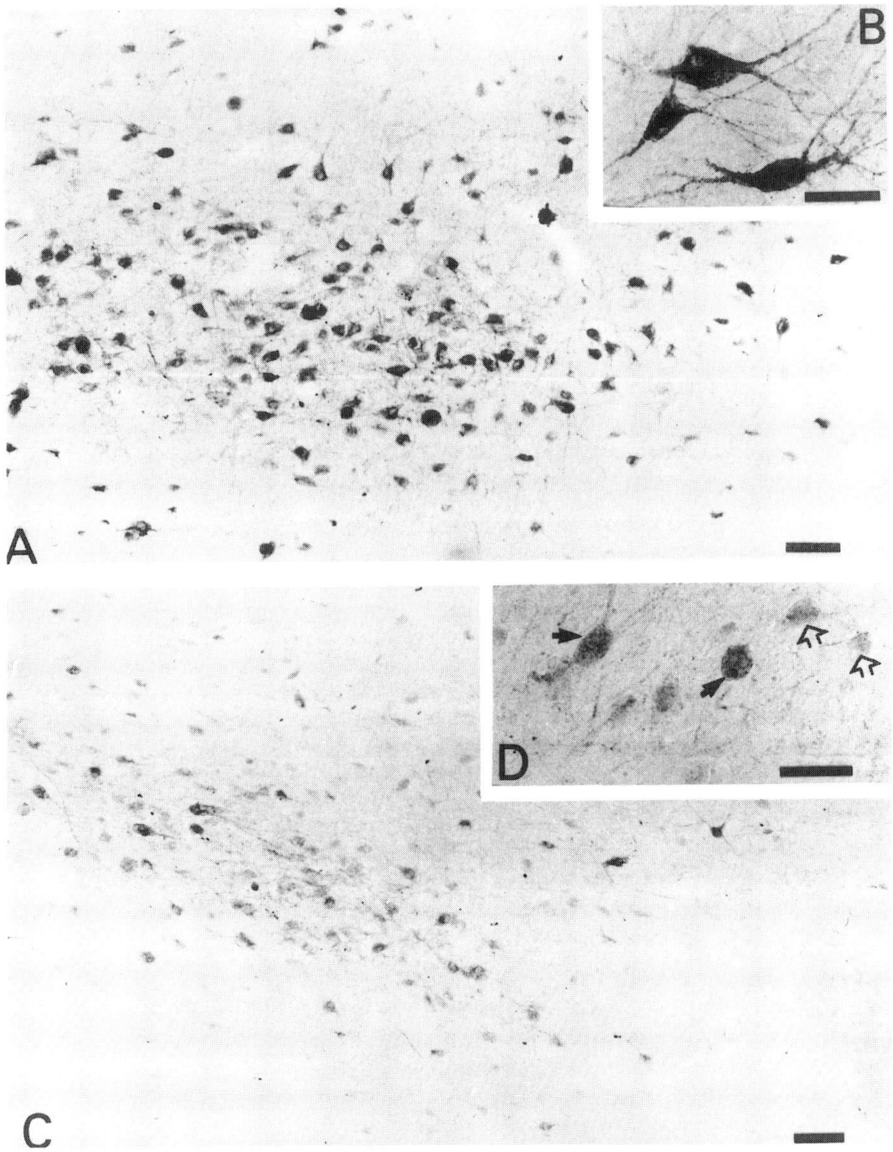


**Figure 22.** Comparison of trkA immunostaining with the anterolateral portion of the nucleus basalis in an aged control (A) and an AD case (C). Note in A that virtually all neurons are trkA-positive and exhibit a classic multipolar appearance (B). By contrast, in C and D (black arrows) many of these neurons are shrunken and express decreased trkA immunoreactivity (open arrows) (D). Bars = 100 μm (A, C) and 50 μm (B, D).

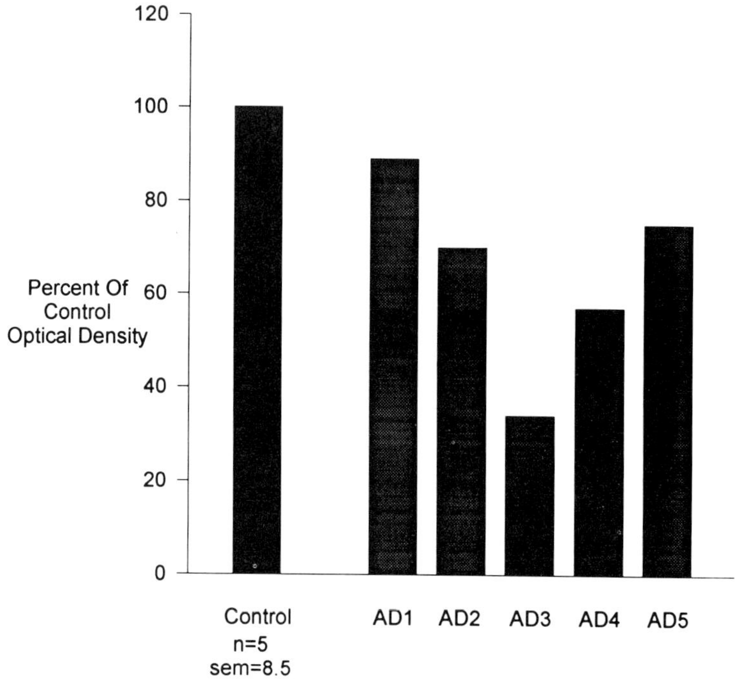


**Figure 23.** Histogram showing difference in optical density measurements of trkA-immunostained nucleus basalis neurons between normal aged controls and AD cases (see text for details). From Mufson *et al.* (1997) with permission.

## 13.3. Is Nerve Growth Factor Off trk in Alzheimer's Disease?

The above findings demonstrating *decrease* in retrogradely transported NGF and trkA mRNA and protein within CBF neurons as well as an *increase* in cortical NGF protein levels have led us to hypothesize that CBF degeneration may be due in part to a defect in the high-affinity trkA NGF receptor in AD. Figure 26 summarizes known changes in NGF and its receptors in AD. Further support for our hypothesis is derived from observations that both the protein and mRNA for $p75^{NTR}$ are stable or increased in AD (see Goedert *et al.*, 1986; Kordower *et al.*, 1989; Mufson *et al.*, 1989b; Ernfors *et al.*, 1990). These findings indicate that at its production site, $p75^{NTR}$ is not defective in AD. In addition to normal neuronal production of the $p75^{NTR}$ protein and mRNA, we also found that residual basal forebrain axons continue to immunostain for $p75^{NTR}$ in AD (Mufson *et al.*, 1989b), suggesting that the $p75^{NTR}$ may transport normally down the axon to its cortical target regions in the diseased brain. This indicates that general transport defects *per se* may not explain the observed reduction in NGF transport.

Although the effect of diminished trkA receptor upon CBF hypofunction in AD is unknown, recent animal studies have begun to shed light on this issue. Recently an age-related reduction in trkA receptor mRNA was seen in aged rats (Cooper *et al.*, 1994) similar to that seen in AD (Mufson *et al.*, 1996). In aged rats displaying atrophic basal forebrain neurons, there is a significant (43%) decrease in trkA, but not p75$^{NTR}$ mRNA, within CBF neurons (Cooper *et al.,* 1994). This experimental study found a reduction in the number of CBF neurons that retrogradely transported iodinated NGF, further supporting our hypothesis (Mufson *et al.,* 1995) that reduced NGF transport may be secondary to diminished trkA, resulting in degenerative processes within CBF neurons. However, it remains to be determined whether there is also an alteration in the binding of trkA with NGF at the level of its cortical production sites in AD (Fig. 26).

Although the exact mechanisms underlying this diminished trkA protein and its role in NGF trophism in AD are unknown, it is important to note that NGF exerts trophic influences upon CBF neurons via a series of complex events, which include trkA receptor synthesis, anterograde transport, NGF binding, receptor autophosphorylation, retrograde transport of the NGF–receptor complex, and other aspects of signal transduction. Defects at any of these steps could result in CBF degeneration secondary to impaired trkA gene expression (Kordower *et al.,* 1989). For example, decreased levels of trkA immunoreactivity could be either primary or secondary to structural alterations within CBF neurons in AD. It is well known that CBF neurons stain intensely for Alz50, an abnormally phosphorylated tau cytoskeletal protein (Mufson *et al.,* 1993). Because tau is a major factor in the maintenance

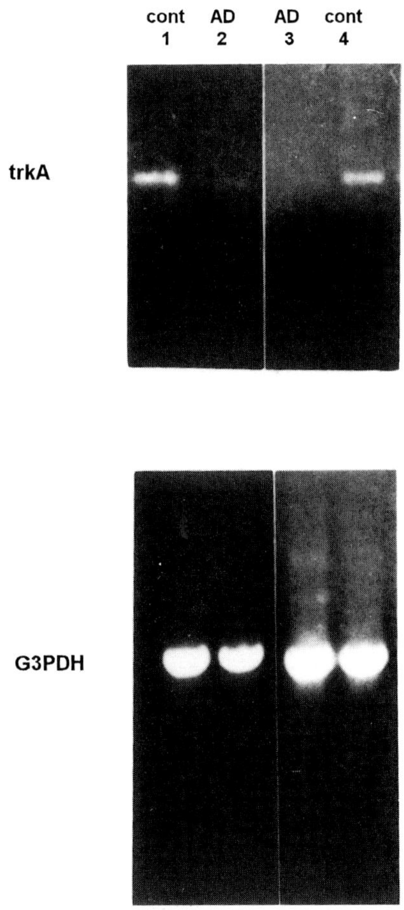


**Figure 24.** RT-PCR amplification of human brain cDNA using oligonucleotides specific for the extracellular domain of the trkA receptor within the anterior subdivision of the nucleus basalis. Note the reduction in trkA expression in AD (lanes 2 and 3) as compared with aged controls (lanes 1 and 4). Bottom PCR blot shows equivalent loading of the housekeeping gene G3PDH within the tissue sample of the nucleus basalis between aged control and AD cases. From Mufson *et al.* (1996) with permission.

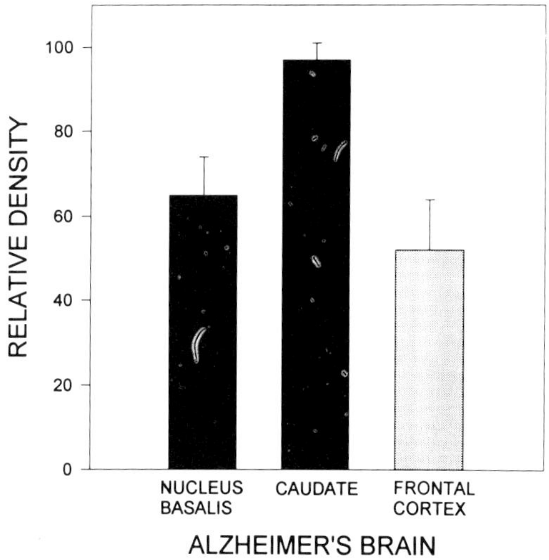


**Figure 25.** TrkA protein expression is decreased in AD basal forebrain and frontal cortex but not in caudate. The indicated regions from AD brains and normal aged control brains were analyzed by Western blotting with mAb 5C3. More than three independent blots were done for each brain sample. The density of trkA bands in X-rays was analyzed by densitometry relative to control samples of aged brain from the same regions (normalized to 100%). One of the AD samples (labeled AD1 in Fig. 7) expressed high trkA levels (similar to control aged brains) and accounts for the relatively large SEM. Percentage of optical density ± SEM AD nucleus basalis and frontal cortex were significantly different from control ($p < 0.02$). From Mufson *et al.* (1997) with permission.

of neuronal morphology and axonal transport, abnormal phosphorylation of this protein could trigger dysfunctional axonal transport. A trafficking defect could lead to diminished trophic support and ultimately a down-regulation of trkA gene expression and protein synthesis in AD (Mufson *et al.,* 1996, 1997; Salehi *et al.,* 1996; Boissière *et al.,* 1996). Alternatively, down-regulation of the trkA message within neurons of the nucleus basalis may be the pivotal factor underlying the atrophy and degeneration of CBF neurons in AD. Thus it remains to be determined whether a defect in axonal transport may partially underlie the neuronal atrophy of CBF neurons by altering trafficking of the NGF–trkA complex in AD.

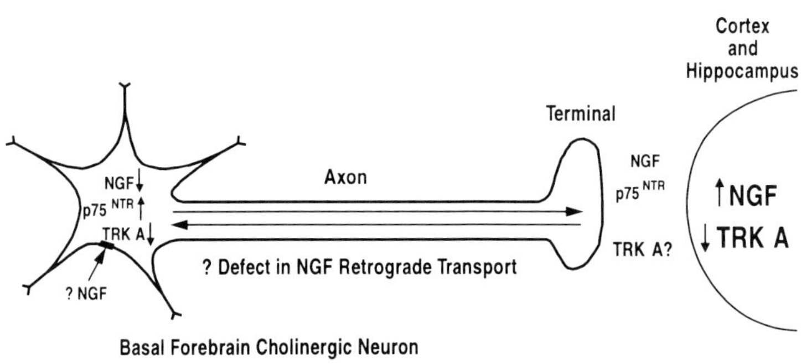


**Figure 26.** Summary diagram of changes to the NGF system associated with the cholinergic basal forebrain (CBF) neurons and their target sites in AD. CBF neurons show a decrease in the protein level and gene expression for the high-affinity trkA receptor (downward arrow), but the low-affinity receptor remains stable or increases (upward arrow). In addition, there is a reduction in protein levels (downward arrow) in AD cortex. Although there is an increase in NGF protein levels in cortical and hippocampal target sites (upward arrow), there is a decrease in NGF retrogradely transported by the CBF neurons, leading us to hypothesize that CBF degeneration may be due in part to a high-affinity trkA receptor defect in AD. It is unknown whether there is a defect in the binding of trkA (?) with NGF, a defect in NGF retrograde transport, alterations in the interaction of NGF and $p75^{NTR}$, or alterations in the complexing of all three proteins in AD. For clarity, these relationships are shown outside of the neurons' target sites.

The present findings of reductions in trkA mRNA (Mufson *et al.*, 1996) and protein (Salehi *et al.*, 1996; Mufson *et al.*, 1997) within CBF neurons in AD suggest that this neurotrophic defect may be the primary signal underlying neurodegenerative changes within CBF perikarya in this disease. These observations, in conjunction with data demonstrating defects in trkB and trkC protein expression within nucleus basalis neurons in AD (Salehi *et al.*, 1996), suggest that these cells may display a uniformly diminished responsiveness to several members of NGF neurotrophin family. Therefore, future research should be directed at understanding the regulation of trk gene expression. Studies aimed at developing novel pharmacologic therapies for AD need to address trk defects in the trk receptor system in addition to alterations in the endogenous neurotrophic ligand. Moreover, it is crucial to consider whether changes in trkA expression (a) occur as a function of other genetic factors (i.e. apoE genotype), (b) occur in sporadic versus familial AD, and (c) are affected by age at onset of the disease process.

## 14. Neuroplasticity of the Nerve Growth Factor Receptor System in Alzheimer's Disease

The increased expression of NGF seen within the neocortex and the reduction of trkA gene and protein expression within CBF neurons in AD must be considered along with the observations that there are regenerative changes occurring in the AD brain associated with the NGF–CBF system. For example, hippocampal fibers putatively derived from the NGF consumer neurons of the septal/diagonal band subfield of the CBF reorganize in AD (Geddes *et al.*, 1985), suggesting that the neurotrophic factors mediate aspects of plasticity in aging and disease. Recently, we described two plasticity responses related to the NGF system in AD concerned with a novel expression of $p75^{NTR}$ in the cortex and with the hyperinnervation of galaninergic fibers upon remaining CBF/$p75^{NTR}$-containing neurons in this disorder.

### 14.1. *De Novo* Expression of $p75^{NTR}$-Containing Cortical Neurons

Using a monoclonal antibody against the $p75^{NTR}$, we observed a novel population of cortical neurons containing $p75^{NTR}$ only in AD patients and in one very old (>90 years) normal adult (Figs. 27, 28) (see Mufson and Kordower, 1992; de Lacalle *et al.*, 1994). These cells were preferentially distributed within the temporal association cortex, a region that displays severe degeneration in AD. A few such neurons were also located in the hippocampal and amygdaloid complexes. These neurons appeared healthy, with bipolar, fusiform, and multipolar morphologies, and they did not contain neurofibrillary tangles (Fig. 29A). Interestingly, we have also observed $p75^{NTR}$-immunoreactive neurons in the developing human cortical subplate zone (Fig. 29B) (see Mufson and Kordower, 1992; Kordower and Mufson, 1992), many of which were morphologically similar to those found in AD and in the very old. In this regard we have suggested that the expression of $p75^{NTR}$ by cortical neurons in extreme old age and in AD reflects a return to a fetal pattern of expressing trophic factors in response to neuronal injury (Mufson and Kordower,

1992; Kordower and Mufson, 1992). On the other hand, it is well established that NGF and its receptors have a "feedforward" relationship (Hefti *et al.*, 1989); infusion of NGF into the striatum (Gage *et al.*, 1989) or lateral ventricle (Koliatsos *et al.*, 1994) results in the expression of $p75^{NTR}$-immunoreactive neurons (but see Bartus *et al.*, 1996). Thus, the increased synthesis of cortical NGF observed in AD (Crutcher *et al.*, 1993; Scott *et al.*, 1995) may provide trophic support for those cortical neurons, which then (re)express $p75^{NTR}$ immunoreactivity within cortical neurons in AD. It would be interesting to determine whether levels of NGF are elevated in the cortex of very old normal people and in people with AD, in whom there are $p75^{NTR}$-immunoreactive cortical neurons; older individuals lack these neurons.

## 14.2. Galanin Fiber Hypertrophy of Remaining $p75^{NTR}$ Basal Forebrain Neurons

Numerous investigations have linked the neuropeptide galanin (GAL) to CBF function in normal states and in neurodegenerative states including AD (see Crawl-

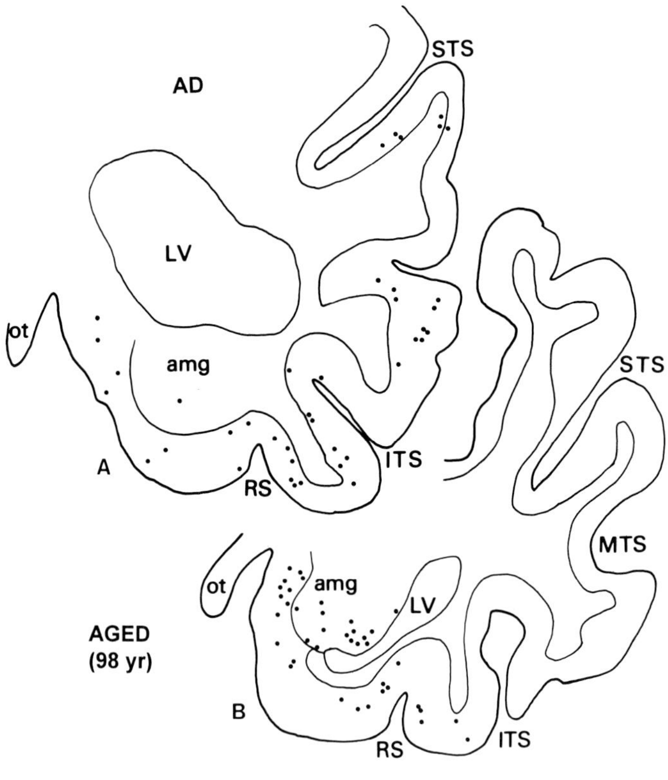


**Figure 27.** Schematic drawing illustrating $p75^{NTR}$-positive neuron distribution within limbic and temporal neocortex of an AD patient (A) and a 98-year-old normal control (B). Each black dot represents a single neuron. amg, amygdala; LV, lateral ventricle; ot, optic tract; ITS, inferior temporal sulcus; MTS, medial temporal sulcus; RS, rhinal sulcus; STS, superior temporal sulcus. From Mufson and Kordower (1992) with permission. Copyright 1992, National Academy of Sciences, U.S.A.

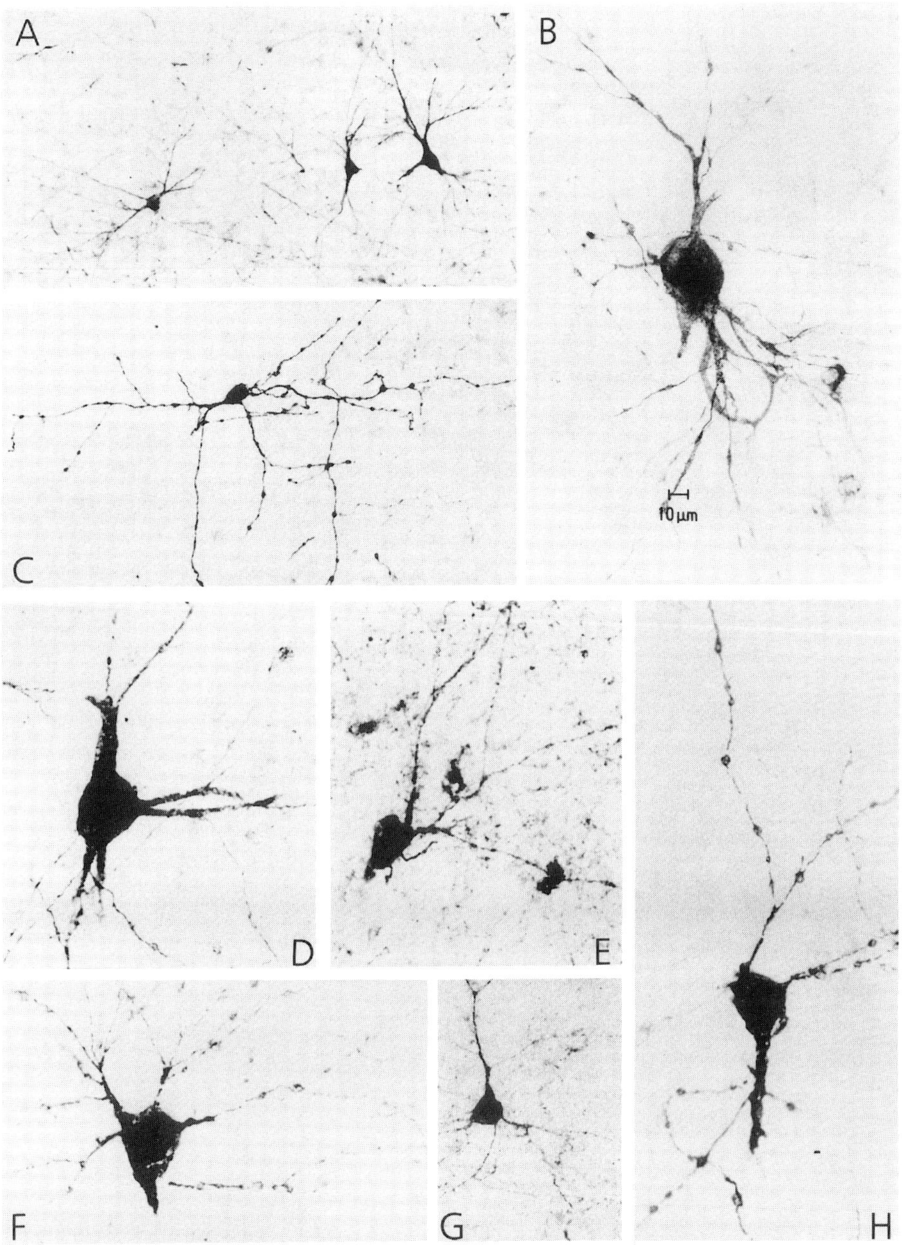


**Figure 28.** Morphological profiles of multipolar p75$^{NTR}$-immunoreactive neurons within AD cortex (A–C; F, H) and a 98-year-old normal control (D, E, G). (A) Cluster of neurons within layer V of inferior temporal cortex, ×80. (B) Neurons within basolateral amygdala, ×100. (C) Stellate-like neuron with beaded dendritic processes in the inferior temporal cortex, ×200. Neurons within superior temporal (D) and entorhinal cortex (E). (F) Neurons with superior temporal cortex. (G) Small layer II entorhinal cortex perikarya. (H) Neuron within the inferior temporal cortex. Bar in B represents the magnification for B and D–H. From Mufson and Kordower (1992) with permission. Copyright 1992, National Academy of Sciences, U.S.A.

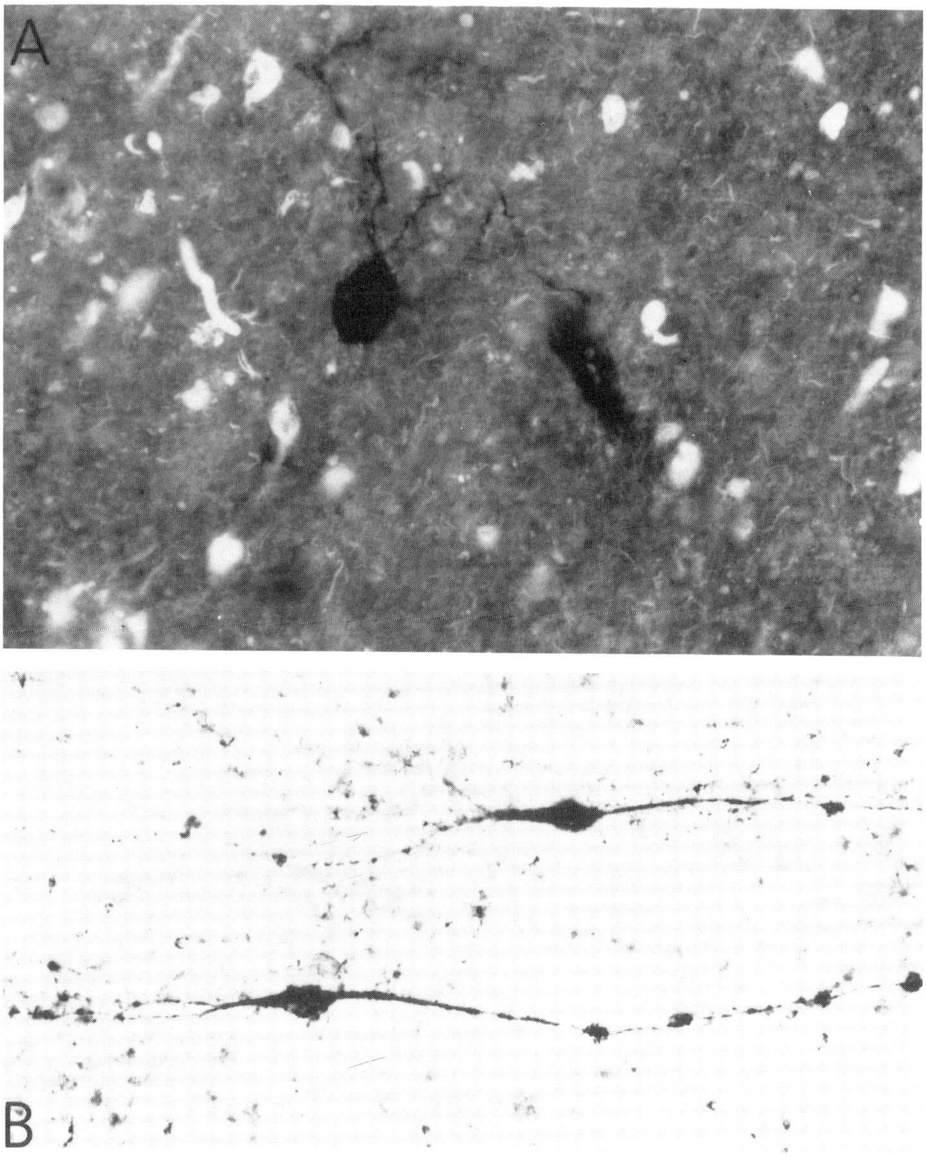


**Figure 29.** (A) p75NTR-immunoreacted section counterstained with thioflavin S to demonstrate the lack of colocalization between white fluorescent NFTs and black immunoreactive cortical neuron, ×350. (B) p75NTR-immunoreactive neurons within neocortex of a 21-week-old human embryo, ×350. From Mufson *et al.* (1992) with permission.

ey, 1996, for review). Galanin is a novel 29- (in humans, 30)-amino acid peptide isolated from porcine small intestine (see Crawley, 1996, for review). Although it has been demonstrated both immunohistochemically and biochemically that GAL has a widespread distribution within the CNS (Melander *et al.*, 1986; Kordower *et al.*, 1992; see Crawley, 1996, for review), its chemoanatomy in the CBF exhibits a species-specific expression within the mammalian basal forebrain (see Kordower and Mufson, 1990; Benzing *et al.,* 1993; Crawley, 1996, for reviews). In the human brain, GAL does not colocalize with the magnocellular cholinergic neurons of the basal forebrain as it does in rats and monkeys (Melander *et al.,* 1986; Kordower and Mufson, 1990; Benzing *et al.,* 1993), but it *is* found within small GAL-immunoreactive interneurons as well as within a widespread fiber network that travels through the basal forebrain (Fig. 30A) (see Kordower *et al.,* 1992; Mufson *et al.,* 1993; Benzing *et al.,* 1993). The finding that GAL inhibits acetylcholine (Fisone *et al.,* 1987) and disrupts working memory in rats (see Crawley, 1996, for review) has prompted the suggestion that this neuropeptide plays a crucial role in the regulation of cholinergic/p75$^{NTR}$ basal forebrain neuron function in AD. Also, the hyperinnervation of remaining CBF/p75$^{NTR}$ neurons by galaninergic fibers (Fig. 30B) in AD suggests that this plasticity response may exacerbate cholinergic deficits, and in turn enhance cognitive decline (Chan-Palay, 1988; Mufson *et al.,* 1993), by altering acetylcholine neurotransmission within the remaining cholinergic neurons, a notion supported by the findings that loss of cortical choline acetyltransferase is more severe than neuronal loss within the CBF in AD (Perry, 1980). Alterations may also occur at the synaptic level between GAL fibers and CBF neurons. In fact, using dual-label confocal microscopy, which can analyze serial optical sections to accurately track galaninergic fibers and identify putative contacts between these fibers and CBF neurons, we have observed that putative GAL synaptic contacts also hypertrophy in patients with AD (Bowser *et al.,* 1997).

These findings indicate that NGF-related systems exhibit regenerative events that may be a response to the extensive neurodegeneration seen in the AD brain. The observation of a *de novo* expression of p75$^{NTR}$ within cortical neurons is part of the initial evidence of neurotrophin-related trophism in AD. The GAL hyperinnervation of remaining CBF/p75$^{NTR}$/trkA-expressing neurons in AD suggests that this peptide may antagonize putative trophic effects of NGF by altering NGF receptor-mediated events associated with cholinergic function. In this regard, NGF is known to induce the expression of ChAT (Heumann *et al.,* 1984) and GAL (Kaplan *et al.,* 1991) in cultured pheochromocytoma (PC12) cells and also following its intraventricular infusion (Planas *et al.,* 1997). Possibly, GAL regulates not only the cellular expression of ChAT but also components of the NGF system *in vivo.* Therefore, understanding whether GAL affects NGF or its receptors will provide greater insight into the pathologic process(es) affecting cholinergic cell dysfunction in AD. Such regenerative events need to be considered when devising therapeutic approaches associated with the use of NGF as a treatment for CBF degeneration in AD.

## 15. Treatment Approaches for Preventing Cholinergic Cell Degeneration in Alzheimer's Disease

For several years NGF has been suggested as a primary or adjunctive therapy for the treatment of the cholinergic deficit in AD. The first AD patient to receive intraventricular infusions of mouse NGF was reported to display modest improvement on a task of episodic memory, although the patient remained stable on other behaviors (Olson, 1993). Furthermore, patients have experienced sustained hyper-

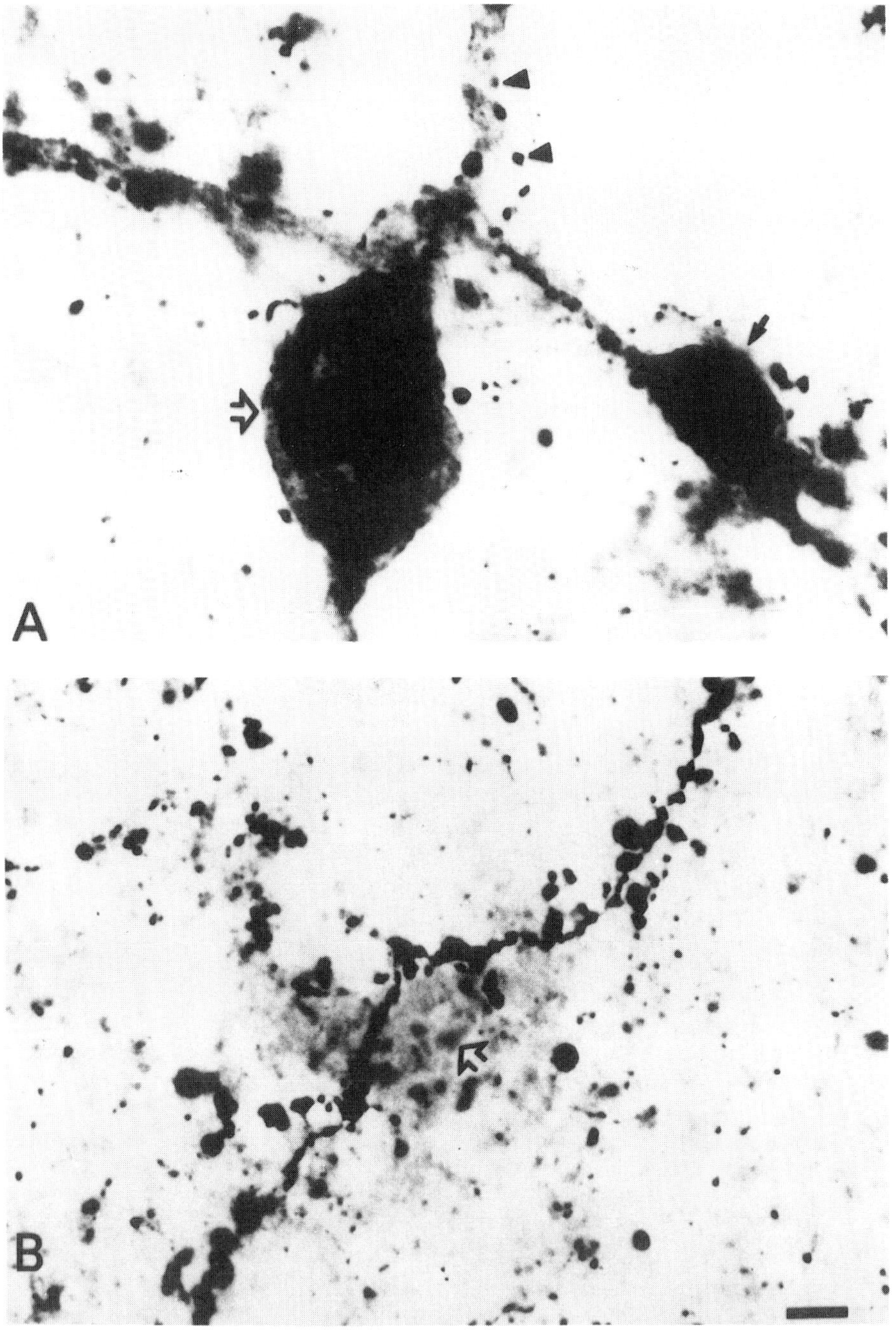


**Figure 30.** Photomicrographs showing the relationship between galaninergic profiles and $p75^{NTR}$-containing neurons in human aged and AD basal forebrain. (A) Septal/diagonal band complex $p75^{NTR}$-containing cholinergic neuron (open arrow) decorated by fine galanin-immunoreactive fibers (arrowheads) in an aged control case. Note the galanin-immunoreactive neuronal process emanating from a galanin-containing neuron (small arrow), which reaches the dendrite of the $p75^{NTR}$-positive neuron. (B) Hypertrophied galanin-immunoreactive fibers surrounding a $p75^{NTR}$-positive neuron (open arrow) within septal/diagonal band complex in AD. Bar in B = 15 μm.

algesia in association with NGF treatment (Olson, 1993). Because human NGF does not readily cross the blood–brain barrier, long-term treatment is likely to be required for AD patients, and methods must be devised for chronic delivery. Recently, we have been evaluating two novel approaches in animal models of CBF degeneration; the grafting of encapsulated cells that have been genetically modified to secrete NGF and the peripheral administration of NGF conjugated to an antibody directed against the transferrin receptor.

## 15.1. Grafting Encapsulated Cells Secreting Human Nerve Growth Factor

Grafting of encapsulated cells that secrete human NGF was used in experiments to rescue axotomized basal forebrain neurons in young (Emerich *et al.*, 1994) and aged (Kordower *et al.*, 1993) monkeys. Figure 31 is a schematic drawing showing the surgical approach employed to intraventricularly transplant polymer-encapsulated baby hamster kidney (BHK) fibroblasts that were genetically modified to secrete hNGF or identical grafts lacking the NGF construct in axotomized monkeys. The findings obtained from the aged monkey will be reviewed, because they are extremely relevant to CBF degeneration in AD. These monkeys showed AD-like pathology replete with numerous β-amyloid-containing immunoreactive plaques within the temporal and parietal cortex, amygdala, and hippocampus (Fig. 32A, B). Interestingly, NGF treatment had no effect on the number or distribution of amyloid plaques as compared with control animals (Kordower *et al.*, 1997). These data indicate that chronic *short-term* administration of NGF does not affect the expression of β-amyloid in young or aged primate brain. However, it is possible that the relatively short intervals between grafting and evaluation were insufficient to detect changes in amyloid deposition. Furthermore, histologic evaluation of the lesion site revealed a complete unilateral transection of the fornix and a loss of AChE-containing fibers within the ipsilateral hippocampus (Fig. 32C, D). Monkeys receiving BHK-control grafts exhibited a significant reduction (53%) of $p75^{NTR}$-positive neurons within the medial septum ipsilateral to the implant (Fig. 33A). Many remaining perikarya within the septum appeared atrophic relative to $p75^{NTR}$-reactive neurons on the contralateral side. By contrast, the ability of grafted NGF-secreting cells to rescue CBF neurons was striking. The lesion-induced degeneration of septal neurons was significantly attenuated ($p < 0.001$) in monkeys receiving grafts of polymer-encapsulated BHK-hNGF cells (Fig. 33A). These animals exhibited only a modest 7–23% reduction in ChAT- and $p75^{NTR}$-immunoreactive neurons within the medial septum ipsilateral to the lesion, as compared with the contralateral side (Fig. 33B, C). In addition to maintaining the viability and/or expression of cholinergic markers in septal cells, BHK-hNGF grafts induced a robust sprouting of fibers expressing the cholinergic markers $p75^{NTR}$, ChAT, and AChE within the ipsilateral septal region (Fig. 34). In contrast, one of the BHK-control-grafted monkeys displayed a cholinergic sprouting response (Fig. 34B).

To examine whether implanted BHK-hNGF-secreting cells continued to produce NGF over the course of the experiment, the grafted capsules were retrieved from the monkeys immediately prior to sacrifice. A large number of BHK cells were found within the capsules, which were then producing hNGF at a rate of 9.6 ng every 24 h/capsule as compared with 44.65 ng of NGF every 24 h/capsule prior to implantation, a lower production level that is biologically relevant, as media ob-

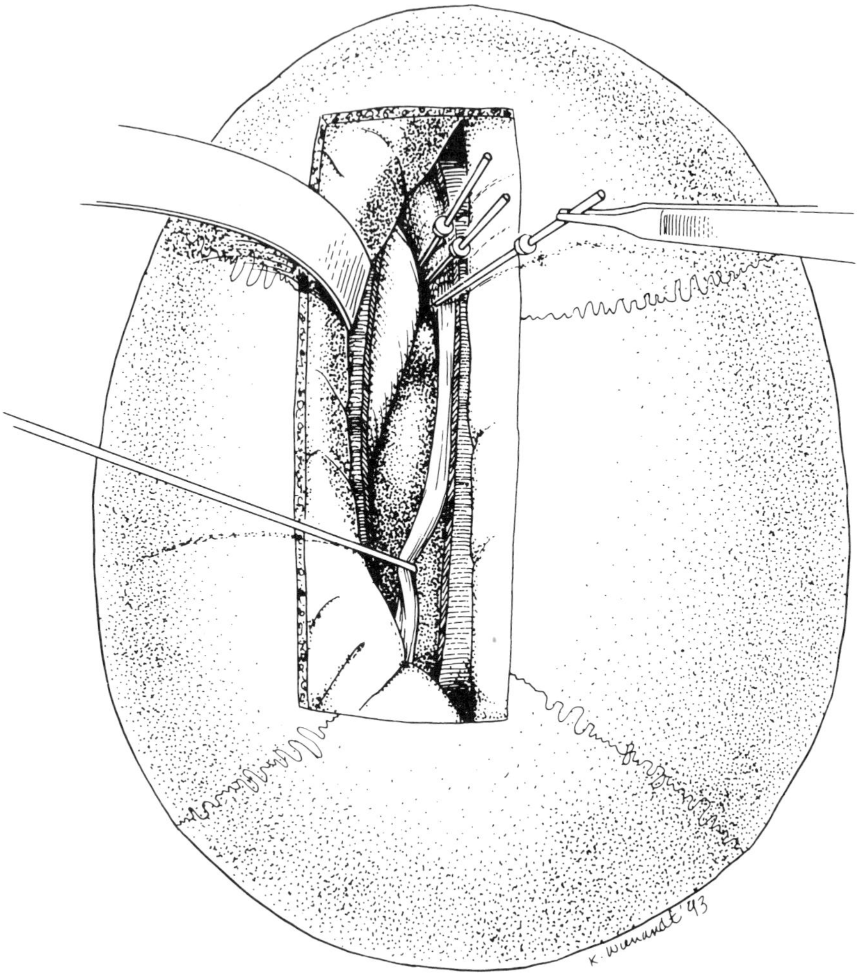

**Figure 31.** Drawing of the dorsal aspect of primate brain illustrates transection of the fornix and placement of polymer capsules containing baby hamster kidney (BHK) cells. After craniotomy, the cerebral hemisphere is retracted and the fornix is unilaterally transected using a microdissector. Microforceps are used to place the polymer capsules within the lateral ventricle, located between caudate (left) and septal (right) areas. The upper portion of the capsules consist of a silastic tether that is clipped to permit closure of the dura. From Emerich *et al.* (1994), Grafts of polymer-encapsulated cells genetically modified to secrete nerve growth factor rescue basal forebrain neurons in nonhuman primates, *J. Comp. Neurol.* **349**:148–165, with permission.

tained from the capsules following grafting for 30 days in aged monkeys induced a robust differentiation of PC12A cells *in vitro* in a manner similar to that observed with exposure to 50 ng of hNGF.

## 15.2. Comments

These observations demonstrate the feasibility of grafting encapsulated cells that have been genetically modified to secrete hNGF in the brain of aged primates.

This encapsulation method permits the use of xenografts without the need for potentially toxic treatment agents that suppress immune function. The pores in the polymer capsule are large enough to allow hNGF and nutrients to move bidirectionally, but are small enough to prevent immune cells from entering and destroying the BHK-grafted cells. Overall, these findings demonstrate that hNGF supports the viability of degenerating aged-primate CBF neurons (Kordower *et al.*, 1994), confirming earlier findings in young adult monkeys (Emerich *et al.*, 1994). Recently, grafted BHK cells modified to secrete NGF were shown to continue producing NGF for at least one year after implantation in rat (Winn *et al.*, 1996). This is additional support for the concept that grafts of genetically modified cells sequestered in a polymer capsule may be a viable way to administer hNGF or other molecules into the human brain for the treatment of the cholinergic deficit arising in dementing illnesses such as AD.

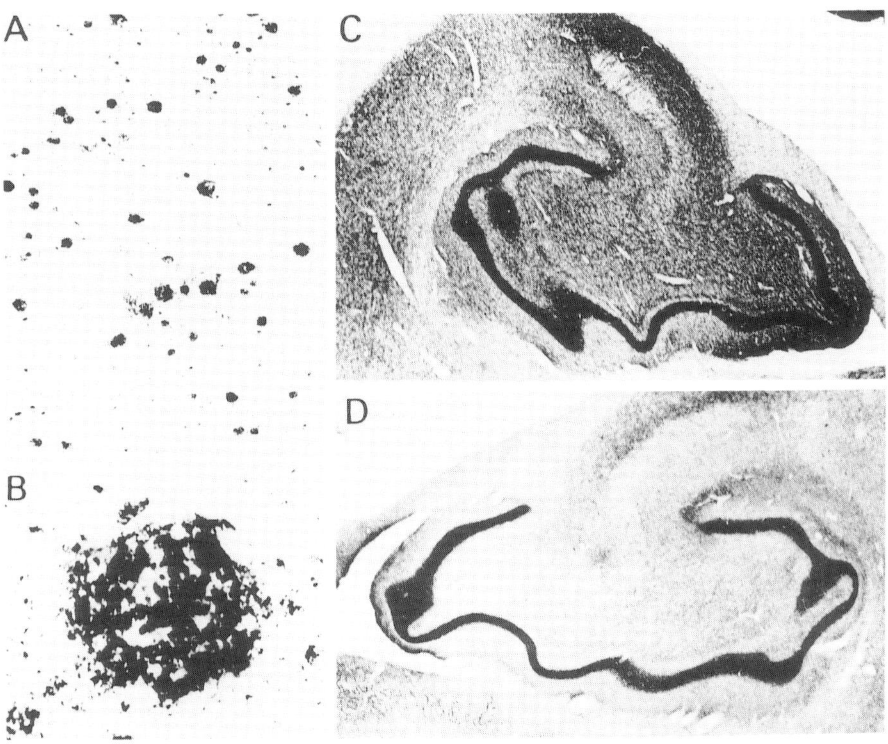


**Figure 32.** (A) Low- and high- (B) power photomicrographs through the temporal neocortex of a 26-year-old monkey, demonstrating the presence and morphology of numerous β-amyloid-immunoreactive senile plaques. (C) The pattern of AChE staining within the hippocampus of a 25-year-old monkey contralateral to the fornix transection. (D) The same section illustrated in panel B, showing a comprehensive loss of AChE-containing fibers ipsilateral to the fornix transection. Bars = 200 μm (A); 50 μm (B); 500 μm (C, D). From Kordower *et al.* (1994c) with permission. Copyright 1994, National Academy of Sciences, U.S.A.

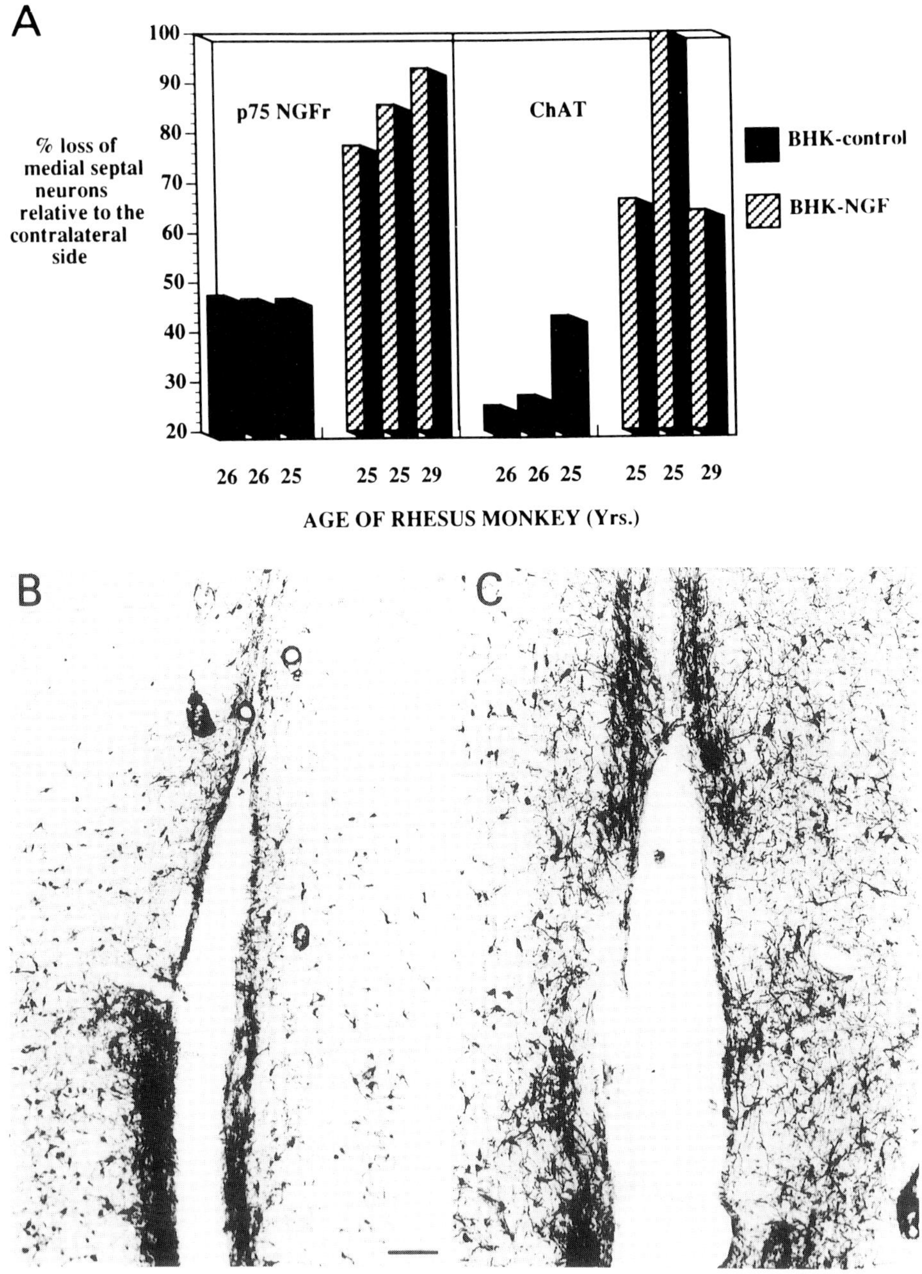


**Figure 33.** (A) Quantification of p75[NTR]- and ChAT-immunostained neuron loss in BHK-control and BHK-hNGF-grafted aged monkeys. (B) p75[NTR]-stained section through the medial septum of a monkey receiving a BHK-control graft. Note the extensive loss of neurons ipsilateral (right) to the lesion relative to the intact (left) side. (C) Monkeys receiving BHK-hNGF grafts displayed a symmetrical pattern of p75[NTR] staining within the medial septum, with minimal neuronal loss. Bar = 100 μm (B, C). From Kordower *et al.* (1994c) with permission. Copyright 1994, National Academy of Sciences, U.S.A.

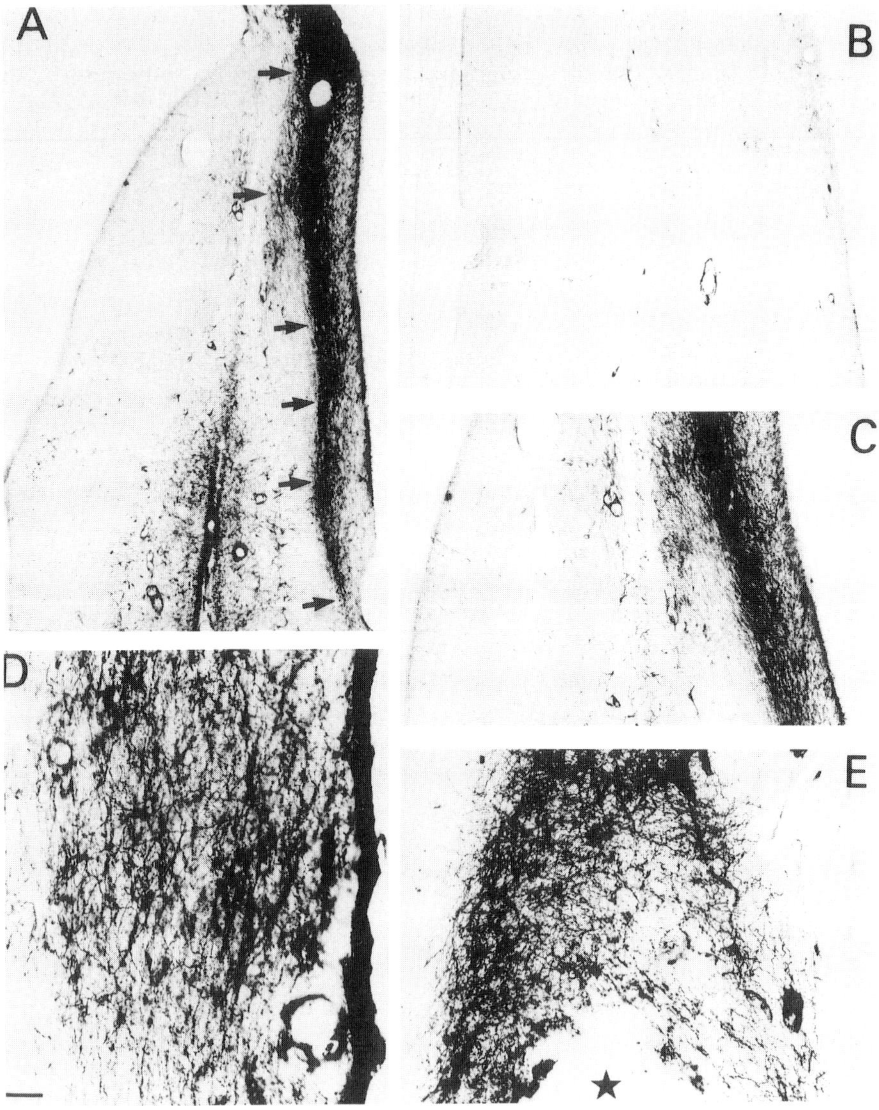


**Figure 34.** (A) Dense unilateral p75$^{NTR}$-immunoreactive sprouting response (arrows) within the septum proximal to a BHK-hNGF graft in a 29-year-old monkey. (B) No such staining pattern is observed within the septum of a 25-year-old monkey receiving a BHK-control graft. (C, D) Morphology of the p75$^{NTR}$-immunoreactive sprouting response, which can be resolved into individual fibers that ramify against the ependyma of the ventricle. (E) In one animal the capsule penetrated into the underlying parenchyma (star). A dense network of p75$^{NTR}$-immunoreactive fibers was also seen in this location adjacent to the implant. Bars = 500 μm (A); 250 μm (C); 50 μm (D, E). From Kordower *et al.* (1994c) with permission. Copyright 1994, National Academy of Sciences, U.S.A.

## 16. Nerve Growth Factor Conjugated to an Antitransferrin Receptor Antibody

Despite the fact that trophic factors may be powerful tools for regulating the viability of specific neuronal populations, their clinical utility is limited by their inability to cross the blood–brain barrier (BBB). Therefore, administration of neurotrophic factors to specific CNS targets requires invasive neurosurgical approaches, such as intraventricular injections (Hefti *et al.*, 1989; Kromer, 1987; Williams *et al.*, 1986) or the grafting of neurotrophin-secreting cells into brain (Schumacher *et al.*, 1991; Stromberg *et al.*, 1990; Kawaja *et al.*, 1992; Emerich *et al.*, 1994; Kordower *et al.*, 1994c). Recently, a novel, noninvasive means of facilitating the transport of macromolecules such as NGF across the BBB has been reported (Friden *et al.*, 1991, 1993; Kordower *et al.*, 1993, 1994c; Granholm *et al.*, 1993, 1994). This delivery system makes use of the endogenous mechanism that actively transports iron into the brain by binding to the transferrin receptor (TfR), which is expressed on brain capillary endothelial cells at a high density. Transferrin, a plasma glycoprotein, binds iron in the blood and functions as the major iron transport protein. Iron is delivered to cells via interaction of the iron–transferrin complex with specific cell-surface receptors (Granholm *et al.*, 1994) and is internalized through receptor-mediated endocytosis. When conjugated to an antibody directed against TfR, proteins normally excluded from the brain can cross the BBB following peripheral administration (Backman *et al.*, 1995; Friden *et al.*, 1991, 1993; Kordower *et al.*, 1994c). Nerve growth factor can be covalently linked to an antibody against the transferrin receptor (termed OX-26) while retaining its biological activity (Friden *et al.*, 1993). This OX-26-NGF conjugate can cross the BBB (Fig. 35) and exert the physiological effects observed previously with direct administration of NGF into brain (Friden *et al.*, 1993; Granholm *et al.*, 1994).

### 16.1. Actions of OX-26-NGF on Cholinergic Systems

The ability of the intravenously administered NGF conjugate to support the viability of central cholinergic neurons has been shown in a number of model

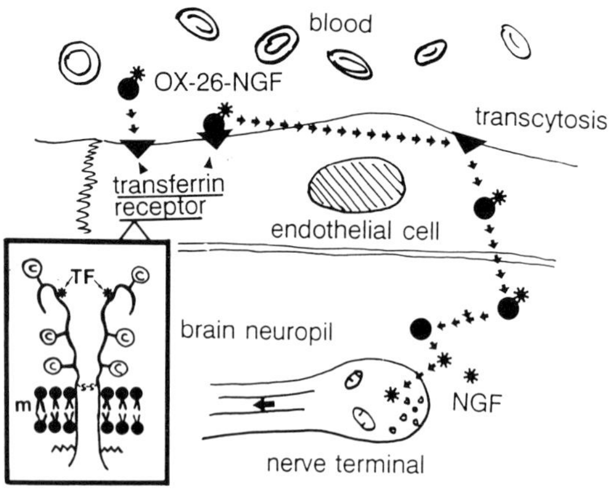


**Figure 35.** Schematic illustration of the hypothesized binding and transfer of the OX-26-NGF conjugate on brain capillary endothelial cells. The conjugate is thought to bind to the transferrin receptor on the endothelial cell surface and then to be transferred by transcytosis into the brain parenchyma. (Inset) Transferrin receptor location in the cell membrane (m) and the two sites for transferrin (TF) binding on the extracellular portion of the receptor. From Granholm *et al.* (1994) with permission.

systems dealing with cholinergic innervation within the CNS. Initial experiments demonstrated that intravenous injections of the NGF conjugate increased the viability and phenotypic expression of CBF perikarya after transplantation into the anterior chamber of the eye (Backman *et al.*, 1995; Friden *et al.*, 1993; Granholm *et al.*, 1994). Within the central nervous system, we recently demonstrated that intravenous NGF-conjugate treatment specifically prevented the degeneration of cholinergic striatal neurons in a rodent model of Huntington's disease (Kordower *et al.*, 1994a), showing that two weeks of NGF conjugate administration prevented the loss of ChAT-immunoreactive, but not somatostatin-immunoreactive or diaphorase-positive neurons that normally occurs following intrastriatal quinolinic acid injections.

### 16.2. OX-26-NGF Prevents Cholinergic Basal Forebrain Cell Degeneration in Animal Correlates of Alzheimer's Disease

In animal experiments mimicking degeneration of the CBF cortical projection system—a hallmark of AD—we tested whether intravenous administration of an OX-26-NGF conjugate could reverse atrophy of CBF neurons following damage to their cortical target zones by epidural application of *N*-methyl-D-aspartic acid (NMDA) (Charles *et al.*, 1996). Seventy-five days post-lesion, the neurons were atrophic ipsilateral to the cortical damage as compared with the contralateral side in control rats receiving intravenous injections of vehicle or a nonconjugated mixture of OX-26 and NGF. By contrast, OX-26-NGF conjugate intravenous injections restored the size of $p75^{NTR}$ and trkA nucleus basalis perikarya to within normal limits relative to the nonlesioned contralateral side (Fig. 36; see also Sofroniew *et al.*, 1983, 1987; Figueriedo *et al.*, 1995). Because the BBB was closed at the time of treatment, this trophic effect did not result from NGF crossing a compromised BBB at the lesion site.

These findings demonstrate that systemic administration of a neurotrophic factor–antibody conjugate, intended to circumvent the BBB, provides trophic influences upon atrophic CBF neurons. The lesion model used does not induce cell death; rather, it shrinks the size of CBF neurons, modeling an early pathological process in AD. These observations support the emerging concept that NGF covalently conjugated to a protein vector delivery system—the OX-26 rat antibody against the transferrin receptor—can facilitate the transfer of impermeable therapeutic molecules (e.g., NGF) across the BBB in animal correlates of AD-like CBF degeneration.

## 17. Conclusion

Although it has been hypothesized that alterations in the expression of NGF may underlie the pathophysiology of CBF neurodegeneration in AD (Appel, 1981; Hefti *et al.*, 1989), the data we have presented here suggest a more complex scenario. Investigations in our laboratories using an antibody that recognizes endogenous NGF (Conner *et al.*, 1992; Mufson *et al.*, 1994) revealed a reduction in transported NGF-like immunoreactivity within remaining CBF consumer neurons in AD

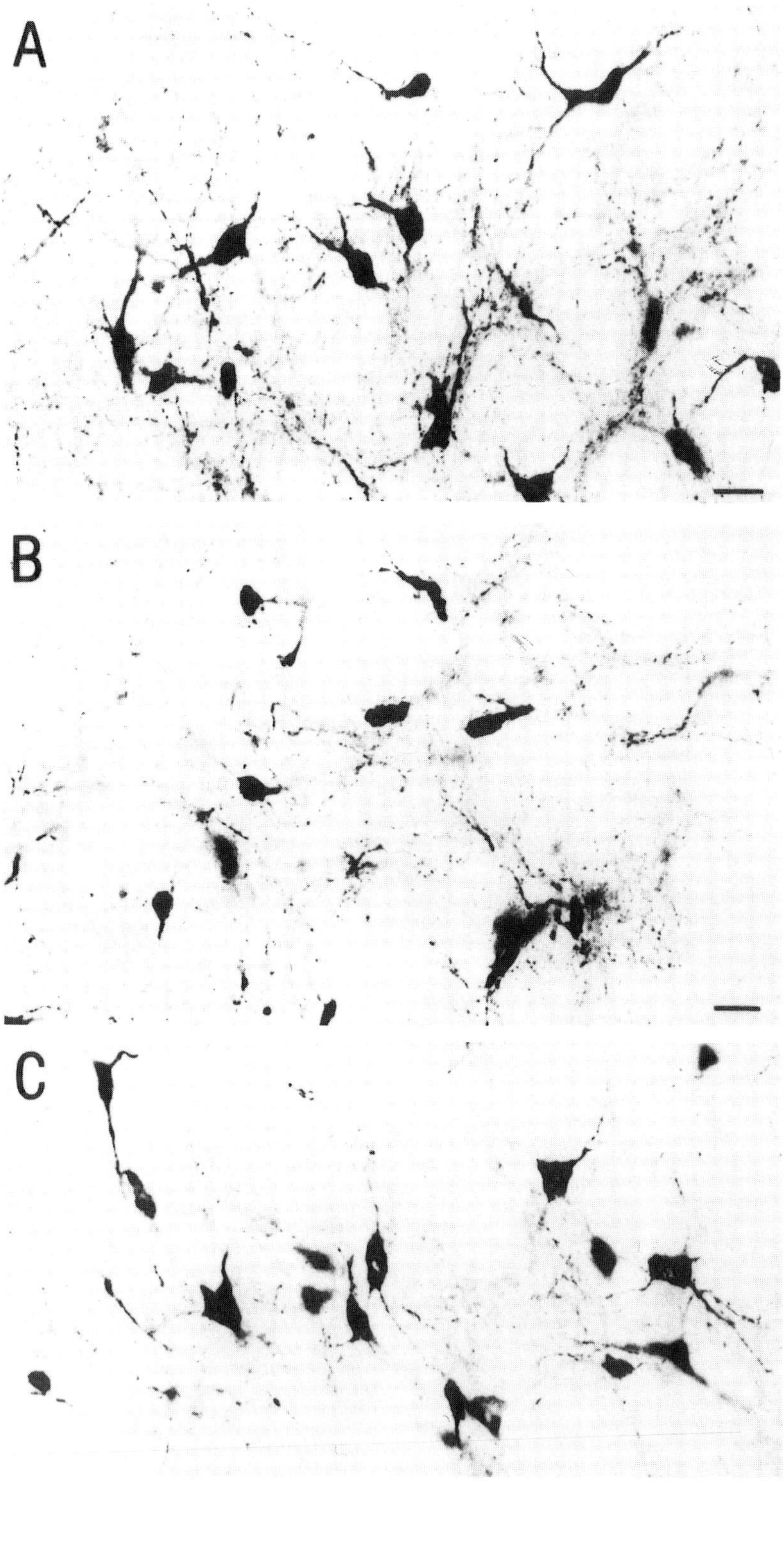
A
B
C

(Mufson *et al.*, 1995). In contrast, NGF levels in CBF cortical target sites are reported to be unchanged (Goedert *et al.*, 1986) or increased (Crutcher *et al.*, 1993; Scott *et al.*, 1995) in AD. Similarly, levels of p75$^{NTR}$ are either stable or elevated in this disease (Higgins and Mufson, 1989; Ernfors *et al.*, 1990), suggesting impaired internalization and/or retrograde transport of target-derived NGF associated with a defect in the complexing of this trophin with its high-affinity trkA receptor. The significant reduction in trkA mRNA and protein in remaining CBF neurons that we observed in AD (Mufson *et al.*, 1996, 1997) supports this concept, and suggests that NGF accumulation in the cortex is due in part to a defect in the production, utilization, and/or binding of NGF with its trkA receptor. Therefore, reductions in trkA message and protein within GBF neurons in AD may signal the crucial neuropathological change that underlies CBF degeneration. Interestingly, activation of trkA alone can prevent degeneration of axotomized CBF neurons in rats (Lucidi-Phillipi *et al.*, 1996), indicating that signaling via the trkA receptor is a pivotal trophic event underlying the viability of CBF neurons in AD. Therefore, a need may exist to develop compounds that do not necessarily regulate endogenous NGF production, but rather its high-affinity trkA receptor, to treat CBF degeneration in AD.

ACKNOWLEDGMENTS. We wish to thank all of our collaborators who participated in the studies reported in this chapter. We would also like to thank Syed Jaffar, Africa Munarriz, Christine Quijano, and Noelen Doogan for histologic, photographic, and secretarial assistance. This work was supported by AG10161, AG09466, AG10668, AG14449, the Alzheimer's Disease Association, the Washington Square Health Foundation, and the Illinois Department of Public Health.

## 18. References

Appel, S. H., 1981, A unifying hypothesis for the cause of amyotrophic lateral sclerosis, Parkinsonism, and Alzheimer's disease, *Ann. Neurol.* **10:**499–505.

Armstrong, D. M., Terry, R. D., Deteresa, R. M., Bruce, G., Hersh, L. B., and Gage, F. H., 1987, Response of septal cholinergic neurons to axotomy, *J. Comp. Neurol.* **264:**421–436.

Backman, C., Biddle, P. T., Ebendal, T., Friden, P. M., Gerhardt, G. A., Henry, M. A., Mackerlova, L., Soderstrom, S., Stromberg, I., Walus, L., and Granholm, A.-C., 1995, Effects of transferrin receptor antibody-NGF conjugate on young and aged septal transplants *in oculo*, *Exp. Neurol.* **132:**1–15.

Barbacid, M., 1994, The trk family of neurotrophin receptors, *J. Neurobiol.* **25:**1386–1403.

Bartus, R. T., Dean, R. L., Beer, B., and Lippa, A. S., 1982, The cholinergic hypothesis of geriatric memory dysfunction, *Science* **217:**408–417.

**Figure 36.** Photomicrographs of trkA-immunoreactive neurons within nucleus basalis. (A) Note the large, healthy-appearing magnocellular perikarya within the intact nucleus basalis. (B) These neurons are shrunken ipsilateral to the lesion in vehicle-treated animals. (C) trkA-immunoreactive nucleus basal neurons are of normal size when treated with the OX-26-NGF conjugate. Bars = 40 μm. Reprinted from *Brain Research* 728, Charles *et al.*, "Atrophy of cholinergic basal forebrain neurons following excitotoxic cortical lesions is reversed by intravenous administration of an NGF conjugate," pp. 193–203, 1996, with kind permission of Elsevier Science—NL, Sara Burgerhartstraat 25, 1055 KV, Amsterdam, The Netherlands.

Bartus, R. T., Dean, R. L., Abelleira, S., Charles, V., and Kordower, J. H., 1996, Dissociation of p75 receptors and nerve growth factor neurotrophic effects: Lack of p75 immunoreactivity in striatum following physical trauma, excitotoxicity, and NGF administration, *Restor. Neurol.* **10:**49–59.

Batchelor, P. E., Armstrong, D. M., Blaker, S. N., and Gage, F. H., 1989, Nerve growth factor receptor and choline acetyltransferase colocalization in neurons within the rat forebrain: Response to fimbria-fornix transection, *J. Comp. Neurol.* **284:**187–204.

Benzing, W. C., Kordower, J. H., and Mufson, E. J., 1993, Galanin immunoreactivity within the primate basal forebrain: Evolutionary change between monkeys and apes, *J. Comp. Neurol.* **283:**189–194.

Biegon, A., Greenberger, V., and Segal, M., 1986, Quantitative histochemistry of brain acetylcholine and learning rate in the aged rat, *Neurobiol. Aging* **7:**215–217.

Bierer, L. M., Haroutunian, V., Gabreil, S., Knott, P. J., Carlin, L. S., Purohit, D. P., Perl, D. P., Schmeidler, J., Kanof, P., and Davis, K. L., 1995, Neurochemical correlates of dementia in Alzheimer's disease: Relative importance of cholinergic deficits, *J. Neurochem.* **64:**749–760.

Boissière, B., Faucheux, Y., Agid, Y., and Hirsch, E. C., 1996, Decrease of trkA mRNA level in cholinergic neurons of the striatum and basal forebrain in Alzheimer's disease, *Soc. Neurosci. Abstr.* **22:**210.

Bothwell, M., 1991, Keeping track of neurotrophin receptors, *Cell* **65:**915–918.

Bowser, R., Kordower, J. H., and Mufson, E. J. (1997), A confocal microscopic analysis of galaninergic hyperinnervation of cholinergic basal forebrain neurons in Alzheimer's disease, *Brain Pathol.* **7:**723–730.

Chan-Palay, V., 1988, Galanin hyperinnervates surviving neurons of the human basal nucleus of Meynert in dementias of Alzheimer's and Parkinson's disease: A hypothesis for the role of galanin in accentuating cholinergic dysfunction in dementia, *J. Comp. Neurol.* **273:**543–557.

Chao, M. V., 1992, Neurotrophin receptors: A window into neuronal differentiation, *Neuron* **9:**583–593.

Chao, M. V., 1995, The p75 neurotrophin receptor, *J. Neurobiol.* **25:**1375–1385.

Chao, M. V., and Hempstead, B. L., 1995, p75 and Trk: A two-receptor system, *Trends Neurosci.* **18:**321–326.

Charles, V., Mufson, E. J., Friden, P. M., Bartus, R. T., and Kordower, J. H., 1996, Atrophy of cholinergic basal forebrain neurons following excitotoxic cortical lesions is reversed by intravenous administration of an NGF conjugate, *Brain Res.* **728:**193–203.

Chen, E.-Y., Mufson, E. J., and Kordower, J. H., 1996, Trk and p75 NGF receptor systems in the developing human brain, *J. Comp. Neurol.* **369:**591–618.

Collier, T. J., and Coleman, P., 1991, Divergence of biological and chronological aging: Evidence from rodent studies, *Neurobiol. Aging* **12:**685–693.

Conner, J. M., and Varon, S., 1992, Distribution of nerve growth factor-like immunoreactive neurons in the adult rat brain following colchicine treatment, *J. Comp. Neurol.* **326:**347–362.

Conner, J. M., Muir, D., Varon, S., Hagg, T., and Manthorpe, M., 1992, The localization of nerve growth factor-like immunoreactivity in the adult rat basal forebrain and hippocampal formation, *J. Comp. Neurol.* **319:**454–462.

Cooper, J. D., Lindholm, D., and Sofroniew, M. V., 1994, Reduced transport of [$^{125}$I] nerve growth factor by cholinergic neurons and down-regulated TrkA expression in the medial septum of aged rats, *Neurosci.* **62:**625–639.

Crawley, J. N., 1996, Galanin-acetylcholine interactions: Relevance to memory and Alzheimer's disease, *Life Sci.* **58:**2185–2199.

Crutcher, K. A., Scott, S. A., Liang, S., Everson, W. V., and Weingartner, J., 1993, Detection of NGF-like activity in the human brain tissue: Increased levels in Alzheimer's disease, *J. Neurosci.* **13:**2540–2550.

de Lacalle, S., Lim, C., Sobreviela, T., Mufson, E. J., Hersh, L. B., and Saper, C. B., 1994, Cholinergic innervation in the human hippocampal formation including the entorhinal cortex, *J. Comp. Neurol.* **345:**321–344.

Emerich, D. F., Winn, S. R., Hammang, J. P., Baetge, E. E., and Kordower, J. H., 1994, Grafts of polymer-encapsulated cells genetically modified to secrete nerve growth factor rescue basal forebrain neurons in nonhuman primates, *J. Comp. Neurol.* **349:**148–165.

Ernfors, P., Lindefors, N., Chan-Palay, V., and Persson, H., 1990, Cholinergic neurons of the nucleus basalis express elevated levels of nerve growth factor receptor mRNA in senile dementia of the Alzheimer's type, *Dementia* **1:**138–145.

Feinstein, D., and Larhammar, D., 1990, Identification of a conserved protein motif in a group of growth factor receptors, *FEBS Lett.* **272:**7–11.

Figueiredo, B. C., Skup, M., Bedard, A. M., Tetzlaff, W., and Cuello, A. C., 1995, Differential expression of p140trk, p75NGFr, and growth-associated phosphoprotein-43 genes in nucleus basalis magnocellularis, thalamus, and adjacent cortex following neocortical infarction and nerve growth factor treatment, *Neuroscience* **68:**29–45.

Fischer, W., Wictorin, K., Bjorklund, A., Williams, L. R., Varon, S., and Gage, F. H., 1987, Amelioration of cholinergic neuron atrophy and spatial memory impairment in aged rats by nerve growth factor, *Nature* **329:**65–68.

Fisone, G., Wu, C. F., Consolo, S., Nordstrom, O., Brynne, N., Bartfai, T., Melander, T., and Hokfelt, T., 1987, Galanin inhibits acetylcholine release in the ventral hippocampus of the rat: Histochemical, autoradiographic, *in vivo,* and *in vitro* studies, *Proc. Natl. Acad. Sci. USA* **84:**7339–7343.

Friden, P. M., Walus, L. R., Musso, G. F., Taylor, M. A., Malfroy, B., and Starzyk, R. M., 1991, Anti-transferrin receptor antibody and antibody-drug conjugates cross the blood-brain barrier, *Proc. Natl. Acad. Sci. USA* **88:**4771–4775.

Friden, P. M., Walus, L. R., Watson, P., Doctrow, S. R., Kozarich, J. W., Backman, C., Bergman, H., Hoffer, B., Bloom, F., and Granholm, A. C., 1993, Blood-brain barrier penetration and *in vivo* activity of an NGF conjugate, *Science* **259:**373–377.

Gage, F. H., Batchelor, P., Chen, K. S., Chin, D., Higgins, G. A., Koh, S., Deputy, S., Rosenberg, M. B., Fischer, W., and Bjorklund, A., 1989, NGF receptor reexpression and NGF-mediated cholinergic neuronal hypertrophy in the damaged adult neostriatum, *Neuron* **2:**1177–1184.

Geddes, J. W., Mongahan, D. T., Cotman, C. W., Lott, I. T., Kim, R. C., and Chui, H.-C., 1985, Plasticity of hippocampal circuitry in Alzheimer's disease, *Science* **230:**1179–1181.

Gibbs, R. B., and Pfaff, D. W., 1994, *In situ* hybridization detection of trkA mRNA in brain: Distribution, colocalization with p75NGFR, and up-regulation by nerve growth factor, *J. Comp. Neurol.* **341:**324–339.

Gnahn, H., Hefti, F., Heumann, R., Schwab, M. E., and Thoenen, H., 1983, NGF mediated increase of choline acetyltransferase (ChAT) in the neonatal forebrain: Evidence for a physiological role of NGF in the brain? *Dev. Brain Res.* **9:**45–52.

Goedert, M., Fine, A., Hunt, S. P., and Ulrich, A., 1986, NGF RNA in peripheral and rat central tissues and in the human central nervous system: Lesion effects in the rat brain and levels in Alzheimer's disease, *Mol. Brain Res.* **1:**85–92.

Granholm, A.-C., Biddle, P. T., Backman, C., Ebenberg, T., Gerhardt, G., Hoffer, B., Mackerlova, L., Olson, L., Soderstrom, S., Walus, L., and Friden, P., 1994, Peripheral administration of nerve growth factor conjugated to an anti-transferrin receptor antibody increase cholinergic neuron survival in intraocular forebrain transplants, in: *Providing Pharmacological Access to the Brain: Alternative Approaches* (T. R. Flanagan, D. W. Emerich, and S. R. Winn, eds.), Academic, New York, pp. 71–92.

Granholm, A. C., Backman, C., Bloom, F., Ebendal, T., Gerhardt, G. A., Hoffer, B., Mackerlova, L., Olson, L., Soderstrom, S., Walus, L. R., and Friden, P. M., 1993, NGF and anti-transferrin receptor antibody conjugate: Short and long-term effects on survival of cholinergic neurons in intraocular septal transplants, *J. Pharmacol. Exp. Ther.* **268:**448–459.

Hefti, F., and Weiner, W. J., 1983, Is Alzheimer's disease caused by lack of nerve growth factor? *Ann. Neurol.* **13:**109–110.

Hefti, F., and Mash, D. C., 1989, Localization of nerve growth factor receptors in the normal human brain and in Alzheimer's disease, *Neurobiol. Aging* **10:**75–87.

Hefti, F., Hartikka, J., Eckenstein, F., Gnahn, H., Heumann, R., and Schwab, M., 1985, Nerve growth factor (NFG) increases choline acetyltransferase, but not survival or fiber outgrowth, of cultured fetal septal cholinergic neurons, *Neuroscience* **14:**55–68.

Hefti, F., Hartikka, J., Salviaterra, P., Weiner, W. J., and Mash, D. C., 1986, Localization of nerve growth factor receptors on cholinergic neurons of the human basal forebrain, *Neurosci. Lett.* **69:**275–281.

Hefti, F., Hartikka, J., and Knusel, B., 1989, Function of neurotrophic factors in the adult and aging brain and their possible use in the treatment of neurodegenerative disease, *Neurobiol. Aging* **10:**515–533.

Heldin, C.-H., 1995, Dimerization of cell surface receptors in signal transduction, *Cell* **80:**213–223.

Hempstead, B. L., Martin-Zanca, D., Kaplan, D. R., Parada, L. R., and Chao, M. V., 1991, High-affinity NFF binding requires coexpression of the trk proto-oncogene and the low-affinity NGF receptor, *Nature* **350:**678–683.

Heumann, R., Schwab, M., Merkl, R., and Thoenen, H., 1984, Nerve growth factor-mediated induction of choline acetyl transferase in PC12 cells: Evaluation of the site of action of nerve growth factor and the involvement of lysosomal degradation products of nerve growth factor, *J. Neurosci.* **4:**3030–3050.

Higgins, G. A., and Mufson, E. J., 1989, NGF receptor gene expression is decreased in the nucleus basalis in Alzheimer's disease, *Exp. Neurol.* **106:**222–236.

Holtzman, D. M., Kilbridge, J., Li, Y., Cunningham, E. T., Jr., Lenn, N. J., Clary, D. O., Reichardt, L. F., and Mobley, W. C., 1995, TrkA expression in the CNS: Evidence for the existence of several novel NGF-responsive CNS neurons, *J. Neurosci.* **15:**1567–1676.

Ibanez, C. F., Ebendal, T., Barbany, G., Murray-Rust, J., Blundell, T. L., and Persson, H., 1992, Similarities and differences in the way neurotrophins interact with the trk receptors in neuronal and nonneuronal cells, *Cell* **69:**329–341.

Ip, N., Ibanez, C., Nye, S., McClain, J., Jones, P. F., Gies, D. R., Belluscio, L., Le Beau, M. M., Espinsosa, R., III, Squinto, S. P., Persson, H., and Yancopoulos, G. D., 1992, Mammalian neurotrophin-4: Structure, chromosomal localization, tissue distribution, and receptor specificity, *Proc. Natl. Acad. Sci. USA* **89:**3060–3064.

Ip, N., Tapley, P., Glass, D. J., Greene, L. A., and Yancopoulos, G. D., 1993, Similarities and differences in the way neurotrophins interact with the trk receptors in neuronal and nonneuronal cells, *Neuron* **10:**137–149.

Jing, S., Tapley, P., and Barbacid, M., 1992, Nerve growth factor mediates signal transduction through trk homodimer receptors, *Neuron* **9:**1067–1079.

Kaplan, D. R., and Stephens, R. M., 1994, Neurotrophin signal transduction by the trk receptor, *J. Neurobiol.* **25:**1404–1417.

Kaplan, D. R., Hempstead, B. L., Martin-Zanca, D., Chao, M. V., and Parada, L. F., 1991, The trk proto-oncogene product: Signal transducting receptor for nerve growth factor, *Science* **252:**554–558.

Kawaja, M. D., Rosenberg, M. B., Yoshida, K., and Gage, F. H., 1992, Somatic gene transfer of nerve growth factor promotes the survival of axotomized septal neurons and the regenerations of the axons in adult rats, *J. Neurosci.* **12:**2849–2864.

Koh, S., and Loy, R., 1988, Age-related loss of nerve growth factor sensitivity in rat basal forebrain neurons, *Brain Res.* **440:**396–401.

Koh, S., Chang, P., Collier, T. J., and Loy, R., 1989, Loss of NGF receptor immunoreactivity in basal forebrain neurons of aged rats: Correlation with spatial memory impairment, *Brain Res.* **498:**397–404.

Koliatsos, V. E., Price, D. L., Gouras, G. K., Cayouette, M. H., Burton, L. E., and Winslow, J. W., 1994, Highly selective effects of nerve growth factor, brain-derived neurotrophic factor, and neurotrophin-e on intact and injured basal forebrain magnocellular neurons, *J. Comp. Neurol.* **343:**247–262.

Kordower, J. H., and Mufson, E. J., 1989, NGF and Alzheimer's disease: Unfulfilled promise and untapped potential, *Neurobiol. Aging* **10:**543–544.

Kordower, J. H., and Mufson, E. J., 1990, Galanin-like immunoreactivity within the primate basal forebrain: Differential staining patterns between humans and monkeys, *J. Comp. Neurol.* **294:**281–292.

Kordower, J. H., and Mufson, E. J., 1992, NGF receptor immunoreactive neurons within the developing human cerebral cortex and hippocampal complex, *J. Comp. Neurol.* **323:**25–41.

Kordower, J. H., Bartus, R. T., Bothwell, M., Schatteman, G., and Gash, D. M., 1988, Nerve growth factor receptor immunoreactivity in the nonhuman primate (*Cebus apella*): Distribution, morphology, and colocalization with cholinergic enzymes, *J. Comp. Neurol.* **277:**465–486.

Kordower, J. H., Gash, D. M., Bothwell, M., Hersh, L. B., and Mufson, E. J., 1989, Nerve growth factor receptor and choline acetyltransferase remain colocalized in the nucleus basalis (Ch4) of Alzheimer's patients, *Neurobiol. Aging* **10:**287–294.

Kordower, J. H., Le, H. K., and Mufson, E. J., 1992, Galanin-immunoreactivity in the primate central nervous system, *J. Comp. Neurol.* **319:**479–500.

Kordower, J. H., Mufson, E. J., Granholm, A. C., Hoffer, B., and Friden, P. M., 1993, Delivery of trophic factors to the primate brain, *Exp. Neurol.* **124:**21–30.

Kordower, J. H., Chen, E.-Y., Sladek, J. R., Jr., and Mufson, E. J., 1994b. Trk-immunoreactivity in the monkey central nervous system. I. Forebrain, *J. Comp. Neurol.* **349:**20–35.

Kordower, J. H., Winn, S. R., Liu, Y.-T., Mufson, E. J., Sladek, J. R., Hammang, J. P., Baetge, E. E., and Emerich, D. F., 1994c, The aged monkey basal forebrain: Rescue and sprouting of degenerating cholinergic neurons following grafts of polymer-encapsulated human NGF secreting cells, *Proc. Natl. Acad. Sci. USA* **91:**10898–10902.

Kordower, J. H., Charles, V., Bayer, R., Bartus, R. T., Putney, S., Walus, L., and Friden, P. M., 1994a, Intravenous administration of an NGF conjugate prevents neuronal degeneration in a model of Huntington's disease, *Proc. Natl. Acad. Sci. USA* **91:**9077–9080.

Kordower, J. H., Mufson, E. J., Fox, N., Martel, L., and Emerich, D., 1997, NGF does not alter the expression of β amyloid-immunoreactivity in the young or aged nonhuman primate, *Exp. Neurol.* **145:**586–591.

Koster, G. R., Winkler, C., Raulf, F., Lottspeich, F., Schartl, M., and Thoenen, H., 1994, Neurotrophin-6 is a new member of the nerve growth family, *Nature* **372:**266–269.

Kromer, L. F., 1987, Nerve growth factor treatment after brain injury prevents neuronal death, *Science* **235:**214–216.

LeSauteur, L., Malirtchouk, S., Lejeune, H., Quirion, R., and Saragovi, H. U., 1996, Potent human p40-trk agonists derived from an anti-receptor monoclonal antibody, *J. Neurosci.* **16:**1308–1316.

Levi-Montalcini, R., and Angeletti, P. U., 1968, Nerve growth factor, *Physiol. Rev.* **8:**534–569.

Lindsay, R. M., Wiegand, S. J., Altar, C. A., and Distephano, P. S., 1994, Neurotrophic factors: From molecule to man, *Trends Neurosci.* **17:**182–190.

Lucidi-Phillipi, C. A., Clary, D. O., Reichardt, L. F., and Gage, F., 1996, TrkA activation is sufficient to rescue axotomized cholinergic neurons, *Neuron* **16:**653–663.

Maisonpierre, P. C., Bellusco, L., Friedman, B., Alderson, R. F., 1990, NT-3, BDNF in the developing rat nervous system: Parallel as well as reciprocal patterns of expression, *Neuron* **4:**501–509.

Martinez, H. J., Dryfuss, C. F., Jonakait, G. M., and Black, I. B., 1985, Nerve growth factor promotes cholinergic development in brain striatal cultures, *Proc. Natl. Acad. Sci. USA* **82:**7777–7781.

Markram, H., and Segal, M., 1990, Regional changes in NGF receptor immunohistochemical labeling in the septum of the aged rat, *Neurobiol. Aging* **11:**481–484.

Meakin, S. O., and Shooter, E. M., 1991, Molecular investigation of the high-affinity nerve growth factor receptor, *Neuron* **6:**153–163.

Melander, T., Hökfelt, T., and Rokaeus, A., 1986, Distribution of galanin-like immunoreactivity in the rat central nervous system, *J. Comp. Neurol.* **248:**475–517.

Mesulam, M. M., Mufson, E. J., Levey, A. I., and Wainer, B. H., 1983, Cholinergic innervation of cortex by the basal forebrain: Cytochemistry and cortical connections of the septal area, diagonal band nuclei, nucleus basalis (substantia innominata), and hypothalamus in the rhesus monkey, *J. Comp. Neurol.* **214:**170–197.

Mobley, W. C., Tennekoon, J. L., Buchannan, K., and Johnston, M. V., 1985, Choline acetyltransferase activity in the striatum of neonatal rats is increased by nerve growth factor, *Science* **229:**284–287.

Mufson, E. J., and Kordower, J. H., 1992, Cortical neurons express the receptor for NGF in advanced age and Alzheimer's disease, *Proc. Natl. Acad. Sci. USA* **89:**569–573.

Mufson, E. J., Bothwell, M., and Kordower, J. H., 1989b, Loss of nerve growth factor receptor-containing neurons in Alzheimer's disease: A quantitative analysis across subregions of the basal forebrain, *Exp. Neurol.* **105:**221–232.

Mufson, E. J., Bothwell, M., Hersh, L. B., and Kordower, J. H., 1989a, Nerve growth factor receptor immunoreactive profiles in the normal aged human basal forebrain: Colocalization with cholinergic neurons, *J. Comp. Neurol.* **285:**196–217.

Mufson, E. J., Cochran, E., Benzing, W. C., and Kordower, J. H., 1993, Galaninergic innervation of the cholinergic vertical limb of the diagonal band (Ch2) and bed nucleus of the striia terminalis in aging, Alzheimer's disease, and Down's syndrome, *Dementia* **4:**237–250.

Mufson, E. J., Connor, J., Varon, S., and Kordower, J. H., 1994, Nerve growth factor immunoreactivity in the basal forebrain and hippocampus in primates and Alzheimer's disease, *J. Comp. Neurol.* **341:**507–519.

Mufson, E. J., Connor, J. M., and Kordower, J. H., 1995, NGF and Alzheimer's disease: Defective retrograde transport to the nucleus basalis, *NeuroReport* **6:**1063–1066.

Mufson, E. J., Li, J.-M., Sobreviela, T., and Kordower, J. H., 1996, Decreased trkA gene expression within basal forebrain neurons in Alzheimer's disease, *NeuroReport* **7:**25–29, 1996.

Mufson, E. J., Lavine, N., Jaffar, S., Kordower, J. H., Quirion, R., and Saragovi, H. U., 1997, Reduction in p140-TrkA receptor protein within the nucleus basalis and cortex in Alzheimer's disease, *Exp. Neurol.* **146:**91–103.

Olson, L., 1993, NGF and the treatment of Alzheimer's disease, *Exp. Neurol.* **124:**5–15.

Parada, L. F., Tsoulfas, P., Tessarollo, L., Blair, J., Reid, S. W., and Soppert, D., 1992, The trk family of tyrosine kinases-receptors for NGF-related neurotrophins, *Cold Spring Harbor Symp. Quant. Biol.* **57:**43–51.

Perry, E. K., 1980, The cholinergic system in old age and Alzheimer's disease, *Age Ageing* **9:**1–8.

Planas, B., Kolb, P. E., Ruskind, M. A., and Miller, M. A., 1997, Nerve growth factor induces galanin gene expression in the rat basal forebrain: implications for treatment of cholinergic dysfunction, *J. Comp. Neurol.* **379:**563–570.

Richardson, P. M,. Verge Issa, V. M. K., and Riopelle, R. J., 1986, Distribution of neuronal receptor for nerve growth factor in the rat, *J. Neurosci.* **6:**2312–2321.

Rodriguez-Tebar, A., Dechant, G., and Barde, Y.-A., 1990, Binding of brain-derived neurotrophic factor to the nerve growth factor receptor, *Neuron* **4:**487–492.

Salehi, A., Verhaggen, J., Dijhuizen, P. A., and Swaab, D. F., 1996, Colocalization of high affinity neurotrophin receptors in nucleus basalis of Meynert neurons and their differential reduction in Alzheimer's disease, *Neuroscience* **75:**373–387.

Saporito, M. S., Carswell, S., 1995, High levels of synthesis and local effects of nerve growth factor in the septal region of the adult rat brain, *J. Neurosci.* **15:**2280–2286.

Schumacher, J. M., Short, M. P., Hyman, B. T., Breakefield, X. O., and Isacson, O., 1991, Intracerebral implantation of nerve growth factor-producing fibroblasts protects striatum against neurotoxic levels of excitatory amino acids, *Neuroscience* **45:**561–570.

Schwaab, M. E., Otten, U., Agid, Y., and Thoenen, H., 1979, Nerve growth factor in the rat CNS: Absence of specific retrograde transport and tyrosine hydroxylase induction in the locus coeruleus and substantia nigra, *Brain Res.* **168:**473–483.

Scott, S. A., and Crutcher, K. A., 1994, Nerve growth factor and Alzheimer's disease, *Rev. Neurosci.* **5:**179–211.

Scott, S. A., Weingartner, J. A., Liang, S., and Crutcher, K. A., 1994, Increased NGF-like activity in young but not aged rat hippocampus after septal lesions, *Neurobiol. Aging* **15:**337–346.

Seiler, M., and Schwaab, M. E., 1984, Specific retrograde transport of nerve growth factor from neocortex to nucleus basalis in rat, *Brain Res.* **300:**33–39.

Shelton, D. L., Sutherland, J., Gripp, J., Camerato, T., Armanini, M., Philips, H., Carroll, K., Spencer, S. D., and Levinson, A. D., 1995, Human trks-molecular cloning, tissue distribution, and expression of extracellular domain immunoadhesions, *J. Neurosci.* **15:**477–491.

Sobreviela, T., Clary, D. O., Reichardt, L. F., Brandabur, M. M., Kordower, J. H., and Mufson, E. J., 1995, TrkA immunoreactive profiles in the central nervous system: Colocalization with neurons containing p75 nerve growth factor receptor, choline acetyltransferase, and serotonin, *J. Comp. Neurol.* **350:**587–611.

Sofroniew, M. V., Pearson, R. C. A., Eckenstein, F., Cuello, A. C., and Powell, T. P. S., 1983, Retrograde changes in cholinergic neurons in the basal forebrain of the rat following cortical damage, *Brain Res.* **289:**370–374.

Sofroniew, M. V., Pearson, R. C. A., and Powell, T. P. S., 1987, The cholinergic nuclei of the basal forebrain of the rat: Normal structure, development, and experimentally induced degeneration, *Brain Res.* **371:**310–331.

Soltoff, S. P., Rabin, S. L., Cnatley, L. C., and Kaplan, D. R., 1992, Nerve growth factor promotes the activation of phosphatidylinositol-3 kinase and its association with the trk tyrosine-kinase, *J. Biol. Chem.* **267:**17472–17477.

Springer, J. E., Koh, S. H., Tayrien, M. W., and Loy, R., 1987, Basal forebrain magnocellular neurons stain for nerve growth factor receptor: Correlation with cholinergic cell bodies and effects of axotomy, *J. Neurosci. Res.* **17:**111–118.

Steininger, T. L., Wainer, B. H., Klein, R., Barbicid, M., and Palfrey, H. C., 1993, High affinity nerve growth factor receptor (trk) immunoreactivity is localized in cholinergic neurons of the basal forebrain and striatum in the adult rat brain, *Brain Res.* **612:**330–335.

Strömberg, L., Wetmore, C. J., Ebendal, T., Ernfors, P., Persson, H., and Olson, L., 1990, Rescue of basal forebrain cholinergic neurons after implantation of genetically modified cells producing recombinant NGF, *J. Neurosci. Res.* **25:**405–411.

Thoenen, H., and Barde, Y. A., 1980, Physiology of nerve growth factor, *Physiol. Rev.* **60:**1284–1335.

Vogels, O. J., Broere, C. A. J., Ter Laak, H. J., ten Donkelaar, H. J., Nieuwenhuys, R., and Schulte, B. P. M., 1990, Cell loss and shrinkage in the nucleus basalis Meynert complex in Alzheimer's disease, *Neurobiol. Aging* **11:**3–13.

Weskamp, G., and Reichardt, L. F., 1991, Evidence that biological activity of NGF is mediated through a novel subclass of high affinity receptors, *Neuron* **6:**649–663.

Whitehouse, P. J., Price, D. L., Struble, R. G., Clark, A. W., Coyle, J. T., and DeLong, M. R., 1983, Alzheimer's disease and senile dementia: Loss of neurons in the basal forebrain, *Science* **215:**1237–1239.

Whittemore, S. R., and Seiger, A., 1987, The expression, localization, and functional significance of B-nerve growth factor in the central nervous system, *Brain Res. Rev.* **12:**439–464.

Will, L. R., and Hefti, F., 1985, Behavioral and neurochemical effects of chronic intraventricular injections of nerve growth factor in adult rats with fimbrial lesions, *Behav. Brain Res.* **17:**17–24.

Williams, L. R., Varon, S., Peterson, G. M., Wictorian, K., Fisher, W., Björklund, A., and Gage, F. H., 1986, Continuous infusion of nerve growth factor prevents basal forebrain neuronal cell death after fimbria-fornix transection, *Proc. Natl. Acad. Sci. USA* **83:**9231–9236.

Winn, S. R., Linder, M. D., Lee, A., Haggett, G., Francis, J. M., and Emerich, D. F., 1996, Polymer-encapsulated genetically modified cells continue to secrete human nerve growth factor for over one year in rat ventricles: Behavioral and anatomical consequences, *Exp. Neurol.* **140:**126–138.

Yan, Q., and Johnson, E. M., Jr., 1988, An immunohistochemical study of the nerve growth factor receptor in developing rats, *J. Neurosci.* **8:**3481–3498.

# 20

# Vascular-Related and Mediated Alterations in Alzheimer's Disease

D. LARRY SPARKS

## 1. Introduction

Alois Alzheimer and his contemporaries recognized a vascular component of the disease that now bears his name: Alzheimer's disease (AD) (1987 translation). In the initial publication, Alzheimer reported the case of a 51 year old with progressive dementia and characteristic neuropathology. The neuropathology detailed after autopsy included: neuron loss, gliosis, neuronal fibrils (paired helical filaments), tangled bundles of fibrils where a neuron had once been (neurofibrillary tangles), and miliary foci of drusen (senile plaques). He also noted vascular changes, including cerebral arteriosclerosis and focal lesions of vessel endothelium. The same year (1987) Fischer suggested that it was not surprising to find neuropathological similarities between demented individuals with arteriosclerotic brain atrophy and those with what is today known as AD (1987 translation). Three years later, Emil Kraepelin (1987 translation) analyzed his own populations grouped according to the presence of presenile dementia, senile dementia, or arteriosclerotic disease, but he was unable to separate the possible roles of arteriosclerotic and senile changes. Among his presenile subjects, most of whom died of heart disease, neuropathologic

D. LARRY SPARKS • Sun Health Research Institute, Sun City, Arizona 85351; and Division of Medical Examiner's Service, Kentucky Justice Cabinet, Frankfort, Kentucky 40601.

*Cerebral Cortex,* Volume 14, edited by Peters and Morrison. Kluwer Academic/Plenum Publishers, New York, 1999.

changes were similar to those reported by Alzheimer, except for the absence of fiber formation. Some subjects exhibited fat in the vessel walls; others did not. Among his senile subjects—most of whom also died of heart attack—arteriosclerotic alterations, enlarged heart, and high blood pressure were common. Included within this senile population were the individuals most severely affected, and who were similar to the subject studied by Alzheimer. Kraepelin (1987 translation) reported that even among these probable AD patients, "most cases are characterized by more or less pronounced vessel alterations" and therefore "we must add this disease pattern to the arteriosclerotic group of senile dementia" (1987 translation, pp. 78–79). These descriptions suggest that senile plaques (drusen) were a common finding in subjects with arteriosclerotic disease.

In 1907 Alzheimer further implied heterogeneity in the outcome of disease in general, stating, "We must reach the stage in which the vast well-known disease groups must be subdivided into smaller groups, each one with its own clinical and anatomical characteristics" (1987 translation, p. 3). Fifty years later, Corsellis (1962) investigated neuropathologic alterations in a population of 207 psychiatric patients subgrouped according to the presence of vascular disease, senile change, mixed vascular and senile change, affective disorder, organic disorder, and functional disorder. He noted that abundant senile plaques (SPs) and neurofibrillary tangles (NFTs) were most prevalent in the senile and mixed vascular and senile groups, but that SPs and NFTs were most abundant in the vascular group as compared with all remaining groups (Corsellis, 1962). This increase is consistent with the findings of Kraepelin, and may relate to the increased incidence of SPs and NFTs in the cortex of nondemented heart disease individuals.

Extending Alzheimer's suggestions to include all disease—not just disease leading to dementia—10 years ago I began subgrouping an elderly nondemented control population according to the presence or absence of autopsy-confirmed heart disease. After excluding from study individuals with grossly observable cerebrovascular disease, stroke, or CNS trauma, I compared the neuropathologic consequences of critical coronary artery disease (cCAD; >75% stenosis of epicardial artery), hypertension, and normal aging (non-heart-disease controls) with those of Alzheimer's disease. Individuals with cCAD, but without cognitive impairment, exhibited cortical SPs identical to those found in AD, and at substantially younger ages than is usual in normal aging (Sparks *et al.,* 1990)—sometimes even at levels diagnostic of AD (Khachaturian, 1985). The SPs in cCAD subjects are deposited in the cortex, with a laminar distribution indistinguishable from AD (Fig. 1) (Sparks and Liu, 1992), and without demonstrable cerebrovascular alterations (Fig. 2) (Sparks *et al.,* 1993). I have also observed that cognitively intact subjects with hypertension often develop both cortical NFTs and SPs well in excess of the levels found in normal aging (Sparks *et al.,* 1995). According to Braak, the more severely affected hypertensive subjects achieve stage III of his scoring system for NFTs (Braak *et al.,* 1993), which is approximately the threshold for the appearance of dementia. These findings indicate that heart disease and hypertension may contribute to conditions that predispose an individual to AD. At the very least, the hypertensive disorder causes derangement(s) conducive to the production of early-AD-like neuropathology, albeit less rapidly or severely. It is not yet determined whether cCAD or hypertension ever progresses to cause discernable cognitive change, but these findings suggest the vascular system may contribute to the development of AD, possibly by initiating the changes associated with the disorder.

The relationship of cCAD, hypertension, and AD presents a quandary in that hypertension and arteriosclerosis are prime risk factors for stroke (Kannel *et al.*, 1983; Wolf *et al.*, 1991), multi-infarct dementia (MID), and ischemic vascular dementia (Metter and Wilson, 1993), which are generally considered distinct entities from AD. This interaction also poses a problem in confirming the diagnosis of AD, particularly if AD-like neuropathology is a common occurrence in vascular dementias, and arteriosclerotic cerebrovascular and cardiovascular changes are frequent in AD.

To further define the interaction between arteriosclerotic vascular disease and AD, this chapter presents current findings indicating a myriad of significant vascular alterations in AD, and ends by presenting new data suggesting the likelihood that vascular-mediated, blood-borne agents contribute to the initiation of AD-like neuropathology in a whole-animal model. It appears that blood vessel alterations and dysfunction occur in both vascular dementia and AD, but at varying levels. Factors that militate vascular dementia may act by inducing a generalized microvessel disorder, thus rendering the large arterioles vulnerable to acute occlusions. Such occlusions locally arrest downstream perfusion in circumscribed regions of the brain. Once there are sufficient numbers of focal disruptions or a critical region of brain is affected, acute onset of multi-infarct dementia ensues, with stable or slightly improved symptomology until the next episode. Those individuals with vascular factors not exhibiting rampant large-vessel disturbances and widespread focal infarctions—but showing microvessel disease—are considered to be experiencing ischemic vascular dementia without the obvious stepwise progression of MID. Such individuals often exhibit AD-like neuropathology, and they frequently obtain a final diagnosis of AD after autopsy, regardless of antemortem clinical impressions. As noted above, nondemented individuals with risk factors for dementia of vascular origin exhibit AD-like neuropathology (Sparks *et al.*, 1990; Sparks *et al.*, 1993; Sparks *et al.*, 1995). Thus it is possible that AD subjects experience a slow, regionally generalized capillary dysfunction in the absence of large arteriolar disease or overt microvessel disorder that might classify them as "of vascular origin."

## 2. Imaging the Vasculature

White-matter lesions demonstrable by MRI or CT scans have been related to reductions in the luminal cross-sectional area of small penetrating arteries and arterioles (Lotz *et al.*, 1986; Roman, 1987). Such white matter lesions are twice as prevalent in all dementing disorders, including AD and vascular dementia, as they are in controls (Englund *et al.*, 1988; Skoog *et al.*, 1994). The white-matter lesions frequently found in AD are believed to represent incomplete infarctions based on comparisons of biochemical and imaging data, and are opined to be the consequence of "an independent white matter disorder of cerebrovascular hypoperfusional/hypoxic origin" (Englund *et al.*, 1988, p. 1425). Although increased numbers of white-matter lesions are reported in AD as compared with controls, they do not

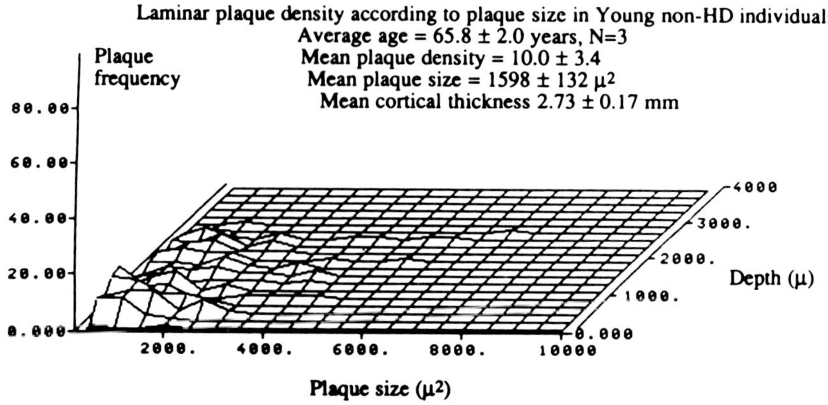


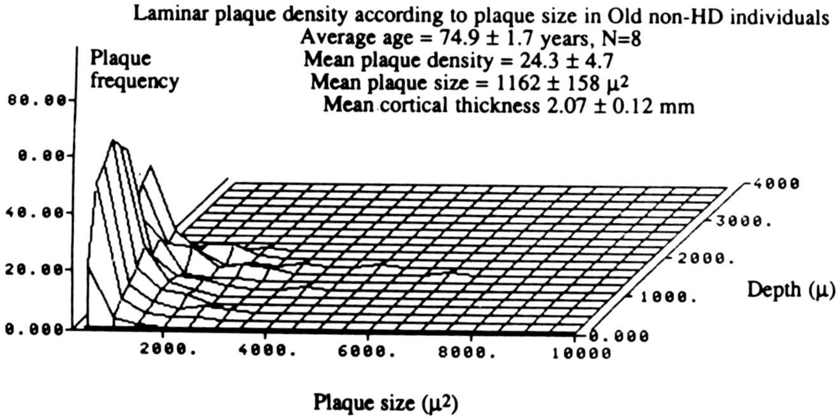


**Figure 1.** Senile plaque frequency (SP; z-axis), according to size (*x*-axis) and position in cortical ribbon (*y*-axis) of presenile and senile nondemented, non-heart-disease subjects (non-HD; top left and top right, respectively), nondemented critical coronary artery disease patients (cCAD; middle left and middle right, respectively), and Alzheimer's disease patients (AD; bottom left and bottom right, respectively). Subjects were grouped by age at death (under 70 or 70 and over). The number of SPs, grouped by size, was determined in five non-overlapping columns throughout the depth of the cortical mantle using Kontron image analysis. Signal-to-noise studies the smallest quantifiable SPs to be 200 $\mu^2$; the largest were nearly 10,000 $\mu^2$. SPs were grouped according to size in 500-$\mu^2$ increments between 200 and 9700 $\mu^2$ for regional density determination. Cortical thickness was determined for each individual. Mean thickness was significantly reduced in the presenile, but not the senile, AD patients as compared with both age-matched cCAD and non-HD groups, because of comparable age-related thickness reductions in cCAD and non-HD, but not AD, subjects. Mean SP size was calculated for each group. An age-related reduction in the average SP size occurred in the non-HD and cCAD groups, but not the AD patients. Mean SP density was assessed for each individual. Presenile density significantly doubled in cCAD as compared with the non-HD controls, doubling again in AD as compared with cCAD. Senile cCAD and AD groups did not differ significantly, but had approximately double the density of age-matched non-HD controls. Non-HD and cCAD, but not AD, populations showed an age-related increase. The regional size distribution of SPs was identical in non-HD, cCAD, and AD; only the absolute numbers were different, with consistently more SPs in the upper half of the cortical ribbon. Presenile non-HD subjects had low SP levels, mainly 200–5000 $\mu^2$ and largely confined to layers II/III. With age, non-HD numbers increased to 200–2000 $\mu^2$ in frontal layers II, III, and V. Younger cCAD subjects showed premature SPs mainly in layers II, III, and V, with 200–4000 $\mu^2$ area, in this respect most closely resembling older controls. SP deposition pattern in older cCAD subjects and both AD groups was similar, with SPs of 200–2000 $\mu^2$ more prevalent in layers II, III, and V.

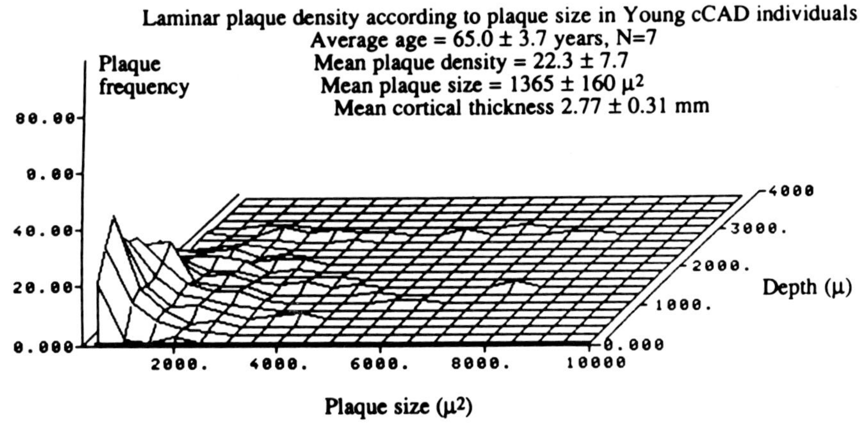


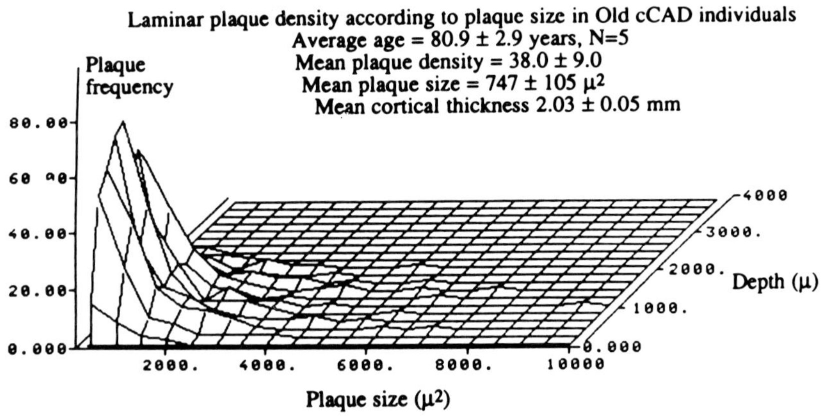


**Figure 1.** (*Continued*)

seem to increase in number or change in character during the clinical progression of the disorder (Obara *et al.,* 1994; Meyer *et al.,* 1995). However, after excluding individuals with cerebrovascular risk factors (i.e., hypertension, cardiovascular disease) from investigation, no differences in white-matter lesions are found between AD and age-matched controls (Kozachuk *et al.,* 1990). It is of note that hypertension is a prime risk factor for producing white matter lesions (Roman, 1987; Ferrer *et al.,* 1990; Hachinski, 1990). Furthermore, the main peptide precursor of β-amyloid (β-amyloid precursor protein; βAPP) has been colocalized with white-matter lesions in human brain (Suenaga *et al.,* 1994), and β-amyloid deposits with associated microglia have also been identified in AD white matter (Uchihara *et al.,* 1995). Thus, although these vascular factors have been associated with white-matter changes, the interaction between the factors and the white matter changes is unclear. Further, the relation between vascular factors, white-matter changes, and dementia is also uncertain (Ashford *et al.,* 1996).

Reduced cerebral blood flow in AD has been observed by MRI, SPECT, and PET scans. Cortical atrophy (gyral widening and thinning of the cortical ribbon) correlates with decreased blood flow on a regional basis in AD brain, particularly in the temporal and parietal lobes (Obara *et al.,* 1994; Meyer *et al.,* 1995, 1996). Reduced blood flow in AD temporal lobe is reported to be the first neocortical

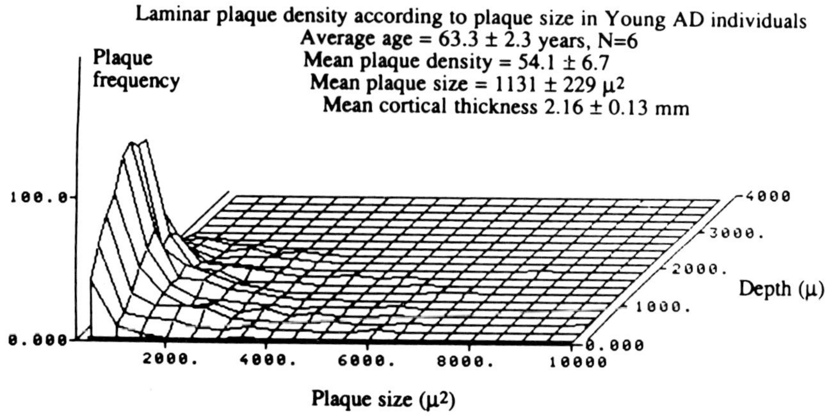


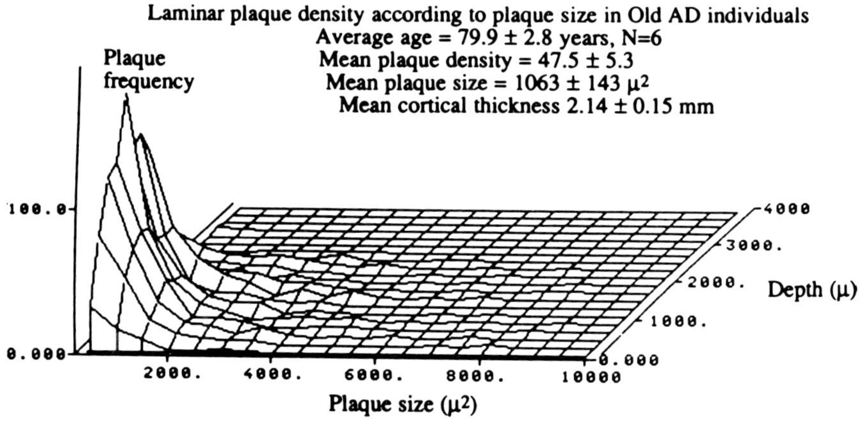


**Figure 1.** (*Continued*)

region affected; other regions of cortex become involved as the disease progresses (Eberling *et al.,* 1992; Harris *et al.,* 1996). In addition, increasingly severe reductions in blood flow in AD temporal lobe have also been correlated to the progression of cognitive deterioration in the disorder (Wolfe *et al.,* 1995). These reductions follow a regional pattern similar to that of the characteristic neuropathology (Harris *et al.,* 1996). In contrast to a number of previous studies (Deutsch and Tweedy, 1987; Mielke *et al.,* 1994), a recent report suggests that blood flow in the temporal and

**Figure 2.** Reproduction of previously published photomicrograph (Sparks *et al.,* 1993) indicating that β-amyloid is deposited in parahippocampal cortex of individuals with critical coronary artery disease (cCAD) before it is demonstrable in the vasculature. Matched views in adjacent sections, stained with: ALZ50 antibody (a and d); Beilschowsky method (b and e); β-amyloid antibody (c and f) from a 62-year-old (a-c) and a 64-year-old (d-f) cCAD subject. Retrospective investigations found no dementia or overt cognitive deficit in any of the subjects (Sparks and Liu, 1993). Standard Alz50 immunocytochemical methods done on 50-μ vibratome sections (antibody supplied by Dr. Peter Davies, Albert Einstein College of Medicine). Beilschowsky methods used for 8-μ paraffin-embedded sections. Standard β-amyloid-ABC immunohistochemical methods used on 8-μ paraffin sections (antibody supplied by Dr.

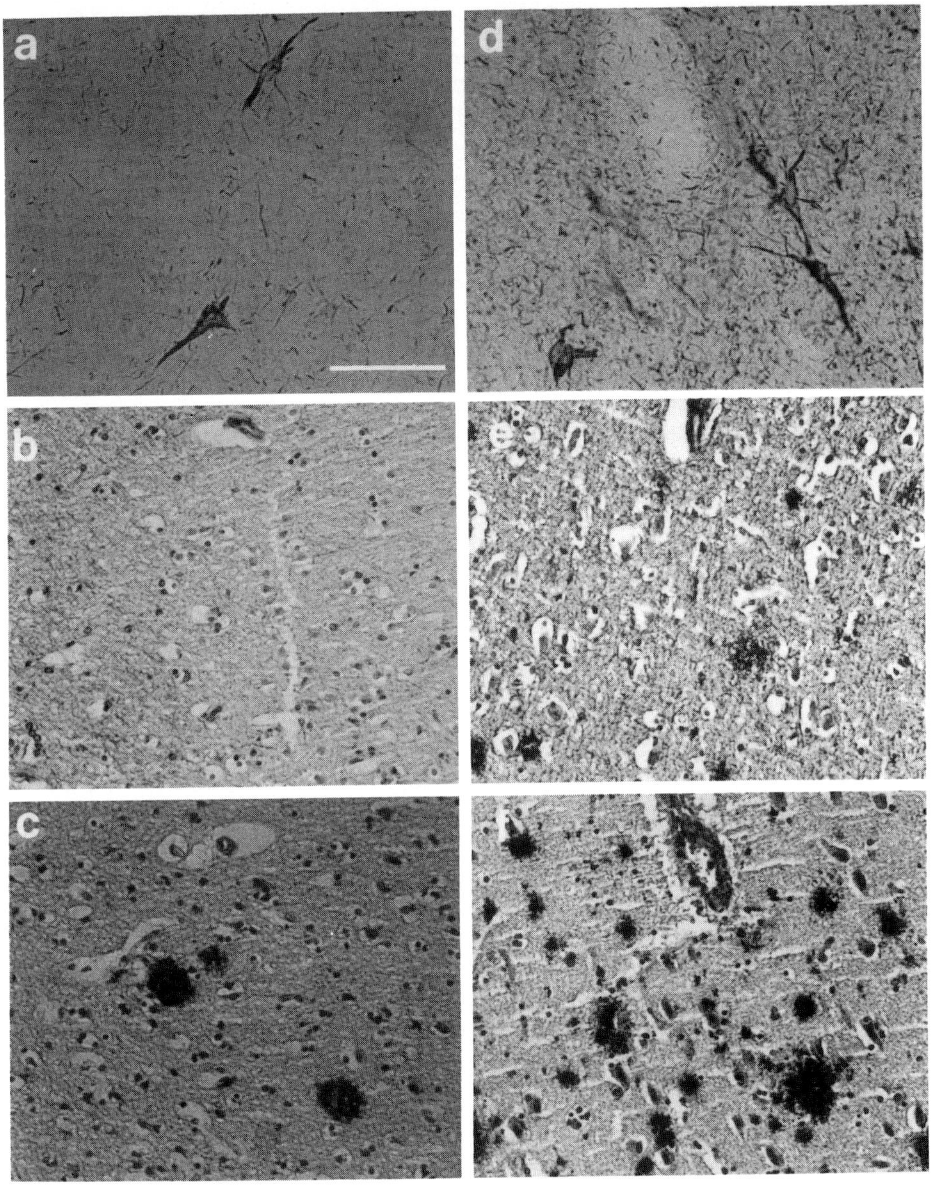


**Figure 2.** (*Continued*) Dale Schenk, 10D5, Athena Neurosciences). Similar staining patterns were found in a 79-year-old non-heart-disease subject and four cCAD subjects (29, 52, 62, and 80 years old). The pattern included abundant Alz50-immunoreactive neurons and neuropil threads (a). β-amyloid immunoreactivity in neuropil, but not vasculature, were present (c) in the absence of argyrophilic features (b). According to the proposed temporal sequence of SP formation (Sparks *et al.*, 1993), the next staining pattern also includes argyrophilic β-amyloid-immunoreactive deposits in parenchyma but not blood vessels (not shown). The next, more-advanced pattern of pathologic staining (64 year-old cCAD subjects) includes abundant Alz50-immunoreactive neurons and neuropil threads (d); argyrophilic and β-amyloid-immunoreactive deposits in neuropil (e, f); β-amyloid immunoreactivity associated with some blood vessels (f). Vascular β-amyloid immunoreactivity was more consistently demonstrable with increased extracellular deposit density and age in cCAD. Bar = 50 μm (a–f). Reproduced with permission from the *Journal of Neuropathology and Experimental Neurology.*

parietal lobes of individuals with vascular dementia is not significantly different from flows observed in AD (Starkstein *et al.*, 1996). Blood flow reductions in frontal cortex were found to be significantly greater in vascular dementia than in AD, and they occurred in the absence of structural damage (Starkstein *et al.*, 1996). These regional flow differences between AD and vascular dementia were correlated to neuropsychologic differences in patients matched for age, sex, and dementia severity (Starkstein *et al.*, 1996). Blood flow of the middle cerebral artery, which feeds the temporal and parietal lobes, is reduced in nondemented hypertensive individuals (Eckland *et al.*, 1995). Consequently, the changes in cerebral blood flow in AD and the dementias related to vascular disease also show overlap.

Previous reports suggest that reduced blood flow becomes progressively more severe during the course of AD (Eberling *et al.*, 1992; Harris *et al.*, 1996). Diminished blood flow is demonstrable in patients with very mild clinical symptoms, and is consistent with the observation that dementia usually appears at Braak stage II or III of neuropathologic alterations in AD (Braak and Braak, 1991; Blancher *et al.*, 1993). On one hand, this is in contrast to a previous report suggesting that symptoms of AD occur prior to decreased cerebral blood flow, whereas reduced flow is known to precede symptoms of MID (Rogers *et al.*, 1986). On the other hand, the premise gains partial support in that reduced blood flow observed in individuals with age-associated memory impairment was halfway between values for controls and for subjects with probable AD (Parnetti *et al.*, 1996).

Cerebral blood flow can be affected by the vessel's ability to accommodate changes in pressure or to respond to chemical inducers, and is well known to accommodate subtle information processing demands. Therefore, there are mechanisms to finely autoregulate cerebral blood flow. The AD cerebrovasculature retains its ability to respond by vasodilation to increase the flow of blood, whereas the vasculature in vascular dementia apparently cannot respond (Meyer *et al.*, 1996). The cerebrovasculature in chronic hypertension exhibits increased medial thickness whether or not luminal stenosis is observed, suggesting a reduced accommodation (Johansson, 1984, 1992; Faraci and Heistad, 1990). This is supported by a reduced ability to dilate cerebral vessels in chronic hypertension (Tamaki *et al.*, 1995).

In ischemic vascular dementia, control of risk factors stabilizes cerebral blood flow and cognitive decline (Meyer *et al.*, 1996). Similarly, treatment of AD patients aimed at increasing cerebral cholinergic activity also increases cerebral blood flow and arrests further slippage in cognition, whereas patients given a placebo continue to experience progressive deterioration (Minthon *et al.*, 1995).

Positron-emission tomography (PET) scans are also used to determine metabolic function in the brain, particularly glucose utilization and oxygen extraction across the vasculature. Early studies revealed reduced glucose consumption in AD patients on a regional basis as compared with age-matched controls (de Leon *et al.*, 1983). Subsequent studies indicated that the regional reductions in glucose metabolism correlated with regional neuron loss in AD cortex (McGeer *et al.*, 1986). This decrease in glucose consumption becomes progressively more severe (Smith *et al.*, 1992), and it has been localized to the temporal and parietal lobes by most investigators (Rapoport, 1991; Burdette *et al.*, 1996). Decreased glucose use is reported to precede the onset of symptoms in AD (Polinsky *et al.*, 1987; Jagust *et al.*, 1988). The spread of reduced glucose metabolism in the temporal lobe has also been related to the severity of dementia in AD (Jagust *et al.,* 1993). Only middle temporal cortex is affected in mildly demented patients, whereas essentially the entire lobe consumes less glucose in moderately demented subjects with AD (Jagust *et al.,* 1993). Interestingly, a recent PET study suggests that the antemortem regional decrease of glucose metabolism does not correlate well with regional SP and NFT densities in autopsy-confirmed AD patients (Mielke *et al.,* 1996).

Further PET studies suggest reductions of forward glucose transport in AD temporal and frontal cortex as a mechanism of decreased consumption (Jagust *et al.,* 1991). Biochemical studies of AD vasculature seem to confirm this observation; however, several mechanisms may explain the reduction of glucose utilization. The density of the glucose transported in AD microvessels is reported to be half of that found in age-matched controls (Harik and Kalaria, 1991), suggesting that glucose transport is impeded directly, though this finding could be a secondary adaption to decreased demand. Other investigators support alternate mechanisms of decreased glucose use in nervous tissue. One line suggests reductions in neuronal uptake of glucose. Decreased glucose transporters GLUT1 and GLUT3 were found in six regions of AD brain as compared with age-matched controls. Reductions of the putative neuronal glucose transporter GLUT3 were more severe than decrements in GLUT1 in all regions of AD cortex (Simpson *et al.,* 1994). It is not clear how decreases in the GLUT3 transporter relate to demonstrably less-severe synapse loss in the same region (Simpson *et al.,* 1994). Alternatively, a metabolic backlog may limit the use of glucose in AD; once within the neuron, anabolic mitochondrial oxidation of glucose is reduced (Blass, 1993). Significant reductions in the activity of the pyruvate and ketoglutarate dehydrogenase complexes have been consistently demonstrated in AD (Blass, 1993), accompanied by alterations in the microscopic appearance of mitochondria (Blass, 1993). Decreased enzymatic use of glucose has been proposed as the cause of β-amyloid production and deposition as SPs (covered below) in AD cortex (Meier-Ruge *et al.,* 1994a, b), and increased numbers of altered mitochondria have been shown to occur prior to the β-amyloid accumulation (Blass, 1993).

Oxygen utilization has been shown by most investigators to be reduced in AD brain (Grubb *et al.,* 1977; Frackowiak *et al.,* 1981; Hoyer, 1986). Decrements of markers for availability and use, without a difference in their ratio (the extraction index), suggests diminished oxygen supply and demand as coupled parameters (Frackowiak *et al.,* 1981). Decreased cortical oxygen use parallels cognitive dysfunction and atrophy of the anterior corpus callosum in AD (Yamauchi *et al.,* 1993). Increased dementia severity seems to be accompanied by greater deficits in oxygen use in both vascular and AD patients; focal derangements are not apparent in either group (Frackowiak *et al.,* 1981). Regional analyses indicate that occipital lobe is less affected than other cortical regions in both of the dementing disorders, and also indicates that increased dementia severity correlates with consonant decrements of oxygen use in all but the temporal cortex in AD (Frackowiak *et al.,* 1981). A near-maximal decrease in oxygen use in less-severely demented AD patients may mean that temporal lobe is involved early in the disorder.

One-quarter of the blood's oxygen is utilized by the brain, though it is only perfused by one-seventh of the body's blood (Schmidt, 1964). A reduction in usable oxygen supply would shift the brain's metabolic machinery from anabolic production of ATP to catabolic processing of energy sources, which would probably lead to oxidative stress, increased free radicals, and conditions incompatible with prolonged neuronal survival in affected regions of brain.

Long-term decrements in glucose result in gliosis and neuronal loss in diabetic patients also (Malamud and Grosh, 1938). By analogy, abundant senile plaques have been observed in the cortex of patients dying after prolonged diabetic coma (David B. Clark, personal communication). Long-term decreases in oxygen cause pyknosis, gliosis, and neuron loss (Malamud and Hirano, 1974). It has recently been hypothesized that the clustering of senile plaques in AD frontal cortex is associated with the anatomic distribution of the vasculature (Armstrong, 1995). Furthermore, immunoblots of unfixed tissue have identified ischemic stress-related deposits of β-amyloid occurring exclusively in the middle layers of cortex (Jendroska *et al.,* 1995), distributed predominantly in the "watershed" between posterior and middle cerebral arteries and correlated to the antemortem presence of hypoxia or ischemia. Such stress-induced deposits of β-amyloid could not be visualized in paraffin-embedded tissue and were more susceptible to proteinase digestion than were SPs. The deposits lacked immunoreactivity for both C-terminal and N-terminal aspects of β-APP, suggesting that they represent a highly labile form of β-amyloid (Jendroska *et al.,* 1995). A role for these lesions in the pathogenesis of AD was proposed because of their prevalence in subjects with neuritic plaques, indicating that AD-like neuropathology or characteristic AD-lesion formation can accompany vascular disturbances and decreased supply of ingredients required for anabolic metabolism.

## 3. Macroscopic Changes in the Vasculature

A variety of alterations in AD vasculature could decrease the supply of all essential nutrients to the brain, including oxygen and glucose, measurable by *in vivo* scans. At the gross level, these alterations would include changes in the number or physical appearance of capillaries; therefore a description of the spatial arrangement of cerebral microvessels is warranted.

Akima and colleagues (1986) described the spatial orientation of normal cortical small-vessel distribution and age-related changes using soft X-ray and scanning electron microscopy techniques compared with traditional paraffin section stains. Small vessels penetrate the cortex, invariably at right angles to the surface, from the leptomeningeal arteries. There are three types of cortical and two types of subcortical supply. *Superficial cortical* microvessels ramify and supply the outer two layers of cortex. *Middle cortical* microvessels send arterioles through layers I and II, which then ramify into "fountainlike" capillary structures that perforate layers III and IV (Akima *et al.,* 1986). *Deep cortical* microvessels branch into fountainlike ramifications of capillaries in middle, and again in deep, layers of cortex (Akima *et al.,* 1986). The *medullary subcortical* vessel does not perfuse the cortex, but rather proceeds directly through it to the periventricular region. The second type of subcortical vessel proceeds in the white matter, but branches into "fountainlike" plexuses in the upper and lower aspects of the bottom half of the cortical ribbon (Akima *et al.,* 1986).

Recent morphometric studies suggest that there is a reduction in the density of the capillary web in AD (Fischer *et al.,* 1990; Buée *et al.,* 1994). Early reports indicated a possible increase in the number of capillaries in AD hippocampus, but these measurements were based on vessel length (Bell and Ball, 1981). A decrease in capillary density has been shown in many dementing disorders, whereas AD-related decrements may be regionally specific (Buée *et al.*, 1994). Reduced capillary number is observed in temporal and frontal cortex in AD (Fischer *et al.*, 1990), but not in occipital cortex (Bell and Ball, 1990; Fischer *et al.*, 1990), and therefore seems to regionally correlate with neuropathologic lesion formation and brain-scanning data. At the same time, the capillary web decrease has been localized to layers III and VI in all regions of AD cortex (Buée *et al.*, 1994), which correlates with the laminar densities of SP in AD (Mandybur, 1975; Rogers and Morrison, 1985) and heart disease (see Fig. 1).

For decades, many AD-related distortions in blood vessels have been known. Most studies investigating the vasculature in AD and senile brain have noted blood vessel tortuosities ("wickerworks," or twisted vessels) and loops (kinks) (Hassler, 1965; Ravens, 1978; Scheibel and Duong, 1986; Scheibel *et al.*, 1987; Fischer *et al.*, 1990; Buée *et al.*, 1994). Of note, looped vessels are not observed in another dementing disorder, Pick's disease (Buée *et al.*, 1994). Direct investigation separating the two vascular deformations indicated that the number of loops was increased in AD cortex, but the number of tortuosities was the same as in age-matched controls. In contrast, the increased tortuosities reported with age appear to be ropelike, fountainlike, twisted ramifications (Akima *et al.*, 1986). As with decreased vessel density, loops in AD microvasculature have been localized to layer V, and to a lesser extent, to layers II and III (Buée *et al.*, 1994). A correlation between the co-incidence of loops and deposition of β-amyloid as SPs has been noted (Robitaille *et al.*, 1987). Similar arteriolar changes have been observed with increasing frequency in normal aging without change in hypertension (Spangler *et al.*, 1994), and they are not Charcot-Bouchard aneurysms (Challa *et al.*, 1992). It is probably that as the cortical ribbon thins, vessels of normal length coil as a consequence of atrophy (Ravens, 1978; Akima *et al.*, 1986), and probably earlier in AD than in normal aging or in heart disease (Fig. 1).

Scanning electron microscopy has shown beautiful images of loops, and has revealed pitting, outpouching, and denervation of the abluminal surface in AD microvasculature (Scheibel and Duong, 1986; Scheibel *et al.*, 1987). Many of these observations were confirmed by a subsequent study (Kimura *et al.*, 1991). The divits in the thickened basement membrane observed by Scheibel and colleagues proceeded up to the inner lamina, but did not appear to breach the vessel wall, therefore, the blood–brain barrier remained intact. Kimura *et al.* (1991) did not note this finding, but did report that AD vessels had a "swiss-cheese" appearance. Capillary outpouchings observed in AD varied from bulges to sacular structures, occasionally with an apparent stalk. The sacs were thought to be pericytes engorged with cellular remnants, but the bulges were shown to be inclusions within the vessel (covered in the next section). The "lumpy-bumpy" appearance of the abluminal vascular surface varied with vessel diameter in AD; the smaller the caliber, the more deformations (Scheibel and Duong, 1986; Scheibel *et al.*, 1987).

The above-mentioned vessel deformations could alter the downstream blood flow and material transfer between brain and circulation. Hassler (1965) suggested these spatial alterations could impede blood flow, whereas Ravens (1978) suggested that they could create a nidus for changes in the blood–brain barrier. Scheibel and colleagues (1987) suggested that alterations in AD microvasculature could attenuate oxygen uptake by underlying tissue. Many investigators accept that changes in the AD vasculature are a result of the disease process and that these changes likely potentiate progression of the disorder.

Recently, altered blood hemodynamics due to brain capillary distortion have been implicated in the pathogenesis of AD (de la Torre, 1994). This is consistent with the possibility that regionally specific defects in the microvasculature could contribute to the cause of neurologic disorders and AD (Roberts, 1986). Physical changes in AD microvessels are proposed to alter normal laminar flow and produce microturbulence (de la Torre and Mussivand, 1993; de la Torre, 1994). Thereafter, disturbed flow would interfere with normal supply of oxygen, glucose, and essential nutrients to the brain while simultaneously disrupting the clearance of its metabolic by-products (de la Torre and Mussivand, 1983; de la Torre, 1994).

The absence of perivascular plexuses of nodule-bearing nervous fibers on the abluminal surface of AD vessels has been related to altered control of blood flow in the disorder (Scheibel *et al.,* 1987). Pathologic alterations in the locus coeruleus (noradrenergic input) have been related to loss of nerve plexuses on the vasculature, to altered blood flow, to glucose use, and to blood–brain permeability in AD (Scheibel *et al.,* 1987). Increased adrenergic receptors in cerebral vessels of AD patients as compared with age-matched controls may suggest denervation supersensitivity and may support a reduced central adrenergic innervation of the cerebrovasculature in the disorder (Harik and Kalaria, 1991). The recent demonstration of acetylcholinergic receptors on cortical microvessels (Kalaria *et al.,* 1994) supports the hypothesis that a dysfunctioning nucleus basalis of Meynert (nbM, cholinergic vascular input) in AD could alter the cholinergic influence on blood flow, vasodilation, and oxygen use (Roberts, 1986).

Decreases in the cross-sectional luminal area of AD microvessels may reduce cerebral blood flow (Bell and Ball, 1981). Apparent thinning or atrophy of the vasculature and the presence of "string" vessels in AD cortex would support this observation (Buée *et al.,* 1994). Direct morphometric investigation indicated that the reduction of capillary diameter in AD was nearly significant ($p = 0.06$) (Stewart *et al.,* 1992) and translates into a 25% reduction in luminal cross-sectional area.

Luminal narrowing, particularly in the smallest capillaries, could produce susceptibility to microembolic events in AD, but no one has yet demonstrated evidence of microemboli in AD with transient ischemia, although occlusion of capillaries by deposition of β-amyloid has been identified (Roher *et al.,* 1993). Thrombosis in stenotic vessels has been shown to result in minute foci of cortical infarction (Mandybur, 1975). It is noted here for later correlation that microcapillary embolic dilations occur in cardiac patients undergoing cardiopulmonary bypass (CPB) (Challa *et al.,* 1993), many of whom experience cognitive difficulties after surgery, which persist in some patients (Kolkka and Hiberman, 1980; Shaw *et al.,* 1986; Sotaniemi *et al.,* 1986; Forsman *et al.,* 1990). Similar embolic findings have been observed in dogs undergoing experimental CPB (Challa *et al.,* 1993). These findings may be strictly due to cardiac surgery (Moody *et al.,* 1990), but cerebral emboli are a common occurrence with a variety of cardiac sources (Sliwka *et al.,* 1995). Nevertheless, each of the aforementioned gross alterations observed in the AD vasculature could disrupt functionally normal blood flow and localized bidirectional transfer across the blood–brain barrier.

## 4. Microscopic Changes in the Vasculature

A basic description of the normal microscopic appearance of blood vessels is appropriate in order to make comparisons with the changes observed in AD. Normal vessel microanatomy has been investigated in cross sections and in longitudinal sections through a vessel. The smallest capillaries have luminal endothelial cells and a basement membrane, and the larger-diameter arteriolar vessels have an additional abluminal smooth-muscle sheath.

Endothelial cells on the luminal surface of the brain's vasculature maintain tight junctions thought to comprise and maintain the blood–brain barrier. Occasional gaps in endothelial continuity, and therefore in tight junctions, have been found in the longitudinal sections (Stewart *et al.,* 1992). Cerebrovascular endothelia have five times more mitochondria than vessels in any other part of the body. These vascular cells also exhibit cytoplasmic vesicles. It is thought that this increase in mitochondria is necessary to produce the energy to actively transport components from the blood in vesicles across the cerebrovascular endothelial cell for utilization by underlying brain tissue, thereby establishing the blood–brain barrier via the endothelial cells.

Directly beneath the endothelial cell layer occasional pericytes are identifiable. These pericytes are thought to engulf and process material that has eluded exclusion or vesicular processing by the endothelial cells. Pericytes are sandwiched between the abluminal tight-junction surface of the endothelial cell layer and the more abluminally located basement membrane.

The basement membrane is arranged in laminae. The inner lamina is on the luminal side of the vessel and the outer lamina is on the abluminal side. Between these extremes, laminar leaflets are arranged like the rings of a tree. The basement membrane has three main constituents: laminin, collagen type-IV, and heparan sulfate proteoglycan (HSPG).

As noted above, larger vessels have a sheath of smooth muscle on the abluminal surface. The thickness of the muscle layer seems proportionate to the diameter of the vessel. Vessels with and without a muscle layer have a smooth appearance by scanning EM, except for the occasional neuroplexus protuberance (Akima *et al.,* 1986; Scheibel and Duong, 1986; Scheibel *et al.,* 1987).

From the outset, focal lesions in the vascular endothelium have been reported in AD (Alzheimer, 1987). Recently, this has been confirmed with immunocytochemical markers for endothelial cells. Conspicuous lack of endothelial cells in microvessels retaining their basement membrane was observed in AD three times as often as in age-matched controls (Kalaria and Hedera, 1995). The number of normally occurring gap junctions in the endothelial cell layer is reported to increase in AD, and has been interpreted to indicate a "leaky" blood–brain barrier (Stewart *et al.,* 1992). The increased numbers of gaps between, or the focal lack of cells putatively representing the barrier between, blood and brain could easily establish regional microenvironmental conditions deleterious to neuron survival. This presumed intermittent physical breach of the blood–brain barrier in AD has been associated with increased deposition of β-amyloid as a neurotoxic agent (SP, "drusen") (Kalaria and Hedera, 1995). For future discussion, it is of note that enhanced blood–brain barrier permeability has been identified in hypertensive individuals and animals (Mueller and Luft, 1982), and hypertension is suggested to play a role in the etiology of microvascular endothelial damage (Caceres *et al.*, 1995).

Degeneration of endothelial cells in AD has been confirmed by ultrastructural studies (Yamashita *et al.*, 1991). Most investigators have sampled regions of AD brain laden with characteristic lesions, gliosis, and neuronal degeneration. The general conclusion has been that vascular change is secondary to the more prevalent, severe, and widespread neuropathologic change. In a unique study that also investigated the vasculature in regions of AD brain devoid of characteristic lesions (e.g., SPs and NFTs), degeneration of endothelial cells and hypertrophic basement laminae still accompanied neuron death and gliosis in adjacent cortex (Yamashita *et al.*, 1991). These authors suggested that in AD, the "primary change occurs in blood vessels . . . followed by nerve cell degeneration" (pp. 83–84). Consequently, relations among vascular change, neuron loss, AD pathology, and dementia remain unclear.

A change in the representative organelle profile of endothelial cells is reported in AD (Stewart *et al.*, 1992; Claudio, 1996) and a decrease in the number of mitochondira in the endothelial layer has been a consistent finding (Stewart *et al.*, 1992; Claudio, 1996). The latter observation has been associated with a metabolic deficiency in the active transport of essential nutrients to the brain. The number of vesicles in endothelial cells of the AD microvasculature is reported to be unchanged (Stewart *et al.*, 1992), or increased (Claudio, 1996). In the latter study, an inverse relationship between an increased number of vesicles and a decreased number of mitochondria in AD patients suggests that a metabolic deficiency induces a backlog (buildup) of transport carriers (microvascular endothelial vesicles) and an inefficient blood–brain barrier.

Another hypothesis proposes that the regional surface characteristics of endothelial cells could instill selective vulnerability in the AD vasculature (Roberts, 1986), based on the premise that the environment in which endothelial cells develop determines their characteristics. Accordingly, regional alterations in the blood–brain barrier could precipitate regional brain dysfunction and neuropathologic lesion formation, which in turn may initiate the progressively more-severe cognitive deficits in AD (Roberts, 1986).

Proceeding in an abluminal direction, identification of increased numbers of pericytes in the AD cerebrovasculature has been a consistent observation, starting with the early morphologic descriptions of the disorder (Kraepelin, 1987 translation; Scheibel *et al.*, 1987; Stewart *et al.*, 1992). Proliferation or enlargement of the pericytes, which are functionally similar to microglia of CNS proper, is proposed as evidence for a breakdown of the blood–brain barrier (Scheibel *et al.*, 1987; Stewart *et al.*, 1992). Increased numbers of pericytes in AD microvessels, which are enlarged with progressively increasing volumes of cellular debris and blood-borne components, may be considered a defense mechanism and probably contributes to the outpouching of AD cortical capillaries demonstrable by scanning EM (Scheibel and Duong, 1986), as well as causing decreased luminal diameter (Stewart *et al.*, 1992). As noted above, engorgement of pericytes with debris may progress to the point that saclike aneurysmal protrusions can be created on the abluminal surface (Scheibel and Duong, 1986). This proliferative response supports the hypothesis that there is a breakdown of the blood–brain barrier in AD, coupled with a functional decrease of essential nutrients in the brains of affected individuals.

Thickening of the vessel wall in senile dementia of the Alzheimer's type was first mentioned by Kraepelin in 1910 (1987 translation). Such thickening, particularly of the vascular basement membrane (VBM), has been confirmed in the AD cerebrovasculature by most investigators (Mandybur, 1975; Mancardi *et al.*, 1980; Robitaille *et al.*, 1987; Scheibel *et al.*, 1987; Yamashita *et al.*, 1991; Perlmutter *et al.*, 1994; Vinters *et al.*, 1994). Early studies noted pools of glycogen-like material trapped between the layers of the VBM. Scheibel and colleagues (1987) described ovoid structures between the internal and external lamina, suggesting that they contributed to the lumpy appearance of the abluminal surface of AD microvessels. They speculated that some of these structures may be cells containing "lakes" of lipids. Studies of the VBM by Perlmutter's group suggest that AD-related thickening is characterized by inclusions of debris-filled vacuoles and redundant patterns of basement membrane (Perlmutter *et al.*, 1994). Their ongoing investigations indicate that thickening of the VBM is a regional phenomenon that correlates to the pattern of lesion deposition in AD brain, which may explain findings that suggest no significant difference in vessel wall thickness between AD and age-matched controls (Stewart *et al.*, 1992). It is of note that the foregoing alterations in the VBM in AD often occur in the absence of amyloid deposition in the vessel walls (Perlmutter *et al.*, 1994) (covered later).

A subset of perivascular mononuclear phagocytic system (MPS) cells sharing many cell surface markers with microglia (Streit *et al.*, 1989; Graeber *et al.*, 1992; Perlmutter *et al.*, 1992) is also observed in the AD vascular basement membrane. Kraepelin, in his early descriptions, mentioned both pericytes and monocytes associated with AD cerebrovasculature (1987 translation), which is supported by more-recent studies (Perlmutter, 1994). These cells are thought to be involved in the deposition of amyloid in the AD vasculature (covered later).

The three main components of the VBM—laminin, collagen type-IV, and HSPG—have been identified as materials co-deposited with SP β-amyloid in AD brain (Snow and Wight, 1989; Buée *et al.*, 1994; Perlmutter *et al.*, 1994). Identical to AD, it is herein shown that cortical SPs occurring in nondemented individuals with heart disease are immunolabeled by HSPG antibody, and the cerebral vessels of heart disease subjects exhibit many of the changes identified in AD (Fig. 3). Other proteoglycans have also been shown to deposit with SPs in AD (Snow *et al.*, 1992; DeWitt *et al.*, 1993), and occasionally with specific SP types (Snow *et al.*, 1996). Thus production of the subtotal perforations of the VBM in AD that cause the "swiss cheese" abluminal surface appearance described by Scheibel *et al.* (1987), and again by Kimura *et al.* (1991), may create the VBM components that are subsequently deposited in SPs.

Larger vessels containing a media tunica have shown muscle cell degeneration in AD (Mandybur, 1975; Kawai *et al.*, 1993). Thickening of the media has also been noted, whereas more-advanced vascular lesions have been associated with pronounced thinning of the muscle layer (Mandybur, 1975); in some cases sufficient death of smooth muscle cells occurs such that only intrinsic connective tissue remains. This vascular muscle degeneration is associated with the pathologic deposition of β-amyloid in AD (Kawai *et al.*, 1993).

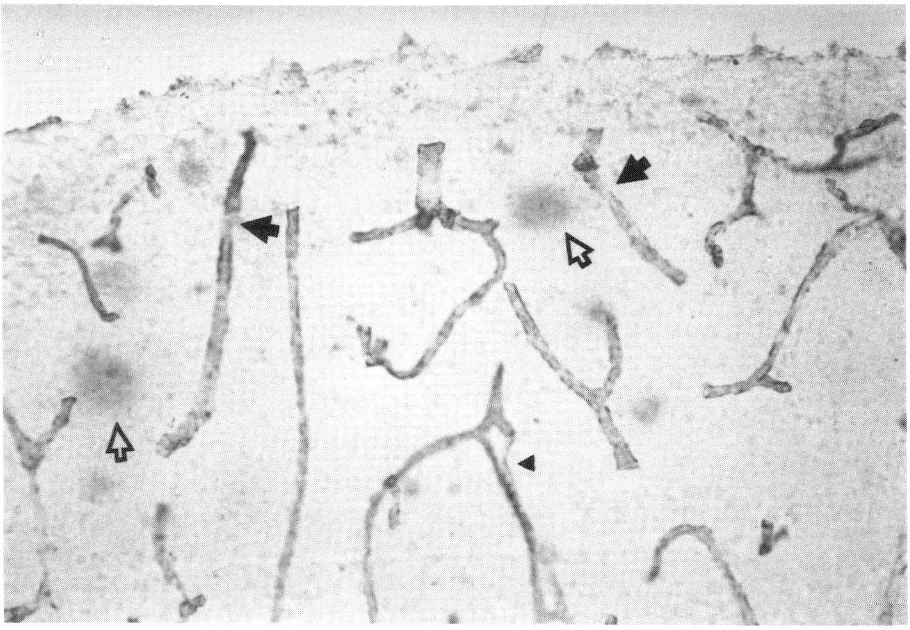

**Figure 3.** Heparan sulfate proteoglycan (HSPG) immunostaining in frontal cortex from a 63-year-old nondemented individual with critical coronary artery disease (cCAD). Paraffin-embedded 8-micron sections were immunostained with antibody supplied by Dr. Howard Fillit, and standard ABC (Vector elite) methods (Buée *et al.,* 1994; Perlmutter, 1994). Frontal cortex shows abundant β-amyloid-immunoreactive SPs, some of which were HSPG-immunoreactive (open arrowheads), further evidence that cCAD SPs are similar to those in Alzheimer's disease (AD). Identical to observations in HSPG-immunostained AD vasculature (Perlmutter, 1994), microvessels in this cCAD subject have a ragged, irregular abluminal surface, and there is a loss of "interconnected" vessel appearance (closed arrowheads); string vessels may be present (small arrow). ×20.

## 5. Pathologic Deposition in the Vasculature

Long before the advent of specific antibodies for what is now known as β-amyloid, the pathologic deposition of proteins reactive with congo red and thereby identified as "amyloid" was recognized in AD vasculature, along with consistently observable collagen accumulations (Mandybur, 1975). Early on it was known that the form of "amyloidosis" depended on the type of vessel affected, whereas the presence of hypertension was proposed as a contributing or accelerating factor (Mandybur, 1975). Larger-bore vessels with media were found to exhibit congophilic amyloid within the outer segments of the muscle layer. Pathologic deposition in the media, and smooth muscle degeneration, occurred with no overt change in the endothelial layer, whereas pathologic deposition in abluminal basement membrane in capillaries was accompanied by focal loss of endothelial cells (Mandybur, 1975). Mandybur suggested that deposition in the muscle constituted the congophilic amyloid angiopathy (CAA) often observed in AD, whether or not there was similar deposition in the basement membrane. Recent studies tend to support this, although deposition of collagen in the absence of β-amyloid has been noted (Vinters *et al.,* 1994).

It is now accepted that pathologic accumulations in the AD blood vessel wall include β-amyloid protein, characterized by Glenner and Wong (1984). The specifics of β-amyloid deposition in the parenchyma is detailed elsewhere in this volume, but there are recognized biochemical similarities requiring brief introductory comment.

The β-amyloid protein is a peptide fragment of a larger membrane-spanning protein (β-APP), the function of which remains elusive. Predominant alpha metabolism of β-APP cleaves the protein in the transmembrane stretch of the β-amyloid sequence to produce the secreted form of β-APP, whereas beta metabolism leads to the production of β-amyloid-containing by-products. The initial step in beta metabolism likely produces a C-terminal 102-amino acid fragment of β-APP (c-100), while liberating the N-terminal of the β-amyloid sequence. Unique to the vasculature, a stable C-terminal form of β-APP approximately 225 amino acids long (22kDa) may serve as the major precursor for β-amyloid (Tomaoka *et al.,* 1992). The final β-amyloid peptide sequence varies in length from 39 to 43 amino acids, with the longer species probably being the earliest form deposited as diffuse SPs (Iwatsubo *et al.,* 1994; Tekirian *et al.,* 1996). Subsequent metabolism, or later production of shorter, more fibrillar-prone β-amyloid sequences, is prominent in lesions accepted as older forms of SP (Iwatsubo *et al.,* 1994). Microglia are thought to oversee the metabolic modification and deposition, but not production, of β-amyloid in the neuropil (Scott *et al.,* 1993). Microglia are also known to participate in the attempted clearance of β-APP by-products (Perlmutter *et al.,* 1994; Sparks, 1996), a function presumably analogous to that of the cerebrovascular pericytes described above.

Peptide with 42(43)-amino acid residues has been shown to be the predominant β-amyloid form in microvessels (Roher *et al.,* 1993; Shinkai *et al.,* 1995). This form comprises three-quarters of the total β-amyloid in AD; the 40-amino acid species comprise the remainder. No 39-amino acid peptide has been observed (Roher *et al.,* 1993). The lack of post-translational modifications to the β-amyloid peptide in AD capillaries, as compared with brain, prompted the opinion that "vascular amyloid is younger" (Roher *et al.,* 1993, p. 10836). Subsequent combined aging-and-AD studies suggest that β-amyloid-42(43) is the form initially deposited in blood vessels, as it can be observed in the absence of the 40-amino acid type (Shinkai *et al.,* 1995). Precipitous age-related increases in vascular β-amyloid-42(43) immunoreactivity make the blood vessels in elderly controls indistinguishable from those in AD. Though the levels of 40-amino acid β-amyloid do increase with age, they never approach the numbers observed in AD (Shinkai *et al.,* 1995). These investigators often identified higher levels of β-amyloid-40 than β-amyloid-42(43) in the AD vasculature, and suggested that it is accumulation of the former that contributes to congophilic amyloid angiopathy (Shinkai *et al.,* 1995).

The presence of β-amyloid in AD cerebrovasculature, identical to that found in the CNS, initially indicated a possible blood-borne origin (Mandybur, 1975; Joachim *et al.,* 1988). One study suggested that all centrally deposited β-amyloid is derived from blood vessels themselves, and that all SPs have associated capillaries (Araki *et al.,* 1991). Other studies suggest that there is no relationship between localized SP deposition in AD neuropil and β-amyloid accumulation in the vasculature (Lippa *et al.*, 1993). My studies of early SP formation in the human heart disease model of AD suggest that significant numbers of SPs—albeit predominantly the diffuse form—are demonstrable at much younger ages and presumably in advance of vascular deposits (see Fig. 2). In these studies, as older forms of SP became more prevalent in older heart disease individuals, the likelihood increased that vascular β-amyloid could be identified, although older forms of SPs were often observed in the absence of any vascular accumulation.

Today the exact origin of β-amyloid in brain retains the burden of controversy, and the situation is not much better for the origin of β-amyloid in vasculature. Most researchers propose that β-amyloid in the vasculature is locally produced, but impeded clearance from brain to blood, with subsequent deposition in the vessel wall, remains a possibility. The latter proposition gains support from observations that β-amyloid injected into the brain is eventually cleared from the neuropil but seems to accumulate in the parenchymal blood vessel walls (Frautschy *et al.*, 1992).

The β-amyloid accumulates routinely in the outermost layer of muscle cells surrounding AD blood vessels (tunica media) (Mandybur, 1975; Kawai *et al.*, 1993; Vinters *et al.*, 1994), but it also invades the basement membrane, producing thick circular or semicircular deposits in vessels with media (Wisniewski and Wegiel, 1994), whereas in vessels without media, it appears to interdigitate between outer laminae of the basement membrane (Perlmutter *et al.*, 1994). It has been suggested that vascular muscle cells containing the precursor protein β-APP produce β-amyloid in association with degeneration (Kawai *et al.*, 1993). Another line of evidence suggests that the media contributes the β-APP for processing by microglial-like cells in basement membrane into fibrillar form β-amyloid (Wisniewski and Wegiel, 1994), where it is then further metabolized to a granular form via enzymes locally released by degenerating muscle cells (Wisniewski and Wegiel, 1994). In capillaries β-amyloid deposition is also thought to be processed by microglial-like perivascular cells or pericytes (Perlmutter, 1994; Wisniewski and Wegiel, 1994), although only a subset of AD capillaries is affected by β-amyloid deposits (Perlmutter *et al.*, 1994).

Although most would accept that β-amyloid deposition in the AD vasculature is part and parcel of the generative AD processes, it is possible that the apparent pathologic events are actually defensive mechanisms. Based on data to be presented, we suggest that an unknown blood-borne agent initiates a cascade of dysfunction and induces production of AD-like neuropathology in the parenchyma. As this continues, thickening of the basement membrane could be a generalized effort to prevent the agent(s) from crossing the blood–brain barrier, while at the same time unfortunately attenuating normal nutrient transport. Thereafter, to alleviate the generalized reduction of normal blood–brain permeability, the pits observed in the basement membrane are created, with subsequent deposition of constituent components in senile plaques. As the process becomes more pronounced, the last line of defense may be deposition of insoluble β-amyloid as a "band-aid" rather than a lesion between blood and brain.

## 6. Evidence for Vascular-Mediated Alterations

The foregoing speculation is based on the premise that one or more blood-borne agents are integral to the process that leads to neuropathologic lesion formation, neurodegeneration, and neurochemical derangement that culminates in cognitive dysfunction in AD. Support for this can be drawn from studies in cultured hippocampal neurons treated with human serum (Brewer and Ashford, 1992), which demonstrated that unknown agents in the blood were able to induce cellular markers characteristic of AD pathology. Changes included elevated neuronal expression of Alz50, MAP2, and β-amyloid and ubiquitin immunoreactivity (Brewer and Ashford, 1992). It is possible that factors normally circulating in the blood at levels considered normal or above are enabled by a primary breakdown of vascular integrity to opportunistically affect surrounding central nervous tissue. Alternatively, elevated steady-state levels of a natural factor could act primarily by first inducing vascular pathology, then producing alterations in central nervous tissue. Another possibility is that a pathologic blood-borne agent, which acts directly or after inducing vascular compromise and breakdown of the blood–brain barrier (BBB), induces pathologic changes in the CNS of AD patients.

Studies by Wisniewski and Kozolowski (1982) first suggested that there was breakdown of the BBB and an invasion of the parenchyma of blood-borne agents in AD brain. Senile plaques (corona) and neurons with and without fibrillary changes were immunodecorated by anti-albumin and anti-IgG antibodies in AD brains to a much greater extent than in age-matched controls, and immunoreactivity was also observed in the neuropil and as perivascular halos. It was concluded that these observations provided indirect evidence of altered permeability of blood–brain barrier microvessels in AD (Wisniewski and Kozolowski, 1982). A subsequent study suggested that there was perivascular staining by anti-IgG and anti-albumin antibodies, which decreased in intensity with increasing distance from the blood vessel (halo) in both demented and nondemented subjects (Alafuzoff *et al.,* 1987). This study also indicated that there was anti-IgG, but not anti-albumin, labeling of SPs only. It was concluded there was compromise of the blood–brain barrier in AD and in nondemented elderly controls, although as a confounding factor all but one of the investigated controls had documented heart disease, and all of the study subjects had nonocclusive atherosclerotic plaques of the circle of Willis (Alafuzoff *et al.,* 1987). More recently, Mori *et al.* (1991) suggested that extravasation of the serum proteins IgG and albumin may be an artifact of fixation, coupled with a postmortem phenomenon. They demonstrated a correlation between extent of vascular leakage (intensity of albumin and IgG immunoreactivity) and length of the postmortem interval (time between death and brain fixation), although the neuronal uptake of these markers in human brain may depend on antemortem damage to neurons expressing immunoreactivity (Mori *et al.,* 1991). The foregoing evidence effectively disputed the hypothesis of a blood–brain barrier breakdown suggested by the flow of albumin and IgG into AD brain.

The presence of cytokines in AD brain, along with increased circulating levels in living patients, requires acknowledging the possibility, however remote, that a breakdown of the BBB may allow a flow of cytokines across the vasculature into the brain tissue. Increased serum levels of interleukin-1 (Cacabelos *et al.,* 1991) and increased production of interleukin-6 by AD mononuclear blood cells have been observed (Huberman *et al.,* 1995), although circulating levels of tumor necrosis factor are reported to be both increased (Fillit *et al.,* 1991) and decreased (Cacabelos *et al.,* 1994) in AD. Increased levels of immunoreactivity for interleukin-1 (Griffin *et al.,* 1989) and interleukin-6 (Huell *et al.,* 1995) have been localized predominantly to the SPs (Huell *et al.,* 1995; Sheng *et al.,* 1995). Because of the intimate relationship between microglia and SP formation in AD, the ability of microglia to produce and secrete interleukin-1 (Frei *et al.*, 1988; Colton *et al.*, 1991), interleukin-6 (Rydel *et al.*, 1992), and tumor necrosis factor (Frei *et al.*, 1988) provides evidence for a central source of cytokines rather than a circulating origin. It has even been suggested that the increased circulation levels of cytokines in AD may be due to flow into the blood after overproduction by microglia in AD brain (Fillit *et al.*, 1991).

Recent studies provide compelling evidence for a breakdown of the BBB in AD, and the data presented strongly suggest that one or more blood-borne agents, or altered hemodynamics alone, are capable of producing AD-like neuropathology in a whole-animal model. In contrast to cytokines, it has not been shown that hemoglobin can be produced in the CNS, and hemoglobin and its protolytic by-products are normally confined by the cerbrovasculature to the blood. It has been shown that nine different hemoglobin by-products have significantly increased concentrations in AD cerebellum (Slemmon *et al.*, 1994) and similar findings have been made in AD cortex (Dr. Dorothy Flood, personal communication, 1996). Immunohistochemically, this presents as increased intensity of neuronal hemoglobin staining in AD cerebellum (Slemmon *et al.*, 1994), whereas in AD cortex, neuropil immunoreactivity has been identified in addition to increased neuronal intensity (Dr. Dorothy Flood, personal communication, 1996). These authors note the neurotoxic character of these hemoglobin by-products, but have not determined whether they are produced in the blood and then traverse a compromised vascular system, or if hemoglobin itself enters and is thereafter metabolized in the brain (Slemmon *et al.*, 1994). Nevertheless, these data persuasively indicate a breakdown of the BBB in AD and strongly support the likelihood of a vascular-mediated component in the pathogenesis of the disease.

Producing early AD-like neuropathology in a whole animal experiencing alteration in the blood only would clearly implicate a vascular-mediated, blood-borne influence in the pathophysiology of AD. Formation of neuropathologic lesions similar to those found in AD have been identified in the brain of an adolescent pig used as the sole support of an isolated heart preparation (Fig. 4). This experimen-

---

**Figure 4.** β-amyloid and Alz50 immunoreactivity in adolescent pig hippocampus. The pig, used as the sole support of an isolated heart preparation, was anesthetized and intubated for positive ventilation, and the femoral vein and artery were exposed and attached to the isolated heart preparation to supply oxygenated blood and venous return for three hours, after which the pig's heart was collected and its brain perfused with buffered paraformaldehyde (4%) (Sparks, 1996) within 5 minutes of death. Fifty-micron vibratome sections of hippocampus were obtained for Alz50 immunohistochemical analysis (antibody supplied by Dr. Peter Davies, Albert Einstein College of Medicine) and visualized with diaminobenzadine (DAB). After paraffin-embedding tissue block, the remaining adjacent 8-micron sections were removed for routine β-amyloid immunohistochemistry using antibody recognizing the N-terminal of peptide sequence (aa 1-16, 10D5, Athena Neurosciences, South San Francisco). β-amyloid immunoreactivity visualized with DAB was assessed by light microscopy. Immunoreactivity highlighted with FITC assessed by confocal laser fluorescent microscopy. (A) Most granule layer cells and many large neurons in hilus displayed intense β-amyloid immunoreactivity, identified with DAB. ×4. (B) Many neurons through the full thickness of cortex (excluding layer I) exhibited β-amyloid immunoreactivity, identified with DAB. ×4. (C). Sequential one-micron confocal fluorescence images (stored on floptical disks) confirmed extracellular β-amyloid immunoreactivity in pig parahippocampal cortex, appearing as approximately 60-μ solid structures in neuropil, and closely resembling diffuse-form SPs observed in

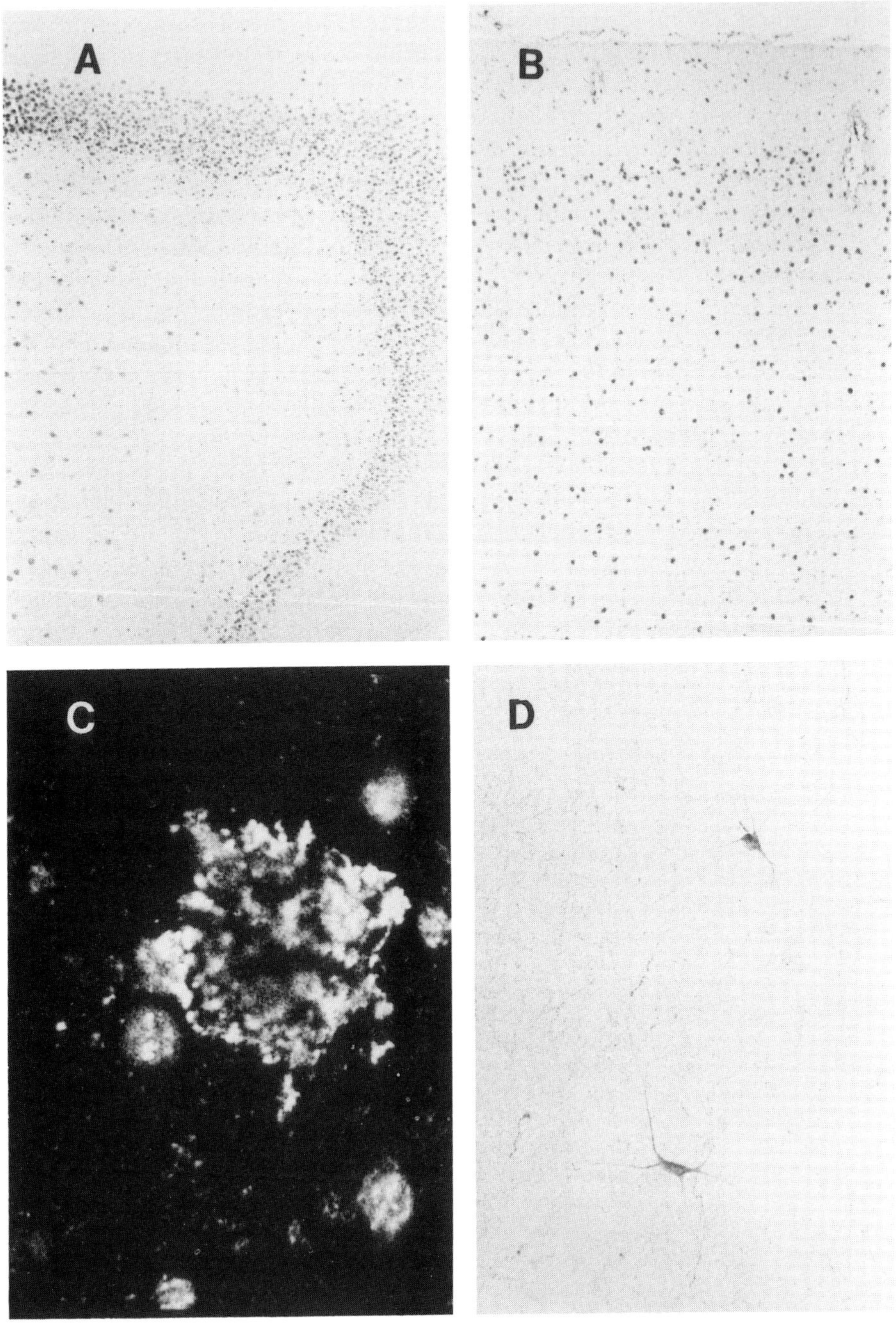


**Figure 4.** (*Continued*) humans. ×80. Similar analysis indicated that nearby neuronal immunoreactivity occurred as hollow spheres surrounding the nuclear compartment of affected cells. (D) Numerous Alz50-immunoreactive neurons, beaded processes, and neuropil threads in animal parahippocampal cortex. ×20, essentially identical in appearance to those in young Down's syndrome patients (Sparks and Hunsaker, 1992) and individuals with critical coronary artery disease (Sparks *et al.*, 1993).

tal procedure lasted just three hours. Lesions observed in hippocampal cortex included: (1) extracellular deposition of β-amyloid closely resembling diffuse-form SP; (2) abundant Alz50-immunoreactive neurons, beaded processes, and neuropil threads; and (3) β-amyloid-immunoreactive neurons distributed in a pattern similar to that previously reported (Sparks, 1996) in hippocampus of subjects with AD, Down's syndrome, and heart disease, and of rabbits fed a high cholesterol diet (Fig. 5). Control rabbits, control pigs and unaffected humans were devoid of any appreciable β-amyloid immunoreactivity. These observations are compared with findings in previously published surgical controls prepared for–but not undergoing–cardiopulmonary bypass (CPB) surgery. In the hippocampus of the controls, there were essentially no β-amyloid-immunoreactive features and no Alz50-immunoreactive neuropil threads or beaded processes, and only one animal exhibited even a limited number of Alz50-immunoreactive neurons (Sparks, 1996). The only surgical procedure experienced by the support animal (other than anesthesia and intubation to provide positive ventilation) was exposure of the left femoral artery and vein to provide oxygenated blood and venous return to and from the isolated heart. Granted this is a "dirty" model and not easily controlled or reproduced because of the myriad of alterations concomitantly produced in the blood, including, but not limited to, full-blown contact activation of the blood, introduction of numerous drugs to maintain and promote function of the isolated heart, and secretion of whatever factors the isolated heart itself may have contributed to the circulation. In spite of these shortcomings, it is clear from this single animal that creating drastic

**Figure 5.** Neuronal β-amyloid immunoreactivity pattern induced in hippocampus of experimental animals and occurring in humans with disorders that cause premature SP formation. In each group eight-micron paraffin-embedded sections were processed by standard ABC (Elite) immunocytochemical methods using β-amyloid antibody recognizing the N-terminal 1–16-amino acid sequence of the peptide (10D5, Athena Neurosciences) and visualized with diaminobenzidine. No traditional cresyl violet, or hemotoxylin, counterstain was done because it obscures intracellular β-amyloid immunoreactivity. *Rabbit panel:* (Left) A rabbit fed 2% cholesterol for 8 weeks exhibits intense β-amyloid immunoreactivity in fascia dentata and adjacent cortex. ×4. (Center) Most cells of the granule cell layer and numerous cells in hilus are densely stained. ×10. (Right) Many cells in the cortex adjacent to hippocampus are immunoreactive. *Pig Panel:* (Left) An adolescent pig used as the sole support of an isolated heart preparation shows intense neuronal β-amyloid immunoreactivity in fascia dentata and adjacent cortex. ×4. (Center) The entire granule cell layer and many hilus cells display intense β-amyloid immunoreactivity. ×10. (Right) Abundant cortical neurons show intracellular immunoreactivity. ×10. *cCAD panel:* (Left) Faint-to-moderate neuronal β-amyloid immunoreactivity in fascia dentata and parahippocampal gyrus in half the nondemented cCAD subjects exhibiting extracellular β-amyloid deposits as SPs. ×4. (Center) Many granule layer cells and some hilus cells are β-amyloid-immunoreactive. ×10. (Right) Immunoreactive neurons are interspersed with SPs in hippocampal cortex. ×10. *Down's panel:* (Left) Light-to-moderately intense neuronal β-amyloid immunoreactivity in fascia dentata and parahippocampal gyrus in half the subjects with Down's syndrome exhibiting SPs. ×4. (Center) Numerous granule layer cells and some hilus cells are β-amyloid antibody-immunoreactive. ×10. (Right) Immunoreactive neurons interspersed with SPs in parahippocampal cortex. ×10. *AD panel:* (Left) Moderately intense neuronal βA4 immunoreactivity is observed in fascia dentata and parahippocampal gyrus from half the subjects with AD. ×4. (Center) Strikingly similar to cCAD and Down's syndrome, granule layer cells and some hilus cells are β-amyloid immunoreactive. ×10. (Right) Although not abundant, immunoreactive neurons are demonstrable among SPs in parahippocampal AD gyrus. ×10. Control animals and unaffected humans were devoid of appreciable β-amyloid immunoreactivity. Reproduced with the permission of Elsevier Science Inc.

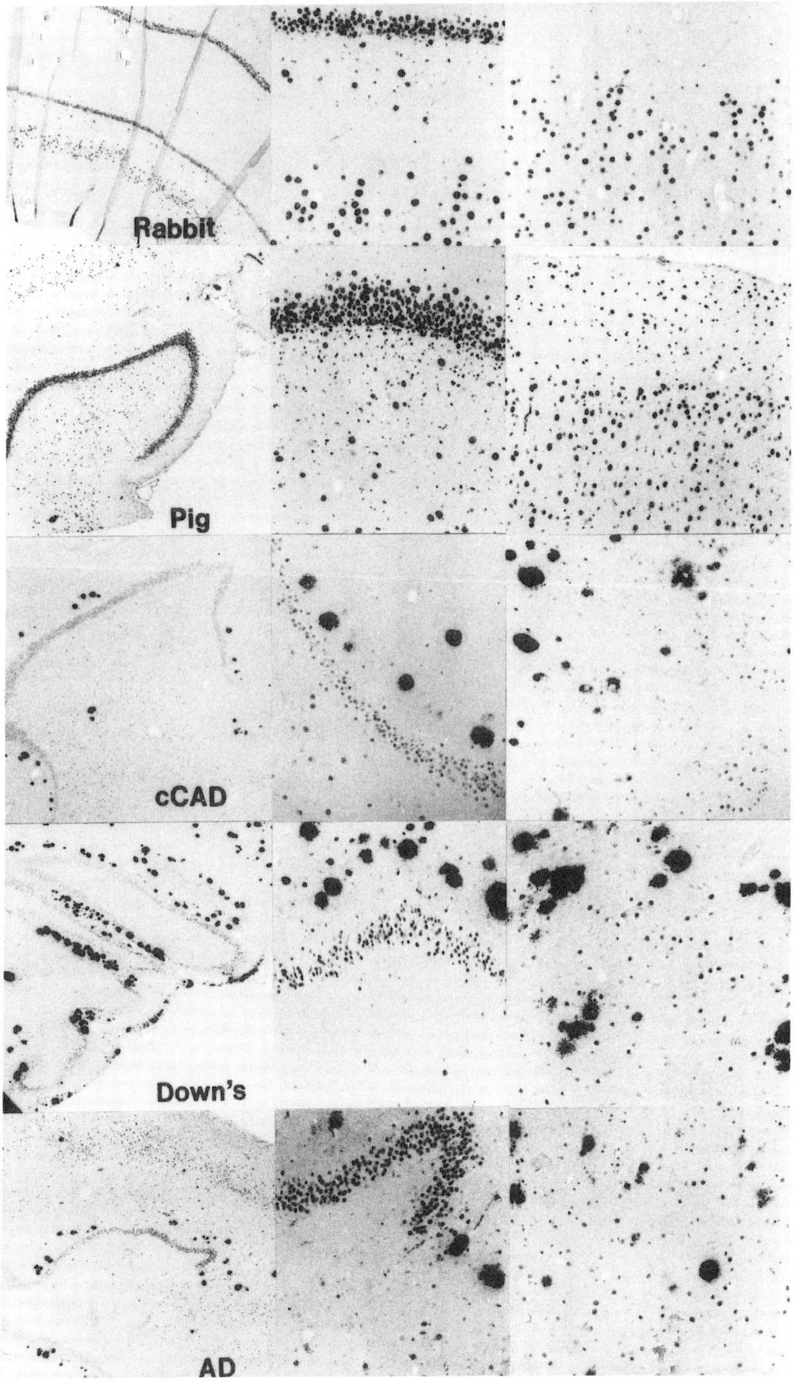
Rabbit
Pig
cCAD
Down's
AD

**Table I. Mean Densities of Immunoreactive Neurons (±SEM) in Adolescent Pig Frontal Pole and Hippocampus after Cardiac Surgical Manipulation***

| Group | Antibody | | | | | | | |
|---|---|---|---|---|---|---|---|---|
| | β-amyloid | | Alz-50 | | | SOD | | |
| | CA | PHG | CA | PHG | FP | CA | Tz | PHG |
| Anesthetic controls (N = 2) | 0 ± 0 | 0 ± 0 | 0 ± 0 | 0 ± 0 | 0 ± 0 | ND | ND | ND |
| Surgical controls (N = 3) | 11 ± 7 | 30 ± 9 | 0 ± 0 | 3.7 ± 3.7* | 0.7 ± 0.7* | 41 ± 12 | 79 ± 4 | 83 ± 9 |
| LAD ligation (N = 3) | 45 ± 16[a] | 129 ± 25[a,c] | .33 ± .33* | 8.7 ± 3.2[a] | 1.0 ± 1.0* | 52 ± 14 | 71 ± 20 | 72 ± 11 |
| CPB good protection (N = 8) | 53 ± 14[a] | 83 ± 16[a] | .75 ± .42[a] | 5.4 ± 1.5[b] | 3.7 ± 1.6[b] | 35 ± 17 | 77 ± 40 | 46 ± 7[a] |
| CPB poor protection (N = 7) | 71 ± 15[a] | 126 ± 17[a,c] | 3.6 ± 0.9[a,c] | 15.0 ± 3.6[a,c] | 12.3 ± 1.9[a,c] | 85 ± 9[a,c] | 127 ± 12[a,c] | 107 ± 9[a,c] |
| Support animal (N = 1) | 78 | 172 | 11 | 48 | 23 | 140 | 210 | 146 |

*Mean numbers of β-amyloid (1–16), Alz-50, and CU/Zn superoxide dismutase (SOD) immunoreactive neurons in hippocampal formation and frontal pole of adolescent pigs experiencing an acute ligation of the left anterior descending (LAD) coronary artery, undergoing cardiopulmonary bypass (CPB) surgery resulting in either good or poor myocardial protection, utilized as the sole support of an isolated heart preparation, compared to surgical controls and anesthetic controls in which diminished cardiac output was produced intentionally. Mean values ± SEM for β-amyloid and SOD immunoreactive neurons were blindly determined from three random 10× objective fields within each specific region of the hippocampal formation. The total number of Alz-50 immunoreactive neurons in an anatomic region was determined blindly for each animal; the mean ± SEM per section is reported. Group differences were determined by two-way ANOVA analysis, followed by independent t-tests when warranted.

* = only one animal affected.

a = $p < 0.05$ as compared with either control group.

b = $p < 0.05$ as compared with anesthetic controls and mean values for both control groups combined.

c = $p < 0.05$ as compared with CPB pigs with good protection.

Abbreviations: ND, not determined; CA, Ammon's horn; PHG, parahippocampal gyrus; FP, frontal pole; Tz, transition zone between CA and PHG.

alterations in the blood can alone induce the neuropathologic changes characteristics of AD, and in only three hours.

Further support for the possibility of a vascular-mediated, blood-borne factor inducing AD-like neuropathology comes from more-reproducible manipulations in adolescent pigs. Compared with the above-noted surgical control pigs, increased numbers of Alz50-immunoreactive neurons and appreciable accumulations of neuronal β-amyloid immunoreactivity in the hippocampus were the consequences of an acute ligation of the left anterior descending (LAD) coronary artery (Table 1). Although no extracellular β-amyloid was observed, the intensity of neuronal immunoreactivity seemed to correlate with the severity of reduced cardiac output in

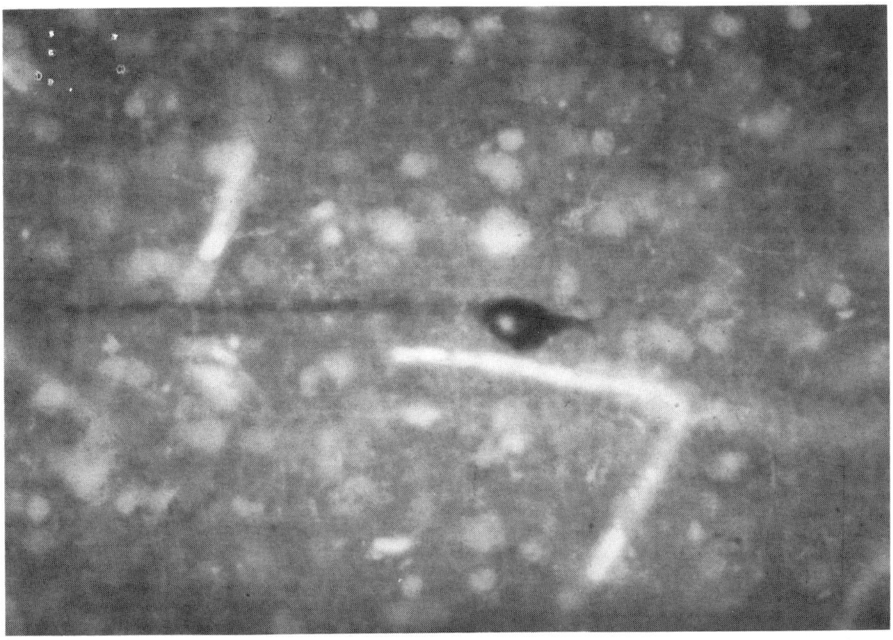

**Figure 6.** Apoptotic neurodegeneration induction in adolescent pigs. Morphologic evidence of apoptosis is presented in an animal undergoing cardiopulmonary bypass surgery. After heart removal, the brain of each animal was perfused with 4% paraformaldehyde within three minutes of sacrifice, removed, and immerse-fixed in perfusate. Without further tissue processing, 50-micron vibratome sections were cut from blocks of hippocampus for analysis, after being first reacted with Alz50 antibody supplied by Dr. Peter Davies (Albert Einstein College of Medicine) using published immunohistochemical methods. After histologically "setting" Alz50 immunoreactivity, sections were treated with commercially available RNAase to digest all nacient RNA, after which each section was reacted with commercially available propidium iodide (PI) to stain the remaining DNA. In each animal the number and appearance of Alz50-immunoreactive neurons were the same as those observed in adjacent sections not undergoing the RNAase and PI treatment. With fluorescent microscopy (×20), Alz50-immunoreactive neurons appeared as fluorescent dark structures backlit by nearby PI-stained DNA. Most of the fluorescent DNA appeared as round-to-ovoid granular deposits approximately the size of the nuclear compartments in Alz50-immunoreactive neurons. At least one pinpoint of intense PI fluorescence was associated with the nuclear region of every Alz50-immunoreactive neuron observed, and nowhere else in the tissue section. Pinpoints are thought to be condensed DNA and deemed apoptotic bodies—the defining morphologic feature of apoptosis—indicating that Alz50-immunoreactive neurons are dying via an apoptotic mechanism.

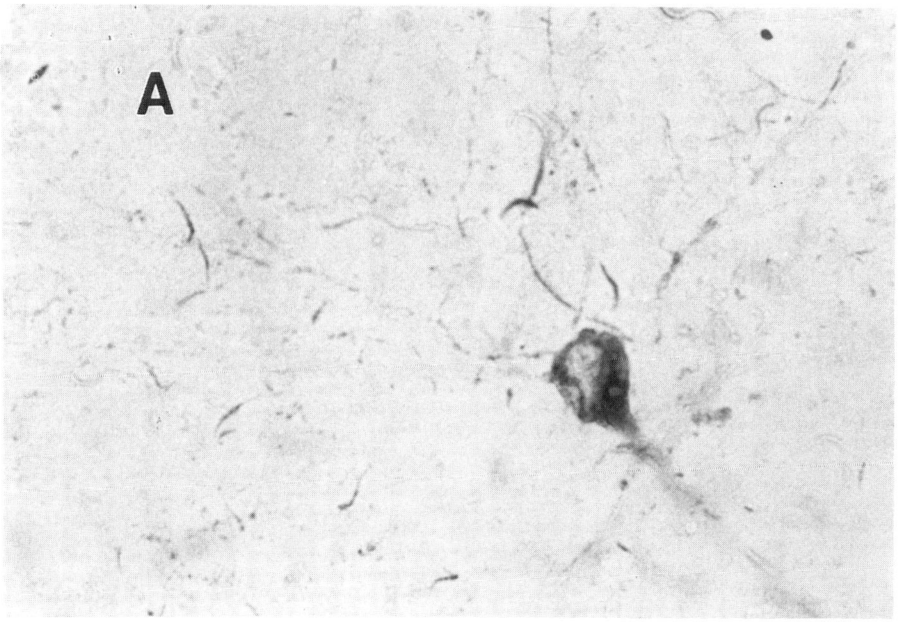


**Figure 7.** Identification of apoptotic neurodegeneration in Down's syndrome. Matched views (natural and fluorescent light) are shown for a nondemented 25-year-old. At autopsy, hippocampal tissue was immerse-fixed in 10% formalin. With no further processing, 50-micron vibratome sections were cut from blocks of hippocampus after being reacted with Alz50 antibody supplied by Dr. Peter Davies (Albert Einstein College of Medicine) using published immunohistochemical methods. Subsequent to histologically "setting" Alz50 immunoreactivity, sections were treated with commercially available RNAase to digest all nacient RNA, after which each section was reacted with commercially available propidium iodide (PI) to stain the remaining DNA. Number and appearance of Alz50-immunoreactive neurons in this and other individuals with Down's syndrome were the same as those in adjacent sections not undergoing the RNAase and PI treatment. Increased numbers of Alz50-immunoreactive neurons

duced in this group by the ligation (Sparks, 1996). In contrast, no Alz-50- or β-amyloid-immunoreactive features were observed in the hippocampus of two adolescent pigs experiencing anesthetically induced reductions of cardiac output similar to that caused by LAD ligations (Sparks, 1996). This clearly suggests that physically mimicking coronary artery disease produces an ill-defined cardiac stress—other than reduced output—that induces neuropathologic alterations similar to those occurring prematurely in humans with critical coronary artery disease (cCAD) (Sparks *et al.*, 1990, 1993, 1995).

Additional support for vascular-mediated, blood-borne induction of AD-like neuropathology comes from studies of pigs undergoing CPB surgery, a procedure often employed in humans when arterial grafting is done to circumvent life-threatening coronary artery disease. It is worth noting again that many cardiac patients undergoing CPB experience subsequent cognitive difficulties (Kolkka and Hiberman, 1980; Shaw *et al.*, 1986; Sotaniemi *et al.*, 1986; Forsman *et al.*, 1990; Challa *et al.*, 1993), which are not associated with reduced cerebral blood flow during surgery (Slogoff *et al.*, 1982) or with alterations observable by MRI (Vik *et al.*, 1991; Sellman *et al.*, 1992). Adolescent pigs undergoing experimental CPB were grouped according to the level of myocardial protection achieved during the surgery. Recov-

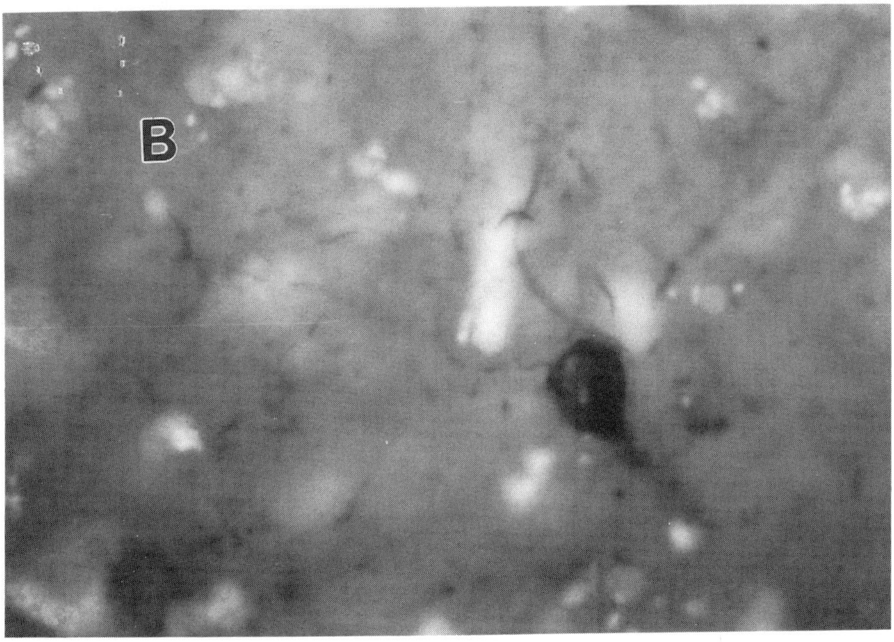


**Figure 7.** (*Continued*) occur in young Down's syndrome subjects as compared with age-matched controls, presumably early in the disorder's progression (Sparks and Hunsaker, 1992). (A) Normal light microscopy; Alz50-immunoreactive neurons and neuropil threads are commonly encountered in parahippocampal cortex of young Down's syndrome individuals. −100. (B) Fluorescent microscopy shows Alz50 neurons as fluorescent dark structures backlit by nearby PI-stained DNA. Most fluorescent background DNA appeared to be diffuse granular deposits. At least one pinpoint of intense PI fluorescence was associated with the nuclear region of every Alz50-immunoreactive neuron observed, and nowhere else in the tissue section (two are seen in this neuron). Pinpoints are thought to be apoptotic bodies—the defining morphologic feature of apoptosis—suggesting that the Alz50-immunoreactive neurons are dying via an apoptotic mechanism.

ery of cardiac function within 30 minutes after weaning an animal from bypass was used as the determinant of myocardial protection. An eventual 80% or greater recovery of cardiac function as compared with pre-bypass baseline values was considered good protection; animals recovering less than 60% function were deemed to have experienced poor protection. Both groups of CPB pigs displayed β-amyloid-immunoreactive neurons in a pattern identical to that observed in the previously presented support pig (Fig. 5), but no animal exhibited extracellular deposits. The mean number of β-amyloid-immunoreactive neurons observed in those animals with good myocardial protection was significantly elevated as compared with both control groups, and was similar to the levels seen in pigs with ligation of the LAD (Table 1). In those animals with poor myocardial protection, a further increase in the number of β-amyloid-immunoreactive neurons occurred, and at a level similar to that found in the support pig (Table 1). Similar to the LAD group, each animal undergoing CPB exhibited Alz50-immunoreactive neurons in the hippocampal formation. Many of the animals with good myocardial protection, and all of those with poor protection, displayed Alz50-immunoreactive neurons in frontal cortex also. The number of Alz50-immunodecorated neurons was significantly increased in the poorly protected CPB pigs in comparison to pigs with good myocardial

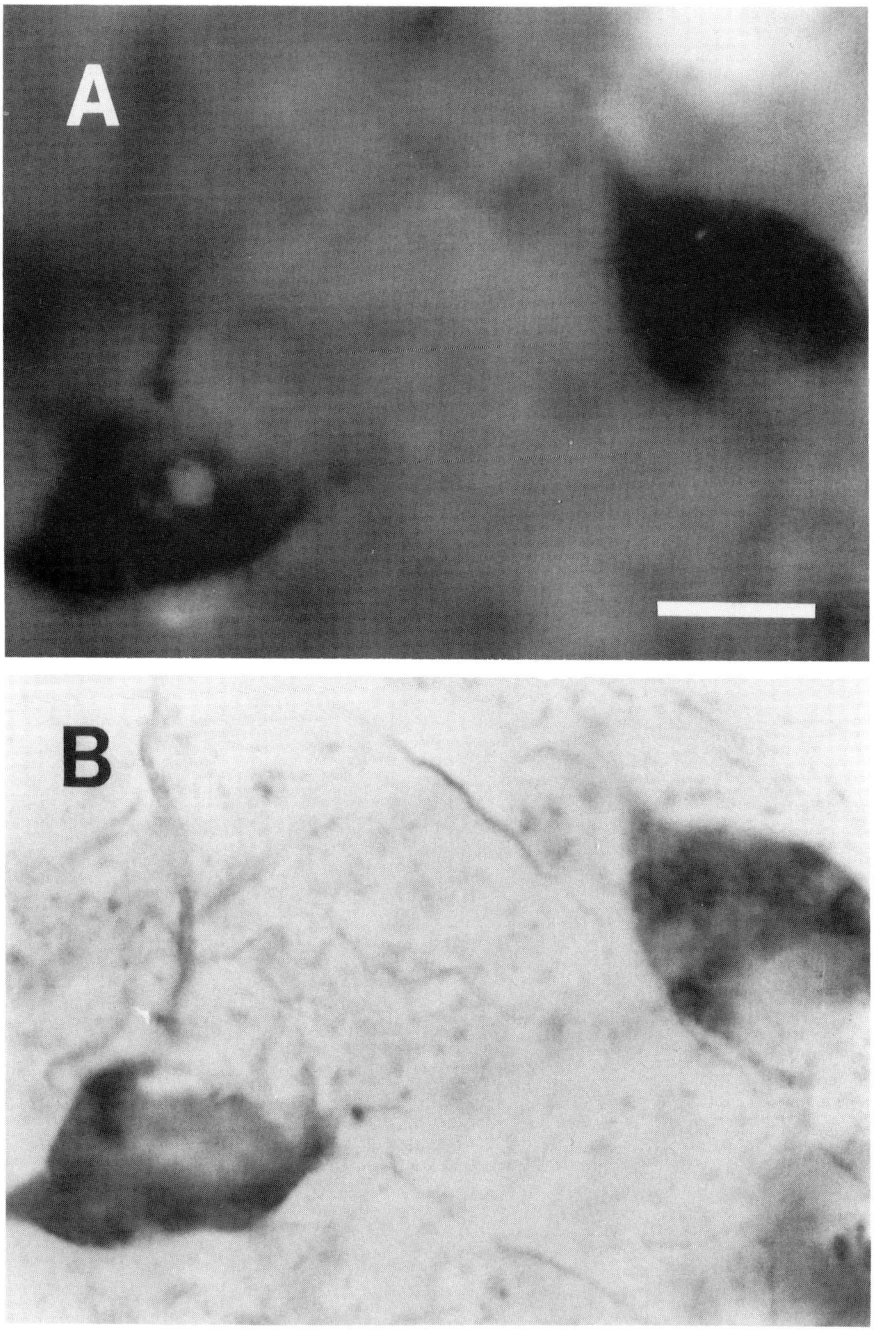


**Figure 8.** Identification of apoptotic neurodegeneration in critical Coronary Artery Disease (cCAD). Matched fields of view (natural and fluorescent light) are presented for a nondemented 71-year-old individual with cCAD. Tissue was processed as described in Figure 7. Increased number of Alz-50 immunoreactive neurons have been shown to occur in advance of senile plaque formation in cCAD (Sparks *et al.*, 1993). Using fluorescence microscopy (A) an apoptotic body is identifiable in the Alz-50 immunoreactive neuron on the left. Utilizing normal light microscopy in the matched field of view (B), besides the neurons exhibiting an apoptotic body the only Alz-50 immunoreactive residues observed were neuropil threads and neurons without apoptotic bodies. Calibration bar = 20 micron.

protection, but never approached the number observed in the sole-support animal (Table 1). Neurons expressing Alz50 immunoreactivity are thought to be dying Wolozin *et al.,* 1988; Al-Ghoul and Miller, 1989; Price *et al.,* 1991), and some are opined to eventually become NFTs (Vickers *et al.,* 1992). Each Alz50-immunoreactive neuron observed in the experimental animals exhibited within the nuclear envelope one or more apoptotic bodies (Fig. 6), the identifying morphologic feature of apoptotic neurodegeneration (Oppenheim, 1991). Just as in experimental pigs, Alz50-immunoreactive neurons in young Down's syndrome patients display apoptotic bodies within the nuclear compartment (Fig. 7). Likewise, many Alz50-immunoreactive neurons in cCAD exhibit apoptotic bodies (Fig. 8), but only in rare instances are apoptotic body containing Alz50-immunoreactive neurons identifiable in AD (Fig. 9). It is possible that increased accumulation of Alz50-immunoreactive cellular remnants in cCAD and AD is an indication of diminished capacity to

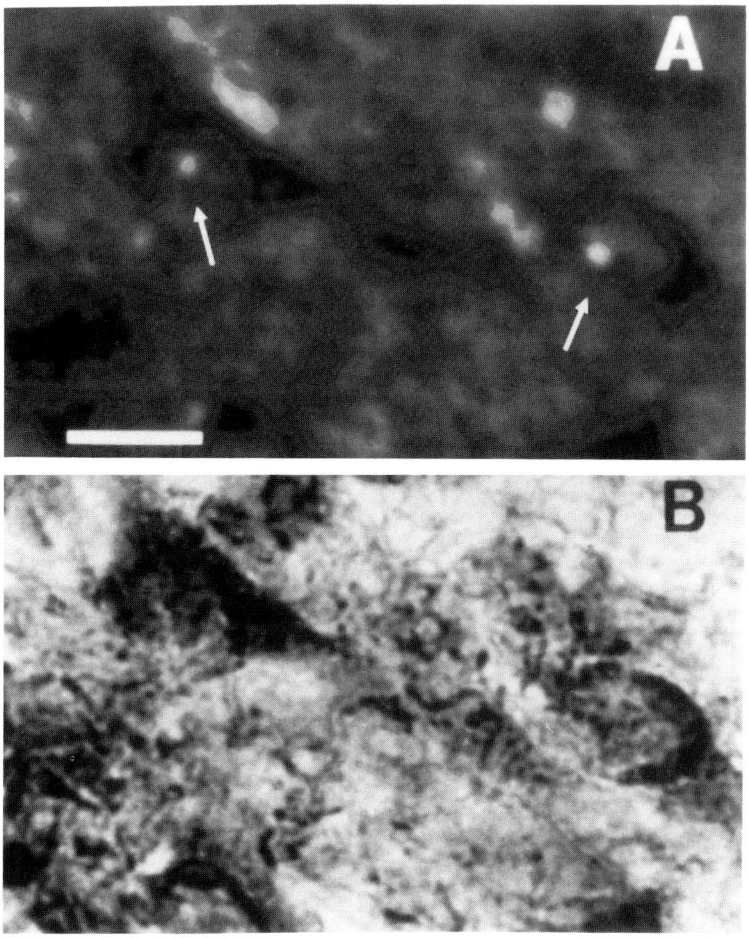


**Figure 9.** Identification of apoptotic neurodegeneration in Alzheimer's disease. Matched fields of view (natural and laser fluorescent light) are presented for hippocampus from a 67-year-old individual with AD. Tissue was processed as described in Figure 7. Laser fluorescence (A) and light (B) images of matched fields of view demonstrate that two neurons (arrows) immunostained with Alz-50 antibody exhibit apoptotic bodies. Calibration bar = 20 micron.

remove apoptotic neuron debris from the neuropil. It may be of note that β-amyloid can induce apoptosis in cultured CNS neurons (Loo *et al.*, 1993).

The influence of previously noted microembolic events associated with CPB can only account for a portion of the neuropathologic change identified because contact activation of the blood (Rinder *et al.*, 1991; Heller and Wendel, 1993; Taggart *et al.*, 1994), increased levels of atrial natriuretic factor (Northridge *et al.*, 1992), brain natriuretic factor (Horio *et al.*, 1993), vasoactive substances (Downing and Edmunds, Jr., 1992), endotoxins and cytokines (Jansen *et al.*, 1992; Butler *et al.*, 1993; Kawamura *et al.*, 1993), and production of embolic factors undoubtedly occurred at equivalent levels in both groups of CPB animals. Thoracic surgery's contribution to the etiology of AD-like neuropathology is obviously negligible. Once again the severity of resultant neuropathologic change seemed to correlate with the level of cardiac dysfunction induced by CPB. It is therefore likely that introduction of new or increased levels of factors into the bloodstream by such an experimentally compromised heart may be involved in causing the observed brain changes. Levels of a variety of circulating factors are known to be altered in humans with severe cardiac disease, including lipid peroxides (Stringer *et al.*, 1989). Besides being increased after CPB (Hoshikawa-Fujimura *et al.*, 1989; Prasad *et al.*, 1992), increased circulating concentrations of free radicals (FR) in coronary artery disease are related to increasing severity of stenosis (Bridges *et al.*, 1992), and FR can be directly produced by the heart under conditions of ischemia (Barsacchi *et al.*, 1992).

Due to the deleterious effects of FR on vascular endothelial integrity (Sellke *et al.*, 1993) and neuronal function, the effect of the aforementioned experimental paradigms on the expression of one of the major FR scavenger enzymes, Cu/Zn superoxide dismutase (SPD), was studied. Increased neuronal SOD immunoreactivity has been noted in AD brain (Ceballos *et al.*, 1991; Pappolla *et al.*, 1992) and oxidative stress has been implicated in the pathogenesis of the disorder (Pappolla *et al.*, 1992). Neurons expressing Cu/Zn SOD epitope in surgical control animals did so at very low intensity, and likely represented maintenance levels of the enzyme (Fig. 8). In each of the experimental groups (LAD, CPB, and the support animal) there was increased intensity of neuronal SOD immunoreactivity as compared with surgical controls (Fig. 10). Increased numbers of SOD-immunoreactive neurons

---

**Figure 10.** Induction of Cu/Zn SOD immunoreactivity in hippocampal cortex of adolescent pigs surgically prepared for experimental cardiopulmonary bypass (CPB). Animals maintained without further manipulation were used as experimental controls; those undergoing CPB were grouped according to post-CPB cardiac outcome (poor $<60\%$ recovery of function; good $\geq 80\%$ recovery of function). Photomicrographs of a representative surgical control animal (A, $\times 10$) and an animal exhibiting poor cardiac function recovery after CPB (B, $\times 10$) are presented. Paraffin-embedded hippocampus was sectioned at 8 microns, and immunocytochemistry was performed using standard ABC methods, with Cu/Zn SOD antibody provided by Dr. Naoyuki Taniguchi (Osaka University Medical School, Japan). Neurons expressing Cu/Zn SOD immunoreactivity were counted in three random fields in each hippocampal region, and values/mm$^2$ were recorded for a particular animal blind to group designation (see Table 1). Significantly increased numbers of SOD-immunoreactive neurons were seen in only CPB pigs with poor outcome and in a pig used as the sole support of an isolated heart preparation. In contrast, increased intensity of neuronal SOD immunoreactivity was observed in each experimental group as compared with surgical controls. (A) Very low intensity of neuronal SOD immunoreactivity in hippocampal cortex of surgical control pigs, probably representing normal levels of free radical scavenger enzyme. $\times 10$. (B) Intense neuronal SOD immunoreactivity induced in adolescent pig hippocampal cortex by experimental CPB, resulting in poor cardiac outcome. $\times 10$.

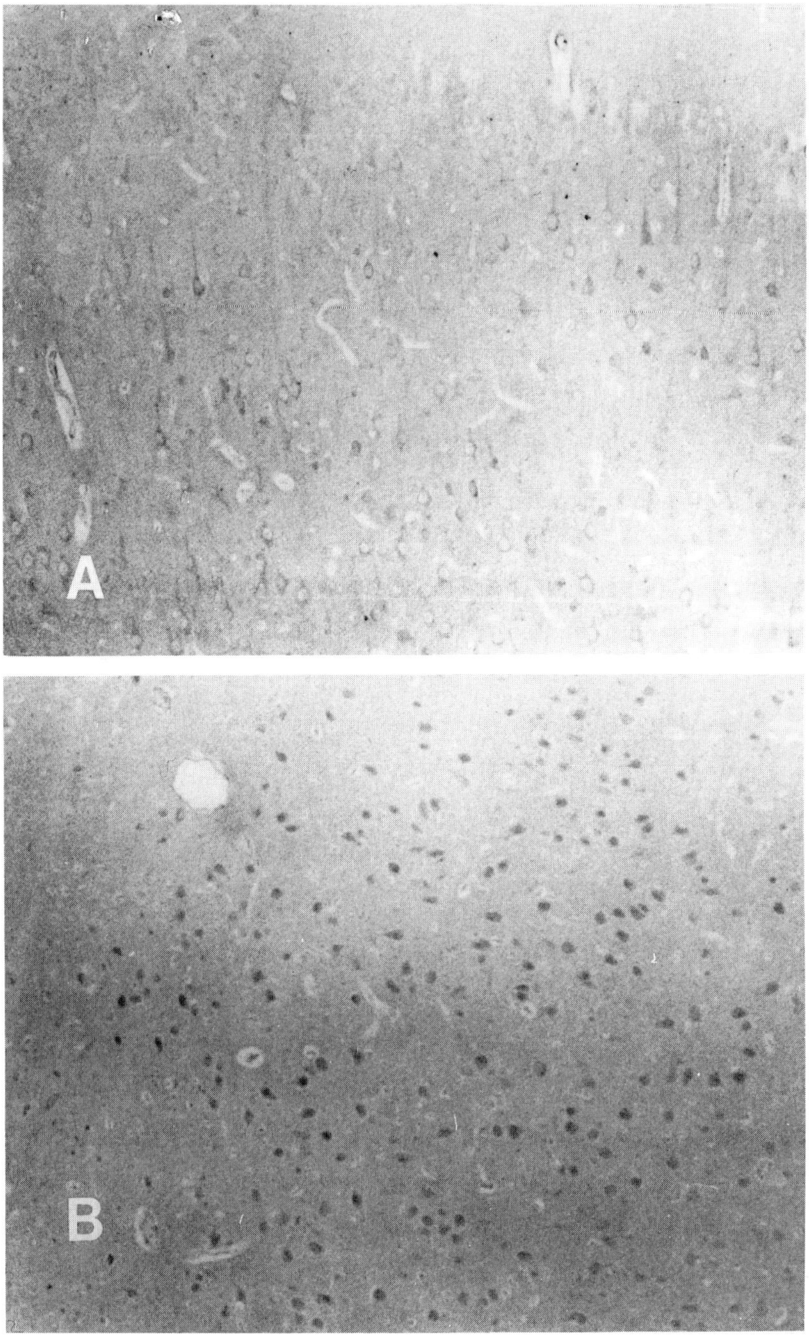
A
B

with accentuated intensity were limited to the CPB animals with poor protection and to the single animal used to support an isolated heart preparation (Table 1). Based on these additional data, there may be a relationship between increased FR activity in the brain and induction of AD-like neuropathology, and possibly such free radicals originate in the bloodstream as a result of increased production due to cardiac stress. This is not a far-fetched concept, because it has been shown that as the steady-state concentrations of FR in the circulation increase, increasing levels of SOD activity occur in the vasculature (Sharma *et al.,* 1992). Vascular SOD immunoreactivity was limited to the luminal surface at low levels of FR in the blood. Immunoreactivity was observed through the full thickness of the vasculature at higher levels of FR, and at the highest circulating levels of FR the underlying tissue on the abluminal surface of the vasculature eventually expressed SOD immunoreactivity (Sharma *et al.,* 1992), clearly suggesting that FR flow from the blood across the vasculature and into the brain is a distinct possibility, and that this factor is among the many candidate agents that could induce the onset of AD-like neuropathology.

It is important to note that experimentally producing dysfunction in the heart—which communicates directly with the brain solely via the blood—induces neuropathologic alterations essentially identical to those occurring prematurely in humans with heart conditions mimicked to cause the cardiac dysfunction in the first place. In these animal studies the factor(s) that induced the most pronounced changes must have been transmitted to the brain by the blood. The effect of surgery, contact activation of essentially every blood-borne system possible, and reduced cardiac output (delivery of fully oxygenated blood to the brain) were of minor or no influence in producing the observed neuropathology. Such a vascular-mediated blood-borne agent is probably not the panacea of AD etiology and likely requires a number of other conditions (vascular, biochemical) or factors (genetic, environmental) acting in unison, but such an agent apparently can produce the neuropathology envisioned to occur early in AD, and shown to occur well in advance of cognitive decline in Down's syndrome (Sparks and Hunsaker, 1992; Hof *et al.,* 1995). Yet the neuropathology induced by the maximal effect of such a blood-borne agent may never become sufficient to produce the severity of change associated with cognitive decline in AD. This is supported by the fact that the neuropathology seen in nondemented humans with HD is never as severe as it is in AD. Alternatively, because heart disease is a prevalent finding in AD (Aronson *et al.,* 1990; Breteler *et al.,* 1994; Prince *et al.,* 1994; Sparks *et al.,* 1995), the presence of which has been correlated with the risk of eventual dementia of the Alzheimer's type (Launer *et al.,* 1995; Skoog *et al.,* 1996), it is possible that the previously studied nondemented subjects died of their heart disease before they displayed an underlying cognitive deterioration (Sparks *et al.,* 1996).

ACKNOWLEDGMENTS. I will eternally thank Dr. John C. Hunsaker III for enabling the human studies that precipitated all subsequent animal investigations. I would like to especially thank Dr. Wes Ashford for his suggestions and invaluable assistance in preparing this chapter, and I thank Drs. Gross, Liu, and Huggle for their expert assistance in completing the animal experimentation. I would also like to warmly thank my friend—a direct descendant of the first Spanish Marquis de Cuba—Captain George Cabanas and his lovely wife, Sheri, for allowing me to hide out in paradise on their back porch in Key West to complete the reading for and organization of this chapter undisturbed.



## 7. References

Akima, M., Nonaka, H., Kagesawa, M., and Tanaka, K., 1986, A study on the microvasculature of the cerebral cortex: Fundamental architecture and its senile change in the frontal cortex, *Lab. Invest.* **55(4)**:482–489.

Al-Ghoul, W. M., and Miller, M. W., 1989, Transient expression of ALZ-50 immunoreactivity in developing rat neocortex: A marker for naturally occurring neuronal death, *Brain Res.* **481**:361–367.

Alafuzoff, I., Adolfsson, R., Grundke-Iqbal, I., and Winblad, B., 1987, Blood-brain barrier in Alzheimer dementia and in non-demented elderly, *Acta Neuropathol. (Berlin)* **73**:160–166.

Alzheimer, A., 1987, A characteristic disease of the cerebral cortex, in: *The Early Story of Alzheimer's Disease: Translation of the Historical Papers*, Liviana, Florence, Italy, 1–3.

Araki, K., Miyakawa, T., and Katsuragi, S., 1991, Ultrastructure of senile plaques using thick sections in the brain with Alzheimer's disease, *Jpn. J. Psychiat. Neurol.* **45**:85–89.

Armstrong, R. A., 1995, Is the clustering of beta-amyloid (A beta) deposits in the frontal cortex of Alzheimer patients determined by blood vessels? *Neurosci. Lett.* **195(2)**:121–124.

Aronson, M. K., Ooi, W. L., Morgenstem, H., Hafner, A., Masur, D., Crystal, H., Frishman, W. H., Fisher, D., and Katzman, R., 1990, Women, myocardial infarction, and dementia in the very old, *Neurology* **40**:1102–1106.

Ashford, J. W., Schmitt, F. A., and Kumar, V., 1996, Diagnosis of Alzheimer's disease, *Psychiat. Ann.* **26(5)**:262–268.

Barsacchi, R., Pelosi, G., Maffei, S., Baroni, M., Savatore, L., Ursini, F., Verunelli, F., and Biagini, A., 1992, Myocardial vitamin E is consumed during cardiopulmonary bypass: Indirect evidence of free radical generation in human ischemic heart, *Int. J. Cardiol.* **37**:339–342.

Bell, M. A., and Ball, M. J., 1981, Morphometric comparison of hippocampal microvasculature in aging and demented people: Diameters and densities, *Acta Neuropathol. (Berlin)* **53**:299–318.

Bell, M. A., and Ball, M. J., 1990, Neuritic plaques and vessels of visual cortex in aging and Alzheimer's dementia, *Neurobiol. Aging* **11**:359–370.

Blancher, C., Braak, H., Fischer, P., and Jellinger, K. A., 1993, Neuropathological staging of Alzheimer lesions and intellectual status in Alzheimer's and Parkinson's disease patients, *Neurosci. Lett.* **162**:179–182.

Blass, J. P., 1993, Pathophysiology of the Alzheimer's syndrome, *Neurology* (Suppl 4) **43**:S25–S38.

Braak, H., and Braak, E., 1991, Neuropathological staging of Alzheimer-related changes, *Acta Neuropathol. (Berlin)* **82**:239–259.

Braak, H., Braak, E., and Bohl, J., 1993, Staging of Alzheimer-related cortical destruction, *Eur. Neurol.* **33**:403–408.

Breteler, M. M., Claus, J. J., Grobbee, D. E., and Hofman, A., 1994, Cardiovascular disease and distribution of cognitive function in elderly people: The Rotterdam Study, *Br. Med. J.* **308**:1604–1608.

Brewer, G. J., and Ashford, J. W., 1992, Human serum stimulates Alzheimer markers in cultured hippocampal neurons, *J. Neurosci. Res.* **33**:355–369.

Bridges, A. B., Scott, N. A., Pringle, T. H., Mcneill, G. P., and Belch, J. J., 1992, Relationship between the extent of coronary artery disease and indicators of free radical activity, *Clin. Cardiol.* **15**:169–174.

Buée, L., Hof, P. R., Bouras, C., Delacourte, A., Perl, D. P., Morrison, J. H., and Fillit, H. M., 1994, Pathological alterations of the cerebral microvasculature in Alzheimer's disease and related dementing disorders, *Acta Neuropathol. (Berlin)* **87**:469–480.

Burdette, J. H., Minoshima, S., Vander Borght, T., Tran, D. D., and Kuhl, D. E., 1996, Alzheimer disease: Improved visual interpretation of PET images by using three-dimensional sterotaxic surface projections, *Radiology* **198**:837–843.

Butler, J., Pillai, R., Rocker, G. M., Westaby, S., Parker, D., and Shale, D. J., 1993, Effect of cardiopulmonary bypass on systemic release of neutrophil elastase and tumor necrosis factor, *J. Thorac. Cardiovasc. Surg.* **105**:25–30.

Cacabelos, R., Franco-Maside, A., and Alvarez, X. A., 1991, Interleukin-1 in Alzheimer's disease and multi-infarct dementia: Neuropsychological correlations, *Meth. Find. Exp. Clin. Pharmacol.* **13:**703–708.

Cacabelos, R., Alvarez, X. A., Franco-Maside, A., Fernandez-Novoa, L., and Caamano, J., 1994, Serum tumor necrosis factor (TNF) in Alzheimer's disease and multi-infarct dementia, *Meth. Find. Exp. Clin. Pharmacol.* **16(1):**29–35.

Caceres, M. J., Schleien, C. L., Kuluz, J. W., Gelman, B., and Dietrich, W. D., 1995, Early endothelial damage and leukocyte accumulation in piglet brains following cardiac arrest, *Acta Neuropathol. (Berlin)* **90:**582–591.

Ceballos, I., Javoy-Agid, F., Delacourte, A., Defossez, A., Lafon, M., Hirsch, E., Nicole, A., Sinet, P. M., and Agid, Y., 1991, Neuronal localization of copper-zinc superoxide dismutase protein and mRNA within the human hippocampus from control and Alzheimer's disease brains, *Free Rad. Res. Comm.* **12/13:**571–580.

Challa, V. R., Moody, D. M., and Bell, M. A., 1992, The Charcot-Bouchard aneurysm controversy: Impact of a new histologic technique, *J. Neuropathol. Exp. Neurol.* **51(3):**264–271.

Challa, V. R., Moody, D. M., and Troost, B. T., 1993, Brain embolic phenomena associated with cardiopulmonary bypass, *J. Neurol. Sci.* **117:**224–231.

Claudio, L., 1996, Ultrastructural features of the blood-brain barrier in biopsy tissue from Alzheimer's disease patients, *Acta Neuropathol. (Berlin)* **91:**6–14.

Colton, C. A., Yao, J., Taffs, R. E., Keri, J. E., and Oster-Granite, M. L., 1991, Abnormal production of interleukin-1 by microglia from trisomy of 16 mice, *Neurosci. Lett.* **132:**270–274.

Corsellis, J. A. N., 1962, *Mental Illness and the Aging Brain: The Distribution of Pathological Change in a Mental Hospital Population,* Oxford University Press, London.

de la Torre, J. C., 1994, Impaired brain microcirculation may trigger Alzheimer's disease, *Neurosci. Biobehav. Rev.* **18(3):**397–401.

de la Torre, J. C., and Mussivand, T., 1993, Can disturbed brain microcirculation cause Alzheimer's disease? *Neurol. Res.* **15(3):**146–153.

de Leon, M. J., Ferris, S. H., George, A. E., Reisberg, B., Christman, D. R., Kricheff, I. I., and Wolf, A. P., 1983, Computed tomography and positron emission transaxial tomography evaluations of normal aging and Alzheimer's disease, *J. Cereb. Blood Flow Metab.* **3:**391–394.

Deutsch, G., and Tweedy, J. R., 1987, Cerebral blood flow in severity matched Alzheimer and multi-infarct patients, *Neurology* **37:**431–438.

DeWitt, D. A., Silver, J., Canning, D. R., and Perry, G., 1993, Chondroitin sulfate proteoglycans are associated with the lesions of Alzheimer's disease, *Exp. Neurol.* **121:**149–152.

Downing, S. W., and Edmunds, H., Jr., 1992, Release of vasoactive substances during cardiopulmonary bypass, *Ann. Thorac. Surg.* **54:**1236–1243.

Eberling, J. L., Jagust, W. J., Reed, B. R., and Baker, M. G., 1992, Reduced temporal lobe blood flow in Alzheimer's disease, *Neurobiol. Aging* **13(4):**483–491.

Eckland, A., Saveland, H., Rommer, B., and Brandt, L., 1995, Transcranial doppler ultrasound in hypertensive versus normotensive patients after aneurysmal subarachnoid hemorrhage, *Stroke* **26:**2071–2074.

Englund, E., Brun, A., and Alling, C., 1988, White matter changes in dementia of Alzheimer's type, *Brain* **111:**1425–1439.

Faraci, F. M., and Heistad, D. D., 1990, Regulation of large cerebral arteries and cerebral microvascular pressure, *Circ. Res.* **66:**8–17.

Ferrer, I., Bella, R., and Serrano, M., 1990, Arteriosclerotic leucoencephalopathy in the elderly and its relation to white-matter lesion in Binswanger's disease, multi-infarct encephalopathy, and Alzheimer's disease, *J. Neurol. Sci.* **98:**37–50.

Fillit, H., Ding, W., Buée, L., Kalman, J., Altstiel, L., Lawlor, B., and Wolf-Klein, G., 1991, Elevated circulating tumor necrosis factor levels in Alzheimer's disease, *Neurosci. Lett.* **129:**318–320.

Fischer, O., 1987, Miliary necrosis with nodular proliferation of the neurofibrils, a common change of the cerebral cortex in senile dementia, in: *The Early Story of Alzheimer's Disease: Translation of the Historical Papers,* Liviana, Florence, Italy, 6–18.

Fischer, V. W., Siddiqi, A., and Yusufaly, Y., 1990, Altered angioarchitecture in selected areas of brains with Alzheimer's disease, *Acta Neuropathol. (Berlin)* **79:**672–679.

Forsman, M., Olsnes, B. T., Semb, G., and Steen, P. A., 1990, Effects of nimodipine on cerebral blood flow and neuropsychological outcome after cardiac surgery, *Br. J. Anaesth.* **65:**514–520.

Frackowiak, R. S. J., Pozzilli, C., Legg, N. J., Du Boulay, G. H., Marshall, J., Lenzi, G. L., and Jones, T.,

1981, Regional cerebral oxygen supply and utilization in dementia: A clinical and physiological study with oxygen-15 and positron tomography, *Brain* **104:**753–778.

Frautschy, S. A., Cole, G. M., and Baird, A., 1992, Phagocytosis and deposition of vascular β-amyloid in rat brains injected with Alzheimer β amyloid, *Am. J. Pathol.* **140:**1389–1399.

Frei, K., Siepi, C., Groscurth, P., Bodmer, S., and Fontana, A., 1988, Immunobiology of microglial cells, *Ann. N.Y. Acad. Sci.* **540:**218–227.

Glenner, G. G., and Wong, C. W., 1984, Alzheimer's disease: Initial report of the purification and characterization of a novel cerebrovascular amyloid protein, *Biochem. Biophys. Res. Commun.* **120:** 885–890.

Graeber, M. B., Streit, W. J., Buringer, D., Sparks, D. L., and Kreutzberg, G. W., 1992, Ultrastructural location of major histocompatibility complex (MHC) class II positive perivascular cells in histologically normal human brain, *J. Neuropathol. Exp. Neurol.* **51(3):**303–311.

Griffin, W. S. T., Stanley, L. C., and Ling, C., 1989, Brain interleukin 1 and S-100 immunoreactivity are elevated in Down syndrome and Alzheimer disease, *Proc. Natl. Acad. Sci. USA* **86:**7611–7615.

Grubb, R., Raichle, M., Gado, M., Eichling, J., and Hughes, C., 1977, Cerebral blood flow, oxygen utilization, and blood volume in dementia, *Neurology* **27:**905–910.

Hachinski, V. C., 1990, The decline and resurgence of vascular dementia, *Can. Med. Assoc. J.* **142:**107–111.

Harik, S. I., and Kalaria, R. N., 1991, Blood-brain barrier abnormalities in Alzheimer's disease, *Ann. N.Y. Acad. Sci.* **640:**47–52.

Harris, G. J., Lewis, R. F., Satlin, A., English, C. D., Scott, T. M., Yurgelun-Todd, D. A., and Renshaw, P. F., 1996, Dynamic susceptibility contrast MRI of regional cerebral blood volume in Alzheimer's disease, *Am. J. Psychiatry* **153:**721–724.

Hassler, O., 1965, Vascular changes in senile brains, *Acta Neuropathol. (Berlin)* **5:**40–53.

Heller, W., and Wendel, H. P., 1993, Clivarin and other LMWHs in an *ex vivo* cardiopulmonary bypass model: Studies on contact activation, *Blood Coag. Fibrinolysis* **4:**s45–s54.

Hof, P. R., Bouras, C., Mehal, N., Sparks, D. L., Perl, D. P., and Morrison, J. H., 1995, Age-related distribution of neuropathologic changes in the cerebral cortex of Down Syndrome cases: Quantitative regional analysis and comparison with Alzheimer's disease, *Arch. Neurol.* **52(4):**379–391.

Horio, T., Shimada, K., Kohno, M., Yoshimura, T., Kawarabayashi, T., Yasunari, K., Murakawa, K. I., Yokokawa, K., Ikeda, M., Fukui, T., and Takeda, T., 1993, Serial changes in atrial and brain natriuretic peptides in patients with acute myocardial infarction treated with early coronary angioplasty, *Am. Heart J.* **126:**293–299.

Hoshikawa-Fujimura, A. Y., Auler, J. O., Jr., Da Rocha, T. R., Brandizzi, L. I., Pascual, J. M., Chamone, D. A., and Jatene, A. D., 1989, PAF-acether, superoxide anion, and beta-glucuronidase as parameters of polymorphonuclear cell activation associated with cardiac surgery and cardiopulmonary bypass, *Braz. J. Med. Biol. Res.* **22:**1077–1082.

Hoyer, S., 1986, Senile dementia and Alzheimer's disease; brain blood flow and metabolism, *Prog. Neuropsychopharmacol. Biol. Psychiatry* **10:**447–478.

Huberman, M., Sredni, B., Stern, L., Kott, E., and Shalit, F., 1995, IL-2 and IL-6 secretion in dementia: Correlation with type and severity of disease, *J. Neurol. Sci.* **130(2):**161–164.

Huell, M., Strauss, S., Volk, B., Berger, M., and Bauer, J., 1995, Interleukin-6 is present in early stages of plaque formation and is restricted to the brains of Alzheimer's disease patients, *Acta Neuropathol. (Berlin)* **89:**544–551.

Iwatsubo, T., Odaka, A., Suzuki, N., Mizusawa, H., Nukina, N., and Ihara, Y., 1994, Visualization of Aβ42(43) and Aβ40 in senile plaques with end-specific Aβ monoclonals: Evidence that an initially deposited species is Aβ42(43), *Neuron* **13:**45–53.

Jagust, W. J., Friedland, R. P., and Budinger, T. F., 1988, Longitudinal studies of regional cerebral metabolism in Alzheimer's disease, *Neurology* **38:**909–912.

Jagust, W. J., Seab, J. P., Huesman, R. H., Valk, P. E., Mathis, C. A., Reed, B. R., Coxson, P. G., and Budinger, T. F., 1991, Diminished glucose transport in Alzheimer's disease: Dynamic PET studies, *J. Cereb. Blood Flow Metab.* **11(2):**323–330.

Jagust, W. J., Eberling, J. L., Richardson, B. C., Reed, B. R., Baker, M. G., Nordahl, T. E., and Budinger, T. F., 1993, The cortical topography of temporal lobe hypometabolism in early Alzheimer's disease, *Brain Res.* **629(2):**189–198.

Jansen, N. J., van Oeveren, W., Gu, Y. J., van Vliet, M. H., Eijsman, L., and Wildevuur, C. R., 1992, Endotoxin release and tumor necrosis factor formation during cardiopulmonary bypass, *Ann. Thorac. Surg.* **54:**744–748.

Jendroska, K., Poewe, W., Kaniel, S. E., Pluess, J., Iwerssen-Schmidt, H., Paulsen, J., Barthel, S., Schelosky, L., Cervos-Navarro, J., and DeArmond, S. J., 1995, Ischemic stress induces deposition of amyloid β immunoreactivity in human brain, *Acta Neuropathol. (Berlin)* **90:**461–466.

Joachim, C. L., Morris, J. H., and Selkoe, D. J., 1988, Clinically diagnosed Alzheimer's disease: Autopsy results in 150 cases, *Ann. Neurol.* **24(1):**50–56.

Johansson, B. B., 1984, Cerebral vascular bed in hypertension and consequences for the brain, *Hypertension* (Suppl III) **6:**III-81–III-86.

Johansson, B. B., 1992, Vascular mechanisms in hypertensive cerebrovascular disease, *J. Cardiovasc. Pharmacol.* (Suppl 3) **19:**S11–S15.

Kalaria, R. N., and Hedera, P., 1995, Differential degeneration of the cerebral microvasculature in Alzheimer's disease, *NeuroReport* **6(3):**477–480.

Kalaria, R. N., Homayoun, P., and Whitehouse, P. J., 1994, Nicotinic cholinergic receptors associated with mammalian cerebral vessels, *J. Auton. Nerv. Syst.* (Suppl) **49:**S3–S7.

Kannel, W. B., Wolf, P. A., and Verter, J., 1983, Manifestations of coronary disease predisposing to stroke: The Framingham Study, *J. Am. Med. Assoc.* **250:**2942–2946.

Kawai, M., Kalaria, R. N., Cras, P., Siedlak, S. L., Velasco, M. E., Shelton, E. R., Chan, H. W., Greenberg, B. D., and Perry, G., 1993, Degeneration of vascular muscle cells in cerebral amyloid angiopathy of Alzheimer disease, *Brain Res.* **623:**142–146.

Kawamura, T., Wakusawa, R., Okada, K., and Inada, S., 1993, Elevation of cytokines during open heart surgery with cardiopulmonary bypass: Participation of interleukin 8 and 6 in reperfusion injury, *Can. J. Anaesth.* **40(11):**1016–1021.

Khachaturian, Z., 1985, Diagnosis of Alzheimer's disease, *Arch. Neurol.* **42:**1097–1105.

Kimura, T., Hashimura, T., and Miyakawa, T., 1991, Observations of microvessels in the brain with Alzheimer's disease by the scanning electron microscopy, *Jpn. J. Psychiat. Neurol.* **45:**671–676.

Kolkka, R., and Hiberman, M., 1980, Neurologic dysfunction following cardiac operation with low-flow, low-pressure cardiopulmonary bypass, *J. Thorac. Cardiovasc. Surg.* **79:**432–437.

Kozachuk, W. E., DeCarli, C., Schapiro, M. B., Wagner, E. E., Rapoport, S. I., and Horwitz, B., 1990, White matter hyperintensities in dementia of Alzheimer's type and in healthy subjects without cerebrovascular risk factors, *Arch. Neurol.* **47:**1306–1310.

Kraepelin, E., 1987, Senile and pre-senile dementia, in: *The Early Story of Alzheimer's Disease: Translation of the Historical Papers,* Liviana, Florence, Italy, 32–81.

Launer, L. J., Masaki, K., Petrovitch, H., Foley, D., and Havlik, R. J., 1995, The association between midlife blood pressure levels and late-life cognitive function: The Honolulu–Asia aging study, *J. Am. Med. Assoc.* **274:**1846–1851.

Lippa, C. F., Hamos, J. E., Smith, T. W., Pulaski-Salo, D., and Drachman, D. A., 1993, Vascular amyloid deposition in Alzheimer's disease: Neither necessary nor sufficient for the local formation of plaques or tangles, *Arch. Neurol.* **50:**1088–1092.

Loo, D. T., Copani, A., Pike, C. J., Whittemore, E. R., Walencewicz, A. J., and Cotman, C. W., 1993, Apoptosis is induced by β-amyloid in cultured central nervous system neurons, *Proc. Natl. Acad. Sci. USA* **90:**7951–7955.

Lotz, P. R., Ballinger, W. E., and Quisling, R. G., 1986, Subcortical arteriosclerotic encephalopathy: CT spectrum and pathologic correlation, *Am. J. Neuroradiol.* **7:**817–822.

Malamud, N., and Grosh, L. C., 1938, Hyperinsulinism and cerebral change, *Arch. Intern. Med.* **61:**579–582.

Malamud, N., and Hirano, A., 1974, *Articles of Neuropathology,* University of California Press, Berkeley, CA, pp. 138–139.

Mancardi, G. L., Perdelli, F., Leonardi, A., and Bugiani, O., 1980, Thickening of the basement membrane of cortical capillaries in Alzheimer's disease, *Acta Neuropathol. (Berlin)* **49:**79–83.

Mandybur, T. I., 1975, The incidence of cerebral amyloid angiopathy in Alzheimer's disease, *Neurology* **25:**120–126.

McGeer, P. L., Kamo, H., Harrop, P., McGeer, E. G., Martin, W. R. W., Pate, B. D., and Li, D. K. B., 1986, Comparison of PET, MRI, and CT with pathology in a proven case of Alzheimer's disease, *Neurology* **36:**1569–1574.

Meier-Ruge, W., Bertoni-Freddari, C., and Iwangoff, P., 1994a, Changes in brain glucose metabolism as a key to the pathogenesis of Alzheimer's disease, *Gerontology* **40(5):**246–252.

Meier-Ruge, W., Iwangoff, P., and Bertoni-Freddari, C., 1994b, What is primary and what secondary for amyloid deposition in Alzheimer's disease, *Ann. N.Y. Acad. Sci.* **719:**230–237.

Metter, E. J., and Wilson, R. S., 1993, *Vascular Dementias,* Oxford University Press, New York.

Meyer, J. S., Obara,K., Muramatsu, K., Mortel, K. F., and Shirai, T., 1995, Cognitive performance after small strokes correlates with ischemia, not atrophy of the brain, *Dementia* **6:**312–322.

Meyer, J. S., Shirai, T., and Akiyama, H., 1996, Neuroimaging for differentiating vascular from Alzheimer's dementias, *Cerebrovasc. Brain Metab. Rev.* **8(1):**1–10.

Mielke, R., Pietrzyk, U., Jacobs, A., Fink, G. R., Ichimiya, A., Kessler, J., Herholz, K., and Keiss, W. D., 1994, HMPAO SPECT and PET in Alzheimer's disease and vascular dementia: Comparison of perfusion and metabolic pattern, *Eur. J. Nucl. Med.* **21:**1052–1060.

Mielke, R., Schroder, R., Fink, G. R., Kessler, J., Herholz, K., and Heiss, W.-D., 1996, Regional cerebral glucose metabolism and postmortem pathology in Alzheimer's disease, *Acta Neuropathol. (Berlin)* **91:**174–179.

Minthon, L., Nilsson, K., Evdinsson, L., Wendt, P. E., and Gustafson, L., 1995, Long-term effects of tacrine on regional cerebral blood flow changes in Alzheimer's disease, *Dementia* **6(5):**245–251.

Moody, D. M., Bell, M. A., Phil, D., Challa, V. R., Johnston, W. E., and Prough, D. S., 1990, Brain microemboli during cardiac surgery or aortography, *Ann. Neurol.* **28:**477–486.

Mori, S., Sternberger, H., Herman, M. M., and Sternberger, L. A., 1991, Leakage and neuronal uptake of serum protein in aged and Alzheimer brains: A postmortem phenomenon with antemortem etiology, *Lab. Invest.* **64(3):**345–351.

Mueller, S. M., and Luft, F. C., 1982, The blood-brain barrier in renovascular hypertension, *Stroke* **13(2):**229–234.

Northridge, D. B., Jamieson, M. P., Jardine, A. G., MacArthur, J. D., MacFarlane, N., and Dargie, H. J., 1992, Pulmonary extraction and left atrial secretion of atrial natriuretic factor during cardiopulmonary bypass surgery, *Am. Heart J.* **123:**698–703.

Obara, K., Meyer, J. S., Mortel, K. F., and Muramatsu, K., 1994, Cognitive declines correlated with decreased cortical volume and perfusion in dementia of Alzheimer's type, *J. Neurol. Sci.* **127:**96–102.

Oppenheim, R. W., 1991, Cell death during development of the nervous system, *Annu. Rev. Neurosci.* **14:**453–501.

Pappolla, M. A., Omar, R. A., Kim, K. S., and Robakis, N. K., 1992, Immunohistochemical evidence of antioxidant stress in Alzheimer's disease, *Am. J. Pathol.* **140:**621–628.

Parnetti, L., Lowenthal, D. T., Presciutti, O., Pelliccioli, G.-P., Palumbo, R., Gobbi, G., Chiarini, P., Palumbo, B., Tarducci, R., and Senin, U., 1996, 1-H-MRS, MRI-based hippocampal volumetry, and 99mTc-HMPAO-SPECT in normal aging, age-associated memory impairment, and probably Alzheimer's disease, *J. Am. Geriat. Soc.* **44:**133–138.

Perlmutter, L. S., 1994, Microvascular pathology and vascular basement membrane components in Alzheimer's disease, *Mol. Neurobiol.* **9(1–3):**33–40.

Perlmutter, L. S., Barron, E., Saperia, D., and Chui, H. C., 1991, Association between vascular basement membrane components and the lesions of Alzheimer's disease, *J. Neurosci. Res.* **30(4):**673–681.

Perlmutter, L. S., Scott, S. A., Barron, E., and Chui, H. C., 1992, MHC class II-positive microglia in human brain: Association with Alzheimer lesions, *J. Neurosci. Res.* **33:**549–558.

Perlmutter, L. S., Myers, M., and Barron, E., 1994, Vascular basement membrane components and the lesions of Alzheimer's disease: Light and electron microscopic analyses, *Microsc. Res. Tech.* **28:**204–215.

Polinsky, R. J., Noble, H., and DiChiro, G., 1987, Dominantly inherited Alzheimer's disease: Cerebral glucose metabolism, *Neurol. Neurosurg. Psychiatry* **50:**752–757.

Prasad, K., Kalra, J., Bharadwaj, B., Chaudhary, A. K., and Pharm, M., 1992, Increased oxygen free radical activity in patients on cardiopulmonary bypass undergoing aortocoronary bypass surgery, *Am. Heart J.* **123:**37–45.

Price, J. L., Davis, P. B., Morris, J. C., and White, D. L., 1991, The distribution of tangles, plaques, and related immunohistochemical markers in healthy aging and Alzheimer's disease, *Neurobiol. Aging* **12:**295–312.

Prince, M., Cullen, M., and Mann, A., 1994, Risk factor for Alzheimer's disease and dementia: A case-control study based on the MRC elderly hypertension trial, *Neurology* **44:**97–104.

Rapoport, S. I., 1991, Positron emission tomography in Alzheimer's disease in relation to disease pathogenesis: A critical review, *Cerebrovasc. Brain Metab. Rev.* **3:**297–335.

Ravens, J. R., 1978, Vascular changes in the human senile brain, in: *Pathology of Cerebrospinal Microcirculation,* Raven, New York, 487–501.

Rinder, C. S., Bohnert, J., Rinder, H. M., Mitchell, J., Ault, K., and Hillman, R., 1991, Platelet activation and aggregation during cardiopulmonary bypass, *Anesthesiology* **75:**388–393.

Roberts, E., 1986, Senile dementia of the Alzheimer's type: A possible cerebrovascular origin and some therapeutic suggestions, in: *The Biological Substrates of Alzheimer's Disease,* Academic, New York, pp. 241–266.

Robitaille, Y., Metuzals, J., Houghton, S., Leblanc, R., and Gauthier, S., 1987, Disintegration of the blood brain barrier and peliosis in Alzheimer's disease, *J. Neuropathol. Exp. Neurol. (Abstr.)* **46:**397.

Rogers, R., and Morrison, J., 1985, Quantitative morphology and regional and laminar distribution of senile plaques in Alzheimer's disease, *J. Neurol. Sci.* **5:**2801–2806.

Rogers, R. L., Meyer, J. S., Mortel, K. F., Mahurin, R. K., and Judd, B. W., 1986, Decreased cerebral blood flow precedes multi-infarct dementia, but follows senile dementia of Alzheimer type, *Neurology* **36:**1–6.

Roher, A. E., Lowenson, J. D., Clarke, S., Woods, A. S., Cotter, R. J., Gowing, E., and Ball, M. J., 1993, β-Amyloid-(1-42) is a major component of cerebrovascular amyloid deposits: Implications for the pathology of Alzheimer disease, *Proc. Natl. Acad. Sci. USA* **90:**10836–10840.

Roman, G. C., 1987, Senile dementia of the Binswanger type: A vascular form of dementia in the elderly, *J. Am. Med. Assoc.* **258:**1782–1788.

Rydel, R. E., Kurdowska, A., Johnson-Wood, K. L., Schenk, D. B., Lieberburg, I., and Travis, J., 1992, Secreted forms of the β-amyloid precursor protein stimulate IL-6 release in cultures of human fibroblasts, *Neurobiol. Aging* (Suppl 1) **13:**S85.

Scheibel , A. B., and Duong, T., 1986, Microvascular changes in Alzheimer's disease, in: *The Biological Substrates of Alzheimer's Disease,* Academic, New York, pp. 177–192.

Scheibel, A. B., Duong, T., and Tomiyasu, U., 1987, Denervation microangiopathy in senile dementia, Alzheimer type, *Alz. Dis. Assoc. Dis.* **1(1):**19–37.

Schmidt, C. F., 1964, The cerebral circulation: Some gerontological considerations, in: *Cerebral Ischemia,* Charles C. Thomas, Springfield, Massachusetts, pp. 5–20.

Scott, S. A., Johnson, S. A., Zarow, C., and Perlmutter, L. S., 1993, Inability to detect beta-amyloid protein precursor mRNA in Alzheimer plaque associated microglia, *Exp. Neurol.* **121(1):**113–118.

Sellke, F. W., Shafique, T., Ely, D. L., and Weintraub, R. M., 1993, Coronary endothelial injury after cardiopulmonary bypass and ischemic cardioplegia is mediated by oxygen-derived free radicals, *Circulation* **88:**395–400.

Sellman, M., Hindmarsh, T., Ivert, T., and Bjarne, K. H., 1992, Magnetic resonance imaging of the brain before and after open heart operations, *Ann. Thorac. Surg.* **53:**807–812.

Sharma, R. C., Crawford, D. W., Kramasch, D. M., Sevanian, A., and Jiao, Q., 1992, Immunolocalization of native antioxidant scavenger enzymes in early hypertensive and atherosclerotic arteries, *Arteriosclerosis and Thrombosis* **12:**403–415.

Shaw, P. J., Bates, D., Carlidge, N. E. F., French, J. M., Heaviside, D., Julian, D. G., and Shaw, D. A., 1986, Early intellectual dysfunction following coronary bypass surgery, *Q. J. Med.* **58(225):**59–68.

Sheng, J. G., Mrak, R. E., and Griffin, W. S., 1995, Microglial interleukin-1 alpha expression in brain regions in Alzheimer's disease: Correlation with neuritic plaque distribution, *Neuropathol. Appl. Neurobiol.* **21(4):**290–301.

Shinkai, Y., Yoshimura, M., Ito, Y., Odaka, A., Suzuki, N., Yanagisawa, K., and Ihara, Y., 1995, Amyloid β-protein 1-40 and 1-42(43) in the soluble faction of extra- and intracranial blood vessels, *Ann. Neurol.* **38:**421–428.

Simpson, I. A., Chundu, K. R., Davies-Hill, T., Honer, W. G., and Davies, P., 1994, Decreased concentration of GLUT1 and GLUT3 glucose transporters in the brains of patients with Alzheimer's disease, *Ann. Neurol.* **35(5):**546–551.

Skoog, I., Palmertz, B., and Andreasson, L.-A., 1994, The prevalence of white-matter lesions on computed tomography of the brain in demented and nondemented 85-year olds, *J. Geriat. Psychiatry Neurol.* **7:**169–175.

Skoog, I., Lemfelt, B., Landahl, S., Palmertz, B., Andreasson, L.-A., Nilsson, L., Persson, G., Oden, A., and Svanborg, A., 1996, 15-year longitudinal study of blood pressure and dementia, *Lancet* **347:**1141–1145.

Slemmon, J. R., Hughes, C. M., Campbell, G. A., and Flood, D. G., 1994, Increased levels of hemoglobin-derived and other peptides in Alzheimer's disease cerebellum, *J. Neurosci.* **14(4):**2225–2235.

Sliwka, U., Job, P.-F., Wissuwa, D., Diehl, R. R., Flachskampf, F.-A., Hanrath, P., and Noth, J., 1995, Occurrence of transcranial doppler high-intensity transient signals in patients with potential cardiac sources of embolism: A prospective study, *Stroke* **26:**2067–2070.

Slogoff, S., Girgis, K. Z., and Keats, A. S., 1982, Etiologic factors in neuropsychiatric complications associated with cardiopulmonary bypass, *Anesth. Analg.* **67:**903–911.

Smith, G. S., de Leon, M. J., George, A. E., Kliyer, A., Volkow, N. D., McRae, T., Golomb, J., Ferris, S. H., Reisberg, B., Garavino, J., and La Regina, M. E., 1992, Topography of cross-sectional and longitudinal glucose metabolic deficits in Alzheimer's disease, *Arch. Neurol.* **49:**1142–1150.

Snow, A. D., and Wight, T. N., 1989, Proteoglycans in the pathogenesis of Alzheimer's disease, *Neurobiol. Aging* **10:**481–497.

Snow, A. D., Mar, H., Nochlin, D., Kresse, H., and Wight, T. N., 1992, Peripheral distribution of dermatan sulfate proteoglycans (decorin) in amyloid-containing plaques and their presence in neurofibrillary tangles of Alzheimer's disease, *J. Histochem. Cytochem.* **40:**105–113.

Snow, A. D., Nochlin, D., Sekiguchi, R., and Carlson, S. S., 1996, Identification and immunolocalization of a new class of proteoglycan (keratan sulfate) to the neuritic plaques of Alzheimer's disease, *Exp. Neurol.* **138:**305–317.

Sotaniemi, K. A., Mononen, H., and Hokkanen, T. E., 1986, Long-term cerebral outcome after open-heart surgery, *Stroke* **17:**410–416.

Spangler, K. M., Challa, V. R., Moody, D. M., and Bell, M. A., 1994, Arterilar tortuosity of the white matter in aging and hypertension: A microradiographic study, *J. Neuropathol. Exp. Neurol.* **53(1):**22–26.

Sparks, D. L., 1996, Intraneuronal β-amyloid immunoreactivity in the CNS, *Neurobiol. Aging* **17(2):**291–299.

Sparks, D. L., and Hunsaker, J. C., 1992, Down's syndrome: Occurrence of ALZ-50 reactive neurons and the formation of senile plaques, *J. Neurol. Sci.* **109:**77–82.

Sparks, D. L., and Liu, H., 1992, Laminar and size distribution of senile plaques in AD, critical coronary artery disease (cCAD), and normal controls, *Soc. Neurosci. Abstr.* **22:**654.

Sparks, D. L., Hunsaker, J. C., Scheff, S. W., Kryscio, R. J., Henson, J. L., and Markesbery, W. R., 1990, Cortical senile plaques in coronary artery disease, aging, and Alzheimer's disease, *Neurobiol. Aging* **11:**601–607.

Sparks, D. L., Liu, H., Scheff, S. W., Coyne, C. M., and Hunsaker, J. C., III, 1993, Temporal sequence of plaque formation in the cerebral cortex of non-demented individuals, *J. Neuropathol. Exp. Neurol.* **52:**135–142.

Sparks, D. L., Scheff, S. W., Liu, H., Landers, T., Coyne, C. M., and Hunsaker, J. C., III, 1995, Increased density of neurofibrillary tangles (NFT) in non-demented individuals with hypertension, *J. Neurol. Sci.* **131:**162–169.

Sparks, D. L., Scheff, S. W., Liu, H., Landers, T., Danner, F., Coyne, C. M., and Hunsaker, J. C., III, 1996, Increased density of senile plaques (SP), but not neurofibrillary tangles (NFT), in non-demented individuals with the Apolipoprotein E4 genotype: Comparison to confirmed Alzheimer's disease patients, *J. Neurol. Sci.* **138:**97–104.

Starkstein, S. E., Sabe, L., Vazquez, S., Teston, A., Petracca, G., Chemerinski, E., Di Lorenzo, G., and Leiguarda, R., 1996, Neuropsychological, psychiatric, and cerebral blood flow findings in vascular dementia and Alzheimer's disease, *Stroke* **27:**408–414.

Stewart, P. A., Hayakawa, K., Akers, M.-A., and Vinters, H. V., 1992, A morphometric study of the blood-brain barrier in Alzheimer's disease, *Lab. Invest.* **67(6):**734–742.

Streit, W. J., Graeber, M. B., and Kreutzberg, G. W., 1989, Expression of Ia antigen on perivascular and microglial cells after sublethal and lethal motor neuron injury, *Exp. Neurol.* **105:**115–126.

Stringer, M. D., Gorog, P. G., Freeman, A., and Kakkar, V. V., 1989, Lipid peroxides and atherosclerosis, *Br. Med. J.* **298:**281–284.

Suenaga, T., Ohnishi, K., Nishimura, M., Nakamura, S., Akiguchi, I., and Kimura, J., 1994, Bundles of amyloid precursor protein-immunoreactive axons in human cerebrovascular white matter lesions, *Acta Neuropathol. (Berlin)* **87:**450–455.

Taggart, D. P., Sundaram, S., McCartney, C., Bowman, A., McIntyre, H., Courtney, J. M., and Wheatley, D. J., 1994, Endotoxemia, complement, and white blood cell activation in cardiac surgery: A randomized trial of laxatives and pulsatile perfusion, *Ann. Thorac. Surg.* **57:**376–382.

Tamaki, K., Nakai, M., Yokota, T., and Ogata, J., 1995, Effects of aging and chronic hypertension on cerebral blood flow and cerebrovascular CO2 reactivity in the rat, *Gerontology* **41:**11–17.

Tekirian, T. L., Cole, G. M., Russell, M. J., Yang, F., Wekstein, D. R., Patel, E., Snowden, D. A., Markesbery, W. R., and Geddes, J. W., 1996, Carboxy terminal of β-amyloid deposits in aged human, canine, and polar bear brains, *Neurobiol. Aging* **17(2):**249–257.

Tomaoka, A., Kalaria, R. N., Lieberburg, I., and Selkoe, D. J., 1992, Identification of a stable fragment of the Alzheimer amyloid precursor containing the beta-protein in brain microvessels, *Proc. Natl. Acad. Sci. USA* **89(4):**1345–1349.

Uchihara, T., Kondo, H., Akiyama, H., and Ikeda, K., 1995, White matter amyloid in Alzheimer's disease brain, *Acta Neuropathol. (Berlin)* **90:**51–56.

Vickers, J. C., Delacourte, A., and Morrison, J. H., 1992, Progressive transformation of the cytoskeleton associated with normal aging and Alzheimer's disease, *Brain Res.* **594:**273–278.

Vik, A., Brubakk, A. O., Rinck, P. A., Sande, E., Levang, O. W., and Sellevold, O., 1991, MRI: A method to detect minor brain damage following coronary bypass surgery, *Neuroradiology* **33:**396–398.

Vinters, H. V., Secor, D. L., Read, S. L., Frazee, J. G., Tomiyasu, U., Stanley, T. M., Ferreiro, J. A., and Akers, M.-A., 1994, Microvasculature in brain biopsy specimens from patients with Alzheimer's disease: An immunohistochemical and ultrastructural study, *Ultrastruct. Pathol.* **18:**333–348.

Wisniewski, H. M., and Kozolowski, P. B., 1982, Evidence for blood–brain barrier changes in senile dementia of Alzheimer's type (AD/SDAT): Alzheimer's disease, Down's syndrome, and aging, *Ann. N.Y. Acad. Sci.* **396:**119–129.

Wisniewski, H. M., and Wegiel, J., 1994, β-amyloid formation by myocytes of lepomeningeal vessels, *Acta Neuropathol. (Berlin)* **87:**233–241.

Wolf, P. A., D'Agostino, R. B., Belanger, A. J., and Kannel, W. B., 1991, Probability of stroke: A risk profile for the Framingham Study, *Stroke* **22:**312–318.

Wolfe, N., Reed, B. R., Eberling, J. L., and Jagust, W. J., 1995, Temporal lobe perfusion on single photon emission computed tomography predicts the rate of cognitive decline in Alzheimer's disease, *Arch. Neurol.* **52(3):**257–262.

Wolozin, B., Scicutella, A., and Davies, P., 1988, Reexpression of a developmentally regulated antigen in Down syndrome and Alzheimer disease, *Proc. Natl. Acad. Sci. USA* **85:**6202–6206.

Yamashita, K., Miyakawa, T., and Katsuragi, S., 1991, Vascular changes in the brains with Alzheimer's disease, *Jpn. J. Psychiat. Neurol.* **45:**79–84.

Yamauchi, H., Fukuyama, H., Harada, K., Nabatame, H., Ogawa, M., Ouchi, Y., Kimura, J., and Konishi, J., 1993, Callosal atrophy parallels decreased cortical oxygen metabolism and neuropsychological impairment in Alzheimer's disease, *Arch. Neurol.* **50:**1070–1075.

# Index